T0328334

Freshwater Ecology: Concepts and Environmental Applications of Limnology

Freshwater Ecology: Concepts and Environmental Applications of Limnology

Third Edition

WALTER K. DODDS
Kansas State University, Manhattan, KS, United States

MATT R. WHILES
University of Florida, Gainesville, FL, United States

ACADEMIC PRESS
An imprint of Elsevier

Academic Press is an imprint of Elsevier
125 London Wall, London EC2Y 5AS, United Kingdom
525 B Street, Suite 1650, San Diego, CA 92101, United States
50 Hampshire Street, 5th Floor, Cambridge, MA 02139, United States
The Boulevard, Langford Lane, Kidlington, Oxford OX5 1GB, United Kingdom

Notices

Knowledge and best practice in this field are constantly changing. As new research and experience
broaden our understanding, changes in research methods, professional practices, or medical
treatment may become necessary.

Practitioners and researchers must always rely on their own experience and knowledge in evaluating
and using any information, methods, compounds, or experiments described herein. In using such
information or methods they should be mindful of their own safety and the safety of others,
including parties for whom they have a professional responsibility.

To the fullest extent of the law, neither the Publisher nor the authors, contributors, or editors,
assume any liability for any injury and/or damage to persons or property as a matter of products
liability, negligence or otherwise, or from any use or operation of any methods, products,
instructions, or ideas contained in the material herein.

British Library Cataloguing-in-Publication Data
A catalogue record for this book is available from the British Library

Library of Congress Cataloging-in-Publication Data
A catalog record for this book is available from the Library of Congress

ISBN: 978-0-12-813255-5

For Information on all Academic Press publications
visit our website at https://www.elsevier.com/books-and-journals

Working together
to grow libraries in
developing countries

www.elsevier.com • www.bookaid.org

Publisher: Candice Janco
Acquisition Editor: Louisa Hutchins
Editorial Project Manager: Hilary Carr
Production Project Manager: Denny Mansingh
Cover Designer: Matthew Limbert

Typeset by MPS Limited, Chennai, India

CONTENTS

PREFACE

FOR THE STUDENT

This book was written for you. We obtained as much student input as possible by having student reviewers assess the text and the approach used in it. The idea for the text was based on teaching students who were not satisfied with the existing texts. Teaching aquatic ecology and limnology have made it clear to us that most students enter ecological sciences for practical reasons. They often are concerned about conservation of resources from a classical (e.g., fisheries program) or from environmental issue perspectives. Most existing texts limit the applied aspects of aquatic ecology to a section at the end.

The aim of this text was to incorporate discussion of the issues as they arise when we cover the basic materials. This allows you to see the applications of difficult topics immediately and, we hope, provides additional impetus for doing the work required to gain an understanding. We also attempted to use the broadest possible approach to freshwater ecosystems; scale and linkages among systems are important in ecology. Most students in ecological courses had some interest in the natural world as children. They spent time exploring under rocks in streams, fishing, camping, hiking, or swimming, which stimulated a love of nature. This book is an attempt to translate this basic affinity for aquatic ecosystems into an appreciation of the scientific aspects of the same world. It is not always easy to write a text for students. Instructors usually choose a text, giving the students little choice. Thus, some authors write for their colleagues, not for students. We tried to avoid such pressures and attempted to tailor the approach to you. We hope you will learn from the materials presented here and that they will adequately supplement your instructor's approach. When you find errors, please let us know. This will improve any future editions. Above all, please appreciate the tremendous luxury of being a student and learning. You are truly fortunate to have this opportunity and we are grateful for the time you take with this text.

FOR THE INSTRUCTOR

We hope this book will make your job a little easier. The chapters are short, mostly self-contained units to allow the text to conform to a wide variety of organizational schemes that may be used to teach about freshwaters. This will also allow you to avoid sections that are outside the scope of the course you are teaching. However, we integrated environmental applications into the text because we view basic science and applications as integrated issues. We attempt to create instructional synergism by combining these aspects of aquatic ecology. Describing applications stimulates student interest in mastering difficult scientific concepts. We employ a variety of pedagogical approaches in an attempt to engage student interest and facilitate learning. These include highlights, biography boxes, and method boxes. We have advanced sections for areas where you might want the students to get a little more in-depth, but in areas that we often skip in our own lectures for a first class in freshwater ecology. We also include an appendix on experimental design in ecological science and a glossary because we have found many advanced undergraduates have little exposure to the practical side of doing science. These sections help when we ask students to read and report on primary literature. It is always difficult to know what to include and where to go into detail. We supply some detailed examples to enforce general ideas. The choice of example is probably not always the best one, just the best one we could find while preparing the text. Suggestions for improvements in this and any other areas of the text are encouraged and appreciated. Thanks to the instructors using the first two editions that did just that. We apologize for any errors.

Why did we write this? In our experience, teaching limnology/freshwater ecology is more work than teaching other courses because of the breadth of subject, differential preparation of students, and associated laboratories, but always seems to be the most fun. Of course it is fun, it is the best subject! We hope this book facilitates your efforts to transmit what is so great about the study of freshwater ecology.

WHY DID WE WRITE A THIRD EDITION?

Given the money savings of used texts for students, there needs to be a good reason for a new edition. In the near decade since writing of the second edition, many new advances have occurred, and new topics have gained prominence with respect to environmental effects and hot areas of research interest. In our attempt to cover these new areas we have added over 800 new and updated references, 46 new figures, 67 figures updated to color, 1 new chapter, and expanded the length of the text. Particularly, we included more emphasis on toxins, pollution, and large-scale approaches than in the previous edition. Dedicated students in WKD's classes found numerous errors in previous editions (with a potential award of test points for each unique error), as did instructors across the country. We hopefully corrected these without introducing too many new ones. We hope you find this new edition even more useful than the last one.

ACKNOWLEDGMENTS

We thank Dolly Gudder, who was involved in all aspects of the writing and compilation of the first two editions of this book, including proofing the entire text, drafting and correcting all the figures from the first edition, library research, writing the first draft of the index, and obtaining permissions. Alan Covich provided extensive conceptual guidance and proofread the first edition of text; his input was essential to producing this work. Eileen Schofield-Barkley provided excellent editorial comments on all chapters of the second edition. The fall of 1998 Kansas State University limnology class proofed Chapters 1—8 and 11—18 of the second edition. The Kansas State University limnology (freshwater ecology) classes proofread all chapters (especially Michelle Let, Katie Bertrand, and Andrea Severson.). The L.A.B. Aquatic Journal Club, Chuck Crumly, Susan Hendricks, Stuart Findlay, Steve Hamilton, Nancy Hinman, Jim Garvey, Chris Guy, and Al Steinman provided suggestions on the first edition. Early helpful reviews on book concepts were provided by James Cotner, David Culver, Jeremey Jones, Peter Morin, Steven Mossberg, Stuart Fisher, Robert Wetzel, and F. M. Williams. The anonymous reviews (obtained by the publisher) are also greatly appreciated, including those who completed a detailed survey on the prospect of a second edition. Many of the good bits and none of the mistakes are attributable to these reviewers. We appreciate the support of the Kansas State University Division of Biology, the Kansas Agricultural Experiment Station, and Southern Illinois University Carbondale. This is publication # **19-234-B** from the Kansas Agricultural Experiment Station.

For the second edition, thanks to Lydia Zeglin for help on describing molecular methods. Excellent suggestions for corrections also came from Robert Humston, David Rogowski, John Havel, Sergi Thomas, Daniel Welsh, Erika Iyengar, Matt McTammany, Jack Webster, and Robert Humston. Kabita Ghimire produced the global distribution maps of freshwater habitats. Keith Gido, James Whitney, and Joe Gerken helped clarify Chapter 23. Thanks to Erika Martin, Joshua Perkin, and Kyle Winders for detailed review of the entire draft manuscript from the second edition. Thanks to Andy Richford who assisted in the development of the ideas for the second edition while he was editor at Elsevier.

For the third edition, we thank Priscilla Molley, Anne Schechner, Elizabeth Renner, and Lindsey Bruckerhoff for proofing and suggestions. Suggestions for corrections came from several freshwater ecology classes, Charles Booth, Kristen Bouska, Jessica Fulgoni, Justin Murdock, Ishi Buffam, Becky Bixby, and David Locky. Kevin Wilson provided comments on Devils Hole Pupfish. Alex Shepack provided valuable assistance with the declining amphibians box, and Frank Anderson updated the invertebrate phylogeny section.

MRW thanks F.E. Anderson, L.L. Battaglia, J.E. Garvey, J.W. Grubaugh, R.O. Hall, A.D. Huryn, K.R. Lips, R. Lira, and S.D. Peterson for valuable input and advice on this work. His early mentors, M.E. Gurtz, C.M. Tate, G.R. Marzolf, and J.B. Wallace provided guidance at critical points of his life and helped him realize that he could actually make a career out of his interests in ecology. His parents, Jim and Jane, and his sister, Wendy, tolerated and even supported his somewhat odd boyhood interests and laid the foundation for all these. S.G. Baer provided support, encouragement, and advice. For the third edition, his recent and current graduate students (Sophia, Jessica, Kelley, Lucas, Jared Katie, Kelsey, Kasey, and Adam) and L. Hsieh were sources of inspiration who tolerated his preoccupation and absent mindedness. N. Baccus helped the Whiles' laboratory running smoothly, and B. Comer and T. Sherk kept the Cooperative Wildlife Research Laboratory humming along. As always, he is particularly grateful for the ongoing, lifelong support of his parents, Jane Whiles and the late James Whiles.

MRW and WKD's children, Hannah, Joey, Sadie, and Rowland, and grandchildren (Jaxon and Samuel) put it all in perspective; the next generations are the reason this text includes environmental applications. We both deeply appreciate the support and love of our families.

CHAPTER 1

Why Study Continental Aquatic Systems?

Contents

Figure 1.1 Crater Lake, Oregon.

Freshwater Ecology
DOI: https://doi.org/10.1016/B978-0-12-813255-5.00001-6

All life depends on water for existence. Although 71% of our planet is covered by water, only a very small proportion is associated with the continental areas to which humans are primarily confined (Table 1.1 and Fig. 1.2). Over 99% of the water associated with continents is in the form of groundwater or ice and is therefore difficult for humans to use. Humans mostly depend on freshwater streams, rivers, marshes, lakes, and shallow groundwaters; we rely heavily on a rare commodity.

Why study the ecology of continental waters? To the academic, the answer is easy: because it is fascinating and we enjoy learning for its own sake. Thus the field of *limnology*[1] (the scientific study of continental waters) has developed. Limnology has a long history of academic rigor and broad interdisciplinary synthesis (e.g., Hutchinson, 1957, 1967, 1975, 1993; Wetzel, 2001). One of the truly exciting aspects of limnology is the synthetic integration of geological, chemical, physical, and biological interactions that define aquatic systems. No limnologist exemplifies the use of such academic synthesis better than G. E. Hutchinson (Biography 1.1); he did more to define modern limnology than any other individual. Numerous other exciting scientific advances have been made by aquatic ecologists, including refining the concept of an ecosystem, developing ecological methods for approaching control of disease, devising methods to assess and remediate water pollution, understanding how effects of large predators can cascade to primary producers, establishing ways to manage fisheries, restoring freshwater habitats, understanding the deadly lakes of Africa, and conserving unique organisms. Each of these advances will be covered in this book.

Table 1.1 Locations, amounts, and turnover times of water compartments in the global hydrologic cycle

Location	Amount (thousands km³)	Total %	% Inland liquid water	Approximate residence time
Freshwater lakes	125	0.009	1.45	75 years
Saline lakes and inland seas	104	0.008	1.20	
Rivers (average volume)	1	0.0001	0.01	5 months
Shallow and deep soil water	67	0.005	0.77	100 years
Groundwater to 4000 m depth	8,350	0.61	96.56	1,000 years
Ice caps and glaciers	29,200	2.14	—	12,000 years
Atmosphere	13	0.001	—	8.9 days
Oceans	1,320,000	97.3	—	3,060 years

Source: Data from Todd (1970); Wetzel (2001); Pidwirny (2006).

[1] The term "limnology" includes saline waters (Wetzel, 2001) and all other continental waters, but limnology courses traditionally do not cover wetlands, groundwater, and even streams. Thus this book is titled "Freshwater Ecology."

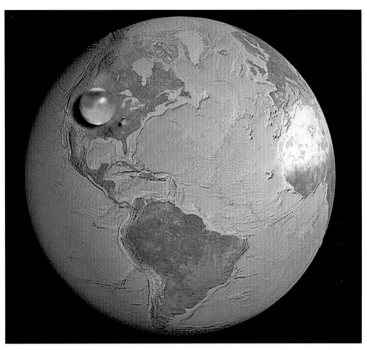

Figure 1.2 An image illustrating the relative amounts of water on Earth. The largest blue sphere represents all the water in groundwater, ice caps, oceans, lakes, and rivers. The medium-size blue sphere is the volume of water in groundwater, and the smallest sphere is the global freshwater in lakes and rivers. *Image courtesy: Howard Perlman, United States Geological Survey.*

We hope to transmit our excitement and appreciation of nature that comes from studying aquatic ecology. Further justification for study may be necessary for those who insist on more concrete benefits from an academic discipline or for those interested in preserving water quality and aquatic ecosystems in the broader political context. There is a need to place a value on water resources and the ecosystems that maintain their integrity, and to understand how the ecology of aquatic ecosystems affects this value. Water is unique and has no substitute. A possible first step toward placing a value on a resource is documenting human dependence on it and how much is available for human use.

Humankind would rapidly use all the water on the continents were it not replenished by precipitation. Understanding hydrologic *fluxes*, or movements of water through the global *hydrologic cycle*, is central to understanding water availability but much uncertainty surrounds some aspects of these fluxes. Given the difficulty that forecasters have predicting the weather over even a short–time period, it is easy to understand why estimates of global change and local and global effects on water budgets are beset with major uncertainties (Mearns et al., 1990; Mulholland and Sale, 1998; Junk et al., 2013). Scientists can account moderately well for evaporation of water into the atmosphere, precipitation, and runoff from land to oceans. This accounting is accomplished with networks of precipitation gauges and measurements

BIOGRAPHY 1.1 George Evelyn Hutchinson

George Evelyn Hutchinson (Fig. 1.3) was one of the top limnologists and ecologists of the 1900s, and likely the most influential of the century. His career spanned an era when ecology moved from a discipline that was mainly the province of natural historians to a modern experimental science. In large part, he and his students were responsible for these developments. Born in 1903 in Cambridge, England, Hutchinson was interested in aquatic entomology as a youth and authored his first publication at age 15. He obtained an MA from Emmanuel College at Cambridge University and worked in Naples, Italy, and South Africa before securing a position at Yale University. He remained at Yale until the end of his career and died in 1991.

Hutchinson's range of knowledge was immense. He was well versed in literature, art, and the social sciences. He published on religious art, psychoanalysis, and history. His broad and innovative view of the world enriched his scientific

Figure 1.3 G. Evelyn Hutchinson. *Courtesy: The Yale Image Library.*

endeavors. Hutchinson published some of the most widely read and cited ecological works of the century. His four volumes of the *Treatise of Limnology* are the most extensive treatment of limnological work ever published. His writings on diversity, complexity, and biogeochemistry inspired numerous investigations. Hutchinson organized a research team to study the Italian Lake Ianula in the 1960s; this multidisciplinary approach has since become a predominant mode of ecological research (Hutchinson and Cowgill, 1970). He reportedly was always able to find positive aspects of his students' ideas and encouraged them to develop creative thoughts into important scientific insights. As a consequence, many of Hutchinson's students and their students are among the most renowned ecologists today.

Hutchinson earned many major scientific awards in his career, including the National Medal of Science. He wrote popular scientific articles and books that were widely distributed. He was a staunch defender of intellectual activities and their importance in the modern world. Hutchinson's mastery of facts, skillful synthesis, evolutionary viewpoint, cross-disciplinary approach, encouragement of students and collaborators, and knack for asking interesting and important questions make him an admirable role model for students of aquatic ecology.

of river discharge. Recently, sophisticated methods for estimating groundwater flow and recharge, including those based on remote sensing, have improved our ability to account for global hydrology.

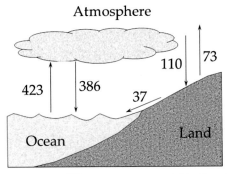

Figure 1.4 Fluxes (movements among different compartments) in the global hydrologic budget (in thousands of km^3 per year). *Data from Berner and Berner (1987).*

The *global water budget* is the estimated amount of water movement (fluxes) among *compartments* (the amount of water that occurs in each area or form) throughout the globe (Fig. 1.3). This hydrologic cycle will be discussed in more detail in Chapter 4, but is presented here briefly to inform our discussion of how much water is available for human use (Fig. 1.4). Total runoff from land to oceans via rivers has been reported as 22,000, 30,000, and 35,000 km^3/year by Leopold (1994), Todd (1970), and Berner and Berner (1987), respectively. These estimates vary because of uncertainty in gauging large rivers in remote regions. So, what are the demands on this potential upper limit of sustainable water supply?

HUMAN USE OF WATER: PRESSURES ON A KEY RESOURCE

People in developed countries generally are not aware of the quantity of water that is necessary to sustain their standard of living. In North America particularly, high-quality water is often used for such luxuries as filling swimming pools and watering lawns. Perhaps people notice that their water bills increase in the summer months. Publicized concern over conservation may translate, at best, into people turning off the tap while brushing their teeth or using low-flow showerheads or low-flush toilets. Few people in developed countries understand the massive demands for water from industry, agriculture, and power generation that their lifestyle requires (Fig. 1.5).

Some of these demands, such as domestic use, require high-quality water. Other uses, such as hydroelectric power generation and industrial cooling, can be accomplished with lower quality water. Some uses are *consumptive* and preclude further use of the water; for instance, a significant portion of water used for agriculture is lost to evaporation. The most extreme example of nonrenewable water resource consumption may be water "mined" (withdrawal rates in excess of rates of renewal from the surface) from *aquifers* (large stores of groundwater) that have extremely long regeneration times. Such *withdrawal* is practiced globally (Postel, 1996; Gleeson et al., 2012)

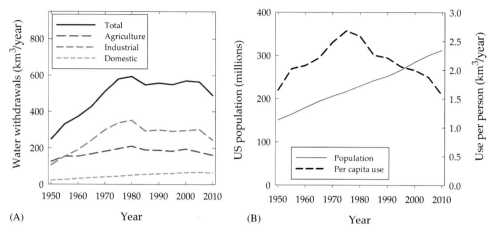

Figure 1.5 Estimated uses of water (A) and total population and per capita water use (B) in the United States from 1950 to 2010. Note that industrial and irrigation uses of water are dominant. Off-stream withdrawals used in these estimates do not include hydroelectric uses. *Data courtesy: The United States Geological Survey.*

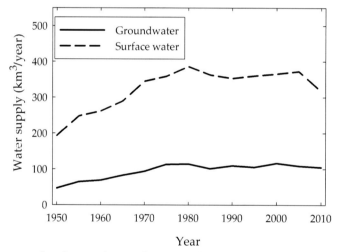

Figure 1.6 Amounts of surface and groundwater used in the United States from 1950 to 2010. These estimates include only withdrawals and not hydroelectric uses. *Data courtesy: The United States Geological Survey.*

and accounts for a significant portion of the United States' water use, particularly for agriculture (Fig. 1.6). In many cases use rates exceed the rate at which the aquifer is replenished, and the groundwater is overexploited. Other uses are less consumptive. For example, hydroelectric power "consumes" relatively less water than agriculture (i.e., evaporation from reservoirs increases water loss, but much of the water moves downstream).

Table 1.2 General ranges of water use with varied socioeconomic conditions on a per capita basis, rates per country are estimates for the year 2000

Society	Range or mean ($m^3 \, y^{-1}$ person^{-1})
Irrigated semiarid industrial countries	3,000–7,000
Irrigated semiarid developing countries	800–4,000
Temperate industrial countries	170–1,200
China	431
Jordan	155
Switzerland	351
Turkmenistan	5,309
Uganda	9
United States	1,688

Source: After la Rivière (1989); Falkenmark (1992); Gleick et al. (2004).

How much water does humankind need? A wide disparity occurs between per capita water use in and among countries, particularly in semiarid countries in which surface water is scarce (Table 1.2). Israel is one of the most water-efficient developed countries, with per capita water use of 500 m^3/year (Falkenmark, 1992), about four times more efficient than in the United States. Increases in standard of living lead to greater water demands (per capita water use) if not accompanied by increases in efficiency of use.

The maximum total water available for human use is the amount that falls as precipitation on land each year minus the amount lost to evaporation. As mentioned earlier, the maximum amount of water available in rivers is 22,000–35,000 km^3/year. However, a large amount of water is lost to floods in populated areas (temporal inaccessibility) or flows occurring in areas far removed from human population centers (spatial inaccessibility), leaving approximately 9,000 km^3/year for all human uses (la Rivière, 1989). Thus only about one-third of the water that flows through the hydrologic cycle is available spatially or temporally. Humans cannot sustain water use rates that exceed this supply rate without withdrawing groundwater more quickly than aquifers can renew their stores, collecting from melting ice caps, transporting from remote areas, or reclaiming (desalinizing) water from oceans. These processes are expensive or impossible to sustain in many continental regions without tremendous energy input, as we discuss later in this chapter.

Predicting future water use is difficult but instructive for exploring possible future patterns and consequences of this use. Total annual off-stream withdrawals (uses that require removal of water from the river or aquifer, not including hydroelectric power generation) in the United States in 1980 were 2,766 m^3/person and have decreased slightly since that time, mostly due to a decrease in total industrial use (Fig. 1.5). If all of the people on Earth used water at the current United States rate (i.e., if global standard of living and water use efficiency were the same as in the United States), over half of every bit of water produced by the hydrologic cycle would be used. Globally, humans currently withdraw about 54% of geographically and spatially accessible runoff (Postel et al., 1996); if all people in the world used water at the per capita rates used in the United States, all the water available (spatially and temporally accessible) would be used.

On a local scale, water scarcity can be severe. Political instability in Africa is predicted based on local population growth rates and limited water supply (Falkenmark, 1992). Similar instabilities are likely to arise from conflicts over water use in many parts of the world (Postel, 1996). Large urban areas have tremendous water demands, and very strong effects on local and regional hydrology, generally decreasing water availability (Fitzhugh and Richter, 2004). For example, in some large cities in arid regions, major river flows are almost completely composed of sewage effluent, and some small streams that were ephemeral have become permanent because of runoff from lawn watering. In the arid southwestern United States, human water uses can account for more than 40% of the total supply (Waggoner and Schefter, 1990). In such cases, degradation of water quality has substantial economic consequences. Cape Town South Africa turned off their taps serving 4 million people in 2018, and this scenario is likely to become more common in cities around the world.

The population of Earth was estimated at 7.4 billion people in 2014 and is predicted to exceed 11 billion people by the end of the century (United Nations, 2015). The global population will have increased more than 10% between the second and third editions of this book. Given the increase in human population and resource use (Brown, 1995a,b), demand for water will only intensify (Postel, 1996). As the total population on Earth expands, the value of clean water will increase as demands escalate for this finite resource (Dodds, 2008). Population growth will increase demand for water supplies, even in the face of uncertainty over future climate (Vörösmarty et al., 2000) with particularly acute effects in rapidly urbanizing areas (McDonald et al., 2011). Increased efficiency has led to decreases in per capita water use in the United States since the early 1980s (Fig. 1.5). Efforts to boost conservation of water will become essential as water becomes more valuable (Brown, 2000).

Despite the existence of technology to make water use more efficient and maintain water quality, the ongoing negative human impact on aquatic environments is widespread. Most uses of water compromise water quality and aquatic ecosystem integrity, and future human impact on water quality and biodiversity is inevitable. An understanding of aquatic ecology will assist humankind in making decisions to minimize adverse impacts on our aquatic resources, and will ultimately be required for the development of policies that lead to sustainable water use practices (Gleick, 1998). Accounting for the true value of water will help policy makers decide how important its use is, and students reading this book understand how water is a vital resource to learn about.

WHAT IS THE VALUE OF WATER?

Accurate accounting for the economic value of water includes determining both the immediate benefit and how obtaining a particular benefit alters future use. This accounting can be essential to planning sustainable development (Garrick et al., 2017). Availability of water is essential, but the quality of water must also be considered. Consumption and contamination associated with each type of use dictate what steps are crucial to maintaining aquatic ecosystems and water quality and quantity. Establishing the direct benefits of water use, including patterns and types of uses, is also necessary. Elucidation of benefits allows determination of economic value of water and how its use should be managed.

Aquatic ecosystems provide us with numerous benefits in addition to direct use. Estimates of the global value of freshwater wetlands ($1.5 trillion/year in 2007) and rivers and lakes ($2.4 trillion/year) indicate the key importance of freshwaters to humans (Costanza et al., 2014, Fig. 1.7). These estimates will increase as researchers are able to assign values to more ecosystem goods and services (Dodds et al., 2008). Freshwater ecosystems have the greatest value per unit area of all habitats. Estimates suggest that the greatest values of natural continental aquatic systems are derived from flood control, water supply, and waste treatment. The value per hectare is greater for wetlands, streams, and rivers than for any terrestrial habitats. The next section details how these values are actually assigned.

We explore values of aquatic ecosystems because monetary figures can influence perceived value to society. Assigning values to ecosystems can provide evidence for advocates of minimizing anthropogenic impacts on the environment. Ignoring

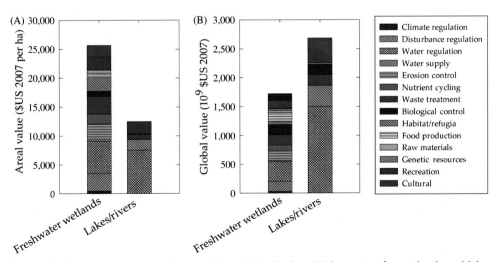

Figure 1.7 Relative values per unit area (A) and global values (B) by service for wetlands and lakes and rivers. *Data from Costanza et al. (2014).*

ecosystem values can be particularly problematic because perceived short-term gain often outweighs poorly quantified long-term harm when political and bureaucratic decisions are made about resource use. For example, restored wetlands may not be as valuable as conserved wetlands (Dodds et al., 2008), so policy decisions should be made accordingly. The concept of "no net loss", when a constructed wetland can be substituted for a wetland that is removed for development, may not capture all the values a natural wetland provides to society.

Management of freshwater resources can lead to conflicting values and trade-offs. For example, increased water supply through reservoirs may not protect biodiversity. Thus valuation of ecosystem services can be used to scale economic consequences of varied management approaches (Dodds et al., 2013). Such approaches can be applied from local to global scales. It can also be used to account for multiple effects of environmental change in specific cases, such as occurs when lakes shift from phytoplankton to macrophyte dominated states (Hilt et al., 2017).

Quantification of some values of water is straightforward, including determining the cost of drinking water, the value of irrigated crops, some costs of pollution, and direct values of fisheries. Others may be more difficult to quantify. What is the value of a canoe ride on a clean lake at sunset or of fishing for catfish on a lazy river? What is the worth of all the species that inhabit continental waters including nongame species?

What is the actual value of water? The local price of clean water will be higher in regions in which it is scarce. Bottled water costs vary by region, but people are willing to pay roughly \$0.50 to \$1.00 L^{-1} of water, or over \$500 m^{-3}. Thus clean drinkable water is very valuable to most people. Highly subsidized irrigation water sold for about \$0.01 m^{-3} in Arizona, but clean drinking water costs \$0.37 m^{-3} in the same area (Rogers, 1986). In the same time period, drinking water costs were between \$0.08 and \$0.16 m^{-3} in other areas of the United States (Postel, 1996). In 2013, average rates for the 50 largest cities in the United States were \$0.13 m^{-3} (retrieved from https://www.saws.org/who_we_are/community/RAC/docs/2014/50-largest-cities-brochure-water-wastewater-rate-survey.pdf, 2 February 2018). At the rate of \$0.01 m^{-3}, and assuming that people on Earth use 2,664 km^3/year for irrigation (according to the Food and Agriculture Organization for 2001), the global value of water for irrigation can be estimated as \$26 billion/year. Global agricultural production was worth \$3.1 trillion in 2015 (retrieved from http://data.worldbank.org/indicator/NV.AGR.TOTL.CD, 26 December 2016). The Food and Agriculture Organization estimates that 20% of crop area is irrigated but 40% of production is from irrigated areas [FAO, 2016. AQUASTAT website. Food and Agriculture Organization of the United Nations (FAO). Website accessed on 26 December 2015].

Use of freshwater for irrigation does not come without a cost. Agricultural pesticide contamination of groundwater in the United States led to total estimated costs of \$1.8 billion annually for monitoring and cleanup (Pimentel et al., 1992). Erosion related to agriculture caused losses of $5.1 billion/year directly related to water quality impairment in the United States (Pimentel et al., 1995). This estimate includes costs for dredging sediments from navigation channels and recreation impacts, but it excludes biological impacts. These estimates illustrate some of the economic impetus to preserve clean water.

The economic value of freshwater fisheries, including aquaculture, worldwide is over $112 billion/year (Table 1.3), and has almost doubled between 2007 and 2014. This economic estimate includes only the actual cash or trade value of the fish and crustaceans. In many countries, sport fishing generates considerable economic activity. For example, in the United States, $41.8 billion was spent on goods and services related to freshwater angling in 2011 (U.S. Department of the Interior and Bureau of the Census, 2011). Similarly, 20 million freshwater anglers were estimated to have spent a collective 10 billion Euros in 2006 (Brainerd, 2010). In addition, 63% of non-consumptive outdoor recreation visits in the United States included lake or streamside destinations, presumably to view wildlife and partake in activities associated with water (U.S. Department of the Interior and Bureau of the Census, 1993). Many of these visits result in economic benefits to the visited areas. Maintaining water quality is vital to healthy fisheries and healthy economies. Pesticide-related fish kills in the United States were estimated to cause $10−24 million/year in losses (Pimentel et al., 1992). Finally, maintaining fish production may be essential to ensuring adequate nutrition in developing countries (Kent, 1987). Managing fisheries clearly requires knowledge of aquatic ecology. These fisheries and other water uses face multiple threats from human activities.

Sediment, pesticide and herbicide residues, fertilizer runoff, other nonpoint runoff, sewage with pathogens and nutrients, chemical spills, garbage dumping, thermal pollution, acid precipitation, mine drainage, hydraulic fracturing for oil extraction, urbanization, damming, and habitat destruction are some of the threats to our water

Table 1.3 Global capture and aquaculture fisheries production relying on freshwater in 2014 (Food and Agriculture Administration, 2015)

Type	Amount (thousand metric tons)	Values (million $ US)
Carps, barbels, and other cyprinids	29,776	42,366
Tilapias and other cichlids	6,036	9,547
Miscellaneous	17,395	21,573
Sturgeon, paddlefish	89	410
River eels	266	1,499
Salmons, trouts, smelts	4,365	23,929
Freshwater crustaceans	2,448	12,840
Freshwater mollusks	513	386
Total	60,888	112,550

resources. Understanding the implications of each of these threats requires detailed understanding of the ecology of aquatic ecosystems. The effects of such human activities on ecosystems are linked across landscapes and encompass wetlands, streams, groundwater, and lakes (Covich, 1993). Management and policy decisions can be ineffective if the linkages among systems and across spatial and temporal scales are not considered (Highlight 1.1). Effective action at the international, national, and local governmental levels, as well as in the private sector, is necessary to protect water and the organisms in it. Success requires whole-system approaches grounded with sound scientific information (Vogt et al., 1997). Productive application of science requires explicit recognition of the role of temporal and spatial scale in the problems being considered and the role of the human observer (Allen and Hoekstra, 1992). We therefore consider scale throughout the book. We give numerous examples in this book that illustrate that the understanding of the mechanisms behind problems such as nutrient pollution, flow alteration in rivers, sewage disposal, and trophic interactions has led to successful mitigation strategies. Many rivers in developed countries are cleaner than they were a half century ago, so there is hope. Future efforts at protection are more likely to be successful if guided by informed aquatic ecologists interested in safeguarding our water resources.

Highlight 1.1 Valuation of Ecosystem Services: Contrasts of Two Desired Outcomes

Ecosystem services refer to the properties of ecosystems that confer benefits to humans. Here, we contrast two types of watershed management and some economic considerations of each. The first case involves the effects of logging on water quality and salmon survival on the northern portion of the Pacific coast of North America and the second involves water supply in some South African watersheds. The preferred management strategies are different, but both rely on understanding ecosystem processes related to vegetation and hydrologic properties of watersheds. When watersheds have more vegetation, particularly closer to streams, runoff can be less and have less sediment. Removal of streamside vegetation is a major concern for those trying to conserve salmon.

Several species of salmon are considered endangered, but they have direct effects on the biology of the streams in which they spawn (Willson et al., 1998). Sport and commercial salmon fisheries have considerable value on the northwest coast of North America. Dams that prevent the passage of adult fish and habitat degradation of streams are the two main threats to salmon survival in the Pacific coastal areas. Logging (Fig. 1.8), agriculture, and urbanization lead to degradation of spawning habitat. The main effects of logging include increased sedimentation and removal of habitat structure (logs in the streams). These factors both decrease survival of eggs and fry. Even moderate decreases in survival of young can have large impacts on potential salmon extinction (Kareiva et al., 2000)

Figure 1.8 A clear cut across a small stream. *Courtesy: The United States Forest Service.*

Economic analysis of efforts to preserve salmon populations includes calculation of the costs of modifying logging, agriculture, and dam construction and operation as opposed to the benefits of maintaining salmon runs. The economic benefits of salmon fisheries were estimated at $1 billion/year (Gillis, 1995). Costs of modifying logging, agriculture, and dam construction and operation probably exceed the direct economic value of the fishery.

The second case involves shrubland watersheds (fynbos) in South Africa that provide water to large agricultural areas downstream and considerable populations of people in urban centers and around their periphery (van Wilgen et al., 1996). Introduced weed species have invaded many of these shrubland drainage basins (watersheds or catchments). The weeds grow more densely than the native vegetation and reduce runoff to streams. Also, about 20% of the native plants in the region are endemic and endangered by the weedy invaders.

Costs of weed management are balanced against benefits from increased water runoff. Costs associated with weed removal are offset by a 29% increase in water yield from the managed watersheds. Given that the costs of operating a water supply system in the watershed do not vary significantly with the amount of water yield, the projected costs of water are $0.12 m^{-3} with weed management and $0.14 m^{-3} without it. Other sources of water (recycled sewage and desalinated water) are between 1.8 and 6.7 times more expensive to use. An added benefit to watershed weed control is protection of native plant species. Weed removal is therefore economically viable.

These two cases illustrate how ecosystem management requires understanding of hydrology and biology. In the case of the salmon, vegetation removal (logging) is undesirable because it lowers water quality and reduces reproductive success. In the South African shrublands, the removal of introduced weeds is desirable because it increases water yield. Knowledge of the ecology of systems is essential to making good decisions.

ADVANCED: METHODS FOR ASSIGNING VALUES TO ECOSYSTEM SERVICES

An entire field of defining ecosystem services has arisen as a branch of ecological economics. In this section, we discuss how values are actually assigned, and how this relates to freshwaters. A simple calculation can illustrate how valuable water is. About 9,000 km^3 of water are available each year to humans through the global hydrologic cycle when and where we need that water. It takes 2,443 J of energy to evaporate each gram of water. This means that 2.19×10^{22} J of energy are required each year to move water from the ocean into the atmosphere. Humans currently use about 5×10^{20} J of energy each year, mostly from fossil fuels. Thus, it would take about 40 times our global energy use to evaporate all the water available, or 16 times to evaporate all the water we actually use. The average elevation on Earth is roughly 800 m, and lifting the 9,000 km^3 of water to that elevation would require about 10% of human energy use (this does not account for frictional loss in pipes or inefficiencies of pumps) that would be required if humans needed to pump the water from sea-level to the areas they inhabit.

Several schemes have been constructed to value ecosystem services. Probably the most widely agreed upon scheme was developed in conjunction with the Millennium Ecosystem Assessment (MEA, 2003). This scheme recognizes utilitarian (provisioning, regulating, and supporting aspects) and nonutilitarian values (ethical, religious, cultural, and philosophical aspects). Utilitarian values can be broken down into direct use values, indirect use values, and option values. Direct use values include consumptive (harvesting fish or aquatic plants) and nonconsumptive (water sports and recreation) use values. Indirect use values include water purification, water supply, and other ecosystem processes that benefit humanity. Option values are values that are not used currently, but may be used in the future (e.g., maintaining a fishery so your offspring can use it). For direct uses, consumptive values can be directly estimated by the market value of the item of interest.

Nonconsumptive values can be determined on the basis of the amount of money people spend to partake in particular activities and surveys to gauge how much they would be willing to pay to maintain that activity (Wilson and Carpenter, 1999). In the previous section, we discussed surveys documenting how much people spend for freshwater angling. One can calculate how much property value would increase if a lake was cleaner, or how many more recreation days would be spent on a clean lake compared to a polluted lake coupled with the average amount of money spent per recreational visit. In an example of determining a relationship between an economic benefit and an ecosystem attribute, Michael et al. (1996) demonstrated that a single meter increase in lake clarity translated into increased property values of $32-656 m^{-1} of frontage. This general approach was taken to assess decreases in value of freshwater ecosystem services caused by nutrient pollution (eutrophication). These analyses for the United States indicated that the largest losses of value were related to declining property values and decreased recreational use (Dodds et al., 2009). Thus, it can be established how much people are willing to pay for some esthetic values.

Indirect uses can also be quantified. For example, if water pollution is of concern, one could calculate the cost required to bring water up to usable quality. Another example would be wetlands, where we rely on flood regulation as an ecosystem service allowing water to spread out and cause less damage downstream. The cost of protecting downstream areas from floods if the wetlands were removed can be used to assign values for wetland flood protection.

The idea of quantifying ecosystem services is controversial in the field of economics (e.g., Bockstael et al., 2000). Critics argue that discounting (the lost cost of investment opportunity) is ignored. The argument is fundamentally philosophical, with proponents for conserving economic values of ecosystem services arguing that benefits are accrued to society rather than individuals, so benefits are worth as much if they are used immediately or withheld as option values for future generations.

The second critique of valuation of ecosystem services is that they do not have exact or directly accounted value (e.g., using willingness to pay survey methods, Keeler et al., 2012). Economists who argue this are more comfortable assuming that values of ecosystem services are externalities, meaning they are economic values that are not readily assigned a monetary value. This critique is based on the argument that a poorly constrained estimate of value is worse than no estimate at all.

Nonutilitarian values are more difficult to quantify. Still, people are willing to pay to protect aspects of the environment that are important to them for ethical, religious, cultural, or philosophical reasons. For example, many people are willing to allot tax dollars to help protect parks and natural areas even if they are unlikely to ever visit those areas. Díaz et al. (2018) suggest that the way that all ecosystem services (utilitarian and nonutilitarian) influence human society and culture should be taken into account. This approach would be less focused on economic benefit and more directly on human wellbeing.

The fact remains that humanity values many aspects of freshwater. Rivers and lakes are some of the most desirable places to live and recreate, and we have multiple dependencies on them as a source of water. As the field of ecological economics matures, we will be able to assign more concrete values to freshwater ecosystem services.

THE ANTHROPOCENE: CLIMATE CHANGE AND WATER RESOURCES

The influence of humans is now global and is causing geological changes radical enough that we are in a new geological era, the *Anthropocene* (Crutzen, 2006; Zalasiewicz et al., 2011). There is now consensus among scientists that human-induced climate change is warming the planet (Intergovernmental Panel on Climate Change (IPCC), 2013), though some of the details remain uncertain. Knowledge of the basics of freshwater ecology will be essential if future scientists, such as the students reading this book, hope to be able to deal with the consequences of global climate change.

There have been large alterations in the Earth's surface waters globally between 1984 and 2015 as verified by remote sensing at high resolution (Pekel et al., 2016). Overall, the amount of surface water has increased because of reservoir construction, with 184,000 km^2 new surface water formed; about twice the surface area of Lake Superior. However, in some regions of the Earth a total decrease in about 90,000 km^2 has occurred due to diversion and withdrawal. This loss is mostly concentrated in the Middle East and Central Asia related to human activities, but long-term drought in Australia and the United States have also led to losses.

The Earth's warming climate is expected to negatively affect the quality and quantity of freshwater resources. Along with direct effects of warming temperatures, climate models forecast changes in regional precipitation patterns and overall higher variability in precipitation that will lead to increased frequency, magnitude, and unpredictability of flooding and droughts in many regions. Reductions in snow pack in mountain regions, which are already shrinking at an alarming rate, will cause reduced stream flows (Giersch et al., 2017). Warming will also result in earlier melting of snow pack causing changes in the seasonal hydrology of receiving streams (Barnett et al., 2008). Climate change will decrease freshwater availability in many areas because of increased losses to evaporation and human use, and warming water temperatures are expected to have both additive and synergistic effects with other stressors such as nutrient pollution and the spread of exotic species. Smaller bodies of water will likely be affected first by climate change because they have less thermal and hydrologic buffering capacity and are more highly influenced by local precipitation patterns (Heino et al., 2009).

Warming temperatures will exacerbate current water pollution problems because increased water loss will reduce the volumes of water in lakes, streams, and wetlands, effectively concentrating pollutants and reducing the flushing of these materials (Whitehead et al., 2009). Along with the obvious ecological consequences, increasing levels of pollutants will increase water treatment costs accordingly. Freshwater habitats in many regions are already stressed from increasing nutrient inputs from human activities; this coupled with reduced water volumes and warming water temperatures will fuel growth of undesirable algae (Wrona et al., 2006). Algal growths will lead to changes in community structure and reduced dissolved oxygen availability, further harming freshwater organisms.

Predicting exact responses of different freshwater habitats to the combined effects of warming and other human impacts is difficult. For example, increased nutrient levels and water temperatures might initially increase biodiversity in a cold, nutrient-poor lake, whereas similar changes to a nutrient-rich lake would likely result in a decrease in biological diversity because of reduced oxygen storage capacity (Heino et al., 2009). Warming may also directly influence the availability of nutrients and other materials. For example, in arctic regions, predicted melting of upper layers of permafrost will liberate phosphorus, which can lead to cascading effects on the productivity of regional streams and lakes (Hobbie et al., 1999).

Climate change will alter biotic conditions for many organisms. For example, seasonal patterns in lakes will change with extended ice-free seasons and periods of summer stratification of lakes in higher latitudes. As temperatures warm, lake mixing patterns will change, with extended periods of summer stratification. Extended stratification will increase the probability of anoxia in cooler, deep water habitats that cold-water species need to survive. In streams, climate change will alter flow regimes globally, altering the conditions which aquatic species need to survive (Pyne and Poff, 2017). These changes will leave aquatic biota vulnerable to climate change. Markovic et al. (2017) document many vulnerable species across Europe, and such vulnerability certainly exists in many other parts of the Earth.

Freshwater habitats are expected to face some of the greatest losses in biodiversity, and thus ecosystem function and stability, as a result of climate change and other human impacts (Xenopoulos et al., 2005). Growth rate, adult size, and ultimately fecundity of most aquatic organisms are all influenced by temperature. Even subtle changes in average or maximum water temperatures can have measurable effects on the biota; in some cases, these effects are sublethal, but can still be serious. For aquatic insects such as many mayflies, warming water temperatures just $2°C-3°C$ above optimal can greatly reduce the number of eggs produced by females (Vannote and Sweeney, 1980; Firth and Fisher, 1992), which has important implications for future mayfly populations, predatory fish production, and overall ecosystem health. Warming can also desynchronize life cycles and seasonal phenologies of consumers and their resources, a pattern that has been documented in flowers and pollinators (Memmott et al., 2007), as well as in freshwater and marine food webs (Thackeray et al., 2010).

Global climate change has led us into a no-analog world, where predicting hydrology on the basis of past patterns is difficult or impossible (Milly et al., 2008). A mechanistic understanding of hydrologic process coupled with downscaled global circulation models is therefore necessary to predicting the future availability of water in various regions as well as patterns of drought and flooding that will influence human society.

POLITICS, CITIZENS, SCIENCE, AND WATER

Given the undeniable importance of water to human affairs, the fact that water issues are political is scarcely surprising. Globally, water could well become the first limiting resource for humanity (Dodds, 2008). More than two billion people on earth live in areas where water is scarce (Oki and Kanae, 2006) and more than a billion lack adequate safe drinking water (Pimentel et al., 2007). Locally, many communities are limited by water. Most major cities on Earth are located near water because people need an adequate supply, and rivers and lakes served as primary transportation avenues before road, rail, and air travel were available. Wetlands also have historic importance; the ancient Chinese fueled one of the world's great civilizations for centuries with rice harvested from wetlands.

Now, particularly in arid regions, politics and water are deeply intertwined. In the arid western United States, water law is convoluted and water rights transcend most other property rights. Communities appropriate water from far away, and subsidize its use to encourage development. However, water is not just limiting in the United States, and conflicts over water have occurred throughout recorded history (Gleick, 2008). Rates of conflicts involving water are increasing globally (Fig. 1.9). Most of these conflicts revolve around controlling or harming water supplies as a military target or in border disputes. Cut-off of water was used as a military tool as early as 2,500 BC, and is still used to this day. Violent clashes have repeatedly occurred over water, including between Ethiopia and Kenya in 2006, between villagers in Kashmir India in 2002, between Chinese farmers in 1999 (with riot damages reaching $1 million), and even between a government and its people, as in 1993 when the Iraqi government started draining the Mesopotamian marshes to punish people living in the area. Political conflict over water occurs across borders including India—Pakistan, Israel—Jordan, and Mexico—United States. The degree of conflict will only increase with a growing population vying for a finite amount of water and water pollution and exploitation decreasing the availability of that which is clean and accessible.

As we harm the quality of water, less is suitable to support native biota and for human uses. When water is a more limited commodity, its market value swells, even more so as the economic values associated with ecosystem goods and services provided

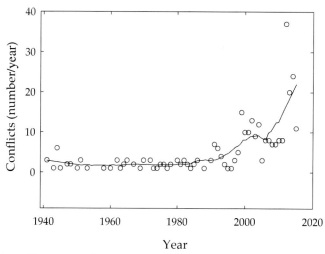

Figure 1.9 Number of human conflicts over water per year. *Data from http://www2.worldwater.org/ conflict/list/, downloaded December 2016.*

by water are increasingly recognized. The United States' Clean Water Act, the European Union's Water Framework Directive, India's system of Water Quality Standards, and the Law of the People's Republic of China on Prevention and Control of Water Pollution are just a few examples of how governments are attempting to protect water quality. Such protection requires sound scientific understanding of how aquatic ecosystems function and how humans influence that function.

So far, the history of water resources for wealthier countries has been to impair and then repair (Vörösmarty et al., 2015). In this model, countries develop by impairing their water, then follow with costly remediation and restoration of their water resources. Less developed countries are able to impair, but often not able to repair damage done to their water resources given the costs. In the best possible outcome, the causes of impairment are treated rather than the symptoms, before the system is damaged. For example, it is less costly for society as a whole to keep organic toxic chemicals from entering water than it is to dredge and remove sediments from contaminated waterways after the pollution enters the system. We address both causes and symptoms of water impairment throughout this book.

One positive development is the global expansion of citizen science as a tool to assess environmental quality (Conrad and Hilchey, 2011). For example, a global network of nutrient collections has allowed comparison of urban streams in South, Central and North America (Loiselle et al., 2016). More detailed analyses have brought in volunteers from corporations to help practice environmental stewardship in Brazil (Cunha et al., 2017). Similar efforts are ongoing in China (Zhang et al., 2017). The application of bringing the public into environmental monitoring has the benefits of not only collecting valuable data, but also educating the public on technical aspects of water quality. Well-trained leaders and programs that are scientifically sound form the backbone of these initiatives.

Managing aquatic resources successfully requires scientific understanding of freshwaters. Through scientific study of freshwaters, students may come to understand the beauty and complexity of a crucial part of the natural world. Part of studying freshwaters is learning the intrinsic value, in addition to the utilitarian value, of water. We hope that this text will serve as a resource for future stewards of our freshwaters.

SUMMARY

1. Clean water is essential to human survival, and we rely most heavily on continental water, including streams, lakes, wetlands, and groundwater.

2. The global renewable supply of water is about 39,000 km^3/year, and humans use about 54% of the runoff that is reasonably accessible. Thus, clean water is one resource that will be limited severely with future increases in human population

size and standard of living. Local problems with water quality and supply may lead to political instability.

3. Economic analysis of the value of clean water is difficult, but factors to consider include the value of clean water for human use, the value of fisheries, and the recreational use of aquatic habitats. The global benefits of these uses translate into hundreds of billions of dollars each year. Intangible benefits include preservation of nongame species and native ecosystems as well as protecting cultural values.

4. The study of the ecology of inland waters will lead to more sound decisions regarding aquatic habitats and provide a solid basis for future research.

QUESTIONS FOR THOUGHT

1. Why are you interested in studying aquatic ecology, and is such study important?
2. What is the difference between fluxes and compartments in water cycles, and what types of units are typically used to describe them?
3. What are some potential economic benefits of maintaining water quality?
4. What are the potential dangers of approaching conservation of aquatic resources from a purely economic viewpoint?
5. Who does freshwater really belong to?
6. List three "trade-offs" that are potentially involved in protecting native species in regulated rivers by attempting to mimic natural water discharge patterns.
7. What types of data do you think that "citizen scientists" can collect that could be valuable to scientists and decision makers, and what types of data collection should be left to more trained individuals?

CHAPTER 2

Properties of Water

Contents

Figure 2.1 Bubbles formed in ice. This image represents several properties of water. Surface tension forces the bubbles to be spherical, and gas that can dissolve in liquid water cannot in solid water. *Courtesy of Steven Lundberg.*

Unique physical properties dictate how water acts as a solvent and how its density responds to temperature. These physical properties have strong biological implications and knowledge of water's characteristics forms the foundation for aquatic science. Physical properties of water are so central to science that they are used to define

Freshwater Ecology
DOI: https://doi.org/10.1016/B978-0-12-813255-5.00002-8

several units of measurement, including mass, heat, viscosity, temperature, and conductivity. Properties of water influence how it controls geomorphology, conveys human waste, links terrestrial and aquatic habitats, and influences evolution of organisms. In this chapter we explore how viscosity and inertia of water vary with scale, temperature, and relative velocity related to aquatic ecology. Movement of water is discussed in the final section, including how flowing water interacts with solid surfaces.

CHEMICAL AND PHYSICAL PROPERTIES

One of the many unusual properties of water is that it exists in liquid form at the normal atmospheric temperatures and pressures encountered on the surface of Earth (Table 2.1). Most common compounds or elements take the form of gas or solid in our biosphere (exceptions include mercury and some organic compounds). *Polarity* of the water molecule and *hydrogen bonding* dictate the range of temperatures and pressures at which water occurs in a liquid state as well as additional vital distinguishing characteristics. Because oxygen atoms attract electrons, the probability is greater that electrons will be closer to the oxygen than the hydrogen atoms. The angle of attachment (104.5°) between the two covalent bonds, one for each of the hydrogen atoms attached to the oxygen, means a slight positive charge near the hydrogen atoms, to one side of the molecule. This unequal distribution of charge leads to each water molecule exhibiting polarity. The negative region near the oxygen attracts positive regions near the hydrogen atoms of nearby water molecules resulting in hydrogen bonding (Fig. 2.2).

Hydrogen bonding not only becomes more prevalent as water freezes but also occurs in the liquid phase (Luzar and Chandler, 1996; Liu et al., 1996); without

Table 2.1 Some properties of water

Property	Comparison with other substances
Density	Maximum near 4°C, not at freezing point, expands upon freezing
Melting and boiling points	Very high
Heat capacity	Only liquid ammonia is higher
Heat of vaporization	Among highest
Surface tension	High
Absorption of radiation	Minimum in visible regions; higher in red, infrared, and ultraviolet
Solvent properties	Excellent solvent for ions and polar molecules. Increases for ions with increasing temperature, decreases for gases with increasing temperature

Source: Adapted from Berner and Berner (1987).

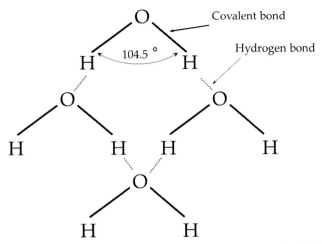

Figure 2.2 Schematic of hydrogen bonding among water molecules. The black lines represent covalent bonds; the dashed lines represent hydrogen bonds. This is an approximate two-dimensional representation. In water, three-dimensional cage-like structures are formed. In liquid water, these structures form and break up very rapidly.

hydrogen bonding, water would be a gas at room temperature. When water freezes, the molecules form tetrahedral aggregates that lead to decreased density. Thus pure ice has a density of 0.917 g/cm^3 at $0°C$, which is less dense than liquid water at any temperature (Fig. 2.3A). Lower density as a solid than as a liquid is another unusual aspect of water; most compounds are denser in solid phase than in liquid phase. There are several forms of crystal ice with varied density, but the ecological relevance of these different forms is minor.

The density of liquid water, which is influenced by temperature and dissolved ions, can control the physical behavior of water in wetlands, groundwater, lakes, reservoirs, rivers, and oceans. Density also influences water flow and viscosity. The differences in density are important because less dense water floats on top of water with greater density. Such density differences can maintain stable layers within bodies of water. Formation of distinct stable layers is called *stratification*. We discuss stratification in detail in Chapter 7 because it can control water movement and distribution of chemicals and organisms in lakes.

Maximum density of pure water occurs at $3.98°C$ (Fig. 2.3B,C). As water cools below $3.98°C$, hydrogen bonding begins to arrange the water molecules into the crystal structure of ice, leading to more space between the water molecules and a less dense fluid. As water is heated above $3.98°C$, it has a continuously greater decrease in density per degree temperature increase above $3.98°C$ (Fig. 2.3). Dissolved ions also increase water density. This density increase that occurs at ionic concentrations found

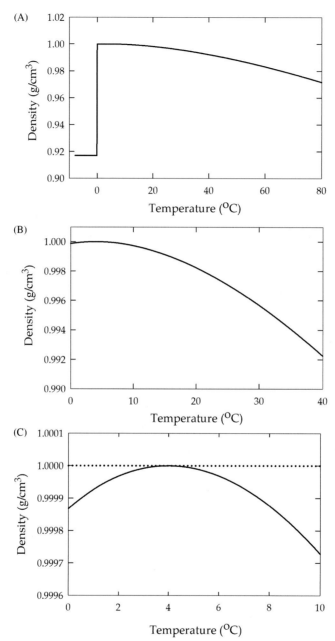

Figure 2.3 The density of water as a function of temperature from freezing to 80°C (A), from 0 to 40°C (B), from 0 to 10°C to focus on the region of maximum density (C). At 0°C, ice forms with a density of 0.917 g/mL. *Data from Cole (1994).*

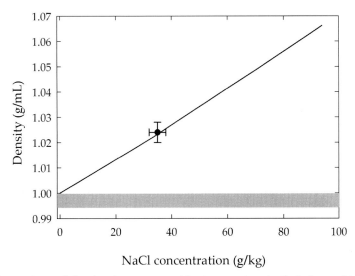

Figure 2.4 Comparison of density change caused by temperature to that changed by salinity. The range of variation in density with temperature over about 40°C is represented by the grey box. Seawater has an approximate salinity of 3.5% (indicated by the point bounded by error bars); saline lakes can exceed this value many times. *Data from Dean (1985).*

in some natural saline lakes, can easily overcome or enhance temperature effects on stratification (Fig. 2.4).

Water is also one of the best *solvents* known and can dissolve both gases and ions. The solvent properties of water have greatly influenced geologic *weathering* of the earth's surface by dissolving ions from rocks. Weathering is the natural source to the biosphere of all nutrients that do not have gaseous form under normal conditions. Weathering also alters geomorphology. For example, about 20% of the continental land is karstic terrain (White et al., 1995), a geological formation caused by rainwater dissolving limestone and leaving very rough land topography.

Most solids dissolve in water more readily as temperature increases (i.e., the rate at which a compound dissolves into water). For instance, this temperature effect on dissolved ions causes sugar to dissolve more quickly in hot tea than in iced tea. Most compounds, but not all, have higher solubility (saturation concentrations) at greater temperatures as well. An exception to this generality is that many carbonate minerals are less soluble in hot water than in cold water.

The *solubility* of gases in water tends to decrease when temperature increases (see Fig. 12.7). Decreased solubility at high temperature is why effervescence is enhanced when a warm beer is opened compared with a cold beer. Temperature's influence on gas solubility can have significant biological consequences; aquatic animals such as fishes are more likely to die of low dissolved oxygen (O_2) stress when water temperatures are elevated because O_2 is held in warm water and the animal's metabolic requirements for O_2 increase as temperature increases.

Additional properties of water include high heat capacity, heat of fusion (freezing), heat of vaporization, and surface tension. Water has a high *heat capacity*, that is, it takes a relatively large amount of energy to increase the temperature of liquid water. To illustrate, specific heat capacities (in calories required to change the temperature of 1 g of a substance by 1°C) are 1, 0.58, and 0.21 for water, ethanol, and aluminum, respectively. Similarly, *heat of fusion* and *vaporization* are high for water compared with other liquids (Table 2.2). A high heat capacity and heat of fusion means that a considerable amount of solar energy is required to heat a lake in the summer, and a long period of cold weather is required to freeze the surface of a lake. The slow heating and cooling of water is evident in the 5- to 7-day lag time that is often evident between air temperatures and stream water temperatures in many regions. High heat capacity buffers water against rapid changes in temperature. Thus aquatic organisms generally do not experience the rapid temperature swings terrestrial organisms are subjected to (and generally are not exposed to temperatures <0°C or >100°C). Many aquatic organisms are very sensitive to even small changes in temperature relative to terrestrial organism because they evolved in thermally buffered environments. For example, many aquarium fishes are not adapted to temperature swings and must be acclimated to new tanks slowly, so the rapid temperature change does not shock and kill them. Temperature sensitivity could also make many aquatic organisms highly susceptible to predicted climate changes (Firth and Fisher, 1992; Heino et al., 2009); global climate change is predicted not only to increase global mean temperatures but also to increase climatic (temperature) variability as well.

A high heat of vaporization means that a considerable amount of energy is needed to evaporate water. Humans and other terrestrial animals take advantage of the heat of vaporization by perspiring; the evaporation of the moisture cools the skin (takes away energy). Surface waters are also cooled by evaporation. Because of its high heat capacity and heat of vaporization, water, in any form, is an important determinant of local,

Table 2.2 Heats of fusion, vaporizations, heat capacities, and surface tensions of various liquids (Keenan and Wood, 1971; Weast, 1978)

Substance	Heat of fusion (cal/g)	Heat capacity at 25°C (cal g^{-1} C^{-1})	Surface tension at 20°C (dyne/cm)	Viscosity (cP)	Heat of vaporization (cal/g)
Water	79.7	1.00	73	1.00	539.6
Benzene	30.3	0.41	40	0.65	94.3
Mercury	2.8	0.03	435	1.55	67.8
Oxygen	3.3	—	—	—	50.9

regional, and global climate. Global climate models must take heat of vaporization and heat capacity into account.

Another important aspect of water is *surface tension*. The high surface tension of water results from hydrogen bonding, which pulls water into a tight surface at a gas—water interface. Several organisms, such as water striders, take advantage of this surface tension to walk on the surface of water (Fig. 2.5). Some lizards (*Basiliscus* and *Hydrosaurus*) also run across the surface of water using the support of surface tension (Vogel, 1994). Special adaptations to walking on the water surface have evolved in different organisms and will be discussed in Chapter 15 further (Hu and Bush, 2010). Surface tension is also important to animals that live below the water surface; the respiratory siphons that many aquatic insect larvae use to breathe atmospheric air require surface tension to function, nor would it be possible for aquatic beetles and true bugs to use air bubbles pulled from the surface as a source of oxygen without surface tension. These structures function because of the tendency of water molecules to stick to each other (*cohesion*) and to other surfaces (*adhesion*).

The influence of surface tension also comes into play in many unexpected ways, such as when water droplets form spheres. Surface tension determines the size of the largest possible drop of water. Most rain drops are 1—2 mm diameter, but a few as big as 10 mm have been observed. Under microgravity, drops as large as 3 cm are stable (Pennisi, 2014). Water droplets can be important in transmission of disease. Finally, surface tension leads to *capillary action*, the ability of water to move up narrow tubes. Capillary action is central to forming the capillary fringe (the moist zone in sediments

Figure 2.5 *Gerris lacustris*, a water strider, using surface tension to stand on the water. Image by Henk Monster, Wikimedia commons.

immediately above groundwater) because water creeps up the narrow spaces between sediment particles. Wetland plants with leaves above the water surface also use capillary action to move moisture up their stems to their leaves.

An interesting adaptation related to capillary action occurs in small surface-walking insects when they move from the surface of the water to land. In areas where water meets solid objects, the surface tension and capillary action of the water causes a *meniscus* (an up-sloping surface where water meets a hydrophilic solid). This meniscus can be observed by looking closely at a clear glass of water at the edge where the surface of the water meets the glass. The meniscus is a difficult barrier to overcome for very small organisms trying to leave the water surface because it is an up slope with nothing solid to push against. However, they can position their bodies so that capillary action pulls them up the meniscus without them providing propulsive force from their legs (Hu and Bush, 2005). These animals essentially pull up on the water sloping up in front of them with their forelegs or front part of their body, and push down on the surface of the water in their middle, and capillary action moves them forward up the meniscus.

Finally, pressure increases with depth, and solubility of gases is a function of pressure. Scuba divers must pay attention to high gas solubility at higher pressure because rapid ascent from depth allows gas bubbles to form from gas dissolved in the blood. This is known as the bends and can be fatal. Fish and other animals are also susceptible to the bends, and this is part of the reason why most fish caught in deep water will die even if released carefully. The fact that gases can dissolve at much higher concentrations under pressure explains how carbon dioxide can become supersaturated at depth in lakes (important in the Lake Nyos disaster explained in Chapter 13) and why fish that are subject to water released from deep waters in reservoirs can be harmed by the supersaturated gases (relative to atmospheric pressure). Pressure increases linearly with depth because the pressure is caused by the mass of water above the object.

$$P_x = P_{atm} \, \rho \, x \, g$$

where P_x is the pressure at depth, P_{atm} is the atmospheric pressure, x is the depth, ρ is the density, and g is the force of gravity. In practical terms, a body is at approximately twice atmospheric pressure at a depth of 10 m below the surface, pressure will double again (4 times atmosphere) at 30 m.

ADVANCED: THE NATURE OF WATER

The physical chemistry of water has advanced considerably because of the instrumentation that is available to study it. Properties of water are based in large part on how the individual molecules interact with their nearest neighbors, but also on the fact that hydrogen ions in acidic solutions move rapidly from one molecule to another (Thämer et al., 2015; Wolke et al., 2016). The nature of liquid water is determined by the hydrogen bond and van der Waals forces (Schmid, 2001). Apparently several types of hydrogen bonding are present simultaneously in liquid water, and there is rapid fluctuation among the types. Hydrogen bonds fluctuate on temporal scales between 10^{-14} and 10^{-11} seconds (Tokmakoff, 2007). The bonds are seemingly traded among water molecules (like handing off a baton in a relay race) rather than broken and reformed at a later time. The true nature of a hydrogen bond is still not yet completely understood (Tokmakoff, 2007). X-ray data suggest that in liquid water, most molecules are found in chains or rings formed by relatively strong hydrogen bonds within more diffuse networks of weaker hydrogen bonds (Wernet et al., 2004). The regions flip back and forth from strong to weaker bonding, and there are regions of tight hydrogen bonding that are maintained even at water temperatures close to boiling (Huang et al., 2009).

Hydrogen bonding forms small hexamers. These six-molecule objects can take any of three isomeric forms (Pérez et al., 2012). Eventually the hope is to create a predictive model of water in all its forms if the small-scale structure of water can be understood. These hexamers may be particularly important when water is confined in very small spaces (e.g., in layers only a few molecules thick such as might occur in fine sediments) and require understanding of quantum properties of water (Richardson et al., 2016).

The hydrophobic effect, the inability of some molecules to dissolve in water, is a function of the small size of the molecule and how the solutes interact with the hydrogen bonding. The exact mechanism behind this effect is still not well understood (Granick and Bae, 2008). The concept of hydrophobicity is extremely important in materials science, but has obvious biological importance as well. For example, protein folding and dissolving depends upon this effect. The capillary effect also depends upon hydrophobicity. A more complex definition of polarity is also suggested by study of what will and will not dissolve in water (Schmid, 2001). The definition used by advanced physical chemists is more complex than what is taught in introductory chemistry classes. These properties may seem esoteric to aquatic ecology students, but they do determine how processes like weathering and transport of nonpolar pollutants occur.

RELATIONSHIPS AMONG WATER VISCOSITY, INERTIA, AND PHYSICAL PARAMETERS

Viscosity is the resistance to change in form, or a sort of internal friction. *Inertia* is the resistance of a body to a change in its state of motion. Water viscosity increases with smaller spatial scale, greater water movement, and lower temperature. Inertia increases with size, density, and velocity. These facts are underappreciated but are very biologically and physically relevant to aquatic ecology. Consequences of these physical properties include, but are not limited to (1) why fish are streamlined, but microscopic swimming organisms are not; (2) why the size of suspended particles captured by filter feeding has a lower limit; and (3) why organisms in flowing water can find refuge near solid surfaces. Many aspects of these features of life in aquatic environments can be discussed conveniently using the *Reynolds number* (*Re*). This number can quantify spatial- and velocity-related effects on viscosity and inertia. The effects of viscosity and inertia and other properties of water on organisms have been described eloquently and in greater detail (Purcell, 1977; Denny, 1993; Vogel, 1994). Here, we attempt to describe water's physical effects briefly, using the Reynolds number as the basis of the discussion.

Relative viscosity increases and inertia decreases as the spatial scale becomes smaller. Viscosity increases because the attractive forces between individual water molecules become more important relative to the organism. Thus the influence of individual water molecules is greatest when organisms are small or the space through which water is moving is small. We discuss the individual components of *Re, inertial force* and *viscous force*, and then provide an example calculation using these relationships. Mathematically, the ratio of inertia to viscosity is the Reynolds number:

$$Re = \frac{F_i}{F_v}$$

where F_i is the inertial force and F_v is the viscous force.

The equation for inertial force (F_i) is

$$F_i = \rho \, S \, U^2$$

where ρ is the density of the fluid, S is the surface area of the object, and U is the velocity of the fluid moving past the object (or the object moving through the fluid).

Inertial relationships can be stated in familiar terms: the faster the object, the greater the inertial force. A slowly pitched baseball does less damage to a batter than a 44 m/s (100 mph) fastball. Of course, larger objects have more inertia; a splash of water from a cup imparts less force than the splash from a bucket of water propelled at the same velocity, and a moderate sized wave can easily knock a person down.

The properties of inertia constrain aquatic organisms. For example, at small scales inertia is generally unimportant. A bacterium will coast one-tenth the diameter of a hydrogen atom if its flagellum stops turning (Vogel, 1994). In contrast, a large fish can coast many body lengths if it stops swimming. Low inertia at small scales also means that turbulence is less likely (i.e., individual parcels of water have less inertia).

The other part of the equation to calculate Reynolds number is viscous force (F_v):

$$F_v = \frac{\mu \, S \, U}{l}$$

where μ is the *dynamic viscosity* of the fluid, a constant that describes the intrinsic viscosity of a fluid (e.g., corn syrup is intrinsically more viscous than water), and l is the length of the object.

Again, certain aspects of this relationship are intuitive. The viscous force can be thought of as a frictional force. Increasing velocity increases viscosity. Water feels more viscous to a person wading up a stream than one wading in a still pool. Dynamic viscosity related to properties of a fluid (i.e., μ) may also be important. Swimming completely immersed in tar would be much more difficult than swimming in water (although maybe not as difficult as it would seem as determined by comparing people swimming in water and syrup in TV's "Myth Busters"). Surface area also influences viscous force. Pulling a large object (with a large frontal surface area) through water is more difficult (takes more force) than pulling one with a small frontal surface area. An object with a large frontal surface area and short length is more difficult to pull through water than one with a small frontal surface area and long length; pulling a boat sideways through the water is more difficult than pulling the same boat bow-first.

Viscous force is dependent upon spatial scale. One way to envision this force is to think of the hydrogen bonds in water as a spider web. A very small object such as a fly is heavily influenced by the web, and a larger object such as a human can barely perceive any resistance when walking through the web. Similarly, small particles such as motile bacteria experience substantial viscosity when moving through water and smaller particles take longer to settle out of water because they experience greater viscous force than do larger particles.

Environmentally related variation in dynamic viscosity can have major effects on aquatic organisms because dynamic viscosity (μ) is greater when temperature is colder (Fig. 2.6). Thus, it requires more energy for a fish or a protozoan to swim in cold water than in warm water, and it is more difficult for animals to filter out small particles at lower temperatures (Podolsky, 1994). As a result, many temperate fishes become inactive in winter and small filtering invertebrates such as zooplankton enter resting stages. Again, thinking about the spider web analogy helps here; if the water is colder, the hydrogen bonding is more prevalent. More hydrogen bonding would be equivalent to a stronger, more densely woven spider web that would be more difficult for a small object to pass through.

Effects of dynamic viscosity and viscous force are diverse and include: constraints on (1) how aquatic organisms collect food, (2) how fast organisms swim (a bacterium with a cell length of 1 μm experiences viscous forces in water similar to a human

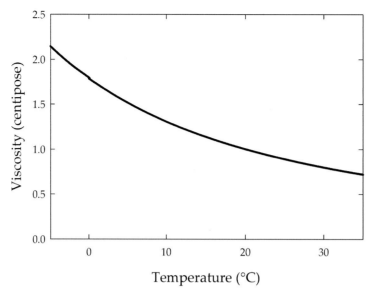

Figure 2.6 Viscosity as a function of temperature. Note that viscosity doubles when temperature drops from 30°C to 0°C (i.e., a range of temperatures across seasons in temperate surface water). *After Weast (1978).*

swimming in tar), (3) when natural selection favors streamlined organisms, (4) how quickly particles settle in water, and (5) how fast groundwater flows. For example, when groundwater is moving through two sediment types that have the same surface area of channels through which fluid can flow but one has more 1-μm diameter pores and the other has fewer pores of 5-μm diameter, the water velocity is slower through the sediment with the 1-μm pore diameter holes. The velocity is lower because the water is subject to greater friction while flowing through the smaller pores. Mesh sizes of filter-feeding organisms are constrained, in part, by this physical fact (Nielsen et al., 2017). The first three constraints can be used to functionally classify planktonic organisms (Kruk et al., 2017).

If we put the equations for inertia and viscosity together and cancel, we get the equation for Reynolds number:

$$Re = \frac{F_i}{F_v} = \frac{\rho \, S \, U^2}{\mu \, S \, U/l} = \frac{\rho \, U \, l}{\mu}$$

The Reynolds number is greater at large spatial scales (a large fish) than at small scales (a bacterium or protozoan) (Fig. 2.7). Calculations of this number (Example 2.1) reveal the wide variations in viscosity and inertia experienced between large and small organisms. A summary of the effects of scale related to *Re* is provided in Table 2.3. Reynolds numbers will be considered again when we discuss filter feeding by

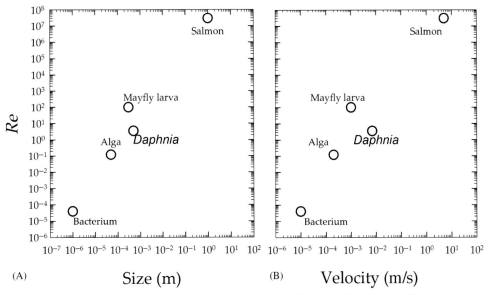

Figure 2.7 Reynolds number as a function of size (A) and velocity (B) for a variety of aquatic organisms. Note the log scales. *Data from Vogel (1994).*

Example 2.1 Reynolds number calculations

Calculate viscous force (F_v), inertia (F_i), and Reynolds (Re) number for two cubes, one 1 μm and the second 1 cm on a side, each moving at 2 lengths/s, given that $\rho = 1 \times 10^6\,\text{g/m}^3$ and $\mu = 1\,\text{g m}^{-1}\,\text{s}^{-1}$. If turbulent flow is more likely to occur above $Re = 100$, which cube would have its hydrodynamic properties altered more by streamlining?

Parameter	Comment	1 cm cube	1 μm cube
L	length of cube	10^{-2} m	10^{-6} m
S	square of length	10^{-4} m^2	10^{-12} m^2
U	2 lengths/s	2×10^{-2} m/s	2×10^{-6} m/s
F_i	from equation	4×10^{-2} g m s^{-2}	4×10^{-18} g m s^{-2}
F_v	from equation	2×10^{-4} g m s^{-2}	2×10^{-12} g m s^{-2}
Re	F_i/F_v	200	2×10^{-6}

Note that inertia is very, very small for the small cube, leading to a much smaller Reynolds number. The 1 cm cube is above $Re = 100$, and would have its hydrodynamic properties altered more than the 1 μm cube if both were streamlined.

Table 2.3 Contrasting effects of scale and Reynolds number on aquatic organisms

Parameter	Small organism (<100 μm)	Large organism (>1 cm)
Re	Low	High
Viscous force (F_v)	High	Low
Inertial force (F_i)	Low	High
Flow	Laminar or none	Turbulent
Body shape	Variable	Streamlined
Diffusion	Molecular	Transport (eddy)
Particle sinking rates	Low	High
Relative energy requirement for motility	High	Low

organisms in lakes and streams, microbial food webs, production of aquatic macrophytes, and flow of water in streams and groundwaters.

MOVEMENT OF WATER

At the very smallest scale, molecules move independently in the process called *Brownian motion*. The warmer the water, the more rapidly the molecules move. The average instantaneous velocity of individual water molecules is extremely rapid (>100 m/s), but because they continuously collide, individual molecules move from any location slowly (50×10^{-9} m/s, Denny, 1993).

We experience this movement of water molecules as water temperature. The faster the water molecules move, the warmer the water is. As such, many chemical reactions occur more and more quickly when water warms because solutes move through the water to react with each other more quickly and when they do interact have more energy and are more likely to have enough energy to overcome the activation energy. Roughly, chemical interaction rates double for every 10°C. This leads to an exponential rate of reaction increase (i.e., 2 times greater for first 10°C, but 20°C leads to 4 times greater). All organism metabolic rates depend on temperature, and most aquatic organisms are isothermal with the water they live in. Among other things, this means that the effects of climate warming will not be linear.

Water flows (bulk movements of many molecules) on larger spatial scales and many methods have been used to measure water velocity (Method 2.1). Flow can be either laminar or turbulent. *Laminar flow* is characterized by flow paths in the water that are primarily unidirectional (parallel to each other). *Turbulent flow* is characterized by eddies, where the flow is not as unidirectional. Turbulent flow (mixing) decreases at small scales because viscosity dampens out turbulence as the Reynolds number decreases (below a value of approximately 1). Richardson (1922, p. 66) noted this cascade of turbulence into the viscous molecular realm in a more poetic way "We realize

METHOD 2.1 Methods used to measure water velocity

The simplest, albeit somewhat inaccurate, way to measure water velocity on a large scale is to place an object that barely floats into moving water and measure the amount of time it takes to move a known distance. For example, an orange is often used because it floats just at the surface of the water and is a bright visible color. Experiments in our freshwater ecology classes confirm that apples and coconuts move at the same velocity as oranges. Velocity estimates obtained with this method can be corrected for streams with a rough bed (multiply the estimate by 0.8) or smooth bed (by 0.9). However, this method is less accurate than methods described below because it is sensitive to wind speed and surface water velocity is only loosely related to velocity of the entire water column.

For flows in water >5 cm deep, propellers are often used to estimate velocity. The more rapidly water is moving, the more rapidly a propeller spins. Given an electronic method to count the revolutions per unit time of a propeller and suitable calibration constants, water velocity can be estimated. Electromagnetic flow meters measure the electrical current that is induced when a conductor is moved through a magnetic field. Because water is a conductor, a flow meter can be constructed to create a magnetic field, and the electrical current increases proportionally as water velocity increases.

For smaller scale (several centimeters or less) measurements of water velocity, other methods are more useful. Very small particles or dye can be suspended in flow, and movement can be timed along a known distance. Particle movement can be measured by photographing the moving particles using a flash of known duration. The length of the particle path on the photographs can be related to the flash duration, and water velocity can be calculated. Pitot tubes are small tubes with one end extending above the surface of the water and that have a 90° bend, which allows the open end of the tube that is under water to be positioned facing upstream. As water velocity increases, the pressure at the end of the tube increases, and the height of the water in the tube above the external water level increases proportionally to water velocity. Pitot tubes are inexpensive but are not sensitive to low water velocity and can foul easily. Head rods or head sticks use this same principle and are very inexpensive and easy to use. These devices are essentially meter sticks with blunt and sharp ends; the difference between depth readings taken with the blunt and sharp sides facing into the current is the hydraulic head and is proportional to the velocity. As with pitot tubes, simple equations are applied to field measurements to convert head to velocity.

Hot film, wire, or thermistors can be used to measure water velocity. These devices operate on the principle that moving water carries heat away from objects. Electronic circuitry can be used to relate water velocity to the cooling effect on the heated electronic sensor. These heat-based devices are most useful for biological and ecological investigations of the effects of water velocities on organisms, when low velocity and small spaces are most common.

Acoustic Doppler velocity (ADV) and laser Doppler velocity (LDV) methods have been used to measure water velocity. These methods rely on reflection of sound and light waves, respectively, from small particles in the flowing water. ADV and LDV are very useful because they allow determination of velocity in all three directions. They also allow very short-duration measurements (many per second) on very fine spatial scales (sub mm). Drawbacks include inability to sense velocity in confined spaces and high equipment costs. Modern units are often combined with global positioning units and can be used to map velocity with depth and create two-dimensional plots of this velocity (Fig. 2.8).

(Continued)

METHOD 2.1 (Continued)

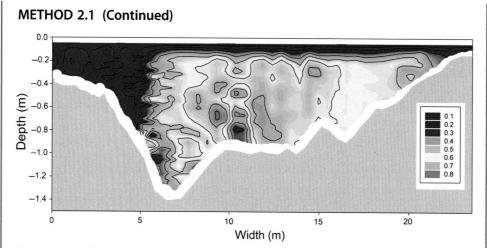

Figure 2.8 A flow velocity profile contour of the Kansas River measured with acoustic Doppler velocity equipment. The white area is too close to the sediment for the equipment to measure velocity. Velocities are in m/s.

thus that: big whirls have little whirls that feed on their velocity, and little whirls have lesser whirls and so on to viscosity."

Surfaces interact with flowing water. Water velocity slows, turbulence is damped and flow becomes laminar near solid surfaces. Eventually, within a few microns of a surface, the primary movement of water is Brownian motion. The equation for F_v indicates that viscous force increases closer to surfaces (as spatial scale decreases). Thus friction with the solid surface is transmitted more efficiently (adhesion is transmitted by cohesion) through the solution at small spatial scales near the surface (Fig. 2.9). Water also flows more slowly, and flow becomes more laminar, at the bottom and sides of stream channels and pipes or near any object that is moving through water. The outer edge of the region where water changes from laminar to turbulent flow is called the *flow boundary layer*.

The thickness of the flow boundary layer increases with decreased water velocity, increased roughness of the surface, increased distance from the upstream edge of an object, and increased size of the object (Fig. 2.10). The flow boundary layer is not a sharp, well-defined line below which no turbulence occurs, even though it is convenient to conceptualize the layer in such a fashion. Rather, the outer edge of the boundary layer represents a transitional zone between fully turbulent flow and laminar flow, and is characterized by patches of turbulence (Barkley et al., 2015). The relationships among flow boundary layer thickness and other factors have many practical aspects. For example, algal growth can have significant influences on hydrodynamic

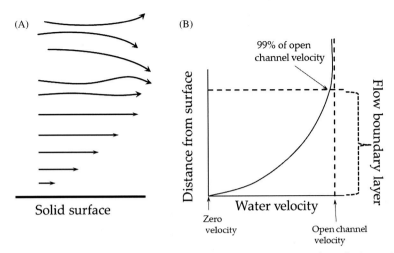

Figure 2.9 The concept of a flow boundary layer. (A) Arrows represent the velocity and direction of water flow. Inside the flow boundary layer, flow is approximately laminar and slows near the surface; outside the layer, turbulence increases. (B) The outer region of the flow boundary layer is where velocity is 99% of that in the open channel. Very close to the solid surface, water velocity approaches zero. *Modified from Vogel (1994).*

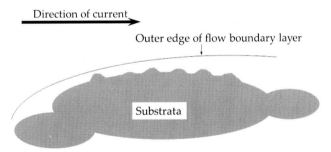

Figure 2.10 Schematic of thickness of the flow boundary layer as a function of surface roughness and distance from leading edge. Water is flowing from left to right. Picture the substrata as a rock in a stream. The thickness of the boundary layer increases with distance from the leading edge and is shallower over bumps and deeper over depressions.

conditions several centimeters from the bottom and edge of a stream (Nikora et al., 1997). Many aquatic animals and organisms can find refuge from high water velocities in cracks in rocks because the cracks occur inside the flow boundary layer and have very slow water velocities. Aquatic invertebrates found in fast flowing waters often have characteristic flattened body forms that allow them to take advantage of the slower velocity microhabitats of the boundary layers.

The effects of scale on flow or movement through water have substantial practical implications. Very small objects experience laminar flow, and larger objects experience

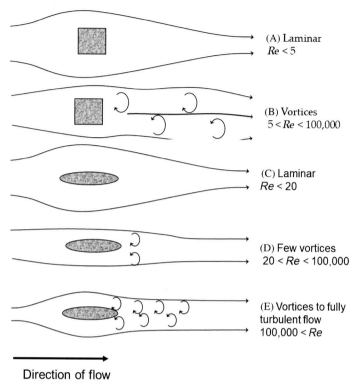

Direction of flow

Figure 2.11 Patterns of flow behind two differently shaped solid objects at three different ranges of Reynolds numbers. When the Reynolds number is low (A), turbulence is minimal. Vortices start to form with increased Reynolds number; vortices and turbulence are more prevalent with the cubic object (B and C) than with the streamlined object (D and E). Compare to Fig. 2.12.

turbulent flow. Large organisms that swim through the water (such as fishes) benefit from being *streamlined* (shaped to avoid turbulence); small organisms (bacteria sized) do not need to streamline (Fig. 2.11) because turbulence increases drag but there is no turbulence at low Reynolds numbers. The breakpoint at which streamlining becomes useful is approximately 1 mm (depending on velocity and temperature). Greater water velocities also increase turbulence (Fig. 2.12). When a large organism moves through water, turbulent flow acts opposite to the motion of the organism and exerts a force that slows the organism. Spatial scale related to flow is also important for organisms that live very close to large surfaces because they experience reduced water velocities (within approximately 0.1 mm, depending on flow velocity and surface geometry).

An interesting aspect of movement through water at low Reynolds number is the effect a wall has on microbes swimming near solid surfaces. A 1 μm-diameter bacterium swimming 50 body widths from a wall is slowed because friction with the wall is transmitted through fluid. Thus, when microbes are observed swimming in a drop of

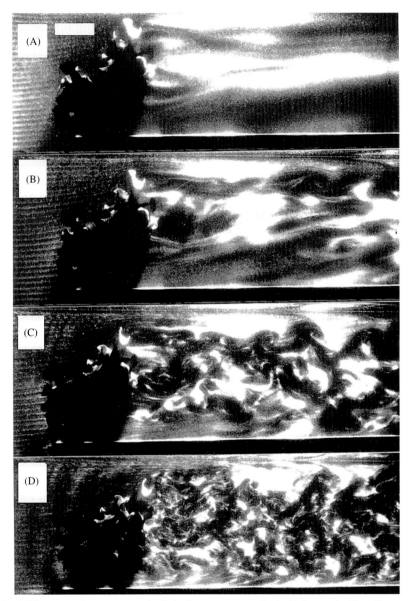

Figure 2.12 Water moving past an algal thallus at progressively higher velocities. Tracer particles allow visualization of turbulence. Velocities are 0.5 cm/s (A), 1.5 cm/s (B), 2 cm/s (C), and 3.5 cm/s (D). *From Hurd and Stevens (1997); reproduced with permission of the Journal of Phycology.*

water confined between a microscope slide and a cover slip, they are moving slower than if they were swimming freely. Also, an organism swimming in the interstices of fine sediments will have a lower velocity than it will in open water. In contrast, a fish swimming in a river is only modestly influenced by the solid surfaces it swims past.

When viewing rapidly moving protozoa through a microscope, one can artificially increase the effects of the wall and viscosity by adding a cellulose solution that further increases the viscosity, slows the organisms, and makes them easier to observe.

In contrast to very small organisms swimming, large organisms can take advantage of turbulence and vortices (Liao et al., 2003). Vortices are produced by an upstream swimming fish or can be shed by a stationary object in flow. Fishes are able to move sinuously through these vortices and use less energy than if they swim through water without vortices flowing at the same average velocity. Very rapid and sensitive detection of hydrostatic pressure is required to identify these vortices and react appropriately; this is a reminder that aquatic animals have a long evolutionary history and sense and experience water very differently than humans.

Reception of hydromechanical signals by small swimming crustaceans (e.g., copepods) is also related to friction transmitted through water at small scales. Copepods can sense small moving prey particles and move toward them, and they can sense predators (larval fishes) and move away from them (Kiørboe and Visser, 1999). Copepods can react to moving predators slightly less than 1 cm away by making several long "jumps" away from the source of the hydrodynamic disturbance (Kiørboe et al., 1999). The actual response distances of the copepods to predators and prey are a function of velocity and relative size; complex models have been proposed to describe the effect (Kiørboe and Visser, 1999).

ADVANCED: EQUATIONS DESCRIBING PROPERTIES OF MOVING WATER

Engineers and hydrologists have developed a sophisticated body of knowledge of the mechanics and influences of moving water or organisms moving through water. We discuss the fundamental metrics of water movement and their ecological relevance (Table 2.4) and some equations that describe moving water and other important aspects (Table 2.5) in this section. Most of these equations can be related, in part, to the properties discussed with respect to Reynolds numbers.

Shear stress (tractive force per unit area) can be used to calculate the force on attached organisms or sediments on the bottom of a flowing river or stream.

$$\tau = \rho \frac{U}{l}$$

where τ is the shear stress, ρ is the density of the fluid, U is the velocity, and l is the distance from the object into the flowing water where full velocity is reached. The steeper the velocity gradient or the greater the density of the fluid, the more shear force is present. Organisms that live in fast flowing rivers are subjected to high shear stress, which has necessitated the evolution of adaptations such as flattened body forms,

Table 2.4 Overview of some additional parameters used to describe properties of fluids

Parameter	Symbol	SI units	Aquatic ecological relevance
Discharge	Q	m^3/s	Downstream loading, variance characterizes flood and drought
Velocity	U	m/s	Any time an organism or the water it inhabits moves
Force	F	N (kg m s^{-1})	What causes object to change direction of motion
Pressure	p	N/m^2	Important in determining the solubility of gases, and the effect of depth on organisms
Shear stress	τ	N/m^2	Force on attached benthic organism
Dynamic viscosity	μ	N m s^{-1}	Intrinsic viscosity of a fluid
Density	ρ	kg/m^3	Alters stratification, sinking rates
Kinematic viscosity	ν	m^2/s	Describes how easily fluids flow
Surface tension	σ	N/m	Capillary rise, spread of floating materials, ability to use water surface, size of bubbles

Table 2.5 Some numbers used to characterize fluid dynamics

Parameter	Symbol	Relevance for aquatic ecology
Reynolds number	Re	Allows scaling of viscosity and inertia[a]
Capillary rise	H	Calculate the height a liquid will move up through soil as a function of water density, degree of wetability of solid phase, and density of water[b]
Sinking velocity	U	Stokes law for small spherical objects[b]
Darcy's law	U	Velocity of fluid through porous media, groundwater flow[b]
Dalton's law	E	Rate of evaporation, for heat and water budgets
Drag		Amount of force required to move an object through a fluid[b]
Lift		Amount of lift a fin or attached organism experiences[b]
Froude number	Fr	Wave behavior at high velocity, floating objects moving across water surface[a]
Sherwood number	Sh	Advection to sheer across diffusion boundary layer[b]
Péclet number	Pe	Ratio of the convection to molecular diffusion of heat[b]
Schmidt number	Sc	Ratio of the convection to molecular diffusion of mass[b]

[a]Vogel (1994).
[b]Denny (1993).

attachment hooks, or suction discs; larval net-winged midges (family Blephariceridae) have six ventral suction discs located on their abdomen that allow them to persist in high shear stress habitats including cascades and waterfalls (Fig. 2.13).

Calculation of the sinking rates of particles in water provides a good example of the ramifications of properties of water and scale in aquatic habitats. In general, larger

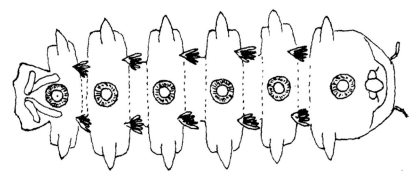

Figure 2.13 Ventral view of a larval net-winged midge (Order Diptera, family Blephariceridae) showing the suction discs that allow them to live in high water velocities in streams. *Reproduced with permission from Thorp and Covich (2001).*

and denser particles sink more rapidly. This relationship can be calculated for small spherical objects using Stokes law:

$$U = \frac{2gr^2(\rho' - \rho)}{9\mu}$$

where U is the velocity, g is the gravitation acceleration, r is the radius of the sphere, ρ' is the density of the sphere, ρ is the density of the liquid, and μ is the dynamic viscosity. The relationship between sinking rate and density of spheres as a function of size can be seen in Fig. 2.14. This relationship can be used to predict how long particles will remain suspended in water (Example 2.2). Denser, larger objects sink more rapidly.

Objects that deviate from spherical form can sink more slowly. *Melosira* is an alga that lives suspended in water. It has cylindrical cells and increases in size by adding the cylindrical cells end to end (like stacking barrels end-to-end). Thus the volume of the colony can amplify with a much greater increase in surface area than if the colony was spherical. The relationship for viscous force (F_v) predicts that adding cells to a filament will lead to an increase in viscous force and thus a slower sinking rate relative to a sphere (Fig. 2.14). Slower sinking allows *Melosira* to stay in the lighted water column and grow. However, longer chains might also decrease losses to predation (Bjærke et al., 2015), lending ecological complexity to the problem.

Drag and lift are related aspects of life in moving water or life moving in water. The equation for drag is:

$$\text{drag} = \frac{1}{2}\rho\, U^2 S\, C_d$$

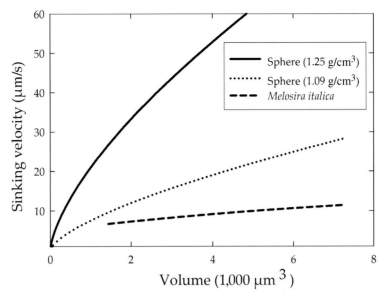

Figure 2.14 Sinking velocities in water of spherical particles with two different densities and of a filamentous diatom, *Melosira*, as a function of volume. The diatom can be found suspended in water and has a density of approximately 1.2 g/cm³. *Data from Reynolds (1984).*

Example 2.2 Calculating the sinking rates of two spheres through 10 m of water

There are two spherical objects, one with a radius of 1 μm (about the size of a bacterium) and the other with a radius of 100 μm (the radius of a large spherical alga). Both have a density of 1,100 kg/m³. Assume the density of water is 998 kg/m³, gravitational acceleration is 9.8 m/s², and viscosity is 0.001 kg m⁻¹ s⁻¹. Calculate the time it would take for each object to sink through 10 m of still water. First, calculate the velocity of each sphere using Stoke's equation. The small and large spheres have velocities of 2.2×10^{-7} and 2.2×10^{-3} m/s, respectively. At this velocity it would take the bacteria-sized sphere 1.44 years and the larger sphere 1.26 h to sink 10 m. Considering that there are few if any bodies of natural water that have no turbulent mixing, it is highly unlikely that a bacterium would ever sink out of the water.

---

where ρ is the density of the fluid, U is the velocity of the fluid, S is the frontal area of the object, and C_d is the drag coefficient. As velocity, frontal area, and density of the fluid increase, overall drag increases. The value for C_d is determined empirically, but is a function of the streamlining of the object (when flow is turbulent) and will vary with U.

If a fluid moves more rapidly on one side of an object than the other, a force of lift is generated. This can be important both for organisms that swim, and for organisms attempting to stay attached to solid surfaces with water flowing past.

$$\text{lift} = \frac{1}{2} \rho \, U^2 S \, C_l$$

where the terms of the equation are almost identical to the equation for drag except the area term, S, is the projected area perpendicular to the direction of flow, and C_l is the coefficient of lift. Larger objects have more lift, as do those in fluid that is flowing at a greater velocity and those in denser media. A larger fin can have more lift. An organism with a larger surface area attempting to stay attached to the bottom of a stream needs to be attached more strongly than a smaller one assuming that both protrude the same distance into the stream channel (i.e., experience the same water velocity over the top of the organism).

The Froude number is useful for relating the propagation of waves through water to the velocity of the water or an object moving on its surface. This number describes forces when objects will move on the surface of the water (a duck or a beaver), when boats will hydroplane, and when rapidly flowing water will exhibit hydraulic jumps (e.g., standing waves that form in rapids). Standing waves form when the wave propagation is slower than the water velocity. The Froude number is calculated as:

$$Fr = \frac{2 \prod U^2}{g \lambda}$$

where U is the velocity, g is the gravitation force, and λ is the wavelength.

Darcy's law allows for calculation of flow rates through porous media. This calculation is important in many areas of groundwater and soil science. For example, calculating the rate a pollutant will spread through groundwater requires calculation of water flow rates.

$$U = \frac{k_p \, \Delta p}{\mu \, l} = \frac{J}{S}$$

where U is the velocity, k_p is the permeability (depending on the size, shape, and volume of pores), p is the pressure, μ is the dynamic viscosity, l is the length, J is the flux, and S is the surface area. The smaller the average size of the pores, the more viscous the fluid, and the lower the pressure, the lower the transport velocity. Pressure generation will be discussed more in the chapter on groundwater.

When water flows down an open channel, the expected average velocity (V) can be related to physical characteristics of the channel by the Manning equation (also referred to as the Gauckler–Manning equation).

$$V = \frac{k}{\eta} R_h^{2/3} S^{1/2}$$

where k is used to convert units ($=1$ in metric calculations), η is an empirically derived roughness coefficient, R_h is the hydraulic radius (the ratio of the cross-sectional area divided by the wetted perimeter), and S is the slope of the channel. Each of these factors that influence velocity makes sense. The rougher the channel bottom is, the more velocity is slowed. Shallow wide channels have a low hydraulic radius (lots of wetted perimeter per unit cross sectional area). Steeper slopes lead to greater velocities.

FORCES THAT MOVE WATER

Solar heating and *evaporation* of water are central to water movement. This solar energy input drives the hydrologic cycle by evaporating water from the ocean that is deposited subsequently as precipitation on land. When water flows downhill in rivers or groundwater, it releases the potential energy it gained against *gravity* when evaporated by the sun. The generation of hydroelectric energy illustrates the massive amount of power released as water flows back to the ocean. The erosive power of water is a direct effect of this release of potential energy. The Grand Canyon of the Colorado River is an impressive example of how much geological change the energy of flowing water can accomplish.

A significant portion of the energy used by all humans on Earth is to provide the water that is geographically and temporally available to humans via the hydrologic cycle. We discussed this calculation in the last chapter and discuss the physics of flowing water and will discuss its effects (hydrology) on diffusion in Chapter 3, and the physiography of rivers and streams in Chapter 6. Gravity plays a further role in water movement by causing density-induced currents that may be important in lakes (Chapter 7).

Wind causes surface waves and mixing in lakes, reservoirs, and ponds, and this water movement can have strong influences on organisms. Much of the water movement of a passing wave is simply a rotation of individual parcels of water. As depth increases, this rotation decreases (Fig. 2.15). Consequently, wind-induced mixing also decreases with depth. We discuss how wind partially controls the stratification and currents in lakes in Chapter 7.

The *Coriolis Effect* can influence large lakes (area greater than 100 km^2) by causing rotational currents. The effect is an apparent deflection when a moving object is viewed from a moving point of reference. As applied to currents in lakes, it is not exactly a force that moves water; rather, the momentum of rotation of Earth and large-scale water currents resist this motion. For example, if you are standing on the

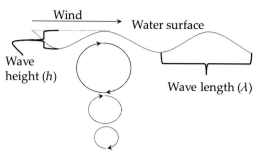

Figure 2.15 Schematic of wind-induced water movement in a body of water illustrating the decrease in movement with depth from the surface, wave length, and wave height. The three-dimensional flow paths are more complex than in this simple diagram, and we discuss them further in Chapter 7.

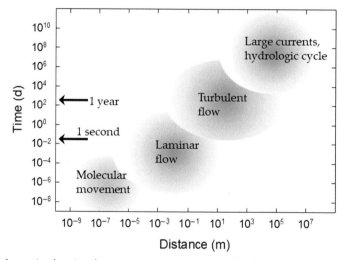

Figure 2.16 Schematic showing how water movement is related to spatial and temporal scales. The x axis ranges from the size of a protein to the size of the earth, and the y axis from the time frame of molecular events to the age of the earth. This figure is not meant to imply that sharp boundaries exist between the adjacent regions. Rather, the regions should be viewed as fuzzy overlapping regions. Forces that drive the water movement are discussed in the text.

surface of the earth looking north and watching a large flow of water moving toward you from north to south, Earth is rotating under the flow from west toward east, so the water would appear to move toward the west, while a current moving away from you would appear to move toward the east. This effect is most evident in large-scale ocean currents, and rotation of large storms, but it can also cause counterclockwise currents in lakes in the Northern Hemisphere and clockwise rotation in the Southern Hemisphere.

Finally, *organisms* can move water on smaller spatial scales as a result of locomotion or attempts to change the amount of water movement. In some cases, the actions of animals can alter the movement of water through sediments and increase the exchange of materials between sediments and the water column. For example, clams and worms that live in sediments effectively mix mud and pore water, beavers and humans slow water in streams by building dams, and humans increase water movement with channelization of streams. The relationship of temporal and spatial *scale* and forces on types of water movement are summarized in Fig. 2.16.

SUMMARY

1. The uniqueness of water is related in large part to the polarity of the molecule and associated hydrogen bonding.
2. Water is an excellent solvent. Ions generally become more soluble and gases become less soluble as water temperature increases.
3. Water is most dense at 3.98°C. Ice is about 9% less dense than liquid water. The variation of density with temperature is characterized by nonlinear relationships.
4. Density decreases above 3.98°C, with a greater density decrease per degree temperature rise as temperature increases.
5. Dissolved ions can increase density. A part per hundred difference in salinity causes a greater density difference than a 50°C temperature variation.
6. Hydrogen bonding of water leads to additional properties, including high heat of fusion, heat of vaporization, heat capacity, and surface tension, all of which are biologically very noteworthy.
7. Reynolds number (*Re*) is the ratio of inertia to viscous force and describes how the properties of water vary with spatial scale and water movement. Small organisms have low values of *Re* and little inertia relative to the viscous forces they experience; the opposite is true for larger organisms.
8. Biological activities, such as swimming, filter and suspension feeding, sinking, and many other aspects of aquatic ecology are constrained by properties of water that can be described by Reynolds numbers.
9. Water movement can be molecular, laminar, or turbulent. Molecular movement is also referred to as Brownian motion. Laminar flow is caused by physical processes that predominate at smaller Reynolds numbers and is more common close to solid surfaces or in the pores of sediments. Turbulent flow commonly occurs in open water at Reynolds numbers greater than 5.
10. Numerous equations have been developed by hydrologists and physicists to describe the properties of water, moving through water, and how these vary with temporal and spatial scale.

11. Evaporation and cooling of water vapor (the hydrologic cycle and gravity), wind, density differences, Coriolis effects, and activities of organisms (particularly animals) are all factors that can move water.

12. Water movements occur across a wide range of temporal and spatial scales and this forms part of the abiotic template that aquatic organisms have adapted to.

QUESTIONS FOR THOUGHT

1. Water is central to humans and science in many ways. For example, think of units of measurement. Which of these are based on physical properties of water?

2. Why is less of a temperature difference required for stratification (stable layers of different density) to occur in tropical lakes than in temperate lakes?

3. Why do bacteria generally sink more slowly than dead fish, even when both have approximately the same density?

4. If you wanted to simulate a 1-μm-diameter spherical bacterium swimming at 10 μm/s with a 1-cm-diameter sphere, how fast would you have to move the sphere to achieve the Reynolds number that the bacterium experiences?

5. Flow is slower in a small pipe than in a large pipe, given the same water pressure gradient. Explain this difference with respect to Reynolds number and viscous force.

6. Why are small swimming crustaceans not nearly as streamlined as larger fish?

7. Why might a lake with a large surface area mix more often and more deeply than a smaller lake?

CHAPTER 3

Movement of Light, Heat, and Chemicals in Water

Contents

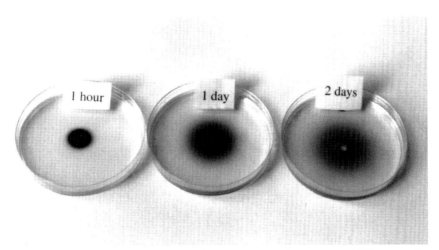

Figure 3.1 Diffusion of dye into a 0.5% agar solution in 10-cm-diameter petri dishes as a function of time. The agar prevents turbulent mixing so the outward spread of the dye is indicative of the rate of molecular diffusion. Length of time since dye was added to a small depression in the middle of each agar plate is labeled.

The movement of chemicals in water is a key factor for the survival and growth of most aquatic organisms and is central to the understanding of water pollution and material transport. Light is the ultimate energy source for most life on our planet. Without light, energy would not flow through ecosystems and most biogeochemical cycles would stop. Light heats water, which ultimately leads lakes to stratify. The temperature and movement of water are tied closely to the rate at which chemicals move through water.

DIFFUSION OF CHEMICALS IN WATER

An insect detecting prey, an algal cell acquiring nutrients, a microbe sensing its environment, a fish breathing, a contaminant moving through an aquifer, and many other aspects of freshwater ecology are all subject to the effects of diffusion of substances through water. Diffusion of heat and dissolved materials can be described in a similar manner, so we will describe the basics in this section and details of the movement of heat will be described more specifically in the next section.

Diffusion of chemicals is affected by factors including (1) the concentration gradient between two points (distance and concentration difference), (2) *advective transport* of water (water currents that move the chemicals), (3) temperature, (4) size of molecules, (5) the presence and structure of sediments or polymers excreted by organisms, and (6) any direct movement of the chemicals by organisms. The rate of diffusive flux between two points is positively related to the difference in concentration and inversely related to the distance between the two points. The basic equation used to describe *chemical diffusion* is *Fick's law* (technically Fick's second law):

$$J = D \frac{C_1 - C_2}{x_1 - x_2}$$

where diffusion *flux* (J, the amount of a compound diffusing per unit area per unit time) is a function of the *diffusion coefficient* (D, the intrinsic rate of diffusion independent of concentration and distance), the difference in concentration (C), and the distance (x) between points 1 and 2. The rate of diffusion is greatest with large concentration gradients (difference between C_1 and C_2) and small distances ($x_1 - x_2$). Fig. 3.2 provides a visual representation of this equation. The diffusion coefficient is a function of fluid identity, size of the diffusing molecule (larger molecules diffuse more slowly), temperature, obstruction of diffusion by pore structure in sediments or other materials, and rate of mixing. Fick's law can be applied to many problems related to diffusion, such as calculating the rate at which pollutants spread (Example 3.1).

Thermal movement of molecules (*Brownian motion*), as discussed in Chapter 2, causes molecules to mix randomly throughout the fluid and leads to *molecular diffusion*. The importance of Brownian motion is related inversely to spatial scale. For example, many molecules randomly collide with a human swimmer from all sides, but because of the size of the human body relative to that of water molecules, the force is the same from all directions. A bacterium is small enough that molecules will collide unevenly around its cell surface causing it to jiggle when viewed under a microscope.

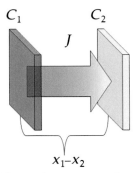

Figure 3.2 Schematic illustrating diffusion between two flat surfaces at different concentrations (C_1 and C_2). The rate of diffusion (J) is described by Fick's law (see text). The concentration at the red plane C_1 is greater than at the yellow plane C_2, so the net diffusion is toward C_2. Diffusion is less rapid as distance (x_1-x_2) between the two planes increases and as the difference between the concentrations at the two planes (C_1-C_2) decreases.

Example 3.1 Calculating diffusion flux into a stream from a groundwater source

Nitrate (NO_3^-) is a common contaminant of drinking water that when present in excessively high concentrations leads to serious health problems. Groundwater is present below a feedlot and contains 100 mg/L NO_3^--N (i.e., the number of milligrams per liter of nitrogen in the form of nitrate) at a distance of 10 m from a stream, where the concentration is 10 mg/L NO_3^--N and the diffusion coefficient for NO_3^-, D, has been measured as 1.85×10^{-3} cm²/s. Assume that D does not change from the feedlot to the stream. What is the daily flux of nitrate into the stream per square meter of stream bottom?

First, convert the nitrate concentration and distance into units consistent with the diffusion coefficient, so 100 mg/L NO_3^--N = 0.1 mg/cm³, 10 mg/L NO_3^--N = 0.01 mg/cm³, and 10 m = 1,000 cm.

Then,

$$
\begin{aligned}
J &= 1.85 \times 10^{-3}\ \text{cm}^2/\text{s} \times (0.1 - 0.01\ \text{mg/cm}^3)/1{,}000\ \text{cm} \\
&= 1.67 \times 10^{-7}\ \text{mg cm}^{-2}\text{s}^{-1} \\
&= 0.14\ \text{g } NO_3^-\text{-}N\ \text{m}^{-2}\text{d}^{-1}
\end{aligned}
$$

The probability of unequal collisions is greater for smaller particles or ions, and the smaller relative difference in mass means that smaller molecules are more influenced by collisions with water molecules. Molecular diffusion is more rapid for smaller molecules (i.e., the diffusion coefficient, D, is greater) because small particles are more likely to be moved by collisions with water molecules. Thermal energy is a property of the average velocity of dissolved ions and water molecules. Consequently, molecular diffusion rates are greater as temperature increases (Fig. 3.3).

Molecular diffusion, in the complete absence of turbulent mixing, is so slow that it is not often observed directly at macroscales (Fig. 3.1). For example, a cube of sugar would take days to dissolve completely and disperse evenly throughout a glass of water in the absence of any movement of the water. However, it is difficult to control water

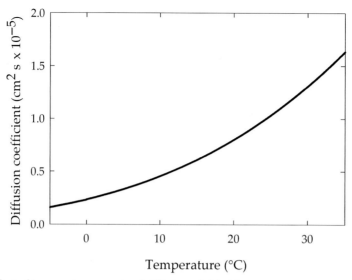

Figure 3.3 Effect of temperature on rate of diffusion of chloride.

movement because evaporation causes cooling and this creates convection currents that slowly mix the water. We further accelerate the mixing process by mixing with a spoon.

Water currents often occur at spatial scales exceeding those of individual molecules, and such currents can dominate material transport. Molecular diffusion is many orders of magnitude slower than diffusion with water movement (variously referred to as *transport diffusion*, *advective transport*, or *eddy diffusion*). As discussed in Chapter 2, overall molecular movements are on the order of several nanometers per second, whereas velocity in streams and rivers can exceed 1 m/s (a billion times more rapid). *Turbulent flow* causes transport diffusion and relatively high diffusion coefficients. Thus, if there is any appreciable movement of water, transport diffusion will dominate over molecular diffusion. With very small spatial scales, Reynolds number is small, viscosity is large, turbulence is not present, and molecular diffusion is more important. Transport diffusion overrides molecular diffusion in the water column of lakes and parts of wetlands, groundwaters, streams, and rivers. There is little or no water movement near solid surfaces or within fine sediments, and molecular diffusion is the dominant mode of diffusion.

The ratio of molecular to transport diffusion is described by the dimensionless *Sherwood number*

$$Sh = \frac{Ul_c}{D}$$

where U is the velocity, l is the length, and D is the molecular diffusion coefficient. When Sh is large, advective transport dominates. When it is small, molecular diffusion predominates. However, D is very small in water ($\approx 10^{-5}$ m^2/s) so an organism must be very small ($<20\,\mu$m) before molecular diffusion predominates (Denny, 1993). Thus, molecular diffusion rules for bacteria and other microscopic organisms, but transport diffusion controls exchange of materials between the individual and the environment that most multicellular organisms inhabit.

In sediments, such as in groundwater or on lake bottoms, the rate of chemical diffusion is also influenced by the mean path length and the size of pores and channels within the sediment. Short path length and large channels lead to high permeability. When channels are long or have many dead ends, diffusion rates are slowed because molecules must take a longer path to diffuse between two points, and transport diffusion is limited because water velocity is slow. Determining mean channel length in sediment requires direct empirical measurement of diffusion of a tracer through the sediment.

Solutes can interact with sediments (both the inorganic particles and the organic materials) and further lower diffusion rates. For example, removing organic contaminants such as DDT (a pesticide banned in the United States and many other countries in the 1970s) and some other pesticides from groundwater is very difficult because they have low solubility (are hydrophobic) and are adsorbed onto the surface of sediment particles, so they have low diffusion rates. Thus, bioremediation efforts (the use of microbes to clean groundwater) and other methods used to remove organic contaminants from groundwaters often employ detergents. Detergents increase diffusion rates by releasing hydrophobic molecules from sediments, increasing the biological availability of organic contaminants, and speeding the decontamination process.

Water velocity nears zero as a solid surface is approached because water molecules adhere to the surface and cohesion transmits the effect of the surface. Transport diffusion slows and molecular diffusion predominates when water movement is sufficiently slow close to the surface. The region near a solid surface where molecular diffusion predominates is called the *diffusion boundary layer*. This is similar to, but thinner than, the flow boundary layer discussed in Chapter 2. However, the same considerations apply to the thickness of diffusion and flow boundary layers; depressions in solid surfaces lead to thicker diffusion boundary layers, and thinner layers occur where a solid surface protrudes into flowing water and as water velocity and turbulence increase. Like the flow boundary layer, the diffusion boundary layer is a fuzzy zone of gradual rather than abrupt transition because particularly strong turbulent eddies will occasionally disrupt the zone.

The diffusion rate is so much slower across the region where molecular diffusion predominates than in regions with eddy diffusion that it represents a rate-limiting boundary layer to the passage of biologically active chemicals. Compounds required for metabolic activity must cross the diffusion boundary layer, so metabolic rate can be limited by the thickness of this layer. The diffusion boundary layer constrains evolutionary pressures that shape chemically mediated interactions among organisms (Dodds, 1990; Brönmark and Hansson, 2000). For example, the dilution associated with transport diffusion makes it unlikely that biologically active compounds will be released into turbulent waters. The evolutionary cost of purposely releasing chemicals that are costly to synthesize is too great if chemicals are rapidly diluted to concentrations so low that they are ineffective.

Chemical diffusion rates are related to the size and geometry of organisms. *Surface area to volume* relationships are essential in the evolution of aquatic organisms. More surface area means more space for materials to move across (and more area to have proteins on a membrane to actively transport materials into cells), and smaller volume means less average distance to the surface. Large objects have a greater volume relative to their surface area compared with smaller objects. Subsequently, geometry can affect relative rates of diffusion (of nutrients inward and metabolic waste products outward), which can limit metabolism. One way to conceptualize this is to think of the average distance of all points inside a sphere to the outside edge. On average, the distance to the edge of a larger sphere is greater than in a smaller sphere. Fick's law states that greater distance translates into less diffusion flux. Thus, metabolic processes in large organisms are more likely to be limited by diffusion.

Changes in form that increase surface area to volume ratios are one way to overcome limitations imposed by diffusion. Form is known to alter diffusion rates in small organisms suspended in water (Highlight 3.1). Surface area to volume relationships also explain why large organisms (more than a few cells wide) have vascular systems. These systems move liquids through the organisms and thus promote transport diffusion; otherwise, the relatively low rates of molecular diffusion would limit the maximum possible metabolic rate. Fungi and plants with large cells use cytoplasmic streaming to move materials within cells and overcome limitations by molecular diffusion. Specialized organs associated with vascular systems such as the gills of fishes and aquatic invertebrates have evolved to increase effective surface area and promote inward diffusion of O_2 and outward diffusion of CO_2. Many insect larvae that live in high velocity stream habitats have evolved relatively small gills because the diffusion boundary layer is so thin at high water velocity, combined with the fact that larger gills increase drag and probability the animal will be swept downstream. Such insect larvae can die if kept in still water because their respiration rate exceeds diffusion rates when advective transport is limited.

Movement of organisms can also increase movement of chemicals, although this movement is often not treated as a diffusion process. A swimming organism emits a plume of excreted compounds that is a function of its size, shape, and swimming speed (Bearon and Magar, 2010). Animals in benthic sediments can disturb the sediments in which they live (*bioturbation*) and increase diffusion of materials by increasing turbulent mixing. Motile bacteria can absorb and then move organic contaminants through

Highlight 3.1 Does diffusion select for morphology of small planktonic organisms?

Factors crucial to the survival of planktonic organisms (organisms suspended in water) include: (1) avoiding sinking into regions where light is insufficient for survival (Reynolds, 1994), (2) a surface area to volume ratio that maximizes nutrient and gas diffusion and increases competitive ability (Reynolds, 1994), and (3) development of defensive appendages that lower the probability of capture or ingestion by predators. The morphology (shape and size) of single- or few-celled planktonic organisms can influence all these factors.

Sinking can be affected by particle size (see formula for F_v and discussion of Stokes law in Chapter 2). Small particles experience high viscosity and sink more slowly. However, increasing surface area also increases viscosity F_v. Therefore, spines or projections (e.g., Figs. 8.8, 9.4, and 19.4) lower sinking rates.

Increasing the surface area with spines or projections also increases the surface area to volume ratio. Assuming that the spines or projections can allow materials to move through them to and from the cell, they also can increase the biologically active surface area. A greater biologically active surface area increases the diffusion rates of incoming nutrients and outgoing waste products. Thus, there can be natural selection for planktonic organisms to increase their biologically active surface area.

Sinking and diffusion rates may not be the only selective factors acting on morphology. Predation risk is also decreased by defensive appendages. Spines or projections can increase handling times and unwieldy organisms can clog mouthparts of planktonic predators. Finally, growth varies nonlinearly as cell size increases due to changes in nutrient acquisition and metabolic activity (Ward et al., 2017). In conclusion, four selective pressures converge, and there are a wide variety of shapes of planktonic organisms. The relative importance of these three factors may vary with the organisms considered and their habitats. Considering the relationship between morphology and diffusion is still necessary to describe the selective pressures on shape of small planktonic organisms.

---

groundwater sediments more rapidly than expected given permeability and water flow. Motile ciliates can increase O_2 transport through sediments up to 10 times above molecular diffusion rates (Glud and Fenchel, 1999). Migration of planktonic organisms from the bottom to the top of lakes (Horne and Goldman, 1994) and migration of fishes upstream to spawn (Kline et al., 1990) can result in significant redistribution of nutrients in aquatic systems. Emerging adult aquatic insects and amphibians can cause significant fluxes of nutrients from aquatic to terrestrial habitats (Gray, 1989; Nakano and Murakami, 2001; Regester et al., 2008).

The ideas concerning movement of materials in water as a function of scale (distance and time) are similar to the concepts for water movement discussed previously (see Fig. 2.16). On the smallest scales, low Reynolds numbers occur, turbulent mixing is not possible, and molecular diffusion dominates. Diffusion rates increase significantly at greater spatial scales because transport diffusion is likely to override molecular diffusion. Turbulent flow can occur at scales ranging from rivulets to large rivers and lakes. Finally, large currents in very large lakes and the hydrologic cycle move materials on continental and global scales.

MOVEMENT OF GASES BETWEEN ATMOSPHERE AND WATER

Movement of gas across the water–atmosphere interface is a special case of diffusion. The process of gas equilibration with the atmosphere is called *reaeration* or *aeration*. This rate is important because it determines how quickly gasses such as O_2 can re-equilibrate to the demands of respiration and production by photosynthesis, as well as how quickly greenhouse gasses can enter the atmosphere. The equation is similar to Fick's law of diffusion, except that the distance term disappears since it is movement across an extremely small barrier (the water–atmosphere interface), where flux F is:

$$F = k([\text{gas}_{\text{dissolved}}] - [\text{gas}_{\text{equilib}}])$$

where k is the reaeration coefficient, $[\text{gas}_{\text{dissolved}}]$ is the concentration of the gas in solution, and $[\text{gas}_{\text{equilib}}]$ is the concentration at equilibrium with the atmosphere. The greater the difference between the actual concentration of the dissolved gas and the equilibrium concentration of the gas in water, the greater the diffusion rate. If the gas concentration in the water is less than it would be at saturation, as can occur when respiration consumes O_2 in the water, then the flux will be from the atmosphere into the water. When photosynthesis supersaturates O_2, the flux will be from water to atmosphere.

Estimating the reaeration coefficient is difficult. It can depend upon factors such as turbulence in the water, wind, temperature, chemicals on the surface of the water, and properties of the gas being considered (Brutsaert and Jirka, 2013). The best methods for determining rates are empirical and based on the use of inert tracers such as sulfur hexafluoride, radon, or argon. The flux rates of these tracers then need to be corrected for the gas of interest since different molecules have slightly different reaeration rates (Raymond and Cole, 2001). Other methods have included modeling and equations on the basis of the physical characteristics that might influence reaeraton, such as turbulence, wind, and temperature. In some cases, modeling gives reasonable estimates, but general equations often do not match estimates from tracers or modeling (e.g., Riley and Dodds, 2012).

Ebullition, loss of gas bubbles from aquatic systems, can be an important mode of transport of gasses from sediments that are otherwise dominated by molecular diffusion of dissolved gasses. This is particularly important when considering the rates of loss of greenhouse gasses such as CO_2 and methane (CH_4) from sediments (Higgins et al., 2008; Crawford et al., 2014; Wilkinson et al., 2015). These rates can be considerable, particularly in the case of CH_4, which can then bypass aquatic organisms that would consume it (see Chapter 13). Measurement of these rates is difficult but can be accomplished by floating chambers on the water surface that capture gas as the bubbles leave the water, or by hydroacoustic methods that can measure size and frequency of gas bubbles moving through the water column (Huttenen et al., 2001; DelSontro et al., 2015).

LIGHT IN WATER

The interaction between light and water is important for at least three reasons: (1) light is needed for photosynthesis, (2) organisms with eyes or light sensors use light as a sensory cue, and (3) light heats water and ultimately leads to stratification in lakes (see Chapter 7). Feinberg (1969) provided a basic account of the properties of light and all limnologists should be familiar with the common names associated with wavelength ranges of the electromagnetic spectrum (Table 3.1), and the basic units of heat and light (Table 3.2).

Earth's atmosphere alters the intensity and composition of solar irradiance that reaches aquatic systems. Some atmospheric components remove specific wavelengths of light (Fig. 3.4). Others, such as dust and clouds, may scatter or absorb light less selectively. One essential aspect of atmospheric influence on irradiance is the absorption of ultraviolet (UV) light by atmospheric ozone. Releases of chlorofluorocarbons

Table 3.1 Names and various ranges of wavelengths of light and their biological relevance

Name	Wavelength (m)	Biological relevance
Radio	$2 \times 10^4 - 1 \times 10^{-4}$	Little effect
Infrared	$10^{-4} - 10^{-6}$	Heats water, some bacteria can use shorter wavelengths for photosynthesis
Visible red-orange	$7 \times 10^{-7} - 5.7 \times 10^{-7}$ (700−570 nm)	Visible, photosynthetic energy, heats water
Visible green	$5.7 \times 10^{-7} - 4.9 \times 10^{-7}$ (570−490 nm)	Visible, photosynthetic energy (some algae), heats water
Visible blue	$4.9 \times 10^{-7} - 4.0 \times 10^{-7}$ (490−400 nm)	Visible, photosynthetic energy, heats water
Ultraviolet UVA	$4 \times 10^{-7} - 3.2 \times 10^{-7}$ (400−320 nm)	Mutagenic, cell damage
Ultraviolet UVB	$3.2 \times 10^{-7} - 2.8 \times 10^{-7}$ (320−280 nm)	Mutagenic, cell damage, but also critical for vitamin D3 production in semiaquatic reptiles
Ultraviolet UVC	$2.8 \times 10^{-7} - 2.0 \times 10^{-7}$ (280−200 nm)	Mutagenic, cell damage, most filtered out by atmosphere before reaching water
Extreme UV, X-rays, gamma rays	$10^{-8} - 10^{-13}$	Mutagenic, cell damage, not normally present at high levels in natural environments

Table 3.2 Units and names of some common light and energy units

Name	Unit	Comment
Joule (J)	N/m	
Watt	J/s	
Calorie	Cal	Energy to heat one gram of water 1°C, 1 cal = 4.184 J
Hertz (Hz)	s^{-1}	Greater frequency is more energy
Wavelength (λ)	M	Shorter wavelengths have greater energy
Photosynthetically available radiation (PAR)	μmol quanta $cm^{-2} s^{-1}$	Flux per unit area of photons between 400 and 700 nm

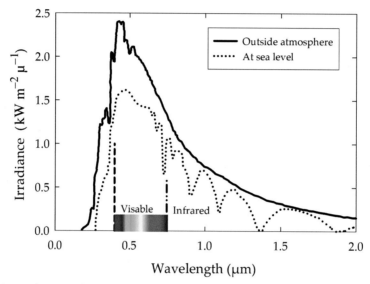

Figure 3.4 Spectral energy distribution of solar radiation outside the earth's atmosphere and inside the atmosphere at sea level. Note how the atmosphere changes the spectral distribution of light. *After U.S. Air Force (1960).*

(freons), bromide-containing compounds, and nitrous oxide have caused destruction of ozone in the upper atmosphere and led to significant increases in the solar UV that reaches the surface, particularly at higher latitudes. Some of the consequences of increased UV to aquatic ecosystems are discussed in Chapter 12. The absorption of light in the atmosphere is wavelength specific; the same is true once light enters water. Light can be intercepted before reaching the water by clouds, canopy, and topography.

Annual total irradiance reaching the Earth's surface is about two times greater at the equator than at high latitudes. The greatest irradiance values are in tropical desert habitats. For each 3 km of elevation gain, there is about a 50% increase in photosyn-thetically available radiation (Barry and Chorley, 2009). Patterns of UV are similar to total irradiance and are also influenced by altitude, latitude, and cloud cover (Beckmann et al., 2014). Light can also be intercepted by features near the surface of the water such as tree canopy or strong topographic relief. For example, Yard et al. (2005) modeled the amount of light entering the Colorado River at the bottom of the Grand Canyon to account for being deep within a winding canyon.

When light reaches the water surface, it can either be reflected or enter the water (Fig. 3.5). If the critical angle of the light reaching the atmosphere/water interface is too great, then it will all be reflected. The amount of light reflected is highly variable because it can be altered by waves on the water surface, the incident angle of the sun, the type of waves (e.g., whitecaps and size), and materials (e.g., dust, pollen, oils) on the surface of the water. If the surface is snow-covered ice, almost all the light is reflected away. Clear ice cover transmits most of the light that enters it.

As light enters the water, refraction causes a bending of the light as it moves from less dense air to denser water. This refraction causes objects observed in water from above the surface to appear to be shallower or larger than they really are (contributing to "fish stories"). The refractive indices of water and ice are very similar, so only minor refraction is expected as light passes between solid and liquid water in an ice-covered body of water.

Once light enters a parcel of water, it can be *absorbed, reflected,* or *transmitted.* Almost all light that is absorbed by water, suspended particles, or dissolved materials is converted to heat. Some of the light is reflected back into the atmosphere, giving lakes their characteristic color as observed from above.

Biological activities related to light and water column heating are constrained by the amount of light transmitted to different depths. Light intensity decreases logarith-mically with depth in a water column (*attenuation*). Attenuation rate is related to reflection and absorption by water, dissolved compounds, and suspended particles. In

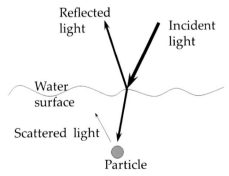

Figure 3.5 Schematic of light striking the water surface where it can enter or be reflected back, scatter off of a particle, or be absorbed in the water column.

more productive water columns with a large biomass of suspended photosynthetic organisms (*eutrophic*), or those with large amounts of suspended inorganic materials or high concentrations of dissolved colored materials, the water contains more material to absorb or reflect out the light; thus, it is not transmitted as far. In less productive (*oligotrophic*) lakes with low amounts of suspended particulate or dissolved colored material, light is transmitted to greater depths (Fig. 3.6).

Logarithmic attenuation (Fig. 3.6) is a physical characteristic that can be explained using the following example. Assume that 1/10 of the full sunlight entering the water column is transmitted to 2 m below a lake surface. If attenuation is constant, only 1/10 of the light remaining at 2 m is transmitted to the second 2 m (by 4 m); thus, only 1/100 remains. By 6 m only 1/1,000 remains (1/10 of 1/100), and so on. We present this example in units convenient to a log base 10 scale, but a plot of the light remaining with depth will be linear for a homogeneous water column using a logarithmic scale of any base.

Several equations describe light attenuation, and coefficients of attenuation are useful indices for aquatic ecologists. If I_1 is the light intensity at depth z_1, and I_2 is the intensity at a deeper depth z_2, then the *attenuation* or *extinction coefficient* (η, identical to the absorption coefficient used in chemistry) is:

$$\frac{\ln I_1 - \ln I_2}{z_2 - z_1} = \eta$$

Alternatively, when the equation is rearranged and raised to the power of e to remove the natural logs,

$$I_2 = I_1 e^{-\eta(z_1 - z_2)}$$

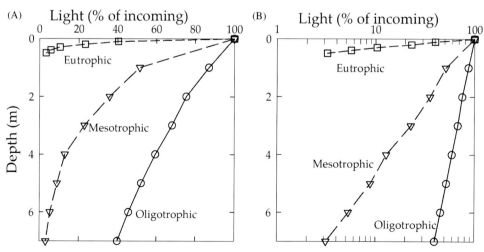

Figure 3.6 Light as a function of depth in three lakes—Waldo Lake (oligotrophic), Triangle Lake (mesotrophic), and a sewage oxidation pond (eutrophic), Oregon—plotted on linear (A) and log (B) scales (R. W. Castenholz, unpublished data).

Notice that the equations for light attenuation are based on natural logs (ln) and that logs to the base 10 will give a different number that is not comparable with most published values without conversion. The exponential nature of the equation describes a very rapid decline in light intensity with depth, a concept that is not intuitive to some beginning students. A more intuitive equation describes light in terms of *percentage of transmission* over a specific depth:

$$\frac{100 \cdot I_2}{I_1} = percentage\ transmission$$

where I_1 and I_2 are light intensity values measured at any two points. If the two points are 1 m apart vertically in the water column exactly then it is the percentage transmission per meter, which is the common way to express the value. Various methods are used to measure light in water (Method 3.1) to characterize attenuation.

METHOD 3.1 Equipment used to measure light in aquatic habitats

Many people who study aquatic habitats use a Secchi disk to estimate transparency. This simple method was first used in 1865 by an Italian astronomer, Father Pietro Angelo Secchi, to test ocean clarity. The Secchi depth is the depth at which a weighted, black-and-white disk, 20 cm in diameter, disappears from view. Black and white provide the maximum contrast regardless of the color of the light transmitted by the lake. The instrument provides the most consistent results in sunny, midday, calm water conditions off the shaded side of a boat or dock to minimize reflection off the water. Secchi depth corresponds to the depth at which approximately 10% of the surface light remains (Wetzel, 1983). The relationship between transmission of photosynthetically available radiation (or the extinction coefficient) and Secchi depth is complex and nonlinear because it depends on ambient light, scattering and absorptive properties of the water, and the measurer. However, Secchi depth can serve as an estimate of light attenuation (Fig. 3.7) and the method is inexpensive and easily performed by minimally trained observers. This method is useful for volunteer lake monitoring groups.

Several other methods are available for measuring light, depending on the reason for the measurements. If an investigator is interested in the total light that is heating the surface of a lake, then the total energy entering the lake should be measured. If photosynthesis is of interest, then the energy associated with wavelengths of light that support photosynthetic activity needs to be quantified. For other applications, specific wavelengths are of interest (e.g., determining the influence of UV-B radiation on biological activity).

To measure *irradiance*, which is the radiance flux density on a given surface, it is necessary to correct for light that is coming at an angle to the sensor as opposed to a parallel beam coming straight toward the collector. The appropriate correction is a cosine curve, so

(Continued)

METHOD 3.1 (Continued)

cosine collectors are the sensors of choice to measure radiation from above. In contrast, some biological processes such as photosynthesis are dependent on total light received from all directions. In this case, a *scalar* or *spherical* (360 degrees) response collector is the sensor of choice (Kirk, 1994).

To measure the total energy entering a water body per unit area per unit time, a pyrheliometer is used which compares the temperatures of a reflective metal surface and one that absorbs all incoming radiation. These measurements are useful for determining the heat budgets of water bodies.

A meter that is used to estimate light available for photosynthesis should measure the number of photons that are available to excite chlorophyll. These sensors measure photons between 400 and 700 nm and the results generally are reported as *photosynthetically available radiation* (PAR) or photosynthetic photon flux density. Units are in μmol quanta m^{-2} s^{-1}, or sometimes μEinsteins (m^2 s) [the Einstein is not an internationally recognized unit for a mole (6.02×10^{23}) of photons]. Spherical sensors are often used for these measurements when considering suspended algae.

The relative absorption of specific wavelengths could also be of interest. In this case, selective filters can be fitted over sensors, or a spectral radiometer can be used. Spectral radiometers have diffraction gratings that allow photodetectors to sense the intensity of specific wavelengths of light.

Problems arise if light is measured in algal mats or sediments because very small sensors are needed (Jørgensen and Des Marais, 1988). In such cases, spherical tips on fiber optic collectors have been used (Dodds, 1992). Photodetectors or spectral radiometers can be used to analyze light collected by such sensors.

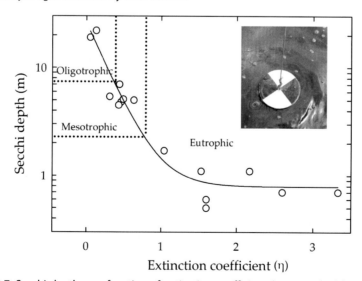

Figure 3.7 Secchi depth as a function of extinction coefficient (measured with a quantum meter, 400−700 nm) for 13 Oregon lakes. Boundaries between trophic states for Secchi depth set according to the classification of OECD (1982) (R. W. Castenholz, unpublished data) (United States Army Corps of Engineers photo by Tim Beauchene).

Not only does total light intensity change with depth but also there is variation in the attenuation of relative amounts of different wavelengths. Pure water has maximum absorption in the visible wavelengths of red light and maximum transmission of blue (Fig. 3.8). As mentioned previously, attenuation is a function of the sum of the absorption of light by the water, by the particles in the water, and by the compounds dissolved in the water (Fig. 3.9). Attenuation coefficients are largest in eutrophic lakes and smallest in oligotrophic lakes (Example 3.2). As the productivity of a lake increases, so does the attenuation, because lakes with greater productivity have more

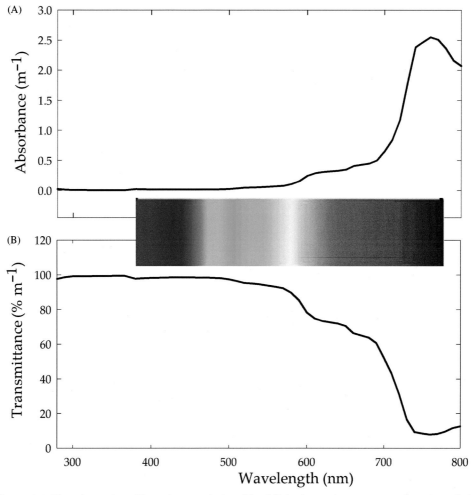

Figure 3.8 The absorption (A) and transmission (B) of light by pure water as a function of the wavelength of light. *Data from Kirk (1994).*

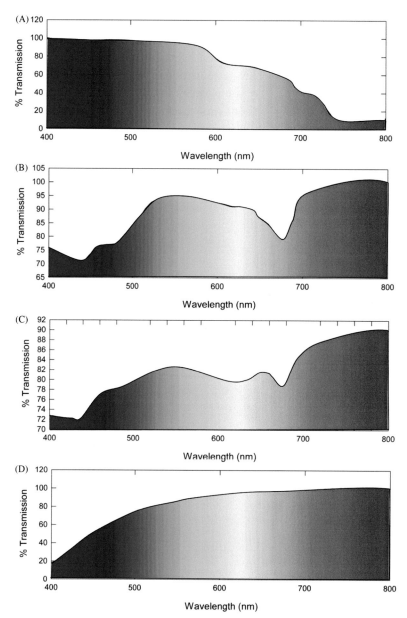

Figure 3.9 Percentage light transmission as a function of wavelength for (A) pure water, (B) a green alga, *Chlorella*, (C) a cyanobacterium, *Microcystis*, and (D) humic substances. Note that water absorbs much of the red light, the green alga allows more green light through, the cyanobacterium lets more blue light through relative to green, and the humic substances remove blue and green light. Thus, an oligotrophic lake has blue water, a lake with green algae appears green, a lake with cyanobacteria appears blue green, and a lake with high levels of dissolved humic substances appears reddish-brown. *Data from Hakvoort (1994) and Wetzel (2001).*

Example 3.2 Use of light attenuation equations

Calculate the attenuation coefficient and percentage transmission per meter at the surface of two lakes, one that is productive and one that is less productive. In both lakes, light is 1,500 µmol quanta m^{-2} s^{-1} at the surface. In the productive lake, the light is 1 µmol quanta m^{-2} s^{-1} at 1 m, and in the unproductive lake it is 1,200 µmol quanta m^{-2} s^{-1} at 1 m.

Calculation	Unproductive	Highly productive
I_0 (ln I), intensity at first depth	1,500 (7.31)	1,500 (7.31)
I_1 (ln I), intensity at second depth	1,200 (7.09)	1 (0)
% transmission m^{-1}	80	0.06
η	0.22	7.31

Note that transmission is greater and attenuation coefficient is less for the unproductive lake.

suspended particles and dissolved organic compounds to absorb light. Other factors that can increase light attenuation include large *sediment* loads (typical of reservoirs in agricultural areas and shallow lakes) and lakes with high concentrations of humic compounds. *Humic compounds* result from the decomposition of plant material and lead to brown-colored water with large absorption coefficients. Blackwater swamps, other wetlands, rivers, and some small lakes can have high concentrations of humic compounds.

In clear oligotrophic lakes, blue wavelengths are more likely to be reflected back out before they are absorbed than are longer (redder) visible wavelengths. In contrast, in more eutrophic lakes green is transmitted furthest (Fig. 3.10). There are always some suspended particles in lakes that reflect light back out of the surface into the eye of the observer. Therefore, clear (oligotrophic) lakes with very pure water appear blue and eutrophic lakes appear green (Fig. 3.11).

Different materials dissolved or suspended in waters influence the transmission of light and alter the spectral characteristics of light as it passes through the water (Fig. 3.9). Pigments of suspended algae (phytoplankton) absorb light in specific regions. The most important of these is chlorophyll *a*, a photosynthetic pigment that absorbs blue and red light. As a lake becomes more eutrophic, more blue is absorbed and relatively more green is transmitted and reflected. Algae that contain chlorophyll remove blue and red light in the water column above, but enough green light is transmitted to depth to allow photosynthesis for organisms, that are adapted to use this wavelength range.

Cyanobacteria (blue—green algae) and red algae have evolved specialized light-collection pigments called *phycobilins* that allow them to use green light (Fig. 3.9). This fact, combined with an understanding of optical properties including wavelength-

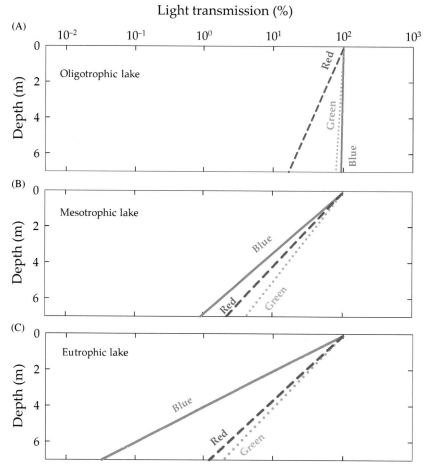

Figure 3.10 Light transmission as a function of color for an oligotrophic lake (Waldo Lake, 1984; A), a mesotrophic lake (Munsel Lake, 1984; B), and an eutrophic lake (Siltcoos Lake, 1983; C) Oregon (R. W. Castenholz, unpublished data).

specific attenuation, can be useful for describing some of the ecology of cyanobacteria. There is a greater relative transmission of green light compared with blue and red light in eutrophic lakes. At a meter or less of depth in an eutrophic lake, the red and blue wavelengths used by most photosynthetic organisms are depleted. Cyanobacteria use their specialized pigments to harvest green light and transfer it to chlorophyll *a* for photosynthesis. Dense populations of cyanobacteria are sometimes found well below the surface of lakes (Fig. 3.12). Hence, the spectral quality of light at depth is a function of the absorptive properties of the lake. Spectral transmission influences perceived colors of lakes and the objects within them (Highlight 3.2).

Figure 3.11 Contrasting color of lakes. An eutrophic pond in Oklahoma with a floating cyanobacterial bloom (left) and ultraoligotrophic Crater Lake (right).

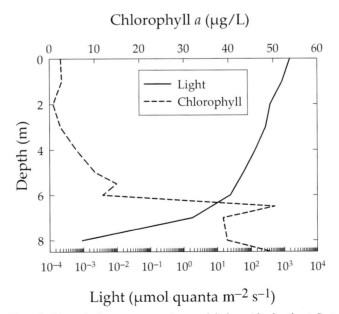

Figure 3.12 Profiles of chlorophyll *a* concentration and light with depth at Pottawatomie State Fishing Lake No. 2, Kansas. Deep chlorophyll peaks are attributable to the presence of large populations of cyanobacteria (*Oscillatoria*). The high biomass of algae occurs in a region with 1%–0.001% of surface sunlight. Also, note how the attenuation of light increases (shallower slope of the light curve) because of the dense algal populations.

Highlight 3.2 Why are lakes the color they are, and how are colored fish lures and other objects perceived under water?

The color of a lake can tell us much about its biological status. Several factors are involved in determining how an observer perceives lake color. Very unproductive (oligotrophic) lakes are a deep blue color because pure water transmits blue and absorbs green and red light. The blue light goes more deeply into the lake, but some of it is scattered by suspended particulate material and returns to the surface. Scattered light that is blue is more likely to leave the lake than scattered green or red light.

In productive (eutrophic) lakes, blue and red wavelengths are absorbed by algal pigments, and green wavelengths travel relatively further. The probability is greater that green light will be scattered back out of the lake than red or blue light, giving highly productive lakes their green color. These predictable relationships between lake color and productivity allow for large-scale assessments of lake productivity using remote sensing techniques and satellite images (Lillesand Johnson et al., 1983).

Dissolved organic materials such as tannin and lignin can impart a brown color to a lake, pond, or stream. Observation of such lakes may reveal little suspended material, but the dissolved material absorbs all wavelengths of light, yielding a brownish color. Sediments can also color lakes. A lake that is very turbid with reddish clay will look red. High-altitude lakes near glaciers often contain very fine sediment particles (glacial flour) created by glacial action. These particles can lend a milky turquoise appearance to otherwise unproductive blue lakes.

Organisms can impart additional colors to lakes and ponds. Species of water ferns can be bright purple and reach very high densities on the surface of certain ponds. Photosynthetic bacteria can reach high densities, particularly in saline ponds, and yield purple, brown, yellow, or blue appearances. Dissolved metals can also color ponds. For example, high concentrations of copper can lead to metallic blue ponds or lakes.

The alteration of spectral quality with depth means that some wavelengths are never present in deeper waters, and colors are perceived differently there than under full sunlight. This may be important to the way that fish are able to perceive color. For example, under full sun, red fish lures appear red because they absorb green and blue light and reflect red. However, red light is not available deep in a lake. To a diver or fish some 5 m deep in a mesotrophic lake a red lure would appear black. Fish lures appear to be different colors deeper in oligotrophic lakes than in eutrophic lakes (Fig. 3.13). A white lure would appear blue 10 m under the surface of an oligotrophic lake but green 1 m deep in an eutrophic lake because of the predominant color of light found with depth in each lake. Certainly, the color patterns of many fish lures are designed to appeal to anglers rather than fishes.

Aquatic ecologists should also remember these spectral properties in other contexts. For example, a deep-water fish may be colored bright red, but this actually may

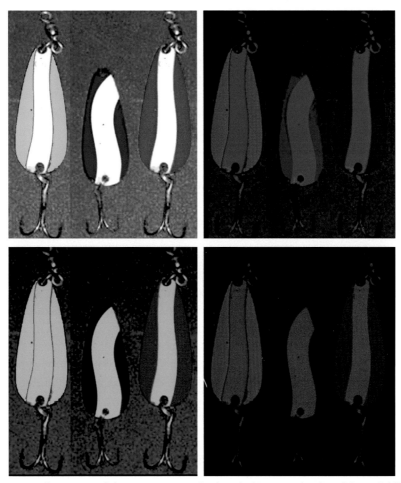

Figure 3.13 Illustration of the appearance of colored objects at depth in lakes of different trophic status. Upper left, full light; upper right, a blue filter simulating light deep in an oligotrophic lake where white looks blue, and blue looks black. Lower left, a green filter as at moderate depth in a mesotrophic lake where white looks green and blue looks black. Lower right, a red filter as at shallow depth in an eutrophic lake where the contrast between red and white is strongly decreased.

represent cryptic coloration when the fish is in deep water because the red appears black. Yet, if the fish moves into shallow water, the coloration may be visible and attractive to mates.

HEAT BALANCE IN WATER

The heat balance in water determines the water temperature, which in turn controls rates of biological activity. Heating, cooling, and movement of heat through water determine temperature of the water at any point in time. Changes in heating can lead to water currents (e.g., water sinks from the surface on cold nights or as it cools on the surface by evaporation). Physical limnologists study heating and mixing that set up density differences that can lead to stratification and specific water currents (more detail in Chapter 7).

Heat energy in water is gained by solar radiation, diffusion of heat from the atmosphere and ground, and warmer water flowing into a particular area (from precipitation, condensation, groundwater, streams, or rivers). Heat energy is lost by reflection, back radiation, evaporation, diffusion to cooler ground, and flow of cooler water into an area. Change in temperature is accounted for by the balance of heat into and out of any specific area. Solar, back radiation, and atmospheric interactions dominate heat budgets of most waters (Wetzel, 2001). Aquatic habitats such as wetlands with emergent vegetation can be influenced by interactions of the atmosphere with emergent vegetation (e.g., modification of the atmospheric boundary layer, condensation on plant leaf surfaces, shading). Once heat enters a water body, distribution through the water begins by processes of diffusion and advection.

The movement of heat can also be viewed as diffusion. An equation similar to that presented for chemical diffusion (Fick's law) can be used to describe diffusion of heat. In this case, J_Q is heat flux per unit area, and the concentration difference is replaced by the temperature difference $(T_1 - T_2)$.

$$J_Q = K \frac{T_1 - T_2}{x_1 - x_2}$$

where J_Q is heat flux (W/m^2), T_1 and T_2 are temperature at points x_1 and x_2, respectively, and K is the coefficient of thermal conductivity. Heat is transferred more rapidly across steep gradients of temperature (shorter distances or greater temperature differences).

As with molecular diffusion of dissolved materials, molecular diffusion of heat predominates when advective transport is minimal. Transport of hot or cold water by flow can redistribute heat rapidly. In contrast, heat is transferred more slowly by molecular collisions in the absence of transport. Thus, molecular diffusion of heat is considerably slower than transport diffusion of heat. Consequently, when lakes stratify, they generally maintain that stratification until processes at the surface bring the surface layer temperature equal to that of the deeper layers because diffusion of heat across layers of stratification is so slow. Stable stratification can result from heating of surface waters and the depth of the heated surface layer will be determined by how deeply light can penetrate and heat the water and how deeply the surface can be mixed by the wind or other water currents. Below the mixing depth the molecular diffusion of heat equilibrates disparate temperatures between the bottom and the top of the lake very slowly.

As with diffusion of chemicals, the relative rates of advective and molecular diffusion of heat can be captured with the Péclet (*Pé*) number

$$Pé = \frac{lU}{K}$$

where *l* is the characteristic length, *U* is the velocity, and *K* is the coefficient of heat conductivity.

The annual heat budget of a lake (not actually a "budget" but a heat storage capacity) indicates its heat content and how it changes over the course of a year. In temperate lakes, this budget is the total amount of energy required to increase the temperature of a lake from its winter minimum to its summer maximum. Not surprisingly, the annual heat budgets of tropical lakes are much smaller because their seasonal temperature differences are much smaller. Analyses of a large number of temperate lake heat budgets (Wetzel, 2001) lead to several generalizations: (1) most heat gain occurs in the spring—summer period, (2) some periods lead to relatively large changes in heat content (10%) in a few days, and (3) heat absorbed by sediments and heat of fusion of ice can have considerable influence on annual heat budgets.

In some ways, heat budgets of rivers and streams are more complex than those of lakes, particularly in smaller streams where groundwater temperatures have a large influence, or where surface runoff (floods, snowmelt) can change heat input over periods of days. Rivers with substantial influence from reservoirs can further complicate heat budgets; shallow and wide reservoirs can artificially warm rivers. If reservoirs are deep and water is released from cool areas, they can artificially cool rivers, as is the case in the Grand Canyon, where the naturally warm Colorado River is cold because of reservoir releases. Characteristic heat budgets of wetlands would fall somewhere between those of flowing waters and lakes.

SUMMARY

1. Diffusion of materials in water can be described by Fick's law. In this relationship, diffusion flux is related positively to a diffusion constant, the concentration gradient, and negatively to the distance between the source and location to which it is diffusing.
2. Diffusion constants can be affected by many factors. The most important factor is the movement of water. Moving masses of water result in diffusion constants many orders of magnitude greater than those for molecular diffusion in still water. Factors that increase diffusion constants also include increased temperature and

bioturbation. Sediment properties, such as affinity of the diffusing molecules for the sediments and low permeability, can be associated with low diffusion constants.

3. Diffusion properties can constrain natural selection with respect to relationships among chemicals and organisms and the body shape of organisms. This constraint includes uptake of nutrients and interactions between organisms that are mediated by chemicals.

4. The quantity and quality of light entering water can be altered by the atmosphere, the surface of the water, and any ice or snow cover over water.

5. Light is attenuated logarithmically when it enters water.

6. Red light is absorbed directly by water, and blue light is transmitted deepest in oligotrophic lakes. Phytoplankton chlorophyll pigments absorb red and blue light. Cyanobacteria also can absorb green light. Other dissolved and particulate materials in waters can also absorb and reflect light.

7. Movement of heat can be described with equations similar to those used to describe diffusion of dissolved materials in water.

8. Heat balance accounting can characterize dynamics of temperature and stratification in aquatic ecosystems.

QUESTIONS FOR THOUGHT

1. Why would you expect transport diffusion to occur in a still glass of water, if evaporation was occurring at the surface, and how would you keep such transport diffusion from occurring?

2. Why do many small invertebrates have a heart but no system of blood vessels?

3. Calculate the surface area to volume ratios for a sphere, a cube, and a right circular cylinder (1 cm high), each with a volume of 1 cm^3. Relate these to diffusion of materials to cells of different shapes (r = radius, h = height, S = surface area, and V = volume, for a sphere $S = 4\pi r^2$, $V = 4/3\pi r^3$, and for a right circular cylinder $S = 2\pi rh + 2\pi r^2$ and $V = \pi r^2 h$).

4. Why might large rivers generally have greater light attenuation coefficients than do large lakes?

5. If objects at the surface of a lake are blue, black, red, or white, what color would they appear if they were viewed at 20-m depth in an oligotrophic lake?

6. Use the data plotted in Fig. 3.6 to demonstrate that you obtain a straight line if the natural log (ln) of light is plotted against depth.

CHAPTER 4

The Hydrologic Cycle and Physiography of Groundwater Habitats

Contents

Figure 4.1 A stream leaving a limestone cave on the South Island of New Zealand.

Identification of aquatic habitats is based, in part, on landscape geomorphology and hydrology. The hydrologic cycle describes the movement of water from the oceans into the atmosphere and across land. In combination with other geological processes, the hydrologic cycle determines many of the physical characteristics of aquatic habitats. This chapter provides some details about the hydrologic cycle and then discusses the

Freshwater Ecology
DOI: https://doi.org/10.1016/B978-0-12-813255-5.00004-1

physical geology of groundwaters. Chapters 5–7 continue this theme with respect to wetlands, streams, rivers, lakes, and reservoirs. The reader should be aware of the many linkages among different types of aquatic habitats across the landscape, even though we discuss them in separate chapters. An understanding of physiography[1] provides a starting point for examining abiotic factors that drive aquatic ecosystems. While there are not clear lines between what is considered a stream, a wetland, or a lake, scientists have classified freshwater habitats as such based on gradients across axes that are based on physical factors. This chapter forms the starting for understanding how these habitats differ.

HABITATS AND THE HYDROLOGIC CYCLE

The definition of aquatic habitats can be based on geology and the *hydrologic cycle*, which describes the way water moves through the environment. Temporal and spatial variations in movement and distribution of water are called *hydrodynamics*. Parts of hydrodynamics were considered in Chapter 2; here, we talk about how water moves across the surface of the land. Links between terrestrial and aquatic ecosystems, as well as links among different aquatic ecosystems, are central to understanding how water moves across the surface of the continents. Chapter 1 included a brief description of a global water budget with respect to water available to humans. This chapter provides a more detailed consideration of the hydrologic cycle and hydrodynamics. The importance of hydrology to ecology is part of the relatively new field of study referred to as *ecohydrology*.

Aquatic habitats span a wide range of spatial and temporal scales; the organism or process of interest dictates the scale chosen for study. For example, proximity to a grain of sand can influence a microbe, but ecosystem processes dominated by microbes can be altered by position in a watershed. Changes in small-scale microbial communities can occur over periods of hours, but changes in ecosystems can take decades to millions of years. We present an indication of the range of habitats and link hydrodynamics and hydrology to aquatic ecology in Table 4.1. We already discussed basic physical properties of water movement; now we will explain how water moves across the landscape.

Climate results in varied quantities of precipitation around the world (Fig. 4.2). Precipitation, in combination with factors that influence return of water to the atmosphere, dictates how much water enters aquatic systems (Fig. 4.3). Precipitation is either intercepted by vegetation or falls directly on nonvegetated surfaces. Water can then return to the atmosphere by direct evaporation or by transpiration through plants.

---

[1] Physiography is an older term for physical geography, and refers to the "natural" rather than social part of the geography of a region.

Table 4.1 Habitat classification by time and distance scales

Habitat	Time range	Distance range	Examples
Microhabitat	1 second–1 year	1 μm–1 mm	Fine particles of detritus, sand and clay particles, surfaces of biotic and abiotic solids in the environment
Macrohabitat	1 day–100 year	1 mm–1,000 m	Riffles and pools in streams, rivers, and underground rivers. Logs, macrophyte beds, pebbles–boulders, position on lakeshore
Local habitat	1 month–1,000 year	1 mm–100 km	Lake, regional aquifer, stream or riffle reach, shallow lake bottom
Watershed	$1–10^6$ year	1–10,000 km	Areas feeding small streams to the basins of large rivers, including lakes, aquifers, and streams within boundaries
Landscape	$10–10^7$ year	10–10,000 km	A mosaic of local habitats or watersheds
Continent	$10,000–10^9$ year	>10,000 km	A composite of large drainage basins and aquifers

This classification is only one of many ways to divide a natural continuum.

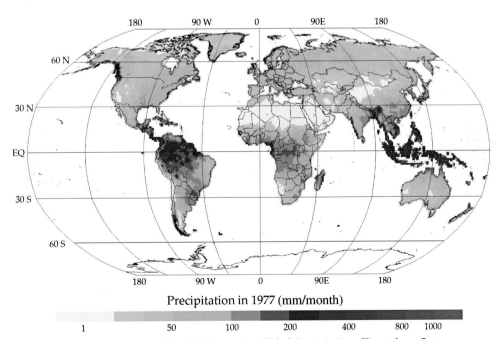

Precipitation in 1977 (mm/month)

1   50   100   200   400   800   1000

Figure 4.2 Global precipitation for 1997. From the Global Precipitation Climatology Center.

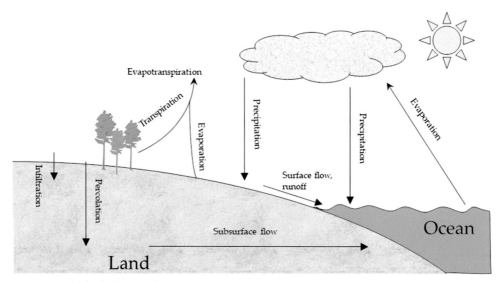

Figure 4.3 The hydrologic cycle.

Transpiration and evaporation together are *evapotranspiration*. If water is not lost to evapotranspiration, it infiltrates (flows down into) the soil or flows across its surface. Very heavy rain or impermeable surfaces can cause water to flow across the surface (*sheet flow*), which enters streams rapidly and can cause flooding.

Water that infiltrates can be stored as soil moisture (or as ice in polar or high-altitude areas). Soil moisture in incompletely saturated soils adheres to soil particles with capillary force. Water that is not stored *percolates* down through the soil layers into groundwater (*infiltration*), or flows through shallow surface soils to streams. This water is stored underground or eventually flows into rivers and streams (*runoff*).

Many factors combine to determine the amount of precipitation that ends up as runoff (*water yield*). The amount of precipitation required to cause runoff is greater when the temperature is high because *potential evapotranspiration* (the maximum possible water loss) is high. Low temperatures and rates of evapotranspiration characterize polar regions. Thus, wetlands, lakes, and streams can form with moderate amounts of precipitation at high latitudes. Impermeable soils or surfaces have high water yield (including rocks, packed soils, and desert "pavements"). Dry soils can hold considerable amounts of moisture, so saturated soils will have greater water yield per unit precipitation. Human alterations to the landscape that compact soils and increase the area of *impervious or impermeable surfaces* (surfaces that do not allow infiltration, such as paved streets and parking lots) can greatly alter patterns of surface flow and infiltration (Arnold and Gibbons, 1996).

Once water enters the freshwater systems, it can either be stored, flow away, evaporate, or in shallow habitats with plant roots in the water be transpired. Habitats in dry climates can have considerable evaporative losses, an important factor in reservoirs used to store water in dry areas, and a factor that causes formation of saline lakes. Surprisingly, plants growing close to streams do not always use stream water (Evaristo et al., 2015).

ADVANCED: PREDICTION OF AMOUNT AND VARIABILITY OF RUNOFF WITH GLOBAL CLIMATE CHANGE

Human activities can alter global patterns of precipitation and the hydrologic cycle in unpredictable ways. As the earth warms in response to increases of greenhouse gases (primarily CO_2), evaporation and precipitation will likely increase worldwide, but variability and distribution will also change. Global circulation models predict the influences of global warming and global climate change. These complex models divide the column of atmosphere above each segment of land into layers, and simulate interactions with the vegetation, soil, or water below. Each model gives somewhat different results on the basis of varying assumptions, requirements for data, and calculation approaches. The current general approach, as taken by the Intergovernmental Panel on Climate Change, is to look at the predictions of a wide variety of models (Pachauri et al., 2014). Understanding the implications of such models for continental aquatic systems requires understanding the movement of water once it reaches terrestrial habitats and moves into the aquatic realm.

Researchers downscale these models to make more local predictions. One study predicts that 100-year floods will occur twice as frequently over 40% of the globe due to a more energetic climate (Arnell and Gosling, 2016). Another study documents that climate change has already altered flood patterns with earlier snow-melt floods and earlier winter floods caused by more soil moisture (Blöschl et al., 2017). How such changes will influence local weather patterns is uncertain in many cases. Predicting the impacts of the greenhouse effect and global change on specific habitats is complicated because the balance between precipitation and evapotranspiration influences freshwater habitats greatly, and we do not completely understand the feedbacks among climate, vegetation, and runoff. The strongest effects will likely occur in areas that are currently arid or where precipitation is equal to or less than potential evapotranspiration (Schaake, 1990) and in areas currently dominated by permafrost (Liljedahl et al., 2016).

Predicting the amount of runoff as a function of precipitation is difficult enough when one assumes that conditions are similar over the long run (e.g., when past trends can be used to predict the future). With global climate change, the earth is shifting to states that were not necessarily common previously; we are living in a "no analog" hydrologic world (Milly et al., 2008). For example, water for many areas is supplied by summer snowmelt. Up to 60% of hydrologic change in the Western United States from 1950 to 1999 may be related to human influence on climate (Barnett et al., 2008), and is related to earlier snow melt and perhaps less snow accumulation. In another example, the Tibetan Plateau provides water for a substantial portion of the world's population. While there is strong indication that water availability will decrease to downstream areas overall with climate change, specific predictions are difficult because detailed regional data are not available in this part of the world (Immerzeel et al., 2010).

A warmer world means a more active (energetic) hydrologic cycle, so the total amount of precipitation falling should increase, as should the total amount of cloud cover. However, clouds and snow can amplify back-radiation of heat, making modeling difficult. A more active hydrologic cycle is also a more variable hydrologic cycle, creating variance that can make predictions more difficult. For example, 10 cm of rain that falls during five storms over a period of a month may lead to very different magnitude and frequency of runoff than 10 cm that falls in one storm over a month. The 10 cm falling in five storms may all evaporate before reaching groundwater or streams, and the 10 cm falling in one storm may runoff very quickly in sheet flow and result in flooding.

Humans have also increased evaporation globally (Jaramillo and Destouni, 2015). These increases are related to irrigation and larger surface area of water through creation of reservoirs. Global warming will further increase evaporation rates and will lead to a more active hydrologic cycle.

Increased temperature coupled with increased atmospheric CO_2 makes predicting the rate of evapotranspiration tricky. Evaporation should increase with temperature, as should transpiration, but with increased CO_2 availability plants can keep their stomata closed longer and still maintain substantial rates of photosynthesis. Thus, transpiration may actually decrease with increased availability of CO_2 despite warming temperatures. Precipitation that is more variable may alter vegetation as well. For example, more pronounced dry periods could lead to a greater probability of fire, leading to altered plant communities and a corresponding shift in whole-system transpiration, as well as interception of precipitation.

Specific local conditions also make understanding the effects of global change on hydrology difficult. For example, local warming in the Himalayas linked to global climate change is accelerating the melting of glaciers and snowfields. This melting threatens downstream areas by lowering water storage for the Indus and Brahmaputra basins, potentially influencing water security for over 60 million people (Immerzeel et al., 2010). These glaciers and their moraines impound freshwater lakes. More rapid melting of glaciers increases the probability of catastrophic failures of these natural impoundments, leading to destructive floods downstream known as Jökulhlaups. Such catastrophic flooding will probably become more likely with global warming (Ng et al., 2007).

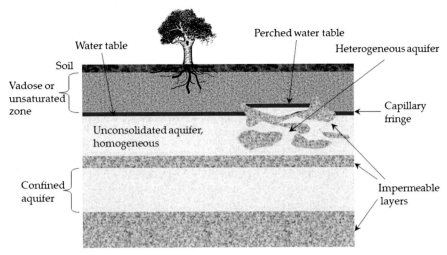

Figure 4.4 Various subsurface habitats.

MOVEMENT OF WATER THROUGH SOIL AND AQUIFERS

Several regions below the surface of soil receive infiltration (Fig. 4.4). The dry or moist sediments below surface soil layers form the *unsaturated zone* (also called the *vadose zone*). The depth of the unsaturated zone can vary from zero (where groundwater reaches surface water) to more than 100 m in some deserts. *Capillary fringe* is the area within the unsaturated zone where capillary action draws groundwater up into the pores or spaces in the sediment above the area completely saturated by water. This zone is generally 1 m or less above the *water table*, which is defined as the top of the region where the pore space is filled with groundwater.

Below the water table is the groundwater habitat. Methods are available for sampling groundwater (Method 4.1), but the spatial extent (vertical and horizontal) of samples that can be collected is limited, relative to those from surface water habitats. The extent of groundwater habitats and the adjacent capillary fringe and vadose zone often vary with precipitation patterns and can be difficult to define because they may not have distinct boundaries.

In this book, we use *aquifer* to describe continuous groundwater systems, but some use the term only for groundwater reservoirs that are useful to humans. The aquifer is also referred to as the *phreatic* zone. A tremendous amount of water is stored in aquifers below Earth's surface, but much of it is not available for human uses. Most of the water is too deep to access, unsuitable for human use because of high salinity, or tied up with mineral materials (many molecules and elements bind to water strongly).

Water stays in aquifers for variable lengths of time. Some aquifers exchange with surface waters in weeks or months. These include karst aquifers with relatively large

METHOD 4.1 Sampling subsurface waters

Sampling groundwater in unsaturated areas requires methods to extract the water that is in the soil. Lysimeters are used to sample water from the vadose zone (Fig. 4.5). This sampler has a ceramic cup on the end that absorbs water from the surrounding soil when it is placed under vacuum (Wilson, 1990).

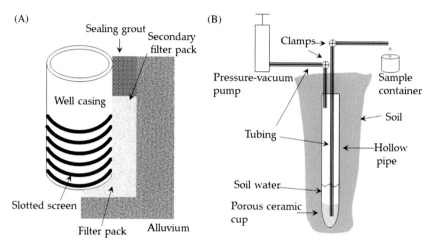

Figure 4.5 Some equipment used for sampling soil water and groundwater. (A) A slotted well casing packed with filtering materials allows sediment-free water to be sampled. (B) A vacuum sampler (lysimeter) relies on negative pressure to extract pore water from soil. The lysimeter is put under vacuum for sampling, and then it is put under pressure so the sample flows out of the sampling tube.

Groundwater is generally sampled through wells, but they can cover only a small part of the habitat. Shallow, temporary wells may be installed by hand where the water table is close to the surface and there are unconsolidated sediments. Deeper sampling requires well drilling machinery. When wells are drilled, samples of the pore water can be collected and sediments can be removed from the drilling apparatus. A split-spoon sample, in which the drill bit takes a core in its center as it cuts downward, is commonly used. The drill bit is then removed and split, and the core can be analyzed.

After a well is drilled, a casing is inserted through the length of the well with slots or screens placed in the region from which water is to be removed (Schalla and Walters, 1990). The outside of the well is then packed with sand fine enough to keep the sediments from the aquifer from entering and plugging the well when water is removed (Fig. 4.5). The fine packing materials used at the base of wells cause problems for researchers interested in groundwater animals because organisms larger than those able to pass though the packing material cannot be sampled. As a result, unless they have specially designed wells (e.g., such as those in Fig. 4.5 but without screens or fine packing materials), groundwater ecologists

(Continued)

METHOD 4.1 (Continued)

may miss significant components of the groundwater fauna. Material is packed in the hole outside the well casing above the slotted portion to form a seal. Otherwise, water could move vertically into the aquifer from the surface or between aquifer layers and contaminate the sample from the desired depth. Bentonite (a type of clay) is commonly used for this sealing because it is relatively chemically inert and swells when wetted.

Once a well is installed, it must be developed. Developing entails removal of a large volume of water and sediments in the water to ensure that the well flows clearly and supplies water representative of the aquifer. The well should be sampled regularly, with enough water removed so water will not become stagnant. During sampling, several volumes of water in the casing must be removed before the actual sample is collected to ensure that the water sampled is from the aquifer outside the well.

Various pumps and bailers are available for sampling groundwater. The type of analysis to be performed on the samples collected is determined before selecting the system. For trace metal analysis, pumps without metal parts that may contaminate samples are used. When organic materials are to be analyzed, pumps that do not use oil are essential because the oil may contaminate the water samples.

conduits for water to flow in (even underground "rivers" flow in some) and the shallow groundwaters connected to rivers in valleys dominated by coarse substrata such as boulders and cobbles. On the other end of the spectrum are deep aquifers like the Ogallala in the central United States that hold water for thousands or even millions of years; water in these systems is often referred to as *fossil water*. Because of slow replenishment rates, fossil water is a nonrenewable resource. Scientists know how old water is because water can be dated by measuring its isotopic composition. A large spike of radioisotopes was dispersed globally during nuclear atmospheric testing from the mid-1940s to the 1960s and entered the hydrologic cycle. Tritium is commonly used as an indicator of surface water entering an aquifer from this time. Proportions of the ^{14}C isotope in dissolved substances in water can be used to determine age up to tens of thousands of years, and isotopes of other elements can age waters to more than 1 million years old (Geyh, 2005).

Knowledge of groundwater flows and processes is integral to the study of all aquatic systems; it influences many aspects of streams (Allan, 1995; Jones and Holmes, 1996; Brunke and Gonser, 1997), wetlands (Mitsch and Gosselink, 1993; Batzer and Sharitz, 2006). and lakes (Freckman et al., 1997; Hagerthey and Kerfoot, 1998). We discuss these examples more specifically in the following three chapters. In some areas such as the southeastern United States, there are also large discharges of groundwater into marine waters (Moore, 1996).

Soil texture and composition determine how rapidly water percolates into groundwater habitats. Impermeable layers, such as intact layers of shale or granite, do not

Table 4.2 Representative particle sizes and hydraulic conductivity of various aquifer materials (Bowen, 1986)

Material	Particle size (mm)	Hydraulic conductivity (m/day)
Clay	0.004	0.0002
Silt	0.004–0.062	0.08
Coarse sand	0.5–1.0	45
Coarse gravel	16–32	150

allow water to flow deeper. In very fine clays or soils with large amounts of organic material, the rate of percolation can be very low. In contrast, gravel and sand allow for relatively rapid water flow (Table 4.2). Infiltration capacity partially determines the proportion of water that flows off the surface, and the quantity that enters groundwater. The rate at which water percolates into groundwater is the rate of *recharge*.

Infiltration rates have important practical consequences. For example, when sewage sludge, mineral fertilizers, or pesticides are applied to cropland, contaminants will enter groundwater if infiltration rates are high and crop uptake is low. Thus, infiltration rate is an important aspect of determining fertilizer application levels (Wilson et al., 1996). Extensive areas with impermeable surfaces associated with urban development can decrease flow into groundwaters, reducing recharge and increasing flooding problems when increased sheet flows occur, as we discuss in Chapter 6.

Permeability determines the potential rate of flow (*hydraulic conductivity*) through aquifers. This flow is variable and dependent on geology. Water will flow slowly in fine sediments and more rapidly where large pores and channels exist (e.g., in limestone aquifers with channels and unconsolidated sediments with large materials such as cobble). Hydraulic conductivity is partially dependent on the Reynolds number (see Chapter 2) because viscosity is high and flow is slow when Reynolds numbers are small (i.e., when sediment particles are small). Darcy's law (Chapter 2) can express the rate at which water moves through aquifers. This law states that flow rate in porous materials climbs with increased pressure and decreases with longer flow paths. Darcy's law is used to mathematically describe flow of groundwater and infiltration through the vadose zone (Bowen, 1986).

The amount of water held in sediment is determined by its *porosity*, or the volume fraction of pores and/or fractures. If sediment is saturated with water, there are two components: (1) water that will drain away from sediment (yield) and (2) water that is held in sediment (retention). As pores get smaller, retention increases because surface tension has more influence at smaller spatial scales. Greater flows are often found in more porous sediments because more porous materials tend to have more channels through which water can pass. For example, gravel and sand pack with relatively large spaces left between the particles for water to flow through. This packing results in large connected channels. Exceptions to this relationship exist; high-porosity sediments may have a low hydraulic conductivity when a large proportion of the pores are dead ends and not involved in flow. Porosity may not be related directly to flow rates because of uneven distribution of pore sizes and the *tortuosity*, or average length of the flow path between two points, which varies as a function of type of material (Sahimi, 1995).

An example of high porosity but low conductivity material is the carbohydrate material excreted by microbes. These extracellular products have a high proportion of water and many microscopic pores, but allow little if any flow through them because the pores are small and the Reynolds number precludes flow at such high viscosity. Microbial secretions can decrease flow through sediments (Battin and Sengschmitt, 1999), making microbial ecology of groundwater and soils significant to hydrology. A specific example of this importance occurs when microbial growth plugs flows from wells or septic systems.

Groundwaters are difficult to sample, and characterize fully. Scientists have recently developed large-scale methods that do not require drilling to understand depth to groundwater. For example, detailed monitoring of global positioning systems documented that loss in mass of groundwater in the Western United States during a time of intense drought in 2012 caused an uplift of 5 mm of the land surface. This uplift is due to the decrease in water mass that depresses the surface and corresponds to a loss of 10 cm of water (Borsa et al., 2014). Döll et al. (2014) used GRACE satellites, which monitor gravitational force, to verify modeled losses in global groundwater. The more water, the more mass, and the stronger the gravitational force that can be detected. They found evidence for global depletion of groundwater, and the greatest rates of loss occurred in India, United States, Iran, Saudi Arabia, and China.

The most difficult smaller scale water movements occur in heterogeneous aquifers with complex flow paths. However, understanding groundwater hydraulics is essential to describing its connection with surface habitats. This can extend to understanding ecology of groundwaters, potential yields of aquifers, and the potential for pollutants to spread in them. In groundwater ecology, for example, much of the carbon in many shallow groundwater systems comes from terrestrial sources. If it takes hundreds or thousands of years for water to reach the habitat from the surface, then the habitat is likely to be extremely oligotrophic. In contrast, many karst aquifers have very brisk flow and contaminants from the surface such as nutrients and pesticides can spread rapidly through them. Determining the velocity and direction of groundwater flow can be tricky. Several methods have been developed to measure flow direction and rate in groundwaters (Method 4.2). In all cases these methods are less straightforward than measurements of surface water, and heterogeneous aquifers require substantial sampling to understand their complexity.

METHOD 4.2 Sampling flow rates and directions in groundwaters

Piezometers are used to determine the depth of groundwater; these are tubes with open bottoms that are drilled into groundwater at various depths so the depth of water in the tube can be determined. Direction of flow is determined in both vertical and the horizontal directions. Nested piezometers (a group that is drilled to various depths in one location) are used to assess vertical flow, and piezometers spaced apart from each other can be used to determine horizontal direction of flow, as well as a vertical component.

The nest of piezometers is used to indicate pressure, with greater pressure leading to greater elevation of the top of the water inside the piezometer. If pressure differs between depths, then water will flow from the region of higher to lower pressure. Nested piezometers indicate an upward flow if the elevation of the top of the water in the piezometer tube that penetrates the aquifer to the deeper point is greater than the elevation of the water in the shallower tube (Fig. 4.6). The water is moving down if the elevation top of the water in the shallowest piezometer is greater than that in the deepest. If all elevations are the same, then the vertical flow is neither up nor down. If the flow is up, then water must be flowing into that location from somewhere else (either another part of the aquifer or through infiltration) and if the flow is down, then it must be replenished or the depth of the water table will decrease.

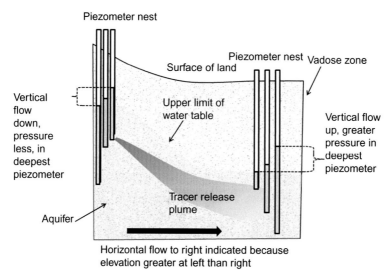

Figure 4.6 Ways to use small wells (piezometers) to estimate direction of groundwater flows. A nest (locally distributed) of piezometers can be used to determine vertical direction of flow, and spatially distributed piezometers can be used to determine lateral direction of flow based on tracers or water table elevation.

When piezometers are placed apart from each other, they can be used to estimate the local depth to the water table. If the elevation of groundwater at one site is lower than at another, and the two are hydraulically connected, then we know that water is flowing from

(Continued)

METHOD 4.2 (Continued)

the higher to the lower site. The difference in elevation between the two sites is the *hydraulic head*. Releasing a tracer at the upper site and monitoring its appearance at the lower site can indicate water velocity directly.

Several additional methods are used to assess flow rates through aquifers (Fitts, 2002; Kalbus et al., 2006). Slug tests consist of rapid water additions or removals from a well and monitoring of how long it takes the water level to reach its original depth. Equations can be used to analyze water depth data and relate them to flow. Longer periods of pumping can be used and the stable lower water table depth in the pumped well and nearby wells can be used to characterize flow rates. In tracer tests, water is added to the well with an inert tracer, and the rate of dilution can be used to calculate flow rates. Flow meters can be installed in wells to determine precisely where in the depth profile the water is flowing from (to characterize heterogeneity of flow). A more advanced method makes use of isotopic composition of water to determine water source and infer flow pathways (Maxwell and Condon, 2016).

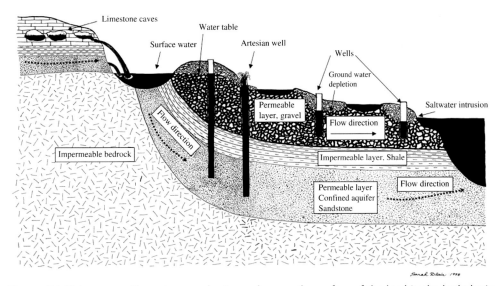

Figure 4.7 Water moves through groundwater and across the surface of the land in the hydrologic cycle. *After Leopold and Davis (1996); drawn by Sarah Blair.*

GROUNDWATER HABITATS

Groundwater habitats can be divided into a variety of subhabitats (Fig. 4.7). For example, groundwater can flow through regions of continuous *homogeneous* substrata (even distribution of permeable substrata such as sand, clay, or gravel) with little obstruction. Aquifers can occur where many alternate flow pathways exist because *heterogeneous* distribution of impermeable materials in the subsurface results in variable flow patterns and directions (e.g., aquifers with large rocks embedded in fine sediments or with patches of low hydraulic-conductivity material interspersed among high hydraulic-conductivity materials). Some rocks such as sandstone are permeable and have flow through the entire rock, while others such as granite only transmit water along cracks or fractures in the rock, and this leads to very complex flow patterns. An aquifer between two impermeable layers is *confined*. Heterogeneous flow patterns are of concern when considering groundwater contamination problems because such patterns interfere with assessing both the extent of the problem and attempts to clean up contaminants.

Groundwater is located worldwide, but the depth below the surface and the amount in the aquifer varies across the globe. In porous substrata, wells can yield a large amount of water. In some areas, such as those underlain by solid rock, groundwater yields can be very low. Examination of the distribution of large aquifers demonstrates heterogeneity in the types of aquifers found in the United States (Fig. 4.8). Some areas have extensive continuous aquifers (e.g., lower Mississippi Valley and High Plains) and others have sparser, localized aquifers (Rocky Mountain region). Humans are depleting groundwater in many of these aquifers at rates faster than the rate of recharge (Gleeson et al., 2012). A famous example of this is the huge Ogallala or High Plains Aquifer (Highlight 4.1). Groundwater depletion is commonly

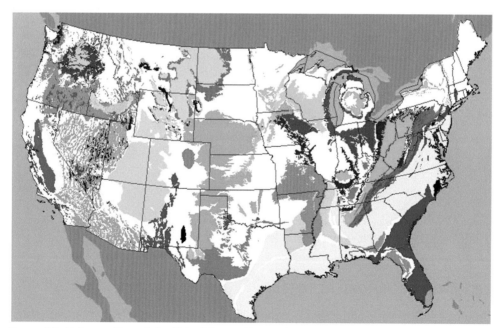

Figure 4.8 Major aquifers of the United States. Blue and turquoise areas are unconsolidated sand and gravel aquifers, yellow are semiconsolidated sand aquifers, green are sandstone aquifers, purple are sandstone and carbonate aquifers, brown and reddish brown are carbonate aquifers, and red and pink are igneous and metamorphic rock aquifers. *Image courtesy of the National Atlas, United States Department of the Interior.*

Highlight 4.1 Mining the Ogallala Aquifer

Groundwater depletion is a global problem (Konikow and Kendy, 2005), but we focus on one highly studied example to illustrate the problem. The High Plains or Ogallala Aquifer stretches from Nebraska to the southern tip of Texas (Fig. 4.9). The aquifer underlies 450,000 km^2 and has an estimated thickness of up to 300 m and an estimated water volume of 4,000 km^3. The volume is comparable to some of the larger lakes in the world. By the late 1980s, the aquifer supplied 30% of all irrigation water in the United States, and one-fifth of the irrigated cropland in the United States is fed by the aquifer (Kromm and White, 1992a; Sophocleous, 2005). Extraction of water from the Ogallala is ongoing.

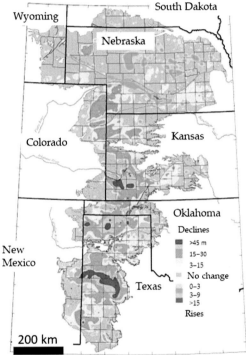

Figure 4.9 The extent of the High Plains Aquifer depletion from predevelopment to 2015. *Data from the U.S. Geological Survey, redrawn from McGuire (2017).*

Mean recharge rate is 1.5 cm/year, and withdrawal rates average about 10 times this rate. Between 1980 and 1994 large regions of the aquifer lost 10 m or more in depth. Precipitation to land above the aquifer is less than that required to support the crops that are irrigated from the aquifer (i.e., potential evapotranspiration exceeds precipitation). Annual withdrawals exceed the total annual discharge of the Colorado River, another important source of irrigation water in the United States (Kromm and White, 1992b). Some regions of the aquifer are very thick and can support withdrawals for decades. In many regions the water table has dropped far enough that it is not economically feasible to use the groundwater for irrigation when grain prices are low and fuel prices are high (Kromm and White, 1992b). Water is being withdrawn at unsustainable rates, so the withdrawals can be referred to as "mining" the aquifer. Rosenberg et al. (1999) predict that local effects of global climate change will increase the rate of decline of the water table.

In addition to loss of economic uses, there are ecological impacts as the water table is drawn deeper underground. Depletion of the groundwater has caused decreased water supply, and stream and river flow has disappeared in many regions (Kromm and White, 1992b). For example, the Arkansas River loses water to the aquifer because agricultural activity has lowered the water table, and a large section only flows during floods (Fig. 4.10). Loss of flow has negative impacts on migrating waterfowl that use the river, and decreases the ability of the river to dilute and remove pollutants. A number of small stream fishes in the region are endangered because most of the small streams no longer flow and perennial flow is not possible (Dodds et al., 2004; Perkin et al., 2015a,b). In the mid-2000s, severe drought substantially compromised surface flow for several years, probably leading to extirpation of several fish species.

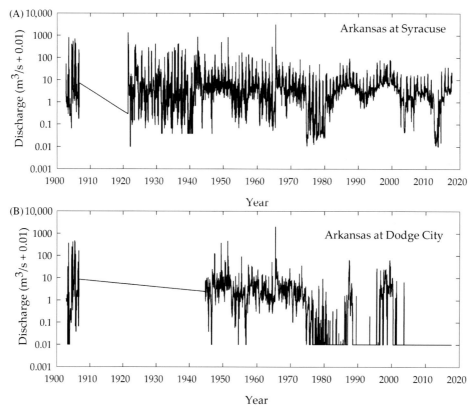

Figure 4.10 Discharge of the Arkansas River as it flows through a region of the High Plains Aquifer that has been used heavily for center-pivot irrigation since the 1960s. The Syracuse station (A) is immediately upstream of the High Plains Aquifer, and the Dodge City station (B) is downstream. Note that the logarithmic scale of discharge is modified so that it reads 0.01 m³/s when the river is dry. The river downstream of the aquifer has flowed only during periods of flooding since the early 1970s; it was almost never dry prior to that. *Data from the U.S. Geological Survey.*

Conserving the remaining water makes good economic and ecologic sense. It remains to be seen if more efficient irrigation technology and dryland farming will allow the same level of economic productivity as was made possible in the region during the past few decades by exploiting the High Plains Aquifer for irrigation water. As crop prices increase (for example as occurred when the US government offered tax breaks for producing ethanol for fuel from corn), the incentive to pump deeper and deeper groundwater increases. When the cost of the energy to pump water up exceeds the profits obtainable from crops, dryland farming will replace irrigation unless subsidies encourage farmers to keep pumping.

---

associated with irrigated land worldwide. Detailed satellite measurements of gravity indicate how overuse of water associated with agriculture could ultimately endanger food security (Famiglietti, 2014).

One of the major types of groundwater habitat is found in karstic limestone regions (White et al., 1995). Understanding specifics of karst aquifer hydrology is important for assessing the impacts of humans on groundwaters (Maire and Pomel, 1994). Large channels can form in these habitats because the water dissolves the limestone. If the water subsides, caves are left (Figs. 4.1 and 4.7). Pools and streams in limestone caves provide a type of groundwater habitat in which the geological formation allows humans to directly interact with and sample the subsurface habitat. *Sinkholes*, depressions in the land surface typical of karst regions, are often used for dumping household and farm wastes in rural areas. Sinkholes are generally closely connected to groundwater and cave systems, and they can be conduits for the rapid movement of pollutants into these underground systems. There have been some rather spectacular occurrences of sinkholes forming rapidly and swallowing houses and other surface features (Fig. 4.11). Sinkholes can become lakes or wetlands in many areas. Hydrology of karst aquifers is very complex, in part because it is difficult to predict the pathways of limestone dissolution (Mangin, 1994).

INTERACTION OF GROUNDWATERS WITH SURFACE WATERS

When groundwater impinges on the surface, a stream, lake, or wetland forms (Fig. 4.7). The dynamic zone of transition where both surface water and groundwater influences are found is referred to as the *hyporheic zone*. This zone forms a transitional habitat (*ecotone*) where there is a change between groundwater and surface water organisms. A hyporheic zone can occur between groundwater and wetlands, streams, or lakes (Gibert et al., 1994).

Figure 4.11 A sinkhole that formed in Bartow, Polk County, Florida on May 22, 1967 was 158 m long, 38 m wide, and 18 m deep. *Photograph courtesy of the U.S. Geological Survey.*

The hyporheic zone represents the interface through which materials exchange between surface and groundwater. This zone may include interstitial water of sediments below lakes and wetlands, gravel bars in rivers, sand below streams, and many other benthic habitats in aquatic systems. Saturated fine sediments located on the fringes of water bodies are often referred to as the *psammon*. As with any habitat classification, the distinction between the hyporheic zone and groundwater is unclear because the zone is transitional, varies over space and time, and depends on whether material transport or habitats of organisms are the response variable (Gibert et al., 1994a,b). For example, hyporheic zones formed by river action can be extremely complex because of erosion and deposition and associated sorting of particles into different sizes that naturally occur in the stream channel (Creuzé des Châtelliers et al., 1994). These zones can be extensive (stretch many kilometers from the river) in larger valleys with substantial amounts of large cobble and boulder (e.g., glacial valleys with large amounts of rocky material in the unconsolidated sediments (Stanford and Ward, 1993)). In the Flathead River system in Montana, stonefly nymphs (aquatic insects that have a terrestrial adult stage) occur in well water drawn from over 2 km away from the surface flowing river channel (Stanford and Ward, 1988), and the food webs are driven by biogeochemical processes deep in the hyphorheic zone (DelVecchia et al., 2016). The ecological importance of hyporheic zones worldwide has become increasingly apparent (Danielopol et al., 1994; Gounot, 1994; Allan, 1995; Stanley and Jones, 2000).

Methods exist to quantify connections between hyporheic zones and surface water (Harvey and Wagner, 2000). These methods include those already discussed for determining groundwater flow rates as well as use of natural or added tracers to detect the source of water. The rate at which groundwater interacts with surface water can be biologically important. Streams are often fed by subsurface flow, and lakes and wetlands can have extensive interaction with groundwater. There are several methods for determining the influence of groundwater on surface water (Kalbus et al., 2006). Groundwater is often a different temperature than surface water, so temperature gradients can reflect rates of groundwater input. Biologically conservative salts (their concentration not influenced by organisms) are often dissolved in groundwater at different concentrations than in surface waters, and the input/dilution of these salts can be used to assess rates of interactions of groundwaters with surface waters. Similarly, radon concentrations are often greater in groundwaters relative to surface water, and these concentrations can indicate inputs of groundwater.

Connections between groundwater and surface water are extremely important because most streams have at least some groundwater input during parts of the year. Thus, groundwater/surface water interactions are very important to water quality as well as dissolved material transport. One area where groundwater contamination reaching surface waters is of great concern is the process of high pressure treatment to fracture oil-bearing sediments to extract oil (fracking, see Highlight 4.2).

Highlight 4.2 Fracking

Fracking, or hydraulic fracturing to increase oil and gas yield from wells, has expanded tremendously since the 1990's. Shale oil and shale gas deposits targeted with the practice occur around the world, so this is indeed a global environmental issue (Burton et al., 2014). In this practice, companies drill vertically into shales or sandstones that contain previously unrecoverable oil that is locked into the sediment. Once drillers encounter a formation that contains oil or gas, they drill horizontally through the layer. The horizontal layer can be as thin as 30 m vertically. After drilling, water ($8,000-80,000$ m^3), chemicals, and sand are pumped into the formation at high pressure ($\sim 70,000-140,000$ kPa) to fracture or crack the layer. This fracturing allows oil and gas to flow through the formerly impermeable sediment through the well up to the surface. Hydraulic fracturing has created an economic boom, lowered fuel prices, and led to shifts toward more use of natural gas. It has also led to concerns about greenhouse gas escape, increased seismicity, surface mining of sand, increased ground and surface water contamination, and massive water demand for the process (Jackson et al., 2014). Here, we focus on ground and surface water contamination.

Water quality impacts range from sediment contamination associated with construction and operation of the facilities to contamination from the chemicals used in the process and those freed by contact with the deep sediments. There is a positive correlation between the number of wells and the degree of sediment pollution in streams in watersheds where fracking occurs (Entrekin et al., 2011). Many of these impacts can probably be mitigated to some degree by careful construction and site management procedures.

Of major concern is the over 750 chemicals, including 29 known to be toxic or carcinogenic, added to water used to fracture the shale and sandstone deposits (Entrekin et al., 2011). Additionally, chemicals are released into the water from the deposits during the process because high pressure, temperature, and the chemical additives all can cause dissolution of harmful chemicals that have been locked underground. Some of the water flows back up the well with the gas, and this water can be very saline in addition to containing other harmful chemicals. Disposal of this water is a major challenge for the industry because it is toxic (He et al., 2017). If this water enters streams and lakes, it can cause considerable environmental damage. In addition, the process causes seismic shifts that could potentially alter groundwater flow paths and allow contaminated underground waters to reach the surface.

The massive demand for water used in fracking is one additional environmental and social concern. If large-scale water use dries streams, both organisms that live in the streams and people that depend on them for water are negatively affected. While industry practices have been adopted to lessen the negative impacts of fracking, there is recent concern in the United States as the current administration in 2017 is decreasing regulations on the industry. In a bigger sense, the increased availability of oil and gas coupled with lower prices have decreased demand for fuel efficiency and subsequently increased release rates of greenhouse gases.

SUMMARY

1. Freshwater habitats vary in scale from individual sediment particles to continental watersheds. The appropriate scale of investigation depends on the question asked.

2. Precipitation falls unevenly across the earth and evaporates or is transpired (evapotranspiration) at different rates depending on a variety of factors, including global weather circulation patterns, geography, and landscape-level influences. Water that is not lost to evapotranspiration either flows across the surface of the land to streams and rivers or infiltrates the soil, becoming groundwater.

3. Characteristics of the medium between soil and groundwater alter the rate at which water flows below the surface and into aquifers. Generally, water flows more slowly through fine-grained sediments. Once water enters an aquifer, the rate at which it moves through is also dependent on slope and the materials that make up the aquifer. Water flows very slowly between the pores of fine-textured sediments such as silts and clays or those with large amounts of organic materials, and relatively rapidly in coarse gravel or limestone with large channels and pores.

4. The hydrodynamics of groundwater dictate its use as a water resource, the ecology of its unique biota, and interactions with other aquatic habitats. Efforts to clean up groundwater pollution also depend on knowledge of groundwater and soil characteristics, particularly flow characteristics.

5. Groundwater interacts with surface water and in many cases dictates important properties of that water. Many streams are primarily fed by groundwater, as are many lakes and wetlands. Surface water generally occurs where the top of the water table is above the elevation of the land surface. Hyporheic zones are at the interface between surface and groundwater habitats, and they can be extensive and ecologically significant components of many freshwater systems.

QUESTIONS FOR THOUGHT

1. Why would understanding the hydrodynamics of groundwater be important when a large oil spill occurs on land (e.g., leakage of a gasoline or oil storage tank)?

2. Darcy's law as presented in the text is one-dimensional (the equation in Chapter 2). When would a three-dimensional model be more useful?

3. Why might understanding the application of Darcy's law to sediments under a wetland be important when calculating a water budget for a wetland?

4. Why are groundwater temperatures one good indicator of annual air temperatures and ultimately regional warming?

5. How could temperature alter flow rates of groundwater?

6. What differences might you expect between the biota inhabiting hyporheic and groundwater habitats?

CHAPTER 5

Hydrology and Physiography of Wetland Habitats

Contents

Figure 5.1 A flock of wintering northern pintail ducks (*Anas acuta*) takes flight from the banks of a wetland in Northern Honshu, Japan. *Photo courtesy: The United States Geological Survey.*

Freshwater Ecology
DOI: https://doi.org/10.1016/B978-0-12-813255-5.00005-3

Wetlands are critical habitats for many plants and animals, including numerous threatened and endangered species. They provide vital and valuable ecosystem services such as flood control and the maintenance of water quality (van der Valk, 2012). Wetlands often receive and process large amounts of runoff from the landscape, and in some cases they are used to treat wastewater. In addition, wetlands are globally important in biogeochemical cycles as they are natural sources of methane (see Chapter 13), a trace gas that plays an important role in the regulation of climate (Schlesinger, 1997). Wetland sediments are valuable because they preserve a long-term record of environmental conditions, and sediments in peat bogs are mined for fuel and use in gardens (Fig. 5.2). Many coal deposits that humans use for energy today came from deposition in wetland habitats over millions of years. Studies of wetlands are less common than studies of lakes and streams because they fall between the traditional disciplines of limnology and terrestrial ecology. However, there is ever increasing interest in wetlands and research on them because of increasing recognition of their ecologic, economic, and esthetic value.

Many wetlands harbor high biological diversity, including numerous terrestrial and aquatic species along with wetland specialists because wetlands are often *ecotones* (transitional habitats); they are frequently adjacent to terrestrial habitats and streams, lakes, or oceans. Some wetlands such as ephemeral pools and acidic bogs host a limited suite of species specially adapted to harsh conditions that can include short periods of inundation, low O_2 levels, or low pH. Although wetlands are generally considered highly productive ecosystems, productivity is actually quite variable; estimates of net primary production in freshwater wetlands range from 800 to 4,000 $g\,C\,m^{-2}\,year^{-1}$, and some tropical and temperate wetlands are among the most productive systems on the planet (Lieth, 1975).

Figure 5.2 Peat moss mining from a bog in Ireland for use as fuel. *Photo courtesy: Charles Ruffner.*

DEFINITION OF WETLANDS

One of the problems with studying and managing wetlands is defining them. Although distinct delineation is difficult between a wetland and a very shallow pond, or a slow, shallow side channel of a stream, the problem of finding a definition for wetlands lies more in deciding what is a wetland versus what is a terrestrial habitat. Wetland definitions and delineations are generally based on the plants that are present (often water-loving plants, called *hydrophytes*) or absent (flood intolerant species), and distinct soils with characteristics related to frequent inundation. The *hydric soils* of wetlands have characteristic colors and textures (often with dark, highly organic surface soils overlying gray mineral layers), and they are anaerobic even in upper layers during the growing season. Some wetland soils are very nutrient poor, resulting in unusual assemblages of plants adapted to low nutrient conditions. Carnivorous plants such as pitcher plants and sundews, which lure and trap insects to gain nutrients (see Chapter 16), can be abundant in nutrient poor wetlands.

The legal definition of wetland status has particular importance with respect to the requirements for wetland preservation. Policymakers are increasingly realizing the central importance of wetlands as wildlife habitat and key providers of valuable ecosystem services. Environmentalists have pressed for more inclusive definitions of wetlands. Many people want more freedom to develop and drain both seasonally wet regions and permanent wetlands and in the United States, the degree of connection of wetlands to other waters has become a major issue in the ability to regulate them. Consequently, numerous definitions of wetlands have been developed by scientists, policymakers, and others (Highlight 5.1). However, there is still no single, indisputable, ecologically sound definition for wetlands because wetland types are so diverse (Sharitz and Batzer, 1999).

Wetlands have been drained, filled in, or otherwise modified worldwide. In the United States, 70% of the *riparian* (near rivers) wetlands were lost between 1940 and 1980, and more than half of the *prairie potholes* (shallow glacial depressions in the northern high plains that form vital habitat for waterfowl), as well as large portions of the Florida Everglades, have been lost since pre-European settlement (Mitsch and Gosselink, 2007). This degree of loss is typical for all types of wetlands in the United States. Twenty-two states have lost more than half of their wetlands in the past 200 years (Fig. 5.3). The amount of loss and interconversion among wetland types has been great (2.5% lost over a period of 10 years), and is driven primarily by agriculture. The creation of ponds, lakes, and reservoirs from wetlands is also an important human activity that has negative consequences for many wetland species (Table 5.1). Peatlands are under increasing pressure throughout the world as a source of fuel and peat moss for gardening. In Southeast Asia, people modify existing wetlands and create many artificial wetlands to allow for rice culture (Grist, 1986); in China half of the current

Highlight 5.1 Definitions of wetlands

Mitsch and Gosselink (1993) and the Committee on Characterization of Wetlands (1995) chronicled several definitions of wetlands. The definition used often depends on the requirements of the user.

United States Fish and Wildlife Service: Wetlands are lands transitional between terrestrial and aquatic systems where the water table is usually at or near the surface or the land is covered by shallow water. Wetlands must have one or more of the following three attributes: (1) at least periodically, the land supports primarily hydrophytes; (2) the substrate is predominantly undrained hydric soil; and (3) the substrate is nonsoil and saturated with water or covered by shallow water at some time.

Canadian National Wetlands Working Group: Wetland is defined as land having the water table at, near, or above the land surface or which is saturated for a long enough period to promote wetland or aquatic processes as indicated by hydric soils, hydrophytic vegetation, and various kinds of biological activity which are adapted to the wet environment.

Section 404 of the 1977 United States Clean Water Act: The term "wetlands" means those areas that are inundated or saturated by surface or groundwater at a frequency and duration sufficient to support, and that under normal circumstances do support, a prevalence of vegetation typically adapted for life in saturated soil conditions. Wetlands generally include swamps, marshes, bogs, and similar areas.

1985 United States Food Security Act: The term "wetland," except when such term is part of the term "converted wetland," means land that (1) has a predominance of hydric soils; (2) is inundated or saturated by surface or groundwater at a frequency and duration sufficient to support a prevalence of hydrophytic vegetation typically adapted for life in saturated soil conditions; and (3) under normal circumstances does support the prevalence of such vegetation.

1995 Committee on Wetlands Characterization, United States. National Research Council: A wetland is an ecosystem that depends on constant or recurrent shallow inundation or saturation at or near the surface of the substrate. The minimum essential characteristics of a wetland are recurrent, sustained inundation or saturation at or near the surface and the presence of physical, chemical, and biological features reflective of recurrent, sustained inundation or saturation. Common diagnostic features of wetlands are hydric soils and hydrophytic vegetation. These features will be present except where specific physicochemical, biotic, or anthropogenic factors have removed them or prevented their development.

Some other words historically used to delineate wetlands or specific types of wetlands: The term wetland has only been in regular use by scientists since the mid-1900s. Terms used prior to this, or to indicate specific types of wetlands, include bog, bottomland, fen, marsh, mire, moor, muskeg, peatland, playa, pothole, reed swamp, slough, swamp, vernal pool, wet meadow, and wet prairie.

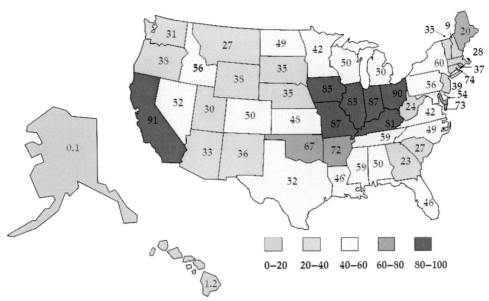

Figure 5.3 Wetland loss in United States since 1780, labeled with percentage lost. *Data from Dahl et al. (1991).*

Table 5.1 Conversion and losses of various wetland types including agricultural and urban uses from the mid-1970s to the mid-1980s in the United States

Wetland type	Amount in mid-1970s	Amount in mid-1980s	Total loss (gain)	Agricultural	Urban land use	Deepwater	Other	Conversion to other wetland types
Swamps	223,000	209,000	−14,000	−4,000	−240	−200	−4,360	−5,200
Marshes	98,000	99,000	1,000	−1,500	−150	0	−350	3,000
Shrubs	63,000	62,000	−1,000	−1,000	0	0	−1,700	1,700
Ponds	22,000	25,000	3,000	900	0	0	1,800	300
Total	406,000	396,000	−11,000	−5,600	−390	−200	−4,600	−200

Deepwater represents conversion to lakes or reservoirs and conversions are placed in the other category if they are not to agriculture, urban, or deepwater. Values are given in thousands of km^2. Positive values indicate a net gain.
Source: Data from Dahl et al. (1991).

wetland area is artificial (Junk et al., 2013). The *rice paddies* feed billions of people and have done so over the centuries. These paddies are the foundation of some of the world's great civilizations, the Chinese dynasties, as well as many others in India and throughout monsoonal Asia. The decline in wetlands is global (Dugan, 1993); large percentages of wetlands have been lost in the United States (54%), Cameroon (80%), New Zealand (90%), Italy (94%), Australia (95%), Thailand (96%), and Vietnam (99%).

WETLAND CONSERVATION AND MITIGATION

In response to decades of wetland destruction, and the realization of the ecological and economic values of wetlands, laws in the United States and other countries increasingly mandate protection of wetlands and, in cases where wetlands are "unavoidably" disturbed or destroyed, some form of compensation. Mitigation, or compensatory mitigation, is a procedure for offsetting wetland loss by creating another wetland, restoring or enhancing an existing wetland, or providing funding to conservation groups or agencies for wetland preservation. Wetland mitigation programs have slowed the net loss of wetland habitats, but there are questions regarding how closely some created or restored wetlands approximate natural systems in terms of their biological diversity and ecosystem functioning (Fig. 5.4) (Galatowitsch and van der Valk, 1995; Brooks et al., 2005; Meyer et al., 2008). We present a more complete discussion of wetland restoration later in this chapter.

On a global scale, the Ramsar Convention on Wetlands is one of the most successful conservation programs in the world. Named because it was originally signed in Ramsar, Iran in 1971, the Ramsar Convention is an intergovernmental treaty that requires participating entities to designate and protect wetlands of international significance, as well as promote wetland conservation. As of 2017, there were 169 signatories to the Convention, with 2,231 wetland sites totaling 2.1 million km^2 designated for inclusion in the Ramsar List of Wetlands of International Importance. Of these wetlands, there are 18 international transboundary sites, and 15 regional initiatives involving multiple countries. Originally developed to protect critical habitats for migratory

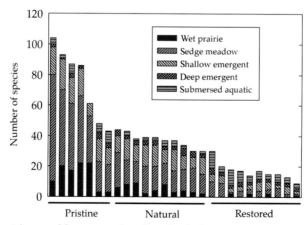

Figure 5.4 Species richness of five categories of wetland plants in pristine, natural (natural systems receiving agricultural runoff), and restored prairie pothole wetlands. The restored wetlands were formerly cropped and natural revegetation was allowed. *Redrawn from Galatowitsch and van der Valk (1995).*

birds, the Ramsar Convention is the only international conservation program we are aware of that explicitly targets wetlands.

Section 404 of the United States Clean Water Act requires permits for discharging any fill materials into any national water bodies including wetlands. According to the Clean Water Act, any impacts to wetlands must first and foremost be avoided. Second, remaining impacts must be minimized. Finally, any unavoidable impacts must be compensated for through mitigation procedures. The food security act of 1985, often called "Swampbuster," further strengthened wetland protection in the United States; it includes a variety of provisions to discourage the conversion of wetlands for agricultural production. Under Swampbuster, farmers can be denied federal farm program benefits if they alter wetlands on their property.

Opportunities for mitigation procedures in the United Sates were expanded in the 1996 Federal Agricultural Improvement and Reform Act, which also began the process of developing wetland mitigation "banks." Wetland mitigation banks are wetlands that have been set aside or restored for the purpose of providing compensation for unavoidable impacts to other wetlands. The value of a bank is defined in compensatory mitigation credits. Mitigation banks represent third party compensation, in that the responsibility for compensatory mitigation is assumed by someone other than the permit holder that impacts a wetland. Mitigation banking is becoming increasingly popular; the Society of Wetland Scientists wrote a position paper in 2004 endorsing mitigation banking as a sound practice that will help achieve the goal of no net loss of wetland habitats. However, most wetland scientists agree that site selection, design, monitoring procedures, and long-term management of mitigation banks and other forms of compensatory mitigation are still evolving and in need of further refinements.

In the European Union, the Water Framework Directive indirectly includes protection of wetlands. This directive requires protection and preservation of all surface waters with respect to species diversity and ecosystem functions. In contrast to the current legal framework in the United States, small streams and isolated wetlands are considered connected to surface waters and are protected under this framework. However, human influence in Europe has been prolonged, and land-use intensive. Many European wetlands have been drained and lost forever.

WETLAND TYPES

Wetlands of various types are distributed worldwide, with particularly large areal coverage in northern Europe, northern North America, and South America. The processes that form these wetlands vary (Table 5.2), but are generally similar to the processes that form lakes (which we discuss in more detail in Chapter 7). Large wetland complexes are found on most continents, with fewer in Australia and none in Antarctica, and in general are associated with large rivers or high latitude glacially influenced areas (Table 5.3). The traditional wetland classification system was

Table 5.2 Some major world wetland types

Type	Description	Distribution	Geomorphology	Vegetation	Ecosystem importance
Tidal salt marsh	A halophytic grassland or dwarf brushland on riverine sediments influenced by tides or other water fluctuations	Mid to high latitude, on intertidal shores worldwide	Form where sediment input exceeds land subsidence in regions with gentle slopes	Salt-tolerant grasses and rushes/ periphyton	Highly productive, serves as nursery area for many commercially important fish and shellfish
Tidal freshwater marsh	Wetland close enough to coast to experience tidal influence, but above the reach of oceanic saltwater	Mid to high latitude, in regions with a broad coastal plain	Area with adequate rain or river flow, with a flat gradient near coastline	High plant diversity including algae, macrophytes, and grasses	Highly productive, many bird species; often close to urban communities and susceptible to human impacts
Mangrove	Tropical and subtropical, coastal, forested wetland	25° north to 25° south worldwide	Forms in areas protected from wave action including bays, estuaries, leeward sides of islands, and peninsulas	Halophytic trees, shrubs and other plants; generally sparse understory	Exports organic matter to coastal food chains, physically stabilizes coastlines; may serve as a nutrient sink
Freshwater marsh	A diverse group of inland wetlands dominated by grasses, sedges, and other emergent hydrophytes; includes important types such as prairie potholes, playas, and the Everglades	Worldwide	Widely varied	Reeds such as *Typha* and *Phragmites*; other grasses such as *Panicum*, sedges (e.g., *Cladium*, *Cyperus*, and *Carex*); broad-leaved monocots (*Sagittaria* spp.); and floating aquatic plants	Wildlife habitat, can serve as nutrient sink
Northern wetland	Bogs and peatlands characterized by low pH and peat accumulation	Cold temperate climates of high humidity, generally in Northern Hemisphere	Forms in moist areas where lakes become filled in or where bay vegetation spreads and blankets; often a terrestrial ecosystem	Acidophilic vegetation, particularly mosses, but also sedges, grasses, and reeds	Low productivity system
Deepwater swamp	Freshwater most or all of the season, forested	Southeast United States	Varied	Bald cypress—tupelo or pond cypress—black gum	Can be low nutrient or high nutrient; can serve as a nutrient sink
Riparian wetland	Wetland adjacent to rivers	Worldwide	In floodplains of rivers in regions with high water table	High diversity of terrestrial plants	Can provide key wildlife habitat and productivity particularly in more arid regions; can act as an essential nutrient filter

Source: After Mitsch and Gosselink (1993).

developed as part of the National Wetlands Inventory (Cowardin et al., 1979). Wetlands can be classified by geomorphology, hydrology, climate, nutrient input, and vegetation (Table 5.4). An alternative classification system for wetlands allows assessment of wetland functions (Brinson et al., 1994). Wetlands can be coarsely divided into two general types, inland and coastal. Four broad geomorphic classifications are riverine, depressional, coastal, and peatland.

Coastal or *fringe wetlands* include *tidal marshes*, which can be freshwater or saline, and *mangrove wetlands*, which are saline and can be quite extensive along tropical coastlines. There are over 11 million hectares of coastal wetlands in the United States (Field et al., 1991), but these systems are often highly modified or completely destroyed by coastal development. People are becoming increasingly aware of the economic and ecological significance of coastal wetlands, and lessons from hurricane Katrina in 2005 illuminated the importance of these systems as buffers from storms. Regions of New Orleans bordering intact coastal wetlands experienced less damage from Katrina than areas where wetlands had been drained and developed, and studies after the storm indicated that the overall destruction in the region would have been reduced if more extensive coastal wetlands had been present to buffer the storm (Costanza and Farley,

Table 5.3 The largest wetland complexes globally

Rank	Location (continent)	Description	Area (thousand km^2)
1	West Siberian Lowland (Eurasia)	Bogs, mires, fens	2,745
2	Amazon River Basin (South America)	Savanna and riverine forest	1,738
3	Hudson Bay Lowland (North America)	Bogs, fens, swamps, marshes	374
4	Congo River Basin (Africa)	Swamps, riverine forest, wet prairie	189
5	Mackenzie River Basin (North America)	Bogs, fens, swamps, marshes	166
6	Pantanal (South America)	Savanna, grassland, riverine forest	160
7	Mississippi River Basin (North America)	Bottomland hardwood forest, swamps, marshes	108
8	Lake Chad Basin (Africa)	Grass and shrub savanna, shrub steppe, marshes	106
9	River Nile Basin (Africa)	Swamps, marshes	92
10	Prairie Potholes (North America)	Marshes, meadows	63
11	Magellanic Moorland (South America)	Peatlands	44

Source: After Fraser and Keddy (2005) and Keddy et al. (2009).

Table 5.4 Some examples of expected ecosystem functions of wetlands based on hydrodynamic characterization

Primary water source	Climate	Geomorphological aspects	Important quantitative attributes	Functions that can relate to ecological properties	Significance of function or maintenance of characteristic
Precipitation	Humid	Poor drainage	Precipitation exceeds evapotranspiration during most times of year so soils waterlogged	Soil constantly waterlogged leading to peat formation and sediment anoxia	Low plant productivity related to anoxic sediment keeping plants from soil sources of nutrients, plants rely upon nutrients in precipitation only
Surface flow from flooding river	Mesic–humid	Floods occur at least annually	Frequency and height of floods and position of wetland an index of connectivity to river	Overbank flow creates influx of nutrients and moves sediments (changes physical structure)	Allows continued high production and high habitat heterogeneity.
Groundwater influx	Mesic	Groundwater springs and seeps often at bottom of slopes or stream margins, some sediments must be permeable to allow influx	Aquifer permanence, yield of springs and seeps dominates hydrologic throughput	Groundwater supplies nutrients and flushes habitat, habitat often very stable	High plant production, stable plant community

Source: Based on Brinson et al. (1994).

Figure 5.5 Researchers working on plant diversity plots in a *Spartina alterniflora* dominated tidal saltwater marsh on the Gulf Coast of Mississippi. *Photo by Loretta Battaglia.*

2007). In 2004, a giant tsunami in the Indian Ocean killed about 230,000 people. Areas with intact mangrove wetlands were more protected against damage than those where the wetlands had been removed for development (Kamthonkiat et al., 2011).

Tidal salt marshes occur throughout the world. These wetlands are brackish and influenced by ocean tides. In the United States, these systems are usually dominated by *Spartina* grasses and *Juncus* rushes (Fig. 5.5). Wet and dry cycles, salinity, and large temperature fluctuations make these challenging habitats for biota and limit diversity, but high abundances of some types of crabs, snails, and other marine invertebrates live in these habitats and are critical food resources for waterbirds and other predators. Further inland, tides influence *freshwater tidal marshes* but these wetlands are not saline, and often have higher biological diversity and are more productive than brackish wetlands closer to the ocean that experience wide swings in salinity. A variety of plants, including grasses, sedges, cattails, and wild rice grow in freshwater tidal marshes. Like their saline counterparts, these systems are critical foraging habitat for birds that feed on the abundant invertebrates, amphibians, and seeds.

Mangrove wetlands (or mangrove swamps) are coastal wetlands dominated by *halophytic* (plants able to grow in saline water) mangrove trees, which include 12 genera distributed across tropical and subtropical coastal areas. These wetlands develop in areas where wave action is minimal and sediments accumulate. Sediments in mangrove swamps are generally anoxic, and thus many species of mangrove trees such as black mangrove (*Avicennia germinanas*) have *pneumatophores*, which are aerial roots that take up atmospheric oxygen for transport to the roots in the sediments. Mangrove swamps

Figure 5.6 Above and below the water structure of mangrove trees. *Photo courtesy: US National Oceanic and Atmospheric Administration.*

show characteristic gradients of species composition or zonation from the open water to land. The above and below water structures associated with mangrove trees provide important habitat for a variety of invertebrates and vertebrates, including commercially important shrimps and oysters (Fig. 5.6). Mangrove swamps and other types of coastal wetlands export much of their plant production to adjacent open water habitats as detritus, which is important fuel for some coastal marine food webs.

Wetlands known as *floating marshes* are common in some coastal areas. These systems are characterized by vegetation that forms thick mats of roots and associated organic matter (e.g., peat) that float on the water. Because they develop on top of the water, the elevation of the plant community changes with fluctuating water levels. Within the normal range of water level fluctuations, water does not inundate floating marshes, which are therefore resistant to flooding disturbances. Floating marshes in the Mississippi delta of the United States are often dominated by maidencane (*Panicum hemitomum*), a grass that can grow over 2 m tall and forms thick, dense mats of roots and rhizomes.

Historically, extensive floating marshes formed in Iraq near Basra above the confluence of the Tigris and Euphrates rivers. These marshes formed on mats of cane material and served as home to the "Marsh Arabs" who built reed houses on floating islands. The lush area provided habitat for fish and wildlife, as well as human inhabitants for all of recorded history. However, by the mid-1980s, a third of the marshes were drained by Iraq during the Iran—Iraq war to facilitate movement of supplies. Saddam Hussein drained more of the marshes because Iran-backed rebels were using them for shelter. By the mid-1990s, most of the marshes were drained and the majority of the Marsh Arabs no longer lived in the area. There have been recent efforts to restore the marshes, in particular the Eden Again project (Alwash et al., 2003), founded by Dr. Azzam Alwash, the founder of Nature Iraq Foundation, and his spouse, Dr. Suzie Alwash, a geologist from the United States. However, despite these efforts, human activities continue to endanger the region and impede restoration efforts. Turkey has dammed water upstream, and agricultural irrigation and other abstractive uses of river water have further decreased flow downstream. Decreasing discharge allows saline water to move into the wetland areas, and the brackish water is not drinkable or suitable for growth of the native plants (Lawler, 2016). It now appears unlikely that the ecosystem structure and function, or the historical way of life of the Marsh Arabs, will ever be restored.

Inland wetlands include *depressional* and *fringe* (those on the edges of other habitats) formations such as *marshes*, *swamps*, *riverine or riparian wetlands*, and *peatlands*. We describe depressional formation processes more fully in the context of lakes in Chapter 7, and the formation of riverine wetlands in Chapter 6. In the United States, marshes and swamps are generally defined by the dominant vegetation; *marshes* are dominated by emergent herbaceous vegetation and *swamps* are dominated by trees. However, this generalization does not extend to some other parts of the world such as Europe and Africa, where wetlands dominated by herbaceous plants are often called swamps. Inland marshes and swamps are some of the most extensive wetlands on the planet and can vary tremendously in timing, frequency, and magnitude of inundation as well as in morphology and size (Fig. 5.7). Extensive marshes such as the Florida

Figure 5.7 An inland freshwater marsh in Louisiana with abundant floating and emergent vegetation including water lilies (*Nymphaea* spp.) and American lotus (*Nelumbo lutea*). *Photo by Loretta Battaglia.*

Figure 5.8 A cypress swamp in Louisiana showing the buttressing and "knees" of the bald cypress trees. The water is covered with duckweed (*Lemna minor*), a floating aquatic plant that is abundant in wetlands. *Photo by Loretta Battaglia.*

Everglades, the fringe marshes associated with the Great Lakes, and the Pantanal in South America are recognized globally as critical habitats for birds and other wildlife.

In the southeastern United States, extensive swamps such as the Okefenokee in southern Georgia and adjacent Florida and the Great Dismal Swamp of southern Virginia and North Carolina are dominated by trees such as bald cypress (*Taxodium distichum*), tupelo (*Nyssa* spp.), and others tolerant of saturated soils and frequent inundation. Trees growing in swamps often have different growth forms than even the same species growing in drier habitats. For example, bald cypress trees in swamps grow "*knees*" and *buttressed* bases (Fig. 5.8), which are absent or not as well developed on bald cypress trees growing in upland areas. Knees and buttressed bases may increase stability in the soft, saturated soils; bald cypress are one of the most resistant trees to windfall during storms. Buttressing is common among many species of trees found in wetland habitats.

Swamps and marshes are often associated with river *floodplains*. Much of the bottomland hardwood forests associated with large rivers in the southeastern United States (e.g., Mississippi), Africa (e.g., the River Nile, the Congo River), and South America (the Pantanal) form globally important wetlands. In humid regions, floodplain forests can extend tens or even hundreds of kilometers (e.g., the Amazon basin) from the main channel of the river, whereas in arid climates they tend to be more constrained to narrow ribbons along the river. These *riparian* systems can range from ephemeral to nearly permanent inundation depending on local topography and connectivity to the river. Natural flood cycles facilitate the exchange of nutrients and energy between floodplain wetlands and the river channel (Junk et al., 1989). Impoundments, levees, channelization, and other human modifications disrupt or eliminate these natural flood cycles. When inundated, floodplain wetlands also serve as important spawning and nursery habitats for many invertebrates, fishes, and amphibians that cannot complete their life cycles in the main channel. In turn, through the process of *ichthyochory*, frugivorous fishes can serve as important dispersers of seeds they consume while foraging in inundated floodplain wetlands (Chick et al., 2003; Correa et al., 2007, Highlight 24.1). The seeds of many other floodplain plants disperse passively through the water, a process known as *hydrochory*. As with fringe wetlands, riparian wetlands often show distinct gradients of vegetation types linked to soil moisture and elevation gradients.

Riparian wetlands around the world are severely diminished and degraded for agricultural purposes because floodplain soils are fertile and in close proximity to water. However, in some less developed tropical regions, extensive river-floodplains are still mostly intact. For example, parts of the Amazon River basin still have a natural, seasonal flooding regime. During the wet season, the river channel and inundated floodplain can be nearly 200 km wide along some reaches, and some floodplain forests remain inundated for over half the year. This system has served as a model for studies of the physical and biological interactions between rivers and riparian wetlands (Junk et al., 1989).

The prairie pothole region of North America is an example of an extensive and critical marsh ecosystem formed through depressional processes (Fig. 5.9). Thousands of shallow wetlands called potholes were formed by retreating glaciers in this 776,000 km^2 region extending from Iowa to central Alberta. The individual wetlands in this region vary from ephemeral to nearly permanent; salinity is highly variable. The wetlands can be high salinity in areas where they have no outlet and receive runoff every year that subsequently dries. This physicochemical diversity among the individual

Figure 5.9 Prairie potholes in northwestern Minnesota. *Photo courtesy: US Fish and Wildlife Service.*

wetlands provides a template for a rich diversity of wetland species such as aquatic invertebrates (Euliss et al., 1999). Much of this region is heavily agricultural, and thus farmers destroyed the majority of the wetlands decades ago. Unfortunately, some losses continued when the US 2010 Farm Bill removed some incentives for protection and high grain prices drove more production. The US 2014 Farm Bill restored some protections. Nonetheless, this wetland area is critical to migratory waterfowl and other flora and fauna in the region and is under constant threat.

Various small, temporary water bodies can be classified as wetlands. These include ephemeral and intermittent vernal pools, and other small wetlands. Although often small and inundated for short periods, these systems can harbor unique species and are critical habitats for many invertebrates and amphibians. The short hydroperiods of these systems limit the species inhabiting them, thus those species adapted to live in them are exposed to less predation pressure and less interspecific competition. Human modifications that alter the hydroperiods of these habitats often eliminate the specialist species that inhabit them. We discuss the ecology of temporary waters and small pools in Chapter 15.

Peatlands are wetlands that accumulate decaying organic matter from mosses, sedges, and other wetland plants over time because the production of organic matter exceeds the rate of decomposition. Extensive peatlands occur at high latitudes of North America and Europe, but they can be found worldwide from the tropics to the Arctic. Peat is a significant carbon storage that currently represents over one-third of soil carbon on the planet (Gorham, 1991), even though peatlands represent only 3% of the land area of the planet. As such, peatlands are carbon sinks that help ameliorate increasing carbon dioxide levels in the atmosphere. However, with warming temperatures and reduced water availability predicted for many regions as a result of climate change, these systems may shift to carbon dioxide sources because of increased decomposition, thereby exacerbating climate change (Voigt et al., 2017).

Bogs are a common type of peatland with no significant water inflows or outflows and are thus *ombrotrophic*, meaning precipitation is the only water source. Bogs are usually nutrient poor habitats inhabited by acid tolerant plants such as sphagnum moss. Humic acids released during the slow decomposition of senescent mosses accumulate and make these systems acidic enough to suppress processes such as decomposition. Although generally inhospitable habitats for many wetland animals, they are an excellent source of well-preserved remains of animals and humans from even thousands of years ago. The so-called "bog bodies" of Northern Europe are human bodies beautifully preserved in the cool, anoxic, acidic conditions of the bogs (Fig. 5.10, Highlight 13.2).

Fens are similar to bogs in that they accumulate peat, but they are fed by additional sources of water such as groundwater and surface runoff and thus they are *minerotrophic*. The pH of the water in fens is usually neutral to alkaline, and nutrient concentrations are higher than in bogs. As a result, fens tend to support a wider range of plant species,

Figure 5.10 A bog body, the naturally mummified corpse of a man who lived during the 4th century BC. He was found in 1950 buried in a peat bog in Denmark. Low pH, cool temperatures, and anaerobic conditions result in remarkable preservation of soft tissues. *Photo courtesy: creative commons.*

including sedges, grasses, spike rushes, reeds, and others. Fens with low nutrient availability and acidic water are known as *poor fens*, which are intermediate between fens and bogs; poor fens are generally dominated by short vegetation such as sedges. *Rich fens* have higher nutrient availability, are neutral or alkaline, and support more diverse plant assemblages. Fens and bogs often occur as the remnants of old lake basins that were originally created from glacial activity and gradually filled by sediments.

WETLAND HYDROLOGY

Regions where the water table is close to the surface are areas where wetlands are found globally (Fan et al., 2013). Hydrologic regimes of wetlands can range from highly variable to constant, and hydrology is probably the most important feature of the abiotic template that influences wetland ecology (Wissinger, 1999). Wetlands can receive any of three sources of water: precipitation, surface water, or groundwater. Hydrodynamic characteristics include fluctuations in water level and direction of water flow. Compared with lakes and streams, water loss by plant transpiration is usually more important to the hydrology of wetlands because they tend to be relatively shallow and densely vegetated. Tidal action can have a strong influence on hydrology of coastal wetlands such as coastal estuaries and mangrove swamps, and tides ultimately result in bidirectional water flow. When rivers flow through wetlands, water moves unidirectionally. Sporadic flooding with unidirectional flow, followed by extended periods of stagnation characterize hydrologic regimes of riparian wetlands.

Important characteristics include permanence, predictability, and seasonality. For example, permanence controls the ability of large aquatic predators such as some fishes, amphibians, and larger invertebrates to inhabit a wetland. The range of permanence results in a *predator-permanence gradient*, because wetlands inundated for longer periods tend to harbor more predators. The presence or absence of these predators, in turn, structures the assemblages of lower trophic levels and can ultimately influence dynamics of basal resources such as plants and detritus. For example, many species of tadpoles, as well as large crustaceans, cannot survive predatory fishes.

On a larger scale, diversity of hydrologic regimes (timing, amount, and source of water) among individual wetlands in a region can greatly influence wetland communities through the process of *cyclic colonization*, whereby individuals from wetlands that are relatively permanent are able to colonize intermittent or ephemeral wetlands when they become inundated (Batzer and Wissinger, 1996). Hydrologic diversity among individual wetlands in a landscape also enhances the diversity of plants and animals in a region, as some species are better adapted to permanent water and others to intermittence.

Hydrologic regimes are critical factors underlying the success of wetland restoration efforts (Hunter et al., 2008; Meyer and Whiles, 2008). For example, in riverine wetlands, hydrodynamic characteristics related to links to the river channels and geomorphology are important components of conservation because many species use connected wetlands as reproductive habitat (Bornette et al., 1998; Galat et al., 1998). Degree of intermittence is another key aspect for success of restoration.

Climate interacts with hydrology to constrain hydrodynamics and ultimately the plants found in wetlands. Given hydrogeomorphic properties, wetlands can be classified by function (Table 5.4). Those wetlands that are fed by constant flows of groundwater may be influenced little by seasonal factors that control surface water flows. At the other extreme, seasonal wetlands such as playas or some riparian wetlands can fill during wet seasons and remain dry throughout the rest of the year.

Wetlands can be further categorized by nutrient input (eutrophic or oligotrophic), salinity, pH, and other components of water chemistry. Because of the variety of hydrogeomorphologic, climatic, and other factors, vegetation can range from dense forest to tundra or from macrophytes to trees (Table 5.4). A variety of different subhabitats can also occur within a wetland, depending on the degree and duration of inundation and the water depth (Fig. 5.11). In forested wetlands, hydrology, light availability, and subtle topographic gradients interact to influence forest composition and regeneration (Battaglia et al., 2000).

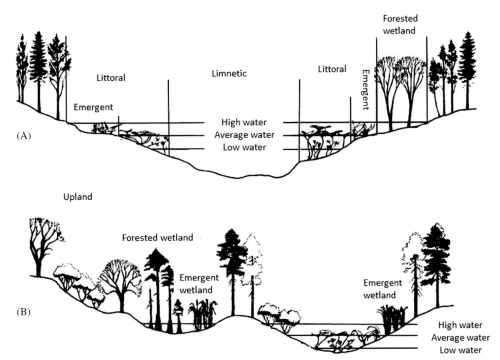

Figure 5.11 Classification of some subhabitats in two wetland types; associated with still open water (A) and associated with a slow-moving stream (B). *From Cowardin et al. (1979).*

Biotic factors can influence wetland hydrology in addition to physical processes such as precipitation and evaporation. Of all animals, humans have the strongest influences on wetland hydrology. While we have discussed destruction of wetlands, large-scale efforts to manage wetland hydrology in efforts to preserve wetlands are becoming more common. A large-scale example of this is the effort to manage the Everglades in Florida (Highlight 5.2).

Other animals such as beavers, alligators, and some other freshwater vertebrates can also alter wetland hydrology and other biotic and abiotic features of wetlands and are known as *ecosystem engineers*. Beavers can alter the geomorphology of entire valleys (Naiman et al., 1994) and are thus one of the more commonly cited examples of eco-system engineers. Extirpation of beavers in drier regions has fundamentally altered the connectivity of aquatic habitats by decreasing wetlands and water retention (Hood and Larson, 2015). Alligators in the Everglades construct holes that keep the wetland from becoming completely dry during times of low precipitation; these deeper areas serve as refugia for fishes, snails, turtles, and other aquatic species.

Highlight 5.2 Managing hydrology of the Florida Everglades

The Everglades and adjacent Big Cypress Swamp are parts of a large wetland area that covered more than 10,000 km^2 of southern Florida prior to massive human modification during the past century (Gleason and Stone, 1994). Slowly flowing freshwater from the Kissimmee–Okeechobee–Everglades watershed characterize the area. Movement of clean water through the wetlands is an ecosystem characteristic that is required to support the native flora and fauna.

Over the years, humans have constructed canals and dikes to drain large areas in the watershed for agriculture, development, and supplying water to the Miami metropolitan area. For example, in one decade in the late 1800s, millionaire Hamilton Disston drained 20,235 ha for agriculture. By 1917, the U.S. Army Corps of Engineers had enlarged four large canals of 380 km total length. Because of these flood control and drainage practices, agricultural production and population increased dramatically in the region (Light and Dineen, 1994). In 1947, Everglades National Park opened, making official a desire to conserve at least part of the wetland.

Today, the South Florida Water Management District in large part controls the drainage systems erected during the past 100 years. The system is extremely complex and includes more than 700 km of canals, 9 large pump stations, 18 gated culverts, and 16 spillways (Light and Dineen, 1994). These control structures are managed to ensure delivery of freshwater for agriculture and Miami drinking water, to control flooding, and to provide enough clean water to maintain the ecological systems of the Everglades National Park. Variations in climate that need to be considered in this management include dry and wet seasons and extreme weather such as hurricanes (Duever et al., 1994). To complicate matters, agricultural runoff has had detrimental effects on the native sawgrass. The input of nutrient-enriched water to sensitive areas must be managed (see Chapter 18), and animal communities respond variably to different management approaches (Rader, 1999).

Plans and actions to mitigate problems associated with altered hydrology and pollution in the Kissimmee–Okeechobee–Everglades are varied. The largest restoration project (up to the 1990s in North America) involves reversing the effects of channelization in the Kissimmee River. We discuss biological effects of this restoration in Chapter 22. In other parts of the watershed, land is being purchased and water running from agricultural areas is being treated to assist with nutrient removal. We describe nutrient pollution problems and solutions in Chapter 18.

The tremendous economic stakes (billions of dollars) conflict with preservation of what is left of the natural environment. Given the large tax base of the region, this has led to a situation in which numerous hydrologists, modelers, and aquatic ecologists, among others, make decisions that minimize the human impact on the Everglades while attempting to maximize the human benefits. In 2000, there were 68 separate projects included in the Comprehensive Everglades Restoration Plan. In 2009, the US government put over $200 million into hydrologic restoration. The problems are complex, but if they are not solved in the near future, the Everglades may be lost forever (Harwell, 1998). Although much important scientific information has been generated, still more is needed to provide the basis for rational management decisions about this important system. Important problems remain including rising sea level and changing hydrology as climate changes (Obeysekera et al., 2015), and understanding linkages to developing urban areas nearby (Sullivan et al., 2014).

Terrestrial animals can also influence and even create wetlands; American bison wallow in certain areas that, through repeated use, become small, depressional wetlands that serve as important habitats for some plants, invertebrates, and amphibians (Knapp et al., 1999; Gerlanc and Kaufman, 2003; see also *temporary waters and small pools* in Chapter 16).

RESTORATION ECOLOGY AND WETLAND RESTORATION

Restoration ecology involves the application of ecological concepts and theory to develop guidelines for ecological restoration. The field was defined in 2004 by the Society for Ecological Restoration International Science and Policy Working Group as "the process of assisting the recovery of an ecosystem that has been degraded, damaged, or destroyed." Since developing as a field in the late 1980s, restoration ecology has grown tremendously. Between 1990 and 2010 there was a 250-fold increase in the number of searchable scientific publications on restoration ecology compared with the two preceding decades (Baer, 2013).

The field of restoration ecology has contributed substantially to basic ecology. Restoration experiments have increased our understanding of community assembly processes, relationships between species diversity and ecosystem function, the dispersal abilities and ecological roles of individual species, successional processes, and competition (Young et al., 2005; Falk et al., 2006). Restoration ecology experiments span all levels of biological organization, from studies of genetics of individuals and populations to biogeochemical cycling and landscape-level dynamics. The field of restoration ecology has also generated considerable scientific debate. This debate includes (1) regarding how we assess and define success of restorations, (2) the source and genetic composition of individuals used to initiate restoration, (3) development of realistic restoration targets in light of global change and other human impacts across the planet, and (4) selection of reference systems or conditions to compare to when assessing the effectiveness or success of a project.

Restoration ecology and conservation biology are similar in that both are goal-oriented, but the intent of restoration is to improve a system that has been damaged by anthropogenic activities, rather than just conserve it. Goals of wetland restoration vary; projects may be focused on restoration or enhancement of ecosystem services such as nutrient uptake and flood control, creating or restoring habitats to promote biological diversity, or enhancing waterfowl habitat for hunters. These goals need not be mutually exclusive. For example, restored riparian wetlands along the central Platte River valley of the central United States enhance carbon and nitrogen uptake and storage (Meyer et al., 2008), provide habitat for aquatic invertebrates and wetland plants, and serve as important foraging habitats for myriad waterfowl and wading birds that migrate along the central flyway (Meyer and Whiles, 2008; Meyer et al., 2010).

However, in many cases, restored wetlands may perform some functions better than others; restored wetlands are often more effective for enhancing ecosystem services such as flood abatement and nutrient uptake than for maintenance of biodiversity (Batzer and Sharitz, 2006).

Restoration methods vary widely according to many factors such as wetland type, degree of degradation, local geomorphology, and restoration goals. Projects can range from small-scale excavations to create vernal pools, to massive landscape-level reconfiguration and reconnection of river floodplain wetlands. Once a project is complete, colonization by wetland communities can be passive (build it and they will come) or organisms may be introduced from natural systems or other restored areas. Most reintroduction efforts are focused on plants, as they are central to wetland structure and function. The source area and genetic makeup of plants and other organisms used in restorations is controversial because the genetic composition of individuals and related adaptations to the particulars of their native habitat may influence their survivorship and competitive interactions in the restoration. Generally, the closer the source of colonists, the better, but sometimes local populations are not available, and in some cases, cultivars may be used.

Cultivars are varieties of plant species that are produced through cultivation and selective breeding. Cultivars can be desirable for restoration because researchers can breed them for traits that may accelerate the restoration process such as rapid growth, resistance to disease, drought tolerance, and high reproductive output. However, the use of cultivars may be discouraged in restoration because they can be aggressive, outcompeting other species and hindering development of diverse plant communities (Schröder and Prasse, 2013). Lesica and Allendorf (1999) suggested using local ecotypes from nearby habitats for restorations of any size that are not attempts to restore highly degraded habitat, the use of genotypic mixtures from multiple populations that may include further locations for restoration of large highly disturbed areas, and to reserve use of cultivars for small severely degraded areas.

Assessing whether a restoration is successful is often difficult because in many cases different individuals or entities desire different outcomes (Zedler and Kercher, 2005). For example, a civil engineer may define success on the basis of hydrology, a local politician may focus on esthetics, birders will focus on waterfowl, and a natural resource manager or ecologist may be more interested in achieving some level of biological diversity or population size of an endangered species. Rosińska et al. (2017) suggested that plant macrophyte communities are good indicators of successful restoration in urban wetlands. Society also tends to put considerably more resources into implementing restoration projects rather than research and postproject monitoring of restorations. Further, research and monitoring of restorations is often focused on easily measured physical variables such as water chemistry and hydrology, which are important, but may not reflect ecosystem function. Research and monitoring are critical because they can guide modification of restoration procedures, accurately assess success or lack thereof of projects, and justify further restoration efforts if they demonstrate success (Galatowitsch and Zedler, 2014). Analysis of the recovery of the value of ecosystem goods and services indicates that many wetland functions can be recovered in restored wetlands over 10 years (Dodds et al., 2008), but that conserving existing wetlands preserves the most value.

Restoring wetlands to historical (e.g., prehuman impact) conditions is becoming increasingly difficult as we continue to alter the planet. Postproject research and monitoring can show which targets and goals may be unachievable. In cases where restoration to historical conditions is unrealistic, goals may be focused on restoring some level of ecosystem function such as nutrient uptake or flood control, and the restoration may be a novel ecosystem. Zedler et al. (2012) recommended landscape scale approaches to wetland restoration. With this approach, wetlands in upper parts of watersheds that are relatively less degraded could more likely be restored to near historical conditions, whereas restoration of highly degraded systems in the lower watershed could focus on restoring some level of ecosystem function such as nutrient uptake.

WETLANDS AND GLOBAL CHANGE

Global change can take several forms including increased temperature caused by greenhouse gases, alterations in precipitation, widespread pollutants, and land use and land cover change. Global climate change is strongly influencing wetlands because water levels therein are the product of a delicate balance between rates of precipitation and evapotranspiration. Changes in either of these processes may directly affect input and output of water, or may indirectly affect wetland water levels by altering depths to groundwater. Predicted warmer temperatures in many regions will produce higher evapotranspiration rates, which can lower water levels even if precipitation rates remain constant. Submerged plants will be most influenced by changes in temperature, and emergent plants by changes in climate (Short et al., 2016).

Given the complex and variable hydraulic characteristics across the different types of wetlands, the magnitude and direction of the changes in these habitats are not easy to predict. For example, deforestation can actually increase water reaching wetlands (although increasing sediment loads) and convert ephemeral wetlands to permanent (Woodward et al., 2014). Detailed modeling approaches scaled regionally can predict the fate of wetlands under climate change. For example, climate modeling linked to hydrologic data suggest declines in density of wetlands in the North American prairie potholes (Sofaer et al., 2016).

Increased demand for food will probably also decrease wetlands (Junk et al., 2013). People commonly drain wetlands to expand areas for crop production, and dry existing wetlands by exploiting surface and groundwater for irrigation. Both human population and standard of living are increasing, leading to expanding demand for production. Pressure to drain wetlands will only increase over the next century. Additionally, reservoir construction and stream channelization threaten riparian wetlands throughout the world.

Predicted warming temperatures and reduced water levels will affect important ecosystem processes in wetlands. Drier, warmer conditions enhance decomposition rates and thus oxidative losses, altering the balance of systems such as bogs that normally accumulate organic carbon. Enhanced decomposition could create a positive feedback, whereby production of the greenhouse gasses CO_2 and CH_4 increases, exacerbating the process of climate change (Bridgham et al., 1995). Further, changes in the hydrologic regimes of wetlands will also disrupt the life cycles of wetland animals, many of which are adapted to specific hydrologic patterns of the wetlands they inhabit (Batzer and Wissinger, 1996; Whiles et al., 1999; Wissinger, 1999). More wetland drying will lead to more intense fires in wetland areas further harming species that inhabit them (Junk et al., 2013).

Rising sea levels associated with warming temperatures will greatly influence coastal wetlands; predictive models based on Intergovernmental Panel on Climate Change (Bernstein et al., 2008) estimates for 2,100 indicate the area of coastal salt marshes will decline by 20%−45%, resulting in significant losses of ecosystem services such as nutrient uptake and flood control (Table 5.5). However, more current analysis indicates that other human pressures are more destructive to coastal wetlands (Wong et al., 2014).

Table 5.5 Predicted changes in tidal marsh areas and associated plant biomass and nutrient uptake and processing on Atlantic coast along Georgia using the mean (52 cm) and maximum (82 cm) estimates of sea level rise from the Intergovernmental Panel on Climate Change Special Report on Emissions Scenarios (Meehl et al., 2007)

Coastal wetland type	Change in area (km²)		Change in plant biomass (t/year)		Change in N sequestration (t/year)		Change in potential denitrification (t/year)	
	52 cm	82 cm	52 cm	82 cm	52 cm	82 cm	52 cm	82 cm
Tidal freshwater	+1	−32	+1,400	−44,800	+8	−262	+7	−211
Brackish	+41	−4	+70,200	−6,800	+307	−30	+184	−18
Saltwater	−226	−496	−225,100	−494,000	−542	−1,188	−384	−843
Cumulative	−184	−532	−153,500	−545,600	−227	−1,480	−193	−1,072
Overall % change	−11%	−33%	−8%	−28%	−4%	−23%	−4%	−25%

Changes in plant biomass and N sequestration and potential denitrification are based on mean values for the three types of wetlands. Negative values indicate a reduction in area and/or process.
Source: Data from Craft et al. (2009).

WETLANDS AS KEY HABITAT FOR WILDLIFE

Wetlands are major focal points for wildlife. Not only the species that inhabit the wetland, but also species that live adjacent to wetlands depend upon the wetlands for food and water. Wetlands are particularly important for hydrophytic plants, waterfowl, and amphibians. It is not just the presence of wetlands, but also how they are connected as a network. For example, some habitats can serve as substitutes for wetlands (ditches, dammed ponds), and help populations persist in a region even if they are lower quality habitats (Uden et al., 2014). On a continental scale, wetlands serve as networks for survival and reproduction of migratory waterfowl, and their management and conservation should be considered holistically (Beatty et al., 2014).

There is increasing interest in how wetlands may serve as important sources of limiting nutrients such as essential fatty acids, particularly docosahexaenoic acid and eicosapentaenoic acid, which are synthesized by algae and are essential dietary elements for vital physiological processes. Essential fatty acids produced by algae in wetlands are fed on by invertebrates and amphibians, and eventually larger consumers foraging in wetland or riparian habitats or from nearby terrestrial habitats. Fritz et al. (2017a) demonstrated that amphibian emergences can transport significant amounts of essential fatty acids from wetlands to adjacent forests in the form of amphibian biomass, and Fritz et al. (2017b) showed that wolf spiders living adjacent to wetlands had higher immune function compared with the same species collected from adjacent upland habitats, presumably because they were feeding on prey with higher concentrations of fatty acids that are linked to immune function.

Wetlands are most important for wildlife in areas transitional between very arid and very moist habitats (Junk et al., 2013). Thus aquatic habitats in deserts are essential to the survival of many animal species. Wet habitats in such dry areas are also under tremendous pressure for development and are often compromised by extraction of groundwater that lowers the water table. For example, the city of Las Vegas, Nevada withdraws groundwater and this is endangering regional biodiversity (Deacon et al., 2007).

SUMMARY

1. Wetlands are distributed worldwide, provide important habitat for wildlife and vital ecosystem services such as flood control and maintenance of water quality, and are important regulators of greenhouse gases.
2. The types of wetlands are extremely variable and generally defined by their morphology and formational process, the length of time they contain water, their vegetation, and the degree of marine influence. The geology of wetlands and their formation varies in different parts of the world.
3. Rice paddies constitute a human-created type of wetland that feeds a large portion of the world's human population.

4. The hydrology of many wetlands has undergone major changes due to human activity. Many wetlands have been drained and lost, and others are compromised severely. Thus, wetlands are among the most endangered habitats throughout the world.

5. Wetland protection efforts in the United States and globally have steadily increased over the past few decades. These generally focus on conservation of existing wetland habitats or, if impacts are inevitable, mitigation procedures. Wetland mitigation involves the conservation or restoration of wetland habitats in order to compensate for modification of wetlands.

6. Wetlands can be coarsely divided into coastal and inland systems, with four broad geomorphic classifications including riverine, depressional, coastal, and peatlands. Coastal wetlands can be heavily influenced by tides, whereas rivers influence the hydrology of riparian wetlands. Depressional wetlands can be influenced solely by local precipitation or various combinations of precipitation and surface and groundwater inputs.

7. Hydrology has a pervasive influence on wetland processes and communities. Biotic interactions such as predation are generally more important in wetlands with long hydroperiods. At the landscape scale, hydrologic diversity among individual wetlands facilitates regional biodiversity and colonization processes.

8. Wetland restoration has the goal of improving wetland habitats impacted by humans. Specific goals of restorations range from complete restoration to historical conditions including biological diversity, to restoring some degree of ecosystem function such as nutrient uptake or flood abatement.

9. Wetlands can serve as important hotspots for production of important nutrients such as essential fatty acids, which are synthesized by the algae inhabiting them.

10. Wetlands are particularly susceptible to predicted climate changes because they are the product of a delicate balance between precipitation and evapotranspiration. These changes may further alter biogeochemical cycles and climate because globally wetlands store tremendous quantities of carbon.

QUESTIONS FOR THOUGHT

1. Do you know of any local wetlands that are endangered or have been drained in your lifetime?

2. How can temporary wetlands in arid habitats be extremely important to wildlife?

3. How might wetland mitigation and restoration procedures be improved?

4. Why do extensive wetlands exist in the high Arctic, even though annual precipitation is similar to that in many temperate or tropical deserts?

5. When people create new wetlands to replace those destroyed by development, what ecosystem services might be easy to replicate and which ones might be difficult to recreate?

CHAPTER 6

Physiography of Flowing Water

Contents

Figure 6.1 Salt Creek Falls, Oregon.

Freshwater Ecology
DOI: https://doi.org/10.1016/B978-0-12-813255-5.00006-5

Rivers and streams are central to life. Small streams are dominant interfaces between all other aquatic habitats and land. Streams and rivers move materials from land to sea through lakes and estuaries, forming a vital link in global biogeochemistry. Streams and rivers have been well characterized by hydrologists because of interest in flooding, erosion, and water supply (for a basic description of river geology and hydrology, see Leopold, 1994; for practical aspects of special interest to ecologists, see Gordon et al., 1992). To comprehend the importance of streams in aquatic ecology, it is necessary to understand their morphology and geology. In this chapter, we discuss ways to describe streams, characteristics of stream flow, geology, and how streams move materials.

CHARACTERIZATION OF STREAMS

We first describe characterization using watershed features such as discharge, number of upstream branches, and area. Then we describe stream classification with respect to water velocity and changes in discharge. One way to characterize a stream is by the size of its drainage area. Drainage area is the land area drained by all of the tributary streams above a chosen point in the main channel. In North America the term *watershed* is synonymous with drainage area but Europeans use the word *catchment*; the word *basin* is also used by some. Watershed in European terminology is the boundary of the catchment (i.e., the ridge that divides catchments). Even though the ridge "sheds" water and the basin "catches" it, we use the American definition of watershed.

Globally, more permanent streams and rivers occur in regions where there is more precipitation and a greater number of intermittent streams are found in drier regions. Over half of the world's river networks are intermittent (Datry et al., 2016). The *discharge*, or volume of water passing through a channel per unit time, is positively correlated to the area of the watershed (Fig. 6.2). The amount of discharge produced per unit area has an approximate upper bound. Many drier areas have lower discharge per unit area than wetter areas, creating a 100-fold variance in the relationship between watershed size and discharge for a watershed of any given size. The world's largest rivers occur in large drainage basins with significant amounts of precipitation. A list of the 15 largest rivers in the world is presented in Table 6.1. The Nile is the longest river at 6,758 km, but historically it only ranked 36th in discharge because it drains a relatively arid landscape. Now, the Nile ranks even lower because in many years it is almost dry by the time it reaches the Mediterranean Sea because of such heavy human use of the river.

Stream order is another way to characterize streams. The most common method for ordering streams is the *Strahler classification system*, often used by ecologists when describing basic stream characteristics (Fig. 6.3). In this method, we assign the smallest

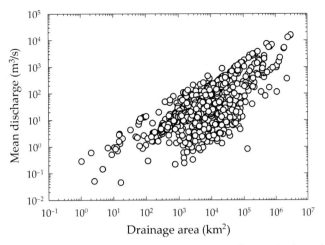

Figure 6.2 Discharge as a function of area for a large number of watersheds in the United States. *Data courtesy the US Geological Survey.*

Table 6.1 Average discharge of the 15 largest rivers of the world, in order of rank

Rank	River	Country	Drainage area (km^2)	Length (km)	Av. Ann. Disch. (m^3/s)	Rank order length	Rank order drainage area
1	Amazon	Brazil	5,950,000	6,597	176,000	3	1
2	Congo	Congo	3,700,000	4,586	41,000	8	2
3	Yangtze	China	1,940,000	5,744	33,100	5	9
4	Orinoco	Venezuela	980,000	2,735	23,000	20	16
5	La Plata	Uruguay	3,110,000	3,894	22,000	14	4
6	Brahmaputra	Bangladesh	930,000	2,896	20,000	17	19
7	Yenisei	Russia	2,610,000	5,937	20,000	4	6
8	Ganges	India	1,000,000	2,510	19,000	24	13
9	Mississippi	U.S.	3,210,000	6,693	18,000	2	3
10	Lena	Russia	2,490,000	4,312	16,000	10	7
11	Mekong	Indochina	790,000	4,248	16,000	11	23
12	Irrawaddy	Burma (Myanmar)	430,000	2,011	14,000	26	—
13	Ob	Russia	2,450,000	5,567	12,000	6	8
14	Tocantins	Brazil	910,000	2,639	11,000	21	20
15	Amur	Russia/China	1,850,000	4,344	11,000	9	10

Source: After Leopold (1994).

streams as first order. Order only increases when two streams of the same order join. Others have proposed modifications to this method of stream ordering, each with its own benefits and drawbacks (see discussions in Allan, 1995; Allan and Carillo, 2007). Some of the practical problems with determining and using stream order include: (1) it is often difficult to determine the smallest stream in a network. Some people might

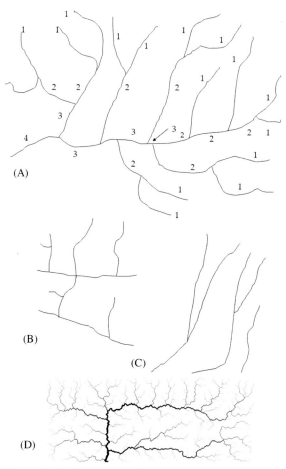

Figure 6.3 The Strahler method of stream ordering on a dendritic stream system (A). Order increases only when two streams of equal order meet. Other types of drainage patterns include rectangular, which may be found in Karst systems (B), and parallel, which occur mainly in deeply eroded areas (C), and a mathematically generated drainage pattern on the basis of minimization of energy dissipation *(D). A-C modified from Strahler and Strahler (1979), and D courtesy Andrea Rinaldo based on the mathematics in Rinaldo et al. (2014).*

consider the smallest permanent stream as the first channel, others might consider the smallest channel as that where the geomorphic action of the stream is evident; (2) researchers cannot always rely upon maps for accurate hydrologic information because blue line and dashed blue line features (permanent and intermittent flowing waters) are not determined consistently; and (3) stream order does not always correlate closely with discharge, water chemistry, or other important abiotic factors, although low-order streams can have some predictable differences from higher orders. For example, a larger order river in a wet climate will usually have a greater discharge than a lower order river, but in deserts a stream may lose discharge as it flows downstream. Definition of "what is a stream" has legal implications similar to those discussed in the last chapter for wetlands; the jurisdictional waters that can be regulated and considered part of a stream are highly controversial (Doyle and Bernhardt, 2011).

More recently, computer-based tools have been used for more accurate (or at least more automated) assessments of stream order and drainage patterns. Analysts can use Geographical Information Systems (GIS) with detailed digital elevation maps to identify stream networks and delineate watersheds. Analysts can use the mapping methodology combined with models of water yield specific to soil types and different regions to allow prediction of where there should be flowing water. To a lesser degree, these tools have some of the same problems as traditional maps, but they can provide fast, automated estimates of order for many streams in a region. Stream order will continue as a primary descriptor characterizing streams at hierarchical levels of organization, but students should not put too much emphasis on the exact order of a stream.

In general, a greater number of low-order streams occur in a watershed. Although streams of higher order have a greater length per stream, the total length of low-order streams is often greater (Fig. 6.4). The high relative abundance of small streams means that processes that occur where small streams interact with land or groundwater usually dominate interactions between aquatic and terrestrial systems.

Stream drainage systems can also be classified by the pattern of the stream channels (Fig. 6.3). Various patterns develop in response to geological factors and can give rise to very different segment lengths of each order. Patterns range from highly reticulated to almost parallel/linear in shape. In general, streams form branching networks; it appears that this even occurred on Mars when surface water was present there (Seybold et al., 2017). This fractal nature of the channels (Fig. 6.3D) is a result of energy minimization as rivers flow (Rinaldo et al., 2014). The formation of branching networks is a function of rock weakness and the amount of precipitation, with branching related to stream valleys that widen and capture smaller neighboring valleys, and the side slopes of these widening valleys that become more susceptible to incision (Perron et al., 2012). Factors such as tectonics, coastal advancement, and geological age determine if watersheds are static or changing rapidly over geological time (Willett et al., 2014). Generally, wider branching angles are characteristic of drier areas (Seybold et al., 2017).

Another useful way to characterize stream hydrology is by discharge and water velocity. Discharge is related to, but different from, water velocity. Discharge is a volume of water passing through a channel per unit time, whereas *water velocity* is the speed of water at any point in a channel (also referred to as current). Flow is a general term for movement that can mean discharge, water velocity, or both. We have discussed the strong biological effects of water velocity on organisms in Chapter 3 and will describe effects of both discharge and water velocity in streams in subsequent chapters.

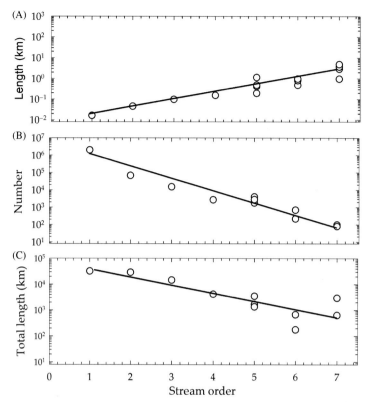

Figure 6.4 Relationships between stream order, average lengths of each order (A), number of streams of each order (B), and total length of streams of each order (C) for several watersheds in the southwestern United States. On average, streams of lower order are shorter, but they can also be more numerous; individual watersheds can have a greater total length of low-order streams. *Data from Allan (1995) and Leopold et al. (1964).*

Discharge is related to water velocity through the cross-sectional area

$$Q = V\,A$$

where Q is the discharge, V is the velocity, and A is the cross-sectional channel surface area.

$$A = w\,h$$

where w is the average width and h is the average depth. The way these features are measured varies with stream size. In larger streams and rivers, it is reasonably accurate to measure average depth and width and multiply them to get cross-sectional area. Average water velocity can be determined with numerous measurements across the channel with depth. Generally, 20 measurements are made across the channel with measurement at 0.6 of the depth if the depth is <0.75 m and at 0.2 and 0.8 of depth when depth >0.75 m (Platts et al., 1983).

In smaller streams, it can be difficult to determine actual depth and much of the discharge may actually occur under rocks or other larger objects. In this case, an alternative is release of an inert tracer like sodium chloride or a dye at a known rate and calculation of discharge from the amount of dilution. For example, a dye tracer can be used to calculate discharge when added at a constant rate

$$Q_{stream} = Q_{tracer}[C_{tracer}] / [C_{stream}]$$

where Q_{stream} is the discharge of the stream, Q_{tracer} is the pumping rate of the tracer, $[C_{tracer}]$ is the concentration of the tracer, and $[C_{stream}]$ is the concentration of tracer in the stream. This method also can provide an estimate of average velocity as calculated by the average time it takes the tracer to reach a point downstream divided by the distance downstream from the release point. The time is determined by the point when the tracer reaches half the concentration it attains at plateau (equilibration of release).

Another common method for measuring discharge in a small stream is with a *weir*, which is a constructed constriction in the stream channel with a known geometry (Fig. 6.5). In this case water height can be related to discharge by making repeated measures of depth and discharge at different depths and constructing a *rating curve*, which is used to predict discharge from depth.

Figure 6.5 A weir used to measure discharge from water height at Konza Prairie Biological Station, viewed from downstream. There is a depth sensor and a lateral pipe system to log water depth continuously in this particular weir. The depth sensor and data logger are housed in an enclosure behind the bison to the left.

A plot of discharge against time is a *hydrograph*. Discharge can vary from fairly constant in rivers fed primarily with groundwater (Fig. 6.6A and B) to intermittent in headwater streams of drier regions (Fig. 6.6C and D). Steep watersheds and intense storms can lead to great variability in discharge (Fig. 6.6E). Damming changes the natural hydrograph in several ways. Generally, dams moderate the magnitude and duration of floods, and may cause low discharge periods to become rarer (Fig. 6.7). With power generation reservoirs, downstream discharge can fluctuate daily because release rates change to meet power demand throughout the day, but much of the seasonal variation in flow can be decreased.

Streams can be classified as *perennial*, flowing all the time or at least at all times except extreme droughts; *intermittent*, flowing some of the time and receiving water from groundwater; or *ephemeral*, flowing rarely and not receiving input from

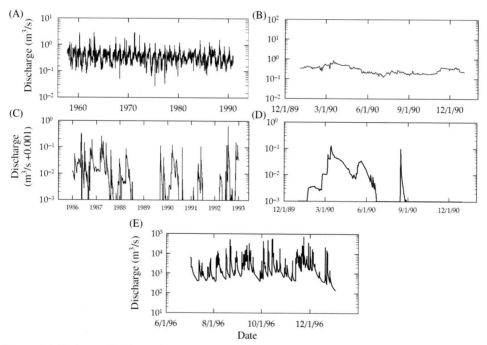

Figure 6.6 Hydrographs from three river systems plotted on log scales. The Niobrara River in Nebraska is mostly spring fed and shows relatively little variation in discharge among (A) and within years (B; note only about a 10-fold difference in each year, whereas two or three orders of magnitude are covered in the remaining hydrographs). Kings Creek in Kansas is a small, intermittent, prairie stream, with alternating periods of wet and dry over the years (C). A typical year in Kings Creek includes both times of no flow and floods (D; note 0.001 = 0 discharge in C and D). A stream in a steep watershed (Slaty River on the west coast of New Zealand) with frequent rainstorms exhibits approximately weekly floods (E). *Data for A and B courtesy US Geological Survey; data for C and D courtesy Konza Prairie Long-Term Ecological Research project; and data for E courtesy of Barry Biggs and Maurice Duncan.*

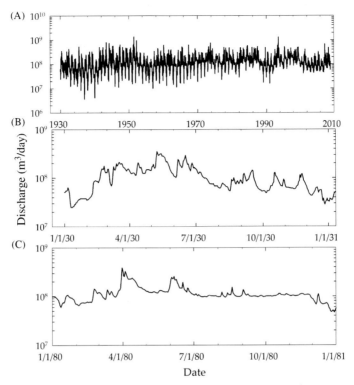

Figure 6.7 Discharge of the lower Missouri River. Prior to the 1950s, discharge was more variable than after dams were installed (A). A typical year before regulation (B) reveals a period of very low discharge in the winter, a spring peak, and a gradual decrease after early summer. After regulation (C), discharge is about the same throughout much of the year, except in winter when it is allowed to decrease after barge traffic halts. *Data courtesy US Geological Survey.*

groundwater. Small, headwater ephemeral stream channels are sometimes difficult to identify except during wet periods when actively flowing channels develop as soils become saturated. This dynamic nature of headwater stream networks is related to the concept of *variable source area hydrology*, which is the expansion and contraction of saturated areas in a watershed during wet and dry periods (Hewlett and Hibbert, 1967).

Fluctuation in water discharge can characterize streams and it can have a profound influence on the community structure of stream organisms (Poff and Ward, 1989). In this approach, discharge variability, flooding patterns, and extent of drying are used to create a classification system (Table 6.2). Classification on the basis of discharge patterns is one approach used by stream ecologists to describe stream characteristics and relate them to the organisms that inhabit them. This understanding of the natural flow regime and how it relates to ecology can serve as a guide to management of flowing waters. For example, mitigating effects of dams can be partially offset by using controlled releases that mimic natural flow patterns of the rivers they are on (McManamay

Table 6.2 A method of classifying streams by discharge patterns and relationship to aquatic communities

Drying frequency	Flood and discharge frequency/ predictability	Stream type	Effect on biota
Often	Rare-frequent	Harsh intermittent	Strong
Low	Frequent	Intermittent flashy	Strong
Low	Infrequent	Intermittent runoff	Strong
Rare	Frequent unpredictable floods, low discharge predictability	Perennial flashy	Strong
Rare	Frequent predictable floods, low discharge predictability	Snow and rain	Strong—intermediate
Rare	Infrequent floods, low discharge predictability	Perennial runoff	Strong—intermediate
Rare	Infrequent floods, high discharge predictability	Mesic groundwater	Weak
Rare	Infrequent predictable floods, high discharge predictability	Winter rain	Seasonally strong
Rare	Infrequent predictable floods, high discharge predictability	Snowmelt	Seasonally strong

Source: After Poff and Ward (1989).

et al., 2016). Understanding the effects of climate change also require prediction of the biological effects of flooding. For example, winter floods are expected to become more common in Europe, and this will influence stream macrophyte communities (Garssen et al., 2017).

We can also characterize streams by their surrounding landscape and the associated vegetation. Thus, scientists speak of desert streams, forest streams, or arctic streams. This classification method can be useful because the terrestrial vegetation in the landscape that the streams drain may drive the biological processes that occur in streams. For example, many stream invertebrates rely on leaves and woody debris from terrestrial vegetation for food. This concept is discussed more detail in Chapter 25.

STREAMFLOW AND GEOLOGY

Important links occur between groundwater and streams. Groundwater feeds most streams the majority of the time. Consider the regularly flowing headwater streams with which you are familiar; it rarely rains hard enough for water to flow across the surface of the land (*sheet flow*). Rather, infiltration through soil and subsurface sediments feeds into the stream to maintain flow. This constant level of discharge in streams is *base flow*. Streams with increased discharge from groundwater as they move downstream are *gaining* streams. Streams that lose flow because they are at a level above groundwater are *losing* (influent) streams. Streams can be gaining or losing over shorter reaches depending upon slope and groundwater flow paths.

Increased or prolonged rain or snowmelt can cause rapid increases in discharge, called *floods* or *spates* (Fig. 6.8). These events often occur randomly, but a probability can be calculated that an individual event of a specific magnitude will occur given a certain amount of time (Fig. 6.9). Thus, when hydrologists speak of a 10-year flood, they are referring to an event that on average will occur once every 10 years. In other words, in any year there is a 1 in 10 chance of such a flood, even if such a flood has not occurred for the past 20 years or one occurred the previous year. The recurrence interval (*RI*, also called return period) is calculated as

$$RI = \frac{n+1}{m}$$

where *n* is the number of floods less than some discharge and *m* is the total number of years of record. To create this calculation for floods you would find greatest flood in each year over *m* years. Then rank each of the floods against each other (from 1 to *m*). If you plot *RI* on the *x*-axis and flood discharge on the *y*-axis (Fig. 6.9), then that plot can be used to place any subsequent flood recurrence interval, as long as that flood is

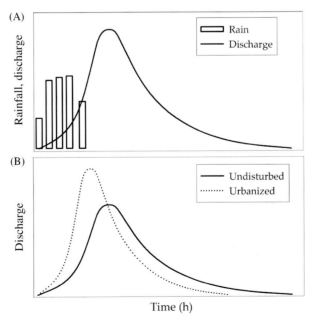

Figure 6.8 A hypothetical hydrograph of a storm event with precipitation and runoff in a natural area (A) and hypothetical comparison of watershed responses before and after urbanization (B). *After Leopold (1994).*

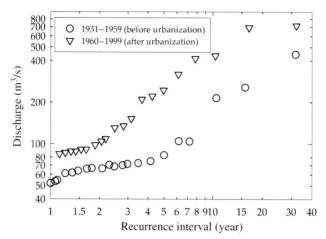

Figure 6.9 Flood frequencies plotted as recurrence intervals as a function of discharge for all recurrence intervals more than 1 year for a watershed before and after urbanization (Seneca Creek in Maryland) (peak discharge data from the US Geological Survey; computed as in Leopold, 1994).

not greater than any others from the prior record. Extrapolating past the record is questionable.

This approach has practical application in defining acceptable flood risk. Thus, if an insurer decides they will not insure or will charge a much higher rate of insuring property that has more than a 1/100 chance of flooding in a particular year, they can use the *RI* and a map of topography to find the elevations that roughly correspond to that flood. Humans can alter the *RI* because watershed characteristics as well as the intensity of precipitation events control severity of floods. Where runoff is more rapid and infiltration is less, floods are often more severe. Channelization and increases in impermeable surfaces (pavement and buildings) associated with urbanization cause increased flooding (Figs. 6.8 and 6.9).

Water velocity also varies within stream channels. As you move down a small stream, you will usually find some shallow areas where the influence of the bottom can be seen at the surface of the flowing water. These turbulent, shallow areas are *riffles*. Deep areas with relatively low water velocity are *pools*, and areas with rapidly moving water but a smooth surface are *runs* (Figs. 6.10 and 6.11). Pools tend to accumulate fine sediments during periods of base flow, and runs have coarser substrata. A section of river with several runs, pools, and riffles is a *reach*. Several reaches together, or sometimes the length between major tributaries, make up a *segment*.

In streams with more permeable substrata, a substantial portion of the flow can go under the riffle (Fig. 6.11). This zone of interaction between the surface water and the groundwater is the *hyporheic zone*. The amount of flow into this zone can be mediated by biofilms on the surface (Aubeneau et al., 2016). The zone has biological

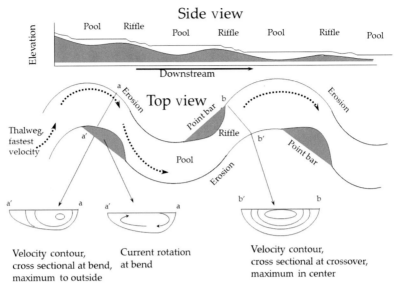

Figure 6.10 Conceptual diagrams of stream meander formation. (Top) The side view is a cross-sectional lengthwise view showing pool and riffle sequence. (Middle) The top view shows a meandering stream, the thalweg (line of maximum velocity), and zones of erosion and deposition (point bars). (Bottom) The water velocity contours (cross-sectional across the channel) show how the maximum velocity is outside of the bend and the lateral current direction. When the thalweg crosses the channel, the maximum velocity is in the center of the channel.

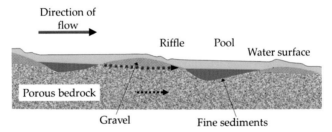

Figure 6.11 Cross-sectional diagram of a stream showing the riffle pool sequence, accumulation of fine sediments, and water flow through the shallow subsurface (hyporheic, indicated by green).

significance in part because it can influence dissolved oxygen, which alters nutrient transformations and how fishes place their nests in river gravel (Cardenas et al., 2016).

The distance downstream between successive riffle and pool areas is approximately five to seven river widths in many streams (Leopold, 1994). The alternating riffle and pool pattern is absent when the bottom material consists of fine sand or smaller particles, is present when it is gravel and small rocks, and occurs sometimes when large boulders make up the stream bottom. In steep mountainous regions with large rocks, pools alternate with small falls. These cascades depend on geographical structure and do not exhibit the typical riffle and pool sequence. The riffle and pool sequences can be tied to stream flow and meandering patterns.

A cross section of the straight portion of a stream shows that the water velocity is often maximum in the center (termed the *thalweg*) and minimum near the sides and bottom (Fig. 6.10). In areas where the stream or river curves, the maximum velocity occurs nearer to the outside of the bend. Furthermore, because the water on top is moving more rapidly than the water on the bottom of the bend, the direction of flow tends to move downward and cut into the outside bank. Higher velocity and down cutting lead to erosion on the outside bend. The slower water velocity on the inside of the bend allows deposition on the inside of the curve and formation of a point bar (Fig. 6.10) and these processes lead to stream meandering.

Streams *meander* (wander in "s"-shaped patterns) unless they are constrained by outcrops or bedrock or conditions are conducive to forming braided channels. Water flowing across surfaces of glaciers, currents flowing in oceans, and rivers flowing into reservoirs or oceans can all meander. The process is a self-organizing procedure that can be characterized by the mathematical tools of fractal geometry (Stølum, 1996). Meandering is characterized by erosion and deposition, which exaggerate the meander over time (Fig. 6.10). In general, erosion occurs on the outside of the curve in the downstream portion of the bend because sediment concentrations are low in the water and it is cutting down into the bank. Water moves along the bottom of the channel downstream and toward the inside of the bend. Because the water has the highest concentration of sediment as it moves toward the inside of the bend and it slows and drops the sediments as it reaches the inside of the bend, it forms a point bar. A point bar is the zone of deposition on the inside of the curve. Eventually, the meander cuts itself off. In a small stream the cut off meander can lead to a short-lived pond, in a large river it forms an *oxbow lake*. Meander formation occurs in similar manner in rivers of all sizes; the wavelength (the distance to meander out, back, out the other way, and back again) averages about 11 times the channel width, and the radius of curvature of a channel bend is generally about one-fifth of the wavelength (Leopold, 1994). Thus, meander size is a function of discharge (Fig. 6.12). While meander formation seems a general property of a flowing channel, researchers have had difficulty experimentally inducing meander formation in laboratory settings mimicking some natural conditions, and this is still and area of active research (van Dijk et al., 2012).

The idea that there are regular relationships between stream width, distance between pools and riffles, and meander lengths is captured by the concept of a *dynamic equilibrium* (Leopold, 1994). This concept is useful because it describes the dynamic nature of rivers and streams, some properties of streams that can be predicted independent of scale, and the possibility of an ideal morphology that dissipates energy most efficiently. Humans alter the flow and morphology of rivers, and deviation from this general idealized morphology may be problematic. A stream

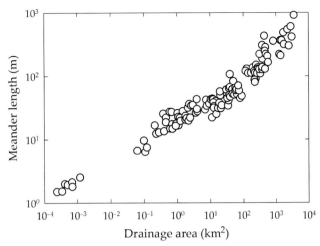

Figure 6.12 Relationship between drainage area and meander length. *Redrawn from Leopold et al. (1964).*

may not bend to the wishes of engineers, eventually regaining its morphology via dynamic equilibrium in ways that humans do not want. Even concrete channels eventually give way to the natural process of streams. Managers tasked with stream restoration need to consider the dynamic nature of stream channels for the greatest chance at success.

A detailed review of river meandering suggests that a variety of factors influence the nature of meandering (Callander, 1978). The *sinuosity* is the distance that water travels between two points divided by the direct distance between the two points. A straight channel has a sinuosity value of 1. Meandering rivers have sinuosity values 1.3 or greater. Sinuosity is a function of surface roughness and slope, with greater sinuosity at higher values of the ratio of resistance to slope (Lazarus and Constantine, 2013). Sinuosity also increases with the ratio of mean width to depth and as the percentage of silt and clay in the river increases.

In addition to meandering, some rivers also flow in a *braided* manner. This pattern generally occurs when water flows in broad sheets across noncohesive sediments such as sand, when slope is relatively steep, and when flow is variable. Individual channels combine and split, form and disappear, sometimes over relatively short periods of time. Braiding is a basic physical process that can be modeled using multiple cells where sediment is transported from one cell to the next (Murray and Paola, 1994). Rivers like the Platte River of the US Great Plains, which have highly variable flows and sand substrata, and other rivers on the plains below mountains such as the Canterbury Plains on the South Island of New Zealand, become highly braided and this natural process is associated with the formation of a great diversity of freshwater habitats within the floodplain. Braided rivers can be found around the world in areas with high sediment transport, often below mountainous areas such as the Himalayas. Desert rivers are also often braided.

Meandering over time and deposition of materials by a river leads to a *floodplain* that is relatively flat across the river valley (Fig. 6.13A). Floodplains are often inundated seasonally and provide numerous wetlands and other types of habitat for many species of plants and animals. Over geological time, a river moves back and forth across its valley. Any traveler on an airplane can observe the related landscape heterogeneity; always ask for a window seat. The complexity that can develop over geologic time is impressive (Fig. 6.14). Human activities that alter flows and modify channel structure greatly reduce the habitat heterogeneity of these systems.

The aggregate effect of these different aspects of channel structure, in addition to the effects of fallen trees and large rocks, leads to a highly heterogeneous system on the scale of tens to hundreds of meters (Fig. 6.13B). The degree of heterogeneity in

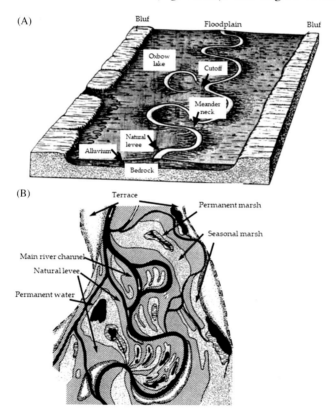

Figure 6.13 (A) General features of a floodplain (from Strahler and Strahler, *Elements of Physical Geology*, Copyright ©1979, reprinted by permission of John Wiley and Sons, Inc.) and (B) a diagram of heterogeneity of a tropical floodplain. *From Welcomme (1979); reprinted by permission of Addison Wesley Longman Ltd.*

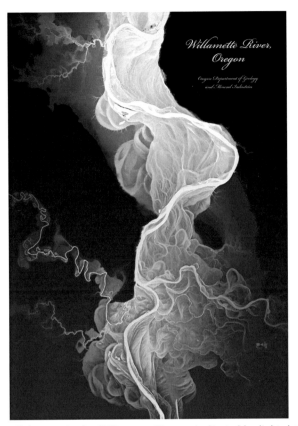

Figure 6.14 Geological changes in the Willamette River as indicated by light detection and ranging (LIDAR) to measure detailed differences in elevation. *Courtesy Oregon Department of Geology and Mineral Industries; Lidar imagery by Daniel E. Coe.*

the floodplain includes raised levees that naturally deposit along stream channels and old scoured oxbows. In large rivers, depressions caused by river actions may form important wetlands, ponds, or lakes. This heterogeneity alters response to floods and greatly influences the ecology of rivers and the riparian zone's function as an interface between terrestrial and aquatic habitats (Naiman and Décamps, 1997).

An active natural river creates a habitat mosaic for organisms to inhabit along its margins (Lorang and Hauer, 2007). The flowing channel is referred to as the *fluvial zone*. The zone where the river regularly floods is *parafluvial*, and the remaining floodplain the *orthofluvial*. Movement of materials in a network can cause an instream habitat mosaic as well. For example, where side channels enter a river, the geomorphology of sediment loading creates an alluvial fan as one stream enters a larger one. The fan has effects that cascade upstream, leading to lower water velocity and a wider stream channel (Benda et al., 2004).

HUMAN INFLUENCES ON PHYSICAL ASPECTS OF RIVERS

Few relatively pristine river systems remain. Damming, channelization, and excessive water use have resulted in major human impacts on the geomorphology of rivers throughout the world (Highlight 6.1). Other organisms, such as beavers, hippopotamuses, crocodiles, and elephants, alter channel morphology and riparian areas (Naiman and Rogers, 1997), but not to the degree that humans have (Highlight 6.2).

Most large rivers are not in their natural state. Globally, 172 of 292 large river systems are impacted by dams (Nilsson et al., 2005). For example, irrigation, diversion, or reservoirs alter 70% of the discharge from the 139 largest rivers in North America, Europe, and the former Soviet Union. Most of the unaffected river systems that remain in these regions are in the far north (Dynesius and Nilsson, 1994). In the United States, researchers estimate that there are 2.5 million dams (National Research Council, 1992).

Reservoirs allow sediments to settle from rivers, and rivers downstream become "sediment starved" and significantly more erosive. Natural sandbars are less likely to form in a starved river. In an extreme case, construction of the Aswan Dam led the River Nile to become sediment starved. The lack of sediment, in turn, has led to erosion of the Nile Delta, where the River Nile enters the Mediterranean, and a subsequent loss of valuable agricultural land that has served as Egypt's breadbasket for millennia (Milliman et al., 1989). Similar processes are occurring in many large river deltas of the world. The starving of rivers combined with sea level rise and extraction of materials such as oil from under the deltas is threatening many deltas that are important to human civilization as well as being vital centers of biodiversity (Syvitski et al., 2009).

Reservoirs can also alter the riparian habitat by interfering with flooding; in dry, sandy rivers they allow establishment of more riparian vegetation and reduce the width of the channel (Friedman et al., 1998). Dams also fragment riparian habitat, leading to distinct changes in plant communities by altering patterns of dispersal and recruitment (Nilsson et al., 1997; Jansson et al., 2000). Perhaps surprisingly, effects of reservoirs and other disturbances can extend to organisms upstream (Pringle, 1997). For example, interruption of salmon runs reduces the transfer of nutrients into small streams via spawning salmon that eventually die in the stream channel.

Reservoirs can also disrupt natural temperature patterns. When relatively cool rivers flow into wide, shallow reservoirs and the reservoirs release their water downstream, it can lead to artificially higher water temperatures in the river below. Alternatively, deep reservoirs can always have cold, deep water (see Chapter 7 for an explanation of temperature stratification). If water is released from deep in the reservoir, artificially low temperatures can occur. For example, the Colorado River is naturally a warm river that would not normally support salmonid populations.

Highlight 6.1 Human impacts on rivers and streams from damming, channelization, and flood control measures

People have channelized many rivers globally, and riparian (streamside) vegetation has been removed to allow rapid boat travel, increased drainage, and allow agricultural and urban expansion. This has dramatic influences on the shoreline habitat. For example, a 25-km stretch (as the crow flies) of the Willamette River in western Oregon had 250 km of shoreline in 1854 (Fig. 6.15A), but human activity decreased it to 64 km by 1967 (Sedell and Froggat, 1984). There was a concurrent loss of at least 41% of the riparian wetlands during this time (Bernert et al., 1999). This removal of virtually all slow-moving portions and straightening of meanders is common in rivers in areas where humans live. Additional measures include construction of projecting short

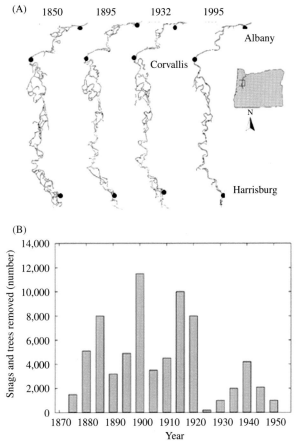

Figure 6.15 Changes in the Willamette River related to channelization from 1850 to 1995 (A, image courtesy of Ashkenas, Gregory, and Minor, Oregon State University) and (B) snag and tree removal. *Data from Sedell and Froggat (1984).*

dykes that move flow from the edge to the middle of the channel (wing dykes) to decrease bank erosion and move deeper faster water areas into the main channel. These dykes are common and further alter river discharge. Such alteration has been well documented on the Missouri River (e.g., Hesse et al., 1989). In addition, removal of large instream obstructions, such as logjams and stumps (often called snagging operations), to facilitate navigation is common and this destroys vital habitat for aquatic organisms. Animals in many large river systems are limited by the amount of stable substrata as habitat and much of the fish and invertebrate diversity and productivity is associated with accumulations of large woody debris (e.g., Benke, 1984). Even historic factors come into play in human alterations of rivers and streams; the more than 10,000 mill dams that were erected in the Northeast United States from the 1700s to early 1900s could have substantially altered stream geomorphology by burying wetlands in sediments and creating deeply incised stream channels throughout the region (Walter and Merritts, 2008).

Flood control measures on many large rivers include levees to contain high discharge. When floods do occur, the levees constrain the water, making it move faster and deeper in the main channel instead of spreading out across the floodplain and flowing with a lower average velocity as it would naturally. If a flood does breach the levy suddenly, it causes considerable damage because of the rapid current velocity when the levee breaks. In addition, such levees constrain flows and act like dams to upstream regions that do not have levees. In the Mississippi River basin, mean annual flood damage has increased by 140% during the past 90 years. This increase is probably attributable to increases in numbers of levees and removal of riparian wetlands (Hey and Philippi, 1995).

Landowners often channelize smaller streams flowing across their property so they can develop or plant crops closer to the edge of the stream and drain their land more quickly. This channelization causes the water to move downstream at higher velocity, increasing erosion in the channelized reach and upstream. Channelization also causes the stream to have stronger flooding impact downstream, leading more landowners to channelize their stream banks. Removal of riparian vegetation creates an even worse situation because the natural retention of sediment and slowing of floodwater does not occur, thus increasing the severity of floods.

Human alteration of stream and river hydrology has effects across all spatial scales. At the smallest scales (Paragamian, 1987), it removes habitat for aquatic organisms. At intermediate scales, alterations of stream hydrology are associated with increased erosion and more severe flooding. At the largest scales, human activities lead to global changes in the transport of materials by rivers. The economic impact of all these human influences is likely very great. Effective management requires understanding of dynamics of natural rivers (Poff et al., 1997).

However, there are now large populations of trout in the Grand Canyon because managers continuously release cold water from the upstream Glen Canyon Dam. These altered conditions and the predatory trout are harming endangered populations of native fishes.

RIVER AND STREAM RESTORATION

River restoration has become important as people have come to realize that alteration of hydrology, water chemistry, and biology of rivers has unintended consequences in urban (Bernhardt and Palmer, 2007) and other areas (Bernhardt et al., 2005a, b). Such efforts require restoration of the natural hydrology and understanding of how the dynamic equilibrium of geomorphology can be restored to lead to long-term stability of the system. One form of river restoration that has become more common, is removing dams to bring the stream back to natural conditions (Highlight 6.2). Restoration targets need to consider geomorphology, life history of species that managers are interested in (Jansson et al., 2007), and desired ecosystem functions (Groffman et al., 2005). The most complete restoration will also consider the natural heterogeneity, landscape connectivity, and linkages across ecological scales (Peipoch et al., 2015). This is required to allow rivers to provide ecosystem services such as flood protection (Nilsson et al., 2018). New high frequency sensing techniques allow managers to assess restoration conditions of flow and water chemistry with much greater ability to detect differences among restored and unrestored systems (Rode et al., 2016). While managers frequently attempt to restore morphology, it is less likely that flow dynamics are restored (Kondolf et al., 2006). Still, restoration of their morphology is essential to recovering ecosystem services associated with flowing waters. This requires classification of existing natural and altered morphology.

One of the most popular river classification systems is the Rosgen system. The main objectives of this system are to (1) predict river behavior from existing morphology, (2) develop specific hydraulic and sediment relationships appropriate for the type of stream, (3) provide a way to assess reference reaches, and (4) provide a common framework for characterizing stream morphology and condition. The first level of classification characterizes the channel and valley shape, slope, and pattern (e.g., confined or wide valley). The second level of classification is a determination of field classification of stream type. For example, surveyors measure width, depth, channel entrenchment, sinuosity, slope, and substratum type in the stream (e.g., sand, gravel, cobble). This description of the Rosgen system is very cursory, and a 2-week training course is generally required to learn the classification methodology and system; it is a complex, multitiered method for surveying streams, and a number of government agencies have adopted it.

The Rosgen method is controversial. Criticisms (Simon et al., 2007, 2008) refuted by Rosgen (2008) are that the method is descriptive and does not appropriately account for temporal evolution of stream morphology on the basis of current geomorphologic principles. The critique suggests that peer-reviewed publication is required to establish the method as scientific and that some restoration projects have failed because

Highlight 6.2 Continued large dam construction and efforts to remove dams

A number of very large dam projects continue to move forward globally. These projects can be attractive as they can provide hydropower, water diversions, and control downstream floods. On the negative side, they destroy river habitat, cause disruption to natural connectivity of rivers, alter downstream flow, and displace people living along a river above a dam site. In spite of the negative aspects, these projects move forward. The Amazon, Congo and Mekong rivers are all experiencing large surges in dam construction and this construction is endangering these rivers with high animal diversity (Winemiller et al., 2016). Many other projects are completed, in progress, or being planned in most areas of the world, particularly in developing countries. In just one example, the Belo Monte dam on the Xingu River in the Amazon basin of Brazil exemplifies the ecological damage that can potentially occur. This reservoir takes advantage of a large elevation drop on the river to create one of the world's megadams. The dam will cover some of the largest rapids in the world. These rapids are home to many endemic fishes and are a hotspot of river biodiversity. The reservoir will destroy this unique natural treasure (Pérez, 2015).

Many dams will and have been built without plans for what to do when, inevitably, the reservoirs behind them fill with sediment, or the dams themselves lose structural integrity. Society has only begun to confront the concept of how to decommission dams that are outdated or causing more ecological harm than their benefits. Dam removal is not an easy thing to do, both with respect to the physical removal and the environmental effects of their decommission. Potential problems include the massive release of sediments downstream and release of any associated toxins deposited when the dam was operational. Benefits include restoring river connectivity. There has been particularly strong focus on removal of dams in the US Pacific Northwest to reestablish runs of anadromous salmonids that had been disrupted by the dams.

By 1999, at least 467 documented dam removals have occurred in the United States since 1912 for environmental, safety, economic, or other reasons (American Rivers, 1999). By 2015 the same organization reported 1,300 removals in the United States, indicating a large acceleration in the rate of removals. Removals are occurring globally with more in developed countries; we know of know of comprehensive data on removals worldwide. These removals have had some beneficial effects, but some of the degradation from dams to aquatic habitats is irreversible (Middleton, 1999) and dam removal may even have some negative aspects such as allowing upstream movements of exotic species, and the release of large quantities of sediments stored behind the dam (Orr et al., 2008). Removal effects on macroinvertebrates can even extend upstream (Pollard and Reed, 2004). Analysis of a small dam removed from a Pennsylvania stream indicated that most effects of the removal downstream were gone once the sediments that were washed from the old reservoir had been flushed from the downstream channel, but some effects lingered a year afterwards (Thomson et al., 2005).

The largest scale dam removal to date occurred on the Elwha River in Washington state in 2011 when two large reservoirs were taken down. This removal had very large geomorphic effects with 7.1 million m^3 of sediments mobilized that had accumulated in the two reservoirs. This sediment moved downstream and led to substantial sediment accumulation (1 m in pools) and cause the lower river to change to a braided state from a meandering state. Part of these changes were probably related to the fact that the river downstream of the reservoirs had been sediment starved for so many years. River flow started to down cut through the massive wave of sediment and by 2 years after removal the researchers estimated that 10% of the released sediment had already been released at the river mouth. The large amount of sediment had obvious effects on biota, from invertebrates to fish and riparian vegetation (East et al., 2015; Foley et al., 2017).

---

of inappropriate application of the classification method. Rosgen (2008) asserts that the critique improperly represents his classification method and its application. Those adopting the Rosgen approach should be aware of this controversy.

Attempts at restoration of hydrology and channel complexity have become common (Bernhardt et al., 2005a, b). Unfortunately, such restorations are not always effective (Alexander and Allan, 2007) and not based on appropriate design standards (Palmer et al., 2005). Complete restoration of large rivers impacted by humans is unlikely, but partial rehabilitation may be possible (Gore and Shields, 1995), particularly with respect to macrohabitat variables such as logs, deep pools, root wads, and cut banks (Poppe et al., 2016). More holistic views of restoration may be necessary. For example, channel restoration can occur more rapidly in some watersheds if beaver (*Castor canadensis*) and natural riparian vegetation are reintroduced (Pollock et al., 2014).

The French River Rhône is one example of large-scale river restoration. In this example 47 km have been restored since 1999. This heavily regulated river produces 25% of the French hydropower energy and is heavily used for industry. This restoration focused on both main channel flow and floodplain connectivity. There were four long stretches where the main channel was bypassed for human uses, and minimum flows in the former main channel were not maintained. The project sought to restore the minimum flows in the natural channels. In addition, improved sewage treatment and riparian forest restoration were done to improve water quality and naturalize flows. In the decades following the restoration, monitoring of fish and invertebrate communities, in addition to habitat assessments indicated that the restoration improved conditions and moved the main channel sites closer to the expected natural condition. The project illustrates that long-term monitoring is necessary to establish restoration success and that restoration methods that integrate multiple ecosystem characteristics (flow, habitat, water quality, and geomorphology) are necessary for successful river restoration (Lamouroux et al., 2015).

The longest-term view of river restoration has been termed "freedom space," which takes into account the natural tendency for rivers to wander across their flood plain. The first level of this is protecting roughly 1.7 times the channel width, which is the frequently flooded area with rapid changes (over about 50 years). The second level is the level across which meanders move across the valley. The third level is the width that is impacted by unusual floods (Biron et al., 2014). It is not likely that all three levels will be implemented in areas that are already highly developed. A long-term view of restoration may be necessary as it may take decades for rivers to return to their native states, and the restoration process itself may be a disturbance (Nilsson et al., 2017).

TRANSPORT OF MATERIALS BY RIVERS AND STREAMS

Rivers can exist for a substantial amount of geological time. There are many cases where mountain ranges have rivers that cut through them. This generally means that the river is older than the mountains, and the river could erode down faster than uplift could raise the mountains. Rivers are the primary source of terrestrial materials to the oceans and over time, they move mountains to the bottom of the ocean.

Rivers carry materials dissolved from land to the sea. In addition, they move larger particles by *erosional processes*. Human activities have substantially altered movement of dissolved materials (Fig. 6.16) to the sea, another hallmark of the Anthropocene (Table 6.3). Dissolved materials in rivers have increased, particularly nutrients such as nitrogen (Vitousek, 1994), phosphorus, and sulfur, because activities such as the burning of fossil fuels and use of fertilizers have increased the actively cycling phases of these materials. Ultimately, the increased transport of nutrients in rivers may alter productivity of coastal marine systems (Downing et al., 1999); increased export of

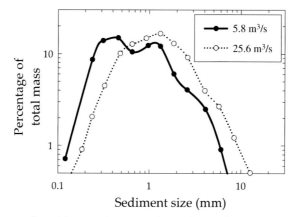

Figure 6.16 Movement of particles as a function of particle size for maximum and minimum flows in the East Fork River. Note that larger particles move more readily at greater discharge rates. *Reprinted by permission of Harvard University Press from Leopold (1994). © 1994 by the President and Fellows of Harvard College.*

Table 6.3 Average chemical composition of river water throughout the world

Attribute	Current concentration	Natural concentration	Pollution	% increase
Ca^{2+}	14.7	13.4	1.3	9
Mg^{2+}	3.7	3.4	0.3	8
Na^+	7.2	5.2	1.3	28
K^+	1.4	1.3	0.1	7
Cl^-	8.3	5.8	2.5	30
SO_4^{2-}	11.5	6.6	4.9	43
HCO_3^-	53.0	52.0	1.0	2
SiO_2	10.4	10.4	0.0	0
Total dissolved solids	110.1	99.6	10.5	11
Dissolved nitrogen	21.5	14.5	7.0	32
Dissolved phosphorus	2.0	1.0	1.0	50

Concentrations in mg/L.
From Berner and Berner (1987) and Meybeck (1982).

Table 6.4 Total organic carbon export by rivers in different terrestrial environments (Meybeck, 1982)

Environment	Average total carbon export ($g\,m^{-2}\,y^{-1}$)	Total carbon load ($10^{12}\,g\,C\,y^{-1}$)
Tundra	0.6	4.5
Taiga	2.5	39.6
Temperate	4.0	88.0
Tropical	6.5	241
Semiarid	0.3	4.6
Desert	0.0	0
Total	3.8	378

nutrients from agricultural regions of the central United States to the Mississippi River and ultimately the Gulf of Mexico are the primary factor contributing to the large "dead zone" (hypoxic, or low oxygen area that can cover $>22,000\,km^2$) in the Gulf of Mexico (Rabalais et al., 2002). Such dead zones caused by nutrient runoff are becoming more common globally because of continuously increasing nutrient inputs. The severity of these dead zones is further exacerbated by global warming (Altieri and Gedan, 2015).

Streams and rivers play a part in the global carbon cycle by moving dissolved and suspended organic materials (e.g., woody debris and leaf fragments) from terrestrial habitats to the sea (Table 6.4). In this context, tropical rain forests, where a large proportion of continental runoff originates, are extremely important. Desert and semiarid habitats are not as important because of their low runoff. Understanding global cycles of materials and the effects of global change requires knowledge of how rivers move materials through the environment.

Greater discharge creates more power that moves more and larger particles downstream (Fig. 6.16). Mobile material can be divided into two categories, suspended load and bed load. The *suspended load* is the fine material suspended in water under normal flows. This suspended load can also be referred to as *turbidity* or *total suspended solids*. The *bed load* moves along the bottom by sliding, rolling, and bouncing. It never moves more than several particle diameters above the bottom. The relative movement of both types of particles and the size of the particles that can remain suspended depend on the water velocity and turbulence (Fig. 6.17). Streams become turbid after a rain. This turbidity is partially attributable to increased sediment inputs from land, but also to the higher water velocity that can resuspended and transport materials from within the channel.

Turbidity is unevenly distributed in rivers. The total concentration of suspended materials is greater near the bottom (Fig. 6.18A). However, the trend of greater concentration with depth does not hold with the finest particles (such as clay), which remain in suspension throughout the water column. Their Reynolds number is small enough that the sinking velocity is never less than the river turbulence. The effect of greater concentration with depth becomes more pronounced with larger particles such as sand (Fig. 6.18B).

Understanding the movement of solid materials can be very important to understanding the ecology of streams. Floods cause erosion and can alter habitat. Some

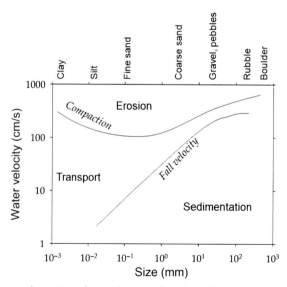

Figure 6.17 Transport and erosion of particles as a function of water velocity. Note that the lines delineating transitions between erosion, transport, and sedimentation represent fuzzy rather than abrupt transitions. *After Allan (1995) and Morisawa (1968).*

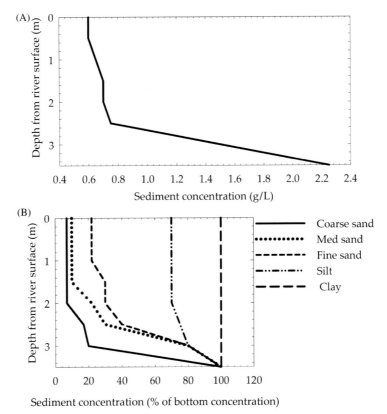

Figure 6.18 Distribution of particles with depth in the Missouri River expressed as total particle concentration (A) and as percentage concentration of bottom sediments for different sized particles (B). *Data from Wilber (1983).*

species of fish and invertebrates are dependent upon entrained organic particles as a food source (Wallace and Merritt, 1980), and some cannot survive when the concentration of suspended particles is too great. Movement of rocks on the bottom of the streams can also have direct ecological effects on the organisms that live on or under those rocks. On land, a rolling stone gathers no moss; in a stream, a rolling stone will not gather much algae or many invertebrates (McAuliffe, 1983), and moving particles can scour animals and plants from solid substrata.

In rivers with contaminated sediments, transport of those sediments during high flows can lead to pollution events because many pollutants such as pesticides and metals adsorb to the surface of sediment particles. For example, in the Clark Fork River in Montana, years of mining have led to contaminated sediments in the basin (Johns, 1995). Each time it rains hard, local fishermen hold their breath; massive trout kills result if enough toxic sediment is resuspended (see Highlight 15.1). Another way that streams are seriously impacted by sediments is from "hill topping" or mountain top removal coal mining (Fig. 6.19). With this mining technique, the top of a mountain containing coal is removed, and the waste rock is disposed of by filling the stream valleys skirting the mountain, obliterating the streams. Movement of dissolved and suspended toxins into the areas downstream from the fills has negative effects on the animals in the streams (Hartman et al., 2005). The mining alters invertebrate secondary productivity and sensitive species can be lost (Voss and Bernhardt, 2017). Bernhardt et al. (2012) found that negative impacts on stream invertebrates from mining occurred when only 5% of the watershed was mined, and that in West Virginia, 22% of the stream length of the state was impacted by mountaintop mining.

Movement of materials is also a key aspect of erosion. Very small particles such as clays are less susceptible to erosion than slightly larger particles (Fig. 6.17). Due to the relationship between particle size and transport, gravel is common in riffles and fine sediments deposit in pools. Under base flow, riffles are erosional habitats and pools are depositional habitats. This may lead the reader to ask, why are pools deep and riffles shallow? The answer is that pools become erosional habitats while riffles are depositional during floods. So, floods scour out pools and deposit the materials in riffles and at baseflow the materials are moved slowly from riffles into pools.

Lateral movement of materials is also important in forming habitat for organisms. For example, when a tributary enters a river it tends to deposit materials leading to

Figure 6.19 Hill top removal mining for coal. *Image courtesy Ohio Valley Environmental Coalition, Vivian Stockman.*

damming of the main channel (Benda et al., 2004). This damming slows flow in the river above the tributary and can lead to a more rapid elevation drop below the tributary (Fig. 6.20).

Retention and transport of particulate materials can be important in describing the long-term effects of sediment pollution. For example, sediments that enter stream channels from erosion caused by watershed disturbance can be retained for decades, prolonging the recovery time from sediment pollution events (Trimble, 1999). The length of time that sediments can be stored in stream networks is an active area of hydrologic research. Many factors, including the presence of macrophytes, can influence rates of sediment trapping and transport (Rovira et al., 2016).

The effects of change in temporal and spatial scales on stream habitats can be linked to processes of erosion and habitat change, leading to a hierarchical classification of stream habitats (Fig. 6.21). Such classification provides a useful framework with which to approach the links between river hydrology and aquatic ecology, and a template for the interaction of organisms with habitat (Gregory et al., 1991). Consideration of scale is a vital component for understanding patchiness in stream ecosystems (Stanley et al., 1997); scale controls chemical transport (Dent et al., 2001), and may be useful in the study of water quality (Hunsaker and Levine, 1995).

Figure 6.20 The Grand Canyon at Hance Rapids. Note the stream entering from the left deposits materials that cause a pool behind the confluence, and a sharp elevation drop leading to the rapids. Also note the river downcutting through sediment layers. *Photograph courtesy US Geological Survey, by L. Leopold.*

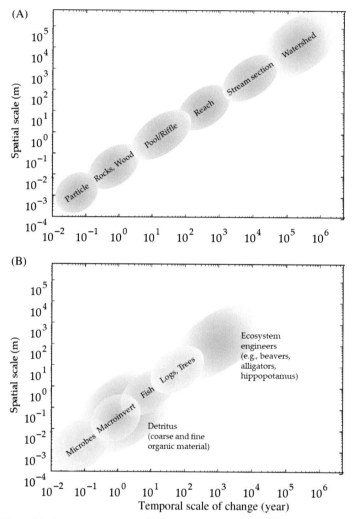

Figure 6.21 Hierarchical representation of the geomorphological changes in stream channels (A) and related biotic processes (B). *After Allan (1995); Frissell et al. (1986); Mitsch and Gosselink (1993).*

ADVANCED: CHARACTERIZING THE MOVEMENT OF DISSOLVED MATERIALS IN RIVERS AND STREAMS

Dissolved materials move down streams as a function of discharge and exchange with the biotic and abiotic components of stream channels. If a chemical is added in a defined pulse to a stream, the pulse will become less coherent as it moves downstream (Fig. 6.22). Streams have areas with relatively slow velocity and others with high water velocity and

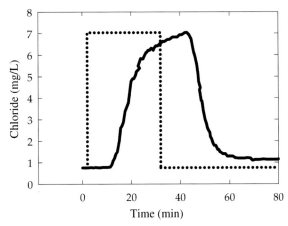

Figure 6.22 Movement of a pulse of chloride through a stream channel with considerable areas of slack flow (transient storage zones). In a perfect channel, the pulse released upstream is indicated by dashed line, the downstream samples by solid symbols. *After Webster and Ehrman (1996).*

this causes spreading of the pulse. In addition, dissolved materials can interact with the sediments or biota as they flow downstream. Departures from the ideal transport of dissolved materials (i.e., a coherent pulse moving through a frictionless, inert channel) are used by researchers to characterize the retentive properties of the channel and the effect of the hyporheic zone (Webster and Ehrman, 1996; Baker and Webster, 2017). The hyporheic zone can play a significant role in retention of nutrients (Hill and Lymburner, 1998).

Characterizing the influence of slower regions of flow including groundwaters, backwaters, pools, and even water in algal mats requires release of nonreactive tracer substances and modeling to match observed transport patterns. Two major processes, *advection* and *dispersion*, describe the pulse of dissolved materials as they flow downstream. Advection is the average velocity of the stream. Dispersion is the diffusion of material (mainly via turbulent transport) from the main pulse. Regions of slower flow are referred to as *transient storage zones*. As a pulse of material moves downstream, some of it moves into the slower zones of flow, and is gradually released back into the main channel. If a solute is released in a coherent pulse, downstream the pulse broadens out, and the extended tail after the main pulse is from material that is rereleased from transient storage zones into the main channel. The equations describing these processes are complex partial differential equations, and in practice are not solved mathematically, but can be modeled with readily available software (Bencala and Walters, 1983; Runkel, 2002).

The common two-compartment model that there is an average cross-sectional surface area of the main channel (A) and a second compartment that is connected to some degree to the main channel that has the cross-sectional area A_s. The computed value of A_s/A then is an indicator of the relative size of the two compartments. With larger values of this ratio, more transient storage area and more dispersion are expected. The parameter α indicates a coefficient of exchange between the transient storage zone and the main channel, and dispersion is noted by the factor D. From these parameters, a number of additional characteristics can be calculated, including the average distance a molecule travels downstream before entering a transient storage zone, the average amount of time a molecule resides in the transient storage zone, and the Peclet number (Pe) which indicates the relative importance of advection and dispersion.

This two-compartment model describes solute dynamics relatively well. However, it is clearly a simplification. Some transient storage zones exchange very quickly with the main stream, and others much more slowly. Thus, there is a range of different transient storage zone types that can occur in a single stream reach. More advanced models view transient storage more as a distribution of different areas of retention, which is probably a more realistic approach (Harman et al., 2016).

Characterizing the movement of dissolved materials has practical applications. As more streams become urbanized, researchers are striving to understand the implications of the hydrologic changes associated with waters that flow thorough areas of dense human population. Nutrient uptake is impaired in urban streams and assessing these changes requires characterization of hydrology and movement of dissolved materials (Meyer et al., 2005). Self-purification of rivers by processes (such as denitrification that removes nitrate, see Chapter 14) can be enhanced in transient storage zones.

SUMMARY

1. Permanent rivers are more common where precipitation is greater; intermittent rivers are characteristic of regions with low or sporadic precipitation.
2. Rivers can be classified by watershed area, stream order, variations in discharge over time, and vegetation.
3. Stream habitats can be characterized into riffles, pools, runs, and falls. These habitats are mostly scale independent.
4. Streams naturally meander; the water flow patterns alter erosion and deposition in such a way that meanders increase in size until they eventually close off and form oxbow lakes. This meandering is scale-independent and occurs in rivers and streams regardless of size.
5. Humans have drastically altered the morphology of stream and river channels throughout the world with activities such as channelization, construction of dams, construction of levies, and alterations of flow from water use.
6. Rivers transport dissolved materials and are thus important in global biogeochemical cycles. The average concentrations of sulfur, nitrogen, and phosphorus have increased 30%−50% because of human activities.
7. Discharge and particle size influence erosion and transport of silt and larger materials on river bottoms.

QUESTIONS FOR THOUGHT

1. Why might some regions in deserts serve as runoff sinks (i.e., have more water flowing in than out)?
2. How can levees act as dams to upstream areas during floods and increase erosion in the channel where they are present?
3. Why does channelization increase the movement of bed load during floods?
4. How can the amount of suspended materials in streams alter the rates of photosynthesis of organisms attached to the stream bottom?
5. How might vegetative characteristics in a watershed relate to stream flow in terms of total amount and in terms of how many floods occur per year?
6. How does frequency and predictability of flooding relate to the possibility that stream organisms are adapted to flooding?

CHAPTER 7

Lakes and Reservoirs: Physiography

Contents

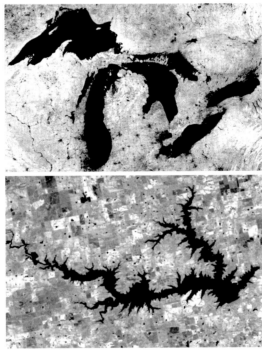

Figure 7.1 Satellite images of the Great Lakes (left) and Smithville Reservoir (Missouri) (right). The reservoir is 16 km long. Note the dendritic pattern of the reservoir and the relatively smooth shorelines of the glacially formed Great Lakes. The numerous black dots around the reservoir are farm ponds. *Images from the US Geological Survey.*

Freshwater Ecology
DOI: https://doi.org/10.1016/B978-0-12-813255-5.00007-7

Lakes cover 3.7% of nonglaciated terrestrial area (Verpoorter et al., 2014), but are disproportionately important to ecological processes and human needs relative to their area. Lakes of all sizes provide us with fisheries, recreation, drinking water, and scenic splendor. Having a clean lake nearby increases property values (Poor et al., 2001). Large lakes (Table 7.1) have played a part in the history, economy, and culture of many nations. Lakes also provide excellent systems for ecological study. The boundaries of the lake community and ecosystem often appear distinct, the water is frequently well mixed, and the bottom is often relatively homogeneous, making lakes tractable systems for ecologists. Much effort has been made to study the physical and biological aspects of lakes (e.g., some of those in Table 7.2 have been studied intensively for over a century) and to manage pollution. The foundation of these studies is an understanding of the geomorphology of the lakes. Different lake morphologies give rise to different levels of productivity and physical effects of water retention, circulation, currents, and waves. For example, the fates of toxins and nutrients in lakes depend partly on lake circulation, which is a function of lake physiography. In this chapter, we describe formation, morphometry, the process of stratification, and water movement in lakes.

FORMATION: GEOLOGICAL PROCESSES

We define a *lake* as a very slowly flowing or nonflowing (*lentic*) open body of water in a depression that is not in contact with the ocean. This definition includes saline lakes but excludes estuaries and other mainly marine embayments. The distinction between a small shallow lake or pond and a wetland is not always clear, and neither is that between a very slow, wide spot in a river and a lake or reservoir with high water throughput; all aquatic habitats occur across a continuum of physical attributes, such as depth and water velocity.

Permanent lakes are common where more precipitation occurs and where geology allows formation of water-retaining basins. Some areas have geological histories that result in more lakes. For example, if we compare the distribution of wetlands to the distribution of freshwater lakes, relatively more lakes than wetlands occur in northern North America, whereas wetlands are relatively important in northern Asia and northeast Europe. Intermittent lakes (those that dry sometimes) are distributed sparsely throughout the world, with greater numbers in arid areas. The western United States, South Australia, India, central Asia, and central Africa all have abundant intermittent lakes.

Humans have made many lakes and ponds and, in doing so, have greatly altered the distribution of lentic habitats. Most regions inhabited by humans with few natural lakes and even moderate precipitation have significant numbers of ponds and reservoirs. At least 2.6 million small impoundments are found in the conterminous United States, and they account for about 20% of the lentic water in the country

Table 7.1 Some properties of the 10 largest lakes by either depth, area, or volume, globally arranged by maximum depth

Lake	Continent	Formation	Mixis	Area (km²)	Maximum depth (m)	Mean depth (m)	Volume (km³)	Length (km)	D_L	Retention (year)
Baikal	Asia	Tectonic	Meromictic	31,500	1,741	730	23,000	2,200	3.4	323
Tanganyika	Africa	Tectonic	Meromictic	34,000	1,470	572	18,940	1,900	3.1	5,500
Caspian	Asia/Europe	Tectonic	Meromictic	436,400	946	182	79,319	6,000	2.6	295
Nyasa	Africa	Tectonic	Meromictic	30,800	706	273	8,400	1,500	2.7	
Issyk Kul	Asia	Tectonic	Meromictic	6,200	702	320	1,732	760	2.8	305
Great Slave	N. America	Glacial	Dimictic	30,000	614	70	2,088	2,200	3.6	
Crater	N. America	Volcanic	Monomictic	55	608	364	20	35	1.3	4.9
Matano	Asia	Tectonic		164	590	240	39	80	1.8	
Toba	Asia	Volcanic–tectonic	Monomictic	1,150	529	216	249	100		
Hornindalsvatn	Europe	Glacial	Dimictic	508	514	237	12	65	2.6	
Great Bear	N. America	Glacial	Dimictic	29,500	452	81	2,381	2,100	3.3	124
Superior	N. America	Glacial	Monomictic	83,300	307	145	12,000	3,000	2.9	184
Michigan	N. America	Glacial	Monomictic	57,850	265	99	5,760	2,210	2.6	104
Huron	N. America	Glacial	Monomictic	59,510	223	76	4,600	2,700	3.1	21
Victoria	Africa	Tectonic	Polymictic	68,800	79	40	2,700	3,440	3.7	23
Aral	Asia	Tectonic	Meromictic	62,000	68	16	970	2,300	2.6	Sink

Data in this table and Table 7.2 are from several sources including Hutchinson (1957), Gasith and Gafny (1990), Herdnedorf (1990), Horne and Goldman (1994), and Messager et al. (2016).

Table 7.2 Selected lakes of historical or research interest that are not included in Table 7.1

Lake	Continent	Formation	Mixis	Area (km²)	Maximum depth (m)	Mean depth (m)	Volume (km³)	Length (km)	D_L	Retention (year)
Biwa	Asia	Tectonic	Monomictic	618	46.2	45	28	46	2.6	5.4
Erie	N. America	Glacial	Monomictic	25,820	64	21	540	1,200	2.1	3
Eyre	Australia	Tectonic	Polymictic	0−8,583	4	2.9	23	70	4.9	Sink
Geneva	Europe	Glacial	Monomictic	580	310	153	89	70		
Kinneret	Asia	Tectonic	Monomictic	1.7	43	26	4.3	2.2	1.16	7.3
Loch Ness	Europe	Glacial	Monomictic	56.4	230	133	7.5	39	3.2	2.8
Mendota	N. America	Glacial	Dimictic	39.8	25.3	12.8	0.47	9.1	1.6	4.6
Ontario	N. America	Glacial	Monomictic	18,760	225	91	1,720	1,380	2.8	8
Tahoe	N. America	Tectonic	Monomictic	499	501	249	124	125	1.6	700
Titicaca	S. America	Tectonic	Monomictic	7,700	280	106	820	176	3.46	1,343
Vanda	Antarctica	Glacial	Amictic	5.2	67	33.8	0.17	5.6	2.28	75 (ablation[a])
Windermere (both basins)	Europe	Glacial	Monomictic	14.3	64	23	0.39	17	1.2	

Note, in English lake naming is inconsistent. For example, Lake Mendota and Geneva Lake. Larger surface area seems to correlate with the "Lake" being first in the name (Beisner and Carey, 2016).
[a]Ablation is evaporation directly from the ice that covers the lake.

(Smith et al., 2002). Similar patterns occur in Europe and all other developed areas on earth. The large number of rivers in the Northern Hemisphere that have been impounded to form reservoirs was discussed in Chapter 6.

Globally there are more small lakes, and the total surface area of large lakes summed is similar to that of smaller lakes. However, large lakes are considerably deeper and contain substantially more water (Fig. 7.2). On a regional level, lake size has important consequences for the chemistry and productivity of lakes (Hanson et al., 2007) because small lakes have biological and physical characteristics that are distinct from larger lakes in the same region. A variety of geological processes lead to the formation of these lakes (Table 7.3). Hutchinson (1957) described these processes in detail; we give a brief summary here.

Tectonic movements of Earth's crust (Fig. 7.3) form some of the largest and oldest lakes. For example, warping of Earth's crust formed the Great Rift Valley in Africa and gave rise to Lakes Edward, Albert, Tanganyika, Victoria, Nyasa, and Rudolf. This group contains some of the oldest, deepest, and most ecologically and evolutionarily interesting lakes on Earth. Although small tectonic lakes are more numerous than large ones, large tectonic lakes cover an area greater than that covered by the small ones on a global scale (Fig. 7.4). *Graben* lakes are tectonic lakes formed where multiple faults allow a block to slip down and form a depression. In half-graben lakes, the blocks tilt and leave a depression that can be filled by water (Fig. 7.3). Lake Baikal of Siberia, the deepest and oldest lake on Earth, is a graben lake. About 3 km of sediment has accumulated on the bottom of Lake Baikal over 16 million years (Fig. 7.5).

Damming by natural processes can form lakes. Examples of these processes include landslides (Fig. 7.6), lava flows, drifting sand dunes, and glacial moraines. In addition, beaver ponds, damming by excessive plant growth, log jams, flows of rivers at deltas, glacial ice dams, and pools formed at the edges of large lakes by shore movement are classified into this general category. Lakes formed by natural dams are usually not large, but some exceptions exist (e.g., Lake Sarez, a large landslide lake in Russia, is 500 m deep). Landslide lakes pose considerable danger to people downstream because of the possibility of catastrophic failure. On May 12, 2008, an earthquake registering 8.0 on the Richter scale killed at least 69,000 people in Sichuan China and caused a landside that created the large, potentially unstable Tangjiashan Lake on the Jianjiang River. Engineers successfully created a stable spillway and the lake was stabilized without further loss of human life. The notching and stabilization caused a large flood but avoided harming any of the 1.3 million people living downstream (Cui et al., 2012).

Glacial activity is responsible for the formation of more lakes than any other process, especially in temperate regions (Fig. 7.7). Areas far enough north or south to have historically had glaciations tend to have far more lakes than areas closer to the

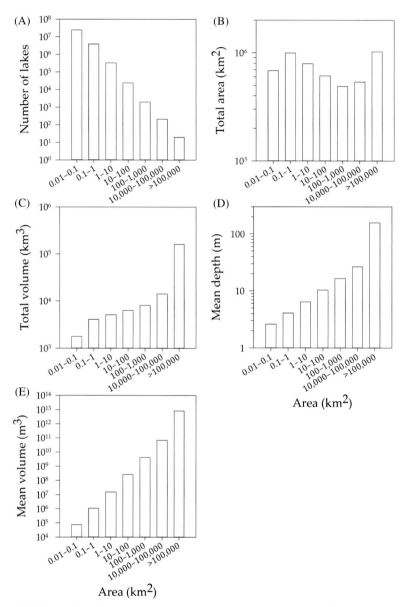

Figure 7.2 Global numbers (A), total area (B), total volume (C), mean depth (D), and mean volume (E), for different size ranges of surface area of lakes. *Data from Cael et al. (2017).*

equator. Several processes associated with glacial activity lead to lake formation. Glaciers scour as they move down valleys. The ice flow of these glaciers creates basins. Lakes occur where glaciers have scoured more deeply, leading to the formation of

Table 7.3 Some ways that lakes form and essential characteristics of each type

Lake type	Formation process	Essential characteristics and examples
Tectonic	Basin formed by movement of earth's crust (Graben: a block slips down between two others, Horst: diagonal slippage).	Can be very old and very deep. Lake Baikal, Asia; Lake Tanganyika, Africa
Pothole or Kettle	Formed when ice left from retreating glacier is buried in till (solid material deposited by glacier) and then melts.	Small lakes/wetlands, prairie pothole region in Alberta and North and South Dakota.
Moraine	Glacial activity deposits a dam of rock and debris.	Narrow, fill valley
Earthslide	Movement of earth dams a stream or river.	Similar to reservoirs, Quake Lake, Montana
Volcanic – Caldera	Volcanic explosion causes hole that is filled with water.	Often round and deep, Crater Lake, Oregon
Dissolution Lake	Limestone dissolves and lake forms.	Small, steep sides
Oxbow	River bend pinches off, leaves lake behind.	Shallow, narrow, may be seasonally flooded

Many more types are possible (e.g., Hutchinson, 1957).

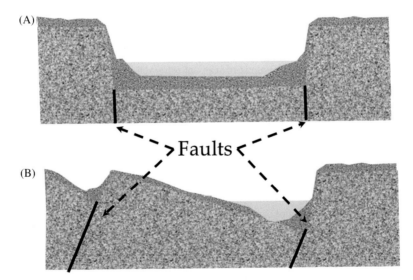

Figure 7.3 Diagrams of two modes of lake formation associated with tectonic processes: (A) graben, a block drops below two others; and (B) tilt block lakes, blocks tip and a lake forms along a single fault line.

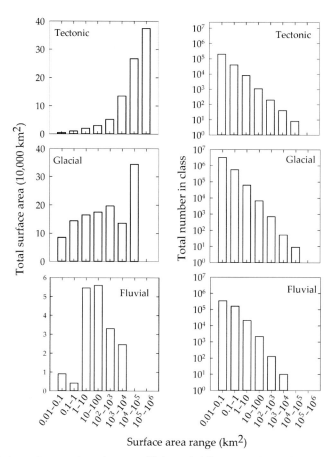

Figure 7.4 Global numbers and total areas of lakes of different geological origins by surface area size class. *Data from Meybeck (1995).*

cirque (also called tarn) lakes in the "amphitheaters" at the heads of the valleys. The glacier forms chains of *paternoster* lakes as it flows further down the valleys (Figs. 7.7 and 7.8).

Glacial scour can lead to formation of extremely large lakes. The Great Slave Lake in Canada was carved to a depth of 464 m below sea level by the massive weight of the continental ice sheet, and is the deepest lake in North America. The Laurentian Great Lakes of North America (e.g., Superior, Huron, and Erie) were partially formed by glacial action. *Fjord* lakes such as Loch Ness (home of a legendary creature) are long glacial lakes formed in steep valleys. One of the strangest lakes associated with glaciers is a gigantic lake below the ice in Antarctica (Highlight 7.1).

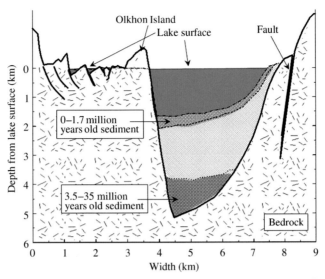

Figure 7.5 Cross section of Lake Baikal, the south basin, in the region of maximum depth (1,637 m). *Redrawn from Belt (1992) and Mats (1993).*

Figure 7.6 Earthquake Lake forming after a large earthquake in 1959 in Montana caused a massive landslide damming the Madison River. This picture taken before the lake was full shows debris still in the water. *Image courtesy US Geological Survey.*

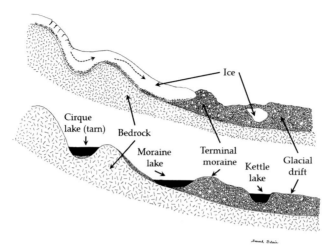

Figure 7.7 Formation of some types of glacial lakes. (A) A cross section of a glacier moving down a valley. (B) After the glacier has retreated, it leaves cirque, moraine dammed, and pothole lakes. *Diagram by Sarah Blair.*

Figure 7.8 Paternoster lakes (a string of glacial lakes) in Grinnell valley, Glacier National Park. *Photograph courtesy the United States National Park Service.*

Highlight 7.1 A large lake beneath the ice in Antarctica

In 1974 and 1975, an airborne radio-echo survey of Antarctic ice depths led to the discovery of a lake under the ice. The ice sitting on the lake's surface is flat relative to the surrounding ice on land, and remote satellite measurements of ice elevation determined the size of the lake to be close to the area of some of the North American Great Lakes (Kapitsa et al., 1996). The lake covers about 15,000 km^2, is about 250 km long, 50–80 km wide, and 200–400 m deep, and rests below 3.5–4 km of ice (Leitchenkov et al., 2016). Preliminary calculations suggest that the residence time of the water in the lake is tens of thousands of years, and that the lake basin is about 1 million years old. However, there is significant water exchange between the lake and the ice sheet (Siegert et al., 2000). Scientists have taken cores through the ice sheet to 3,950-m depth (about 120 m above the lake). The ice at 3,310 m is about 420,000 years old and was formed by refrozen lake water (Jouzel et al., 1999). Analyses of the refrozen lake water from the ice cores indicate the presence of a microbial community (Karl et al., 1999; Priscu et al., 1999), and some of these bacteria may still be viable (Karl et al., 1999). Data suggest that these lakes below the Antarctic icecap are pressurizing and may quickly release their contents under the ice into the ocean leading to a rapid drop in ice elevation (Wingham et al., 2006). Sampling the lake without contaminating is technically difficult. The first successful drilling into the lake occurred at 3,726 m in 2012, but the water immediately rose over 300 m into the borehole and was probably microbiologically contaminated. The lake was drilled into again in 2015 and the water was again probably contaminated, although unique microbial DNA was recovered. Future research is directed toward retrieving uncontaminated microbial samples from the water (Bulat, 2016). Continued research suggests that there are more than 400 lakes below the Antarctic ice sheet and that channels and groundwater may connect some of these lakes (Siegert et al., 2016).

As glaciers move, they entrain rocks and sediments into the ice. Where glaciers melt at the edges and front, they deposit these materials. As the glaciers retreat, they leave this material, called glacial till, behind. If large blocks of ice remain in this till, they melt and eventually leave lakes, ponds, or wetlands called *kettles* or potholes (Figs. 7.7 and 5.9). This process formed the many lakes and ponds that provide vital habitat to waterfowl in the northern prairies of North America, although many have been filled for agricultural purposes.

If the forward flow of a glacier is approximately equal to its backward melting rate, a wall of deposited material called a *terminal moraine* is formed, which can impound water flow and lead to formation of lakes. Materials deposited along the sides of glaciers form lateral moraines, and these also can dam streams and create lakes. Glacial lakes tend to be smaller than tectonic lakes, but a few very large glacial lakes (e.g., the North American Great Lakes) make up a considerable area when considered on a global basis (Fig. 7.4).

The release of large volumes of water from behind glacial ice dams is a catastrophic mode of lake loss followed by downstream formation of smaller lakes. These outbursts happen on a moderate scale now and probably more commonly than thought based on analyses or remote sensing images from the Himalayas (Veh et al., 2018). Huge ones occur during ice ages. These outbursts occurred because of pooling of glacial water as large ice sheets receded, followed by collapse of the ice dam. Such outbursts created massive floods that scoured out existing lakes and created new lakes below any spillways that existed in the channels. Kehew and Lord (1987) suggest that such outbursts established the courses of most major rivers in the mid-continental United States and Canada. Lake Missoula was formed in western Montana behind the retreating ice sheet and was responsible for several catastrophic floods downstream in the Columbia River basin. The lakes in the Grand Coulee in eastern Washington State are actually plunge basins from the giant waterfalls that resulted from this flooding (Fig. 7.9). With a greater discharge of any river currently flowing on earth, and a waterfall five times wider than Niagra Falls, water dropped over cliffs and gouged out plunge basins that were so big that we see them as lakes now.

Volcanic activities can lead to formation of lakes. Explosions of volcanoes or pockets of steam can leave behind depressions in craters that fill with water (*caldera* or *maar* lakes). An example of a volcanic lake is the exceptionally clear and deep Crater Lake in Oregon (Fig. 1.1). The lake was formed when the volcano Mount Mazama exploded about 6,000 years ago. This eruption must have been a catastrophic event for Native Americans living in the region at the time because several meters of ash

Figure 7.9 Lakes formed from glacial outburst flooding as plunge basins of gigantic waterfalls, at Dry Falls, Washington. *Photo taken by Steven Pavlov in 2011, courtesy Wikimedia Commons.*

were deposited throughout western North America. The resulting crater filled with water and a subsequent eruption formed a volcanic cone in the lake known as Wizard Island. Similarly, Lake Yellowstone in Wyoming formed as the result of a very large volcanic explosion, and such an explosion could occur again in this location leading to devastation in the Western United States and severe global cooling because of ash ejected into the atmosphere.

Water can dissolve sedimentary rocks and lead to depressions that form lakes. In karst regions, *sinkholes* form where limestone is dissolved and the cavity collapses (Fig 4.11) and in cases where the bottom is sealed after the collapse, a lake can form. Sometimes these sinkholes simply drop down below the level of groundwater and a lake or pond is created at the bottom of a very steep-walled hole. Such pools, called cenotes, are common in the Yucatán in Mexico; the most famous is probably the Cenote of Sacrifice in Chichén Itzá where numerous Mayan artifacts, including gold objects, have been recovered as well as human bones from sacrificial offerings. Sinkholes can also occur where old subterranean salt deposits dissolve or where sandstone washes away. These dissolution lakes are generally small.

Activities of rivers can form *fluvial* lakes, including oxbow lakes where meanders pinch off (see Chapter 6). A levee lake is another fluvial type that is formed next to rivers where periodic floods scour and fill depressions parallel to river channels. The lowland areas surrounding the Amazon River contain many of these lakes. They are connected to the Amazon during times of high flow. Fluvial lakes provide an important habitat for many organisms and are involved intimately with the ecology of the river. Fluvial lakes tend to be smaller than either glacial or tectonic lakes and are less abundant globally than the other two lake types (Fig. 7.4).

Additional processes that can form lakes include erosion by wind (aeolian lakes), crater formation by meteoric impacts, formation of depressions by alligators (*Alligator mississippiensis*) or bison (*Bison bison*), dam building by beavers (*Castor* spp.), and accretion of corals leading to isolation from the ocean and lakes in the centers of coral atolls.

LAKE HABITATS AND MORPHOMETRY

We classify lakes into several subhabitats or zones. Lake habitats in general are referred to as *lentic* or *lacustrine* (i.e., habitats with deep, nonflowing waters). The open water of a lake, particularly the water column above the depth that does not receive enough light to maintain benthic photosynthetic organisms, is the *pelagic* zone. The *profundal* zone is the benthic habitat below the pelagic waters with light so low that photosynthesis is negligible. Materials that settle from the pelagic waters influence the profundal zone, and the sediment below is usually composed of fine silt or mud. The shallow zone of a lake, where enough light reaches the bottom to allow the growth of

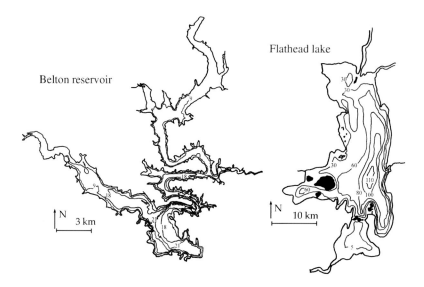

Figure 7.10 Bathymetric maps of Belton Reservoir, Texas (left) and Flathead Lake, Montana (right). The reservoir is dendritic and shallow, with the deepest portion near the dam (lower right). Flathead Lake was formed by a combination of tectonic and glacial processes and is deep with a regular shoreline and shallow outlet (lower left).

photosynthetic organisms, is referred to as the *littoral* zone. The size and shape of a lake determines the relative occurrence and extent of these different subhabitats.

Morphometry, or the shape and size of lakes and their watersheds, is one of the first ways that people classify lakes. The *bathymetric map* (a depth-contour map of a lake bottom) of a lake provides important information on geomorphologic properties (Fig. 7.10). Generally, the first measurement made is of the area of the lake (A) and the second is of depth (z). The maximum depth (z_{max}), mean depth ($\bar{z}$), and volume (v) are also of interest. The volume is the product of area and mean depth:

$$v = A\bar{z}$$

In general, lakes with a shallow mean depth are more productive than deeper lakes. Greater productivity of shallow lakes is a consequence of wind mixing the nutrients up from the bottom more readily, more extensive littoral habitat for primary producers that use the lake bottom, and other morphometric considerations. In general, natural lakes tend to increase in depth as their area increases, but the relationship is quite variable. Still, over about six orders of magnitude of increase in surface area, mean lake depth only increases about one order of magnitude (Fig. 7.11).

Given the volume of a lake and the amount of water entering and leaving the lake, the *retention time* or *water residence time* of the water in the lake can be determined.

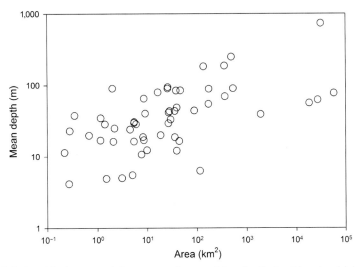

Figure 7.11 Relationship between lake area and mean lake depth for 54 natural lakes. *Data from Hutchinson (1957).*

The average retention time (*R*) can be generally calculated as follows if volume (*v*) and water loss (*L*) from evaporation and outflow is known:

$$R = v / L$$

This equation assumes inflow and outflow are equal. However, in some cases, it is more difficult to calculate water retention because closed basins gain water and increase in volume until evaporation balances inputs. Water retention time can vary widely from several hours for a small pond with a large inflow to thousands of years for very large lakes. The average residence time is less than 1.12 years for lakes in the United States as determined by water isotopic analyses and 60% of the lakes were flow-through lakes (Brooks et al., 2014). Water residence time is important in determining the residence time of pollutants in a lake, how quickly the biota can be washed out, and the general influence of tributaries entering a lake. For example, Lake Tahoe, a beautiful, clear lake in the Western United States, has about a 700-year residence time (Table 7.2) so it is very sensitive to nutrient pollution (see Chapter 18).

Another important aspect of morphometry is the irregularity or degree of convolution of the shore. The *shoreline development* index (*D*_L) quantifies convolution. This index compares the minimum possible circumference of the lake, given a specific surface area (i.e., a circle), to the actual circumference of the lake and its surface area. A value of 1 for shoreline development is a circle with the same area of the lake and a larger value means that the shoreline is highly dissected. High shoreline development is generally related to small values of mean depth and mode of formation, and is indicative of a high degree of watershed influence. Shoreline development is calculated as follows:

$$D_L = \frac{L}{2\sqrt{\pi\, A_0}}$$

where L is the length of the shore and A_0 is the surface area of the lake.

Lakes with high shoreline development are often naturally productive relative to those with low shoreline development of the same size (Example 7.1). The idea of a dissected shoreline resulting in increased productivity leads to consideration of the watershed of a lake as a determinant of productivity. A lake with a relatively large

---

Example 7.1 Compare the morphology of two lakes—Crater Lake and Milford Reservoir

Crater Lake, Oregon and Milford Reservoir, Kansas are lakes of contrasting properties, despite the fact that they are similar in surface area. Crater Lake (Fig. 1.1) is a deep oligotrophic lake in the crater of a large volcano; its scenic grandeur has earned it national park status. Milford Reservoir is a typical Midwest US lake; it is eutrophic but widely used for recreation, including a vibrant fishery. Much of the difference in levels of productivity in these two systems can be related to their contrasting morphometric properties, but the level of agricultural activity in the watershed of Milford Reservoir is also much greater. The following are vital characteristics of the two lakes:

	Crater Lake	Milford Reservoir
Area (km^2)	55	65
Mean depth (m)	364	7.4
Watershed area (km^2)	81	64,465
Shoreline length (km)	42	122
Inflow (m^3/s)	4.3	27.2

Calculate volume, water replacement time, and shoreline development. Also calculate the ratio of the area of the watershed to the lake volume and speculate about how these features may alter trophic state.

Volume for Crater Lake = area × mean depth = 55 × 0.364 = 20.3 km^3
Volume for Milford Reservoir = 65 × 0.0074 = 0.48 km^3
Water replacement for Crater Lake = volume/discharge = 20.3/0.136 = 150 years
Water replacement for Milford Reservoir = 0.48 km^3/(0.858 km^3/year) = 0.56 years
Shoreline development for Crater Lake = D_L = 42/(2$\sqrt{\pi}$55) = 1.6
Shoreline development for Milford Reservoir = D_L = 122/(2$\sqrt{\pi}$65) = 4.3
Watershed area/lake volume for Crater Lake = 4.0
Watershed area/lake volume for Milford Reservoir = 134,000

All these parameters but one suggest that the watershed will have a much greater influence on the water in Milford Reservoir and that nutrients and light should be greater in Milford Reservoir. The only caveat is that the water replacement is so slow in Crater Lake that once a nutrient enters the system, it could be recycled for some time. The high shoreline development index, watershed area to volume, and low mean depth suggest that Milford Reservoir should indeed be more eutrophic than Crater Lake.

---

watershed will have more land area to contribute nutrients. Such a lake is likely to be more productive than a lake with a small watershed. Land use practices also play a major role in determining nutrient inputs.

As lakes get larger, they tend to have smaller shoreline length per unit area of lake. This occurs even though larger lakes tend to have more developed shorelines (Fig. 7.12). So larger lakes tend to be deeper, have more volume per unit surface area,

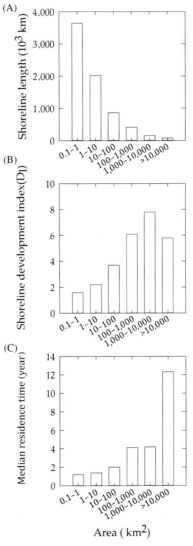

Figure 7.12 Morphometric properties of lakes worldwide including (A) total shoreline length, (B) mean shoreline development, and (C) water residence time as a function of surface area size. *Data from Messager et al. (2016).*

and tend to have longer water residence times. Even though the total volume of water stored is greater in the larger lakes on Earth, the total amount of shoreline habitat is greater associated with smaller lakes.

UNIQUE PROPERTIES OF RESERVOIRS

Reservoirs are man-made lakes, and common features of today's landscape. They differ in several important ways from natural lakes. Damming can also form natural lakes, and presumably, reservoirs are not much different, except that natural lakes usually do not release deep waters downstream, which is sometimes the case with reservoirs. Hybrids between lakes and reservoirs are created when the outlet of a natural lake is dammed to raise the level of the lake.

Unlike most natural lakes, reservoirs are deepest near the dam and generally become shallower near the deltas of the rivers that feed them. Reservoir shape is controlled by the stream network structure of the surrounding topography. The reservoirs therefore often have a smaller relative mean depth than many natural lakes in the region. However, some high-dam reservoirs are 200–300 m deep. Shallow mean depth can lead to increased mixing and associated suspended solids, as well as decreased likelihood of stratification.

Reservoirs fill the drainage basins of rivers and streams, and each arm of a reservoir moves up into a former stream channel. Thus, a typical reservoir has a dendritic or tree-like shape (Figs. 7.1 and 7.9). A dendritic shape results in a high value for the shoreline development index. Shallow mean depths and high shoreline development indices indicate that many reservoirs are very productive, unless turbidity limits light for photosynthetic production.

Larger reservoirs often have more lake-like biological and physical characteristics as one moves from the inlet rivers toward the dam. In the shallow inlets, river flow dominates, but as the water velocity and turbulence decreases with greater width and depth, a lentic character is established. Thus, species commonly associated with rivers are gradually replaced with those more commonly found in lakes as one moves closer to the dam.

Rivers deposit sedimentary material as velocity slows, creating river deltas at the upstream end of reservoirs. Reservoirs fill with sediments from the top down toward the dam. All reservoirs will eventually fill with sediments without dredging, with the useful life of some reservoirs only a few decades. Controlling upstream sediment inputs can prolong the life of a reservoir. Few reservoirs are constructed with plans for being decommissioned, and we discussed problems with dam removal in the last chapter.

Reservoirs have global importance. Even though the amount of sediment erosion from the land has increased worldwide, the amount of sediment that reaches the oceans has decreased because sediments settle in reservoirs, and there are ever more reservoirs, particularly in tropical areas (Syvitski et al., 2005). Organic materials also settle in reservoirs leading to significant retention of organic carbon (Friedl and Wüest, 2002) that would otherwise be exported to the ocean. Reservoirs can increase the export of greenhouse gasses to the atmosphere (Louis et al., 2000) and large reservoirs have increased the total global combined lake and reservoir surface area by at least 10% over that of natural lakes (Messager et al., 2016).

Most lakes have relatively stable water levels because their outflow stream is at a single elevation, except in cases where outflow is severely constricted and flooding can cause increases in lake elevation. In contrast, most reservoirs have the capacity to raise and lower water levels as a practical consequence of their function of water storage and power generation. Thus a relatively unnatural condition occurs where large areas of shoreline are inundated for part of the year. Managers often tightly control water level to allow for recreational boating on the reservoir, to meet downstream navigation requirements, to meet hydroelectric power generation requirements, to meet downstream species conservation goals, or a combination of these. This mode of reservoir operation can lead to a ring of vegetation that cannot withstand inundation around the reservoir at its maximum elevation; a barren area is often seen below this ring when water level is low. The level of alteration can affect the ability of submerged plants to survive, and thus the nature and extent of the littoral zone. This alteration, can increase sediment erosion problems, as well as influence the reproductive behavior and success of fishes and invertebrates in the reservoir.

GEOMORPHOLOGICAL EVOLUTION OF LAKES AND RESERVOIRS

Lakes and reservoirs have lower water velocity than rivers and streams and are more likely to retain sediments. This feature leads to sediment build-up and the gradual filling of lakes and reservoirs. The sediments form from materials entering the lake from tributaries feeding into it, from windblown particles, and from organisms in the lake dying and sinking. As sedimentation fills the lake, the bottom becomes more and more homogenous and regular. The salt flats in Utah formed as sediment on the bottom of a large lake that dried and deposited salts into an extremely large flat area. This former lake bottom currently is used to set land speed records for motorized vehicles.

The standard model of lake evolution is starting with a modest-to-small lake, and over time, the sediments fill the lake, which eventually becomes a wetland and then a meadow. This is only a generalization as many lakes have different fates. For example, oxbow lakes might fill and become wetlands or the river may meander and recapture the oxbow before it does. In tectonically formed lakes, they will not become much shallower if the rate of increase in depth exceeds the sedimentation rate.

Sedimentation particularly influences reservoirs, and this has practical implications as it limits the life of the reservoir. Reservoirs often represent slow parts of relatively large rivers, and once the water slows, the larger sediments drop out. If upstream areas are disturbed, the reservoir can receive a substantial amount of sediment and fill relatively quickly. Reservoir management worldwide, in part, is based on management practices that will maximize the lifetime of the reservoir by routing sediments around the reservoir or flushing sediments from the reservoir during high flow periods (Kondolf et al., 2014).

This sedimentation can lead to improvements in downstream water quality (Fantin-Cruz et al., 2016) by lowering suspended sediment and total phosphorus concentrations. However, the reservoirs can fundamentally change the nature of the particulate material in the river to finer sediment and smaller particulate organic material. This shift away from larger sediment can lead to downstream erosion and change downstream biological communities (Whiles and Dodds, 2002). The sedimentation also creates new habitat at the head of the reservoir. These deltas are ecologically distinct and unique habitats. Little study has occurred on their ecology and function (Volke et al., 2015).

STRATIFICATION

Factors influencing density of water that were discussed in Chapter 2, and heating effects of light discussed in Chapter 3 have profound effects on mixing in lakes. These effects influence the biogeochemistry, biology, and physical geology of lakes. A primary factor creating stratification of lakes is the difference in density resulting from temperature or salinity variation. The classical understanding of lake stratification is based on consideration of cold-temperate lakes, so this seasonal sequence of stratification is considered first.

During the early spring in a cold-temperate lake, the water is *isothermal*, or approximately the same temperature from top to bottom (Fig. 7.13). Wind can completely mix an isothermal lake, leading to *spring mixing*. The entire lake will continue to mix as long as the wind continues to blow. Solar energy warms the surface of the water as the spring season progresses. Surface waters of the lake heat the most because infrared radiation (heat) is absorbed quickly with depth. If you have ever swum in cold water on a calm, sunny spring day, you are familiar with the phenomenon of the top several centimeters of the water being much warmer than the deeper water. Such stratification is only temporary because wind can mix a shallow layer of warm water into the lake.

When a series of calm, warm days occurs, the lake stratifies. Surface waters of the lake heat enough so that the wind cannot completely mix the warmer, less dense surface water into the cooler water below. The top of the stratified lake is the *epilimnion*. The zone of rapid temperature transition just below the epilimnion is the *metalimnion* or *thermocline*. The bottom of the lake is at fairly constant

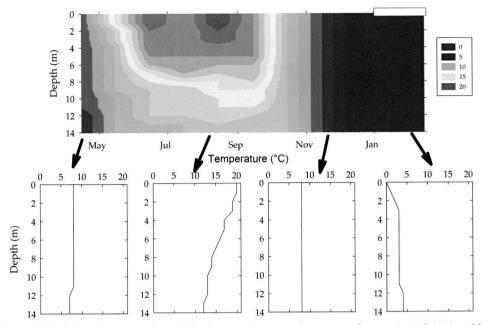

Figure 7.13 A depth contour plot of lake temperature over the course of a year in a dimictic cold-temperate lake (Esthwaite Water, an English lake). The thick white box at the top right corner of the contour plot indicates ice cover. Two-dimensional representations of the temperature versus depth at each phase of stratification are plotted on the bottom. *Data from Mortimer (1941).*

temperature and is called the *hypolimnion* (Fig. 7.14). The stratified layers will stay distinct until a prolonged period of cool weather occurs. The period with distinct layers in the summer in temperate lakes is called *summer stratification*. Prolonged summer stratification is a combined function of the very slow rate of diffusion of heat across the metalimnion because turbulent mixing no longer occurs, and the continued heating of the epilimnion. Because there is minimal mixing across the metalimnion, no eddy diffusion of heat occurs; only molecular diffusion occurs. We explained slow rates of molecular diffusion in Chapter 3.

The epilimnion is very stable relative to the ability of the wind to mix a lake. There can be some mixing of the top of the hypolimnion (*entrainment*) with extreme winds, but even hurricane-force winds will not fully mix a well-stratified lake (Fig. 7.15) because the mixing power of the wind declines exponentially with depth (see next section). Stratification will break down only when the autumn weather can cool the epilimnion to approximately the same temperature (*isothermal*) as the hypolimnion. Cool air coupled with continued heat losses from surface evaporation decreases the temperature of the surface. Cooled surface water is denser than the water immediately below, so it sinks. The wind can mix the lake once the entire lake is isothermal, and the *fall*

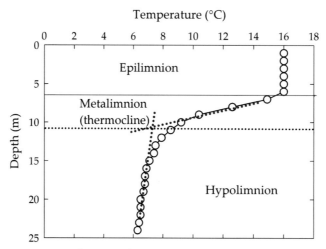

Figure 7.14 Temperature as a function of depth for Triangle Lake, Oregon, on October 1, 1983, and positions of epilimnion, metalimnion, and hypolimnion. *Data from R. W. Castenholz.*

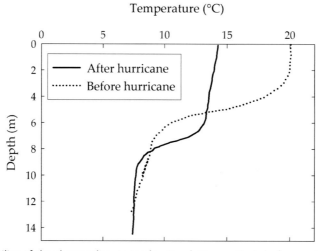

Figure 7.15 Stability of the thermocline in Linsley Pond, Connecticut, before and after a hurricane on September 21, 1938, with wind speeds up to 100 km/h. *From Hutchinson (1957). Copyright © 1957 John Wiley & Sons, Inc. Reprinted by permission of John Wiley & Sons, Inc.*

mixing period begins. The lake will potentially continue to cool and mix until formation of an ice cover on the surface of the lake. In very deep lakes, the wind still cannot mix the bottom waters with the top, even if the lake is isothermal.

The surface of the lake can freeze when the temperature of the entire lake is below 3.9°C. If the lake is warmer, cool water from the surface will continue to sink and mix with the less dense water below it. If the lake temperature is 3.9°C, the water is at its densest, so cooler water will sit on the surface of the lake as long as the wind does not mix it. The surface of the lake can freeze if there is a cold, calm night. Low-density ice and cold, low-density water will sit on the surface of the lake once the surface has frozen. No more wind reaches the surface of the water once ice forms, so no more mixing occurs. *Winter stratification* is the second period during the year when a cold-temperate lake does not mix (Fig. 7.13) and it lasts as long as cold weather maintains the ice cover.

Duration of ice cover provides one of the best long-term records of the effects of global warming on freshwater systems. Detailed data on dates of formation and breakup of ice cover for many lakes and rivers for the past 100 years are available, and records exist for Lake Suwa in Japan almost continuously since 1450. These data suggest that freeze dates have shifted 5.8 days later over the past 100 years and breakup dates 6.5 days earlier (Magnuson et al., 2000).

Limnologists have created many terms to describe the mixing regimes of lakes (Table 7.1). Lakes that mix twice a year are *dimictic*. Lakes that only mix once during the year are *monomictic* and are common in temperate and subtropical regions where winters are not cold enough to freeze lake surfaces but are cool enough to allow the lakes to become isothermal. Lakes that never mix are *amictic* or *meromictic*. *Polymictic* lakes stratify or mix several times a year and are found mainly in tropical regions.

Several conditions cause meromictic lakes. Seasonal temperature regimes can be constant enough (mainly in tropical areas) that lakes rarely mix. The temperature difference in tropical lakes does not need to be as great to form a stable stratification as in the temperate zone because the water temperatures are higher and a greater relative difference in density occurs for each degree difference in water temperature (Fig. 2.3). For example, the density difference is 1.92 times greater between 20°C and 25°C than between 10°C and 15°C water.

Some lakes rarely if ever mix if they contain dissolved compounds in the hypolimnion that stabilize density layers. Salinity differences can lead to extremely stable stratification. In this case, water that is more saline can sit below cooler surface waters when the salinity-caused density difference is greater than the temperature-related differences. Several conditions can cause such salinity differences. In tropical lakes that are stratified for long periods, nutrients enter the surface waters from rivers. These nutrients enter the biomass of the planktonic food web, and when organisms die, they sink. The sinking organisms slowly release dissolved nutrients as they decompose and a portion is transported to the hypolimnion. Slowly, the salinity of the hypolimnion increases and the stratification is stabilized. Saline spring or groundwater inputs can also stabilize lakes. Some of the dry-valley lakes in Antarctica have such stable layers (see Chapter 15).

In arid regions, evaporation can lead to increases in dissolved salt concentrations. Fresh river water flowing into a lake will remain on top of the denser saline water. A fresh surface lake is a common occurrence in closed basins where saline lakes form.

Such was the case in the Dead Sea, where a stable stratification was maintained for nearly 300 years until water diversions for human uses reduced inflow to the lake and the dilute surface layer disappeared in the 1970s (Gavrieli, 1997).

Fjord lakes near marine habitats can also have saline waters below freshwater. In this case, a glacial valley is formed where the glacier carves the valley to an elevation below sea level. As the glacier recedes, saline marine water floods the valley. The floor of the valley rebounds from the weight of the glacier and if a raised portion exists at the end of the valley (e.g., a terminal moraine), the saline ocean water can be isolated. Freshwater flows on top of the saline water and a lake forms with saline water on the bottom and freshwater on the top.

The biological and biogeochemical effects of stratification on lake organisms are strong. Molecular diffusion rates that dominate movement of dissolved materials across the metalimnion are slow enough that a significant depletion of O_2 in the hypolimnion will lead to anoxia during the summer. In turn, O_2 loss from the hypolimnion alters biogeochemical cycling and lake productivity, as discussed in detail in Chapters 12—14, Nitrogen, Sulfur, Phosphorus, and Other Nutrients. Given the very complex chemical and physical characteristics in many stratified lakes, several methods for sampling lake waters from different depths have been developed (Method 7.1).

METHOD 7.1 How do limnologists sample water and sediments from lakes?

The main consideration in determining how to sample water from a lake is what information investigators are interested in. If the information can be obtained with a submersible sensor (such as for temperature, dissolved oxygen, pH, and conductivity), it is easiest to take measurements without removing samples. However, for many chemical and biological parameters, researchers need to remove water from known depths with a minimum of contamination or turbulence.

Several devices are available that allow water to be sampled from specific depths. The Van Dorn bottle consists of a tube with covers over each end. The messenger releases a trigger that releases a stretched rubber connector that pulls the covers onto the tube, sealing the water into it. Kemmerer bottles operate similarly, but rather than using rubber connectors to pull ends onto a tube, gravity is used (Fig. 7.16C). Other devices include pumps to remove water from a given depth, bottles with strings attached to stoppers that can be unplugged at a desired depth, and pipes that allow water to flow up to containers that displace surface waters (Fig. 7.16B). Samples are obtained by lowering the equipment into the water using ropes or cables. The devices are generally triggered with a weight that is dropped down the line to the sampling equipment. This weight is called a messenger.

Lake substrata and associated sediments, the waters contained in them, and organisms that inhabit them can also be sampled with devices called dredge or grab samplers that snap shut when a messenger is dropped down an attached rope (Fig. 7.17). Alternatively, depending on the type of sample and information that is desired, various types of coring devices can be used; these range from small, handheld corers to large, boat-mounted

(Continued)

METHOD 7.1 (Continued)

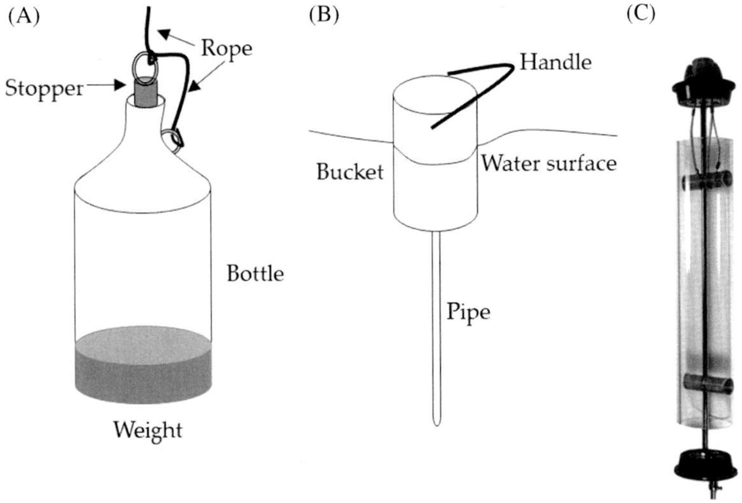

(A)

Rope

Stopper →

Bottle

Weight

(B)

Handle

Bucket

Water surface

Pipe

(C)

Figure 7.16 (A) A simple water sampler made from a weighted bottle and stopper, (B) a sampler that collects water from depth by displacing water at the surface, and (C) a Kemmerer sampler (photograph courtesy of Wildlife Supply Company). Only samplers (such as type C) that close at depth are suitable for collecting dissolved gas samples.

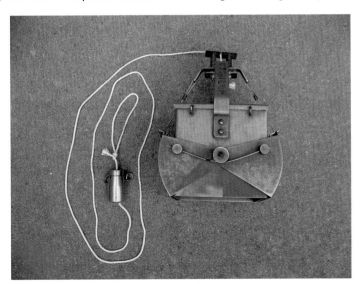

Figure 7.17 An Ekman grab sampler for sampling sediments from the bottom of a lake. These types of grab or dredge samplers come in a variety of shapes and sizes. Most have a spring-loaded trigger (visible on the top of the sampler), which is tripped by a messenger (cylindrical weight on the rope) that is sent down the attached rope; when the messenger trips the trigger, the sampler snaps shut around a parcel of sediment. *Photograph by S. Peterson.*

(Continued)

METHOD 7.1 (Continued)

oceanographic drilling devices, depending on the specifics of the habitat and type of sample needed. Researchers have developed a variety of suction or hydraulic dredge samplers for collecting lake and ocean sediments. Suction samplers require a power source and, like coring devices, can range from handheld samplers to large, boat mounted machines.

The choice of device for water or sediment sampling depends on the type of sample that is required, with many factors to consider before sampling. Toxic materials will poison biological samples, and samplers that have metal parts can contaminate samples used for the chemical detection of metals. The violent closure of some samplers can harm organisms that are susceptible to pressure shock. Zooplankton may avoid an opaque sampler more than a clear one because of their predator avoidance behaviors. Some sampling devices can also be expensive or produce samples that require considerable time and resources to process.

ADVANCED: HEAT BUDGETS OF LAKES

The amount of energy required to heat a lake from its coldest to its warmest point in a year is referred to as the *heat budget* or *heat balance*. This is an index of how much the temperature of a lake changes over the year. Temperature drives rates of metabolism, growth, and activity of organisms. Heat extremes limit distributions of organisms. Water can store a substantial amount of heat, and large lakes can moderate local climate (e.g., keeping areas near lakes warmer in winters, cooler in summers, and creating phenomena like lake effect snow). Understanding how lakes heat leads to deeper understanding of factors that are influenced by temperature. This understanding is attaining greater urgency because of the rapid warming of lakes globally, leading to large changes in the temperature regimes of lakes world-wide (O'Reilly et al., 2015). Wetzel and Likens (1991) note that the term "heat budget" is slightly incorrect as it is not so much of a budget as a measure of the heat capacity of the lake.

The heat content of energy stored in water (or any other substance, *H*) depends on the product of the mass (M), temperature (*t*), and specific heat (*s*). The change in heat content over a year gives the heat balance, which is defined as:

$$Q_R + Q_E + Q_L + Q_V + Q_S + Q_B = 0$$

The energy added (positive) equals the energy lost (negative) over the year, and Q_R is the net radiation or the total flux of energy in from the sun, Q_E is the latent heat exchange, Q_L is the latent heat of evaporation which is the heat lost to the process of evaporation, Q_s is the sensible heat exchange which is the exchange of energy with the atmosphere, Q_V is the net advective exchange including that which is flowing in and out of lakes via streams and groundwater, and Q_B is the conductive heat exchange with the sediments, all in cal cm^{-2} day^{-1} (Ragotzkie, 1978; Wetzel and Likens, 1991).

The net radiation flux in the summer is the total radiation minus the amount reflected back. In the winter, Q_R can be lowered considerably by reflection from ice and snow relative to ice-free water. Since ice and snow have a different density than water, they also store heat differently. In addition, heat entering the lake from the surface goes into melting ice and snow and does not cause an increase in measurable temperature. Strangely, condensation of fog may be a considerable heat input in some large lakes (Hutchinson, 1957).

Methods for measuring heat balance in lakes are based on integrated temperature change through the lake which is done using repeated temperature profiles, and weighting for bathymetry of the lake (Wetzel and Likens, 1991). In general, lakes that are deeper can have a greater annual heat balance as there is more mass of water below each unit surface area to store energy (Fig. 7.18). Additional factors that can influence heat balance include altitude (although it is possible this is due to failure to account for the latent heat of ice melting (Hutchinson, 1957)). Annual budgets can vary up to twofold from year to year in some lakes depending on interannual climate variation, and sediments can account for 7%—38% of the heat budget for the few lakes where this determination has been made (Wetzel, 2001). Reservoirs are generally more heavily influenced by rivers and streams, and as such the effects of inflow and outflow on heat budgets of reservoirs and high turnover lakes should be greater than in larger natural lakes.

Heat budgets of lakes will be influenced by global climate change. A more variable climate will lead to greater year-to-year variance in budgets. The altitudinal effect may lessen and the effect of ice on heat budgets is decreased. Groundwater generally reflects annual temperature, and as this will lag behind atmospheric energy inputs, heat loss to groundwater and sediments may increase modestly as the Earth's climate warms.

WATER MOVEMENT AND CURRENTS IN LAKES

Movement of wind is generally the main cause of waves across lakes, although motorboat activity can cause significant wave action. Wave action is important partially because it is associated with surface mixing and erosion of the shoreline. Lakeshore erosion can lead to habitat destruction and large financial losses associated with property damage; many environmental engineering firms specialize in controlling erosion. Wave action can influence which species can successfully inhabit the different depths of the shallow benthic zone (littoral zone). The two main determinants of wave height are the strength (speed and duration) of the wind and the length of lake on which the wind acts. The influence of the wave also is dependent on the geometry and materials

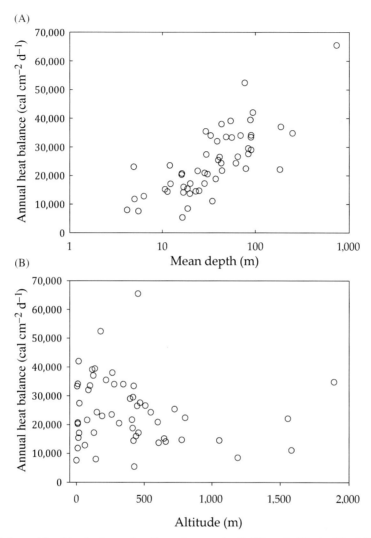

Figure 7.18 Annual heat budget as a function of mean depth (A) and altitude (B) of 52 lakes. *Data from Hutchinson (1957).*

that make up the shoreline. A long shallow beach can dissipate the force of breaking waves across a larger area and is less prone to erosion than is a steep lake shoreline.

The length of the lake on which the wind acts is the *fetch* (Fig. 7.19). The longer the fetch, the higher the waves (Fig. 7.19). Wind from any direction will influence a perfectly round lake similarly throughout. On an irregularly shaped lake, certain wind directions lead to the largest waves. Features such as hills or trees can prevent the wind from making large waves in smaller ponds and lakes. For example, deforestation can

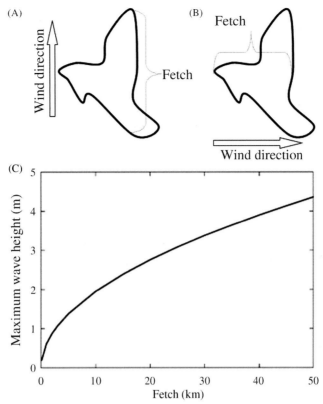

Figure 7.19 How fetch of an irregularly shaped lake varies with wind direction (A, B) and relationship between maximum wave height and fetch (C) (equation for C from Wetzel, 1983).

lead to a deeper epilimnion (by about 2 m in some small Canadian lakes) because of increased mixing during spring warming (France, 1997a). Even though the processes of surface wave formation are the most apparent to human observers, the wind causes other water movements in lakes that can also be important.

As wind moves water on the surface of a lake forward, it must be replaced by water from below. This process leads to spiral circulation patterns called *Langmuir circulation* cells (Fig. 7.20). This process sets up zones of upwelling and downwelling water. The lines of upwelling and downwelling occur parallel to the direction of the wind, and set up spiral movement of water parcels. The spirals rotate in alternating directions compared to the ones next to them. Floating materials aggregate at the water surface along the downwelling lines and form streaks in the same direction as the wind is blowing. These circulation cells are several meters wide, and streaks can commonly be observed on lakes and reservoirs under a constant wind.

In addition to the smaller scale waves and Langmuir cells, movement of water can also occur within the whole lake's volume. When a sustained wind occurs, it causes

Figure 7.20 Langmuir circulation cells on a lake. Note zones of convergence where spiral currents meet and go down into the lake. *Drawing by Pete Else.*

water to pile up on the downwind side of the lake, and when the wind suddenly ceases, the surface of the lake can rock. This rocking of a lake's entire surface is a *seiche*. The surface seiche in Lake Erie can be several meters leading to destructive storm surges and stranded watercraft. Seiches can have periods of a few hours to days, and can also be formed by earthquakes.

An interesting phenomenon, an internal wave, occurs in stratified lakes that are subjected to a sustained unidirectional wind. The force of the wind causes the water in the epilimnion to move across the lake to the downwind side (Fig. 7.21) and the depth of the epilimnion is greater downwind than upwind. Under extreme winds, the hypolimnion can come to the surface on the upwind side. When the wind ceases, the less dense water of the epilimnion moves back across the surface of the lake, and the hypolimnion moves back toward its original position. Like a pendulum, the surface of the hypolimnion rocks back farther than its original position. Thus, winds create an *internal seiche* where the surface of the lake appears still, but the plane that forms the top of the hypolimnion continues to oscillate for hours or days after the wind ceases. Aside from the intrinsic elegance of the seiche as a physical phenomenon, this type of water movement has an important biological implication.

Even though the hypolimnion is very stable, seiches can lead to a moderate amount of mixing of hypolimnetic and epilimnetic water. The movement of this water up to the epilimnion is *entrainment*. Entrainment causes nutrient-rich water from the hypolimnion to reach the epilimnion (Fig. 17.21C), often causing stimulation of primary production (Giling et al., 2017). A study of phytoplankton responses in deep

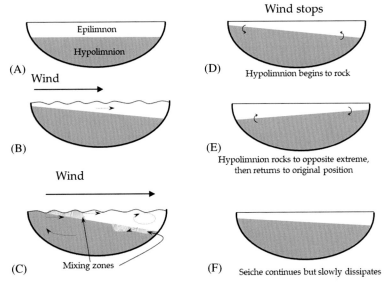

Figure 7.21 Formation of an internal seiche and entrainment associated with wind. Dashed arrows show water flow. (A) The lake under calm conditions; (B) the wind deepens the epilimnion on the right; (C) a strong wind mixes some of the epilimnion with the hypolimnion; (D) the wind stops and the hypolimnion begins to oscillate; and (E and F) the amplitude of the seiche diminishes over time.

tropical lakes to typhoons illustrates that there can be strong effects of such mixing on production and biomass (Ko et al., 2017). Nutrient mixing can be significant biologically because the mixing rate far exceeds the rate of molecular diffusion that usually predominates between the hypolimnion and the epilimnion. Seiches can also influence rooted plants and benthic invertebrates by altering temperature and nutrient regimes. The lake model exercises discussed by Wetzel and Likens (1991) are highly recommended for students who want a demonstration of the processes of stratification and seiches.

Very large lakes, as discussed in Chapter 3, can also be subjected to Coriolis forces causing a counter-clockwise rotation in the Northern Hemisphere. Other forces that move water in lakes can include human caused disturbances, earthquakes, and floods.

SUMMARY

1. Processes that form lake basins include tectonic, glacial, fluvial, volcanic, and damming. Glacial lakes are the most numerous worldwide, but tectonic forces form some of the largest, deepest, and oldest lakes. Fluvial lakes can be very important to riverine ecology and can occur in areas where few other types of natural lakes are common.

2. Lake basin morphology is described with various parameters, including mean depth, area, maximum depth, volume, shoreline development (D_L), and watershed area relative to lake surface area. Shallow lakes with large watersheds and highly dissected shorelines are generally the most productive.

3. Reservoirs are unique habitats that have characteristics of both streams and lakes. They tend to be more lake-like near the dam, and more river-like upstream near the inflowing waters. Depending upon how they are constructed and managed, reservoirs can have profound impacts on upstream and downstream habitats and communities.

4. Waves are greatest where the wind has the longest length of lake (fetch) to act on.

5. Stratification can alter the water circulation in lakes and thus alter biogeochemical, ecosystem, and community properties. Mixing can occur often (polymictic), once a year (monomictic), twice a year (dimictic), or rarely (amictic or meromictic), depending on climate and type of stratification.

6. Thermal stratification occurs when warm surface water sits above denser, cooler waters. The warm surface layer of a thermally stratified lake is the epilimnion, the zone of steep temperature transition is the metalimnion, and the deepest stable zone is the hypolimnion.

7. High concentrations of dissolved substances can also lead to stratified layers in lakes. Such chemically-driven stability can exceed temperature-driven stability because density differences can be greater than are possible with natural temperature differences.

8. Wind causes Langmuir circulation patterns, which lead to streaks of floating material on the water surface while they mix the lake to depth.

9. A sustained wind that suddenly stops can cause oscillation of the lake surface (a surface seiche) or the hypolimnion (internal seiche). This rocking can lead to breakdown of stratification. Mixing of deeper waters into the surface is entrainment.

QUESTIONS FOR THOUGHT

1. Why is it sometimes difficult to assign a single geological explanation to a lake's origin?

2. Why is a lake with a high value for D_L likely to have smaller waves than a lake of comparable surface area with a D_L close to 1?

3. Why do more lakes occur farther from the equator?

4. In which order (from greatest to least) should lakes be ranked with respect to the overall average ratio of maximum depth divided by the mean depth: tectonic, glacial, and fluvial?

5. Langmuir circulation cells concentrate particles slightly denser than water below the surface of the water: where will these particles be concentrated and why?

6. Under what conditions would thermal stratification lead to anoxia in the hypolimnion?

7. Explain why some rivers flowing into lakes flow down into the hypolimnion, some flow across the surface, and others flow into the metalimnion.

8. How do you expect latitude to influence heat budgets of lakes?

CHAPTER 8

Types of Aquatic Organisms

Contents

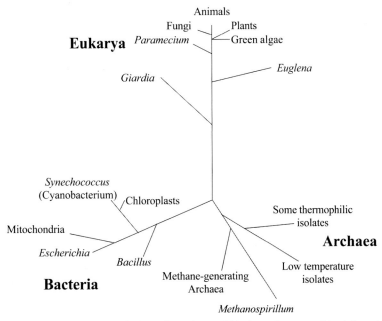

Figure 8.1 A simple phylogenetic diagram based on rRNA data. *Data combined from Fenchel and Finlay (1995) and Pace (1997).*

Correct identification of freshwater organisms is essential to understanding their ecology. The identification of organisms is part of *systematics*, which consists of both *taxonomy* (the identification and naming of life on earth) and *phylogeny* (the understanding of the evolutionary relationships that make up the tree of life). Understanding and determining the identification and relatedness of organisms is important for several reasons. Different species of plants, animals, and microbes interact with the environment in different ways to alter water quality and perform ecological services. Identifying species in food webs is a critical part of managing fisheries. The presence, absence, and overall diversity of invertebrates, microbes, algae, plants, and fishes can indicate water quality and pollution problems through various *biological assessment* (*bioassesment*) procedures. Such assessments rely upon the identification of pollution sensitive and tolerant taxa to indicate water quality, and thus proper identification of these organisms is essential. Some species of algae are toxic, so identification of these species may be important in maintaining safe water quality. Tracking the invasion and influence of pest organisms also requires taxonomic expertise. Finally, endangered species can have a major influence on management decisions, again requiring taxonomic information as well as an understanding of their phylogeny. This chapter provides very basic taxonomic principles, introduces different ways to classify organisms, outlines the habitats they frequent, and examines the ways they can interact with other organisms. This chapter also provides background for the following two chapters on microbe, plant, and animal groups found in freshwaters, and introduces some essential terminology related to habitats and species interactions.

THE SPECIES CONCEPT

Biologists have traditionally considered the species the fundamental unit of taxonomic division and base the scientific system for naming organisms on distinguishing species. Species are then classified into ever-broader groups (Table 8.1). The traditional biological species definition is "a genetically distinctive group of populations whose members are able to interbreed freely under natural conditions and are reproductively isolated from all members of other such groups" (McFadden and Keeton, 1995). However, many aquatic organisms (especially the microbes, but also many plants and animals) do not reproduce sexually. Other species may be able to reproduce with organisms that taxonomists consider separate species (e.g., many species of trout are able to hybridize). Bacteria have no sexual reproduction but can pass genes within or among different taxonomic groups. Even for those organisms that do reproduce sexually, it is often difficult to test if two individuals will interbreed successfully or to determine the degree of genetic similarity. We have limited data on the reproductive biology of most aquatic organisms in their habitat except for some game fish and emergent wetland

Table 8.1 Botanical and zoological naming scheme for organisms, name endings associated with each taxonomic group, and examples of naming of a cattail and a North American beaver

Taxonomic classification (sub or supergroup)	Botanical name endings	Example	Zoological name endings	Example
Super Kingdom or Domain	-a	Eukarya	-a	Eukarya
Kingdom	-ae	Plantae	-a	Animalia
Phylum or Division	-phyta	Anthophyta	-a	Chordata (subphylum Vertebrata)
Class	-opsida	Liliopsida	-a	Mammalia
Order	-ales	Typhales	-a (-formes, birds; -oidea, some mammals)	Rodentia
Family (Subfamily, Tribe, Subtribe)	-aceae	Typhaceae	-idae	Castoridae
Genus		*Typha*		*Castor*
Species (Subspecies)		*latifolia*		*canadensis*

plants. Finally, there is no objective measure of how "genetically distinctive" an organism must be from another before it qualifies as its own species.

Systematists traditionally differentiated between most nonbacterial species based on morphological characteristics, but molecular techniques are becoming increasingly common. The operational species definition uses an older, formal definition of a species as "a group of organisms that more closely resemble each other, with respect to their physical appearance (morphology), physiology, behavior, and reproductive patterns, than they resemble any other organisms" (McFadden and Keeton, 1995). However, no hard and fast line delineates the amount of morphological differentiation that is necessary for organisms to be considered distinct species. Difficulty arises because of the natural morphological variation found in the same species (Fig. 8.2), especially for those in different environments that lead to development of divergent morphologies. Thus, a large sample of individuals from different environmental conditions is often useful to ensure correct identifications. Generally, systematists that specialize in a specific group of organisms have clearly defined the characteristics used to differentiate distinct species.

Lacking a single system to define species that works well across all taxonomic groups leads us to adopt a utilitarian species definition for those who are not taxonomists: we consider a species distinct if the majority of the systematists studying the group of organisms agree that it is a distinct species. This approach allows aquatic ecologists to use systematic data without information about sexual reproduction of the

Figure 8.2 *Eichhornia azurea* with different morphology of floating versus submersed leaves within the same plant. *Courtesy Steve Hamilton.*

species they are studying, or other complex methods of analysis. Scientists can then communicate the taxonomic identity within the bounds of the best current scheme of identification, providing a solid basis for ecological information. A reliable scientific name provides a method to obtain information on a species as well as a way to contribute to scientific data.

A particularly difficult situation arises with microbes that do not have enough morphological variation to classify with traditional techniques, so genetic sequences are best used to determine "species." Some widely adopted approaches classify microbes as taxonomically distinct if DNA sequences vary by more than 30% or if ribosomal rRNA sequences vary by more than 5% (Buckley and Roberts, 2006). These approaches to classification are purely for convenience, and 30% or 5% are mostly subjective. For this reason, microbial researchers classify different microbes under these criteria as *"operational taxonomic units"* (OTU's) rather than species. The methods are empirical, and not based on a theory of microbial evolution (e.g., there is not, as far as we know, analytical evidence that a genome is a coherent group of genes over evolutionary time at some threshold of rRNA similarity). There seems to be some coherence in some groups, particularly obligate pathogens. More general groups do not cluster well (Konstantinidis et al., 2006). The currently used methods do allow investigators to define the species complement of an environment that contains many species, a large proportion of which have not yet been cultured and characterized. We will discuss such methods in more detail here.

CHEMICAL TAXONOMIC METHODS

Taxonomists traditionally based systematics on morphological characteristics of macroscopic organisms, but chemical taxonomy has almost as long of a scientific tradition. For example, differing chemical structure of pigments has allowed classification of algae into major groups, and chemical analyses of bacteria have historically been used to identify species. One of the most fundamental differences between bacteria, gram negative or gram positive, relates to chemical properties of their cell walls with the original staining method developed in 1884. The identification of bacteria, before modern molecular methods, relied heavily upon chemical characteristics including enzyme and metabolic capacities of cells.

More recently, scientists have used protein sequences, lipid identities, and most of all, genetic sequences as taxonomic tools for classifying organisms. Methodology behind sequencing of DNA has expanded rapidly, and the technology is leading to an explosion of ecological research on groups or organisms that scientists could not previously identify in their natural environment (Method 8.1). Other advanced chemical methods such as mass spectrometry combined with gas chromatography can separate complex mixtures of molecules. Some of these compounds are specific to single organisms, groups of organisms, or single processes. For example, volatile fatty acids provide unique chemical fingerprints that depend upon identity of microorganisms (Sasser, 1990). The phospholipid fatty acids can provide estimates of microbial community structure, the biomass of microbes in the community, and microbial stress (Peacock and White, 2017). Furthermore, many multicellular animals cannot synthesize all their fatty acids and need to get them from their food. Subsequently, fatty acid analyses of tissues of higher organisms can reflect their food sources (e.g., Taipale et al., 2014).

A common problem with chemical methods relates to the fact that genes can transfer laterally among organisms. Thus, the existence of a particular chemical in an organism does not necessarily indicate its taxonomic position. For this reason, rRNA is often preferable because the probability of transfer of rRNA genes is lower than it is for many other genes. However, scientists have even called the interpretation of

METHOD 8.1 Nucleic acid sequence methods for taxonomists and ecologists studying freshwater organisms

Methods based on sequences of DNA and RNA have exploded and moved into many ecological research programs. They can be employed to label-specific organisms in their environment with known genetic sequences, can be sampled to obtain taxonomic identity of organisms, or used to gain information on which "species" are present or how many of their genes are being expressed. The general steps associated with their use involve amplifying sequences from tissue samples of organisms or bulk samples from the natural environment, sequencing them, and analyzing the sequences against known sequences or reporting them as unknown.

The first step for getting sequences from the environment is extraction. A major problem with this step is collecting uncontaminated samples (e.g., the researcher has their own DNA that can contaminate the sample as well as the bacteria that are ubiquitous in a laboratory or associated with any sampling gear). Obtaining uncontaminated samples is relatively simple for tissue samples from large multicellular organisms because the DNA to be analyzed dominates the sample, and contamination is not as much of a problem. These samples can be taken from the natural environment, and for larger organisms they can be taken without killing the plant or animal the sample is taken from. Samples can even be taken from museum specimens to gain historical perspective or analyze extinct species.

Taking good samples from an environment with many species in it is more difficult. This is necessary where analyses of microscopic organisms are required, or there is a need to analyze genetic material found in the environment. The goal is to extract all the targeted DNA or RNA without damaging it. Incomplete extraction or damage to sequences can lead to methodological bias. Extracting expressed mRNA as an indicator of gene activity is substantially more difficult than obtaining unbiased samples of DNA because RNA is more labile (it has to be so cells can control rates of protein expression by continuously breaking down and resynthesizing mRNA). For mRNA extraction, all the other types of nucleic acids (DNA, rRNA) are removed from the sample chemically, then the complimentary strand of DNA (cDNA) is synthesized for amplification and sequencing.

The second step to analyze environmental sequences is to amplify to get enough of a signal. The polymerase chain reaction (PCR) exponentially increases the number of copies of the targeted sequences. The PCR reaction is based on breaking (denaturing) apart the two joined DNA strands with heat, and resynthesizing the complimentary strand to each half of the two-stranded molecule. Repeated cycles of this breaking and synthesis exponentially increase the number of copies from a very small sample. While there may be problems with the generality of the method including taxonomic biases, contamination, and the lack of truly universal primers with sequences common to all organisms (Forney et al., 2004), this approach is very useful and forms a cornerstone method of modern molecular approaches.

Once enough DNA is extracted to be sequenced, the sample is generally sent to a laboratory for sequencing. High-throughput sequencing methods, which refer to a family of methods that have been developed for rapid sequencing of many base pairs, provide the data. The specialized sequencing equipment is difficult to use and very expensive so most

(Continued)

METHOD 8.1 (Continued)

laboratories contract out this work. The field is developing very rapidly and sequencing has become relatively inexpensive. The method of choice depends upon the exact data needed for the project. Goodwin et al. (2016) review the common methods, and how much more powerful they have become over the last decade.

The third step is analyzing the sequences. The analysis of large amounts of data produced by sequencing requires advanced bioinformatics. The reported sequences are generally short as many methods break up the DNA laterally to sequence it. These sequences need alignment to represent continuous reads of sequences from a single organism. Ultimately, combining all the aligned sequences produces the entire sequence for a chromosome or an organism's genome. Problems include errors in sequencing and natural variation in a population of sequences. The specific application used to analyze the data depends on the information needed (e.g., the genome of a single organism, the community composition, or identification of specific organisms in the sample). Computer programs for such analyses are currently available, many are open-source and free.

Once researchers know sequences from the natural environment, they can use this information to label organisms in their natural environment. Labeling specific organisms in the environment requires a known target sequence of DNA, then the complimentary strand (the sequence that would base pair with the known coding region) is bonded with a label. The label is generally a fluorescent molecule, but other approaches (e.g., radioactive labels) also provide detectable labels. Researchers analyze the fluorescently labeled sample with a fluorescence microscope to see labeled cells.

All of these methods are developing rapidly because of their application in all areas of biology, from disease research to taxonomy, as well as practical applications such as forensics and identification of genetic disorders. The methodology is certain to improve and change over the next few years, but students should have at least a passing familiarity with the general approaches.

rRNA into question because of possible transfer of genetic material among organisms over evolutionary time (Williams and Embley, 1996; Doolittle, 1999).

MOLECULAR APPROACHES FOR ASSESSING TAXONOMY AND DIVERSITY IN NATURAL ENVIRONMENTS

Molecular ecologists have developed a variety of techniques to assess biodiversity of microbes in their natural environment. Molecular tools are now making it possible to determine which species are present as well as taxonomic relationships among organisms. Aquatic ecologists have applied these tools widely (Hughes et al., 2014). Researchers can extract gene sequences for small subunit ribosomal RNA (SSU rRNA, 16S rRNA for Bacteria and Archaea, and 18S rRNA for Eukarya) from the environment from very small samples, collected organisms, or cultures. All organisms use ribosomes to make proteins, and the genetic code for ribosomal structure is very similar ("highly conserved") across the whole tree of life. It is necessary to amplify the genes for rRNA against a tremendous diversity of genes naturally present in any environment, so a primer (complimentary sequence to a known part of the rRNA gene) allows researchers to amplify the desired sequences selectively.

Taxonomists increasingly use rRNA sequence comparison as a taxonomic tool for all organisms. The method is particularly useful for microorganisms because their phenotype (morphology) does not differ enough to distinguish among species, and many cannot be cultured. It is also useful for macroscopic organisms, as many still look similar despite belonging to different species. The technique is based on comparing sequences of rRNA molecules from different organisms. The greater the time since evolutionary divergence of species, the more the sequences will vary. There are several advantages of the use of rRNA; some portions of the sequence are very conservative, meaning they retain some sections that all organisms have in common. Generally, these parts of the sequence are essential for function of the organism, and mutations in these regions of the rRNA genes are fatal. The rRNA molecules also contain regions that change more rapidly over evolutionary time (those that are less essential to function of the ribosome) and allow detailed analysis at finer taxonomic levels. The literature refers to these regions as "slow clock" and "fast clock" regions, respectively, as their sequences diverge at different rates over evolutionary time. The sections of the rRNA (Fig. 8.3) that are essential can be used by a taxonomist needing to make kingdom-level comparisons, and one distinguishing among closely related species would choose a section that accumulates mutations more rapidly.

In cases in which the technique can be compared with the taxonomic trees generated by morphology (e.g., cyanobacteria), reasonably good agreement occurs between results from conventional techniques and rRNA analysis (Giovannoni et al., 1988; Taylor, 1999). The fossil record of larger animals and plants can corroborate inferences from rRNA taxonomic trees. Thus, scientists are confident that the method also applies to bacterial organisms that cannot be distinguished based on morphology.

In situ identification of species is possible when an rRNA sequence is known for a particular species. A complementary DNA strand is synthesized that matches a unique section of the DNA that codes for the rRNA, and a label such as a fluorescent or radioactive molecule is attached to the end of the complementary strand. The labeled probe then will attach selectively to target organisms in the natural environment with the complementary sequence and allows rapid *in situ* identification.

The use of molecular techniques based on rRNA will likely become more common and more feasible for aquatic ecologists. Decades ago, scientists first successfully determined RNA sequences from environmental samples taken from open ocean

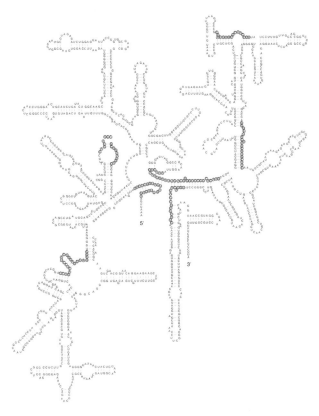

Figure 8.3 A map of a bacterial 16S rRNA molecule from a bacterium. Colored areas are locations of universal sequences that can be used for primers for sequencing the molecule. These are areas that mostly do not vary over evolutionary time. *Image courtesy of The Microbiome Portal Era7 Bioinformatics, Creative Commons (http://themicrobiome.com/en/ downloaded from the viewer on August 29, 2017).*

(Giovannoni et al., 1990) and hot spring organisms (Ward et al., 1990). Now, the method has become relatively rapid and commonly used by aquatic taxonomists and ecologists.

Interestingly, many of the rRNA sequences of bacteria found in natural environments are from organisms that researchers have not yet grown successfully in the laboratory. As such, we can now identify new species without traditional microbiological methodology. The methodology has been employed by researchers in an astonishing number of ways. For example, a known rRNA sequence was used to label a bacterium from the natural environment. Following labeling, the researchers used "optical tweezers" (lasers used to manipulate microscopic particles) to isolate and culture individually marked cells of known bacterial species (Huber et al., 1995).

The sequencing of rRNA from the natural environment also allows for taxonomic identification. In this method, researchers collect and homogenize bulk samples and then determine the presence or absence of all species. As sequencing becomes cheaper and faster, and larger databases of taxonomy based on rRNA are established, it will be possible to determine species presence much more quickly than we can sample some groups with traditional methods. Bacteria clearly fall into this category, but even invertebrates collected in freshwaters can be taxonomically difficult to distinguish and it can be time consuming to collect and process these samples. For example, chironomid midge larvae are difficult to identify without clearing and mounting head capsules of individuals. The ability to put a bulk sample into a machine and see which species of midge larvae are present would make the jobs of some aquatic ecologists substantially easier.

MOLECULAR METHODS FOR GENERAL AQUATIC ECOLOGY

We now have the ability to sequence all DNA collected from a specific environment, an approach generally termed "metagenomics." This allows for determination of the total genomic diversity (a somewhat different concept than the idea of total species diversity, see Chapter 11). In this case DNA is chopped (partially digested) with enzymes into lengths that are readily sequenced. Analytical programs can then assemble the sequence fragments to recreate the entire sequence of the many genomes of microorganisms in the environment. The ability to handle huge amounts of data and rapidly sequence massive numbers of base pairs has just begun to allow researchers to apply this approach in natural habitats, though the cost remains high for extensive application. These methods also work for eukaryotic species and are showing great promise in determining taxonomic patterns in groups that were traditionally difficult or time-intensive to assign a taxonomic identity.

The combination of targeted PCR of environmental genes and fingerprinting techniques can separate functional genes as well. For example, Tsiknia et al. (2015) characterized the distribution of genes associated with nitrogen cycling in a Mediterranean watershed. The genes chosen can be common with all organisms (e.g., genes for DNA synthesis) or restricted to a taxonomic or functional group (e.g., genes for photosynthetic machinery) depending upon the question of interest. Instead of investigating diversity of all microbes in a certain phylogenetic group, the functional gene approach investigates diversity of all microbes with a certain metabolic function (e.g., the ability to use a particular form of nitrogen). Taking this approach further, the abundance of expressed RNA molecules indicates relative gene activity in a sample.

An additional application of molecular methods is using environmental DNA (*eDNA*) to detect organisms in their environment. This is also referred to as *DNA barcoding*, using sequences to identify species. Barcoding techniques are useful for identification of *cryptic species*, (species that are morphologically similar but genetically different enough to be considered distinct). It is also useful in detecting rare species in the environment, serving as an early warning of invaders or presence of rare and endangered species. This type of sequencing can also detect the presence of disease causing or water-quality impairing organisms (Otten et al., 2016).

These approaches are somewhat like using trace DNA as a forensic tool. Even if an animal or plant is not at a specific location at a specific time, they shed DNA into the environment. Thus if we are interested in the presence or absence of an invasive species, scientists may be able to determine if the species is present simply by extracting all the DNA from a water sample and sequencing for known targeted genes for that species. This is a developing method, and it is difficult to know how useful such information is. For example, the rate at which most organisms shed DNA and how quickly shed DNA breaks down in the environment is not well documented. In one case, Eichmiller et al. (2016) documented increased rates of degradation of common carp (*Cyprinus carpio*) eDNA with increased temperature, and with lower dissolved organic carbon. Even these two environmental factors could lead to a wide range of degradation rates. The mechanisms of the spread of eDNA are poorly understood at this point (Jerde et al., 2016). Application of appropriate statistical techniques can increase the success of the method (Shelton et al., 2016).

The eDNA method has much promise. Doi et al. (2017) combined stream fish snorkeling with eDNA collections in Japanese streams. They found a good correlation between presence and abundance of species and the corresponding amount of eDNA matching each species. Fujiwara et al. (2016) found that eDNA was a reliable way to indicate if an invasive macrophyte was present in Japan. Simmons et al. (2015) used this method to detect invasive fish [bighead carp (*Hypophthalmichthys nobilis*) and Northern snakehead (*Channa argus*)] in an Ohio river. Larson et al. (2017) successfully detected invasive crayfish using eDNA in large Nevada and California lakes. This list is incomplete and the method is certain to see much greater adoption.

Molecular techniques can also be useful for paleolimnological studies. Frisch et al. (2017) took sediment cores from two lakes in Minnesota and isolated resting stages of the zooplankton, and then linked the ages of sediment with depth to isotope dates. They analyzed the genetic structure of populations in both lakes. One of the lakes was pristine and the structure stayed the same as it was before European settlement. The other was stable for about 1,000 years, but upon European settlement, phosphorus in the sediment increased more than 20 times and there was a dramatic shift in zooplankton genetic structure.

MAJOR TAXONOMIC GROUPS

Three major groups of organisms occur at the broadest level of classification, the Eukarya (eukaryotes), the Bacteria, and the Archaea (Woese et al., 1990). The Bacteria and Archaea were formerly the Prokaryota. Before microscopic and chemical recognition of the unique cellular composition of bacteria, organisms were classified into animals (mobile) or plants (sedentary and green). After light microscopy became established, the classification dividing organisms into animal and plant groups became difficult because many microbes are photosynthetic, motile, and exhibit simple behaviors (e.g., attraction to light or food). The observation of diverse microbial lifestyles blurred traditional distinctions between animals and plants. Electron microscopy allowed definitive differentiation between organisms with complex inner architecture (eukaryotes) and those with simpler cells (then called prokaryotes). Then, analysis of rRNA and other biological molecules revealed that Archaea split from Eukarya shortly (relative to the 4-billion-year-old Earth) after they diverged from the Bacteria (Fig. 8.1). Such analyses have also revealed that the Bacteria, Archaea, and Eukarya should be assigned to domains, not to the traditional kingdoms (Woese et al., 1990). If Eukarya is assigned a kingdom-level designation, then it retains the name Eukaryota. The most basic tree of life means that the term prokaryote has little meaning in modern systematics because it includes the widely divergent Archaea and Bacteria

Even in the eukaryotic groups, where morphology is often distinct, phylogenetic relationships do not always follow traditional taxonomic lines (Fig. 8.4). For example, the group protozoologists use to classify amoebae actually falls into three widely divergent groups, and those groups are more different from each other than are humans and fungi. Molecular data, combined with the fact that there is lateral transfer of genes, and lack of sequence information for many organisms lead to many phylogenetic questions that currently have no definitive answers. Phylogeny is currently a field in a state of rapid flux, so many relationships are provisional. Students should also realize that this is a very exciting time to be involved in phylogenetic research, as the most basic evolutionary questions of origins of species are being solved in this field on a daily basis.

One aspect of the most basic issues in origin of diversity is the concept that eukaryotic cells formed organelles by ingesting and appropriating bacterial cells. The cells were enveloped and became endosymbiotic, living inside of the Eukaryotic hosts (Sapp, 1991). This idea is the *serial endosymbiosis theory* and was developed through the work of Dr Lynn Margulis. Most biologists accept this as a theory for the origin of chloroplasts and mitochondria (by "theory" we mean scientific truth such as the Theory of Electricity or the Theory of Gravity). Mitochondria and chloroplasts found in eukaryotes were initially cells of purple photosynthetic bacteria and cyanobacteria (blue-green algae), respectively. This evolutionary innovation occurred several billion years ago and vastly increased the complexity of organisms. Molecular evidence strongly supports independent origins of mitochondria and chloroplasts. The numerous less tightly integrated intracellular associations that exist in aquatic ecosystems provide additional support (albeit not as strong) for the serial endosymbiosis theory. For

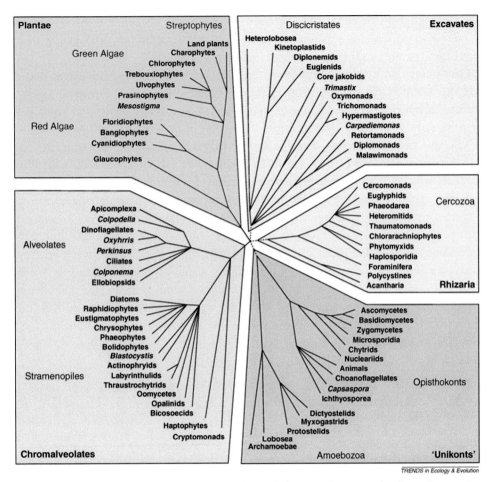

Figure 8.4 A phylogenetic diagram of relationships of the major groups of eukaryotic organisms. *From Keeling et al. (2005).*

example, inclusion of photosynthetic algae in protozoa and small animals is functionally similar to chloroplasts in plants. Such associations occur in freshwaters and we cover them in this book, including *Chlorella* in *Hydra*, *Paramecium bursari*, and *Spongilla*.

There are six or more kingdoms of organisms, if one accepts molecular taxonomy that indicates the three major domains of organisms. The kingdoms include the Plantae, Animalia, Fungi, Archaea, and Bacteria. The Protista likely will be divided into several kingdoms, given the broad spread in molecular trees (Figs. 8.1 and 8.4) and to maintain a classification scheme consistent with evolutionary origins (Hickman and Roberts, 1995). The classification of the protists is currently in flux (Patterson, 1999), although the Catalog of Life currently splits what were formerly the protists into the kingdoms Chromista and Protozoa (Roskov et al., 2016). The Archaea probably contains two kingdoms as well (Woese et al., 1990), although the Catalog of Life assigns these to the phyla Crenarchaeota and Euryarcheota.

One point that becomes clear when analyzing the molecular evidence is that the divergence within the Bacteria and Archaea exceeds that of the Eukarya. Thus the diversity of morphology and behavior that is so evident in the plants and animals has arisen recently. Molecular diversification is the forte of the Bacteria and Archaea.

The question often arises (even with the fuzziness of the species concept), how many species are there on Earth? As of 2017, there were about 1.5 million named species (Larsen et al., 2017). Recent estimates range from 2 million to 1 trillion, with most projections around 11 million species. Using genomic methods increases counts of species that are morphologically cryptic. The Bacteria and Archeae certainly fall into this category, but many invertebrates and prostists do as well. For example, DNA sequencing indicates that there are 200,000 to 250,000 species of freshwater microbial eukaryotes. This well exceeds the number that researchers have identified with traditional taxonomy (Debroas et al., 2017). The revised total estimates for all organisms on Earth suggest that there are 1 to 6 billion species and that about 70%−90% of those are bacteria (Larsen et al., 2017). This suggests that scientist have named less than 0.1% of species on Earth!

CLASSIFICATION OF ORGANISMS BY FUNCTION, HABITATS, AND INTERACTIONS

For the aquatic ecologist, taxonomy of organisms is based not only on phylogenetic relationships but also on their functional roles in communities and ecosystems. Such classifications include how the organisms acquire carbon, what habitat they occupy, and how they interact with other organisms. We introduce these concepts here because they are used in the following chapters to characterize organisms.

Organisms can be *autotrophic* ("self-feeding") and rely on carbon dioxide (CO_2) as the primary source of carbon to build cells. The other option is to be *heterotrophic* ("other feeding") and acquire carbon for cells from organic carbon (Table 8.2). Some organisms are able to use both autotrophic and heterotrophic processes to obtain carbon. We refer to these species as *mixotrophic*. This mode of life is probably more common than previously thought, particularly among microbes (Selosse et al., 2017).

Most autotrophs use light as an energy source to reduce CO_2 to organic carbon and are classified as *photoautotrophic* (i.e., photosynthetic organisms). Some microbes are able to use chemical energy, as opposed to light, as a source of energy to reduce CO_2 to organic carbon; these organisms are *chemoautotrophic*. The chemoautotrophs are less

Table 8.2 Classification of organisms by energy source and nutrient requirements

Carbon source	Mode	Energy source/ electron donor	Electron acceptor	Organisms (example)
CO_2	Photoautotroph	Light/H_2O	O_2	Cyanobacteria/ eukaryotic algae/ macrophytes
CO_2	Photoautotroph	Light/H_2S	Organic C	Green and purple sulfur bacteria (anaerobic)
CO_2	Photoautotroph	Light/H_2, organic C	Organic C	Purple nonsulfur bacteria (anaerobic)
CO_2	Chemoautotroph	Reduced inorganic compounds (e.g., NH_4^+, H_2S, Fe^{2+}, H_2)	O_2	Bacteria (e.g., *Nitrosomonas*, iron bacteria, hydrogen bacteria)
Organic C	Heterotroph	Organic C	Organic C	Fermentative bacteria (anaerobic)
Organic C	Heterotroph	Organic C	O_2	Aerobic bacteria, protozoa, and animals
Organic C	Heterotroph	Organic C	Oxidized compounds such as NO_3^-, SO_4^{2-}, Fe^{3+}	Anaerobic bacteria that respire organic C

See Chapters 9–10, Chapters 12–14, for further discussion of types of organisms. Not all types are shown.
After Yanagita (1990).

important to most carbon budgets than photoautotrophs, but they dominate some unusual environments and play a key part in several nutrient cycles (see Chapter 14).

Organisms that acquire carbon from other living organisms (predation, herbivory, parasitism, etc.), dissolved or particulate organic compounds, or dead organisms (decomposers and carrion eaters) are heterotrophic. Heterotrophs that decompose organic carbon are sometimes called *saprophytes* or *detritivores*.

A variety of additional classifications are used to describe the functional roles of organisms in aquatic food webs (*functional feeding groups*; Cummins, 1973). Organisms that sieve particles from the water column are *filterers*; those that build nets or have morphological features that filter particles out of flowing waters are *passive filterers*, whereas those that actively pump water or create currents are *active filterers*. Organisms that acquire their nutrition from small organic particles in the benthos are *collectors*.

Shredders break up larger organic materials (like decaying leaves) for their nutrition, and *scrapers* remove biofilms from hard benthic substrata. Predators, a functional group that is self-explanatory, are sometimes further grouped by how they feed, such as *engulfing predators*, which swallow prey whole or bite off chunks, versus *piercing predators*, which pierce their prey and suck bodily fluids from them.

Functional feeding groups are somewhat similar to *guilds*, which are often used in terrestrial studies of consumers and communities, except that the guild concept is based on organisms using the *same resource* in the same fashion (Root, 1967). Members of the same functional group may use different resources. For example, some shredders may feed on decomposing leaves, whereas other shredders may feed on wood or living aquatic vascular plants, but they all shred large organic materials. Functional and guild analyses can be more relevant than taxonomic considerations for relating organisms to ecosystem processes like nutrient cycling and energy flow.

Consumers are often classified further by their position in the food web. For example, *grazers* or *herbivores* (primary consumers) eat algae and plants (primary producers), or sometimes bacteria. *Carnivores* or secondary consumers eat other consumers, and *top carnivores* eat consumers but larger animals do not generally eat them. Thus, classification schemes based on modes of obtaining nutrition are one of many ways to classify organisms.

Additionally, organisms may be classified by the habitat they occupy and some of the special terminology for this purpose is presented in Table 8.3. Such classification can be useful because it allows an investigator to make predictions about abiotic and biotic conditions important to organisms. For example, an epilithic alga (living on rock) in a rapidly moving stream may experience a relatively high water velocity. An epiphytic alga on a macrophyte may have competitive or facilitative interactions with the macrophyte. This basic terminology is necessary to understand the literature of aquatic ecology.

Organisms can also be classified by how they interact with other organisms (*interspecific interactions*). Many different types of interspecific interactions are possible, and we adhere to the interaction scheme shown in Table 8.4. There are *direct interactions* and *indirect interactions*. Direct interactions occur between individuals of two species and involve no others; those mediated by additional species are indirect interactions. The terms for interaction types presented here are not all standard, but they allow for a very general classification scheme. *Exploitation* is a general term for an interaction that harms one species and helps another. This term is not widely accepted yet, but includes interactions that are not necessarily *predation* or *parasitism*. For example, an epiphyte that harms a macrophyte but receives benefit from living on its leaves is exploiting the plant. *Mutualism* denotes any positive reciprocal interaction. Others have used various terms for mutualism, including symbiosis, synergism, and protocooperation. *Symbiosis* refers to organisms that live close together (but not to how they

Table 8.3 Some terms used to classify aquatic organisms by habitat

Habitat	Description
Benthic	On the bottom
Emergent	Emerging from the water
Endosymbiotic	Living within another organism
Epilithic	On rocks
Epigean	Above ground
Epipelic	On mud
Epiphytic	On plants
Epipsammic	On sand
Hygropetric	In water on vertical rock surfaces
Hyporheic	In groundwater influenced by surface water
Lentic	In still water
Littoral	On lake shores, in shallow benthic zone of lakes
Lotic	In flowing water
Neustonic	On the surface of water
Pelagic	In open water
Periphytic (aufwuchs, biofilm, microphytobenthos)	Benthic, in a complex mixture including algae
Profundal	Deep in a lake
Symbiotic	Living very near or within another organism
Stygophilic	Actively use groundwater habitats for part of life cycle
Stygobitic	Specialized for life in groundwater
Torrenticole	Adapted to live in swiftly moving water

Table 8.4 Classification of interactions between two species (A and B)

Effect of A on B	Effect of B on A	Name of interaction
Positive	Negative	Exploitation (includes predation and parasitism)
Negative	Negative	Competition
Positive	Positive	Mutualism
None	Positive	Commensalism
None	Negative	Amensalism
None	None	Neutralism

are interacting), and synergism and protocooperation have not received widespread use outside of studies of animal behavior.

Of all the interaction types found in macroscopic ecological communities, *commensalism* (positive on one, none on the other) and *amensalism* (negative on one, no effect on the other) are likely the most common, followed by exploitation and then competition and mutualism (assuming that positive interactions are as likely as negative interactions; Dodds, 1997b). In general, commensalism and amensalism have received almost no attention in the ecological literature; predation has received the most, followed by competition and mutualism. It is up to future ecologists to study amensalism and commensalism more intensively. We discuss these interactions in detail in Chapters 19—22.

ORGANISMS FOUND IN FRESHWATERS

Freshwater habitats contain representatives of most of the groups of organisms on Earth. Several books are available to introduce students to freshwater organism diversity in general (e.g., Reid and Fichter, 1967; Needham and Needham, 1975), and there are also numerous readily available guides covering various taxonomic groups (Table 8.5). Detailed taxonomic analyses of many groups (e.g., genus and/or species-level identifications of algae or invertebrates) often require keys available in specialized books or in the primary literature, and there are no universally accepted taxonomic keys for problematic groups such as some water mites and some groups of protozoa.

The Archaea and Bacteria are difficult to distinguish unless they can be brought into culture and metabolic characteristics can be used as taxonomic characteristics (Holt et al., 1994). Algae are the primary autotrophs in many aquatic ecosystems and are well represented in freshwaters (South and Whittick, 1987; Graham and Wilcox, 2000; Wehr and Sheath 2002). We cover the classification and ecology of microbes in Chapter 9. The incredible diversity of the algae and protozoa make

Table 8.5 Some initial resources for identifying various groups of aquatic organisms

Group	References
General	Reid and Fichter (1967), Needham and Needham (1975)
Algae	Prescott (1978), Dillard (1999), Wehr et al. (2015)
Protozoa	Jahn et al. (1979), Lee et al. (2000), Thorp and Rogers (2014)
Nonvascular plants	Conrad and Redfearn (1979)
Aquatic plants	Riemer (1984), Cook (1996), Broman et al. (1997)
Aquatic invertebrates	Lehmkuhl (1979), Thorp and Rogers (2014), Smith (2001), Voshell (2002)
Aquatic insects	Merritt et al. (2008), McCafferty (1988)
Reptiles and amphibians	Powell et al. (2016), Stebbins (2003)
Fishes	Eddy and Underhill (1969), Page and Burr (2011)

observation of aquatic samples under the microscope fascinating; we can imagine the excitement when the first microscopes were invented and they revealed an entirely new world.

Taxonomy of the Eukarya is relatively well defined at least at the family levels for most groups. Protozoa are common in all freshwater habitats and can often be identified to the family or genus level if a good microscope is available. All major phyla of invertebrates, with the exception of the Echinodermata, have some freshwater species and some groups such as the insects and arachnids are much more diverse in freshwaters compared to marine environments. Students are often surprised to learn that some groups of organisms that they traditionally associate with marine habitats, such as the sponges and jellyfish, can be found in abundance in some freshwaters.

Some freshwater invertebrates, such as the aquatic insects and arachnids, are secondarily aquatic in that their evolutionary paths involved marine ancestors that evolved into terrestrial forms, which subsequently colonized and adapted to freshwater habitats. Others, such as the freshwater cnidarians and decapods, evolved directly from marine ancestors. Secondarily aquatic forms often have morphological or physiological features that reflect their terrestrial ancestry. For example, many aquatic insects use air bubbles or breathing tubes to breathe atmospheric air, even though they spend the bulk of their time under water.

Many beginning students find the invertebrates the most fascinating of freshwater organisms; take a child to a stream and turn over rocks and you will see how early such fascination can begin. Invertebrates are abundant, diverse, and ecologically important in just about every habitat on the planet, and freshwaters are no exception. The diversity of body forms, life cycles, and behaviors among the invertebrates is astonishing. Invertebrates constitute the bulk of animal abundance and diversity in freshwater habitats. Because of their abundance, diversity, and relatively fast individual and population growth rates, invertebrates are ecologically very important in freshwater habitats, as well as in other systems (Wallace and Webster, 1996; Thorp and Rogers, 2014). Identification of many invertebrate groups to the species level, particularly for immature forms (e.g., aquatic fly larvae and mite larvae), is difficult, but taxonomic keys are available for adult and immature forms of most invertebrate groups. We consider major groups of invertebrates in more detail in Chapter 10 and aspects of their ecological significance in Chapters 19—21.

Vertebrates are also abundant and diverse in freshwater habitats, although the bony fishes (superclass Osteichthyes) make up bulk of this diversity. Aside from the fishes, most vertebrates associated with freshwater habitats or are only partially aquatic (e.g., birds and mammals) or have biphasic amphibious life cycles (e.g., frogs and salamanders). Identification of vertebrates is generally easier than invertebrates because fewer, larger, and better-studied organisms are represented, although some groups of smaller-bodied fishes (e.g., some minnows and darters) and immature forms of fishes and amphibians can be challenging and require an expert for accurate species-level identification. Many freshwater vertebrates are already familiar to students and have common names. We introduce the major groups of freshwater vertebrates in Chapter 10 and fish ecology and fisheries management in Chapter 23.

As with vertebrates, many plants in aquatic systems have been well characterized. Emergent wetland species generally are included in traditional plant taxonomic references (Smith, 1977). For the more obscure mosses and liverworts, identification is more difficult, and even some larger groups such as the sedges can be problematic to identify. Truly aquatic plants are only moderately diverse; Riemer (1984) provides a good introduction to their ecology, and we explore major groups of freshwater plants in Chapter 9.

SUMMARY

1. Several species definitions are available. The most utilitarian approach is to define a species by criteria established by the taxonomists of a particular group.
2. Traditional taxonomic schemes have distinguished among organisms using behavior, metabolic characteristics, and morphology. Recently, scientists have adopted molecular techniques.
3. Molecular methods can distinguish species, detect expression of genes in the environment, and assess biodiversity including the presence and absence of species.
4. Traditional taxonomic classifications at the broadest level (e.g., kingdom and phylum) are probably not completely natural, and more research is necessary to untangle these evolutionary relationships.
5. Bacteria and Archaea are two groups with the greatest amount of metabolic diversity. Behavioral and morphological diversity are greatest in the Eukarya.
6. Organisms can be classified by their mode of obtaining nutrition and by their habitat in addition to their evolutionary relatedness.
7. Organisms that use CO_2 as their primary carbon source are autotrophic; those that use organic carbon are heterotrophic. Autotrophic organisms include those that obtain energy from light (photoautotrophic) and chemicals (chemoautotrophic). Heterotrophic organisms include predators, detritivores, and organisms that live on dissolved organic compounds. Some organisms are mixotrophic and have autotrophic and heterotopic carbon sources.
8. Organisms can be classified by their direct interactions (competition, mutualism, exploitation, commensalism, amensalism, and neutralism) with other organisms.

9. Most groups of organisms found on earth include species that live in freshwater habitats, some of which evolved directly from marine ancestors and others that are secondarily aquatic and evolved from terrestrial species.

QUESTIONS FOR THOUGHT

1. Why might legislation designed to prevent extinction of species require a precise definition of a species?
2. Why did the inclusion of mitochondria and chloroplasts in cells of Eukarya represent a sudden large increase in complexity of cells and how does this observation damage the creationists' "blind watchmaker" arguments?
3. Why do freshwater invertebrates have fewer species as a whole than marine invertebrates?
4. Why does lateral transfer of genetic material among widely disparate organisms (e.g., plants and bacteria) cause difficulties for molecular taxonomists?
5. Can you think of more specialized habitats than those listed in Table 8.3?
6. Is taxonomy fixed when a species is described or do perceived taxonomic relationships among organisms change over the years as more information becomes available?
7. Evolution can occur at much more rapid rates than previously thought (over a few generations); how could this influence taxonomy?
8. How would morphological plasticity interfere with taxonomic identification?

CHAPTER 9

Microbes and Plants

Contents

Freshwater Ecology
DOI: https://doi.org/10.1016/B978-0-12-813255-5.00009-0

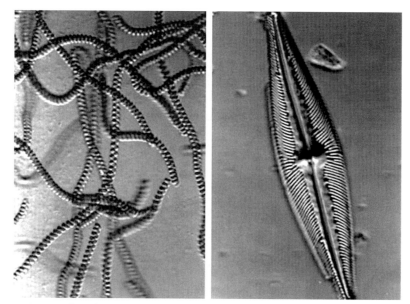

Figure 9.1 Micrographs of a spiral cyanobacterium (*Spirulina*) and the frustule (silicon shell) of a diatom (*Navicula*). *Spirulina* spirals are 5 μm wide, *Navicula* is 30 μm wide.

Primary producers capture much of the energy that flows through freshwater food webs, and microbes are responsible for the bulk of the biogeochemical transformational fluxes (including decomposition and nutrient recycling) in aquatic systems. Some of this freshwater biogeochemistry (e.g., production of methane in wetlands) is important on a global scale because methane and carbon dioxide are greenhouse gasses. Knowledge about these organisms is essential for those involved with water quality issues as well as general ecological studies. In this chapter, we consider the microbes and plants found in freshwaters. This placement of microbes and plants into a single chapter is certainly an artificial classification; the organisms considered here span taxonomic groups from viruses to complex plants including Bacteria, Archaea, and Eukarya. For practical purposes, we mostly follow traditional taxonomic protocol because many determinations are based on obvious morphological characteristics, and this method is readily available to researchers who are not specialists in taxonomic identification via molecular methods. Still, where the traditional taxonomy does not match molecular methods at deep phylogenetic levels, we point out the differences.

VIRUSES

All organisms have viruses. Viruses are not really organisms because they cannot survive without a host and are not capable of basic metabolic function. Nonetheless, they are important in population dynamics of aquatic organisms, aquaculture, and public health (Table 9.1). Taxonomy of viruses is possible with modern molecular methods and this is a rapidly developing field, both with respect to epidemiology and for viruses that are common in the environment but do not infect humans, their crops, or their livestock. For example, Short and Short (2008) found a low of 6 to more than 20 algal viruses in each of 6 sampled North American lakes. Viruses also transfer genetic material among microorganisms, so they can be important in issues related to the release of genetically engineered microorganisms into the environment. Over longer periods, the lateral transfer of genetic material, mediated in part by viruses, causes difficulty in taxonomy because genes or groups of genes "jump" across phylogenetic lines.

Table 9.1 Some organisms causing human diseases transmitted by water or wastewater

Group	Organisms	Disease/symptoms
Bacteria	*Salmonella* spp.	Typhoid fever, paratyphoid fever, gastroenteritis
	Shigella spp.	Gastroenteritis-dysentery
	Vibrio cholerae	Cholera
	Escherichia coli	Gastroenteritis
	Leptospira icterohaemorrhagiae	Weil's disease
	Campylobacter spp.	Gastroenteritis
	Yersinia enterocolotica	Gastroenteritis
	Mycobacterium spp.	Tuberculosis/respiratory illness
	Legionella pneumophila	Legionnaire's disease/acute respiratory illness
Virus	Hepatitis A	Liver disease
	Norwalk agent, Rotaviruses, Astroviruses	Gastroenteritis
	Poliovirus	Polio
	Coxsackievirus	Herpangia/meningitis, respiratory illness, paralysis, fever
	Enteroviruses (68–71)	Pleurodynia/menengitis, pericarditis, myocarditis
Protozoa	*Giardia lamblia*	Diarrhea, abdominal pains, nausea, fatigue, weight loss
	Entamoeba histolytica	Acute dysentery
	Acanthomoeba castellani, Naegleria spp.	Meningoencephalitis
	Balantidium coli	Dysentery
	Cryptosporidium spp.	Dysentery
Helminths	Nematodes (*Ascaris lumbricoides, Trichuris trichiura*)	Intestinal obstruction in children
	Hookworms (*Necator americanus, Ancylostoma duodenale*)	Hookworm disease/gastrointestinal tract
	Tapeworms (*Taenia* spp.)	Abdominal discomfort, hunger pains
	Schistosoma mansoni	Schistosomiasis (liver, bladder, and large intestine)

Source: After Bitton (1994).

Viruses are generally very small particles (25–350 nm) that remain in suspension and are too small to sink relative to turbulence and Brownian motion at very small scales. Larger viruses do occur, however, as big as smaller bacteria with as many genes. These viruses can infect aquatic eukaryotic microorganisms and more are being discovered. Some scientists claim that they are a fourth domain of life (Abergel et al., 2015).

When lake water is filtered and the filtrate is analyzed with scanning electron microscopy, particles that look like viruses are seen (Fig. 9.2). It is difficult to determine if particles that look like viruses are actually infectious and to determine their host. Generally, an assay for the numbers of infectious units per volume water based on exposure of organisms is the definitive test for active viruses. For this method, a researcher makes a series of dilutions containing viruses and exposes organisms to the dilution series. The number of infections caused by the various dilutions indicates the number of active viruses. This protocol is used in testing for viruses that infect humans (with a surrogate infective host) as well as those that infect other organisms. Molecular probes and labeled antibodies are available for the more common human infectious agents and can be synthesized for any virus for which the RNA or DNA sequence is known.

Viruses are classified by their possession of DNA or RNA, by occurrence of single- or double-stranded nucleic acids, and by the molecular weight of nucleic acids. Classification of the capsid (protein coat around the nucleic acid), including the

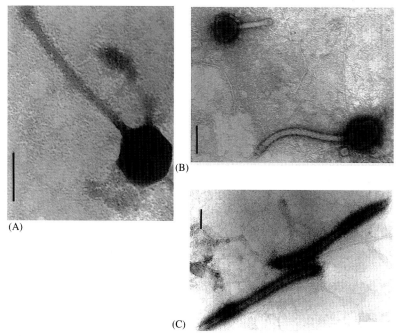

Figure 9.2 Electron micrographs of aquatic virus-like particles from two high-mountain lakes. Scale bars = 100 nm. *Reproduced with permission from Pina et al. (1998).*

number of subunits, the shape or symmetry, and where in the host cell the capsid is assembled, is also important. Some viruses also have a lipid or lipoprotein coat. Characterization schemes differ among viruses of plants, animals, and microorganisms. Animal viruses are recognized by the diseases they cause, plant viruses by the disease and plant species that serve as host, and microbial viruses by the organism they infect. Additionally, prions (infectious proteins) and viroids (naked RNA) may be significant parasites. Prions are evolutionarily ancient and infect both bacteria and eukaryotes, so presumably are taxonomically widespread (Yuan and Hochschild, 2017). The ecology of virus assemblages in their natural environment is complex, as they evolve very rapidly to stay ahead of defenses evolved by their hosts in an evolutionary "arms race" (Anderson and Banfield, 2008).

Viruses can be specific for one species or strain of organisms or more widely infective. Those of the greatest human interest cause disease, and water transmits many of these (Table 9.1). Viruses that infect unicellular organisms are often fatal if an infection proceeds because the reproductive virus lyses (bursts) the cell. Understanding the dynamics of microbial communities requires knowledge of virus transmission.

A successful virus in an aquatic habitat must make contact with the correct type of host cell. Viruses only are active for so long in the environment, and need to remain active long enough to randomly encounter the appropriate host. The spread of viral infections is greater when the density of host cells in the environment is high and the length of time that a virus can remain viable outside the host is long. Many things may inactivate viruses when outside of their hosts, including UV light, absorption onto cells or remains of cells that are not proper hosts, attaching to bits of dead host cells, and predation by microflagellates that can ingest very small particles. Inorganic particles such as clay can enhance or impede infection rates. The particles can increase the time of viral survival by limiting the previous factors, but can also lower infection rates because tight association with the inorganic particles lowers the probability of contact with a host cell. Some of these factors influence how long viruses can survive in groundwater, an important public health issue (Highlight 9.1).

Highlight 9.1 Survival of human pathogenic viruses in groundwater

Pathogenic viruses can enter groundwaters through many different sources. Some of the most common sources are land disposal of sewage, overflow from septic systems, livestock waste, and landfill leachate. Noroviruses, Hepatitis A, Hepatitis E, adenovirus, astro-virus, enteroviruses, and rotaviruses are common waterborne viruses that lead to millions of infections around the world each year (Gibson, 2014). Contamination of groundwater used for drinking water is most difficult to deal with because most treatments do not remove all viruses. For example, human infections were traced to viral contamination of groundwater in Georgetown, Texas (coxsackievirus and hepatitis A) and Meade County, Kentucky (hepatitis A; Lipson and Stotzky, 1987), and 1,500 people were infected with Norwalk virus (norovirus) from a contaminated spring in

Rome, Georgia (Bitton, 1994). Contamination of drinking water wells with hepatitis A, polio, or enteroviruses occurs throughout the world. Knowing how long these viruses can remain infective is important to allow estimation of the probability that groundwater flows will move them to drinking water wells while they are still active.

Infective viruses can travel over 50 m (depth) from septic tanks into drinking water wells. Controlled studies have demonstrated movement up to 1.6 km horizontally through soils (Gerba, 1987). Clearly, viruses could travel even greater distances in aquifers with rapid water velocity, such as karst systems or alluvium with coarse cobbles (Sinton et al., 1997). For example, poliovirus moved at least 20 m in a cobble aquifer with less than 1% virus mortality (Deborde et al., 1999).

Factors that influence the movement of viruses into and through groundwater include the rate of water flow through the sediment, the retentive properties of the sediments, and the survival time of the virus. Factors that lower viral survival times in sediments include high temperatures, microbial activity, drying, lack of aggregation with other particles, and low organic matter. Inactivation of viruses can be very rapid, but poliovirus can remain active for up to 416 days in sandy soils (Sobsey and Shields, 1987).

Understanding the hydrology of soils and sediments is necessary to assess the problems that may be related to sewage contamination of groundwaters. Aquatic microbial ecologists are only beginning to elucidate the mechanisms of deactivation of viruses related to microbial infections. Because little research has been conducted on community dynamics of groundwater microbes, this is a potentially valuable and exciting field for future study.

ARCHAEA

The *Archaea* are single celled organisms with no nucleus or organelles. They reproduce asexually by fission or budding and no known species produces spores. Analysis of ribosomal RNA sequences has led to a classification of these organisms as separate from the Bacteria (Fig. 8.1). Archaea are more closely related to the Eukarya than to the Bacteria as indicated by similarity of genes for the fundamental cellular processes of transcription and translation.

The Archaea were originally isolated from extreme environments and thus scientists thought that they mainly predominated in environments such as anaerobic waters, hot springs, and hypersaline environments including salt lakes. Molecular methods have revealed that Archaea are common in all environments, although not as dominant as in some extreme habitats. These organisms are common in the ocean and probably in many freshwater habitats as well. The study of Archaea and their importance in freshwater habitats, as well as the general phylogeny of the group is a rapidly developing and very exciting area of aquatic microbial ecology.

The Archaea are broadly diverse and include members that can create methane from carbon dioxide and hydrogen (methanogens), some that can use organic carbon (particularly in very cold habitats with low carbon availability), and others that are capable of photosynthesis, but do not generate oxygen gas. Some members of the group are chemoautotrophic and combine inorganic compounds such as ammonium or sulfide with oxygen as a source of energy. Some groups such as the methanogens have global biogeochemical importance, especially those populations found in wetlands. The identification of species or strains of archaebacteria generally is based on metabolic characteristics (Table 9.2) and molecular analyses.

Table 9.2 Major groups of bacteria from Bergey's manual (Holt et al., 1994)

Major group	Group	Genera represented in aquatic systems (# of described genera)	Importance
Gram-negative with cell wall	Spirochetes	*Spirochaeta* (8+)	Free living in aquatic waters, some pathogens
	Aerobic/microerophlic, motile, hilical/vibriod	*Campylobacter, Bdellovibrio* (16)	In freshwaters, some denitrifiers, includes the predatory *Bdellovibrio*
	Nonmotile curved	*Ancylobacter* (8)	In freshwater
	Aerobic/microaerophilic rods and cocci	*Azotobacter, Psuedomonas, Francisella, Legionella* (84)	Aerobic nitrogen fixation (e.g., *Azotobacter*), some disease organisms, very diverse group
	Facultative anaerobic rods	*Escherichia, Vibrio* (45)	Contains many waterborne disease organisms
	Anaerobic straight, curved and helical rods	*Thermotoga* (47)	Common from anoxic muds and animal intestinal tracts, also a number of thermophiles and halophiles
	Dissimilatory sulfur reducing	*Desulfomonas* (18)	Reduces oxidized sulfur compounds to sulfide
	Anaerobic cocci	*Megashaera* (4)	Mainly animal parasites
	Rickettsias and chlamydias	2 subgroups	Parasites
	Anoxygenic photosynthetic	*Rhodospirillum* (28+)	Able to use sulfide as electron donor for photosynthesis
	Oxygenic photosynthetic	*Cyanobaceteria* (37)	Important photosynthetic and nitrogen fixers
	Aerobic chemilithotrophic	*Thermothrix, Nitrobacter* (28)	Important biogeochemically, including nitrifiers, iron oxidizers, and sulfur oxidizers
	Budding or appendaged	*Caulobacter* (25)	*Caulobacter* indicative of oligotrophic conditions
	Sheathed	*Clonothrix* (7)	Some important in iron and manganese cycles
	Gliding, nonphotosynthetic, nonfruiting	*Beggiatoa* (28)	Mostly aquatic, some important in sulfur cycling
	Myxobacteria	*Polyangium* (12)	Decomposers, predominantly in soils but some in freshwaters

(Continued)

Table 9.2 (Continued)

Major group	Group	Genera represented in aquatic systems (# of described genera)	Importance
Gram positive with cell walls	Cocci	*Streptococcus, Trichhococcus* (24)	Some pathogens, also found in aquatic habitats
	Endospore forming	*Bacillus* (10)	Widespread species, some sulfur oxidizers
	Nonsporing regular rods	*Lactobacillus* (8)	Widespread, some fish pathogens
	Nonsporing irregular rods	*Microbacterium* (36)	An artificial group, some pathogens, mainly in soil, some aquatic thermophiles
	Mycobacteria	*Mycobacterium* (1)	Widely distributed in soil and water, some pathogens
	Actinomyctes	*Dactylosporangium, Streptomyces* (49)	Fungi-like morphology, important in decomposition in soils
No cell wall, Bacteria	Mycoplasmas	*Mycoplasma* (6)	Smallest known self-reproducing organisms
Archaea	Methanogens	*Methanobacterium* (18)	Generates methane
	Sulfate reducers	*Archaeoglobus* (1)	Deep sea vents
	Extreme halophyles	*Halobacterium* (6)	Require at least 1.5M NaCl for growth
	No cell wall	*Thermoplasma* (1)	Found in mine waste
	Extreme thermophiles	*Sulfolobus* (14)	Optimum growth from 70°C −105°C

The number of described genera is only included as a rough guide to relative diversity in the group.

BACTERIA

Bacteria are single-celled organisms without organelles or a nucleus. They are ubiquitous and may have greater active biomass (protoplasm) than any other group of organisms on Earth (Whitman et al., 1998) particularly because they dominate deep groundwaters that extend the biosphere thousands of meters into the Earth. Bacteria are in every habitat and they regulate cycling of nutrients and flows of energy through aquatic ecosystems.

The Bacteria are the most metabolically diverse group of organisms on Earth. These unicellular organisms include heterotrophs that obtain energy from oxidizing organic carbon, as well as predators and parasites. Bacteria can also be autotrophic (chemoautotrophy and photoautotrophy) or even mixotrophic (a mixture of heterotrophic and autotrophic activities). Some of the most crucial biogeochemical fluxes mediated by these organisms will be discussed in Chapters 12—14. Bacteria are essential partners with animals and plants, and the bacteria and other microbes associated with larger organisms is termed their (the larger organism's) *microbiome*. In general use, a microbiome is the entire microbial community in a specific area of a larger organism;

bacteria tend to be most important in animal microbiomes, whereas fungi are more important in plants (Christian et al., 2015).

Microbiome research is one of the fastest growing fields in the biological sciences, fueled mainly by rapid developments in molecular taxonomic techniques. There is increasing evidence of the importance of the microbiome of an individual organism to its nutrition, metabolism, physiology, immunology, behavior, and ultimately fitness (Ottman et al., 2012). Microbiomes can vary considerably among and within species and individuals. Warne et al. (2017) found that gut microbiomes differ among species of amphibians and change through development, with large intraspecific differences among tadpoles, metamorphic individuals, and juvenile frogs. Experiments demonstrate that gut microbiota can be transplanted among species and developmental stages, and that eggs and egg jellies are important stages for microbial colonization of tadpoles (Warne et al., 2017), similar to the vertical transmission of important microbes from human mothers to infants during birth.

Some of the most important human pathogens regularly transmitted by water are bacterial (Table 9.1) and can be highly specific or facultative pathogens to many aquatic plants and animals. Over a half million children under 5 years of age die from dysentery each year in developing countries, and it is a tragedy that clean water is not available to every person on Earth. The human bacterial pathogens are not restricted solely to developing countries, with outbreaks of waterborne bacterial illnesses occuring in parts of the world with a high standard of living (Young, 1996). Finally, bacteria are the most common organisms used in bioremediation, which is the process of using organisms to clean up (break down) pollution (see Chapter 16).

The methods used to determine bacterial groups are often based on morphology or simple metabolic characteristics and likely do not accurately represent evolutionary relationships among all bacteria. A common identification scheme is presented in *Bergey's Manual of Determinative Bacteriology* (Holt et al., 1994), but this scheme is gradually being replaced by molecular methods. Classically, differentiation among bacterial species or strains is based on reaction to staining compounds, morphology, motility, production of extracellular materials, color, and metabolic capabilities. Metabolic capabilities have the greatest utility for discrimination among taxa because the other attributes vary little among species. Most bacteria are only 1 or 2 μm in diameter, but larger and smaller examples exist and a modest variety of morphologies occur (Fig. 9.3). We list the major groups of bacteria, on the basis of standard bacteriological techniques, in Table 9.2.

The number of species of bacteria is unknown, partially because of a fundamentally different species concept. Most vertebrates and many of the plant species have been described, but less than 1% of bacterial species had been described by the late 1900s (Young, 1997). Locey and Lennon (2016) used biodiversity scaling laws and estimated

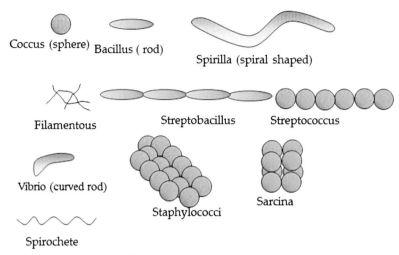

Figure 9.3 Some bacterial morphologies.

there are as many as 1 trillion bacterial species globally; Larsen et al. (2017) reached a similar number. Microbial diversity has received limited study because of difficulties associated with identification in natural samples. Furthermore, obtaining representative samples in some habitats, such as groundwaters, is difficult (Alfreider et al., 1997). The gut of each aquatic invertebrate species could harbor several unique microbial species. Each milliliter of water or gram of sediment has species that have never been cultured, so it is possible that there are more species of Bacteria than any other type of organism.

Early analysis of intensively studied hot spring communities suggests that only a fraction of the viable bacteria in any habitat can be cultivated successfully with current techniques (Ward et al., 1990). In a study of Octopus Spring in Yellowstone National Park, rRNA sequences were obtained from natural samples. Few of these sequences matched known sequences, even though numerous microbes have been cultured from the same study site and hot springs likely contain simple microbial communities. Subsequent work with substantially greater data sets and more efficient methods has verified the original finding by Ward et al. (1990) that only a portion of the bacteria in an environment has ever been cultured, and novel sequences continue to be identified from this one spring (Colman et al., 2016).

Zwart et al. (2002) searched published rRNA data and found 34 species clusters that occur exclusively or mostly in freshwater; 23 of these clusters included organisms that have never been cultured. A survey of lakes indicated that larger lakes had more species (distinct rRNA bands) than smaller lakes, lakes near each other tended to have fairly unique groups of bacteria, and there were 10−100 dominant species in each lake (Reche et al., 2005).

Cyanobacteria, the blue-green algae

The *Cyanobacteria* are a particularly important group of photosynthetic bacteria. Aquatic ecologists are very concerned with the cyanobacteria because of their tremendous impact on water quality; they form *blooms* or extremely high cell densities in eutrophic waters. Most researchers have adopted the more modern terminology "cyanobacteria" rather than "blue-green algae" or "cyanophytes" to clearly delineate the fact that this group is bacterial. The taxonomy of these organisms has been studied more completely than that of other bacteria because of their economic importance and because most are large and morphologically distinct under the light microscope.

Most cyanobacteria are O_2-producing photosynthetic bacteria. Fossils similar to extant cyanobacteria are the oldest known records of life (Schopf, 1993). Cyanobacteria have been successful for billions of years and are currently able to exploit some of the most extreme habitats on Earth, including very cold, very hot, and extremely saline environments.

Cyanobacteria are found in most habitats and can range from 1 μm in diameter to several hundred μm (Fig. 9.4). Shapes range from simple spheres to complex branching structures. They have specialized cells including *akinetes* (resting cells) and *heterocysts* (the site of most nitrogen fixation). Nitrogen fixation is the acquisition of gaseous N_2 into cellular nitrogen and we discuss this more thoroughly in the context of nitrogen cycling (see Chapter 14).

Proteinaceous vacuoles called *gas vesicles* lend buoyancy to the cyanobacteria and lead to formation of surface scums on water bodies under calm conditions. These surface scums can be up to 1 m thick, can have very objectionable odors, and can render a lake useless for swimming and other forms of recreation. The gas vesicles give the cyanobacteria a competitive advantage in eutrophic conditions by allowing them to compete well for light at the surface and shade out the phytoplankton below. Halting vesicle synthesis allows cells to sink to deeper, more nutrient-rich waters.

Cyanobacteria are excellent competitors for light because they have *phycobilins*, pigments that absorb light in the green region (where chlorophyll does not absorb). The ability to use green light allows some species of cyanobacteria to inhabit very deep waters and remain photosynthetically active because green light is not absorbed by most other photosynthetic organisms and thus does not attenuate in the water column as rapidly as other parts of the spectrum such as reds and blues. We first introduced the ecological consequences of phycobilins in Chapter 3 (Fig. 3.12).

Cyanobacteria are often difficult for herbivores to consume, partly because of their gelatinous coatings and because many of them produce toxins (Highlight 9.2). These toxins have likely evolved to limit grazing, but are very broad spectrum and may also have adverse effects on fishes, humans, crop plants, and many other organisms. Cyanobacteria also synthesize compounds with "earthy" or "musty" flavors and odors that can lead to undesirable qualities in drinking water and poor taste of fishes

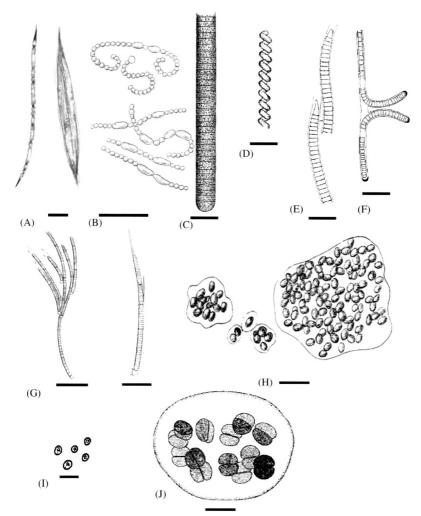

Figure 9.4 Selected genera of cyanobacteria, with length of scale bar: (A) *Aphanizomenon*, 30 μm for bundle, 5 μm for individual filament; (B) *Anabaena*, 20 μm; (C) *Oscillatoria*, 20 μm; (D) *Spirulina* 10 μm; (E) *Phormidium* (distinguish from C by mucous sheath), 20 μm; (F) *Scytonema* (with false branching), 30 μm; (G) *Rivularia*, habit view 30 μm, single trichome 15 μm; (H) *Microcystis*, 20 μm; (I) picocyanobacteria (indeterminate genus), 3 μm; (J) *Chroococcus*, 10 μm; (A, Sadie Whiles, B, Coupin (1911) courtesy Wikimedia Commons, C and H, Brotherus (1903), D, E, F, J, Wehr and Sheath, (2003), G and H Conn and Webster (1908) courtesy of Wikimedia Commons).

inhabiting waters with high population densities of cyanobacteria (Schrader and Dennis, 2005). These taste and odor problems lead to considerable costs for cleaning drinking waters and cause people to purchase bottled water (Dodds et al., 2009).

Benthic species of cyanobacteria inhabit diverse habitats, including wetlands, streams, and temporary waters, and are somewhat more common in oligotrophic

Highlight 9.2 Cyanobacterial toxins

Cyanobacteria are among several groups of toxic primary producers found in fresh-waters. Cyanobacteria produce several general types of toxins, neurotoxins, hepatotox-ins, and cylindrospermopsin. Both planktonic and benthic species can produce toxins (Gaget et al., 2017).

These toxins can be responsible for a variety of problems, including illness of humans who drink water containing the toxins, death of dialysis patients dialyzed with water containing the toxins, dermatitis from skin contact, potential long-term liver damage from contaminated water supplies, and animal deaths from drinking water con-taining cyanobacterial blooms (Falconer, 1999; Codd, et al., 1999). Cyanobacterial tox-ins could facilitate cancer by enhancing melanoma invasion of other cells (Zhang et al., 2012), and toxin concentrations also correlated with rates of nonalcoholic liver disease (Zhang et al., 2015). Cyanobacterial toxins can accumulate in fish and contaminate some of the largest freshwater fisheries in the world (Poste et al., 2011). At least 25 genera containing 40 species of cyanobacteria have members that produce toxins (Codd, 1995; Carmichael, 1997).

The neurotoxins produced by cyanobacteria act very rapidly (also known as very rapid death factors) and are responsible for the deaths of domestic animals that drink from water containing the toxins (Carmichael, 1994). The neurotoxins are lethal at very dilute concentrations; the notorious toxin dioxin is $10-60$ times less toxic than the cyanobacterial aphantoxin (Kotak et al., 1993). The neurotoxins include anatoxin-a, anatoxin-a(s), saxitoxin, and neosaxitoxin (the first two are unique to cyanobacteria). Some cyanobacterial genera that contain species that produce neurotoxins include *Anabaena*, *Aphanizomenon*, and *Oscillatoria*. It is difficult to know if a species is produc-ing a toxin in a particular lake because different strains of each species can produce dif-ferent amounts of toxins.

Cylindrospermopsins inhibit protein synthesis and are potentially problematic because they can reach appreciable concentrations dissolved in natural waters. These toxins can harm the human liver, kidneys, small intestine, and white blood cells (Falconer and Humpage, 2006).

Hepatotoxins kill animals by damaging the liver, including the associated pooling of blood. These toxins are in a family of at least 53 related small peptides. A complex series of factors induce expression of the genes for hepatotoxins (Pineda-Mendoza et al., 2016). This group of toxins has some that are more toxic than others, and some that are more prone to bioconcentration (Pick, 2016). There is concern that these compounds lead to increased rates of liver cancer (Carmichael, 1994). Microcystins remain in the blood serum of people exposed to the toxins for more than 50 days after initial exposure (Hilborn et al., 2007).

The Canadian government implemented a recommended water quality guideline of 0.5 μg/L microcystin-LR (the most common hepatotoxin) as a result of this threat, and other countries will likely follow suit (Codd et al., 1999; Fitzgerald et al., 1999). The World Health Organization recommends a maximum level of 1 μg/L microcystin-LR and a number of European countries as well as Australia, New Zealand, Denmark, Spain, France, Czechoslovakia, Poland, Belgium, and Brazil have adopted regulations at or near this recommended level. Countries all over the world are battling problems with blooms. For example, Lake Taihu, the third largest freshwater lake in China had such severe blooms in 2007 that 2 million people were without drinking water for 2 weeks, and the blooms continue in this lake (Xu et al., 2017)

There are 15 genera with species known to produce hepatotoxins include *Microcystis* and *Nodularia* (Paerl and Otten, 2013). These genera pose a threat to drinking water quality because they commonly form large blooms in nutrient-rich drinking water reservoirs during summer. In the treatment of algal blooms in lakes, methods that lyse cells and release toxins should be avoided (Lam and Prepas, 1997). Copper treatments commonly used on algal blooms release most toxins present within 3 days, but lime (calcium hydroxide) will remove algae without immediate release of toxins (Kenefick et al., 1993). The toxins are remarkably stable once they enter drinking water and can be removed only by chlorination and activated charcoal. Chlorination of drinking water rich in organics may be problematic because it may form chlorinated hydrocarbons (known carcinogens). We cover methods for controlling cyanobacterial blooms in Chapter 18.

Given the intense blooms of cyanobacteria that can form in some lakes, the ecological importance of these toxins in terms of ecosystem and community properties is likely underappreciated. The cyanobacterial toxins affect food crop (bean) photosynthesis when they are present in irrigation water (Abe et al., 1996). They can also modify zooplankton assemblages (Hietala and Walls, 1995; Ward and Codd, 1999), reduce growth of trout (Bury et al., 1995), kill turtles (Nasri et al., 2008), interfere with development of fish and amphibians (Oberemm et al., 1999), as well as rotifers (Beyer and Hambright, 2016), have caused massive flamingo die-offs (Krienitz et al., 2003), and certainly affect many other organisms. The toxins have been found in mayflies that emerge from tainted waters (Moy et al., 2016) and to be highly concentrated in bats living near a lake with *Microcystis* blooms because they feed on insects that emerge from the lake (Woller-Skar et al., 2015). However, some animals may actually prefer water containing toxic algae even though it is toxic to them (Rodas and Costas, 1999), though the reasons for this are not clear. The toxins can also be bioconcentrated by clams (Prepas et al., 1997), mussels, aquatic insect larvae, and oligochaetes (Chen and Xie, 2008).

Students should heed a note of caution related to cyanobacterial toxins. Some companies provide dietary supplements made from cyanobacteria (blue-green algae). A screening of 87 samples of cyanobacterial nutritional supplements found toxins in 85 of them. Even *Spirulina*, once thought to be nontoxic, has some strains capable of producing toxins (Dittman and Wiegan, 2006). If the genera in the product have strains known to produce toxins, the product should not be consumed without verification of routine negative test results (Schaeffer et al., 1999).

Prediction of when toxic blooms will occur has remained elusive. An extensive survey of 241 lakes in the Midwestern United States measured numerous chemical and morphological factors. Statistical analyses only accounted for 50% of the variance in toxin concentrations, and many relationships predicting toxins were nonlinear (Graham et al., 2004). Low nitrogen to high phosphorus ratios correlate positively to microcystin concentrations in Canadian lakes (Orihel et al., 2012). Many factors are involved, including wind mixing, temperature, grazing, iron, water residence time, and light

(Paerl and Otten, 2013; Ibelings et al., 2016). Exposure to grazing zooplankton may cause more toxicity; the toxin-producing *Microcystis* can upregulate toxin genes in the presence of *Daphnia* (Harke et al., 2017). Climate change may exacerbate harmful blooms (Paerl, 2017; Wood et al., 2017) by increasing water temperatures and altering mixing by wind. Even very large lakes, such as the Great Lakes of North America, are susceptible to cyanobacterial toxin problems. Genomic analyses have indicated that toxic strains of *Microsystis* replace nontoxic strains under conditions that were not constant across three studied lakes (Kardinaal et al., 2007).

Other groups of algae (dinoflagellates and diatoms) have toxic species or strains but cause problems more rarely in freshwaters. Dinoflagellate blooms (similar to the marine red tide) in freshwater lakes or reservoirs can poison fish. The factors that lead to blooms of these toxic algae also need more study.

---

waters (in contrast with planktonic species that dominate eutrophic waters). We discuss benthic cyanobacteria in hot springs and hypersaline habitats, including their motility and behavior, in Chapter 15.

Cyanobacteria tend to do well in warmer waters, and some species are capable of photosynthesis at higher temperatures than any eukaryotic algae. Tropical areas have substantial problems with cyanobacterial blooms and have been studied for years by researchers such as Dr Maria do Carmo Calijuri (Biography 9.1). Given that cyanobacterial blooms are undesirable, an additional concern is that such blooms are more probable with warmer weather. With predicted climate warming, cyanobacterial problems will probably increase (Jöhnk et al., 2008; Visser et al., 2016).

PROTOCTISTA

The protists include a wide variety of organisms from single celled to multicellular, including algae (other than cyanobacteria) and the heterotrophic protozoa. Together, the various groups of protists are responsible for much of the primary production and nutrient recycling that occurs in aquatic habitats. Traditionally these were split into the protozoa (heterotrophic and unicellular) and the protophyta (autotrophic) groups, but the difference between many of these is a matter of absence or presence of chloroplasts, with all other characteristics similar. These diverse and often elegant organisms are often the first microbes that students see and have doubtlessly inspired numerous careers in microbiology and aquatic ecology. Those whose imaginations have not yet been captured may want to examine the excellent photomicrographs of algae by Canter-Lund and Lund (1995). As discussed in Chapter 8, our understanding of the phylogenetic relationships among protists is in a state of flux. We present the following groups roughly in conventional rather than modern phylogenetic terms, mainly because this is how taxonomic books that students can use are organized.

BIOGRAPHY 9.1 Maria do Carmo Calijuri

Professor Maria do Carmo Calijuri (Fig. 9.5) is an internationally respected researcher on cyanobacterial blooms and one of the top water quality researchers in Brazil. She received a bachelor's degree in Biological Sciences (1982) and Master's degree in Ecology and Natural Resources (1985) from the Federal University of São Carlos (SP, Brazil) and obtained PhD in Hydraulic and Sanitation Engineering (1988) from the University of São Paulo where she is now a Distinguished Professor of the Department of Hydraulics and Sanitation. She has worked in Limnology and Environmental Sanitation, and supervised 42 Masters and 28 Doctoral students. She published over 80 scientific articles in national and international periodicals, authored two books, and was an editor of two more.

Figure 9.5 Maria do Carma Calijuri.

During graduate school, she began her studies in aquatic ecology and limnology. She participated intensively in fieldwork, and learned about the development and execution of large research projects. Her project on reservoir primary production considered dam operation systems, retention time, depth, morphometry, circulation, currents, and eutrophication. Her work was pioneering in tropical limnology. She found that in large Brazilian reservoirs, storms cause pulses of particulate and dissolved material that interfere with phytoplankton photosynthesis, nutrient cycling, and biomass loss. Her research has shown that in contrast to seasonal temperate reservoirs, tropical systems have frequent and irregular changes of smaller amplitudes. She recommends that young researchers pay attention to shorter time scales to understand process and functioning of tropical aquatic systems.

As her national and international research fame increased, she also assumed the coordination of the Graduate Program in Environmental Engineering Sciences and the Direction of the São Carlos School of Engineering at USP. She found she had an aptitude and passion for administration, and notes that she could never have imagined reaching such a high level of achievement in Engineering as a biologist and a woman.

She eventually became the head of 10 programs in different areas of Engineering. She appreciates that this gave her the opportunity to work with people with different backgrounds, many of whom became good friends. As she reflected on what her career can teach others, she noted that there were moments of hard work, others of great anxiety, of pure joy, and of pride in duty accomplished. These moments developed her maturity and allowed her continued success.

Eukaryotic algae

A wide variety of algal species occurs in freshwaters, but we only discuss the most common here. The *algae* are nonvascular eukaryotic organisms that are capable of oxygenic photosynthesis and contain chlorophyll *a*. Some algae may be considered protozoa because they have colorless forms that survive by ingesting other organisms. Table 9.3 and the following text summarizes the characteristics of selected groups of eukaryotic algae. Eukaryotic algae are globally important in freshwaters because they form the base of many food webs. They are also used for bioenergy production and removing nutrients from waste water. Understanding their taxonomy and ecology will be essential for future developments in these areas (Smith and Mcbride, 2015).

Rhodophyceae, the red algae

Red algae are abundant in marine habitats but are relatively rare in freshwaters; in freshwaters, they usually inhabit lotic waterbodies. For example, *Batrachospermum* (Fig. 9.6C) is a red alga found in streams and springs throughout the world. The algae are red because of their phycoerythrins, which impart a red hue. Phycoerythrins are phycobilin pigments similar to those found in the cyanobacteria that allow red algae to use blue-green wavelengths of light. Because of their ability to use wavelengths of light that penetrate further into the water, which other species are unable to use, red algae are often abundant in heavily shaded or deep habitats.

Chrysophyceae, the golden algae

The Chrysophyceae, sometimes called golden algae, are common components of the plankton in oligotrophic lakes. They have two flagella and, interestingly, most species are able to shift between photosynthesis and ingesting smaller organisms or particles for food (are mixotrophic). Alternatively, these organisms can ingest inorganic particles for their nutrient content (e.g., nitrogen and phosphorus). Many members of the Chrysophyceae are thus considered facultative heterotrophs.

Dinobryon is a common genus that forms chain-like colonies (Fig. 9.6). The large size of *Dinobryon* probably makes them difficult for herbivorous zooplankton to consume. *Dinobryon* are common in the summer in lakes when zooplankton populations are elevated. It is interesting that the colony is able to swim in a directed fashion well enough to stay in the photic zone of lakes, given that it consists of individual cells inside of shells (loricas) that are not in contact with each other. Like diatoms, members of the Chrysophyceae contain significant quantities of silica, but it is in the form of small scales outside the plasma membrane rather than a solid wall that surrounds the cell.

Bacillariophyceae, the diatoms

Diatoms are extremely important primary producers in lakes, streams, and wetlands. They are often dominant in plankton tows during the spring in oligotrophic—mesotrophic lakes and in the benthic zone of lakes, streams, and wetlands year-round.

Table 9.3 Characteristics of major groups of freshwater algae

Group (common name)	Dominant pigments	Cell wall	Habitats	Estimated # of species (% freshwater)	Ecological importance
Cyanobacteria	Chl a, chl d, phycobilins	Peptidoglycan	Oligotrophic to eutrophic, benign to harsh environments	8,000 (50%)	Some fix nitrogen, some toxic, floating blooms characteristic of nutrient rich lakes
Rhodophyceae (red algae)	Chl a, phycobilins	Cellulose	Freshwater species in streams	14,000 (5%)	Rare in freshwaters except *Batrachospermum* in streams
Chrysophyceae	Chl a, chl c, carotenoids	Chrysolaminarin	Freshwater, temperate, plankton	300–1,000 (80%)	*Dinobryon* a common dominant in phytoplankton
Bacillariophyceae (diatoms)	Chl a, chl c, carotenoids	Silica frustule	Plankton and benthos	20,000 (20%)	An essential primary producer, both in freshwaters and globally
Dynophyceae	Chl a, chl c, carotenoids	Cellulose	Primarily planktonic	3,000 (7%)	Some toxic, some phagotrophic, involved in many symbiotic interactions
Euglenophyceae (Euglenozoa)	Chl a, chl b	Protein	Commonly in eutrophic waters, associated with sediments	3,000	Can be phagorophic, indicative of eutrophic conditions
Chlorophyceae (green algae)	Chl a, chl b	Naked, cellulose or calcified	Oligotrophic to eutrophic, planktonic to benthic	6,500–20,000 (87%)	Very variable morphology, very important primary producers. Filamentous types in streams, unicellular in plankton
Charophyceae	Chl a, chl b	Cellulose, many calcified	Benthic, still to slowly flowing water	690 (95%)	Often calcareous deposits

See Figs. 9.4–9.7, and 9.9 and 9.10 for representative genera, and some morphological characteristics.
Source: After Larkum et al. (2016), South and Whittick (1987); Vymazal (1995) and estimates of total numbers by Guiry (2012).

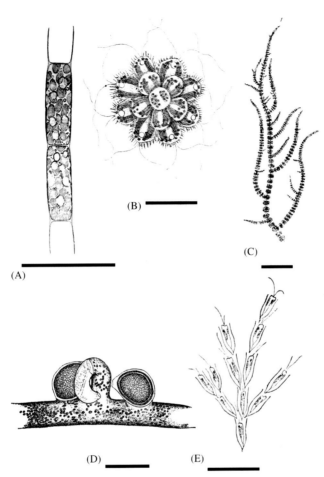

Figure 9.6 Selected algal genera, with scale bar length: (A) *Tribonema* (a Xanthophyte), 40 μm; (B) *Synura* (a Chrysophyte), 50 μm; (C) *Batrachospermum* (a red alga), 1 cm; (D) *Vaucheria* (a Xanthophyte), 200 μm; and (E) *Dinobryon* (a Chrysophyte), 20 μm (A Wehr and Sheath (2003, B, Brotherus (1903). C, Coupin (1911), D, Atkinson (1905), E, Algen (1914)).

The key defining characteristic of diatoms is the silica opalescent—glass cell wall called the frustule. The silica frustule makes them one of the few groups of organisms (along with the chrysophytes) that can be limited by silicon. This frustule has two halves, and the halves fit together to make an elongate, pennate (Fig. 9.7), or circular centric form. Centric forms are common in planktonic species, and pennate forms are common in benthic species. The frustules can attach to each other to form chains or filaments of many cells. The frustules are resistant to dissolution, so they may remain in the sediments for some time. This attribute makes them a valuable tool in forensic medicine with respect to drowning victims (Highlight 9.3).

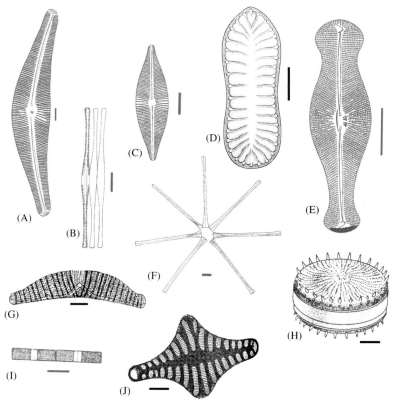

Figure 9.7 Common genera of diatoms with scale bar length: (A) *Cymbella*, 10 μm; (B) *Fragilaria*, 10 μm; (C) *Navicula*, 10 μm (D) *Surirella*, 10 μm; (E) *Didymosphenia*,10 μm; (F), *Asterionella* 10 μm; (G) *Epithemia*,10 μm; (H) *Stephanodiscus*, 2 μm; (I) *Aulacoseira*, 10 μm. (J) *Staurosirella*, 3 μm; (A, C and E, reproduced with permission from Patrick and Reimer 1975; B, C and F reproduced with permission from Patrick and Reimer 1966; D, courtesy Wikimedia Commons).

Highlight 9.3 Diatoms in forensics

Diatoms are one tool used to determine if drowning is a cause of death and where a drowning or other crimes occurred. When a person dies by drowning, one of the last things they do is take a breath of water. The water enters the lungs and bursts some of the alveoli (site of contact between blood and atmosphere). Diatoms in the water enter the bloodstream and the victim's circulatory system moves them through the body until the heart stops. When a forensic scientist searches for the cause of death, she digests a tissue sample with strong acid or with enzymes so the diatom frustules will remain behind (Timperman, 1969). If diatom frustules are in the tissues (particularly the bone marrow), the person likely died from drowning (Ludes et al., 1996). If the person was not breathing when they entered the water (i.e., he or she was already dead), no diatom frustules will be found deep within the organs. The technique is useful enough that the establishment of routine monitoring programs for diatoms has been recommended in areas where frequent drowning cases occur (Ludes et al., 1996).

Diatoms can establish location of drowning or trace suspects to a particular place because specific diatoms occur in known areas. In one case, a Finnish man was assaulted and thrown into a ditch. Five years later, the corpse was discovered and diatoms found in the lungs and bone marrow were the same species as those occurring in the ditch. Investigators concluded that he died by drowning in the ditch where the body was found (Auer, 1991). In another case, a group of teenagers assaulted two boys who were fishing in a pond and attempted to drown them. The boys escaped and the authorities apprehended the teenagers. Investigation confirmed that the teenagers had been at the pond because the residue found on their shoes contained the same diatom community as the pond mud (Siver et al., 1994).

Diatoms are also useful in paleolimnological studies because they sink and accumulate in the sediments and leave a record of the community structure of planktonic diatoms. Sampling the sediments by sectioning with depth and identifying diatom frustules in each depth layer can produce a story of which diatom assemblages dominated in the lake over long periods. Changes in diatom assemblages over time correlate to changes in climate, water chemistry, nutrient availability. Isotopic analyses can date sediments from various depths so that ecological changes inferred from diatom frustules correspond to a specific period in the past. These techniques give us information on the history of lakes, such as long-term temporal patterns of salinity and trophic state, and showed that acid precipitation was the result of industrialization (Flower and Battarbee, 1983). Ruth Patrick, one of the leading environmental researchers in the United States, made diatoms and their use in environmental studies her specialty (Biography 9.2).

Dinophyceae, the dinoflagellates

The dinoflagellates are common in lakes, wetlands, and ponds, and occasionally occur in streams. They are unicellular and free swimming, and usually planktonic. They can have cellulose plates or armor covering their body (Fig. 9.9). One flagellum encircles the cell, and another trails behind. Many members of this group are able to ingest other organisms (are mixotrophic). Some have no photosynthetic pigments, and some exist as predators, ingesting smaller cells.

Some dinoflagellates have complex life cycles and are able to assume a variety of forms, including spores, ameboid forms, and flagellated cells (Burkholder and Glasgow, 1997). In addition, some species of dinoflagellates ingest small unicellular algae and use them as chloroplasts. The dinoflagellates form a group that does not fit comfortably in the old classification system of plants or animals, because there are closely related autotrophic and heterotrophic species.

The toxic dinoflagellate, *Pfiesteria piscicida* occurs in estuaries. It has caused fish kills in the Chesapeake Bay. *P. piscicida* may harm humans and authorities publicize

BIOGRAPHY 9.2 Ruth Patrick

Dr Ruth Patrick (Fig. 9.8) was one of the leading diatom systematists in the world. She used her taxonomic expertise to extend the general theory of how aquatic microorganisms colonize new habitats and for over 50 years assessed the condition of stream ecosystems from the structure of biological communities. Her publication list spans 62 years and includes over 200 works, 143 as sole author and 38 as the first author; she is also coauthor of the authoritative monograph on diatom systematics in North America. Active in her 90s, Patrick completed the third book of a series on rivers and estuaries. She exhibited a continued dedication to pollution control in aquatic systems, where she pioneered the use of diatoms as indicators of chronic pollution. Dr Patrick died in 2013 at 105 years old.

Figure 9.8 Ruth Patrick.

Patrick served as president for major scientific societies and served on committees for several U.S. presidents, Congress, the National Academy of Sciences, and others. Patrick was the recipient of the prestigious National Medal of Science, conferred by President Clinton. This is among a long list of awards that includes 25 honorary doctorates and election to the National Academy of Sciences.

Given the period that her career spanned, she overcame tremendous obstacles to become a leading scientist when women were not typically scientists, an environmentalist when few were concerned about human impacts on the environment, and only the 12th woman in 100 years when she was elected to the National Academy of Sciences. A hallmark of Patrick's career was her insistence on making a positive difference. Her father, who allowed her to climb onto his lap to look through a microscope when she was 4 or 5 years old, guided her. Patrick notes that her father would get up from the dinner table every night and say, "Remember, you must leave this world a better place." She has.

swimming advisories with blooms. Nutrient pollution transported via freshwaters to the estuary probably exacerbates blooms of this toxic alga (Burkholder and Glasgow, 1997), although there is some controversy over the causes of blooms and the biology of the organism. Freshwater dinoflagellates may also be toxic to humans and some harm fishes, but they have not received as much study as their marine and estuarine counterparts (Carty, 2014).

Euglenophyceae

The euglenoids can be pigmented or colorless, and thus can be autotrophic or heterotrophic. Pigmented members of this group (hence the suffix "phyceae" indicating "plant") often have pigments similar to those of the green algae, but are always unicellular and generally motile. Molecular analyses suggest that they are not closely related to the green algae, in spite of their similar pigment content (Fig. 8.4). They occur most commonly in eutrophic situations, including shallow sediments. Euglenoids are capable of ingesting particles. A flexible protein sheath covers the cell, and amoeboid cell movement can occur. Additionally, many cells have a single flagellum for locomotion. Characteristic features include a red photosensitive spot in one end and numerous chloroplasts in the cell (Fig. 9.9).

Chlorophyceae and Charophyceae, green algae and relatives

These algae range from simple, single-celled organisms to complex, multicellular assemblages (Figs. 9.10 and 9.11). They are found in all surface aquatic habitats from damp soil and wetlands to the benthic zone of rapidly flowing streams and the plankton of large lakes, and they are the most diverse group of freshwater algae.

Some species mainly inhabit oligotrophic habitats, whereas others are common in eutrophic habitats. Unicellular types are most common in lake plankton but occur in benthic habitats. There is a tremendous variety of unicellular morphologies (Fig. 9.10). Species with filamentous morphologies (Fig. 9.11) generally attach to the benthic substrata in streams and lakes. One species, *Basicladia*, almost exclusively attaches to the backs of aquatic turtles. Filamentous green algae are often the most obvious algae in nutrient enriched streams and along shorelines of nutrient enriched lakes, in some cases with massive populations that degrade water quality and hinder recreational activities.

The Charophytes (stoneworts) are related closely to the Chlorophyceae, but are more complex (Fig. 9.9). The ancestors of the stoneworts were the evolutionary precursors to land plants. These plant-like algae are commonly called stoneworts because they precipitate calcium carbonate on the exterior of their cell walls and feel rough, or stone-like to the touch. They are not vascular, but have multicellular reproductive structures more like land plants than the other algae. The charophytes can sometimes cause problems because of immense biomass that impedes water flow or navigation on rivers. Charophytes may also be an important component of more productive wetlands. *Chara* can be abundant in the benthic zone of some oligotrophic lakes. Many charophyte species are sensitive to nutrient enrichment and distribution of the stoneworts serves as an indicator of nutrient pollution. Species found only in very clean waters may be endangered and some species are now protected in Europe.

Additional algal groups

Additional groups of algae occur in freshwaters, including the Cryptophyceae, the Tribophyceae, and the Phaeophyceae. Members of these groups can dominate

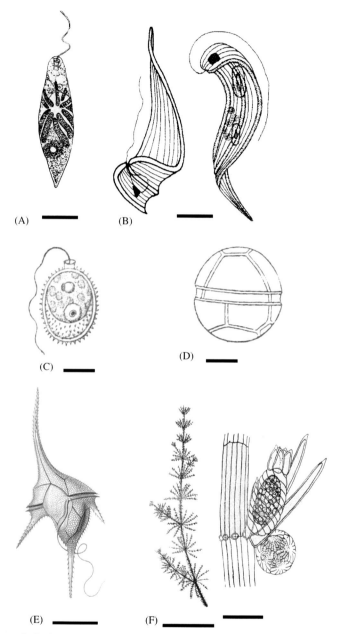

Figure 9.9 Selected algal genera, with scale bar length: (A) *Euglena* (a Euglenophyte), 20 μm; (B) *Phacus* (a Euglenophyte), 20 μm; (C) *Trachelomonas* (a Euglenophyte), 20 μm; (D) *Peridinium* (a Dinoflagellate), 20 μm; (E) *Ceratium* (a Dinoflagellate), 20 μm; and (F) *Chara* (a Charophyte) large view 2 cm, close-up 500 μm; A, C, D, E, F, Wikimedia Commons, B Wehr and Sheath (2003).

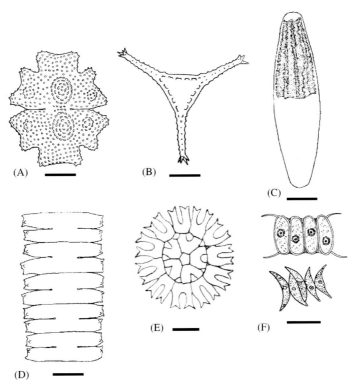

Figure 9.10 Some colonial and single celled green algae and scale bars (A) *Euastrum*, 30 μm; (B) *Staurastrum*, 25 μm; (C) *Netrium*, 35 μm; (D) *Spondylosium*, 25 μm; (E) *Pediastrum*, 20 μm; (F) *Scenedesmus*, 20 μm (A–D, Wehr and Sheath, 2003, E, Pasher (1914), F Coupin (1911)).

freshwater systems at times. However, detailed description is left to phycology courses and the comprehensive phycological texts (South and Whittick, 1987; Graham and Wilcox, 2000).

Protozoa

Protozoa are found in nearly every habitat on earth, and can be very abundant and diverse in freshwaters. This group includes autotrophs and heterotrophs. Many types are heterotrophic and survive by ingesting particles or absorbing dissolved organic carbon. They are very important predators of bacteria in aquatic environments. There are also many ecologically and economically important parasites in this group. Some of the smallest protozoa are only slightly larger than bacteria and can ingest virus-sized particles. The largest are visible to the unaided eye. This is a very diverse group (Fig. 9.12) and includes the most complex single-celled organisms known.

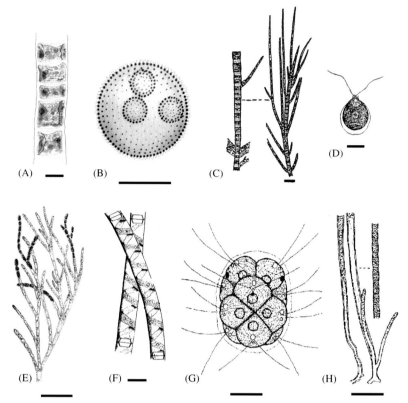

Figure 9.11 Common genera of filamentous and flagellated green algae, with scale bar length: (A) *Ulothrix*, 20 μm; (B) *Volvox*, 10 μm; (C) *Stigeoclonium*, 20 μm; (D) *Chlamadymonas*, 10 μm; (E) *Cladophora*, 50 μm; (F) *Spirogyra*, 20 μm (G) *Pandorina*, 10 μm; and (H) *Basicladia*, 30 μm (A and B, Sadie Whiles, C, G, H Wehr and Sheath, 2003, D, E, F Wikimedia Commons).

Life histories are generally simple. Sexual reproduction is widespread but not universal. Many protozoa form cysts that are resistant to drying and other environmental extremes, allowing them to persist in harsh habitats and quickly exploit moisture when it becomes available. Other morphological variations among life cycle stages can also occur, such as differentiation between forms that search for food and those that consume it (Taylor and Sanders, 1991).

Various classifications of the protozoa have been proposed (Taylor and Sanders, 1991), and molecular analyses indicate that there should be several phyla of protozoa (Figs. 8.1 and 8.4). This is because they were historically lumped into everything that was not plant, animal, or fungi. For the sake of simplicity, we will discuss the major functional groups of protozoa, following the groupings of Taylor and Sanders (1991).

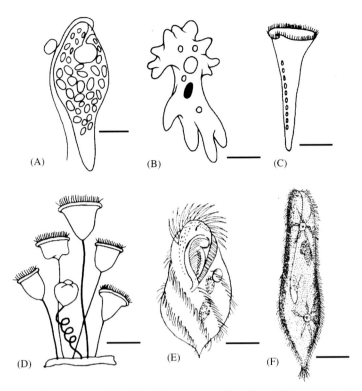

Figure 9.12 Selected protozoa: (A) *Khawkinea*, a zooflagellate, 20 μm; (B) *Amoeba*, 50 μm; (C) *Stentor*, a solitary ciliate, 200 μm; (D) *Vorticella*, a colonial ciliate 75 μm; (E) *Hypotrichidium*, a ciliate, 30 μm; and (F) *Paramecium* a ciliate, 60 μm. *Reproduced with permission from Thorp and Covich (1991b).*

The flagellated protozoa (those with one or more flagella) can be divided into two major functional groups. The phytoflagellates includes groups that are autotrophic, heterotrophic, and mixotrophic, some of which we discussed in the section on algae (e.g., dinoflagellates, chrysophytes, euglenoids, and flagellated green algae). The zooflagellates (Fig. 9.12) are all heterotrophic. The phylum Zoomastigina includes several important human parasites (e.g., trypanosomes and *Leishmania*) and many free-swimming forms. The heterotrophic nanoflagellates are very small zooflagellates that are often the most important consumers of pelagic bacteria and can serve a vital role in nutrient cycling.

The amoeboid protozoa, including phyla Karyoblastea and Rhizopoda, constitutes the largest group of described species of protozoa. The amoeboid protozoa include those that move by protoplasmic flow and use of pseudopodia (extensions of the protoplasm; e.g., *Amoeba*, Fig. 9.12B). The flowing movements of amoebae illustrate this process. These groups of protozoa are more often associated with benthic habitats and sediments than open freshwaters, but some floating forms exist. Some amoeboids live in shells (called tests) that they secrete or construct from sediments. Two species from two genera (*Naegleria* and *Acanthamoeba*) can cause meningitis in human swimmers. Many of the amoeboid protozoa are also important microbial predators.

The ciliated protozoa (phylum Ciliophora, including 10 classes) are characterized by having more than four cilia, which are primarily used for locomotion. Nuclear dualism is a unique feature of the ciliates; individuals have one or more macronuclei and one or more micronuclei. This group contains the familiar *Paramecium* and other free-swimming genera (Fig. 9.12F). Some members of this group are floating or sessile predators with ciliated larval stages for dispersal. The common attached protozoan, *Vorticella*, is also a member of the Ciliophora (Fig. 9.12C). Most of the ciliates feed on bacteria or other protozoa, but a few species feed on algae and detritus. *Balantidium coli*, an intestinal parasite of pigs and occasionally humans, is the only member of the Ciliophora known to cause human disease.

FUNGI

Fewer species of fungi occur in aquatic habitats than in terrestrial habitats. Nonetheless, they are very important in the degradation of detritus (leaf litter and woody debris) that enters streams and lake margins, and as a source of nutrition for detritivores. Fungal colonization softens detritus and increases its nutritional value. Without the activity of fungi and bacteria, significant detrital carbon sources would remain unavailable to invertebrates (Arsuffi and Suberkropp, 1989; Suberkropp and Weyers, 1996). The role of fungi in lakes is not as well characterized as it should be, given how common and abundant fungi are in lentic ecosystems (Wurzbacher et al., 2010).

Aquatic fungi

More than 2,500 species of fungi occur in freshwaters (Wong et al., 1998; Shearer et al., 2007), including those of the Labyrinthulomycetes (slime molds), Ooprotista (algal fungi), Ascomycetes (filamentous fungi), Basidiomycetes (column fungi), and Deuteromycetes (imperfect fungi). Most fungi are *saprophytes*, meaning they live on dead organic matter. The systematics of the fungi are based on their reproductive features and morphology, except for the Dueteromycetes, in which no reproductive structures have been found. The taxonomy of the Deuteromycetes will probably be resolved more satisfactorily upon application of more molecular methods.

Of the groups of fungi, only the Ooprotista (formerly the Phycomycetes) are predominantly aquatic. In general, they are unicellular. The Ooprotista includes many parasitic species that are pathogens of planktonic algae, small animals, and the eggs of crustacean larvae and fish (Rheinheimer, 1991). The Chytridioprotista (chytrids) live in aquatic habitats and moist soils. While members of this group are considered fungi, they have a motile zoospore with one whip-like posterior flagella, more like the protists. The chytrids can be important parasites and the chytrid, *Batrachochytrium dendrobatidis*, which we will discuss in Chapter 11, is the primary cause of many of the catastrophic amphibian declines occurring around the planet (Lips et al., 2006; James et al., 2015).

The Ascomycetes and Deuteromycetes (particularly the aquatic Hyphomycetes) are often abundant on decaying leaves and wood (Fig. 9.13). The Hyphomycetes are divided into the Ingoldian fungi (with branched or radiate conidia) and the Helicosporous (with helical conidia) fungi (Alexopolus et al., 1996). Yeasts (in the group Ascomycetes) can occur in rivers and lakes, particularly in polluted waters.

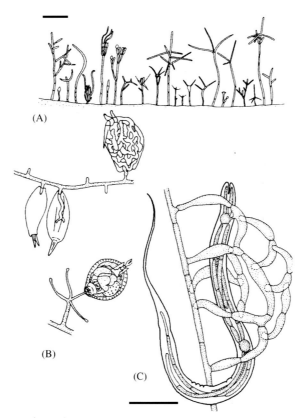

Figure 9.13 Selected aquatic fungi: (A) Aquatic deuteromycetes, (B) *Zoophagus* species with trapped rotifers, and (C) *Arthrobotrys oligospora* and a trapped nematode. *Reproduced with permission from Rheinheimer (1991).*

"Sewage fungus" associated commonly with organic-rich pollution is actually composed predominantly of sheathed bacteria, not fungi.

Fungi are not abundant in pristine groundwater because of low concentrations of organic matter (Madsen and Ghiorse, 1993). Likewise, pristine spring water rarely has significant numbers of fungi. Fungi may be locally abundant where pollution or natural inputs enrich groundwater with organic compounds, but when the ground waters are anoxic, the fungi are less abundant than in oxic water.

An interesting mode of nutrition for some aquatic fungi is predation on rotifers or nematodes. The fungi that prey on rotifers have sticky appendages that trap the organisms and then rapidly grow into them. The nematode trapping fungi inhabit soils and aquatic sediments and can form a net of loops or snares that trap the nematodes as they crawl through (Fig. 9.13).

Fungi often associate with plant roots (mycorrhizae), particularly in terrestrial habitats. However, these fungi are also important in aquatic environments. In some wetlands, they can alter the dominant plant species, depending upon wetland hydrology. Experiments with plants in Northeastern United States calcareous fens suggests that the importance of mycorrhizal fungi decreases with increased inundation (Wolfe et al., 2006). Macrophytes in lakes have mycorrhizal symbionts that are most prevalent in low phosphorus/high redox lakes, with many species of macrophytes associated with arbuscular mycorrhizal fungi (Baar et al., 2011).

Aquatic lichens

Lichens form with ascomycete fungi, a basidiomycete, and a green algal and or a cyanobacterial partner. The basidiomycete (yeast) partner was just recently described (Spribille et al., 2016). The fungi create protection for the alga, attaches it to a sold surface, and can disperse the alga. In exchange, the algae provide photosynthetically acquired organic carbon, and in the case of the cyanobacterium, can fix atmospheric N_2. The taxonomy of lichens traditionally is based on external morphology. The morphological differentiation among lichens requires determining if the form is foliose (leaf like), fruticose (finger-like projections), or crustose (appressed to a solid surface). Taxonomists now use molecular techniques to distinguish lichen species that they could not differentiate in the past.

Lichens do not occur in groundwaters, but are present occasionally in the benthic zone of lakes and rivers. There are approximately 200 species of freshwater lichens, and there will probably be more described as tropical streams and rivers are studied more closely (Hawksworth, 2000). Dense growths of the lichen *Dermatocarpon fluviatile* grow in benthic habitats of streams and some lakes. Lichens can be abundant in wetlands, particularly those in northern temperate, boreal, or polar regions. In one northern European wetland, lichens and mosses were responsible for 9% of the carbon input to the bog (Mitsch and Gosselink, 1993). Lichens on rocks near the waterline of lakes have received some study (Hutchinson, 1975; Hawksworth, 2000). Lichen species change with distance above the water as their tolerance of submergence decreases and tolerance of desiccation increases.

PLANTAE

Plants dominate in many shallow waters. Plants in water are *macrophytes*. They are the dominant organisms in wetlands and many lake margins and streams. They can play essential roles in biogeochemistry and ecology. For example, macrophyte beds can provide important spawning habitat and shelter for small fishes. Macrophytes can also be the dominant photosynthetic organisms in small or shallow lakes. Thus, they can even be central to food webs in some lakes and wetlands. Riparian plants are very important in biogeochemistry because they intercept materials (e.g., pollutants, nutrients, and sediments) that are entering aquatic habitats from the land. Furthermore, much of the diversity of plants in many terrestrial landscapes is associated with riparian zones or wetland areas. Aquatic plants can cause nuisance conditions, and there are aquatic or riparian species that are invasive and cause problems.

Nonvascular plants

Bryophytes (mosses and liverworts) are abundant in some freshwaters. Traditionally, they received little study relative to their importance in some systems, but appreciation of their importance is increasing (Arscott et al., 1998). Bryophytes can be important components of streams, particularly in alpine and arctic regions, where they can be productive and provide important structural habitats for invertebrates (Bowden, 1999). Aquatic mosses can be divided into three orders (Hutchinson, 1975): the Sphagnales, the Andreales, and the Bryales. The Sphagnales and Bryales have numerous aquatic representatives, and the Andreales has few. The Sphagnales contains only one genus, *Sphagnum*.

Species of the genus *Sphagnum* are often a dominant component of the vegetation in the shallow acidic waters of peat bogs and can be very important in many high-latitude wetlands. The total global biomass of *Sphagnum* is greater than that of any other bryophyte genus (Clymo and Hayward, 1982). Carbon deposition in these peat bogs may be important in the global carbon cycle. The moss promotes acidic conditions because microbial breakdown of organic material produced by *Sphagnum* produces organic acids. The acidity leads to a stable dominance by the acid-tolerant moss and slows breakdown of organic material. Thus, peat accumulations are significant in the bogs where *Sphagnum* dominates. We discussed peat and bogs in more detail in Chapter 5.

The Bryales includes several interesting aquatic genera, including *Fontinalis*, which can be collected up 120 m deep in Crater Lake, and *Fissidens*, which occurs to 122 m deep in Lake Tahoe (Hutchinson, 1975). Thus, among plants, the mosses can inhabit some of the deepest habitats.

Vascular plants

The angiosperms, the true flowering plants, are the dominant aquatic vascular plants, with representatives of both monocots and eudicots. In addition, some of the ferns and fern allies are associated with aquatic habitats. Of the angiosperms, the monocots are relatively more abundant in aquatic habitats than they are in terrestrial habitats (Hutchinson, 1975). Vascular plants represent important structural habitat for other organisms (e.g., fishes, invertebrates, and epiphytes), food for some herbivores and detritivores, and, as with algae, their photosynthesis and respiration can greatly influence the local chemical environment, including availability of dissolved oxygen. While we often think of aquatic plants as more significant to food webs in wetlands and shallow lakes, they can be quite important in rivers as well. For example, the federally endangered West Indian Manatee, an herbivore that can consume up to 15% of its massive body weight in aquatic plants per day, feeds on floating and submersed macrophyte beds found in many of the warm, spring fed streams of Florida. These slow moving, clear water springs and rivers serve as thermal refugia during winter when coastal habitats are cold, and also provide the manatees abundant aquatic plants to graze.

Aquatic ecologists tend toward classifications of plants based on functional roles and habitats (Fig. 9.14). Distinguishing aquatic plants on the basis of morphology can be difficult because there is tremendous variation in morphology of plant structures within genera. In addition, leaf size and shape can change appreciably in the same species grown under different environmental conditions (Fig. 9.15) or even in the same plant above and below water (Fig. 8.2).

The traditional categories include: (1) *floating unattached* macrophytes (roots not attached to substratum, Fig. 9.16), (2) *floating attached* plants (leaves floating at the

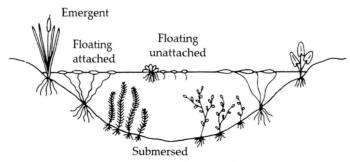

Figure 9.14 Growth habit types of aquatic plants. *Reproduced with permission from Riemer (1984).*

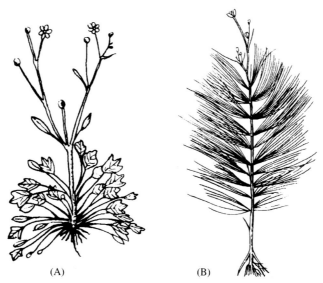

Figure 9.15 Two morphologies of the buttercup *Ranunculus polyphyllus* growing on land (A) and submersed (B). *From Hutchinson copyright © 1975. Reprinted by permission of John Wiley & Sons, Inc.*

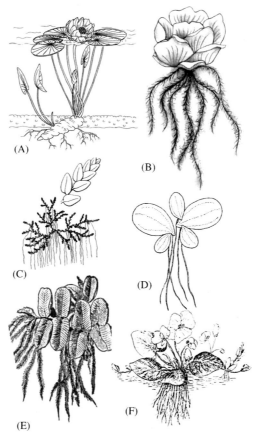

Figure 9.16 Some species of floating attached (A), floating unattached (B–F) macrophytes, (A) *Nymphaea*, water lily; (B) *Pistia*, water lettuce; (C) *Azolla*, water velvet; (D) *Spirodela polyrhiza*, duckweed; (E) *Salvinia*, water fern; (F) *Eichornia crassipes*, water hyacinth. Scale bar = 2 cm. A, C, E, E, F Wikimedia Commons, B, by Sadie Whiles.

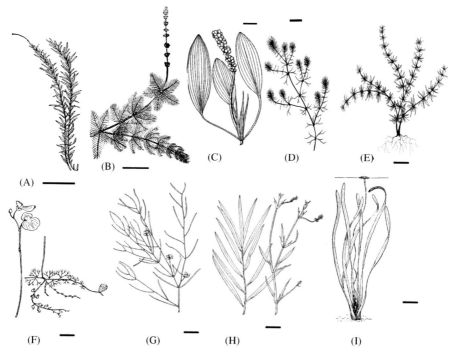

Figure 9.17 Some species of submerged aquatic plants. (A) *Elodea pusillis*; (B) *Myriophyllum spicatum*, water milfoil; (C) *Potomogeton nodosus*, pondweed; (D) *Ceratophyllum*, coontail; (E) *Hydrilla verticillata*, hydrilla, (F) *Utricularia*, bladderwort; (G) *Potamogeton pussilus*, pondweed, compare with C; (H) *Potamogeton robbinsii*, pondweed, compare with C and G; and (I) *Vallisneria Americana*, wild celery; Scale bar = 4 cm A, C, F, G, H, I, Wikimedia Commons, B, D, E Courtesy Sadie Whiles.

surface, and roots anchored in the sediments), (3) *submersed* or *submerged* plants (entire life cycle, except flowering, under water; generally attached to sediment, Fig. 9.17), and (4) *emergent* (growing in saturated soils up to a water depth of 1.5 m and producing aerial leaves, Fig. 9.18).

A wide variety of plant groups have given rise to aquatic species; some of the representative genera are listed in Table 9.4. Trees associated with wetlands are important in defining wetland types (Table 9.5). The genera of submersed plants tend to be confined to aquatic habitats, but emergent genera also have many representatives in terrestrial habitats (Hutchinson, 1975). The aquatic plants form a vital part of the ecosystem, but some have become serious invaders or pests. Invasion of North American wetlands by purple loosestrife (*Lythrum salicaria*) (Highlight 9.4) is one of many examples of nuisance exotic plants that are altering freshwater ecosystems.

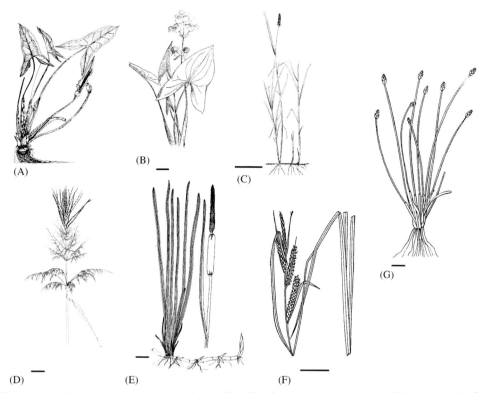

Figure 9.18 Some emergent aquatic plants. (A) *Peltandra virginica*, arrow arum; (B) *Sagittaria latifolia*, arrowhead; (C) *Scirpus validus*, great bulrush; (D) *Zizania aquatica*, wild rice; (E) *Typha*, cattail; (F) *Carex aquatilis*, water sedge; (G) *Eliocharis obtuse*, spikerush Scale bar in A, B, D, and E 10 cm; C, F and G 5 cm. *Images courtesy Wikimedia Commons.*

Table 9.4 Some common genera of aquatic plants, with their diversity, habit, and distribution

Habit	Genus	Common name	Species	Habitat/distribution/comments
Emergent	*Oryza sativa*	Rice	1	Tropical temperate food crop, most important in world
	Peltandra	Arrow arum	3	Shallow water eastern North America
	Sagittaria	Arrowhead	20	New World, tuber edible by waterfowl
	Scirpus	Bulrush	150	Worldwide, mostly North America
	Sparganium	Bur reed	20	Temperate and arctic Northern Hemisphere
	Typha	Cattail	10	Worldwide, common in monospecific stands in swamps, marshes, and along streams
	Phragmites	Giant reed	3	Worldwide
	Pontederia	Pickerelweed	5	New World, shallow muddy areas
	Juncus	Rushes	225	Northern Hemisphere, few submerged species
	Carex	Sedges	1,000	Worldwide, common in damp areas to shallow waters

(Continued)

Table 9.4 (Continued)

Habit	Genus	Common name	Species	Habitat/distribution/comments
	Eliocharis	Spike rush	200	Worldwide, may be completely submerged, but will be sterile if so
	Zizania	Wild rice	2	North America, is a prized food
Floating attached	*Nuphar*	Spatterdock, yellow water lilies	25?	Large yellow flowers
	Brasenia	Water shield	1	Scattered world wide
	Nymphaia	White water lilies	40	Worldwide, introduced in many places, large flowers, mostly white
Floating unattached	*Lemnaceae*	Duckweed	30	Worldwide, can cover small pond surfaces
	Salvinia	Water fern	12	Tropical
	Eichornia crassipes	Water hyacinth	1	Tropical and subtropical, one of the worst weeds in the world, also used for tertiary sewage treatment
	Pistia stratiodes	Water lettuce	1	Tropical and subtropical, can be a pest
	Azolla	Water velvet		Worldwide, contain N fixing cyanobacteria, may be important in traditional rice culture, can be red or purple on surface of water
Submersed	*Utricularia*	Bladderwort	150 (30 aquatic)	Seed-like bladders trap and digest aquatic animals
	Ceratophyllum	Coontail	30	Worldwide
	Elodea	Elodea	17	North and South America, but introduced elsewhere as an escapee of aquaria
	Cabomba	Fanwort	7	Tropical to temperate New World
	Hydrilla verticillata	Hydrilla	1	Similar to *Elodea*
	Najas	Naiad	50	Worldwide
	Potomogeton	Pondweeds	100?	Worldwide, found in most types of fresh surface waters, important food sources for wildlife, some pest species
	Myriophyllum	Water milfoil	40	Africa, some species introduced pests elsewhere
	Vallisneria	Wild celery	10	Worldwide, warm areas

Source: After Riemer (1984).

Table 9.5 Some important trees associated with wetlands in North America

Common name	Scientific name	Distribution
Black spruce	*Picea mariana*	Boreal wetlands
Tamarack	*Larix laricina*	Boreal wetlands
Red maple	*Acer rubrum*	Temperate wetlands
Northern white cedar	*Thuja occidentalis*	Northeast temperate North America
Atlantic white cedar	*Chamaecyparis thyoides*	Southeast United States
Cypress	*Taxodium* spp.	Southeast United States, deepwater swamps
Tupelo	*Nyssa aquatica*	Southeast United States, deepwater swamps
Cottonwood	*Populus* spp.	Riparian wetlands
Willow	*Salix* spp.	Riparian wetlands
Red mangrove	*Rhizophora* spp.	Brackish tropical waters
Black mangrove	*Avicennia* spp.	Brackish tropical waters

Highlight 9.4 Invasion of wetlands by purple loosestrife

Purple loosestrife (*Lythrum salicaria*) is a perennial plant that is invading many North American wetlands (Fig. 9.19). It is an emergent plant with purple flowers that could be considered beautiful if they did not belong to an aggressive invader. Purple loosestrife can reach some of the highest levels of biomass and annual production reported for freshwater vegetation (Mitsch and Gosselink, 1993). Unfortunately, the plant is a poor food source for most waterfowl, and large stands with a high percentage of cover possibly lower the numbers of nesting sites for ducks and other water birds as well as provide additional cover for predators. Purple loosestrife can outcompete native plants and lower biodiversity (Malecki et al., 1993). In wet areas used for hay, it lowers the forage value. However, Anderson (1995a,b) suggested that the effects of purple loosestrife have been overestimated and more research should be done to quantify its impacts on native ecosystems. Lavoie (2010) reiterated this point.

Figure 9.19 Purple loosestrife (*Lythrum salicaria*). *Image courtesy Pixabay.*

L. salicaria occurs naturally in Europe from Great Britain to Russia (Mal et al., 1992). People introduced it to eastern North America in the ballast of ships and as a medicinal and decorative plant (Malecki et al., 1993). Loosetrife has since become a serious pest species around the Great Lakes of North America and has spread across Canada and the United States to the west coast. Introductions have also occurred in Australia, New Zealand, and Tasmania (Mal et al., 1992).

Several control strategies have been attempted (Malecki et al., 1992), including herbicides (which also harm other wetland species), physical removal, burning, manipulation of water levels to favor native species, use of native insects, and introduction of exotic insects to control the plant. There are concerns that insects introduced as control agents will harm native species. After tests for host specificity, managers released three insect species in the United States as control agents (Piper, 1996). Their efficiency in loosestrife control was strong, but nontarget species were also harmed (Hinz et al., 2014). As always, an understanding of wetland ecology is crucial to assessing the impact of and control options for this exotic invader.

SUMMARY

1. Viruses are common in freshwater habitats and have important consequences in terms of diseases of aquatic organisms and human health, as well as population dynamics of microorganisms.

2. Archaea are important in extreme habitats and for some types of biogeochemical cycling, particularly the formation of methane. They are probably more important in most aquatic habitats than was appreciated previously.

3. The biomass, metabolic diversity, and species diversity of bacteria probably exceeds that of any other group of organisms on Earth. Understanding the role of bacteria is central to attempts to understand the aquatic environment, including organic carbon decomposition, energy flow, and nutrient cycling.

4. Cyanobacteria are a significant ecosystem component of many lighted aquatic habitats. They can cause problems when they form large blooms. Many varieties produce toxins and foul the water.

5. Algae constitute an important and diverse group found in freshwaters and form the basis of many aquatic food webs. The importance of algae also includes an intimate role in water quality. Diatoms are used in paleolimnology to document historical biological patterns over thousands of years.

6. Protozoa are a very diverse and taxonomically problematic group. Protozoa are generally the main consumers of bacteria in aquatic systems and can represent important food for larger consumers.

7. Fungi are responsible for much of the degradation of particulate organic material that occurs in freshwaters, and some groups are also commonly parasitic on aquatic organisms.

8. The bryophytes can be important in some shallow aquatic habitats. Formation and maintenance of high latitude wetland communities by *Sphagnum* in peat bogs is of global significance.

9. Plants can be classified as submerged, floating, or emergent, and they represent important structural habitat for other organisms. Vascular plants are dominant contributors to organic matter in many shallow aquatic ecosystems and are also an important food source for aquatic herbivores. The types of flowering plants present define many wetlands.

QUESTIONS FOR THOUGHT

1. Do more types of viruses exist than species of organisms on Earth?
2. Should separate taxonomic definitions of species be used for microbes than those used for animals?
3. Why are there no known fish-pollinated aquatic angiosperms?
4. How does stability of the benthic substrata partially determine if benthic systems are dominated by microalgae or by macrophytes?
5. Should molecular taxonomy methods be more useful for aquatic angiosperms or unicellular algae?
6. Why are floating-leafed macrophytes relatively rare on large lakes?
7. Why are bryophytes relatively more important in alpine and high latitude freshwater habitats?

CHAPTER 10

Multicellular Animals

Contents

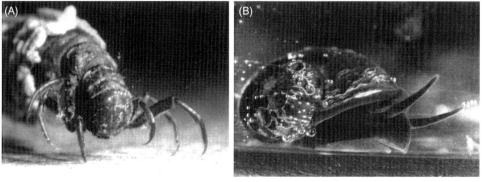

Figure 10.1 A caddis larva, *Psychoglypha subborialis*, and snail, *Vorticifex effusa*. The caddis larva is 1 cm long, and the snail is 0.5 cm long. Both are from Mare's Egg Spring, Oregon.

Animals are some of the most fascinating organisms found in aquatic habitats because of their diversity and behavior. Many animals play critical roles in freshwater ecosystems (Wallace and Webster, 1996; Vanni, 2002) and can be important indicators of freshwater ecosystem health because they integrate stresses and sensitive species cannot tolerate polluted habitats (Palmer et al., 1997; Covich et al., 1999). Food webs are a major pathway of energy flow through ecosystems and animals generally dominate them. Unfortunately, freshwater animal communities, particularly those found in streams, are some of the most imperiled in the world because of myriad human impacts on freshwater habitats (Strayer and Dudgeon, 2010; Vorosmarty et al., 2010). We have already discussed the economic importance of commercial and sports fisheries (Chapter 1). This chapter considers each group of organisms approximately in order of evolutionary origins.

INVERTEBRATES

Phylum Porifera

Sponges are generally thought of as marine animals but they can be abundant in freshwaters. There are about 25 species of freshwater sponges in North America and about 300 species worldwide (Frost, 1991). They grow in a variety of lentic and lotic habitats and some species have small ranges, whereas others are widespread. Sponges are among the most primitive animals.

Sponges feed on particles by filtration with sizes ranging from several hundred micrometers to smaller than bacteria (<1 μm). They feed selectively and do not digest unsuitable particles and eject them instead. Many species of sponges have algal endosymbionts that provide photosynthate and give them a bright green color. Most of these green species harbor a green alga, *Chlorella*, but some contain algae from other classes. Freshwater sponges are important food for some predators, including spongilla-flies, insects whose larvae are specifically adapted to pierce and suck fluids from sponge cells. Some freshwater sponges have an apparent mutualistic relationship with caddisfly larvae in the genus *Ceraclea*, which feed on the sponges and use their spicules to build their cases, but also incorporate living pieces of sponge in the case that are dispersed as the caddisfly moves about the habitat (Corallini and Gaino, 2003).

Macroscopically, colonies can appear as round clumps, as flattened encrusting bodies, or as finger-like growths (Fig. 10.2E). All freshwater sponges are composed of collagen and silicaceous spicules (Fig. 10.2F). Sponges have no multicellular organs, but they do have a variety of specialized cells, including epithelial cells, flagellated cells that pump water through a canal system in the sponge, and digestive cells that break down ingested particles and transport nutrients to various parts of the sponge. Sponges reproduce both sexually and asexually. Asexual reproduction can range from simple

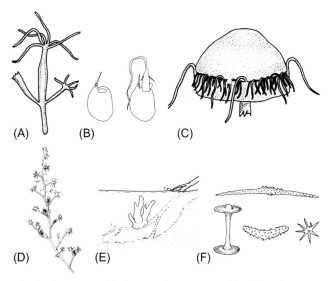

Figure 10.2 Example freshwater cnidarians and structures (A–D) and sponges and structures (E and F). Organisms and structures and their approximate lengths are as follows: (A) *Hydra*, ~5 mm; (B) discharged and undischarged *Hydra* cnidocysts, the microscopic stinging structures that contain venom; (C) *Craspedacusta* medusa, ~ 1 cm; (D) *Cordylophora* colony, 20 mm; (E) a sponge colony growing on a stick, 20 cm; (F) spicules made of silicon from several species of sponges (about 50 μm each). *A and C–F reproduced with permission from Thorp and Covich (2001); B original.*

fragmentation to formation of specialized resistant stages called gemmules that are resistant to adverse conditions such as drying and can aid in survival in marginal habitats as well as serve as a dispersal stage.

Phylum Cnidaria

The Cnidaria (formerly called Coelenterata), which includes jellyfish, anemones, corals, and hydroids, are mostly marine, but a few species of small jellyfish (Fig. 10.2C), polyps, and colonial hydrozoa occur in freshwaters, all of which are members of the class Hydrozoa (Slobodkin and Bossert, 1991; Smith, 2001). These animals are radially symmetrical, and all have cnidocytes, specialized cells that contain cnidocysts (nematocysts) that can fire variously modified filaments to capture prey and may contain strong toxins (Fig. 10.2B). Most cnidarians exhibit obvious alternation of generations between sessile (hydroid or polyp) and free-swimming (medusa) forms. Cnidarians can reproduce asexually through budding or fission, or sexually. Freshwater species often have a drought-resistant stage that allows for widespread dispersal and persistence in temporary and marginal habitats.

Hydra (Fig. 10.2A) may be the most commonly observed freshwater cnidarian occurring in streams, wetlands, and lakes. These small polyps are 1−20 mm long with 10−12 tentacles crowning a tubular body. *Hydra* species can float in the plankton or neuston, but are most commonly observed attached to hard substrata, detritus, or macrophytes in benthic habitats. *Hydra* can move across solid surfaces by "cartwheeling"—attaching tentacles, releasing the posterior end, and flipping it over the body to reattach on the other end. They can also move vertically in the water column by secreting a gas bubble that allows them to float to the surface. These small predators feed on a variety of smaller invertebrates such as copepods and cladocerans, which they paralyze with nematocysts.

As noted for sponges, *Hydra* can also appear bright green from the endosymbiotic green alga, *Chlorella*. This relationship is usually optional for *Hydra*. The alga produces photosynthate in lighted habitats but is ejected or digested during an extended period of low light when they are not useful to the *Hydra*. *Hydra* species with algal symbionts can have higher growth rates in the light than those without algae (Slobodkin and Bossert, 1991).

Cordylophora (Figs. 10.2D and 10.3) are the familiar colonial polyps that can be very abundant in brackish and freshwater habitats. Large colonies resemble moss on rocks and can reach up to 70 mm above the substrata. *Cordylophora*, along with several other hydroids and all anthozoans (corals and related marine cnidarians), have no true

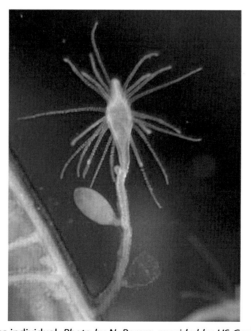

Figure 10.3 *Cordylophora* individual. *Photo by N. Rorem, provided by US Geological Survey.*

medusoid generation. *Cordylophora* are likely native to the Black Sea—Caspian Sea region, but are now considered globally invasive. In some regions, dense growths can clog industrial water intake systems. The filamentous structure of *Cordylophora* colonies may also enhance the establishment of another invasive group, dreissenid mussels (Folino-Rorem, 2015).

A unique jellyfish (*Mastigias*; Fig. 10.4A) occurs in a saline, marine-influenced lake in Palau in the West Caroline Islands (Hamner et al., 1982). The lake is stratified and has an anoxic, high nutrient hypolimnion. This medusa contains an endosymbiotic dinoflagellate. The medusa moves down at night to follow the copepods it consumes for food and it moves up during the day to allow the endosymbionts to photosynthesize. During the day, the medusae migrate up to 1 km horizontally to maximize exposure to light.

The freshwater jellyfish, *Craspedacusta sowerbii* (Figs. 10.2C and 10.4B), can also be important in linking the lower and upper strata of water in temperate freshwater lakes,

Figure 10.4 Lake jellyfish (A) *Mastigias*, a jellyfish abundant in a saline lake in Palau, and (B) *Craspedacusta* an invasive freshwater jellyfish. *Images from Creative Commons, (A) taken at the Monterey Bay Aquarium and (B) courtesy Alexander Mrkvicka.*

at least for short periods of time when the medusae are abundant. Their role in food webs can be substantial when they aggregate in swarms of more than 1,000 medusae/m^3 (Angradi, 1998; Spadinger and Maier, 1999). *Craspedacusta* occur in freshwater habitats throughout the world, and are somewhat unusual among the freshwater cnidarians in that the medusa is the more obvious, commonly observed life stage. The native range of *Craspedacusta* is the Yangtze River basin in China, but like *Cordylophora* they have spread globally and are invasive pests in many regions.

Phyla Platyhelminthes and Nemertea

The Platyhelminthes includes three classes: the Turbellaria (free-living flatworms), the Trematoda (flukes), and the Cestoda (tapeworms). This diverse phylum includes some species that reproduce only sexually, others that reproduce only through asexual means including budding and fission, and some that can reproduce either way. The phylum Platyhelminthes includes groups of significant ecological and economic importance.

The Turbellaria (Fig. 10.5H–J) are common in freshwaters, with about 400 species found throughout the world (Kolasa, 1991). Most turbellarians are predators or scavengers of dead animals, but some smaller forms are omnivores that feed on detritus, microbes, and living or dead smaller invertebrates. Some predaceous forms use secreted mucus to entrap and subdue their prey. Turbellarians do not have an anus or closed circulatory system, but they do have an intestine and a ciliated epidermis over the entire body. The Turbellaria are divided into two groups—the microturbellarians, with about 300 species, and the macroturbellarians, or Tricladida. Microturbellarians usually have cosmopolitan distributions, whereas the triclads are distributed less widely. Microturbellarians inhabit rivers, ponds, lakes, and subsurface habitats. *Microstomum* and a few other types of microturbellarians have the interesting adaptation of ingesting *Hydra* and retaining their undischarged cnidocysts, which they keep on their body surface to serve as a defense (Smith, 2001).

Triclads in the families Dugesiidae and Planariidae are probably the best known turbellarians to beginning students, as they are common in freshwater habitats and common subjects of study in introductory biology laboratories because of their amazing regenerative capabilities. These free-living flatworms are distinguished by the extension of a muscular pharynx from the mouth that ingests food. Caves and underground waters have many unique and endemic triclad species; diversity of triclads is particularly high in karst regions where species with no pigments flourish in subterranean habitats.

The flukes represent one of the major groups of animal parasites of humans. Many flukes have complex and fascinating life cycles that include multiple hosts; many rely on freshwater hosts such as snails or fishes for at least part of their life cycle. One of the more well-known flukes, *Schistosoma*, is the cause of schistosomiasis. This parasite uses pulmonate snails as the intermediate host, so understanding the epidemiology of

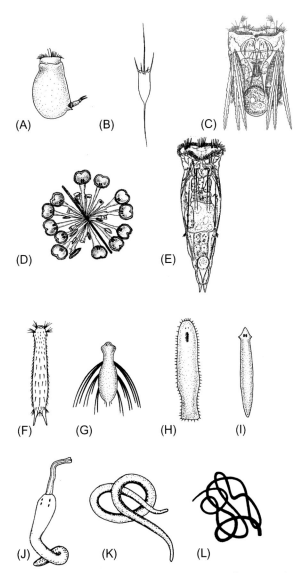

(A)   (B)   (C)

(D)   (E)

(F)   (G)   (H)   (I)

(J)   (K)   (L)

Figure 10.5 Representative rotifers (A—E), gastrotrichs (F and G), flatworms (H—I), and a nemertean (J), a nematode (K), and a nematomorphan (L). Organisms shown and their approximate lengths are as follows: (A) *Gastropus*, 0.3 mm; (B) *Kellicottia*, 0.6 mm; (C) *Limnias*, a tube-building rotifer, ~0.15 mm; (D) a colony of *Sinantherina*, colony diameter 2 mm; (E) *Epiphanes*, 0.6 mm; (F) *Chaetonotus*, 0.4 mm; (G) *Stylochaeta*, 0.4 mm; (H) *Macrostomum*, 2 mm; (I) *Girardia* (*Dugesia*), 10 mm; (J) *Protostoma* with proboscis extended, ~20 mm; (K) a nematode, 1 mm; and (L) a horse-hair worm, Nematomorpha, 10 cm. *A, B, D—H, and K—L reproduced with permission from Thorp and Covich (2001) and C and J from Smith (2001).*

this parasite requires knowledge of the ecology of snails. Humans are a definitive host of *Schistosoma*, meaning the flukes reproduce within them. Humans pass the eggs in feces. Thus, management of snails and sewage treatment are both central to controlling schistosomiasis. Schistosomiasis is one of the most significant and devastating human diseases on the planet; the United States Centers for Disease Control estimated that in 2012, 200 million people had the disease worldwide and about 20 million of those had severely debilitating infections, making it second only to malaria as the most devastating human parasitic disease globally. Swimmers' itch is caused by a related fluke that is normally parasitic on birds—in this case, the specialized cercariae fluke larvae burrow into bathers' skin, leaving an irritating, but usually harmless rash.

The nonsegmented proboscis worms (Fig. 10.5J) in the Nemertea are distinguished from the Turbellaria by having an anus and closed circulatory system. Twelve freshwater species and many more marine species have been described. All known species are benthic predators and freshwater forms generally occur in warm, shallow, heavily vegetated habitats (Kolasa, 1991; Smith, 2001). Nemerteans actively hunt; when they encounter a prey item, they repeatedly stab it with stylets located in their proboscis and further subdue the prey with sticky mucus secretions from the proboscis. Nemerteans are generally small, but large forms can exceed 100 mm in length.

Phylum Gastrotricha

The gastrotrichs can be tremendously abundant in freshwaters ($10,000-100,000$ m^{-2}), but are poorly studied (Strayer and Hummon, 1991). About 250 species in 22 genera have been described from the freshwaters of the world (Smith, 2001). Morphologically, they are about $50-800\,\mu$m long and bowling pin shaped. Gastrotrichs usually have a distinct head with sensory appendages and a cuticle covered with scales or spines (Fig. 10.5F and G).

Gastrotrichs are mainly benthic and are common in the shallow waters including littoral zones of lakes and ponds, as well as many wetland habitats including temporary pools. Gastrotrichs lay two types of eggs, tachyblastic and opsiblastic eggs; the latter has a heavier shell and can withstand drying, heat, and freezing for years, allowing them to persist in ephemeral habitats. Gastrotrichs are among the few animals that can withstand extended anoxia and concomitant exposure to sulfide (Strayer and Hummon, 1991). Gastrotrichs feed on bacteria, protozoa, algae, and detritus. Most genera have cosmopolitan distributions.

Phylum Rotifera

About 2,000 species of rotifers occur in freshwaters, and members of the phylum generally have cosmopolitan distributions. Rotifers can be found in all freshwater habitats.

They are more diverse in fresh than marine waters, and some species inhabit saline lakes (Wallace and Snell, 1991).

Rotifers are small ($\sim 60-250$ μm long) and distinguished by a ciliated head region (corona) that moves water in a circular fashion. Rotifer body forms range from worm-like to vase-shaped and some are colonial (Fig. 10.5A–E). They have a well-developed digestive system that includes a mastax to grind food, a stomach, an intestine, and an anus. Rotifers have a small brain (around 15 cells) and eyespots that allow them to respond to environmental stimuli with moderately complex behavior. Rotifers can also have a "foot" that extends ventrally and several appendages called "toes" that can be used for movement or attachment.

Some rotifers reproduce only sexually, others asexually, and yet others mostly by asexual parthenogenesis with occasional sexual reproduction. Amictic generations from parthenogenic females have no recombination of DNA. Sexual reproduction often results in formation of a thick-walled, highly resistant resting egg. Thus, sexual reproduction may be a strategy to avoid undesirable environmental conditions. It is noteworthy that bdelloid rotifers are the largest metazoan taxon completely without sexual reproduction; males have never been seen (Welch and Meselson, 2000). This verifies that sexual reproduction is not a requirement for multicellular animals to maintain stable species over the tens of millions of years that the bdelloid rotifers have been a distinct group. The bdelloid rotifers have been the recipients of many genes via horizontal transfer across broad taxonomic lines, so evolution of genomes can occur in the absence of sexual reproduction (Gladyshev et al., 2008).

Several species of rotifers are adapted to drying. In some cases, they can survive decades of desiccation and revive within minutes to hours after rewetting. Such species are important animal components of temporary waters and are well adapted for dispersal via wind or movement of waterfowl.

Many species of rotifers are omnivorous filter feeders. These species use their cilia to filter large volumes of water and the rotating motion of the water from their feeding gives the group their name. Some benthic species act as "sit and wait" predators that engulf prey when it swims near (Wallace and Snell, 1991). Rotifers feed selectively and reject unsuitable food particles. Rotifers are important consumers of bacteria in aquatic habitats and are therefore important links between the microbial community and higher trophic levels. Rotifers are a major food source for zooplanktivores, including many small crustaceans and insects, but they are generally too small for planktivorous fish to capture. Some species produce long spines in response to predation risk; consequently, body form can vary considerably (Fig. 10.6). Other species have adaptive behavioral responses to avoid predators such as making rapid jumps when touched.

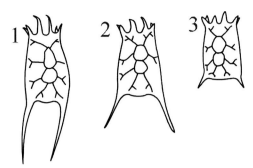

Figure 10.6 Change in body form (cyclomorphosis) of the planktonic rotifer *Karatella quadratica* in successive generations in laboratory culture after Hutchinson (1967).

Phylum Nematoda

Nematodes or roundworms inhabit freshwaters, soils, and marine habitats. Probably because of their small size and complex taxonomy, they have not received much attention in freshwater habitats, but their importance in freshwaters and all other habitats is undisputed. Two thousand freshwater species were reported in 1978 (Pennak, 1978), and about 2,500 by 2008, which was 7% of the total estimated number of species (Abebe et al., 2008). Nematodes are usually benthic but occur in most aquatic habitats, including those as extreme as hot springs and snowmelt pools (Poinar, 1991). Nematodes often reach densities of 1 million/m^2 in freshwater benthic habitats.

The nematodes are nonsegmented, worm-like, cylindrical in cross section, and possess a complete alimentary tract and a body cavity (Fig. 10.5K). They also have a well-developed nervous system, excretory system, and musculature. Most freshwater nematodes are less than 1 cm in length. Reproduction is parthenogenic in some species, and sexual reproduction can occur as well.

A wide variety of feeding strategies occur among the nematodes. Some feed on detritus, some feed on algae, many feed on aquatic plants, and some species are carnivorous. Predatory nematodes may be the biggest consumers of other species of nematodes in their natural habitats. Nematodes can be important parasites of humans, other animals, and plants. In freshwater habitats, many species are endoparasites in other invertebrates as well as vertebrates. Species that are parasitic on humans use mosquitoes, black flies, or other dipterans as intermediate hosts.

The nematode that causes river blindness (onchocerciasis) is transmitted by black flies (Simuliidae); the World Health Organization estimated that in 2008, 17.7 million people were infected, 270,000 were blind and about a half million had visual impairment, most living in Africa. There has since been considerable progress in lowering disease rates with using aerial spraying of larvicides to kill the blackfly hosts and treatment of infected individuals. The goal of complete eradication seems possible. For example, the disease occurred in Columbia, probably introduced in the slave trade, but was declared eliminated in 2013.

Phylum Nematomorpha

Members of the Nematomorpha are known as horsehair worms or Gordian worms (Fig. 10.5L) and are parasites. Some species are parasitic on humans, but invertebrates and other vertebrates generally serve as hosts. The free-living adults are several centimeters to 1 m long and about 3 mm wide. They often appear like thin sticks or thick hairs and nonliving until they start to move. Their typical life cycle includes adults that reproduce sexually and lay eggs. The eggs hatch and the larvae then encyst on vegetation where they may be ingested by a host animal (commonly an invertebrate, although not always an aquatic host; grasshoppers and crickets are common hosts). Sometimes, another predator engulfs this host, and the parasitic larva infects the predator.

The horsehair worm host must be in contact with water for the mature adult to emerge from the host's body cavity. At this stage of development, the horsehair worm can be almost as large as its host. Horsehair worms are one of a few types of parasites that can alter the behavior of their host. Hosts infected with horsehair worms exhibit water seeking behavior and thus the chance that a parasitized host ends up in a water body is much higher than for uninfected individuals (Thomas et al., 2002). The effects of the parasite on host behavior are apparently reversible and some hosts can survive the emergence of the adult from its body, but the host often dies from drowning once it ends up in the water (Ponton et al., 2011). Although reproductive, the adult never ingests food and has no functional digestive tract.

Phylum Mollusca

The freshwater mollusks include two classes, the Gastropoda (snails and limpets) and the Bivalvia (clams and mussels). Mollusks are widespread, conspicuous, and often abundant. They are soft-bodied, unsegmented animals. Their body has a head, a muscular foot, a visceral mass, and a mantle that often excretes a calcareous shell. They form an important part of the biodiversity and food webs of many aquatic ecosystems.

The gastropods constitute the most diverse class of the phylum Mollusca, with about 75,000 species of marine and freshwater snails worldwide. Freshwater snails are very diverse in North America, with about 500 species present, and 3,700 species globally (Strong et al., 2008). The gastropods have a univalve (one-piece) shell and a file-like radula that scrapes surfaces while feeding (Brown, 1991). Shell geometry can range from simple and conical as in the freshwater limpets (Fig. 10.7F) to spiral and flat (*Planorbella*; Fig. 10.7E) or spiral and elevated (Fig. 10.7D). Most snails reproduce sexually, but a few are parthenogenic. Many freshwater species are hermaphroditic (both male and female in same organism); hermaphroditic species usually reproduce

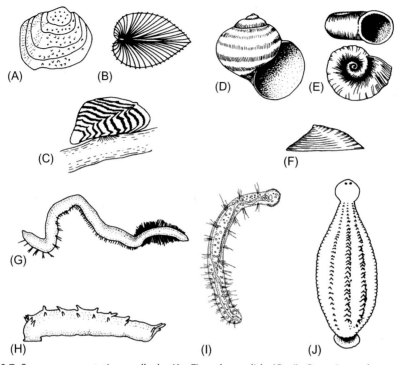

Figure 10.7 Some representative mollusks (A—F) and annelids (G—J). Organisms shown and their approximate lengths are as follows: (A) a unionid mussel, *Quadrula*, 7 cm; (B) the Asiatic clam, *Corbicula*, 3 cm; (C) a zebra mussel, *Dreissena* on a stick, 3 cm; (D) *Pomacea*, 4 cm; (E) *Planorbella*, 3 cm; (F) the freshwater limpet, *Ferrissia*, 4 cm; (G) *Branchiura*, 10 cm; (H) *Ceratodrilus*, 3 mm; (I) *Aeolosoma*, ~6 mm; and (J) *Placobdella*, 16 mm. A—H and J reproduced with permission from Thorp and Covich (2001). I reproduced with permission of Smith (2001).

sexually by exchanging sperm with other individuals, although some species are capable of self-fertilization. Reproductive activity ranges from once per year to continuous, depending on the species and the geographic region.

Snails generally feed on detritus, periphyton and biofilms, macrophytes, and occasionally carrion. They usually prefer periphyton to macrophytes (Brönmark, 1985) and can be major consumers of periphyton in aquatic systems. In the process of grazing, snails can greatly alter the composition and structure of periphyton assemblages and can even stimulate growth by reducing light limitation and excreting nutrients and mucus that fertilize periphyton and biofilms (Steinman, 1996a,b). Despite the protection of their shell, there are many important predators of snails. Some, such as crayfish and some species of sunfish, crush the shells. Other predators, such as leeches, flatworms, and some aquatic insect larvae, invade the shells (Brown, 1991). Because shells get thicker and stronger as snails grow, predation by predators that crush the shells, such as crayfish, is often size dependent, with smaller individuals more vulnerable (Crowl and Covich, 1990). Freshwater snails are intermediate hosts for some important parasites such as the human lung fluke (*Paragonimus*) and blood flukes (*Schistosoma*).

A shell with two halves and enlarged gills with long ciliated filaments used for filter feeding (McMahon, 1991) characterize the freshwater bivalves. Most of the native North American bivalves burrow in sediments. The bivalves inhabit the benthic zone of streams, lakes, and rivers. Several bivalve species have recently invaded North America, including the Asiatic clam *Corbicula fluminea* (Fig. 10.7B), the zebra mussel *Dreissena polymorpha* (Fig. 10.7C), and the quagga mussel *Dreissena bugensis*. The zebra mussel has caused significant economic and ecological damage (Highlight 10.1).

Highlight 10.1 Invasion by the zebra mussel

The zebra mussel was established in the Volga drainage in Europe from its native Caspian Sea drainages before 1800. It spread from the Volga throughout most of the major river systems in Europe through the canal transportation system following introduction to some far western European countries in the early 1800s (Hutchinson, 1967). This invasion represents the first major expansion of its range caused by human activity. In 1986, a ship with water from a European port, probably taking on cargo at the St. Claire River, dumped ballast water and released the zebra mussel into North America. Since that time, the population has increased its distribution to cover much of the Mississippi drainage (Fig. 10.8). This explosive spread has occurred partly because females can produce more than 1,000,000 eggs each reproductive cycle that give rise to easily transportable veliger larvae (de Vaate, 1991; Sprung, 1993) and partly because of the many ways the mussel can be transported. Carlton (1973) reports two natural transport methods (currents and animals other than humans), and 20 human-caused movements, including water traffic, fisheries activities, and navigation. Ultimately, this species probably will spread through much of North America (Strayer, 1991). These mussels attach tightly to any solid surface with byssal threads, can reach tremendous densities (Fig. 10.9), and overgrow native species. This invader has many potential effects on food webs and ecosystems (Table 10.1).

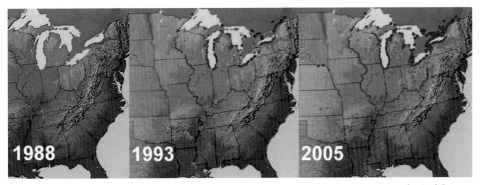

Figure 10.8 Spread of the zebra mussel from 1988 to 2005. *Images courtesy US Geological Survey.*

Figure 10.9 Zebra mussels foul a current meter that has just been removed from a lake. *Image courtesy US National Oceanic and Atmospheric Administration.*

The Hudson River provides a good case study for examining the effects of the zebra mussel on ecosystem properties (Strayer et al., 1999). The zebra mussels were first observed in the Hudson River in 1991 and by 1993, densities were high enough that the mussels filtered the entire water column every 1.2—3.6 days. The adult mussels are filter feeders and can remove suspended particles including phytoplankton, small zoo-plankton, and detritus. Amounts of all these suspended particles decreased concomitantly with mussel increases. This consumption ultimately may influence fish by lowering the amount of food available to animals that feed on suspended particles and serve as food for piscivorous fishes. The increase in zebra mussels leads to a decrease in native bivalves, particularly native unionid mussels. The zebra mussels will probably cause extirpation of two of the species of unionid mussels in the Hudson River. Dissolved phosphorus and water clarity have increased since the invasion of the mussels, probably leading to increases in macrophytes. The various ecosystem effects can be viewed as positive or negative. For example, an increase in water clarity may be good, but the effects are obviously disastrous for some native species.

Another example from the North American Great Lakes illustrates the massive shifts in ecosystem structure that can occur with introduction. In this case, the

Table 10.1 Some community, food web, and ecosystem-level effects of the zebra mussel, *Dreissena polymorpha*

Effect	Citation
Lower phytoplankton biomass, greater water clarity	Fahnenstiel et al. (1995), Heath et al. (1995), Holland et al. (1995), and Makarewics et al. (1999)
Decreased dissolved oxygen from mussel metabolism	Caraco et al. (2000) and Effler et al. (1998)
Increases in total benthic biomass	Higgins and Vander Zanden (2010)
New food source for waterfowl; effective conduit for biomagnification of organic toxicants	Mazak et al. (1997)
Increased variability in phytoplankton biomass and phosphorus concentrations	Mellina et al. (1995)
Increased abundance and depth of macrophytes related to decreased turbidity	Skubinna et al. (1995)
Declines in native unionid bivalves, and clams	Nalepa (1994), Strayer et al. (1998)
Increases in periphyton biomass and productivity related to increased water clarity, about equal to phytoplankton decreases	Fahnenstiel et al. (1995) and Lowe and Pillsbury (1995)
Increased numbers of small planktonic heterotrophic bacteria, decreased numbers of predatory protozoa	Cotner et al. (1995) and Findlay et al. (1998)
Increased nitrogen regeneration rates	Gardner et al. (1995) and Holland et al. (1995)
Decreases in planktonic protozoa	Lavrentyev et al. (1995)
Moderate effect on cladoceran grazing rates in open, deep waters of a large lake	Wu and Culver (1991)
Large initial impacts on ecosystem structure, followed but weaker and less predictable ecosystem effects	Karatayev et al. (2015)

planktonic production decreased from 5% to 45% after introduction of the mussels, and benthic production increased up to 190%. This caused a large shift in the location and type of production in these lakes, which will also shift the food web in favor of benthic species (Brothers et al., 2016).

Zebra mussels have caused chronic problems in water intakes in both Europe and the United States (Kovalak et al., 1993). Layers of attachments up to 30 cm thick can clog pipes and screens. Removal efforts are time intensive and costly, usually involving dewatering (shutting down and drying) water systems and cleaning with high-pressure water hoses (Kovalak et al., 1993). Chemical controls such as treatment with chlorine and other toxic chemicals are effective in confined areas (e.g., water pipes), but toxic when released to the environment. The economic impact was estimated to be $5 billion in the Great Lakes region alone by the year 2000 (Ludyanskiy et al., 1993), and other surveys put actual costs at a minimum of $10 million/year (O'Neill, 1997). Biological controls are being investigated, but no inexpensive method is available that has been widely adopted. Biological controls may be somewhat effective; crayfish (Perry et al., 1997), waterfowl (de Vaate, 1991), and some fishes prey upon zebra mussels (in particular introduced species, but also native drum, sunfishes, and redhorses; French, 1993). There is no good way currently to control zebra mussels at the ecosystem level.

Unionid mussels are the most diverse group of bivalves found in freshwaters of North America (Fig. 10.7A). Unionids are unique in that they produce specialized larvae called glochidia, which are important for dispersal. These larvae attach to host fish species, mainly in the gill region (but some attach to fins), and encyst for 6–60 days. After encystment, the larvae drop off the host, settle into the substrata, and develop into sedentary adult forms with shells. Some adult female mussels attract potential hosts with elaborate lures that are muscular extensions of the mantle. These lures may look like small fish or other probable prey items, and when potential host fish attack them the glochidia are released forcibly (Fig. 10.10). The unionid mussels are also unique because of their long life span. Individuals can live from 6 to 100 years; other bivalves live less than 7 years. The long life span and infrequent reproduction make unionids vulnerable to human impacts on streams. Of the 297 native species and subspecies of North American mussels, 19 are extinct, 62 are federally listed in the United States as endangered or threatened, and 130 need further study to determine their conservation status. These species mediate many ecological processes through their role as filter feeders (Vaughn and Taylor, 1999; Hoellein et al., 2017). We discuss conservation of mussels and other species in Chapter 11. At one time, unionids were harvested in large quantities for the button industry (Fig. 10.11), and they are still the basis of a historically important pearl and shell fishery that continues at modest levels.

Phylum Annelida

Annelids are segmented worms with a tubular body and a specialized digestive system with a terminal mouth and anus. Their body cavity has thin transverse septa that delineate the segments. They generally reproduce sexually by cross-fertilization and are often hermaphroditic, but many also reproduce asexually by budding. The freshwater annelids include the oligochaetes, the leeches, and several other less diverse groups.

Aquatic oligochaetes (Fig. 10.7G) are similar to their terrestrial analogs (earthworms). They usually have four bundles of chaetae (hairs) on each segment (Brinkhurst and Gelder, 1991). Most of these worms burrow through sediment in lotic

Figure 10.10 A red-eye bass (*Micropterus coosae*) "attacking" the lure of a freshwater mussel (*Lampsilis cardium*). (A) View of the gravid mantle that serves as a lure. The fish approaches the lure (B). After the fish bites the mantle (C), the glochidia are released in a cloud and the fish rapidly leaves (D). The mussel is about 6 cm long. *From Haag and Warren, Jr. (1999). Images courtesy of Wendell R. Haag, U.S. Forest Service.*

Figure 10.11 Pearl shell buttons and a *Megalonaias nervosa* mussel shell from the Mississippi River that was drilled for buttons. *Photograph by J. W. Grubaugh.*

and lentic habitats and ingest organic particles, but some are important algal feeders or predators. The aquatic oligochaetes have a thinner muscular layer around their body, so fisherman use terrestrial oligochaetes (earthworms) for bait more often as they will not disintegrate on a hook. Some oligochaetes, particularly the tubificid (Naididae) worms such as *Tubifex*, are highly resistant to low dissolved oxygen concentrations and high levels of organic pollution. Thus, they indicate polluted waters and can be components of biotic indices to assess ecosystem health. Oligochaetes can also be vectors for important parasites such as whirling disease, which infects trout in North America, Europe, South Africa, and New Zealand.

Members of the annelid class Polychaeta mostly inhabit marine habitats, where they can be very abundant and tremendously diverse in size and form. In contrast, only a few species of polychaetes make freshwater habitats their home, including *Aeolosoma* (Fig. 10.7I). Members of the genus *Aeolosoma* grow in freshwater habitats throughout the world, including open water habitats of large lakes, stream sediments, and groundwater habitats, where they feed on fine organic particles. Many polychaetes are predators or omnivores, but some have specialized appendages for filter feeding.

Leeches (Hirudinea) are mostly predators that feed on midge larvae, amphipods, oligochaetes, and mollusks, but some are parasites that feed on the blood of vertebrates, including humans. Species of leeches that use blood are being investigated for the pharmacological value of the anticoagulants used during feeding (Davies, 1991; Salzet, 2001). They are used medically as a good option for removing congealed blood around a wound.

Leeches are distinguished by several characteristics, including dorsal—ventral flattening (Fig. 10.7J), an oral sucker and usually a posterior sucker, usually 34 true body segments, and a muscular body. Leeches commonly inhabit shallow, warm waters, including slow-moving streams and rivers, lakes, and wetlands (but generally not acid peat bogs). Some species live in moist terrestrial habitats and can be quite large, reaching nearly 30 cm in length.

Leeches can sense moving animals in the water and swim vigorously toward them. When parasitic leeches attach to prey for a blood meal, they attach with the posterior sucker and explore for a suitable feeding spot with the anterior end. They then attach their oral sucker, make three painless cuts with the jaws, and inject anticoagulants. The leech eats its fill and then drops off the host. Frequent meals are not necessary, and specimens can live without feeding for more than 2 years (Smith, 2001).

Members of one annelid order, the Branchiobdellida (sometimes called crayfish worms; Fig. 10.7H) are exclusively commensals and parasites of crustaceans, particularly crayfish. Branchiobdellids range from 1 to 10 mm in length, and an individual crayfish can harbor hundreds of individuals representing multiple species. Most branchiobdellids feed on organic materials and smaller organisms on the surface of their host, or food particles from the host's feeding, and actual parasitism in this group is rare.

Phyla Ectoprocta and Entoprocta: Bryozoans and kamptozoans

Bryozoans (Ectoprocta) and kamptozoans (Entoprocta) are generally *sessile* (attached to substrata) colonial invertebrates that use ciliated tentacles to capture suspended food particles. These groups are primarily marine, with more than 4,000 species worldwide, about 50 of which live in freshwater habitats (Smith, 2001). Colonies are composed of many individual zooids, each of which are approximately tubular and have a crown of tentacles (Fig. 10.12A−D). Although sometimes lumped together, these two groups are actually distantly related phyla; the Entoprocta (Fig. 10.12D) have zooids with external segmented stalks and lack a coelom, and the Ectoprocta (Fig. 10.12A and B) are coelomates that lack external stalks. Both can reproduce sexually or asexually through budding and fission, and individual zooids are hermaphroditic. Statoblasts (Figs. 10.12C and 10.13), the small, resistant structures of bryozoans that are formed through budding, are important for dispersal and surviving harsh conditions. Bryozoans and kamptozoans are generally restricted to warm water and dwell in both lentic and lotic habitats (Wood, 1991). They require solid substrata such as rocks or wood for attachment. Colonies can reproduce asexually by formation of encapsulated dormant buds, and most can reproduce sexually once a year. Members of both phyla are resistant to predation. Large, gelatinous *Pectinatella* colonies, which can exceed the size of a softball, are a common and spectacular sight in some lakes and reservoirs in the southeastern United

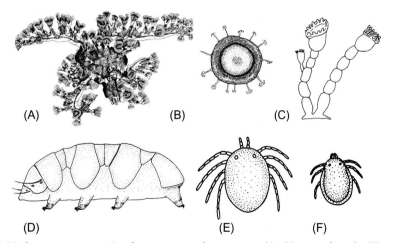

(A)   (B)   (C)

(D)   (E)   (F)

Figure 10.12 Some representative bryozoans and structures (A−D), a tardigrade (E), and water mites (F−H). Organisms and structures shown and their approximate lengths are as follows: (A) *Hyalinella* colony, ∼2 cm wide; (B) bryozoan statoblast, 1.4 mm; (C) *Urnatella* colony, 5 mm; (D) heterotardigrade, 0.3 mm; (E) generalized adult water mite, 1 mm; (F) generalized larval water mite, 0.5 mm; *A reproduced with permission of Smith (2001). B−F reproduced with permission from Thorp and Covich (2001).*

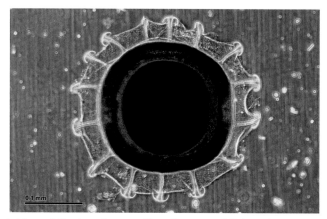

Figure 10.13 Statoblast of *Pectinatella magnifica. Courtesy of Wikimedia Commons, photo by Josef Reischig.*

States where they are sometimes called "dinosaur snot." Bryozoans can be very abundant in lentic habitats and, in some cases, can be a nuisance because they clog pipes in water treatment facilities.

Phylum Tardigrada

Water bears (tardigrades) are microscopic animals with a cosmopolitan distribution (Nelson, 1991); about 900 species are reported worldwide, but only 62 of these are truly aquatic (Garey et al., 2008). They have a bilaterally symmetrical body with four body segments, each with a pair of legs (Fig. 10.12E). Most adults are tiny, ranging from 250 to 500 μm long. Some tardigrade species reproduce through parthenogenesis, while others have sexual reproduction; many are hermaphroditic.

Tardigrades are well known for their ability to withstand drying, even for periods up to 4−7 years with no apparent biological activity, a characteristic related to their unusual habitat preferences. Tardigrades rarely are planktonic; aquatic species mainly inhabit benthic habitats in which they can reach very high densities, notably in the capillary water in wet sand of beaches. Most species are semiaquatic and are associated with droplets of water and water films on terrestrial mosses and liverworts. They are some of the only animals found in extreme habitats, such as Arctic and Antarctic lakes, streams, and pools on ice sheets (McInnes and Pugh, 1998; Pugh and McInnes, 1998).

Phylum Arthropoda

Arthropods are one of the most successful groups of organisms on the planet and are well-represented in all continental surface waters. They are important components of aquatic biodiversity and central to ecosystem function. They have an exoskeleton and stiff jointed appendages (including legs, mouthparts, and antennae). There are two subphyla common in freshwater habitats: Chelicerata (water mites and aquatic spiders) and Pancrustacea, a clade comprising the hexapods (Insecta and Collembola), and all groups that are often referred to as crustaceans (crayfish, shrimp, amphipods, isopods, mysids, fairy shrimp, cladocerans, and copepods).

Subphylum Chelicerata: Water mites and spiders

Water mites are a taxonomically diverse and poorly studied group including several families. More than 6,000 species of water mites have been described worldwide (Di Sabatino et al., 2008); about 1,500 occur in North America, but only half of them are named. They can be extraordinarily diverse, with as many as 75 species representing 25 genera/m^2 in plant beds in eutrophic habitats, and 50 species from 30 genera in a single stream riffle (Smith and Cook, 1991).

Water mites are quite small, generally between 0.5 and 1.5 mm, and have a mouth region and a body that has a fused cephalothorax section (head and thorax) and an abdomen (Fig. 10.12F—H). Six pairs of appendages are present; the last four pairs, the legs, are the most conspicuous. The appendages can have setae, or spines, which taxonomists use as characteristics for identification.

Water mites inhabit a variety of benthic habitats, including springs, riffles, interstitial habitats, lakes (with mostly benthic but a few planktonic forms), and temporary pools. The majority of water mites are carnivorous or parasitic (primarily on aquatic insects). Species that are more sedentary may feed on carrion or possibly detritus.

Among the true spiders, no North American species are completely aquatic. However, several species live near water and are able to run on the water surface and even dive beneath it. Many spiders, particularly members of the genus *Tetragnatha* (long-jawed spiders) build their webs close to water on emergent vegetation and other structures to catch adult aquatic insects. Spiders can be important consumers of insects in riparian habitats and wetlands. Fishing spiders, members of the genus *Dolomedes*, are common in and along lake and stream margins where they are active predators on aquatic insects and small vertebrates such as small fish and amphibians. They trap air in fine hairs on their bodies and subsequently can remain submerged up to 30 minutes. The European water spider *Argyroneta aquatica* (Clerck) and a closely related Japanese species spends its entire life under water. These spiders keep an air bubble around their body for breathing and build silk under-water air containers. The spiders use the containers for shelter and nests (Schutz et al., 2007).

Subphylum Pancrustacea: Hexapods: insects and collembolans

Many insect orders contain aquatic or semiaquatic species (Merritt et al., 2008). Insects are found in most freshwater habitats, where they are arguably the best studied group of freshwater invertebrates. The majority of the aquatic insects spend most of their immature lives in the water, and the adults emerge from the aquatic environment to mate and disperse. Insects have three major body regions (head, thorax, and abdomen), one pair of antennae, compound eyes (as adults), and specialized mouthparts. The thorax has three segments, each with a pair of legs, with each leg divided into five parts. The immature forms of aquatic insects are variously referred to as *larvae*, *nymphs*, or *naiads*. The term larva is generally applied to immature forms of groups with complete metamorphosis and nymph is generally used to describe immatures of those with incomplete metamorphosis, although opinions on this terminology vary. We summarize some characteristics of the orders of insects with aquatic or semiaquatic species in Table 10.2 and discuss each of the orders in turn.

The Collembola, or springtails, are no longer considered insects by most taxonomists. There are about 525 water-dependent species (Deharveng et al., 2008). They are small, eyeless and wingless arthropods (usually less than 6 mm long) that possess a

Table 10.2 Characteristics of orders of aquatic insects and Collembola

Order	Common name	Approximate number of aquatic and semiaquatic species	Primary habitats	Functional feeding groups
Collembola	Springtails	50	Lentic, shallow	Collectors
Ephemeroptera	Mayflies	2,250	Lentic and lotic	Scrapers, collectors, few predators
Odonata	Dragonflies and damselflies	5,500	Lentic and lotic	Predators
Orthoptera	Grasshoppers and crickets	~100	Mostly lentic	Shredders
Plecoptera	Stoneflies	2,140	Lentic mainly	Shredders, collectors, predators
Trichoptera	Caddisflies	7,000	Lentic and lotic	Predators, scrapers, collectors
Megaloptera and Neuroptera	Fishflies, alderflies, and spongillaflies	300		Predators
Hemiptera	Bugs, plant hoppers, and others	3,800	Lentic and lotic	Predators
Lepidoptera	Aquatic caterpillars	100	Lentic and lotic	Shredders, scrapers
Coleoptera	Water beetles	5,000	Lentic and lotic	Predators, scrapers, collectors, shredders
Hymenoptera	Ants, bees, and wasps	100	Lentic and lotic	parasites
Diptera	Flies and midges	>30,000	Lentic and lotic	Predators, scrapers, collectors, shredders

characteristic ventral tube (collophore) that functions in respiration and osmoregulation and can also be adhesive (Fig. 10.14A). The name springtail refers to the ability of some collembolans to jump using a specialized bifurcate structure on the abdomen called a furcula. They differ from insects by having only six abdominal segments and the mouthparts are withdrawn into a pouch in the head capsule. Most species are terrestrial or semiaquatic and occur in lentic habitats, but they can also be abundant along stream margins. Some occur in ice or snow habitats. Collembola are commonly neustonic, where their hydrophobic integuments allow them to rest on the surface tension. Their biology and ecology in freshwaters are not well-known, but they are generally considered primary consumers that feed on detritus, algae, fungi, and/or biofilms.

Mayflies (Ephemeroptera) are often dominant components of the aquatic invertebrate communities of streams and benthic habitats of some lakes. Mayfly nymphs are distinguished from other aquatic insects by long filaments on their posterior end (generally three) and the presence of conspicuous gills on the abdominal segments (Fig. 10.14B and C). The abdominal gills of mayfly nymphs range from plate-like, to feathery, to long filaments. Another unique feature of mayflies is the subimago stage. This "extra step" in the mayfly life cycle follows the larval stage and precedes the adult stage. The subimago stage is the only winged preadult stage known in insects. Most mayfly nymphs feed on periphyton or biofilms, but some filter organic particles from the water and a few are predaceous. Mayfly nymphs generally crawl about on the substratum, but some species are rapid swimmers and others construct burrows. Mayflies are diverse in well-oxygenated, unpolluted streams and generally indicate good water quality and ecosystem health. Adult mayflies do not feed and live for only a few days while they attempt to reproduce. Emergences of mayflies (and other insects) indicate to fly fishers which fly patterns are most likely to attract fish. Massive emergences of millions of mayfly adults in spring and summer in some regions are a spectacular site; densities can be so high that they stop traffic and can be detected with Doppler weather radar (Fig. 10.15).

The Odonata (dragonflies and damselflies) are fantastic predators as both aquatic nymphs and terrestrial adults. About one-third of the nymphs are lotic and two-thirds are lentic (Hilsenhoff, 1991). The nymphs can be distinguished from other aquatic insects by the long, hinged labium that has been modified to eject rapidly and seize moving prey. Nymphs have large compound eyes and short antennae (Fig. 10.14D−F) and move about the substrata by crawling; some stalk their prey, which can include small fish and tadpoles along with other invertebrates. To escape predators, nymphs can forcefully eject water from the rectum as a form of jet propulsion. Adults are able to fly at 25−35 km/hour, are significant predators of mosquitoes and other insect, and are thus beneficial. Adult odonates are some of the most familiar insects associated with freshwaters and many have beautiful colors and patterns

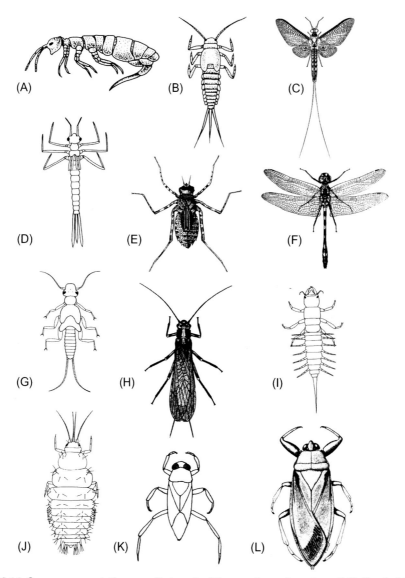

Figure 10.14 Some representative aquatic insects: (A) a semiaquatic springtail (Collembola), 1 mm; (B) *Baetis* mayfly nymph (Ephemeroptera), 1 cm; (C) adult *Hexagenia* mayfly (Ephemeroptera), ~3 cm; (D) a damselfly nymph (Odonata), *Calopteryx*, 1 cm; (E) a dragonfly nymph, *Macromia* (Odonata), ~5 cm; (F) an adult dragonfly, *Macromia*, ~8 cm; (G) a stonefly nymph, *Isoperla* (Plecoptera), 0.7 cm; (H) an adult stonefly, *Clioperla*, ~3 cm; (I) an alderfly larvae, *Sialis* (Megaloptera), 2 cm; (J) a spongilla fly larvae, *Climacia* (Neuroptera), 0.5 cm; (K) an adult backswimmer, *Notonecta* (Hemiptera), ~2.5 cm; (L) an adult giant water bug, *Lethocerus* (Hemiptera), ~7 cm. *A, B, and J reproduced with permission from Thorp and Covich (2001). C, E, F, H, and L reproduced with permission from Borror et al. (1989) and D, G, I, and K reproduced with permission of Hilsenhoff (1991).*

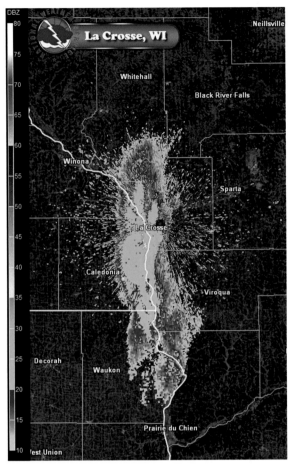

Figure 10.15 A dense mayfly hatch from the Mississippi River near La Crosse, Wisconsin indicated by Doppler weather radar. *Image courtesy US National Weather Service.*

(Fig. 10.14F). Mating systems of odonates are fascinating and involve intense sexual competition; males of some species have copulatory organs designed to destroy or displace sperm from males that have mated with a female before them (Cordoba-Aguilar et al., 2003).

The order Orthoptera (grasshoppers, crickets, and others) does not include any truly aquatic forms, but some species are adapted for a semiaquatic existence. Semiaquatic crickets and grasshoppers general associate with the emergent aquatic plants that they feed on, or sometimes with damp marginal substrata. Many have the ability to swim on the surface of the water. Some tropical forms can swim under the water surface, remain submerged for a few minutes, and lay their eggs in underwater portions of plants.

The Plecoptera (stoneflies) are important in streams as food for fish and other vertebrates. Some species are predators of other invertebrates, while others are important leaf shredders, which facilitate the decomposition and recycling of materials. Stoneflies are associated primarily with pollution-free, cool, highly oxygenated running waters. This preference for clean habitats has led to their use in biotic indices for stream water quality and biotic integrity. Plecopterans are unusual in that although they generally inhabit streams with high water quality, they are somewhat tolerant of acidification (Herrmann et al., 2003). The nymphs are similar in body form to mayflies, but can be distinguished from other aquatic insect nymphs by the presence of two long cerci (appendages) on the posterior end of the abdomen and their elongate, often flattened bodies (Fig. 10.14G and H). The gills of stonefly nymphs are generally less conspicuous than those of mayflies or, if they are obvious, usually appear as small tufts or finger-like projections.

The order Hemiptera includes the suborders Heteroptera (true bugs; Fig. 10.14K and L) and Homoptera (plant hoppers, aphids, and others), most of which are terrestrial. About one-third of the aquatic or semiaquatic species live on the water surface and two-thirds live in the water. All breathe air, so those that live underwater generally use siphons or air bubbles to breathe. This group is well represented in lentic and lotic habitats, particularly in highly vegetated areas, and they are tolerant of low dissolved oxygen concentrations in the water. The Hemiptera are distinguished by mouthparts that are modified to form a sucking and piercing beak, which is used to pierce plant tissues in herbivorous forms or feed on the blood of animals in carnivorous species. This group includes some of the most familiar aquatic insects, including the giant water bugs and water scorpions (big enough to prey on small fishes and tadpoles), the water boatmen (often seen in shallow littoral zones), and the gerrids (water striders). Water boatmen can be extremely abundant in lentic habitats, where they represent a very important food for fishes. Humans collect, dry, and eat water boatmen in some regions of Mexico and Egypt.

The order Neuroptera used to include what is now considered a separate order, the Megaloptera (fishflies, dobsonflies, alderflies, and hellgrammites). Neuropterans are mostly terrestrial, except for spongillaflies, which live in and feed on freshwater sponges (Fig. 10.14J). Megalopteran larvae have seven or eight pairs of lateral filaments and large mandibles (Fig. 10.14I). All megalopteran larvae are predators, mostly on other invertebrates. Larval forms with abdominal gills generally inhabit cool, oxygen-rich streams, but some species with respiratory siphons live in wetlands and other lentic habitats. Some megalopteran larvae (*Corydalus*, the hellgrammites) can be more than 6 cm long and live for 2—5 years before emerging. Megaloptera crawl from the water to pupate under logs or leaf litter. The pupae are unusual in that they are fully motile and have fully functional mandibles that they can use to defend themselves from predators (Fig. 10.16).

Figure 10.16 Exarate pupa of an eastern dobsonfly (*Corydalus cornutus*). *Courtesy University of Florida, photo by D.W. Hall.*

Caddisflies (Trichoptera; Figs. 10.1 and 10.17D—F) are best known for their ability to build cases (Fig. 10.18), retreats, and nets that they construct using silk and materials from the environment such as sand grains or parts of leaves. Some species are free living, highly mobile, and lack cases or nets. Most species occur in running waters, but some occur in lakes and wetlands. Along with the mayflies and stoneflies, a diverse caddisfly community indicates a clean stream or river. The free-living caddisflies are mostly predators, those that build cases are primarily grazers of periphyton or eat dead leaves, and species that spin nets are filter feeders. Net-spinning caddisflies use stream flow to filter organic materials from the water column onto their nets. Adult caddisflies look like moths, which reflects the close relationship between caddisflies and the Lepidoptera (butterflies and moths), along with the ability of the larvae to produce silk. Caddisflies can reach very high densities in some freshwater habitats, particularly filter feeding species in streams with abundant stable substrata for them to build nets on. Although massive emergences of adults can be a nuisance to humans, emergences can signal anglers that fish may be actively foraging and lures that approximate adult caddisflies could be successful.

The order Lepidoptera (butterflies and moths) is not diverse in freshwaters, but can be abundant in some stream and lentic habitats. Those that live in streams generally graze on periphyton, whereas those that live in lakes and wetlands often feed on aquatic vascular plants and may bore into stems and leaves. Larvae are similar in form to terrestrial caterpillars, but some forms that live in streams have prominent filamentous gills. Like the caddisflies, this group has the ability to produce silk, which they use to build retreats in stream-dwelling forms and cases in lentic forms. Like their terrestrial counterparts, aquatic caterpillars pupate and emerge as aerial adults.

The Coleoptera (beetles) species with aquatic larval and/or adult stages (Fig. 10.17G—I) represent only about 3% of this mostly terrestrial order. However, there are so many species of beetles in the world that they are significant components

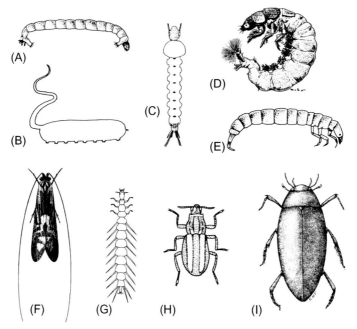

Figure 10.17 Some representative aquatic insects (A) a midge larva, *Chironomus* (Diptera) 0.5 cm; (B) a syrphid fly larvae, *Eristalis* (Diptera) with respiratory siphon extended, ∼5 cm; (C) a mosquito larvae, *Anopheles* (Diptera), ∼1.5 cm; (D) a hydropsychid caddisfly larvae, *Hydropsyche* (Trichoptera), ∼2 cm; (E) a caddisfly larva, *Polycentropus*, 1.5 cm; (F) an adult caddisfly, *Macronemum*, ∼2.5 cm; (G) a whirligig beetle larva, *Dineutus* (Coleoptera), 1 cm; (H) a riffle beetle *Stenelmis*, 0.5 cm; and (I) a hydrophilid beetle, *Hydrophilus* (Coleoptera), ∼2.5 cm. *A−C, E, and H reproduced with permission from Thorp and Covich (2001). D, F, and I reproduced with permission of Borrer et al. (1998); and G reproduced with permission of Hilsenhoff (1991).*

of the biodiversity of both lentic and lotic habitats. The group includes the riffle beetles (Elmidae, with aquatic larvae and adults), water pennies (Psephenidae larvae, attached to rocks), whirligig beetles (Gyrinidae, with adults found on the surface), and predacious diving beetles (Dytiscidae) and water scavenger beetles (Hydrophilidae), both of which have aquatic larvae and adults. Some adult Coleoptera, along with some Hemiptera, transport air bubbles or air films (plastrons) under water by means of specialized hairs. Underwater, these air bubbles can function as a physical gill. As oxygen levels decline in the bubble, dissolved oxygen diffuses in, replenishing it from the surrounding water. The aquatic larvae of some Curculionidae and Chrysomelidae are equipped with specialized spines that pierce plant tissues and extract dissolved oxygen from submerged portions of emergent macrophytes. Because of the many adaptations for using atmospheric oxygen, coleopterans are often abundant and diverse in poorly oxygenated wetlands. Many coleopterans are important predators, whereas others are adapted to scrape periphyton, graze aquatic plants, or consume detritus.

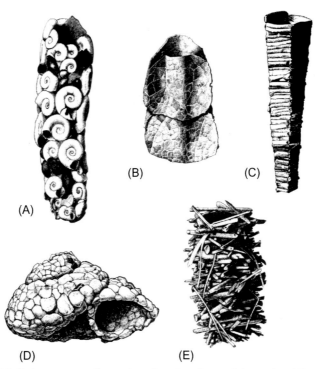

(A) (B) (C) (D) (E)

Figure 10.18 Caddisfly larvae cases illustrating diversity of materials used and form of construction, all about 1 cm long: (A) *Philarctus*, (B) *Clostoeca*, (C) *Brachycentrus*, (D) *Helicopsyche*, and (E) *Platycentropus. Reproduced with permission from Wiggins (1995).*

The order Hymenoptera (bees, wasps, and ants) does not include any known truly aquatic species, but some small wasps parasitize aquatic insects, and in some cases, the adult wasp actually swims or crawls underwater to lay eggs in the host. Most "aquatic" Hymenoptera parasitize larval or pupal stages of aquatic insects such as aquatic flies, caterpillars, and caddisflies. The nearly featureless larvae are endoparasites in the host, and like their terrestrial counterparts, most are host specific. One spider wasp (family Pompilidae) specializes on the semiaquatic spider, *Dolomedes*. These wasps can crawl under water, where they hunt the spiders as they crawl on the surface or dive. Once stung, the wasp carries the paralyzed spider to a nest on the shore where the female wasp lays eggs on it (Bennett, 2008).

Flies and midges (Diptera) constitute a large group, to which about 40% of all aquatic insects belong (Hilsenhoff, 1991). This group is dominated by the family Chironomidae (Fig. 10.17A), which includes about one-third of all species of aquatic Diptera, can reach densities of well over 10,000 individuals/m^2 in productive freshwater habitats, and represents an important food for a variety of fishes. The Diptera also includes many aquatic larvae with adults that are nuisance species and important disease vectors, such as mosquitoes (Culicidae, Fig. 10.17C), black flies (Simuliidae), biting midges (Ceratopogonidae), and horseflies and deerflies (Tabanidae). Some mosquitoes are very important vectors of diseases such as dengue fever and malaria in many parts of the world, and black flies transmit onchocerciasis (river blindness). Although some are considered pests, this group includes members that are a primary food source for fishes, waterfowl, and other aquatic and riparian predators. This group also includes the phantom midges (Chaoboridae), which are important predators on zooplankton in many lakes. Dipterans show a great degree of diversity and adaptation to different types of aquatic habitats. The "blood worms," which are actually chironomid midge larvae, have hemoglobin in their blood that imparts a red coloration and allows them to survive in low-oxygen sediments. Bloodworms can be very abundant in some habitats and pet stores sell them as aquarium fish food. *Eristalis* (Syrphidae, Fig. 10.17B) larvae have elongated respiratory siphons that allow them to breathe atmospheric oxygen and thrive in anaerobic waters such as sewage lagoons. Shorefly larvae (Ephydridae) can be abundant in extreme environments, such as hot springs and saline lakes.

The field of forensic entomology uses insect evidence to examine crimes and, in some cases, aquatic insects have helped solve crimes. One of the more common approaches is to use information on colonization patterns of scavengers on bodies to determine the time of death. In a fascinating case involving blackflies (Simuliidae), Richard Merritt, an aquatic entomologist at Michigan State University, examined a car recovered from the Muskego River that contained a woman's body. The husband of the deceased woman claimed she had left in the car after an argument they had a few weeks earlier in June. He indicated it was a foggy night and suggested she might have accidentally driven in to the river while she was upset and visibility in the fog was poor. However, Richard found that blackfly pupal cases that were attached to the car were from a species that pupates and emerges in April, and thus the car must have entered the river many weeks before the man claimed he last saw his wife. The husband was obviously lying, and thanks to the blackfly pupal cases and other evidence, he was eventually found guilty of murdering his wife and then staging the accident (Merritt and Wallace, 2000).

Crustaceans

Crustaceans are a paraphyletic group, but we still use the term for convenience to refer collectively to crayfish, shrimp, amphipods, isopods, cladocerans, copepods, mysids, and fairy shrimp. Crustaceans are abundant and diverse in freshwater habitats; many species are critical in food webs and ecosystem processes. Cladocerans and copepods are key primary consumers in many lakes and ponds, and decapods (crayfish and others) are important omnivores in benthic food webs. Crustaceans exhibit a wide diversity of body forms, but are characterized by respiration through gills or across the body surface, a hard chitinous exoskeleton that is often reinforced with calcium, two pairs of antennae, and most body segments with paired and jointed appendages. The taxonomic groups considered here are the Ostracoda, Copepoda, Branchiopoda, Decapoda, Mysidaceae, Isopoda, Amphipoda, and Bathynellacea.

The Ostracoda (seed shrimp; Fig. 10.19F) are benthic species that are covered by a carapace often composed of chitin and calcium carbonate that is modified into two shells or valves, which cover the body. Ostracod shells are particularly well preserved in sediments; paleolimnologists make use of them as indicators of ancient environments and these fossils are the oldest known microfauna (Delorme, 1991). Ostracods (sometimes spelled ostracodes) occur in most aquatic habitats, including temporary pools (even in very small pools in the bracts of bromeliads), ponds, lakes, rivers, and groundwaters. They even live in hot springs and hypersaline habitats. Superficially, they resemble small bivalves, but the body concealed by the shells has the typical crustacean array of variously modified segmented appendages. Ostracods mostly eat detritus and algae; a few are scavengers or predators. Ostracods rarely exceed 3 mm in total length, but they can reach tremendous densities in some benthic habitats and may

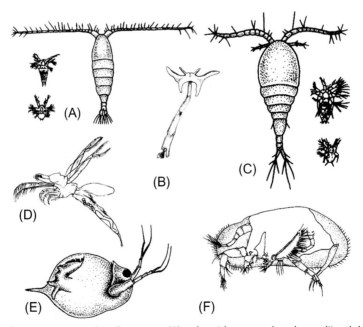

Figure 10.19 Some representative Crustacea: (A) calanoid copepod and nauplii, adult 0.5 mm; (B) the parasitic copepod, *Lernaea*, ~7 mm; (C) cyclopoid copepod and nauplii, adult 0.5 mm; (D) the cladoceran, *Leptodora*, 3 mm; (E) *Daphnia*, 0.5 mm; and (F) the ostracod, *Candona* with left valve carapace removed, 0.5 mm. *A and C−F reproduced with permission from Thorp and Covich (2001). B reproduced with permission from Food and Agriculture Organization of the United Nations and Kumar (1992).*

serve as an important food source for predators that specialize on benthic microfauna. Sexual and asexual (parthenogenesis) reproduction are evident in this group, with some species capable of both modes.

The Copepoda is a group of microcrustaceans that can be important component of zooplankton in both freshwater and marine systems, but most freshwater species are associated with the benthic zones of streams, wetlands, and groundwaters. About 500 freshwater species have been described worldwide (Williamson, 1991). They have a cylindrical body (Fig. 10.18A—C) that is 0.2—4 mm long, with numerous segmented appendages on the cephalothorax and abdomen. They have conspicuous first antennae and a single, simple, anterior eye. Most copepods are drab in coloration, but some forms are brilliant shades of red or orange. The copepods can be herbivorous, detritivorous, carnivorous, or parasitic. Parasitic forms such as *Lernea*, a common ectoparasite on fish, are highly modified and do not resemble free-living forms whatsoever (Fig. 10.19B).

Reproduction in copepods is usually sexual. The fertilized eggs hatch into a larval stage called a *nauplius*. Some copepods can produce eggs that are resistant to drying, and copepods are often among the first invertebrates to appear in freshly inundated habitats such as intermittent streams and wetlands. Six naupliar stages proceed six copepodite stages (the last being the adult stage). Plankton tows from lakes often reveal numerous nauplii. Behavior of the copepods can be complex, with predator avoidance, mating behaviors, and foraging behaviors all well developed.

The Branchiopoda includes the Cladocera (water fleas) and the fairy, tadpole, brine, and clam shrimps. This order is generally composed of small crustaceans with flattened, leaf-like legs. They occur in a wide variety of freshwater habitats but usually are not abundant in flowing waters. There are about 400 species worldwide (Dodson and Frey, 1991). The fairy, tadpole, and clam shrimps are relatively large crustaceans and are susceptible to predation by fish and large invertebrates. Many have desiccation-resistant stages that allow them to live in temporary pools where predators are scarce. Brine shrimp (*Artemia salina*) are tolerant of very saline waters and can establish dense populations in lakes that are too saline to contain fish. Brine shrimp are commonly used as food for aquarium fish and in aquaculture.

Cladocerans (Fig. 10.19D and E) are often extremely important zooplankters in lakes, and they can occur in wetlands and lotic environments as well. Researchers have developed many methods to sample the cladocerans and other zooplankton from lakes (Method 10.1). Cladocerans have a small, compound eye in the center of their head and a carapace on their back that serves as a brood chamber. Adults range from 0.2 to 18 mm in length and body forms in this group are quite variable. The thorax has four to six pairs of legs that beat continuously and create water currents that bring in algae, protozoa, and detritus. They use the many fine setae on their legs to filter food, and then move it toward the mouth with the legs. Planktonic species use the

---

METHOD 10.1 Sampling zooplankton

Scientists have devised a variety of zooplankton nets and sampling methods. Techniques depend on the species of interest. Zooplankton populations are patchy over space and time, so it is necessary to sample at a variety of depths in a lake to obtain adequate population estimates and to ensure that sampling captures all species.

For very small species, such as rotifers, that are able to pass through larger nets, samples are collected with a Van Dorn or similar sampler (see Chapter 7, preserved, and concentrated by settling or use of a fine filter in the laboratory. For larger species, researchers must process more water and have used a variety of net configurations with standardized mesh sizes.

Wetzel and Likens (1991) present detailed methods for obtaining a volumetric sample. One approach is to filter a volume of water through a net in the lake, and another is to remove a set volume of water from the lake and filter it at the surface. Metered nets can obtain quantitative zooplankton samples. These nets have a flow meter on the mouth and are towed horizontally at a set depth and then retrieved. The meter indicates the volume sampled. Alternatively, vertical tows indicate all species in the water column over a range of depths. Nets are available that close when a messenger is sent down the line that is used to pull the net through the water. These nets can sample vertically over a fixed range of depths. If the area of the mouth of the net and the distance sampled are known, then the volume of water filtered can be calculated.

Several techniques are available for removing water from a lake and then filtering it. In all cases, the collection devices are clear to reduce the chance that predator avoidance behaviors related to evading large, dark objects do not decrease the amount of zooplankton collected. One method is to take a Van Dorn sample and pour the water sample through a filter to concentrate the zooplankton. Another method involves lowering a box (Schindler trap) that encloses a known amount of water at a specific depth. The box has an exit net where the water flows out as the operator pulls it above the surface.

Researchers generally immediately preserve the samples with formalin; they then return them to the laboratory for counting and identification. Quick preservation is necessary to prevent predatory species from consuming their prey before the sample is processed.

---

large, second antennae for swimming, and the animals move with a jerky motion. The jerky motion leads to the common name "water fleas" for the cladoceran *Daphnia*.

Reproduction in many cladocerans is parthenogenic for most of the year. Thus, females are most common in nature. At times, often when conditions are suboptimal for growth, females produce both males and sexual females, and sexual reproduction occurs. The fertilized eggs are encased in a resistant *ephippium* that can remain dormant until conditions once again are favorable for growth and reproduction. These ephippium can be dormant for centuries, and clones have been resurrected from sediments dated through isotopes to have been deposited 600 years ago

(Morton et al., 2015). After hatching, they go through several juvenile instars that generally appear similar to adults.

A change in body form across seasons called *cyclomorphosis* is one aspect of cladoceran biology that has received considerable attention, and is also evident in rotifers. The variation in shape is diverse (Fig. 10.20) and can be related to temperature, varied predation pressure, other abiotic factors, or synergistic effects of several factors (Yurista, 2000). Cladocerans exhibit a variety of interesting behaviors including swarming,

Figure 10.20 Cyclomorphosis of adults of the cladoceran *Daphnia retrocurva* over a season in Bantam Lake, Connecticut, during 1945. Only body shape was traced. *From Brooks (1946).*

predator avoidance, and mating and feeding selectivity. The most obvious behavior is probably vertical migration in lakes, with the populations moving deep during the day to avoid sight-feeding predators and moving to the surface at night to feed on phytoplankton.

Cladocerans are frequently studied aquatic invertebrates and reproduce asexually in laboratory settings (allowing production of numerous clones). Thus, *Daphnia*, and the amphipod *Hyalella*, are common subjects of toxicology studies. Cultures of each can be easily then exposed to compounds to assess the toxicity of various pollutants at different concentrations.

The Decapoda includes some of the most familiar crustaceans such as crayfishes, crabs, and shrimps. Most decapod species live in marine or estuarine habitats, but hundreds of freshwater crayfish, shrimp, and crab species have been described, and they can be very abundant and diverse in the tropics. Decapods live in both lentic and lotic environments; some species have evolved to live in caves and groundwaters, and others live in swamps and intermittent wetlands. They are the largest and longest-lived crustaceans in most freshwater systems. The bodies of shrimp and crayfish are cylindrical (Fig. 10.21A and C) with a carapace that encloses at least a branchial chamber. Chelae, or pincers, are well developed in most adult decapods and are the feature most commonly associated with this group. While many are drab colored, or even un-pigmented (e.g., many cave-dwelling species), some crayfish and freshwater shrimps are brightly colored.

The Tasmanian giant crayfish (*Astacopsis gouldi*) is the largest freshwater invertebrate in the world. It can live up to 60 years and reach a mass of 6 kg and a length of 80 cm. However, overharvesting, pollution, and habitat destruction threaten this species. Freshwater decapods can provide a food source for humans; therefore,

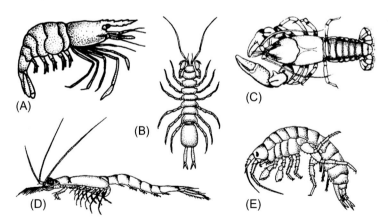

Figure 10.21 Some representative Crustacea: (A) a cave shrimp, *Palaemonias*, 2 cm; (B) the isopod, *Caecidotea*, 2 cm; (C) the crayfish, *Cambarus*, 10 cm; (D) the opossum shrimp, *Mysis*, 1 cm; and (E) the amphipod *Gammarus*, 1 cm. *Reproduced with permission from Thorp and Covich (2001).*

aquaculture of freshwater decapods such as crayfish and shrimps is an important economic activity in some regions and is a multimillion-dollar industry worldwide. Some freshwater decapods can be serious pests in rice fields, where they consume plants and disturb the water (Schmitt, 1965).

Many of the freshwater shrimps in the tropics are amphidromous; adult females release planktonic larvae that drift downstream to estuaries where they further develop before migrating back into freshwater streams (March et al., 1998). Impoundments have serious negative impacts on these ecologically and economically important crustaceans (Benstead et al., 1999).

Crayfish are omnivorous and can be dominant consumers (Hobbs, 1991). Freshwater shrimp are more often grazers or detritivores, but some such as the large *Macrobrachium* shrimps, are predators. The freshwater decapods can be important prey for fishes and other large animals. Some species of crayfish have become nuisance invaders and may outcompete native species. All decapods were thought to reproduce sexually, in one form or another, until the discovery of the marbled crayfish in aquaria in Germany. Marbled crayfish (or Marmorkrebs) reproduce through parthenogenesis (Scholtz et al., 2003) and while they are common in the aquarium trade, the natural range and ultimate origins of this crayfish remain virtually unknown. The rapid reproduction of this animal, and its widespread adoption as an aquarium pet, has led to considerable concern about its spread as an invasive species, with escapees establishing themselves in Madagascar and Europe. They are not legal to possess, trade, or transport in the European Union.

Many students will be familiar with Isopoda (isopods) as pill bugs or sow bugs in terrestrial habitats. There are many marine and terrestrial species, but freshwater forms also exist and can be very abundant in some habitats. The isopods are strongly dorsal—ventrally flattened (Fig. 10.21B) and range in length from 5 to 20 mm. Many of the species in North America inhabit springs, spring brooks, groundwaters, or streams; those living in subterranean habitats often lack pigments. A few make the littoral zones of lakes their home. Most isopods are thought to be detritivores and scavengers because they are often observed eating dead aquatic animals. The isopods *Caecidotea* (formerly *Asellus*) and *Lirceus* have received some study because of their roles in processing organic matter and as an indicator of water quality in both surface and subsurface waters.

The Amphipoda (scuds and sideswimmers) are similar in size to the isopods, but their bodies are laterally flattened (Fig. 10.21E). There are about 800 freshwater species worldwide (Smith, 2001). Amphipods usually have well developed eyes (except in the subterranean species). The amphipods are omnivorous and often considered scavengers. They are mainly nocturnal benthic species and can be present in densities up to 10,000 m^{-2}. Amphipods are very important as fish food. The amphipod *Hyalella* is cultured widely and is one of the most commonly used test organisms for freshwater toxicology studies. Freshwater amphipods, as well as isopods, reproduce sexually.

Discovery of pesticide resistant populations of *Hyalella* captured a lot of attention because this was one of the first documented cases of toxin resistance in a nontarget aquatic invertebrate (Weston et al., 2013). Resistant forms tolerate pyrethroid insecticides, which people use in agricultural systems and yards to combat common pest insects. This resistance may have far-reaching consequences because the process that imparts their resistance facilitates bioaccumulation of pyrethroids in their bodies, which are then passed on to the many aquatic and semiaquatic vertebrates that feed on *Hyalella* (Muggelberg et al., 2017).

The Mysidaceae (opossum shrimps; Fig. 10.21D) and Bathynellaceae are usually minor components of the freshwater fauna. The mysid shrimps are planktonic zooplanktivorous species that are several millimeters long. Twenty-five species occur in freshwaters, whereas 780 occur in marine habitats. Mysids can be an important food source for fishes and strong competitors of other planktivores in some temperate lakes. The bathynellids are microscopic to macroscopic invertebrates that occur in the hyporheic zones of some streams and other groundwaters. Because of their cryptic habitat, taxonomists have historically paid them scant attention, and new species are actively being described.

PHYLUM CHORDATA, SUBPHYLUM VERTEBRATA

Many people consider the fishes the most important freshwater animals. Other types of vertebrates also live in or rely on freshwater habitats. We will only briefly discuss them here. Still, vertebrates make up a major component of the diversity of freshwaters, second only to the insects (Fig. 10.22).

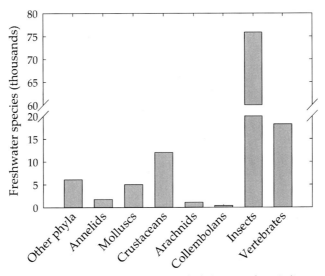

Figure 10.22 Numbers of freshwater species in several phyla. *Data from Balian et al. (2008).*

Fishes

Fishes are the most diverse aquatic vertebrates, with more than 24,000 described species, and this number may rise to 28,000 as taxonomy progresses (Moyle and Cech, 1996) (Table 10.3). Slightly less than half of these species live in freshwater and more than 3,000 inhabit Amazonian waters alone (and there are still many

Table 10.3 Some orders of fishes that have freshwater representatives

Order	Common name of freshwater species	Main region of dominance	Approximate number of freshwater species
Petromyzontiformes	Lampreys	Worldwide, coastal	40
Myliobatiformes	SRrays	Coastal	
Lepidosireniformes	Lungfishes	AF, SA	
Polypteriformes	Bichirs	AF	10
Acipenseriformes	Sturgeons, paddlefishes	PA, NA, AS	28
Lepisosteiformes	Gars	NA	7
Amiiformes	Bowfin	NA	1
Osteoglossiformes	Elaphantfishes, bonytongues	AF	200
Anguilliformes	Eels	Worldwide	26
Clupeiformes	Herrings, shads	Worldwide, some freshwater	80
Gonorynchiformes	Milkfishes	AF, AS	29
Cypriniformes	Minnows, carps, algae eaters, suckers, loaches	Worldwide	2,600
Characiformes	Characins	SA, AF	1,300
Siluriformes	Catfishes	Worldwide	2,280
Gymnotiformes	Knifefishes	SA	62
Esociformes	Pikes, mudminnows	NA, PA	10
Osmeriformes	Smelts, galaxiids	Worldwide, AU	71
Salmoniformes	Salmon, trout, whitefishes, chars, graylings	NA, PA	66
Atheriniformes	Silversides, rainbow fishes, blueeyes	AU, AS	160
Beloniformes		AS	45
Cyprinodontiformes	Top minnows, killifishes, pupfishes	Worldwide	805
Synrannchiformes	Swamp eels, spiny eels	AF, AS, SA, US	87
Scorpaeniformes	Sculpins	NA, PA	62
Perciformes	Basses, perch, sunfish, darters, cichlids, gobies, gouramis	Worldwide	2,200

Regions are African (AF), Neotropical (Central and South America, SA), Asia (Southeast Asia and peninsular India, AS), Palearctic Region (northern Eurasia, PA), Nearctic Region (North America, NA), and Australian Region (AU).
Source: From Moyle and Cech Jr. (1996) and Matthews (1998).

undescribed Amazonian fish species). Fishes provide the major economic impetus to conserve and protect many freshwaters because of their importance as food and for recreation. Ecologists have done a considerable amount of research on fishes. We discuss the major groups of fishes from freshwaters here. The lampreys (Fig. 10.24A) are the most primitive and are jawless; the jawed fishes include the Chondrichthyes (sharks and rays) and the Osteichthyes (bony fishes). Of these groups, the bony fishes are by far the most diverse in contemporary fresh and marine waters (Matthews, 1998).

Fishes have a variety of adaptations that allow them to specialize in various ecological roles and to display complex behavior. Body form is variable (Figs. 10.23 and 10.24). The streamlined bodies of active predators such as the salmon (Fig. 10.24E) allow them to pursue prey. Lie-and-wait predators with pointed snouts (Fig. 10.24B), torpedo-shaped bodies, and large posterior fins are able to generate sudden thrust and acceleration. Fishes that live near the surface and obtain food from it generally have upturned mouths, flattened heads, and large eyes (Fig. 10.23B). Bottom fishes generally have flattened bodies and often special adaptations, such as the lack of a swim bladder or the presence of barbels or paddles for sensing prey in benthic habitats (Figs. 10.23E and 10.24C). Deep-bodied fishes have narrow bodies and large fins (often spiny; Fig. 10.23D), and are adapted to maneuvering in tight quarters such as dense macrophyte beds. Eel-like fishes (Fig. 10.23C) have reduced fins and elongated bodies, and they are adapted to move through narrow spaces, soft sediments, or holes.

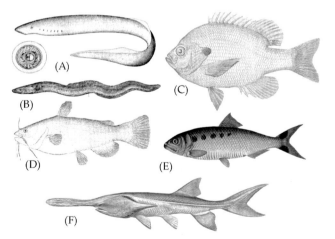

Figure 10.23 Some representative fishes: (A) lamprey, *Petromyzon*, 60 cm; (B) eel, *Anguilla*, 1 m; (C) bluegill, *Lepomis*, 20 cm; (D) bullhead, *Ictalurus*, 50 cm; (E) paddlefish, *Polyodon*, 2.5 m; and (F) alewife, *Alosa*, 40 cm. *Figures from Wikimedia Commons. C and D courtesy of Freshwater and Marine Image Bank, University of Washington.*

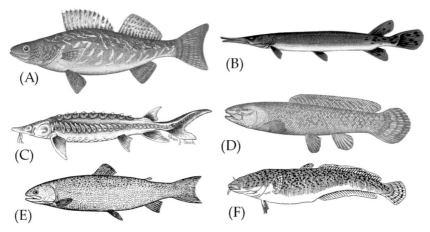

Figure 10.24 Some representative fishes: (A) walleye, *Stizostedion*, 90 cm; (B) gar, *Lepisosteus*, 1.5 m; (C) sturgeon, *Scaphirhynchus*, 20 cm; (D) bowfin, *Amia*, 60 cm; (E) salmon, *Salmo*, 1 m; and (F) burbot, *Lota*, 90 cm. *Images courtesy of Wikimedia Commons. C US Fish and Wildlife Service Karen Couch. D courtesy of Freshwater and Marine Image Bank, University of Washington.*

Fishes also have extensive sensory adaptations. The ability for chemoreception is acute; chemical concentrations as low as 10^{-13} moles/Liter can stimulate behavioral responses (Moyle and Cech, 1996). Sounds and vibrations can also be sensed by the inner ear and lateral line system. This includes the ability to sense remote objects by their hydromechanical signals. Fishes also have a series of pit organs that allow sensing of weak electrical currents generated by prey. Many fishes have well-developed eyes that facilitate sight feeding and sexual displays.

Freshwater fishes can be divided into three geographic types. A few marine fishes can enter freshwaters for extended periods and enter the lower regions of coastal streams. Obligatory freshwater species must inhabit freshwaters for at least part of their life cycles and many cannot tolerate marine waters for any significant length of time. *Diadromous* fishes spend part of their life cycles in marine systems. These species include fishes that are *catadromous* and move from freshwater to saltwater to spawn (e.g., some eels) and those that are *anadromous* and move from saltwater to freshwater to spawn (e.g., salmon).

Tetrapods

Fishes gave rise to several groups of vertebrates that are prominent inhabitants of some freshwater systems. These include amphibians, reptiles, birds, and mammals. The total diversity of these species is not high relative to that of the fishes or the arthropods, and many are only semiaquatic, but they draw public interest. Several species such as beavers and crocodilians have extensive effects on ecosystems, which we discuss later in the text.

Amphibians are divided into three groups: the salamanders (Urodela or Caudata), the caecilians (Gymnophiona), and the frogs and toads (Anura). The anurans account for 4,100 of the more than 4,600 species of amphibians (Pough et al., 1998). Many of these species spend part or all of their time in freshwaters, particularly as eggs or larvae. Adult amphibians are predators but some larvae, such as many tadpoles, consume algae and detritus. Amphibians can be associated with lentic and lotic habitats. Many species of tadpoles are highly susceptible to predation by fish and other large predators, although some are toxic or distasteful and are resistant to predation (Alford, 1999). Amphibians are consumed by a wide variety of animals, including fishes, other amphibians, reptiles, birds, and mammals, and even some of the larger invertebrates. Many amphibians are sensitive to environmental change and are thus good indicators of habitat disturbance and pollution. Amphibians are among the most imperiled groups of animals on the planet; their diversity is declining catastrophically in many parts of the world (see Highlight 11.2).

Aquatic reptiles, birds, and mammals are familiar to most people. A list of some of the more important or charismatic species includes some that can be abundant in wetlands, rivers, and lakes (Table 10.4). Some of these species can alter hydrology or nutrient cycles. Others are of interest because of their rarity, unusual nature, or

Table 10.4 Some reptiles, birds, and mammals of interest in freshwater ecosystems

Taxonomic group	Common name	Habitat/trophic position	Comments
Testudines	Turtles	Rivers, wetlands, ponds, lakes/predators and herbivores	About 260 species worldwide
Eunectes	Anaconda	Rivers, wetlands, ponds/ predators	Largest snake in world, up to 10 m long
Nerodia and Natrix	Water snakes	Rivers, wetlands, ponds, lakes/predators	
Crocodilians	Crocodiles, alligators, caimans	Rivers, wetlands, ponds, estuaries/predators	All 21 species endangered or threatened
Podicipediformes	Grebes	Lakes, estuaries/ predators	21 species
Pelecaniformes	Pelicans, cormorants, anhingas	Lakes, estuaries, marine/ predators	
Anseriformes	Swans, geese, ducks, screamers	Lentic and lotic waters/ predators and herbivores	Temporary ponds and wetlands important to many species
Phoenicopterifromes	Flamingos	Shallow lagoons and lakes/filter feeders	Five species, all tropical

(Continued)

Table 10.4 (Continued)

Taxonomic group	Common name	Habitat/trophic position	Comments
Coconiiformes	Herons and storks	Shallow water/predators	
Falconiformes	Osprey, hawks, eagles	Surface waters/predators	Mainly take fish from surface
Gruiformes	Cranes and rails	Shallow waters/ predators	
Cinculus	Dipper	Mountain stream/ insectivore	Dive into clear, running mountain streams to forage
Charadriiformes	Shorebirds and gulls	Shallow waters/ predators	
Gaviiformes	Loons	Lakes/predators	
Ornithorhynchus anatinus	Duck-billed platypus	Lakes, rivers/predators	An Australian monotreme
Noctilio	Fishing bat	Lakes, rivers, wetlands/ predators	
Phoca siberica	Baikal seal	Lake Baikal/predators	Only freshwater seal
Lontra	River otter	Streams, lakes/predators	
Delphinidae	Dolphins	Coastal rivers/predators	Asia and South America
Platanistidae	River dolphins	Large rivers in Asia and Indopacific/predators	Generally blind, locate prey by echolation
Sirenia	Dugong, manatee	Tropical, estuarine, coastal rivers/ herbivorous	
Hippopotamus amphibious	Hippopotamus	Lakes, rivers/herbivorous	
Castor	Beaver	Lakes, rivers, wetlands/ herbivorous	Northern Hemisphere
Myocastor coypus	Nutria	Rivers, wetlands/ herbivores	South American native introduced into Europe and North America

perceived dangerousness. They can be important predators of aquatic species. Most of these species are associated with shallow habitats, such as the littoral zone of lakes, ponds, rivers, streams, and wetlands.

SUMMARY

1. The major groups of animals in freshwaters are the Porifera, Cnidaria, Turbellaria, Nemertea, Gastrotricha, Rotifera, Nematoda, Mollusca, Annelida, Bryozoa and Kamptozoa (Entoprocta and Ectoprocta), Arthropoda, and Chordata.

2. The taxonomy of the smaller bodied taxa is less completely resolved than that of larger organisms.

3. Sexual reproduction is often sporadic or nonexistent in more primitive organisms. It is usually required in larger organisms such as the vertebrates, with notable exceptions.

4. Body form can vary with season or exposure to predation in several groups, including the rotifers and cladocerans. A seasonal change in body form is called cyclomorphosis.

5. Arthropods are the most diverse animal groups in freshwater systems and have adapted to all major aquatic habitats.

6. Aquatic insects are particularly diverse in rivers and streams.

7. Fishes assume an important role in aquatic food webs. Their body shape is related to their place in the food web and their habits.

8. Many of the mammalian, amphibian, reptilian, and avian species that use freshwater habitats are endangered; some have become extinct.

QUESTIONS FOR THOUGHT

1. What constrains arthropods from reaching the size of humans?

2. What groups of invertebrates do you think are most likely to contain undescribed species? Why?

3. Why do most aquatic insects leave the water as adults to mate?

4. Why has behavior evolved to be more complex, but biochemistry less complex, in animals in comparison to microorganisms?

5. Why are aquatic insects rare and not diverse in marine habitats relative to freshwater habitats?

6. Why are animal assemblages of lakes generally less diverse than those in streams?

7. How have animals adapted to subsurface habitats?

8. Why is lifespan tied to the utility of various freshwater animals for use in biological assessments?

9. Why are there more species of insects than mammals and birds?

10. Why do fishes that are ambush predators have a different morphology than piscivorous species that forage in the pelagic zone?

CHAPTER 11

Evolution of Organisms and Biodiversity of Freshwaters

Contents

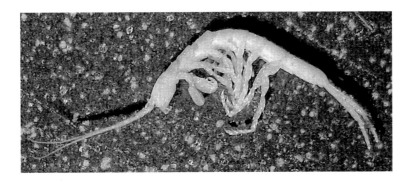

Figure 11.1 A groundwater-dwelling isopod (*Caecidotea tridentata* previous page) and a green heron (*Butorides striatus*). *Heron photo courtesy: Steve Hamilton.*

Describing the distribution patterns of organisms and factors that control them is central to the study of ecology and historically has been the focus of many limnological studies. An entire field, conservation biology, is now dedicated to the science of conserving organisms. We are living in a time of unprecedented rates of human-caused extinction, particularly in freshwater habitats (Jenkins, 2003; Vaughn, 2010). Efforts to understand, manage, or conserve species require knowledge of distribution of organisms in their habitats, how they evolved, and how they maintain populations. In addition to causing many species to go extinct and threatening many more, people are introducing exotic species into habitats at very high rates. Species introductions can lead to direct negative effects on humans and often harm the ecosystems that exotic species invade.

In this chapter, we discuss measures of diversity and how and why diversity varies among and within habitats, and we describe some particularly spectacular cases of high diversity. We also discuss extinctions caused by humans and some of the associated consequences, as well as threats of invasive species.

MEASURES OF DIVERSITY

The simplest measure of diversity is richness. This is the number of taxonomic units per unit area. Most commonly, researchers report *species richness*, the total number of species found in an area. Another important aspect of diversity is how equally represented the numbers of each species are, termed *evenness* or *equitability*. For example, consider two habitats, both with 10 species. If one habitat has approximately even numbers of each species, it will have a high evenness. If the other is dominated by one species and has one or a few representatives of the rest, it has the same species richness, but lower evenness. This can be visualized in Fig. 11.2 where ponds B and F have the same richness, but pond F has a greater evenness. Some diversity indices include both evenness and richness.

One commonly used measure of diversity that includes both evenness and richness is the Shannon index,[1] H':

$$H' = - \sum_{j=1}^{s} p_j \ln(p_j)$$

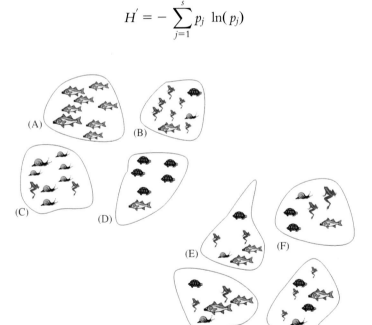

Figure 11.2 A diagram of diversity in two sets of ponds. When A–D are considered versus E–H, both have approximately the same overall diversity. When α diversity (within-habitat diversity) is measured, ponds in the A–D group have lower diversity than those in the E–H group. When β diversity (between-habitat diversity) is measured, ponds in the A–D group have higher diversity than those in the E–H group.

---

[1] Often incorrectly referred to as the Shannon–Weiner or the Shannon–Weaver index.

where there is an assemblage of organisms with S species and p_j is the proportion of species j (i.e., the number of individuals of species j divided by the total number of organisms in the assemblage). The summation sign at the beginning of the equation means that the proportions of all the species in the community are summed. The equation is most often calculated with the natural log, but others use $\log_{10}$ or $\log_2$. Here we use the natural log, but care should be taken when comparing results to other studies to be certain that the index was calculated the same way. H' increases with more species and with greater evenness. One index used to indicate evenness is to divide H' by the diversity at maximum evenness for a given number of species. The following equation can be used to create an index of evenness using this approach:

$$E = \frac{H'}{\ln S}$$

where E is the evenness and S is the number of species. Extensive sampling is required for this index to be comparable among different communities, because measures of species richness are highly sensitive to sampling effort (Pielou, 1977). *Rarefaction* is a method ecologists use to estimate the number of species based on sampling effort. In this method, researchers take repeated samples and the number of new species found with each subsequent sample is recorded. Then statistical methods are used to estimate the total number of species (Fig. 11.3). The method allows researchers to compare species richness across studies with unequal sampling effort. Calculation of diversity and evenness is demonstrated in Example 11.1.

Within-habitat diversity (α *diversity*) and between-habitat diversity (β *diversity*) are additional aspects of diversity. These two aspects link spatial scale and diversity. Each habitat has a variety of subhabitats and each has its characteristic species. For example, a wetland may contain habitat types ranging from open water to damp soil, and contain phytoplankton to emergent vegetation. Fig. 11.2 illustrates how spatial arrangement relates to determinations of diversity. The two groups of four ponds (A–D and E–H) each have the same number of total species and approximately the same evenness if all the ponds in each group are considered. However, each individual pond in the first group (A–D) has lower diversity (evenness and richness) than each pond in the second group. The ponds in the first group have low within-habitat diversity (α diversity) but high between-habitat diversity (β diversity).

We have considered only three indices of diversity here—species richness, Shannon diversity, and evenness. Ecologists have used several other indices of diversity throughout the years. Some mathematical benefits and drawbacks of the various indices are discussed by Pielou (1977) and Magurran (2004) but are beyond the scope of this text. Once diversity is estimated, the next issue considered is the major patterns of β diversity.

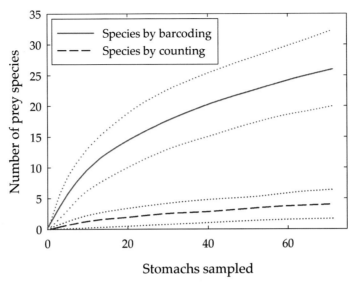

Figure 11.3 Examples of species rarefication curves. In this example, each sample was a fish gut taken from one of eight predatory fish species collected on the Canadian Shield, and the number of new species found with each additional sample was recorded. Two methods were used, (1) traditional taxonomy based on morphology of animal remains in the fish gut and (2) DNA barcoding to indicate the species composition in each stomach. The solid lines give the curves of sampling effort for each method and the dashed lines the 95% confidence intervals around those lines. *Figure after Bartley et al. (2015).*

One approach to classification of habitats is to use the distributions of organisms (β diversity) and abiotic characteristics, termed the *ecoregion* concept. Ecoregions are relatively large areas of land or water that contain a geographically distinct assemblage of natural communities. Many scientists use the term ecoregion interchangeably with *biome*; however, biomes in terrestrial usage are not phylogenetically based (e.g., a desert is a desert regardless of the phylogeny of xeric-tolerant plants that are found in it) whereas ecoregional designations often have a phylogenetic component. For example, the freshwater habitats of North America have been divided into 76 ecoregions from 8 major regions based on biological distinctiveness with regard to fishes, amphibians, crayfish, unionid mussels, and aquatic reptiles (Abell et al., 2000). It is not yet clear how well these ecoregions extend to aquatic insects, freshwater plants, and algae. The next issue we will consider is the processes that lead to observed distributions of diversity given that freshwaters can be biologically distinct.

Example 11.1 Calculating diversity

A graduate student takes samples of stream insects from both pool and riffle habitats. Find and calculate the mean species richness, the Shannon diversity index, and the evenness for both habitats. For this exercise, assume the genera without species identifications represent one species.

Sample	Species	#/m²	p_j	$p_j \times \ln p_j$	H'	E	S
Pool	Perlesta placida	10	0.24	−0.34			
	Zealeuctra claasseni	20	0.49	−0.35			
	Stenonema femoratum	5	0.12	−0.25			
	Baetis sp.	6	0.15	−0.28			
	Total pool	41			1.22	0.88	4
Riffle	P. placida	17	0.21	−0.33			
	Z. claasseni	18	0.23	−0.34			
	S. femoratum	5	0.06	−0.17			
	Baetis sp.	3	0.04	−0.13			
	Tipula sp.	21	0.26	−0.35			
	Pseudolimnophila sp.	16	0.20	−0.32			
	Total riffle	80			1.64	0.91	6

The diversity in the riffle is higher because more species are present and evenness is greater. Statistical experimental designs including replication followed by statistical techniques such as Analysis of Variance should be used to determine if these differences are indeed significant (see Appendix 1). These data are made up and are presented only to demonstrate calculation of diversity.

TEMPORAL AND SPATIAL FACTORS INFLUENCING EVOLUTION OF FRESHWATER ORGANISMS

The ultimate source of biological diversity is evolution. Two overarching factors that influence the evolution of new species are time available for evolution and reproductive isolation. These factors, coupled with what Hutchinson (1959) called the mosaic nature of the environment (what we now call spatial and temporal aspects of habitat heterogeneity) have resulted in millions of species. We discuss these factors with regard to lakes, streams, groundwater habitats, and wetlands. Consideration of spatial and temporal scale is important in describing the biological processes. We describe specific examples of situations in which evolution has resulted in high numbers of species, and explore some short-term determinants of biodiversity in the next section.

One of the most striking demonstrations of the importance of time to evolution of biological diversity is the high number of species found in geologically ancient habitats that have had ample time for new species to arise. Typically, such habitats contain many *endemic* species, those species with a restricted distribution. Additionally, such habitats provide some of the clearest examples of *adaptive radiation* (evolution of many species from a single or few founder species).

Ancient lakes contain a large proportion of freshwater biodiversity (Cohen, 1995a, 1995b). These tectonic lakes have unique assemblages of vertebrates and invertebrates compared to the Great Lakes of North America (Fig. 11.4). The Great Lakes were covered by ice during the last ice age and others, and are young in an evolutionary sense since most species could not survive this. This high degree of endemism occurs even though the surface area (i.e., potential habitat) of the Great Lakes is almost 10 times greater than that of the largest of the ancient lakes. We discuss the fact that we expect more species in a greater area in the next section; this makes the diversity of old tectonic lakes relative to the Great Lakes even more remarkable.

Lakes that have existed continuously for millions of years generally have changes that lead to geographic isolation. Variations in water level, formation and isolation of various sub-basins, and fluvial processes allow for divergence and evolution of new species (Brooks, 1950). Divergence and evolution of species can occur because of reproductive isolation of populations. Reproductive isolation permits organisms to evolve to be different enough that they become distinct species that can no longer interbreed if their populations become contiguous again.

Lake Baikal, located in Russia, may be the most studied of the ancient lakes. It has 377 endemic crustacean species, 291 of which are amphipods. Fig. 11.5

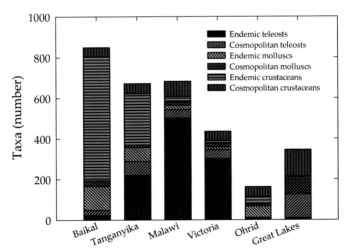

Figure 11.4 Number of endemic species for some large lakes of the world. The area of the North American Great Lakes is approximately 10 times greater than that of any of the other lakes shown. *Reproduced with permission from Cohen (1995).*

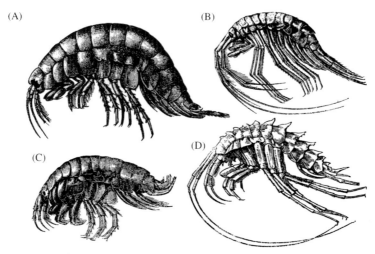

Figure 11.5 A variety of gammarids from Lake Baikal, demonstrating diversity of body form: (A) *Ommatogammarus albinus*, body length up to 25 mm; (B) *Abyssogammarus sarmatus*, body length up to 63 mm; (C) *Crypturopus pachytus*, body length up to 18 mm; and (D) *Garjajewia cabanisi*, body length up to 80 mm. *Reproduced with permission from Kozhov (1963).*

demonstrates a small part of the diversity in body form that can be found in these amphipods; this diversity in body form is consistent with differences in rRNA sequences (Sherbakov et al., 1998). In addition, there are 86 endemic species of turbellarians, 98 endemic mollusks, and 29 endemic species of fish [of which 26 species are sculpins (Cottoidei); Sherbakov, 1999]. About 80%−100% of the species in these groups are endemic. Slightly less than half of the 679 species of diatoms are endemic (Brooks, 1950). As with most large lakes, the incredible biodiversity of Lake Baikal is threatened by human-caused pollution (Galaziy, 1980).

The African Rift Valley has formed a deep depression in the continent of Africa that has held lakes for millions of years. Lake Tanganyika is one of these and has been in existence for 9−12 million years. It is renowned for its large number of endemic fishes, dominated by the perch-like tropical cichlid fishes. Greenwood (1974) reported that the family Cichlidae has 126 species, all of which are endemic. An additional 67 species of fishes from other families live in this African lake, 47 of which are endemic. Coulter (1991a) reported 287 species and subspecies of fishes, 219 of which occur only in Tanganyika. The Coulter and the Greenwood numbers do not exactly match because of changes in taxonomy and inclusion of subspecies by Coulter.

Lakes Victoria and Malawi, also found in the African Rift Valley, have more species of Cichlidae than Lake Tanganyika but not more endemic species in other families, and they have fewer endemic genera. The variety of adaptations that have evolved in the fishes from the Rift Valley lakes is astounding (Meyer, 1993). Various species of fishes in this one family live by eating zooplankton, phytoplankton, gastropods, benthic algae, macrophytes, detritus, or fishes. Among the piscivores, some eat the scales of other fishes and have evolved to look like their prey; their victims do not realize they are predators (Fryer and Iles, 1972). The diversity of body form in just the one family, Cichlidae (Fig. 11.6), can exceed that seen in all families found in some temperate habitats. This group of fishes is evolving very rapidly and, as such, has proved an excellent subject for studies of evolution (e.g., Malinsky et al., 2015). There are a number of species in Tanganyika that exhibit convergent evolution, looking more like more distantly related lineages than closer relatives (Fig. 11.7).

Behavioral diversity in the Cichlidae in these lakes is also high (Hori et al., 1993), and this family has been used in animal behavior studies (e.g., Alcazar et al., 2016). Wide variation in coloration (Axelrod, 1973) probably results from sexual selection that leads to sympatric speciation (i.e., evolution of a species within one habitat;

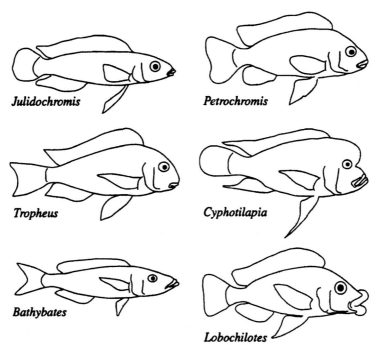

Figure 11.6 Some of the many varieties of fishes in the family Cichlidae from Lake Tanganyika. *Julidochromis*, omnivorous; *Petrochromis*, herbivorous; *Tropheus*, herbivorous; *Cyphotilapia*, ambush predator, gastropods; *Bathybates*, piscivorous; *Lobochilotes*, insectivorous. *From Gillespie et al. (2001).*

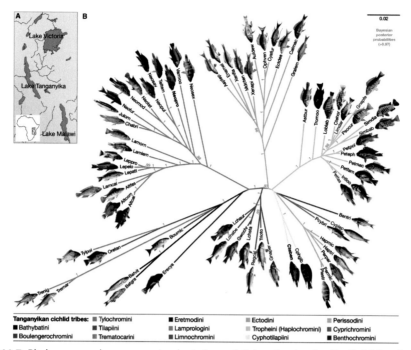

Figure 11.7 Phylogeny and convergent adaptation of fishes in the family Cichlidae from Lake Tanganyika. *Reproduced with permission from Muschick et al. (2012).*

Seehausen and van Alphen, 1999). Many cichlids exhibit parental care, which ranges from guarding eggs and fry from predators, to incubating eggs and protecting fry in the oral cavity (Keenleyside, 1991). Two species can brood in the same region and mutually defend against predators. In another case of evolved cooperation, mixed groups of predators can cooperate to increase success (Nakai, 1993). These complex behavioral adaptations are more likely to arise in a diverse animal assemblage.

Tanganyika also supports a high proportion of endemic animals in other groups, including 12 of 20 leech species, 37 of 60 gastropods, 33 of 69 copepod species, 74 of 85 ostracods, and 22 of 25 decapod species (Coulter, 1991b). As in Lake Baikal, the millions of years this lake has existed has allowed for diversification of many animal groups. Unfortunately, the biological diversity of Tanganyika is increasingly threatened by overfishing for food and the aquarium trade, as well as other human impacts such as deforestation in the watersheds that drain into the lake, and the inevitable pollution that human activities cause. Warming has also caused declines in endemic mollusks and production of commercially important fishes in Lake Tanganyika (Cohen et al., 2016).

An obvious question when considering the various groups that have produced large numbers of endemic species in ancient lakes is the following: Why does a particular group diversify more than another (Gillespie et al., 2001)? For example, the teleost fishes are very diverse in the African Rift Valley lakes like Tanganyika, but the amphipods have the greatest number of endemic species in Baikal. Amphipods occur in the African lakes, and teleosts occur in Baikal, so why not extensive speciation in both groups in both lakes? Chance plays a large part in evolution. If the right group already happens to be present and already is diverse, it may be predisposed to amplify the existing variation when conditions occur that are favorable to evolution of new species. The evolution of greater diversity can then be accelerated in particular related groups. For example, the diverse open water fish assemblages found in Lake Malawi probably arose from multiple independent lines of species that inhabit the shoreline (Fryer, 2006). Thus, a diverse shoreline community led to a diverse community in the pelagic (open water) areas of the lake. Likewise, the diverse cichlid fish assemblages found in the rivers of southern Africa probably arose in a lake that is now dry (Joyce et al., 2005).

Advances in molecular evolutionary ecology also are providing some answers. Molecular sequences from genomic analyses indicate a number of characteristics that correlate with cichlid groups from Africa that have diverged rapidly. These include variation in many noncoding elements, rapid coding sequence evolution, changes associated with transposable element insertions, and novel microRNAs (Brawand et al., 2014).

Certain biological traits may favor rapid speciation in a group. The gastropods provide an example of some characteristics in a specific group that can lead to evolution of high numbers of species. Gastropod taxa that are depth tolerant, brood their young, and have poor dispersal ability can diverge more rapidly. All these characteristics increase the chances of reproductive isolation and, consequently, evolution of independent lineages, as is known from study of evolutionary mechanisms (Michel, 1994).

River basins can also have a continuous existence over long geological periods, leading to evolution of high numbers of unique species. The Mississippi River drainage has been geologically stable, the river has flowed continuously, and little Late Cenozoic extinction has occurred (e.g., in the last 10 million years). Much of the diversity in North American fishes has radiated from this drainage basin (Briggs, 1986); there are 300 fish species found in Tennessee, a tributary of the Mississippi, alone (Etnier, 1997). The mussel assemblages in the superfamily Unionacea have evolved 227 native species in North America (McMahon, 1991), most of which are found in the Mississippi River basin. However, many of these mussel species are now endangered by human activities.

Geographic isolation is another factor that leads to high levels of endemism. Islands have relatively high numbers of endemic species (but lower total species richness). Islands are an extreme case of geographic isolation. Such isolation occurs in continental habitats as well; many freshwater habitats are essentially islands in a terrestrial "sea." Geographic isolation into different habitat types led to species divergence in three species of sticklebacks (*Gasterosteus* spp.), but divergence did not occur where isolated habitats were similar (Rundle et al., 2000). Dry landscapes can lead to more isolation of aquatic habitats and greater divergence of amphipod species (Thomas et al., 1998) and leeches (Govedich et al., 1999). Nine species of ostracods in the genus *Elpidium* evolved in Jamaica (Little and Hebert, 1996). These ostracods inhabit the small pools of water that form in bromeliads (terrestrial epiphytic plants). Thus, segregation of habitats that leads to evolution of new species can occur over very small spatial scales.

Groundwater habitats in aquifers with channels large enough to allow movement of animals have given rise to many unique endemic species. The organisms found in groundwaters can be accidental wanderers from surface waters, can occur mostly in groundwaters and have some adaptations to the subsurface life (*stygophile*), or can be specifically adapted to life in groundwaters (*stygobite*) (Gibert et al., 1994a,b). Adaptations to life in groundwater habitats include loss of eyes and pigmentation, increased length of sensory appendages, slowed metabolic rates, long life histories, and production of fewer and larger eggs.

Apparently, dispersal ability is low in groundwater fauna (Strayer, 1994), probably contributing to endemism. For example, little recovery of some North American hypogean (subterranean) invertebrates occurred following the last glaciation. In some cases, the species found in groundwaters can vary over small spatial scales. Diversity of ostracods in the shallow groundwater (hyporheos) near the Sava River (Croatia) varies considerably across a 135-m transect moving away from the river channel (Rogulj et al., 1994). Studies of subterranean crustacean species found in subsurface streams in a Virginia karst aquifer indicated the importance of spatial scale (Culver and Fong, 1994; Fong and Culver, 1994). These authors demonstrated unique species assemblages that varied at scales from individual rocks to differences among pools and riffles. Greater differences occurred among different branches of the streams, and the greatest differences arose among adjacent subsurface drainage basins.

Rivers with extensive hyporheic systems in very permeable or channelized aquifers can have a unique biota associated with the subsurface water (Marmonier et al., 1993; Popisil, 1994; Ward and Voelz, 1994; Dole-Olivier et al., 1994; Boulton, 2000; Hakenkamp and Palmer, 2000). In a fascinating series of studies on groundwater communities in a gravel-bed river system, numerous aquatic insect larvae and other invertebrates occurred up to 2 km from the river channel (Stanford and Ward, 1988). Some of these species are unique stoneflies that spend their entire nymphal life cycle underground and then return to surface water habitats to emerge as adults and mate (Gibert et al., 1994a,b; Stanford and Gaufin, 1974). Hyporheic habitats may be constant enough to provide refugia for ancient taxa of cladocerans over evolutionary time (Dumont, 1995).

The data available on biodiversity of hypogean animals suggest that a tremendous number of undescribed species exist. The difficulties encountered in sampling these organisms, coupled with the documented spatial variation in distributions of the animals, make it highly probable that species will not be collected unless areas are sampled intensively. Despite our ignorance about the distribution of many of these species, many are highly susceptible to human-caused pollution of groundwaters. Therefore, the use of groundwater invertebrates as bioindicators of contamination has been suggested (Malard et al., 1994), particularly as indicators of low dissolved oxygen conditions (Malard and Hervant, 1999).

Some temporary pools are also centers of endemism. Such pools are isolated from other aquatic habitats and require special adaptations in the form of desiccation resistance. The vernal pools of California are examples of this because they provide habitat for plant communities with substantial endemism. In addition, when 58 of these vernal pools were sampled, 67 species of crustaceans were found and about half these were endemic, mostly Anostraca, the fairy shrimp (King et al., 1996). Such pools with endemic plant and crustacean communities occur in dry habitats throughout the world (Belk, 1984; Thorne, 1984).

Several cases of endemism have yet to be explained. For example, benthic copepods show high levels of endemism in South American wetlands (Reid, 1994). However, science has not yet explained why such endemism occurs in these wetlands and not in those of North America, such as the vernal pools in California, in which the Anostraca are diverse.

SHORT-TERM FACTORS INFLUENCING LOCAL DISTRIBUTION OF SPECIES

Numerous factors, including species interactions, productivity, habitat type, and colonization control how many and what types of species occur in a specific environment. Here, we discuss colonization and habitat type. Later, we focus on species introductions. We discuss species interactions in Chapters 19–22, and productivity in Chapters 17 and 18. Factors related to colonization are dispersal ability, competitive ability, and distance from sources of colonists when new habitats form. Several areas of ecological theory relate to these concepts.

Habitat type and diversity are interrelated. The habitat serves as a template on which evolution can occur. The organisms best matched to the particular conditions in the habitat will reproduce more successfully, reinforcing specialization. Thus, heterogeneity of habitats will lead to increased diversity. For example, a lake has several distinct types of habitats that different groups of organisms dominate

(Fig. 11.8). Changes in these habitats can influence biodiversity. For instance, decreases in woody debris on lakeshores associated with residential development can lead to decades-long decreases in biodiversity (Christensen et al., 1996). Similarly, decreases of benthic habitat diversity in streams led to decreases in macrophyte diversity (Baattrup-Pedersen and Riis, 1999), and macrophytes can control heterogeneity of sediments (Schneider and Melzer, 2004). Within taxonomic groups, different species also specialize in specific habitat types, which has been well documented for many plants and animals, even those as simple as rotifers (Pejler, 1995). Scale can also be an important determinant of diversity patterns, ranging from microbes being associated with the smallest particles to animals such as beavers actually creating and responding to changes at the watershed level. Fig. 6.22 illustrates how scale can affect biodiversity in streams and rivers.

How habitats link to each other is also a central determinant of biodiversity. Interfaces among habitat types usually have high numbers of species. High diversity in these edge habitats occurs because species that are adapted to both habitat types can occur there (species ranges overlap), in addition to species that specifically exploit the transitional zone. The relationships among biodiversity and riparian zones of streams are a prime example of high diversity associated with an ecosystem interface (Tockner and Ward, 1999). Maintenance of riparian zones can preserve both diversity and ecosystem function in a watershed (Naiman et al., 1993; Spackman and Hughes, 1995; Patten, 1998) and in the stream (Vuori and Joensuu, 1996). Likewise, managers should consider that hyporheic zones are an edge habitat between groundwater and stream water when biodiversity of river systems is of concern (Stanford and Ward, 1993).

Dispersal ability is also important in determining distributions of species and biodiversity. Widely distributed freshwater species (cosmopolitan species) often have the following characteristics: (1) life history stages that can survive transport (often

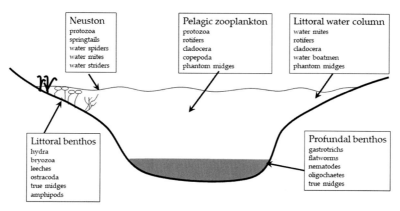

Figure 11.8 Representative invertebrate groups as a function of habitat in a small lake. *Redrawn from Thorp and Covich (1991b).*

desiccation resistance); (2) production of large numbers of individuals in the transportable life stage form; and (3) the ability to survive, including competing successfully and avoiding predation, in new habitats (Cairns, 1993). Therefore, some species distribute very broadly. For example, many protozoa have cosmopolitan distributions; about 8% of all protozoan species ever described on Earth can often be found in a single sample (Fenchel et al., 1997). Protozoa have all the characteristics listed above and thus are very successful at colonizing new habitats. Consequently, identification keys for protozoa that have been written anywhere in the world are useful in most other locations (Cairns, 1993). However, many more endemic protozoa could be described once molecular sequencing of natural samples becomes more common.

Some of the natural modes of species transport include swimming or being moved by currents through connected habitats, transport on the feet or in the guts of animals (including insects, birds, and mammals), windborne dispersal, and having life stages specialized for dispersal (e.g., many amphibians and aquatic insects). Transport of algae in waterfowl guts has been documented (Proctor, 1959, 1966) and is probably common on the local scale (Figuerola and Green, 2002). A few larger zooplankton species can be wind transported, but are not as likely to be transported by waterfowl (Jenkins and Underwood, 1998). Waterfowl can also regurgitate seeds, spreading aquatic plants (Kleyheeg and van Leeuwen, 2015).

Resting stages of organisms are particularly important in some cases—not only in dispersal but also in the ability to respond to long-term changes in the environment. For example, copepods produce eggs that can remain viable but physiologically inactive in sediments for long periods of time (*diapause*). This long-term survival is an adaptation to changing environmental conditions and allows the copepods to establish in environments that only sporadically support a reproductive population (Hairston, 1996). Hairston et al. (1995) studied sediments of two small freshwater lakes in Rhode Island. They dated the sediments of the lakes with isotopes and hatched eggs taken from sediments of different ages. In one lake, the mean age of eggs that hatched was 49 years, and the maximum was 120 years. In the second lake, the mean was 70 years and the maximum was 332 years.

A consistent ecological pattern relating spatial scale and diversity is the positive relationship between the size of an island of habitat and the number of species that are found in that habitat (Preston, 1962). If S is the number of species, then

$$S = cA^z$$

or, in a form easier for calculations,

$$\log S = \log c + z(\log A)$$

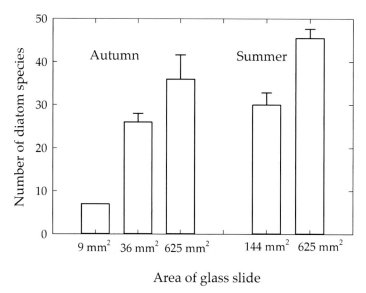

Figure 11.9 Number of diatom species on glass plates of various surface areas after 1 week in Ridley Creek, Pennsylvania. *Data from Patrick (1967).*

where c is a constant measuring the number of species per one unit area and z is the slope of the line on a plot of the log of the area versus the log number of species. The relationship between size of habitat and area holds well for many types of organisms. The species−area relationship applies broadly (Rosenzweig, 1995) and includes diatoms colonizing glass slides (Fig. 11.9), invertebrates on stones in streams (Fig. 11.10A), and fishes in lakes (Fig. 11.10B). This relationship can be used to estimate the effects of habitat reduction on species diversity (Example 11.2). However, caution is in order; the relationship does not always hold. There is a small island effect, where islands below a minimum size show no further decrease in species richness. This effect was documented for a series of small ponds in Hungary; diatom communities had a low and fairly constant richness up to about a surface area of $1,000 \text{ m}^2$ and diversity increased with pond size beyond that threshold (Bolgovics et al., 2016).

Spatial considerations also determine diversity of a given habitat. The distance from a source of colonists paired with the area of the habitat allows researchers to estimate a relative expected diversity (MacArthur and Wilson, 1967). Although the specifics of these models are somewhat complex to treat here, they include two important points related to biodiversity and success of invaders. First, lowering the connectivity of habitats (i.e., sources of natural colonists) can disrupt the natural diversity. Second, human transport of organisms can vastly reduce the effective isolation of some habitats.

Succession, the sequence of species colonizing newly formed habitats, is a dominant and long-standing ecological principle. The central idea related to diversity is that

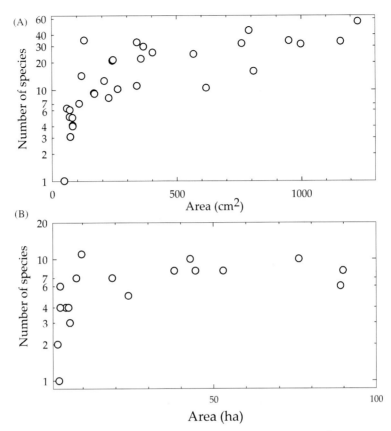

Figure 11.10 Number of invertebrates as a function of stone size in an Australian stream and number of fish as a function of lake size in small Wisconsin lakes. *(A) From Douglas and Lake (1994); (B) Data from Tonn and Magnuson (1982).*

Example 11.2 Calculating expected numbers of species after a habitat reduction

A developer wants to decrease the area of a small lake in Wisconsin from 90 to 10 ha. Given that data from Tonn and Magnuson (1982) can be used to calculate values of $c = 2.211$ and $z = 0.3325$ for lakes of similar size, calculate the expected number of fish species for each size of lake.

For a 90 ha lake, $\log S = \log 2.211 + 0.3325 \times \log(90) = 0.994$, so $S = 9.9$ species of fish

For a 10 ha lake, $\log S = \log 2.211 + 0.3325 \times \log(10) = 0.677$, so $S = 4.8$ species of fish

Thus, we can expect that half the fish species will be supported in the smaller lake. This calculation ignores other possible factors such as what type of habitat will be present after alteration of the lake and the high variance in the predictive relationship. It also does not tell us which species will survive. However, such relationships are useful for illustrating the idea that the size, as well as the quality of habitat, is an important parameter in conserving biodiversity.

---

species that occur early in succession reproduce rapidly and disperse widely. Species that occur late in succession are better competitors and displace the early colonists. Diversity is low early in succession because only the early colonists are present. Diversity is also low late in a successional sequence because only the competitive dominants survive. Diversity is maximal in the intermediate successional stages because both early and late successional species coexist. Despite these general patterns, chance can play a large part in the actual species that become established in a successional sequence in new aquatic communities (Jenkins and Buikema, 1998). We discuss specific examples of succession in Chapter 22.

The *dynamic equilibrium model* of species diversity suggests that disturbance, such as drying or flooding, can decrease diversity if it exceeds the ability of species to recover. However, it can increase diversity if it removes dominant species that outcompete others (Huston, 1994). Pollutants and other environmental pressures can decrease diversity accordingly. For example, species diversity in lakes and rivers decreases dramatically with phosphorus pollution (Azevedo et al., 2013). We will also discuss this more thoroughly in Chapter 22.

The *intermediate disturbance hypothesis* (Connell, 1978) relates diversity to disturbance frequency and intensity. This hypothesis suggests that maximum diversity occurs at intermediate levels of natural disturbance, and links to the concept of diversity as a function of succession discussed in the paragraphs above. At high rates of disturbance, only early successional species, which are generally good colonizers but poor competitors, are able to survive. At very low levels of disturbance, late successional species, which are strong competitors, dominate. At intermediate levels of disturbance, both types of species co-occur, and diversity is greatest. The problem with the intermediate disturbance hypothesis is that it provides no *a priori* way to establish relative disturbance intensity among communities. Again, we provide more detail in Chapter 22.

Species interactions can alter diversity. For example, Leibold et al. (2017) found that grazing zooplankton doubled phytoplankton diversity in experimental ponds. Disturbances by humans can also cause variations in species diversity, and human alterations of natural disturbance regimes can alter diversity. For example, impoundments that reduce the frequency and magnitude of downstream flooding events in rivers are linked to reductions of native fish species that are adapted to the natural disturbance regime (Poff et al., 2007; Geist, 2011). Assessing changes in types and numbers of species present forms the basis of several indices of biotic integrity. We describe use of such indicators in Chapter 16.

GENETICS AND POPULATIONS OF SPECIES

Genetic diversity, within and among species, is a key component of biological diversity. Using modern molecular techniques, ecologists can now analyze sediment and water samples for microbial genetic diversity by amplifying and sequencing DNA extracted from environmental samples. Maps have even been created of global genetic diversity (Miraldo et al., 2016). Within species, genetic diversity is critical for long-term viability because it determines an organism's ability to adapt to dynamic environments. Genetic diversity can increase the probability that populations and species can survive adverse periods (e.g., prolonged drought or climate changes), disease outbreaks, and other causes of mass mortality, because more genetically diverse populations are more likely to include individuals with differential susceptibility to these stressors.

Genetic diversity includes the proportion of polymorphic loci across the genome (gene diversity), numbers of individuals within a species with polymorphic loci (heterozygosity), and the number of alleles per locus. Ultimately, mutation is the source of genetic diversity, and it is maintained by continued mutations, emigration and immigration, and environmental variability. Environmental variability can influence genetic diversity at multiple scales over relatively short-time periods. For example, genetic diversity of the European freshwater snail, *Radix balthica*, decreased locally in response to a severe drought (within a single study site; alpha diversity), but genetic diversity across sites (beta diversity) simultaneously increased at larger scales (Evanno et al., 2009). These patterns of alpha and beta diversity were the same for snail species diversity within and among the multiple study sites.

Conservation efforts for endangered species increasingly focus on maintaining or enhancing genetic diversity because it inherently links to species survival (Vrijenhoek, 1998). The same is true for fish and invertebrate aquaculture programs, where genetic diversity is key to long-term viability and productivity, as well as for retaining genes for disease resistance. Genomic methods will become more important, for example Macdonald et al. (2017) applied sequencing methods to guide conservation of genetic diversity in three stream invertebrates, two stonefly species and a mayfly.

Conservation efforts for rare species strive to keep population sizes above the *minimum viable population*, which is the lowest calculated number of individuals required for a population to survive in the wild given the acceptable risk of extinction. Within a population, only a portion of the members contribute genes to succeeding generations, and these individuals constitute the *effective population size*, which can be a fraction of the actual population, depending on a variety of demographic factors and life history specifics.

Below the minimum viable population size, the probability a population will survive stochastic (random or chance) environmental or demographic events is low. Accurate estimates of minimum population size are difficult, and they vary greatly depending on the life cycles and natural histories of individual species of interest. Using life history information and a criterion of less than 10% probability of extinction in 100 years, minimum viable population size for the threatened Rio Grande cutthroat trout was estimated at 2,750 individuals (Cowley, 2008).

A review of published estimates of minimum viable population sizes for a variety of terrestrial and aquatic endangered species found a median value of 4,169 individuals using a standardized population persistence probability of 99% (Traill et al., 2007).

The *extinction vortex* (Gilpin and Soulé, 1986) suggests that as populations become smaller, their susceptibility to extinction accelerates exponentially. Because of both environmental and demographic stochasticity, smaller populations have much greater probabilities of going extinct. As populations decline, they experience degraded genetic diversity and individual fitness, and numbers of individuals fluctuate widely as a function of environmental variability (Fagan and Holmes, 2006). The extinction vortex illustrates the "slippery slope" of extinction and the importance of maintaining viable population sizes for effective conservation efforts.

The controversy over endemism and microbial species

There are two major views of microbial diversity, the first is that "everything is everywhere and the environment selects" (Whitfield, 2005), and second is that there is dispersal limitation of microbes. The first view is supported by a large amount of work on protozoa. These studies have found few endemic species of free living protozoa (although patristic protozoa can have a limited host range), and taxonomic keys from one part of the world work as well in another distant location (Finlay, 2002). The fact that microbes have such huge populations suggests that even if the probability of transport of any individual cell is low, the probability that at least one cell out of billions will be transported is high. Of course, the organisms must survive the transport, so if an aquatic species is not resistant to desiccation, it will not usually survive terrestrial transport. Extreme weather events like hurricanes and tornadoes can move water and soil high into the atmosphere. Birds migrate long distances, some from the North to the South Poles, and are known to transport organisms. Furthermore, sediments taken from ponds can yield species of protozoa that are not currently present in the ponds, but can be induced to grow by altering environmental conditions such as temperature, salinity, and food sources.

Still, there are microbes that do not have good dispersal stages. Diatoms are an example of this. Diatoms are easily identified taxonomically because of the distinctive ornamentations on their frustules. A substantial number of endemic freshwater species are found in New Zealand (Kilroy et al., 2007), a location that is geographically very distant from many other freshwater sources. While this high endemism appears related to geographic isolation, these diatoms might not have been found elsewhere because of incomplete taxonomic coverage, or that the distinctive ornamentation is not indicative of distinct species. The second argument, that the diatoms are not actually different species than those found in other parts of the world, could be settled with molecular analyses such as DNA barcoding.

The argument between the two world views has yet to be settled. A study of 16 regional data sets from North America and Europe demonstrated that assemblages of lake diatoms responded regionally to changes in pH. These data suggest that dispersal rates are slow enough that it might be possible for genetic differentiation and endemism to develop (Telford et al., 2006). As more habitats are sampled, and molecular techniques allow for distinction of morphologically similar species, a definitive answer should emerge as to what groups have truly global populations and what other groups are geographically isolated. Until we have this answer, Griffith (2012) argued that we need a strategy to conserve potentially endemic microbes.

GLOBAL CHANGE AND SHIFTS IN BIODIVERSITY

Changes in temperature, precipitation, and annual timing and intensity of climatic events are already altering the distribution of species. Additional changes, such as nutrient inputs, land use disturbance, damming rivers, and altering flow regimes, also occur globally and are altering biodiversity. The current global biodiversity crisis related to these and other causes is most pronounced in aquatic habitats, where extinction rates are highest (Vaughn, 2010). Managing biodiversity of aquatic species in the Anthropocene will be challenging given the lack of understanding of all responses to change and no prior analog to such changes during times when scientists directly studied freshwater systems (Kopf et al., 2015). Global change effects on biodiversity is a reoccurring theme throughout this book. For example, we consider effects of changes in network structure in watersheds on diversity in Chapter 25, and influence of nutrients and pollutants on diversity in Chapters 17 and 18.

Climate change will probably increase numbers of invasive species in aquatic habitats as conditions become more favorable for generalist species with relatively wide thermal tolerances and good dispersal capabilities (Rahel and Olden, 2008). Species intolerant of colder water temperatures will also be able to spread to higher latitudes and altitudes as these regions warm. These invaders will likely be mobile, weedy species, such as the American Bullfrog (*Rana catesbeiana*) and Asian Carp (*Hypophthalmichthys* spp.), while less mobile and more sensitive species decline.

Microbial species and some invertebrates may be able to adapt quickly enough to keep pace with climate change because of their rapid generation times, and thus may not need to shift their distributions to survive. Life cycles of some lentic invertebrates are already shifting with climate change, and some species are experiencing decreases in body size as water temperatures warm, a common response observed among ectothermic vertebrates and invertebrates (Stoks et al., 2014). Other documented responses to warming include increases in growth and developmental rates, the latter of which should facilitate adaptation as conditions change.

One of the key issues with global warming is the ability of species to expand their distributions into areas that were formerly too cold for their continued existence. While thermal refugia may be available for many terrestrial systems for organisms to expand their distributions to greater latitudes, this avenue is not available for aquatic organisms that are confined to drainage networks without significant north–south gradients. In this case, mountain streams may be good refugia for cold-water species as long as abiotic and biotic characteristics other than temperature are appropriate (Isaak et al., 2016). However, not all drainage systems have elevation-based temperature refugia.

Climate change will interact with other human impacts to alter diversity. For example, warming as well as nutrient enrichment has altered plant communities in the lower Mississippi River delta (White and Visser, 2016). Bush and Hoskins (2017) modeled range shifts of 527 freshwater fish species in New South Wales, Australia and demonstrated that substantial decreases in range sizes of these species will occur with warming. Distribution models from various areas of Earth suggest that regions that are currently too cool for some species will experience increases in diversity, while regions that are already warm and dry will lose species.

Climate change effects will be difficult to predict without understanding the ecology of specific systems. For example, warming will encourage blooms of the toxic cyanobacterium *Microcystis*. However, zebra mussels facilitate blooms of the cyanobacterium, and they may not survive warming episodes. Therefore, prediction of the influence of warming on toxic blooms in some lakes may not be straightforward (White et al., 2017). Further, warming temperatures will likely interact with other stressors such nutrients, metals, and pesticides, rendering the ultimate responses of individual species difficult to predict. Many studies suggest that aquatic animals will be more susceptible to pollutants and other stressors as temperatures warm. For example, Laetz et al., (2014) demonstrated that the susceptibility of juvenile Coho Salmon to organophosphate insecticides increased with warming temperatures within the range they normally experience.

One ingenious study on streams circumvented some of the problems associated with short term warming experiments at less than the ecosystem level. Nelson et al. (2017a,b) sampled two cold Iceland streams for a year, and then water in one of the streams was warmed by heat exchange with a nearby geothermally warm stream. They found warming decreased invertebrate abundance but total biomass did not change as larger species became more common in the experimental (warmed) stream over two years. Their results indicated greater energy flux through the food web occurred with warming.

NONNATIVE SPECIES

Invasion by species introduced by humans, either intentionally or inadvertently, is probably the most permanent form of pollution. Dispersal of species is one of the less recognized forms of global change instigated by humans (Vitousek et al., 1996). Once an invader becomes established, eradicating it is almost impossible. We do not know of any examples of eradication of an aquatic pest species once established in a large area (although some species may be successfully controlled to low numbers). The best way to control invasive aquatic species is to regulate their initial introduction, as illustrated by efforts to control invasive aquatic plants (Hussner et al., 2017). Once species are introduced, the changes they lead to may never reverse. For example, the peacock bass (*Cichla monoculus*) was introduced into Lake Gatun, Panama a half century ago. Even 45 years after the introduction the dramatic fish community shifts still persist (Sharpe et al., 2017).

Understanding the ecology of ecosystems is essential to managing invasive species. Aquatic invaders occur world-wide and have broad effects on the communities and ecosystems that they invade (Gallardo et al., 2016; Tricarico et al., 2016). Movement of materials may, in part, be responsible for increased diseases in some freshwater organisms (Johnson and Paull, 2011).

As humans increase speed and frequency of global travel and move more materials across international boundaries, the probability for transport of undesirable species also increases dramatically. Humans have created a world without borders for many species (Mack et al., 2000), and rates of human-caused species invasions are far greater than natural rates, although natural invasions of species influence how evolution of new species relates to community formation. Numerous case studies illustrate potential problems associated with unwanted species introductions; we will present only a few here.

Why do species invade successfully? Predicting which species will invade and the effects of the invasion is difficult (Fuller and Drake, 2000). A sound knowledge of the natural history of the invader and the community it has entered allows for the best predictive assessment of the influence of an invading species. Qualitative techniques may be used to assess the risks associated with freshwater species introductions (Li et al., 2000). Freshwater species with traits that encourage growth and consumption, while avoiding predators, are more likely to be established according to a broad meta-analysis (McKnight et al., 2017).

Moyle and Light (1996) proposed a conceptual model of species invasions and establishment based on case studies of species of invading fishes. Their points appear useful and are generalized here with additional observations provided by Kolar and Lodge (2001). As with any conceptual ecological model, exceptions exist, and the model provides one way to conceptualize the problems. First, most invaders fail to establish. Most failed invasions are never documented; thus, the success rate of invasions is not well-known. Second, most successful invaders establish without major effects on the ecosystem or community, although some have major effects. Third, all aquatic systems can be invaded. Fourth, major community effects are observed most often in low-diversity systems, including island and highly disturbed habitats. Fifth, top predators that invade successfully are more likely to have strong community effects than successful invaders at lower trophic levels. Sixth, species must have physiological and morphological characteristics suited to the environment to invade successfully. Seventh, invaders are most likely to establish when native assemblages are disturbed. Natural or anthropogenic disturbance increases susceptibility to invasion. Eighth, success of invaders can depend on environmental variability. Invasion and establishment in an extremely harsh and variable environment may be difficult unless invaders are preadapted to the variation (see the sixth point). Variation can also play a role in the seventh point. Ninth, even stable systems may be susceptible to invasion. Tenth, the greater the number of invading individuals and times they are introduced, the greater the probability they will become established. Finally, species with a history of being invasive are most likely to invade other habitats. Thouvenot et al. (2017) added to this list and noted that some invasive species can facilitate invasions of others.

The previous model does not follow more traditional treatments of community assembly theory that depend on the sequence of invasion and species interactions. Many treatments of invasion success have revolved around competition and predation as determinants (Roughgarden, 1989). For example, early attempts at understanding success of invasions led to suggestions that alien species are more likely to colonize cropland and disturbed habitats, and by extension, more diverse communities are less susceptible to invasions (Elton, 1958). This idea of diverse communities being resistant to invasion has since been challenged (e.g., Moyle and Light, 1996; Levine, 2000). Still, disturbance and alteration of habitat can influence the success of species invasions. For example, reservoirs can harbor lentic species in river networks that historically had little lake habitat and receive many introduced fish species. These fishes that were rare or absent historically can invade rivers and streams upstream from the reservoirs (Falke and Gido, 2006).

The Great Lakes of North America have experienced a tremendous number of invasive species (Highlight 11.1), including the zebra mussel (*Dreissena polymorpha*), which has had considerable economic and ecological impacts. The spread of the zebra mussel is one of the more pronounced examples of the effects that a nonnative species can have, in terms of both explosive invasion and economic consequences (see Highlight 10.1). This increase in invasive species is of broader concern because the Great Lakes now connect to the Mississippi River basin via the Chicago waterworks (Grippo et al., 2017).

Highlight 11.1 Invaders of the North American Great Lakes

The Great Lakes have a long history of invasions of exotic organisms caused by humans (Table 11.1). As of 1990, more than 139 alien species had become established, including plants, fish, invertebrates, and algae (Mills et al., 1994). By 1998, 145 alien species had been documented (Ricciardi and MacIsaac, 2000), and by 2009, the National Center for Research on Aquatic Invasive Species reported 181 nonnative species in the Great Lakes. Some of these introductions have had massive economic impacts, such as the effect of the zebra mussel on water intakes for municipalities and industries and the crash of the large lake trout fishery caused by the sea lamprey. Damages to reverse negative effects on water quality from invasion of a single species, the spiny water flea (*Bythotrephes longimanus*) would cost between $86 and $163 million. Many of the other introductions have had moderate impacts or impacts that do not directly affect humans in an economic sense.

Table 11.1 Some invaders of the North American Great Lakes

Organism	Year	Source	How introduced	Effects
Sea lamprey, *Petromyzon marinus*	∼1830	Atlantic	Shipping canals	Decreases native lake trout
Purple loosetrife, *Lythrum salicaria*	1869	Europe	Ship ballast	see Highlight 9.4
Alewife, *Alosa pseudoharengus*	1873	Atlantic	Shipping canals	Suppresses native fish species; new prey for salmon
Chinook salmon, *Oncorhynchus tshawytscha*	1873	Pacific	Intentional	New piscivore; important sport fish
Common carp, *Cyprinus carpio*	1879	Europe	Intentional	Destroys habitat for waterfowl and fish
Brown trout, *Salmo trutta*	1883	Europe	Intentional	New piscivore; important sport fish
Coho salmon, *Oncorhynchus kisutch*	1933	Pacific	Intentional	New piscivore, important sport fish
White perch, *Morone americana*	∼1950	Atlantic	Shipping canals	Competes with native fish
Eurasian watermilfoil, *Myriophyllum spicatum*	1952	Eurasia	Not known	Competes with native plants
European ruffe, *Gymnocephalus cernuus*	1986	Europe	Ballast water	Competes with native fish and eats eggs
Zebra mussel, *Dreissena polymorpha*	1988	Europe	Ballast water	Biofouling; outcompetes native species

Source: After Mills et al. (1994).

Routes of introduction include intentional releases, movement through the aquarium trade and bait buckets, release of ship ballast water taken from other areas, and connection of the Great Lakes with the Atlantic Ocean by a system of shipping canals (the Erie Canal and the St. Lawrence Seaway) and other canals with the Mississippi basin.

Measures to limit the number of new introductions include requirements that ships from overseas originating in freshwater ports exchange ballast water while at sea to avoid additional introductions of freshwater species. However, species that are tolerant to salinity changes will not be kept out by these methods (Ricciardi and MacIsaac, 2000). It is likely that new species will continue to be transported into the Great Lakes. Control of new species introductions has been attempted since the first edition of this book was written (data from 1999), yet the number of invasive species continues to climb. The voluntary measures and the mandated open ocean ballast change regulations have had little influence to date, the rates of invasion still continue to increase (Holeck et al., 2004). The damage associated with some of these species and the existing nonnative organisms will continue to be considerable.

A recent example of efforts to halt invasion of two species includes bighead (*Hypophthalmichthys nobilis*) and silver (*Hypophthalmichthys molitrix*) carps. These two species of Asian carps are present in the Mississippi River basin and threaten to move into the Great Lakes via canals that connect the Great Lakes to the Mississippi. Some methods that are being tried or under development include electric barriers and elevated CO_2 concentrations in connecting waters. Managers are also considering targeted toxin particles as well as using algal attractants as bait for traps. One of the major barriers to understanding if the fish have entered the Great Lakes is that it is very difficult to capture and verify their presences with traditional techniques when carp numbers are very low early in the invasion process. One approach to provide early warning for their presence is to use eDNA to indicate presence of species in waters without their direct capture. Initial testing suggests that eDNA approaches will be useful (Jerde et al., 2013).

Another particularly difficult case of a species that has recently become a nuisance, even in areas where it is native, is the diatom *Didymospenia gemenata* (Fig. 11.11). It is native to the northern hemisphere and is considered an invasive species in New Zealand, Argentina, Chile, and Australia, although it may have been native in South America, just previously undetected (Taylor and Bothwell, 2014). It was possibly moved to these locations by anglers with diatom materials on their equipment traveling to fish in other areas. This stalked diatom will bloom in cold-water streams, covering the entire bottom of the stream with a brownish slimy layer a few centimeters thick. The slime is composed of the mucopolysaccharide stalks of the diatom and leads to the common name "rock snot." This slick layer covering the stream bottom inhibits angling by creating slippery conditions and fouling fishing gear. The diatom cover can

Figure 11.11 The diatom *Didymospenia gemenata* causes dense mats that completely cover the bottom of some streams (A) and has become abundant in many streams were it previously was not found. It forms dense mucilaginous mats (B) and is referred to commonly as "rock snot." *Images courtesy: Justin Murdock.*

alter the food web by shifting invertebrate community composition, although the exact response is not predictable (Larned and Kilroy, 2014). The apparent "blooms" of the diatom are correlated with low soluble reactive phosphorus concentrations. The low phosphorus induces formation of the stalks, leading to the nuisance conditions. The recent apparent increase in cases of blooms might be caused by climate change-altered P inputs into streams and increased atmospheric N deposition leading to less transport of P to streams (Bothwell et al., 2014). Experiments in New Zealand streams indicate that elevated P concentrations discourage *D. gemenata* blooms (Kilroy and Larned, 2016). Other local factors such as low nitrate and scouring floods may also be important (Jackson et al., 2015; Bray et al., 2016). Keeping anglers from transporting cells into new areas is an obvious method of control, but once the blooms are already established it is difficult to know how to achieve control. Addition of phosphate may discourage the blooms as James et al. (2015) demonstrated in whole-stream fertilization experiments, but downstream transport will cause other unwanted algal blooms.

Many other introductions of nonnative species to freshwaters have occurred (Table 11.2) and many of these were intentional. Introduction of fish species has led to homogenization of fish communities; many of the same species are found across the continent now (Rahel, 2000). Some fishes are introduced for sport or harvest, but another common reason for fish introductions is for mosquito control. At least 10 species of fishes have been released for this purpose, but their effectiveness in mosquito control is doubtful (Azevedo-Santos et al., 2016). Hopefully, this practice of purposely stocking exotic species will decrease in popularity as it becomes clear that unintended consequences are common with species introductions.

Table 11.2 Some species introductions not covered elsewhere in the text

Species	Source	Habitat introduced to	Effects	Citation
Trout, *Salmo* spp.	Intentional introduction	Introduced world-wide, pariculary strong effects in New Zealand lakes and streams	Extinction of native galaxiid fishes in New Zealand	Flecker and Townsend (1994), Townsend (1996)
Carp, *Cyprinus carpio*	Intentional introduction	North American lakes and streams	Habitat destruction, increased turbidity, competition with native fishes	Lever (1994) and Bajer et al. (2016)
Asiatic clam, *Corbicula fluminea*	Probably deliberate	Sandy sediments of lakes rivers and streams in North America	Possible competitive effects on native bivalves, high rates of filtration	Strayer (1999) and McMahon (2000)
Water hyacinth, *Eichornia crassipes*	Intentional introduction	Tropical and subtropical rivers worldwide	Interferes with boat traffic and waterflow; outcompetes native macrophytes	
Opossum shrimp, *Mysis relicta*	Intentional introduction	Flathead Basin	Collapse of salmon fishery	Spencer et al. (1991)
Daphnia lumholtzi	Unknown, probably human caused	Southeastern and south–central, United States	Possibly more resistant to predation by zooplanktivores	Havel and Hebert (1993) and Kolar and Wahl (1998)
Bythotrephes cederstroemi	Ballast water from Europe	Northeast United States	Possibly more resistant to predation by zooplanktivores	National Research Council (1996)
Crayfish, *Pacifastacus leniusculus*	Intentional introduction	Europe	Transmission of disease and competitive removal of native crayfish	Vorburger and Ribi (1999)

This list contains only a small portion of the introduced species that have entered freshwaters (e.g., Benson, 2000).

EXTINCTION

Man has been reducing diversity by a rapidly increasing tendency to cause extinction of supposedly unwanted species, often in an indiscriminate manner. Finally, we may hope for a limited reversal of this process when man becomes aware of the value of diversity no less in an economic than in an esthetic and scientific sense.

—Hutchinson (1959).

Humans likely will cause extinctions of at least half of all the approximately 10 million species on Earth in the next 50–100 years (May, 1988), despite the fact that the problem has been recognized clearly for more than 70 years. May (1988) calculated that worldwide extinction rates are currently 1 million times greater than rates of evolution, and the majority of taxonomists and ecologists understand that we are causing the sixth mass extinction event over the 3.5 billion years life has existed on Earth (Pimm et al., 1995; Wake and Vredenburg, 2008).

In an even bleaker assessment of the situation, when area–diversity relationships (discussed previously) are used to estimate the ultimate effects of habitat destruction, the habitat destruction caused by humans to date will ultimately result in the loss of 95% of the earth's species (Rosenzweig, 1999). Freshwater habitats are probably the most impacted by humans because they integrate the landscapes they drain. Consequently, aquatic systems suffer corresponding effects from all extensive upstream terrestrial disturbances. Humans live near water and discharge their wastes into it. Control of hydrology by humans destroys habitats for many species. Consequently, losses of freshwater species have been substantial (Warren et al., 2000; Jenkins, 2003; Strayer, 2006), and the outlook for sensitive aquatic species is bleak (Folkerts, 1997; Vaughn, 2010).

Rates of species extinction caused by humans far exceed the rates of evolution of freshwater organisms. More than 300 species of endemic fishes may have evolved in Lake Victoria, Africa, during the past 12,000 years (Johnson et al., 1996), one of the highest rates of species evolution ever documented. Many of these species have disappeared during the past two decades from overfishing, introduction of nonnative predators, and eutrophication (Barel et al., 1985; Kaufman, 1992; Seehausen et al., 1997; van Rijssel et al., 2016). About 200 of these species disappeared in a single decade, primarily because of introduction of *Labrus niloticus*, the predatory Nile perch (Goldschmidt et al., 1993); the pharyngeal jaws of the native fauna evolved in a way that this group of species consisted of inferior competitors to the perch (McGee et al., 2015). The Victoria fishes may represent one of the most rapid speciation events, but species accrued 500 times more slowly than humans destroyed them.

The Amazon basin is another area where large-scale extinctions could occur. There are over 277 planned dams and already 155 existing dams in the basin. About 20% of the watershed has been developed for cattle grazing or crop production, and in those areas 50% of the riparian zones are impacted (Castello and Macedo, 2016). Given the over 800 species of freshwater fishes documented from the basin (and likely many more undescribed species), the potential for biodiversity loss is high. There have already been local extinctions (extirpations) related to fishing pressure (Castello et al., 2015).

Scientists and resource managers generally assume that endemic species are most likely to go extinct and deserve the most attention for conservation. Some cosmopolitan species also could go extinct because they generally rely on habitats that are temporarily available for colonization and act as fugitives that move across the landscape from one habitat to another. The massive alteration of habitats associated with human activities may cause extinction of such species, and this may have a broader geographic effect on community formation and ecosystem function than extinction of localized endemics (Cairns, 1993). The blackfin cisco (*Coregonus nigripinnis*) was once so abundant in the North American Great Lakes that it was fished commercially. Overfishing and predation by the exotic sea lamprey (*Petromyzon marinus*) led to catastrophic declines. The last individual was observed in Lake Michigan in 1969, and it was listed as extinct in 1996.

Many endangered species are aquatic or require aquatic habitat. In 2016, the US Fish and Wildlife Service listed hundreds of aquatic species or subspecies on their threatened and endangered list, including 163 fishes, 89 bivalves, 52 snails, 35 amphibians, and 28 crustaceans. All these numbers have increased, by about 20%, since 2009, and increased ~30% between 1999 and 2009. In addition, mammals and birds that require aquatic habitat are listed, such as the whooping crane (*Grus americanus*), the Florida panther (*Puma concolor coryi*; the only remaining habitat for this animal is wetland), and the piping plover (*Charadrius melodus*).

Globally, the IUCN (World Conservation Union Red List, searched March 2017) lists 26,070 extinct, extinct in wild, regionally extirpated, endangered, or vulnerable freshwater species. It lists 156 species as already extinct. Many more may be endangered, but because they are inconspicuous they have not been listed (Strayer, 2001). Many wetland plants, particularly those associated with the vernal pools of California, are also listed as endangered and threatened. About half of all the crayfish species in United States and Canada are threatened. Pressures on native crayfish include habitat destruction, limited natural ranges, and introduced competitors (Taylor et al., 1996).

Some of the most endangered fishes are the very large species found mostly in tropical waters (Stone, 2007), as they are susceptible to overharvesting, habitat fragmentation, and live in large rivers where pollution can be considerable. The largest freshwater fish can be up to 500 kg and 700 cm long (the Chinese paddlefish *Psephurus gladius*). Of the 20 largest species, 6 are endangered. This does not include the endangered sturgeon species that move between fresh and salt water. Most of these species are endangered by overharvesting, but pollution and habitat fragmentation (by dams) are also of concern. Interviews with elders who have fished the waters where these giants live for decades confirm that numbers and sizes of these massive fishes are decreasing.

The very largest species, the freshwater megafauna, have been proposed as flagship species for aquatic conservation (Carrizo et al., 2017). These are classified as animals larger than 30 kg as adults, and this includes large mammals such as hippopotamuses and freshwater dolphins, reptiles such as crocodilians, and fishes. These species are assessed by the Conservation of Nature (IUCN) Red List and 58% of them are endangered. A preponderance of these species are found in South America, Central Africa, and South and Southeast Asia.

Just because a group of species does not appear on the list, it does not mean that no species in the group are threatened with extinction. We know much less about smaller organisms than larger ones. New species of aquatic insects, protozoa, algae, and bacteria are described daily, whereas new species of mammals and plants are described more rarely. Pollution, dewatering, and habitat destruction have negative effects on all freshwater organisms from fishes to bacteria. Therefore, small organisms may not be listed as endangered simply because their existence or the state of their populations are not known.

Many species are *extirpated* (locally extinct). For example, the pearly mussels (Unionacea) of North America are among the most imperiled groups of animals on Earth. These mussels are long-lived and most inhabit mid to large rivers, and are thus influenced by any disturbances occurring in the watershed above (Strayer et al., 2004). It is particularly easy to document species disappearances because old mussel shells can be found on river banks, particularly in archeological sites of indigenous peoples that harvested them (Miller et al., 2014). Of the 44 species of unionid mussels ever found in Kansas, 4 are extinct, 6 are endangered statewide, and an additional 16 are listed as threatened (Obermeyer et al., 1997). Not all these species are rare or extinct in other parts of their historical range. The numbers of extinct, endangered, and threatened mussel species are similar in all states throughout the Mississippi drainage. Between 34% and 71% of the mussel species found in the southeastern United States are threatened or endangered. Thirty-six species that are thought to be extinct in the United States were found in the Tennessee River basin alone (Neves et al., 1997).

Some states in the western United States have high proportions of threatened endemic species. For example, in the Great Basin, Klamath and Sacramento basins, 58% of the native fishes are endemic. All the native fish species are threatened in Nevada (Warren and Burr, 1994). One species, the Devils Hole Pupfish (*Cyprinodon diabolis*) is highly endangered and emblematic of many of the challenges in protecting rare species (Highlight 11.2).

Highlight 11.2 Conserving the highly endangered Devils Hole Pupfish (*Cyprinodon diabolis*)

The Devils Hole Pupfish is one of the most endangered species on the planet and was one of the first species listed in 1967 under the Endangered Species Preservation Act of 1966. It was grandfathered into the United States Endangered Species Act in 1973. This species occurs in only one warm-water filled cavern limnocrene (water filled cavern with no outflow) in the Nevada Desert. This small warm-water spring is less than a 100 m^2 in area (Fig. 11.12) but is over 124 m deep (the bottom has not been found so

Figure 11.12 The Devils Hole Pupfish (A), the only small spring where it is found with water monitoring gear installed (B), and Ash Meadows Fish Conservation Facility (C). *Photographs A and B courtesy: United States National Park Service.*

the true depth is not known). The fish are only found down to about 24 m and are entirely dependent upon a 4 by 6 m shelf for food (algae and invertebrates growing on the shelf) and reproduction.

The fish were threatened in the 1960s and 1970s from groundwater withdrawals for agriculture that were decreasing the water levels enough to expose the shelf. This continued until the federal government won a Supreme Court lawsuit in 1976 limiting groundwater pumping. Between the 1970s to 1990s the average population was 300 fish, although the population fluctuates because of low food availability during the winter and spring when little light reaches the shelf and algal growth is limited.

Since the 1990s the population has dropped to around 35 fish several times. It is not clear why the population has dropped. Several changes have occurred that may be responsible. One potential recent threat is that Devils Hole has been colonized by a dytiscid beetle, *Neocyeodytes cinctellus*. This relatively recent colonist might eat the pupfish eggs, compromise their viability, or might be competing for food. Ostracods that used to be found in the pupfish guts are rarely found in guts since the beetle colonized. The fish live only a year, lay only one egg at a time, and can cannibalize their young. While the population would double during good growth periods, recently the population has been stable at around 100 fish.

The United States Fish and Wildlife service is establishing an additional population in case natural or human-caused events eradicate the wild population. As such, the Ash Meadows Fish Conservation Facility was established to rear an independent population (Fig. 11.12). This was a daunting task, prior efforts to propagate the species had been unsuccessful. The facility is built to replicate the natural habitat with a lighted shelf and a deep cavern. Temperature and chemistry are tightly controlled to match that in the Devils Hole. The facility started with eggs collected in Devils Hole, and now has successfully bred fish. They have about 30 fish now.

The future of this species is in question. It is not clear if the population will survive given the population is likely below the minimum viable population size. The expense and effort to conserve this species is high, and if the species is destined for extinction, this could all be for naught.

Ongoing amphibian declines are resulting in mass extirpations and extinctions of much of the amphibian diversity on the planet (Collins and Crump, 2009; Highlight 11.3). An estimated 200 species of frogs have gone extinct over the century and 100s more are expected be lost over the next century (Alroy, 2015). Many of the amphibian species that we are losing inhabit freshwater habitats for at least part of their lives, and declines are homogenizing amphibian species diversity (Smith et al., 2009). Unfortunately, regulatory barriers and delays are problematic in protecting endangered amphibians in the United States (Walls et al., 2017) and presumably elsewhere.

Efforts to conserve species are facilitated by protecting areas with high diversity so as many species can be protected simultaneously as possible. Collen et al. (2014) analyzed global patterns of diversity of mammals, amphibians, reptiles, fishes, crabs, and crayfish. They found the freshwater representatives were at greater threat to extinction. They analyzed three measures of diversity that are used to assess conservation priority: endemism, species richness, and richness of threatened species. They found that these three indicators were not congruent with each other, and that the degree to which the indicators agreed was not consistent with the group of organisms considered. Efforts are rarely made to conserve freshwater invertebrate species, and efforts to conserve vertebrates will not necessarily protect invertebrates.

Highlight 11.3 Global amphibian declines

Amphibian populations are declining around the planet at alarming rates (Collins and Crump, 2009). Estimates indicate that 43% of all amphibian species have experienced some degree of population decline, 33% are globally threatened, and over a hundred may be extinct (Stuart et al., 2004). Further, most of these losses have occurred in the last three decades. The causes of some amphibian declines are still not entirely understood, but climate change, ultraviolet-B radiation, fungal pathogens, pesticides, habitat destruction, overexploitation, or combinations of these factors have all been implicated (Kiesecker et al., 2001; Davidson, 2004; Lips et al., 2006; Pounds et al., 2006).

In cases of severe habitat destruction and pollution, declines of amphibians and other freshwater organisms are not surprising. However, in many regions, amphibians are disappearing from seemingly undisturbed habitats such as high mountain cloud forests in the tropics, national parks in the United States, and other protected areas around the world. Scientists originally labeled these declines "enigmatic," but investigations have since linked many of these mysterious events to the chytrid fungus *Batrachochytrium dendrobatidis* (Bd) (Skerratt et al., 2007; Lips et al., 2008). Since its discovery in the 1990s, Bd has now been associated with amphibian declines on every continent inhabited by amphibians. Bd causes the disease chytridiomycosis, which spreads rapidly through populations and can be lethal to most species under appropriate conditions. In many cases infected frogs become lethargic, slough their skin, and die, but otherwise show few symptoms.

Sudden, catastrophic amphibian declines associated with chytridiomycosis have been documented in remote, mountainous regions of Central America (Fig. 11.13), resulting in extirpations of $\sim 1/2$ the species and an overall 50% decrease in the abundance of ampibians (Lips et al., 2006). Extinctions of some African, Australian, Mexican, and South and Central American frogs have now been linked to chytridiomycosis. Unfortunately, chytridiomycosis outbreaks are most devastating in cool, moist habitats at high altitudes, where amphibian species diversity and endemism are highest.

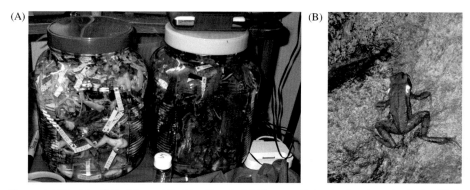

Figure 11.13 A jar full of frogs killed by the chytrid fungus *Batrachochytrium dendrobatidis* in Panama (A) and a dead *Bufo haematiticus* (B) in a Panamanian stream. *Photographs courtesy: Scott Connolly and Forrest Brem.*

Studies in Central America indicate that species more closely associated with streams are most susceptible (Lips et al., 2003; Brem and Lips 2008), and thus the ecological impacts of these losses may be greatest in streams (Whiles et al., 2006). Scientists have now linked many amphibian declines to chytridiomycosis, but there are still unanswered questions about the disease, such as the details of how it spreads, how it survives in areas once amphibians have declined, and why it is so lethal to so many species in so many habitats.

There has been some hope in finding ways to resist the disease. Most promisingly, it appears that some innate defenses found on amphibian skin, like bacteria or the frogs own antimicrobial peptides, may confer resistance to infection (Antwis et al., 2015; Voyles et al., 2018). Some species of frogs, once exposed to the disease and cleared of it, display immunity either from immunological or behavioral adaptations (McMahon et al., 2014). It also appears that some resistance may be genetically inherited, primarily through the major histocompatibitily complex of the genome (Savage and Zamudio, 2011).

Scientists agree that Bd is an exotic species, but there is no consensus on why it is now killing amphibians in so many regions. There are several lineages of the Bd fungus now identified, including those from Korea and Brazil; however, most current declines have been attributed to the Global Pandemic Lineage (James et al., 2015). Some studies implicate climate changes that have made the disease more virulent and stressed amphibians, but there is also evidence that Bd may have simply spread around the planet with the global transport of amphibians or other animals and plants. Two species, the African clawed frog (*Xenopus laevis*) and the American bullfrog (*Lithobates catesbeianus*), have received considerable attention because they have been moved around the planet by humans in large quantities, they are very hardy, and they appear capable of carrying the disease with them (Daszak et al., 2004; Weldon et al., 2004). Massive quantities of *X. laevis* were shipped from Africa around the world for decades for use in human pregnancy tests, scientific studies, and the pet trade, and today there are established feral populations of *X. laevis* in Europe, the United States, and South America. Large numbers of American bullfrogs have also been exported from their native habitats in eastern and central North America for food, and bullfrogs have also established feral populations around the world.

Humans may be responsible for moving Bd and other diseases through the amphibian trade. Importation data for live amphibians into three major US ports of entry (Los Angeles, San Francisco, and New York) indicated that nearly 28 million amphibians of various species entered the country from 2000 to 2005. Testing with polymerase chain reaction (PCR) techniques indicated that over half of these imported amphibians were infected with Bd, and $\sim 9\%$ with ranaviruses, another disease agent linked to amphibian declines (Schloegel et al., 2009). An international group of scientists recently examined the DNA of hundreds of Bd samples collected from every continent and found strong evidence that it originated from Asia, specifically Korea in the 20th century (O'hanlon et al., 2018). The international group attributed the global spread of all lineages of Bd to globalization and the amphibian trade.

Some species are beginning to recover from Bd driven declines. These recoveries, first noticed in Australia and Panama, seem to be increasing in frequency (Newell et al., 2013; Perez et al., 2014). The recovering populations appear to have developed strategies to cope with the continuing threat of chytridiomycosis (Voyles et al., 2018). These rebounding populations provide hope for the future, but it is still uncertain how these amphibians will fit back into the ecosystems they were briefly absent from.

While some species appear to be recovering from infections caused by Bd, a new and closely related pathogen has recently been discovered. This salamander chytrid fungus, *Batrachochytrium salamandrivorans*, is causing dramatic population declines in European mountain salamander populations (Martel et al., 2013). While the ecological impact of these declines has not been assessed, research in the United States has shown that salamanders play important ecological roles (Regester et al., 2006; Semlitsch et al., 2014).

This extinction event is unprecedented in geographic or taxonomic scope in recent history. There are undoubtedly ecological consequences beyond the inherent tragedy of this catastrophic loss of biodiversity. Loss of amphibians results in declines in predators of amphibians and changes in the structure and function of freshwater habitats, including primary production, invertebrate communities, and nutrient uptake dynamics in tropical streams (Whiles et al., 2006, 2013; Rantala et al., 2015).

Habitat destruction may well be the most important factor in global declines. Most of the world's amphibian species reside in tropical areas, and in particular tropical rain forests are very speciose, with hundreds of species found in the Amazonian basin alone. The rate of rainforest clearing is high and continues unabated.

Regardless of the ultimate causes for declines and extinctions, there are now some focused efforts to save remaining populations and species by establishing captive breeding programs, as well as researching ways to stop disease transmission. However, if habitat destruction is not stopped, the only place to see many of these species in the future may be in zoos and museums.

WHAT IS THE VALUE OF FRESHWATER SPECIES DIVERSITY?

Processing of human wastes by aquatic microbes has been estimated to be worth more than $700 billion/year worldwide, and bioremediation of hazardous waste will save 85% of the costs over nonbiological methods of treatment (estimated at $135 billion worldwide over the next 30 years; Hunter-Cevera, 1998). We documented some of the economic benefits of fisheries in Chapter 1 and biodiversity is important to the growing aquaculture industry; species diversity provides genetic diversity and animals and plants for culture. Biological diversity is also important because of the reliance of aquaculture on healthy biological systems (Beveridge et al., 1994). Inherent with species diversity is biochemical diversity, which has already provided myriad medicines and other important chemicals that benefit humans: this is still a rich area of scientific exploration.

Organisms can contribute to ecosystem structure and processes in a variety of ways, and thus biological diversity is important for sustained ecosystem function and integrity. Various species of microbes, plants, and animals in aquatic systems govern vital ecosystem processes such as decomposition and nitrogen cycling, and thus a reduction in the numbers of species in a system can be expected to negatively influence these processes at some point, as we will see in the remainder of this text.

Various stressors are increasing the probability of species extinctions (Fig. 11.14). These include, but are not limited to, invasive species, pollution, habitat destruction,

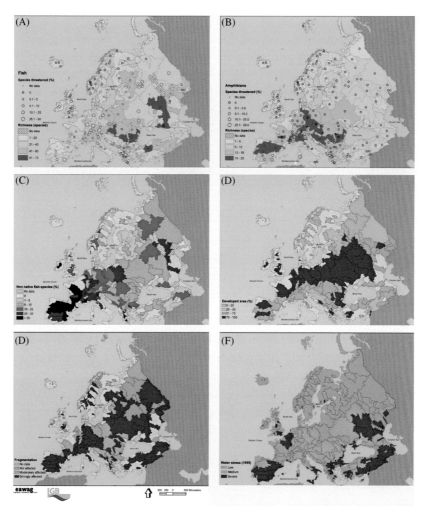

Figure 11.14 Percentage threatened species, numbers of native species, and stressors in Europe. (A) % fishes threatened (larger red dots mean more threatened) and fish species (darker green areas have more native diversity), (B) % amphibians threatened (larger red dots mean more threatened) and amphibian species (darker green areas have more native diversity), (C) nonnative fishes (darker colors mean more species), (D) % developed areas, (E) fragmentation, and (F) degree of water stress. *Image from Tockner et al. (2009).*

and global climate change. Developed areas, such as Western Europe and North America, are attempting to conserve some species, but freshwater species are particularly vulnerable because watersheds integrate large areas. Subsequently, another reason for concern over extinction of species is that the species we live with are analogous to the canary in the coal mine; when the bird dies, it is time to get out. When numerous species go extinct, this could indicate that the ability of Earth to support life is endangered (Lawton, 1991). Unlike the canary in the coal mine, we cannot "leave the mine."

Although important considerations, basing arguments for conserving species purely on economic and utilitarian grounds is dangerous. Many species may exist that have no economic value and are not essential to ecosystem function. Many species are not attractive to the public, unless they have big eyes or flashy coloration. Do these species deserve to go extinct? It is common in public debate over listing a species as endangered to hear the question, "Aren't people more important than animals?" We suspect what is usually meant in this case is that an individual believes his or her right to make money the way he or she wants is more important than the right of a species to exist. If the immediate past is any gauge, humans will continue to drive many species to extinction, regardless of their value.

SUMMARY

1. Several aspects of diversity are used by ecologists, including the number of species (richness), evenness, within-habitat (α) diversity, and between-habitat (β) diversity.
2. Evolution ultimately drives diversity. Aquatic habitats that have been in existence for long geological periods, such as some tectonic lakes and old river basins, tend to have high diversity and many endemic species. Spatial segregation of habitats also influences species evolution.
3. Over short-term timescales (decades and centuries), factors other than evolution may influence diversity, including size of habitat, connectivity of habitats, habitat disturbance, sources of new species, species interactions, productivity, and species introductions.
4. Genetic diversity within populations and species is important for long-term survival. Conservation efforts and aquaculture programs are increasingly focused on maintenance of genetic diversity.
5. Population size is a critical factor in the survival of species. As populations decline below the minimum viable population size, the probability of extinction increases exponentially.
6. Human-caused species introductions are common forms of pollution in aquatic environments. Some of these invading species have serious economic and

ecological impacts on the communities that are invaded. Most successful invasions are difficult or impossible to reverse.

7. Climate change is altering biodiversity in various ways, including altering the geographic ranges of some species and increasing susceptibility to toxins.

8. Humans are causing extinction of existing species at rates from hundreds to millions of times greater than the evolution of new species is occurring. Many of us alive today, particularly those the age of many students reading this text, will likely witness the disappearance of half or more of the biodiversity on Earth, including many species in aquatic habitats.

QUESTIONS FOR THOUGHT

1. Why are temperate lakes generally relatively species poor?
2. What are the evolutionary advantages of very long diapause periods?
3. Are there more species of large than small animals?
4. Should conservation be based on local diversity (e.g., if a species is rare in one state and common in another, should it be conserved where it is already rare)?
5. What is the benefit of an ecosystem-based method of conservation?
6. Should DNA samples of endangered species be stored?
7. How could warming temperatures increase the susceptibility of a species to toxins such as insecticides?
8. Will captive breeding programs allow us to preserve amphibian diversity?
9. How might genetic diversity be different than species diversity, particularly in microbial species?
10. Why might long-lived species appear to be doing ok because there is a large number of large adults, but then abruptly go extinct?
11. Should we spend millions of dollars to project a species that has a reasonably high probability of extinction even with protective measures?

CHAPTER 12

Aquatic Chemistry and Factors Controlling Nutrient Cycling: Redox and O_2

Contents

Figure 12.1 The macrophyte *Elodea* with O_2 bubbles from photosynthesis.

Chemistry controls physiology, biogeochemistry, behavior of pollutants, and many aspects of how organisms interact with their environment and each other. The way chemicals transform and move through the environment is an essential aspect of

Freshwater Ecology
DOI: https://doi.org/10.1016/B978-0-12-813255-5.00012-0

ecosystem function. Nutrients often control primary production, and understanding nutrient cycling is central to understanding the influence of nutrient pollution. On a global scale, nutrient cycling ties to patterns of climate and production. Understanding nutrient cycling requires careful study of the concept of oxidation—reduction state (redox). The dynamics of the element oxygen and how it reacts with other compounds central to life are related closely to redox. We discuss both in this chapter. In following chapters, we discuss additional nutrients (carbon, nitrogen, phosphorus, sulfur, and iron). The material in this chapter forms a conceptual foundation for treatment of all biogeochemical discussions in this book as well as those on toxins, saline systems, and chemical interactions among organisms.

CHEMICALS IN FRESHWATERS

Among the chemical parameters of interest to the aquatic ecologist, some of the most important are the concentrations of all inorganic ions (charged molecules) and concentrations of specific ions and gasses dissolved in solution. The total amount of dissolved and suspended organic material (carbon containing) and concentrations of specific organic molecules are also of interest. The chemistry of natural waters is an extremely complex field of study given the tremendous number of chemical compounds and all their possible interactions (Stumm and Morgan, 1981; Brezonik, 1994).

Chemical materials in aquatic systems can be either *dissolved* or *particulate*. The distinction between dissolved materials and particulate materials is arbitrary and commonly based on the ability to pass through a filter of a specified size. For example, researchers operationally define dissolved metals and dissolved organic carbon based on methods to analyze water passed through a 0.45-μm filter, but total dissolved solids are determined using material that passes through a filter with slightly greater than 1 μm retention (Eaton et al., 1995). A more natural definition scheme has been proposed (Gustafsson and Gschwend, 1997) in which chemicals fall into three classes: dissolved, *colloidal* (particles not settled by gravity), and *gravitoidal* (particles that will settle) (Fig. 12.2). A colloidal particle is any particle whose movement is not affected significantly by gravitational settling and can bind dissolved chemicals. Most aquatic ecologists have not adopted this definition scheme since its proposal because definitions based on analytical methods are more practical. Still, the scheme allows more realistic consideration of the physical and chemical state of materials associated with natural waters.

Total mass of ions relates to *total dissolved solids* (total mass of material dissolved in water, often noted as TDS), *salinity*, and *conductivity*. Conductivity and salinity are easier to measure. Conductivity is proportional to the relative amount of electricity conducted by water. The more dissolved ions present, the higher the conductivity. Conductivity correlates approximately to system productivity because high nutrient

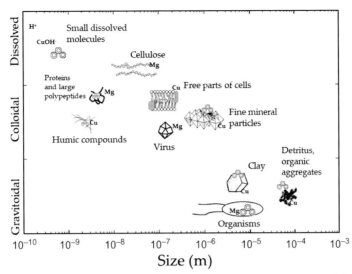

Figure 12.2 Chemical associations with dissolved, colloidal, and gravitoidal particles. The association is depicted with three trace substances, inorganic copper and magnesium ions and the three-ringed organic compound phenanthrene. Dissolved chemicals depicted include a variety of organic chemicals. Colloidal particles include membrane pieces, viruses, inorganic precipitates, and aggregations of organic chemicals. Gravitoidal particles include clays, planktonic cells, and larger aggregates of organic and inorganic materials. *Redrawn from Gustafsson and Gschwend (1997).*

waters have high conductivity, but other factors including concentration of non-nutrient salts also influence conductivity. For example, saline lakes have extremely high conductivity, but do not necessarily have correspondingly high productivity if nutrients such nitrogen and phosphorus are low.

The value for total dissolved solids does not always correlate to that of conductivity. Only ionic compounds are included in conductivity, whereas uncharged molecules (such as many dissolved organic compounds) also contribute to total dissolved solids. Also, various salts lead to different conductivity. For example, the ions ferrous iron (Fe^{2+}) and ferrous lead (Pb^{2+}) influence conductivity similarly because each have the same charge, but iron and lead have molecular weights of 55.85 and 207.2 g/mol, respectively, roughly a fourfold difference of mass. Salinity is the mass of dissolved salts per unit volume. Complete chemical analysis is necessary to determine true salinity, but conductivity is a common surrogate.

As water moves through terrestrial ecosystems, materials dissolve or *weather* from land. Chemical weathering releases dissolved matter, whereas mechanical weathering releases particulate matter that may react to form dissolved matter at some point. Mechanical weathering can include freezing and thawing as well as erosional activity of wind and water that moves particles against each other. Microbes can enhance rates of weathering (Bennett et al., 2001), but their activity requires water and time. Thus, the total concentration of dissolved matter relates inversely to the amount of runoff because the higher the runoff, the less time water has to dissolve ions. However, the relationship between runoff and total dissolved solids is variable because of differences in geomorphology, geology of the parent material (e.g., the relative abundance and solubility of the ions in the native sedimentary and igneous rocks and soils), and area of runoff. Hence, the amount of dissolved materials associated with a set amount of precipitation can vary over an order of magnitude (Fig. 12.3). Furthermore, some ions, in particular nutrients such as nitrate, can decrease in concentration, as they are assimilated (incorporated into biomass) by terrestrial and aquatic biota. Therefore the relative abundance of ions in water flowing from terrestrial habitats also depends on chemical interactions with biota. Much of the material that enters watersheds probably does so through smaller rivers and streams (Hynes, 1975; Alexander et al., 2000).

Solubility and relative abundance of elements in the earth's crust lead to some general patterns of average relative abundance of dissolved ions in river waters. For example, about 75% of the earth's crust is composed of silicate, but it is relatively insoluble so silicate does not dominate in natural waters. Aluminum (abundant but not soluble) and manganese (low abundance and solubility) are found at low concentrations relative to calcium or sulfate. Concentrations of some ions are constant across many rivers. Silicate is very abundant, but with relatively low weathering rates. Silicon has moderate biological importance (in some plants and diatoms) but concentrations are generally constant. Concentrations of other elements (e.g., sodium and iron) vary over many orders of magnitude, depending on regional geology (Fig. 12.3). Concentrations of

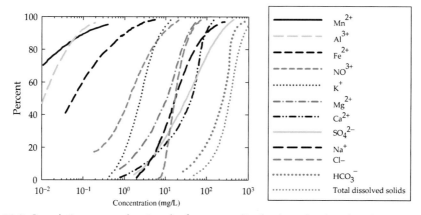

Figure 12.3 Cumulative curves showing the frequency distribution of various ions in water running off land. Each line shows the percentage of water samples with less than the corresponding concentration. *Redrawn from Davis and De Wiest (1966).*

nitrate and phosphate vary over several orders of magnitude because plants have high affinity for these nutrients when they limit primary production, but agricultural fertilization or weathering of natural deposits can lead to very high concentrations. Variance in nitrogen and phosphorus concentration is large because biological demands can lead organisms to deplete concentrations to very low values in relatively pristine waters, and human activities have boosted nutrient concentrations far above natural levels in many rivers.

Solubility of compounds in water can be characterized by solubility constants. Solubility of most compounds increases with temperature, but several important compounds ($CaCO_3$ and $CaSO_4$) become less soluble at higher temperatures. Equilibrium solubility constants indicate what concentration of a material will dissolve in water when it is saturated. When concentrations are below saturation, the material will continue to dissolve, while above saturation the compound will precipitate.

Concentration of hydrogen ions (protons) is also central to maintaining biological activity as well as defining the chemical nature of the environment. This concentration, or acidity, is *pH*, which is expressed on a logarithmic scale, corresponding to the following equation:

$$pH = -\log_{10}\{H^+\}$$

where $\{H^+\}$ is the hydrogen ion activity, which is closely related to concentration expressed in moles per liter. Hydrogen ions interact with other ions except when they are present in very low concentrations, so concentration does not exactly correspond to hydrogen ion activity. The smaller the pH value is the greater the hydrogen ion activity. On this log scale, each change of one unit of pH corresponds to a 10-fold change in hydrogen ion activity. Pure water has a pH of 7.0 (1.0×10^{-7} mol H^+ L^{-1}). Vinegar and beer have a pH of about 3, stomach acid has a pH of 2, and household ammonia has a pH of about 11. The actual range found in most aquatic ecosystems is near neutrality (several orders of magnitude in activity; $pH = 7 \pm 1$). Acid precipitation causes devastation in many aquatic systems because it both increases the H^+ concentration a 100- to 1000-fold and it mobilizes and increases solubility of toxic metals. We discuss this problem in Chapter 16. Increases of global atmospheric CO_2 concentrations are also increasing hydrogen ion activity because carbonic acid forms when CO_2 gas dissolves in water. Runoff from mining waste can also be very acidic. Some natural habitats such as acid springs (pH < 1) or alkaline lakes (pH > 10) have extreme pH values. The value of pH is easy to determine because reliable hand-held probes are readily available, easy to use, and inexpensive.

Alkalinity and *acidity* indicate the capacity of water to react with strong acids and bases, respectively, and indicate the suitability of domestic water and the relative content of inorganic carbon. *Hardness* is the sum of magnesium and calcium ions present and an indicator of the ability of water to precipitate soap; another property of water that is of practical interest for domestic water supplies. Hard water causes scaling (buildup of precipitates) and interferes with the ability of soaps and detergents to clean properly. Scaling can reduce the efficiency of hot water heaters, and some vegetables lose flavor when cooked in hard water. For these reasons, hard water is not desirable for domestic uses and people treat (soften) water to reduce hardness. Water treatment plants typically use chemical precipitation methods to remove hardness, and in individual homes, ion exchange resin systems are common. Many companies claim magnetic treatment can lower hardness of water. There is currently no credible scientific evidence to back up the viability of these products on the basis of magnetism, nor an adequate scientific explanation for the mechanism of magnetism lowering water hardness. Such products could lead consumers to "pour their money down the drain."

Additional bulk chemical parameters that characterize water chemistry or quality include color, taste, odor, suspended solids, and *turbidity*. Turbidity is the light absorption or the light scattering of water. Light absorption increases with more suspended particles such as clays or suspended algae, but other factors can also influence absorption of light, such as tannins and lignins from decomposition of organic material. Turbidity correlates significantly with total suspended solids. Total suspended solids are determined by filtering water onto a preweighed filter, and then drying and weighing the filter. Material is filtered onto noncombustible glass fiber filters and heated at high temperature (e.g., 450°C) to burn off organic material. The loss in mass following combustion can be used to calculate total suspended organic material and total suspended inorganic material. The mass lost to burning indicates *volatile suspended solids*.

These bulk parameters are traditionally the first indicators used to characterize general water quality. We will discuss other dissolved materials in more detail in this and other chapters. Researchers use a wide variety of analytical methods to measure concentration of specific dissolved ions, other dissolved compounds, and suspended particulate material (Method 12.1), and these methods continue to evolve as technology develops.

REDOX POTENTIAL, POTENTIAL ENERGY, AND CHEMICAL TRANSFORMATIONS

Some fundamental chemical principles are required to understand chemical interactions and transformation. Students are encouraged to review their basic chemistry. Some key concepts are chemical equilibrium, activation energy, free energy, and solubility constants. We will mention these concepts, but in-depth discussion is beyond the scope of this text. However, we will discuss some concepts, such as redox

METHOD 12.1 Analysis of concentration of specific dissolved ions and other dissolved and particulate materials in natural waters

Many methods can detect the concentrations of specific ions in natural waters; we only discuss a few here. For a more definitive treatment of a wide variety of methods, the reader is referred to *Standard Methods for the Determination of Water and Wastewater* (Eaton et al., 2005). Probably the simplest method of analysis is the use of ion-specific probes. The probes are calibrated in solutions with known ion concentrations. The calibrated probes are then immersed in the unknown solution and the concentration of the ion read directly from the meter. Many biologically important ions occur in concentrations too low for analysis with current-generation ion-specific probes in unpolluted habitats; so, other methods of analysis are required.

Analysts use many colorimetric methods to analyze concentrations of specific ions. These methods are based on formation of a specific colored compound from the ion of interest. The concentration of the colored compound can then be determined with a spectrophotometer (a machine that measures the absorption of light at specific wavelengths). Beer's Law (the Beer–Lambert Law) states that the relationship between absorption of light by a colored compound and concentration of that compound is linear at intermediate concentrations.

$$A = C * l$$

where A is the absorbance, C is the concentration of a chemical that absorbs light, and l is the distance the light needs to travel through the solution. Beer's Law allows the laboratory worker to make a standard curve from which the unknown concentration in natural waters can be calculated directly. Some analysts do these assays manually, but autoanalyzers are used routinely in laboratories that need to analyze a large number of water samples. These machines automatically take a small sample, add the appropriate chemical mixtures for a colorimetric reaction to occur, measure absorption at the appropriate wavelengths, and use a computer program to calculate concentration automatically from absorption.

Ion chromatography is also used in many laboratories. This technique uses the idea that dissolved ions will pass at different rates through specific materials. A column filled with an appropriate material is used and a carrier solution passes through the column. A small amount of the sample is added to this carrier solution, and the sample is carried through the column. The retention time (time needed for specific ions to pass through the column) is known for ions of interest. The machine detects the ions in the sample after they pass through the column and separates them according to their retention times. A nonspecific detector for ions (e.g., a sensitive conductivity meter) can be used to estimate the amount of material passing out the end of the column. The method is rapid and allows analysis of a large number of ions simultaneously. However, it is generally less sensitive than colorimetric methods.

Atomic absorption spectrometry is used to estimate concentrations of dissolved and particulate metals. This method uses the idea that individual elements emit light at very specific wavelengths when their electrons are excited. This is the same concept used to

(Continued)

METHOD 12.1 (Continued)

detect the elemental composition of distant stars. For the analysis, the water sample is injected into a chamber, where it is subjected to high energy that causes excitation of electrons. The intensity of the light at the wavelength specific to the ion is directly proportional to the amount of the ion. The signal is compared to a standard concentration, thus giving the concentration of the element in the sample.

Bulk dissolved and particulate organic carbon concentrations are usually analyzed by conversion to CO_2 followed by analysis of concentration of this gas. Some controversy exists regarding the efficiency of different methods (e.g., UV, high temperature, or persulfate oxidation digestions) used to decompose organic carbon to CO_2 (Koprivnjak et al., 1995). Given the complex chemical composition of dissolved and particulate organic carbon, it is not surprising that analysis is difficult. Several methods are available for analyzing particulate materials. One involves degradation of particulate material into dissolved ionic forms and then analysis by methods mentioned previously for dissolved ions. Another method involves combustion at high temperature and analysis of the resulting gasses (e.g., organic nitrogen is converted to N_2 gas). As with organic carbon, problems can occur with the efficiency of degradation.

Organic materials can be analyzed more specifically with gas or liquid chromatography and by a number of advanced analytical methods such as spectral fluorometry, mass spectrometry, and other techniques. Many of these methods rely on a principle similar to ion chromatography as discussed earlier; molecules of different mass or charge pass through separation columns at various specific rates. With appropriate detectors, very small concentrations (nanomoles per liter) can be sensed. This ability to sense vanishingly small concentrations is important because many toxins are highly active even at very small concentrations. If the output of a chromatograph then flows into to a mass spectrometer, the exact mass of compounds that are separated by the chromatography can be determined to identify specific compounds more exactly.

potential, in more detail because experience has shown that students require substantial review on this topic to understand chemical interactions.

The relative availability of electrons for chemical reactions in a solution is the *oxidation—reduction potential* or *redox potential*. The electrons are not floating around free; they are associated with dissolved elements and molecules to various degrees, and redox of a solution is a measure of how tightly associated the electrons are. This parameter is important because it quantifies the general chemical environment of the water. This environment dictates which chemical reactions will occur spontaneously given the numbers of electrons associated with reactants and products. By way of review, a compound that is oxidized loses electrons, and one that is reduced gains electrons. In a chemical reaction, one molecule or element gains electrons and another loses them. The term oxidation comes from oxygen, which as a high affinity for electrons, and tends to oxidize other compounds.

Redox potentials determine the potential energy requirements or yields during biotic or abiotic processes that transform chemical compounds in the environment. Stated differently, redox potential is a large determinant of what chemical reactions will occur without an input of energy, what reactions will require energy, and what products will be favored in the environment. Even though the concept of redox can be difficult for some students to grasp, it is worth the effort because it provides an understanding of the way that nutrients cycle and the behavior of pollutants in the environment. Building a rational framework based on principles of chemistry is essential to comprehending the cycles that underlie ecosystem function as well as metabolic constraints that organisms face.

Redox potential of natural systems is simple to measure with electrodes that assess availability of transferable electrons relative to the availability of electrons in hydrogen gas. Such sensors read in millivolts, with low values (e.g., below 100 mV) denoting large numbers of transferable electrons and high values of redox denoting few transferable electrons. If the electrons are more available than in hydrogen gas, then the reading is negative. If there are many oxidized ions in solution, the reading is positive. Even though redox potential is easy to measure, prediction of an exact redox potential for natural aquatic habitats based on chemical composition is difficult because of myriad chemical compounds that co-occur, because the different compounds may not be at equilibrium, and due to the complex interactions among many dissolved organic and inorganic chemicals.

The central idea that links the concept of redox potential to biogeochemical cycling is that chemical compounds have *potential energy* when they have a redox significantly different from the average redox of the rest of the compounds dissolved in the water. *A compound can have a high potential energy if it is a low redox compound in a high redox environment or if it is a high redox compound in a low redox environment.* This is a difficult concept for many students to grasp and bears close thought. Gibbs free energy diagrams are the best way to visualize the potential energy (Fig. 12.4). The redox potential of different chemical transformations varies relative to which pair of chemicals is considered (Fig. 12.5). The energy yield or cost of a chemical transformation is the distance between the head and tail of the arrow in Fig. 12.5. The transformation will yield energy if the redox potential of the solution is less than the head of the arrow for the reductions and greater than the head of the arrow for the oxidations.

Redox reactions require energy; this energy is the activation energy (Fig. 12.4). The reaction will not occur rapidly if the activation energy is high. For example, ammonium can exist in an oxidized solution containing dissolved oxygen without spontaneously combining with the dissolved oxygen gas and converting to nitrate, even though ammonium has a significantly higher potential energy than nitrate under oxidizing conditions. The activation energy of this conversion is too high for the

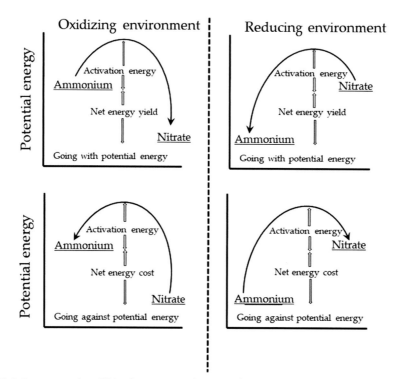

Figure 12.4 Representative Gibbs free energy diagrams for ammonium and nitrate ions in oxidizing and reducing environments. The activation energy is the energy required to move from a state of high to low potential energy, and the energy yield is what is released when the transformation occurs. The activation energy plus the energy yield are required to accomplish a transformation from low to high potential energy.

reaction to proceed at significant rates under normal conditions found in natural aquatic environments.

One of the primary determinants of redox is the concentration of dissolved oxygen gas (O_2). O_2 concentration is a major determinant of biogeochemistry and redox because of the O_2 molecule's tremendous affinity for electrons. When O_2 is present, redox values must be high—generally in excess of 200 mV. This high redox potential indicates an oxidizing environment, with very few available electrons, such that only specific chemical reactions (those that release electrons) will proceed without a net input of energy.

Iron concentrations in a dimictic lake are an example of how O_2 concentrations regulate redox potential to control the concentration of a chemical in the aquatic environment (Fig. 12.6). In this case, ferrous iron (Fe^{2+}), the reduced form of iron, is soluble, but ferric iron (Fe^{3+}), the oxidized form of iron, forms an insoluble precipitate in water. Ferrous iron converts readily to ferric iron in the presence of O_2

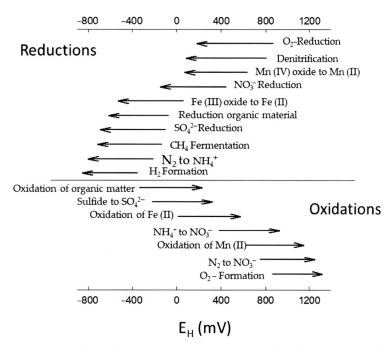

Figure 12.5 Microbe-mediated chemical transformations plotted to show energy yield as the difference between the tail and the head of the arrow and redox potential required to complete transformation. *Redrawn from Stumm and Morgan (1981).*

(i.e., the activation energy for the reaction is low) and then precipitates and settles to the sediments if O_2 is present through the water column and mixing rates are low. When wind mixes the lake fully, such that eddy diffusion allows full contact with the atmosphere, iron concentrations are low throughout the water column. As biological activity depletes O_2 from the deeper stratified layers, the dissolved iron concentration increases as ferrous iron diffuses out of the sediment. In this case, low redox potential and high dissolved iron concentrations correlate closely.

Organisms can promote chemical reactions that would not otherwise occur by lowering the activation energy, and in some cases providing energy to force a chemical reaction that goes against potential energy. Catalytic enzymes that lower activation energy promote chemical reactions. In the previous example, in which ammonium is stable to abiotic transformation in aquatic habitats containing O_2, microorganisms can lower the activation energy required to oxidize ammonium to nitrate. This reaction releases energy because nitrate has a lower potential energy than does ammonium in the presence of O_2. Bacteria can direct this energy toward cellular growth; the process is nitrification, and we discuss it in detail in Chapter 14.

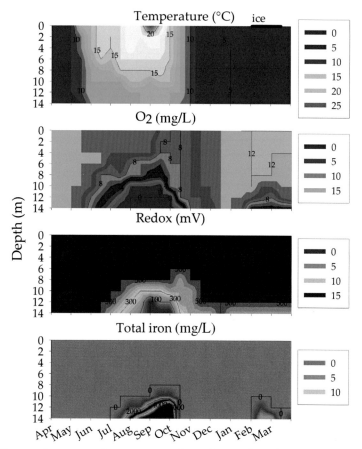

Figure 12.6 Temporal patterns of temperature, O_2, redox, and total iron in Esthwaite Water (an English lake) over a year. This type of figure is common for representing time series in lakes. The contours represent the boundaries of the values, with depth on the *y*-axis and time on the *x*-axis. *Redrawn from Mortimer (1941).*

Organisms can drive chemical reactions against potential energy (create more energetic chemical compounds). Such reactions require more input of potential energy than is stored in the products. Photosynthesis is an excellent example of a reaction that goes against potential energy; CO_2 is transformed to sugar (with a higher potential energy) using the energy of sunlight to accomplish the energy-requiring transformation. Many forms of nutrient assimilation (incorporating nutrients into biochemical compounds that make up the cell) require energy for uptake and transformation to forms that enter central biochemical pathways. For example, nitrate requires more energy for use than does ammonium, and nitrogen gas more energy than nitrate (we provide more detail in Chapter 14), but if nitrogen is limiting, then the energy will be used to transform the nutrients so they can be assimilated.

OXYGEN: FORMS AND TRANSFORMATIONS

The element oxygen occurs in many forms in the natural environment including water. The predominant form in the atmosphere is oxygen gas, O_2, at about 21% of atmospheric gas. Oxygen is a major component of organic compounds and biologically relevant inorganic compounds. As a result, it is important to understand the behavior and distribution of oxygen in the natural environment in order to appreciate its impact on aquatic ecosystems.

The amount of O_2 dissolved in water is a function of many factors, including metabolic activity rates, diffusion, temperature, and proximity to the atmosphere. This amount can be expressed in several related concentration units, including mg/L, mol/L, and percentage saturation. The percentage saturation is the concentration of O_2 relative to the maximum equilibrium concentration for that solution. *Dissolved oxygen* refers to the O_2 dissolved in water (as opposed to oxygen that is part of other chemical compounds).

The *saturation concentration* of O_2 is determined as the equilibrium concentration when pure water is well mixed and in contact with the atmosphere for an extended period. The amount of O_2 that can be dissolved in water is a function of temperature; the lower the temperature, the greater the concentration of O_2 under equilibrium conditions (Fig. 12.7). In addition, the greater the atmospheric pressure, the greater

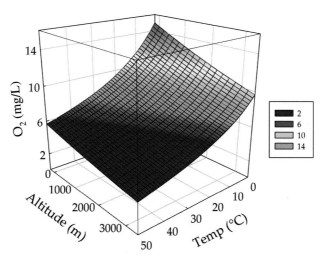

Figure 12.7 Saturation concentrations of dissolved O_2 as a function of temperature and altitude. The equation that describes the curve is: $\ln(O_2) = 2.692 - 1.27 \times 10^{-4}$ (alt) $- 6.15 \times 10^{-10}$ (alt)^2 $- 0.0286$ (temp) $+ 2.72 \times 10^{-4}$ (temp)^2 $- 2.09 \times 10^{-6}$ (temp)^3, where $O_2 =$ mg/L, alt is altitude in meters, and temp is temperature in °C. *Equations modified from Eaton et al. (1995).*

the saturation O_2 concentration. Atmospheric pressure is a function mainly of altitude (Fig. 12.7), but variation of O_2 saturation with barometric pressure can also be important. A 10% change in barometric pressure at one location is extreme, but air pressure on a very tall mountain (5,400 m) can be half the value observed at sea level. The elevation effect has ecological impacts; Jacobsen (2008) sampled stream invertebrate communities on an elevation gradient from 2,600 to 4,000 m above sea level in Ecuador. They observed a strong relationship between the saturation of O_2 and invertebrate diversity. Finally, O_2 saturation increases with increasing water pressure (i.e., depth in the water) and decreases with increasing salinity.

A point of confusion for some students is the concept that O_2 concentrations can exceed the level of saturation. If O_2 becomes highly supersaturated, it will form bubbles and come out of solution. However, at concentrations greater than saturation, O_2 can remain dissolved and slowly equilibrate with the atmosphere. Likewise, when O_2 is below saturation in solution, it will slowly come to equilibrium with exposure to the atmosphere. Both supersaturation and subsaturation are common in natural waters because they can be caused by photosynthesis and respiration, respectively. Supersaturating O_2 concentrations can also occur in very high energy streams, waterfalls, and below reservoir outputs.

The presence or absence of O_2 is also an important aspect of freshwaters because it determines whether and what type of organisms can live in a given ecosystem. Habitats that have any O_2 in the water are referred to as *oxic* or *aerobic*, and those without detectable O_2 are *anoxic* or *anaerobic*.[1] The absolute concentration of O_2 is also important as different organisms have different tolerances for low O_2, and several methods can measure O_2 concentration (Method 12.2).

The main cause of biological consumption of O_2 in most environments is aerobic respiration. All organisms must metabolize, and the oxidation of organic carbon with molecular O_2 gives a greater energy yield than does oxidation using other molecules. The general reaction is

$$CH_2O + O_2 \rightarrow CO_2 + H_2O + \text{chemical energy}$$

where CH_2O is a general stoichiometric representation of the formula for sugar (not the structure of formaldehyde), and chemical energy is in the form of ATP. However, the ratios of C to H and O vary in different organic compounds (e.g., phospholipids have relatively low oxygen concentrations relative to sugars). This is why the equation is not presented with glucose as the energy source and exact chemical balance; the formula should be considered only an approximate representation of respiratory metabolism.

---

[1] The older terminology (aerobic and anaerobic) is less precise than the newer terminology (oxic and anoxic) because the prefix aer refers to air, which also contains other gasses.

METHOD 12.2 Measuring O_2 concentration

Winkler titration was one of the first methods developed for measurement of dissolved O_2. This method is many years old and is very reliable. In this method, reagents cause iodine to react with O_2 molecules, forming iodate, IO_3^-. The iodate is then titrated back to iodine, and the amount of titrant necessary to react with the iodate is directly proportional to the O_2 concentration (Eaton et al., 1995). The Winkler method is more time-consuming, but can be more accurate and more precise than use of standard O_2 electrodes when performed by an accomplished analyst. The Winkler method requires collecting samples without introducing O_2. Van Dorn samplers (see Chapter 7) and other methods provide techniques for collecting samples and not exposing them to the atmosphere.

The most commonly used method for measuring O_2 concentration in the field as well as laboratory or industrial applications is an O_2 electrode. The older type of electrode has a gold-plated cathode and a reference electrode (anode). When a voltage is applied across the cathode and the anode, the cathode reacts with O_2 and causes an electrical current to flow. The more O_2 that reacts with the cathode, the higher the current that registers on the meter, corresponding with a higher O_2 concentration. This type of electrode must be calibrated regularly for variations in temperature and atmospheric pressure. A more recently introduced electrode takes advantage of a fluorescent compound impregnated in a membrane that fluoresces more brightly when exposed to more O_2. The membrane is excited optically and the resulting wavelength of fluorescence detected. The signal is proportional to the O_2 concentration. These sensors are more stable than the cathode-style electrodes.

These two types of electrodes are useful because they can have long lead wires and be lowered into lakes or wells to assess in situ O_2 concentrations. These electrodes can also be attached to a data logger in a submersible housing and used to monitor dissolved O_2 unattended over days or weeks. Very small-scale cathode tips (microelectrodes) have been constructed to allow for determination of O_2 with <0.1 mm spatial resolution (Revsbech and Jørgensen, 1986).

Dissolved oxygen is difficult to detect at very low concentrations (Fenchel and Finlay, 1995), so it is problematic to determine whether an environment is strictly anoxic or not. Strict anoxia is required for some biogeochemical processes. Some organisms are extremely sensitive to low O_2 concentrations and must live in stringently anoxic conditions. Specialized dyes and sample handling methods are used to sample for strict anoxia.

PHOTOSYNTHESIS

Photosynthesis is the process that ultimately provides energy to run most ecosystems. Light, temperature, and nutrients all control photosynthetic rates. Light is generally the predominant factor over the short term. While scientists most fundamentally account for photosynthesis and respiration in units of carbon because this most closely relates to energy, researchers commonly use changes in O_2 for estimation of rates of photosynthesis and respiration. We discuss carbon in the next chapter, but introduce photosynthesis here with O_2 for practicality.

Photosynthesis can contribute a significant amount of O_2 in lighted surface environments in addition to the O_2 that dissolves in surface water from the atmosphere. A generalized equation for oxygenic photosynthesis is

$$CO_2 + H_2O + light\ energy \rightarrow CH_2O + O_2$$

The photosynthesis equation is essentially the reverse of the equation for respiration. Again, the equation for photosynthesis is a general equation because the exact composition of the organic molecules (represented as a general stoichiometry for sugars, CH_2O, in the equation) varies depending on the biological molecules an organism synthesizes.

Photosynthesis—irradiance (P—I) relationships describe the effects of light on photosynthetic rate. Several parameters are generally used to describe this relationship (Fig. 12.8A). These *P—I* parameters include the respiration rate (O_2 consumption in

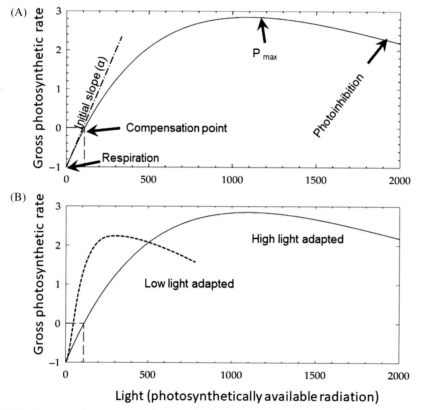

Figure 12.8 Diagram of a representative relationship between net photosynthetic rate and irradiance (A) and comparison of high-light- and low-light-adapted species (B). The low-light species has a steeper α, lower compensation point and P_{max}, and greater photoinhibition. Irradiance is in photosynthetically available radiation, and photosynthetic rate is in arbitrary units. A single species can acclimate to light in a similar fashion.

the dark), the *compensation point* (where gross production equals respiration), α (the increase in photosynthetic rate per unit increase light at low irradiance prior to any light saturation of photosynthesis), P_{max} (the maximum photosynthetic rate), and β (a parameter describing the deleterious effects of high light or *photoinhibition*). Understanding this curve provides initial insight into how light alters photosynthetic rates and the strategies photosynthetic organisms use to compete successfully in their environment.

Organisms living in low–light habitats (e.g., deep in lakes, in high sediment or colored waters, or in a small shaded stream flowing through a forest) have several characteristics allowing them to survive and compete. $P-I$ curves describe these characteristics. They have a relatively low compensation point so respiration is equal to photosynthetic O$_2$ production at very low light. Such organisms also have a rapid increase in photosynthetic rate as light increases (a steep α). These organisms tend to be photoinhibited at relatively low irradiance (Fig. 12.8B).

The light field can vary spatially and temporally. Clouds, vegetation, and waves cause variation in the light intensity reaching photosynthetic organisms (as discussed earlier in Chapter 3). Some species of algae are adapted (have appropriate evolutionary characteristics) to compete more effectively in a variable light field (Litchman, 1998). Steep spatial light gradients (e.g., from full light to less than 1% sunlight over cm or mm) can occur in natural environments; periphyton and sediments attenuate light rapidly in benthic habitats. Unless turbidity is high, benthic sediments attenuate light more rapidly than in the water column of lakes and rivers. Primary producers can either have permanent physiological characteristics (be adapted) to specific light conditions, or can alter their physiology (acclimate) to maximize production under various light conditions.

Mixing also influences the light regime of phytoplankton in lakes. If turbulent mixing is deep enough, the average light experienced can be too low to support growth. The mixing depth below which growth does not occur is the *critical mixing depth*. Phytoplankton is most likely to mix below the critical depth during winter when a lake is not stratified and ambient light is low. High levels of inorganic turbidity or extended periods of very cloudy weather can also lead to lakes where phytoplankters mix below the critical mixing depth.

Organisms that live in intense sunlight near the water's surface have special characteristics that allow them to succeed in this harsh habitat. The strongest effect of high light is photoinhibition caused by the direct or indirect damaging effects of high-intensity light on the molecules in the cells. The most likely site of damage in photoinhibited cells is in the photosynthetic apparatus at photosystem II (Long et al., 1994). Most studies of photoinhibition in the natural environment have been on phytoplankton (Long et al., 1994). Photosynthetic organisms can protect themselves from high light by synthesizing special protective pigments, and if they are motile, by moving into areas with lower irradiance. Primary producers often synthesize carotenoids with a red color to protect themselves from high light. These compounds can be concentrated by zooplankton and fish and lead to an orange or red color of their tissues. The red tissue is common in high altitude lakes where irradiance is intense, but also in the pink flesh of salmon. Photosynthetic organisms that are acclimated to intense light have low values for α as well. The compensation point of organisms acclimated to high light occurs at relatively high irradiance. Thus, an organism acclimated to high light will not compete well in low light. The inhibitory influence of ultraviolet radiation on photosynthetic rates is of particular concern given the increased amounts reaching Earth's surface with the thinning of the atmospheric ozone layer (Highlight 12.1).

Highlight 12.1 The influence of UV radiation on aquatic photosynthetic organisms

Human activities have led to significant decreases in concentrations of ozone (O_3) in the stratosphere, allowing more UV radiation to penetrate the atmosphere and causing substantial biological effects (Bancroft et al., 2007). The two types of UV radiation that are of concern are UV-A (320–400 nm) and UV-B (280–320 nm). They have increased significantly at high and low latitudes, with the greatest increases near the poles. Even though phase-out of chemicals that cause ozone depletion began seriously in the early 1990s, UV levels continued to rise with the largest Antarctic ozone hole occurring in 2006. Finally, in 2016 scientists could that detect long-term increases in the ozone hole had halted and the hole was starting to heal (Solomon et al., 2016). UV should gradually decrease over the next 50 years (Madronich et al., 1995), but it could take considerably longer for Earth's atmosphere to recover. Unfortunately, ozone destruction from human-caused increases in nitrous oxide emissions are increasing, and this is the current largest cause for ozone destruction (Ravishankara et al., 2009). Understanding the ecological effects of this increased UV irradiance requires knowledge of how it influences photosynthetic organisms (Häder, 1997).

Photoautotrophs must be in light to survive, grow, and reproduce, so consequently also are exposed UV. A variety of atmospheric factors can alter incoming UV, including seasonal variation in O_3 depletion, amounts of UV-scattering particulate material (including pollutants) in the air, increased cloudiness, and altitude. Yearly variation in incident UV irradiance can be considerable (Leavitt et al., 1997). Once UV enters water, the depth of the water and concentration of UV-absorbing compounds control the amount that ultimately reaches aquatic organisms. Dissolved organic carbon is the predominant material that absorbs UV, and rapid attenuation occurs with moderate dissolved organic carbon concentrations (i.e., more than several mg C/L). Given these considerations, another important determinant of UV exposure is the position of organisms in the aquatic habitat. Organisms inhabiting shallow waters, such as those in wetlands, open streams, and littoral regions of lakes, may have very high exposures. The epilimnia of lakes, particularly those in high altitudes, may receive high levels of UV. Finally, species of primary producers can protect themselves by synthesizing mycosporine-like amino acids (in the diatoms), scytonemen (in cyanobacteria), or flave-noids (in green algae and higher plants). However, synthesis of these compounds costs energy and may lower overall production (Karentz et al., 1994).

The question remains; what is the effect of increased UV on aquatic primary pro-duction? Increased UV has a wide variety of influences in aquatic systems as mediated by primary producers (Table 12.1). However, the most complicated of the UV effects relate to community interactions and ecosystem influences (Karentz et al., 1994). In artificial stream experiments, natural levels of solar UV drastically reduced grazing midge larvae. In these experiments, UV had a negative direct influence on the periphy-ton over weeks, but the release from grazing pressure with higher UV eventually led to more luxuriant periphyton growth over months (Bothwell et al., 1994). In contrast, experiments on grazing snails did not support a general prediction that UV always relaxes grazing pressure (Hill et al., 1997), but production of grazing tadpoles may be negatively affected (Licht, 2003).

Table 12.1 Some possible effects of increased UV on primary producers

Effect of increased UV	Reference
Harms cyanobacteria	Bebout and Garcia-Pichel (1995) and Castenholz (2004)
Decreases photosynthetic rates of phytoplankton and periphyton in high altitude tropical lake	Kinzie et al. (1998)
Decreases nitrogen uptake rates of plankton	Behrenfeld et al. (1995)
Lowers populations of consumers of the producers	Bothwell et al. (1994) and Häder et al. (1995)
Damage to DNA	Jeffrey et al. (1996)
Damage to ability of cyanobacteria to fix nitrogen	Kumar et al. (1996)
Minimal affects with increased dissolved organic carbon	Morris et al. (1995)
Damages photosynthetic apparatus	Nedunchezhian et al. (1996)
Harms stream mosses	Rader and Belish (1997a)
Selects for tube building or mucopolysaccharide producing diatoms	Rader and Belish (1997b)
Has selective effects on competitive ability of different periphyton species	Francoeur and Lowe (1998), Hodoki (2005), Tank and Schindler (2004), and Vinebrooke and Leavitt (1999)
Alters response to nutrient enrichments	Bergeron and Vincent (1997)
Decreases growth of *Sphagnum* in a bog	Gehrke (1998) and Robson et al. (2003)
Alters phytoplankton species composition	Laurion et al. (1998) and Xenopoulos and Schindler (2003)
Exacerbates the biotic effects of acid precipitation on lakes	Yan et al. (1996)

In lakes, UV can have a variety of effects on microbial components of the food web, and complex interactions can occur among UV, dissolved organic C, and nutrient supply (Bergeron and Vincent, 1997; De Lange et al., 2003). Initially, increased concentration of dissolved organic carbon may shield primary producers from UV, but as concentrations further increase, photosynthetic light can be absorbed leading to light limitation. These effects translate up the food web to production of trout (Finstad et al., 2014). In addition, all primary producers have viruses that may infect and destroy cells. With increased UV, the survival of viruses decreases, and consequently rates of transmission and cell mortality are lower. A global meta-analysis of UV effects on freshwater animals indicated that effects on zooplankton were the greatest followed by fishes and amphibians (Peng et al., 2017). Given that zooplankton are key grazers of phytoplankton in lakes, there is strong indication that lake food webs can be influenced by increases in UV.

More research is necessary if we are to understand the influence of UV on aquatic communities and ecosystems, especially in the context of climate warming and lake acidification. Climate warming and lake acidification will both lower dissolved organic C, leading to increased UV penetration into aquatic ecosystems (Schindler et al., 1996).

Other factors in addition to light influence photosynthetic rate. Temperature has a distinct influence (Fig. 12.9). As with other metabolic rates, the rate of photosynthesis approximately doubles with each $10°C$ increase in temperature up to a species-specific threshold (DeNicola, 1996). Above this threshold, further increases in temperature harm the photosynthetic organisms and lower the rate; eventually, with a great enough increase in temperature, death occurs (see Chapter 15). The amount of nutrients available also has an influence on photosynthetic rates. Nutrient-starved cells will lower their rates of photosynthesis when nutrients become available so cellular metabolism can be directed toward acquisition of nutrients (Lean and Pick, 1981), but ultimately (after a day or two), nutrients will stimulate photosynthetic rate as cells synthesize the required molecular machinery from the newly available nutrients.

Water velocity can alter photosynthetic rates (Fig. 12.10). As the diffusion boundary layer thickness decreases with greater water velocity, the transport of material across the layer increases. The influx rate of CO_2 across this diffusion boundary layer can be an important determinant of photosynthetic rate of macrophytes (Raven, 1992), as can diffusion controls over the influx of nutrients. High levels of dissolved O_2 can inhibit photosynthesis. Increased water velocity can increase transport of O_2 and other potentially inhibitory chemicals. Therefore, increases in water velocity often increase photosynthetic rates. However, very high velocities may stress photosynthetic organisms and lead to decreases in photosynthetic rates (Stevenson, 1996).

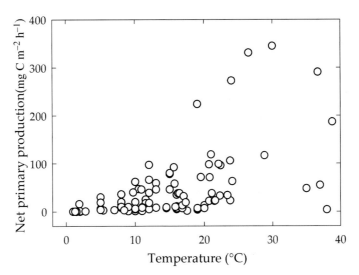

Figure 12.9 The relationship between photosynthetic rate and temperature based on 94 measurements of stream epilithon production in 14 studies. *Reproduced with permission from DeNicola (1996).*

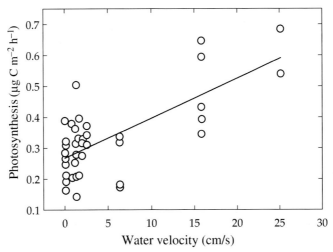

Figure 12.10 The relationship between water velocity and photosynthetic rate of the benthic cyanobacterium *Nostoc*. *From Dodds (1989), with permission of the* Journal of Phycology.

Organic compounds, particularly herbicides that target molecular photosynthetic machinery, can have strong negative effects on photosynthetic rates. The herbicide Atrazine is common in waters of the United States (it is banned in Europe) and acts by interfering with photosystem II. Chemicals designed to control macrophytes often act by interfering with photosynthesis (Murphy and Barrett, 1990). Organic chemicals that lower photosynthetic rates can also occur naturally; macrophytes can release organic compounds that inhibit photosynthetic rates of epiphytes that grow on their surface (Dodds, 1991).

Determining rates of primary production over extended periods in entire ecosystems is difficult given the large variation observed in photosynthetic rates and the wide variety of factors that influence rates. However, researchers must estimate annual rates for their work to have ecosystem relevance (Wetzel, 2001), and a variety of methods are available for photosynthetic rate measurements (Method 12.3). Therefore, measurements should be made under the array of conditions that occur over the full seasonal cycle to achieve accurate estimates of rates of primary production.

METHOD 12.3 Techniques for measuring photosynthetic rates in freshwaters

Researchers use several methods to measure photosynthetic rates, with the best method often determined by the system of interest and the research question. Two general classes of methods are available: those that make use of chemical tracers and those that measure bulk change in dissolved products or substrates of photosynthesis. Numerous scientific articles and books detail various methods; Wetzel and Likens (1991) present one of the clearest accounts for lakes, and Hall and Hotchkiss (2017) for streams.

Changes in dissolved O_2 concentration form the basis of the most widely used methods. For this technique, a sealed system containing the photosynthetic organisms is usually used, and the O_2 change in light versus dark enclosures is compared following a known incubation time. This technique requires moderately high biomass and is difficult to apply to oligotrophic phytoplankton because O_2 production and consumption rates are low relative to background O_2 concentrations. However, with great care, O_2 methods may still be suitable for oligotrophic lakes (Carignan et al., 1998).

Several enclosed systems allow determination of photosynthetic rates with O_2 exchange. Researchers try to mimic in situ conditions when rates are measured. Glass bottles are commonly used for phytoplankton studies. UV-transparent plastic or quartz bottles allow the investigator to account for UV effects. It is necessary to duplicate water movement found in the natural habitat for investigation of photosynthetic rates of benthic algae. Enclosed recirculating systems constructed of transparent plastic are generally used for these measurements (e.g., Dodds and Brock, 1998; Rüegg et al., 2015).

Open-water methods can also indicate system O_2 production and consumption. Odum (1956) pioneered such methods for use in streams. The technique relies on measuring the increase in O_2 as light increases or the decrease in dark. The exchange rate with the

(Continued)

METHOD 12.3 (Continued)

atmosphere must be known for these measurements. The gas exchange rate is referred to as *aeration* or *reaeration*. This exchange rate is difficult to measure as various factors such as wind, turbulence, and surface slicks can alter the rates (Brutsaert and Jirka, 2013). While equations exist to estimate aeration rates, their application in small streams is questionable. Trace gases such as propane or sulfur hexaflouride can be used to estimate this exchange rate (Marzolf et al., 1994, 1998). Modeling methods have been refined to allow for modeling of aeration and error estimation of parameters (e.g., Grace et al., 2015). With the whole-stream O_2 method, the stream communities remain under their natural conditions and the method integrates across large areas (stream reaches). The production and consumption of CO_2 can be used to measure respiration and photosynthesis, respectively, in a fashion similar to that outlined previously for O_2. The main problem with this approach is that background concentrations of CO_2 and associated dissolved forms (discussed in Chapter 13) are considerably greater than even dissolved O_2 concentrations. Measuring changes in CO_2 is primarily useful for emergent plants or macrophytes with very high biomass.

For any methods that track changes in dissolved O_2, results can simply be expressed as change in O_2 per unit time, or converted to actual carbon fixed by photosynthesis or released by respiration using the following relationship:

$$mgC = mgO_2 \times \frac{1}{PQ} \times \frac{12}{32}$$

where PQ is the *photosynthetic quotient* (moles of O_2 released during photosynthesis per unit CO_2 incorporated $= 1.2$ for algae at moderate light levels; Strickland and Parsons, 1972), 12 is the atomic weight of C, and 32 is the molecular weight of O_2.

Photosynthetic organisms also consume protons (increase pH) while they photosynthesize. Some investigators have used this fact to estimate photosynthetic rates in the natural environment. This method is not highly sensitive and may not work well in waters that are resistant to pH changes (buffered).

The radioactive isotope of carbon (in $^{14}CO_2$) has seen broad application in measurements of photosynthetic rates. If the ratio of $^{14}CO_2$ to ambient unlabeled $^{12}CO_2$ is known, then the rate of uptake of radioactive carbon into plant carbon can be used to calculate total photosynthetic rate. This approach is useful in very oligotrophic waters where O_2 methods fail. However, the practical difficulties of using radioactive isotopes (in laboratory and particularly in field settings) hamper this method. Also, separating net from gross photosynthetic rate with $^{14}CO_2$ techniques is difficult because some of the carbon fixed by photosynthesis can be respired immediately. Careful planning is necessary before using $^{14}CO_2$ methods.

Additionally, a fluorescence method, pulse amplitude-modulated fluorescence, has been applied to measure relative photosynthetic rates. This method relies on the biophysics of photosystem II and how it responds to pulses of saturating light over time. The problem with this method for determination of absolute photosynthetic rates of complex assemblages is that different species of primary producers respond differently to fluorescence (Juneau and Harrison, 2005), and it should mainly be used for short-term comparisons of relative physiological responses and carefully calibrated against more traditional methods.

RESPIRATION

Respiration is the primary route of O_2 consumption in most environments. Other biotic and abiotic activities can lead to consumption of O_2 such as oxidation of ammonium, sulfide, or iron (discussed above with respect to iron and in more depth in Chapter 14), but these rates are usually low relative to respiratory consumption in most environments. Similar factors to those that influence photosynthetic rates can influence respiration rates.

All organisms must respire. If there is a lot of living biomass, then respiration rates will be substantial in an ecosystem. A common error of beginning students is to discount the fact that photosynthetic organisms respire. An example of the importance of respiration by producers occurs when algal blooms lead to anoxia because the algae continue to respire at night or during cloudy periods when they are not photosynthesizing.

Quality of organic carbon available for consumption is a primary determinant of rate of respiration. Easily metabolized molecules such as simple sugars can lead to greater rates of respiration than a source that is less bioavailable (such as cellulose). We cover details of carbon use in Chapter 13. Other nutrients such as nitrogen and phosphorus can increase availability of carbon, biomass of heterotrophic organisms, and ultimately respiration rate.

Exchange rates with the environment can also alter rates of respiration; diffusion controls the rate of transport of O_2 and other compounds required for respiration. As with photosynthesis, any other compound that influences physiology can alter respiration rate (such as toxins, extreme pH, and salinity). Animals can have rapid variation in respiration rates based upon their levels of activity.

METABOLIC BALANCE OF PHOTOSYNTHESIS AND RESPIRATION, AND TEMPERATURE EFFECTS

Respiration and photosynthesis are primarily responsible for maintaining a constant concentration of O_2 in the Earth's atmosphere. At smaller scales, O_2 concentrations are not constant. When photosynthesis dominates, O_2 concentrations exceed saturation, but when respiration dominates, a habitat could become anoxic in the absence of any O_2 input from the atmosphere.

We initially explain some of the terminology used to describe photosynthetic processes and the metabolic balance that occurs in individual autotrophic organisms as well as for whole ecosystems. Because respiration and photosynthesis are both occurring simultaneously, even within individual photosynthetic organisms, we can distinguish between net primary productivity (*NPP*) and gross primary productivity (*GPP*) as follows:

$$NPP = GPP - respiration$$

Net primary production is the photosynthesis that occurs in excess of the respiratory demand. *Gross primary production* is the total amount of photosynthesis that occurs before accounting for losses to respiration. Aquatic ecologists use the equation relating *NPP*, *GPP*, and respiration frequently (Example 12.1). The value for *NPP* (positive, negative, or close to zero) determines if organisms in the environment supply or consume O_2 overall. When the production of an entire ecosystem (all primary producers and heterotrophs in an area) is being spoken of, the terms *net ecosystem production* (*NEP*), *gross ecosystem production* (*GEP*), and *ecosystem respiration* (*ER*) are often used and the same equation applies:

$$NEP = GEP - ER$$

In part, temperature can control the metabolic balance of GPP and ER. Temperature controls respiration and photosynthetic rates; some of the earliest physiological observations were on temperature effects on respiration. *Homeotherms* (warm-blooded animals) generally have higher rates of respiration than *poikilotherms* (cold-blooded animals). The Van't Hoff-Arrhenius equation describes the general relationship

---

Example 12.1 Calculating net and gross photosynthetic rate and respiration from lake water samples

A researcher wanted to determine the photosynthetic rate of phytoplankton in a lake at midday. She sampled the water, and determined that the initial O_2 content was 8.0 mg/L. She filled three clear and three dark bottles with this lake water and suspended them in the lake at the depth of collection. Following 1 hour, she analyzed the water in the bottles for O_2 content. Water in the clear bottles had 9.5, 10, and 10.5 mg O_2/L and the dark bottles had 7.7, 7.5, and 7.3 mg O_2/L. Calculate net and gross photosynthetic rate and respiration.

Net photosynthetic rate refers to the photosynthesis that occurs in excess of respiratory demand and is calculated by the increase in O_2 in the light bottles compared to the initial O_2 concentration. The average final concentration was 10 mg O_2/L, so the net photosynthetic rate = $(10-8$ mg $O_2 L^{-1} h^{-1} = 2$ mg $O_2 L^{-1} h^{-1}$.

The respiration calculation is similar; the rate is the difference between the dark bottle and the initial O_2 concentration. Therefore, respiration rate = $(8 - 7.5$ mg $O_2 L^{-1} h^{-1} = 0.5$ mg $O_2 L^{-1} h^{-1}$. Note that when respiration is calculated in the same way as net photosynthesis, a negative flux rate (negative rate of O_2 production) is obtained. Finally, gross photosynthetic rate is net photosynthetic rate + the oxygen consumed by respiration. Gross photosynthetic rate = $2 + 0.5$ mg $O_2 L^{-1} h^{-1} = 2.5$ mg $O_2 L^{-1} h^{-1}$. This value can also be calculated by subtracting the concentration value for the dark bottles from that of the light bottles and dividing by time.

between temperature and rate of metabolism of a particular organism adapted to a particular range of temperatures,

$$R = e^{-E/kT}$$

where R is the rate, E is the activation energy, k is the Boltzman's constant, T is the absolute temperature (degrees Kelvin), and e is the mathematical constant approximately equal to 2.718.

A central issue in understanding the effects of temperature on net metabolic rate is how the activation energies of ecosystem respiration and gross primary productivity vary. This will be particularly important to understand how climate change will influence carbon transformations in freshwaters as a function of global warming. While some research has been done in this area (e.g., Demars et al., 2016), more is necessary to understand broad patterns among freshwater systems. Song et al. (2018) made numerous metabolic measurements across six biomes and assessed the temperature sensitivity (E_a) for GEP and ER. They found that with global warming ER will increase more than GEP and that this should lead to greater rates of carbon dioxide emission.

CONTROLS OF DISTRIBUTION OF DISSOLVED OXYGEN IN THE ENVIRONMENT

As mentioned previously, the distribution of O_2 over time and space in aquatic habitats is a function of O_2 transport (influx and efflux), production by photosynthesis, and consumption by respiration and some reducing chemicals. Given the natural variation in the rates of these different processes, and differences driven by the relative inputs of organic C, habitats can be either anoxic or oxic. In this section, we describe spatial and temporal variations of O_2 in lakes, sediments, groundwaters, and small particles.

Measurement of vertical patterns of dissolved O_2 in lakes and relation to thermal stratification is a common exercise in limnology courses and limnological research. Such measurements illustrate the processes leading to production and consumption of dissolved O_2 and the biological importance of density stratification in lakes. Movement of O_2 across the metalimnion is slow because it depends mostly on molecular diffusion. In addition, the hypolimnion is usually deep enough that little light reaches it. With little or no photosynthesis, respiration predominates in the hypolimnion as organic carbon rains down from above in the form of settling planktonic cells and other organic particles. Given a high enough rate of carbon input and associated heterotrophic activity, heterotrophs can consume O_2 completely during a summer season of a dimictic or monomictic lake (Fig. 12.11A).

The hypolimnia of eutrophic lakes tends to have O_2 deficits by the end of summer, and oligotrophic hypolimnia of monomictic or dimictic lakes retain at least some O_2.

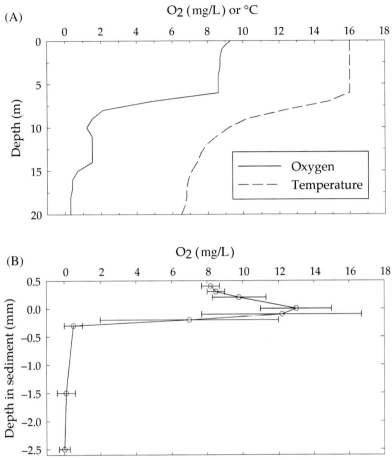

Figure 12.11 O$_2$ profiles from the stratified Triangle Lake, Oregon, on October 1, 1983, mid-morning (A) and from the South Saskatchewan River in an active algal mat, midday on June 9, 1993 (B). Note the lack of O$_2$ in the hypolimnion and the possible deep photosynthetic activity (at 10−14 m) that causes a slight increase in O$_2$ in (A) and the supersaturating O$_2$ concentration at the sediment surface (B). *Data for (A) courtesy of R. W. Castenholz; data for (B) from Bott et al. (1997).*

The greater deficit of O$_2$ in eutrophic lakes occurs because the amount of organic carbon sinking per unit time into the hypolimnion from the epilimnion is greater in eutrophic lakes than in oligotrophic lakes. Understanding anoxia of hypolimnia is important because taste and odor problems in drinking water become more acute in anoxic conditions and as trophic state increases (see Chapter 18) and because the presence or absence of O$_2$ can determine distributional patterns of organisms. For example, fish will generally avoid an anoxic hypolimnion.

The O_2 can also disappear from the hypolimnion of an amictic lake, even if the lake is oligotrophic. Such anoxia occurs because the O_2 is depleted gradually in the hypolimnion over time. Even though the consumption is gradual, the lake never mixes. Lake Tanganyika in Africa (discussed in Chapter 7 and Chapter 11) is an example of an oligotrophic lake with an anoxic hypolimnion; even though the lake is 1,470 m deep and is oligotrophic, only approximately the top 100 m are oxygenated.

As mentioned previously, dissolved O_2 can exceed saturation with respect to the atmosphere in lakes. When there is a high biomass of phytoplankton at the metalimnion, an O_2 peak may result. This deep chlorophyll maximum can be associated with enough photosynthesis that O_2 concentration can build in the metalimnion where mixing and contact with the atmosphere is limited.

Significant daily changes in the O_2 concentration in the epilimnion of a lake related to the balance between photosynthesis and respiration can also occur. Even though there can be considerable exchange with the atmosphere, during relatively calm days an O_2 supersaturation can build up in the epilimnion, particularly in littoral zones (Fig. 12.12B). The observation of relatively high O_2 production in the littoral highlights the importance of benthic primary producers in lakes, a subject that Robert Wetzel had researched extensively (Biography 12.1).

When lakes mix completely, dissolved O_2 mixes throughout. In some cases, when lakes are hypereutrophic and wind (i.e., mixing) and light are low (under extended cloudy conditions or early in the morning), the heterotrophic demand for O_2 can be great enough to cause the water to become anoxic. Ice cover can also lead to anoxia by limiting photosynthesis and O_2 transport into lakes. Anoxia can cause fish kills in lakes, wetlands, and streams (Highlight 12.2).

Anoxic conditions are very common in sediments of lakes, streams, and wetlands, even if there is ample O_2 in the water above. Anoxia develops because the sediments retard mixing, and diffusion is generally molecular. Furthermore, sediments serve as a store of organic carbon, and heterotrophic activity is relatively high. Even though a highly active photosynthetic community can lead to supersaturated O_2 concentrations at the surface of lighted sediments, anoxic conditions often still develop within depths of millimeters or centimeters below the sediment surface (Fig. 12.11B).

Plants rooted in anoxic sediments must often cope with a lack of O_2 for their roots and wetland plants often have specific adaptations to living in saturated soils. These adaptions of aquatic plants include the ability to actively or passively transport O_2 to their roots or exhibit fermentative metabolism. Vascular systems that transport O_2 down to the roots also serve to transfer methane, CH_4, an important greenhouse gas, to the atmosphere (as we will discuss in Chapter 13).

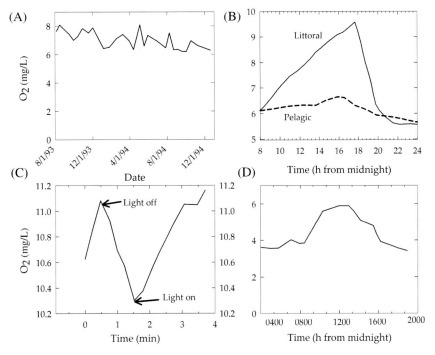

Figure 12.12 Temporal variation in O_2 in Kansas groundwater (A), the pelagic and littoral zones of an Indiana lake (B), a periphyton assemblage (C), and a stream (D). *Data in (A) courtesy of Konza Prairie long-term ecological research site; data in (B) from Scott (1923); (C), original data; (D), Odum (1956).*

In sediments that receive sunlight and are not toxic, there is invariably a photosynthetic community associated with the sediment surface. The production of these communities is generally high, leading to steep gradients in O_2 over depth (Fig. 12.11B) or time (Fig. 12.12C). The shape of the O_2 curve with depth in sediments is similar to the shape of the O_2 curves with depth from hypereutrophic lakes, but the vertical scale is in millimeters instead of meters. Distribution of O_2 across these sediments can be an important factor controlling biogeochemical cycling. Thus, factors that alter O_2 distribution, such as animal burrows, can have strong ecosystem effects.

O_2 concentration in streams can also vary over time (Fig. 12.12D) allowing for in situ measurement of photosynthetic rates (Method 12.2). O_2 variation over time can provide an index of the relative degree of system productivity, with large diurnal swings in O_2 occurring in the more productive systems. O_2 dynamics provide information useful for ecosystem analyses or as a management tool to assess effects of river pollution (Auer and Effler, 1989). In rivers that receive a great deal of untreated sewage (such as commonly occurs in developing countries), O_2 can disappear completely (Highlight 12.2).

BIOGRAPHY 12.1 Robert Wetzel

Robert Wetzel (Fig. 12.13) had a major influence on the field of aquatic ecology. One of his many contributions was to demonstrate that primary production by periphyton and macrophytes in lakes is often greater than that by phytoplankton. Many researchers incorrectly assumed that phytoplankton photosynthesis dominated in lakes because of the large volume of water containing phytoplankton (see Chapter 24). In addition to establishing the importance of benthic producers, his research documented the fate and cycling of carbon in ecosystems.

Figure 12.13 Robert Wetzel. *Courtesy Robert Wetzel.*

Perhaps his greatest influence has been through his textbook, *Limnology* (Wetzel, 1983), which has been used to train innumerable aquatic ecologists. The most recent edition, *Limnology: Lake and River Ecosystems* (2001), is an essential reference text for advanced limnologists.

Dr. Wetzel became interested in ecology because an inspiring high school biology teacher took the time to show him the virtues of nature. From this start, he became one of the most respected aquatic ecologists in the world. He published over 400 papers and 23 books, and received numerous awards and honors. He was deeply involved in international limnological pursuits and was elected member of the Danish, Russian, and Hungarian National Academies of Science.

When Wetzel was interviewed for the first edition of this book, he said he found an academic career extremely rewarding, and he deeply appreciated the freedom to satisfy his intellectual curiosity. He had adventures related to limnology as well. For example, he recalled a time sampling alone on Borax Lake, in a rubber raft, when a youth on shore with a rifle tried to shoot the raft out from under him! Fortunately, he survived that experience.

Wetzel cautioned new students of aquatic ecology not to put too much stock in simple explanations and to remember that biology is a complex and sophisticated field of study. He suggested that an important area for future study is the regulation of growth and the productivity of aquatic organisms by chemically mediated signals among them (Wetzel, 1991).

Robert Wetzel died on April 18, 2005 at the age of 68. His academic career was inspired by that of his hero, G. E. Hutchinson. He carried on the tradition of synthesis and intellectual rigor established by Hutchinson and became a giant in the field of limnology.

Highlight 12.2 Fish kills result from anoxia in streams and lakes

When organic carbon is high and exchange with the atmosphere is low, habitats can temporarily become anoxic or hypoxic (low O_2, generally below 2 mg/L) and fishes and other animals die. While these events are often referred to as fish kills, it is important to consider that many other aquatic animals including amphibians and invertebrates may be killed or adversely affected. This condition occurs when total respiratory demand of all organisms in the ecosystem exceeds input of photosynthetic and atmospheric O_2. High temperatures often exacerbate the problem because the metabolic rates are greater and O_2 solubility is lower in warmer water (Cooper and Washburn, 1949).

Fish kills can occur during the summer in eutrophic lakes. Kills occur in eutrophic and hypereutrophic lakes when productive systems experience a series of calm, cloudy days. Under these conditions, algal blooms have high total respiration rates. With little input of atmospheric O_2, anoxia develops leading to fish kills. Such summer kills can be common in many areas and similar kills can occur year-round in the tropics.

Fish kills also occur in lakes in the winter when ice cover prevents O_2 transport into the water. If snow covers the ice, light transmission and photosynthetic O_2 production are low. However, a eutrophic lake contains a significant amount of biomass, so respiration continues and O_2 concentrations decrease. Fish kills occur in such situations unless the fish can find an inlet stream or O_2 is bubbled into the lake.

Fish kills from anoxia are relatively common in many areas. For example, in the state of Missouri, from 1970 to 1979 there were more than 40 known winter kills and at least 100,000 fish deaths. During the same period, there were about 20 summer kills resulting in the death of more than 200,000 fishes (Meyer, 1990). Anoxia causes very large-scale fish kills in Lake Victoria, Africa. In this case, storms suspend sediments and wash organic material from surrounding wetlands into the lakes. The resulting anoxia kills metric tons (hundreds of thousands) of fishes (Ochumba, 1990). These kills are problematic because local people rely on the fishes for food.

An interesting case of fish kills occurs in pools with hippopotamus. They graze in surrounding terrestrial areas at night and return to the pools during the day, where they rest and defecate. During dry periods, the pools they inhabit become anoxic as microbes break down the defecated materials. During wet periods, the higher discharge flushes the pools and low oxygen water is transported downstream killing fish (Dutton et al.,2018).

Hypoxic conditions can exacerbate other stresses that fishes experience such as high temperature and exposure to toxins (either man-made or resulting from toxic algal blooms). Thus, keeping O_2 concentrations at reasonable levels in naturally oxic waters is key for maintaining the biotic integrity of fish assemblages. A meta-analysis indicated that when O_2 is below 4 mg/L, many species of fish exhibit reduced growth and food consumption (Hrycikd et al., 2017).

Fish kills from O_2 depletion occur in wetlands (Calheiros and Hamilton, 1998) and flowing waters as well. People release untreated sewage into wetlands, rivers, and streams throughout the world. Such releases were widespread in the United States and Western Europe until the 1970s, when environmental laws were enacted requiring reductions in the amount of organic carbon (i.e., biochemical oxygen demand) in sewage discharges. When the load of organic carbon in untreated sewage stimulates respiration and consumes O_2 at a rate in excess of the rate of atmospheric exchange, a river can become anoxic and the fishes die. Such problems have become rare in developed countries since municipalities have been required to lower the organic carbon in the sewage that they release. Consequently, fish species that are less tolerant of low O_2 are reestablishing in areas where they were absent for many years.

Variation occurs in O_2 over time in wetlands. This fact has been used to estimate rates of primary production and respiration in the Everglades (McCormick et al., 1997). The variation of O_2 was extreme in areas that received phosphorus pollution in the Everglades. The water column was completely anoxic at night and fully saturated during the day.

Groundwaters can be oxic or anoxic depending on the relative supply of organic carbon to heterotrophs, the scale of observation, the residence time of the water in the aquifer, and the dissolved O_2 concentration in the incoming water. Many pristine aquifers are oxygenated, but human activities can lead to anoxia. However, row-crop agriculture actually may decrease organic C input into aquifers leading to higher O_2 concentrations (Fig. 12.14A). Septic systems, feedlots, spills of organic chemicals, and subsurface disposal of sewage effluent can lead to anoxic groundwaters by stimulating respiration with addition of organic carbon (Madsen and Ghiorse, 1993). When groundwater is deep enough, it is generally anoxic because of very low supply of O_2, high temperatures associated with geothermal heating, O_2 consumption by organisms, reactions with inorganic chemicals, and long turnover times. If groundwaters become anoxic, biogeochemical cycling changes and many species of invertebrates and microbes cannot exist in the aquifer. Consequently, O_2 can control biodiversity of groundwaters.

Additional small-scale anoxic habitats exist that are important in a variety of aquatic environments. The interior of organic particles may have anoxic zones associated with them even if they occur in oxygenated waters. For example, decaying leaves in oxic streams and lakes may have anoxic zones at their surface and inside individual leaves or entire aggregates of leaves (Fig. 12.14B). The digestive systems of many animals are anoxic, and a distinct microflora forms in these locations. Researchers have only given modest attention to the role of aquatic animal digestive tracts in biogeochemical cycling. In lighted sediments, the algal community that is present can cause great

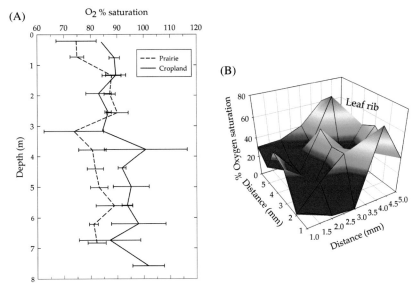

Figure 12.14 O_2 profiles in unconsolidated sediments and groundwater below cropland and prairie (A) and an O_2 concentration map at the surface of a leaf particle from groundwater (B). Statistical analysis showed that the O_2 concentration was significantly lower for prairie in (A) than under cropland. The groundwater was at 4.2 m in the prairie profile and 5.3 m in the cropland profile. *(A) Reprinted from Dodds et al. (1996), with permission from Elsevier Science © 1996; (B) Reproduced with permission from Eichem et al. (1993).*

variations in O_2 concentration with time. Upon darkening, respiration rapidly consumes O_2; there is a measurable decrease within one second (Fig. 12.12C). Organisms living in such a habitat must be adapted to rapid changes in O_2 concentration.

SUMMARY

1. Materials in water can be dissolved, colloidal, or particulate (gravitoidal).

2. Conductivity (total dissolved ions) and pH are chemical properties of water and important descriptive parameters used by aquatic ecologists because these properties can control the distribution and activity of organisms.

3. Redox potential is another important parameter that controls chemical and biochemical processes in aquatic ecosystems. The way chemicals are transformed in the environment is determined partially by the redox potential of the environment. Redox potential is an estimate of the relative concentration of the available electrons in the environment. O_2 concentration is a primary determinant of redox potential.

4. Potential energy drives chemical and biochemical processes. Potential energy is positive when the redox potential of the reactants is very different from the redox potential of the surrounding environment. A chemical reaction will release energy if the products of the reaction have less potential energy than do the reactants.

5. Organisms can promote chemical reactions by lowering the activation energy with enzymes. In the case of reactions that go against potential energy, another source of energy must be present. For example, organisms can shunt chemical energy into a chemical reaction. Photosynthetic organisms convert light into chemical energy and use this chemical energy to drive carbon fixation (converting CO_2 to sugar).

6. A chemical reaction will not occur spontaneously if the activation energy is too large, regardless of the potential energy released.

7. The presence of O_2 is very important because of its role in redox and respiration. The concentration of O_2 varies with space and time in aquatic environments. Habitats with O_2 are termed oxic or aerobic, and those without O_2 are called anoxic or anaerobic.

8. Photosynthesis produces O_2 in aquatic environments, and aerobic respiration consumes it.

9. Photosynthetic rates are controlled mainly by the amount of light, temperature, and nutrient availability. Organisms are able to acclimate to low light. High light (especially UV) can cause photoinhibition, and adaptations such as special protective pigments can provide protection.

10. Respiration rates are mainly controlled by carbon availability and quality, temperature, and nutrient availability.

11. The balance of photosynthesis and respiration, the degree of contact with the atmosphere, and transport processes determine actual O_2 concentration.

12. Some common anoxic habitats include the hypolimnia of eutrophic lakes; organic-rich sediments in lakes, streams, and wetlands; organic-rich groundwaters; digestive tracts of animals; decaying vegetation in water; and particles with high rates of associated microbial activity.

QUESTIONS FOR THOUGHT

1. Why are nutrients taken up by cells often preferred in a reduced form, even when organisms inhabit oxidized environments? (Hint: think of the conditions under which life evolved.)

2. Why do midge larvae that live in the profundal benthos of lakes often turn bright red when brought to the surface?

3. Why are aquifers in karst regions often oxic?
4. Why is oxidation of organic carbon by O_2 more efficient than anaerobic respiration?
5. Dense periphyton mats can detach from the bottom of a lake during the day, float to the surface, and then sink again at night. Why might this happen?
6. Winter fish kills from anoxia can occur in Arctic lakes that are not very productive and do not freeze to the bottom. How could this happen?

CHAPTER 13

Carbon

Contents

Carbon cycling is central to the way ecosystems operate from local to global scales because organic carbon is the currency of energy exchange in ecosystems, as well as a primary building block of life. Comprehending carbon cycling is central to understanding food webs and aquatic community dynamics. Inorganic carbon in water is involved in the bicarbonate equilibrium, which connects intimately to pH control and responses to acid precipitation. Production of methane, a greenhouse gas, by wetlands is one example of the direct impact of the carbon cycle of freshwater ecosystems on global biogeochemistry and climate. In this chapter, we discuss forms of organic and inorganic carbon, as well as transformations and fluxes of carbon in the environment (including the carbon cycle).

FORMS OF CARBON

Inorganic carbon

Figure 13.1 Lake Nyos, Cameroon, Africa. This is the site of a catastrophic CO_2 release that killed 1,700 people in 1986. *Image courtesy: George Kling.*

Carbon dioxide (CO_2) is the primary form of inorganic carbon in the atmosphere. The first edition of this book was published in 2002, when the atmospheric concentration was approximately 375 ppm; it was roughly 410 ppm in 2018, an increase of about 3.5% per decade. The concentration has been constantly increasing since the industrial revolution, intensifying the greenhouse effect. The greenhouse effect (increases in atmospheric CO_2, other gasses, temperature, and associated climate change) is influencing aquatic ecosystems in a variety of ways (Hutchin et al., 1995; Tobert et al., 1996; Magnuson et al., 1997; Megonigal and Schlesinger, 1997; Doll and Zhang, 2010); we discuss such effects in more detail in Chapter 1 and elsewhere in the book.

When CO_2 dissolves in water, it can exist in a variety of forms, depending on pH. The forms are *carbon dioxide, carbonic acid, bicarbonate,* and *carbonate.* The sum of the concentrations of all these forms is the inorganic carbon concentration ($\sum CO_2$). Under most conditions in aquatic systems, CO_2 rapidly converts to carbonic acid, so we consider them as the same here. The chemical conversions among these forms make up the *bicarbonate equilibrium.* Understanding this series of chemical reactions is necessary for comprehension of how aquatic ecosystems buffer against changes in pH, and how CO_2 becomes available for photosynthesis (Butler, 1991). The bicarbonate equilibrium can be represented as

$$CO_2 + H_2O \Leftrightarrow H_2CO_3 \Leftrightarrow H^+ + HCO_3^- \Leftrightarrow 2H^+ + CO_3^=$$

$$\text{carbon dioxide} \qquad \text{carbonic acid} \qquad \text{bicarbonate} \qquad \text{carbonate}$$

The "$\Leftrightarrow$" symbol indicates a carbonic equilibrium reaction. Adding or taking away chemicals from any part of the reaction can force the reaction to compensate. For example, if we add acid (H^+) to a bicarbonate solution, the equilibrium is weighted too heavily to the right-hand side of the equation, so the bicarbonate will convert spontaneously to carbonic acid or carbon dioxide. You can demonstrate this easily by adding an acid such as vinegar to a solution of the sodium salt of bicarbonate (baking soda). Adding the acid will cause production of CO_2 as the equilibrium is reestablished. Because the CO_2 gas has a limited solubility in acidic water, it will bubble out. Hence, as the pH changes so do the relative amounts of bicarbonate, carbonate, and carbonic acid (Fig. 13.2).

As atmospheric pressure increases, the concentration of dissolved CO_2 increases. Carbonated beverages stored under pressure (such as soda, beer, and champagne) lose CO_2 when opened. These beverages must be slightly acidic, so the equilibrium is forced to the side of CO_2. A similar, but catastrophic, release of CO_2 had disastrous consequences in the African Lake Nyos (Highlight 13.1).

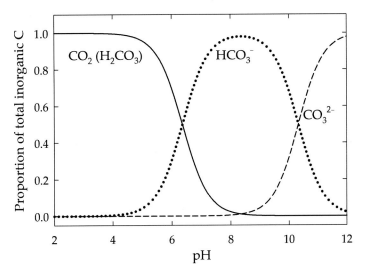

Figure 13.2 The relative concentrations of inorganic compounds involved in the bicarbonate equilibrium as a function of pH.

Highlight 13.1 The Lake Nyos disaster

One thousand seven hundred people died on August 21, 1986, near Lake Nyos (Fig. 13.1) in Cameroon, Africa. At about 9:30 pm, people in the area heard a loud rumbling. One survivor reported viewing a mist rising off the lake and a large water surge (subsequently demonstrated to have washed up to 25 m high on the southern shore). Some survivors reported smelling an odor like rotten eggs, experiencing a warm sensation, and then losing consciousness. When they awoke 6–36 hours later they were weak and confused, and many of their family members were dead. Thousands of livestock and other animals succumbed as well (Kling et al., 1987). Research teams visiting the area later pieced together a picture of what caused this disaster. The deaths resulted from a catastrophic release of more than a million metric tons of CO_2 from the lake; CO_2 is heavier than air, and the large amount released from the lake filled the valley around the lake, displacing O_2 and suffocating people and animals.

Something caused CO_2 bubbles to start coming out of solution, leading to mixing of the lake and catastrophic CO_2 release. The lake went from clear blue to reddish brown; the iron dissolved in the hypolimnion was mixed up and oxidized forming a

Figure 13.3 Water jet in Lake Nyos upon installation of CO_2 releasing pipe into the hypolimnion. *Photograph courtesy: Michel Halbwachs.*

rusty precipitate at the surface of the lake. Evans et al. (1993) suggested that volcanic eruption, alteration in limnological factors, or both initiated the gas release. The limnological explanation revolves around the idea that seasonal mixing allowed for the CO_2 to degas rapidly when the hypolimnion exceeded a threshold CO_2 concentration. Evidence for this hypothesis includes the observation that a similar gas release killed 37 people at nearby Lake Manoun during the same season 2 years earlier (Kling, 1987).

The CO_2 continues to build up in this amictic tropical lake since the disaster (Evans et al., 1993) and could lead to hypolimnetic saturation again within 140 years. However, saturation may be reached near the bottom in <20 years (Evans et al., 1994). The source of CO_2 is most likely volcanic activity occurring below the lake. The lake stratifies, with a chemocline (transitional zone of a lake with stratification stabilized by salinity), currently at about 50 m depth. Therefore, limnological factors prohibit mixing of CO_2-rich waters with the atmosphere to relieve high concentrations deep in the lake. Relatively high concentrations of dissolved CO_2 can build up at depth because of the high pressure under the water.

Several actions were contemplated to avoid future catastrophic releases (Holloway, 2000). The final solution was to use a natural pump-lift pipe to remove excess CO_2 from the hypolimnion. This ingenious solution placed a pipe down into the hypolimnion with the other end above the surface of the water. Water is pumped up from below, and as it decreases in pressure, dissolved CO_2 bubbles out of solution and causes an airlift pump (Fig. 13.3) that continues pumping water to the surface with no further energy input (Halbwachs et al., 2004). The fountain when this pipe was installed shot 50 m into the air. About 20% of the gas is methane, and now plans are underway to harvest this gas and use it as fuel. Never before has the field of physical limnology had such direct involvement in a human health issue. Careful future monitoring will be necessary to avoid repeating the disaster. Currently, there is also concern that the natural dam of volcanic rock that impounds the lake is weakening and could collapse. Rapid downstream flooding would probably lead to thousands more deaths. Research is underway to find how to stabilize the dam.

The equilibrium of inorganic carbon also explains the acid-neutralizing or *buffering* capability of bicarbonate and carbonate. Protons form covalent bonds with carbonate and bicarbonate and ultimately with hydroxyl ions if CO_2 is produced. The protons are not free, so pH changes only a small amount relative to the concentration of H^+ added to solution as long as there is carbonate or bicarbonate to react. Once these forms convert to CO_2, pH can decrease rapidly with further additions of acid. Systems with a significant amount of dissolved bicarbonate (e.g., limestone watersheds) are able to resist the effects of acid precipitation. An opposite response occurs to addition of base (OH^-). The OH^- ions associate with the H^+ ions so the equilibrium balances by moving toward the bicarbonate side. Alternatively, as CO_2 dissolves in water it causes decreases in pH. Thus, rain is naturally slightly acidic, and increases in atmospheric CO_2 are causing acidification of surface waters.

The acid- and base-neutralizing capacity of bicarbonate and the predominance of bicarbonate ions in many systems have led to using *alkalinity* and *acidity* titrations to estimate $\sum CO_2$. For alkalinity titrations, acid can be added with little initial change in pH. After all the bicarbonate and carbonate have reacted with the added acid, further additions cause proportionally greater decreases in pH per unit of acid added. The alkalinity is the amount of acid needed to cause these greater decreases in pH, which is measured with pH-sensors or dyes while a sample is titrated with a known concentration of acid.

The dependence of the bicarbonate equilibrium on pH yields plots that show the relative concentrations of each of the forms of inorganic carbon at various levels of pH (Fig. 13.2). Such data are useful because CO_2 is the form of inorganic carbon required for photosynthesis. Many photosynthetic organisms can convert bicarbonate to CO_2, but CO_2 is still the most easily used form. Knowing the alkalinity and pH of a solution allows an investigator to calculate the amount of inorganic carbon that is immediately available for photosynthesis.

Calcite ($CaCO_3$) is an important precipitate of the bicarbonate equilibrium. The precipitate can form spontaneously when CO_2 is removed from solution (as a way to balance the equilibrium). This precipitation can occur when photosynthesis removes CO_2, when physical factors remove CO_2 (e.g., degassing of spring waters), or when organisms such as mollusks build their shells (Wetzel, 2001). The resulting precipitate can build up impressive concretions of whitish calcium bicarbonate on the stems of the green alga *Chara* (hence the name stoneworts), in terraced outflows of hot springs, and in some benthic habitats with photosynthetic microorganisms. Mammoth Hot Springs in Yellowstone National Park is composed of massive carbonate terraces.

Organic carbon

Organic carbon (carbon compounds bonded with hydrogen, usually oxygen, and sometimes other elements) takes a tremendous variety of forms. The broadest classifications are *dissolved organic carbon* (DOC) and *particulate organic carbon* (POC). Stream ecologists further divide the particulate fractions into *fine particulate organic matter* (FPOM) and *coarse particulate organic matter* (CPOM). A dividing line of 0.45 µm has been proposed for the difference between DOC and POC, and a line of 1 mm has been proposed for the division between FPOM and CPOM (Hutchens et al., 2017). Some investigators have used further divisions of the particle sizes (e.g., ultrafine particles or large debris). CPOM and FPOM contain both living and dead organic materials. Nonliving material derived from terrestrial plants often dominates in streams and benthic sediments but living algae and macrophytes can be present. Living material dominates CPOM and FPOM suspended in many lakes.

The amount of organic carbon in ecosystems that is biologically available is particularly important in habitats dominated by heterotrophic organisms (e.g., groundwaters, sediments, and forested streams) but is important in all aquatic environments (Dodds and Cole, 2007). Given the huge number of organic compounds synthesized by organisms, and all the possible pathways of degradation leading to by-products, even determining the total organic chemical content of water can be difficult. Some of the organic carbon compounds may be very resistant to degradation (*recalcitrant*), and others are *labile* and have high biological availability (Wetzel, 2001), so simply lumping them together may not make ecological sense unless the aim is to characterize total export from or storage by an environment. However, one common method to estimate the total available organic carbon for heterotrophs is based on the concept of *biochemical oxygen demand* (BOD) or the total demand for oxygen by chemical and biological (respiration) oxidative reactions. Change in concentration of BOD is a cornerstone of understanding the effects of sewage effluent on aquatic ecosystems (Method 13.1).

METHOD 13.1 What is biochemical oxygen demand and how is it measured?

Biochemical oxygen demand (BOD) is a simple and practical indicator of the total organic content that is available to organisms plus any chemicals that spontaneously react with O_2. The procedure is straightforward: an analyst incubates water in sealed bottles and monitors the decrease in O_2 over time. If all the O_2 is consumed over the time of the incubation, the original water sample must be diluted and analyzed again. During the incubation, naturally present heterotrophic organisms use O_2 to respire the organic carbon that is biologically available and any chemicals that spontaneously react with O_2 (e.g., sulfide) will also do so, allowing analysts to assess total BOD in wastewaters (Eaton et al., 1995).

When humans release sewage into natural waters, it provides heterotrophs with additional organic carbon and creates a demand for dissolved O_2. High sewage influx leads to anoxic conditions in the waters, particularly during summertime low flows. During summer low-flow at high temperatures, dilution is at a minimum and dissolved O_2 concentrations are low. Regulations for the degree of sewage treatment are based on the amount of BOD released into the receiving waters. The regulated facilities calculate a permissible loading rate based on dilution, aeration rate in the receiving water, and BOD concentration of the sewage effluent. Sewage treatment facilities often make daily BOD measurements to assess the treatment efficiency.

The dissolved pool of organic carbon can be divided into two major classes—humic and nonhumic substances. *Humic compounds* are large molecular weight compounds and lend a brownish color to water. The nonhumic fraction includes sugars and other carbohydrates, amino acids, urea, proteins, pigments, lipids, and some additional low-molecular-weight compounds. Such a classification is ecological; heterotrophic processes break down some nonhumic substances to yield humic compounds as a degradation byproduct (Stumm and Morgan, 1981).

Microbial activities result in formation of humic substances as the byproducts of the breakdown of still larger molecular weight compounds such as *celluloses, tannins,* and *lignins.* Tannins and lignins leach from bark and leaves of plants. The microbes use the more easily metabolized parts of the larger compounds; humic substances are the remaining resistant portions of the molecules and are resistant to microbial use (except see Lovley et al., 1996). Humic compounds can be classified into three groups—the *humic acids* (soluble in alkaline solutions and precipitate in acid), the *fulvic acids* (remain in solution in acidic solutions), and the *humin* (not extractable by acid or base) (Stumm and Morgan, 1981). The compounds form from a variety of different source molecules, ultimately derived from either terrestrial or aquatic sources (Reemtsma and These, 2005).

Both the humic substances and the tannins have the following important features (Stumm and Morgan, 1981): (1) they attach to many other organic substances (i.e., they can be important in transport and fate of organic pollutants), (2) they form complexes with metal ions (which can be particularly important in keeping iron in solution), (3) they form colloids including large organic flocs, (4) they color the water brown or tan (absorbing light that could fuel photosynthesis) when in high concentrations such as in blackwater swamps, (5) they absorb ultra violet radiation, (6) they are resistant to biological degradation, and (7) they can interfere with water purification for human use.

Advanced analytic methods are allowing better characterization of carbon compounds found in natural waters. A study of dissolved organic matter in Lake Ontario using nuclear magnetic resonance indicated that the majority of the compounds were aliphatic (did not contain carbon rings), with heteropolysaccharides, carboxyl-rich and acyclic molecules (Lam et al., 2007). Methods such as size exclusion chromatography, mass spectrometry, and scanning fluorescence spectroscopy all allow analyses of specific types of molecules to determine characteristic patterns that serve as "fingerprints" for dissolved organic matter. For example, fluorescence spectroscopy can separate general classes of dissolved organic carbon by microbial or terrestrial origin (McKnight et al., 2001).

TRANSFORMATIONS OF CARBON

We discussed photosynthesis and aerobic respiration in Chapter 12. These two fluxes are central to carbon cycling on a global scale and in surface freshwater habitats, and understanding freshwater fluxes will be essential to balancing the global terrestrial carbon budget (Butman et al., 2016). There are also bacteria that are capable of *anoxygenic photosynthesis* (photosynthesis with no O_2 production) that we discuss later. In general, the complexity in carbon cycling transformations lies in anaerobic cycling and in use of complex organic compounds in oxic and anoxic habitats.

Organisms evolved the ability to use most types of organic molecules (the exception so far being many human-synthesized plastics). This evolution is evident in the rapid appearance of strains of microbes able to use novel organic carbon compounds produced by humans and released into the environment. The process of *in situ* bioremediation is based on this phenomenon (see Chapter 16). Similarly, organisms are perpetually participating in an evolutionary arms race, with one species evolving chemical defense or chemicals to lower activities of competitors, and other species evolving resistance to the chemicals.

Most naturally occurring compounds can be broken down, and numerous metabolic pathways exist given the millions of distinct organic compounds occurring in the environment. The general metabolic approach to deal with large organic compounds (e.g., proteins, cellulose, tannins, and fatty acids) is to modify the compound by energy-requiring hydrolysis reactions into a more readily used compound that can be metabolized with energy yielding reactions (Fig. 13.4). Consequently, cellulose is broken down into its component sugars at an energy cost, but metabolism of the sugars

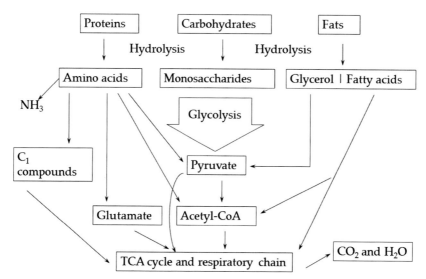

Figure 13.4 A general diagram of aerobic breakdown of organic carbon by organisms. *Modified from Rheinheimer (1991).*

provides energy greater than this initial cost. Complex organic compounds can be degraded in the cell, or enzymes can be excreted outside the cell (*exoenzymes*) to break down larger organic carbon compounds into compounds that can be taken up (Sinsabaugh et al., 1991). A full understanding of carbon cycling will require elucidation of cycling of complex organic carbon compounds (Hobbie, 1992; Wetzel, 2001).

Breakdown rates of detrital organic carbon

Empirical measurements of organic carbon decay suggest that mass generally declines at rates adequately described by a negative exponential relationship:

$$M_t = M_0 \; e^{-kt}$$

where t is the time, M_t and M_0 are mass at times t and 0, respectively, and k is the decay coefficient. This relationship is common because heterotrophs use the most labile organic material first, and the more recalcitrant material, with slower absolute decomposition rates, later. Generally, scientists determine these decomposition rates by incubating organic materials such as leaves, wood, dead algae, or macrophytes in natural environments and determining the change in organic mass over time (e.g., Benfield et al., 2017).

When organic material of just one type is plentiful, rates should be constant over time (e.g., when there is a high concentration of glucose, the breakdown rate is roughly a constant mass per unit time). When concentrations become low, then rates decrease because of decreased uptake rates of materials. When organic materials such as leaves enter aquatic ecosystems, heterotrophs use the simple sugars, proteins, and starch quickly, whereas the more complex structural materials such as cellulose and lignin break down more slowly. Bacteria and fungi colonize the leaves rapidly and begin to break them down (Gessner and Chauvet, 1994). These organisms possess enzymes necessary to break down cellulose and lignin. As discussed later in Chapter 19, animals that eat low-quality organic materials, like senescent leaves, are actually deriving much of their nutrition from the microbial community associated with decay of these materials (Cummins, 1974).

Different species of terrestrial vegetation invest variable amounts of their resources to produce recalcitrant compounds in their leaves to protect against herbivory and other mechanical damage; so, decomposition of leaves of different species can be highly variable (Jackrel et al., 2016). Rapidly growing species often have leaves that break down very quickly, whereas slower growing types have much tougher leaves that are more difficult to degrade (Fig. 13.5). Portions of leaves such as ribs and veins, which are tougher, break down more slowly than other parts of the leaf tissue (Fig. 13.6).

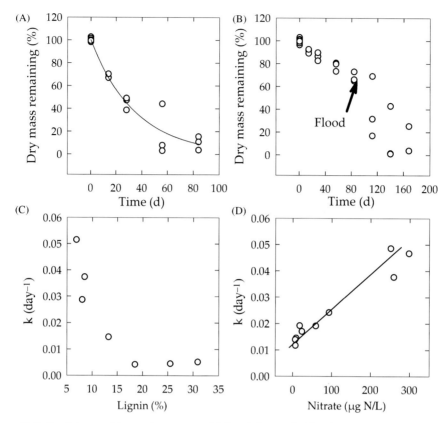

Figure 13.5 Breakdown of leaves in streams. (A) breakdown of alder follows a typical exponential decay, (B) decomposition of beech leaves increases following a flood (at around 80 days), (C) rates of various species as a function of lignin content, and (D) rates as a function of nitrate concentration in the stream. *(A, B, and D) Data from Gessner and Chauvet (1994). (C) Data from Suberkropp and Chauvet (1995).*

Biochar, or black organic carbon, may be one of the most important recalcitrant compounds. This carbon results from burning terrestrial vegetation and detritus, and is an important component of marine dissolved organic carbon (Jaffé et al., 2013). We are only just beginning to understand the processing of this material in freshwaters. Effects on stream biota are unclear. Biochar can impede growth of some aquatic invertebrates, but it can also adsorb metal and organic contaminants, leading to positive effects on stream biota (Clements et al., 2015).

Other factors, such as temperature, mechanical effects of animals, and inorganic nutrients, can also control rates of decay of organic matter. One of the concerns over global warming is that increased temperatures will stimulate organic material

Figure 13.6 A leaf in Valium Stream, Ihla Grande, in the Atlantic Rainforest of Brazil. Only the recalcitrant portions of the leaf that have not been degraded remain.

breakdown, returning CO_2 to the atmosphere more quickly and not allowing for deposition and burial that would sequester the organic carbon from the atmosphere. The most basic approach to this issue is to assume that rates roughly double with every 10°C increase, as discussed earlier (Q_{10} and biological or chemical rates). However, Follstad Shah et al. (2017) did a global analysis of 1,025 measurements of leaf breakdown rates and found the increase of rate of decomposition was about half as much as expected. This still translates into increased CO_2 release, both from increased rates of microbial activity and decreased efficiency of invertebrates that consume the detritus and the microbes associated with it (Mas-Martí et al., 2015).

Researchers often using mesh bags in decomposition experiments to exclude or allow animals to access the leaves. These experiments are somewhat difficult to assess, because the finer mesh bags, while excluding animals, also impede water exchange, which could slow decomposition rates. Nonetheless, numerous studies have shown that decomposition rates are greater when animals can access the decomposing leaf material (Follstad Shah et al., 2017).

Leaves and wood are very low in nitrogen and phosphorus, and nutrients can stimulate decomposition rates (Gulis et al., 2004). A whole-stream nutrient enrichment experiment in an Appalachian Mountain stream documented that leaf breakdown rates were greater with nutrient addition, with concurrent increases in the microbial respiration rates and ultimately greater biomass of macroinvertebrates associated with leaves

(Greenwood et al., 2007). Human land use can alter rates of leaf breakdown; in agriculturally influenced streams, increased nutrients expedite breakdown rates and in urbanized streams, more frequent hydrologic disturbance increases fragmentation of leaves leading to quicker decomposition (Paul et al., 2006).

Oxidation of organic carbon with inorganic electron acceptors other than O_2

Organic carbon can release the most energy to organisms if it is oxidized with O_2 through the process of aerobic respiration. In the absence of O_2, the next best energy yield is from other electron acceptors (such as nitrate, sulfate, and oxidized iron) to oxidize organic carbon. The relative efficiency of these oxidations depends on the oxidation state of the compound. In other words, O_2 is the most oxidized compound abundant in the natural environment that organisms can use to react with the reduced organic carbon, so they obtain the greatest amount of energy through aerobic respiration. Then, organisms turn to other compounds as oxidants of organic carbon in order of redox potential (see Fig. 12.5). The most efficient process is not always used (e.g., a bacterium that can only use nitrate to oxidize organic carbon may continue to do so even in the presence of O_2). However, if the transformation is less efficient and carbon is limiting, the organism that relies on the less efficient mode of oxidizing carbon will ultimately be outcompeted, unless it is able to switch to the more efficient mode. In variable environments, conditions can change rapidly so an organism that does not rapidly change metabolic strategy can be successful, even if it is using a less efficient mode of respiration for part of the time. For example, it may be more energetically expensive to turn off the biochemical machinery for anaerobic respiration than to retain it until the next anoxic event.

In anoxic habitats, this series of organic carbon oxidations can be distributed across a redox gradient (across millimeters or centimeters in sediments, micrometers or millimeters in decaying organic particles, or meters at the interface of an anoxic hypolimnion in a lake, reservoir, or wetland). Aerobic respiration will dominate in the locations in most contact with the oxic zone. After O_2 is used up, then each successive type of oxidant is used up according to the maximum potential energy yield, leading to areas with progressively lower redox (Fig. 13.7). As such, NO_3^- is used first, followed by Mn^{4+}, Fe^{3+}, and SO_4^{2-}. All these oxidations are more efficient than acetogenesis and methanogenesis, which we discuss in following sections.

Fermentation

In addition to oxidizing organic compounds with inorganic electron acceptors, heterotrophic organisms in anoxic environments can use organic carbon by *fermentation*, a process of rearranging the organic molecules to yield simpler organic and inorganic

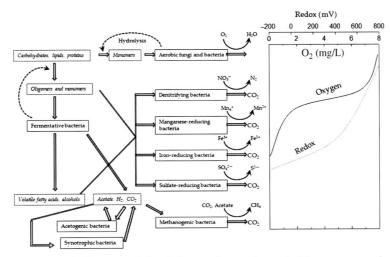

Figure 13.7 A general diagram of breakdown of organic material across a redox gradient. Organisms are shown in solid boxes, and chemical pools are shown in dashed boxes. Curved arrows with solid lines indicate alternative electron acceptors for oxidation of organic carbon, and the preferred redox of each of these transformations approximately corresponds with the redox curve drawn on the right. *Redrawn from Westermann (1993).*

compounds (e.g., acetate, ethanol, CH_4, CO_2, H_2, and H_2O) and energy. A wide variety of these reactions occur because there is a tremendous number of different structures of organic molecules that can be used; examples are presented in Table 13.1. Many of these reactions are of enormous commercial benefit (e.g., fermentation to produce alcohol), and they are central to the carbon flux of anoxic aquatic habits.

A general feature of these fermentation processes is that many yield organic acids. These acids lower pH and generally decrease rates of degradation. No individual species of fermenter is able to metabolize organic polymers (such as cellulose, proteins, and lipids) completely to CO_2 and H_2. In contrast, individual species are able to degrade polymers to CO_2 and H_2O in the presence of O_2 (Fenchel and Finlay, 1995). Consequently, complex communities of *"syntrophic"* microorganisms are required to continue energy cycling in anoxic systems. Individual species from these complex groups of heterotrophic anoxic microbes cannot grow in isolation without a supply of very specific metabolic substrates, so they tend to "cooperate" to break down organic materials.

As complex organic compounds produced by terrestrial plants undergo microbial degradation in the absence of O_2, humic compounds form. Degradation of the organic compounds leads to lowered pH and creation of phenolic and acidic organic compounds that are inhibitory to further microbial activity. Lowered pH encourages precipitation of the humic compounds. As these build up in the sediments, further degradation slows. For the most part, humic substances are difficult for microbes to degrade, but some microbes can

Table 13.1 Some representative fermentative, methanogenic, and acetogenic transformations that occur in anoxic communities

Reaction	Name	Comment
Fermentation		
Glucose $\rightarrow$ 2 ethanol $+ 2CO_2$	Ethanol fermentation	Formation of alcohol
Glucose $\rightarrow$ 2 lactate $+ 2H^+$	Lactate fermentation	
Glucose $\rightarrow$ ethanol $+$ acetate $+ CO_2 + H_2$	Mixed acid fermentation	Produces variable amounts of products
Glucose $\rightarrow$ butyrate $+ 2CO_2 + 2H_2$	Butyrate fermentation	
3 lactate $\rightarrow$ 2 propionate $+$ acetate $+ CO_2$	Propionate fermentation	Gives Swiss cheese flavor
Alanine $+ 2$ glycine $\rightarrow$ 3 acetate $+ 3NH_3 + CO_2$	Paired amino acid fermentation	Important when proteins being broken down
Acetogenesis		
$2CO_2 + 8H^+ \rightarrow$ acetate $+ 2H_2O$	Heterotrophic acetogenesis	
$2CO_2 + 4H_2 \rightarrow$ acetate $+ 2H_2O$	Autotrophic acetogenesis	
Methanogenesis		
$CO_2 + 4H_2 \rightarrow CH_4 + 2H_2O$	Autotrophic methanogenesis	Uses CO_2 and H_2 as a source of energy
Acetate $\rightarrow CH_4 + CO_2$	Acetoclastic methanogenesis	Disproportionation of acetate

Acetogenic and methanogenic processes require very low redox (Fig. 12.5). The listed processes represent a small proportion of the fermentative processes that can occur in anoxic communities.

use them slowly (Lovley et al., 1996). Anoxic conditions lead to slower breakdown of organic materials because of reduced efficiency of organic carbon oxidation and inhibition of microbial activity by metabolic by-products. Accumulation of organic compounds under anoxic and acidic conditions has led to formation of peat bogs as we discussed in Chapter 5. These wetlands are sites of organic accumulation, and meters of material can accumulate over many years. Such bogs occur worldwide, notably in boreal regions of the Northern Hemisphere. Similar processes lead to the formation of blackwater swamps. Humic compounds in blackwater swamps lead the water to appear brown or black in color, and the acidity allows buildup of organic carbon. Peat bogs are important in archeology and paleolimnology (Highlight 13.2).

Methanotrophy

Fermentation reactions can produce methane and carbon monoxide. These compounds are simple but energetically inefficient for most microorganisms to metabolize. Some specialized aerobic bacteria, the *methylotrophs*, can harvest energy by oxidizing simple compounds containing methyl groups and carbon monoxide. Bacteria that

Highlight 13.2 Acid bogs in archeology, paleolimnology, and palynology

Bogs lead to preservation of organic materials for thousands of years. Human bodies are the most spectacular finds of archeological interest (see also Chapter 5 and Fig. 5.10). Several circumstances characteristic of carbon cycling in bogs are required for preservation of bodies: (1) the water must be deep enough to prevent carrion-eating animals from consuming the body when it is first deposited; (2) the water must remain anoxic to inhibit microbial growth and oxidation (preserved materials will decay within days when re-exposed to O_2); (3) sufficient concentrations of tannic acids must be present to cause tanning of the skin (in nonacidic waters, only the bones are preserved); and (4) water must be cold, generally below $4°C$, to inhibit microbial growth. Above this temperature, flesh rots and acids attack and degrade the bones (Coles and Coles, 1989). Bodies 2,000 years old are so well preserved that the color of the hair and eyes could be determined, as well as the last meal eaten. Many human artifacts preserved in bogs have been uncovered as well.

Scientists also use the preservation of organic materials in anoxic bogs to indicate changes in plant communities over time. Isotopic analyses allow for dating the depth layers in sediments. The researchers can use preserved pollen, pigments, or other plant parts to indicate the local plant community. Such information may be useful for detecting environmental trends over centuries or longer periods (Taylor and Taylor, 1993). For example, isotopic composition of sedges and mosses preserved in peat bogs reflects atmospheric CO_2 content with a temporal resolution of about a decade (White et al., 1994). These data are useful to climate modelers who are interested in rapid climate change events that have occurred during the past 12,000 years.

---

specifically oxidize methane are *methanotrophs*. These bacteria are central to the global carbon cycle because their activity is a major reason that methane does not build to large concentrations in Earth's atmosphere. Methanotrophs generally live in close proximity to anoxic habitats from which there is a constant diffusive flux of methane.

Methanogenesis, acetogenesis, and disproportionation of acetate

Some Archaea can produce methane under anoxic and extremely low redox conditions in a process called *methanogenesis*. Under these conditions, CO_2 has more potential energy than the reduced methane at low redox (Figs. 13.7 and 12.5). We present some of the reactions that produce methane in Table 13.1. Microbes can use CO_2 and H_2 or acetate to produce methane. Using acetate is also referred to as a disproportionation, because one of the two carbons ends up as methane, and the other as CO_2 (i.e., proportioned into two different redox states). Microbe-mediated reactions similar to methanogenesis also include acetogenesis (making acetate). Acetate production can be more important than methane production in some conditions, particularly when pH is lower. We present some examples of equations for these reactions in Table 13.1.

Methanogenesis drives biogas production as an alternative energy source and was responsible for the formation of natural gas deposits stored underground across geological time. Methane is a greenhouse gas, and methane concentration in the Earth's atmosphere is increasing because of human activities. Anoxic habitats are common in wetlands, and freshwater habitats contribute significantly to the global methane budget.

The production of methane can fuel some food webs. Methanotrophic bacteria intercept methane and use it for growth, and then larger organisms intercept these bacteria. In some lakes, isotopic analyses indicate that *Daphnia* received 50% of their carbon from methanogenesis. In a stream, macroinvertebrates can derive up to 30% of their carbon from methanogenic sources, and in a lake, fish species that feed on chironomid midge larvae received up to 20% of their carbon via methanogens (Jones and Grey, 2011). Methanotrophs can also fuel portions of wetland food webs (van Duinen et al., 2013).

GLOBAL EMISSION OF METHANE AND CARBON DIOXIDE RELATED TO INLAND AQUATIC HABITATS AND CLIMATE CHANGE

Global methane concentrations have more than doubled in the past 200 years. The absolute level is less than that of CO_2 in the atmosphere, but the rate of increase is greater. Increasing concentrations of methane cause concern because methane absorbs about 20 times as much heat per molecule as CO_2 and acts as a potent greenhouse gas. While some details of the increase are not well established (Schlesinger, 1997), it is clear that human activities put a significant amount of methane into the atmosphere. Such human activities include burning, gas and coal production, methane release from landfills, and ruminant livestock production.

Wetlands are the predominant sources of methane, even more so when we take rice paddies (agricultural wetlands) into account (Fig. 13.8). Wetlands and rice paddies make up 54% of the total methane produced globally. Methane escapes wetlands because anoxic processes lead to methanogenesis. Methanotrophic bacteria intercept and oxidize most of the methane, and we do not include this in the budget in Fig. 13.8. However, the stems of plants form a conduit from anoxic sediments to the atmosphere, bypassing the populations of methanotrophs found at the oxic–anoxic interface in the sediments (Joabsson et al., 1999). The extent and importance of methane release from tropical wetlands is probably substantially underestimated (Gumbricht et al., 2017).

Many factors play into how much methane a specific wetland produces. The differences in types of wetlands relate to rate of methane release, as does hydrology of those wetlands. For example, wet periods following long dry periods lead to increased

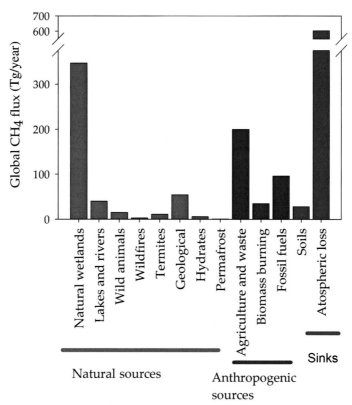

Figure 13.8 Global sources (anthropogenic and natural) and sinks of methane. Estimates are from 2000 to 2009. *Data from Kirschke et al. (2013).*

methane emissions (Turetsky et al., 2014). Thus, climate change models predicting more severe drought and flood could suggest increased methane release associated with increased climate variability. Different plant species transfer different amounts of methane, and the grasses seem to transfer the most (Kao-Kniffin et al., 2010). Rice is a grass, as are many other emergent wetland species such as sedges.

Increasing complexity of the process, damage to plants by herbivores can increase rates of methane export through plants to the atmosphere from roots (Petruzzella et al., 2015). Winton and Richardson (2017) excluded herbivorous waterfowl from wetlands in North Carolina and found increased rates of methane emission with exclusion, and greater plant growth. They found that the herbivory clipped plants and inhibited O_2 transport into the sediments as well as methane transport out. Their data suggested that the decreased O_2 into the sediments was the controlling factor because the anoxic habitat did not allow methanotrophic bacteria to consume the methane.

Rates of methane and CO_2 formation can be so great that gas bubbles containing methane move directly through the water column from anoxic sediments (*ebullition*). This process is an important source of methane to the atmosphere from freshwaters, but it is difficult to quantify. Floating chambers can catch the released gas, but the episodes of gas release are sporadic and heterogeneous, and difficult to sample well. To make things more difficult, the largest of the gas bubbles (the upper 10%) could account for 60% of the methane production from lakes, as determined by hydroacoustic sampling (DelSontro et al., 2015). Release of these large bubbles is even more heterogeneous and sporadic than small bubbles. Increased temperature also links to increased ebullition in the German Saar River (Wilkinson et al., 2015).

Understanding carbon cycling in wetlands will aid future efforts to predict atmospheric methane concentrations. For example, it is unknown what will happen to methane production in high latitudes if the permafrost thaws and significant numbers of new wetlands are formed. This thawing could result in a "runaway greenhouse effect" as more gas release warms the Earth even further. Analyses indicate significant increases in methane from high latitude wetlands will further exacerbate greenhouse warming (Zhangd et al., 2017). One study for the Barrow Peninsula in Alaska estimated a 60% increase in methane emissions from tundra ponds in the last 40 years (Andresen et al., 2017). Increased CO_2 can also stimulate methanogenesis in wetlands (Megonigal and Schlesinger, 1997), as well as phytoplankton production in lakes (Vogt et al., 2017), complicating the relationships among climate change, CO_2, and methane.

Large amounts of peat, as well as organic carbon, are stored in tundra regions. There is concern over how much C from peat will enter the global atmospheric stores of CO_2 and CH_4 as global warming melts permafrost that is preserving large amounts of peat in the northern regions of the planet. When soils melt, the organic carbon locked in the permafrost can mobilize. Some is respired immediately but some washes into wetlands and ponds. This organic carbon pushes down the redox and ultimately leads to increased methanogenesis (Koven et al., 2011). Though the potential for this accelerated carbon release via wetlands had long been understood (Bridgham et al., 1995), research is still needed to understand how strong a feedback will occur (Kirschke et al., 2013; Schuur et al., 2015). The concern is that release of the radiative forcing gases will accelerate global warming and lead to further loss of high latitude stored carbon.

One unexpected observation from tundra ecosystems is that light oxidizes organic carbon that enters waters draining to the ocean in addition to microbial oxidation. The light oxidation accounts for about 75% of the total oxidation and 30% of the CO_2 production (Cory et al., 2014). Hence, potentially even more of the carbon released with permafrost melting will return as greenhouse forcing gas than was predicted on the basis of microbial activity measures. This effect will be most pronounced in shallow habitats that are fully exposed to the sun.

High latitude areas are not the only regions that will release more C with warming. Lakes in tropical areas could also increase emission rates, with CO_2 release rates increasing from 9% to 20% and methane from 20% to 58% by 2100. Thus, carbon that washes from terrestrial habitats will be less likely to be stored in sediments. A further unknown is the amount of inundated tropical habitat in areas such as the Amazon basin (Marotta et al., 2015).

Streams and lakes also release methane and carbon dioxide to the atmosphere. Global releases of CO_2 from streams and rivers are about 1.8 PG/year, and about 0.32 PG/year from lakes; streams and rivers are particularly important "hot spots" of release across the landscape (Raymond et al., 2013). Global releases of methane from streams are about 15% of those from wetlands, and lakes are about 40% less than wetlands (Stanley et al., 2016). The gasses can either be produced in the stream sediments, or be carried into the stream from groundwater where the processes are ongoing. Watershed disturbance by humans can increase methane production by streams by increasing organic sediments (Crawford and Stanley, 2016), even in the presence of increased nitrate, which could to decrease rates of methanogenesis. In addition, modification of stream channels by humans tends to decrease rates of organic carbon storage, leading to greater release of terrestrially derived materials to the atmosphere (Wohl et al., 2017). Measurements of diffusive flux from streams also do not account for ebullition; when accounted for, it doubles the methane production from streams (Crawford et al., 2014).

Lakes can release methane as part of the process of phosphate uptake from methylphosphonate (Yao et al., 2016). Lakes build up CO_2 in hypolimnetic waters (Weyhenmeyer et al., 2012) that presumably can be released in fall mixing over relatively short periods that routine sampling may miss. However, shallow lake habitats are probably the most important sources of methane (Encinas Fernández et al., 2016), and warming of even tropical lakes will increase emission rates (Marotta et al., 2014). Ebullition in lakes can be a dominant pathway of gas release from anoxic habitats into the atmosphere (Beaulieu et al., 2016; Miller et al., 2017). There is substantial regional variability in how much CO_2 lakes and reservoirs produce (Lapierre et al., 2017). As temperature increases, so will rates of respiration. However, photosynthetic rates will also increase with temperature at about the same rate. Therefore, some CO_2 release with increased temperatures may be mitigated (Demars et al., 2016).

A CONCEPTUAL INTRODUCTION TO NUTRIENT CYCLING

In this section, we introduce how to diagram a complete nutrient cycle, rather than considering individual fluxes specific to individual elements. Such cycling is a key feature of all compounds used by organisms. All the fluxes that occur in an environment make up the *cycle*. The *budget* is an account of the relative magnitude of the fluxes. If any chemical form in a cycle is not recycled, it builds up in the environment. Evolution has led to organisms that can metabolize most naturally occurring organic carbon compounds (including novel compounds synthesized by humans). Such adaptations can occur in days or weeks in the case of most novel organic compounds. Likewise, almost any inorganic chemical with potential energy in an environment can be converted by organisms to yield energy. Thus, we have complete nutrient cycles.

Nutrient cycles have general features that we, your attendant authors, have used to build a conceptual framework of the way that materials move through the environment. Some of this framework is built on the requirements of organisms and some on the constraints that redox exerts on chemical reactions. The general features make it easier for students to understand nutrient cycles and ultimately link the cycles together in a larger, more comprehensive view of aquatic ecosystems.

All organisms require nutrients, and they can acquire them as dissolved inorganic forms, organic forms, or both. For example, humans obtain carbon in the form of organic molecules, but plants can obtain carbon from either CO_2 or organic carbon sources. As we discussed in Chapter 12, the acquisition of nutrients is *assimilation*, and generally occurs regardless of the redox state of the environment. Likewise, organisms excrete inorganic nutrients; the general name for this process is *remineralization, mineralization,* or *regeneration*. Remineralization of organic carbon to CO_2 is respiration or fermentation in the carbon cycle. When reduced inorganic chemicals are in an oxidized environment (e.g., an oxic region), microbes can oxidize them, and metabolic energy can be gained. Energy is gained because the flux goes with potential energy. Similarly, oxidized compounds can release energy when reduced in an anoxic environment.

Here, we introduce a method for diagraming nutrient cycles to assist the reader in understanding and remembering complete nutrient cycles. It is important to keep in mind the idea of potential energy in environments of differing redox (see Chapter 12). A generalized nutrient cycle is presented to illustrate the point (Fig. 13.9). In this diagraming method, we list inorganic compounds from left to right as they become more oxidized, oxic processes are placed in the top half of the diagram, and anoxic processes are placed in the bottom half of the diagram.

The boundary between anoxic and oxic that passes through the horizontal center of the diagram can be considered similar to (1) the dividing line between an oxic epilimnion and an anoxic hypolimnion, (2) the difference between an oxic sediment surface and an anoxic portion deeper in the sediments, or (3) the difference between the oxic habitat outside a decaying leaf and the anoxic habitat inside. Assimilation and heterotrophy generally occur regardless of the presence of O_2, so we place them across the oxic/anoxic boundary.

In general, processes that yield energy move from left to right on the top (oxic) part of the diagram and from right to left on the bottom (anoxic) part of the diagram.

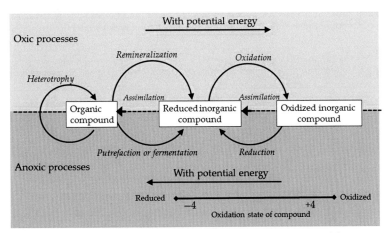

Figure 13.9 Diagram of a hypothetical nutrient cycle. This will be the general format used to represent nutrient cycles. Oxic processes are above the center horizontal line and anoxic processes are below. Those that move on the center line are required, independent of O_2 concentration. Inorganic forms are listed from left to right, from reduced to oxidized. The redox state of the inorganic compounds varies with the elements being considered, and are arbitrarily set to +4 and −4. Hence, transformations are generally occurring with potential energy if they move from left to right in the top half of the diagram or from right to left in the bottom half of the diagram.

Some processes move against potential energy. For example, photosynthesis uses carbon dioxide to synthesize organic carbon in oxic environments. Each of the nutrient cycles has its own idiosyncrasies related to the different chemical properties of the individual nutrient and the conventions of researchers, but this general diagraming method should assist in learning the generalities across nutrient cycles.

THE CARBON CYCLE

We diagram the carbon cycle using the technique presented for general nutrient cycles (Fig. 13.10). We list CH_4 separately from other organic forms because of its crucial role in global carbon cycling. In addition, photosynthesis is an assimilatory flux, but the processes of oxic and anoxic photosynthesis are different, so we separate them in this chart. All processes occur in freshwater.

We covered the global importance of methane production in wetlands earlier in this chapter. Wetlands are also globally important in the process of carbon deposition or *sequestration*. Most of the coal on Earth formed from trees that died and accumulated in wetlands and other freshwater systems, and were buried.

Carbon transport through freshwaters is also of global importance. Far more heterotrophic activity fuels food webs in marine systems than previously thought, and

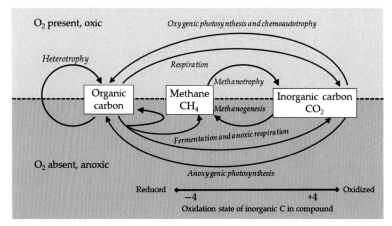

Figure 13.10 A diagram of the generalized carbon cycle. See Fig. 13.9 for a general explanation of the diagram layout.

many marine ecosystems are net heterotrophic (Dodds and Cole, 2007). Extensive impoundment of water in reservoirs has altered how much organic carbon moves from terrestrial to aquatic systems; reservoirs intercept about 20% of global carbon transport from rivers into the ocean (Syvitski et al., 2005). Deforestation, urbanization, and hydraulic modification all have the potential to influence carbon transport.

Advanced: hydrogen and carbon cycling

Hydrogen gas (H_2) production is closely linked to anaerobic carbon cycling. Syntrophic assemblages of bacteria use hydrogen cycling coupled with carbon cycling to complete fermentation cycles (Fenchel and Finlay, 1995). For example, pyruvate, ethanol, or butyrate fermentation pathways can lead to excretion of H_2 as a way to dissipate reducing power (electrons). If H_2 builds up, then the fermentation can proceed no further. However, if methanogens are present to consume excess H_2, or if other pathways occur (such as using sulfate to oxidize hydrogen gas) then fermentation can proceed. The relationship between hydrogen-consuming and hydrogen-producing organisms forms the basis of syntrophy. Whereas syntrophic associations are primarily composed of Archaea and Bacteria, some Eukarya are capable of fermentation that yields H_2. Some protists have hydrogenosomes that convert pyruvate to acetate and H_2. This source of H_2 is probably only moderately important in aquatic ecosystems in most cases.

Hydrogen gas also mediates control over methanogenesis by sulfate. In this case sulfate reducers (covered in more detail in the next chapter) oxidize acetate and H_2 while converting sulfate to sulfide. These organisms are better competitors for acetate and H_2 because redox relationships dictate greater energy yield occurs when the oxidant is sulfate (used by sulfate reducers) rather than CO_2 (used by methanogens).

Essentially, coupling of the anoxic C cycle with H_2 dictates that methanogenesis will only occur at very low rates when sulfate is present in the environment. Maintaining a balance of electrons and the wide variety of chemical interactions that microorganisms can mediate leads to complex interactions among the various nutrient cycles; considering hydrogen gas is an important part of understanding these anoxic interactions.

SUMMARY

1. Inorganic carbon is found in the form of carbon dioxide (CO_2) gas in the atmosphere at about 400 ppm (it reached this level for the first time in recorded history in 2013), with continuous increases of about 2 ppm/year. This gas readily dissolves in water, where it can take the form of CO_2, carbonic acid (H_2CO_3), bicarbonate (HCO_3^-), and carbonate (CO_2^-).

2. The equilibration between the different ionic forms of inorganic carbon is the bicarbonate equilibrium. One of the most important factors driving equilibrium is the pH of the water. Under low pH, CO_2 predominates; under high pH, the equilibrium moves toward carbonate. The equilibration can buffer natural waters against changes in pH.

3. Organic carbon can be dissolved or particulate; the particulate fractions traditionally are divided into fine and coarse components. Dissolved organic material can be divided into humic and nonhumic substances. Nonhumic substances generally (but not always) have low-molecular weight and are metabolized easily by aquatic microbes. Humic substances have high-molecular weight and are more resistant to breakdown.

4. Oxygenic (O_2-producing) and anoxygenic photosynthesis produce organic carbon, as do chemoautotrophic microorganisms. Organic carbon is oxidized to CO_2 by a variety of metabolic pathways, including aerobic respiration, fermentation, and anaerobic respiration using oxidized inorganic compounds to release energy from organic molecules in the absence of O_2.

5. Methanotrophic organisms rely on oxidation of methane as a primary energy source and generally live close to anoxic habitats with methane production.

6. Methanogenic organisms produce methane from CO_2 and H_2 or from acetate under extremely low redox conditions. These organisms in freshwater habitats have a significant input to the global atmospheric methane pool.

QUESTIONS FOR THOUGHT

1. Why could global warming alter deposition rates of organic carbon in peat bogs?

2. Should we preserve wetlands, even though some are a net source of greenhouse gasses?

3. Why can the microlayer at the surface of lakes be important in determining influx and efflux rates of lake water CO_2?

4. Do the processes of manganese reduction and iron reduction by bacteria (using these elements to oxidize organic C) have to occur in regions of different redox in the environment?

5. Why is CO_2 more likely to limit production of benthic organisms than phytoplankton?

6. Why is it important to know the pH when conducting experiments to measure photosynthetic rate with $^{14}CO_2$?

7. Why is alkalinity generally high in limestone watersheds?

CHAPTER 14

Nitrogen, Sulfur, Phosphorus, and Other Nutrients

Contents

Freshwater Ecology
DOI: https://doi.org/10.1016/B978-0-12-813255-5.00014-4

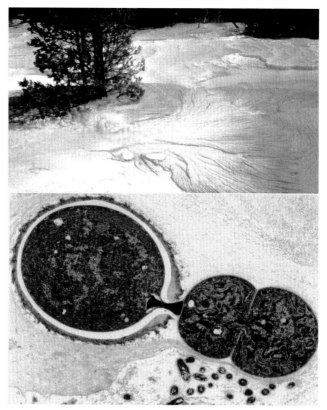

Figure 14.1 Streamers composed of the sulfur-oxidizing bacterium *Thermothrix* at Mammoth Terrace, Yellowstone National Park (courtesy of R.W. Castenholz) and a transmission electron micrograph of a heterocyst (the site of nitrogen fixation in *Nostoc* and other cyanobacteria) attached to a smaller dividing vegetative cell with a diameter of approximately 8 μm.

Other nutrients, in addition to carbon, are central to aquatic ecosystem function. Some forms of these nutrients, such as sulfide, ammonium, and nitrate, can cause water quality problems if they are present in large quantities. In addition, these nutrients (most commonly phosphorus and nitrogen) can limit ecosystem processes, so knowledge of their forms and cycling is an important part of understanding ecosystem function. In this chapter, we discuss cycling of nitrogen, sulfur, phosphorus, iron, and silicon. Nutrient cycles are all interrelated; none stand alone. Consequently, the chapter concludes with a discussion on how the cycles interact and revisits how redox controls nutrient cycling.

NITROGEN

Nitrogen forms

The most common form of nitrogen (N) in the biosphere is N_2 gas. The atmosphere is composed of about 78% N_2. Water generally contains N_2 as a dissolved gas. Even though N_2 is less soluble in water than is O_2, the higher atmospheric concentration of N_2 leads to dissolved concentrations at equilibrium with the atmosphere similar to those observed for O_2. The N_2 molecule is very difficult for organisms to use directly because it has a triple covalent chemical bond that requires a significant amount of energy to break. Thus, forms of organic and inorganic nitrogen other than N_2 are *combined nitrogen*.

The two most important forms of dissolved inorganic N in natural waters are *nitrate* (NO_3^-) and *ammonium* (NH_4^+). Ammonium is the ionic form found in neutral to acidic waters. Under high pH conditions, the ion is converted to *ammonia* gas (NH_3) which can move between the atmosphere and the water and is toxic to aquatic organisms. Roughly, less than 10% of ammonia is in gaseous form below pH 8, and at pH 9 about half the ammonia is in the gas form. *Nitrite* (NO_2^-), an additional form of dissolved inorganic nitrogen, occasionally occurs in significant concentrations in natural water, especially when sewage is present, and can be problematic because of its toxic nature. The sum of nitrate, nitrite, and ammonium is *dissolved inorganic nitrogen*. Other forms of inorganic N that also dissolve in water, such as N_2 gas, are generally not considered as part of dissolved inorganic nitrogen because they are not so biologically available. Dissolved nitrous oxide (N_2O) gas also occurs in low concentrations in most waters and in the atmosphere. Nitrous oxide has the common name laughing gas, but also absorbs heat 298 times per molecule more efficiently than CO_2 and is an important contributor to the greenhouse effect.

Organic N can take many forms, including amino acids, nucleic acids, proteins, and urea. *Urea* may be particularly important because many organisms excrete it (although most aquatic organisms excrete far more ammonium) and may be a crucial form in nitrogen cycling. Organic N in aquatic habitats can be in the form of *dissolved organic nitrogen* and *particulate organic nitrogen*; usually a 0.45-μm filter is used to separate dissolved from particulate.

Nitrogen transformations

Organisms must assimilate nitrogen in some form because it is a required component for many biological molecules. Many animals can assimilate nitrogen only in the form of organic molecules in the tissues of organisms they consume, such as amino acids and nucleic acids. In general, cells cannot take whole proteins across cell membranes, and they must be broken into amino acids. Hence, some organisms excrete enzymes that cleave proteins into the surrounding water (Billen, 1991) or into their guts if they are multicellular organisms. Some protozoa can ingest particles through their cell membranes, and the breakdown of proteins then happens within the cell.

Primary producers, fungi, archaea, and bacteria generally can use nitrate, nitrite, or ammonium, which they can take up from the water. The assimilatory pathways for nitrogen acquisition in such organisms require the nitrogen to be in the form of

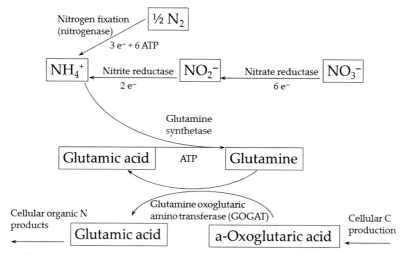

Figure 14.2 Nitrogen assimilation. This figure illustrates that nitrogen must be assimilated in the form of ammonium, and energy requirements for assimilation are $N_2 > NO_3^- > NO_2^- > NH_4^+$

ammonium before entering biochemical pathways in the cell (Fig. 14.2). The enzyme nitrate reductase converts nitrate to nitrite, and nitrite reductase transforms nitrite to ammonium, which cells can then assimilate. These reductase enzymes require energetically costly reducing equivalents, in the form of NADH, NADPH, or ferredoxin. Nitrate reductase requires molybdenum to function properly, so a limited supply of molybdenum may lead to an inability to use nitrate. In oxic environments, ammonium has higher potential energy than nitrate, so energy is required to convert nitrate to ammonium before assimilation. Most aquatic bacteria and primary producers prefer ammonium because it requires less energy to use, and affinity for ammonium is relatively high.

Some bacteria, including some cyanobacteria (Young, 1992) and some Archaea, have the capacity to assimilate N_2. This capacity is *nitrogen fixation*. The transformation does not occur spontaneously because it requires extremely high activation energy. The enzyme nitrogenase catalyzes the conversion of N_2 to ammonium, and requires molybdenum as an essential component. The process is one of the most energetically expensive metabolic reactions, requiring at least six ATP molecules and three electrons (reducing equivalents) for each ammonium molecule produced. Importantly, O_2 inactivates nitrogenase. Organisms either have to inhabit anoxic habitats to fix nitrogen or protect the enzyme from exposure (Bothe, 1982).

Some groups of cyanobacteria have heterocysts, specialized cells that protect nitrogenase from O_2 (Figs. 14.1 and 14.3). These cells are unique in appearance under the microscope and have a variety of adaptations that allow for nitrogenase activity in the heterocysts and photosynthetic O_2 evolution in adjacent cells. Adaptations include

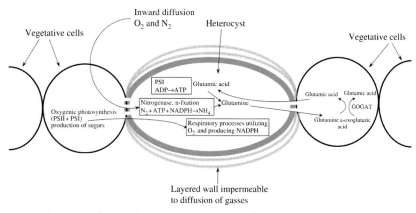

Figure 14.3 A diagram of cyanobacterial vegetative cells, a heterocyst, adaptations to protect nitrogenase from deactivation by O_2, and mode of N transport from the heterocyst into vegetative cells.

high respiratory rates to consume O_2, thick gels or mucilage around the heterocysts to retard inward diffusion of O_2, and loss of O_2 evolution in photosynthesis with retention of cyclic photophosphorylation (i.e., generation of ATP by photosystem I) in heterocysts (Haselkorn and Buikema, 1992). Other groups of cyanobacteria have no heterocysts but form aggregations in which their respiration can deplete O_2. Heterocysts and patterns of aggregation allow low O_2 concentrations while promoting high-energy availability through respiration for the costly process of nitrogen fixation (Paerl and Pinckney, 1996). The nif genes control nitrogen fixation (Böhme, 1998) and the chemical signaling controlling heterocyst formation occurs in response to limitation by other forms of nitrogen (Yoon and Golden, 1998). Chemical signaling controlling heterocyst formation is particularly interesting because it is the first documented case of intercellular signaling with a peptide among the Bacteria.

Some wetland and riparian species of plants are associated with nitrogen-fixing bacteria. Species of alder (*Alnus*) trees can harbor symbiotic bacteria, *Frankia alni*, that fix N_2. The fixed nitrogen can enter aquatic habitats through leaching, or when leaves fall into the water. Some wetlands can also have significant populations of nitrogen-fixing cyanobacteria, which may be free-living or associated with other organisms. For example, the aquatic fern, *Azolla*, has endosymbiotic, nitrogen-fixing cyanobacteria. *Azolla* growth is encouraged in traditional rice culture; farmers drain the field, which causes the plants to die and to release the fixed nitrogen. Then they reflood the field and plant it with rice. Cyanobacterial crusts associated with sediments in wetlands can also fix substantial amounts of nitrogen.

High-energy reactions also fix nitrogen in the atmosphere when lightning produces enough energy to cause N_2 and O_2 to combine and form nitrate. Thus, rainwater naturally contains modest amounts of nitrate. Additional nitrogen occurs in rainfall and particulates in the atmosphere that are suspended from terrestrial systems. Burning fossil fuels introduces nitrogen oxides into the atmosphere. Volatilization of ammonia from livestock operations and fertilizer application also enters the atmosphere and converts to nitrogen oxides. In aquatic systems with low amounts of nitrogen, atmospheric deposition can be a significant source of nitrogen pollution. Industrial production of fertilizers also fixes atmospheric nitrogen. This process has approximately doubled worldwide rates of nitrogen fixation and led to widespread nitrogen contamination of waters (Vitousek, 1994; Vitousek et al., 1997). This process has also resulted in damage to marine ecosystems worldwide as nitrogen stimulates primary production and when the cells eventually decompose, O_2 levels become low enough to harm marine life (Breitburg et al., 2018).

When ammonium is in water with dissolved O_2, it has potential energy relative to the more oxidized forms nitrite and nitrate. Some chemoautotrophic bacteria use this stored potential energy through the process of *nitrification*. This two-step process yields energy to allow for carbon fixation:

$$\underset{\text{Ammonium oxidation}}{NH_4^+ \longrightarrow NO_2^- + energy} \underset{\text{Nitrite oxidation}}{\longrightarrow NO_3^- + energy}$$

Different genera of bacteria are responsible for ammonium oxidation (e.g., *Nitrosomonas*) and nitrite oxidation (*Nitrobacter*). Nitrification is a dissimilatory process because the nitrogen form changes but cells do not assimilate it. Nitrification rates are minimal in anoxic habitats because nitrate has higher potential energy than nitrite and ammonium at low redox. Ammonium and nitrite oxidation are important processes for several reasons: (1) nitrification forms a vital link in the nitrogen cycle [the only other processes that produce nitrate are weathering of rocks (generally occurring slowly) and lightning], (2) nitrite formed by this process can have negative health implications, and (3) nitrifying bacteria compete for ammonium with primary producers. Initially, researchers had assumed that nitrifying bacteria occur only where ammonium is high. However, these bacteria occur commonly in oligotrophic, low ammonium environments and are able to compete successfully for ammonium with other microbes (Ward, 1996).

Denitrification forms a key link in the nitrogen cycle because it leads to the conversion of inorganic combined nitrogen (nitrate) to the relatively unavailable N_2 gas. Researchers pay much attention to this process given its importance in agriculture and water quality (Payne, 1981). Denitrification uses nitrate as an electron acceptor in a fashion analogous to the use of O_2 for aerobic respiration. Denitrification converts the forms of nitrogen in the following sequence:

$$NO_3^- \rightarrow NO_2^- \rightarrow NO \rightarrow N_2O \rightarrow N_2$$

During this process, cells convert organic carbon to CO_2 and cellular energy. The process of denitrification is less efficient than aerobic respiration because nitrate is less oxidized than O_2 and yields less free energy when reacting with organic carbon.

The process of denitrification is an important route of nitrogen removal from aquatic ecosystems and can occur in anoxic portions of streams, wetlands, and lakes. Understanding the loss through denitrification helps us understand responses to pervasive nitrogen pollution, and how we can mitigate that pollution. Effects of pollution include fertilization of aquatic habitats downstream and increased production of nitrous oxide; when the denitrification process is incomplete, the greenhouse gas escapes into the atmosphere. Numerous factors control the rate of denitrification including temperature, oxic state, carbon supply, and nitrate supply. Wetland plants encourage denitrification (Alldred and Baines, 2016). Increases in phosphorus availability might stimulate denitrification in lakes (Finlay et al., 2013). Increased nitrate stimulates denitrification in streams (Mulholland et al., 2008), leading to substantial rates of nitrous oxide generation globally (Beaulieu et al., 2011). These rates might be underestimates, as they do not include ebullition of nitrous oxide, which can be important in streams and rivers (Higgins et al., 2008). Water column denitrification can be important in some rivers (Reisinger et al., 2016; Marzadri et al., 2017), and depends upon suspended particle size, which presumably controls occurrence of microscale areas of anoxia (Xia et al., 2017). Geomorphology dictates sediment structure, which can also play a role in denitrification rates (Tatariw et al., 2013).

Finally, excretion of nitrogen by organisms is an important flux. Excess nitrogen must be excreted if an organism ingests more nitrogen than is needed. If water is limiting, animals convert nitrogenous waste into urea because ammonium requires more water to produce and is toxic at high concentrations. However, more chemical energy is necessary to produce urea than ammonium. Ammonium is often the primary form of excreted nitrogen because water rarely limits aquatic organisms, although some organisms (e.g., some fishes) excrete urea. Urea excretion is more important when water ammonium concentrations are high or in alkaline waters where ionic ammonium will spontaneously convert into toxic ammonia gas. When organisms remineralize ammonium, the process is referred to as *ammonification*. The study of ammonification is particularly important in aquaculture because excessive ammonium produced by high densities of confined fishes can be toxic (Jana, 1994).

Nitrate is generally the predominant form of dissolved inorganic nitrogen in oxidized (oxic) waters, and ammonium is the predominant form in anoxic waters. The absolute amounts of each are highly variable in many freshwaters (see Fig. 12.3; nitrate). The biotic processes that lead to reduced and oxidized forms of ammonium (e.g., nitrification, remineralization, and denitrification) can vary over time and space in freshwater habitats. Also, soils have a greater affinity for nitrate than ammonium, but different soils have varying degrees of affinity. In areas where combustion of fossil fuels contributes most to atmospheric nitrogen content, rainwater generally has more nitrate than ammonium. Ammonium can be more prevalent when agricultural inputs dominate. Consequently, nitrogen concentrations in precipitation are geographically patchy, dependent in part on human activities. Dissolved nitrogen enters freshwaters from sewage and agricultural activities. The observed concentrations of dissolved inorganic nitrogen in freshwaters are spatially heterogeneous as a result.

In a lake with an anoxic hypolimnion, ammonium dominates the hypolimnion, and nitrate is mainly confined to the epilimnion during stratification (Fig. 14.4). Several processes drive this spatial pattern of nitrate and ammonium concentrations: In the epilimnion, nitrification transforms ammonium to nitrate in the presence of O_2, but in the hypolimnion, denitrification removes nitrate under anoxic conditions, and cells continue to excrete nitrogen in the form of ammonium. A similar pattern occurs in oligotrophic lakes and wetlands, except that high ammonium occurs mainly in the anoxic zone within the sediments.

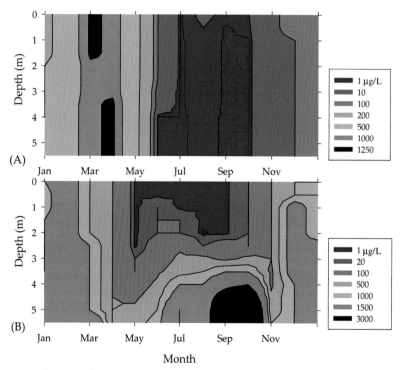

Figure 14.4 Distribution of nitrate (A) and ammonium (B) in hypereutrophic Wintergreen Lake, Michigan as a function of depth and time. Ice cover occurred from January to March. Darker colors represent higher concentrations. Contours are in μg/L. *Reproduced with permission from Wetzel (1983).*

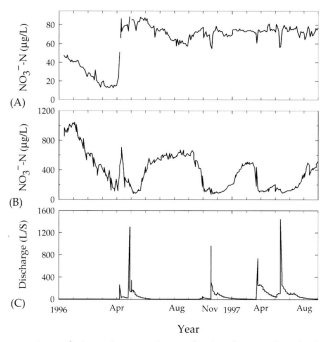

Figure 14.5 Concentrations of nitrate in groundwater flowing from undisturbed prairie (A), nitrate in Kings Creek, Kansas (B), and discharge in Kings Creek (C). High nitrate is related to input of groundwater from below fertilized cropland that dominates during periods of low discharge. Note nitrate variations in groundwater from under prairie. *Data courtesy of Konza Prairie Long-Term Ecological Research site.*

High ammonium in streams is often associated with input of anoxic groundwaters or pollution (Duff and Triska, 2000). The temporal variation of nitrate in streams can be large, given the complex interactions between precipitation and groundwater sources (Fig. 14.5), whereas concentrations in groundwaters may be more stable. Understanding processes that control nitrate in streams generally requires understanding nitrogen dynamics in the surrounding terrestrial ecosystem. For example, riparian zones decrease inorganic nitrogen and increase the ratio of ammonium: nitrate in watersheds of the Amazon (McClain et al., 1994). Wetlands can also have high ammonium concentrations in the water because of the anoxic sediments and the fact that nitrogen deposition in the form of organic materials is converted to ammonium as the organic compounds are used for energy.

Advanced: Alternative pathways of anoxic N cycling and nitrous oxide production by freshwater environments

Several pathways compete with denitrification and other processes that create N_2, yielding ammonium instead. Under some conditions, cells reduce nitrate to ammonium instead of N_2 in the process *dissimilatory nitrate reduction to ammonium* (DNRA). This is a fermentative pathway where nitrate serves as an electron acceptor to oxidize organic carbon. DNRA can account for 0%—60% of the total of nitrate removal that is not accounted for by assimilation (Burgin and Hamilton, 2007). There is an additional pathway where cells use nitrate to oxidize sulfide to yield energy and produce ammonia.

Pathways in the nitrogen cycle can remove nitrogen from the ecosystem by conversion to N_2 in addition to denitrification. For example, nitrate can be reduced N_2 using reduced forms of iron or manganese and yielding energy. Nitrite can also react with ammonium, yielding energy and N_2 in a process referred to as anaerobic ammonium oxidation (*anammox*). Anammox can dominate N_2 production in some groundwaters (Smith et al., 2015).

All of these processes form N_2 gas and are potential nitrogen removal routes from ecosystems. Scientists established these pathways more recently than the other pathways, and we are only beginning to understand their importance in freshwaters; in marine waters anammox can be responsible for up to 60% of N_2 production (Burgin and Hamilton, 2007). Given the relative paucity of information on DNRA, anammox, and other pathways that yield N_2 gas in freshwaters, it is an exciting time to be researching the nitrogen cycle.

Several of these pathways have nitrous oxide as an intermediate product (interestingly annamox has the toxic nitrogen compound hydrazine as an intermediate instead). Nitrous oxide produced by these pathways is important because of its role as a greenhouse gas and contribution to the depletion of the atmospheric ozone and to subsequent increases in UV radiation. The general idea is that nitrous oxide is lost during the processes of denitrification or nitrification as the gas diffuses away before the nitrous oxide can be converted to the final product. Conditions under which these losses occur in aquatic environments need more study. Along with methane, global warming and thawing of permafrost can also stimulate nitrous oxide release by high latitude wetlands (Voigt et al., 2017). Processes occurring in stream hyporheic zones are important (Quick et al., 2016), and rivers and estuaries could account for a considerable portion of nitrous oxide production that is stimulated by human use of nitrogenous fertilizers (Kroeze et al., 2005).

Nitrogen cycle

The nitrogen cycle (Fig. 14.6) can be represented in the format provided for carbon in Chapter 13. Remember that the diagram is divided into oxic and anoxic components, with more oxidized forms to the right of the diagram, and fluxes that go with potential energy occurring from left to right on the top of the diagram and from right to left on the bottom of the diagram. An anoxic habitat could be the hypolimnion of a

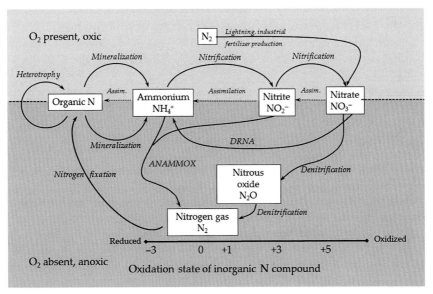

Figure 14.6 A conceptual diagram of the nitrogen cycle.

stratified eutrophic lake; sediments in a lake, stream, or wetland; or a decaying organic particle as we discussed in Chapter 13, for the carbon cycle. This generalized cycle illustrates some of the complexities of the nitrogen cycle related to varied redox potential of the different forms of inorganic nitrogen.

Understanding the complex fluxes in the nitrogen cycle is crucial because nitrogen can limit primary production in lakes, streams, wetlands, groundwaters, riparian zones, and marine habitats. In addition, nitrate and nitrite can be toxic (Highlight 14.1) and understanding the nitrogen fluxes and factors controlling them may allow for mitigation of some pollution problems. For example, if methods of management can maximize denitrification, then nitrate pollution that enters surface waters will return to the atmosphere as benign and inert N_2 gas.

SULFUR

The sulfur cycle is more complex than either the nitrogen or the carbon cycle. The greater number of redox states sulfur can assume drives the complexity. The cycle is of interest because it illustrates organisms evolved to use many of the common compounds with potential energy in the biosphere. The sulfur cycle is also important because some water quality problems revolve around sulfide contamination. Importantly, sulfur is tightly coupled to the inorganic metal cycles such as iron and manganese, and thus indirectly to phosphorus.

Highlight 14.1 Problems with excessive nitrate contamination

The United States and many other countries have enacted limits on the amount of nitrate that can be present in drinking water. In the United States, the allowable level is 10 mg NO_3^- -N/L (note the −N signifies it is the mass of N per liter, not the total mass of nitrate). The primary reason for this limit is methemoglobinemia, a medical condition that occurs because nitrite binds more strongly to hemoglobin than to O_2. This condition can result when humans drink water with large amounts of nitrite or nitrate. In infants, denitrifying bacteria in their guts produce nitrite. The condition is most often fatal in infants fed with formula made with high nitrate water. Methemoglobinemia causes the infant's skin to appear blue as it suffocates, and blue baby syndrome is the common name for this problem.

People can also convert nitrate to carcinogenic nitrosamines in the stomach. Researchers have documented a relationship between gastric cancer rates and fertilizer use (Fig. 14.7). The relationship is only a correlation; other factors such as lifestyle, pesticide and herbicide use, and diet may explain the correlation. However, the interdependence of cancer and nitrate should be investigated more thoroughly. Brender et al. (2013) found increased rates of birth defects in infants as a function of maternal nitrate intake in drinking water. They found significant increases in rates of spina bifida and oral clefts with nitrate intake of 5 mg/day or more.

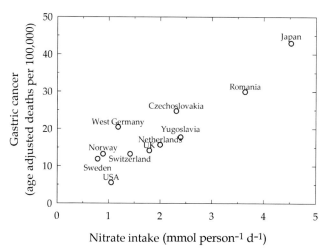

Figure 14.7 Correlation between nitrate intake and rates of gastrointestinal cancer. *After P. E. Hartman.* © *1983, reprinted by permission of Wiley–Liss, Inc., a subsidiary of John Wiley and Sons, Inc.*

The problem of nitrate contamination is widespread and worst in highly agricultural areas. In France, researchers sampled 266 groundwater sources, and 98% of those exceeded the recommended levels of 10 mg N L^{-1}. Worldwide, 10% of rivers exceed 9 mg N L^{-1} (Meybeck et al., 1989). Some of the large contributors to nitrate in groundwater and streams include agricultural fertilization and feedlots (confined livestock facilities). Some sewage treatment plants release large amounts of nitrate as well.

Once nitrate contaminates an aquifer, treating the problem is very difficult. One possibility is to fertilize with organic carbon to cause respiratory consumption of O_2 by microbes and promote denitrification. The nitrogen then is lost as N_2. The problem with this treatment is that it leads to anoxic groundwater. Anoxia can result in other problems, such as high iron content and bad tastes and odors related to sulfide and other chemicals.

The best way to control the problem of nitrate contamination is to do so at the source. Options include lining livestock feedlots and treating all runoff before it reaches the groundwater or streams, only adding as much fertilizer to crops as is necessary, leaving intact riparian buffer zones (Hedin et al., 1998), better design of septic drain fields, and treating sewage to remove nitrogen (discussed in later chapters). All the treatment and control options require a basic understanding of nitrogen cycling and cycling of other nutrients.

Sulfur forms

Inorganic sulfur can take a wide variety of forms. Some of the forms that occur include:

$$\underset{\text{Sulfide}}{S^{2-}} \qquad \underset{\text{Elemental sulfur}}{S^{\circ}} \qquad \underset{\text{Thiosulfate}}{S_2O_3^{2-}} \qquad \underset{\text{Sulfate}}{SO_4^{2-}}$$

These are listed above in order of increasing redox (more oxidized) from left to right. Hydrogen sulfide dissolves in water as an ion (S^{2-}) under basic conditions or converts to a gas (H_2S) under acidic conditions. Hydrogen sulfide gas is toxic in high concentrations but humans can detect it by odor in exceptionally small concentrations. Hydrogen sulfide is partially responsible for the distinctive smell (an odor reminiscent of rotten eggs) of anoxic sediments in lakes and wetlands.

Sulfur is a component of many organic molecules. This includes some amino acids (e.g., methionine and cysteine) that are central to protein structure. Dimethyl sulfide is a significant gaseous product of phytoplankton metabolism that forms an important component of the global sulfur budget. Other organic compounds give rise to putrid smells when sulfur-containing proteins are degraded.

Fundamental sulfur transformations

Numerous transformations are possible because of the many redox states sulfur can take. We illustrate a few of these transformations in Table 14.1. The general fluxes of heterotrophy and remineralization (aerobic or anaerobic) are present, as for all other nutrient cycles.

Table 14.1 Some sulfur transformations

Equation	Conditions	Name	Classification
$H_2S + \frac{1}{2}O_2 \rightarrow S^{\circ} + H_2O$	Oxic	Sulfide oxidation	Abiotic and dissimilatory biotic
$S^{\circ} + H_2O + 1\frac{1}{2}O_2 \rightarrow H_2SO_4$	Oxic	Elemental sulfur oxidation	Abiotic and dissimilatory biotic
$4NO_3^- + 3S^{\circ} \rightarrow 3SO_4^{2-} + 2N_2$	Anoxic	Inorganic sulfur oxidation	Dissimilatory biotic
$2CO_2 + 2H_2S + 2H_2O + light \rightarrow 2(CH_2O) + H_2SO_4$	Anoxic	Photosynthetic sulfur oxidation (anoxygenic photosynthesis)	Biotic
$CH_3COOH + 2H_2O + 4S^{\circ} \rightarrow 2CO_2 + 4H_2S$	Anoxic	Acetate oxidation	Dissimilatory, reduction
$2(CH_2O) + H_2SO_4 \rightarrow 2CO_2 + 2H_2O + H_2S$	Anoxic	Anaerobic respiration, sulfate as the electron acceptor	Dissimilatory, reduction
$4H_2 + H_2SO_4 \rightarrow 4H_2O + H_2S$	Anoxic	Anaerobic hydrogen respiration	Dissimilatory, reduction
$S_2O_3 \rightarrow SO_4^{2-} + S^{2-} + H^+$	Anoxic	Disproportionation	Dissimilatory
$SO_4^{2-} \rightarrow S^{2-} \rightarrow cysteine$	Oxic/ anoxic	Assimilation	Biotic
$H_2S + Fe^{2+} \rightarrow FeS + H_2$	Anoxic	Iron pyrite formation	Abiotic, spontaneous

Note: Others occur, but this table partially illustrates the complexity of the sulfur cycle.

Reduced forms of sulfur can combine with O_2 for a net release of potential energy (Table 14.1). *Abiotic sulfur oxidation* occurs spontaneously, but more slowly than *biotic sulfur oxidation*. Therefore, the biotic oxidation of reduced sulfur compounds by chemoautotrophic organisms generally occurs in areas at the interface between oxic and anoxic habitats where the supply of reduced sulfur compounds is high, and there is not much time for abiotic oxidation as the reduced sulfur compounds have just diffused into oxic conditions. In unique cases, the inflow of sulfide-rich water into caves with airspace or water containing O_2 can lead to an entire cave ecosystem supported by sulfur-oxidizing bacteria (Highlight 14.2).

Some bacteria can use oxidized sulfur as an electron acceptor to respire organic carbon in a process similar to denitrification. This *dissimilatory sulfur reduction* is a primary source of sulfide found in anoxic sediments and waters. Sulfate forces the redox to remain moderately high and can indicate conditions under which methanogenesis will not occur.

Several unique transformations exist in the sulfur cycle. One is *disproportionation*, an anoxic transformation in which cells convert thiosulfate to sulfate and sulfide, yielding

Highlight 14.2 A Romanian cave ecosystem fueled by sulfide-oxidizing primary producers

In 1986, researchers found a groundwater ecosystem within Movile Cave, which has a considerable influx of springwater rich in H_2S. The cave has a terrestrial fauna with 30 obligate cave-dwelling invertebrates, of which 24 are endemic. The aquatic fauna includes 18 invertebrates, with 9 endemic species. Air pockets in the cave have floating microbial mats (composed of bacteria, fungi, and five nematode species; Riess et al., 1999) and other mats are attached to walls. Bubbles of methane float the mats, presumably produced by methanogenic bacteria. Scientists hypothesized that sulfide-oxidizing bacteria form the base of the food web in these caves.

Subsequent studies demonstrated that the microbial communities were able to assimilate ^{14}C-labeled bicarbonate into microbial lipids, verifying that the assemblages growing in the cave were chemoautotrophic. The ratios of stable isotopes of carbon ($^{13}C/^{12}C$) and nitrogen ($^{15}N/^{14}N$) indicated that the carbon fixed by microbial autotrophs was entering the food webs. This analysis demonstrated that the microbial biofilms, the invertebrate consumers of the microbes, and the invertebrate predators had isotopic ratios more similar to each other than to those found in nearby surface aquatic or terrestrial samples (Sarbu et al., 1996). Furthermore, isotopic signatures were consistent with the hypothesis that invertebrates subsisted on microbes or other invertebrates that feed on microbes in the cave. Analyses of the microbial community by molecular methods suggested that the methane bubbles in the microbial mats support methanotrophic microbial biomass (Hutchens et al., 2004). This unusual cave ecosystem and a similar one discovered in Israel (Por, 2007) are unique, but perhaps more widespread than now known because of the difficulty of finding them. One could even consider them a distinct biome.

---

energy (Table 14.1). Another transformation is the use of sulfide as an electron donor for *anoxygenic photosynthesis* that yields sulfate in a process analogous to using water as an electron donor for oxygenic photosynthesis. The production of sulfate in anoxic habitats leads to the possibility of complete autotrophic and heterotrophic components and a complete sulfur cycle with no O_2 present. Such processes could have been central to early life in the anoxic habitats of primordial Earth.

Sulfide can combine with metals to form *pyrites*, the most common being iron pyrite (fool's gold). These precipitates have very low solubility under anoxic conditions and can represent an important loss of iron to aquatic ecosystems. The precipitate is black and gives anoxic sediments their characteristic black color. Sulfur-oxidizing bacteria can hasten breakdown rates of pyrite when exposed to O_2 because they can use it chemoautotrophically. Mining operations take advantage of this microbial activity where metals are bound to pyrites; miners mix bacteria that oxidize the pyrite compounds with the ore, and the bacteria release the metals into solution as they oxidize the pyrite. Regardless of whether this oxidation occurs biotically or abiotically, the end product is sulfuric acid, so acid mine waste can be a substantial pollution problem in surface waters influenced by mine drainage.

Advanced: More complex sulfur transformations

Like methane, N_2, and nitrous oxide, dimethyl sulfide is a gaseous by-product of organisms. Gas can enter the atmosphere and lead to a net loss of sulfur from some ecosystems. The significance of this production has not been well studied in freshwaters; it is of interest to those who link global sulfur budgets to marine primary producers (Malin and Kirst, 1997), and production by fermentations in freshwater sediments may be important as well (Lomans et al., 1997; Yoch et al., 2001). Dimethyl sulfide oxidizes to sulfate in the atmosphere. The gas is important globally because it provides nuclei for formation of clouds and could help regulate global climate.

Bacteria and Archaea have evolved a dazzling array of ways to cycle sulfur, several of which we already discussed in this chapter. For example, some chemoautotrophs use a variety of oxidized compounds, including nitrate, Fe^{2+}, and Mn^{2+} to oxidize reduced sulfur compounds such as sulfide and elemental sulfur and yield energy. The bacterium *Beggiatoa* mediates an interesting adaptation to oxidizing sulfur (Hinck et al., 2007). These large filamentous gliding bacteria have big vacuoles and can use them to store nitrate taken up from oxic areas in hot springs. The *Beggiatoa* can then glide down into the anoxic sediments where sulfide concentrations are high, and use the nitrate to oxidize the sulfide and generate energy.

Additionally, some microbes can use elemental sulfur to oxidize acetate, others react sulfate with hydrogen gas to yield energy, and denitrification can occur using elemental sulfur and nitrate. Sulfur plays a pivotal role in the cycling of other compounds because it removes iron from solution, which would otherwise precipitate phosphate. Sulfide can precipitate with many other metals, and when sulfate is present, methanogenesis is less energetically favorable than using sulfate to oxidize organic carbon.

Sulfur cycle

We study the cycling of sulfur because of interactions with other nutrients in addition to its basic biogeochemical elegance. The sulfur cycle in aquatic systems is indirectly important to ecosystem productivity because sulfur is rarely a limiting nutrient for primary producers; the requirement is low and the levels available are generally relatively high. Sulfide can be high enough in some extreme ecosystems to be toxic. The sulfur cycle can be diagrammed using the method presented previously (Fig. 14.8).

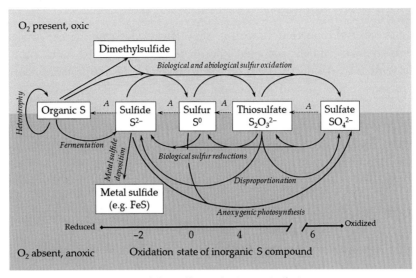

Figure 14.8 A conceptual diagram of the sulfur cycle. A = assimilation.

PHOSPHORUS

Phosphorus is the key nutrient limiting primary production in many aquatic habitats. We discuss nutrient pollution and use by organisms extensively in Chapters 17 and 18. Unlike carbon, nitrogen, and sulfur, phosphorus mainly occurs in only one inorganic form (phosphate), so most research has centered on organic transformations and interaction of other inorganic chemicals with phosphate.

Phosphorus forms

Phosphate (PO_4^{3-}) is a dominant form of inorganic phosphorus in natural waters, but concentrations are often near or below detection in pristine waters (about $1-10\,\mu g/L$). Determining the precise level of phosphate is difficult because standard methods of analysis also detect a variable and poorly defined group of phosphate compounds (Rigler, 1966), even though people routinely think such assays represent phosphate concentrations. These assays more accurately represent the soluble reactive phosphorus (reactive to the assay) and the assays often to not give results that are biologically relevant (Dodds, 2003a). Cells can also store phosphate as a polymer (polyphosphate). Organic phosphorus occurs in compounds including nucleic acids and lipids. The crudest classification found in the literature denotes dissolved organic phosphorus (DOP) and particulate phosphorus (PP). The sum of DOP, PP, and soluble reactive phosphorus is total phosphorus.

Phosphorus transformations

The interaction of phosphate with iron is important in determining the availability of phosphorus in many aquatic systems. Phosphate will precipitate with some metals, including ferric iron (Fe^{3+}, an oxidized form of iron). The precipitation occurs only in the presence of O_2, and the complex dissociates under anoxic conditions. Precipitation of ferric phosphate leads to the deposition of phosphorus in sediments with oxygenated surface water. When the precipitate settles into an anoxic zone (such as an anoxic hypolimnion of a lake) the phosphate dissociates. Transport processes, such as eddy diffusion, move dissolved materials including the dissociated phosphate. Thus, phosphate mixes to the surface when fall mixing breaks down an anoxic hypolimnion (Fig. 14.9). This precipitation process can be important in deposition and retention of phosphorus in streams and rivers as well (Small et al., 2016; Smolders et al., 2017).

Another important precipitation of phosphate occurs with calcium to form calcium phosphate (apatite). This reaction is favored at high alkalinity, high phosphate, and high pH (Stumm and Morgan, 1981). Carbonate rich waters will commonly exhibit these conditions. Photosynthesis by periphyton or macrophytes will raise pH and encourage precipitation of apatite. This process of deposition can strip phosphorus from the water column and lead to long-term sequestration of phosphorus in the sediments. The process is part of the reason for strong phosphorus deficiency in the

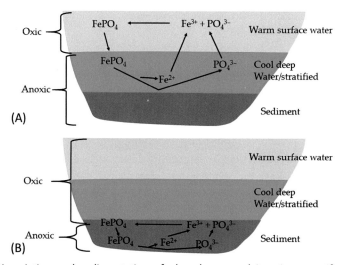

Figure 14.9 Dissociation and sedimentation of phosphorus and iron in a stratified lake with an anoxic hypolimnion (A) and one with an oxic hypolimnion over anoxic sediments with an oxidized microzone right at the top of the sediments (B).

Everglades (Dodds, 2003b). Over geological time, the process can lead to formation of marine sedimentary rocks that are very rich in phosphate. Phosphate that is mined for fertilizers and industrial uses largely originates from this source. Similarly, carbonate deposition in streams can limit phosphorus availability (Corman et al., 2016).

Cells can assimilate phosphate at vanishingly small concentrations under phosphorus-limited conditions. Billions of years of natural selection under phosphorus limiting conditions have created a selective pressure for very efficient uptake mechanisms. We cover uptake of nutrients Chapter 17. This uptake results in extraordinarily low (nanomoles per liter) concentrations of phosphate, and these concentrations are far below those normally detected by conventional methods (Hudson and Taylor, 2005). Natural concentrations of phosphate in very phosphorus-limited waters are also far lower than conventional reagent grade ultrapure laboratory water; organisms are better at removing phosphate from water than are standard water purification methods created by humans. When phosphate is in excess, cells can take it up and store it as polyphosphate (Rier et al., 2016).

Organisms have *phosphatase* enzymes that cleave dissolved organic phosphorus DOP compounds to liberate phosphate. The phosphatases are common to all life as they are necessary for normal cell function. Toxic microcystins synthesized by cyanobacteria inhibit phosphatase activity and are thus toxic to most organisms.

Phosphatases can also be excreted outside the cell (extracellular) or be associated with the exterior cell surface (Chróst, 1991; Olsson, 1991). These enzymes increase the availability of phosphate to cells, so the excretion of extracellular phosphatases increases when phosphorus becomes scarce to cleave organic phosphorus compounds that cannot be taken across the cell wall. The ubiquitous nature of these compounds in natural waters leads to rapid turnover of many organic phosphorus compounds. Assays for phosphatase activity can gauge the degree of phosphorus deficiency in aquatic habitats; under limiting conditions, organisms tend to produce and excrete ample amounts of phosphatase.

Heterotrophy results in rearrangement of organic phosphorus compounds as in all other cycles (Fig. 14.10). Many animals cannot use phosphate directly, but must consume organic phosphorus to meet their needs (e.g., lipids, nucleic acids). Organisms excrete excess phosphorus as phosphate or organic phosphorus in both oxic and anoxic environments.

In contrast to other nutrient cycles, phosphate cannot serve to oxidize organic carbon (e.g., denitrification and dissimilatory sulfate reduction). Reduced phosphorus compounds are extremely toxic, and natural phosphate concentrations are low, which explains why organisms have not evolved the ability to use phosphate as an electron acceptor in anaerobic respiration (although we may still discover organisms that are capable of this transformation).

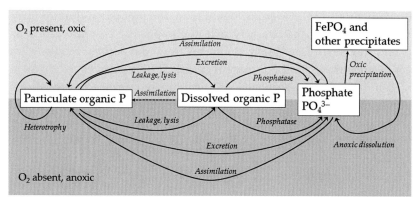

Figure 14.10 A diagram of the phosphorus cycle.

Given the importance of phosphorus in nutrient pollution, it is common for people studying lakes to enumerate a phosphorus budget. In Chapter 18, we explore the effects of phosphorus in lakes, and how water quality engineers can use wetlands to remove phosphorus from water.

SILICON AND IRON

Silicon cycles drive the dominance of diatoms in many surface waters and can be a primary determinant of algal community structure in lakes and streams. Iron is intimately tied to the sulfur and phosphorus cycles in addition to being a water quality variable of concern on its own. Manganese, molybdenum, and many other nutrients have their own cycles as well.

Silicon

Aluminosilicate minerals are the most abundant in the earth's crust (Schlesinger, 1997). However, silicate is generally not the major dissolved ion in natural water because of the limited solubility of silicon-containing compounds (Stumm and Morgan, 1981). Microbes contribute to weathering rates of silicates, which can be important in chemistry and geology of groundwaters (Hiebert and Bennett, 1992). Silicon can take several forms, including *silica* (SiO_2) and *silicic acid* (H_4SiO_4). Silicic acid is generally the form dissolved in waters. The greatest concentrations are in waters draining volcanic watersheds, and the lowest from limestone watersheds.

Clays are composed mainly of silicon-containing compounds and are a primary component in turbid aquatic systems and sediments. Clays can be very important as sites for microbial activity and areas where chemicals can adsorb. For example, clays can adsorb and hold organic pollutants making it difficult to clean spills in groundwaters. Sand, composed mainly of silicates, also forms a major component of the benthic substrata of many aquatic systems.

Diatoms rely on silicon as a component of their frustules (specialized cell walls) where it is deposited as opal. The opal is relatively insoluble, so it sinks to the sediments and persists after the diatom dies. The burial of these frustules allows for paleolimnological determinations of the diatom species present over time in lakes. Diatomaceous earth, which is soft, sedimentary rock formed from ancient accumulations of diatom frustules, is used widely in water filtration systems, as a mild abrasive, and as a natural insecticide. Silica from diatoms and sponge spicules forms chalk. Emergent grasses also use silicate as a structural material and as a defense against herbivores, and freshwater sponges make their spicules from silicate for similar reasons.

Diatoms can deplete silicon in the epilimnion of both eutrophic and oligotrophic lakes during summer. Depletion occurs because of the relatively slow cycling of silicon, its incorporation into diatom frustules, and subsequent sinking of frustules before silicon can be redissolved. Dissolved silicon concentrations build near the sediments during summer and winter stratification (Fig. 14.11) because diatom frustules sink to

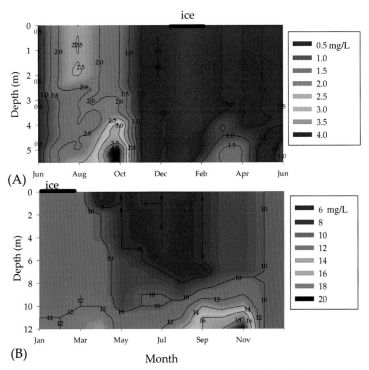

Figure 14.11 Concentration of silica as a function of depth and time in hypereutrophic Wintergreen Lake, Michigan (A), and oligotrophic Lawrence Lake, Michigan (B). Concentrations are given in mg/L. Note start of time series is different in the two panels. *Reproduced with permission from Wetzel (1983).*

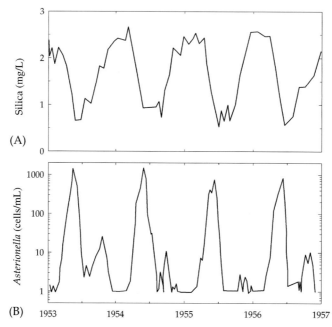

Figure 14.12 The relationship between epilimnetic silicon and biomass of the diatom, *Asterionella*, in Lake Windermere, England. Note how decreases in silicon correspond with high densities of diatoms. *Data from Lund (1964).*

the sediments and slowly dissolve. The temporal and spatial variation in silicon availability may be important in the successional processes of phytoplankton, as diatoms deplete silicon seasonally (Fig. 14.12).

In contrast, concentrations of silicon in rivers are remarkably constant over time and across continents (Wetzel, 2001); hence, periphytic diatom communities in flowing waters experience less drastic seasonal changes in silicon concentrations than plankton in lakes. However, increased nitrogen loading to world rivers is apparently stimulating diatoms, leading to greater silicon uptake and lower concentrations of silicate exported by rivers to marine habitats (Turner et al., 2003).

Iron

Iron is a key element in many biological molecules, including electron transport proteins, hemoglobin, enzymes used in synthetic pathways for chlorophyll and proteins, and enzymes used for nitrate assimilation, photosynthesis, and other essential metabolic processes. The total demand for iron is not high relative to nitrogen, carbon, and phosphorus because iron serves as a cofactor in these reactions (one or a few iron atoms for each entire protein). However, iron is required as a minor nutrient and serves as a key link among many biogeochemical interactions, such as the phosphorus and sulfur cycles.

Iron occurs as a dissolved ion in several forms. Under high redox (oxic conditions), iron takes the form of *ferric* (Fe^{3+}) ions. At low redox, iron is in the *ferrous* (Fe^{2+}) form. O_2 converts ferrous to ferric iron by accepting electrons from the ion, but the process occurs over many minutes (Stumm and Morgan, 1981). Ferric ions react with hydroxyl ions to form the flocculent precipitate ferric hydroxide [$Fe(OH)_3$, more commonly known as rust] under oxic conditions and near-neutral pH. Those who obtain water from anoxic wells with iron-rich groundwater can directly observe this precipitation. The water appears clear when it is withdrawn, but exposure to O_2 converts the ferrous iron to ferric iron, which then forms a flocculent precipitate as ferric hydroxide; a clear glass of water looks a bit rusty after a few minutes. Modern water treatment plants that use anoxic groundwater rich in iron as a water source often aerate drinking water and allow the ferric hydroxide to settle in a pool.

Inorganic iron concentrations vary seasonally in stratified lakes, with especially pronounced variation in concentrations related to anoxic events such as summer stratification in eutrophic lakes (see Fig. 12.6). These patterns link closely to the redox of the lake.

Chelators maintain higher concentrations of iron in oxic solutions. These compounds briefly bind iron and other metals and do not allow precipitation of ferric hydroxide. Most chelators in natural waters are organic compounds, although inorganic compounds can also serve as chelators. Some algae produce and excrete organic chelators of iron to keep iron available for uptake. Other substances, such as humic materials formed by plant decomposition, are also chelators.

Large concentrations of humic materials found in natural water can complex with iron so successfully that they make iron unavailable to organisms. As a result, some dystrophic wetlands and lakes are unproductive not only because of high concentrations of humic materials that absorb light but also because they make iron less available. At intermediate concentrations, these chelators make iron more available by interfering with precipitation but they are detrimental at high concentrations.

Ferric iron can also precipitate phosphate, as mentioned previously. These flocculent precipitates will settle to the sediments in an oligotrophic lake or wetland. If the hypolimnion is anoxic as expected in a eutrophic lake, the $FePO_4$ precipitate will dissociate into phosphate and ferrous iron. This process is a key interaction between elemental cycles and has a major impact on algal production.

Iron pyrite forms when ferrous iron reacts with sulfide. This precipitate has limited solubility and forms black deposits in anoxic sediments. Thus, in wetlands or lakes with O_2 in the water column immediately above the sediments, the first few centimeters of sediment are light colored, but in the deeper anoxic portions, the dark pyrites color the sediments dark brown or black. This reaction and its indirect

---

Highlight 14.3 Sulfur and iron dynamics increase wetland phosphorus and associated growth of noxious plants

Large areas of the Netherlands are composed of peaty lowlands. These wetland areas are fed with river water to compensate for groundwater withdrawals for agriculture (Smolders and Roelofs, 1993). The inflow of river water was hypothesized to cause deterioration in water quality (increases in turbidity, filamentous algae, and duckweed) and a shift away from characteristic emergent wetland plants. Managers attempted to halt this deterioration by stripping incoming river water of phosphate, but the water quality did not improve.

Further investigation revealed that the sulfate and iron contents in the river water were altering the biogeochemical cycling. People treated the river water entering the wetlands by stripping the phosphate, resulting in high sulfate, and low iron. The sulfate diffused into the anoxic sediments where microbes reduced it to sulfide. Increased sulfide concentrations in the anoxic sediments led to greater precipitation of ferrous iron as iron pyrite into the sediments. Because deposition removed the iron from the oxic portion of the system, it could no longer precipitate with phosphate in the oxic waters of the wetland. The lowered amount of phosphate precipitation allowed higher concentrations of phosphorus to be maintained in the surface waters and led to stimulation of algal growth. Concurrent increases in the sulfide levels in the pore waters of the wetland sediments combined with low iron concentrations could have harmed the native macrophytes via sulfide toxicity and growth limitation by low iron.

This European case study is an excellent example of the importance of understanding nutrient cycling and how interactions among the cycles influence primary producers. The management of the system appeared to be simple, but complex interactions among nutrient cycles conspired against the feasibility of a "dilution solution to pollution."

---

mediation of the phosphorus cycle can influence management responses to nutrient pollution (Highlight 14.3).

Ferrous iron has potential energy in the presence of O_2 and can serve as an energy source for microbes able to oxidize it before spontaneous conversion occurs. This reaction is confined to areas with O_2 that are close to an anoxic habitat that provides ample diffusive flux of iron in the reduced (ferrous) form. This oxidation is one route for dissociation of iron pyrite (FeS) in oxic environments.

Iron-oxidizing bacteria can cause problems in wells in anoxic groundwaters high in iron. In these situations, O_2 dissolved in the well water moves down the well. The aquifer immediately outside the bottom of the well has a zone in which O_2 and Fe^{2+} both occur. The iron oxidizers can use this potential energy source to incorporate CO_2 and grow. Bacterial growth can eventually plug the well intake screens and the porous materials surrounding the well intake. Removing the bacterial growth can be difficult, and measures such as back-flushing wells provide only temporary help.

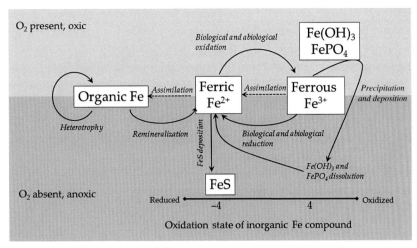

Figure 14.13 A conceptual diagram of the iron cycle.

Iron cycling (Fig. 14.13) is similar to nitrogen, carbon, and sulfur cycling because the concepts of redox and potential energy can be used to order and understand the processes. The iron cycle was one of the first nutrient cycles to receive extensive study (Mortimer, 1941).

CYCLING OF OTHER ELEMENTS

Cells need a number of minor elements including copper, manganese, selenium, zinc, and molybdenum. Sodium, potassium, chloride, and boron are also required. Most of these occur at low concentrations. These elements can also act as pollutants at greater concentrations.

Minor nutrient cycles, such as molybdenum, selenium, and copper, have not been well studied; even though organisms require these micronutrients, the requirement is small and the chances of these elements limiting system production are relatively low (but see Goldman, 1972). Potassium can limit plant production in some wetlands. An example of minor nutrients limiting productivity is the case of molybdenum limitation in Castle Lake, California. Molybdenum is a requirement for nitrogen assimilation via nitrate, as this enzyme (nitrate reductase) reduces nitrate to ammonium before assimilation into biochemical pathways. Alder trees in the watershed scavenge the molybdenum, creating apparent molybdenum limitation in the lake (Goldman, 1960). Ammonium assimilation does not require molybdenum, so presence of ammonium overcomes limitation. Similar limitation by molybdenum may occur in some groundwaters (Thorgersen et al., 2015).

Manganese cycling is similar to iron cycling in many ways (Wetzel, 2001). Some chemoautotrophs can combine reduced manganese with O_2 to yield energy. Some microbes can use oxidized manganese as an electron acceptor for anaerobic respiration. There are some differences between the iron and manganese cycles. For example, iron oxidation is spontaneous in the oxic waters of lakes, but manganese oxidation requires the high pH and redox associated with photosynthesizing organisms (Richardson et al., 1988). Manganese nodules occasionally form on freshwater sediments. These nodules form when bacteria oxidize reduced iron and manganese (Chapnick et al., 1982; Atlas and Bartha, 1998).

Selenium and copper are both required for proper cell function, but they are far more commonly studied as pollutants. Some agricultural practices can lead to selenium occurring at toxic levels, and some forms of mine waste can lead to copper pollution.

GRADIENTS OF REDOX AND NUTRIENT CYCLES AND INTERACTIONS AMONG THE CYCLES

Redox is a major factor controlling biological and abiological nutrient transformations as we stressed in previous discussions of biogeochemical cycling. Redox potential in a specific area allows us to predict the metabolic activities that predominate. Redox decreases with increasing depth in anoxic sediment or across the metalimnion of a eutrophic stratified lake. Changes in redox also occur over time, giving rise to an orderly sequence of preferred nutrient transformations. The more oxidized a compound is, the more energy that can be generated when using it as an electron donor to oxidize organic carbon with anaerobic respiration. We presented the order in which compounds serve as electron acceptors (their relative redox state) in Fig. 13.7.

Redox gradients are sites of high rates and diverse types of metabolic activities. These microbial "hot spots" occur because of the dependence of many aquatic microbial geochemical processes on either reduced chemicals in oxidized environments or oxidized chemicals in reduced environments. Given Fick's law of diffusion, the steeper the gradient, the greater the diffusive flux and, redox gradients are locations where reduced compounds are diffusing rapidly into oxidized conditions and vice versa.

The complex interrelationships of nutrient cycling in the vicinity of a redox gradient is illustrated by considering the hypothetical distribution of some of these activities and populations of microbes (Fig. 14.14). The fact that multiple chemicals interact with each other and organisms transform them, leads to substantial complexity. Additional complexity arises from the variety of fermentative processes that can occur and the fact that many different microbial strains or species can be responsible for each of the processes.

Gradients of redox can occur in a variety of habitats (discussed in Chapter 12). Consideration of redox adds more complexity to the view of biogeochemical cycling

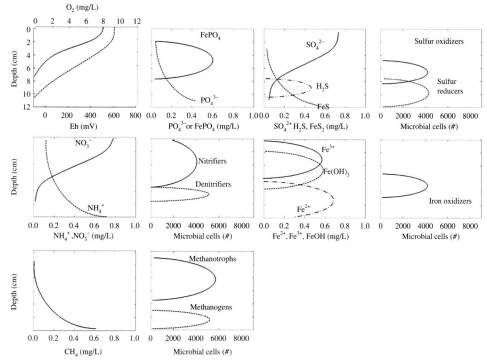

Figure 14.14 Relationships among redox gradients, dissolved oxygen, nutrient concentrations, and functional groups of microorganisms responsible for biogeochemical fluxes. The figure illustrates the steep gradients that occur at oxic/anoxic interfaces, and how such interfaces are a hot spot for biogeochemical activities.

because of the series of redox potentials. In the absence of O_2, redox gradients still provide a habitat gradient with different metabolically driven fluxes depending on the amount of potential energy that organisms can extract. Because many of these organisms have a long evolutionary history, natural selection dictates that organisms relying on inefficient metabolic pathways will be outcompeted and are unlikely to dominate in areas where other processes are more efficient. A very interesting group of bacteria inhabits freshwater associated with redox gradients. Known as "cable bacteria," this group uses redox gradients for energy by directly transporting electrons into oxic conditions and harvesting energy, bypassing the need to wait for reduced compounds to diffuse into oxic environments (Risgaard-Petersen et al., 2015).

One of the influences of humans is alteration of the natural distribution of redox heterogeneity. For example, urbanization can lower heterogeneity, and decrease anoxic carbon-rich habitats leading to lower rates of denitrification. Restoration of stream channel heterogeneity can increase the rates of denitrification by creating numerous small low-redox areas that are conducive to denitrification (Groffman et al., 2005).

The complex interactions among organisms are constrained by redox but also link to organic carbon because carbon provides energy and materials for organism growth. For example, chemoautotrophic organisms, such as those that oxidize ammonium, sulfide, or ferrous iron, use the energy obtained to assimilate CO_2. Assimilatory processes can provide new nutrients to a habitat, as in the case of nitrogen fixation. The redox series of metabolic interactions discussed previously is in large part a series of increasingly less efficient ways for organisms to oxidize organic carbon to CO_2 and release energy.

Some unexpected interactions can occur among nutrient cycles that make sense based on redox. For example, nitrate addition can protect some wetlands from eutrophication (Lucassen et al., 2004) because nitrate has a higher redox and is used as an electron acceptor before oxidized Fe^{3+}. Oxidized iron can then precipitate the phosphate. The phosphate deposits in the sediments and is less available to wetland plants. Understanding this effect requires a thorough understanding of redox and interactions among chemicals in freshwaters.

In addition, key links occur between the cycles as mediated by inorganic chemical reactions. These include interactions between ferric iron and phosphate, ferrous iron and sulfide, and between calcium and phosphate.

SUMMARY

1. The major forms of inorganic nitrogen are N_2 gas, nitrate, nitrite, and ammonium. Organic nitrogen occurs in many forms, including amino acids, proteins, nucleic acids, nucleotides, and urea.
2. The major fluxes in the nitrogen cycle include denitrification (using nitrate to oxidize organic C, yielding N_2), oxidation of ammonium to nitrate by chemosynthetic bacteria, assimilation of ammonium, fixation of N_2 by bacteria, and excretion of ammonium by heterotrophs.
3. Forms of inorganic sulfur include sulfide, thiosulfate, sulfate, elemental sulfur, and metal sulfides. Organic sulfur is part of proteins and amino acids.
4. The most important fluxes in the sulfur cycle are biological and abiological sulfur oxidation, biological sulfur reduction (a form of anoxic respiration), production of sulfide by fermentation, disproportionation, and metal pyrite precipitation and deposition.
5. Phosphorus is a key element that determines the productivity of many aquatic ecosystems and occurs as phosphate and organic phosphate. Phosphatase enzymes can cleave organic phosphorus to phosphate. Phosphate forms a low-solubility

precipitate with ferric iron in the presence of O_2 that can cause its removal from oxic environments.

6. Silicon is a vital component of the cell walls of diatoms and can be a key factor in controlling composition of phytoplankton communities. It is redissolved slowly; thus, when frustules sink out of the photic zone in lakes, it takes months for the silicon to become available again. Under some conditions, frustules accumulate over time and form silica-rich sedimentary rock.

7. Iron occurs as ferric and ferrous ions in oxic and anoxic habitats, respectively. It can also occur as a metal pyrite (FeS) in anoxic habitats and a flocculent precipitate [Fe(OH)$_3$] in oxic habitats. Iron is an important component of many proteins, including those for electron transport, nitrate assimilation, and chlorophyll synthesis.

8. Bacteria in oxic environments can oxidize ferrous iron. Ferric iron combines with OH^- or PO_4^{3-} to form low-solubility precipitates. Chelators can keep ferric iron in solution (prevent it from precipitating with OH^-). Chelators are organic molecules that form complexes with iron.

9. Gradients of redox that occur in natural waters allow for complex populations of microbes that are capable of a wide variety of nutrient transformations in localized hot spots. There is a predictable order in which each process will occur across the redox gradient, given the relative potential energy of each chemical reaction at each redox point.

10. Nutrient cycles do not occur in isolation. Complex interactions occur among all of them, in part because organisms have similar nutrient requirements. Interactions also occur because different chemicals interact with each other in the absence of organisms.

QUESTIONS FOR THOUGHT

1. How might acid precipitation dominated by sulfuric acid increase the availability of phosphate, aside from pH effects?

2. Why are isolated chemosynthetic communities, such as those found where sulfide enters oxic cave waters or methane enters sediments, still ultimately dependent on photosynthetic organisms?

3. When people control eutrophication by only removing nitrogen, cyanobacteria can dominate. Why?

4. What was the earth like before oxygenic photosynthesis and how did the sulfur cycle likely drive the redox processes of ecosystems in aquatic habitats at that time?

5. Why are areas downstream from hyporheic zones where water re-enters streams often sites of very high algal production?

6. Why do sewage treatment plant operators generally want to avoid denitrification in ponds designed to settle solids?
7. Why does it make sense to remove nitrogen from sewage with denitrification?
8. How might phosphate-rich water be treated with iron to remove phosphate (including O_2 concentrations and forms of iron used)?
9. Why does silicon often disappear from the epilimnion of lakes more rapidly than phosphorus or nitrogen?
10. Why is it difficult to reduce the effects of excessive phosphorus pollution in lakes when O_2 disappears from the hypolimnion?

CHAPTER 15

Adaptations to Extreme and Unusual Habitats

Contents

Figure 15.1 Lake Bonney (the lake is under the flat ice in the center of the valley), a permanently ice-covered lake in Antarctica (top), and salt pillars that are several meters high at hypersaline Mono Lake, California (bottom). *(Top) Courtesy: John Priscu; (Bottom) courtesy: Dave Herbst.*

Abiotic extremes create unusual environments that are engaging to aquatic ecologists because of their novelty. Microorganisms that can live in almost boiling water and animals that can live in near-freezing water are fascinating. In addition to piquing academic interest, these habitats can provide insight into how organisms may tolerate pollution and other human-caused environmental perturbations. For example, studies on thermal pollution and global warming may require an understanding of how organisms cope with high temperatures (Sand-Jensen et al., 2007), and organisms from saline lakes may provide clues to species' responses to salinization by agricultural and urban runoff. Furthermore, aquatic microbial ecologists and biotechnologists have isolated useful microbes from extreme habitats, such as those that produce the enzymes essential for the polymerase chain reaction, an essential tool in modern molecular biology. In this chapter, we discuss how organisms adapt to different extremes and the environments in which the extremes occur. The extremes considered here include high and low temperatures, periodic drying, high salinity, surface layers of water (experiencing damagingly high light), and ultraoligotrophic waters. In the next chapter, we will discuss how human stressors, such as pollution, can have detrimental effects on organisms, and adaptations discussed in this chapter will be very relevant to those discussed in the next.

ADAPTATIONS TO EXTREMES

Adaptation can be physiological alterations of behavioral, cellular, or molecular characteristics within an organism, or can refer to evolutionary differences in the ability of organisms to specialize or tolerate specific environments. We consider both in this section, so the student needs to be aware of the context in which we use the word adaptation.

Shelford's law of tolerance suggests that there is an optimal range in which fitness is greatest; as conditions deviate from the optimum, fitness declines, and beyond some limit, the organism cannot survive and reproduce. With many potential environmental extremes, this is true. However, with some extremes it is not a "law" (see Dodds, 2009 for a discussion of what might qualify as a law in ecology). For example, ionizing radiation probably has no intermediate level of radiation exposure where fitness is optimal. Still, for many environmental drivers that influence organisms, such as temperature and pH, species adapt to optima and are less fit outside the optimum range.

Many of the adaptations we discuss in this section are cellular or molecular, and most of the organisms that inhabit extreme environments are microorganisms. Microbes probably dominate because higher plants and animals have complex multicellular systems that cannot evolve to compensate for extremes, such as particularly high temperatures, salinity, and variations in pH—although notable exceptions occur and we mention them where appropriate.

Understanding the influence of extremes in pH, salinity, and temperature requires knowledge of the structure and function of biological molecules. We begin with a discussion of adaptation to temperature. The main influences of temperature relate to protein structure, DNA and RNA structures, and lipid fluidity. Metabolism at high temperatures presents particular difficulties when molecules involved are unstable at high temperatures, or when they freeze at low temperatures. In order to function, proteins must maintain structure and have ample thermal energy. An enzyme not only needs to maintain an active site in a very specific configuration, but also must be able to translate thermal energy into making or breaking chemical bonds. Thus, enzymes have a temperature range in which they are able to maintain structure and activity, and within that range, they have an optimum temperature for activity. Enzyme activity increases with temperature up to a point, beyond which the enzyme starts to break down (denature) and rapidly loses function. Proteins in organisms in high-temperature environments need enhanced thermal stability. Factors that are correlated with thermal stability of proteins include a hydrophobic core, reduced glycine content, more alpha helical conformation, high ionic interactions, reduced surface area to volume ratio (better core packing), more stabilizing bonds (hydrogen bonds and salt bridges), and numerous related adaptations (Madigan and Oren, 1999; Kumar et al., 2007).

Biological membranes need to maintain an optimum degree of fluidity to function properly. If the membranes are excessively fluid they will not maintain cell integrity, and if the membranes are solid they lose biological activity. As temperature increases, the melting point of the lipids in an organism that can function at those temperatures increases as well. Lipid melting points increase with greater proportions of single bonds (increasing saturation), increased branching, greater length, and, in the Archaea, ether lipids (Russell and Hamamoto, 1998). For lipids to remain fluid at low temperatures, the opposite properties (unsaturated, short, and unbranched) are required.

DNA and RNA molecules also have a specific temperature range in which they function optimally. RNA molecules of organisms adapted to high temperatures may have more C−G bonds than those growing at lower temperatures because C−G pairs have three hydrogen bonds, whereas A−T pairs only have two bonds. The C−G pairs, increased cation concentrations, and specialized proteins can all stabilize DNA (Stetter, 1998; Madigan and Oren, 1999; Grosjean and Oshima, 2007). High temperatures can damage DNA and RNA, and thermophilic cells have greater rates of DNA repair and RNA turnover. RNA can have modified nucleotides that stabilize in high temperatures (Grosjean and Oshima, 2007). Messenger and ribosomal RNA need portions of the molecule held in specific configurations (secondary and tertiary structures) to function properly. If these secondary (Fig. 8.3) and tertiary structures are not maintained, protein synthesis (translation) will not occur. In organisms adapted to high temperatures, RNA has extended regions of base pairing (C−G or A−U) to stabilize secondary and tertiary structures.

The requirement for these extended regions is less stringent or even puts an organism at a disadvantage in low-temperature environments. Some multicellular organisms can withstand freezing, but a variety of molecular adaptations to low temperatures can also aid in survival (Rodrigues and Tiedje, 2008). Many of the molecular adaptations to cold are opposite to thermophilic adaptations (e.g., membranes more fluid at low temperatures, fewer C−G pairs).

To survive freezing, organisms must have the ability to avoid the damaging effect of ice crystal formation in cells (Sakai and Larcher, 1987). Ice crystals rupture the plasma membrane and destroy the integrity of the cells. Organisms that live in high latitudes or altitudes use a variety of strategies to avoid or survive freezing including *supercooling*, which is the lowering of the freezing temperature of the water in their cells (Storey and Storey, 1988). For example, some fishes, amphibians, and invertebrates produce glycerol, which inhibits freezing down to −100°C in a 60% solution with water, and concentrate it in critical body parts. Tardigrades almost completely replace their water with the sugar trihalose and can survive decades of freezing. Some animals freeze nearly completely, but they use ice nucleators, proteins or other substances in the blood and tissues that facilitate formation of ice crystals to control freezing and localize it to places that damage will not occur. This process slows the rate of freezing, allowing time for physiological and metabolic adjustments as the body freezes. Many invertebrates simply enter a resting stage called a *diapause*, in which they are more tolerant of freezing or temperatures close to freezing. Diapauses often occur during life stages of reduced biological activity, such as an egg, pupae, or cyst, and can be used to survive other adverse conditions including drought, high temperatures, or anoxic conditions.

Different adaptations are required for organisms living under extremes of salinity. In these situations, ions outside the cells (such as magnesium compounds) become highly hydrated, and water becomes limiting. Animals and microbes can survive moderate levels of salinity by excreting excess salt. Osmoregulation of fishes has received considerable study (Eddy, 1981), and fish physiologists have documented the mechanisms to control salt balance. Only organisms able to withstand high intracellular concentrations of salt (above about 10% salinity) can survive in high-salinity environments. Hypersaline microbes have multiple sets of genes coding for sodium and potassium transport (Vellieux et al., 2006). Osmotic pressure will collapse cells and drain their water if they do not maintain an internal concentration of dissolved materials approximately equal to the external ion concentration. For example, *Dunaliella*, a green alga that thrives in saline waters, synthesizes high concentrations of glycerol to counteract the effects of increased salinity (Javor, 1989). Many other species of algae and fungi also use glycerol to counteract osmotic pressure in aquatic environments. Some halobacteria can accumulate up to 5 molar KCl in their cells and can remain isotonic with the external solution (Grant et al., 1998). Salinity also alters proteins by increasing hydrophobic interactions, and hypersaline organisms have high amounts of negatively charged amino acids on their external surface to counteract ion effects (Vellieux et al., 2006).

A number of habitats have extremely high or low acidity. For example, basic hot springs or alkaline lakes can have a pH as high as 12, and acid hot springs can have a pH as low as 1.5. Some contaminated habitats, such as acid mine drainage, can also be highly acidic with negative pH values when acid mine water is concentrated by evaporation. Acidophilic microorganisms have membranes that are impermeable to hydrogen ions (Johnson, 2007). Cells must maintain their internal pH close to neutral to allow for normal metabolism, but extracellular proteins must function at the pH of the environment.

Lack of water is a particularly severe condition for aquatic organisms. Temporary pools, wetlands, and intermittent streams all have periods of drying. Many groups of aquatic organisms have the ability to withstand desiccation. An impressive example is the cyanobacterium *Nostoc*, which accumulates sucrose to maintain biological molecules during drying, and can withstand 107 years of desiccation (Dodds et al., 1995). Similarly, crustaceans that inhabit temporary pools in arid regions produce resting eggs that can remain viable without water for decades (Smith, 2001).

Finally, habitats with high light intensity can be detrimental to many organisms because solar radiation harms cells. Some species that live in shallow, clear water bodies and those that live on the surface need some protection from damaging UV radiation. Solar irradiance causes formation of free radicals that can react with biological molecules. High-energy light, particularly UV, causes the most damage. Compounds such as carotenoids (many types of organisms), mycosporine-like amino acids (diatoms), flavenoids (green algae and higher plants), and scytonemin (cyanobacteria) absorb damaging light. These compounds protect organisms from damage by preventing formation of the harmful free radicals that would react with essential molecules such as DNA and protein (Long et al., 1994). The drought resistant eggs of some invertebrates that inhabit temporary waters protect themselves from damaging UV radiation by protective coatings. Planktonic cladocerans and copepods that live in clear, high latitude lakes use protective pigments and behavioral responses, which allow them to feed in the water column during the day when UV exposure is high (Hansson et al., 2007).

SALINE LAKES

There is roughly as much water in saline lakes on Earth as is contained in freshwater lakes, but these lakes are generally not important water supplies, so they have received substantially less study than freshwater lakes. Saline lakes are often the only surface water in dry regions of the world. Some large, permanent saline lakes include the Great Salt Lake, the Dead Sea, Issyk Kul, the Caspian Sea, the Aral Sea, and Lake Balkash. Many saline lakes, such as the Australian Lake Eyre, are intermittent.

Saline lakes and ponds occur in closed basins in which water leaves primarily through evaporation. In these situations, salts weather from the watershed and flow to the lowest point. Water then accumulates until the water body is large enough for evaporation to equal inflow. As the water evaporates, it leaves behind the salts. Saline lakes can have concentrations of salts 10 times or more in excess of those found in marine waters (oceans are approximately 3.5% salt by weight). The relative proportion of ions in the inflowing water and varied solubility of different ion pairs as the salts are concentrated determine the chemistry of the lake.

One interesting feature of some salt lakes is the formation of salt or tufa (carbonate mineral) columns. These are formed as ion-rich water moves up through elevated columns of deposited salts by capillary action and evaporates off of the top, leaving the salt behind. Carbonate deposition by photosynthetic organisms can enhance the process. Such a process forms the bizarre landscape with tufa pillars on the shore of parts of Mono Lake (Fig. 15.1), and similar pillars of salt along the Dead Sea may have been mistaken for Lot's wife.

Depth of salt lakes varies considerably over the years because their levels depend on a balance between evaporation and inflow. During wet years, the depth of the lake increases until the surface area is great enough to allow evaporation to equal inflow. In these cases, salinity decreases, which can have ecological effects on the aquatic community. For example, a series of wet years led to flooding of the Great Salt Lake in Utah in the 1980s and led to shifts in the lake's food web. Decreases in salinity allowed the predaceous insect, *Trichocorixa verticalis*, to invade the pelagic zone of the lake. This predator caused decreases in the brine shrimp, *Artemia franciscana*, and subsequent increases in protozoa and three species of microcrustacean zooplankton (Wurtsbaugh, 1992).

There is a general decrease of diversity of animals and plants as salinity increases and the upper tolerance limit of various organisms is exceeded (Table 15.1). As with other extreme environments, the Bacteria and Archaea dominate in the harshest habitats. Some animals such as brine shrimp (the anostracan *Artemia salina*) can withstand more than 30% salt. The upper salinity limit of some animals may actually be a lower limit for O_2; solubility of O_2 decreases with increased salinity. However, the majority of freshwater species disappear under only moderate salinity levels, presumably because of an inability to osmoregulate (Bayley, 1972).

Salt lakes have scientific, economic, cultural, recreational, and ecological values. They can contain unique microbial species that could be useful to humans (Paul and Mormile, 2017). These lakes are very sensitive to decreases in flow. Human appropriation of freshwaters in the dry regions where they are located can have devastating effects. For example, prior to 1960, the Aral Sea in Russia was the fourth largest lake in the world, with a surface area of 68,000 km^2 and volume of 1,090 km^3. By 1993, the lake area had decreased almost by half and the volume decreased to 340 km^3. In 2013, the lake area was down to 1,000 km^2 and volume decreased correspondingly

Table 15.1 Upper salinity tolerances of some members of selected groups of organisms

Organism	Upper salinity range (%)
Fishes	11
Nematodes	12.5
Ostracods	13
Gastropods	15.9
Rotifers	16
Isopods	16
Copepods	17.6
Diatoms	20.5
Chironomids (Diptera)	28.5
Ephydra cinerea (Diptera)	30
Artemia salina (brine shrimp, Anastroca)	33
Cyanobacteria	35
Ciliate protozoa	35
Green algae (*Dunaliella*)	35
Parartemia salina (Anostroca)	35.3
Phototrophic bacteria	40
Extreme halophilic bacteria	Saturated

Source: Data from Javor (1989).

(Shi et al., 2014). Agricultural uses reduced water inflow from 50 to 7 km^3/year. The former lakebed is a source of salt and dust storms that have negative impacts on human health. The increase in salinity and numerous introductions of animals have caused sharp decreases in biodiversity (Williams, 1993). Other saline lakes, notably Mono Lake in California, are subject to similar pressures. Such lakes are unique and we should conserve them.

Saline lakes can serve as important sources of food for waterfowl. The exclusion of fish by high salinity allows relatively large crustaceans such as brine shrimp to attain high densities. This can provide an ample food source for birds. Flamingos use saline lakes for this reason, consuming both crustaceans and algae. The pink color of flamingo feathers is from the birds retaining the carotenoids that are synthesized by the algae and concentrated by the zooplankton to protect against ultraviolet radiation. Additionally, red colored Archaea (*Halococcus* and *Halogeometricum*) actually live in the plumage of some flamingos and lend red color to the birds (Yim et al., 2015). Many other species of waterfowl depend upon saline lakes world-wide.

Advanced: Chemistry of saline lakes and salt production

Most modern nonmarine lake brines are dominated by the anion chloride, followed by carbonate and sulfate. Marine-influenced brines are also dominated by chloride. Sodium and potassium are the most common cations, but some lakes are dominated by magnesium or calcium instead (Hardie, 1984). As concentration of salt increases, some salts become insoluble and precipitate.

Of the major anions, chloride salts are generally more soluble than sulfate salts, which in turn are more soluble than carbonate salts. For the cations, sodium salts are soluble to higher concentrations than magnesium, while calcium salts are least soluble (Table 15.2). The exact concentration at which chemical precipitation occurs depends upon concentration of salts, interactions among the salts (those in this table and others), pH, and temperature. Table 15.2 describes how different salts precipitate as saline water dries and concentrates.

Saltworks (salt production facilities) use the known series of deposition; using a series of evaporation ponds, the salts become more concentrated as the water passes through the sequence of ponds. The saltworks "harvest" different salts in each pond as they are concentrated by evaporation. Saltworks are common near oceans, but people also build them near saline lakes. Since lakes have variable chemistries, the saltworks near various saline lakes will yield different types of salts than those near the oceans, or next to lakes with relict marine waters. Biology complicates this process, as some species of halophytic organisms can influence the precipitation of salts or cause them to precipitate in less unpredictable orders. Therefore, the aquatic microbiology of saline systems has practical applications.

Table 15.2 Saturation concentrations, ordered from least to most soluble, of various salts found in saline lakes

Ion	Concentration [molar]
$CaCO_3$	0.00015
$MgCO_3$	0.0019
$CaSO_4$	0.015
Na_2SO_4	0.33
$NaCO_3$	0.67
$MgSO_4$	2.1
$MgCl_2$	5.7
$CaCl_2$	6.7
$NaCl$	16.4

Source: Data from Weast (1978).

HOT SPRINGS

Hot springs associated with geothermal activity have piqued the interest of many scientists. Visitors to thermal areas such as Yellowstone National Park might not be aware that many of the beautiful colors they see in the pools and streams formed by the hot springs are actually living microorganisms. These colors on the bottoms and walls of the hot springs are highly organized microbial mats (Castenholz, 1984). Many *thermophilic* (heat-loving) organisms not only tolerate but also require the presence of elevated temperatures to grow successfully. Bacteria and Archaea mainly inhabit areas above 55°C. These habitats have lower biotic diversity and hence form an attractive system for ecological research.

The chemistry of hot springs is variable and the differences influence the biology of the organisms in the spring; they can range from highly acidic (as low as pH <1) to very basic (pH >11). The distribution of pH values is generally bimodal, with acidic springs dominated by sulfates, basic springs dominated by carbonates or silicates, and few neutral springs (Brock, 1978). This difference depends upon the geology that forms the hot spring. Water percolates from the surface to deep underground, where under high temperatures and pressures rocks dissolve into the water.

Anoxic waters high in sulfide generally feed acidic springs. The sulfide is oxidized either physically or biologically by sulfur-oxidizing bacteria (Fig. 14.1), with both processes leading to formation of sulfuric acid. At the highest temperatures, chemoautotrophic bacteria that oxidize sulfide are the dominant primary producers. The biota of hot springs is usually less diverse as temperature increases, although some groups such as cyanobacteria apparently prefer warm temperatures (30°C–40°C) (Fig. 15.2). The thermal tolerance limits of an increasing number of phylogenetic groups are exceeded

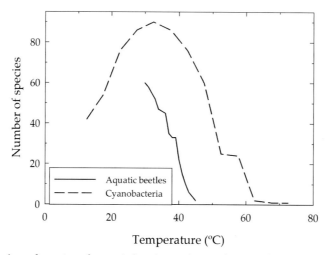

Figure 15.2 Number of species of aquatic beetles and cyanobacteria found in springs of different temperatures. *Data from Brock (1978).*

Table 15.3 Upper temperature tolerances for various groups of organisms

Group	Approximate upper limit (°C)
Fishes	38
Vascular plants	45
Insects	50
Ostracods	50
Mosses	50
Protozoa	56
Algae (eukaryotic)	60
Fungi	62
Cyanobacteria	73
Photosynthetic Bacteria	73
Extreme thermophilic Bacteria and Archaea	122

Source: From Brock (1978) and Takai et al. (2008).

as temperature increases above 25°C (Table 15.3). Multicellular plants and animals generally cannot withstand temperatures greater than 50°C, but some single-celled Eukarya and filamentous Fungi can withstand temperatures up to 62°C. Photosynthetic bacterial primary producers occur up to 73°C.

Strains of individual microbial species can also be distributed along a temperature gradient (Fig. 15.3). In this case, unicellular cyanobacteria that appear identical under

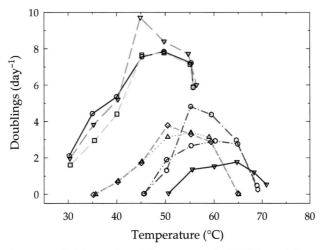

Figure 15.3 Growth curves of eight strains of *Senecococcus* isolated from different temperatures in one hot spring. *Reprinted with permission from Peary and Castenholz (1964).* © *1964 Macmillan Magazines Limited.*

the microscope have adaptations to different temperature optima and can dominate in a narrowly defined habitat. This range of temperature optima of various strains clearly illustrates that biochemical specialization is necessary for a strain to compete success-fully at individual temperatures. The extremely stable nature of hot spring tempera-tures allows these strains to dominate in the narrow regions of their temperature optima.

Wickstrom and Castenholz (1985) described an interesting case of organism dis-tribution related to temperature in Hunter's Hot Spring in Oregon. The spring leaves the ground at slightly less than boiling and the water cools as it contacts the atmosphere and ground, creating a gradient of decreasing temperature downstream. The cyanobacterium *Synechococcus* dominates from 74°C to 54°C because other pri-mary producers are unable to survive (Fig. 15.4). As the stream cools, the motile fila-mentous cyanobacterium *Oscillatoria terebriformis* dominates, covering the surface of the mat at moderate light levels and contracting to the margins under very high light. As the stream cools further, the herbivorous ostracod *Potamocypris* is able to survive. This algivorous thermophilic ostracod can crop down *Synechococcus* and *Oscillatoria*, allowing for the development of a mixed leathery mat community of two cyanobacteria (*Pleurocapsa* and *Calothrix*), which are poorer competitors but are resistant to grazing (Wickstrom and Castenholz, 1978). Such clear-cut effects of competition and predation on community structure across a physical gradient are not often observed in nature.

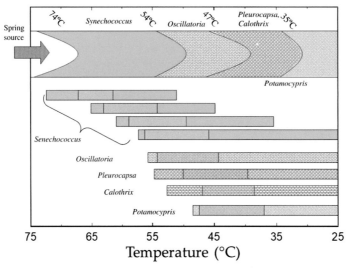

Figure 15.4 Distribution of cyanobacterial genera and strains and the grazing ostracod *Potamocypris* in Hunter's Hot Spring, Oregon. Dominant species on top with distribution limits. Bars give temperature ranges of each organism, with shaded portion of each bar representing the temperature range for optimum growth. *Adapted from Wickstrom and Castenholz (1985).*

COLD HABITATS

Cold habitats include ice, snow, polar, and high altitude lakes, streams, and wetlands. These habitats can be present for part of the year in temperate areas or much of the year in polar or high-altitude regions. Organisms that live in these habitats have to be able to function at low temperatures. There are two general groups of organisms. Most commonly, organisms found in low-temperature habitats also occur in more moderate habitat. They have much greater rates of metabolism in the moderate habitat. However, some organisms are *psychrophilic*, meaning they require cold temperatures (generally below 5°C) to grow and/or reproduce. The psychrophiles are less common, but more interesting physiologically.

Data on these organisms may be of applied use in the food industry with respect to organisms that spoil refrigerated food as well as a source of enzymes that work well at low temperatures. Biotechnology researchers have focused less on psychrophilic than thermophilic organisms. Such organisms produce proteins that are active at low temperatures. These compounds could be useful in cold food preparation and in detergents for washing in cold water (Russell and Hamamoto, 1998).

Many temperate systems are frozen or very cold for part of the year, and freshwater organisms need to be able to survive these periods. Ice cover on temperate lakes does not necessarily mean that all organisms are static. A global survey of 101 temperate lakes found that chlorophyll concentrations under ice, on average, were 43% of values in the summer but biovolume was about 15%, indicating that per-cell chlorophyll was greater und the low-light conditions under the ice. Zooplankton abundances were about ¼ of those in summer (Hampton et al., 2017).

Microbial communities also occur in the slush and snow on the surface of alpine lakes (Felip et al., 1995). These communities include bacteria and autotrophic and heterotrophic ciliates. The production of these communities can be higher than planktonic production in the ice-covered water below. Many of the species present in the slush are either derived from the plankton in the lake below or from the snow pack above. Apparently, some of the species are adapted to the icy habitats because they occur mainly in the slush and not in the adjacent lake or snow.

Lakes in the dry valleys of Antarctica (Fig. 15.1) provide a unique permanently cold habitat. In general, the same biogeochemical processes occur in these habitats as in temperate systems, but extreme environmental conditions control rates (Howard-Williams and Hawes, 2007). These lakes have several meters of ice cover year-round, so they receive very low levels of light. The primary producers (planktonic algae) found in the lakes are adapted to compete for light (steep α and low compensation points for the photosynthesis—irradiance curves; see Chapter 12. Some primary producers are able to consume small particles as well as photosynthesize (are mixotrophic), and this may allow them to survive the long winter with no light (Roberts and Laybourn-Parry, 1999). The communities are simple, with no fishes or large invertebrates. Some of the lakes have warmer regions fed by saline, geothermally heated warm springs (Fig. 15.5). These warmer regions are anoxic, have high nutrients (Green et al., 1993), and have an enhanced population of primary producers located at the chemocline (Fig. 15.5B).

A unique community is associated with liquid inclusion in the ice layers on the surface of the dry-valley lakes (Priscu et al., 1998). Particles from the terrestrial habitat blow onto the ice surface. The particles absorb heat in the summer and melt down into the ice cover. The liquid water surrounding the particles supports a community of algae and bacteria. Similar communities also occur on glaciers (Bagshaw et al., 2016); meltwater, ice-marginal habitats on glaciers can provide a home for a diverse assemblage of microbes (Ambrosini et al., 2017).

There are also lakes and channels under the Antarctic ice sheet. Christner et al. (2014) sampled subglacial Lake Whillans under 800 m of ice. This lake is part of a network of subglacial drainage channels. An active microbial community was detected with molecular sequencing and the community evidently subsists on chemoautotrophic carbon fixation.

Streams feed the dry-valley Antarctic lakes. These streams flow only during a few months of the year when the sun is warm enough to melt the glaciers. The channels

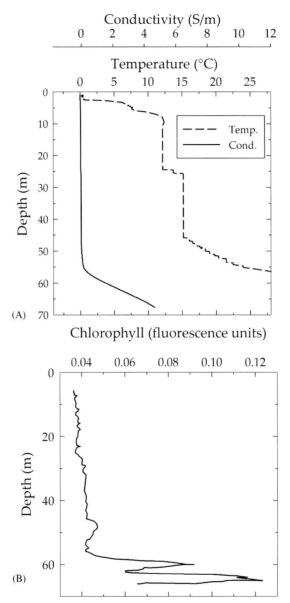

Figure 15.5 Vertical profiles of temperature and conductivity (A) and phytoplankton (B) (chlorophyll fluorescence) from Lake Vanda, Antarctica. *Reproduced with permission from (A) Spigel and Priscu (1998). (B) Howard-Williams et al. (1998).*

dry frequently. Generally, any remaining water freezes when the channel is not flowing. Amazingly, the channels have significant biomass of algal primary producers, mostly cyanobacteria. These organisms can be freeze-dried for much of the year, but they are able to actively photosynthesize minutes after being wetted (Vincent, 1988). The diatom communities in these streams are composed of many endemic species (Stanish et al., 2011). The composition of the bacterial communities correlate to the composition of the diatom communities (Stanish et al., 2013). These streams also provide water to the nearby soils, substantially increasing microbial activities in the "riparian" soils (Zeglin et al., 2011).

Arctic lakes, ponds, and wetlands are also generally very cold. Ponds and lakes can freeze to the bottom; if they freeze completely, they will not contain fishes or many macrophytes. Many fishes can withstand and compete well at temperatures down to 0°C; however, most species have optimum growth above 8°C (Elliott, 1981). Aquatic mosses are often the only macrophytes found in Arctic lakes and streams. These mosses grow slowly and are 7−10 years old, which is greater longevity than has been documented previously for any rooted freshwater macrophyte (Sand-Jensen et al., 1999). The cyanobacteria *Nostoc commune* is also a common primary producer found in aquatic high Arctic habitats (Sheath and Müller, 1997).

High-altitude ponds or lakes are similar to polar habitats because they are ice-free for only a few months a year. The lakes and ponds in high altitudes at temperate or tropical latitudes experience very high levels of light in the summer, and the zooplankton in these habitats are often red or orange because they contain carotenoids that protect against damage by UV-B. If fishes feed on the zooplankton, they retain the carotenoids and have pink flesh.

Simple microbial communities can be associated with snowfields that occur in high mountainous or polar regions. Aristotle first noted snow algae around 300 B.C., and detailed study began in the early 1800s. The microbial primary producers in snowfields can include chlorophytes, euglenoids, chrysophytes, cyanobacteria, and diatoms. The primary producers can support a community of fungi, bacteria, rotifers, protozoa, and some invertebrates (Hoham, 1980). Increased photosynthesis by the snow algae leads to greater bacterial productivity (Thomas and Duval, 1995). This productivity can be transferred to the terrestrial food web that includes small mammals and birds (Jones, 1999).

The most common algae in snow generally are single-celled green algae (chlorophytes). The most obvious sign of these algae is the pink ("watermelon") snow associated with the psychrophilic *Chlamydamonas nivialis*, a green alga that can acquire a strong reddish color produced by high levels of carotenoids. *C. nivialis* and other species of *Chlamydamonas* produce these pigments to protect the cells from ultraviolet irradiance (Bidigare et al., 1993). The irradiance is extremely high at the snow surface because of the high altitude and the reflective properties of snow.

A problem in the life cycle of *C. nivialis* is how to inhabit the upper, lighted portions of snowfields when they start buried deep under the snow each winter. The evolutionary solution to the problem is that the spores of the alga rest in the soil over the winter, then hatch and swim to the surface when the snow starts melting. The motile cells then reproduce sexually and produce more resting spores.

TEMPORARY WATERS AND SMALL POOLS

Drying is probably the most extreme disturbance that can occur in an aquatic system. However, organisms colonize temporary or ephemeral habitats within days or weeks of rewetting. These habitats include temporary pools (Fig. 15.6), streams, lakes, and wetlands. For some organisms, these represent marginal habitats, and for others temporary systems are the only habitats where they are found. For example, ephemeral pools can be important sources of mosquitoes in many areas and are important in urban areas because of disease transmission.

Temporary pools can be categorized according to the permanency of their water. One scheme (Williams, 1996) divides pools and streams into those with periods of drying that occur every several years, systems with regular drying that occurs during specific seasons of the year, and systems with very unpredictable drying (few or several times per year). Such classifications have been useful for predicting life histories and invertebrate community structure. Another approach considers streams across gradients of permanence from always flowing to mostly dry (Feminella, 1996). Again, some species occur only in permanent waters and others prefer temporary stream habitats.

(A)   (B)

Figure 15.6 A temporary pool formed in granite by freezing and thawing of water (A) and a temporary pool (B) formed by bison activity. *Bison image courtesy: N. Gerlanc.*

Obviously, given a gradient of stream and pool types, and cyclic climatic variation, there are not hard and fast definitions of intermittence and no hard line that delineates ephemeral streams. Even a normally perennial stream or pool could dry during a period of prolonged drought.

Intermittent streams are receiving increased attention, which makes sense because they occur in dry areas with occasionally heavy precipitation where they are the main aquatic habitat, in spite of their lack of permanence. Half of the world's rivers have channels that dry at times; some from natural causes and some from human abstraction or a combination of both. A new view of these systems describes what is unique about them, particularly how they alternate having characteristics of terrestrial and aquatic habitats (Datry et al., 2016).

Most fishes are unable to exploit temporary habitats unless there is a refuge nearby that serves as a source of colonists. Where fishes do not occur, large invertebrates can be found that would otherwise be absent because of their susceptibility to predation. In the absence of predatory fishes, these invertebrates, such as fairy shrimp and tadpole shrimp, can successfully develop and reproduce during wet periods (Fig. 15.7). Life histories of organisms in temporary pools usually feature resting stages (that are resistant to desiccation) and/or life stages conferring the ability to fly, crawl, or be blown into pools.

(A)

(B)

Figure 15.7 Tadpole shrimp [*Triops australiensis*, this species can be as large as 7 cm (A)] and fairy shrimp [*Eubranchipus grubii*, ~10 mm (B)] typical of temporary pools. *Photo A by Stijn Ghesquiere courtesy: Wikimedia Commons, photo B by Christian Fisher courtesy: Creative Commons.*

Temporary pools serve as important habitats for amphibians because of the lack of fish predation in such sites. In these systems, amphibians such as ambystomatid salamanders can be top predators that structure prey communities (Wissinger et al., 1999). Temporary pools also allow amphibians with adults that use terrestrial habitats to reproduce in areas without year-round water. The life history of tadpoles is often linked intimately to water permanence, and species interactions are also related to permanence (Skelly, 1997). The competition-predation-gradient hypothesis suggests that there is a tradeoff between competitive ability and predator avoidance among tadpole species, such that strong competitors are more vulnerable to predation, and poor competitors are better at avoiding predation. Strong competitors should thus be more abundant in less permanent habitats where predators such as fishes are absent, whereas tadpole species that are better at avoiding predation should be more abundant in permanent sites with large predators. Demonstrating the importance of interspecific competition among tadpoles has proven difficult, but the importance of predators is clear in many systems.

Another tradeoff involves development time and foraging activity among tadpole species; fast developing species tend to be very active, foraging frequently to fuel rapid development. This activity renders them vulnerable to predators. As such, habitats that dry frequently and lack large predators are generally inhabited by active, rapidly developing tadpole species such as chorus frogs (*Pseudacris* spp.), compared to permanent habitats with fishes that are inhabited by slow-developing, inactive species such as bullfrogs (*Rana catesbeiana*) (Skelly, 1997).

Ephemeral pools or wetlands form an important habitat for other organisms, including waterfowl. Prairie potholes, many of which only hold water for the wet part of the year, are key habitats for many of the ducks that live in the central Midwestern United States (Batt et al., 1989; see Chapter 5). The waterfowl are mainly migratory and use the pools during wet times of the year. The pools have large populations of invertebrates (as fish that would eat them cannot survive in habitats that dry) that serve as a food source for many of the waterfowl. Ephemeral pools and wetlands as habitat for waterfowl and other organisms are important globally (Roshier et al., 2001), although little is known about this function of wetlands in many parts of the world (Williams et al., 2009).

Mediterranean temporary ponds are important because they occur in high diversity areas and have many unique species (Bagella et al., 2016). The vernal pools of California are unique systems in that they contain endemic plant assemblages. Georgia, Texas, Mexico, Chile, South Africa, and Australia also have temporary pools with endemic plant assemblages (Thorne, 1984). Likewise, the fairy shrimps (Anostraca) have adapted to vernal pools throughout the world, with some genera that are distributed broadly and others that are endemic to local regions (Belk, 1984). In a good example of endemism in temporary pools, when 58 vernal pools in California were sampled by King et al. (1996), they recorded 67 species of crustaceans. They suggested that 30 were likely new species.

Rock pools form another habitat that is widely distributed and unique; they occur on all continents in all biomes. They are some of the oldest habitats, and in spite of being intermittent, evolution has had much time to work on the organisms inhabiting them. As such, there are a number of endemic species associated with rock pools. The pools are small and tend to have broad fluctuations of chemical and physical characteristics. Pools on tops of granite outcrops are common but pools can form in limestone or sandstone. At least 460 animal species have been described that are found in rock pools (Jocque et al., 2010).

Impermanent streams and rivers are the dominant lotic habitat in many arid regions of the world (Davies et al., 1994). Their unique characteristics demand different management approaches and sound understanding of their ecology (Datry et al., 2017). These streams are becoming more common as people deplete water in formerly permanent streams and rivers. In addition, global climate change will increase intensity of drought in many areas, increasing intermittence. Most information on the ecology of intermittent streams comes from Mediterranean streams and those from Australia, although such streams occur on all continents (Datry et al., 2014).

The level of permanence controls invertebrate community structure (Miller and Golladay, 1996; Bogan et al., 2013), and drying probably has stronger effects than flooding (Boulton et al., 1992). Primary production by periphyton in streams is resilient to desiccation (Dodds et al., 1996), recovering in days to weeks following rewetting. Invertebrate grazers put little initial pressure on primary producers following rewetting, and then various groups of grazers colonize in the weeks following as grazer communities reassemble (Murdock et al., 2010). The top predators in many intermittent systems are commonly relatively small insectivorous fishes or large invertebrates (Sabo et al., 2010). The degree and frequency of connectivity with permanent waters is a key factor influencing community structure and maintaining biodiversity in these systems (Larned et al., 2010). Intermittent streams are also areas of high plant diversity because they are often the wettest areas in drier landscapes. Disturbances such as flooding and drought likely allow some species to establish that may not be competitive under more stable conditions (Katz et al., 2012).

Another specialized aquatic habitat that has received attention from ecologists is the small pools formed in pitcher plants, tree holes, and the leaves of bromeliads; these pools of water found in terrestrial plants are referred to as *phytotelmata*. Humans also create small aquatic habitats such as those associated with abandoned car tires, birdbaths, and water bowls for pets. Many different insect larvae exist in these small pools, and the larvae partition the environment so they will not compete for the same resources. Tadpoles of some amphibians also inhabit these small pools. The pools are attractive study systems because they form a well-defined ecosystem in which all members of the community can be identified, and pools can be easily replicated and sampled. These habitats occurring in urban areas allow for greater populations of disease-carrying mosquitoes (including *Aedes aegypti* and other *Aedes* species). Diseases of concern associated with these habitats include Dengue, Yellow Fever, Zika, and Chikungunya.

Pool size can be very important in these small habitats. Not only does pool size relate to the probability of drying, but also to the ability of adults to find the habitat and lay their eggs. Larger pools are more complex and predators such as damselfly larvae are less effective in larger, more complex habitats (Srivastava, 2006).

Pitcher plants (Fig. 15.8) form small, deep wells with slippery sides, and the pool of water that collects at the bottom serves as an insect trap as well as a habitat for aquatic organisms, including bacteria, protozoa, and aquatic invertebrates. These plants have a unique band of tissue surrounding the rim of the plant with epidermal cells overlapping in a step-like fashion that are extremely slippery for insects when wetted by rain or nectar produced by the plant (Bohn and Federle, 2004). More than 17 invertebrate species are obligate associates of pitcher plants in the southeastern United States (Folkers, 1999). Pitcher and other carnivorous plants typically grow in nutrient-poor wetlands and use the trapped insects as a nutrient source.

Invertebrates that inhabit the pitcher plant *Sarracenia purpurea* have a positive influence on the plant. The inhabitants include a chironomid (*Metriocnemus knabi*) and a culicid (*Wyeomyia smithii*) that populate the small pools in the plants. They accelerate breakdown of trapped prey, making nutrients and CO_2 more available (i.e., stimulating primary production) than in the absence of the two invertebrates (Bradshaw and Creelman, 1984). These two invertebrates partition the habitat spatially with a third species (*Blaesoxipha fletcheri*), and this allows for their coexistence (Giberson and Hardwick, 1999).

A very large pitcher plant (30 × 16 cm pitchers) grows in the Philippines (Robinson et al., 2009). The plant, *Nepenthes attenboroughii* (named after the famous naturalist David Attenborough) contains mosquito larvae, and probably other species. The pitchers are mostly full of liquid; the upper layer is composed of clear water and the lower a cloudy viscous liquid that contains digestive enzymes. The plant is interesting in that it is large enough to be able to consume small rodents. Once they fall into the pitcher, they cannot escape and the plants digest them. One species of pitcher plant (*Nepenthes rafflesiana*) in Borneo grows in trees and provides habitat for bats. The bats roost in the plant to receive protection and the plant receives over 40% of its nitrogen from the bat's waste products (Grafe et al., 2011). This is a mutualistic interaction.

Figure 15.8 *Darlingtonia*, a pitcher plant that contains small pools. It is inhabited by some insects and preys upon others.

Small pools with several hundred milliliters of water form in the bracts of the tropical monocot *Heliconia*, supporting a complex community of invertebrates, protists, and bacteria. Studies of insect community interactions have demonstrated that positive and negative interactions in the pools occur among the residents (i.e., competition is not the only structuring force in the community). These studies also provide some of the early direct measurements of interspecific interaction strengths (Seifert and Seifert, 1976).

Mosquito larvae can inhabit tree holes and other small pools. Eisenberg et al. (2000) investigated communities dominated by larvae of *Aedes sierrensis* for community effects of larval feeding. The larvae reduced numbers of planktonic protozoa. When biofilms of bacteria and fungi increased, the predation pressure on planktonic protozoa decreased.

ULTRAOLIGOTROPHIC HABITATS

Aquatic systems with very low amounts of available nutrients are also extreme environments. Morita (1997) suggested that the normal state of bacteria is one of depletion and starvation with respect to supplies of organic carbon. If this is the case, most bacteria must experience the stress of oligotrophy at least occasionally. Other organisms are subject to the influence of oligotrophy as well when production of photosynthetic organisms and heterotrophs is low, and food webs are severely energy limited.

Physiological adaptations to such habitats include slow growth and resting or static stages. Some lakes (e.g., Lake Tahoe and Crater Lake) and many groundwater habitats are extreme oligotrophic environments (we discuss these in the section Deep Subsurface Habitats later in this chapter). Picocyanobacteria compete well in ultraoligotrophic Andean lakes with high regional diversity (Caravati et al., 2010). Presumably, they do well because their small size results in a high surface area relative to volume allowing them to take up nutrients more efficiently. Similar adaptations are required for survival in the ultraoligotrophic lakes of Antarctica (Laybourn–Parry et al., 1995).

HYPERTROPHIC HABITATS

In sharp contrast to ultraoligotrophic habitats, hypertrophic habitats have excess nutrients and are thus highly productive. However, a few species that are tolerant of the harsh conditions that characterize these systems generally dominate. Inorganic forms of limiting nutrients (mainly phosphorus and nitrogen) are readily available in hypertrophic systems, and this facilitates dominance by a few microbial species that are best able to exploit a constant supply of nutrients. Livestock and human sewage lagoons are good examples of hypertrophic habitats, but some temperate zone and tropical wetlands and shallow lakes are naturally hypertrophic. We discuss human nutrient pollution in detail in Chapter 18.

Hypertrophic systems are generally light limited because of dense growths of algae and associated high *bioturbidity*. Along with light limitation, dissolved oxygen is very limited; dissolved oxygen supersaturation is common during daylight hours where light can penetrate and photosynthesis proceeds, but high respiration at night leads to frequent anoxia. Even during daylight, only the uppermost layers contain dissolved oxygen because of limited light penetration beyond the surface. Blooms of filamentous or colonial cyanobacteria are common in these systems, and planktonic invertebrate communities are generally limited (Scheffer, 1998). Vascular plants are generally poorly represented, although some floating species can be abundant.

Invertebrate communities in hypertrophic habitats are limited to a characteristic community dominated by small forms with low oxygen demands as a result of severe oxygen limitation. These include annelids and flatworms (Figs. 10.5 and 10.7), air breathing insects [e.g., true bugs (Hemiptera; Fig. 10.14) and dipteran larvae with respiratory siphons such as the rat-tailed maggots (Syrphidae; Fig. 10.17)], and pulmonate snails (Fig. 10.7). Other specializations for low oxygen environments include those of the "bloodworms" (chironomid midge larvae with high concentrations of hemoglobin that makes them appear blood red in color). Depending upon the degrees of nutrient enrichment, fishes and amphibians that rely on dissolved oxygen in the water are rare or absent.

DEEP SUBSURFACE HABITATS

Deep subsurface habitats are extreme not only because energy sources are limited in these deep areas, but also because temperature and pressure increase and dissolved oxygen decreases. Scientists viewed deep subsurface groundwaters as essentially sterile until recently. Such a view is incorrect because bacteria, fungi, and protozoa can now be cultivated from subsurface samples (Fig. 15.9), and bacteria have been found as deep as 400—500 m (Balkwill and Boone, 1997). A study in Finland documented bacteria at a depth of 940 m (Haveman and Pedersen, 1999). These deep microbial communities include a moderately diverse group of bacteria capable of many common nutrient transformations (e.g., denitrifiers, sulfate reducers, and nitrogen fixers) and a somewhat diverse assemblage of heterotrophic microorganisms (Sinclair and Ghiorse, 1989). Temperature tolerances could control depth limits of organisms because geothermal heating increases temperatures with depth (Ghiorse, 1997).

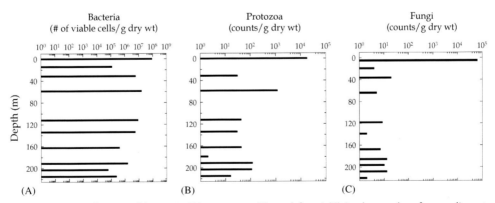

Figure 15.9 Distribution of bacteria (A), protozoa (B), and fungi (C) in deep subsurface sediments from near the Savannah River. © 1989 from Geomicrobiology Journal *Fig. 2, p. 22, and Fig. 3, p. 23, by Sinclair and Ghiorse (1989). Reproduced by permission of Taylor and Francis Inc.*

Animals (stygobionts) inhabit deep waters. Amphipods and fishes have been found in wells up to 500 m deep, amphipods and dipterans in caves up to 2,200 m deep, and nematodes to a depth of 3,600 m in mines (Fišer et al., 2014). Since nematodes have been found as deep as Bacteria and Archeae, how much deeper will we find bacteria? There are obvious questions that arise upon finding such organisms: How long have they been there and what are they living on? The answer to the first question is surprising. Many of these sediments were deposited millions of years ago and pore water ages of 1,200 years (as determined by isotopic dating) have been measured where active microbes have been isolated (Kieft et al., 1998). The microbial communities inhabiting at least some groundwaters are likely derived from the microbes present when the sediments were deposited (Amy, 1997). Apparently, some communities in deep groundwaters have been isolated from the surface for more than 10,000 years. Such isolation raises additional questions: What are they eating? Why hasn't it been depleted? How much have the microbes evolved since their isolation?

In some cases, where subsurface hydrocarbon deposits occur, organic material is sufficient to support an active microbial community (Krumholz et al., 1997). In other cases, organic C is limited and the communities must be adapted to a very oligotrophic way of life. Consequently, rates of respiration in the deep subsurface are generally extremely slow relative to those in most other aquatic sediments (see Chapter 24).

One interesting study of basalt rocks that formed 6–17 million years ago in the Pacific Northwest of the United States suggested that chemoautotrophic processes supported the microbial community present from 200 to 1,000 m deep. In this case, Stevens and McKinley (1995) suggested that microbes used H_2 gas with CO_2 to produce methane and energy. Scientists studying deep wells in Sweden (Kotelnikova and Pedersen, 1998) have made similar claims, but studies in Finland found no such autotrophic activity (Haveman and Pedersen, 1999). If these studies are correct that hydrogen and carbon dioxide can support an entire ecosystem, this is the only known ecosystem on Earth that is not ultimately dependent on O_2 derived from photosynthesis or photosynthetic products. However, Anderson et al. (1998) suggested that the production rates of H_2 are too low in the environment to support microbial growth. Since then, researchers have isolated other Archaea from deep groundwater habitats that are capable of using H_2 gas with CO_2 to produce methane, indicating that the subsurface ecosystems could well be fueled in such a fashion (Shimizu et al., 2013, 2015).

A provocative paper suggests that one groundwater habitat exists dominated by a single microbe (Chivian et al., 2008). These researchers sampled water in fractures 2.8 km deep in an African gold mine. They found that over 99% of the genetic material in the sample was from a single bacterium. The microbe can use organic carbon or live by oxidizing carbon monoxide. It also has the capacity to fix nitrogen. This could be the only environment on Earth where a single organism lives in isolation.

Why is it important to understand the ecology of these deep ecosystems? Subsurface disposal of highly radioactive materials and other waste is common. An active microbial community at these depths could alter transport and containment of such wastes. Bagnoud et al. (2016) studied clay deposits 300 m under the surface that were being considered as a location for radioactive waste disposal based on the assumption of no microbial activity that could degrade containment. They found an active microbial community at that depth subsisting on organic materials included in the clay. Subsurface communities can also alter oil deposits and have global geochemical effects (Stevens, 1997). Furthermore, given microbial biomass and the depth at which it has been located, bacteria could have a greater total biomass of active cells than any other type of organism on Earth (Whitman et al., 1998).

THE WATER SURFACE LAYER

The air—water interface is often not studied but represents a distinct habitat that includes organisms with specialized adaptations (Fig. 15.10). Microorganisms living at the surface are *neustonic* and surface macroorganisms are *pleustonic*. Those organisms found above the surface are *epineustonic*, and those below are *hyponeustonic*. One of the key characteristics of this habitat is the water surface tension. The force at the interface is considerable, and it is quite difficult for a small organism to escape once it has entered (Vogel, 1994). Accordingly, coming in contact with a lake surface may spell death for some species of *Daphnia* and other zooplankton. Other organisms, such as water striders and whirligig beetles, require the water tension to function. Addition of substances to the water that interrupt the surface tension, such as detergents, renders these insects helpless in the water.

The surface layer of water (within 100 µm) represents a unique chemical environment (Napolitano and Cicerone, 1999). Biogenic surfactants, primarily humic and fulvic acids, accumulate here. Lipids, metals of environmental concern, nutrients, and some microorganisms can accumulate in this layer. Bubbles can interact with the chemicals on the surface leading to production of foams. Lipids and other organic molecules, both natural and human produced, stabilize the foams. Even pristine mountain streams can accumulate foam on their surface because the steep gradient entrains bubbles; natural organic compounds trap them and create the foam.

Organisms that specialize in the surface layer must be able to withstand very high levels of light. Such high light must lead to increased energetic costs associated with repair of cellular damage from free radicals formed by high-energy UV irradiance. The

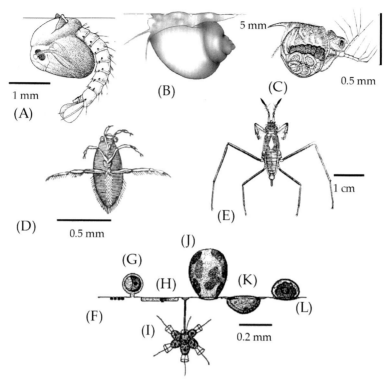

Figure 15.10 Some organisms adapted to use the water surface as a habitat. (A) A mosquito pupa, *Anopheles maculipenis;* (B) a snail, Physa; (C) the cladoceran, *Scapholeberis mucronata;* (D) Notonecta, a water boatman; (E) a water strider, *Rhenmatobates rileyi;* (F) Lampropedia hyalina; (G) the chrysophyte *Ochromonas vischerii;* (H) the diatom Navicula; (I) Codonosig botrytis; (J) *Botrydiopsis;* (K) Arcella; (L) Nautococcus. [A, D, D, E from Wikimedia Commons, B, Sadie Whiles, and (F–L) Ruttner (1963)].

constant influx of nutrients and organic carbon from the air above offset this disadvantage. The bacterial group *Betaproteobacteria* dominate this layer in at least one high altitude Austrian lake (Hörtnagl et al., 2010) and presumably they can withstand this harsh habitat.

Animals have several adaptations to locomotion on the surface. They tend to have waxy hairy legs that repel water. Some brush the surface of the water to move forward, but with more force, they deform the surface and capillary action causes the water to act as a "trampoline" (Hu and Bush, 2010). Water striders can actually jump off the surface of the water using the surface tension (Koh et al., 2015).

Surface-dwelling organisms can also alter the properties of the habitat. For example, they can manipulate surface tension by exuding organic compounds that spread across the surface. An interesting form of locomotion occurs this way; the veliid, *Velia caprai*, and beetles in the genus *Stenus* excrete material that lowers the water tension behind them, so the surface tension in front pulls them forward. The beetle *Dianous coerulescens* can move using water tension in this way at speeds up to 70 cm/s (Hynes, 1970).

SUMMARY

1. Organisms have special adaptations to extreme habitats, allowing them to use a tremendous range of extremes. Bacteria and Archaea dominate in extreme habitats.
2. Physiological adaptations to high temperature include lipids with higher melting points and stabilizing features of proteins and nucleic acids.
3. Organisms in high-salinity habitats need to regulate osmotic pressure, as do those that can withstand drying.
4. Diversity decreases as habitats become more extreme.
5. Hot springs have served as attractive communities for study because of their stable nature, low diversity, and the adaptations of the organisms that are able to live in near-boiling water.
6. Temporary pools are colonized quickly by organisms that are able to withstand desiccation or those that can move in from nearby sources.
7. Communities in intermittent and perennial habitats differ because of life history tradeoffs of the species that inhabit each.
8. Active microbial communities grow in regions of melted water in ice and snow, in habitats ranging from ultraoligotrophic to hypertrophic, and in groundwater up to 1,000 m below the earth's surface.
9. The air–water interface is an extreme environment. High surface tension and high irradiance are characteristics of this habitat.

QUESTIONS FOR THOUGHT

1. Can extreme habitats serve as models for early life on Earth or possible life on other planets?
2. Should efforts be made to conserve the biodiversity of unusual habitats such as hot springs?
3. Should companies be able to patent and take full profit from gene sequences taken from organisms collected in national parks without remuneration to the government?
4. Are "extreme" habitats actually extreme for organisms adapted to live in them?
5. Why can the depth of a saline lake be highly variable from year to year and from decade to decade, and how may global climate change influence such lakes?

6. Why might saltworks that precipitate brines be interested in the microbiology of saline waters?

7. What features allow certain characteristic species to dominate habitats with excess nutrients?

8. How much (%) is the estimated thickness of the biosphere increased by the understanding that organisms can inhabit up to 500-m depth?

CHAPTER 16

Responses to Stress, Toxic Chemicals, and Other Pollutants in Aquatic Ecosystems

Contents

Freshwater Ecology
DOI: https://doi.org/10.1016/B978-0-12-813255-5.00016-8

Figure 16.1 Organic pollutants burn on the Cuyahoga River in 1952. *Courtesy Cleveland State University,* The Cleveland Press *collection.*

The modern aquatic environment has suffered greatly from physical disturbance as well as organic and inorganic toxic pollution. Although scientists recognized the negative effects of pollutants in the 1950s and before, it was not until Rachel Carson's book *Silent Spring* was published in 1962 that it became common public knowledge that organic and inorganic pollutants can have strongly negative, far-reaching, and unpredictable influences on human health and ecosystems (Biography 16.1). Furthermore, acid precipitation, thermal pollution, acid mine wastes, and increases in suspended solids all cause environmental damage. More recently, scientists have developed concerns over the plastics released into the environment, the release of chemicals with endocrine activity, and nanomaterials. The relative importance of various types and causes of lake and river pollution has been determined in the United States from state reports (Fig. 16.3). These data suggest that 36% of the river and stream miles and 37% of lakes were impaired in the last decade. The US Environmental Protection Agency (1997) defines impairment as having evidence of damage to aquatic life, unsuitability for drinking, production of fish that are not safe to eat, or being unsafe for swimming. Large improvements in water quality have not occurred in the United States over the last decades; developed countries around the world had the largest improvements in the 1960s and 1970s but are subject to slow continued degradation of many aspects of water quality since then.

Pristine aquatic habitats no longer exist, except perhaps the deepest, most isolated groundwaters. The atmosphere transports pollutants throughout the world in our atmosphere (Ramade, 1989). Climate change is ubiquitous. Most major watersheds are disturbed. The question is no longer if the pollutants are present, but rather in what quantities, and what are their effects?

BIOGRAPHY 16.1 Rachel Carson

The positive influence of Rachel Carson (Fig. 16.2) may exceed that of any academic aquatic ecologist. In 1962, she published a book titled *Silent Spring* that became a bestseller and had a tremendous impact on public awareness of the pollution caused by pesticides. Her ability to take a technical subject and make it accessible to the public led to some of the first laws enacted to control the release of pesticides into the natural environment. Lear (1997) chronicles her life in an informative biography.

Carson's undergraduate studies in biology at the Pennsylvania College for Women (now Chatham College) prepared her to pursue a master's degree at Johns Hopkins University. Her research on the developmental biology of catfish eventually led to a job writing for the Bureau of Fisheries. She wrote her first book in 1941 (*Under the Sea-Wind*), followed by two critically acclaimed books and numerous popular articles that trans-

Figure 16.2 Rachel Carson. *Image courtesy of the National Oceanic and Atmospheric Administration.*

lated scientific concepts into lay terms. She then published *Silent Spring*, in which she chronicled the wanton use of pesticides and some of their effects on the environment, including biomagnification, death of wildlife (including the loss of bird life that leads to a silent spring), and potential influence on human health (toxicity and carcinogenic properties of toxic pollutants).

The completion of *Silent Spring* was a tremendous professional and personal accomplishment. While writing the book, Carson tended to her dying mother, and after her sister died, she became a single mother to her orphaned nephew. She also began the battle with breast cancer that claimed her life a few years later. Her careful attention to scientific detail was crucial because her book became the focal point of the debate over pesticide use, and she was called to testify to inform of the Senate on the issue. The exceptional popular response to her book led to strong backlash from many chemical companies, entomologists, and government officials; the detractors generally had a financial or professional stake in maintaining indiscriminant pesticide use. Carson documented her facts so well that her critics turned to personal attacks in their attempts to discredit *Silent Spring*.

The life and work of Rachel Carson prove that aquatic ecologists can make a difference in the world. She demonstrated that traditional academic and management careers are not the only ways to have a positive impact, and that combining two disparate strengths (in her case, excellent popular writing and science) can yield impressive results.

We do not know about much chemical pollution given all the chemicals released into the environment. For example, of the more than 72,000 chemicals in commercial use in the 1990s, toxicologists have only screened about 10% for toxicity and screened only 2% of the total for carcinogenicity. In the United States, federal and state governments only regulate about 0.5% of these chemicals (Miller, 1998). The drug database, drug bank, lists over 40,000 known drugs, and many of these end up in the waste stream that enters freshwaters. The Chemical Abstract Service lists over 133 million unique organic and inorganic compounds. In addition, humans create and release novel materials such as microplastics and nanomaterials into the environment with little knowledge of their effects. There are many impacts of humans on freshwaters, many of which are not well studied. For example, light pollution has substantial effects on stream-associated invertebrates (Meyer and Sullivan, 2013) as well as frogs (Hall, 2016), but this will not be covered in detail as little is known about it at this time. Here we discuss some general concepts of toxicology, causes and effects of pollution by inorganic and organic contaminants, and thermal pollution. We also cover mitigating solutions. Nutrient pollution has had a strong influence on aquatic systems; however, we will discuss this in Chapter 18.

BASIC TOXICOLOGY

Exposure to toxic substances can either come in large pulses over a short period of time (*acute*) or in low doses over a long time (*chronic*). Responses can be *lethal* or *sublethal* (not causing death) and can result from *instantaneous* or *cumulative* (a response to numerous events) reactions to exposure. Studies of toxicity require accounting for variability in responses of organisms. Thus, the *lethal dose*, the amount ingested that causes death, is labeled with a subscript that indicates the percentage of animals killed (e.g., LD_{50} is the lethal dose for 50% of the animals tested). Organisms can also be exposed to toxic substances through the water, including absorption across cell membranes, gills, and skin. Therefore, toxicologists also report lethal concentrations (e.g., LC_{50}). *Effective concentration* is the concentration that causes some effect other than death (e.g., on reproduction, growth, behavior); a subscript also denotes the percentage showing the effect (Mason, 1996).

Some toxicants have negative effects on reproduction while having little influence on the general health of the adult organism. These compounds can cause complete extinction of a population, but the effects may be difficult to demonstrate with standard laboratory tests (i.e., the LD_{50} is much higher than environmental concentrations). An example of deleterious effects on reproduction is the response of certain waterbirds to 1,1-dichloro-2,2-bis(p-chlorophenyl)ethylene (DDE), which is a

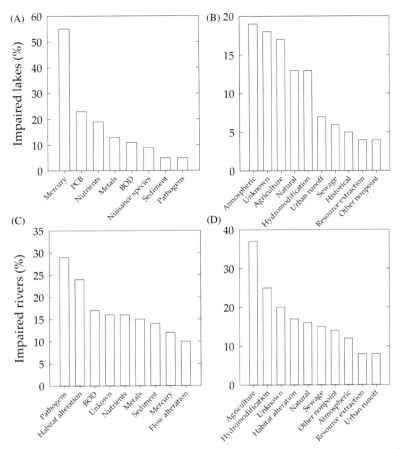

Figure 16.3 Percentages of impaired lakes (A), causes of impaired lakes (B), percentage of streams impaired (C), and causes of stream impairment (D) in the United States by type. *Data from the US Environmental Protection Agency for 2004 National Water Quality Inventory.*

metabolite of dichlorodiphenyltrichloroethane (DDT). DDE causes the birds to lay eggs with thin shells, leading to reproductive failure (Laws, 1993) and extirpation of local populations. This effect almost led to the extinction of the bald eagle; it still threatens many migratory birds, particularly in parts of the world where DDT is still used.

Some chronic effects do not occur immediately. Such delays are particularly the case with mutagenic substances in which prolonged exposure increases the chance of deleterious mutations. If these mutations lead to formation of cancerous cells, a substance is termed carcinogenic.

Several additional issues are important with regard to estimating the influence of pollutants on aquatic organisms and humans. Extrapolating effects to low concentrations of pollutants can be a problem. A threshold below which contaminants are not harmful may occur. A minimum threshold of toxicity is expected to be the case if an organism can repair a limited amount of damage caused by a toxicant, if it can be excreted up to some limited rate, or if the compound does not interact with biological molecules below some concentration. Scientists can have difficulty detecting such a threshold for toxic chemicals because the effects might be subtle, require long periods of exposure, or are sensitive to other environmental parameters. Identifying thresholds is particularly important in regulating human carcinogens in the environment. If no threshold exists, then scientists can extrapolate to very low concentrations of materials to predict the number of deaths or illnesses with exposure of a human population over many years. If there is a threshold, then exposure to levels below the threshold should not cause problems. Understanding human effects is more difficult as controlled exposure experiments are not ethically acceptable, so researchers need to use proxy organisms to study toxicity.

Low concentrations of toxicants may actually stimulate biological activity (Calabrese and Baldwin, 1999). This stimulation could be a negative stressor. The fact that stimulation or depression of biological activities can occur, and that the consequences are context dependent, further complicates regulation of a toxic substance and estimation of long-term effects. Such effects mean extensive testing of each suspected toxic substance with large sample sizes and extended exposure, while considering the life history of target organisms, is necessary before release into the environment. Nontoxic factors such as benign chemicals and temperature can alter toxicity, adding another level of complexity. Obviously, it is difficult to predict toxicity of a compound when it is a function of several other variable environmental factors. For example, zinc toxicity is greater for fishes in high temperatures and in low conductivity water (Fig. 16.4). Extrapolating laboratory results such as those from Fig. 16.4 to real world effects may yield inaccurate results; hence, a combination of field and laboratory approaches may be best for assessments of toxic substances (Blus and Henny, 1997).

It is difficult to know what the influence of two toxicants will be on one another. In some cases they may alleviate the influence of each chemical alone (*antagonism*), but in others the effects may be strictly *additive*. In the worst-case scenario, the sum of the effects is greater than simply adding the individual toxic effects (*synergistic*). Direct testing is generally necessary to establish an interactive effect. Additionally, toxic materials can alter sensitivity to other factors. For example, metal toxicity can increase susceptibility to ultraviolet radiation (Kashian et al., 2007).

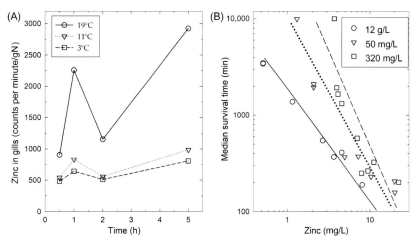

Figure 16.4 Effects of temperature on uptake of zinc into salmon gills (A) and influence of calcium carbonate (mg/L) on mortality of trout exposed to various concentrations of zinc (B). *Redrawn from Hodson (1975) and Lloyd (1960).*

Assessing and predicting the effects of any single substance in freshwater habitats is increasingly complex because pollutants rarely occur singly and may interact in antagonistic, additive, or synergistic ways with myriad other contaminants or natural environmental gradients. Aquatic species often suffer from *multiple stressors*. For example, fish and invertebrate communities inhabiting streams draining agricultural fields experience altered hydrology, degraded physical habitat structure, pesticide runoff, excess nutrients, and numerous other stressors, which may interact in an overwhelming array of ways. One study tested the interactions of increased sediments, nutrient addition, and temperature in stream mesocosms. The results were complex and not always predictable; effects of toxin interactions tended to be synergistic for populations and antagonistic for communities (Piggott et al., 2015). Toxicologists generally avoid this problem by performing laboratory studies on one or a few pollutants at a time. While this reductionist approach is necessary and critical for understanding potential toxic effects of substances, the environmental relevance of these approaches is decreasing as freshwater habitats receive increasing numbers of pollutants and other stressors. A review of 88 papers suggests that additive, synergistic, and antagonistic responses all occur in freshwater environments (Jackson et al., 2016). Somewhat comfortingly, they found that the most common response was antagonistic, indicating that there is a tendency for some stressors to cancel each other out.

Biota can concentrate toxicants. The first step in this process is *bioconcentration*, or the ability of a compound to move into an organism from the water. *Bioaccumulation* refers to the bioconcentration plus the accumulation of the compound from food. This process can lead to toxicological effects even if environmental concentrations are low. *Biomagnification* refers to the entire increase in concentration from the bottom to the top of the food web. Biomagnification is a particular concern with lipid-soluble organic contaminants and some metals. Less water-soluble organic compounds are more concentrated by organisms in laboratory studies (Fig. 16.5). However, the compounds that concentrate the most are only moderately hydrophobic and animals metabolize them slowly, as determined by inspection of 1,500 records of biomagnification (Walters et al., 2016).

Bioconcentration and bioaccumulation factors can be difficult to determine for animals and plants in their natural environment. Factors influencing uptake and retention of a contaminant (such as metabolic rate, rate of assimilation of contaminated food, heterogeneous distribution of the pollutant, and rate of excretion of the contaminant) can all depend on a variable environment. Some compounds that do not bioconcentrate in fully aquatic organisms can bioconcentrate efficiently in air-breathing organisms living in or near the aquatic environment because such compounds have volatile phases than can be transmitted in the air but are nonpolar so they dissolve poorly in water (Kelly et al., 2007). Despite the uncertainties, biomagnification is a well-documented problem and pollutants can be concentrated many millions of times, even if the range of concentrations and bioaccumulation factors is wide (Table 16.1).

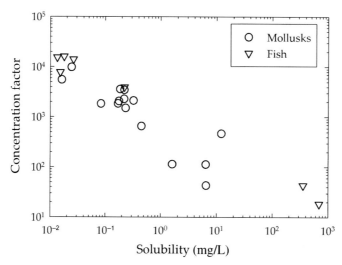

Figure 16.5 Relationships between water solubility and bioconcentration factors of various organic compounds in fishes and mollusks. *Adapted from Ernst (1980).*

Table 16.1 Range of concentrations (parts/trillion) and approximate biomagnification factors calculated for DDT and PCBs in Lake Ontario

Chemical	DDT	PCBs	DDT bioconcentration	PCB bioconcentration
Water	0.3–57	5–60	1	1
Benthic sediments	25,000–18,000	110,000–1,600,000	4,200	26,000
Suspended sediments	40,000	600,000–6,000,000	1,400	100,000
Plankton	63,000–72,000	110,000–6,100,000	2,300	94,000
Fish	620,000–7,700,000	1,378,000–7,000,000	143,000	278,000
Herring gull eggs	7,700,000–34,000,000	41,000,000–204,000,000	719,000	3,710,000

Source: Ranges from Allan (1989).

BIOASSESSMENT

Aquatic organisms, particularly invertebrates and fishes, are very useful for assessing the acute and chronic effects of pollutants because the diversity of organisms present and their characteristics correlate with the degree of pollution and environmental gradients (Loeb and Spacie, 1994). For instance, data on many stream invertebrate species (Fig. 16.6) can be used to demonstrate two possible responses to environmental extremes. This evaluation based on diversity of organisms is *bioassessment*. In the case of O_2 and pH, diversity is maximal at intermediate values (pH about neutral, O_2 about 8 mg/L). This is an example of Shelford's "Law of Tolerance" we discussed in the last chapter. In contrast, chloride and turbidity are always harmful as diversity is greatest at the smallest concentrations. These data illustrate that biodiversity can serve as an indicator of environmental conditions.

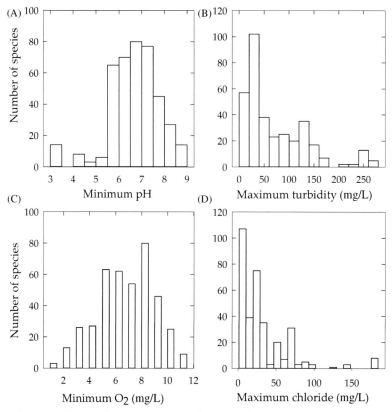

Figure 16.6 Number of invertebrate species as a function of pH (A), turbidity (B), minimum O_2 (C), and chloride (D). *Data from Roback (1974).*

Specific indices based upon more refined taxonomic characteristics are most reliable. Some species or groups are commonly found in eutrophic situations (e.g., cyanobacteria dominate eutrophic lakes, and *Tubifex* worms inhabit sewage outfalls) and others are sensitive to specific environmental factors (e.g., amphibians are susceptible to many types of pollution, and salmonid fishes are limited by water temperature and O_2 concentrations). Bioassessment methods are a common tool for resource managers monitoring water quality and overall ecosystem health. Researchers and managers have developed standard methods that use algae, plants, invertebrates, fish, and even riparian birds for use in streams, lakes, and wetlands (e.g., Karr, 1981; Danielson, 1998; Gerritsen et al., 1998; Barbour et al., 1999; Hilsenhoff, 2017).

A basic community-level indicator of stream health and water quality is the total number of insect taxa in the groups Ephemeroptera, Plecoptera, and Trichoptera (EPT). More species of EPT taxa usually inhabit cleaner waters because these groups are generally intolerant of pollution and low dissolved oxygen. The Index of Biotic Integrity provides a detailed rating for freshwaters using fishes as indicator species. It is a measure of stream quality composed of 12 indicators, including total number of fish species, pollution-tolerant species, food web structure, and fish condition (Karr, 1991). Such indices are useful for determining the suitability of habitats for supporting aquatic life and discerning chronic effects of pollutants, and become more useful when calibrated for specific regions.

Many indices include some measure of tolerance, whereby individual taxa are assigned tolerance values based on their distributions across pollution gradients (Hilsenhoff, 2017). For example, a species only found in pristine waters would be assigned a score of 0—1 (low tolerance for pollution) and a species found in highly polluted waters would be assigned a tolerance value in the range of 8—10. These values are applied to abundance and diversity data from field samples of fishes, invertebrates, or other groups to calculate an average tolerance value for the community in a habitat.

A more comprehensive approach combines several indices into multimetric indices for bioassessment. Here, individual metrics that assess diversity, community structure, and tolerance are tallied and combined into an overall rank or score of biotic integrity that is compared to other regional water bodies, historical data, or modeled predictions of biotic integrity for a given region. Habitat quality data aid in interpretation; if the physical habitat of a stream is degraded (e.g., a channelized stream with a concrete bottom and little natural habitat), poor biotic index scores might not be related to water quality *per se*.

Ecological function can also indicate the ecological health of freshwater systems, particularly litter decomposition (Gessner and Chauvet, 2002) and ecosystem metabolism (Fellows et al., 2006; Young et al., 2008). Decomposition and metabolism are functions governed by many physical and biological features and processes in a system and could relate more closely to specific ecosystem services. Most methods of biological assessment focus on elements of ecosystem structure that we refer to as biotic integrity (e.g., community composition and habitat quality). Wallace et al. (1996) studied invertebrate community and ecosystem responses to an experimental chemical removal of most invertebrates from a headwater stream. They found that both the EPT metric and North Carolina Biotic Index (based on taxa tolerance values similar to the Hilsenhoff index) tracked leaf litter decomposition and seston generation closely. As standard, efficient, and cost-effective methods are developed, future assessments of freshwater habitat health may include more functional measures along with standard structural metrics.

ORGANIC POLLUTANTS

There are millions of known organic compounds; more than 100,000 have been created and used by humans, and billions of years of evolution have led to many more. Humans synthesize at least several hundred new chemicals each year. The large number of compounds makes regulation difficult. Modern society has a consistent record of releasing toxic organic compounds into the environment, only to determine afterwards that they have negative effects on ecosystem and human health. We know almost nothing about how complex mixtures of these compounds at low concentrations will influence human health (Schwarzenbach et al., 2006). The effects of unregulated release of pollutants into a large ecosystem are exemplified by the experiences in the Great Lakes of North America. Problems associated with pollution of these lakes peaked in the 1960s, and the slogan "Lake Erie is dying" served as a rallying point for concerned citizens (Highlight 16.1). Fortunately, our society has mitigated the problems to some degree, though organic contaminants linger in the system and eutrophication and species introductions remain as problems.

The use of organic compounds in agriculture is widespread (Nowell et al., 1999). Worldwide, about 2.3 million metric tons of pesticides are used yearly, and in the United States about 630 different active compounds are employed. Corn, cotton, wheat, and soybean crop management accounts for about 70% of the insecticide use and 80% of the herbicide use in the United States. About 25% of the pesticide use occurs in urban settings, such as on lawns and golf courses (Miller, 1998). Use of pesticides by individuals in suburban areas is generally unregulated beyond recommendations on product packages, which can lead to overuse. The amounts of pesticides found in the waters of the United States have not decreased since the 1990s, and the

Highlight 16.1 "Lake Erie Is Dying"

Until the 1960s, most municipalities and industries surrounding the lake dumped sewage and other wastes into Lake Erie or rivers feeding it without treatment. The lake seemed so large as to be unaffected by such releases. As the population grew, the problems associated with the releases, such as organic chemical contamination, pathogenic bacteria, and eutrophication, worsened. Such problems led to public pressure to clean up the lake (hence the slogan "Lake Erie is dying") and confrontation between citizens, state and federal government officials, and entities causing the pollution (Kehoe, 1997).

Total loads of phosphorus increased fivefold from 1900 to 1970, leading to eutrophication problems. In this sense, the lake was not dying but actually becoming more productive as the phosphorus and nitrogen inputs stimulated algae. This stimulation of algae led to undesirable accumulations of the benthic filamentous green alga *Cladophora* that fouled beaches (Burns, 1985). Some areas of the lake became anoxic and taste and odor problems developed because of algal blooms.

Loading of mercury, lead, cadmium, copper, and zinc increased greatly, with sediment contents 12.4, 4.4, 3.6, 2.5, and 3 times greater, respectively, than in presettlement times (Burns, 1985). Mercury contents of fishes became so high that they were not healthy for human consumption. Inputs of toxic metals from industry have decreased in the past few decades, but contaminated sediments continue to cause problems.

The Great Lakes Water Quality Agreement of 1978 listed 22 hazardous or potentially hazardous organic compounds that were polluting the lake. Of these, polychlorinated biphenyls (PCBs), DDT, and dieldrin caused the greatest concern. DDT use was restricted in 1970, and the concentrations in the smelt taken from the lake decreased from 1.59 to 0.04 μg/g between 1967 and the late 1970s. Low levels of DDT contamination continue because DDT is sequestered in the sediments and slowly reenters the food webs. Manufacture and use of PCBs has been illegal in the United States since 1976; in 1978, PCBs were entering Lake Erie at about 0.9 metric tons/year, with the majority coming from atmospheric deposition. A decade later, fishes in Lake Erie had enough PCB content that consumption of more than 5 kg of fish per year was unsafe (Burns, 1985). The invasive zebra mussels now bioconcentrate PCBs and pass them on to the waterfowl that consume them (Mazak et al., 1997).

Human activities did not kill Lake Erie, but severe damage was done and it still has problems. Runoff from agriculture continues to cause eutrophication problems leading to large toxic blooms. In 2014, a toxic bloom forced people in Toledo Ohio to use bottled water out of concerns about toxic substances in their drinking water. Such blooms continue to plague the lake. The system is a good example of how multiple human impacts on a lake can decrease the value for recreation, fisheries, and drinking water. With careful stewardship, the water quality in the lake will improve, but slowly.

amounts found in urban streams have increased substantially since that time (Stone et al., 2014). A survey of 38 streams across the United States in 2014 found pesticides were the most common contaminant (Bradley et al., 2017). Annual costs estimated over 20 years ago associated with the use of pesticides include $1.8 billion for cleaning groundwater, $24 million in fishery losses, and $2.1 billion in losses of terrestrial and aquatic birds (Pimentel et al., 1992). There is no reason to think these costs have decreased since this time, though we are aware of no more recent estimates.

Pesticides have clear negative effects on stream animals. Effects may extend to microbial communities (DeLorenzo et al., 2001). Stream invertebrate larvae can bio-concentrate pesticides, and adults move the contaminants to riparian areas after they emerge. The riparian predators, such as spiders, ingest the toxic substances (Walters et al., 2008; Laws et al., 2016), and insectivores eat the contaminated animals. A broad survey of pesticide effects on stream invertebrates across Australia and Europe indicated widespread impacts on diversity. The authors detected significant decreases of diversity even when pesticide concentrations were within the regulatory allowed limit (Beketov et al., 2013).

There is an increasing array of chemical types used as pesticides. In particular, widespread use of organophosphates, carbamates, pyrethrins, and organochlorines, which are all neurotoxins used to control a variety of pest species. These contaminate many freshwater habitats and species, even in relatively undisturbed regions. Agricultural insecticides are found in freshwaters globally, are rarely monitored, and when they are monitored, concentrations exceeding regulatory limits are common (Stehle and Schulz, 2015). Amphibian population declines in the Sierra Nevada Mountains correlate to the pesticides used in the agricultural Central Valley of California. Wind and precipitation transport pesticides used in the valley into freshwater habitats in the mountains (Sparling and Fellers, 2009).

Eliminating use of pesticides in the near future is unrealistic, and so careful regulation of their use and disposal is critical. Despite their negative consequences for freshwaters, pesticides and herbicides have had tremendous positive effects on humanity. Increased agricultural production to feed the world's human population is possible, in part, because of these compounds. Pesticides also aid in control of a variety of insect vectors of important diseases. Control of *Anopheles* mosquitoes has led to eradication of malaria in many parts of the world. Pesticides used on aquatic snails have been effective at controlling river blindness, with only moderate effects on non-target aquatic species (Resh et al., 2004).

Although biomagnification of toxic organic compounds is a serious problem, compounds that do not biomagnify as readily can still be of concern. Atrazine is a commonly used chemical for control of weeds in croplands in the United States and elsewhere. It is fairly water soluble (Nowell et al., 1999), persists 6—9 months, and bioconcentrates much less than many other pesticides do, although bivalves used in a laboratory study did bioaccumulate atrazine (Jacomini et al., 2006). The chemical properties of atrazine lead to efficient transfer through the environment (Pang and Close, 1999). Atrazine has seen widespread use in the Midwestern United States, with 32 million kg applied annually, with less use in the European Union where it was banned in 2004. Unfortunately, it is carcinogenic and harms aquatic life (particularly photosynthetic organisms) at levels of $2 \mu g/L$ (Carder and Hoagland, 1998). If agricultural producers discontinue the use of atrazine, they may use worse compounds (Vighi and Zanin, 1994). Given its persistence and water solubility, better management practices are necessary to keep atrazine from entering the surface waters in many agricultural regions.

The glyphosate herbicide (N-(phosphonomethyl)glycine), Roundup, has seen significant use, and has been the number one selling herbicide in the world since 1980. The use of this herbicide has increased because biotechnologists genetically modified crops to be resistant to the chemical (Roundup-ready). Commercially produced Roundup also contains a surfactant that is toxic to wildlife in addition to the glyphosate. Roundup is toxic to many aquatic wildlife species, and is particularly toxic to amphibians (Relyea, 2005).

Genetically-modified crops have also been engineered to produce toxins to deter herbivores. A particularly widespread crop with a genetic modification is Bt modified maize (corn). Bt corn is a variant of maize that has been genetically altered to express proteins from the bacterium *Bacillus thuringiensis*. These proteins are toxic to pests such as the European corn borer, which can cause serious damage to crops. The potential negative effects of these modified crops on aquatic organisms that come in contact with the pollen and residues are well established and have generated a substantial amount of controversy (Highlight 16.2).

Some of the toxic organic compounds found in aquatic systems readily move through the atmosphere. Volatile persistent organochlorine compounds occur worldwide (Simonich and Hites, 1995). The compounds condense from the atmosphere depending on temperature, with the most volatile organics condensing in the polar regions (Wania and Mackay, 1993) or at higher elevations. Atmospheric transport, in combination with biomagnification over a long food chain, can account for the unusually high concentrations of the toxic organic compound toxaphene in fish collected from a remote subarctic lake (Kidd et al., 1995). One would assume this lake is a pristine habitat because it is far from civilization. The fact that a toxic organic compound contaminates fishes in the lake illustrates the pervasive impacts of humans on aquatic environments.

Highlight 16.2 Transgenic crops and freshwater habitats

Considerable controversy has occurred regarding transgenic crops and potential adverse effects on the environment and human health. Although humans have been genetically modifying plants and animals for agriculture and other purposes through selective breeding for centuries, we now have the technology to make changes that are more radical over shorter periods, including transfer of genes across broad taxonomic lines.

Corn has been genetically modified to express crystalline protein toxins initially derived from the genes in the bacterium *Bacillus thuringiensis* (*Bt*). The proteins expressed by transgenic *Bt* corn plants are endotoxins, which are toxic to many common agricultural pests because they bind to receptors in the gut and cause lethal septicemia. These genetically engineered crops have been at the center of controversy for decades.

The bacterium *Bacillus thuringiensis* and its associated endotoxins have been used since the 1930s in the form of dried spores and crystal toxins to control agricultural pests. Its use increased in the 1980s as resistance to synthetic insecticides increased among agricultural pests. Ironically, *Bt* was developed primarily through organic farming because it is naturally occurring and different forms of the toxin affect specific insects, mostly those related to the Lepidoptera. It was not until *Bt* endotoxin genes were integrated into plant genomes to create genetically modified *Bt* corn and other crops that controversy began. Genetically modified *Bt* crops generally express the endotoxins in all plant tissues and can be used to control pests ranging from the European corn borer to corn rootworm, depending on variety.

In the late 1990s, a study by a group of scientists at Cornell University sparked heated debate over the potential adverse environmental effects of *Bt* crops. Losey et al. (1999) found that monarch butterfly larvae that consumed milkweed leaves with *Bt* corn pollen on them had much higher rates of mortality and lower growth than those fed milkweed leaves with non- *Bt* corn pollen. This study was highly publicized and subsequently heavily criticized for being somewhat preliminary and unrealistic in terms of the amounts of pollen to which the caterpillars were exposed. Debate over potential environmental impacts has continued since, with some subsequent studies showing adverse effects, and others showing no effects.

In 2007, a group of researchers at Midwestern universities began looking at whether *Bt* crops could influence food webs and ecosystem processes in streams draining Midwestern agricultural fields with ever increasing proportions of *Bt* crops. *Bt* corn represented over 75% of the corn planted in the US in 2017. Rosi-Marshall et al. (2007) examined detritus and pollen from *Bt* corn with toxins that target the European corn borer (Lepidoptera). The endotoxin protein is detectable in crop detritus from crop fields for at least 240 days (Zwahlen et al., 2003), and some of this material makes its way into streams via wind and water movements (Griffiths et al., 2017). Rosi-Marshall et al. (2007) and Chambers et al. (2010) found that significant amounts of crop residues, including corn leaf detritus and pollen, entered streams bordering agricultural fields. They also found that these materials still had active *Bt* endotoxins in them, the toxins could be transported significant distances downstream, and that caddisfly larvae, which are closely related to the target lepidopteran pest, were adversely affected in laboratory feeding studies using pollen and leaf material. Some of these studies were criticized for being preliminary and the authors were criticized for overstating the results. However, much of this criticism came from individuals whose research was funded by the industry producing and marketing *Bt* crops.

A recent study found that *Bt* toxins were widely detected in agricultural streams in the Midwestern US, but that the toxins degraded rapidly. However, despite its rapid degradation, the toxin was commonly found in stream water because of chronic inputs from crop fields. As such, Griffiths et al. (2017) considered *Bt* toxins *pseudo-persistent* in the environment, a term applied to chemicals with short have lives but chronic inputs.

Bioassays using *Daphia magna* showed toxic effects at far lower concentrations than expected (de Souza Machado et al., 2017). However, the possible effects of *Bt* toxins in aquatic ecosystems remain controversial and relatively poorly studied. *Bacillus thuringiensis* var. *israelensis* (B.t.i.) has been used to control black fly larvae in streams for years and studies indicate few effects on non-target aquatic insects (Jackson et al., 1994). The USEPA has asserted that not enough *Bt* toxin could enter the water to cause harmful effects on aquatic invertebrates (USEPA, 2005), but results of Rosi-Marshall et al. (2007) seem to contradict this. Further, the USEPA's stance that there should be no adverse affects in freshwater habitats is primarily based on 48-hour toxicity tests on *Daphnia* performed by scientists employed by the company developing *Bt* crops.

As research on possible negative impacts of *Bt* crops on freshwater habitats progresses, the costs and benefits need careful consideration. Recent field examinations of invertebrate communities from streams draining *Bt* and non-*Bt* corn fields showed no patterns; agricultural streams and the communities that inhabit them are subjected to myriad stressors including nutrients, sediments, hydrologic alterations, and channelization, and thus Bt toxins may be relatively inconsequential (Chambers et al., 2010). Further, a comprehensive assessment of the environmental impacts of transgenic *Bt* crops should consider the environmental and economic consequences of the alternatives, ranging from traditional pesticide applications to yield reductions that would occur with no pesticide use.

Petroleum products are another common source of aquatic contamination in many parts of the world. Urban runoff is a significant source of oil contamination, with about 1 g per person per day (Laws, 1993). Multiplying this by the U.S. urban population of 200 million yields 7.3×10^{10} g (about 14 million gallons) of oil entering aquatic habitats per year. Urban runoff is now treated in some developed countries, so this source of contamination is decreasing in these areas. In countries with increasing automobile use, the source is increasing. An unknown portion of the oil that enters freshwaters is consumed by microbes or flows to the ocean; the absolute damage to freshwater aquatic habitats is unclear. Another common source of contamination is leakage from underground gasoline storage tanks into groundwater. Cleaning spills from such leaks has cost billions of dollars. In addition, there are numerous pipelines used to transport oil across land and freshwater. These pipelines inevitably leak, particularly as they get older and can cause substantial environmental and economic damage, particularly if they enter freshwaters. These incidents are happening worldwide from the tropics to the Arctic (Jernelöv, 2010).

Hydraulic fracturing (Highlight 4.2) for oil recovery can cause release of organic compounds if the fluids used to pressurize the oil formations are released into the environment (He et al., 2017) or with leakage of the petroleum products that are being extracted (Vengosh et al., 2014). It is difficult to study release of the industrial fluids because many of the companies do not disclose what chemicals they use. One study demonstrated lethal and sub-lethal effects of the fluids on *Daphnia magna* (Blewett et al., 2017). Trout bioaccumulated organic compounds from fracturing fluids (He et al., 2017).

Oil and gas also leak into aquatic ecosystems from outboard engines used on watercraft. Visible slicks of oil and gas are common around busy marinas. Engine exhaust also pollutes water. Two-stroke engines release more pollution than four-stroke engines. The organic compounds in the exhaust of both engine types can kill zooplankton and bacteria. A 15-kW (20-hp), two-stroke engine that operates for 1 hour makes $11,000 \, \text{m}^3$ of water undrinkable by causing bad taste and odor. Expensive treatment is required to reverse these effects (Jüttner et al., 1995).

Chlorinated hydrocarbons such as polychlorinated biphenyls (PCBs) are of concern in aquatic systems because of their persistence in the environment and their toxicity. Production of PCBs, which were used in a variety of industrial applications as coolants, lubricants, and liquid insulators, was banned for most purposes in the United States in 1979. However, PCB residues persist in sediments of freshwater habitats and organisms, particularly near urban areas. Analyses of biological samples in 2009 from across the United States indicated that PCBs were the most abundant, occurring in 93% of the samples (Batt et al., 2017). Before regulations banned their use, large-scale production and industrial use of PCBs resulted in severe pollution of some water bodies. The Hudson River in New York had significant PCB contamination, and citizens and local governments are now paying a high price because of closure of commercial and recreational fishing, and massive ongoing cleanup operations. Many regions of the world have shut down fishing and issued fish consumption warnings because of the presence of PCBs in fish tissues. Because they bioaccumulate readily, PCBs make their way into riparian predators such as birds, spiders, and amphibians, which feed on insects emerging from contaminated water bodies (Maul et al., 2006; Walters et al., 2008).

Additionally, many municipal sewage plants used to treat their final effluent with chlorine to kill pathogens. The chlorine reacts with dissolved organic materials and forms chlorinated hydrocarbons. These compounds are known carcinogens and toxins. Many municipalities are switching to ultraviolet radiation treatment schemes instead.

A common way to clean up spills of organic materials in the environment is *biore-mediation* (Anderson and Lovley, 1997). Bioremediation uses organisms that can break down or inactivate pollutants. In some cases, people add organisms to do the job, and in other cases, native bacteria have the ability to degrade the organic pollutant. Some bacteria can metabolize novel organic compounds, an adaptation that arises because evolution favors microbes able to use unique carbon. Evolutionarily, this would have mostly been metabolites synthesized by other microorganisms. The number of individual bacteria is high, and their generation times are short. The probability that an individual microbe will have a mutation that allows use of a unique source of organic carbon is low. However, the probability that one of the millions of bacteria found in each mL of water will have a beneficial mutation that helps metabolize the compound is substantial. These features of bacteria lead to the rapid establishment of new genotypes capable of using pollutants.

Bioremediation is probably most important in cases of contaminated groundwater because spills of any size are extremely difficult to remove from underground, particularly if the compounds are not water-soluble and are associated with sediments. Several strategies for bioremediation exist, including pumping the water and treating it at the surface, using plants that bioconcentrate the compound taken into their roots, addition of engineered microbes to the aquifer to consume the pollutants, and use of *in situ* microbial activity to eradicate pollution. In most cases, workers release surfactants (compounds that decrease the ability of organic compounds to associate with solid surfaces) to dissociate the compounds from the sediments. They also commonly add nutrients and dissolved oxygen to groundwaters to stimulate microbial activity. An understanding of the ecology of groundwaters is useful in optimizing rates of bioremediation. For example, protozoan populations can decrease rates of bioremediation by consuming bacteria that would break down the pollutants (Kota et al., 1999).

The ability to metabolize or inactivate toxic substances is often coded upon plasmids (small circular pieces of DNA that are free in the cytoplasm), which can move within and among microbial species and allow transfer of genetic information. This lateral transfer is of concern in relation to genetically engineered microbes but also may be helpful in bioremediation efforts. In many cases, bacteria capable of metabolizing an organic compound disappear quickly upon release into a contaminated site, but the indigenous bacteria acquire the plasmid that codes for proteins that can degrade the pollutant. Movement of plasmids among natural populations of bacteria is well established.

We use so many organic chemicals that it is difficult to predict what their effects will be and where they will occur. For example, flame-retardants are commonly used on clothing and household materials. When researchers studied the Columbia River, they found 21 different flame retardant chemicals in the water (Schreder et al., 2014). How these compounds will affect the biota of the river is unknown, particularly given how many of them we release. A study of pesticide and personal care contaminants in Spain indicated complex interactive effects on structure of the periphyton community (Ponsatí et al., 2016).

Pharmaceuticals, personal care products, and endocrine disruptors

Humans produce and use numerous chemicals in their daily lives that have biological activity, and many of these eventually end up in freshwater habitats. Humans use at least 80,000 organic chemicals (Pimentel, 1996) and there is increasing concern over the widespread occurrence and potential environmental effects of these emerging contaminants in freshwater habitats (Ternes, 1998; Kolpin et al., 2002). These substances wash directly into the environment during storms when wastewater systems are overwhelmed, pass through sewage treatment plants, and run off from farms where livestock are treated with antibiotics and other drugs. A global analysis of the literature suggests that antibiotics, painkillers, antidepressants, blood lipid regulators, and other cardiovascular drugs are the most commonly found pharmaceuticals (Hughes et al., 2012). Significant amounts of antibiotics, hormones, disinfectants, fragrances, caffeine, and other substances are excreted or dumped by humans, and then enter wastewater treatment facilities that are not designed to remove them (Daughton and Ternes, 1999).

Researchers also routinely detect illicit drugs in wastewater (Petrie et al., 2015). For example, Lee et al. (2016) found amphetamines at stream sites near Baltimore, MD. They also documented that amphetamines in streams alter bacterial and diatom communities and negatively influence invertebrates.

The average residence time for a given compound in a wastewater treatment facility ranges from less than 1 hour to a few days, which is shorter than the degradation half-lives of many of them (Halling-Sørenson et al., 1998; Xia et al., 2005). Even natural estrogens, which are easy for organisms to degrade, can pass through sewage treatment plants (Liu et al., 2016). In addition to the known compounds, pharmaceuticals break down in the environment due to biological activity or abiotic factors such as UV light; there are few studies on the compounds formed as the pharmaceuticals decompose (Petrie et al., 2015). Aymerich et al. (2016) studied attenuation of 8 pharmaceuticals and 11 transformation products in wastewater plants and their receiving waters. They found only 5 of the 19 compounds were reduced to below 10% of their original concentrations.

A survey of US surface waters found numerous pharmaceuticals and personal care products, including hormones, caffeine, antacids, and painkillers at detectable levels (Barnes et al., 2002; Buxton and Kolpin, 2002; Kolpin et al., 2002). While some of the increased attention to these substances has likely resulted from improved analytical procedures for detecting their presence, there are likely ecological and human health consequences, particularly in urban streams where sewage effluent can dominate discharge.

Most measured concentrations of pharmaceuticals and related contaminants are relatively low (e.g., less than 1 part/billion). The products can bioconcentrate; Arnnok et al. (2017) found pharmaceutical concentrations up to 3,000 times greater than background in Great Lakes fishes. However, different products vary greatly in their observed bioaccumulation (Chen et al., 2017). They found antidepressants and their metabolic byproducts at the greatest concentrations in the brains of the fishes. Bioaccumulation of pharmaceuticals also occurred in fishes from a stream in Czechoslovakia (Grabicova et al., 2017).

Chronic exposure to low concentrations may result in sublethal effects including changes in behavior, growth, or reproductive capacity (De Lange et al., 2006). Richmond et al. (2016) exposed stream organisms to low concentrations of antidepressant drugs. They found that the drugs suppressed primary production by 29% and eplilithic respiration also decreased. Additionally, the chemical exposure altered development of dipteran midge larvae.

Some of these compounds could persist in the environment for many years; a Swedish study documented that the antianxiety/insomnia drug oxazepam persisted in lake sediments in active form for 40 years (Klaminder et al., 2015). Science knows little about the potential ecosystem-level effects of these compounds (Rosi-Marshall and Royer, 2012).

While many consequences of pharmaceuticals and other emerging contaminants in freshwaters are not well established, there is evidence that organism assimilate them and the compounds can have deleterious effects (Kinney et al., 2008; Vajda et al., 2008). Some emerging contaminants are *endocrine-disrupting compounds* in that they act as biological signals. The presence of endocrine disruptors in the environment is of great concern, as these substances can seriously influence the development and reproduction of organisms (Highlight 16.3).

The presence of antibiotics in freshwater is a concern because of their potential role in development of antibiotic resistant bacteria, particularly pathogenic forms, and possible negative impacts on microbially mediated ecological processes (Halling-Sorensen et al., 1998; Maul et al., 2006). Researchers have detected numerous antibiotics in streams receiving sewage effluent and those located below confined animal operations such as cattle feedlots. Intensive farming with manure additions leads to widespread occurrence of antibiotic resistant bacteria in New Zeland (Winkworth-Lawrence and Lange, 2016). Bacteria resistant to multiple antibiotics are now common in aquatic environments (Leff et al., 1993; McKeon et al., 1995). Bacteria resistant to human-synthesized antibiotics have been isolated from many rivers and billabongs in remote rural areas of Australia (Boon, 1992); both are regions with low human densities.

Highlight 16.3 Ecoestrogens: compounds that mimic natural hormone activities

Numerous organic compounds can mimic natural metabolic compounds, leading to endocrine disruption (Sonnenschein and Soto, 1997; Stahlschmidt-Allner et al., 1997). An example of this form of pollution is the release of compounds that mimic estrogen (variously called oestrogens, ecoestrogens, or environmental estrogens). These compounds include pesticides and even ingredients in sunscreens (Schlumpf et al., 2001; Klann et al., 2005).

Exposure to the pesticide DDT correlates positively to nonfunctional testes in male alligators, and other reports of feminized wildlife have begun to surface. In this case, DDT behaves like estrogen; this adds a new dimension to the documented effects of organic compounds intentionally released into the environment (McLachlan and Arnold, 1996). Endocrine-disrupting compounds influence reproduction of fishes, birds, mollusks, mammals (Colborn et al., 1993), and reptiles (Crain et al., 1998). Other possible cases of influence of environmental estrogens include male fishes in polluted waters that produce abnormal amounts of the egg yolk protein normally produced by female fishes and sex reversals of turtles when exposed to estrogenic chemicals. Ecoestrogens can bioaccumulate and be passed to offspring (Crews et al., 2000).

Water samples from 34 of the 35 steam sites sampled across the United States reacted with estrogen receptors in laboratory assays (Conley et al., 2017). An examination from 1995 to 2004 of fishes in 111 US water bodies by the US Geological Survey found that 33% of smallmouth bass and 18% of largemouth bass were intersex, in that they had both male and female reproductive structures (Hinck et al., 2009). While the exact cause for the intersex condition of so many fishes has yet to be determined, estrogens from human birth control pills and other sources are the most likely cause because rates were highest in densely populated regions and examinations of museum specimens collected decades ago show no intersex individuals. However, naturally occurring materials (oak leaves) and salt pollution can alter sex ratios of tadpoles (Lambert et al., 2016), so determining exact causes of these changes in the environment can be challenging. On the positive side, the number of intersex fishes decreased downstream of a sewage treatment plant that had been upgraded to improve efficiency (Hicks et al., 2017).

Some researchers attribute the highly controversial reports of reduced human sperm counts to environmental chemicals, and others report no effects. Meta-analysis is inconclusive on the effects of ecoestrogens on human sperm counts and more research is needed before conclusive results are available (Perry, 2008). Endocrine-disrupting compounds have also been linked to formation of human cancers (Gillesby and Zacharewski, 1998). Apparently, combinations of organic chemicals can also activate the estrogen receptor (Arnold et al., 1996). Such inadvertent biological signaling may have far-reaching and unpredictable effects in aquatic habitats. Fortunately, standard water purification techniques can remove ecoestrogens (Fawell et al., 2001). Regardless of the strength of an individual claim, the topic of ecoestrogens illustrates that wholesale release of organic contaminants into the environment can have unintended effects.

A study found significantly higher numbers of tetracycline-resistant bacterial gene types in water from wastewater treatment plants as compared to nearby lakes (Auerbach et al., 2007), indicating that human antibiotic use can stimulate the prevalence of antibiotic resistance. Antibiotic resistance genes were associated with both human sewage and animal feeding operations in a large study on the Platte River (Pruden et al., 2012). Antibiotics are used as a routine addition to livestock feed in the United States because they increase growth in healthy animals. In 2006, the European Union banned the feeding of all antibiotics and related drugs to livestock for growth promotion (but allowed them to treat sick animals). Increasing reliance on aquaculture for fish production around the world is further contributing to the problem; fish farms can use large quantities of antibiotics and other drugs, and aquaculturists apply them directly to water. The escalating prevalence of antibiotics or microbes exposed to antibiotics (e.g., animal feedlot runoff) that enter freshwater environments increases the probability that microbes that cause human disease will possess plasmids coding for resistance to antibiotics used to treat those diseases. This is because harmless bacteria in the environment can develop and transmit these plasmids into disease-causing bacteria. Furthermore, viruses that attack bacteria can transmit the genes for antibiotic resistance, moving resistance from benign bacteria into disease-causing bacteria (Calero-Cáceres and Muniesa, 2016).

Plastics

Plastics are very resistant to microbial degradation, and can occur in the environment as large pieces (>5 mm, macroplastics), as microparticles of plastics (1 μm—5 mm, *microplastics*), or as even smaller sub μm particles (*nanoplastics*). Smaller pieces form when larger pieces break down or as additives to commercial products (exfoliants in personal care products, cleaning products, and for industrial cleaning processes). Exposure to UV or high temperatures makes plastic more fragile and more likely to break down. An additional source of microplastics is microfibers, which are released from synthetic clothing when it is washed and dried, with most released during drying (Pirc et al., 2016). Most plastic is disposed of on land, but some moves out of the terrestrial system into freshwaters or is directly deposited there. Commonly used plastics, such as polystyrene or polyethylene are highly resistant to biodegradation and Horton et al. (2017) claimed that essentially all the plastic produced by humans is still in the environment in one form or another. There have been calls for bans on using microbeads in personal care products and elsewhere (Rochman et al., 2015). The United States banned them in 2017, and a few other countries worldwide have as well. Bans on plastic bags are also becoming more common.

Research on microplastics is most advanced in marine systems (e.g., Green et al., 2017), but microplastics are common in freshwater environments (Eerkes-Medrano et al., 2015; Horton et al., 2017). Microplastics in rivers are most common near urban environments. Levels are variable in the water column but concentrations in freshwater sediments are similar to those found in marine sediments. Microplastics pass through animal guts, but there is some evidence they can cross the gut out of the digestive tract into the animal. Studies on copepods revealed that nanoplastics were more lethal than microplastics. In freshwaters, aged nanostyrene plastics were more toxic to *Daphnia* than were unaged particles. Terrestrial earthworm studies have also revealed negative effects, suggesting that aquatic oligochaetes could have similar responses (Horton et al., 2017). There are pigments associated with microplastics, so they are optically active along with paint chips that enter freshwaters (Imhof et al., 2016). These pigments could alter the optical properties of lakes, leading them to appear less clear and potentially intercepting light that would fuel primary production.

Microplastics also serve as habitat for microbial organisms. McCormick et al. (2014) studied a highly urbanized river in Chicago and found very high concentrations of microplastics originating in sewage treatment plants. High-throughput sequencing indicated microbial communities associated with the microplastics were less diverse and distinct from other habitats within the river. They also found evidence for plastic-degrading microbes and pathogens. Hoellein et al. (2017) studied microplastics downstream from a sewage input. They found the bacterial communities associated with microplastics became more similar to those associated with natural particles in the river when sampled further from the sewage input. Microplastics can also alter the growth rate and community composition of microbial primary producers (Yokota et al., 2017).

Microplastics can be sources or transporters of pollutants. Many of the chemicals used to create plastics (plasticizers) are toxic. These chemicals leak into the environment from the plastics as they break down or are exposed to environmental factors. Microplastics bind hydrophobic compounds, and can concentrate pollutants such as PCBs, dioxin, and heavy metals. If animals, including people, ingest these plastics there is risk of exposure to toxic compounds (Horton et al., 2017). However, one review suggests that biomagnification from eating other organisms is still more important than exposure from ingested plastic particles with bound harmful organic compounds (Koelmans et al., 2016).

ACID PRECIPITATION

Contamination by acid precipitation has had enormous environmental and economic impacts on aquatic systems. Loss of fisheries and concomitant loss of many tourist dollars are common in affected areas. Here, we discuss sources of acid, distribution of the problem, biological effects, and potential solutions to problems associated with acid precipitation. There are additional sources of acid contamination that are not specifi-cally covered, such as mine drainage (Gray, 1998) and natural acidic systems, but the generalities of the following discussion apply to pH effects regardless of source.

A recent study demonstrated some surprising, long-term impacts of chronic acidifi-cation on streams. Somewhat contradictory to what one might predict, long-term effects of acidification are actually increasing stream water alkalinity. Kaushal et al. (2013) examined long-term water chemistry patterns in 97 rivers in the eastern United States and found that 62 of them showed significant increases in alkalinity. These increases in alkalinity are a result of the long term accelerated weathering of bedrock in watersheds from acidic rain water, and the trend has been referred to as "rivers on Rolaids" by the researchers who documented the pattern. The ultimate ecological impacts of these changes in stream water chemistry are not completely understood, but the increased alkalinity stimulates growth of some algae and bacteria, and this increased production can lead to decreased dissolved oxygen levels. When ammonia is present in the water, such as near confined animal operations and sewage outflows, the increased alkalinity also drives conversion of ammonia to ammonium, which is less toxic to animals. Although acid rain has decreased since the Clean Air Act, precipita-tion in many regions still remains acid enough to continue the weathering and increas-ing alkalinity.

Sources and geography of acid precipitation

Acid precipitation has far-reaching effects on aquatic ecosystems. It is primarily associ-ated with industrial activity and vehicle use that combust hydrocarbons. Acid rain has acidified lakes and streams in all industrialized regions of the world. The US Environmental Protection Agency surveyed more than 1,000 lakes and 211,000 km of streams during the 1980s. About 75% of the lakes and 50% of the streams surveyed suffered the effects of acid precipitation. The areas most impacted were the Adirondacks, the mid-Appalachian highlands, the upper Midwest, and the high-elevation West. In the worst case, 90% of the streams in the New Jersey Pine Barrens were acidic. In mid-Appalachia, there were 1,350 acidic streams. Furthermore, the Canadian government estimates that 14,000 lakes in eastern Canada are acidic. The Norwegian government sampled 1,000 lakes and found that 52% of the lakes were endangered. In the southern part of Norway, 60%—70% of the lakes had lost their fishes (Henriksen et al., 1990). Although scientists identified the problem in the 1970s and 1980s, and regulators established emission control regulations, biological and chemical recovery has been slow and incomplete (e.g., Arseneau et al., 2011; Hesthagen et al., 2011).

The proximate cause of acid precipitation is sulfuric and nitric acids in rain, snow, and fog. Combustion of coal and petroleum products forms these acids in the atmosphere, which dissolve in the water droplets of clouds, ultimately reaching the ground. Acid precipitation is concentrated downwind from industrial and urban areas because they have more concentrated emissions from factories and automobile exhaust. Acid deposition in the United States has historically been greater in the heavily populated and industrialized northeast. Acid deposition correlates most closely with sulfate deposition, but also with nitrate deposition. The control of acid precipitation is an environmental victory in the United States (Fig. 16.7). The Clean Air Act controlled acid precipitation and recovery has accelerated in the last decade (Strock et al., 2014). While acid deposition has decreased, nitrogen deposition in the form of ammonium (instead of nitrate) has increased (Li et al., 2016), which could have other effects.

When acid precipitation reaches the ground, it can react with the terrestrial ecosystem. If sufficient base is present, it will neutralize the acid. The ability of a soil or water body to absorb acidity without a change in pH is called *buffering capacity*. The most common material that confers the ability to resist changes in pH is the

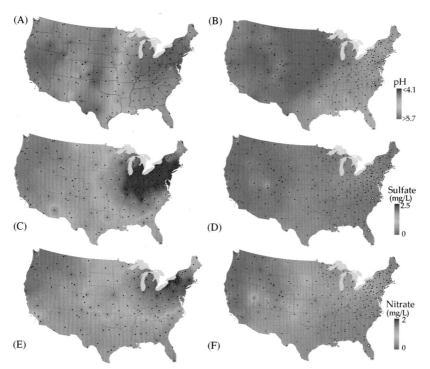

Figure 16.7 Concentrations of acid, sulfate, and nitrate in US precipitation from 1985 (A, C, and E, respectively) and 2015 (B, D, and F, respectively). *Images courtesy of the National Atmospheric Deposition Program (NRSP-3). 2017. NADP Program Office, Illinois State Water Survey, University of Illinois, Champaign, IL.*

bicarbonate in limestone. The bicarbonate equilibrium (discussed in Chapter 13) leads to neutralization of the acid and release of CO_2. Watersheds and aquatic systems that have a significant amount of limestone have a high buffering capacity and are able to resist the effects of acid precipitation. Responses have been variable since regulatory control; anion concentrations such as chloride have decreased in some areas, cation concentrations remain high, and organic acids have replaced inorganic acids to lower pH (Futter et al., 2014). As such, regional recovery across Europe and North America is variable following regulations (Stoddard et al., 1999).

Biological effects of acidification

Acid rain has major effects on biological systems ranging from altered microbial activity to the impaired ability of fishes to survive and reproduce (Table 16.2). Naturally acidic habitats include acid peat bogs (*Sphagnum* bogs) and blackwater swamps (Benner et al., 1989). Lakes and streams in watersheds dominated by such bogs or swamps can be relatively acidic ($\sim$pH 3–4). Some geothermal springs are very acidic and have very distinct microbial communities associated with them. Much of our understanding of the effects of long-term acidification on aquatic ecosystems derives from the study of naturally acidic habitats. Amazingly, an iron-oxidizing archaeal isolate from acid mine drainage can grow at pH 0 (Edwards et al., 2000).

One of the basic ecosystem influences of acidification is the lowered rate of decomposition mediated by microbes. Microbes from the naturally acidic Okefenokee Swamp are less efficient at metabolizing low-molecular-weight carbon compounds relative to those in nearby neutral wetlands (Benner et al., 1989). The microbes in acidic habitats are less able to metabolize recalcitrant cellulose and lignin, although some

Table 16.2 Influences of decreasing pH on several groups of aquatic organisms

Organism or process	Approximate pH value
Bacterial decomposition slows/fungal decomposition takes over.	5
Phytoplankton species decline/green filamentous periphyton dominate.	6
Most mollusks disappear.	5.5–6
Most mayflies disappear.	6.5
Beetles, bugs, dragonflies, damselflies disappear.	4.5
Caddis flies, stoneflies, Megaloptera disappear.	4.5–5
Salmonid reproduction fails, aluminum toxicity increases.	5
Most adult fishes harmed.	4.5
Most amphibians disappear.	5
Waterfowl breeding declines.	5.5

Source: After Jeffries and Mills (1990).

degree of adaptation to the acids does occur (Fig. 16.8). Inhibition of microbial activity by low pH leads to greater rates of deposition of organic material and may partially explain the stable existence of acidic depositional wetlands (i.e., once a wetland sediment becomes acidic, microbial activity maintains acidity and carbon continues to accumulate). Rates of microbial decomposition of leaves are also lower in acidified streams (Clivot et al., 2013, Fig. 16.9), which may increase carbon accumulation and alter the associated food webs.

Acidification alters algal populations. Filamentous green algae characteristically bloom in the littoral zones of acidified lakes. When these blooms collapse, the resulting O_2 depletion can have negative impacts on animals (Turner et al., 1995). Diversity of planktonic and benthic algae decreases with lower pH (Dickman and Rao, 1989). Similar decreases in algal diversity and replacement with filamentous green algae occur in acidified streams (Meegan and Perry, 1996).

Shifts in algal communities in lake sediment cores resulting from pH changes verify historical trends in acidification (Mallory et al., 1998). Such verification is required before politicians are willing to enact stringent and potentially costly emission controls. In this technique, existing lakes serve as a baseline to create an index that correlates current water column algal communities with pH. This index

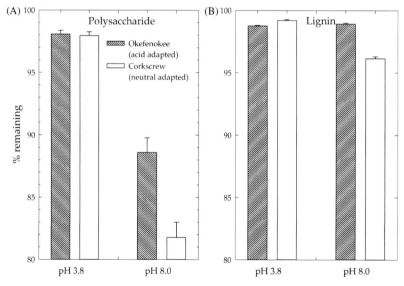

Figure 16.8 Percentage of remaining polysaccharide (A) and lignin (B) compounds after degradation by microbes from a naturally acidic swamp (Okefenokee Swamp in Georgia, pH 3.4—4.2) and a neutral swamp (Corkscrew Swamp in Florida, pH 6—8) after incubation at different pH levels. *Modified from Benner et al. (1989).*

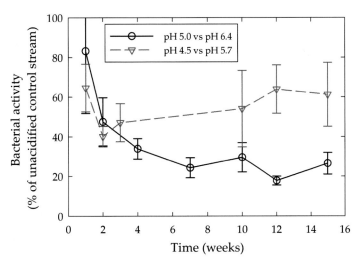

Figure 16.9 Microbial activity on leaves placed in acid and neutral streams as measured by the rate of thymidine incorporation into nucleic acids in two acidified streams compared to two nearby neutral streams. *Modified from Palumbo et al. (1989).*

indicates pH using species with parts that remain preserved in the sediment, such as diatom frustules or chrysophyte scales. Analysis of sediment cores can establish changes in the community over time. The deeper in the sediments, the longer ago the algae were deposited. This index, coupled with isotope analyses to date specific depths of sediments, yields a record of pH in a lake over time (Fig. 16.10). In the case of Big Moose Lake, New York, some chrysophyte species are dominant in low pH, whereas others occurred only at the higher pH values associated with preindustrial conditions.

Diversity of plants and animals also decreases as aquatic systems become more acidic. Macrophyte diversity decreases in low pH lakes and streams (Thiébaut and Muller, 1999). Fungal diversity is reduced in acidic streams (Fig. 16.11). Invertebrates exhibit a wide range of acid sensitivities. Perhaps the most sensitive invertebrates are those that require calcium bicarbonate for shells (e.g., Mollusca). These shells dissolve or are unable to form when pH decreases. However, a few species from these groups are adapted to survive in waters with pH < 5 (Freyer, 1993). As aquatic systems become acidified, biomass and diversity of crustaceans (Fig. 16.12) and other invertebrates decreases. Among aquatic insects, mayflies are particularly sensitive to low pH (Herrmann et al., 1993). In lakes, not only does the diversity of zooplankton decrease with increased acidification, but also the efficiency of energy transfer up the food web is lowered (Havens, 1992a,b).

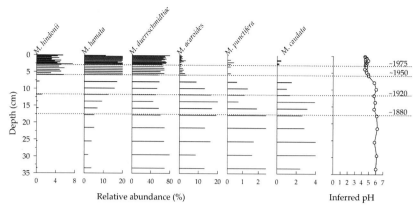

Figure 16.10 Distribution of *Mallomonas* spp. scales (a chrysophyte) with depth and reconstructed pH from Big Moose Lake (New York). Sediments were dated by ^{210}Pb content. *From Majewski and Cumming (1999) with kind permission from Kluwer Academic Publishers.*

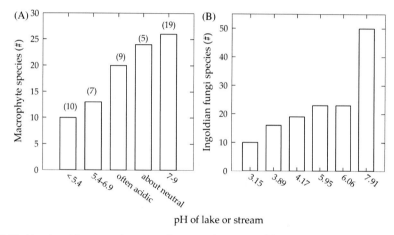

Figure 16.11 Number of macrophyte species as a function of lake water pH (A) and number of species of Ingoldian fungi as a function of stream water pH (B). The numbers of lakes sampled are shown in parentheses. *Data from Hutchinson (1975). (A) and Dubey et al. (1994).*

Fishes are susceptible to acidification. Many examples of pH damages to fishes come from salmonids because they are of the greatest economic importance in the areas most polluted by acid precipitation. One study suggests that economic benefits of fisheries drop from $38 per angler day in lakes with quality trout fisheries to <$4.50 day^{-1} in lakes with pH < 4.5 (Caputo et al., 2017). Acidification increases the concentration of aluminum (Fig. 16.13), which causes damage to fish gills. The low pH increases the toxicity of the aluminum (Gensemer and Playle, 1999). Subsequently,

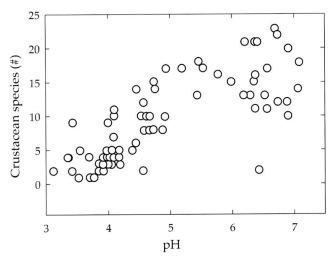

Figure 16.12 Crustacean species diversity as a function of pH. *Reproduced with permission from Freyer (1980).*

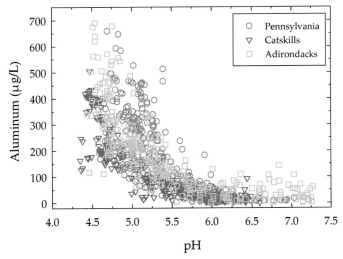

Figure 16.13 Relationship of aluminum concentrations to pH. *Reproduced with permission from Wigington et al. (1996).*

the number of sensitive fishes decreases in acidified waters (Fig. 16.14), and the most acidified waters have no fishes.

Treatments to reverse acid precipitation effects

Several treatments are available to counter the effects of acid precipitation. The most obvious is stopping the source by burning low-sulfur fuels for industry and power

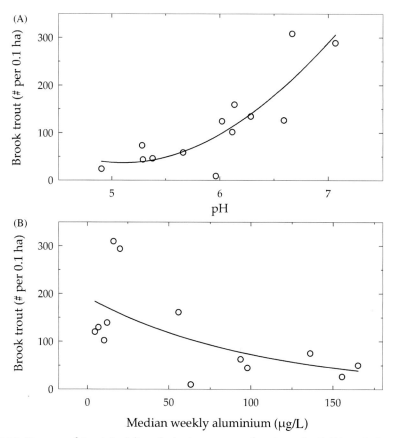

Figure 16.14 Biomass of trout in Adirondack streams as a function of pH (A) and aluminum (B). *Reproduced with permission from Baker et al. (1996).*

production, and decreasing the emission of nitrogen compounds in automobile exhaust. The most common local treatment is to add lime (calcium carbonate) to lakes and watersheds to neutralize the effects of the acid (Fairchild and Sherman, 1990). Directly adding calcium carbonate to the lake causes short-term (years or less) increases of pH and recovery of some biota (Hörnström, 1999). Adding calcium carbonate to the entire watershed may have longer lasting effects but is costlier (National Research Council, 1992). Unfortunately, even if acidification is reversed, losses of calcium and magnesium from soils and internal sulfate cycling may lead to long-term changes in water chemistry (Likens et al., 1996; Miles et al., 2012). Research by Gene Likens has illustrated that knowledge of biogeochemical cycling is important in understanding causes and effects of acid precipitation (Biography 16.2). Another possible solution is to fertilize the lake and allow the biota to reverse the problem

(Davison et al., 1995), but as discussed in Chapter 18, eutrophication has its own problems.

Emission controls have led to decreases in acid deposition in North America and Europe and associated reversals in surface water acidification (Stoddard et al., 1999). The biological community has recovered in some areas (e.g., Arseneau et al., 2011). However, some areas have not recovered in North America; those watersheds received so much acid deposition that they were not able to respond to decreased sulfate loading.

Another cause of acidification of surface waters is mine drainage. Acid mine drainage and metal contamination are related in many instances. Metal pyrites weather when exposed to oxygenated surface waters and metals dissociate with concurrent formation of sulfuric acid. This acid mine drainage results from coal mining and metal mining operations. Treatment options include neutralization with limestone (Hedin et al., 1994), oxidation in wetlands, and various combinations of these treatments (Robb and Robinson, 1995).

METALS AND RADIOACTIVE POLLUTANTS

A wide variety of metals and some other inorganic materials act as toxic pollutants in aquatic ecosystems (Table 16.3). Arsenic, chromium, lead, zinc, mercury, cadmium, and other metals are naturally occurring, but human activities create situations where concentrations are higher than they would be naturally and are toxic to aquatic life. Metals can bioaccumulate in many organisms starting at the base of the food web (Kassaye et al., 2016) and can be bioconcentrated in food chains. Bioconcentration has led to problems such as excessive lead contamination of freshwater fishes throughout all areas with cities. Complex pelagic food webs with many lateral links transfer less metals up the food chain (Stemberger and Chen, 1998), an additional argument for maintenance of biodiversity. Atmospheric deposition and industrial waste releases, particularly mining (Table 16.4), are common sources of metal contamination. Such mining activities have had extensive negative impacts in some aquatic habitats (Highlight 16.4).

Chemical conditions can alter the bioconcentration and toxicity of metals. For example, cadmium, silver, nickel, and zinc uptake by invertebrates is highly influenced by reactive sulfides in sediments (Lee et al., 2000). High-sulfide sediments bind the metals and render them less toxic. Also, the redox state of metals can influence toxicity; hexavalent chromium (oxidized) is much more toxic than trivalent chromium.

Lead toxicity in waterfowl has been a particular concern in freshwater systems because of the historical use of lead shot pellets for hunting. Waterfowl such as ducks, geese, and coots ingest the pellets as grit for their crops. Less than 10 lead pellets will

BIOGRAPHY 16.2 Gene E. Likens

The study of biogeochemistry is argu- ably the most related area of freshwa- ter science to water quality, the links between aquatic and terrestrial habi- tats, and the influence of aquatic pol- lutants. Dr Likens (Fig. 16.15) is one of the foremost contemporary scientists specializing in the biogeochemistry of ecosystems. He has received numer- ous honorary degrees and awards, including the Tyler prize (a World Prize for Environmental Achievement), elec- tion to the US National Academy of

Figure 16.15 Gene Likens. *Courtesy Gene Likens.*

Sciences, and top awards from the American Society of Limnology and Oceanography, the Ecological Society of America, and many other international societies. He has more than 330 publications, including 12 books.

Likens grew up on a farm in northern Indiana, where he fished, collected aquatic organ- isms, and generally enjoyed exploring aquatic habitats. Likens maintains that a love of natu- ral history is the single best predictor of success for an aquatic ecologist. He attended a small liberal arts college (Manchester) and obtained a PhD from the University of Wisconsin. Following a lecture on the conservation of aquatic resources, he told the professor he was interested in the subject and wondered if he could get paid for that type of work. Obviously, the answer was yes.

Fortunately for the aquatic sciences, Likens did not follow his other career goal; he also wanted to be a professional baseball player and played for 2 years in the rookie league in Kansas, a league that also gave rise to baseball great Mickey Mantle. Likens was a most valuable player, but he decided that the life of an academician was preferable to that of a professional athlete.

Likens says he feels lucky to have been able to travel to and study some of the most beautiful places in the world, including Hubbard Brook, where he conducted important research on the influence of logging on nutrient transport by streams (Likens et al., 1978). Hubbard Brook is also the site of much of his research on acid precipitation effects; it has been the site of experimental watershed acidification and produced crucial insights into the long-term impacts of acid leaching of soils (Likens et al., 1996). Likens predicts that a major future challenge in aquatic ecology will be to understand the implications of com- plexity. He suggests we currently do a good job at assessing the influence of one or two fac- tors, but that to truly understand ecosystems we need to account for the simultaneous influence of multiple biotic and abiotic factors.

Table 16.3 Maximum allowable concentrations of some toxic metals in natural waters used by humans for the United States, and potential human health problems associated with each (after Budavari et al., 1989; Laws, 1993)

Metal	Chemical symbol	Maximum conc. (μg/L)	Some responses to acute poisoning	Some chronic effects
Mercury	Hg	0.144	Death within 10 days, severe nausea, abdominal pain, blood diarrhea, kidney damage	Loss of teeth, kidney damage, muscle tremors, spasms, depression, irritability, birth defects
Lead	Pb	5	Anorexia, vomiting, malaise, convulsions, brain damage	Weight loss, weakness, anemia
Cadmium	Cd	10		Cancer, throat dryness, headache, vomiting
Selenium	Se	10		Nervousness, depression, liver injury (this is an essential element in small amounts, but toxic at higher concentrations)
Thallium	Tl	13	Nausea, vomiting, diarrhea, tingling pain in extremities, weakness, coma, convulsions, death	Weakness and pain in extremities
Nickel	Ni	13.4		Cancer, dermatitis, nausea, vomiting, diarrhea
Silver	Ag	50		Bluish color of skin, skin and mucous membrane irritation
Manganese	Mn	50		Languor, sleepiness, weakness, emotional disturbances, paralysis
Chromium	Cr	50		Cancer, skin and respiratory irritation, renal damage (chromium III is not toxic; chromium II is)
Iron	Fe	300		
Barium	Ba	1,000	Excessive vomiting, violent diarrhea, tremors, death	

kill a bird, but marshes frequented by hunters may have 6 or 7 pellets/m^2 in the sediments. For this reason, the US Fish and Wildlife Service has phased out use of lead shot in favor of steel shot (Laws, 1993). In upland areas of the United States, it is still legal to use lead shot and this shot can move into aquatic habitats or water birds, such as geese, that use upland areas part time and can ingest these poisonous pellets. Lead fishing weights are still in use and can cause wildlife deaths. In addition, atmospheric lead deposition increases lead concentration in lakes throughout the

Table 16.4 Some effects on aquatic environments related to various mining activities

Type of mining	Potential impacts
Sulfide ores (copper, nickel, lead and zinc)	Acidification by sulfuric acid, possible arsenic contamination, sediments
Gold and silver	Same as sulfur ores, possible mercury, cyanide or arsenic contamination
Uranium	Acid tailing drainage; runoff of radioactive materials, toxic metals, sediments, and organic compounds
Iron	Heavy water demand; runoff of sediments, toxic metals
Coal	High water demand, high sediment load in runoff, acid runoff from high sulfur deposits
Salt	Salinization of waste water

Source: After Ripley et al. (1996).

world. Analysis of peat bog sediments in Switzerland indicated that anthropogenic inputs increased lead contamination starting 3,000 years ago, and in 1979 deposition rates were 1,570 times the natural background values found prior to 1000 BC (Shotyk et al., 1998). A similar sediment analysis in Michigan indicates lead contamination from copper mining by indigenous people 7,000—8,000 years before the present day (Pompeani et al., 2013).

Mercury is a neurotoxin and microbes transform it to its most toxic state, methylmercury, in anoxic conditions. It is a worldwide contaminant influencing all environments (Driscoll et al., 2013; Ozersky et al., 2017). This is an old problem; mercury accumulation in a Spanish peat bog increased about 2,500 years ago, at a time when mercury mining began in the region (Martínez-Cortizas et al., 1999). In some countries, people use mercury indiscriminately to extract gold in mining operations and can heavily contaminate freshwater systems (Cursino et al., 1999).

Mercury contamination of fish is a problem that has beset many areas. About half the streams with large piscivorous fishes in the Western United States have fishes with mercury concentrations exceeding safe limits (Peterson et al., 2007). Organisms assimilate and concentrate methylmercury in aquatic food webs. Mercury most commonly enters aquatic systems from atmospheric fallout from coal burning, trash incineration, and industrial emissions; once methylated under anaerobic conditions in water-saturated sediments, it readily moves into the food web. Periphyton mats can be an important site of mercury methylation and its entry into the food web (Cleckner et al., 1999). Biomagnification then can cause levels of mercury in fish high enough to warrant consumption advisories. Such restriction on consumption may be problematic for people that use fishes as a large component of their diet (Egeland and Middaugh, 1997).

Highlight 16.4 Massive contamination of the Doce River (Brazil) and the Clark Fork River (United States) by mining waste

In 2015 a dam containing mining wastes burst, releasing a wave of toxic mud that spread down the Doce River in Brazil and killing 20 people (Garcia et al., 2017). This toxic waste decimated the biological communities of the river leading to a loss of $521 million/year of ecosystem services. The pollution moved down the river and eventually contaminated Atlantic coastal waters. The fines assessed to the mining company did not cover the estimated costs of all losses. Furthermore, experience from the United States indicates these costs will accrue for many decades to come. In January 2019 another mine dam collapsed in Brazil in the state of Minas Gerias, leading to deaths of hundreds of people and massive environmental damage.

Over 100 years of mining (primarily copper) in the region of Butte, Montana, has resulted in numerous contamination problems in the Clark Fork River. Mines discharged waste from the mines directly into the river or allowed it to wash into tributaries. The operations discharged an estimated 99.8 billion kg of waste into the system prior to 1959, and 2 or 3 million m^3 of contaminated sediments is present in the floodplain. Contamination has affected the upper 200 km of the river.

Mine operations installed treatment ponds to trap wastes over the years, and initiated liming to precipitate metals in the waste in 1959. However, cadmium, copper, lead, and zinc in the water column continue to exceed criteria for protection of aquatic life. Metal waste that drains into the groundwater influences microbial communities (Feris et al., 2004). This extensive contamination has led to designation of the upper Clark Fork River as a Superfund site. Cleanup of the site started in the late 1980s and continues, costing US$123 million as of 2016.

Historic fish surveys in 1950 showed no fishes in regions of the upper Clark Fork River. With improved water quality, fisheries managers have reintroduced trout into the upper river. However, thunderstorms cause episodic contamination events during high discharge, and significant fish kills happened in 1983−1985 and 1988−1991. Physiological abnormalities of fish and concentration of the contaminants in the food chain still occur. The data suggest that metal contamination problems are likely to defy attempts at remediation for significant periods after contamination (Phillips and Lipton, 1995). The Milltown dam, upstream of Missoula, Montana, is filled with contaminated mine waste that accumulated in 1908, shortly after it was built. The dam is currently being removed and sediment is being removed from the river. These two examples illustrate the potential impacts of mining on water quality.

Eutrophic systems can have less severe problems with production and concentration of methylmercury in the food web (Gilmour et al., 1998), but generalizations may be difficult since the relationships between organic C and methylmercury concentrations are complex (Hurley et al., 1998). A global synthesis of mercury magnification in food webs suggested that biomagnification was greatest in cold, low productivity high-latitude systems; additionally, it was greater when dissolved organic C concentrations were higher (Lavoie et al., 2013).

Mercury is one of the major problems influencing the Everglades, where public health officials have issued fish consumption advisories because methylmercury concentrations in fish tissues exceed 30 ng/g (Cleckner et al., 1998). A landscape-level study found that concentrations of mercury in tissues of most species of stream invertebrates increased significantly with stream size, and this pattern was attributed to in-stream production of methylmercury along the stream network. This production was greatest in *Cladophora* mats (Tsui et al., 2009).

Selenium has caused severe problems in some wetlands. Irrigation mobilizes selenium naturally found in soils and concentrates the selenium as the water evaporates. In Kesterson Reservoir, a National Wildlife Refuge in central California, selenium contamination caused congenital deformities and mass mortality of waterfowl. Although selenium is a required nutrient in trace levels, it bioaccumulates and becomes toxic at higher concentrations. The US Geological Survey has identified about 500,000 km^2 in the western United States that are susceptible to similar problems. The worst cases occur where irrigation runoff is reused for irrigation, and water ends up in terminal wetlands or lakes. Such lakes and wetlands have no outlets and concentrate selenium by evaporation.

Arsenic can cause problems because it can be present in high concentrations naturally or as runoff from industrial uses. Historically, arsenic was a common pesticide and subsequently contaminated aquatic systems. There are widespread problems with arsenic in drinking water in China; an estimated 19.6 million people are at risk for contaminated drinking water (Rodríguez-Lado et al., 2013). In a particularly horrible case, thousands of drinking water wells in West Bengal, India, contain naturally occurring arsenic (Bagla and Kaiser, 1996). An estimated 200,000 people in this area have arsenic-induced skin lesions and hardened patches of skin that may become cancerous. More than 1 million Indians may be drinking this contaminated water. Successful management includes providing safe drinking water and uncontaminated food (groundwater-irrigated crops concentrate the arsenic). However, in a study where clean water and food were provided for 5 years, moderate cases of skin lesions were reversed, but severe cases of lesions and chronic lung disease were not (Bhowmick et al., 2018).

The West Bengal problem could be related to recent large-scale withdrawal of groundwater for agriculture, leading to rapid fluctuations in groundwater level and input of O_2, which allows for release of the arsenic from sulfides in the pyrite-rich rocks of the area. Phosphorus from fertilizers also increases arsenic release rates. Microbes can oxidize arsenic (Oremland and Stolz, 2003) and could control arsenic release, which starts after iron is oxidized from the system (Islam et al., 2004). The microbial effect provides a potential partial explanation for the phosphorus effect since phosphate tends to bind with iron. More research by groundwater geochemists and hydrologists is needed to study this problem and find solutions.

We mine rare earth elements for various technical uses. We know little about how they concentrate in the environment and their biological effects. Amyot et al. (2017) studied food webs in 14 lakes in Quebec that were not subject to mining in their watersheds. They summed the rare earth elements analyzed in the biota, including Y, La, Ce, Pr, Nd, Sm, Eu, Gd, Tb, Dy, Ho, Er, Tm, Yb, and Lu. They found clear trends of biomagnification and bioconcentration with concentrations in fishes several orders of magnitude greater than in the water, but concentrations in lake sediments greater than in biota.

Radioactive compounds can be contaminants of water. These usually occur naturally. The primary contaminants are isotopes of radium, radon, and uranium. Radium contaminates approximately 1% of drinking water supplies with radium above acceptable levels, and radium and uranium occur in significant concentrations in many surface and groundwaters. Numerous human deaths are attributed anually to exposure to radium (6−120), uranium (2−20), and radon (80−800) in the United States (Milvy and Cothern, 1990).

The effects of natural radioactive materials on aquatic habitats are difficult to gauge. Laws (1993) states a made up example to illustrate why we do not know much about biological effects of many pollutants. "While the deaths of 250,000 Americans out of a population of 250 million (i.e., 0.1%) might seem an alarming statistic to many persons, the loss of 0.1% of a population of crabs or tunicates would not be likely to cause much public alarm." Consequently, we do not know the effects of many contaminants on aquatic organisms, given the limited information on the effects of most chemicals on humans. Most of the research on the influence of radioactive compounds on aquatic habitats has occurred downstream from nuclear power-generation plants. Perhaps the greatest concern is with biomagnification; body tissues retain many radioactive isotopes, and concentrations increase with each increase in trophic level.

NANOMATERIALS

Nanomaterials are very small molecules or clusters of molecules (on the nanometer scale), often of materials that usually occur in large aggregations. Nanomaterials can be organic or inorganic. While these compounds can occur naturally, people are now engineering these particles and the can ultimately enter freshwaters. The products have many benefits due to their unique properties, but environmental research on their influences is in its infancy (Wilson, 2018). Small clusters of molecules behave differently from larger aggregates because of quantum mechanical properties. For example, silver particles are more toxic at nanosize because the silver is more likely to release into solution, and titanium oxide nanoparticles (but not larger titanium oxide particles) react with UV light to form hydroxyl radicals that are harmful to cells. We discussed nanoplastics earlier in this chapter. While the unique properties of these materials make them useful in many applications, and humans are using them, their effects on the freshwater environment are unclear. There is some concern about the effects of these products on human health and how they will enter our drinking water supplies (Troester et al., 2016). Understanding their influences in freshwaters is difficult because there are so many types, they have different modes of action, they can interact with other pollutants, they can be difficult to work with in environmentally realistic ways, and they interact with the environment (e.g., light), which alters their toxicity (Bundschuh et al., 2016).

Some commonly used nanomaterials include titanium dioxide (TiO_2), silver particles, zinc oxide (ZnO), carbon nanotubes, fullerenes, and cerium dioxide (CeO_2). These particles generally occur in freshwaters in $<1\,\mu g/L$ except for TiO_2, which is commonly found in the 10s of $\mu g/L$ (Bundschuh et al., 2016).

Titanium dioxide is used as a whitener in products such as paints, sunscreens, foods (e.g., powdered sugar doughnuts), and many other applications. The use of standardly ground TiO_2 powders is being rapidly replaced by nanosized particles. Bacteria exposed to nanosized TiO_2 take up the particles and have lower metabolic capacity (Combarros et al., 2016). A pulse of TiO_2 to an experimental stream system caused an immediate decrease of bacterial abundance with recovery over three weeks. Analysis of 16S rRNA genes suggested no influence on the composition of the bacterial community, indicating that the effects were very broad spectrum and did not target specific groups of bacteria (Ozaki et al., 2016). Additionally, TiO_2 exposure can increase accumulation of toxic arsenate in *Daphnia magna* (Li et al., 2016).

Silver is one of the most toxic metals in freshwaters, and the use of nanosilver particles as an antimicrobial agent has increased dramatically. Nanosilver is synthesized in particles less than 100 nm diameter, and the tiny particles are more toxic than larger particles. Hundreds of products, such as child toys and textiles, contain silver nanoparticles and much of the silver will eventually enter aquatic environments. While silver nanoparticles are undoubtedly reaching freshwater environments, and they are strongly toxic to microbes, researchers know little about how large their effects are (Zhang et al., 2016). These particles can come in several forms, all of which accumulate in duckweed (Stegemeier et al., 2017). Silver particles at environmentally relevant concentrations can alter behavior of the snail, *Physa acuta* (Bernot and Brandenburg, 2013). Particles in short term laboratory experiments can have deleterious effects on bacteria, but one lake mesocosm study found only modest effects of nanosilver (Blakelock et al., 2016).

Carbon materials (e.g., bucky balls, carbon nanotubes, graphene compounds) are becoming more and more prevalent. While these compounds are relatively unreactive, the consequences of their release into the environment are a mystery. Carbon nanotubes can bioconcentrate in *Daphnia* (Petersen et al., 2009). Graphene nanomaterials are probably more toxic than carbon nanotubes to green algae (Zhao et al., 2017). Influence of nanomaterials in freshwaters is certainly a ripe area for future research, as people tend to release compounds into the environment first and worry about the effects later.

SALT POLLUTION

Salinization of freshwaters by human activities is widespread and has strong negative consequences for biota (Cañedo-Argüelles et al., 2016). A variety of human activities can increase the salinity of freshwater habitats. Applications of salt to roadways and sidewalks can directly influence nearby freshwater habitats when snow and ice melt (Kaushal et al., 2005), and has salinized many lakes in North America (Dugan et al., 2017). A study in Ontario, Canada indicated that ~50% of the total salt applied to a major highway eventually washed away with overland flow and the remainder entered shallow groundwater habitats, resulting in degradation of groundwater resources (Meriano et al., 2009). Studies in the northeastern United States indicate that the salinity of many urban and suburban streams has been increasing for decades and already exceeds lethal limits for many freshwater organisms (Fig. 16.16) (Kaushal et al., 2005). Retention ponds, often used to control floods in urban areas, exacerbate the salt effects by prolonging a slow release of runoff from snowmelt from salted roads (Snodgrass et al., 2017). If these trends continue, many surface waters in the northeastern United States will be unfit for human consumption within the next century.

Direct applications of salt are not the only cause of increasing salinity of freshwaters. Agricultural and urban runoff, as well as point sources such as sewage effluent, can increase salinity of receiving waterways by increasing concentrations of chloride and other ions. Dewatering of rivers, groundwaters, and coastal wetlands facilitates salt intrusion into formerly freshwater habitats in coastal regions (Fig. 4.7). Industrial activities, such as production of soda ash, generate large quantities of waste products that

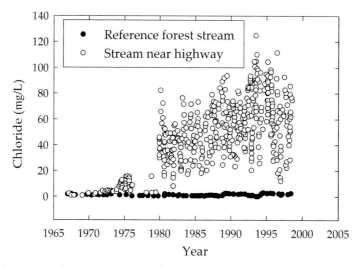

Figure 16.16 Long-term data showing a significant increase in baseline chloride concentration in a stream following highway construction and a nearby control stream with no roads and no change in chloride concentrations over the same time period. *Adapted from Kaushal et al. (2005).*

can greatly increase salinity in regional freshwater habitats if not carefully contained. Groundwater extraction activities, including coalbed methane extraction, pump deep groundwaters to the surface that are often naturally saline, which can contaminate surface and shallow groundwater habitats.

Agricultural irrigation can increase salinity of downstream waters. Continued application of water will cause concentration of salts as evaporation leaves salt behind. A slight excess of irrigation water is added to wash away concentrated salts. Runoff from such practices leads to higher concentrations of dissolved salts. Such problems affect many drier irrigated areas; most of the irrigated lands in California's Imperial and San Joaquin Valleys have salinity problems, and salts and agricultural chemicals contaminate runoff.

Climate change may also increase salinity of freshwater habitats in some regions as increased frequency and magnitude of droughts and warmer temperatures enhance evaporation of surface waters, concentrating solutes. In some regions, such as the Murray-Darling River Basin of Australia, which has been experiencing increasing salinity for decades, identification of specific causes has proven difficult because of complex interactions, mostly among weather, geology, and human activities. However, natural weathering processes are contributors to some degree (Morton and Cunningham, 1985; White et al., 2009).

Regardless of the ultimate cause, increased salinity has detrimental effects on aquatic life for reasons discussed above; osmotic stress can result in slower growth, reduced reproduction, and ultimately death, depending on the degree of pollution and tolerance of individual species. Unlike most marine habitats, salinity of freshwater habitats can be quite variable in time and space, and thus tolerances of individual species vary greatly. Above certain levels (e.g., above 2.0 g NaCl/L), few species can persist in salt-polluted habitats. Chronically salt-polluted habitats have reduced biological diversity and ecosystem functions such as primary production and decomposition rates. Even modest chronic additions lower stream microbial leaf decomposition rates (Tyree et al., 2016).

SUSPENDED SOLIDS

Turbidity and suspended solids or sediments are natural parts of all freshwater environments. Some habitats are naturally highly turbid, but human activities have increased levels of suspended solids in many habitats (Fig. 16.3). Agricultural and urban runoff, watershed disturbance (e.g., logging, construction, and roads; Forman and Alexander, 1998), removal of riparian vegetation, alteration of hydrodynamic regimes, and introduction of species such as common carp that stir up sediments all can lead to anthropogenic increases in total suspended solids. In addition, alterations in flow can increase sediment loads, and understanding the sources and causes of excess sedimentation is complex (Belmont et al., 2011). Arctic warming and melting of permafrost will also increase sediment pollution (Chin et al., 2016), and clearing channels of macrophytes to increase water flow or navigability can lead to high sediment concentrations (Greer et al., 2017).

Sediments can have different biological and physical effects depending on the type of suspended solids (Table 16.5). High concentrations of suspended solids can lower primary productivity of systems by shading algae and macrophytes, at times leading to almost complete removal. Suspended solids can also have negative effects on aquatic animals by interfering with reproduction, respiratory O_2 transport, filter feeding, and habitat availability. The negative impact of excessive sediments on stream biota is clear (Hynes, 1970; Waters, 1995). Such sediments lower incoming light and primary production, increase scour, harm sensitive invertebrate species, reduce biodiversity, and lower the aesthetic values of streams and lakes. Excess sediments can also decouple the food web in streams. For example, Louhi et al. (2016) found that sediment additions to artificial stream channels reduced the effects of predators on grazing invertebrates, altering a top—down trophic cascade. Probably the strongest negative effect of sediments is filling and embedding gravel and cobble habitat through deposition, leading to reduced water and oxygen flow through substrata and increased anoxia. Sediment

Table 16.5 Classification of suspended solids and their possible impacts on freshwater systems

Type of solid	Physical and chemical influences	Biotic influences
Fine inorganic particles	Decrease light, impede flow, increase scour, metals, and other ions can associate	Clog respiratory organs, reduce metabolite flux, light limitation, alteration of habitat
Fine particulate organic matter (natural)	Clogging sediments, decreasing oxygen flux. Lower redox	Food source for heterotrophs, stimulation of respiration
Fine particulate organic matter (sewage)	High nutrient, carbon, and potential metal content. Lower redox	Food source for heterotrophs, stimulation of respiration
Fine particulate organic matter (microplastics)	Potential light absorption, disruption of flow	Carry toxicants, filter feeders may be deceived
Toxicants on particles	Movement of hydrophobic compounds, survival of viruses	Toxicity and disease

Source: After Wilber (1983).

clogging can harm interstitial invertebrates and eggs of many spawning fishes. Lowering flow through gravels can cause O_2 levels to decrease below levels necessary for eggs to develop successfully. Salmonids are sensitive to sediments (Wilber, 1983), but some other species are tolerant (e.g., many catfishes and carps).

Light attenuation in lakes may comprise a large part of the influence of suspended solids on the biota. A highly turbid lake or reservoir may have limited rates of primary production. However, if there is sufficient organic material in the suspended particles, a productive food web based on microbial use of the suspended particulates can occur.

THERMAL POLLUTION

Human alterations to natural temperature regimes of freshwater habitats cause thermal pollution. Thermal pollution is any deviation from the natural temperature in a habitat and can range from elevated temperatures associated with industrial cooling activities to discharges of cold water into streams below large impoundments. Given that the metabolic rates of ectotherms are controlled by temperature, and that the vast majority of freshwater organisms are ectothermic, thermal pollution can strongly affect freshwater communities. Alterations to normal water temperature regimes have myriad biological effects, including interfering with temperature cues for spawning fishes, facilitating establishment of exotic species, and altering growth and development of aquatic organisms (Langford, 1990). Further, aquatic organisms evolved in relatively thermally buffered environments, and thus they are generally more sensitive to temperature fluctuations than are their terrestrial counterparts.

Temperature tolerances among species of freshwater organisms are highly variable, but all have an optimal range and low and high limits within which they can survive. Increases in temperature cause an increase in growth rate up to a point. Above some threshold, damage occurs. Because temperature governs rates at multiple levels of biological organization (e.g., from enzymatic reactions to metabolism of whole organisms), changes in temperature associated with thermal pollution ultimately influence rates of ecosystem processes and functions such as nutrient cycling and decomposition.

Most forms of thermal pollution involve temperature increases, and while the effects of extreme temperature increases are obvious, relatively small changes can also be biologically significant. We discussed molecular adaptations to temperature in Chapter 15. Temperature increases as little as 1°C−2°C can alter communities because they are lethal to some species and can affect growth and reproduction of others. Raising water temperatures just 2°C−3°C above the optimal for some aquatic insects can greatly reduce the number of eggs produced by females because more energy is used to support higher metabolic rates and less is available for egg production (Vannote and Sweeney, 1980; Firth and Fisher, 1992).

Thermal pollution associated with sewage effluent is linked to differences in leaf litter decomposition rates in Lake Titicaca in South America (Costantini et al., 2004). Trees can die when thermal pollution alters wetlands. As temperatures increase, cyanobacteria dominate over green algae and diatoms. One of the key issues in thermal pollution is the replacement of cold-water fishes with warm-water fishes.

Power plants and industrial factories are major point source contributors to thermal pollution. In this case, plant operators withdraw cool water from streams, use it to cool generators and other machinery, and then return it to the stream at elevated temperatures. Rapid changes in temperature associated with power plant operations can kill fishes by thermal shock (Ottinger et al., 1990). Mitigating the thermal effects of power plant effluent obviously has a significant financial cost. Temperature regimes of small lakes or near-shore portions of lakes are also altered by human activities including effluent from municipal facilities and industry.

Along with industrial sources, urban and suburban runoff can contribute to thermal pollution, particularly during short, intense thunderstorms in watersheds with high amounts of impervious surfaces such as asphalt (Herb et al., 2008). Depending on local groundwater inputs, discharge, and other factors that influence thermal regimes, even small municipal discharges can alter stream temperatures for considerable distances downstream.

Reductions in stream flows or lake volumes alter temperature regimes by reducing the thermal buffering capacity of the water body. Water withdrawals for irrigation, hydroelectric, and other human uses reduce stream flows and lead to significantly increased water temperatures during warm periods. This warming can ultimately result in increased fish kills associated with high temperatures (Caissie, 2006).

Impoundments that release water from the surface can result in higher stream temperatures during warm periods because water velocity is decreased and solar penetration enhanced in the impounded water. Along with the direct effects of warmer temperatures on aquatic life, the solubility of O_2 in water decreases with increasing temperature, so O_2 stress increases as temperatures rise. During cold periods when stream water temperatures are normally near freezing, hypolimnetic releases can artificially warm streams.

Deforestation is also a major contributor to thermal pollution in streams, small ponds, and wetlands, as removal of riparian vegetation greatly increases solar penetration and temperature (Beschta et al., 1987). Small streams and ponds in forested regions are particularly vulnerable because they are normally shaded during warm months and have less thermal buffering capacity. Studies in forested headwater streams show increases in summer maximum temperatures of $5°C-8°C$ after logging, and recovery periods to normal thermal regimes can take $5-15$ years (Caissie, 2006).

Anthropogenic cooling of freshwaters can also have strong effects on aquatic life. Releases of cold water from the hypolimnia of large reservoirs alter stream thermal regimes for long distances. Hypolimnetic releases can greatly alter seasonal temperature patterns, lower maximum temperatures by more than $10°C$, and dampen normal diurnal temperature fluctuations (Baxter, 1977; Ward and Stanford, 1979; Ward, 1985). The overall effect is a dampening of normal diel and seasonal temperature fluctuations. Water temperatures below large impoundments (like the Glen Canyon Dam on the Colorado River in the southwestern United States) are now much cooler in the summer than they were historically. Fisheries managers often stock non-native cold-water fishes such as salmonids in temperate zone streams below large impoundments. This occurs in regions where natural summer stream temperatures exceed the upper limits for most cold-water species. Although important for recreation, these tailwater fisheries (fisheries downstream from dams) differ substantially from natural river systems and are often detrimental to native species.

While many forms of thermal pollution originate at point sources, climate change (discussed throughout this book) represents a nonpoint source of thermal pollution that is already influencing a wide range of freshwater habitats. The long-term impacts of climate change on freshwater habitats remain to be seen, but predictions indicate significant shifts in structure and function of streams, lakes, and wetlands, particularly in higher latitudes. Responses will vary; in a specific example, wetland plant cover will expand on the Mississippi delta with increased temperatures associated with global climate change (White and Visser, 2016).

ANTHROPOGENIC INCREASES IN UV RADIATION

Freshwater organisms living in high light environments have adapted defenses against damaging radiation. However, human activities are increasing UV-B radiation in freshwater habitats through a variety of mechanisms, including reducing atmospheric ozone, which absorbs incoming UV radiation; reducing water depths and thus attenuation of UV radiation; and altering amounts of suspended particles and dissolved organic matter in water columns, which absorb UV radiation. Freshwater habitats in polar regions and at high latitudes are most susceptible, as they have higher exposure.

Ultraviolet radiation harms primary producers (see Highlight 12.1). Increased UV-B radiation adversely affects oceanic bacterioplankton activity (Herndl et al., 1993), lake phytoplankton productivity (Harrison and Smith, 2009), and lake zooplankton communities (Williamson et al., 1994). Increased UV-B exposure also causes to abnormalities and death of amphibian larvae (Bancroft et al., 2008). Exclusion of UV radiation from stream mesocosms in New Zealand increased the abundance of aquatic insects by 54% (Clements et al., 2008). Ultraviolet radiation can interact with other toxic substances such as metals and pesticides, and nutrient levels can determine the level of damage from UV radiation (Carrillo et al., 2008). These interactions can be *synergistic*, in that effects are more than simply additive.

URBANIZATION

We have covered some of the areas where urbanization can influence aquatic habitats, but not in a cohesive fashion. Recently, the concept of *urban stream syndrome* has emerged in the literature as a pressing area of research given the fact that over half of the world's population lives in urban areas. Urban areas also have detrimental influences on lakes and wetlands. Urban stream syndrome includes flashier hydrographs, simplification of stream channels, concrete-lined channels, urban pesticide contamination, runoff of petrochemicals and salt from roadways, increased frequency/intensity of algal blooms, and reduced species richness (Walsh et al., 2005; Wenger et al., 2009; Cunha et al., 2017). Most of the research so far occurred in developed countries, but developing countries have their own set of problems (Booth et al., 2015; Capp et al., 2015; Loiselle et al., 2016). For example, cities do not treat their sewage in many developing countries, so streams become open sewers with the potential for disease transmission.

Urbanization can cause unnatural changes in streams, and understanding them requires comparison to local native conditions of streams. For example, in the desert Southwestern United States, watering lawns leads to runoff and formation of permanent streams where none existed before (Kaye et al., 2006). Stream channels can have open riparian canopy where they would naturally have closed canopy, and vice versa.

With respect to wetlands, the most common result of urbanization is draining and filling for development. However, some cities construct small retention ponds or wetlands to mitigate the flashy runoff caused by impermeable surfaces. Many of these wetlands are not hydrologically or biogeochemically similar to their natural counterparts (Stander and Ehrenfeld, 2009).

Urban lakes are often shallow, hypereutrophic, and have unnatural hydrology and morphology (Birch and McCaskie, 1999; Schueler and Simpson, 2001). Most people never experience a natural lake. These lakes are often used for runoff control and are exposed to many contaminates associated with cities. It is common for people to dam small streams to provide artificial lakes, which people find enticing and can increase residential property values. People want lakes that are crystal-clear and produce many big fish. This is very rare considering the anthropogenic pressures on such lakes and their drainage basins.

SUMMARY

1. Humans influence all surface waters on Earth.
2. Toxicologists are concerned with the acute and chronic effects of pollutants. Natural variations in uptake, sensitivities, concentration, additive effects of different pollutants, and effects of other environmental factors all complicate predictions of how a particular toxicant will influence a specific organism.
3. Biomagnification of pollutants can cause high concentrations of toxic compounds in the tissues of organisms at the top of food webs. The most lipid-soluble compounds usually magnify to the greatest degree. Metals, organochlorine pesticides, and PCBs are examples of highly persistent pollutants that readily bioaccumulate.
4. Bioassessment protocols are tools to assess the impacts of pollutants and other habitat alterations on aquatic communities. Indices for bioassessment are generally constructed using data on invertebrates or fishes that are present and the state of their habitat, but methods have been developed for all types of aquatic organisms and habitats.
5. Acid precipitation is mainly caused by humans burning fossil fuels, leading to increased sulfuric and nitric acid in the atmosphere. Acid mine drainage is also of concern in many regions of the world.
6. Acidification of aquatic ecosystems impacts all aquatic organisms.
7. Under acidic conditions, microbes degrade complex organic compounds more slowly, cyanobacteria and diatoms are selected against, filamentous green algae are selected for, invertebrates that use calcium carbonate become rare, aluminum has

a greater impact on fish gills, and reproduction of many animals is impacted negatively.

8. Metals can have a broad array of negative impacts on aquatic ecosystems. Mining wastes often cause contamination, but runoff from industrial applications and naturally occurring sources can cause problems as well. Mercury contamination of freshwater habitats has resulted in fish consumption warnings in many regions of the world.

9. Tens of thousands of organic compounds are discharged by humans into aquatic habitats, including pesticides, oil, and materials in urban runoff. Only a small percentage of these compounds have been tested for toxicity. In some cases, microbes can break down these compounds (bioremediation) given enough time.

10. Endocrine disruptors act as biological triggers and are linked to reproductive deformities in a variety of freshwater species.

11. Increasing prevalence of antibiotics from human and livestock waste in freshwater habitats leads to development of antibiotic resistant bacteria.

12. Freshwater organisms experience multiple stressors, which may interact in antagonistic, additive, or synergistic ways. This makes identification of the effects of one particular type of pollutant difficult.

13. Nanomaterials are becoming common contaminants with little understanding of their influences in fresh waters.

14. Suspended solids can cause harm to aquatic organisms. Generally, sediments interfere with photosynthesis from increased light attenuation and with respiration by clogging water flow. In streams, fine solids fill up and destroy gravel and cobble habitats, can increase scour associated with high flow, and reduce movement of water into subsurface habitats.

15. Thermal pollution can cause shifts in community structure. These shifts may allow for establishment of exotic species and local extinction of native species.

16. Urban areas have heavily altered freshwater habitats. Lakes, streams, and wetlands in urban areas receive substantial amounts of pollutants and are hydrologically altered; they generally do not represent natural habitats nearby.

QUESTIONS FOR THOUGHT

1. Why is biomagnification worse with lipid-soluble compounds that are resistant to abiotic and biotic deactivation?

2. Are there any habitats on Earth that have not been influenced by human activities?

3. What political conditions have led to a world in which toxicants are released routinely into the aquatic environment before even cursory testing of their effects on organisms, including humans, has been conducted?

4. Why can controls on emissions of greenhouse gasses ultimately decrease acid precipitation?

5. Why can acid precipitation lead to lower iron availability and greater phosphorus in some lakes and wetlands with anoxic sediments?

6. Why is bioremediation of metal contamination more difficult than that of contamination by organic compounds?

7. Why are microbes able to more rapidly evolve ways to inactivate toxicants and develop resistance than are fishes?

8. Under what conditions may suspended solids have positive influences on aquatic ecosystems?

9. Should aquatic scientists assume a role of advocacy with regard to issues of aquatic pollution, or should their role be primarily to provide data for informed decisions to be made by managers and policymakers?

CHAPTER 17

Nutrient Use and Remineralization

Contents

Figure 17.1 Carboys used for an in situ bioassay of nutrient limitation at Milford Reservoir, Kansas. The first four experimental additions (from right to left) are control, N, P, and N + P. One week after addition, the nitrogen + phosphorus treatment had the most chlorophyll.

Nutrients can be critical basal resources in aquatic systems because they often limit primary production or heterotrophic activity. Understanding ecosystem production, nutrient pollution, and interactions among heterotrophs, autotrophs, and their environment requires an elucidation of nutrient dynamics. In this chapter, we discuss how organisms acquire and assimilate nutrients, the relative amounts needed in different systems, availability of nutrients, and the crucial concept of nutrient limitation. How nutrients are made available (recycled) by heterotrophs as the nutrients cycle through the food web of aquatic systems is also discussed. We also cover the ecological ramifications of ratios of nutrients (*stoichiometry*); as we discussed in Chapter 14, nutrient cycles do not happen in isolation.

Organisms need to take nutrients in from the water surrounding them or in food that is consumed (*uptake*) and then incorporated into organic molecules used for growth (*assimilation*). Generally, each of these steps requires energy. In animals that ingest food, uptake and assimilation are not the same; animals excrete many parts of food that are used. Loss of nutrients is excretion and more specifically if it is in inorganic form, *remineralization* (return to the inorganic state). *Mineralization, remineralization*, and *nutrient regeneration* are terms that ecologists use interchangeably.

USE OF NUTRIENTS

In the broadest sense, a *nutrient* is any element required by organisms for growth, but many ecologists use this term primarily in reference to elements that are limiting. For example, even though organisms are composed of large amounts of oxygen and hydrogen, rarely do aquatic ecologists consider these as nutrients because they seldom limit primary production. When oxygen limits ecosystem activity, this is due to lack of availability for use as an oxidant of organic carbon, not a lack of availability as material to build cells. Likewise, primary producers can use CO_2 directly, and many researchers think carbon supply ultimately does not limit ecosystems driven by photoautotrophs, and should not be considered a "nutrient." Stated differently, while carbon is the energy currency for ecosystems, the availability of carbon to build cells is rarely considered limiting for autotrophs. Nitrogen, phosphorus, silicon, and iron are the nutrients most often studied by aquatic ecologists and referred to as nutrients. Organic carbon compounds are, at times, referred to as nutrients needed for heterotrophic growth, and the concept of organic carbon limitation is a central idea in food web and ecosystem ecology.

The major classes of biological molecules require different amounts of nutrients. Lipids are phosphorus and carbon rich. Amino acids and proteins require relatively more nitrogen than most molecules. Nucleotides and nucleic acids require significant amounts of nitrogen and phosphorus relative to carbohydrates, such as starch and sugars, which are composed entirely of carbon, oxygen, and hydrogen. Other elements such as iron are required in very small amounts as cofactors in enzymes. Specialized requirements include calcium for mollusks and silicon for diatoms. Large amounts of phosphorus are required for vertebrate bones. Each organism must acquire nutrients to survive and grow.

Organisms acquire nutrients in a variety of forms. Most primary producers and many bacteria use inorganic nutrients, such as nitrate or phosphate. In a case that stretches the definition "autotroph," algae such as the dinoflagellates, chrysophytes, and euglenophytes can ingest particulate material such as bacterial cells to meet requirements for nitrogen, phosphorus, and some carbon. Most animals and a good proportion of the heterotrophic eukaryotic unicellular organisms acquire their nutrients in organic form (e.g., they must assimilate nitrogen as proteins or amino acids, and carbon as organic molecules).

Nutrients in organic form dissolve in the surrounding water or occur in particulate material (including organisms). Algae can use some of these dissolved organic forms directly (Berman and Chava, 1999). Individual bacteria and archaea have different degrees of ability to use inorganic or organic compounds, but the bacterial assemblage found in natural waters is generally able to use most organic and inorganic nutrient sources (with individual species specializing on recalcitrant or less easily assimilated forms). Here, we consider the general principles of assimilation and uptake of dissolved nutrients. Uptake of materials in the form of particles (e.g., living and dead organisms) will be considered in later chapters. We present three common equations to describe the functional relationships among nutrient concentrations, uptake, and assimilation.

The *Michaelis–Menten* relationship describes the influence of nutrient concentration on uptake rate:

$$V = V_{max} \frac{[S]}{K_s + [S]}$$

where V is the uptake rate of substrate, V_{max} is the maximum uptake rate, $[S]$ is the concentration of substrate (nutrient),[1] and K_s is the concentration of S where $V = 1/2$ V_{max}. The shape of the curve (Fig. 17.2A) reveals that as concentration increases, uptake increases rapidly only at low concentrations. At high substrate concentrations,

---

[1] The term substrate, in most of biology with biochemical or microbiological basis, refers to materials that organisms assimilate for growth (e.g., nutrients and organic carbon) and in chemistry the chemical that is being acted upon in a reaction. In limnology, this creates confusion because substrate is often used to denote solid surfaces that organisms are associated with. For this reason, we follow Wetzel and refer to nutrients as substrate or substrates, and solid habitat as substratum or substrata. Also, the square brackets around [S] indicate the concentration of the substrate S, generally in moles per liter. Moles per liter can be converted to grams per liter by multiplying by molecular weight. Units of molecular weight are in grams per mole.

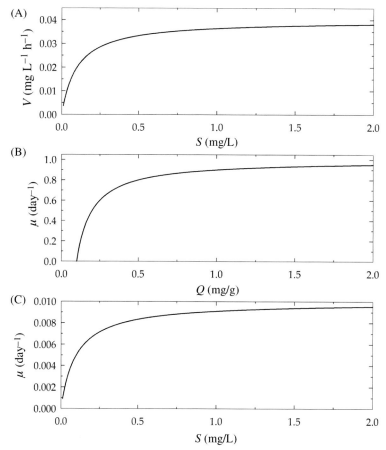

Figure 17.2 Graphical representation of equations used to describe nutrient uptake and assimilation: (A) Michaelis–Menten, (B) Droop, and (C) Monod relationships.

the maximal rate of uptake is approached. Values for K_s vary widely among and within algal taxa (Table 17.1) and may even vary among individual cells as they become adapted to surrounding nutrient concentrations. Values of V_{max} can be a useful physiological indicator (Zevenboom et al., 1982). For example, organisms existing in low-nutrient environments tend to have low values of K_s and those in areas with high nutrient availability have high values of V_{max}. Larger cells have greater values for V_{max} and K_s than do small cells (Suttle et al., 1988). We present a calculation with this equation in Example 17.1.

Organisms cannot always immediately used nutrients taken up for growth. Organisms that are under nutrient stress are able to take up nutrients at very high rates upon transient exposure to high nutrient concentrations but are not able to sustain growth rates proportional to this uptake. This high uptake is a selective advantage

Table 17.1 Some values for Monod nutrient uptake and Droop equations for phosphate (after EPA, 1985); means are followed by ranges in parentheses

Algal type	K_s (mg/L)	μ_{max} (day^{-1})	Q_0 (see sub-title)
$NO_3^- $-$N$ or NH_4^+-N			
Phytoplankton	0.114 (0.0014–0.2)	0.039 (0.0024–0.15)	0.024 (0.015–0.04)
Diatoms	0.051 (0.001–0.13)	0.07 (0.015–0.125)	3×10^{-7} (0.5–6 $\times 10^{-7}$)
Green algae	0.016 (0.0014–0.030)	0.0925 (0.06–0.125)	1.7×10^{-6} (0.5–34.2 $\times 10^{-7}$)
Cyanobacteria	0.50 (0.03–0.98)	0.0825 (0.04–0.125)	1.93×10^{-7} (0.52–4.3 $\times 10^{-7}$)
PO_4^{3-}-P			
Phytoplankton	0.0357 (0.0028–0.07)	0.222 (0.0014–2.95)	0.0014 (0.001–0.003)
Benthic algae	0.125	0.045	0.0005
Diatoms	0.065 (0.0002–0.06)	0.262 (0.024–0.5)	5.06×10^{-7} (0.01–7 $\times 10^{-7}$)
Green algae	0.28 (0.001–1.5)	0.317 (0.133–0.5)	1.7×10^{-9} (1.7–4.5 $\times 10^{-9}$)
Cyanobacteria	0.5 (0.007–0.98)	0.585 (0.042–0.5)	2.91×10^{-9} (0.58–5.66 $\times 10^{-9}$)

Units of Q_0 are in mg nutrient/mg cell for phytoplankton and in µmoles/cell for benthic algae. Values obtained from literature review of one to five studies for each parameter.

Example 17.1 Uptake and growth calculations

Two species of phytoplankton have the following characteristics:

Phytoplankton species	K_s (mg/L)	V_{max} (mg L^{-1} h^{-1})	μ_{max} (day^{-1})	Q_0 (g/g)
Anabaena	0.5	3	0.5	0.05
Scenedesmu	0.1	1	0.1	0.01

Problem 1: Which species will be the best competitor for nutrients (i.e., have the highest nutrient uptake rates) when nutrient concentrations are 0.01 and 0.5 mg/L?

The Michealis—Menten equation yields the following results:

Phytoplankton species	V (at $S = 0.01$ mg/L)	V (at $S = 0.5$ mg/L)
Anabaena	0.059	1.5
Scenedesmus	0.091	0.83

The *Anabaena* will have a competitive advantage at the higher nutrient concentration. The *Scenedesmus* will have the advantage at the lower nutrient concentration. We present the curves for this and the following two problems in Fig. 17.3.

Problem 2: Which species will have the highest growth rate at $[S] = 0.01$ and 0.5 mg/L?

Phytoplankton species	μ (at $S = 0.01$ mg/L)	μ (at $S = 0.5$ mg/L)
Anabaena	0.0098	0.250
Scenedesmus	0.0091	0.0833

The *Scenedesmus* has the disadvantage in growth at both substrate concentrations.

Problem 3: Which species will have the highest growth at a Q of 0.04 and 0.1 g/g?

Phytoplankton species	μ (at $Q = 0.04$ g/g)	μ (at $Q = 0.1$ g/g)
Anabaena	0.0	0.25
Scenedesmus	0.075	0.09

The *Scenedesmus* will have a competitive advantage at the lower cell quota because the *Anabaena* cannot grow. The *Anabaena* will have the advantage at the higher cell quota.

when nutrients are scarce because pulses of nutrients can be stored. Such consumption in excess of growth requirements is *luxury consumption*. In many microorganisms, enough phosphorus can be stored in the form of the polymer polyphosphate through luxury consumption for several cell divisions (Healey and Stewart, 1973) and macrophytes may luxury consume phosphorus as well (Yan et al., 2016). Stream biofilms can accumulate phosphorus as polyphosphate during storm-related pulses of phosphate (Rier et al., 2016).

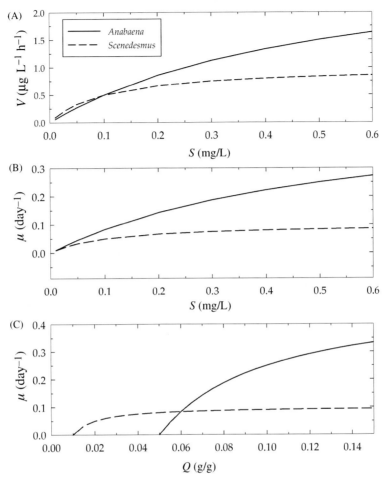

Figure 17.3 Graphical representations of the equations used in Example 17.1 for uptake and growth as a function of nutrients for two algae: (A) Michaelis−Menten uptake, problem 1; (B) Monod growth, problem 2; and (C) Droop growth, problem 3.

Nutrient supply, temperature (Fig. 15.3; Eppley, 1972), light (see discussion on photosynthesis−irradiance relationships in Chapter 12), pH, and other factors can control maximum growth rates (μ_{max}). Logically, growth rate relates more directly to nutrient concentration inside ([Q]) than outside the cells. The relationship between internal nutrient concentrations and growth is the *Droop equation*:

$$\mu = \mu_{max}\left(1 - \frac{Q_0}{Q}\right)$$

where μ is the growth rate, μ_{max} is the maximum growth rate, Q is the cell quota (concentration inside the cell), and Q_0 is the minimum cell quota [the concentration in the cell below which no growth occurs (Fig. 17.2B)]. Typical values for the constants in the Droop equation are in Table 17.1. This equation is useful if internal nutrient concentrations can be determined. The final step is to link growth to external nutrient concentration. This is accomplished using the *Monod equation*:

$$\mu = \mu_{max} \frac{[S]}{K_s + S}$$

where μ is growth rate, μ_{max} is the maximum growth rate, $[S]$ is the substrate concentration, and K_s is the concentration at which $\mu = 1/2 \ \mu_{max}$ (Fig. 17.2C). The equation is in the same form as the Michealis−Menten equation. This equation works well for single species in culture and moderately well for phytoplankton assemblages. The actual physiological bases for K_s and μ_{max} are complex and can vary over time and with nutrients, even within single cells (Ferenci, 1999). We present typical values for these two constants for nitrate, ammonium, and phosphate in Table 17.1.

Nutrient concentration is a prime determinant of uptake, but other factors, such as light, temperature, diffusion (Grant et al., 2018), and metabolic characteristics, control uptake rate as well. For example, low temperature decreases affinity for limiting substrates (Nedwell, 1999). The greater energy requirement for use of nitrate than that of ammonium under oxic conditions provides another example (see Chapter 14). The energy requirement translates into a greater effect of light on nitrate than on ammonium uptake by phytoplankton (Fig. 17.4).

This energy requirement has led to interactions among nitrogen forms with respect to uptake preferences. For example, ammonium can inhibit nitrate uptake (Fig. 17.4). This adaptation is advantageous because using nitrate takes more energy, so using ammonium when it is available is an efficient strategy. Similarly, energetically expensive nitrogen fixation will not occur at high rates when nitrate or ammonium are plentiful.

Macrophytes have the ability to obtain dissolved nutrients from the sediments through their roots in addition to using nutrients from the water column (White and Hendricks, 2000). Uptake kinetics are less clear for macrophytes because nutrient uptake from the water column can be important for some submerged macrophytes, and partitioning water column uptake from that of the sediments can be difficult. Scientists have debated the relative importance of nutrient supply from sediments compared to supply from the water column for many years (Sculthorpe, 1967). O'Brien et al. (2014) measured stream metabolism and phosphorus uptake before and after removing macrophytes from the channel of a lowland stream. They found that primary production of the stream dropped dramatically with removal, but phosphorus

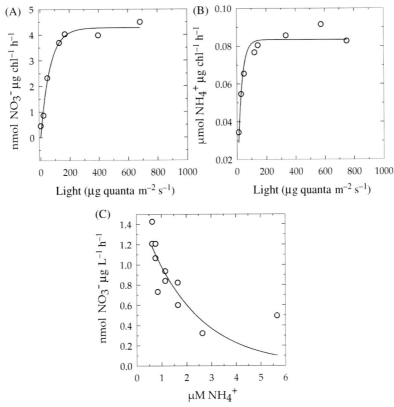

Figure 17.4 Uptake rates of nitrate (A) and ammonium (B) as a function of light. (reproduced with permission from Dodds and Priscu, 1989) and nitrate (C) as a function of ammonium (reproduced from Dodds et al., 1991, by permission of Oxford University Press) for epilimnetic plankton in Flathead Lake Montana, July 1987.

uptake did not. Their results do not support the idea that stream macrophytes may rely more heavily on nutrients in the water column than do lake macrophytes (Pelton et al., 1998). Emergent vegetation in wetlands probably acquires the majority of its nutrients from the sediments.

NUTRIENT LIMITATION AND RELATIVE AVAILABILITY

Relative availability of nutrients

Exploring nutrient requirements and how they translate into limitation of growth by nutrients is the next step after understanding basic uptake. *Nutrient limitation* is the control of growth or production by a nutrient or nutrients (in contrast to limitation by other factors, such as light, predation, or temperature). The concept of nutrient limitation is central to aquatic ecology because it allows determination of which nutrient or nutrients control primary production or heterotrophic activity of the ecosystem.

All organisms have approximately the same nutrient requirements because every one of them is composed of the same major types of molecules. The typical composition of algal cells with balanced growth (Table 17.2) reveals that various nutrients are required in general relative proportions. The stoichiometry of carbon, nitrogen, and phosphorus at balanced growth (with all nutrients supplied at rates faster than needed for growth) is usually taken as 106:16:1 (C:N:P by atoms or moles) and is referred to as the *Redfield ratio* (Redfield, 1958). Variations in stoichiometry can relate to physiological state or taxonomy. For example, diatoms require much more silicon than other organisms. The plasticity of nutrient contents varies among individual organisms. Many primary producers can build up significant amounts of starch, lipids, or cellulose to store carbon and alter their C:N:P accordingly. Therefore, stoichiometry can be very useful for determining and understanding nutrient limitation.

Some nutrients are readily accessed by most cells. Oxygen and hydrogen are easily available to photosynthetic organisms able to split water. Carbon is generally available to autotrophs in the form of CO_2. Ultimately, most other nutrients must come from weathering of the Earth's crust. The relative abundance of materials dissolved in rivers

Table 17.2 Elemental composition of algae and plants compared to availability in world rivers

Element	Plants and algae	World rivers	Average demand/supply
H	13,400,000	3,520,000,000	≪ 1
O	5,880,000	1,780,000,000	≪ 1
C	2,750,000	31,900	86
N	689,000	525	1,312
Si	163,000	7,390	22
K	34,900	1,880	18
P	24,400	10	2,440
Ca	23,300	12,000	2
Na	17,900	8,340	2
Mg	17,100	5,260	3
S	13,700	3,980	3
Fe	7,240	401	18
Zn	314	5	63
B	264	296	1
Cu	102	5	20
Mn	82	9	9
Mo	1	1	1

All composition data are in moles or atoms relative to molybdenum. Data for algae from Healey and Stewart (1973) combined with data for plants and algae. Data for rivers from Vallentyne (1974). Average demand/supply is algae and plants divided by rivers.

provides an approximate guide to the comparative supplies of nutrients available to aquatic organisms (Table 17.2). Some exceptions exist; for example, nitrogen-fixing organisms can use N_2 from the atmosphere. It is also very important to understand that standing stock and supply rates are not exactly the same thing. For example, carbon, nitrogen, and phosphorus are difficult to obtain relative to other elements, given relative concentrations in the world's rivers. However, CO_2 readily enters water from the Earth's atmosphere (has a high supply rate) and thus can be replenished quickly in most surface waters. Since inorganic carbon is more readily available (and becoming increasingly so from human activities), nitrogen and phosphorus are most likely to limit growth of algae and aquatic macrophytes.

Factors such as chemical characteristics, geology, and human land use alter the availability of nutrients in aquatic systems. For instance, phosphate tends to bind to clays, so it is transported slowly into aquatic systems, and can be transported in rivers at times of flooding (Banner et al., 2009). Nitrogen is lost from anoxic systems by denitrification but can be gained by nitrogen fixation, so the relative supply of nitrogen in individual systems may be difficult to predict. Geology and land use in particular areas can alter relative nutrient availability. For example, phosphorus can be amply available in watersheds with phosphorus-rich volcanic ashes or sedimentary deposits of apatite (mainly calcium phosphate deposits associated with limestone) and nitrogen can weather from nitrogen-rich rocks (Houlton et al., 2018). Variation in surface reference concentration of total phosphorus and nitrogen across US rivers shows areas that are naturally higher in phosphorus than other places (Fig. 17.5). The idea that fundamental characteristics of watersheds control nutrients has led to development of the concept of *nutrient ecoregions*. These regions were proposed for the United States in the 1980s and continue to be refined (Omernik and Griffith, 2014). Developing similar classifications globally will help us understand the natural distribution of nutrients and how human activities alter those distributions. Current concentrations of both nitrogen and phosphorus are at least 10 times greater than reference conditions, mostly related to high densities of human populations or intensive agricultural activity. Nutrients continue to increase in spite of some efforts to control them, but the reasons why are not currently understood in some cases; for example, phosphorus is increasing in streams in undeveloped watersheds (Stoddard et al., 2016). Hence, knowledge of nutrient supply in a specific groundwater aquifer, wetland, lake, or stream requires detailed knowledge of the geological and land-use characteristics of the watershed, including point and nonpoint sources of nutrients.

Nutrient limitation

The factor or factors present in the lowest relative rate of supply that is required for synthesizing the cellular constituents limits growth of organisms. The initial application

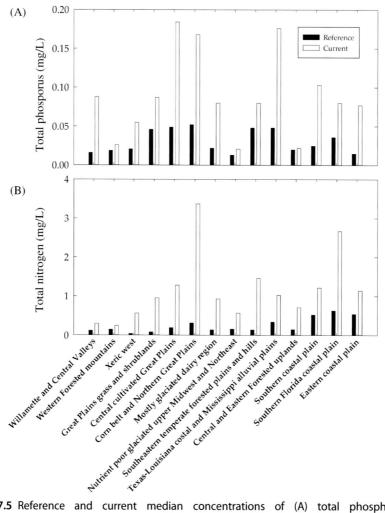

Figure 17.5 Reference and current median concentrations of (A) total phosphorus and (B) total nitrogen for rivers in the major ecoregions of the United States. *Data from Dodds et al. (2009).*

of this idea to aquatic sciences used the concept of *Leibig's Law of the Minimum:* An overall process rate is limited by the rate of its slowest subprocess. Leibig popularized this idea by referencing fertilization for crop production. He made the argument that a single constituent that was in the lowest relative supply would eventually disappear and leave the land barren if people planted successive crops in one area with no nutrient amendments.

Consider an automobile construction plant as an analogy for Leibig's law. The plant uses parts shipped from suppliers all over the world. Different amounts of various parts are required for each car (e.g., four tires/car, one transmission, and six engine cylinders). If 400 tires, 100 transmissions, and 540 cylinders are supplied each month, only 90 cars (540/6) can be constructed monthly, because supply rate of cylinders is limiting, even though sufficient tires and transmissions are available for 100 cars/month. Similarly, a living cell requires that their environment supplies elements at specific ratios for growth. Liebig's law of the minimum has been applied to nutrient limitation of primary producer assemblages in aquatic systems with the assumption that all producers have equal nutrient requirements and nutrients are evenly distributed in space and time in the environment (i.e., a homogeneous equilibrium condition of nutrients exists). The law, combined with the idea of equilibrium nutrient availability and the assumption of spatial and temporal homogeneity, predicts that only one nutrient will limit primary production of a system.

How closely does this prediction match empirical data on nutrient limitation? Several techniques have been used to assay nutrient limitation (Method 17.1). There is controversy regarding the use of such methods, but the results of the assays indicate some interesting patterns. We consider results of bioassays on primary producers in

---

METHOD 17.1 Bioassay tests for determination of nutrient limitation

The most realistic way to determine nutrient limitation is to test the entire system. However, this approach is not practical in many situations because experimental controls, multiple treatments, and replication are difficult, if not impossible, and nutrient pollution of an entire system is undesirable. The most commonly used alternative is in situ experiments in enclosures or small-scale approaches.

In lakes, mesocosms (limnocorrals) or containers (Goldman, 1962) from 1 to 1,000 liters have been used. Researchers add nutrients and measure the response of the algal biomass after about 1 week (Fig. 17.1). Enclosing the water may lead to artifacts, such as attached algae proliferating on the walls and lack of external nutrient inputs, but these "container effects" may be minimal in short-term studies in large enclosures.

Scientists often test benthic systems with nutrient-diffusing substrata. In these tests, they place the nutrients inside a container, out of which they slowly diffuse across a permeable surface that benthic algae colonize. Unglazed clay pots filled with nutrient-enriched agar are commonly used for such tests, but use of other permeable surfaces is increasing (Johnson et al., 1990). Factors that can interfere with these tests include grazers that crop algae as it grows and the inability to duplicate natural surfaces (e.g., nutrients rarely diffuse out of solid surfaces in nature). Another factor that should be considered is that agar and phosphate when autoclaved together create peroxide, a toxin (Tanaka et al., 2014). Thus, bioassays done without considering this should view phosphorus limitation results with caution. (The solution to the problem is to mix the phosphate and the agar right before they cool and harden!, Tank et al., 2017.)

(Continued)

METHOD 17.1 (Continued)

Another alternative is to take a water sample, return it to the laboratory, and test how well it stimulates production of laboratory cultures of algae (Eaton et al., 1995). In this case, researchers add different nutrients to different incubations to assess which nutrient is limiting algal growth. The drawbacks to this method are that it does not simulate natural conditions and species of phytoplankton used in the test may not be found in the system of interest. Duckweed (*Lemna minor*) also serves as a laboratory test organism (Eaton et al., 1995) because it is widely distributed in nature, its small size makes it easy to collect, and it is simple to grow in specific media.

Finally, a variety of short-term physiological bioassays have been proposed (Beardall et al., 2001). Of these, the most successful has been use of Redfield ratios for indication of nutrient limitation (Example 17.2). Other physiological methods are generally less reliable although quicker than growth-based bioassays (Dodds and Priscu, 1990).

Example 17.2 Using Redfield ratios of primary producers to indicate nutrient deficiency

We list molecular ratios of C, N, and P for phytoplankton assemblages from three separate lakes. Use the ratios to predict nutrient limitations.

Phytoplankton stoichiometry	C:N:P (molar)	C:N (molar)
Balanced growth	106:16:1	6.6:1
Mirror Lake	212:32:1	6.6:1
Deep Lake	106:3:1	35:1
Clear Lake	400:16:1	25:1

You should make the comparison to balanced growth (the Redfield ratio). The C:N ratio is presented to clarify the example. In Mirror Lake, the N:P ratio is greater than 16, so phosphorus is limiting relative to N. The C:N ratio in Mirror Lake is the same as balanced growth, so N likely is not limiting but phosphorus is limited. In Deep Lake, the N:P ratio is lower than balanced growth, the C:P ratio is the same as balanced growth, and the C:N ratio is greater than balanced growth; so, nitrogen is limiting in this lake. Finally, in Clear Lake the N:P ratio is the same as balanced growth, but the C:P and C:N ratios exceed those at balanced growth so both nitrogen and phosphorus are limiting.

lakes, wetlands, and streams; we know little about nutrient limitation of microbial activities in groundwaters.

Broad analyses of nutrient limitation of primary producers indicate nitrogen, phosphorus, or both commonly limit terrestrial, freshwater, and marine waters (Elser et al., 2007). Given that nutrients mostly originate from terrestrial habitats and rivers transport them through freshwaters to marine habitats, the idea that limiting nutrients are similar across habitats is not surprising. Tests of nutrient limitation in lakes indicate that either nitrogen or phosphorus most commonly limits primary production (Elser et al., 1990; Bracken et al., 2015), and that many lakes are colimited by both nitrogen and phosphorus (Fig. 17.6A). Nitrogen may limit production in many tropical lakes and these systems have received far less study than temperate lakes (Lewis, 2002). Furthermore, bioassays in streams indicate that different aspects of streams are limited by different nutrients, or several simultaneously (Beck et al., 2017). An experiment using bioassays in reference, agricultural, and urban streams across the United States indicated that nitrogen, phosphorus, or both could stimulate gross primary production, respiration, and chlorophyll. The bioassays also indicated that these variables did not necessarily respond the same way at the same location, and that responses to nutrients depended on substrata (inert glass or wood) the biofilms were growing on (Johnson et al., 2009).

Colimitation is a common response to nutrients, at least over relatively short timescales. The idea of colimitation has received attention on theoretical grounds (e.g., Kaspari and Powers, 2016; Sperfeld et al., 2016). While some still hold on to the idea that only one nutrient can limit primary production, this view is becoming less common based on both empirical evidence and theoretical treatments. Colimitation could be a result of physiological properties of individual

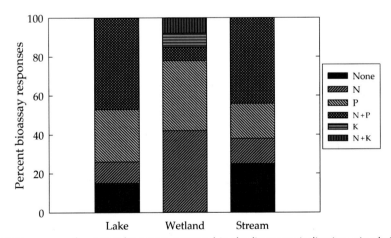

Figure 17.6 Summary of nutrient bioassays reported in the literature indicating stimulation of biomass of phytoplankton, wetland plants, and stream periphyton. *Data for lakes from Elser et al. (1990), data for wetlands from Verhoeven et al. (1996), and data for streams compiled from various sources.*

primary producers or spatial heterogeneity of nutrients, producers, and their herbivores (Marleau et al., 2015).

Nutrients other than nitrogen and phosphorus can limit some lakes. Studies of Lake Erie suggest that phytoplankton are at times colimited by nitrogen, phosphorus, or iron (North et al., 2007), and cross-system analyses suggest that iron might be more commonly limiting in oligotrophic lakes than previously thought (Vrede and Tranvik, 2006). Analyses of background iron levels across the United States support the idea of low iron in some watersheds, particularly those with few wetlands (Larson et al., 2015). CO_2 can limit phytoplankton in some cases. Floating cyanobacteria can maintain dominance in some eutrophic lakes by intercepting CO_2 so that it is not available to other primary producers (Shapiro, 1997), and phytoplankton in lakes with low CO_2 can exhibit lower photosynthetic rates (Hein, 1997). Supply of CO_2 can limit benthic algae in the benthic zone of lakes, particularly in acidified waters with low total inorganic carbon concentrations (Fairchild and Sherman, 1990). Silicon can limit many diatom populations (Schelske and Stoermer, 1972). In some cases, micronutrients such as molybdenum (a cofactor in several enzymes, including those for nitrate uptake) may limit algal growth (Goldman, 1960, 1972; Howarth and Cole, 1985). Early recognition of variable nutrient limitation and use of bioassays to indicate limitation is one of the many contributions by Dr Charles Goldman (Biography 17.1). Not all lakes are nutrient limited (Tilzer et al., 1991); light, particularly in high dissolved organic carbon lakes, can be limiting (Deininger et al., 2017); toxins can also limit primary productivity.

Assays on wetlands indicate that nitrogen or phosphorus limit plant production in many peat mires, but potassium may also limit plant growth (Verhoeven, 1986).

BIOGRAPHY 17.1 Charles Goldman

Professor Charles Goldman (Fig. 17.7) has dedicated his career to researching factors controlling production in lakes. Early contributions include recognition of the potential importance of trace nutrients (see Chapter 14). Goldman has supervised more than 100 graduate students and 30 postdoctoral researchers. He has published 5 books and more than 400 scientific articles, and he has produced four documentary films. He has won many prestigious national and international awards for his scientific contributions, including the Albert Einstein World Award of Science.

Goldman is an adventurer who has traveled to and studied remote lakes throughout the world. In 1967, "Goldman Glacier" was named in Antarctica, reflecting his early involvement in limnological research in polar regions. Once, while taking primary production measurements on a very hot day at Lake Victoria, Goldman decided to take a swim and impressed the African crew with back flips off the bridge. After his sixth flip he noticed that a 6-m long Nile crocodile had eased up to the opposite side of the boat. He did not try a seventh flip.

(Continued)

BIOGRAPHY 17.1 (Continued)

Figure 17.7 Charles Goldman. Photograph courtesy of Charles Goldman.

Goldman credits his father (an amateur ichthyologist) with getting him started in aquatic sciences and for the idea that "reading is the only way to compensate for the shortness of life." Goldman started as a geologist but changed his focus after taking a limnology course from David Chandler and realizing that "limnology is the queen of the ecological sciences." He suggests that students study as broad of a base in sciences as possible and include humanities because being a successful aquatic ecologist requires a multidisciplinary approach.

Goldman and his students have studied Lake Tahoe since 1958. Research on Lake Tahoe and Castle Lake in California has convinced him of the value of long-term data sets; data sets on these lakes contain trends over a few years that are the opposite of long-term trends in the same data. He has also used his research to support social action to protect Lake Tahoe. His experience with mixing social action and research has also led him to become involved in efforts to protect Lake Baikal. Goldman is an example of an exceptional scientist applying basic scientific knowledge to environmental problems.

Phosphorus limitation, or nitrogen and phosphorus colimitation, is common in most wetland types (Bedford et al., 1999). Wetlands with high hydraulic throughput (drained wetlands) can be more prone to potassium limitation (Van Duren et al., 1997). Zinc concentrations can limit wetland rice production (Neue et al., 1998). As with lakes and streams, some cases of multiple nutrient limitations occur in wetlands (Fig. 17.6B).

Nitrogen, phosphorus, nitrogen and phosphorus, or neither can also limit auto-trophs in streams (Fig. 17.6C). It is unwise to assume that any particular nutrient limits primary production in streams or that nutrients cannot be colimiting. CO_2 might also limit some primary producers in streams (Dodds, 1989; Raven, 1992) and silicon can limit epiphytic diatoms at times (Zimba, 1998). Cases of no nutrient limitation occur more commonly in streams than lakes or nonforested wetlands because of the greater influence of riparian canopy cover on light (light limitation) in streams and scouring floods that remove algal biomass. Light patches in forested streams can lead to areas where benthic biofilms are nutrient limited, and other areas where they are not (Warren et al., 2017). Meta-analysis of stream nutrient concentrations and bioassay responses supports the idea that nitrogen or phosphorus can be limiting in streams (Keck and Lepori, 2012).

If Leibig's law of nutrient limitation holds true, this brings in to question the common observations of limitation of primary producers by more than one nutrient in lakes, wetlands, and streams (Dodds et al., 1989). Given that streams are commonly far from equilibrium, the existence of colimitation in at least some streams is not surprising. Likewise, benthic processes structure wetlands, and it is easy to imagine spatial variations in sediments and gradients created by different points of nutrient inflow and outflow lead to significant spatial heterogeneity in nutrient limitation. The concept of several nutrients limiting primary production in lakes is more difficult because lakes are more homogeneous than are streams and wetlands.

Several of the assumptions used to apply Leibig's law may not hold, even in lakes, because all producers do not have equivalent requirements (e.g., diatoms need more silicon than other algae), nutrients may not be equally available in environments (e.g., pulses of nutrients occur), and not all organisms have the same competitive abilities. Models have been proposed for multiple limiting factors (Verduin, 1988), but these have not been well investigated. In contrast, researchers have paid much attention to the links between Leibig's law and phytoplankton diversity and this leads to a famous paradox.

The paradox of the plankton and nutrient limitation

The *paradox of the plankton* was proposed by Hutchinson (1961). Although we could view the paradox as a "straw man" given what is currently known about aquatic ecology, the proposed paradox forms a useful starting point for discussion and perhaps forms the basis for the most common question asked in aquatic ecology graduate qualifying exams in the past 50 years.

Hutchinson based the paradox on applying Leibig's law and the *competitive exclusion principle* (Hardin, 1960) to phytoplankton communities. Competitive exclusion occurs because only one species can be the superior competitor for a single limiting resource. In an environment in which several species are competing for a single resource, the superior competitor eventually will extirpate the others. The application of Leibig's law and the competitive exclusion principle require several assumptions: (1) all nutrients are well mixed over time and space in the environment (equilibrium), (2) all organisms have roughly the same requirements, and (3) competition has time to cause inferior competitors to disappear from the community. Hutchinson argued that because lakes are very well mixed environments, most aspects of these assumptions should hold. The paradox he noted is that a typical mesotrophic or oligotrophic lake has many species (typically 10–100) of phytoplankton present at any one time.

There are a range of explanations for high planktonic diversity proposed by Hutchinson and others. (1) Predation by zooplankton and viruses removes or suppresses dominant competitors (Suttle et al., 1990). (2) Pulses and micropatches of nutrients from uneven mixing and excretion lead to nonequilibrium conditions (discussed later in this chapter). (3) Mutualistic or beneficial interactions promote otherwise inferior competitors. (4) Many lakes are not at equilibrium conditions over timescales greater than a month, and the time required for dominant phytoplankton species to outcompete inferior competitors is more than a month (Harris, 1986). (5) Different competitive abilities lead to different nutrients limiting different species. And (6), chaos (in the mathematical sense) arises when species compete for three or more resources (Huisman and Weissing, 1999). The idea of different competitive abilities led to the resource ratio theory and the determination of how the Redfield ratio links to nutrient limitation.

RESOURCE RATIOS AND STOICHIOMETRY OF PRIMARY PRODUCERS

Primary producers can alter the relative proportions of elements that comprise their cells (their stoichiometry). This adaptation allows producers to acquire and store cellular components when resources are not limiting for use during times when they are limiting. Luxury consumption (discussed previously) alters the stoichiometry of primary producers. When nutrients are limiting, photosynthesis still occurs and leads to accumulation of carbon in lipids or starch (i.e., cellular stoichiometry is shifted toward relatively more carbon). This relationship between nutrient supply and cell stoichiometry has led to the extensive use of deviations from the Redfield ratio to indicate nutrient limitations.

The Redfield ratio derives from nutrient contents of phytoplankton grown with excess concentrations of all nutrients at conditions optimal for maximum growth. Deviations from these ratios indicate nutrient limitation (Example 17.2). A similar approach may be useful in determining nutrient limitation of wetland plants (Boeye et al., 1997; Bedford et al., 1999). For example, nitrogen content of pitcher plant tissues increases when they grow in higher nitrogen environments (Bott et al., 2008). Interestingly, when nitrogen increases, pitcher plants change their morphology to one that favors more light capture and less carnivory (a nitrogen source).

Cellular stoichiometry determines how primary producers will respond to changes in nutrient supply. This can be important in understanding eutrophication and its consequences. For example, cellular stoichiometry might regulate toxin production in cyanobacteria (Van de Waal et al., 2014).

NUTRIENT REMINERALIZATION

Uptake rates of limiting nutrients at natural concentrations in most aquatic habitats are great enough that the dissolved pools of nutrients will disappear rapidly if they are not replenished (Axler et al., 1981; Kilham and Kilham, 1990). Turnover may take hours or days for the nitrate pool or only seconds for the phosphate pool, but without supply of nutrients from some source, uptake cannot continue at rates measured in the environment. In general, external sources of nutrients (*new nutrients*), such as river and groundwater inflow or atmospheric deposition, cannot supply nutrients to lakes and wetlands at measured rates of uptake. Even in unpolluted streams and rivers, uptake depletes nutrients in the water column in a few hundred meters of flow (see discussion on nutrient spiraling in Chapter 24). Therefore, the predominant short-term source of nutrients is from remineralization. Remineralization is an important ecosystem driver, particularly in the epilimnia of large lakes (Fee et al., 1994). As a result, the productivity of a system is a function of both regeneration and supplies of new nutrients. We describe how the remineralization occurs, what organisms are responsible, and the dynamics of nutrient remineralization.

What short-term processes control the levels of dissolved inorganic nutrients such as ammonium and phosphate?

If the water chemistry of a lake, stream at base flow, wetland pool, or groundwater source is sampled each day for several days, the concentrations of ammonium and phosphate generally vary little. This lack of variability over short periods occurs even though uptake rates are often sufficient to remove all dissolved inorganic nutrients in considerably less than one day.

The balance between uptake and remineralization is the reason that nutrient concentrations are moderately stable in the short term (Dodds, 1993). Gross uptake removes nutrients at a variable rate that increases with increased nutrient concentration to a point, and remineralization replenishes them at an approximately constant rate (Fig. 17.8). This replenishment results in a dissolved nutrient concentration that is moderately resistant to perturbation from hours to days. This balance of uptake and remineralization is common in many planktonic systems (Fig. 17.9). Remineralization is important in wetlands, particularly those with limited hydrologic nutrient inputs (Bridgham et al., 1998).

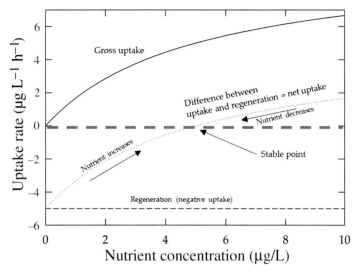

Figure 17.8 Graphical representation of the concept that gross nutrient uptake and regeneration can stabilize dissolved nutrient concentrations. This occurs when net uptake = 0 and gross uptake = regeneration. The graph illustrates a net increase in nutrients (excess regeneration) when nutrient concentrations are low and a net decrease (excess uptake) when nutrient concentrations are high. *Redrawn from Dodds (1993).*

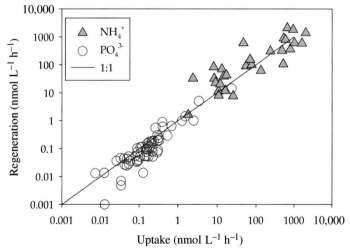

Figure 17.9 Uptake and regeneration of ammonium and phosphate from a variety of surface waters. Note the approximate 1:1 correspondence between the rates. *Reproduced with permission from Dodds (1993).*

The concept of nutrient balance is illustrated with the relationship between net uptake (V_{net}), gross uptake (V_{gross}), and remineralization (M) where:

$$V_{net} = V_{gross} - M$$

If concentration is constant, then net uptake must equal zero. If concentrations are in equilibrium, this equation implies gross uptake is difficult to measure by simply calculating the change in nutrient concentrations when the concentration is near the equilibrium point. This is the reason tracer studies are useful in determining gross uptake rates at ambient nutrient concentrations (Mulholland et al., 2002). Note that the equation is analogous to the equation for net and gross photosynthesis and respiration we present in Chapter 12.

Meaningful discussion about system dynamics is difficult if the only information is on concentrations of dissolved inorganic nutrients (standing stocks) because rates of uptake and remineralization, not the amounts of nutrients, are the important parameters (Dodds, 2003b). For example, algae in a highly eutrophic lake can have a very high nutrient demand and thus keep phosphate and ammonium levels near or below detection. Conversely, if the algal bloom in the same lake is declining and dead cells are decomposing, very high levels of ammonium and phosphate can occur. High values for dissolved inorganic nutrients are not confined to eutrophic systems; in a very carbon-limited aquifer, dissolved inorganic nutrients may be high even though the system has very low levels of microbial activity (e.g., it is relatively "oligotrophic" for a groundwater aquifer). Consequently, understanding the processes leading to supply and consumption of nutrients may be more important than just knowing the concentrations of the inorganic nutrients, and nutrient concentrations alone cannot necessarily be used to determine which nutrients are limiting. This is a common problem of assuming that pool (compartment) size is reflective of flux into or out of that pool. Just as lake size does not necessarily indicate flux of water through it (the Great Salt Lake is very large and has no outflow, the Great Lakes have a large river that empties them), nutrient concentration is not indicative of nutrient uptake or remineralization.

Processes leading to remineralization

Several processes can lead to nutrient remineralization; some remineralized nutrients originate from decomposition of dissolved and particulate organic material and others from living organisms. Dissolved nutrients in organic form are common in aquatic environments because organisms release them with normal metabolic activity, cells break and release their contents, and organic nutrients enter aquatic habitats from land. Even healthy algal cells can release up to half of fixed photosynthetic carbon directly into the dissolved form (Zlotnik and Dubinsky, 1989). These dissolved organic molecules often contain nitrogen and phosphorus. Part of this leakage may be associated with release of extracellular enzymes (such as phosphatase), and part may be unavoidable losses related to inefficiencies or cell lysis. In addition, a proportion of cells break from viral infections, receive damage from feeding activities of algivores, or other causes. These cells release dissolved organic nutrients into solution. In forested streams and wetlands, senescent autumn leaves and other riparian inputs can contribute large quantities of dissolved nutrients when they enter the aquatic environment and leach soluble materials.

Activities of heterotrophic organisms release additional organic molecules containing nutrients. Consumers release organic molecules as excreta and through sloppy feeding (i.e., damaging tissues of other organisms but not ingesting them). These organic molecules become available to other heterotrophic organisms for consumption. When organisms metabolize the organic molecules, they excrete excess inorganic nutrients back into the water. We will discuss processes controlling excretion rates under the framework of ecological stoichiometry near the end of this chapter. Given the variety of processes leading to remineralization, which are the most important? Few studies have partitioned out the relative contributions of different organisms to remineralization. The most common way to partition remineralization into functional groups of planktonic systems is to *size fractionate* (separate into different size classes by filtration). For the most part, these experiments indicate that very small organisms dominate nutrient remineralization in many planktonic systems (Table 17.3). Similar patterns occur for size fractionation of uptake. This route for nutrient remineralization is often called the *microbial loop*, where small, unicellular algae, bacteria, viruses, protozoa (particularly very small flagellates), and rotifers rapidly recycle carbon and nutrients. We discuss the microbial loop in detail in Chapter 19.

Nutrient supply associated with larger organisms can be important in some instances (Capps et al., 2015). Fishes in lakes can contain a substantial portion of the phosphorus in the lake, making remineralization by fishes important (Sereda et al., 2008). One study suggested that as lake productivity increased, the abundant detritivorous fish, gizzard shad (*Dorosoma cepedianum*), was responsible for increasing total available nutrients from 18% in reservoirs with less human influence in their watersheds to 50% in more heavily influenced watersheds (Vanni et al., 2006). Different species of fisher excrete phosphorus at different rates (Gido, 2002). Grazing snails can also remineralize nutrients to stimulate algal biofilms in streams, and pollutants may influence this process by altering excretion rates (Taylor et al., 2016). Activities of large animals such as moose (*Alces alces*) can release substantial amounts of nutrients into water (Bump et al., 2017).

Movements of nutrients and energy from one habitat to another (e.g., from forest to stream) are often referred to as *ecological subsidies* (Polis et al., 1997). As coarse spatial scale approaches to freshwater ecology become more common, there is increasing

Table 17.3 Size fractionation of regeneration of ammonium and phosphate in the epilimnion of pelagic freshwater systems

Lake	Nutrient	High cutoff (μm)	Low cutoff (μm)	% High	% Medium	% Small	References
Lake Calado (Amazon floodplain)	N	20	3	1	39	60	Fisher et al. (1988)
Lake Calado (Amazon floodplain)	P	20	3	0	0–45	55–100	Fisher et al. (1988)
Flathead Lake (Montana)	N	280	3	0–10	10–25	75–100	Dodds et al. (1991)
Flathead Lake (Montana)	P	280	3	0–30	0–45	50–100	Dodds et al. (1991)
Lake Biwa (Japan)	N	98	–	50	–	50	Urabe et al. (1995)
Lake Biwa (Japan)	P	98	–	15	–	85	Urabe et al. (1995)
Lake Biwa (Japan)	N	100	20	3–16	7–18	63–98	Haga et al. (1995)
Lake Kizaki (Japan)	N	100	20	1–62	27–68	40–70	Haga et al. (1995)
Ranger Lake (Ontario)	P	40	0.8	52	30	18	Hudson and Taylor (1996)
Mouse Lake (Ontario)	P	40	0.8	15	65	20	Hudson and Taylor (1996)
Lake Herrensee (Germany)	P	150	–	18	–	82	Hantke et al. (1996)
Lake Bräuhaussee (Germany)	P	150	–	11	–	89	Hantke et al. (1996)
Lake Thaler See (Germany)	P	150	–	2	–	98	Hantke et al. (1996)

The % categories are the % regeneration in the size fraction above the high cutoff (% high), between high and low (% medium) and below low (% small). Where several seasons were studied, ranges are presented for the percentages.

interest in ecological subsidies among aquatic and terrestrial habitats (Polis et al., 2004). Animal inputs can be particularly important when they make nutrients available from an area with spatial segregation from the region of interest. For example, activities of beavers bring terrestrial nutrients into streams (Fig. 24.4; Naiman et al., 1988; Naiman et al., 1994). In lakes, vertical migration of zooplankton from the hypolimnion to the epilimnion (Fig. 20.5) can move nutrients ingested in the hypolimnion into the epilimnion. Movement of benthic organisms, such as the amphipod *Gammarus*, from sediments into the water column can bring as much as 33% of the phosphorus into the water column (Wilhelm et al., 1999).

Other examples of external sources in streams include nutrient input from the rotting carcasses of fishes that swim upstream to spawn and then die (Kline et al., 1990; Bilby et al., 1996). In coastal Alaskan lakes, this input is enough to cause substantial effects on lake ecosystem production and plankton community structure (Finney et al., 2000). Salmon moving into small streams and then consumed by bears provide about one-fourth of the riparian nutrients in some Alaskan streams (Helfield and Naiman, 2006). Like salmon, migrating sea lamprey (*Petromyzon marinus*) can also move marine nutrients into streams (Weaver et al., 2016). Additional examples include excretion into ponds and wetlands by flocks of ducks, egg deposition in wetlands by amphibians (Regester et al., 2006), and movement of hippopotami out to graze terrestrial vegetation and then excreting material into rivers or wetlands (Subalusky et al., 2015). Mass drownings of wildebeest (*Connochaetes taurinus*) can enrich African rivers (Subalusky et al., 2017).

Advanced: Remineralization as a source of nutrient pulses in lentic systems

Nutrient pulses can be important because they can persist for some minutes before diffusing to background concentrations at small scales, low Reynolds numbers, and high viscosity. Pulses could allow microscopic organisms lucky enough to encounter the patch the opportunity to take up large amounts of a limiting nutrient. Pulses might be ecologically important as a partial explanation for large phytoplankton cells doing well in lakes. Large algal cells are poor competitors for nutrients at low concentrations relative to smaller cells (Suttle et al., 1988). Larger cells have a low ratio of surface area to volume and cannot assimilate nutrients as well. The larger cells may be able to use high concentrations of nutrients associated with pulses and thus maintain competitive ability by storing nutrients for use between pulses. A large cell has more room for nutrient storage than does a smaller one.

Zooplankton excretion is a process that could create patches of elevated nutrients in planktonic environments. In an ingenious study, Lehman and Scavia (1982) showed zooplankton could excrete pulses that remained stable for a long enough time to give phytoplankton cells in their vicinity a possible competitive advantage. In their study, they fed a culture of the cladoceran *Daphnia* algal cells labeled with radioactive phosphorus (^{32}P). Then they transferred these radioactive *Daphnia* into bottles containing unlabeled planktonic algae. After a short time, they harvested the algal cells and placed some on microscope slides coated with a photographic emulsion sensitive to the radioactive ^{32}P. After the slides were exposed and developed, they examined the slides microscopically for tracks of beta radiation emission. These tracks could be used to determine if each cell had taken up radioactive phosphorus and how much. Analysis of the distribution of the label in the algal cells revealed that some cells had significantly more label in them than would be expected if the phosphorus excreted by the *Daphnia* was dispersed randomly into the bottle (i.e., the pulses were mixed quickly). Thus, the experiment demonstrated that nutrient pulses could exist in planktonic communities and favor specific cells.

The importance of pulses from zooplankton excretion is uncertain. Artificial nutrient pulses change phytoplankton community composition in algal cultures (Scavia and Fahnenstiel, 1984). However, if it is true that microorganisms smaller than copepods and cladocerans dominate most nutrient remineralization in planktonic communities, then such pulses are less likely to have overall importance. Very small cells can only create small pulses of remineralized nutrients. Given Fick's law (see Chapter 3 on properties of water), we know that diffusion rates are greater at smaller scales. Small pulses probably disperse quickly and are unlikely to be present for long enough to stimulate large planktonic algae. Other factors such as grazer resistance may select for larger cells, and nutrient pulses may not be necessary to explain their presence (see Sidebar 3.1 on factors selecting for morphology of plankton).

STOICHIOMETRY OF HETEROTROPHS, THEIR FOOD, AND NUTRIENT REMINERALIZATION

Ecological stoichiometry is a powerful tool in aquatic and terrestrial ecology (Sterner and Elser, 2002), and has a large influence on how primary producers interact with primary consumers. Heterotrophs rely on organic material and remineralize nutrients such as nitrogen and phosphorus when they are in excess. Primary producers are very flexible in their stoichiometry and can build cells that are rich in carbon relative to nitrogen and phosphorus under conditions of nutrient limitation. In contrast to primary producers, the stoichiometry of heterotrophs is more similar to that of the Redfield ratio, and they are generally much less flexible in their ability to alter these ratios. However, some heterotrophs can alter their stoichiometry, sometimes in response to stressors such as pesticides (Janssens et al., 2017). Because heterotrophic organisms need to meet both their energy and carbon demands for growth from the organic material they consume, the nutrients in the food they eat can frequently exceed the amount needed.

As an example of the stoichiometric effects of the carbon requirement for both growth and respiration, consider a fish that can convert only 10% of the carbon it assimilates into biomass. The fish must use the remaining 90% of the carbon to create energy for metabolism required for survival and reproduction. In this case, the required molar nutrient ratio is 1060:16:1. If food is consumed that has the Redfield ratio of 106:16:1 mol of C:N:P, only one-tenth of the carbon, nitrogen, and phosphorus can be used for growth and the fish will excrete the excess nitrogen and phosphorus. Food for heterotrophs is not always at the Redfield ratio, and requirements of all heterotrophs are not the same as the Redfield ratio. Consideration of stoichiometry has led to much study of the requirements for ratios of nutrients, the stoichiometry of heterotrophs, and the composition of their food.

Most bacterial heterotrophs rely on dissolved organic material for carbon, nitrogen, and phosphorus requirements. Consequently, Hall et al. (2011) argue that linking stoichiometry to microbial ecology is essential to understanding how microbes mediate ecosystem processes. This material ultimately comes from primary producers (either phytoplankton in lakes or benthic algae and terrestrial vegetation in wetlands and streams) and can vary considerably in stoichiometry, as discussed previously. Bacteria can retain nitrogen increasingly as the C:N ratio of the dissolved organic material consumed decreases; hence, net remineralization is high at low C:N ratios (Fig. 17.10).

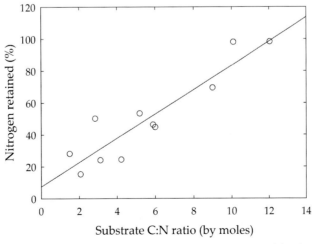

Figure 17.10 Nitrogen retention efficiency as a function of C:N ratio of food source for bacteria. Note that when food is relatively nitrogen rich (i.e., C:N is low), a low percentage of the nitrogen is used and most of the nitrogen taken in is regenerated. *Redrawn from Goldman et al. (1987).*

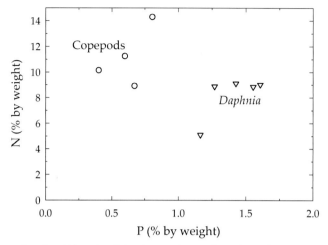

Figure 17.11 Data showing N:P ratio of *Daphnia* is lower than that of copepods, indicating different nutrient requirements for both types of grazers. *Reproduced with permission from Elser et al. (1996).* © *American Institute of Biological Science.*

The dissolved organic carbon available to bacteria may be low in nitrogen and phosphorus (particularly if it is derived from algae growing under strong nutrient limitation or terrestrial detritus), and bacteria may need to meet their requirements for these materials by assimilating (also referred to as immobilizing or incorporating) inorganic forms, such as nitrate, ammonium, and phosphate (Tezuka, 1990). Consequently, bacteria can account for a substantial portion of inorganic nutrient uptake in some lakes and presumably other aquatic habitats (Currie and Kalff, 1984).

Ecosystem processes (e.g., remineralization) can link to stoichiometry of organisms (Elser et al., 1996). For example, copepods have a higher N:P ratio than the cladoceran *Daphnia* (Fig. 17.11). The low N:P ratio of *Daphnia* means they have a relatively high phosphorus requirement for growth (Sterner, 1993). This requirement can lead ultimately to more intense phosphorus limitation in lakes (Elser and Hassett, 1994). The high requirement of *Daphnia* for phosphorus can lead to a shift to stronger phosphorus limitation in phytoplankton (Sterner, 1990; Sterner, Elser, and Hessen, 1992) because of preferential assimilation of phosphorus relative to nitrogen and relatively high ratios of N:P in nutrients remineralized by *Daphnia* (Sterner and Hessen, 1994). Similarly, the high nitrogen requirement of copepods relative to cladocerans such as *Daphnia* can intensify nitrogen limitation and select for nitrogen fixing cyanobacteria (Hambright et al., 2007). The differential temporal response of herbivores and primary producers to nutrients can lead to colimitation at the ecosystem level (Marleau et al., 2015).

While streams and wetlands have been the subjects of less research, the constraints of stoichiometry will hold in all environments. For example, consumers can alter nutrient content of periphyton. In an experiment with snails (molar N:P = 28) and crayfish (molar N:P = 18), grazer identity had various effects on food source nutrient content (Evans-White and Lamberti, 2006). The crayfish, with greater relative phosphorus demand, depressed the phosphorus content of periphyton to a greater degree than snails when external supplies of dissolved inorganic phosphorus were low. Periphyton stoichiometry was altered because the crayfish held more phosphorus and excreted less relative to the snails. When supply rate of dissolved inorganic phosphorus was high, consumers had no influence on phosphorus content. Examination of stoichiometric relationships in forested headwater streams revealed larger elemental imbalances among primary consumers and their detrital resources than has been observed in many other systems (Cross et al., 2003).

Detritivores can also be sensitive to stoichiometry of their food. For example, the shredder *Tipula abdominalis* was fed maple or oak leaves that had the C:P content experimentally manipulated in both species of leaves. The shredder always preferred oak leaves and exhibited greater growth when fed maple leaves. When the C:P ratio was lower in maple leaves, the *Tipula* larvae grew more quickly. The researchers did not observe the same dependence on C:P when the larvae were fed oak leaves (Fuller, et al., 2015). The response is a bit surprising, as oak leaves are supposed to be more refractory; however, the study did indicate that taxonomy of food source can dictate stoichiometric response of consumers.

There are some broad relationships between taxonomic groups and phosphorus content (Fig. 17.12); fishes have the lowest C:P ratios because their bones have high phosphorus content), and invertebrates the greatest and most variable ratios. Detritivores, which generally subsist on poor quality food, usually have greater body C:P. The general hypothesis is that animals with high growth rates depend upon high phosphorus food, and that animals tend to match the food source they have adapted to consuming (Frost et al., 2006).

In a related study, Evans-White et al. (2009) found that diversity of macroinvertebrates decreased as total phosphorus in the water column increased, up to a point, and then leveled out with low diversity at high total phosphorus contents (Fig. 17.13). The authors hypothesized that fast growing organisms could compete well at high phosphorus concentrations and would dominate under polluted conditions. This hypothesis assumed that primary consumers evolved under a natural range of phosphorus concentrations in their food, and that above this range, only some species would be effective competitors. The effect is mediated by the fact that stoichiometry of food for primary consumers (periphyton and detritus) increases in phosphorus content as stream water becomes more phosphorus-rich. The hypothesis suggests that primary consumers should be more highly influenced by stream nutrient concentrations than

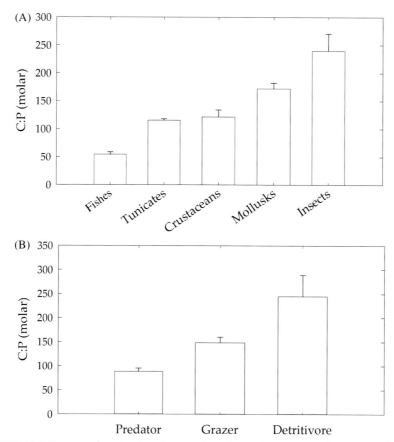

Figure 17.12 Variation in carbon to phosphorus ratios of aquatic consumers by taxonomic group (A) and functional group (B). Fishes have significantly lower C:P than all other taxonomic groups, and detritivores have significantly greater C:P than predators or grazers. *Data replotted from Frost et al. (2006).*

predators because predators rely on animals that have relatively constant stoichiometry. Analysis of the data supported the hypothesis because primary consumer diversity was more closely related to total phosphorus concentrations than predator diversity. In a related study, Manning et al. (2016) experimentally dosed a stream with inorganic nitrogen and phosphorus. They found that nutrient enrichment not only altered microbial breakdown rates, but also increased detritivory. Thus, the concepts of stoichiometry and nutrient limitation have implications for food webs and ecosystem function because remineralization feeds back to primary producers and heterotrophic organisms at the base of the food web.

Requirements for complex organic materials are a potential complicating factor in establishment of stoichiometric constraints on growth and production of animals. Just as humans require some specific vitamins and other organic compounds that they cannot synthesize, other animals have some requirements that are not related to the ratios

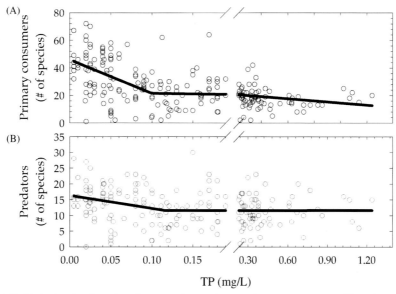

Figure 17.13 Variation in diversity of stream invertebrate primary consumers (A) and predators (B) from rivers and streams in Kansas, Missouri, and Nebraska in spring and autumn samples as a function of water phosphorus concentration. There were significant breaks in the relationships as denoted by the two lines, with the predator break occurring at greater total phosphorus than the consumers. *Data from Evans-White et al. (2009).*

of carbon to nitrogen or phosphorus. Polyunsaturated fatty acids comprise one group of compounds that has received considerable attention for being potential limiting nutrients (Brett et al., 2017). The fact that algae can be particularly rich in these compounds has led some researchers to suggest that polyunsaturated fatty acids might be limiting to animals and that would make algae more important sources of food than reflected by their C:N:P ratios. While algal fatty acids might be important components of stream invertebrate growth and reproduction, further study is required for us to fully understand their role in aquatic ecosystems (Guo et al., 2016). Evidence suggests predators that forage closer to aquatic habitats and thus consume prey that are richer in algal-derived essential fatty acids can have higher immune function (Fritz et al., 2017).

SUMMARY

1. Nutrient uptake is described by the Michaelis–Menten uptake equation. After cells take up nutrients, they must be assimilated (converted to the chemical compounds that make up cells). This process is described by the Droop equation, which links intracellular nutrient concentrations with growth, and the Monod equation, which describes the relationship of external dissolved nutrients to growth rate.

2. Other factors that influence uptake and assimilation of nutrients include the ability to acquire and store nutrients for later use (luxury consumption), temperature, and light.

3. Nutrients are required in known ratios (stoichiometry) for growth. The ratio of C:N:P required for algal growth is approximately 106:16:1 by moles and is known as the Redfield ratio. If the relative availability of a nutrient is lower than its requirement relative to availability of others, it can be limiting.

4. Some aquatic scientists maintain that only one nutrient can be limiting at a time, but others argue that more than one nutrient can limit primary producer assemblages at a time. Empirical evidence suggests that N, P, or both are usually the limiting nutrients in lakes, streams, and wetlands.

5. Nitrogen and phosphorus are generally the limiting nutrients in freshwater habitats, but iron, potassium, and others can also be limiting in some situations.

6. Nutrients can be supplied from outside (new nutrients) or inside the system by the process of remineralization (regeneration). Nutrient remineralization provides the primary source of the nutrients available to primary producers in aquatic ecosystems. Sources of these remineralized nutrients include organic material excreted or lost by producers and processed by heterotrophs, and excretion by predators or consumers.

7. The balance of uptake and remineralization often controls dissolved inorganic nutrient concentrations in aquatic ecosystems.

8. Microbes are responsible for the bulk of remineralization in most ecosystems, but in certain cases, larger organisms can be important.

9. Nutrient pulses created by larger organisms can persist in the environment, but those produced by smaller microbes likely disperse quickly by diffusion.

10. Stoichiometry of heterotrophs feeds back to alter nutrient limitation of primary producers as well as determining some aspects of community structure.

QUESTIONS FOR THOUGHT

1. Why might dissolved nutrient levels be more variable over days to weeks in streams than in large lakes?

2. Nuisance filamentous benthic algae in the Clark Fork River, Montana, are limited by nitrogen in the summer, despite the fact that phosphorus concentrations dissolved in the water at that time are extremely low. Given that dissolved phosphorus concentrations are very high in the spring, what is a potential reason for the lack of phosphorus limitation in the summer?

3. Why isn't nutrient limitation necessarily additive (i.e., why are there generally small responses to additions of nonlimiting nutrients)?

4. What is an evolutionary argument for why nutrient competition should lead to limitation by multiple nutrients?

5. Why may nutrient pulses be more likely to form and persist in groundwater and wetland sediments than in planktonic habitats?

6. Why are large cells more likely to have high maximum rates of nutrient uptake, high half-saturation constants, and the ability for greater luxury consumption relative to small cells?

7. Why do many scientists think that total phosphorus concentrations are more useful indicators of nutrient supply than dissolved phosphate concentrations?

8. Why can it be misleading to use the ratio of dissolved inorganic nitrogen to dissolved phosphate, rather than the Redfield ratio of organisms, to indicate nutrient limitation?

CHAPTER 18

Trophic State and Eutrophication

Contents

Figure 18.1 Whole-lake nutrient additions at Lake 226 in northwestern Ontario. The far part of lake received nitrogen, phosphorus, and carbon, and the near part received only nitrogen and carbon. The algal bloom in the far lake gives the lake a light color. *From Schindler (1974). Copyright 1974 by the American Association for the Advancement of Science. Photograph courtesy D.W. Schindler.*

Humans have caused global increases in the numbers of lakes negatively impacted by nutrient pollution since the 1900s (Jenny et al., 2016). Changes in nutrients can alter ecosystem structure and function in all freshwater habitats. Alteration by increased nutrients is termed *eutrophication*. Anthropogenic actions (*cultural eutrophication*) or natural conditions can increase nutrient input. In this chapter, we describe how comparisons among aquatic systems define *trophic state*, the level of ecosystem productivity, and consider problems that may be associated with eutrophication. Next, we examine the linkages among nutrient loading, nutrients, algal biomass, water clarity, and fish production. Finally, we describe methods for controlling eutrophication and present several case studies. Given the large economic costs associated with improvement of water quality, and the large costs associated with negative effects of nutrient enrichment (Dodds et al., 2009), eutrophication continues to be a very relevant issue in lakes, streams, and wetlands.

DEFINITION OF TROPHIC STATE

Classifications of the trophic state of aquatic ecosystems are useful because they allow people to compare productivity of ecosystems within and among ecoregions, and provide an initial approach for determining the extent of cultural eutrophication. Trophic state is historically signified by the terms oligotrophic, mesotrophic, and eutrophic. Oligo means "few," trophic means "foods," eutrophic means "many foods," and mesotrophic falls between these two categories. The three categories are only one way to characterize a continuum of ecosystem productivity. Over the years, scientists have used several systems to describe the trophic state of lakes; trophic state classifications are not as highly developed for streams, groundwaters, and wetlands.

Traditional definitions of trophic state depend on primary producer biomass as driven by nutrient availability. This makes sense, as one of the most problematic aspects of nutrient pollution is nuisance or harmful algal blooms. However, recognition of the effect of nutrients on heterotrophic activity, with potential influences up to the ecosystem level, has led to a broader view of eutrophication. This view considers *heterotrophic state* as represented by the whole system CO_2 production per unit area and *autotrophic state* the whole system CO_2 fixation per unit area (Dodds and Cole, 2007). Eutrophication then is an increase in a nutritive factor or factors leading to greater rates of whole-system heterotrophic or autotrophic metabolism (Dodds, 2006).

Early limnologists noticed that certain types of phytoplankton and zooplankton were common in high-nutrient lakes and others in nutrient-poor lakes. This observation led to extensive efforts to characterize the trophic state of lakes with regard to their phytoplankton communities (Hutchinson, 1967). Thus, limnologists recognized the links among nutrients, phytoplankton biomass and productivity, and water quality. These links will be described quantitatively in this and subsequent sections.

Current classifications of autotrophic state of lakes generally use water clarity, phytoplankton biomass, and nutrient concentrations (productivity is not as easy to measure, so it is used less in trophic classification). Epilimnetic hypoxia is another potential indicator of autotrophic state of lakes (e.g., Yuan and Pollard, 2015). In general, oligotrophic lakes have low algal biomass, low algal productivity, low nutrients, high clarity deep photic zones, oxygen in the hypolimnion (if the lake is monomictic or dimictic), and may support coldwater fisheries. Eutrophic lakes are characterized by cyanobacterial blooms, high total nutrients, large variation in O_2 concentrations (including potential anoxia in the hypolimnion), and may have frequent fish kills. The trophic state of lakes is usually based on phytoplankton concentrations, but shallow eutrophic lakes can have extensive macrophyte populations. One proposal in such lakes has been to add the nutrient concentration in the water to the nutrient content in the macrophytes to indicate trophic state (Canfield et al., 1983), but this scheme is not widely adopted. A group of researchers working in African lakes recently proposed a system of ranking trophic state based on macrophyte communities (Kennedy et al., 2016).

A large group of limnologists interested in eutrophication (OECD, 1982) constructed one of the commonly used classifications for lakes. This group constructed a classification system by combining data from many lakes. For this analysis, scientists from around the world classified lakes in their region as eutrophic, mesotrophic, or oligotrophic (leading to a certain inherent degree of subjectivity in the scheme). The lakes in these three categories were subjected to two classification approaches: a probability distribution (Fig. 18.2A) and a fixed boundary classification (Table 18.1). There are several fixed boundary classification systems for lakes; for the most part, the boundary levels are consistent (Nürnberg, 1996). Another commonly used method of classification involves calculating a trophic index that places trophic state on an exponential scale of Secchi depth, chlorophyll, and total phosphorus (Fig. 18.2B), where 10 scale units represent a doubling of algal biomass (Carlson, 1977). This index is useful because it more realistically represents the continuous distribution of trophic state and it offers quick relative comparisons among the various factors related to autotrophic state.

Lakes may not clearly fall into an individual category in any of the classification systems. For example, phosphorus could be high enough for a lake to be classified as eutrophic, but light attenuation by suspended sediments could keep chlorophyll levels in the mesotrophic range. Also, total phosphorus and phytoplankton concentrations could be low in a lake with extensive macrophyte biomass and production (Brenner et al., 1999).

Classification of stream autotrophic state can be based on suspended algae, attached algal biomass, or nutrients. Suspended algal mass in streams is usually a function of the extent to which the benthic algal community has entered the water column from the stream bottom, except where water flow is slow enough to allow development of a truly planktonic algal assemblage. Trophic classification is difficult because hydrological variation (flooding) and light limitation by riparian canopies translate into high variability in benthic algal biomass over time (weeks or months) and space. Furthermore, many streams have food webs dominated by input from terrestrial organic material, so biomass of primary producers can be a poor indicator of whole-system productivity, and the distinction between autotrophic and heterotrophic state becomes very important.

A trophic classification for temperate streams has been proposed that uses the data available for reference conditions (Table 18.2). This system assigns the bottom third of

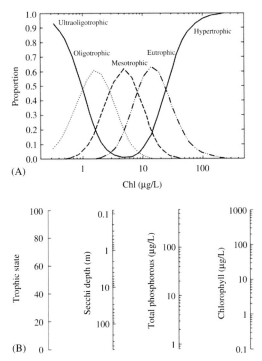

Figure 18.2 Two trophic classification systems for lakes. (A) Probability distribution for chlorophyll related to trophic state. The *y*-axis is the probability that a lake will have a specific trophic state given a set value of chlorophyll. For example, at 10 μg chl/L, the chances are approximately 0% that the lake is ultraoligotrophic, 5% that the lake is oligotrophic, 50% that it is mesotrophic, 45% that it is eutrophic, and 5% that it is hypertrophic (hypereutrophic). (B) A logarithmic scale that allows a continuous index to be derived from Secchi depth, total phosphorus, or chlorophyll *a*. *A, Adapted from* Eutrophication of waters. Monitoring and assessment and control © *OECD, 1982. B, Plotted from data of Carlson (1977).*

Table 18.1 Two fixed-boundary trophic classification systems for lakes

Parameter	Ultraoligotrophic	Oligotrophic	Mesotrophic	Eutrophic	Hypertrophic
OECD (1982)					
Total P (μg/L)	<4	4−10	10−35	35−100	>100
Mean chl (μg/L)	<1	1−2.5	2.5−8	8−25	>25
Maximum chl (μg/L)	<2.5	2.5−8	8−25	25−75	>75
Mean Secchi (m)	>12	12−6	6−3	3−1.5	<1.5
Nürnberg (1996)					
Total P (μg/L)		<10	10−30	30−100	>100
Total N (μg/L)		<350	350−650	650−1,200	>1,200
Mean chl (μg/L)		<3.5	3.5−9	9−25	>25
Mean Secchi (m)		>4	4−2	2−1	<1
O$_2$ depletion rate (mg m^{-2} day^{-1})		<250	250−400	400−550	>550

Table 18.2 Autotrophic state boundaries for streams

Nutrient	Autotrophic state boundary	Nutrient concentration (mg/m^3)	% cases of mean chl <100 mg/m^2	% cases of maximum chl <100 mg/m^2
Total N	Lower 1/3	330	93	73
Total N	Upper 1/3	685	90	71
Total P	Lower 1/3	26	95	83
Total P	Upper 1/3	60	87	75

The lower 1/3 and upper 1/3 of the distribution of stream total nitrogen and total phosphorus are pooled across 14 ecoregions. In each of these ecoregions reference values were determined by Smith et al. (2003) and/or Dodds and Oakes (2004). The relationship of the boundary numbers is translated to the percentage of mean or maximum benthic chlorophyll values expected to be less than 100 mg/m^2 when nutrient values were less than the boundary.
Source: After Dodds (2006).

the streams to the oligotrophic category, the middle third to the mesotrophic group, and the streams with the highest chlorophyll or nutrients in the top third to the eutrophic group (Dodds et al., 1998). Again, this is a method to describe a continuum, but it satisfies the convention of assigning systems to one of three trophic categories. The method has not been applied to macrophyte-dominated streams.

Trophic categorizations for groundwater and wetlands have not been developed extensively, to our knowledge. Trophic state may correlate to biological characteristics of wetlands. For instance, some oligotrophic, nitrogen-limited wetlands contain carnivorous plants that use captured insects as a nitrogen source. Phosphorus fertilization of the Everglades leads to shifts in plant and periphyton communities. Analysis of published data suggest that N:P ratios alter plant diversity, with greatest wetland plant diversity at intermediate ratios. The same analysis also suggests that at high absolute nutrient concentrations, common weedy species tend to dominate (Güsewell et al., 2005). Nutrients clearly influence biotic integrity of wetlands, and can alter fundamental ecosystem properties such as litter decomposition rates (Emsens et al., 2016). Therefore, it should be possible to create biologically meaningful classifications of wetland trophic state.

Productivity of groundwaters is almost solely a function of heterotrophic state, although it is possible that chemoautotrophic state plays a roll. Heterotrophic state should link to the influx of organic carbon and O_2, so more eutrophic groundwaters would have high supply rates of organic C, low O_2, and potentially high microbial productivity. The rates of microbial activity may actually decrease in aquifers that are anoxic because anoxic carbon cycling is less efficient than respiration using O_2. Aquifers that are eutrophic because of high organic carbon input could lack a complex invertebrate community due to anoxic conditions, so trophic state may correlate to biological characteristics.

ADVANCED: DETERMINING REFERENCE NUTRIENT CONDITIONS IN FRESHWATER ENVIRONMENTS

Understanding the natural distribution of nutrient content is of interest for several reasons. (1) These are the conditions under which organisms evolved and could be part of the conditions required for species conservation. (2) Regulating nutrients requires understanding what is the likely best possible condition for a region. (3) Understanding downstream loading requires knowledge of the nonanthropogenic nutrient inputs to a system.

The best way to identify reference conditions is to find minimally impacted sites (lakes, wetlands, or streams) and measure the nutrient contents in these sites. The sites should be representative of the region, and the sampling should be randomly stratified based on the relative occurrence of geology and soil types because nutrient yields from reference watersheds can vary over 100-fold (Smith et al., 2003). An example of this type of stratification of regions is the nutrient ecoregions used by the US Environmental Protection Agency (Omernik, 1995). Unfortunately, there are many areas where all sites experience nutrient pollution. For example, the Corn Belt ecoregion in the Midwestern United States and many agricultural areas in Europe have fertilized cropland in almost every watershed. Atmospheric nitrogen deposition influences aquatic ecosystems in all regions with moderate to high densities of humans.

An alternative method for determining reference nutrient conditions is to use a few small reference watersheds in conjunction with modeling methods to scale up to entire watersheds (Smith et al., 2003). These nutrient yield models can account for atmospheric nutrient deposition. These models indicate that current total nitrogen (TN) concentrations are more than six times greater than reference conditions, while total phosphorus concentrations have doubled.

Yet another method is to plot human influences on the x-axis, nutrient concentration on the y-axis, and use the intercept to extrapolate to reference conditions. In this method, for example, an equation such as the following can be used for extrapolation

$$\log_{10} TP = -0.727 + 0.00668 \times \% \text{ cropland} + 0.01465 \times \% \text{ urban land}$$

where $\log_{10} TP$ is the log to the base 10 of total phosphorus concentration in mg/L and % cropland is the percentage of the watershed in croplands. If the % urban land and % cropland is zero, then this should be reflective of a lack of human inputs and the reference concentration is simply the constant in the regression ($10^{-0.727}$ results in a total phosphorus concentration of 187 µg/L (0.187 mg/L)). This method does not account for atmospheric deposition. The method does generally agree with the modeling method of Smith et al. (2003), with some exceptions (Dodds and Oakes, 2004), and appears to apply well to reservoirs as well (Dodds et al., 2006).

The US Environmental Protection Agency has also suggested that in areas with no reference sites, the lower quartile of the existing distribution could represent reference conditions. In our view, this is an exceedingly arbitrary method of assigning reference conditions. Data from Smith et al. (2003) also suggest that this is a poor way to represent expected background conditions in the absence of reference sites.

WHY DOES ALTERATION OF TROPHIC STATE BY NUTRIENT POLLUTION MATTER IN LAKES?

Problems with algal blooms seem to be worsening as climate warms and population density increases worldwide (Highlight 18.1). The stimulation of algal blooms and creation of anoxic hypolimnia in lakes leads to many problems. As we mentioned in Chapter 1, the monetary value of property on a lake can decrease with eutrophication. Algal blooms are not aesthetically pleasing; they look bad and smell worse. Water taste and odor problems become more acute as lakes become more eutrophic (Fig. 18.3). Both planktonic and attached cyanobacteria contribute to taste and odor problems (Sugiura et al., 1998). These problems related to eutrophication are difficult to solve with standard water purification methods, leading to greatly increased costs for supplying potable water (Wnorowski, 1992). To make matters worse, algal blooms may be toxic; cyanobacteria and dinoflagellates produce neurotoxins and hepatotoxins (see Highlight 9.2). Phosphorus pollution could stimulate algal toxin production (Jacoby et al., 2000; Harke et al., 2016). Toxic algal blooms can harm people and livestock, and chronic exposure to toxins in drinking water could increase probability of liver cancer (Zhang et al., 2015). The probability of harmful or objectionable algal blooms occurring increases with greater cultural eutrophication (Hart et al., 1999).

Fish kills related to anoxic events are common symptoms of eutrophication (see Highlight 12.2). With a series of cloudy days or under ice cover in a eutrophic lake, fish may die. Cold-water fisheries are possible in deep lakes with cool hypolimnia (see Chapter 23). If the hypolimnion is anoxic, temperature-sensitive fishes have no refuge from high temperatures in the epilimnion. High pH associated with algal blooms can also cause fish kills (Kann and Smith, 1999).

Eutrophication can alter biotic integrity. For example, eutrophication can lead to decreases in species richness and diversity of algae, which may have negative consequences for the food web (Proulx et al., 1996). A long-term analysis of Clear Lake in Iowa demonstrated that since the 1950s the number of macrophyte species declined from 30 to 12 by 2004 (Egertson, Kopaska, and Downing, 2004). During this time, the lake became more eutrophic, going from a Secchi depth of approximately 0.8 to 0.4 m. Even naturally eutrophic lakes have shifts in algal communities associated with agriculture (Räsänen et al., 2006). Eutrophication also influenced evolutionary processes of whitefish,

Highlight 18.1 Global increases in eutrophication and unwanted algal blooms under a changing climate and increased anthropogenic pressures

As population density increases, worldwide nutrient concentrations have increased in freshwaters. Cyanobacterial blooms are favored by higher temperatures (Kosten et al., 2012) occurring globally due to global climate change both because they compete better in warmer waters, and thermal stratification is more stable under warmer conditions. Additionally, greater runoff in areas that become wetter will transport more nutrients and increase eutrophication (Sinha et al., 2017). Thus, cyanobacterial blooms are expected to worsen most with global climate change in dimictic lakes (Taranu et al., 2012), creating strong influences in temperate zone lakes. Predictions based on global climate change models (general circulation models) also forecast about three times average duration of harmful algal blooms in the United States by the end of this century (Chapra et al., 2017). Furthermore, increased carbon dioxide might stimulate blooms. As a result, cyanobacterial blooms are expected to get worse in many places (Paerl et al., 2011; O'Neil et al., 2012; Paerl and Paul, 2012; Visser et al., 2016). However, local meteorological conditions will also play a role (Huber et al., 2012).

Several very large blooms bring this point to bear. Lake Taihu, the largest lake in China, experienced a massive bloom in 2007. The lake is a key drinking water source for about 10 million people. The blooms on the lake have become progressively worse with watershed development. The water in the taps became so bad in 2007 that people had to switch to bottled water for fears of toxin and horrible tastes and odors of the drinking water. Two million people lost access to safe drinking water in this event (Qin et al., 2010). Efforts will be required to control nutrient inputs into the lake to hopefully avoid future problems as bad as or worse than the 2007 event. Officials have closed polluting factories and built new sewage treatment facilities since 2007, decreasing nutrient loads. The decreases are likely to lower the probability of cyanobacterial blooms (Xu et al., 2017).

Lake Erie had been recovering from eutrophication that was threating in the 1960s and 1970s because nutrient inputs had decreased. However, since the 1990s blooms were becoming more and more common and in 2011 the lake experienced its largest bloom in recorded history. The bloom was composed of potentially toxic *Microcystis* and *Anabaena*. This bloom was linked to agricultural practices in the watersheds of the lake and warmer than normal conditions as expected with regional climate change (Michalak et al., 2013). It remains to be seen if the nonpoint pollution that led to this event can be controlled or if the problem will continue to intensify.

During the summer of 2016, a harmful algal bloom developed along hundreds of miles of the Ohio River, shutting down drinking water plants (Brooks et al., 2017). The bloom developed because of high temperatures and nutrients as well as sluggish flow that did not wash out the bloom. Similarly, toxic blooms continue to form on the Murray River in Australia (Crawford et al., 2017).

Cyanobacterial blooms will continue to plague humanity as long as nutrients entering freshwaters are not controlled. The potential toxicity of these blooms makes the situation direr. Both science and management help understand steps required to control the problem, from local nutrient control to global decreases in greenhouse gasses that are causing global warming.

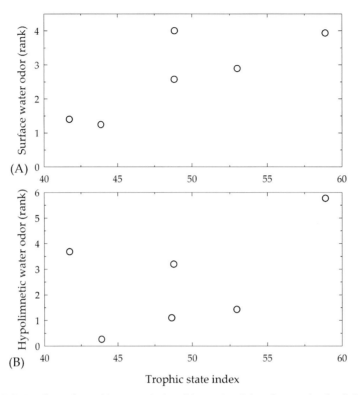

Figure 18.3 Relationship of trophic state index (determined by the method of Carlson, 1977; Fig. 18.2B) and water odor of surface (A) and hypolimnetic (B) samples from six Kansas reservoirs. Human testers ranked water odor with a higher rank indicating lower drinking water quality. *Reproduced with permission from Arruda and From (1989).*

leading to reversal of traits evolved under more oligotrophic conditions (Vonlanthen et al., 2012).

Eutrophication can influence benthic algal communities as well. For example, the filamentous green alga *Cladophora* can attain nuisance levels in the Great Lakes and elsewhere. Invasive zebra mussels that increase water clarity and light penetration through filter feeding increase the depth that *Cladophora* can grow at and exacerbate the problem (Kuczynski et al., 2016). This increase can cause additional problems because *Cladophora* can harbor pathogenic bacteria (Ishii et al., 2006).

Not all the results of eutrophication are bad. Fish biomass may be greatest in eutrophic lakes (Fig. 18.4). Increased productivity can lead to increases in fish productivity, although the fish that dominate in eutrophic lakes may not be desirable species (Lee and Jones, 1991). Some lakes are fertilized artificially to increase fish production

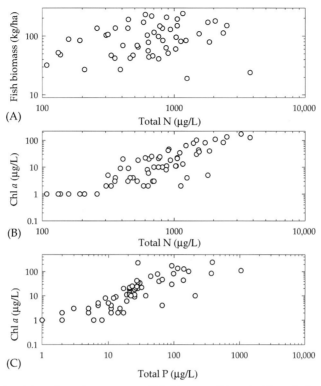

Figure 18.4 Relationships of total nitrogen to fish biomass (A) and chlorophyll to total N (B) and total P (C) in 67 Florida lakes. Note that the relationship between fish and nutrients is weaker than that between chlorophyll and nutrients. *Data from Bachmann et al. (1996).*

(Highlight 18.2). Eutrophic lakes are also more resistant to the effects of acid precipitation because metabolic activities buffer them (Davison et al., 1995) and photosynthesis tends to increase pH.

NATURAL AND CULTURAL PROCESSES OF EUTROPHICATION

The idea that, over thousands of years, a natural developmental ontogeny of lakes occurs from deep and oligotrophic to shallow and eutrophic, then to a wetland, and then a terrestrial meadow has been present in the ecological literature for decades. The filling of lakes with sediments is a natural process because lakes are depressions in the watershed that collect sediments over time. This idea of a succession of lake types is applicable to many small to medium lakes but students should view it with caution, especially with respect to very deep lakes.

Highlight 18.2 Two examples of fertilizing lakes to increase fish production

The following are two examples of situations in which alterations in nutrient regimes led to attempts to fertilize ecosystems to increase fish production. In the first case, managers fertilized lakes on the coast of British Columbia (Vancouver Island) to increase survival of young salmon. As human activity decimated natural salmon runs, the numbers of adult salmon returning to spawn in small streams decreased substantially. The adult salmon die after spawning, and nutrients release from their carcasses and wash from the streams into the lakes as they decay. Historically, coastal lakes in the Pacific Northwest were likely more productive because of this fertilization, leading to increased survival and growth of young salmon. Concern over survival of juvenile salmon led to a project of artificial fertilization of 20 lakes over 20 years to determine if increases in survival and growth of juvenile salmon would result (Stockner and MacIsaac, 1996).

The fertilization led to approximate doublings in bacterial abundance, phytoplankton biomass and productivity, and zooplankton biomass. Growth and survival of juvenile sockeye salmon (*Oncorhynchus nerka*) increased more than 60%. The fertilization and associated costs were about $1 million/year, and calculated benefits in increased returns of adult sockeye were about $12 million/year (Stockner and MacIsaac, 1996). These estimates suggest that fertilizing nutrient-poor coastal lakes to levels similar to those thought to occur historically, or to those in pristine ecosystems, is economically feasible. If salmon runs ever recover to near historic levels, no fertilization will be required because the spent adult spawning salmon that acquired their nutrients in the ocean would die and release the nutrients to the system.

A fertilization project in the southwest United States was less successful. Lake Mead is a large reservoir in Nevada and Arizona that supports a sports fishery valued at approximately $7 million/year. Fish production in Lake Mead has decreased; largemouth bass (*Micropterus salmoides*) harvest has declined more than 90% since the 1960s. Researchers hypothesized that the closure of Glen Canyon Dam in 1963 lowered nutrient input into Lake Mead and led to declines in fish production. Scientists initiated a large-scale fertilization experiment over a 4-year period to assess the effect of increased nutrient input on fish production and water quality. This experiment resulted in a moderate decrease in water quality (increased taste and odor problems and chlorophyll *a*). However, increases in zooplankton and forage fish were not significant (Vaux et al., 1995). In this case, more fertilization might be necessary to stimulate fish production, but could cause degradation of the quality of the drinking water from Lake Mead. Furthermore, there is no guarantee that more fertilization would lead to increased fish production.

Large tectonic lake basins are generally oligotrophic, and likely will remain so for the majority of their histories (in the absence of human intervention). For example, Lake Baikal is millions of years old and about 1.5 km deep with up to 7 km of sediment (Fig. 7.5). The period when the Baikal basin fills and becomes a shallow productive lake would likely be very short relative to the entire geological life span of the lake. In long-lived lakes, long-term changes related to geological processes (e.g., deforestation related to glaciation) may lead to periods when lakes are mesotrophic or eutrophic and others when they are oligotrophic.

Paleolimnological methods using isotopic dating and preserved remains of algae in the sediments (primarily diatoms) can be useful for estimating the history of a lake's trophic state (Anderson, 1993). Such methods often reveal that lakes thought to be naturally eutrophic were more oligotrophic thousands of years ago (Anderson, 1995). For example, researchers analyzed the sediments of Lake Okeechobee, Florida, using lead isotopes to establish age of sediments back to the late 1800s. The dated sediments indicated that the phosphorus increased in the lake along with algal pigments (trophic state), with particular increases from the 1950s that correspond to the expansion of crop fertilization in the watershed (Engstroma et al., 2006).

Natural eutrophication can occur with watershed disturbances. Waters (2016) used paleolimnologic methods to show three periods when Lake Griffin, Florida had toxic blooms over the last 4,700 years. The last period was due to cultural eutrophication, but the earlier two correlated to changes in climate. In an interesting case, Spirit Lake in the northwestern United States was altered greatly following the volcanic eruption of Mount St. Helens. The eruption occurred on May 18, 1980, and was the equivalent of a 10-megaton nuclear explosion leading to massive input of downed timber, volcanic ash, and an abrupt temperature increase from 10°C to 30°C. Spirit Lake was deep and oligotrophic before the blast. After the eruption, Spirit Lake had a shallower basin with a large surface area, ultimately leading to increased macrophyte growth and production (Larson, 1993). Such a rapid and drastic change is rare in most natural lakes on human timescales of observation.

Cultural eutrophication is common on all continents but Antarctica. Cultural eutrophication occurs rapidly (relative to most geological processes) and can be difficult to reverse. Human activities that lead to cultural eutrophication include the use of agricultural fertilizers, livestock practices, watershed disturbance such as deforestation, burning fossil fuels, and release of nutrient-rich sewage into surface waters (Loehr, 1974). Road building also leads to increased erosion and infilling of lakes. Historical examples of eutrophication caused by watershed disturbance include road construction of the Via Cassia by the Romans (Highlight 18.3) and eutrophication caused by agriculture in early Mexico (O'Hara et al., 1993).

Eutrophication control can be costly. Hence, political battles over the relative importance of phosphorus control to solve eutrophication problems caused by humans

Highlight 18.3 Lago Di Monterosi: anthropogenic eutrophication, BC

In the 1960s, G. E. Hutchinson assembled a group of scientists to study the history of the Italian Lago Di Monterosi (Hutchinson and Cowgill, 1970). The scientists included paleontologists who worked on sediment cores from the lake, historians, a geologist, and limnologists. The group members were from Italy, the United States, and Britain.

Lago Di Monterosi formed by a volcanic blast about 35,000 years ago. It has remained shallower than 10 m since that time, and it approached a depth of 1 m during a very dry period about 10,000 years ago. Analyses of pollen and preserved remains of aquatic plants, animals, and microalgae suggest that the lake remained moderately productive until about 2,000 years ago. During approximately the first 30,000 years, the lake slowly evolved into a shallow oligotrophic lake, became slightly acidic, and contained some *Sphagnum*, indicating peat bog formation, at least along parts of the shores.

Historical records and archeological remains suggest that the Roman Empire built the Via Cassia by 171 BC. The romans probably built the paved road to improve speed of transit from Rome to and from the strategically important Tuscany. The road passed through the edge of the lake watershed (Ward-Perkins, 1970), which resulted in settlement and deforestation in the watershed.

Analyses of lake sediments dated with ^{14}C to the time when Via Cassia was built reveal a marked increase in the rate of sedimentation, a decrease of tree pollen, a decrease in the amount of aquatic plant pollen, and increased carbon and nutrient content of the sediments. These and other characteristics are consistent with deforestation of the watershed, increased sedimentation, more nutrient input associated with increased runoff, and greater productivity of the lake. This eutrophic state abated somewhat after the fall of the Roman Empire, and the lake has maintained a moderately eutrophic state since that time.

This study has several important implications. It demonstrates that humans have a long history of causing eutrophication and affecting habitats on a watershed scale. It also demonstrates that lakes do not necessarily undergo a constant succession from oligotrophy to eutrophy over geological time. Finally, this is an early example of study of a limnological problem that was tackled by assembling a team of specialists. It illustrates that limnology is a holistic subject, and that observations from both "hard" and "social" sciences facilitate study of ecologically and environmentally relevant questions. This groundbreaking approach to ecological issues has now been adopted many places to allow more effective study of environmental issues than would be possible with just one investigator.

can be intense (Edmondson, 1991). Perhaps the most important scientific verification of the role of phosphorus in eutrophication was the work headed by David Schindler (Biography 18.1) at the Experimental Lakes Area in Canada. These whole-lake experiments demonstrated that phosphorus and nitrogen additions, and not organic carbon additions, were clearly responsible for nuisance algal blooms (Fig. 18.1). The relative roles of phosphorus and nitrogen in lake eutrophication have been controversial in the scientific community (Highlight 18.4).

BIOGRAPHY 18.1 David Schindler

David Schindler (Fig. 18.5) is one of the most influential scientists who have studied human-caused pollution in aquatic systems. Although he has many publications and honors for his eutrophication work at the Experimental Lakes Area in Canada, he has conducted important research on basic ecosystem processes, acid precipitation, organic carbon contamination, and the influence of global change on aquatic ecosystems. He has also investigated the cumulative effects (multiple stressors) caused by anthropogenic inputs (Schindler, 2001).

Figure 18.5 David Schindler.

Dr. Schindler always loved lakes and ponds, but he entered the aquatic sciences by accident. He began college as a physics major but luckily a limnological laboratory hired him as a technician. After reading some books off the shelf there, he was hooked. For him, the three most influential books were Hutchinson's *Treatise of Limnology*, Vol. 1 (1957), Elton's book on animal invasions (1958), and Tinbergen's book on animal behavior (1951). Schindler has more than 200 publications, many in the top scientific journals. He has honorary PhD degrees from several universities. He has won major awards, including the Stockholm Water prize, and the G. E. Hutchinson Award of the American Society of Limnology and Oceanography. He has also received the Naumann-Thienemann Medal of the International Limnological Society, the Tyler Prize, and the Volvo Environment prize. Schindler is involved in many national and international committees and panels related to human impacts on aquatic systems.

Schindler is proud that all of his three children chose careers involving aquatic systems. His family also joins him in competitive dog sled racing, his favorite hobby and sport.

Schindler suggests that all undergraduates work on writing skills because communicating and publishing scientific discoveries are crucial to a successful scientific career. He sees resurgence in research on eutrophication, particularly on problems related to nonpoint sources of nitrogen and phosphorus. In 1998, for the first edition of this book, Schindler predicted there would be a realization of the problems associated with mercury and organic contaminants. Since that time, mercury contamination has been widely recognized, and problems associated with organic pollutants, including ecoestrogens, have become a matter of increased concern.

Highlight 18.4 Controversies over nitrogen and phosphorus control in lakes

The controversy over limitation by nitrogen or phosphorus has been an ongoing concern of limnologists studying eutrophication. Smith (1982) noted that algal biomass in lakes statistically related to both nitrogen and phosphorus, and that algal biomass was greater when both nutrients were relatively abundant. This early stoichiometric approach implied a revised view that considered colimitation as a possibility. Reanalysis of the data from the Experimental Lakes Area also indicated to some that nitrogen and phosphorus colimitation might be important (Fee, 1979). Researchers also noticed that bioassays in lakes commonly gave greater responses when they added both nitrogen and phosphorus together and nitrogen responses alone were common. These short-term experiments suggested that N-limitation or colimitation by phosphorus and nitrogen is common (e.g., Dodds et al., 1989; Elser et al., 1990; Morris and Lewis, 1988). The simple view that phosphorus limits freshwaters is questionable.

The scientific tensions over nitrogen and phosphorus limitation in lakes and coastal areas came to a head when Lewis and Wurtsbaugh (2008) published a critique of the idea of strict control of freshwater phytoplankton blooms by phosphorus limitation. This idea received direct response in the literature (Schindler, 2012; Schindler and Hecky, 2008; Schindler et al., 2008). The crux of the argument against nitrogen limitation is that N_2-fixation (use of atmospherically derived N_2) can ultimately make up for any nitrogen deficiency, so control of phosphorus inputs alone should be all that is necessary. David Schindler (1974) expressed this view early, but it is still controversial (Conley et al., 2009). The concept that N_2 fixation will make up for any deficiency given abundance of other elements, taken to its logical conclusion, suggests that it is an unnecessary expense to control nitrogen pollution because given enough time, nitrogen will accumulate in the ecosystem to match the level of phosphorus inputs.

Dr Schindler has argued that short-term bioassays simply cannot be relied upon to indicate limitation over the long term (Schindler et al., 2016), as they diverge rapidly, become very unrealistic, and cannot be maintained in a close enough to natural state to allow natural populations of nitrogen-fixing cyanobacteria to develop. Thus, bottle experiments cannot represent the long-term balance of nutrient inputs and outputs from natural lakes. The opposing arguments are that: (1) this balance may not be reached in timeframes relevant for lake management (Scott and McCarthy, 2010), (2) higher chlorophyll yield per unit nitrogen (i.e., greater bloom intensity or probability that bloom conditions will occur with more N availability across natural lakes (Smith, 1982)) is a result that does not depend on bottle experiments, and (3) denitrification continuously causes losses of nitrogen (Paerl et al., 2016). Further concerns over nitrogen pollution include the potential for influences of nitrogen on biotic integrity. Species declines across the United States are linked to nitrogen pollution (Hernández et al., 2016). In addition, the toxin microcystin is a nitrogen-rich molecule and laboratory and field studies suggest that nitrogen control might also lower toxin production of *Microcystis* blooms (Horst et al., 2014). Statistical analyses also tie toxin production to TN in lakes across the United States (Yuan and Pollard, 2017). However, toxicity of blooms correlated to phosphorus concentrations in the Great Lakes (Harke et al., 2016). High phosphorus to nitrogen ratios in Canadian lakes (Orihel et al., 2012) and iron may also play a role in toxin formation by binding phosphorus (Orihel et al., 2016). Off odors caused by cyanobacteria might require both nitrogen and phosphorus increases, but not either alone (Olsen et al., 2016).

The stakes in these scientific battles are real; controlling nutrient pollution can cost money. However, assuming phosphorus alone limits freshwater algal blooms over time-scales relevant to management seems imprudent given the opposing data. At the very least, the precautionary principle suggests that controlling both nitrogen and phosphorus is the prudent approach. Fortunately, many control methods such as using wetlands to remove nutrients, adding only as much fertilizer as cropland needs, maintaining riparian protection, and isolating livestock wastes from water control both nitrogen and phosphorus.

RELATIONSHIPS AMONG NUTRIENTS, WATER CLARITY, AND PHYTOPLANKTON: MANAGING EUTROPHICATION IN LAKES

Vollenweider (1976) convincingly documented the relationships among nutrient loading, algal biomass, and lake clarity. This correlation of nutrients with lake autotrophic state represented a milestone in lake management because it allowed managers to predict the outcome of nutrient control strategies. As such, it is also a great triumph for predictive ecology. The watershed forms the natural unit for this approach to nutrient management (Likens, 2001). The models are based on empirical data relating watershed loading to in-lake nutrients and nutrients to algal biomass (Figs. 18.4 and 18.6)

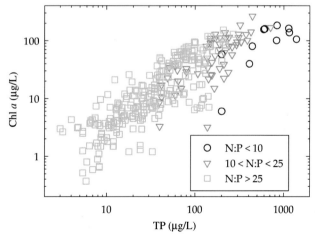

Figure 18.6 Relationship between mean growing season concentrations of total phosphorus (TP) and chlorophyll in 228 temperate lakes coded by N:P ratios (corrected data plotted following Smith, 1982).

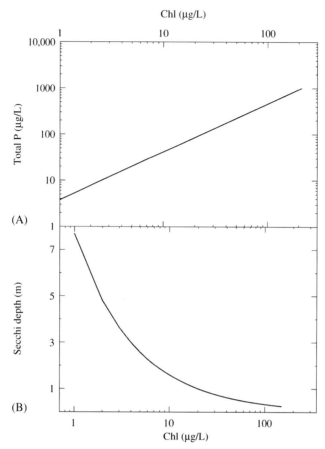

Figure 18.7 A nomogram relating epilimnetic chlorophyll concentration to total phosphorus (A) and Secchi depth (B) (based on equations in OECD, 1982). This graph can be used to estimate changes in clarity related to a known change in total phosphorus.

and provide a conceptual framework that links nutrient supply to lakes with phytoplankton biomass and water clarity (Fig. 18.7).

A simplified view of the sequence of events that can occur to mitigate eutrophication includes: (1) identifying a lake with problems, including determination of uses that interfere with the desirable condition of the lake; (2) characterization of the system, including lake morphology, land use in the watershed, nutrient loading into the lake, lake water retention, and sedimentation rates; (3) identification of feasible strategies for nutrient control considering both point and nonpoint sources; (4) projecting the influence of management actions on nutrient concentrations in the lake;

(5) predicting the response of chlorophyll to lower nutrient concentrations in the lake; (6) assessing the potential effect of decreased chlorophyll on lake clarity; (7) assessing if the projected costs of the nutrient control strategies justify the predicted benefits to the lake (O'Riordan, 1999); and (8) if nutrient control or mitigation strategies are instigated, monitoring the system to determine if the sought-after improvements have actually occurred (Fig. 18.8).

Lowering *external loading* (supply of nutrients from outside the system) is generally necessary to control eutrophication. The lowering of external loading usually incurs some cost; so, lake managers may need to estimate the amount of improvement in water quality that will result from a set amount of nutrient control. This estimation involves calculation of in-lake phosphorus concentrations and subsequent algal biomass.

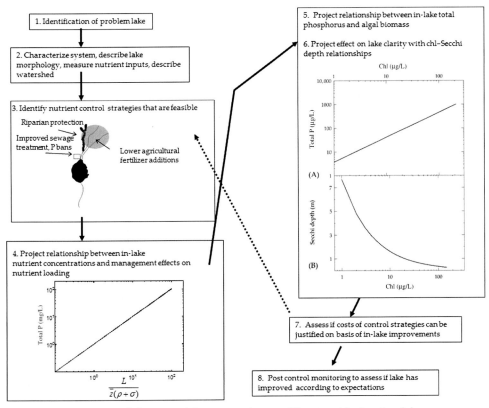

Figure 18.8 A diagram of the potential steps used to modify eutrophication in a lake.

ADVANCED: EMPIRICAL RELATIONSHIPS USED TO PREDICT CONTROL OF EUTROPHICATION

Equations allow prediction of the influence of altered nutrient regimes on productivity for a variety of lake types. We provide a very general description of such equations as an introduction to the method. A more detailed explanation is provided by Cooke et al. (1993) and Ryding and Rast (1989). Managers most often use equations for phosphorus because they generally assume that phosphorus limits primary production in lakes (Correll, 1999; Schindler, 1974). However, some have suggested that nitrogen is more likely limiting in tropical lakes (Golterman and de Oude, 1991), and bioassays (see Chapter 17) indicate that colimitation by nitrogen and phosphorus is likely in many lakes.

The following is a general, steady-state equation to calculate in-lake total phosphorus concentration:

$$TP = \frac{L}{\overline{z}(\rho + \sigma)}$$

where TP is the total phosphorus in mg/m^3 (μg/L), L is the phosphorus loading in mg m^{-2} year^{-1}), $\overline{z}$ is the mean depth in m, ρ is the flushing rate in year^{-1}, and σ is the sedimentation rate in year^{-1}, which is roughly equal to $10/\overline{z}$.

The general steady state equation represents one of the simplest cases. It accounts for sources and losses of phosphorus in the lake. A similar equation can also be used to estimate TN. The sources are loading from rivers, groundwater, and atmosphere. Losses are from washout (flushing and sedimentation). Assumptions include steady-state phosphorus concentration, complete mixing of inputs, constant sedimentation, little fluctuation of loading over time, and limited phosphorus input from sediments (*internal loading*). Relationships that are more complex are available to deal with exceptions to most of these assumptions (Cooke et al., 1993).

In practice, L is determined by measurements of total phosphorus in inflowing streams; atmospheric deposition and groundwater inputs are generally ignored. Nutrient input into streams is often heavily dependent on land-use patterns, which we will discuss in the next section. Groundwater input may be difficult to determine, particularly in heterogeneous geological substrata or where septic inflows create areas with exceptionally high phosphorus influx. Determination of mean depth and calculation of flushing rate require morphological mapping of the lake basin and hydrologic measurements.

Sedimentation rate can be highly variable among and within lakes, depending on characteristics such as fetch, epilimnion depth, and type and phosphorus content of sediment (i.e., considerable variance occurs in the relationship $\sigma = 10/\overline{z}$). Direct determination of sedimentation rates may provide more accurate estimates of phosphorus loss from the epilimnion.

Once the TP in the lake is calculated, the next step is to calculate the chlorophyll that this amount of nutrient could support. A strong relationship exists between total phosphorus and chlorophyll (Figs. 18.6 and 18.7) when values for many lakes are plotted. Equations can be derived from such data sets; the following has been proposed by Jones and Bachmann (1976) using data from 143 lakes:

$$\log_{10} \text{chl } a = 1.46 \log_{10} \text{TP} - 1.09, \quad r^2 = 0.90$$

where chl a is the summer mean chlorophyll in mg/m^3 ($\mu g/L$), TP is the summer mean total phosphorus in mg/m^3 ($\mu g/L$), and r^2 is the proportion of the variance that can be described by the relationship. We demonstrate use of this and the preceding equation in Example 18.1.

Smith (1982) used a larger and more variable data set and demonstrated that he could account for more variance in his data set when total nitrogen was considered in addition to total phosphorus. If a plot of these chlorophyll values versus total

---

Example 18.1 Using loading equations to predict response to nutrient control

If a lake manager is able to lower mean total summer phosphorus inputs to a lake by 50% from an initial loading value of 1g m^{-2} year^{-1}) total P, what will be the expected decrease in chlorophyll given a mean depth of 10 m, a flushing rate (ρ) of 2 year^{-1}, and a sedimentation (σ) rate of 1 year^{-1}? How does this translate into increased Secchi depth?

First, we need to solve for initial chlorophyll (in practice, this will probably be a measured value). To do so, we solve for TP concentration first, then use the equation relating TP to chlorophyll. To calculate TP, do not forget to convert phosphorus loading (L) into mg m^{-2} year^{-1} and that $mg/m^3 = \mu g/L$:

$$\text{TP} = \frac{1000}{10(2+1)} = 33.3 \, \mu\text{g P/L}$$

If we rearrange the equation relating chlorophyll to TP, we get:

$$chl = 10^{1.46 \log_{10} \text{ TP} - 1.09} = 13.6 \, \mu\text{g chl/L}$$

A 50% decrease in loading will reduce the TP by half to 16.7 $\mu g/L$. Using the second equation, this concentration will yield 4.9 $\mu g/L$ chl. Inspection of Fig. 18.7, constructed using slightly different equations, allows the calculated relationships to be checked. If we use the nomogram on Fig. 18.7, we can also see that the Secchi depth should increase from approximately 1.5 to 2.7 m. An additional issue is variance; it is beyond the scope of this discussion, but a significant amount of variance occurs in the empirical relationships and managers and planners should consider this source of uncertainty in an actual management situation to avoid disappointment if the response is not all that they predicted.

phosphorus is divided into categories of TN:TP ratios (Fig. 18.6), we see that chlorophyll per unit phosphorus is lower when the relative amount of nitrogen is low. Smith (1982) proposed a corrected equation ($n = 311$ lakes) equation relating algal biomass to chlorophyll using nitrogen and phosphorus:

$$\log_{10} \text{chl } a = 0.640 \log_{10} \text{TP} + 0.587 \log_{10} \text{TN} - 0.753, \quad r^2 = 0.75$$

Units and variables are the same as in the previous equation. This equation is probably most useful in high phosphorus waters (Cooke et al., 1993) and may not apply to tropical lakes (Sarnelle et al., 1998). Tropical lakes also generally have greater variance in the chlorophyll–total phosphorus relationships than do the more studied temperate lakes (Huszar et al., 2006).

The probability that an algal bloom will occur, particularly a bloom of cyanobacteria, may be more important than average chlorophyll values. A 21-year data set on phosphorus loading to Lake Mendota, Wisconsin allowed calculation of the probability of algal blooms (Lathrop et al., 1998). In this analysis, with no change in current levels of loading, there was a 60% chance of a cyanobacterial bloom on any given summer day. When loading was deceased by half, there was only a 20% chance of a bloom. Beaulieu et al. (2013) analyzed data for over 1000 lakes in the United States and found that high levels of nitrogen and phosphorus (with TN being the most important) as well as temperature were the best factors predicting cyanobacterial blooms, but the model only accounted for 10% of the variability. Beaver et al. (2014) analyzed the same data set and found nitrogen, temperature, and dissolved organic carbon positively correlated with the occurrence of the cyanobacterial toxin microcystin, and linked these to land use in the watersheds of the lakes. Yuan et al. (2014) found that a model with TN and algal biomass was the simplest model to predict toxic cyanobacterial blooms. Prediction accuracy can also improve if wind can be accounted for (Wang et al., 2016); windy conditions do not allow formation of the surface scums that favor cyanobacterial dominance. Understanding local factors is most important in predicting cyanobacterial toxin presence on the basis of statistical analyses of over 1000 US lakes (Taranu et al., 2017). These studies illustrate that managers deal with variable and unpredictable systems, and that nutrient control methods may decrease, but not preclude, the probability of a noxious bloom. The probability of a cyanobacterial bloom in a lake increases sharply above about 30 μg/L total phosphorus (Downing et al., 2001). Newer statistical techniques, such as Bayesian network modeling and regression trees could improve predictive ability for anticipating conditions leading to cyanobacterial blooms (Rigosi et al., 2015), as well as formation of compounds with bad tastes and odors (Kehoe et al., 2015).

MITIGATING LAKE EUTROPHICATION

Eutrophication management can begin with control at the nutrient source (treating the cause) or within-lake treatment (treating the symptom; Table 18.3). Treating causes is usually more effective in stopping negative effects of pollutants than treatment after contamination. Part of the reason that treating causes before eutrophication is important is that recovery is slow; a meta-analysis of lake eutrophication showed that after nutrient control, systems were on average 30% recovered, even decades later (McCrackin et al., 2017). Just as with other pollutants, treating the cause of eutrophication by controlling nutrient sources is generally most cost-effective over the long term.

Nutrients can come from *point sources*, such as sewage outfalls, factory effluents, septic tanks, and waste flowing from the surface of intensive livestock operations. Nutrients can also come from *nonpoint sources*, such as agricultural fields, urban storm runoff systems (lawn fertilizers), disturbance of watersheds, addition of fertilizers to golf courses and pastures, and atmospheric deposition. Even heavy use by waterfowl that feed outside the aquatic habitat then return to the water to excrete can lead to substantial nutrient loading (Olson et al., 2005). Nutrient input from point sources is relatively easy to determine because it is concentrated and sampling is easy.

Establishing rates of input from nonpoint sources is more difficult, but broad ranges of nutrient loss rates from different types of land uses are known (Fig. 18.9). Agricultural and urban uses lead to the greatest degree of runoff, with human population density in a watershed demonstrating a significant positive correlation to nitrogen and phosphorus runoff (Caraco and Cole, 1999). A 50% increase in agricultural and urban land use can result in a doubling of TN runoff (Fig. 18.10). Phosphorus inputs from nonpoint agricultural sources can be considerable (Metson et al., 2017). Urban areas in developing countries where sewage treatment is minimal can cause drastic increases in phosphorus when even a small proportion of the watershed or the riparian zone is developed. In some cases, temporal variance in runoff can be extreme. For example, the first large rains after extended drought in the Midwest United States led to a >30% increase in annual N export and violations of nitrate standards downstream (Loecke et al., 2017). Thus, the human uses in the watershed above the water body of interest need to be characterized to estimate the approximate impact of different land uses on nutrients flowing into the system.

Identifying land uses that can increase nutrient inputs has become substantially easier because of advances in information availability and mapping capabilities. A *geographic information system* (GIS) is a powerful tool for mapping land-use patterns and effects of changes in those patterns as related to eutrophication (Hunsaker and Levine, 1995). GIS systems consist of a series of map layers, in which an analyst assigns different attributes to a spatial grid. The layers can represent different attributes or the

Table 18.3 Methods for controlling causes and symptoms of eutrophication associated with excessive phytoplankton

Method	Explanation	Positive aspects	Negative aspects
Control of cause			
Control of point sources	Bans on phosphorus in detergents, tertiary sewage treatment	Clean up at source	Tertiary treatment can be expensive. Generally ineffective in lakes where hypolimnion goes anoxic
Control of nonpoint sources	Control of watershed disturbance, feedlot effluent, intact riparian zones	Clean up at source, potential for long-term improvement	Can be politically unpopular. Generally ineffective in lakes where hypolimnion goes anoxic
In-lake control of symptoms			
Dilution and flushing	Use of low nutrient water to dilute nutrients and phytoplankton	Where practical can be an easy, inexpensive solution	Requires large supply of low-nutrient water. Usually only practical in smaller lakes
Destratification, mixing	Keeping O_2 in the hypolimnion keeps phosphorus in sediments. Deeper mixing increases light limitation of phytoplankton	Rapid results	Energy required to mix and de-stratify lakes. Not practical on large lakes. Can select against desirable cold water fishes.
Hypolimnetic release	Release nutrient rich water from hypolimnion	Easy to implement in reservoirs with possibility of hypolimnetic release	Costly to pump water, greater nutrient input to downstream systems.
Biomanipulation	Manipulate food web by increasing piscivores or decreasing planktivores to increase numbers of zooplankton that graze phytoplankton	Can lead to rapid increases in water quality with minimal costs	Unpredictable results, does not work on extremely eutrophic systems. May lead to excessive macrophyte growth in shallow systems
Alum	Alum seals phosphate in hypolimnion, flocculates, and settles phytoplankton	Rapid response	May need repeated application, may be cost-prohibitive in large lakes
Copper treatment	Copper kills phytoplankton	Acts within days	Repeated treatment necessary, can lead to sediment contamination and negative effects on nontarget species

Source: After Cooke et al. (1993).

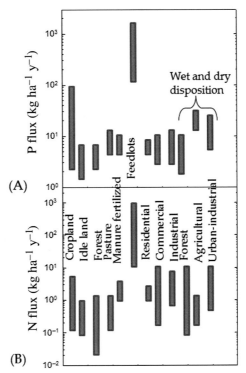

Figure 18.9 Ranges of phosphorus and nitrogen fluxes from different land-use categories and the rates of nitrogen and phosphorus loading from atmospheric deposition. *Adapted from Loehr et al. (1989).*

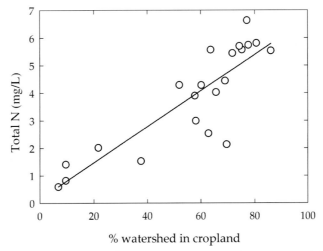

Figure 18.10 Relationship between total nitrogen concentration in streams and the percentage of land in agricultural and urban use from a variety of watersheds in Kansas. *Data courtesy Kansas Department of Health and Environment.*

changes in an individual attribute over time. Data layers include land use, drainage networks, soil types, population densities, and elevations. Data sets that are more detailed continue to become available, as these are useful for many other management purposes. These data layers now pair with the extensive data sets on nutrients that many governments now make available online, and they provide a powerful, open-source tool for landscape-scale analyses of ecological patterns.

Control of nutrient sources

Control of nonpoint sources is often difficult because it requires coordination across watersheds and cooperation of many different types of landowners. Patience is also required when enacting steps to control nutrient loading because recovery generally takes a decade or much longer to achieve after nutrient control measures begin (Bennett et al., 1999; Gerling et al., 2016; Jeppesen et al., 1998). Generally, agricultural land use is the most important source of nonpoint source pollution (Kronvang et al., 1995). Typical ways to control nonpoint nutrient input include lowering fertilizer applications (to lawns in urban regions and to crops in agricultural regions), proper timing of fertilizer application, establishing erosion-control strategies (e.g., maintaining riparian vegetation and minimizing exposed soil), keeping livestock out of streams and ponds with fences and by providing stock tanks, and restoration of natural vegetation. Control of nearshore land use in lakes could be important in controlling cyanobacterial blooms (Doubek et al., 2015). Regulation and education are necessary components to nutrient control programs. Given the variety of sources for nonpoint pollution and the complexity of ecosystem valuation, economic analysis of the costs and benefits of nutrient control can be difficult (Carpenter et al., 1998).

Control of nonpoint sources can have benefits beyond the local effects of lowering lake, stream, and groundwater eutrophication. Since the 1990s, a large anoxic zone has increased in size in the Gulf of Mexico, and the reduced O_2 has damaged fisheries in the region. This anoxic zone results from river-borne nutrients (Turner and Rabalais, 1994) and continues to increase in size. Similar zones have developed in more than 400 near-coastal marine areas (Diaz and Rosenberg, 2008).

Interaction with added nutrients lower silicon availability, which can shift downstream algal communities. There is more nitrogen and phosphorus in rivers worldwide but less increase in silicon. The natural weathering rates of silicon remain constant and humans have had little influence on these rates. Nitrogen and phosphorus can stimulate diatom growth, leading to uptake and burial of silicon. This shift in stoichiometry and nutrient amounts has led to increases in algal productivity and shifts in composition of near-shore marine plankton communities (Justic et al., 1995a).

Therefore, increased nonpoint source nutrient pollution has worsened eutrophication in marine coastal regions throughout the world (Justic et al., 1995b).

One of the first steps toward lowering phosphorus input into watersheds from point sources is generally a ban on phosphate-containing detergents. This restriction can decrease the phosphate entering sewage works by half. In these situations, detergents for automatic dishwashers and automatic car washes are generally exempt because of the reduced efficacy of low-phosphate alternatives. This approach is more cost effective than chemical removal of phosphorus at the sewage treatment plant.

Control of point sources generally puts the majority of the financial burden on fewer institutions (e.g., a municipal sewage treatment plant or a specific factory) than does the control of nonpoint sources. However, regulation of point sources is easier from the regulators' point of view and regulators often write laws specifically to control point sources. Removing phosphorus from waste streams can be costly. One method of removal involves chemical treatment with alum or Fe^{3+} to precipitate the phosphate. The precipitate then settles and the low phosphorus water is released. This method generally can bring effluent concentrations down to $0.2-1$ mg phosphorus/L (Clasen et al., 1989).

Nitrogen can be removed by converting ammonium to ammonia gas by raising the pH. The solution is stripped of ammonia by bubbling with air, then the water is neutralized and released. When the gas enters the atmosphere, it must come down as atmospheric deposition elsewhere. Alternatively, waste can go through an aerobic treatment to convert the ammonium to nitrate, followed by an anoxic phase in which nitrate is used in denitrification. The resulting N_2 gas enters the atmosphere. Finally, wetlands can remove nitrogen through plant uptake and denitrification, as we discuss later in this chapter.

Removing nitrogen and leaving phosphorus in a system may not solve eutrophication problems because many species of cyanobacteria that form undesirable blooms can use N_2 gas via fixation and do not need nitrate or ammonium to bloom. In the Experimental Lakes Area in Canada, phosphorus addition without concurrent nitrogen addition eventually led to large blooms of the nitrogen-fixing cyanobacterium *Aphanizomonon*. This effect continued for several years until nitrogen fixation brought the lake's nitrogen content to match the phosphorus addition rates (Hecky et al., 1994). Even if total biomass decreases somewhat, undesirable species may be selected for when only nitrogen is removed. Because these cyanobacteria have gas vesicles and are buoyant, they concentrate on the surface. They are more apparent as "scum" and visually indicate the high level of nutrients in the lake, especially phosphorus. This example illustrates the importance of limiting factors and indicates how phosphorus limitation can be important in regulating algal communities.

Treatment in the lake

When a lake becomes eutrophic, managers can use any of several methods to treat the symptoms. Treatment becomes more difficult when O_2 has disappeared from the hypolimnion because phosphate that would bind with Fe^{3+} in an oxic hypolimnion can enter from the sediments, drastically increasing rates of internal loading. The $FePO_4$ locked in sediments dissociates into Fe^{2+} and PO_4^{3-} in the anoxic hypolimnion and diffuses into the water column (Fig. 14.9). The phosphate released from the sediments then can become available to phytoplankton when the lake mixes. Phytoplankton use luxury uptake to acquire and store this phosphate, which provides nutrients for future blooms. The problem is most severe in medium-depth lakes and under warmer conditions (Genkai-Kato and Carpenter, 2005), although some lakes recover more quickly than expected (Müller et al., 2014).

Managers have devised several strategies to combat this resuspension of phosphate. We summarize these strategies here and in Table 18.3, Cooke et al. (1993) discuss them in detail elsewhere. In smaller lakes, dredging of sediments combined with other algal control methods may be the most effective treatment (Waajen et al., 2016).

One method to counteract the symptoms of eutrophication is to provide O_2 to the hypolimnion so phosphate remains in the sediments. This *hypolimnetic aeration* requires large amounts of energy; it can be prohibitively expensive in any but the smallest of lakes. If the main goal is to protect a cold-water fishery, then only a small part of the hypolimnion needs to be oxygenated and care must be taken to not break stratification or the entire lake will be too warm for the fishes. This approach provides low-temperature, oxygenated water as a refuge for salmonid species.

Aeration does not always lower algal biomass (Soltero et al., 1994). However, aeration for many consecutive years can successfully mitigate water quality problems in shallow urban lakes (Lindenschmidt and Hamblin, 1997). Destratification can also keep phosphate in the sediments and can have an inhibitory effect on nuisance cyanobacteria by increasing mixing depth and erasing the advantage they gain by floating on the surface (Visser et al., 2016). This mixing can select for green algae and diatoms instead of the toxic cyanobacterium *Microcystis* (Visser et al., 1996).

The use of copper as a method to control algae has been widespread. Copper is particularly toxic to cyanobacteria and removes the objectionable algae. In hard waters copper can precipitate as copper carbonate, so that repeated applications are necessary to achieve results. The copper contaminates the sediments and can eventually poison other aquatic life (such as crustaceans) if pH is acidic (pH is commonly acidic in anaerobic sediments). Furthermore, the copper may break the cells of cyanobacteria and release toxins into the water (Lam et al., 1995). Extended treatment with copper may become more problematic than the condition it was supposed to cure (Cooke et al., 1993).

Lake managers have used several compounds in attempts to bind phosphate and keep it buried in the sediments. Alum is one of the most commonly used chemicals. The expense is high to treat any but small lakes. Lanthanum modified bentonite has been tested to reverse eutrophication. This compound strongly binds phosphate. The compound can be toxic, but appears to control phosphorus concentrations in some lakes and remains below toxic levels except in low alkalinity and potentially saline waters. Its effectiveness is limited in high organic carbon waters. The accumulation problem seen with repeated treatment with copper could also occur with lanthanum (Copetti et al., 2016). Magnetic nanoparticles are also being investigated for in-lake phosphorus control (Funes et al., 2016), but we know little about the environmental effects of such particles.

Hydrologic methods can decrease eutrophication. For example, water drawdowns in reservoirs dilute nutrients and decrease blooms (Pawlik-Skowronska and Toporowska, 2016). Another method of control is to release water from the hypolimnion. This is most easily accomplished in reservoirs that have hypolimnetic release tubes, as lakes and reservoirs with surface release require pumping to remove the water from depths. The nutrient-rich water that builds up in the hypolimnion is removed in this method. This method, if applied correctly (not leading to destratification of lakes), can improve water quality (Nurnberg, 2007). However, it does simply transport the higher nutrient water downstream.

Addition of barley straw may control phytoplankton blooms. Some studies of this method have shown measurable lowering of algal biomass (Barrett et al., 1996; Everall and Lees, 1996; Ridge et al., 1999), and dissolved phosphorus (Garbett, 2005). Repeated treatment with barley straw was effective in lowering cyanobacterial populations and decreasing taste and odor problems in one drinking water supply reservoir (Barrett et al., 1999). The mechanism for this control is not well established, but the microbes degrading the barley could release of inhibitory compounds (Iredale et al., 2012).

Increasing the biomass of phytoplankton grazers could decrease algal biomass. Alterations of the food web (top-down control) to increase zooplankton biomass have been advocated to control algal blooms. We discuss this method in Chapter 20. Some fishes could also lower cyanobacterial blooms. Nile tilapia (*Oreochromis niloticus*) directly eat cyanobacteria and could be useful in controlling blooms in some situations (Salazar Torres et al., 2016).

Macrophyte removal

One symptom of eutrophication in shallow lakes can be excessive growth of macrophyte vegetation (Chambers et al., 1999). Some macrophyte growth is a healthy part of aquatic ecosystems; the plants provide habitat for other desirable species (e.g., fishes) and stands of macrophytes can prevent unwanted sediment suspension (Bachmann et al., 1999). However, macrophytes can interfere with recreation, clog water flow structures, lead to low O_2 conditions, and cause taste and odor problems. Thus, removal of macrophytes is desirable at times. Physical, chemical, or biological controls can remove macrophytes (Table 18.4).

Physical control methods include direct harvesting, sealing aquatic sediments with plastic to prevent establishment of rooted macrophytes, shading, and alteration of water level (dry down; Wade, 1990). The managers use the lack of ability to withstand freezing to control macrophytes in some temperate zone reservoirs. They drop water levels in reservoirs during freezing weather, killing some species (Murphy and Pieterse, 1990).

Chemical control methods require application of herbicide. Preferable herbicide properties include a limited lifetime in the water, toxicity primarily to target plants, and no bioconcentration in the food web. In general, physical methods are more expensive than chemical methods (Murphy and Barrett, 1990). Both methods often require repeated use because the macrophytes can recolonize.

Table 18.4 Methods for controlling macrophytes

Method	Explanation	Positive aspects	Negative aspects
Physical	Dredge or cut out macrophytes	Removal rapid; high control on area treated	Costly; ineffective in large systems; needs to be repeated
Chemical	Apply herbicides	Removal rapid; moderate control on area treated	May affect nontarget species. Needs repeated application. Biomagnification of toxins
Sealing sediment	Seal sediments with plastic film to stop macrophyte establishment	Also keeps phosphate from reentering lake, inhibiting phytoplankton blooms	Expensive, not effective in large systems; plastic lake bottom may be aesthetically unpleasant
Biological control	Find organisms that specifically graze macrophytes	Inexpensive, lasting control	May eat desirable species; may become pest species; may be impossible to eradicate after introduction

Whole-lake control of nutrients (Table 18.3) can be useful in controlling excessive macrophytes.
Source: After Cooke et al. (1993).

Biological control methods include use of fungi, insects, and herbivorous fishes. Selective organisms are the most desirable agents of biological control. Nonselective control agents can harm beneficial species and can be extremely difficult to eradicate after they become established. We discuss some problems associated with unwanted species introductions in Chapter 11.

A common biological control method is the introduction of the herbivorous grass carp (*Ctenopharyngodon idella*). These fish are nonselective and can remove a large amount of both beneficial and unwanted aquatic vegetation. The grass carp are also a potential source of protein. A concern regarding grass carp is that they will escape their area of introduction and remove desirable macrophytes elsewhere. Triploid grass carp that are not capable of reproduction are available for use in control programs. In addition, spawning requires a temperature of at least 17°C, so reproduction is unlikely in some cool-temperate habitats (van der Zweerde, 1990), but this could change with climate warming. The use of grass carp is a relatively inexpensive control method, and therefore a common problem with this method of control is adding too many fish. In this case, the carp strip all the macrophytes from a lake and the lake can switch to an algal-bloom dominated system.

In general, there is a balance between too many macrophytes and complete removal. Macrophytes form a refuge for larval fishes and can stimulate a productive fishery. The macrophytes also tend to lower algal blooms and lead to clearer water; removal of macrophytes can result in undesirable algal blooms (Genkai-Kato and Carpenter, 2005) and shift the fish community to more planktivorous species (Michaletz and Bonneau, 2005). For example, decomposing stalks of the reed *Phragmites* can inhibit cyanobacterial growth (Nakai et al., 2006). Excess addition of grass carp commonly results in complete removal of macrophytes, increases in suspended sediment, and greater probability of phytoplankton blooms.

MANAGING EUTROPHICATION IN STREAMS AND RIVERS

The relationship in streams between benthic algal biomass and water column nutrients is not as strong as that between nutrients and phytoplankton in lakes (Fig. 18.11). This lack of predictability occurs because floods, grazing, and light limitation can lower algal biomass even when nutrients are high. Eutrophication in streams has become a serious issue, and control of nutrients is likely to be the best solution to the problem (Dodds et al., 1997). Problems with eutrophication in streams include: (1) the negative aesthetic impact of excessive algal growth, alteration of food webs, taste and odor problems, low O_2, and high pH (Dodds and Welch, 2000), (2) shifts in macrophyte community structure (Bowden et al., 1994), (3) excessive macrophyte growth (O'Hare et al., 2010) potentially leading to interference with channel flows (Ferreira et al., 1999), and (4) alteration of carbon retention (Rosemond et al., 2015).

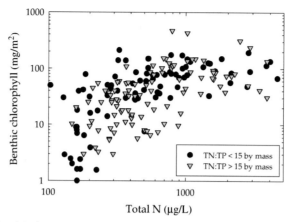

Figure 18.11 Relationship between total nitrogen concentration in the water column and mean benthic chlorophyll from about 200 temperate streams coded by N:P ratios. *Data from Dodds et al. (2006).*

Cyanobacterial blooms can occur in rivers and are expected to be worse under higher temperatures and in areas where climate change is increasing water retention times (Cha et al., 2017). Benthic cyanobacteria growths in streams can be a result of eutrophication and some of these are toxic (Wood et al., 2012; Quiblier et al., 2013; Wood et al., 2017). In addition, there has been increased concern about eutrophication in estuaries and near coastal marine environments caused by nutrients delivered by streams (Conley, 2000; Turner and Rabalais, 1994).

Streams, even more so than lakes, can be driven by heterotrophic processes altered by nutrient addition. Consequently, control of nutrients in streams is important regardless of the natural condition of light and periphyton dominance (Dodds, 2007). Bini et al. (2014) suggest that nutrient pollution in streams is a major cause of decreased diversity and increased proportions of tolerant species. Bumpers et al. (2015) fertilized streams dominated by detrital carbon input, which stimulated salamander production at the top of the food web. Nutrient concentrations in the water stimulate decomposition of leaves in streams (a fundamental energy flow pathway in stream food webs) more than the nutrients in the leaves themselves (Biasi et al., 2017).

Historically, the worst problems with sewage input to streams and rivers were the loss of O_2 and the introduction of pathogenic bacteria (Fig. 18.12A). This sewage input represented an artificial increase in heterotrophic state. In many areas, sewage treatment does not lower total nutrient input much but does lower loads of biochemical oxygen demand and pathogenic bacteria (primary and secondary treatment are common, but tertiary treatment to remove nitrogen and phosphorus are rarer). Older sewage treatment plants can release significant amounts of ammonium,

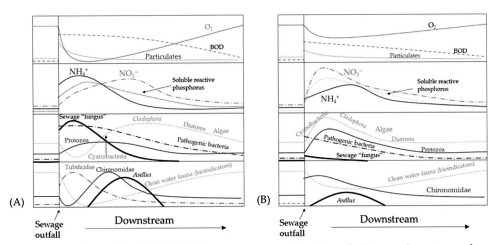

Figure 18.12 Chemical and biological parameters as a function of distance downstream from untreated sewage effluent (A) and from a modern sewage treatment plant (B). *A, Redrawn from Hynes (1960).*

which is toxic to aquatic life. More advanced sewage plants oxygenate treated water, and nitrification converts most ammonium to nitrate, leading to different chemical and biological patterns downstream from the sewage outfall (Fig. 18.12B). Many modern sewage plants remove nitrogen via denitrification, requiring management of the processes described in Chapter 14.

CASE STUDIES OF EUTROPHICATION IN LAKES AND LOTIC SYSTEMS

Examples of pollution control can aid understanding of general ecological concepts and their application to solving eutrophication problems. This section outlines some successes and failures of lake managers that illustrate the complexity of the issues involved and the variety of problems that have arisen.

Lake Washington

The case of Lake Washington (Fig. 18.13) is one of the great triumphs for limnologists, lake managers, and environmentalists. This lake has experienced a strong recovery on the basis of scientific understanding of limnological properties and particularly the efforts of the late limnologist Prof. W. T. Edmondson (Biography 18.2). Edmondson (1991) wrote a fascinating account of the political and scientific aspects of cleaning up this lake.

Lake Washington is a large (28-km long and 65-m deep) lake that forms the eastern border of the city and suburbs of Seattle, Washington. Lake Washington is a monomictic lake (summer stratification) that was historically oligotrophic. As human

Figure 18.13 The highly urbanized Lake Washington (A, Wiki Commons); Lake Trummen, an example of a lake recovered by sediment dredging (B, Wiki Commons); Lake Tahoe, an ultraoligotrophic lake (C); and a toxic algal bloom in Lake Okeechobee (D). *C, Photo courtesy Michael Marfell. D, Photo courtesy Planet Labs.*

population in the lake's watershed grew, pollution of the lake increased (Table 18.5). Dr Edmondson realized that the lake was becoming eutrophic on the basis of species of algae and zooplankton that are characteristic of more polluted lakes and started informing the public of the potential damage to the lake. The city eventually enacted a plan to protect the lake after protracted, and at times ugly, political battles. In 1936, the city of Seattle diverted its sewage from the lake, and by 1963, cities around the lake halted all major sewage inputs to the lake.

The phosphorus input from sewage dumped into the lake caused a decrease in lake clarity in a naturally oligotrophic system. Following the halt of sewage input into the lake, phosphorus levels in the upper waters of the lake decreased significantly (Edmonson and Lehman, 1981), populations of the eutrophic cyanobacterium *Oscillatoria rubescens* decreased (Fig. 18.15), and a species of *Daphnia* typical of oligotrophic waters became abundant again. In this case, removal of the nutrient input from sewage allowed the lake to return to close to its original oligotrophic state.

In Lake Washington, O_2 never completely disappeared from the hypolimnion of the lake, and nutrient control brought about rapid reversal of eutrophication

BIOGRAPHY 18.2 W. Thomas Edmondson

The career of Dr W. Thomas Edmondson (Fig. 18.14) is an excellent example of linking basic scientific principles with successful environmental management. Due to his work, the case of Lake Washington is one of the most visible achievements of modern limnology. Edmondson had a distinguished and scholarly career with numerous publications and awards. He was a member of the National Academy of Sciences of the United States, was academic advisor to some of the top limnologists in the world, and published in the top journals on many aspects of aquatic ecology.

Edmondson got his start in limnology as a high school student studying rotifers in the laboratory of G. E. Hutchinson. Edmondson attended Yale, and by the time he graduated with a bachelor's degree, he had eight publications on rotifers. During World

Figure 18.14 W. T. Edmondson.

War II, he served as an oceanographer making measurements on ocean waves to determine if they conformed to theoretical predictions. He also participated in studies of sound refraction in the deep sea, a dangerous project that required deploying many depth charges by hand at sea.

In 1955, Edmondson was a faculty member at the University of Washington. One of his students discovered that the cyanobacteria, *Oscillatoria rubescens*, had appeared in Lake Washington. The appearance of this alga signaled that eutrophication was beginning, as had been documented in Lake Zurich a half a century before. After studying the problem, Edmondson began to correspond with the chairman of the committee appointed to study the pollution of Lake Washington. Ultimately, there was a public vote to determine if municipalities should divert all sewage effluent from Lake Washington. During the time before the vote, Edmondson was careful not to endorse a specific position but to provide accurate scientific information. Members of the public verbally attacked him for his scientific positions and complained to the university president, but he held fast to the scientific facts as he saw them. One of the attacks took him to task for worrying that the dissolved oxygen would disappear from the hypolimnion of the lake. The attacker said that there is plenty of oxygen in the lake because water molecules contain oxygen. In 1958, voters approved the sewage diversion, leading one prominent political figure to state, "If you explain it well enough, people will do the right thing." The lake made a strong recovery; evidently, Edmondson explained it well enough. It is not so clear in today's world that explaining the science to the public would be enough to sway public policy.

The story of Lake Washington illustrates, in part, the difficult position of an environmental scientist when having to balance advocacy and scientific information. If Edmondson had not brought his scientific findings to the attention of the public, the public may have ignored the eutrophication of the lake or viewed it as unavoidable, the sewage input may have continued, and the lake ended up irreversibly degraded. However, had Edmondson let politics dictate his actions, and strayed too far from his role as a source of scientific information, he may have been discredited. Edmondson's course of action demonstrates one way to be an effective scientist and confront environmental problems.

Table 18.5 Sewage inputs into Lake Washington

Years	Sewage input	Sewage phosphorus input (1000 kg/year)
1891–1936	Sewage from Seattle dumped directly to Lake Washington	?
1936–1958	Suburbs dump sewage directly into the lake	20–40
1958–1963	Population of suburbs grows as individual communities work to divert sewage outfalls from lake	100, then decreasing
1963	All major sewage inputs into lake halted—sewage diverted to Puget Sound with large mixing zone	0

Source: After Edmondson (1991).

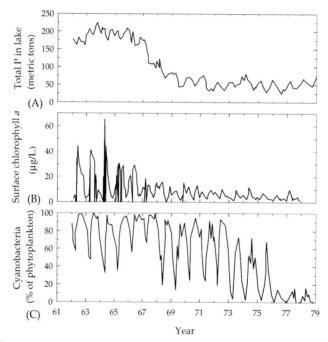

Figure 18.15 Changes over time in phosphorus loading (A), epilimnetic chlorophyll (B), and proportion of cyanobacteria (bluegreen algae; C) in Lake Washington. *Redrawn from Edmonson and Lehman (1981).*

without release of excessive phosphorus from the sedimentary $FePO_4$. The excess phosphorus added to the lake in the past remains buried in the sediments and some has been washed out of the lake. The lake now receives heavy recreational use, supports a salmon fishery, and maintains a reasonable clarity and absence of algal blooms.

Lake Trummen

This is a well documented case of the long-term effects of sediment dredging to mitigate internal loading problems. The 100-ha Swedish lake began receiving sewage in the late 1800s, this continued until 1959, leading to poor lake quality and frequent winter fish kills. Even after stopping the sewage input, the lake quality was so bad that citizens considered filling in the lake. Rich phosphorus deposits in the top layers of the sediments of the shallow lake (mean depth 1.1 m) caused high rates of internal loading. In 1970 and 1971, the lake was dredged to remove sediments, increasing the mean depth from 1.1 to 1.75 m. The dredged sediments were drained and sold as topsoil, partially offsetting the dredging costs (Cooke et al., 1993).

Dredging lowered phosphorus concentrations in the sediments from 2.4 to 0.1 mg/L, leading to a 90% decrease of phosphorus in the water column, an increase in Secchi depth from 0.23 to 0.75 m, and decreases in nuisance cyanobacterial blooms. The decreased phosphorus and increased mixing depth were responsible for lower phytoplankton biomass. The dredging and minimization of nutrient inputs to the lake resulted in better water quality, and fishing, swimming, and wind surfing continue to be popular activities on the lake (Cooke et al., 1993).

Lake Tahoe

Lake Tahoe is one of the most visited large oligotrophic lakes in the United States. It is naturally oligotrophic, with a small watershed (812 km^2) to surface area (501 km^2) ratio and great depth (505 m maximum). As development threatens the lake because of associated nutrient input, considerable study on the primary production has occurred. Separating natural variation in processes controlling production of phytoplankton from the effects of nutrient pollution has been important. Dr Charles Goldman (Biography 17.1) and his research group documented a threefold increase in primary production from 1968 to 1987 (Goldman et al., 1989).

Tahoe is a large tectonic lake on the border between California and Nevada. This lake is ultraoligotrophic, with Secchi depths historically reaching about 40 m, averaging 30 m, and a retention time of 700 years. Tahoe has gone from a regionally popular vacation spot with numerous homes and cabins in the watershed to an international tourist destination. There are gambling casinos on the Nevada side of the lake and several ski areas around the lake.

Part of the tremendous attraction of Lake Tahoe is the steely blue color associated with its ultraoligotrophic nature. Over time, a decrease in clarity occurred related directly to increased nutrient input into the lake. The initial problem was that septic system drain fields were leaking nutrient-rich waters into the lake. The solution to this problem involved installing sewage systems and pumping the treated sewage out of the lake basin. The installation of sewage systems had the unforeseen negative impact of encouraging further construction and development. The associated removal of trees, increases in area of paved surfaces, and road building instigated more nutrient runoff into the lake (nonpoint source runoff).

The nutrient limitation in the lake has switched with pollution input. When septic systems were polluting the lake, an excess of phosphorus was present (primarily from detergents containing phosphate), and nitrogen was limiting. As watershed disturbance and atmospheric deposition became the dominant sources of nutrient pollution, phosphorus pollution became less important, nitrogen additions increased, and the lake passed through a stage of colimitation by nitrogen and phosphorus to a state of phosphorus limitation (Goldman et al., 1993; Jassby et al., 1995).

The future of Lake Tahoe is uncertain. The economic pressures for real estate development along the shore are immense. If development continues, the biomass of the phytoplankton will continue to increase. This increased amount of algae will lead to decreased clarity, and the lake will become a more typical oligotrophic lake instead of one of the purest lakes in the world. The clarity of the lake seems to have stabilized, but 2017 has the lowest clarity recorded (Fig. 18.16). The low clarity was probably related to an unusually high snow pack and runoff into the lake. Additionally, the near-shore benthic areas continue to become more eutrophic, as this is the area where algal communities intercept nutrient inputs from the shoreline. This nutrient pollution has resulted in decreases in endemic benthic taxa (Caires et al., 2013) and changes in the pelagic energetics (Chandra et al., 2005).

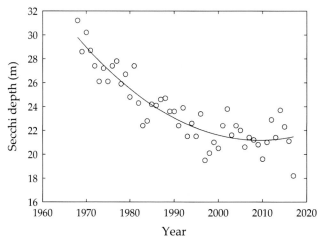

Figure 18.16 Secchi depth in Lake Tahoe. *Data courtesy UC Davis Tahoe Environmental Research Center.*

Lake Okeechobee

Lake Okeechobee in southern Florida is one of the largest lakes in the United States in terms of surface area (1,840 km^2). However, the mean depth is 3 m, which is less than the maximum wave height possible given the fetch. The shallow lake has turbid conditions from wind-induced sediment resuspension. The lake has become more eutrophic because agricultural runoff in the watershed has led to a doubling of total phosphorus (Havens et al., 1996). Paleolimnological evidence supports the idea that development and increases in agriculture have accelerated eutrophication (Engstrom et al., 2006). This eutrophication threatens a recreational fishery (valued at more than $1 million/year) and the lake's value as a domestic and agricultural water supply. Water quality of this lake was a major issue in the 2018 state elections.

Nutrient enrichment of the lake has caused increases in cyanobacteria known to produce toxins and increases in algal biomass. Phosphorus inputs have apparently stimulated nitrogen-fixing species. Predicting when and where cyanobacteria are going to bloom is difficult. Wind induces mixing of the lake and leads to lower probabilities of bloom formation by increasing light limitation (Bierman and James, 1995; James and Havens, 1996). Nutrient concentrations correlate most directly to bloom formation in the shallower western regions (James and Havens, 1996).

Historically, Lake Okeechobee was probably phosphorus limited, but increases in phosphorus loading related to agriculture increased the degree of nitrogen limitation (Havens, 1995). Improvements in agricultural management practices have lowered phosphorus inputs into the lake by 40% compared to those in the 1980s. Nitrogen inputs from water pumped from nearby agricultural areas also decreased, lowering nitrogen inputs by about 50%. However, these reductions have not led to improvements in water quality parameters, probably because of the large amounts of phosphorus stored in the well-mixed sediments of the lake (Havens et al., 1996). Phosphorus values continued to increase until 2004 when further load reductions were attempted (James and Pollman, 2011). However, as late as 2015, phosphorus loads had not declined substantially; diary operations were shut down but human population increased in the watershed at the same time.

Complicating the management scenario, the Everglades receive water from Lake Okeechobee. We discussed the complex hydrological management of the Everglades system in Highlight 5.2 and the eutrophication problems in the Everglades later.

The Clark Fork River

We discussed the problems associated with metal contamination from mine runoff and this river previously (Highlight 16.4). An additional problem is dense algal growth related to point source inputs of nutrients during times of low discharge in the summer (mainly municipal sewage outfall). Benthic chlorophyll values frequently exceeded

100 mg/m^2 (Watson, 1989) and the public perceived this high algal biomass as a nuisance.

Two approaches were used to calculate TN and TP in the water column that would lead to acceptable periphyton biomass (Dodds et al., 1997). The first was to identify reaches where chlorophyll levels were generally acceptable and analyze the TN and TP. The second was to use equations generated from the general relationship between TN, TP, and benthic chlorophyll similar to those used in lakes to predict what level of nutrient should lead to generally acceptable values of benthic chlorophyll. Both methods suggested about 350 and 30 µg/L TN and TP, respectively. A simple model of nitrogen and phosphorus inputs was then used to estimate the influence of sewage effluent controls on water column TN and TP. A voluntary nutrient reduction program was put in place, and by 2008 nutrients had been controlled in parts of the river, but the upper part of the river still had extremely high algal biomass.

Suplee et al. (2012) assessed the success of the voluntary reduction programs up to 2011. This analysis found that advanced sewage treatment to remove nutrients in the city of Missoula had reduced the nutrient content in the river below their sewage outfall. However, nutrients remained high in river reaches above that point and nuisance levels of *Cladophora* blooms continued to form during summer low flow periods. This case study suggests that voluntary regulation of nutrients can be ineffective.

The Murray-Darling River

The Murray-Darling basin drains a seventh of Australia and is over 3,000 km long. Total water discharge is low, and the river dries in many years leading to a series of isolated pools in the larger portions of the river. Extractive uses have increased the probability of the river drying completely. The region produces one-third of Australia's agricultural output and demand for the water is intense.

The relatively high nutrients in the water and the stagnant warm conditions are conducive to algal blooms. The toxic cyanobacterium *Anabaena circinalis* creates water quality problems. Toxic algae commonly occur in samples from the river (Baker and Humpage, 1994). Unlike many other systems, management of hydrology is the primary way to control the problem, including lowering water levels behind some weirs (Maier et al., 2001). Water level also influences macrophyte assemblages (Walker et al., 1994) and invertebrate and attached algal populations (Burns and Walker, 2000).

Control of water level is difficult because the climate is arid, and the demand for water is high. In dry years, the problem is worse. While precipitation could increase with climate change, so will temperature and evaporative demand, potentially putting further demands on the water in the Murray-Darling basin. Intense cycles of flood and drought complicate management of this system (Gale et al., 2014); such extreme variance is expected to increase with global climate change.

MANAGING EUTROPHICATION IN WETLANDS

Wetlands are also susceptible to eutrophication, but researchers and managers have only paid modest attention to eutrophication in wetlands. This is in spite of their crucial role in maintaining diversity, regulating ecosystem function, and providing economic value to humanity. Wetlands can have tremendous benefits in nutrient immobilization and sediment trapping, which eutrophication could compromise.

Many of the same factors that cause eutrophication in lakes and streams also apply to wetlands; Sánchez-Carrillo et al. (2010) argued that similar approaches to eutrophication control in lakes could be useful in wetlands. In addition, the use of wetlands is increasingly viewed as a way to remove nutrient loads from sewage effluent, which is a form of deliberate cultural eutrophication. Furthermore, atmospheric deposition of nitrogen in wetlands can cause eutrophication (Morris, 1991). The Everglades are one wetland where eutrophication and its control are a major management focus, and serves as an example of some of the issues related to causes and control of nutrient pollution in wetlands (Highlight 18.5). We also discussed Eutrophication control strategies in a European wetland in Highlight 14.3.

Nitrogen pollution of wetlands can alter plant communities and lead to invasion by undesirable species of plants (Tomassen et al., 2004). For example, Rejmankova et al. (2008) made experimental nutrient additions in plots inside 15 phosphorus-limited wetlands in Belize of nitrogen or nitrogen + phosphorus. Both additions led to a dramatic shift in primary producer communities from mainly cyanobacterial mats to macrophyte beds (*Eleocharis* and *Typha*).

Consumer responses to wetland eutrophication can be complex because they depend upon plant responses as food and as habitat structure. In a study in the Florida Everglades, invertebrate density increased with phosphorus additions to experimental plots. Periphyton mats decreased or were absent in medium (0.8 g P added $m^{-2} y^{-1}$) and high (3.2 g P $m^{-2} y^{-1}$) phosphorus treatments because increased grazing by invertebrates cropped the mats (Liston et al., 2008). Fertilization of cattle pastures in Florida increased water column nutrient concentrations in adjacent wetlands, altered wetland plant communities, and overall reduced aquatic invertebrate richness and diversity (Steinman et al., 2003).

Protecting wetlands from nutrient inputs is not different from methods used to protect other freshwater habitats, with strategies used such as watershed nutrient control, phosphorus bans, riparian protection, and tertiary sewage treatment. While management of riparian zone vegetation can reduce nutrient inputs associated with overland flow and shallow groundwater transport, reducing nutrient inputs originating from atmospheric deposition represents a bigger challenge. Materials such as limestone, loess, and cinder, which adsorb phosphorus, can be used as substrata in constructed

Highlight 18.5 Eutrophication and the Everglades

The Everglades are an extensive, naturally oligotrophic complex of wetlands on the southern tip of Florida impacted extensively by upstream agriculture and urbanization. This impact has harmed native plant species assemblages by increased nutrient input and decreased hydrologic flushing. Preservation of the native wetland plants is necessary for animal conservation efforts; approximately one-third of the wading bird populations have decreased, and several other species that rely on the wetlands are nearly extinct, including the Florida panther and snail kites (Davis and Ogden, 1994). A natural flow regime that varies seasonally drives the Everglades' food web.

The predominant plant cover in the Everglades was historically sawgrass (*Cladium jamaicense*), but it is being replaced by cattail (*Typha*) in response to eutrophication. A shift in cyanobacterial community is associated with phosphorus enrichment and hydrodynamic alteration (Browder et al., 1994), as well as shifts from *Utricularia*- to *Chara*-dominated communities (Craft et al., 1995). Such community changes have occurred in areas where phosphorus concentrations have increased, mainly from alterations in hydrology (Newman et al., 1998) and increases in agricultural runoff (Doren et al., 1996). Increases in agricultural activities have resulted in almost a threefold increase in input rates of phosphorus compared to historical levels (Davis, 1994). One major agricultural increase was expansion of sugarcane production in the area as fueled by the crisis in relations between the United States and Cuba in the 1950s and 1960s (Harwell, 1998). This illustrates how sociopolitical conflicts can cascade to environmental effects.

Controlling phosphorus inputs from upstream agriculture will be difficult. Control will require advanced water treatment of agricultural runoff, purchase of sugar farming operations, curtailing fertilization, restoration of some wetlands, or some combination of these measures. Complicating eutrophication management, conversion of mercury deposited from the atmosphere to more readily bioconcentrated toxic forms (methylmercury) by microbes occurs at lower rates under more eutrophic conditions, leading to less contamination of fishes (Gilmour et al., 1998). Thus, eutrophication may lead to fewer fish consumption advisories.

The South Florida Water Management District constructed 1,821 ha of treatment wetlands between 1993 and 2007. Engineers designed these wetlands to reduce phosphorus content of storm water before it flows into the Everglades. Human activities still severely threaten the Everglades despite the best efforts to preserve them. The Everglades may retain long-term biotic integrity only if we take a series of steps to control nutrient inputs, provide a natural hydrologic regime, and control atmospheric inputs of mercury. Ultimately, sea level rise associated with global warming will probably lead to inundation of much of the wetland with brackish water.

wetlands to enhance phosphorus inhibition (Guan et al., 2009), and may be useful for remediation of polluted natural wetlands. Oyster shells, which are rich in calcium like limestone, have also been used in constructed wetlands and associated filtration systems for nutrient removal from wastewaters.

Wetlands as nutrient sinks

Wetlands can have major impacts on flows of nutrients, sediments, and water through watersheds. As floodwater moves through a wetland area, it spreads and slows, dropping sediments and surrendering nutrients to the plants growing in the wetlands. Riparian wetlands may be particularly important in this regard. Humans increasingly use constructed wetlands for effective wastewater treatment, but in some cases, natural wetlands may also be used. For example, managers have "sacrificed" sections of the Everglades to retain phosphorus and retard downstream phosphorus transport. Given increasing evidence for adverse effects of nutrient enrichment on wetland communities, managers should give careful consideration before subjecting natural wetlands to high nutrient inputs.

Wetlands commonly provide nutrient removal and general sewage treatment in both North America and Europe (Brix, 1994; Sloey et al., 1978). Newly constructed wetlands have high initial rates of nutrient removal related to nutrient uptake by growing plants and algae (Richardson and Schwegler, 1986). After the first few months of heavy nutrient loads into the wetland, phosphorus removal can decrease significantly, but nitrogen removal can remain at moderate levels. The reason for the difference in nitrogen and phosphorus removal is that denitrification can remove nitrogen in the form of N_2 gas, but when the system saturates with phosphorus, little additional removal occurs. However, some phosphorus removal may continue because sedimentation in wetlands can account for a significant loss of phosphorus from the incoming water (Mitsch et al., 1995). If managers harvest plant biomass continuously from the wetland, plants will add new growth and assimilate additional nutrients. This process is necessary if nutrient removal is to continue. However, finding a use for removed plants, such as mulch or papermaking, may be difficult because such uses are not always economically profitable.

Wetlands that serve as sites of nitrate removal can also release organic carbon (Ingersoll and Baker, 1998). Unfortunately, increased organic carbon export increases the biochemical oxygen demand draining from the wetland. Such demand can lead to water quality problems downstream.

The retention efficiencies of wetlands for various materials vary greatly. Such variation is not surprising given the wide assortment of wetland types that occur naturally. Wetlands can retain from 23% to 91% of sediments coming in, from 12 to 1,370 mg N m^{-2} y^{-1}), and from 1.2 to 110 mg P m^{-2} y^{-1}) (Johnston, 1991). Essentially all studied wetlands retain nitrogen (from 21% to 95% in those in which inputs and outputs have been monitored) and therefore function as *nutrient sinks*. However, 9 of 24 wetland studies reviewed by Johnston (1991) and 1 of 19 studies reviewed by Kadlec (1994) showed that wetlands actually serve as a net source of phosphorus. Hence, scientific management of wetlands for nutrient removal is required to ensure that the wetlands serve as nutrient sinks. Wetlands may assist in removal of nutrients from agricultural waters (Woltemade, 2000). Large areas of wetlands are required for effective nutrient removal. These wetlands are particularly effective if incorporated into riparian buffer strips.

SUMMARY

1. Aquatic systems have different levels of trophic states based on autotrophic or heterotrophic activity in the system.
2. Eutrophic lakes have wide swings in O_2 concentration and pH, anoxia in the hypolimnion, fish kills, and algal blooms including increased abundance of cyanobacteria. Eutrophic streams and wetlands can have high biomass of primary producers, high rates of litter decomposition, and decreased invertebrate diversity. Eutrophic groundwaters can have high inputs of organic carbon and be anoxic.
3. Lakes can naturally become more eutrophic over thousands of years. However, cultural (human-caused) eutrophication is currently far more common.
4. Eutrophication of lakes can lead to taste and odor problems, toxic algal blooms, fish kills, lowered water clarity, and decreased property values.
5. Quantitative equations are available to calculate expected relationships among nutrient loading, nutrient concentration, algal biomass, and Secchi depth in lakes. Lake managers use these equations to make decisions on efforts to alter lake productivity. Similar equations are also available for streams.
6. Solving eutrophication problems generally requires control of point sources and nonpoint sources of nutrients. Control of nonpoint sources includes limiting excessive application of fertilizers, terracing fields, maintaining riparian and nearshore vegetation, and keeping livestock out of water bodies with fences and provision of stock tanks. Control of point sources includes bans on phosphorus-containing detergents and treatment methods such as chemical precipitation and denitrification to remove nutrients from wastewater.
7. Management of eutrophication in lakes can include oxygenation of the hypolimnion, addition of chemicals to precipitate phosphorus to lock it in the sediments, and application chemicals that kill algae.
8. Case studies of eutrophication demonstrate that controlling sources of nutrients and improving water management, rather than treating the symptoms of eutrophication, are the best way to avoid problems associated with excessive nutrients. In lakes, correcting eutrophication problems is much more difficult after the lake has become excessively productive and O_2 disappears from the hypolimnion.
9. Wetlands can also become eutrophic. Eutrophication of wetlands results in changes in microbe, plant, and animal communities, and can facilitate establishment of undesirable exotic species.

10. Wetlands can provide tertiary treatment of sewage because of their ability to retain nutrients. Natural and constructed wetlands can be very effective for nutrient removal, but the use of natural systems for wastewater treatment can adversely affect wetland communities. The Florida Everglades are compromised by eutrophication through mismanagement of the drainage basin.

QUESTIONS FOR THOUGHT

1. Why is a common classification system for trophic state useful for aquatic scientists, even if it is mainly a way to classify a continuous gradient of habitat types?
2. Why is the notion of a slow, constant movement toward a more eutrophic state in natural systems probably naive?
3. Why does a lake manager need to be aware of the variance associated with loading equations when making management recommendations?
4. Why might some ecoregions have lakes that naturally have blooms of heterocystous cyanobacteria?
5. Why would addition of Fe^{3+} to remove PO_4^{3-} from the epilimnion cause only temporary relief from eutrophication when the hypolimnion is anoxic?
6. Why is the relationship between total phosphorus and planktonic chlorophyll in lakes stronger (less variable) than the relationship between total phosphorus in the water column of streams and benthic chlorophyll?
7. Why might eutrophication of wetlands make insectivorous plants, such as sundews and Venus flytraps, less competitive?
8. Some people have treated eutrophication problems in lakes by diluting them with river flow. What conditions are necessary for this solution to work?

CHAPTER 19

Behavior and Interactions Among Microorganisms and Invertebrates

Contents

Figure 19.1 A composite scanning electron micrograph of a cross section of a periphyton community. The area is approximately 1 mm². Diatom species present include *Melosira varians* (long cylinders), *Gyrosigma attenuatum* (sygmoid shaped), and various smaller species of *Navicula* and *Nitzschia*. A layer of debris and extracellular exudates is at the bottom of the mat. *Photo courtesy: Jennifer Greenwood; reproduced with permission from Greenwood et al. (1999).*

Examination of interactions within and among species is central to the study of aquatic ecology, as well as ecology in general. Such interactions are mediated by behavior, physiology, and metabolism, and are shaped by evolution. The behavior and interactions of microbes are simpler than those of multicellular organisms, and provide a basic model for the origins of more complex behaviors of more specialized organisms. Many interactions among microorganisms depend on behavioral responses mediated by motility and responses to chemicals. Methods of determining bacterial species (operational taxonomic units) in their native environment have been developed relatively recently compared to those for multicellular species typically studied in the field of ecology, so we are just beginning to study microbial communities from classical ecological viewpoints (White, 1995). Chapters 20—23 explore the ecology of macroscale plants and animals. In this chapter, we describe aspects of the ecology of microbes. Methodology to understand microbiomes (complete communities of microorganisms in a region) allows us to begin to appreciate their importance in freshwater ecology (Highlight 19.1) We begin with a discussion of behavior and motility, provide a general classification of interaction types, and then discuss species interactions in microbial communities and how macroscopic organisms interact with microscopic organisms.

Highlight 19.1 Microbiomes

The ability to sequence large amounts of environmental DNA has led to the ability to identify all the common bacteria and archaea as well as fungi and protozoa in a microbial assemblage. In any prescribed area, all the living microbes compose the microbiome. We are only just starting to describe microbiomes. However, they can be extremely important for nutrient cycling, animal health, and any other ecological process dominated by microbes. Unique microbiomes live in unique habitats.

Each animal has a microbiome associated with its gut as well as the external surface of its body. Gut microbiomes can help animals digest food and outcompete disease microbes that could harm the animal. Interestingly, carnivorous pitcher plants (*Sarracenia alata*) have microbiomes analogous to animal gut microbiomes (Koopman and Carstens, 2011). Microbiomes alter nutrient uptake of plants. Every stream, lake, wetland, groundwater, and subhabitat within them has a unique microbiome.

The technology to describe microbiomes has created an explosion in information about them. There are many technical hurdles dealing with the huge amounts of data. Unique sequences that do not match those of any known organisms are common, but as sequence data accumulate in online data repositories, more and more sequences become available for comparison, and we will gradually describe the entire world of microorganisms. Just a few examples here illustrate that point.

The gut microbiome of wild freshwater fishes reflects their diets. Analysis of the fish community in Liangzi Lake, China showed unique microbiomes depending on trophic position; herbivorous fishes had cellulose-degrading bacteria and carnivorous fishes had more bacteria with protease (Liu et al., 2016). Microbiomec in grass carp can also vary in the same individual by diet (Wu et al., 2012).

Kolmonen et al. (2011) analyzed the microbial communities of 67 Finnish lakes. They demonstrated considerable variance in these communities. They also found shifts in bacterial community structure related to trophic status of the lakes. Ice-covered lakes have unique biomes from ice-free lakes (Bertilsson et al., 2013). The microbiome of Lake Biakal showed distinct differences between the relatively constant deeper waters and the seasonally variable surface waters (Kurilkina et al., 2016).

Metagenomic sequencing in streams allows researchers to tease apart biotic and abiotic factors influencing stream microbial communities (Lang et al., 2015). For example, understanding of microbiomes across landscapes allows scientists to track human septic inputs into watersheds and indicate the potential presence of fecal bacteria (Verhougstraete et al., 2015).

BEHAVIOR OF MICROORGANISMS

The behavioral ecology of microbes is relatively simple, but important to the survival of many species nonetheless. The ability to control position in the environment is essential to the survival of many microbial species. Clearly, motility is important because evolution has led to several different modes. Here we discuss the cues microorganisms use to move and the modes of motility.

Motility

Several modes of motility occur among microbes. Bacteria and Archaea use simple flagella for locomotion. The flagella of Bacteria and Archaea are similar, but probably arose independently (flagella are an analogous trait). The molecular biology of flagellar motility is well characterized for *Escherichia coli* (Glagolev, 1984; Koshland, 1980), but most other mechanisms of motility are less understood. Cyanobacteria have no flagella, but they are capable of gliding on solid substrata. In addition, gas vacuoles allow bacteria and cyanobacteria to float or sink. The amoeboid protozoa and *Euglena* are capable of moving across solid surfaces by changing the shape of their cells. Other eukaryotic microbes, including diatoms and desmids (a green alga), can glide across solid surfaces; the exact mechanisms for this gliding are poorly understood. Eukaryotic microorganisms also use a variety of strategies to swim through open water, including flagella (protozoa and algae), paddles, other swimming appendages found in larger multicellular organisms such as Cladocera, or by undulating body movements (Copepoda).

In general, copepods are capable of the greatest relative speeds among aquatic organisms (200 body lengths/s), with the flagellated bacteria coming in a distant second (Table 19.1). Although fishes are capable of the greatest absolute speeds, they are less impressive when considering velocity scaled to body size. It is even more amazing that microbial species are so fast given that viscosity is greater at low Reynolds numbers.

Taxis

Moving toward or away from objects or other environmental stimuli is important for many organisms' survival. They have many strategies, with more elaborate strategies common in organisms that are more complex. Movement of organisms in response to

Table 19.1 Approximate velocities of various organisms in water

Organism	Velocity (m/s)	Relative velocity (body lengths/s)
Desmids	1×10^{-6}	0.01
Amoebae	6×10^{-6}	0.03
Bacteria (gliding)	2.5×10^{-6}	0.1
Diatoms	2.5×10^{-5}	0.1
Cyanobacteria (gliding)	1×10^{-5}	0.1
Cyanobacteria (floating)	2×10^{-5}	0.2
Carp	0.4	1
Eels	0.5	1
Trout	4	8
Salmon	6	10
Ciliate protozoan	5×10^{-4}	10
Bacteria (flagellar movement)	20×10^{-5}	20
Copepod	0.1	200

Source: Data from various sources, maximum speed generally presented

stimuli from a distance is *taxis* or *tropism*. Any strategy requires determining the relative change in signal (stimulatory response of light, chemicals, gravity, etc.) and coupling this to passage of time or change in position. *Chemotaxis, phototaxis,* and *magnetotaxis* are movements of organisms stimulated by chemicals, light, and magnetic fields, respectively. A negative tactic response is a *phobic* response; for example, negative phototaxis is also a photophobic response.

The simplest type of taxis involves moving when the environment is not suitable and remaining in place when it is. An example of this is control of the rate of synthesis of gas vesicles used for flotation in planktonic cyanobacteria. Under low light and high nutrients, cyanobacteria synthesize vesicles; when light is high, synthesis stops (Walsby, 1994). This adaptation for movement allows the cyanobacteria to dominate in eutrophic waters. At the surface of a lake, in a dense cyanobacterial bloom, ample light may be available. However, a large, actively growing population of algal cells floating at the water surface will locally deplete nutrient concentrations. In this case, the synthesis of gas vesicles stops, and the cells sink. At greater depth, nutrients are high and light is low. Then they assimilate nutrients and the rate of gas vesicle synthesis increases again. Concentration of CO_2 may also be involved in buoyancy regulation, but the specific mechanism is less clear (Klemer et al., 1996). This "behavior" may allow the cyanobacteria to compete in eutrophic waters by moving up and down to avoid light and nutrient limitation, but field validation of the adaptive value of this behavior is lacking (Bormans et al., 1999).

Geotaxis, movement with respect to gravity, may be useful for benthic organisms. These organisms need to remain in the benthic zone to survive (Hemmersbach et al., 1999). They may seek the bottom by avoiding light, sensing gravity, or using Earth's magnetic field. Magnetotactic bacteria (Fig. 19.2) are examples of organisms that use magnetic fields to move downward (Blakemore, 1982). In the Northern Hemisphere, the electromagnetic field toward magnetic north also has a downward component and in the Southern Hemisphere, the downward component is associated with the south magnetic pole. Thus, bacteria that move toward magnetic north in the Northern Hemisphere tend to move down into sediments. In the Southern Hemisphere, benthic magnetotactic bacteria must move toward the magnetic south to move down into the sediments. These magnetotactic bacteria can be diverse in freshwaters (Wang et al., 2013).

Negative phototaxis can serve as a form of geotaxis. This active movement is a strategy used by some protozoa that prefer anoxic habitats (Doughty, 1991). Photophobic responses allow them to avoid areas where photosynthetic organisms are producing O_2 and where atmospheric O_2 is diffusing into the water.

Chemotaxis is the movement toward or away from chemicals and is the most common tactic response of motile microorganisms. Microorganisms are generally small enough that they cannot sense a chemical gradient across the cell (molecular diffusion

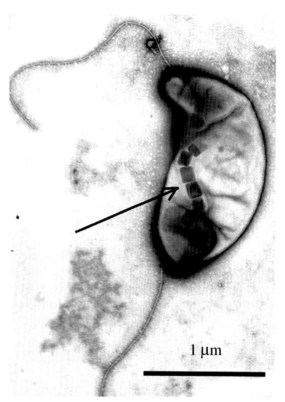

Figure 19.2 A bacterium containing magnetosomes; arrow points to magnetosomes. *Image courtesy: Richard Blakemore.*

is rapid enough over several micrometers to disallow such steep gradients over any but the shortest periods). Only the largest multicellular microbes may be able to sense differences from one side of the organism to the other. However, the ability to find and exploit chemical patches offers a selective advantage even to pelagic bacteria (Blackburn and Fenchel, 1999), so the ability to integrate information on chemical concentrations across time is important to some bacteria. For example, bacteria are attracted to organic compounds released from phytoplankton (Smriga et al., 2016).

Organism use several search strategies to move into regions where the desirable signal originates. Microbes most commonly use the random walk strategy (Fig. 19.3). An example of this is the swim and tumble strategy used by *E. coli* in which the bacterium swims for a short time and then stops and tumbles rapidly. If conditions improve (e.g., higher sugar concentrations), longer runs are taken; when conditions become worse while moving, shorter runs are taken before each tumble. This searching mode allows the bacterium to move toward a chemical signal. The search strategy is not limited to microorganisms; salmon trying to find their home streams exhibit similar behavior (Hasler and Scholz, 1983).

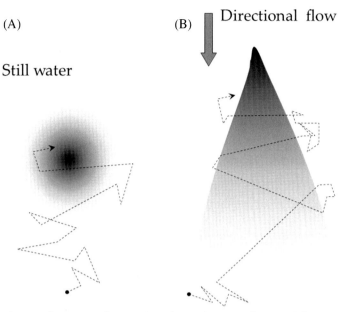

Figure 19.3 Random walk strategy for positive chemotaxis in still (A) and flowing water (B). The organism tumbles and moves a short distance if the concentration of the attractant is not increasing. If the concentration is increasing, the organism continues moving in the same direction. Darker regions represent higher concentrations of a diffusing attractant.

A moderately complex behavioral response is exhibited by the hot spring-inhabiting cyanobacterium, *Oscillatoria terebriformis* (Richardson and Castenholz, 1987a, 1987b). We discussed the community setting of this organism in Chapter 15. This cyanobacterium is positively phototactic under moderate irradiance. This phototaxis causes it to move to the surface of the microbial mat as light increases. Under high irradiance (full sunlight at midday), the cyanobacterium is photophobic and forms self-shading clumps to avoid photoinhibition or moves down into the mat. At night, a secondary behavioral response toward sulfide comes into play. The *Oscillatoria* is attracted to moderate concentrations of sulfide but avoids higher concentrations. Dissimilatory sulfate reducers and fermenters produce sulfide deeper in the mat and the cyanobacterium moves down several millimeters into the dark mat. These deeper portions of the mat are anoxic and rich in organic carbon, and the *O. terebriformis* is able to use the organic carbon, functioning as an anoxic heterotroph. The next morning, phototaxis overrides chemotaxis, and the cyanobacterium moves up into the mat again. The depth of the sulfide concentration also cycles diurnally; oxygenic photosynthesis allows oxidation of the sulfide during the day, but the sulfide gradient is closer to the upper mat surface at night. These behaviors give the cyanobacterium a competitive advantage over some of the other photosynthetic organisms that occur in the mat, but do not move up and down in response to light gradients.

INTERACTION TYPES IN COMMUNITIES

We discuss interspecific (not intraspecific) interactions in the following sections, but we first need to introduce the general framework of describing interactions among organisms, be they microbes or macroscopic plants or animals. It is useful to make the distinction between trait-mediated and density-mediated interspecific interactions. *Density-mediated* interactions refer to the population changes of one species in response to another. *Trait-mediated* interactions are traits evolved in response to nonlethal selective pressure associated with the interaction. The selective pressure is not density dependent, but based on traits. For example, individual behaviors arise based on predation risk, these would be trait-mediated. Most cases of species interactions have been explored through experiments that focus on density-mediated responses; this approach typifies most examples in this and the following two chapters. In general, density-mediated species interactions are important to those interested in questions of how to manage organisms and how to predict the immediate effects of disturbances on organisms. Trait-mediated interactions are important in determining why specific interactions occur and in generalizing about interaction types across communities. The reason that these two interaction types are separate is that population regulation does not necessarily define what is causing natural selection for specific traits.

As discussed in Chapter 8, the interactions that can occur among organisms are exploitation (mainly parasitism and predation), mutualism, competition, commensalism, amensalism, and neutralism. The relative occurrence of each of these interactions in microbial communities is generally not well known. In a study of interactions among the cyanobacterium *Nostoc* and bacteria (Gantar, 1985), positive interactions were about as likely to occur as negative interactions. However, studies of phytoplankton summarized in Hutchinson (1967) show an excess of negative interactions. Twenty-seven species of tropical and subtropical fungi were mostly inhibitory toward each other when grown together in culture (Yuen et al., 1999). If an excessive number of negative interactions occur, amensalism and competition will be more important than commensalism or mutualism. We discuss some possible microbial interactions in the following sections.

PREDATION AND PARASITISM, INCLUDING THE MICROBIAL LOOP

Microbial food webs are central to nutrient cycling and energy transfer in most aquatic systems. Transfer of energy, carbon, and other nutrients through the microbial food web is referred to as the *microbial loop* (Fig. 19.4), and can occur in any ecosystem, including freshwaters. The idea that microbial assemblages have an active role in transfer of energy in aquatic systems has changed the way that science views food webs and energy transfer. The microbial loop is essential to scavenging dissolved organic compounds in water and returning this organic material to the food web. Bacteria release a large variety of enzymes into the environment that break down organic carbon into smaller compounds that they then transport across the cell membrane (Münster and DeHaan, 1998). Microbes process the majority of organic material originating in terrestrial environments that enters aquatic environments. Organisms that eat these bacteria and fungi, or use them in their guts, return the organic carbon (or a portion of it) back into the food web. Without the microbial loop, nonliving dissolved and particulate organic material would contain the bulk of the organic carbon in aquatic ecosystems. The microbial loop can also be important to understanding how pollutants move into food webs (Wallberg et al., 1997); bactivory can increase rates of bacterial activity (Ribblett et al., 2005), increasing rates at which pollutants are processed.

Phytoplankton, bacteria, protozoa, viruses, and rotifers dominate microbial food webs in lakes. Very small flagellated protozoa are important because they are dominant consumers of bacteria. Phytoplankton can be divided into very small cells (e.g., <2 or 3 μm) and larger cells. The very small photosynthetic cells are *picophytoplankton*, which are present in a wide variety of lake types (Søndergaard, 1991). These very small algae can also be common in rivers (Contant and Pick, 2013).

Microbial food webs are a key supplier of carbon in stream food webs (Bott, 1995; Sigee 2005; Findlay 2010); Judy Meyer (Biography 19.1) demonstrated that bacterial

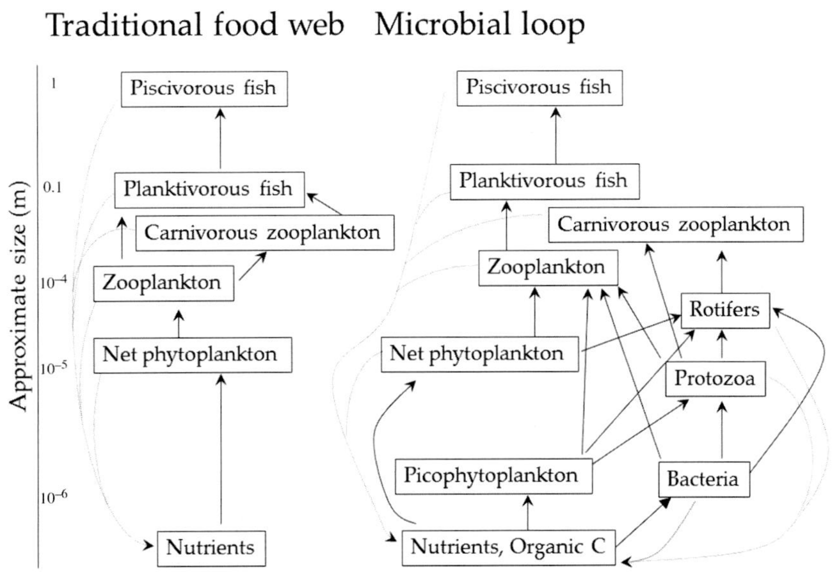

Figure 19.4 Representation of traditional views of a pelagic food web and the microbial loop.

BIOGRAPHY 19.1 Judy Meyer

Dr Judy Meyer (Fig. 19.5) is a leading investigator in the field of stream ecology. She began to study for a master's degree in aquatic biology at the University of Hawaii because of an interest in dolphins but wound up studying the mathematics of nutrient limitation of phytoplankton. When she realized that a PhD would give her control over her career, she switched to studying freshwater because she did not enjoy organizing and participating in large oceanographic cruises. She decided to specialize in streams because she liked collaborating with the stream ecologists she met.

Meyer has more than 100 publications, many in the top scientific journals. She has been president of the Ecological Society of America and the North American Benthological Society (now the Society for Freshwater Science) and has served on numerous

Figure 19.5 Judy Meyer.

scientific boards and committees. She has studied a broad variety of topics from coral reefs to stream food webs. Much of her research is on organic carbon as a food source for stream bacteria and the subsequent use of the microbes as a food source for higher trophic levels. She also has emphasized human impacts on aquatic environments both in recent research and through her involvement in policy issues.

Her suggestions for students are to seek multidisciplinary training to gain a holistic perspective and to remember that everyone starts as a novice. As an undergraduate, Meyer went to work at a premier stream research site, Hubbard Brook, and did not know what a weir (essential to measuring discharge for a stream ecologist) was when she started, even though she worked with them throughout her career after that. Also, she thinks students should concentrate on humans as part of the ecosystem. In stream ecology, she considers linkages between streams and riparian zones a crucial component of future research.

Meyer is an excellent example of a top-level scientist who works collaboratively. She has had an outstanding career and says she also is proud of successfully raising a family. She now devotes her time to working on environmental issues. She is an original thinker, a hard worker, and a high achiever.

consumption of dissolved organic carbon in streams is substantial. Bacteria and fungi break down the wood and leaves that enter streams, making the carbon available to primary consumers. Energy transfer via the microbial loop in streams could even be more efficient than in planktonic systems (Meyer, 1994).

The microbial loop is also responsible for much of the energy flow in wetlands, with amoebae rather than ciliates and heterotrophic flagellates predominating (Gilbert et al., 1998). The microbial loop is the primary energy pathway in many groundwater habitats (Gibert et al., 1994) including hyporheic sediments (Findlay and Sobczak, 2000).

Given their small size and difficulty of identification of individual species, microbial communities in general, and food webs in specific, are difficult to study. Subsequently, there is not the long history of research analogous to that associated with macroscale aquatic food webs. In the following sections, we discuss important aspects of microbial food webs, including viruses, how organisms can prey upon very small particles, and other forms of parasitism.

Viruses

Viral infections occur in all known species and types of organisms where scientists have looked for them. Microbes are no exception, with bacteriophages infecting bacteria, cyanophages infecting cyanobacteria, and so on. Virus-like particles are common in freshwaters (Figs. 9.2 and 19.6)—about 23 per bacterial cell (Maranger and Bird, 1995). Viruses also occur in extreme environments such as hot springs (Rice et al., 2001). Viruses were common in an Antarctic lake, but unlike in temperate lakes, viruses that infect eukaryotic organisms rather than bacteriophages dominated the viral assemblages (Lopez-Bueno et al., 2009). Researchers initially found these viruses by electron microscopy, but are now more commonly turning to genomic sequencing methods. Some of these viruses are active, and broadly infective, but others could be inactive. The reason scientists refer to them as virus-like particles is that they look like viruses when viewed with an electron microscope, but how infective they are and what they infect is not known simply by looking at water samples. Similarly, the existence of a sequence of RNA or DNA that appears to be viral is not necessarily infective.

Viruses must remain viable between hosts to propagate infection. Factors that inactivate viruses include UV radiation and sunlight (Wommack et al., 1996), absorption to organic (Murray, 1995) and inorganic particles, adsorption to nonhost cells, and predation by microflagellates (Gonzalez and Suttle, 1993). Below a certain density, the infection will not spread. As hosts develop immunity to viral attack, the effective density of hosts decreases. Standard ecological predator—prey approaches can be used to model epidemiology of virus infections. The models apply to simple laboratory systems with bacteria and viruses (Bohannan and Lenski, 1997).

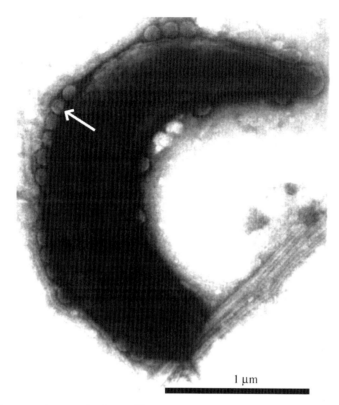

Figure 19.6 Electron micrograph of virus-like particles attached to a bacterium from a freshwater lake. *Reproduced with permission from Pina et al. (1998).*

Rates of viral infection of phytoplankton can be on par with zooplankton grazing in a shallow eutrophic lake (Tijdens et al., 2008). These rates are commonly between 10% and 50% of phytoplankton production, as high as 100%, but may be lower in freshwater benthic habitats (Filippini et al., 2006). Bactivory can be relatively inefficient (see next section), so viral predation on bacteria could be a major route of control of bacterial populations.

Population controls by viral infection can have several community effects. Molecular analyses of 25 French lakes indicated viral infection rates altered bacterial community structure, but just in lakes with relatively high rates of viral lysis (Keshri et al., 2017). Many of these mechanisms are based on the epidemiology of infections. The host cell density needs to be great enough that the virus survives long enough to reach a viable host cell. Algal viruses of freshwater lakes appear to be closely related to marine algal viruses (Short and Short, 2008), so the more extensive work on marine viruses probably applies to freshwaters.

Dramatic collapses of cyanobacterial blooms in lakes can occur when viral infections spread rapidly through the population (Van Hannen et al., 1999; Simis et al., 2005). Viral infections could keep cell densities below a threshold level, above which viral infection becomes much more probable, and may prevent competitive dominants from overrunning less competitive cells. This relationship between host cell density and viral epidemics was invoked by Suttle et al. (1990) to help explain the presence of phytoplankton species that are not competitively dominant (i.e., the "paradox of the plankton"; see Chapter 17). Interestingly, viruses transported by rivers into coastal areas can infect the bacteria there and alter bacterial community structure (Auguet et al., 2009), so the connectivity of aquatic habitats across watersheds could influence bacterial communities of lakes, wetlands, and downstream lotic habitats by transport of viruses as well.

In addition to controlling populations of organisms in the environment, viruses can be important vectors of gene transfer among organisms (Sigee, 2005). Transduction (gene transfer among bacteria by viruses) is one of the three routes of gene transfer. Occasionally both host and viral DNA are included in viral genomes, and viruses then insert this DNA into that of other organisms. In general, the amount of DNA transferred is small, but the process is still of interest with respect to evolution and spread of genetically modified characteristics of organisms. Genes from widely disparate organisms occur in some genomes, and the actions of viruses over evolutionary time explain part of this lateral transfer of genetic material.

Viral infection leads to lysis or breaking of cells and the cell contents may be an important source of dissolved organic compounds that fuel the microbial heterotrophic community. When scientists used radioactive carbon dioxide tracers to label photosynthetic organisms, they demonstrated that a portion of organic carbon fixed by photosynthetic activity was lost to solution. Viral infections may explain some of this loss. Interestingly, the release of dissolved organic matter by healthy planktonic cells may lower rates of viral infection (Murray, 1995). The released substances bind the viruses so they are no longer able to attach to cells. It is likely that these substances are proteins and mucopolysaccharides similar to those found on the planktonic cell membrane. They could be released into the water and viruses that use these cues to detect and infect cells bind to them and are unable to detach. Directly or indirectly, viral activity may be a key feature in providing energy and nutrients to the microbial loop by stimulating release of organic carbon (Bratbak et al., 1994).

Consumption of small planktonic cells

The consumption of small cells (Bacteria, Archaea, heterotrophic nanoflagellates, small algae, and fungi) is necessary if production from the microbial loop is to move into the macroscopic food web. The primary difficulty in consuming very small particles from planktonic systems is the energy required to extract them from a dilute and viscous aquatic environment. Very small particles are physically difficult to remove via filtration because the Reynolds number is very low and the viscosity is very high. A very fine-retention filter will not allow significant fluid flow between the meshes without a tremendous input of energy. Thus, protozoa engulf the smallest particles individually or rely upon random encounter with bacteria for capture (Sigee, 2005). These strategies allow for effective removal of particles as small as viruses from solution (Gonzalez and Suttle, 1993). Marine planktonic ciliates are major consumers of very small bacteria and phytoplankton (Christaki et al., 1998). Although less researched, ciliates can also be important consumers of very small plankton in lakes (Porter et al., 1985); mixotrophic flagellates can be the most important consumers of bacteria in lakes (Saad et al., 2016; Fischer et al., 2017). Flagellates <20 μm in size may be the primary consumers of bacteria in many pelagic environments (Porter et al., 1985; Sherr and Sherr, 1994). Nanoflagellates consume water column bacteria in rivers extensively, altering bacterial community structure (Batani et al., 2016). Protozoa could be the only consumers of any importance in anoxic environments (Finlay and Esteban, 1998).

The rate of particle capture is often described as a clearance rate, essentially the volume of water that can be cleared of particles per unit time. Maximal capture efficiencies for different species of protozoa occur at different particle sizes when plotted as a function of particle size (Fig. 19.7). Specialization for particle size allows for coexistence of protozoa by partitioning the resource base. The physical constraints of

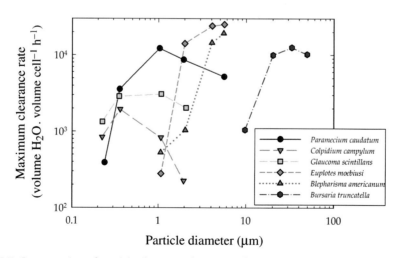

Figure 19.7 Consumption of particles by several species of ciliated protozoa as a function of particle diameter. Note that different species have different maximum clearance rates, and that clearance rates vary with particle diameter within a species. *Reproduced with permission from Fenchel (1980).*

ingesting and handling large particles constrains the upper size limit of particles that cells can consume. The ability to locate and ingest large numbers of smaller particles alongside hydrodynamic constraints related to high viscosity at low Reynolds numbers set the lower limits.

The idea that increased viscosity makes feeding costlier is supported by the observation that rotifer growth decreases with increasing viscosity because ingestion rates decrease (Hagiwara et al., 1998). Temperature has a strong influence on viscosity, so *filter feeding* (filtering fine particles for food) becomes more energetically difficult in colder, more viscous water. Coupled with a decrease in metabolic rate at lower temperature, the relative rates of feeding on planktonic bacterial-sized particles should be low in cold water.

The concentration of particles of a specific size also influences ingestion efficiency. Much energy is required to capture and ingest each particle consumed at low concentrations. It is necessary to clear greater volumes for the same amount of food as concentration of food particles decreases. The predator will not be able to ingest enough food to stay alive if particles are at a low enough concentration. The ingestion rate will increase with concentration up to a point, but then the organism becomes satiated and cannot or does not need to process any more particles per unit time (Fig. 19.8).

Large zooplankton, such as *Daphnia*, may be able to consume bacteria directly at low rates, but they may release organic compounds from sloppy feeding on phytoplankton that stimulate bacteria (Kamjunke and Zehrer, 1999). However, grazing of

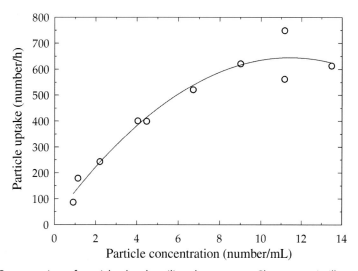

Figure 19.8 Consumption of particles by the ciliated protozoan *Glaucoma scintillans* as a function of particle concentration. *Reproduced with permission from Fenchel (1980).*

phytoplankton by zooplankton may ultimately lower bacterial production by depressing production of organic compounds from a decreased algal biomass.

There are many small particles in aquatic ecosystems, ranging from inorganic particles such as clays to living organisms such as phytoplankton and bacteria. Particle feeders may seem to be filtering or consuming particles indiscriminately, but studies using a variety of methods to trace particle consumption have shown that this is not necessarily the case. Protozoa ingested inert particles such as fluorescent plastic beads, but at reduced rates relative to live bacteria of the same size (Sherr et al., 1987), suggesting that organisms that feed on microbes are at least partially selective. Particle feeders will not grow or reproduce as well when ingesting particles that are of poor nutritional quality (Tessier and Consolatti, 1991; Sterner, 1993), and they prefer "soft" algae (DeMott, 1995). Mixotrophic (photosynthetic cells that also ingest particles) also can feed selectively. Ballen-Segura et al. (2017) tested three taxa (*Rhodomonas*, *Dinobryon*, and *Cryptomonas*) from a mountain lake for contents in their food vacuoles using molecular methods. They found that *Rhodomonas* fed almost exclusively on Archaea, and that all three avoided Actinobacteria. Therefore, there is an evolutionary advantage for selecting high-quality food, even for organisms that eat microbes.

Basic energetics provides a good framework for understanding controls on rates of consumption of very small particles in planktonic systems. More energy is required for survival and reproduction of an organism that consumes small particles when (1) particles are dilute, (2) particles are low quality, (3) particles are very small, (4) temperatures are low, and (5) particles protect themselves (e.g., are toxic). These factors all interact and understanding their interrelationships is necessary for characterizing planktonic food webs.

In contrast to pelagic habitats, benthic habitats contain larger organisms that are able to process considerable amounts of sediment and hence capture a significant number of bacteria per unit time (Fig. 19.9). In this case, invertebrate animals' ingestion rates may equal or exceed the ability of protozoa to ingest bacteria, even if we normalize rates per unit biomass to account for the much larger size of the invertebrates. The following sections cover several strategies used by macroorganisms to ingest microorganisms.

Scrapers

Organisms that scrape biofilms can consume a wide variety of microbial species. Scrapers, also referred to as grazers, include snails, some tadpoles, and many types of immature aquatic insects. Scrapers feed on substrata surfaces, consuming attached algae, heterotrophic components of biofilms, and associated deposited organic sediments. Scrapers can also remove heterotrophic biofilms found in groundwaters, so not all scrapers are grazers. In general, scrapers do not discriminate among individual cells

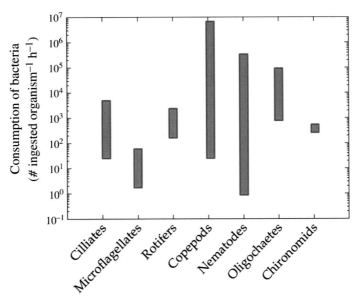

Figure 19.9 Ranges of observed consumption rates of bacteria by several groups of benthic organisms from marine and freshwaters. *Data from 12 laboratory and field studies compiled by Bott (1995). Reproduced with permission.*

because biofilms are generally complex mixtures of many microbial species. However, studies using ^{15}N tracer additions to streams indicate that some scrapers selectively feed on the more biologically active components of biofilms (Dodds et al., 2014). The total nutritional content of the biofilm could alter scraper feeding and movement rate based on the dominant particles in the biofilm (i.e., some biofilms may be more attractive than others) or the efficiency of digestion of various types of particles. In turn, biofilm grazers in Lake Ereken in Sweden altered the microbial community, both autotrophic and heterotrophic, of the biofilms they consumed (Burgmer et al., 2010).

In general, particulate materials derived from the algal components of biofilms are more easily assimilated than is detrital material derived from terrestrial plants because dead leaves and wood tend to be recalcitrant and nutrient-poor. The nutritional ecology of scrapers can be complex because of differential assimilation of ingested materials. Some fishes and tadpoles that have traditionally been considered algivorous scrapers actually derive considerable nutrition from microbes associated with biofilms and detritus, as well as animal material consumed while grazing (Evans-White et al., 2003; Whiles et al., 2010).

Evolution has led a variety of organisms to converge on the scraper lifestyle. Some of the most common scrapers in aquatic environments are snails that possess a radula, a characteristic structure of the gastropods composed of bands of rows of tiny teeth. Grazing fishes (e.g., *Campostoma anomalum* and numerous tropical species) also have subterminal mouthparts that have evolved to optimize scraping. Scrapers that feed on periphyton can greatly reduce algal biomass in freshwater habitats. In some cases, scrapers can enhance biomass-specific production of algae through removal of senescent tissues and reducing light limitation, excretion of nutrients during feeding, and altering algal community structure (Lamberti and Moore, 1984; Steinman, 1996). Scrapers can also be important in *bioturbation*, as their feeding activities produce suspended materials and re-suspend deposited particles.

Shredders

Organisms that eat detritus, either in the form of leaves and wood or as finer benthic organic material, are essentially microbial predators, as much of their nutrition could come from microbes associated with detritus (Cummins and Klug, 1979). Shredders, including many crustacean and insect species in freshwater habitats, feed on coarse detritus such as wood and leaves. In doing so, they contribute significantly to a key ecosystem function, the decomposition of organic materials. For example, experimental removal of invertebrates, including shredders, from a headwater stream in the Appalachian Mountains resulted in significantly slower leaf litter decomposition and a sixfold decrease in fine particle concentrations in the stream water (Cuffney et al., 1990).

Although they feed on coarse detritus, the nutritional ecology of shredders and other detritivores links to microbes. Some animals (insects, mollusks, and crustaceans) have the ability to produce enzymes called cellulases that breakdown cellulose (Watanabe and Tokuda, 2001; Davison and Blaxter, 2005), but it is thought that most cellulose degradation is accomplished by bacteria and fungi, which are subsequently digested by animals. Archaea also produce cellulases, but their role in detritus decomposition is not well understood as of yet.

Microbial colonization of detritus is termed *conditioning*; when given a choice, shredders preferentially feed on more highly conditioned materials (Golladay et al., 1983; Arsuffi and Suberkropp, 1984). Microbes associated with plant detritus have been termed the "peanut butter on the cracker" because shredders feeding on this material gain more protein from the microbes than the detritus "cracker" (Cummins, 1973). While wood has relatively low nutritional value among detritus types, wood biofilms, or *epixylon*, can be more nutritious than leaf detritus to some shredders (Eggert and Wallace, 2007). However, this is somewhat controversial, as microbes associated with detritus may contain very low levels of essential fatty acids (Brett et al., 2017), which are linked to a variety of physiological processes and overall health and fitness of animals.

Shredders tear leaf and wood material up before ingesting it. This increases rates of microbial degradation of cellulose in their guts and stimulates breakdown of the remaining material by increasing surface area for further microbial colonization. Because they feed on relatively low-quality material, shredders have high ingestion rates, which facilitate breakdown of coarse organic materials and generation of fine particulates (Wallace and Webster, 1996). Feeding activities of shredders generate small particles that feed a variety of collectors such as midge larvae and filter feeding caddis-flies and mussels. Shredders that burrow enhance organic matter and associated micro-bial activity in sediments in much the same fashion as earthworms in terrestrial systems (Wagner, 1991). In lotic systems, shredders enhance the entrainment and downstream movement of materials by reducing particle sizes.

The vast majority of research on shredders, microbes, and leaf breakdown has been done in lotic systems, where allochthonous litter inputs and subsequent decomposition are often primary energy flow pathways (Vannote et al., 1980; Webster et al., 1999). Breakdown of leaves is clearly important in wetlands, probably most shallow lakes sur-rounded by trees, and even groundwaters that receive leaves from cracks or interflow.

Collector-gatherers

Freshwater organisms that consume relatively small (<1 mm diameter), deposited organic particles from the surface of substrata are termed collector-gatherers. Many of these can eat entire microbial assemblages or pick out particles that are rich in cellular material and are more biologically active. Oligochaetes and chironomids process large quantities of sediments and assimilate the microbial component and any other digest-ible organic materials in the sediments. As with shredders, this group receives much of its nutrition from microbes associated with the detritus they consume. For example, stable isotope analyses suggest that chironomid larvae can receive a substantial portion of their diet from methane oxidizing bacteria (Eller et al., 2005; Eller et al., 2007). The use of a ^{13}C sodium acetate tracer indicated that chironomids, copepods, and other gatherers in a headwater mountain stream preferentially assimilated bacterial car-bon (Hall, 1995).

Collector-gatherers can contribute significantly to turnover rates of organic parti-cles. In a study of a Sonoran Desert stream, collector-gatherers ingested the equivalent of their body weight in 4–6 hours, and overall ingestion of particles by collector-gatherers exceeded rates of primary production (Fisher and Gray, 1983).

Collector-gatherers such as oligochaetes and chironomids are often the most abun-dant macroscopic organisms in many shallow water habitats (Wallace and Webster, 1996), and thus their conversion of detritus and microbial biomass to invertebrate bio-mass is significant for larger consumers such as fishes and waterfowl, which rely heavily on these groups for food.

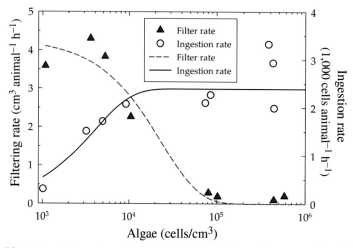

Figure 19.10 Filtering and ingestion rates of algal cells (*Chlamydamonas reinhardtii*) by *Daphnia magna* as a function of algal concentration. Although the filtering rate falls off sharply with increased concentration, the ingestion rate increases up to a point and then becomes constant. *Reproduced with permission from Porter et al. (1982).*

Filter feeders

Organisms such as *Daphnia* can filter particles from water and maintain continuous ingestion (Fig. 19.10). As with protozoa, the ingestion rate is a function of particle size and concentration. Rotifers (Fig. 19.11) and *Daphnia* have many feeding appendages that circulate and filter water. These appendages capture particles and move them toward the mouthparts. Some of the particles are actually filtered, but many others may stick to individual feeding appendages rather than become lodged between them (i.e., impaction rather than filtration is used to capture some particles).

Larger *Daphnia* are more efficient feeders on larger particles than are smaller *Daphnia* (Burns, 1969; Hall and Threlkeld, 1976), and in the absence of fishes, they can dominate planktonic communities. This feature is essential to the trophic cascade in lakes we describe in Chapter 20. The clearance rates of filter feeders can be impressive: zooplankton can filter the entire volume of a pond up to 4.7 times per day. One estimate suggested that more than 100% of the primary production by phytoplankton could be cropped each day (Porter, 1977).

Filter feeders associated with benthic habitats, such as blackfly larvae, freshwater mussels, and net spinning caddisfly larvae can also feed on microbial particles and strongly influence energy flow and nutrient cycling in freshwater habitats. *Passive filter feeders*, those that let the moving water do the work for them, often dominate filter-feeding assemblages in flowing waters. Net-spinning caddisflies in the family

Figure 19.11 The filtering apparatus of a rotifer, *Epiphanes senta. Reproduced from Melone (1998).*

Hydropsychidae are abundant filter feeders found on stable substrata in streams and rivers. Different species of hydropsychids have different mesh sizes of nets, and this alters the optimal captured particle size.

Two species of caddisfly larvae, *Ceratopsyche morosa* and *Ceratopsyche sparna*, were investigated for capture of various sizes (12, 84, and 528 μm diameter) of particles (Edler and Georgian, 2004). *C. morosa* has an average net mesh size about 20% greater than that of *C. sparna*. They captured the larger particles most efficiently overall, but some of the smaller particles were also captured. *C. sparna*, with smaller mesh size, captured the smaller particles more efficiently, but because Reynolds number dictates less flow through the smaller mesh size, the nets were less efficient with larger particles. The natural seston in this system was dominated by particles in the smaller range of those tested, but large particles, though rare, would have far greater energy content. These two co-occurring species could presumably specialize on different sized particles based on the hydrodynamics of their nets.

Blackfly larvae (Simuliidae) can also be very abundant passive filter feeders in streams. Blackflies use specialized cephalic fan structures to filter very fine particles from the water column, including individual bacteria cells. In turn, blackflies produce relatively large fecal pellets that larger filter feeders consume, or that settle out of the water to fuel benthic food webs. Production of fecal pellets by abundant filterers is not necessarily trivial; estimates from large rivers in Sweden indicate that peak loads of blackfly fecal pellets in transport in the water column can reach over 400 metric tons/day (Malmqvist et al., 2001).

Freshwater bivalves are *active filter feeders* (e.g., they pump water through filtering structures) that can filter large quantities of water and remove significant amounts of phytoplankton and suspended organic materials when they are abundant. Depending on their local densities, freshwater mussels and other bivalves can filter 10%–100% of the water column per day. In shallow water systems, bivalves can be the most significant filter feeders, and they often consume particles at rates higher than downstream transport and settling (Strayer et al., 1999). The pearly mussels, one of the formerly dominant groups of freshwater bivalves, have declined precipitously with increasing human impacts on streams and rivers, with ecosystem level consequences (Strayer et al., 2004). On the other hand, exotic bivalve species such as the zebra and quagga mussels (*Dreissena* spp.) have invaded many freshwater habitats, with measurable ecological changes (see Highlight 10.1).

Microbial adaptations to avoid predation

We have already mentioned several adaptations for avoiding predation, including indigestibility, poor nutritional value, and chemical, behavioral, and mechanical defenses. All these adaptations have evolved to be useful in certain circumstances; we briefly discuss each of them.

Some unicellular algae are indigestible. They are able to pass through the guts of grazers because they are coated in thick mucilage (polysaccharide and proteins) that protects them from digestion. In addition to protection, these cells actually gain an advantage if ingested. The zooplankton gut is a high-nutrient environment, and luxury consumption of phosphorus by the algae in the process of passage through the gut translates into an advantage after expulsion (Porter, 1976). We do not know the overall importance of this type of adaptation for most microbes consumed by animals.

Cells may not be ingested or may cause the death of their predators if they have poor nutritional quality or manufacture toxic substances. Poor nutritional quality or toxin production works only for cells that represent a relatively high proportion of the diet of a predator. Toxins are diluted if too few cells are consumed. Also, higher quality food cells make up for low-quality cells. Cells found in mixed assemblages, such as periphyton, a biofilm, or a diverse phytoplankton assemblage, may waste their competitive edge by putting energy into chemical defense. If predators do not eat selectively (as is likely with a more diverse prey assemblage and/or if predators are very large and individual prey very small), the defended cells will be eaten anyway and will grow slower than will those that put energy into growth and reproduction rather than into defense.

A unicellular algal bloom provides a situation in which production of toxins to deter grazers may be advantageous for planktonic organisms eaten by filter feeders. The fact that blooms of one or a few species can form can explain why it is evolutionarily advantageous for planktonic cyanobacteria to produce broad-spectrum toxins that can harm or even kill grazing zooplankton (Hietala et al., 1995; Ward and Codd, 1999). Given the complete dominance of cyanobacteria in many eutrophic habitats, the energy used on chemical defense from predation is not wasted.

When toxic cells dominate, this can leave larger grazers at a disadvantage, generating selection pressure for smaller grazers such as rotifers that specialize on smaller algae and bacteria that are present in lower concentrations (Hansson et al., 2007). Some species of zooplankton are adapted to feed on cyanobacteria, but toxins strongly harm many species (Tillmanns et al., 2008). Presumably, an evolutionary "arms race" has occurred where algae develop toxins, and their grazers then evolve resistance to the toxins.

Chemical defenses may be useful deterrents to protozoa that are predators of planktonic cells. Investments in chemical defenses confer an evolutionary advantage in this case because predators select individual cells. For example, one unicellular marine alga has a concentrated compound (dimethyl sulphoniopropionate) that it converts to dimethyl sulfide and toxic acrylate when the cell is grazed (Wolfe et al., 1997). Presumably, such defenses also occur in freshwater algae, as dinoflagellates can contain significant quantities of dimethyl sulphoniopropionate (Ginzburg et al., 1998) and commonly occur in freshwater habitats.

Spines and large body size are defenses because predators cannot ingest the cells efficiently. The majority of the phytoplankton the epilimnetic waters of a mesotrophic lake in midsummer are large cells or aggregates that are difficult for zooplankton grazers to ingest. Such mechanical defenses are less useful in benthic habitats, where grazers such as snails, insect larvae, crayfish, and fishes are large enough to be unaffected by microscopic spines or large colony size. The cyanobacterium *Microcystis* forms large colonies in the presence of flagellate grazers, rendering them less vulnerable to predation (Yang et al., 2006). Bacterial strains also can alter morphology to avoid predation by small flagellates. Salcher et al. (2016) grew four bacterial strains and subjected them to flagellate grazing pressure. One strain altered its morphology and several strains survived better when grown with other strains as opposed to being the only bacterial strain present.

Chemical cues produced by grazing activity commonly induce defenses, but the specific compounds induced are rarely identified (Van Donk et al., 2011). For example, grazing can lead to development of mechanical defense in the green alga *Scenedesmus* (Lüring, 1998; von Elert and Franck, 1999). When *Scenedesmus actus* was grown in water filtered from a *Daphnia* culture, colonies of four to eight cells formed rapidly. Solitary algal cells were produced when the alga was grown without exposure to *Daphnia* water. This defense against *Daphnia* grazing by formation of large colonies occurred only when the predator was present. Larger colonies settle more quickly, so there is an evolutionary cost associated with larger colonies that is counterbalanced by resistance to predation (Lürling and Van Donk, 2000).

Parasitism

Parasites or diseases afflict all organisms. Naturalists often treat these opportunistic diseases and infections as oddities. It may be difficult to study parasitism because infections can be sporadic and unpredictable. Diseases of fishes have attracted attention (Table 19.2), particularly those associated with aquaria and fish culture (Untergasser, 1989). We know much about human parasites with intermediate aquatic hosts. Scientists have expressed concerns that global climate change will worsen human diseases associated with freshwater, but the issue is complex and definitive statements are difficult (Okamura and Feist, 2011).

Schistosomiasis is one of the most common debilitating human diseases on Earth with an aquatic organism serving as an intermediate host. About 200 million people worldwide have this infection, with about 10% of those suffering from severe clinical disease (World Health Organization, 1993). Five species of *Schistosoma* trematodes can cause human disease, and each requires a different species of snail as an intermediate host. One option for control of this disease is to control the snail species that spread the infection. Molluscicides are generally used to do this, but some snails have developed resistance. Understanding the ecology of the snails may increase the effectiveness of control strategies. Proper design of reservoirs, irrigation canals, and drainage ditches can decrease snail populations and lower disease rates. Restoration of food webs to natural conditions can help control the disease. For example, in the Senegal River

Table 19.2 Some parasitic diseases of freshwater fishes

Taxonomic group of fish parasite	Example of diseases (causative agent)/fish affected
Virus	Lymphocystis/aquarium fishes
Bacteria	Furunculosis (*Aeromonas salmonicida*)/Salmonids
	Fin rot (several species of bacteria)/aquarium fishes
Fungi	(*Ichthyophonus hoferi*)/all known aquarium fishes
Protozoa	Whirling disease (Myxospora)/cold water fishes
	Ich or whitespot disease (*Ichthyophthirius multifiliis*)/many fishes
Trematoda	Hookworms/freshwater and marine fishes
Hirudinea	Leeches/freshwater species
Copepoda	Fish lice/can attack many species
Vertebrata	Lamprey (*Petromyzon*)/salmonids

basin, dam construction blocked the native river prawn (*Macrobrachium vollenhoveni*). Restoration of the prawn populations decreased populations of the snail vectors of the disease (Sokolow et al., 2015). Control efforts have included many attempts to develop a human vaccine (Butterworth, 1988) in combination with biological control of the snails, although several decades of research have yet to produce an effective vaccine as of 2018.

Other human parasites have different aquatic hosts. These include Guinea worm, a disease that is almost eradicated with only 22 cases in 2015 (down from over 3 million cases in 1986). Zebra mussels (*Dreissena polymorpha*) and Asian freshwater clams (*Corbicula fluminea*) can bioconcentrate human enteric bacterial pathogens (Graczyk et al., 2003). Microcrustaceans serve as vectors of cholera. A number of flukes and roundworms are transmitted by consuming undercooked freshwater fishes and crustaceans.

Parasites do not only cause human disease, microbes also have their own diseases. Fungal pathogens can heavily infest blooms of phytoplankton (Van Donk, 1989). The chytrid fungi are the most studied infections, and in particular infections of the diatom *Asterionella* (Sigee, 2005). Fungal infection is a potentially important controlling factor in successional sequences. Such parasitic infections are more likely to spread when algal populations are growing more rapidly and population densities are high. (Osman et al., 2017). About half of the strains of actinomycetes isolated from a lake sediment lysed *Microcystis* cells (Yamamoto et al., 1998). Such predation may offer methods for controlling algal blooms. Members of the bacterial genera Acinetobaceter, Pseudomonas, Stenotrophomonas, and Delftia lyse cyanobacterial cells (Osman et al., 2017).

Understanding the factors controlling disease in aquatic organisms can be complex. For example, the zooplankter *Daphnia dentifera* suffers infection by the fungi *Metschnikowia bicuspidata*. The proportion of infected individuals can be influenced by factors including (1) density of hosts, (2) selective predation on infected animals, (3) sloppy feeding on infected animals that releases propagules, and (4) filter feeders that consume fungal parasites and lower disease transmission rates. Understanding the community context is important in characterizing disease transmission and prevalence (Strauss et al., 2016). To make things even more complex, the presence of cyanobacteria may increase the susceptibility of *Daphnia* to fungal infections (Tellenbach et al., 2016).

Microbial parasites commonly infect frogs (Smyth and Smyth, 1980). We discussed chytrid infection of frogs in Chapter 11. Frogs are also hosts for all major groups of animal parasites: Protozoa, Trematoda, Cestoda, Acanthocephala, and Nematoda. Some of these life cycles are complex. For example, the trematode *Gorgoderina vitelliloba* infects the frog *Rana temporaria*. Adult insects or tadpoles ingest the parasitic trematode, which enters the kidneys. After 21 days, the trematode flukes enter the bladder and deposit eggs, which the frogs excrete into the water. The trematode eggs hatch into a small swimming form that enters the gills of the freshwater fingernail clam *Pisidium*. From here, they emerge as a cercaria form (a small worm-like form) that is eaten by tadpoles or aquatic insect larvae. The cercariae encyst in the body cavities of their hosts. Adult frogs then eat the infected tadpoles or adult aquatic insects, and the cycle of infection begins again (Smyth and Smyth, 1980).

The trematode *Ribeiroia* can cause deformities in frogs because the cercariae larvae of this parasite penetrate and encyst in developing limb buds of larval amphibians (Blaustein and Johnson, 2003). Along with the negative consequences for the infected frog, causing deformities in the host (Fig. 19.12) may be beneficial to the trematode, as deformed frogs are more easily captured by the definitive bird host. Eutrophication that leads to higher densities of the intermediate snail host and can enhance infection rates by *Ribeiroia* (Szuroczki and Richardson, 2009). Negative impacts of *Ribeiroia* can also be synergistic with pesticides (Kiesecker, 2002). While some well-publicized cases of malformed frogs link to trematode parasites, this is by no means the only cause; a variety of pollutants including pesticides, metals, and road de-icing salt can also cause amphibian malformations.

Other exploitative interactions

Other exploitative interactions rarely are appreciated as general processes in aquatic communities, but they undoubtedly are important. These interactions vary widely, so we provide only a few examples to illustrate their potential importance.

Epiphytes often cover macrophytes, and benefit from living on them. The epiphytes compete for light and nutrients with the macrophytes and may increase drag and associated detachment, and therefore have a negative influence on macrophytes. The relationships among epiphytes, shear force, photosynthetic rates, and nutrient

Figure 19.12 A deformed adult Pacific tree frog, *Pseudacris regilla*. This animal shows multiple hind limbs bilaterally, as well as duplicated and fused pelvic elements. *Photograph by Goodman and Johnson, PLOS One 2011 www.ncbi.nlm.nih.gov/pmc/articles/PMC3102088.*

uptake rates are complex (Hansen et al., 2014). Such interactions are widespread in aquatic communities (Gross, 2003). Chemicals released by macrophytes can inhibit epiphytes (Dodds, 1991; Burkholder, 1996; Gross, 1999; Erhard and Gross, 2006; Nakai et al., 2006; Hilt and Gross, 2008). Thus, biofilms on macrophytes may be less diverse than on nearby substrata (Levi et al., 2017). Most evidence for these interactions is not from the field, but derived from laboratory studies because controlled field experiments on the topic are so difficult (Gross, 2003).

Vibrio cholerae associates with crustaceans and can break down their chitinous exoskeletons. The bacterium causes the deadly human disease cholera. This exploitaive interaction between zooplankton and a bacterium has significant human health implications. This relationship was discovered by one of the most accomplished microbial ecologists of our time, Rita Colwell (Biography 19.2).

Another type of exploitative interactions can occur when new habitat is exposed in aquatic systems (e.g., floods expose new rock surfaces), a sequence of colonization occurs. Often, the earlier colonists must condition the habitat before establishment of the later organisms in the sequence. This conditioning may inhibit or ultimately exclude the initial colonists, but it facilitates the later organisms that can colonize the site. We discuss succession more completely in the next section and in Chapter 22.

COMPETITION

Competition can account for many species interactions in aquatic communities and could be a key driving force in evolution. In practice, demonstrating competition directly can be difficult (Connell, 1983; Schoener, 1983). Some competition must occur because many aquatic organisms have similar requirements. Consequently, we might expect to find competition for scarce resources if physical conditions remain uniform for a sufficiently long time. Understanding microbial competition has some immediate practical application. For example, competition with indigenous freshwater bacteria influences how long human fecal bacteria will survive in aquatic environments (Wanjugi et al., 2016).

The actual time required for competition to become important is a function of relative growth and reproductive rates of potential competitors. These, in turn, relate to the relative supply rate of the resources and other environmental factors influencing growth and reproduction. As with all ecological processes, spatial and temporal scales are important considerations.

BIOGRAPHY 19.2 Rita Colwell

Rita Colwell (Fig. 19.13) is a specialist on *Vibrio cholerae*, the causative factor of the disease cholera. The disease causes thousands of deaths each year and infects many more people, mainly in developing countries where people have limited access to clean drinking water. Dr. Colwell established that viable but noncultivable cells of *V. cholerae* are in most estuaries and even rivers. She showed that cholera bacteria are often associated with zooplankton and other microcrustaceans with chitin exoskeletons. She discovered that simple filtration of drinking water through several layers of tightly woven cloth can dramatically reduce occurrence rates of the disease by removing small zooplankton. The concept of viable but noncultivable bacteria has become central in the field of microbial ecology and Dr. Colwell's work brought the idea to the forefront as well as demonstrating that it had implications for public health.

Figure 19.13 Rita Colwell.

Dr. Colwell is one of the top scientists in the world and has been incredibly successful. She is a member of the US National Academy of Sciences, has received the National Medal for Science, and was the director of the US National Science Foundation for 6 years. Rita has been a coauthor on over 700 publications including numerous books, and has served as president of the American Association for the Advancement of Science. Colwell has numerous other awards and honors, and continues, as an active researcher, to produce top-level scientific work.

Rita says that she gained an appreciation for aquatic sciences initially because she grew up near the ocean and spent much time playing on the shore as a child. As an undergraduate, she became fascinated by microbiology, and eventually realized that she could spend her life researching aquatic bacteria, and obtained her doctorate degree working on marine vibrios. Her early work demonstrating *V. cholerae* is common in estuaries was initially met with much skepticism; she had substantial difficulty publishing the research and convincing the public health community of the truth of her results.

Dr Colwell notes that most of her important research results initially were viewed skeptically, but that she trusted the data and ultimately was proven right by others. She learned the hard way to trust the data; she ignored an undergraduate working with her who showed her a species of *Vibrio* that had lateral flagella rather than polar flagella as all were thought to have at the time. A colleague published the same result shortly thereafter.

She claims that growing up in a large family with little money made her competitive and stubborn. Therefore, when others criticize her work, she works harder to get it published. She suggests that students should trust the data they generate and, if possible, publish their results. She thinks the process of writing up research helps crystallize ideas and separates the "wheat from the chaff," in addition to moving the field of science forward.

Now Dr. Colwell, in addition to her research, is working on efforts to improve education for young women. She feels that many of the world's problems arise because women are not empowered to control their lives, and that education changes this and will make the world a better place.

There are two forms of competition. *Exploitation competition* is competition by organisms both exploiting the same resource. Direct negative effects on other organisms are termed *interference competition*. Tilman (1982) has explored the consequences of exploitation competition related to simultaneous consumption of several resources. This theoretical framework helps us consider the possible ramifications of unequal abilities among organisms to use nutrients. Often, a trade-off exists between the ability to grow well with low versus high nutrients. If one organism can grow well under low nutrient concentrations, and the second can grow well under high nutrient concentrations (Fig. 19.14), they will be competitively dominant at different nutrient concentrations. If their competitive abilities are reversed for a second nutrient, then they are able to coexist under certain ratios of nutrients because they are both limited by different nutrients. In

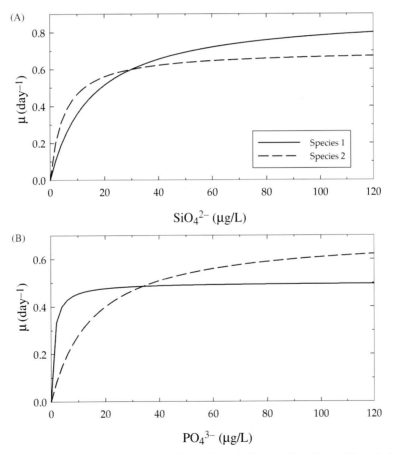

Figure 19.14 Competition of two species of hypothetical diatoms for silicon (A) and phosphorus (B). Species 1 has a relatively higher affinity for low concentrations of silicate but is outcompeted at higher concentrations. The situation reverses for species 2.

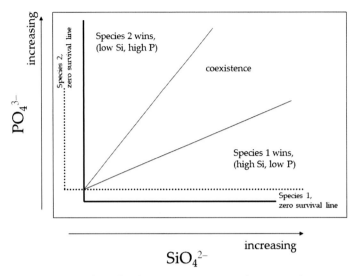

Figure 19.15 Scheme showing how the data in Fig. 19.14 translate into relative success of two species based on nutrient ratios.

the example presented in Fig. 19.14, high Si:P ratio favors species 1, low Si:P ratio favors species 2, and at intermediate ratios both species may coexist (Fig. 19.15). Data suggest that such mechanisms of competition can be important in phytoplankton communities (Tilman et al., 1982; Kilham et al., 1996), but we should be careful when extrapolating from results of laboratory experiments to the effects of nutrient ratios on natural populations (Sommer, 1999).

Interference competition is more difficult to establish for microbes. Interference competition can occur through production of toxic compounds that inhibit a potential competitor. Organic compounds that alter growth rates of other organisms are *allelochemicals* and planktonic (Keating, 1977, 1978) and benthic community members produce them. Production of chemicals by microscopic algae that inhibit other primary producers is common and occurs across many taxa (Gross, 2003). For example, cyanobacterial allelochemicals mediate composition of biofilms in Antarctic Dry Valley lakes (Slattery and Lesser, 2017).

Chemically mediated interactions (allelopathy) can drive successional sequences in phytoplankton (Rice, 1984). Given evolutionary and ecological constraints, such organisms should only excrete these chemicals expressly to act as allelochemicals under certain conditions. The benefits to the organism must exceed the cost of synthesizing the chemicals. In phytoplankton communities, allelochemicals that are synthesized specifically to lower competition are excreted only when cell densities are sufficient to bring total concentration in the water up to effective levels (Lewis, 1986), as may be the case during algal blooms. However, organisms release some chemicals into the water for other reasons (e.g., cell lysis and excretion of exoenzymes) and such chemicals may serve as environmental cues for planktonic species, causing them to increase or decrease growth rates (Keating, 1977, 1978).

In biofilms, in which cells are often in proximity to the same competing cells and molecular diffusion dominates (e.g., diffusion is relatively slow), allelochemicals may be more common. Some bacteria that form biofilms secrete *quorum sensing* compounds that signal cells to settle and form aggregates when they reach high enough concentrations (Elias and Banin, 2012). Understanding of molecular and genomic features of quorum sensing is increasing (Whiteley et al., 2017). Few data are available on chemical interactions among individual organisms living in biofilms (Characklis et al., 1990). There is evidence of allelopathy in some species of cyanobacteria inhabiting biofilms (Leao et al., 2009). Still, given the close proximity of cells in biofilms (Fig. 19.1), numerous and complex chemical inhibitions must occur.

Successional sequences commonly take place in microbial communities, and competitive ability may drive the sequences. Turnover in bacterial communities in intermittent prairie streams was driven less by stochastic processes several weeks after beginning (Veach et al., 2016). Bacteria and protozoa undergo a successional sequence in sewage treatment (Cairns, 1982). In periphyton communities, a successional sequence may occur as the community grows thicker and competition becomes more intense (Hoagland et al., 1982; Fig. 19.16). The community is reset by disturbances such as flooding and grazing. Research on competition in these communities is sparse.

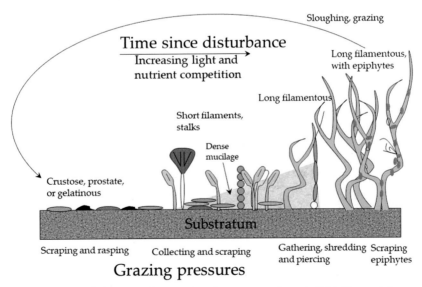

Figure 19.16 Conceptual diagram of successional sequence of an algal biofilm and the associated increases in nutrient and light competition. The type of grazer is indicated at each stage. *After Steinman (1996).*

The picture of microbial succession is more complex than simple competition; autotrophic organisms facilitate colonization of new habitats by heterotrophic bacteria (Roeselers et al., 2007). Pollutants can alter rates of bacterial and fungal growth and succession (Veach et al., 2012; Heikki et al., 2017). Furthermore, as the community changes, the types of grazing activities on the community also change. Given that animals can excrete nutrients, there may be complex feedbacks between microbial producer communities and larger consumers (Bertrand et al., 2009).

MUTUALISM: FACILITATION AND SYNTROPHY

Mutualism is rarely studied in aquatic ecology, possibly because it is rare in aquatic communities relative to other interactions, or because researchers have not widely recognized the need to study it. The most common occurrence of two or more species benefiting each other is probably nutrient cycling. Some argue that the example of nutrient cycling stretches the definition of mutualism, but both organisms benefit (albeit indirectly in many cases), thus fitting the definition used in this book. Interactions involving microorganisms in which both species benefit are highly diverse and we will discuss them by example.

Even different strains of a single species of bacteria cooperate (Rainey and Rainey, 2003). *Pseudomonas fluorescens* in static culture can evolve types that produce polysaccharides that cause them to float and to avoid anoxic conditions in the liquid culture. The bacteria "cooperate" to make a floating mat because evolution selects for cooperative characteristics in these microbes.

Mutualism in phytoplankton communities could be more common than thought. As metagenomics methods are applied to plankton, more facilitative interactions have been described. While we still don't know how pervasive mutualism is in these communities, future research will certainly shed more light on this issue (Kazamia et al., 2016). One area that could represent important mutualism is the exchange of genetic materials to deal with toxins. It is long known that antibiotic resistance can be transmitted in bacterial communities (see Chapter 16). Pollution of lake water with organic toxins led to movement of genes to degrade the compounds from rare species into more common species (Wang et al., 2017), indicating facilitation.

Most animals have microorganisms in their guts, and aquatic organisms are no exception. The gut microbiota can change with life cycle and diet (Li et al., 2017). The most commonly reported genera of bacteria found in guts of aquatic invertebrates include *Vibrio*, *Pseudomonas*, *Flavobacterium*, *Micrococcus*, and *Aeromonas* (Harris, 1993). These microbes can benefit organisms by decomposing food that would otherwise be unavailable and by outcompeting pathogenic organisms. Science is just beginning to characterize interactions with gut microorganisms, but there are documented cases of

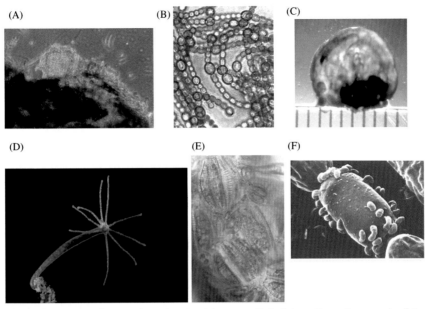

Figure 19.17 Some aquatic organisms involved in mutualistic interactions; the pouch of the water fern *Azolla* (A) that contains endosymbiotic *Nostoc* (B), (C) *Nostoc parmelioides* containing the midge *Cricotopus nostocicola*, (D) *Hydra* with endosymbiotic *Chlorella* (courtesy: Frank Fox, www. mikro-foto.de), (E) the diatom *Epithemia turgida* with cyanobacterial endosymbionts (courtesy: Rex Lowe), and (F) bacteria attached to a heterocyst of *Anabaena flos-aquae* (courtesy: Hans Paerl).

gut microbes aiding aquatic invertebrate digestion and increases in resistance to toxins associated with gut fauna (Harris, 1993).

Association of nitrogen-fixing bacteria or cyanobacteria with plants is probably the most likely interaction involving nutrient cycling to be accepted as mutualistic. Examples from aquatic habitats include the interaction between the water fern *Azolla* (Fig. 19.17A) and the nitrogen-fixing heterocystous cyanobacterium *Anabaena azollae* (Fig. 19.17B). Many wetland or riparian plants may also associate with nitrogen-fixing microbes, including the flowering plant *Gunnera* and the cyanobacterium *Nostoc* (Meeks, 1998). Alder is a common tree in boreal riparian zones that has nitrogen-fixing bacteria associated with its roots and can be a significant source of fixed N to nutrient-limited systems (Rytter et al., 1991). The diatom *Epithemia* contains nitrogen-fixing cyanobacteria (DeYoe et al., 1992), and both organisms may benefit from the interaction (Fig. 19.17E).

Mutualistic mycorrhizal interactions occur in some wetland plants (Søndergaard and Laegaard, 1977; Rickerl et al., 1994; Daleo et al., 2007). Such relationships can increase competitive ability for nutrients in some lake macrophytes (Wigand et al., 1998). Mycorrhizae can alter wetland plant assemblage composition

(Wolfe et al., 2006). Invasions of non-native riparian plants can be facilitated if they invade with their obligate mycorrhizae (McInerney and Rees, 2017).

Syntrophy is complementary metabolism. Anaerobic microorganisms in anoxic habitats commonly exhibit syntrophy. In these interactions, each organism provides the other with an organic carbon source or uses a carbon source that would become toxic and limit the other (Fenchel and Finlay, 1995). A specific example of syntrophy is interspecies H_2 transfer in which hydrogen–generating microbes facilitate activity of methanogens. The removal of H_2 is beneficial to the microbes that produce it because high concentrations of H_2 are inhibitory. Understanding this and other syntrophic interactions is central to describing anaerobic sewage digestion in addition to nutrient cycling in natural anoxic environments. These interactions often involve dense aggregates of organisms in very close proximity to each other.

A general type of mutualistic interaction that is moderately common in aquatic microbial communities involves animals that ingest and maintain algal cells within their bodies to obtain fixed carbon from them. Presumably, the alga receives protection from predation and inorganic nutrients from the animal host. Organisms with this type of interaction include *Hydra* (Fig. 19.17D) and *Paramecium bursuri* with the green alga *Chlorella*. Dinoflagellates have ingested several different types of algae in this fashion, including cyanobacteria and green algae.

Similarly, protozoa ingest chemoautotrophic bacteria. The most studied of these interactions is the sequestering of methanogenic bacteria by protozoa (Fenchel and Finlay, 1995). This is important because the protozoa and their bacterial endosymbionts can encyst and this allows their transmission through oxic habitats (that would kill the methanogens) to colonize other anoxic habitats. Such an association of organisms may be responsible for much of the unwanted methane production that occurs in landfills and it is likely common in anoxic sediments in all aquatic habitats.

Bacteria are often associated with cyanobacterial heterocysts (Fig. 19.17F). In this case, the photosynthetic cyanobacterium provides fixed carbon and nitrogen, and bacteria respire to remove O_2. Lowering the O_2 tension around heterocysts promotes N_2 fixation because O_2 damages nitrogenase. Thus, both microorganisms benefit from the interaction (Paerl, 1990).

Grazer-resistant macrophytes may benefit from organisms that remove algae and bacteria from their surface, and the grazers may benefit from the macrophyte that provides growth substrata for their food and perhaps protection from predation. Examples of this type of interaction include snails that remove epiphytic bacteria and algae from *Nostoc* and possibly epiphyte grazers that remove epiphytes from the filamentous green alga *Cladophora* (Dodds, 1991). Grazing snails commonly remove the biofilm from macrophytes, and this could be a mutualistic interaction.

CHEMICAL MEDIATION OF MICROBIAL INTERACTIONS

Given that most microorganisms evaluate the environment around them by sensing chemicals, it is not surprising that many of the interactions among microorganisms (and among macroorganisms; Dodson et al., 1994) are mediated by chemicals. Many studies describe chemically mediated interactions, including aspects such as attachment cues, toxic chemicals excreted by bacteria, chemicals that alter morphology, and chemicals that attract and repel other microbes (Aaronson, 1981; Borowitzka, 2016). Chemically mediated interactions are documented for all major groups of microorganisms, including the bacteria, protozoa, fungi, algae (Aaronson, 1981), and rotifers (Snell, 1998).

Examples of chemically mediated interactions among phytoplankton species are accumulating slowly in the literature, but such interactions are likely important (Wetzel, 2001). The work of Keating (1977, 1978), discussed previously, shows how specific ecological predictions can be made if the details of chemically mediated interactions are documented. Zooplankton also have a complex chemical ecology, but it is not well studied (Heuschele and Selander, 2014).

An interesting example of chemically mediated interactions involving microbes has been described for *Daphnia* diving in response to fish. Bacteria growing on the fish that prey on *Daphnia* excrete an unidentified compound that causes *Daphnia* to move to deeper water (Ringelberg and Van Gool, 1998). This positive effect on *Daphnia* probably has no effect on the bacteria, but the bacteria have an indirect negative effect on the fish.

Grazers can use compounds released from microbial primary producers to find food. For example, Moelzner and Fink (2015) found that alga compounds above a threshold concentration attracted snails. Furthermore, they found that feeding by snails releases more of these attractant compounds.

The ramifications of directional water flow on chemically mediated interactions are obvious. Generally, such interactions are more likely to be reciprocal among species when they occur within the diffusion boundary layer than when they occur outside it (Dodds, 1990); Reynolds number, flow dynamics, and Fick's law allow predictions about the type of chemically meditated interactions that will evolve in microorganisms.

SUMMARY

1. The behavior of microorganisms is based primarily on factors that control motility. The ability to sense changes in the environment, coupled with control of movement, allows microorganisms to move to habitats that are more favorable.
2. Chemotaxis is movement toward high or low chemical concentrations, phototaxis is movement toward light, and geotaxis is movement in response to gravity.

Different microbes have evolved the ability to use some or several of these taxes. Phobic responses are movements away from stimuli.

3. All the basic interaction types among macroscopic species also occur in microbial species (exploitation, competition, mutualism, commensalism, amensalism, and neutralism).

4. Viruses are important parasites of microbes.

5. The rate of feeding on small cells can depend on their concentration and food quality. Functional feeding groups classify animals based on these relationships.

6. Invertebrates and other macroconsumers in freshwater habitats interact with microbes in a variety of ways. For those that consume detritus, fungi and bacteria can represent the major source of nutrition.

7. Invertebrate functional groups can influence rates of critical ecosystem processes such as primary production and decomposition through their feeding activities.

8. Microbial adaptations to avoid predation include indigestibility, low nutrient content, and chemical, behavioral, and mechanical defenses.

9. Parasitism by microbial species is an important ecological interaction for aquatic organisms.

10. Competition can influence many microbial species in various ways, including changes in successional sequences and determination of what species will dominate under specific nutrient regimes.

11. Mutualism occurs in aquatic habitats; syntrophic assemblages of anoxic microbes, gut microbes, and nutrient remineralization are likely the most common mutualistic interactions.

12. Chemicals excreted into the water mediate many microbial interactions. Examples include successional sequences of phytoplankton and excretion of inhibitory compounds to decrease competition.

QUESTIONS FOR THOUGHT

1. Why do obligate mutualisms appear to be a more important type of interaction in coral reefs than in the benthic zones of lakes, wetlands, and streams?

2. How might shredders and scrapers alter microbial activity on leaves in streams? Specifically, how could they stimulate production of the microbes they consume on the litter?

3. How can what an organism ingests and what it assimilates vary, and how might this influence its ecological roles?

4. How can interactions among organisms change over time or with changes in biotic conditions?

5. Why should individual diatom species found in periphyton assemblages be less likely to produce chemicals to deter scrapers than would macrophytes?

6. Planktonic bacterial populations may respond in a complex manner to temperature. Can you predict if lower growth or decreases in predation rates related to increased viscosity should be more important controls of biomass?

7. What kind of predator was the precursor to eukaryotic cells (before chloroplasts and mitochondria)?

8. Why is chemical sensing of microbial prey so important but visual identification more likely with larger scale prey?

9. Under what conditions might an algal species exhibit photophobic response? Why may an obligate anaerobe be photophobic?

10. Adhesion to collecting appendages may be an important mode of collection of bacterial prey. Why specifically are viscous forces and Reynolds numbers important to consider in such cases?

11. Why are filter feeders found in the water column of lakes but mainly in the benthic zone (not as much in the water column) of rivers?

CHAPTER 20

Predation and Food Webs

Contents

Figure 20.1 A walleye (*Sander vitreum*) consuming shiners (Cyprinidae). *Photograph courtesy: Bill Lindauer Photography.*

Freshwater Ecology
DOI: https://doi.org/10.1016/B978-0-12-813255-5.00020-X

In this chapter we first consider herbivory, detritivory, omnivory, and predation on animals. Second, we discuss adaptations of macroscopic organisms in response to being prey or predator. Third, we cover food webs and their dynamics. A *food web* is the network of predator—prey interactions that occurs in an ecological community. *Food chains* are the most simplistic view of food webs, in which only trophic levels (e.g., a single level of producers and discrete levels of consumers) are considered. Food webs and predation in lakes and streams have received a tremendous amount of attention, less so in wetlands and groundwaters. Some claim that "food webs are a central, if not the central, idea in ecology" (Wilbur, 1997). Though not all ecologists agree with the statement that food webs are "the" central idea in ecology, most recognize that food webs are an essential aspect of ecological interactions and must be considered with other factors, such as abiotic effects and competition (Wilbur, 1987; Garvey and Whiles, 2016).

HERBIVORY

Herbivory is the consumption of primary producers and can be divided into consumption of macrophytes or microscopic algae. Microscopic algae can be consumed as phytoplankton or as periphyton. The adaptations of organisms for consuming small cells were discussed in Chapter 19. Herbivory can be intense, leading to algal populations that have relatively high productivity but have a total mass that is less than the mass of the animals that consume them (Vadeboncoeur and Power, 2017).

Although invertebrates are often the primary consumers of phytoplankton, planktivorous and omnivorous fishes may ingest suspended algae (Matthews, 1998). Gizzard shad (*Dorosoma cepedianum*) can effectively use the mucus on their gill rakers to trap cells as small as 20 μm in diameter. Several species of fishes use similar strategies to filter and retain small particles (Sanderson et al., 1991). Some small *Tilapia* (including species used in aquaculture) are also able to capture and ingest phytoplankton.

Numerous large organisms consume periphyton. The effects of grazers on benthic algae have been extensively studied and can be divided into functional and structural responses. The structural responses of periphyton to grazers include (1) a general decrease in biomass (but not always); (2) changes in taxonomic composition (but prediction of specific general taxonomic shifts is difficult); (3) changes in the form and structure (physiognomy) of communities, with a general decrease in large erect forms; and (4) alteration of species richness and diversity (perhaps with intermediate levels of grazing leading to maximum diversity).

Functional responses of periphyton to grazing include (1) a general decrease in primary production per unit area, (2) constant or increasing production per unit biomass, (3) changes in nutrient content, (4) increased rate of nutrient cycling, (5) increased rates of export of cells from the assemblage, and (6) alteration of successional trajectories (Steinman, 1996).

Many fishes consume periphyton and small macrophytes (Matthews et al., 1987; Matthews, 1998). Herbivorous fishes that consume periphyton are common in tropical waters. The minnow *Campostoma anomalum* (central stoneroller) is an important herbivorous fish in small temperate streams in the United States. This minnow can consume significant amounts of periphyton and will be discussed later with respect to its role in food webs. Many organisms consume macrophytes in aquatic habitats, including crayfish, common carp, grass carp (Highlight 20.1), lepidopteran and trichopteran larvae, moose (*Alces alces*), wild Asiatic water buffalo (*Bubalis bubalis*), muskrat (*Ondatra zibethicus*), manatees (*Trichechus*), ducks (Anatidae), beavers (*Castor*), swans (*Cygnus*), and snails. Many tropical fishes also consume macrophytes (Matthews, 1998).

Highlight 20.1 Using grass carp to remove aquatic vegetation

The grass carp (*Ctenopharyngodon idella*) is a cyprinid that consumes aquatic vegetation as an adult. This herbivorous fish has been introduced into many areas to assist in removal of unwanted macrophyte growth. The fish is cultured for human consumption in China and was first used in Europe to control macrophytes. It was first released in the United States in Arkansas in the early 1960s and has since become widespread.

This species is a very effective herbivore. For example, in Texas, grass carp were stocked at 74 fish/ha in a reservoir with 40% macrophyte cover. All macrophytes were consumed within 1 year (Maceina et al., 1992). There was a concurrent increase in cyanobacterial plankton and a decrease in water clarity.

There is concern regarding how much this species can spread, and some states have completely restricted its use for macrophyte control. The temperature range for successful reproduction is 19°C−30°C (Stanley et al., 1978), and because reproduction occurs mainly in larger rivers, it was thought that the grass carp was unlikely to spread when added to ponds. However, ponds have breached and reproductive fish have escaped. As a result, grass carp larvae have been found in the lower Missouri River, and the species has become established there (Brown and Coon, 1991). As the climate warms with global warming, the potential for reproduction will shift to higher latitudes.

One approach to keep stocked fish from reproducing is to use triploid grass carp produced with thermal or temperature shocks of newly fertilized eggs. The triploids are unable to reproduce and are the only grass carp allowed in some states (Allen and Wattendorf, 1987). Scientific equipment and training are needed to determine if grass carp are actually triploid. It is difficult for the average consumer or manager to know if the grass carp they purchase are truly triploid.

A problem with use of grass carp to control macrophytes, in a management sense, is inappropriate overstocking. Some people perceive macrophytes as a nuisance in lakes and ponds. If many grass carp are added, all vegetation is removed. This clears the way for large algal blooms to occur in the absence of suppression by macrophytes (except see Lodge et al., 1987, for an alternative outcome). A rational approach is to accept a moderate amount of macrophytes as a healthy component of natural ponds and wetlands and control them only if they become so thick as to completely preclude desired uses such as swimming and fishing. In this case, judicious use of grass carp may be warranted. Sufficiently low densities should be used, so that not all macrophytes are removed. Nonfertile triploids should be stocked, so that fish will not multiply and completely remove the macrophytes and, worse, spread to other habitats and become a nuisance. Those who use grass carp should be aware that reduced macrophyte densities in lakes can limit recruitment of desired species such as largemouth bass, and the carp can cause a decrease in water clarity by increasing bioturbation and stimulating algal blooms.

Consumption rates of submerged macrophytes are high in freshwaters, with 40%—48% of plant biomass consumed; this is roughly 5—10 times more than in terrestrial habitats (Bakker et al., 2016). This is probably related to the fact that macrophytes are more nutrient-rich and less recalcitrant. Accordingly, herbivores can have large impacts on ecosystems. For example, water lily leaf beetles control cover of the floating macrophyte *Nuphar lutea* allowing light to reach submerged macrophytes and algae and strongly altering community and ecosystem structure (Stenberg and Stenberg, 2012). Terrestrial grazers also eat emergent macrophytes. This herbivory lowers wetland plant diversity and shapes riparian communities. Sarneel et al. (2014) suggest that this herbivory sharpens the boundary between terrestrial and aquatic habitats.

Some macrophyte species are unpalatable to herbivorous animals because they are chemically defended against consumption. In addition, macrophytes may be too tough for some herbivores to process (Brönmark, 1985). Macrophytes can either synthesize or accumulate toxins (Hutchinson, 1975; Porter, 1977; Kerfoot et al., 1998). Macrophytes can be induced by grazers to produce chemical protectants. The freshwater macrophyte *Cabomba caroliniana* is induced to produce a chemical defense when attacked by either snail or crayfish grazers. Not only does this chemical lower feeding rates and growth rates of grazers, but these chemicals also inhibit bacteria growing on the plant. This inhibition could protect the macrophyte from bacterial infection at the site of grazer damage (Morrison and Hay, 2011).

Some macrophytes, such as the filamentous green alga *Cladophora*, can be a poor food source for herbivores because they have low nitrogen and phosphorus content (Dodds and Gudder, 1992). Fertilization can increase apparent palatability of macrophytes to grazing mallard ducks. In a pond experiment, grazing effects on macrophytes were greater in nutrient-enriched treatments. However, the fertilization also caused a shift in the macrophyte community, so the grazing effects were interactive with competitive effects among the plants (Bakker and Nolet, 2014).

DETRITIVORY

Inputs of terrestrial organic material into rivers, lakes, streams, wetlands, and groundwaters can be substantial. These can include allochthonous materials such as autumn-shed leaves, wood from riparian trees, and dissolved carbon transported from terrestrial habitats. Detritus can also arise from dead primary producers in the water such as macrophytes. In some cases, groundwaters can also receive coarse organic materials from terrestrial vegetation (Eichem et al., 1993). In Chapter 19, we discussed detritivory with respect to microbial communities and feeding strategies used, here we discuss the topic from the point of view of food webs and the consumers. Detrital material is consumed by many organisms and is the major source of energy in cases with high input rates, or where primary production is limited.

Analysis of benthic and pelagic food webs of a subtropical lake suggests that omnivory and detritivory are general features of aquatic food webs (Havens et al., 1996). Most orders of aquatic insects (Cummins and Klug, 1979) and other groups of invertebrates contain omnivorous organisms that consume detritus. Any benthic organism that does not exclusively specialize on macrophytes, periphyton, or predation on animals probably consumes detritus as a significant portion of its diet. For example, in the detailed food web documented for a riffle in an Ontario creek, the majority of the primary consumers were omnivorous and consumed detritus (Fig. 20.2). Many species of tadpoles feed on detritus and some fishes are detritivores as well, notably the carp species used in Asian polyculture we discuss in Chapter 23.

Detritus itself is not particularly nutritious and it often contains significant quantities of cellulose and lignins, which most animals cannot digest. The microbial community generally must first condition allochthonous materials before they can be profitably consumed (Allan, 1995). As we mentioned in the last chapter, detritivores derive much of their nutrition from the microbes that eat the detritus; microbial colonization of detritus is called *conditioning*. Fungi and bacteria are a very important component of the microbial community that conditions leaves and wood entering from the surrounding riparian vegetation. Detritivorous fishes, such as larvae of common carp (*Cyprinus carpio*) and tilapia (*Oreochromis niloticus*), ingest significant quantities of bacteria in the detritus they consume (Matena et al., 1995).

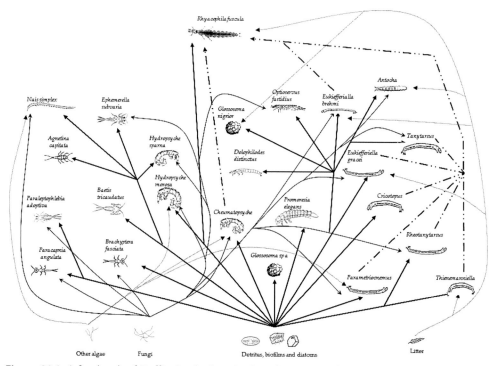

Figure 20.2 A food web of Duffin Creek, Ontario, Canada, in December illustrating the high degree of omnivory of the aquatic invertebrates. *Data from Tavares-Cromar and Williams (1996).*

Detritivores can enhance their nutritional intake by selectively feeding on more nutritious components of available detritus. In laboratory studies, leaf-shredding caddisfly larvae were offered leaf disks conditioned for different amounts of time, and selectively fed on the more conditioned disks (Arsuffi and Suberkropp, 1985). Similarly, wood-feeding detritivores apparently derive much of their nutrition from epixylic biofilms composed of microbes, extracellular polysaccharides, and organic particulates (Couch and Meyer, 1992; Eggert and Wallace, 2007). Different species of detritus-colonizing fungi can also vary in their nutritional quality, and detritivores select materials colonized by more nutritious types of fungi (Arsuffi and Suberkropp, 1989). Data on nitrogen uptake by animals indicate that there is a substantial amount of recalcitrant nitrogen in detritus and that animals are only assimilating the most biologically active fractions, either by selective consumption or inability to assimilate less available forms (Dodds et al., 2014). Consumption of detritus can create positive feedbacks; feeding activities of many detritivores rupture leaf cuticles, reduce particle sizes, and increase surface area for microbial colonization. Egested materials are also readily colonized by microbes, and coprophagy by detritivores is an important process in some freshwater food webs (Wallace and Webster, 1996).

The identity of the food sources also influences its nutritional quality. Several species of riparian tree leaves were fed to the stonefly *Tallaperla maria*. Some leaf species resulted in negative growth rates while others resulted in positive growth. Effects of additions of multiple species of leaves were not additive, suggesting complex nutritional interactions among food-source types, and that identity of riparian vegetation can alter the detritivore community in streams (Swan and Palmer, 2006).

Quality of detritus may be altered by global increases in CO_2. Tuchman et al. (2002) grew *Populus tremuloides* trees under doubled current atmospheric CO_2 concentrations and ambient concentrations. The leaves from trees grown under elevated CO_2 were more carbon rich and had more recalcitrant compounds. The leaves were conditioned by microbial communities and then fed to mosquito larvae (*Aedes albopictus*). Mortality of larvae fed leaves from the elevated CO_2 treatments was significantly greater than of those fed control leaves.

OMNIVORY

Omnivores consume materials from multiple trophic levels of the food web or multiple food web compartments (e.g., detritus and algae). Many, if not most, aquatic animals eat more than one type of food during their life span. Some organisms, such as crayfish, can be predators, herbivores, and detritivores within a life stage. Others, such as many predatory fishes, can be zooplanktivores as fry and piscivores as adults. The prevalence of omnivory greatly complicates food web and functional analyses. It can be difficult to determine the trophic position of organisms given omnivory and technical difficulties associated with determination of ingestion and assimilation. In general, what an organism ingests is more related to its function (e.g., leaf shredding insects facilitating decomposition) and what it assimilates is more related to its trophic status.

Several methods have become available to help identify diets and trophic status of omnivores (Method 20.1). Isotopic N data can be used to establish food chain length. The degree of enrichment of ^{15}N between primary food sources and top consumers can be used as an index of food chain length. Interestingly, omnivory does not alter the ability to estimate chain lengths by using this method, but the method does require assuming a standard amount of enrichment at each level of the food chain. A study involving cross-system analyses using these methods (Vander Zanden and Fetzer, 2007) assumed 3.4 $\delta^{15}N$ fractionation per trophic level, and suggested that food chain lengths in streams are significantly less (3.5 levels) than in lakes (4 levels). The use of these techniques is revealing that omnivory is more the rule than the exception in most ecosystems.

METHOD 20.1 Tracers in food web and nutrition studies

Gut contents are often used to determine what organisms eat, but this method can be misleading because it does not necessarily describe what organisms assimilate. For example, trout in some streams can have considerable amounts of filamentous algae (e.g., *Cladophora*) in their guts. They are not able to digest this material, but simply ingest it when feeding on invertebrates that live in the algae. The trout ingest but do not assimilate the algae. Feeding and growth experiments in the laboratory are one way to assess the use of various food categories. Analysis of natural abundance of *stable isotopes*, primarily the stable isotopes (not radioactive) of carbon and nitrogen, is becoming an increasingly common method for assessing trophic position of animals and what food sources primarily supply food webs. Fatty acids can also be used to trace food sources; both fatty acid and stable isotope analyses reflect what a consumer has assimilated.

Stable carbon and nitrogen isotopes occur naturally and are useful in the determination of food sources in food webs (Peterson and Fry, 1987; Peterson, 1999). Nitrogen and carbon isotopes ^{15}N and ^{13}C are not radioactive but are heavier than their more abundant counterparts (^{14}N and ^{12}C) in the natural environment. These isotopes are fractionated (selected for or against) by physical and biological processes to some degree, causing slight but often consistent variations in natural abundance. *Natural abundance* is the ratio of the trace isotope to the more abundant isotope. Because these ratios are often very small, natural abundance is commonly expressed as a δ value relative to some known standard. The equation used for ^{15}N is

$$\delta^{15}N = \left[\left(\frac{^{15}N_{sample}/^{14}N_{sample}}{^{15}N_{standard}/^{14}N_{standard}} \right) - 1 \right] \times 1000$$

which is simply the $^{15}N/^{14}N$ ratio in a sample to that of a standard (in this case atmospheric N_2 gas). Units are parts per thousand (‰). An equation of the same form is used for carbon. A mass spectrometer is usually used to determine the isotopic ratios.

Typically, ^{15}N is the first choice for determining trophic status. There is generally a 3‰–5‰ increase in $\delta^{15}N$ for each trophic transfer, allowing resolution of the feeding levels (Fig. 20.3). In addition, different food sources (e.g., terrestrial vegetation vs algal material) often differ significantly in their natural abundance signatures.

The use of ^{13}C ratios is more difficult. There is not consistent fractionation of ^{13}C across food webs and the fractionation is not as great (Fig. 20.3). However, ^{13}C ratios can often be used to resolve food web differences where there is overlap in the ^{15}N data. The use of natural abundance of both isotopes simultaneously can provide information that laboratory experiments cannot. Additional isotopes (^{18}O, ^{34}S, deuterium) can further resolve food sources in some cases (Finlay et al., 2010).

(Continued)

METHOD 20.1 (Continued)

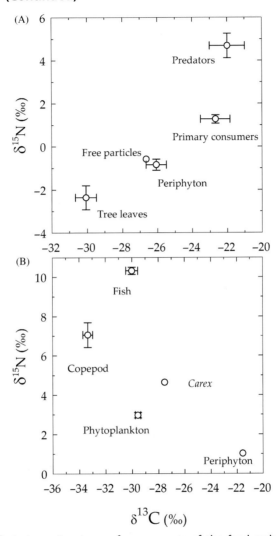

Figure 20.3 Stable isotope signatures of components of the food web in Lookout Creek, Oregon (A) and Toolik Lake, Alaska (B). Errors plotted as standard error where data were available. Note that the use of both isotopes allows for clearer separation of the food web components. The primary consumers in the stream probably rely on periphyton, but those in the lake likely rely on phytoplankton. *Data from Fry (1991).*

There are numerous examples of the utility of isotope analyses; a few are noted here. Littoral fishes and crayfish in Canadian Shield lakes are dependent on terrestrial vegetation (France, 1996, 1997b). Habitat usage varies and food preferences at different life

(Continued)

METHOD 20.1 (Continued)

history stages changed in fishes in a subtropical lake (Fry et al., 1999). Ocean-derived nitrogen has a distinct signal and this signal shows marine nitrogen is transported into streams and lakes by spawning salmon (Kline et al., 1990; Finney et al., 2000). Nitrogen fluxes in stream food webs have been quantified using added labeled nitrogen (Peterson et al., 1997; Hall et al., 1998; Mulholland et al., 2008; Norman et al., 2017; Tank et al., 2018). Isotopes have also been used to quantify how important bacteria are as a carbon source in stream food webs (Hall and Meyer, 1998; Collins et al., 2016).

One problem is identifying the base signal of isotopes in complex mixed food sources such as periphyton. Consequently, the signal reaching the consumers is often very difficult to ascertain from the bulk isotopic composition of their food (Dodds et al., 2014). One method to avoid this problem is compound-specific isotopic analyses. Amino acids are particularly interesting with respect to this because some amino acids follow food sources without fractionating their nitrogen isotopes, yet others in the same protein do fractionate. This interesting method awaits further testing and verification (O'Connell, 2017; Thorp and Bowes, 2017).

Other tracers can be used to examine food web structure in aquatic ecosystems in addition to stable isotopes. Lipids can trace sources of organic matter (Derrien et al., 2017). Fatty acid profiles can be used to assess diets and trophic relations and are gaining increasing attention among aquatic ecologists (e.g., Guo et al., 2017). Fatty acids are essential foundations of energy storage lipids in consumers, but many are not readily synthesized by animals; they must be obtained through feeding and they are conserved. Various food resources such as algae, bacteria, and detritus have unique combinations of fatty acids, which are reflected in the fatty acid profiles of consumer tissues. To assess diets, samples of consumers and possible food resources are collected and lipids are extracted with organic solvents. Analyses involve separation of polar and nonpolar lipids with a solid-phase extraction system, conversion into fatty acid methyl esters, and separation using a gas chromatograph with a flame ionization or mass spectrometer detector.

Fatty acid analyses have been used to assess diets and food web structure in marine and freshwater systems (Muller-Navarra et al., 2000; Stübing et al., 2003; Sushchik et al., 2003), as well as aquatic-terrestrial food web linkages (Koussoroplis et al., 2008). In some cases, fatty acid analyses can provide detailed taxonomic information, including specific types of algae assimilated (e.g., Napolitano et al., 1997). In a study of the food web of a forested headwater stream, fatty acid analyses revealed that the mayfly *Ephemerella* and the caddisfly *Hydropsyche* assimilated primarily autochthonous materials, even during the summer when shading by the riparian canopy would be expected to limit primary production in the stream (Torres-Ruiz et al., 2007). Combination of fatty acid analysis and stable isotope methods allowed better resolution of a food web in an Australian River (Jardine et al., 2015).

Some organisms that have traditionally been assigned to one functional feeding group are more omnivorous than previously thought. For example, *Daphnia* were thought by many to mostly subsist on phytoplankton. However, bacteria may also be an important food source for them (Mcmeans et al., 2015).

Omnivory can have important consequences in aquatic food webs. Wootton (2017) reviewed some of these for freshwaters. He suggested (1) strong omnivory destabilizes food webs, but weak omnivory stabilizes them, (2) omnivory may be stabilized by other food web characteristics, (3) omnivory is most likely in habitats with intermediate productivity and disturbed habitats, (4) omnivory will strengthen tropic cascades (see section later), and (5) omnivorous invaders will be more successful. While omnivory in the community context is complex, understanding it will help manage freshwaters.

ADAPTATION TO PREDATION PRESSURE

Defenses against predation include mechanical, chemical, life history, and behavioral protections. Chemical protection includes both toxic or unpalatable chemicals and low nutritional quality. Mechanical protection includes size (either too large or too small to be eaten) and protective spines or projections. Behavioral responses to predation include avoidance and escape behaviors.

Evolved defenses can be permanent or inducible. It is often energetically efficient to only defend against predation when there is a threat. Thus, inducible defenses (chemical, mechanical, or behavioral) can be selected for. We have already discussed some inducible defenses in the last chapter. In other cases, there is always a threat of predation, so permanent defenses are more common.

One common mechanical defense against predation is for the prey to be too large for the predator to effectively consume. In Chapter 19, we discussed how large colonial algae dominate the plankton of lakes with significant zooplankton populations. In a benthic example involving both size and alteration of life history, a snail, *Physella virgata*, increases its growth rate and delays reproduction in the presence of chemical cues from the predatory crayfish *Orconectes virilis*. The more rapid growth increases the probability the snail will reach a size at which the crayfish can no longer effectively prey upon it. In the absence of this predator, the snail reproduces at a smaller size (Crowl and Covich, 1990). Therefore, evolution favors growth and deferred reproduction in the presence of predators. Fishes also increase body size to limit predation, as demonstrated by the crucian carp (*Carassius carassius*), which increases the height of its body in response to the presence of predators (Brönmark et al., 1999). This adaptation requires predators on the crucian carp to have a larger gape to consume them.

Spiny fin rays of fishes, hard shells of mollusks, gelatinous sheaths of algae, and cases of caddisfly larvae and other aquatic insects are examples of mechanical defense. Growth of spines in *Daphnia* can serve as a mechanical defense, increasing their effective size. Several invasive species of zooplankton in the United States, such as *Bythotrephes cederstroemi* (the spiny water flea) and *Daphnia lumholtzi*, have very long spines (East et al., 1999; Fig. 20.4). This may allow them to resist predation and could be partially responsible for their successful invasion of lakes that do not contain zoo-plankton species with such long spines. The responses may be complex; the planktonic rotifer, *Brachionus calyciflorus*, increases spine length in the presence of a predatory roti-fer. However, it also increases spine length when food is limited or in cold tempera-tures, presumably to lower sinking rates (Gilbert, 2018).

Chemical protection from predation includes toxic or bad tasting chemicals (*allomones*, chemicals that affect another species to the benefit of the producer) and poor nutritional quality. Invertebrates may use visual, hydromechanical, or chemical cues to avoid predators (Dodson et al., 1994). Chemical cues to predators include *kairomones*, compounds produced by predators that affect the behavior, morphology, or life history characteristics of prey species, and *alarm chemicals*, chemicals that are produced by damaged or stressed organisms. Alteration of behavior by alarm chemicals is common in fishes.

(A)   (B)

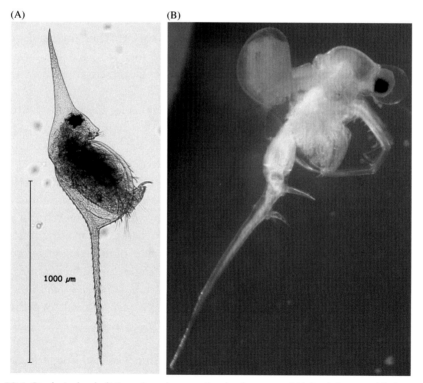

1000 µm

Figure 20.4 *Daphnia lumholtzi*, an invasive species in the central United States with large spines that protect it from predation (A) and *Bythotrephes cederstroemi*, the spiny waterflea (B). *(A) Photograph courtesy: K. D. Hambright; (B) photograph courtesy: The National Oceanic and Atmospheric Administration Great Lakes Environmental Research Laboratory.*

Chemical defenses are common in freshwater organisms. Where such defenses have been sought, they have been well documented; there are numerous examples of toxin production in Coleoptera and Hemiptera (Scrimshaw and Kerfoot, 1987). Whirligig beetles of the genus *Dineutus* produce a volatile chemical that smells like sour apples, which is used for protection from vertebrate predators. In parts of Africa, some young women use adult gyrinid and dytiscid beetles to stimulate breast development; the beetles are allowed to bite their breasts and, in the process, expose the women to defensive chemicals containing hormone-like steroids that enhance breast development (Kutalek and Kassa, 2005). Some stream-dwelling stoneflies use autohemorrhaging, or reflex bleeding, when threatened, whereby they increase their body fluid pressure until hemolymph (blood) exudes from some joints of the exoskeleton (Moore and Williams, 1990). Insect hemolymph can serve as a chemical defense when it is sticky and coagulates quickly, or it can contain foul tasting chemicals.

Poor food quality can also provide some protection against predation. However, the first response to low quality food is to consume more, so this adaptation is only effective if quality is so poor that so much energy is expended to consume enough nutritious food that reproduction and survival are compromised. An example of poor food quality leading to decreased success of a predator is the relationship between highly unsaturated fatty acid content in phytoplankton and growth of their zooplankton predators. These fatty acids are almost exclusively synthesized by photosynthetic organisms but are required for animal growth. Low-phosphorus algae that are high in these fatty acids are a better food source for *Daphnia* than high-phosphorus, low-fatty-acid phytoplankton (Brett and Müller-Navarra, 1997), potentially leading to more effective grazer control of trophic state at lower phosphorus concentrations (Müller-Navarra et al., 2004). Single species of diatoms can be nutritionally unsuitable prey for copepods (Jones and Flynn, 2005). Therefore simple measures of food quality, such as nitrogen and phosphorus content, may not adequately represent nutritional quality.

Animals exhibit numerous behavioral responses to predation. Behavioral defenses may be quite complex and require sensing of predators and behaviors to avoid predation. They can also be context dependent. For example, predator avoidance can be a function of temperature in fishes. Pink and Abrahams (2015) found that predation risk suppressed fathead minnow (*Pimephales promelas*) activity and foraging, except at higher temperatures when metabolic needs apparently superseded predator avoidance. Among invertebrates, behavioral responses range from nocturnal foraging to avoid visual predators, thanatosis (feigning death), and release from the substrata and downstream drifting in stream insects. Behavioral defenses can be combined with chemical defenses, as with reflex bleeding we described earlier.

Probably one of the most studied behavioral responses to predation in aquatic systems is the diel vertical migration of *Daphnia* and other zooplankton in lakes (Fig. 20.5). These zooplankters enter surface waters to consume phytoplankton at

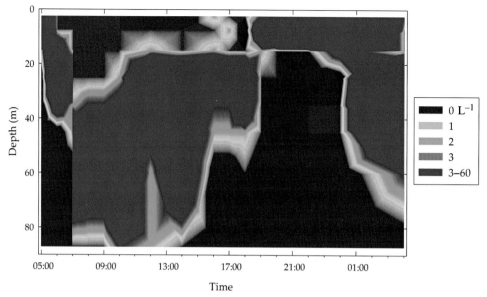

Figure 20.5 Diel vertical migration of young *Daphnia longispina* in Lake Lucerne. *Data from Worthington (1931).*

night when the darkness lowers predation by zooplanktivores that rely on vision to find prey. At daybreak they swim down to darker parts of the lake to avoid the visual feeders. In some cases, larger *Daphnia* that would be more susceptible to predation move deeper than smaller individuals (Hutchinson, 1967; De Meester et al., 1995). Chemical cues excreted by predators or released by predators during feeding may be required to trigger the migration response (Folt and Burns, 1999; Tollrian and Dodson, 1999). Similar migrations can be seen in larger rivers (Jack et al., 2006), probably to avoid sight-feeding predatory fishes.

A chemically mediated behavioral defense of *Daphnia* to predation by fishes involves predator-avoidance behavior triggered by alarm chemicals released when other *Daphnia* are eaten. In this case, *Daphnia* swim downward, form aggregate swarms, or increase avoidance behavior after exposure to water with crushed *Daphnia* (Pijanowska, 1997). These behaviors lower predation rates on *Daphnia* from bream (*Abramis brama*). Similar defenses have been documented for the related cladoceran *Ceriodaphnia reticulata* in response to chemicals produced by green sunfish, *Lepomis cyanellus* (Seely and Lutnesky, 1998). Tadpoles also use chemical cues to avoid predators, and they may be less alarmed by invasive predators than native predators with which they have evolved (Hettyey et al., 2016).

Complex behavioral responses to predation are common. For example, copepod nauplii move quickly to avoid predation. However, due to their small size, they experience high viscosity (low Reynolds number). This viscosity becomes even greater at the lower end of their temperature range. The nauplii alter their swimming strategy by increasing their power stroke at lower temperatures. This allows them to maintain their escape response (Gemmell et al., 2013).

Schooling behavior is a common response to predators, particularly in fishes. Ioannou et al. (2012) used simulated virtual prey for sunfish to study responses to schooling. They projected moving dots and logged what prey behaviors elicited attack responses by the fish. They found that swarms or schools of prey that aligned and moved synchronously were less likely to be attacked. Therefore, these researchers were able to show how predators could apply selective pressures that lead to schooling behaviors in prey.

Inducible responses to predation have evolved in many species. In Midwest United States reservoirs, co-occurring *Dahnia pulicaria* and *Daphnia mendotae* exhibit different evolved responses to predation (Bernot et al., 2006). These species come under increased predation pressure as larval fishes increase in number in the late spring. Exposure to chemicals released by fishes into waters (kariomones) caused *Daphnia mendotae* to reproduce at smaller size. Smaller size at first reproduction aids in avoidance of predation because smaller adults are less susceptible to predation. In contrast, *Daphnia pulicaria* produced resting eggs (ephippia) as temperature increased, simply disappearing from the water column as predation pressure increased.

Hydromechanical cues can be an important component of predator avoidance (Peckarsky and Penton, 1989). These pressure waves are transmitted rapidly at the small spatial scales at which insect larvae and smaller organisms operate. Chemicals released by fishes can increase sensitivity of *Daphnia* to mechanical signals arising from movement of predators (Brewer et al., 1999). Although use of hydromechanical cues may be widespread, much less is known about these than chemical cues (Dodson et al., 1994).

Predator avoidance in time and space is a crucial adaptation to predation in many species. Many species simply hide or swim away when they sense a predator in the vicinity. Timing of reproduction is another approach to avoiding predation. This is the case for some species of zooplankton that produce diapausing eggs in the presence of increased predation pressure (Hairston, 1987). Massive synchronous hatches of aquatic insects (Fig. 10.15) not only increase the chance that adults will be able to find a mate, but also saturate predators, allowing a greater probability that some adults survive and mate.

ADAPTATIONS OF PREDATORS

Predation can be characterized logically by a sequence of events. Prey must be encountered and detected, then attacked and captured, and finally ingested (Brönmark and Hansson, 1998). Adaptations are evident at all stages; we organize this section by the natural sequence of events. The idea that evolution through selection leads to maximization of net energy gained per unit time feeding, led to the concept of *optimal foraging* (Schoener, 1987; Pyke et al., 1977). This simple concept applies to all phases of predation (encounter, detection, attack, capture, and ingestion). We have already discussed the relative merits of sitting and waiting for prey as opposed to active searching. Food quality, quantity, and spatial distribution are often additional considerations for optimal foraging. Optimal foraging has been well established for several fish species (Mittelbach and Osenberg, 1994). A food item may not be preferred if it is high quality but very rare relative to a lower quality food source. All items, even remotely suitable items, may be taken when food is limiting, but only the most profitable may be taken when more is available. For example, bluegill will capture all sizes of zooplankton in equal amounts when the food is at low density but will take large *Daphnia* preferentially at higher zooplankton concentrations (Fig. 20.6).

The behavioral strategy used by organisms to obtain prey can vary from remaining immobile and allowing prey to approach, to active foraging. A species-specific strategy presumably allows for the most efficient harvesting of food given morphological, abiotic, and community constraints. The simplest encounter strategy is to "sit and wait" for prey. For example, pike (*Esox*) lay hidden, waiting for prey to come close. Pike have large tail (caudal) fins and pull their bodies into "S" shapes from which they can accelerate explosively to catch prey. Odonate dragonfly nymphs also wait for prey. They have a hinged labial jaw that ejects very rapidly and snatches prey. *Hydra* capture larval bluegill (*Lepomis macrochirus*) that contact their stinging nematocysts, a form of sit-and-wait predation that can have considerable impact on the populations of the larval fish (Elliott, Elliott, Legett, 1997).

A spectacular method of capturing prey is via electric currents. Electric eels use several strategies. When they are trying to capture free-swimming prey they use high frequency volleys to cause all muscles in the prey body to contract, immobilizing them. When a prey item is hidden, they can emit two or three discharges that cause the prey to twitch and reveal their location. The frequency and intensity of the discharges appears to be tuned to induce contractions of involuntary muscles (Catania, 2014).

Carnivorous plants in wetlands are notable sit-and-wait predators on insects. Most species tolerate or require saturated soils and are wetland species (Juniper et al., 1989). Passive trapping strategies are used, which include the use of adhesive traps such as in sundew (*Drosera*), chambers that can be entered but not left as in the pitcher plants (e.g., *Sarracenia*), snap traps such as the Venus flytrap (*Dionaea*), and triggered chambers as in the bladderworts (*Utricularia*). These plants often attract prey, and adaptations include visual stimuli such as UV patterns visible to insects. Olfactory substances can be produced, including those with nectar scent or the smell of putrefaction, and nectar

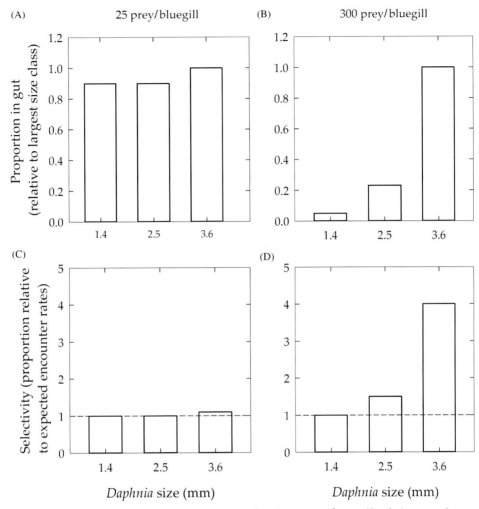

Figure 20.6 Selectivity of bluegill (*Lepomis macrochirus*) on size of prey (*Daphnia magna*) at two prey densities. At low densities, all sizes of *Daphnia* are equally represented in the bluegill guts. At high densities of prey, the larger zooplankton are preferred (selectivity relative to expected encounter rates). *Data from Werner and Hall (1974).*

rewards may be used to lure insects. Tactile stimuli are also important in some cases, such as that of the bladderwort, *Utricularia*, which has filamentous extensions on submerged gas-filled bladders that mimic filamentous algae and attract epiphyte feeders. When the invertebrate contacts the extensions, the bladder fills rapidly with water and sucks in the prey (Juniper et al., 1989). Carnivorous plants are generally found in low-nutrient environments and are photosynthetic, so they use their prey as a source of nitrogen and phosphorus. Carnivorous plants are an excellent example of convergent evolution leading to solution of a problem (nutrient limitation) from divergent plant lineages using a wide variety of capture and attraction mechanisms.

Sensing prey can be accomplished by a variety of adaptations, depending on the organisms and their prey. Invertebrates use visual (in a few organisms with well-developed eyes), mechanical, tactile, and chemical cues (Peckarsky, 1982). In an ingenious demonstration of the importance of the use of mechanical cues, Peckarsky and Wilcox (1989) recorded hydrodynamic pressure wave patterns associated with escaping *Baetis* nymphs. Predatory stonefly nymphs (*Kogotus modestus*) attacked *Baetis* models in greater frequency when the wave patterns were played back than when they were not. Gastropod grazers use volatile organic compounds released from benthic algae as food finding cues (Moelzner and Fink, 2015). Grazing beetles (*Macroplea appendiculata*) use the chemical defense compounds of pondweed (*Potamogeton perfoliatus*) to find the plants in turbid waters. The beetles have evolved resistance to the defense compounds and use them as a cue instead (Röder et al., 2017).

Fishes can sense prey visually, chemically, electrically, or hydrodynamically. Electrical sensory systems in fishes are highly developed; the paddlefish (*Polyodon spathula*) can sense the electrical activity of a swarm of *Daphnia* 5 cm away (Russell et al., 1999). Fishes can sense turbulence trails to track prey as well (Montgomery et al., 2002).

Scientists can also use the concept of optimal foraging to predict how long a predator will remain in a patch of food. If a patch is of low quality, then moving to another patch may be more beneficial. However, if the cost of moving to another patch is high, it can be beneficial to extract more food from the current patch. Obviously, temporal and spatial scales are important considerations when making predictions using optimal foraging models. Similar considerations form the basis of the field of landscape ecology, discussed in Chapter 25.

A problem with using optimal foraging theory to make predictions about behavior of predators is that an investigator may not be able to identify the most important selective forces (Gatz, 1983). For example, one could predict that large prey items are preferable to smaller items because they contain more energy. However, large prey may be encountered infrequently and may take more energy to capture. Taking many small prey items may be more energy efficient; taking many small prey seems to be the strategy used by many predatory freshwater fishes (Juanes, 1994).

The number of prey eaten per unit prey density is referred to as the *functional response*. The number of predators per unit prey density is referred to as the *numerical response*. A functional response curve can take one of three general forms (Fig. 20.7).

In a type I functional response, predation is linearly related to prey density until some saturation is reached; this is the simplest predation model. In a type II response,

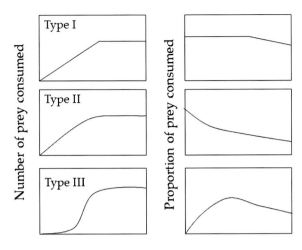

Figure 20.7 Holling's three types of functional response curves and the proportion of prey consumed assuming constant predator numbers.

there is a hyperbolic response to prey density. The form of this response is similar to that seen in the Michaelis—Menten uptake kinetics described in Chapter 17. This form of response is typical of predators that do not have a complex behavioral response to acquiring prey. Just as an enzyme can react with only so many molecules per unit time, a predator can only take a maximum number of prey items per unit time.

The type III response is seen in more advanced predators. At very low prey densities, few or no prey items are taken (optimal foraging theory predicts that there is not enough energy gain to profitably take prey). At intermediate prey densities the rate of capture increases, and eventually the predator is saturated, as in types I and II. Numerical response generally occurs over longer time periods because it requires changes in the predator's populations (from birth, death, immigration, and emigration), unlike functional responses that are primarily limited by behavioral and physiological aspects of the predator and distribution of prey in the environment.

Several strategies are used to consume prey. Many predators need to consume prey whole. These are referred to as *gape limited* predators, and the size of prey they can consume is limited by the size of their mouth or gape (Zaret, 1980). The dominant vertebrate predators in freshwaters are mostly of this type.

NONLETHAL EFFECTS OF PREDATION

The previous discussion assumes the predator will kill its prey; however, other effects can occur, including injury, restriction in habitat use or foraging behavior, and changes in life history (Allan, 1995). Strategies of nonlethal predators include piercing or sucking prey and removing bites or chunks of prey. Such predators include copepods, predatory cladocera, some midge larvae, fishes that are scale predators, and other insect larvae. Parasites are a special class of predators that have a wide variety of ways to attack prey, but often do not kill their hosts.

Prey can spend a large amount of time avoiding predation, and less time feeding or inhabiting food-rich patches. In this case, the reduced consumption of food can lower fitness. The indirect effects of predators on prey behavior are often referred to as the "ecology of fear." This ecology of fear can fundamentally alter food web structure and energy flow by altering the balance of feeding on detritus or primary producers (Sitvarin et al., 2016).

An example of sublethal effects and how they can alter a lake food web was documented by Hill and Lodge (1995). Omnivorous crayfish (*Orconectes*) were exposed to largemouth bass (*Micropterus salmoides*) that were too small to eat them. These crayfish appeared to perceive the fish as a predation risk because crayfish survival and their feeding on macrophytes and invertebrates decreased. Even though the predator did not directly kill the crayfish, the nonlethal effects on the crayfish and its food species were significant. Additional examples of nonlethal predation have been described elsewhere in the book and include herbivores that consume part of a macrophyte, the scale-eating cichlids of Africa, restricted foraging time in the presence of a predator, and most cases of parasitism. In addition, earlier in this chapter we described how the presence of crayfish could lead to delayed reproduction of snails so they more quickly attain a size too large to be consumed. We also described a variety of behavioral responses that alter behavior of prey. Most cases of predator avoidance will be classified as nonlethal effects of predation because of the associated energetic cost.

TROPHIC LEVELS, FOOD WEBS, AND FOOD CHAINS

Food webs can be simplified into trophic levels for ecological analyses (Garvey and Whiles, 2016). These levels were described briefly in Chapter 8. Traditionally, levels include *primary producers* (photosynthetic organisms), *decomposers* or detritivores (consume dead organic material), *primary consumers* (herbivores or grazers), and *secondary consumers* (eat primary consumers). This simplification has been criticized because numerous organisms feed on several trophic levels. Some suggest that species can be assigned fractional trophic levels (e.g., primary producers are assigned level 1, herbivores level 2, predators of the herbivores level 3, and animals that eat half herbivores and half primary producers level 2.5). The idea of trophic levels, regardless of its weaknesses, has proven particularly important in analysis of the effects of top consumers on organisms lower in the food web. We discuss these cascading trophic interactions next.

THE TROPHIC CASCADE

In 1880, Lorenzo Camerano postulated that animals can control the biomass of lower trophic levels and diagramed how a carnivore can control herbivore populations and this can have a positive influence on plants (reprinted in Camerano, 1994). The idea that predation at the upper level of food chains can have a cascading effect down through the food chain is called the *trophic cascade*. Control of primary production by abiotic factors such as nutrients or light is called *bottom-up control*. Control of primary producers from the upper levels of the food chain is referred to as *top-down control*. These ideas entered modern ecology in a key paper by Hairston et al. (1960) that argued that plants dominate terrestrial systems because predators keep herbivores in check. These arguments were extended to the idea that even numbers of links in food chains (Fig. 20.8) will lead to higher biomass of primary producers (Fretwell, 1977). In this section, we discuss how trophic cascades could apply in lakes, streams, wetlands, and groundwaters.

In a highly cited paper, Brooks and Dodson (1965) documented that increased predation pressure by a planktivorous fish led to much smaller zooplankton species. Smaller zooplankton are less efficient grazers and the population size shift resulted in

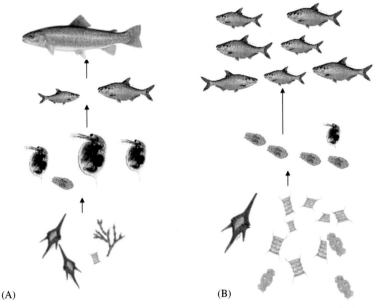

(A)   (B)

Figure 20.8 A conceptual diagram of the trophic cascade in the pelagic zone of lakes. When large piscivores are present, smaller zooplanktivores are uncommon, body size and numbers of zooplankton increase, and phytoplankton decrease (A), when they are not, phytoplankton numbers increase (B). Note, not drawn to scale outside of each tropic level.

increases in chlorophyll associated with increased phytoplankton survival. The idea that trophic cascades operate in lakes received some subsequent attention (Arruda, 1979; Shapiro, 1979), but a series of papers by Carpenter and Kitchell (1987) and Carpenter et al. (1987) stimulated abundant research (Power, 1992). This research was stimulated because managers could possibly use *biomanipulation* of lakes to improve water quality by controlling the fish community. The general approach is to encourage the populations of larger fishes that consume smaller fishes that are zooplanktivores. The trophic manipulation leads to increases in large grazing zooplankton (i.e., average body size of zooplankton increases). The large grazing zooplankton can consume more phytoplankton. Consequently, there is an increase in water clarity related to a decrease in suspended algae. Food web manipulation in concert with nutrient control can lead to significant increases in water quality over decades, as has been demonstrated in Lake Mendota, Wisconsin (Lathrop et al., 1996). However, the approximately 10 decades of research on Lake Mendota illustrate that bottom-up and top-down conditions must both be considered as factors influencing water quality (Kitchell and Carpenter, 1992).

The key issue in propagation of bottom-up or top-down effects in lakes seems to be how well the effects are transmitted between the zooplankton—phytoplankton link (Carney, 1990; Elser et al., 1990; Elser and Goldman, 1991), although the trophic cascade may also break down at the link between zooplankton and fishes (Currie et al., 1999). Empirical analysis suggests that the grazer link is where top-down and bottom-up effects are decoupled (Brett and Goldman, 1997). A study comparing 11 experiments in shallow lakes found that nutrient effects were three times stronger than top-down effects (Moss et al., 2004). The same analysis indicated that the zooplankton-grazer link transmitted top-down effects. In oligotrophic lakes, nutrients may be so limiting that top-down effects cannot occur (Spencer and Ellis, 1998).

Additional factors can influence trophic cascades. For example, concentrations of dissolved organic carbon that are high enough to interfere with sight feeding can affect predation rates and decouple trophic cascades (Bartels et al., 2016). In highly eutrophic lakes, large, grazer-resistant cyanobacterial colonies can dominate, leading to lack of grazer control on phytoplankton and decoupling of top-down effects from the bulk of primary producers. Anoxic hypolimnia of lakes may allow some species to avoid control by trophic manipulation of fish species in the epilimnion. The zooplanktivorous predator *Chaoborus* can withstand low oxygen conditions and find refuge in the anoxic hypolimnion of a eutrophic lake; alteration of the fish component of the zooplanktivorous food web in this case can have little influence (Dawidowicz et al., 2002). In a mesocosm experiment, salt pollution actually accentuated a trophic cascade by making zooplankton more susceptible to fish predation, which led to increased phytoplankton because of reduced grazing (Hintz et al., 2017).

The trophic cascade can extend to physical and chemical aspects of the lake. Low zooplankton grazing rates can lead to greater influx of atmospheric CO_2 related to increased algal biomass with high photosynthetic demand (Schindler et al., 1997). Alternatively, large predators such as fishes in streams can lead to decreased CO_2 emissions from the water because they decrease prey biomass, which leads to increased biomass of detritus and algae (Atwood et al., 2014). Predation by zooplanktivorous fishes can increase nutrient supply by excretion, enhancing algal production both by indirectly lowering grazing pressure and by directly providing nutrients (Persson, 1997; Vanni and Layne, 1997; Vanni et al., 1997). Mazumder et al. (1990) demonstrated that lakes with higher abundance of planktivorous fishes had lower zooplankton, higher phytoplankton, and lower light penetration. The lower light penetration led to less heating of the deeper water in the spring, which led to a shallower epilimnion depth. Similar results occurred both in large enclosures in which the plankton was manipulated and in 27 small Ontario lakes with varied levels of planktivorous fishes. These results demonstrated a clear local effect of organisms on heat content and physical structure of lakes as mediated by the food chain.

The trophic cascade concept in lakes requires simplification of food webs into food chains with discrete trophic levels. Several examples illustrate how community complexity can alter the intended effects of food chain manipulation. In shallow eutrophic ponds, very low rates of zooplankton grazing resulting from intense predation can lead to blooms of unwanted cyanobacteria (Spencer and King, 1984). In the same ponds without fishes, zooplankton flourish, phytoplankton decrease drastically, and macrophytes and periphyton growths can reach nuisance levels. Similar results were seen in shallow eutrophic lakes in the Netherlands. In this case, fish biomass was lowered, zooplankton initially flourished, phytoplankton decreased, and then macrophytes filled the lakes, leading to suppression of both zooplankton and phytoplankton numbers (Fig. 20.9).

Another example of how complexity of food webs interferes with lake biomanipulation is the fact that large fishes reproduce well when their numbers are increased for trophic control of zooplankton. The young of the year fishes that are produced are zooplanktivorous and can reverse the effects of the biomanipulation, at least temporarily (Hansson et al., 1998). Trophic cascades in bromeliad pool communities can be altered by presence of insectivores. When model birds were placed above bromeliads, the top predator damselfly larvae population sizes decreased in the pools, presumably because of a fear response of egg laying adults (Breviglieri et al., 2017). To add even more complexity, the trophic cascade in bromeliads is mediated as much by chemical cues leading to predation avoidance as it is the actual feeding of damselflies (Marino et al., 2016).

In warm reservoirs in the United States, gizzard shad (*Dorosoma cepedianum*) can be the most abundant fish. As young of the year, these fish consume individual

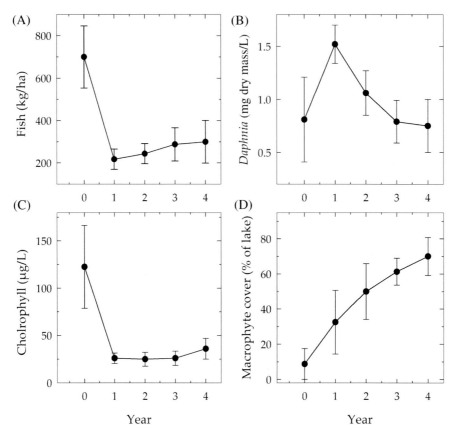

Figure 20.9 Effects of biomanipulation on fish, chlorophyll, *Daphnia*, and macrophyte cover in four shallow eutrophic lakes in The Netherlands. Year 0 is before biomanipulation. Points are means from four lakes and error bars equal 1 SD. *Data from Meijer et al. (1994).*

zooplankton. After growing to several centimeters in length, the fish develop a long gut that includes a grinding chamber and also a filtering structure on their gills. This allows gizzard shad to consume and thrive on phytoplankton, zooplankton, and detritus. Gizzard shad feed on several trophic levels at once and this habit confounds the use of a food chain model to describe trophic dynamics in these reservoirs (Stein et al., 1995). In other shallow lakes, the greatest effect of fishes on algae may not be through the food web but through increased sediment suspension (Havens, 1991) and nutrient mineralization by benthic fishes (Gido, 2002).

Another example in which the idea of a food web rather than a food chain more accurately characterizes aquatic trophic relations is the link between the trophic cascade and the microbial loop. This link is equivocal; some investigators have established the presence of a top-down effect on the microbial loop, and others have not. For example, Pace et al. (1998) demonstrated variable effects on heterotrophic flagellates but increased populations of ciliates and rotifers in lakes with low *Daphnia* populations. Gasol et al. (1995) showed that cladocerans have the greatest effects on heterotrophic nanoflagellates (very small protozoa) during the summer. Pace and Cole (1994) suggested that there was little influence of *Daphnia* populations on bacteria. Simek et al. (1998) described a lake in which a filtering cladoceran controlled bacterial populations in one basin and a ciliate was the primary bacterivore in another basin. Jeppesen et al. (1998) found a weak link between the microbial loop and zooplankton abundance in an 18-year time series from a hypereutrophic lake.

Consideration of temporal and spatial scales can also alter the response of food webs to manipulation. For example, pulses of nutrients can have different effects depending on food web structure and their timing (Cottingham and Schindler, 2000). If zooplankton have time to respond to nutrient enhancement of phytoplankton growth, further pulses of nutrients will have little influence. If predation on zooplankton is high, they may not be able to suppress ephemeral phytoplankton blooms in response to nutrient pulses (Strauss et al., 1994).

Top-down effects can also occur in benthic habitats of lakes. Snails remove littoral periphyton and are susceptible to predation by sunfish. When sunfish are excluded, algal biomass decreases significantly (Brönmark et al., 1992). Since sunfish are prey for large piscivores, high biomass of large piscivores could lead to a decreased biomass of periphyton through the tropic cascade. The alteration of the snails can result in top-down effects on the entire microbial community in the benthic periphyton of lakes (Burgmer et al., 2010). In another case, fishes were removed from eutrophic Lake Ringsjön, Sweden, including the littoral benthic feeding bream *A. brama*, in an attempt to use biomanipulation to improve water quality. Water clarity improved in the pelagic zone and there was an unintended increase in benthic invertebrate populations and concurrent increases in staging waterfowl abundance (Bergman et al., 1999).

Both top-down and bottom-up control of primary production can occur in streams, and in many cases both operate simultaneously (Rosemond et al., 1993; Lourenço-Amorim et al., 2014; Collins et al., 2016). Mary Power has been a leader in the field of understanding the importance of top-down effects of animals in ecosystems, and has particularly specialized on streams (Biography 20.1). In some California streams (Fig. 20.11), exclosure of fishes resulted in decreases in *Cladophora* (a filamentous green alga) biomass because predation on midge larvae was decreased, and the midge larvae suppressed *Cladophora* (Power, 1990a). These results are consistent with the view that odd numbers of trophic levels lead to high producer biomass.

Research on the herbivorous stoneroller (*C. anomalum*) and piscivorous bass (*Micropterus* spp.) suggests that bass have a top-down effect on primary producers in

BIOGRAPHY 20.1 Mary Power

Mary Power (Fig. 20.10) is an aquatic ecologist specializing in fishes and food webs. She has played an important role in defining the idea of keystone species and describing trophic cascades in streams. Her work takes a broad view of ecology and she is comfortable working with algae, nutrients, invertebrates, and fishes. She is one of the preeminent ecologists world-wide and has done important work in tropical and temperate rivers and streams.

Dr. Power has over 100 publications including some of the most cited works in ecology. She has served as society president for the Ecological Society of America and the American Society of Naturalists. Her numerous awards and honors include the Hutchinson Award from the American Society of Limnology and Oceanography, being elected a fellow for the California Academy of Sciences, and fellow of the Ecological Society of America. She is also a member of the National Academy of Sciences of the United States, one of the most prestigious honors a scientist can achieve.

Figure 20.10 Dr. Mary Power.

Mary says she was nearsighted as a child. When she finally got glasses, she could see leaves on trees and the fishes and crabs in the ocean and got hooked on life under water. She was always interested in ecology but at Brown University as an undergraduate, the Biology department shifted strongly toward molecular biology. She felt that they did not value ecology and she had a professor tell her that ethology will rot her mind. This turned her off to biology because she did not like the lab chemicals they were using. Luckily, she taught adult education at Boston University, where the nontraditional students' enthusiasm for ecology reignited her passion for ecology. She took a tropical marine biology course and saw coral reef fishes for the first time and wanted to research fish ecology.

(Continued)

BIOGRAPHY 20.1 (Continued)

Mary entered graduate school and settled on a research project in Panama in 1970s. Her project was to be on fishes in the diverse Rio Frijoles, a river along the Panama Canal. When she got there the canal officials were adding rotenone poison and there was a huge number of dead fishes at her research site. They were poisoning the native fishes because they were worried the natives would compete with the Asian grass carp (*Ctenopharyngodon idella*) they planned to introduce to the canal. The officials were introducing grass carp to eat invasive macrophyte, *Hydrilla*, they thought would clog the canal. So her first days of her field work, she came into a research site full of dead fish. She mentions this as a place where her persistence paid off; some graduate students would have just given up. She investigated and noticed that the canal authorities did not add enough to kill all the fish; some of them crawled out into the riparian litter or hid in deep pools. Therefore she was able to do her research anyway. Later the canal authorities enlisted her to help check if the grass carp were established. They took boats out into the canal and threw in dynamite, sending water 10 m up into the air and killing a lot of fishes but, but no grass carp.

Dr. Power says there are four things a successful researcher needs. First, they need persistence and motivation, because energy and commitment are more important than being brilliant. Second, they need to be good, respectful, thorough, and generous scholars. They should look into the literature to see if idea is new and give credit where it is due. It is important to show sense of gratitude for whose shoulders you are standing on. Third, a student should develop their own world view. This gives them something to test ideas against and help take in and remember things from nature and reading. It is the tree trunk that you can use to build the branches. In her case the view is the importance of food webs in ecological systems. Finally, ecologists should study both process and pattern (local context) and use both to inform each other. This allows confirmation of hypotheses so one can predict the future ecological conditions.

She sees several areas important to the future of ecology. She thinks it is important to study how traits influence populations, communities, and ecosystems. A specific question is how do gene drives influence ecology? For example, if you release a large number of genetically modified mosquitoes for disease control how will that influence local ecology and ultimately evolution of the species? Mary thinks we need to know more about the characteristics of the watershed (climate, soils, geology, and plants) that influence water storage and release. Finally, Dr Power asks, what will freshwater food webs look like in 10–100 years?

small prairie streams. Pools with bass have few stonerollers and high algal biomass, whereas pools without bass have low algal biomass and high numbers of stonerollers. Finally, when a bass was tethered in a pool, algae proliferated only in the areas around the bass, but not outside their reach (Power and Matthews, 1983). The tethering effect may have occurred because bass also can control crayfish in this system, but the effect is still a top-down effect. Grazing by the stonerollers led to lower algal biomass, more cyanobacteria, less heterotrophic bacteria, higher invertebrate density, and less

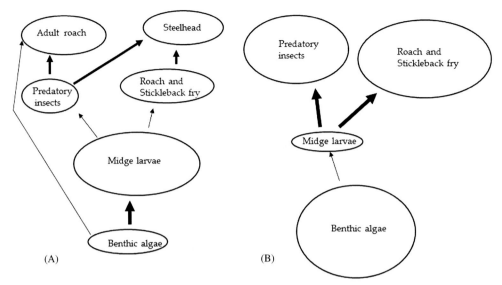

Figure 20.11 Summary of effects of enclosure (A) and exclosure (B) on predatory fishes [roach (*Hesperoleucas symmetricus*) and steelhead (*Oncorhynchus mykiss*)], roach and stickleback (*Gasterosteus aculeatus*) fry, invertebrate predators (lestid damselflies), midge larvae (*Pseudochironomus richardsoni*), and benthic algae (*Nostoc* and *Cladophora*). Size of oval represents amount of biomass at each trophic level. *Data from Power (1990).*

particulate organic material, so the effects of this trophic cascade extended to basic stream ecosystem properties (Gelwick and Matthews, 1992).

Mechanisms driving top-down effects are not always obvious. For example, Alvarez and Peckarsky (2014) added brook trout fish chemicals to naturally fishless streams. They observed increases in palatable diatoms in the streams that received the chemicals, but not in control streams. They hypothesized that predator avoidance behavior by a dominant grazing mayfly in response to the fish chemicals led to lower grazing pressure on the diatoms.

Control of grazers does not always have the anticipated effects in streams. Armored catfish (Loricariidae) are common grazers in the Rio Frijoles in Panama. Exclosure of the catfish led to an initial increase in benthic algal biomass, but after a few weeks, enclosures were smothered with sediments (Power, 1990b). In this case, a secondary effect (sediment resuspension or decrease in sediment trapping) led to results that would not be predicted by a straight trophic view of streams.

Bottom-up factors can also be propagated through the food web to parasites. In New Zealand, parasites infecting fishes and invertebrates were more prevalent and stable when host density was higher. Not all of these were contact-transmitted parasites, so host density effects were not solely epidemiological (Lagrue and Poulin, 2015). Horsehair worms (Nematomorpha) are parasites that can decouple trophic cascades. The parasite alters the behavior of its host, crickets and grasshoppers, causing them to enter the water, where the parasite then leaves the host body. The host crickets and grasshoppers are an important food source to salmonids in some Japanese streams. When these terrestrial prey are not available, the fish concentrate on herbivorous stream invertebrates and algal biomass increases. When the parasitized crickets and grasshoppers are available, stream algal biomass is reduced (Sato et al., 2012).

Understanding trophic cascades and parasites has practical importance; restoration of a native river prawn (*Macrobrachium vollenhoveni*) lowered schistosomiasis by 18% in the human population in Senegal. The prawn was extirpated because a dam blocked their natural migration. Sokolow et al. (2015) reintroduced the prawn, which lowered the intermediate host snail populations by 80%.

Less is known about trophic cascades in wetlands, though the food webs should not differ drastically in structure from other benthic food webs; therefore, there is no reason to believe that the interplay between top-down and bottom-up controls is not important. For example, predaceous hydrophilid beetle larvae can control algae-grazing chironomid midge larvae in a seasonal wetland. In the absence of predators, the midge larvae outstripped their food supply (Batzer and Resh, 1991). Aquatic herbivores can control wetland riparian vegetation (Sarneel et al., 2014) and water lilies (*Naphur lutea*) can be controlled by herbivorous beetles (Stenberg and Stenberg, 2012). An interesting indirect cascade occurred in Florida wetlands where ponds containing fishes were compared to those that did not. Fish predation decreased dragonfly larval survival and subsequently adult populations decreased. The decrease in adult dragonflies led to an increase in pollinators, increasing nearby plant production (Knight et al., 2005). Removal of wolves (*Canis lupus*) in Montana led to increases in grazing by elk (*Cervis elaphus*) on willows (*Salix*) and decreased riparian cover (Ripple and Beschta, 2004), ultimately influencing the aquatic community by altering shading and litter input. The golden apple snail (*Pomacea canaliculata*) has invaded numerous wetlands in Southeast Asia leading to almost complete removal of macrophytes and concurrent increases in phytoplankton (Carlsson et al., 2004).

As with wetlands, the trophic cascade in groundwaters is poorly documented. Grazer control of bacterial production occurs in several groundwater habitats. The isopod *Caecidotea tridentata* stimulated bacterial production in a limestone aquifer in which carbon additions did not (Edler and Dodds, 1996). A case of bottom-up control was demonstrated in which organic contamination of an aquifer stimulated bacterial and protozoan production relative to nearby uncontaminated groundwater (Madsen et al., 1991). Protozoa depressed bacterial abundance, but increased nitrification rates in a laboratory study using groundwater and associated sediments (Strauss and Dodds, 1997). As more groundwater food webs are described, numerous cases of top-down and bottom-up controls are likely to be documented.

A strong caution regarding use of biomanipulation to control eutrophication based on the trophic cascade concept has been put forward by Wetzel (2001). He mentions the numerous compensatory mechanisms that emerge quickly after biomanipulation. These could include predator protection and avoidance by zooplankton, development of an inedible phytoplankton assemblage, and eventual high macrophyte dominance (in shallow lakes). Ultimately, evolution will counter changes in predation pressure; this concept brings up the argument over biotic versus abiotic factors ultimately controlling evolution.

SUMMARY

1. Some species from most major groups of animals inhabiting freshwaters can eat phytoplankton, periphyton, macrophytes, detritus, or other animals. Omnivory and detritivory are common in freshwater animals.
2. Herbivores influence plant and algae in a variety of ways, ranging from reducing biomass to increasing biomass-specific production.
3. Detritivores often feed selectively on microbially conditioned materials and gain much of their nutrition from fungi associated with detritus.
4. Stable isotopes and fatty acid profiles are now commonly used to assess trophic interactions.
5. Prey can respond to chemical, visual, and hydromechanical cues.
6. Protective adaptations to predation include mechanical (size and spines), chemical, and behavioral.
7. Predators forage optimally to maximize their efficiency; this may lead to one of several functional responses to prey numbers.
8. Predators can sense their prey by using chemical, visual, tactile, and hydromechanical cues.
9. Trophic cascades are an important part of aquatic food webs and may occur in all freshwater habitats. Effects can be transmitted through food webs from the top (predators) or the bottom (nutrients and light). The grazer—producer link appears to be crucial in transmission of these effects.

QUESTIONS FOR THOUGHT

1. If a predator is consuming prey at a rate lower than the rate at which the prey is able to replace itself, can the effect of predation be considered significant even if the prey population is increasing?
2. How might the increasing use of fungicides in agricultural systems influence adjacent streams, lakes, and wetlands?

3. Why can increasing turbidity of large rivers cause shifts in types of predators that are successful, and adaptations of prey to those predators?

4. Why might it be more effective to control algal blooms by biomanipulation with removal of all fish than by imposing fishing regulations to increase the number of piscivorous fish?

5. Why are brightly colored organisms less common in freshwaters than in benthic marine systems?

6. What single cosmopolitan species is the top predator in more freshwater systems throughout the world than any other species?

7. Why are chemical cues to prey easier to follow in streams and benthic habitats than in pelagic habitats?

8. How do selective pressures for streamlining of fishes conflict with limitations on gape width?

9. Are trophic levels a valid concept in most freshwater habitats?

CHAPTER 21

Nonpredatory Interspecific Interactions Among Plants and Animals in Freshwater Communities

Contents

Figure 21.1 An experimental stream mesocosm facility used to study interactions among organisms. Each cylinder represents a pool and each rectangular portion a riffle, with water recirculating in each channel. In the panel below, two herbivorous fish species (*Campostoma anomalum* and *Phoxinus erythrogaster*) are used in a behavioral experiment in the same mesocosm.

Freshwater Ecology
DOI: https://doi.org/10.1016/B978-0-12-813255-5.00021-1

653

Competition is central to many aspects of aquatic communities and, thus, has been the focus of much research. Other ways that species interact (including indirect interactions, succession, mutualism, and the effects of keystone species) also have consequences in determining community structure. Some of these concepts were discussed in previous chapters including basic definitions of types of interactions (Chapter 8 classification of interaction types (e.g., trait-mediated vs density-mediated interactions), and interactions involving microorganisms Chapter 19). This chapter focuses on interactions among larger organisms.

COMPETITION

Competition (both species have a negative effect on each other) has been described as a dominant community interaction. It can occur between organisms from any taxon, but it is more likely to occur among organisms in the same functional group or guild. We discuss competition among species of plants and animals in this section. The concept of competition intermingles with that of the *niche*, the ecological role of an organism, particularly as defined by its resource consumption. The idea of niche is somewhat controversial with respect to variability in types of resources used over time and how much overlap occurs in diets of various species. Regardless, species often specialize on resources and partition those resources among themselves based on varied attributes of those resources. Competition occurs within (*intraspecific*) or between species (*interspecific*).

Intraspecifc competition is *density-dependent* in that competition for a given resource such as food becomes more intense as relative availability decreases (*density-independent* factors are not a function of population size). At some level of competition, mortality increases or natality decreases, which results in reduced or negative population growth. This type of interaction is important in natural systems, as well as aquaculture and fisheries management, where the goal is to maximize growth of desirable species.

Intraspecific competition can also reduce individual growth rates. For example, tadpoles can reach very high densities in breeding ponds. In experiments where different total numbers of tadpoles are placed in the same sized rearing tanks, reduced growth occurs with increased densities (Fig. 21.2). In natural ponds in Sweden, experimentally increased densities of *Rana temporaria* tadpoles also resulted in significantly higher mortality rates (Loman, 2004). Density manipulations of other amphibian species have found that tadpoles reared at high densities metamorphose earlier, and at a smaller size, than those reared at lower densities, which has implications for individual fitness (Richter et al., 2009).

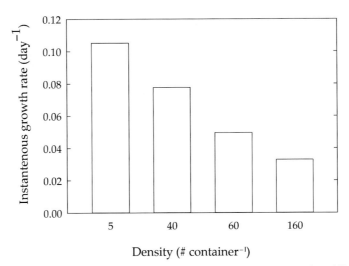

Figure 21.2 Daily instantaneous growth rates of *Rana tigrina* tadpoles reared at different densities. Numbers below bars represent densities of individuals growing in containers of the same size. *Data replotted from Dash and Hota (1980).*

Intraspecific competition can be a cause of disruptive selective pressure as evolution selects for different morphologies or strategies of resource use and the intermediate form is at a disadvantage. For instance, the three-spine stickleback, *Gasterosteus aculeatus*, lives in lakes. Individuals within the same species and the same lake specialize on pelagic or benthic food sources. The morphology of the gill rakers reflects this specialization. In experimental manipulations, there was a greater difference between the morphologies when both were present. The effect was accentuated at higher densities, as would be expected if the morphological selection resulted from intraspecific competition for resources (Bolnick, 2004).

Both intraspecific and interspecific competition fall into two general types. *Scramble competition (exploitative competition)* occurs when all individuals are "scrambling" to acquire as much as they can of the limiting resource; as availability of the resource decreases, each individual gets less, and all are equally affected. This type of competition does not involve direct interactions among members of the population and is quite common when population densities are high. In *contest competition (interference competition)*, some individuals get more than others do by directly interfering with the ability of others to acquire the resource. Examples of intraspecific contest competition include competition among male dragonflies for territories and mates, and competition among male cichlids for breeding territories (Alcazar et al., 2016). Contest competition involves direct interactions among members of the population (e.g., individuals preempting another individual's ability to acquire the resource).

Allelopathy among plant species and territoriality in animals are common forms of interspecific contest competition.

Sexual selection can result from intraspecific competition. For example, male fishes can have broadly different morphology from females ultimately related to competition for mates (Fig. 21.3). In a more complex example, water striders (*Aquarius remigis*) have intense competition for mates. In this case, the water striders congregate in pools and males compete to mate with females. The intense competition to mate actually can harm the females. In conditions where there is only one habitat, the most aggressive males win because they are the most successful at mating. However, when there are several habitats available (e.g., a stream with connected pools, some with many water striders and others with few), females will leave pools with many males to avoid being overwhelmed by mating attempts. If less aggressive males inhabit the pools with

Figure 21.3 Breeding-age adult central stonerollers (*Campostoma anomalum*). The male, in the top panel, has obvious tubercles and fin markings. These could play a part in aggressive interactions among males competing for mates as well as female mate choice. *Photograph courtesy: Garold Sneegas.*

few water striders, then they will also mate successfully when females enter their pools to avoid the higher-density pools. This is a case where population structure can alter the selective pressures of intraspecific competition (Eldakar et al., 2009).

Competition is often difficult to establish directly in the field, in part because evolution leads to decreased overlap in resource use. Intraspecific competition might not be apparent in cases where organisms inhabit defined territories. Interspecific competitors that are usually found in the same habitat at the same time rarely specialize on the same resource. For example, three-spined sticklebacks (*Gasterosteus* spp.) in three lakes of British Columbia have recently evolved from a single marine species. In all these cases, the new pairs of species found in each lake partition the habitat by feeding on benthic or limnetic invertebrates. These species evolved mechanisms of reproductive isolation to ensure that their offspring will inherit their parents' behavioral traits. Limnetic adults prefer to mate with limnetic adults and benthic adults prefer to mate with benthic adults (Rundle et al., 2000). In this case, researchers cannot observe competition observed among species currently inhabiting a lake, even though competition led to diversification of species.

The idea that competition can be difficult to find in extant communities because evolution has led to species partitioning niches and not currently competing is encapsulated in the term "ghost of competition past." It is very difficult to prove that this effect is real (Connell, 1980). Still, the idea provides one good explanation of why it could be difficult to document competition in field experiments.

An additional reason that competition may be difficult to document is that competitive exclusion may lead to conditions where potential competitors do not co-occur. We discussed competitive exclusion in Chapter 17, in the section on the Paradox of the Plankton. The concept behind the competitive exclusion principle is that under equilibrium conditions, the superior competitor should eliminate the inferior competitor. The principle also assumes that both competitors specialize on the same resources. The exclusion may take many generations, but eventually the superior competitor will come to dominate. Our discussion on the Paradox of the Plankton explored how nonequilibrium conditions, predation, mutualism, disease, and multiple limiting factors could explain coexistence of potentially competitive species of phytoplankton. Nonequilibrium conditions are generally more prevalent in benthic habitats, streams, and wetlands than in the pelagic zones of large lakes. Predation and disease are common features of all habitats (see prior chapter) that push systems further from equilibrium. The assumptions of the competitive exclusion principle thus may not always apply, and coexistence of interspecific competitors could be a potential feature of many habitats.

Intraspecific competition can lead to strong selective pressures on organisms. Organisms, in some instances, can avoid these interactions. For example, intraspecific competition can lead to "Alee effects" where organisms self-limit their population growth so as not to crash the entire population. The interplay between success of individuals and success of the group has led to substantial controversy in the field of evolution.

We now document several types of interspecific competition that commonly occur in freshwater environments. Primary producers compete for nutrients and light. Competition among macrophyte species is common (Gopal and Goel, 1993). Macrophytes can form dense stands that shade all primary producers below them (Haslam, 1978). Floating macrophytes, such as water hyacinth (*Eichhornia*) and duckweed (*Lemna*), can blanket the surface of lentic habitats and essentially block all light penetration in to the water. This can give them a strong competitive advantage, even under low nutrient conditions where rooted macrophytes have greater access to nutrients (van Gerven et al., 2015). Macrophytes can intercept 90% of the incident light before it reaches a lake bottom, leading to a 65% decrease in benthic algal production (Lassen et al., 1997). If systems are sufficiently shallow, emergent primary producers intercept light before it reaches the water's surface.

Researchers documented competition for light among macrophytes inhabiting a stream in North Carolina (Everitt and Burkholder, 1991). In this case, they characterized riparian vegetation and macrophyte assemblages in 10 stream segments. The red alga *Lemanea australis* or the aquatic moss *Fontanalis* dominated low-light assemblages in the winter. The alga *L. australis* and the angiosperm *Podostemum ceratophyllum* dominated high light sites. These sites were on the same stream and had similar water velocity and depth, indicating that light availability as influenced by riparian shading was the major abiotic difference between them. Apparently, competition for light structured the macrophyte assemblages.

Macrophytes and phytoplankton compete for light and nutrients in shallow lakes. Several species of macrophytes are capable of interference competition; they release allelopathic chemicals that inhibit phytoplankton species (Gross et al., 2007). In shallow lakes dominated by macrophytes, the deeper portions often have relatively clear water (low phytoplankton concentrations). Hence, allelopathy can stabilize macrophyte dominance in shallow lakes (Hilt and Gross, 2008).

Competition can shape wetland plant assemblages. Competition between two species of wetland plants may occur aboveground (for light) or belowground (for nutrients), and both may be important simultaneously (Twolan-Strutt and Keddy, 1996). Competitive rankings of individual plants can remain stable against changes in nutrients and flooding in some instances (Keddy et al., 1994). However, disturbance can release wetland plants from competition (Keddy, 1989). These studies suggest that competition can determine which plant species dominate in a particular wetland.

An example of competition determining spatial patterns in macrophytes occurs between two species of the cattail, *Typha latifolia* and *Typha angustifolia*. *T. latifolia* has broad leaves and can outcompete *T. angustifolia* in shallow water, but not in deeper water (Grace and Wetzel, 1981). As the season progresses, the two species are almost completely segregated (Fig. 21.4). This segregation occurs even though both species have optimum growth rates at 50 cm depth when occurring in isolation.

Zooplankton species often compete for food. In Chapter 19, we discussed the idea that ratios of nutrients may alter competitive ability or lead to coexistence of phytoplankton species. Similar resource partitioning could lead to coexistence of potentially competing zooplankton species (DeMott, 1995; Lampert, 1997).

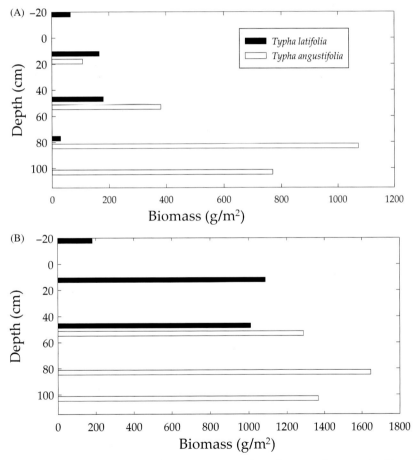

Figure 21.4 Competitive exclusion of two species of *Typha*. Distributions in May (A) and September (B). *Data plotted from Grace and Wetzel (1981).*

Resource ratio theory is a successful way to assess competition for several resources, for example competition by plants for both nitrogen and phosphorus (Sterner and Elser, 2002). Resource ratio theory predicts how tradeoffs associated with using two or more resources that are limiting influence how species coexist (Tilman, 1982). For instance, two species that are differentially limited by two resources will coexist at intermediate ratios of resource supply. At more skewed ratios of supply, the best competitor for the most limiting resource can exclude other species.

In some cases, a trade-off may occur in competitive ability. By way of illustration, one species of rotifer (*Keratella cochlearis*) may compete better at a lower food concentration and the other (*Keratella earlinae*) at higher concentrations (Fig. 21.5). Both species could coexist in a spatially or temporally variable environment. Competitive interactions among zooplankton species probably vary over space and time. In Chapter 20, we discussed the idea that large zooplankton are superior competitors for phytoplankton cells. However, this competitive interaction may not be so simplistic. Large zooplankton may slow or cease feeding in blooms of large inedible algae while smaller zooplanktons continue to feed unhindered on subdominant small phytoplankton and bacteria (Gliwicz, 1980). Similarly, cladocerans may be superior competitors for food compared to rotifers, but suspended clay can decrease the growth of cladocerans while having little effect on rotifers. Thus, the clay releases rotifers from competition with cladocerans (Kirk and Gilbert, 1990).

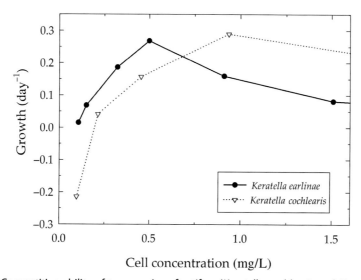

Figure 21.5 Competitive ability of two species of rotifers (*Keratella cochlearis* and *Keratella earlinae*) that feed on the cryptomonad *Rhodomonas*. Note that *K. cochlearis* is able to grow more rapidly at lower concentrations of food, but *K. earlinae* grows better at higher concentrations of food. *Redrawn from Stemberger and Gilbert (1985).*

There are situations in which competitive exclusion should operate but may not be strong enough to drive out weaker competitors. An example occurred with five species of herbivorous zooplankton in a small humic lake (Hessen, 1990). These species coexisted, but pairwise growth experiments in bottles of lake water showed no evidence of significant predation or a single dominant competitor. Weak competition for abundant nutrient-poor food in this lake was hypothesized to allow coexistence of the species. This study and those noted previously are a few of many that suggest that competition is an important, but context-dependent feature of zooplankton assemblages.

Invasive species are often very strong competitors. For example, purple loosestrife (*Lythrum salicaria*) has expanded rapidly in the United States, displacing other wetland species (Blossey et al., 2001). The reed canary grass (*Phalaris arundinacea*) is also an invasive species. In the wetlands where they have invaded, they often become dominant; purple loosestrife makes up about 30% of the total abundance, and canary grass about 8% overall (Fig. 21.6). In most cases where both were present (6 of 7), purple loosestrife was dominant. In 6 of 24 wetlands, reed canary grass was the most abundant species, and in 7 of 24 cases purple loosestrife was dominant (Schooler et al., 2006). These data indicate that invaders can be strong competitors and they can decrease plant diversity of the wetlands they have invaded when present in high abundance.

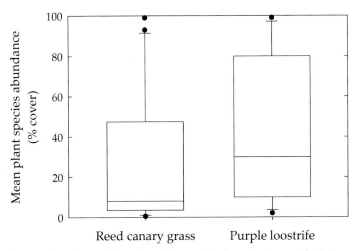

Figure 21.6 Relative abundance of two invasive species (purple loosestrife *Lythrum salicaria* and reed canary grass *Phalaris arundinacea*) in Pacific Northwest US wetlands. Box plot line shows median, box sides the 25th and 75th percentiles, and the whiskers the 90th and 10th percentiles. While overall neither species dominates in all wetlands, there is a high degree of dominance in some wetlands. *Data replotted from Schooler et al. (2006).*

Competition for food has led to clear specialization of benthic freshwater macroinvertebrates over evolutionary time, and invertebrates can be classified into functional feeding groups based on mode of food acquisition (such as collectors, filterers, shredders, and scrapers). Within these groups, competition can still occur. For example, net-spinning hydropsychid caddisflies collect particles from the water column with silk nets of different mesh sizes, with each mesh size characteristic of a species. Smaller mesh nets are more efficient at collecting small particles, whereas large meshes capture large particles more effectively (Loudon and Alstad, 1990). Species with fine nets will likely compete more effectively for fine particles. Such competitive specialization may determine which species dominate.

Although competition among coexisting species using similar resources (such as assemblages of net-spinning caddisflies) might be expected, interactions can be complex and sometimes counterintuitive. Laboratory experiments using different numbers of species of net-spinning caddisflies found that other species enhanced the particle capture efficiency of individual species because of changes in near-bed flow dynamics (Cardinale et al., 2002).

Competition among pond-breeding amphibian species is also of interest. In some ponds, predators remove the competitive dominant frogs, and predation allows less competitive species to coexist with stronger competitors (Morin, 1983). The chytrid fungus *Batrachochytrium dendrobatidis* is causing amphibian declines around the world (Highlight 11.3) and can alter competitive interactions among amphibian species that do not die from infection. In the presence of the fungus, both fowler's toad (*Bufo fowleri*) and gray treefrog (*Hyla versicolor*) tadpoles metamorphosed at smaller body masses when reared together in tanks compared to when they were reared separately (Parris and Cornelius, 2004). The toad tadpoles also had strong negative effects on treefrog tadpole development, but only in the presence of the fungus.

Competition among fishes may be an important consideration for fisheries managers. For instance, Hodgson et al. (1991) demonstrated that introduction of rainbow trout (*Oncorhynchus mykiss*) into lakes with 2- and 3-year-old largemouth bass (*Micropterus salmoides*) can lower the condition (weight to length ratio) of bass. The diet of the bass shifted from *Daphnia* to odonate naiads, and bass condition was lower after the introduction of trout into one lake, particularly when compared to a nearby lake with no trout. Cage experiments suggest that sunfishes (Centrarchidae) compete when predators drive them to take refuge in macrophyte beds (Mittelbach, 1988), and competition may occur with juveniles and translate into lower production of adult fish (Osenberg et al., 1992).

MUTUALISM AND FACILITATION

Mutualisms (both species have a positive effect on each other) are less conspicuous in freshwater than in marine systems; possibly because the continuous time for evolution of mutualisms has been less in freshwaters than in marine systems (i.e., many freshwater habitats have a shorter continuous history than marine or terrestrial habitats). However, some of the conditions for mutualism occur in freshwater. Fish that clean other fishes are common on marine reefs, and this mutualism involves many species of fishes from diverse taxonomic groups. One would expect this behavior to be common in freshwaters but it is rarer, despite comparable benefits to freshwater fishes, and the long geological age of some river systems. Interestingly, a rare example of a freshwater haplochromine cichlid cleaner fish occurs in the highly speciose, but relatively recently evolved fish assemblage in Lake Victoria (Witte and Witte-Maas, 1980). Many of the mutualisms that occur in freshwaters involve microorganisms and we discussed them in Chapter 19.

Some researchers suggest that herbivory can be mutualistic, increasing both primary and secondary production. Herren et al. (2017) found that the chironomids of Lake Mývatn in Iceland stimulated gross primary production by 71% with higher chlorophyll concentrations near the larvae, particularly on the tubes they construct to live in. The effect was greater at higher midge densities, and the growth of the larvae was greater at high densities. These data suggest, at least on a short-term experimental basis, that mutualism between grazers and producers is possible.

Grazers and macrophytes can also form mutualistic interactions. The grazers can remove epiphytes from the macrophytes that would compete with them, or even overgrow them. The macrophytes can provide a surface for the grazers to inhabit and find food on, such as with gastropod grazers and macrophytes in freshwater springs of Florida (Nifong, 2017). A similar relationship occurs with the snail, *Vorticifex effuse*, and giant cyanobacterial colonies of *Nostoc parmeliodes* (Fig. 21.7) in an Oregon freshwater spring (Dodds and Castenholz, 1988a).

Mutualisms based on behavior require coevolved systems and organisms capable of complex behavioral patterns, such as fishes. Cichlids from Lake Tanganyika demonstrate parental care, including guarding eggs and fry from predators, and two species can brood in the same region and mutually defend their broods (Keenleyside, 1991). Fishes in the same lake have evolved cooperation in which predators hunt in mixed groups and this cooperation increases success (Nakai, 1993). Mixed-feeding schools may occur in other fish assemblages and shoaling (mixed schools) are common but we do not know much about the mutualistic aspects of this behavior. Likely, there are benefits; birds have a similar cooperative strategy of mixed-feeding flocks (Sridhar et al., 2009).

Figure 21.7 A large cyanobacterial colony (about 10 cm diameter) of *Nostoc pruniforme* and the scraping snail *Vorticifex effuse*. The snail cleans epiphytes off the surface of the *Nostoc* and when the snail is excluded, the growth rate of the *Nostoc* decreases.

Facilitation is any unidirectional positive effect of one species on another. Facilitation can include mutualism and commensalism, as well as some exploitation relationships. Facilitation may be an important and overlooked aspect of community interactions (Bertness and Callaway, 1994). Plants in stressful environments can facilitate each other by increasing structural stability against disturbance or providing other benefits (Callaway, 1995; Callaway and Walker, 1997). Few macrophyte and wetland plant assemblages have been studied with regard to facilitation, but it could be important in stressful freshwater habitats, as has been demonstrated for estuarine marshes (Bertness and Hacker, 1994). For instance, emergent freshwater marsh plants that are aerenchymous (transport oxygen to their roots) can facilitate other emergent plants living nearby by aerating the sediments (Callaway and King, 1996). As researchers study more aquatic plant assemblages, more examples of facilitation will likely be documented, given the importance of facilitation among terrestrial plants (Brooker et al., 2008).

Indirect facilitation may occur in streams. Crayfish exclude the green alga *Cladophora* from pools. The exclusion of *Cladophora* facilitates the growth of epilithic diatoms. The diatoms in turn support an increased biomass of grazing insect larvae (Hart, 1992; Creed, 1994). Such complex interactions may be a common feature of communities.

Facilitation is also common for nest-building stream fishes in some areas. In this case, the nest builders make stone or pebble nests for their eggs, and other species of fishes rely on these for their own reproduction. These interactions can possibly drive species distributions at large scales (Peoples and Frimpong, 2016a). For example, in a Virginia stream, the bluehead chub *Nocomis leptocephalus* facilitates the mountain redbelly dace *Chrosomus oreas*, which relies on *N. leptocephalus* nests for spawning. *C. oreas* decreases egg predation and increases reproductive output of *N. leptocephalus* if the density of egg predators is low enough. The relationship moved between mutualistic and commensalistic depending upon environmental context (Peoples and Frimpong, 2016b).

Cleaning mutualisms may be more common in freshwaters than ecologists have realized and relationships they classified as parasitic could be mutualistic in some cases. For example, Brown et al. (2012) studied the relationship between the branchiobdellid annelid worm *Cambarincola ingens* and the crayfish *Cambarus chasmodactylus* in field manipulations. They tested the idea that the worm benefits the crayfish by cleaning its gills. They found that crayfish growth was highest at intermediate densities of worms; the relationship shifted to parasitism at high worm densities. Subsequent research demonstrated that the crayfish actually groom and maintain optimal numbers of worms (Farrell et al., 2014). Creed et al. (2015) found that the worm requires the crayfish for reproduction, meaning the relationship is obligate and positive for the worms. Finally, Thomas et al. (2016) established that different worm species could have the same positive cleaning effects, and that species of worms shift as the crayfish grow. This interesting mutualism (Fig. 21.8) suggests that other relationships among animals might be mutualistic if we look at them more closely.

Species invasion does not always have negative effects on native species. For instance, invasive bighead carp (*Hypophthalmichthys nobilis*) facilitates increased midge populations while decreasing zooplankton and production of planktonic fishes (Collins and Wahl, 2017). This facilitation subsequently benefits bluegill (*Lepomis macrochirus*) growth and survival because if the increase in midges that serve as bluegill food (Collins et al., 2017).

An interesting mutualism occurs between the chironomid midge larva *Cricotopus nostocicola* and the cyanobacterium *Nostoc parmelioides* (Fig. 19.17C). The midge receives sustenance from the *Nostoc* and lives inside it until pupation and emergence as an adult (Brock, 1960). The midge is a poor swimmer and highly susceptible to predation without the shelter in the tough leathery *Nostoc* colony. In turn, the midge increases the photosynthetic rate of the *Nostoc* (Ward et al., 1985) by altering its morphology and by attaching it more firmly onto rocks so it can extend into flow and have a smaller diffusive boundary layer (Dodds, 1989). Respiration by the midge lowers O_2 concentration near the larva and enhances nitrogen fixation rates as well as photosynthetic rates (Table 21.1). This example is particularly curious because the eukaryotic animal is endosymbiotic to the bacterial organism.

(A)

(B)

Figure 21.8 The crayfish *Cambarus appalachiensis* (formerly *Cambarus sciotensis*) with approximate carapace length of 45 mm (A), and the branchiobdellid annelid worm mutualist *Cambarincola ingens*, about 11 mm long, attached to its cephalothorax (B). *(A) Courtesy: Bryan Brown, Lauren Krauss, and Erin Spivey; (B) Courtesy: Bryan Brown and James Skelton.*

Snails could facilitate other grazing snails of different species by providing substrata for algae to grow on their shells (Abbott and Bergey, 2007). In this case, the nutrients excreted by grazing snails can stimulate growth of diatoms that attach to their shells. Other snails then graze the diatoms off the snail shells. Lukens et al. (2017) similarly argued that snails from Lake Tanganyika and their epibionts could serve as net autotrophic components of the ecosystem because the primary production on them exceeds the snail's respiration. Grazing tadpoles in tropical streams facilitate small grazing mayflies by removing sediments from substrata and exposing underlying periphyton; grazing mayflies respond by increasing their densities in tadpole grazed patches (Ranvestel et al., 2004; Whiles et al., 2006), and amphibian declines in tropical

Table 21.1 Influence of mutualistic midge larvae on *Nostoc* nitrogen fixation rates, O_2 concentrations, and photosynthetic rates

Measurement	Units	Condition	Mean (95% confidence band)
$^{15}N_2$ incorporation	Del ^{15}N of *Nostoc*	With midge	0.188 (0.005)
		Without midge	0.153 (0.009)
Microscale O_2 concentration with microelectrodes	mmol/L	Over midge	0.67 (0.09)
		Away from midge	0.81 (0.06)
Photosynthetic rate with microelectrodes	$\mu mol\ O_2\ L^{-1}\ s^{-1}$	Over midge	1.4 (0.3)
		Away from midge	1.1 (0.2)

Source: Data from Dodds (1989).

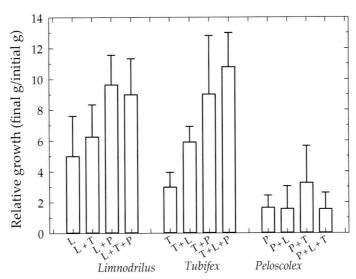

Figure 21.9 Growth of three tubificid oligochaetes alone and in culture with the other species. L, *Limnodrilus hoffmeisteri*; T, *Tubifex tubifex*; and P, *Peloscolex multisetosus*. Note that in two of three experiments, growth was greater in the presence of the other two species. *Data from Brinkhurst et al. (1972).*

mountain streams have resulted in population declines of small grazing mayflies (Rantala et al., 2015).

    Facilitation was demonstrated in a group of three oligochaete species (Fig. 21.9), but the mechanisms were not clear (Brinkhurst et al., 1972). In this case, two of three species tested had greater weight gains when growing with the others than when growing in a single-species culture. Facilitation may also be a common feature of stream invertebrates that process litter. Shredders excrete fine particulate organic material that collectors then ingest. Collectors may remove fine material that interferes with shredders or excrete nutrients that stimulate the microbes and make litter usable for shredders. Such facultative links merit additional study.

Facilitation could be more important for some predators in streams than thought previously. Even though scientists assume that wading birds compete with bass for prey, a cage experiment challenged this assumption. In a prairie stream in Illinois where the dominant predatory wading birds are great blue heron (*Ardea herodias*), green heron (*Butorides virescens*), and great egret (*Ardea alba*), cages were used to manipulate access by birds and smallmouth bass (*Micropterus dolomieu*) (Steinmetz et al., 2008). The birds are only effective predators in shallow waters, whereas bass can consume prey in deeper water. In this case, more small prey fishes were consumed with both types of predators (wading birds and fish) present than either alone, and the effect was multiplicative. This experiment suggests that facilitation occurs among the predators because prey fishes could not find refuge in deep or shallow waters.

OTHER SPECIES INTERACTIONS

In addition to predation and exploitation $(+/-)$, competition $(-/-)$, and mutualism $(+/+)$, other species interactions [neutralism $(0/0)$, amensalism $(-/0)$, and commensalism $(+/0)$] may be important but are rarely studied. We give several examples here to explain potentially important interactions in which one species has an influence on a species that does not have an influence in return.

Tadpoles of a common frog (*Rana temporaria*) have a negative effect on the snail, *Lymnaea stagnalis*, by competing for microalgae (Brönmark et al., 1991). The snail then consumes lower quality *Cladophora* and excretes nutrients that stimulate microalgal growth. The tadpole has a strong negative effect on the snail, but the snail has a weak positive effect on the tadpole.

Macrophytes in lakes and ponds may provide a vital habitat for survival of small fishes. The macrophytes apparently receive little direct benefit from the fishes that live in them. However, the macrophytes could receive some nutrients from fish excretion. In addition, indirect relationships could occur if the small fishes eat invertebrates that remove competitive epiphytes or prey upon herbivorous insects that eat the macrophytes.

Macrophytes also provide important habitat for epiphytes. The specific architecture of the macrophytes controls the amount of epiphytes that grow on them as shown by research in an Australian floodplain. Grasses tend to have lower epiphyte loads than macrophytes with horizontal structure, particularly those growing immediately below the water surface (Pettit et al., 2016).

Macrophytes can also provide nutrients to epiphytes growing on them. Moeller et al. (1988) planted the macrophyte *Najas flexilis* in sediments labeled with radioactive phosphorus. The amount of the radioisotope taken into epiphytes was quantified using track autoradiography. Epiphytes derived a substantial portion of their phosphorus from the macrophytes. Thus, macrophytes provide epiphytes nutrients and a place to live, but they also compete with the macrophytes for light, so this would be an exploitative interaction. Furthermore, different species obtained their phosphorus in varied proportions with the filamentous and more erect forms relying more heavily on the water column for nutrients, which indicated niche separation. Similarly, nitrogen-fixing microbes could leak nitrogen and stimulate nearby organisms that are unable to use N_2. Farmers manipulate this interaction in rice paddies under traditional cultivation as we discussed in Chapter 19.

Most interactions of aquatic organisms with humans are amensal. Humans have negative effects on many aquatic species, but the effects of most of these species on humans are negligible. With some thought, the reader could identify more examples.

SUMMARY

1. Competition occurs in species from a variety of taxonomic groups found in freshwaters.
2. Competition can occur among the members of the same species (intraspecific) and members of different species (interspecific).
3. Competition can be divided into two general types; scramble competition, where individuals exploit the same resource, and interference competition, where individuals interfere with the ability of others to use a resource.
4. Competition can be difficult to establish in field studies.
5. Mutualism, amensalism, and commensalism have been studied much less than predation or competition, but they are important.
6. Facilitation among plant species in harsh environments is an example of positive interactions that may be important in wetland communities.
7. Facilitation may be much more common than previously thought.

QUESTIONS FOR THOUGHT

1. How can disturbance in a habitat act as an agent of natural selection?
2. Are indirect interactions so strong and numerous that they complicate predicting the effects of interactions within an ecological community?

3. Is there a maximum diversity based on competition?

4. How might temporal patterns of disturbance act so that competition becomes unimportant?

5. Are positive or negative interactions more important in natural communities?

6. Why are both per capita interaction strength and the number of species interacting important?

7. Can you think of additional cases of facilitative interactions that are not included above?

CHAPTER 22

Complex Community Interactions

Contents

Figure 22.1 Mt. St. Helens in Washington State erupted and drastically altered the community of Spirit Lake (foreground). This led to documentation of an unusual case of natural succession. *Photograph courtesy US Geological Survey, Lynn Topinka.*

Freshwater Ecology
DOI: https://doi.org/10.1016/B978-0-12-813255-5.00022-3

Communities have complex interplay among abiotic factors and interactions among many species. Community ecologists face a difficult task in making predictions and understanding pattern, and process in complex groups of interacting organisms (Dodds, 2009). Still, we can make some strong generalizations about aquatic communities. Methodological approaches are being developed to contend with this complexity, so this area of research is likely to advance rapidly (e.g., Gao et al., 2016; Ye and Sugihara, 2016; Fernández et al., 2017). Generalizations can be made about temporal (e.g., time of recovery since disturbance) and spatial (e.g., landscape ecology) patterns in aquatic ecosystems. We attempt to cover some of the major areas, delineate where predictions can be made, and describe patterns that may not hold in all systems but are important to consider because they can occur.

DISTURBANCE

Defining *disturbance* is difficult. White and Pickett (1985) suggest that a disturbance is "any relatively discrete event in time that disrupts ecosystem, community, or population structure and changes resources, substrate availability, or the physical environment." This definition is very broad and allows many important aspects of disturbance to be included: spatial distribution, frequency, return interval, predictability, area, and intensity can all be aspects of disturbance. Ecologists must characterize community properties and the associated disturbance at appropriate spatial and temporal scales relative to the life histories of the organisms of interest.

Understanding the meanings of stability is important to classify disturbance. Stability has a specific mathematical definition when we express interactions in a community in a linear matrix form of interaction strengths, but this form of stability has limited practical use. While the concept seems simple on one level, it is a relatively nonspecific term that can take different meanings among managers and researchers (Donohue et al., 2016). In one example, certain diatom species appear in a lake for a short time in the spring every year. Over the period of a year, the population is not stable, but over decades, the pattern may repeat reproducibly. This is exactly the situation in the diatoms of Lake Windermere (Fig. 14.12).

Another thorny issue in disturbance ecology is how to classify natural events. If a species has evolved to take advantage of an event that would be a disturbance for other species, how do we classify an event? For example, a flood in a stream may decimate the diversity of insect larvae, but some benthic stream algae may grow better after a flood because of reduced grazing pressure; hence, is a flood a disturbance for the algae? Likewise, dry periods in intermittent streams and wetlands may seem like disturbances to some species, but others may depend on these events to complete their life cycles. A key issue regarding disturbance is predictability; predictable events such as summer drying in intermittent streams and spring floods in large rivers may not constitute disturbances because they fall within the normal range of conditions for those systems (Resh et al., 1988). Human alteration of natural disturbance regimes (e.g., regulation of rivers to reduce high and low flows) can adversely affect aquatic communities and favor exotic species (Lytle and Poff, 2004; Poff et al., 2007). Managers have extended the idea that flooding is a natural part of flowing water habitats to habitat management, including operation of dams and reservoirs.

Natural disturbance has variable influence across freshwater habitats. In general, groundwaters are least influenced by disturbance. However, karst aquifers may respond rapidly to heavy precipitation and cave streams can flood. Deeper groundwaters in unconsolidated sediments generally have very slow rates of change. Disturbances only moderately affect lakes in most cases; severe weather can cause mixing and floods can strongly influence shallow or small lakes. Streams, rivers, and some wetlands are more prone to disturbance. Flooding is an integral part of many streams and riparian wetlands; recovery from flooding and adaptations to disturbances are important to many aquatic organisms. Along with hydrologic events, wetlands with emergent vegetation can be subject to fires. A nonequilibrium view of streams and wetlands has become essential to understanding their ecology (Palmer and Poff, 1997).

The nonequilibrium view of disturbance and its relationship to competition as a determinant of species diversity has been termed the *dynamic equilibrium model* (Huston, 1994). This model suggests that at high productivity, growth rates—and therefore competition for resources—are high. Thus, only strong disturbances can counteract competition, so in a high productivity system diversity will be low unless disturbance is high. In contrast, at low productivity very little disturbance is necessary to disrupt competition and increase diversity. This model explains some responses to disturbance by stream invertebrate communities. In a study by McCabe and Gotelli (2000), frequency of disturbance did not decrease invertebrate species richness, but intensity and area of disturbance did alter richness. Similarly, the dynamic equilibrium model was consistent with riparian plant richness in riverine wetlands in Alaska (Pollock et al., 1998); in their study plant productivity (a gauge of competition) altered the effect of disturbance on species richness. The response to disturbance can depend upon resource availability as predicted by the dynamic equilibrium model; disturbance of rocks in an alpine river had substantial effects on the macroinvertebrate community under natural nutrient concentrations, but there was much less response to disturbance with nutrient fertilization (Gafner and Robinson, 2007).

A slightly different model, the *intermediate disturbance hypothesis* (Connell, 1978), predicts maximum diversity when disturbance is intermediate. The model predicts that at very high disturbance rate only rapidly growing and colonizing species can remain in the community, and at very low disturbance rates the better competitors drive out other species. The intermediate disturbance hypothesis and the dynamic equilibrium model are not exclusive of each other, but responses to these two concepts do depend upon the type of response variable we use to characterize diversity (e.g., evenness or species richness, Svensson et al., 2012). The intermediate disturbance hypothesis predicts patterns in some lakes (Reynolds et al., 1993), streams (Townsend et al., 1997), and wetlands (Whiles and Goldowitz, 2001).

Scientists have characterized disturbance as a *press* (a continuous disturbance) or *pulse* (a discrete event). Additionally, the overall impact of a disturbance depends on the *resistance* (the ability to withstand the disturbance) and *resilience* (the ability to recover) of the system. Resistance often relates to the magnitude and intensity of the disturbance, as well as features of the system such as the types and amounts of refugia (Lake, 2000). Resilience depends on the ability of colonists to reach an area (Cushing and Gaines, 1989), their ability to reproduce, the severity of the disturbance, and the existence of refugia (Townsend, 1989; Lancaster and Hildrew, 1993).

Recovery of macroinvertebrate species found in intermittent prairie streams following flooding and drought exemplifies disturbance harshness. Fritz and Dodds (2005) found that species richness could be predicted as a function of length of drought, severity of flood, time since flooding or drought, and distance from spring-fed pools that served as refugia from disturbance (Fig. 22.2). They created an index of harshness that incorporated both spatial and temporal aspects related to recovery from disturbance. In this case, there was no strong evidence for an "intermediate" level of disturbance.

Consideration of natural disturbance dynamics is crucial in wetland restoration (Middleton, 1999). Such disturbance includes flood pulses, fires (de Szalay and Resh, 1997), hurricanes, beaver activity, and herbivory. A large-scale example of managing disturbance in wetland restoration is the Kissimmee River project in Florida (Middleton, 1999). This project has continued through multiple phases and includes backfilling a drainage channel with tons of soil and levee removal. In the future, it will also include creation of flood pulses from upstream control structures to mimic natural patterns of discharge. This project has attempted to reverse the effects of channelization and flood control structures put in place between 1964 and 1971 by restoring the system's natural hydrology (Toth, 1996). Disruption of natural floods led to more dryland plant species in the floodplains around the river, more lentic species of macrophytes (including the invading water hyacinth), and more animals in the pools adjacent to and within the river adapted to lentic habitats (Harris et al., 1995; Toth, 1996). Several species of wading birds and waterfowl declined with channelization (Weller, 1995). Restoration to reintroduce flooding and reverse channelization started in 1984 and is planned to involve 70 km of river channel. These efforts have increased flooding, increased the numbers of wetland plants and waterfowl, and restored lotic species to the river. Invertebrate communities colonized the newly restored habitats rapidly (Merritt et al., 1999).

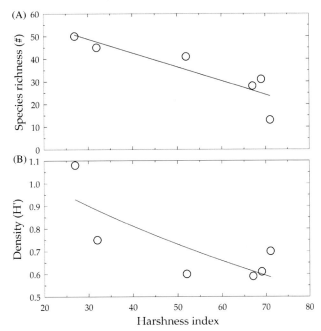

Figure 22.2 The relationship between macroinvertebrate species richness (A), Shannon diversity (B), and a harshness index in Kings Creek, Kansas. The harshness index is a score based on time since flood or drought, intensity of last flood or drought, and the distance from sources of new colonists. A higher index indicates a harsher habitat. The two points on the left are from permanent sites; the rest are from intermittent sites. Both lines are statistically significant ($P < .05$; *data from Fritz and Dodds (2005)*).

The US Army Corps of Engineers used explosives to blow holes in a levee to activate the Bird's Point-New Madrid Floodway during a historic flood in the lower Mississippi River in 2011. The Corps took this action to reduce flooding of Cairo and other cities in the region. The action created an opportunity to examine ecological responses to the sudden reconnection of about 55,000 ha of floodplain habitat that was at the time used primarily for agriculture (Rantala et al., 2015). Water entered the newly connected floodplain at flows >1 m/s through two openings created in the upper portion of the levee and exited through a crevasse at the lower end. Community responses were mixed, but young gizzard shad (*Dorosoma cepedianum*), an important component of the river food web, grew faster in the floodway than in the adjacent river. Water flows through the floodway were rapid and the period of inundation was relatively brief (<1.3 days), and thus some of the expected ecological responses of floodplain reconnection were not evident, but this event nonetheless demonstrated some tangible benefits of floodplain reconnection (Rantala et al., 2015).

Disruption of flooding can cause major changes in a river, from geomorphological to biotic aspects. Such is the case with the Glen Canyon Dam on the Colorado River. Lack of flooding and continuous release of clear, cold water from the hypolimnion of Lake Powell has drastically changed the communities in the river, leading to a depauperate river dominated by fewer algal and invertebrate species than occur naturally (Stevens et al., 1997). The cold-water releases have allowed establishment of introduced trout and have had a negative impact on native warm-water fishes. However, the lack of flooding has also led to increases in fluvial marshes bordering the Colorado River immediately downstream of the dam (Stevens, 1995). These marshes increase wildlife habitat and lead to an increased diversity of terrestrial and wetland species. Dam operators allowed a limited artificial flood in March 1996. The scouring reestablished sandbars but did not completely remove invasive species, such as carp, catfish, and tamarisk (Middleton, 1999). There was a short-term reduction in primary producer and invertebrate biomass followed by recovery (McKinney et al., 1999; Shannon et al., 2001). The cost of the release was $1.8 million of power-generating capacity to the dam.

Since the initial experimental flood in 1996, dam operators allowed experimental high flows in 2004 and 2008. Although these floods were of lesser magnitude than peak predam high flows (i.e., 1,133 vs 5,947 m^3/s; Topping et al., 2003), they did reestablish lateral sandbars. Sandbar habitats throughout the Grand Canyon benefit various aspects of the system, including native fishes, recreational users, and preservation of archeological sites. For these reasons, applications of high flow releases from Glen Canyon Dam that are specifically designed to promote sandbar habitats will likely continue in the Grand Canyon. Restoration of flood regimes is becoming a more common management approach worldwide (e.g., Robinson and Uehlinger, 2008).

Disturbance of terrestrial habitats can cascade to aquatic habitats. For example, riparian disturbance can increase sediment inputs, alter inputs of detritus, decrease the amount of pebble and cobble substrata, and influence invertebrate and algal communities (Stevens and Cummins, 1999). Forest fires can result in large inputs of ash and nutrients into streams, increase sedimentation, and alter light regimes (Minshall et al., 1989; Rhoades et al., 2017). In addition, flooding becomes more common and debris flows can occur with fires. These debris flows can have large effects including altering stream metabolic characteristics (Tuckett and Koetsier, 2016). Terrestrial fires can alter the aquatic community (Reale et al., 2015), leading to strong influences on instream primary producers (Klose et al., 2015) and other members of the community, with changes from population to ecosystem levels (Bixby et al., 2015). Many pollutants enter aquatic habitats from terrestrial sources and influence communities. Watershed disturbance is a prime source of nutrients that often cause eutrophication.

SUCCESSION

Succession is the predictable change in species composition over time following disturbance, ranging from regular harsh conditions associated with seasonality, through stronger disturbances such as flood or drought, to complete obliteration of species or creation of entirely new habitats. The ecological literature differentiates between *primary* and *secondary succession*. In primary succession, disturbance removes all or most of the organisms or conditions cause creation of a new habitat. In secondary succession, disturbance removes only some of the species or substantially reduces populations. In the mildest case, *seasonal succession* occurs on an annual basis because harsh seasons knock back the community that occurs during the growing season, and is the predictable change in species composition and biomass as abiotic conditions change seasonally.

The Phytoplankton Ecology Group created the most complete model of seasonal succession of plankton for temperate lakes with summer stratification. This group of 30 limnologists created a model consisting of 24 steps (Sommer, 1989). We diagram this hypothetical successional cycle in Fig. 22.3 and a simplified version is as follows:

1. Nutrients build in the euphotic zone of lakes during the winter because nutrients mix from the hypolimnion during fall mixing, light is limiting, and there is little demand for nutrients by phytoplankton. Light becomes limiting because of less solar irradiance and wind may mix phytoplankton deeply below the compensation point, or ice cover severely limits light input.
2. During the spring, populations of small, rapidly growing phytoplankton species peak because nutrients and light are high and zooplankton grazing is low.
3. In late spring, reproduction allows zooplankton populations to increase, and their grazing causes a clear water phase that lasts until grazer-resistant species of phytoplankton develop during the summer.
4. Zooplankton populations decrease because of declining food and increased fish predation (particularly by newly hatched fish larvae).
5. A later bloom of cyanobacteria may occur with higher lake temperatures and lower nitrogen availability.
6. Fall mixing can stimulate a second bloom of edible phytoplankton and larger zooplankton.

Fishes can have life histories that allow different coexisting species to hatch at distinct and predictable times in a seasonal sequence. Amundrud et al. (1974) demonstrated an order of appearance of larvae in a small eutrophic lake (Fig. 22.4). In this lake, yellow perch (*Perca flavescens*) larvae appeared first, followed by log perch (*Percina caprodes*), black crappies (*Pomoxis nigromaculatus*), and finally pumpkinseeds and bluegills (*Lepomis* spp.). The sequence suggests that competition has led these fishes to habitat partitioning, and that they specialize on prey that occur at different times during the seasonal successional cycle.

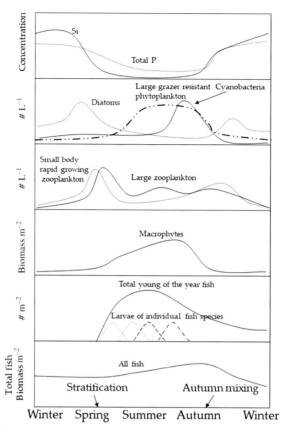

Figure 22.3 A hypothetical successional sequence in the epilimnion of a temperate lake with summer stratification.

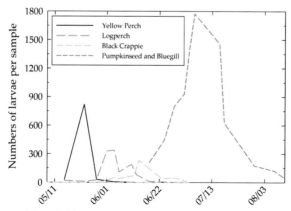

Figure 22.4 Numbers of larval fishes in a small eutrophic lake during a summer successional sequence. *Redrawn from Amundrud et al. (1974).*

On a longer timeframe, succession can occur in newly created lakes or existing lakes that are subject to very strong or newly created disturbance (i.e., primary succession). There are geological periods when new freshwater habitats are created relatively rapidly; following periods of glaciation numerous lakes, streams, and wetlands are created. Creation of new lentic habitat is also a common occurrence in our modern world due to the establishment of new reservoirs.

As glaciers retreat, they leave behind new streams, leading to primary succession. Global warming is now causing retreats of glaciers in alpine regions around the world. The stream invertebrate communities will reflect this succession (Burgherr and Ward, 2001). Milner et al. (2011) found 57 macroinvertebrate and 27 microcrustacean species colonized a stream formed in Alaska within 20 years of glacial retreat. It took salmonids about 10 years to colonize the stream, followed by sticklebacks and sculpin. Robinson et al. (2014) found that the rapidly reproducing chironomid midge larvae were the first colonists in deglaciated streams in the Swiss Alps. Finn et al. (2013) used population genetic methods and found that macroinvertebrate diversity will probably decrease regionally in the Pyrenees (France) as small glaciers and the streams they feed disappear.

In reservoirs, the general view is that an early eutrophication phase is common (Donar et al., 1996). Sequential colonization of the reservoir by invertebrates more adapted to lentic waters also occurs, related to formation of a sediment-covered bottom and establishment of macrophytes (Voshell and Simmons, 1984; Bass, 1992). The sequence of events that occurs following installation of a new reservoir can vary depending on the area being dammed, the morphology of the reservoir, and a variety of other factors. For example, Hall et al. (1999) contrasted community formation in two reservoirs in south Saskatchewan. Paleolimnological techniques were used to analyze algal and chironomid midge assemblages in the two reservoirs over time. One reservoir was formed by damming a river (Lake Diefenbaker) and the other by raising the level of an existing lake (Buffalo Pound Lake). The river reservoir had fluctuations in water level of 6 m/year and the flooded lake had fluctuations of 1−3 m per year. Lake Diefenbaker (500 km^2) exhibited a typical sequence of succession found in newly formed reservoirs; an initial period (4 years) of eutrophic conditions occurred, followed by a decade of mesotrophy, and a recent shift back to eutrophic conditions. Buffalo Pound Lake (50 km^2) had lower phytoplankton biomass after flooding, but midge larvae and macrophyte populations expanded. The different routes of formation, lake level fluctuations, and sizes of reservoirs formed probably led to different successional trajectories.

One of the more detailed studies of succession in a reservoir was on the Gocsalkowice reservoir in Poland. This reservoir filled in 1955 and researchers summarized observations in 1986. The Secchi depth increased from 1−2 m during the first 20 years but then decreased to <1 m because of anthropogenic nutrient inputs in the watershed (Kasza and Winohradnik, 1986). Diatoms and green algae dominated in high numbers in the 1950s and 1960s but cyanobacteria replaced them as dominants in the 1980s (Pajak, 1986). Macrophyte growth was extensive within a year after the reservoir reached full pool, but few species were present. Emergent macrophytes dominated the reservoir margins in later years, and macrophyte community diversity increased and then stabilized after two decades (Kuflikowski, 1986). Species more common in lakes replaced riverine fishes and invertebrate species within the first few years (Krzyzanek, 1986; Starmach, 1986).

External storms can also drive succession in lakes. The planktonic bacterial assemblages of a subalpine freshwater humic lake in Taiwan changed predictably following typhoons that destabilized the water column of the lake (Jones et al., 2008). Following typhoons, the bacterial communities were generally similar to those immediately following other typhoons. As time since disturbance increased, the epilimnetic and hypolimnetic assemblages diverged, and successional trajectories varied following each disturbance. In contrast, the phytoplankton communities did not respond predictably to this disturbance.

An unusual but interesting case of primary succession occurred when Mt. St. Helens in Washington State (Fig. 22.1) erupted and drastically altered Spirit Lake (Larson, 1993). The lake was an oligotrophic mountain lake before eruption. The eruption superheated the lake's waters, killed most of the animals and plants, filled the lake basin with volcanic ash and tree trunks, and a large debris flow blocked the lake's outlet. Soon after the eruption, the water went completely anoxic, leading to a community dominated by heterotrophic and chemoautotrophic bacteria, a few protozoa, and rotifers. By the next spring, O_2 returned to the epilimnion of the lake, and a moderately diverse phytoplankton community developed. Five years later, a diverse phytoplankton assemblage had colonized the lake as well as at least four zooplankton species. Eight years after eruption, macrophytes [milfoil (*Myriophyllum*) and stonewort (*Chara*)] had become established, as had snails and other macroinvertebrates. The future could bring further catastrophic changes to the lake and the Cowlitz River that it feeds—the debris dam is unstable and engineers had to create a tunnel through it to release water. If the tunnel fails, the lake could overtop the dam, drain rapidly, and cause catastrophic flooding downstream (Service, 2016).

Seasonal wetlands may experience a strong successional sequence. For example, in the Pantanal wetlands in Brazil, a seasonal wet—dry cycle leads to large and predictable changes in the aquatic communities (Heckman, 1994). During the wet season, considerable flow can occur in the main channels, but during the dry season, these same channels become lentic. Macrophyte populations develop during the dry seasons and high flows flush them out during the wet seasons. During the dry periods, the large number of fishes trapped in drying ponds attract many waterbirds and caimans (*Caiman* spp., a crocodilian) that consume them. These isolated ponds become hypereutrophic from the nutrients released from the dying fishes and excretion from the waterbirds and caimans, and they exhibit algal blooms. Eventually, all but the deepest pools dry until the following wet season, when the cycle begins again.

Successional patterns of macrophytes in riparian wetlands are commonly associated with any river that exhibits seasonal variability. In the Rhone River in France, successional sequence depends on the degree of connectivity to the main channel. The diversity is highest in frequently flooded habitats because more propagules are available to establish plants (Bornette et al., 1998). In a more spatially restricted study, four species of macrophytes coexisted in a side channel: *Sparganium emersum* grew in the least disturbed areas of the channel, *Hippuris vulgaris* and *Groenlandia densa* inhabited moderately disturbed areas, and *Luronium natans* grew best in the most disturbed areas (Greulich and Bornette, 1999). This pattern illustrates the relationship between competitive ability and successional processes. Without disturbance and succession, *S. emersum* would probably be the only species present because it is the competitive dominant.

Understanding successional patterns may be crucial in wetland restoration schemes (Middleton, 1999). Restoration ecologists have advanced two general lines of thought: The first is that succession depends on the initial inhabitants, and the second is that a natural successional series will restore a wetland if the restoration accurately reproduces abiotic features. Some people assume that the manager must plant the species desired, and others assume that species will naturally populate a restored wetland (Middleton, 1999). The reality probably lies on a continuum between these two points. If a severely disturbed wetland is to recover, reintroduction of lost or rare species may be necessary. Alternatively, in a wetland of any size, it is difficult to control exactly which species become established over the years and where.

Rivers and streams may also undergo succession on seasonal timescales. In small streams in deciduous forests (Fig. 22.5), the light regime and leaf litter inputs vary over the seasons. Increased light and limited litter inputs favor periphyton and species that consume it (scrapers). Increased litter inputs favor species that depend on leaves for nutrition (shredders). Seasonal flooding may also alter the diversity and biomass of species in such a stream. Invertebrates in temperate streams can be categorized into three groups: slow-seasonal, fast-seasonal, and nonseasonal life cycles (Anderson and Wallace, 1984). Many slow-seasonal insects reproduce in the fall and grow through the winter (e.g., winter stoneflies and some Trichoptera). These species can specialize on the large amount of leaf litter entering the stream in fall. Fast-seasonal insects have a prolonged diapause and a short period of rapid growth followed by reproduction. These species may reproduce in spring, summer, or fall. Nonseasonal species may have long life spans or very short and overlapping generations. Data collected from a small Michigan stream suggest that total invertebrate biomass remains approximately constant throughout the year despite the dominance of different

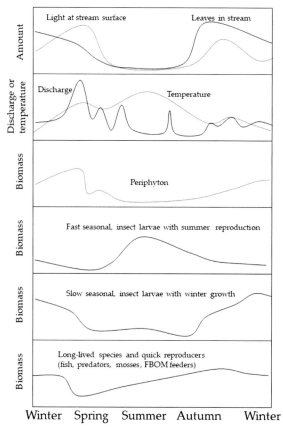

Figure 22.5 A hypothetical seasonal successional sequence in a small temperate stream in a deciduous forest.

species in specific seasons (Cummins and Klug, 1979). For instance, insect larvae that are shredders attain maximum biomass in spring after a winter of feeding on detritus. Seasonal cycles can also influence invertebrate responses to flooding (Robinson and Minshall, 1986).

Seasonal flooding is also an important feature of other streams from tropics to deserts. A predictable sequence of colonization by algae occurs in Sycamore Creek, Arizona (Fig. 22.6). Quick colonizers include diatoms. Filamentous green algae take more time to establish. Finally, cyanobacteria dominate as nitrogen becomes limiting. Many tropical areas are subject to monsoonal weather patterns with flooding common during periods of high rain and more stable flow during the dry seasons. The successional patterns discussed here provide generalized and simplified models, to which exceptions clearly exist, mainly to illustrate the potential for seasonal succession in streams.

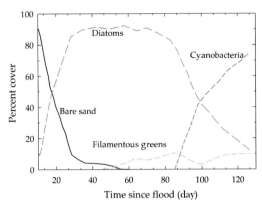

Figure 22.6 A successional sequence of algal groups in a desert stream following a flood. *Reproduced with permission from Fisher and Grimm (1991).*

INDIRECT INTERACTIONS

Indirect interactions are those between two species mediated by one or more other species (Wootton, 1994). The number of indirect interactions in a community is potentially large because of all the linkages among species in each ecosystem. The difficulty with characterizing the influence of indirect interactions is the vast number of interactions that potentially occur. Theoretically, in the case where no species can be involved in one indirect interaction chain more than once, given a group of S species, there are $S \times (S-1)$ possible direct interactions, or interactions with one link. However, the number of possible indirect interaction chains with n links is $S \times (S-1) \times (S-2) \times \ldots \times (S-n)$. Consider a group of three species with six potential direct interactions and six potential indirect interaction chains. When there are 7 species, there are 42 potential direct interactions, 210 with 2 links, 840 with 3 links, 2,520 with 4 links, and 5,040 potential indirect interaction chains with 5 and 6 links. How can an experimental ecologist deal with this complexity? Most all communities have far more than seven members. Luckily, there are weak and zero interactions that can attenuate or truncate indirect interactions, and a community may have a defined size (Dodds and Nelson, 2006). Understanding the complexity of indirect interactions can be daunting, so we first start with some specific examples rather than a general theoretical treatment.

One of the strongest examples of indirect interactions is the trophic cascade, which we discussed in Chapter 20, in which changes in the abundance of a top predator can alter the abundance of primary producers, as mediated by interactions among several other species. While the trophic cascade is a relatively clean example of indirect interactions, a simple food chain is not a completely accurate characterization of trophic interactions; most food webs have omnivory and feeding on more than one trophic level. Even more interactions that are indirect can occur in food webs than food chains. For example, removal of grazers may lead to a decrease in algal production because the grazers remineralize nutrients and increase the growth efficiency of the algae (Darcy-Hall, 2006; Bertrand et al., 2009).

Indirect interactions can alter predictions made by considering only direct interactions. This can be illustrated by the case of one predator with two competing prey species where indirect interactions set up a situation where one of the prey species that is an inferior competitor can coexist with the superior competitor. According to the competitive exclusion principle (we discussed in previous chapters), the inferior competitor should eventually be extirpated by the superior competitor. Additionally, a predator should extirpate its prey if the predator and only one prey species are present. However, if two prey species are present, the predator may switch to prefer the more abundant prey species. This switching can occur because it can be more energy efficient for a predator to forage preferentially for the more abundant prey (see Chapter 20). Because the preferred prey species is not always the competitive dominant, the predator may promote the coexistence of both species. In this case, both prey species benefit each other in the presence of the predator, even though they compete directly.

A potential example of this type of indirect interaction resulted from beaver reintroduction in Scotland. Law et al. (2014) followed a series of ponds for 9 years after introduction of Eurasian beaver (*Castor fiber*). They found that beavers grazed selectively and lowered biomass of macrophytes over the short term using exclusion experiments. After several years, beavers caused a threefold increase in macrophyte species richness and increased both α and β diversity. The selective feeding likely allowed competitive subordinates to survive in the presence of dominants. In a related study, Nummi and Holopainen (2014) found introduction of non-native beaver (*Castor canadensis*) in Finland caused flooding that led to increases in waterbirds, suggesting beavers have many indirect effects in streams.

Indirect interactions can give rise to mutualistic relationships in lake (Lane, 1985). Mathematical analyses of lake food webs confirm this for a variety of pelagic ecosystems. The example from the previous paragraph of a predator allowing an inferior competitor to remain in a community is but one of the possible ways such mutualistic interactions can arise. Although the details of such analyses are beyond the scope of this book, they do suggest that trophic cascades in food webs are not the only important indirect interactions.

The common algal macrophyte *Cladophora*, its epiphytes, and grazers of the epiphytes illustrate a specific example of indirect and complex interactions (Dodds, 1991; Dodds and Gudder, 1992). In this case, the epiphytes generally compete for nutrients and light with the *Cladophora*. However, the *Cladophora* is N limited, and some of the epiphytes fix N and so may ultimately provide N to the *Cladophora*.

Contrary to initial expectations, epiphytes reduce drag on *Cladophora*. This occurs because the epiphytes decrease the amount of space between the filaments, lowering Reynolds number, and decreasing flow through the algal mass. Forcing the water to flow around the algal mass decreases friction and drag. At the same time, forcing water around the mass decreases advective transport through the *Cladophora* filaments potentially limiting nutrient influx and removal of wastes. Consequently, the epiphytes may harm or help the *Cladophora*, depending on the environmental conditions. Invertebrate insect larvae remove sediments and epiphytes from the surface of the *Cladophora* and remineralize nutrients; this may facilitate *Cladophora* growth. At high densities, when other food sources are unavailable to invertebrates, they will eat the *Cladophora*. The *Cladophora* provides habitat and protection from predation for the invertebrates. The three groups of organisms have complex interactions that vary over space and time in the environment, can be mediated by other organisms, and fluctuate between being positive and negative.

An interesting indirect interaction occurs when the snail, *Physa acuta*, suffers from trematode parasites. The parasites often reach biomass of more than 30% of the snail animal tissue. The infection stresses the snails and causes them to graze more rapidly than uninfected snails. Snail populations with 50% infection rates grazed 20% more than uninfected populations. The increased grazing also causes a shift in algal community structure with fewer diatoms and cyanobacteria and increased dominance by the grazing-resistant filamentous alga *Cladophora* (Bernot and Lamberti, 2008).

Another example of complex, and sometimes indirect, interactions was shown in an experiment that was performed on competition between tufted ducks (*Aythya fuligula*) and fishes in a small pond formed by quarry activities in the south of England (Giles, 1994). This shallow pond had a fish community including bream, roach, perch, and pike and was turbid with few macrophytes. Most of the fishes were removed during a 2-year period, after which midge larvae, macrophytes, and snails increased (Fig. 22.7). They reintroduced fishes to enclosures inside the lake, and midge larvae and snail numbers returned to levels found in the entire lake before fish removal. Predation on ducklings by pike is a direct interaction that occurred before the fish removal, but fish removal had indirect effects as well; feeding by ducks, numbers of brooding ducks, and brooding success increased after fish removal because of their dependence on snails and midge larvae. Indirect effects of fishes on the ducks included competition for midge larvae. Indirect effects of fishes also included the removal of macrophytes, which provided habitat for snails. These data are instructive because they compare competition among birds and fishes and document an indirect effect mediated by macrophytes.

Linkages can occur among ecosystems, and indirect interactions can propagate across ecosystems. Many terrestrial animals rely on aquatic margins for resources linking terrestrial processes to aquatic in both directions (Schindler and Smits, 2017). For example,

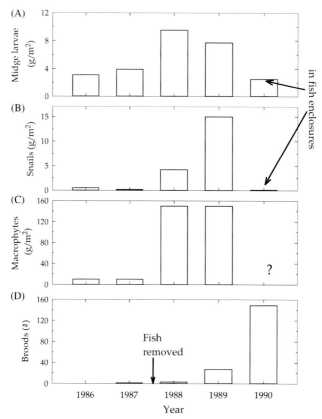

Figure 22.7 Interactions of fishes and tufted ducks in a small English lake. Researchers removed fishes in 1987 and 1988. Peak densities of chironomid midge larvae (A), snails (B), macrophytes (C), and number of duck broods (D) all changed. Researchers added fishes to the enclosures in 1990, and midge and snail densities are reported from inside the enclosures, demonstrating that fishes were able to lower both populations. *Data from Giles (1994).*

predation by fishes can reduce abundance of dragonfly larvae in ponds. The terrestrial areas around ponds with fewer adult dragonflies preying upon aerial insects have more pollinators. More pollinators can lead to changes in terrestrial plant communities (Knight et al., 2005). This chain of interactions indicates that we should not consider aquatic and terrestrial habitats in isolation from each other, and how knowledge of the life history of organisms is essential to understanding complex interactions among them.

STRONG INTERACTORS

Keystone species are those that have a disproportionately large impact on their community or ecosystem relative to their abundance (Power et al., 1996). An alternative form of the same idea is that some species are strong interactors. The ecological literature has used the term "keystone species" loosely, so we should view claims that a species is a keystone species with caution. Identifying keystone species requires knowledge of the particular system of study and the organisms found in it. Stated differently, there is not always a keystone species, and which species will be a strong interactor is difficult to predict in advance, but the potential existence of strongly interacting species can have major consequences for understanding community structure and, by extension, conservation efforts as well as dynamics of introduced species.

An example of a potential keystone species is the detritivorous fish *Prochilodus mariae* found in the Orinoco river basin in South America. The fish removes sediments from the stream bottom through its normal feeding activities (Flecker, 1996). These activities lower algal biomass and invertebrate mass, though some species of invertebrates and algae are stimulated. With experimental removal of this detritivore, large amounts of organic detritus built up, and the carbon flux into the food web and the patterns of carbon flux were substantially altered (Taylor et al., 2006). These results were surprising because of the high diversity in the system (80 species of fishes) and the expectation that another species would take over the role of the large detritivore. This detritivore is an example of a strong interactor that has direct and indirect influences at several levels of the food web.

Bluegill sunfish (*Lepomis macrochirus*) may structure communities in small lakes and ponds throughout the eastern United States (Smith et al., 1999). Frog tadpoles (except the toxic bullfrog tadpoles, *Lithobates catesbeianus*), newts, and predatory aquatic insects are much less abundant when bluegill are present. Since bluegill are common in many small ponds, their presence could alter community structure. Power et al. (1996) provided additional examples of possible keystone species in freshwater systems that included predatory fishes in river and stream food webs, piscivorous fishes in the trophic cascade of lakes, predatory salamanders that prey on anuran tadpoles in temporary ponds, and planktivorous fishes that have strong impacts on zooplankton communities. Finally, a weevil that attacks Eurasian water milfoil (*Myriophyllum spicatum*) could be a keystone species because it stops the milfoil from dominating (Creed, 2000).

Beaver (*Castor canadensis* in North America and *Castor fiber* in Europe) have substantial effects on streams, from altering hydrology, to importing nutrients, and causing strong changes in the aquatic plant community by herbivory (Parker et al., 2007). Beaver activity can alter community structure and life history of fishes by providing deep habitat for larger fishes and spawning habitat in the shallower parts of their pools (Bylak et al., 2014). Selective feeding can hamper competitively dominant macrophytes, and herbivory by beaver can increase macrophyte diversity (Law et al., 2014).

THEORETICAL COMMUNITY ECOLOGY AND AQUATIC FOOD WEBS

Ecologists have detailed many lake food webs, and theoretical community ecologists have analyzed these results in attempts to describe general patterns (Pimm, 1982). Questions considered include the following: What limits the length of food chains? Are complex systems more or less stable? How interconnected are large food webs? Does aggregation of species into trophic groups alter the results obtained from analyses of food webs? How variable are food webs over space and time? Can state changes (regime shifts) be predicted?

It is not clear what limits the lengths of aquatic food chains. The length of the food chain is of particular importance when determining biomagnification of lipid-soluble pollutants. General analyses suggest that the amount of primary production at the base of the food chain is not a good indicator of food chain length (Briand and Cohen, 1990); but these analyses are not necessarily consistent with the idea that energy is lost at each tropic level. However, at least one study suggests that trophic transfer of energy in the Okefenokee Swamp may be very efficient (Patten, 1993), so our concepts of food web efficiency and how it should link to the number of trophic levels may need to accommodate situations with high energy transfer efficiencies. Groundwaters in which the number of large animals is physically limited by their ability to move through aquifers typically have short food chains. Marine pelagic food chains are considerably longer than most pelagic freshwater chains. Stream food webs tend to be short but wider relative to those of lakes (i.e., with more species as primary consumers). Large rivers and lakes tend to have similar food web structure (Briand, 1985). We discuss movement of energy in freshwater ecosystems in more detail in Chapter 24.

The relationship between food web stability and complexity is also unclear (see the next section for more explicit discussion on the meaning of ecological stability). Early ecologists viewed simple systems as less stable (Elton, 1958; MacArthur, 1955). Mathematical analyses of simple linear models of randomly assembled "communities" then suggested decreased stability with more interacting species (May, 1972). Since then, empirical (Frank and McNaughton, 1991) and modeling (Dodds and Henebry, 1996) approaches have suggested that increased diversity begets greater stability. We are not aware of any data from aquatic systems that can resolve the controversy regarding food web complexity and stability.

The degree of connectivity (the number of interaction links per species) and how it changes with community size are also debated. It has been suggested that Connectivity may increase as food webs become more complex (Havens, 1992b), but this analysis is controversial (Havens, 1993; Martinez, 1993). The proportion of species in each trophic category does not seem to change with food web size (Havens, 1992b). However, even this idea of constant proportions of species at each trophic

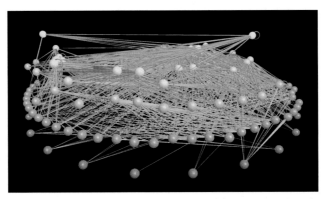

Figure 22.8 The food web of Little Rock Lake as described by Martinez (1991). Producers are on the bottom ring, consumers above. The figure uses a ball to indicate each species, and the sticks connecting the balls indicate which species below are being eaten. Note some of the consumers have loops to themselves representing cannibalism. *Image produced with FoodWeb3D, written by R.J. Williams and provided by the Pacific Ecoinformatics and Computational Ecology Lab (www.food-webs.org, Yoon et al., 2004).*

level independent of food web size was false in one stream community (Tavares-Cromar and Williams, 1996).

As a convenience, scientists conducting community analyses of food webs lump organisms into trophic categories. For example, in benthic food webs, one of the primary food sources is often detritus. In reality, detritus is a complex microbial and invertebrate community associated with decaying organic material. Analysis of a highly resolved food web from Little Rock Lake, Wisconsin (Fig. 22.8) suggests that lumping or aggregation of food webs into functional groups strongly influences properties used by ecologists to describe communities such as the proportion of species at each trophic level and the connectivity of the food webs (Martinez, 1991).

Food webs can have variable structure over space and time. Descriptive characteristics of a detritus-based food web varied over time where the food web was more or less complex across seasons depending on the life cycle of the benthic invertebrates (Tavares-Cromar and Williams, 1996). Temporal variation in stream food webs can link to seasonal hydraulic cycles (Power et al., 1995). Spatial variation also has clear effects; benthic and pelagic food webs in lakes are distinct (Havens et al., 1996), pools and riffles in streams contain different organisms, and there are strong effects of water depth on species composition in wetlands.

THRESHOLDS AND ALTERNATIVE STABLE STATES

One concern about ecological systems is that they will experience regime changes (cross thresholds that are difficult to return from) as humans and other processes influence them. Systems can respond to pressures in a nonlinear manner and it is important to understand these types of responses to manage freshwater ecosystems (Dodds et al., 2010). Often, managers are interested in allowing impact to aquatic systems to a point somewhat before the systems cross a threshold into an undesirable state (Wagenhoff et al., 2017). In general, an *ecological state* can be defined on the basis of measured characteristics of the system. If conditions change such that the measured characteristics exhibit a significant change, then the system has undergone a *state change*. A system that does not return to the initial state after the initial conditions are reestablished exhibits *hysteresis*, and could be in an *alternative stable* state. When a pushed system makes a rapid state change, this is a *threshold* and some say it has gone past a *tipping point*. Such thresholds are, by definition, difficult to predict because they are abrupt and completely change system structure. Predicting these changes based on mechanistic understanding is becoming increasingly important as humans have global directional influences on the environment (Scheffer et al., 2001). Perhaps the most obvious state change is that caused by extinction; once a species is extinct, the probability is essentially zero that it will evolve again.

Making things more difficult, multiple stressors may be involved in changes. Interactive effects of warming, drought, and acidification, rather than a simple summation of individual stressor effects, best explained responses of consumer and primary producer communities in boreal lakes (Christensen et al., 2006). With global climate change and strong local influences, multiple interactive stressors are probably the norm in freshwater ecosystems. Warming, in particular, can have very strong effects leading to altered ecosystem states. Woodward et al. (2010) used natural geothermal streams in Iceland to show that warming can cause radical changes in community structure and food webs of streams. Nelson et al. (2017a,b) report similar results for experimental stream warming.

Another example of alternative stable states that has been the subject of over 1000 papers is two potential states in shallow lakes (Hilt et al., 2017). There are two stable states, phytoplankton dominated and a clear water phase dominated by macrophytes (Phillips et al., 2016). Eutrophication is able to drive these shallow lakes from macrophyte dominance to phytoplankton dominance, partially because of the inhibitory effects of cyanobacteria on some macrophytes (Zheng et al., 2013). Macrophytes can also chemically inhibit phytoplankton, and dominate competition for light at times too, as well as having the additional advantage of being able to take nutrients up through their roots in the sediments. In Lake Veluwe (located in The Netherlands), increases in phosphorus above 20 µg/L led to a shift away from the clear water state. Managers implemented nutrient control, and the lake was flushed with low phosphorus water. The lake did not attain a macrophyte-dominated clear water state until the total phosphorus concentrations fell below 10 µg/L. This system exhibited strong hysteresis (Ibelings et al., 2007).

Internal phosphorus loading, as we discussed in Chapter 18, is a good example of a threshold with an alternative stable state that requires understanding of mechanism. When increases in nutrients are great enough to lead to disappearance of dissolved oxygen from the hypolimnion, there is a fundamental change in biogeochemical processes. In this case, internal loading is much greater if the surface of the sediment has no detectable dissolved oxygen and $FePO_4$ can dissociate and allow phosphate to diffuse from the sediments. Thus, the system is slow in recovering an oligotrophic state. The lake attains a new state with high internal loading and continued algal productivity after it has crossed a threshold. The inability to return easily to the original oligotrophic state indicates hysteresis and that the lake has entered an alternative stable state. A similar eutrophication threshold may also occur in the Everglades, where rapid changes in biotic components occur when total phosphorus concentrations exceed $12-15$ μg/L (Richardson et al., 2007).

Additional thresholds can occur in large lakes with respect to phytoplankton communities. These changes can be lake-specific and depend upon factors such as nutrient retention and presence of zebra mussels. Responses of species richness occur with respect to both nitrogen and phosphorus in the Great Lakes (Kovalenko et al., 2017).

Experimental approaches altering organic material loading to pools inside of pitcher plants (*Sarracenia purpurea*) induced a strong and predictable tipping point. As inputs of detrital material increased, there was an abrupt shift to anoxia in the pools in the plants. These shifts alter the communities in the plants as well as the biogeochemical conditions (Sirota et al., 2013).

One concern is that some systems could be very sensitive to human impacts and that thresholds occur at very low levels of disturbance. King et al. (2010) analyzed macroinvertebrate data from streams in Maryland and found that at only 2% impermeable surface there was a decline in macroinvertebrate species richness. The absolute threshold varied by location (e.g., mountain or lowland). However, one might not expect such sensitivity to development and could incorrectly choose "reference" watersheds that already had passed a threshold and were experiencing strong shifts in community structure.

How do we predict when a system is reaching a point where a threshold will occur? Forecasting is difficult because new states are often outside our range of experience (Groffman et al., 2006). Alterations in system variance may precede regime changes (Carpenter and Brock, 2006). Increased temporal autocorrelation could be a robust indicator of systems that are approaching state transitions (Dakos et al., 2010). Analyses of several early warning indicators proposed to signal critical changes in five freshwater systems suggested only modest agreement among indicators (Gsell et al., 2016).

Carpenter et al. (2011) tested rising temporal variance as an indicator with two lakes, one with large predatory fish addition and the other without. The food web eventually underwent a dramatic shift, but a year before the shift started the variance of zooplankton and algal biomass increased as expected. Similarly, rising temporal variance was a predictor of state change in one whole lake experiment (Pace et al., 2013). In this study, they added large predatory bass (*Micropterus salmoides*) to one lake dominated by small fishes and left a similar lake unmanipulated. The bass led to a trophic cascade where zooplankton density was increased and algal biomass decreased. The shift to a new state was preceded by both rising variance of daily zooplankton counts as well as increased autocorrelation. These shifts did not occur in the reference lake.

Pace et al. (2017) also tested the concept with respect to eutrophication of two lakes. They enriched a small lake until variance in parameters such as pH, dissolved oxygen, and chlorophyll started to rise, and were able to reverse eutrophication. They continued to fertilize a nearby lake, which exhibited a large algal bloom. Belle et al. (2017) analyzed a shift in chironomid midge communities in the benthic zone of a French subalpine lake driven by eutrophication. They found some evidence for rising variance before a strong shift in community composition.

Prediction is difficult, but the fact that thresholds can occur when a system is pushed is important to keep in mind. While new indicators of incipient state change continue to be added, unanticipated thresholds are probably an unavoidable feature of ecosystems (Carpenter et al., 2014). Alternative states, such as those that occur in shallow lakes, can have important implications for ecosystem functions and values (Hilt et al., 2017). There are several important points to managing ecological systems with potential thresholds. First, a mechanistic understanding of the system is required to understand the its specific peculiarities and drivers. Second, the precautionary principle (do not change the system unless you are relatively confident the change will not cause an undesirable outcome) should be taken.

INVASION AND EXTINCTION REVISITED

Some ecologists have termed this the era of homogenization, or *homogocene*, because increasing globalization is spreading species around the planet while leading to extinction of many rare species (Rosenzweig, 2001). While *exotic species* simply refers to species located outside of their natural range, *invasive exotic species* are those that proliferate and cause damage in new habitats; invasive exotics are often weedy, generalist taxa with high reproductive capacities. Humans are transporting organisms ranging from microbes to large herbivores around the world, some with devastating consequences (see Chapter 11 for further discussion on species invasions). At the same time, most ecologists agree that we are causing the sixth mass extinction event in Earth's history (Wake and Vredenburg, 2008), and exotic species can be a major cause. Certainly, humans have invaded many habitats where they have not been abundant, and have had tremendous negative effects on many species. Nonhuman exotic species can outcompete or increase predation pressure on increasingly stressed native species, and subsequently the biodiversity of most regions of the planet is rapidly declining.

One area of controversy is the idea of designer ecosystems. Hobbs et al. (2009) and Martinez and Lopez-Barrera (2008) have made the case that severely degraded areas will never return to their historical native state, and we need to consider the possibility of designing ecosystems to meet human needs. The specific goals would be dependent upon what humans want (e.g., conservation of species, recovery of the value of ecosystem services). This argument has some truth because with global climate change we are moving into an abiotic regime that many species have not experienced over their evolutionary history. However, Wilson (2016) has strongly argued the other side. He reminds us that we have not even named many of the species on Earth, and we have only a rudimentary understanding of the life history of most organisms. Therefore it is impossible to design an ecosystem if you do not know how it actually works. Wilson argues for conservation of remaining areas being the first priority (of half of the Earth, specifically). Regardless, understanding the ramifications of extinctions and species invasions will be necessary to understand the future of freshwater ecosystems.

Invasive exotic freshwater species can cause declines and extirpations of native species through a variety of direct and indirect mechanisms, which are often difficult to identify. For example, rusty crayfish (*Orconectes rusticus*) and other invasive crayfish in the United States are replacing native species in parts of the northeast and Great Lakes where they have been introduced (Thorp and Covich, 2001). Rusty crayfish are large, aggressive crayfish that can displace smaller species from preferred habitats such as rock shelters, and displaced crayfish are more vulnerable to fish predation (an indirect effect of the interaction among crayfish). Other hypothesized mechanisms for crayfish species replacements include reproductive interference, direct predation, and competition for food and habitat. However, definitive experimental evidence for these various mechanisms is generally lacking.

Many species of carp originating from Asia have become invasive in other parts of the world. One recent introduction to US rivers, the Black Carp (*Mylopharyngodon piceus*) is a major concern because this species feeds on mollusks and crayfish, and they get large enough to pose a serious threat to native mussels (Unionidae), one of the most imperiled groups of freshwater animals on Earth. Black carp probably entered the United States in the 1970s as a contaminant in imported grass carp stocks, but aquaculturists subsequently widely used them in the Southern United States for snail control in aquaculture ponds. Predictably, they escaped, most likely during flooding, and have colonized the Mississippi River and some major tributaries (Fig. 22.9). Black Carp are currently expanding their range in the United States from lower portions of the Mississippi to the Upper Mississippi River and major tributaries. We do not yet understand the full implications of this large predator entering the rivers of United States.

Figure 22.9 A large female Black Carp (*Mylopharyngodon piceus*) caught in a gill net in the Middle Mississippi River south of Thebes, IL. Researchers caught the fish in February 2018. It weighed 522 kg and was 1.6 m in total length. *Photo provided by Hudman Evans, a graduate student in the Fisheries and Aquaculture program at Southern Illinois University Carbondale.*

Determining if exotic invasive species are in fact the causes of native species declines is difficult, as the presence of invasive species and habitat degradation are almost inevitably correlated (Didham et al., 2005). As an example, the direct, negative impacts of zebra mussels (*Dreissena polymorpha*) on native mussels and other species seem obvious (see Highlight 10.1). However, some have questioned whether habitat degradation and loss of native unionid mussels opened the door for zebra mussel invasions (Gurevitch and Padilla, 2004), and experimental evidence for either scenario is lacking. Regardless, the result of the spread of exotic invasive species and loss of native species is a biologically less diverse and homogenized planet.

Along with the ecological consequences, invasive species are having major economic impacts. As of 2005, there were an estimated 50,000 exotic species in the

United States, which cost an estimated \$120 billion/year (Pimentel et al., 2005). Major costs associated with exotic species include agricultural losses; recreational and ecotourism losses; and control and eradication efforts.

While some of the consequences of the loss of so much biodiversity remain a mystery, many adverse ecological and economic effects are already evident, including declines and losses of fisheries, reductions in water quality, increased costs of water treatment, and adverse effects on agriculture and aquaculture (Edwards and Abivardi, 1998). Fully understanding these effects requires understanding the complex interactions among organisms. Studies of relationships between biodiversity and ecosystem function (discussed more fully in Chapter 24) suggest that the structure (e.g., community structure), function (e.g., resource use efficiency and productivity), and stability (e.g., resistance and resilience to disturbance) of ecosystems is being compromised as species are lost (Tilman et al., 1997; Loreau et al., 2001). A manipulative mesocosm experiment suggested that higher diversity aquatic food webs could be more resilient to disturbance (Downing and Leibold, 2010). In this study, researchers subjected mesocosms to pulses of acidification as a disturbance. While the richer invertebrate communities did not have increased resistance to perturbation, they did recover more quickly.

The ecological consequences of declining species diversity are less studied in freshwater systems as compared with terrestrial (Covich et al., 2004), but there is ample evidence that loss of species will alter freshwater ecosystem function and reduce stability (e.g., Vanni, 2002; Cardinale et al., 2006). As discussed in the section on strong interactors above, experimental manipulations with the detritivorous fish, *P. mariae*, in a Venezuelan river found that their feeding, egestion, and associated bioturbation accounted for over half of downstream particulate carbon export; removal of the fish from a stream reach resulted in a >400% increase in deposited organic sediments (Taylor et al., 2006). Removal of this fish also resulted in significant increases in respiration and primary production.

Studies of fish assemblages in Lake Tanganyika, Africa, and a South American river provide further evidence that declining fish diversity will affect ecosystem functioning. Modeled excretion rates and extinction scenarios indicate that some species in these assemblages have stronger influences on nutrient cycling rates than others; unfortunately, some of the most important contributors to nutrient recycling are targets of commercial harvest (McIntyre et al., 2007). The ultimate consequences of extinctions depend on which species are lost and if other species compensate (e.g., their populations increase in response to the loss of other species). Either way, the models indicate that declining freshwater fish diversity will negatively influence nutrient recycling. Clearly, declining diversity of fishes and other consumers will have ecological consequences. Given that extinction rates are highest in freshwater habitats, understanding linkages among biodiversity and ecosystem function and stability is an active area of research in freshwater ecology.

SUMMARY

1. Disturbance is an important ecological process. Perhaps the best documented form of disturbance in aquatic ecosystems is river flooding. This is a natural part of the ecosystem, and river restoration projects are beginning to consider this aspect of aquatic ecosystem dynamics.

2. Succession occurs both seasonally and over longer periods in most aquatic habitats. The best examples of seasonal succession are from the series of events that occur in the epilimnion of stratified temperate lakes. Succession after reservoir construction, when a shift from lotic to lentic communities occurs, is also well documented.

3. Streams may also exhibit seasonal succession related to changes in deciduous vegetation or predictable seasonal flooding. Riparian wetlands also exhibit successional patterns related to seasonal flooding.

4. Indirect interactions are likely important in all aquatic habitats. Such interactions may be difficult to predict without sound knowledge of the life history and ecological roles of organisms in a particular habitat.

5. Some species in aquatic ecosystems have disproportionately strong effects on many other species. These are termed strong interactors or keystone species.

6. Complex interactions are common in aquatic ecosystems.

7. Thresholds and alternative stable states complicate understanding and management of aquatic ecosystems. The potential difficulty in returning some systems to their original ecological state suggests that managers should use the precautionary approach until they fully understand system dynamics.

8. Human activities are facilitating biological homogenization through the spread of exotic species and loss of native biodiversity. We are currently experiencing one of the most dramatic extinction events in the history of the planet, and freshwater species are among the hardest hit.

9. The ecological consequences of species losses and introductions is an active area of research in freshwater ecology. These losses will have far-reaching consequences that ecologists do not yet fully understand.

QUESTIONS FOR THOUGHT

1. How predictable are successional trajectories (i.e., can sequences of species colonization be predicted or just general patterns)?

2. How might understanding species interactions be important for predicting the effects of introduced species?

3. How can disturbance make competition less intense?

4. If a species is a keystone species, should greater attention be paid to conservation of that species than others in a habitat?

5. Why would introduction of a keystone species potentially be a threshold in an aquatic environment?

6. How many potential indirect interactions, excluding loops, are there in a community of 10 species?

7. How could you design an experiment to test whether an invasive species or habitat degradation caused the extinction of a native species?

CHAPTER 23

Fish Ecology, Fisheries, and Aquaculture

Contents

Figure 23.1 A large pike (*Esox lucius*). *Photograph courtesy: Chris Guy.*

Much of the economic impetus behind protecting freshwaters comes from the maintenance of productive sport and commercial fisheries. Fishes dominate many freshwater food webs as the top predator in streams and lakes. Fishes are good indicators of water quality and model organisms for physiological and behavioral research. Knowledge of fish ecology and traits is essential to fish conservation (Angermeier, 1995). In this chapter, we discuss the biogeography of fish communities and factors that influence growth, survival, and reproduction of fishes. We briefly describe population dynamics, communities, and production of fishes, as well as harvesting,

Freshwater Ecology
DOI: https://doi.org/10.1016/B978-0-12-813255-5.00023-5

management, and aquaculture methods. We discuss these topics with specific emphasis on principles of freshwater ecology introduced elsewhere in this book.

BIOGEOGRAPHICAL AND ENVIRONMENTAL DETERMINANTS OF FISH ASSEMBLAGE DIVERSITY

Fisheries biologists have become more interested in managing species beyond the traditional focus on those used for harvest or recreation. Efforts to conserve aquatic species have required agencies to get involved with management of nongame species, including maintenance and even restoration of biodiversity and protection of endangered species. Furthermore, fishery *yield* (amount of fish taken per unit time) tends to be greater with more diverse fish communities (Brooks et al., 2016). These aspects of fisheries require consideration of factors that influence diversity and distribution of fishes. We describe patterns from the most general to the smallest scale.

Current estimates of fish diversity indicate similar numbers of species in freshwater and marine systems, although scientists have not yet described all species in either habitat. There are more species of fishes described for the Amazon Basin than in the Atlantic Ocean. This is an interesting observation because marine fishes have had a longer evolutionary history and marine habitat is substantially larger (see discussion of the species area relationship in Chapter 11), so they would be expected to be more diverse. However, continental waters are more geographically isolated, so the possibility for reproductive isolation and subsequent speciation is greater in freshwaters.

Speciation, in large part, occurs within drainage basins, and larger drainage basins tend to have more species (Oberdorff and Hugueny, 1995; Guégan et al., 1998). Islands tend to have small watersheds and lower fish diversity. This diversity pattern is consistent with the theory of island biogeography we covered in Chapter 11. Additionally, consistent with many other animal and plant groups, fish species richness is greater in the tropics than in temperate zones. Data compiled by Matthews (1998) demonstrated that approximately 7% of samples fish sampling efforts from both tropical and temperate streams had no fishes in them (completely unsuitable for fishes), but that maximum possible biodiversity is considerably higher in tropical streams and rivers.

Several explanations could illuminate why there is high diversity in the tropics. Glaciation of higher latitudes that led to local extirpation of fish species partially causes lower diversity in temperate waters. Time for evolution (i.e., total generations) of fish species has also been greater in the tropics because more generations per year can be produced. Tropical areas also have greater terrestrial net productivity, and this has been linked to fish diversity (Oberdorff and Hugueny, 1995; Guégan et al., 1998). Thus, the number of species per sample peaks at 10−20 in many temperate streams, but many samples in tropical streams have 50 or more species (Fig. 23.2). A striking example of tropical adaptive radiation is the diverse and highly coevolved assemblage of fishes associated with the riparian zone in the Amazon basin (Highlight 23.1). We described the spectacular diversity of fishes in some African Rift lakes in Chapter 11.

Within continents, different ecoregions have varied diversity. For example, in North America exclusive of Mexico, there are about 740 fish species. About 300 of these occur in the Mississippi basin. This basin has an area about equal to that of the

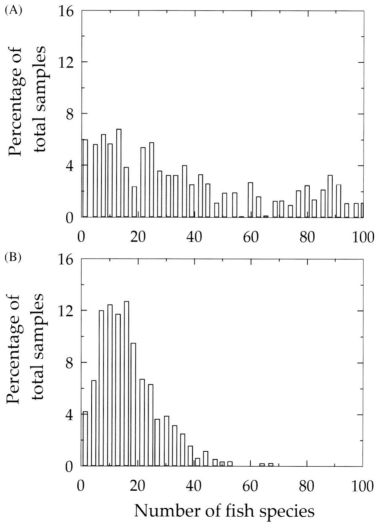

Figure 23.2 The proportion of total samples containing various numbers of fish species from tropical (A) and temperate streams (B). The total numbers of samples were 204 in the tropical streams and 815 in the temperate streams. *After Matthews (1998).*

Highlight 23.1 Fishes in the forest

Many fishes associated with the Amazon River drainage are dependent on annual flooding of the riparian forest (Goulding, 1980). The Amazon and its tributaries flowing through tropical forest annually flood the low-lying areas around the river channels, allowing fishes access to the forest floor, and temporarily connecting many small ponds and lakes with the main river channel. The total number of Amazonian fish species may be between 2,500 and 3,000, and a diverse fish assemblage forages in the floodplains. Some of the species move from the main river into tributaries to spawn, and others move from the flooded forest through the tributaries to the main river channel to spawn. These migrations allow predatory fishes and local fishers to capture spawning fishes as they move predictably in large schools through specific places during specific times.

A most interesting group of fishes has evolved to eat fruits and seeds. Ichthyochory is seed dispersal by fishes and could be more common than thought (Pollux, 2011). These frugivorous fishes occur mostly in South America. In a coevolutionary response, some of the riparian forest plant species have evolved fruits that the fishes disperse. These trees and shrubs fruit during the flooding season, and their fruits are under or near the water surface. Fishers take advantage of the feeding habits of the frugivores and bait their hooks with fruits that are preferred by the fishes.

Other fishes are seed predators and crush, consume, and digest seeds. Seeds form a significant proportion of the diet of some species of piranhas (*Serrasalmus* spp.), which are known to most people as aggressive animal predators. These fishes can have impressive jaws and teeth that are adapted to breaking the hard shells protecting nuts and seeds.

The fishes in the forest provide a clear example of the linkage between riparian wetlands and production of fishes. Many of these fish species have an absolute requirement in their life cycle for the seasonal availability of resources in the flooded riparian zone. Some plants benefit from the interaction as well.

Hudson Bay drainage, which has about 100 species. Consequently, lack of glaciation and a long evolutionary history (over 20 million years since the sea level fell during the Tertiary period) probably led to evolution of more species in the Mississippi basin. However, terrestrial net primary productivity is greater in the basin, leading to more energy entering the river. Systems that are more productive can be more diverse, so strict explanations are difficult.

Within a basin, fish diversity in rivers and streams increases with distance from headwater streams (Fig. 23.3). Larger rivers are less likely to dry. Small streams and rivers are more subject to debris torrents and catastrophic flooding. The floodplain of a large river contains a larger range of habitat types (e.g., wetlands, oxbows, small side channels, small streams entering the main river, and open channels with high water velocity). At times, samples taken in large order streams can have few or no fish species

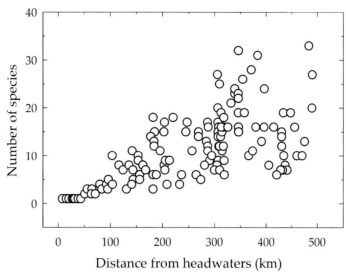

Figure 23.3 Number of fish species as a function of distance from headwaters in the Kali Gandaki River, Nepal. Statistical analysis of the data suggests that the number of fish species increases with river distance up to about 300 km. *Data from David Edds.*

(e.g., if driven anoxic by pollution), but the observed maximum number of fish species per sample is generally higher in larger rivers.

Within basins, factors such as flooding, pollution, turbidity, reservoirs, channel modifications, and exploitation by humans can alter the numbers and types of fishes found. We discussed pollution effects on diversity in Chapter 15. Heavily polluted waters, particularly those in which O_2 levels are very low, support few or no fishes. Increased turbidity disadvantages sight-feeding fishes, and channelization can remove critical habitat (e.g., deep pools or side channels). Reservoirs can exclude fishes that move upstream or downstream to spawn or feed from their breeding or foraging areas (Pelicice et al., 2015). Reservoirs also harm species that require flowing waters to keep their eggs in the water column until they hatch or have floating larvae (Dudley and Platania, 2007). In these cases, lentic fish assemblages replace riverine fishes in the reservoirs. Currently, dams obstruct half of freshwater watersheds, particularly harming some species of lampreys, eels, shads, and other migratory species. Threatened areas from dams include the Murray-Darling (Australia), Southern Italy, the middle and lower Indus basin, West Korea, the Upper Paraná (Brazil), and the Mobile Bay and South Atlantic of the United States (Liermann et al., 2012).

Within a stream segment or reach, or within a lake, the type of habitat present can be very important. Any angler knows about specific areas that fishes are more likely to inhabit. In streams, for example, cut banks, woody debris, and deep pools all are places where fishes may congregate. Humans tend to remove habitat structure, and subsequently have made much effort in many lakes and streams to restore habitat that is essential to maintaining a diverse fish assemblage and productive fishery.

Reservoirs generally have a blend of lotic and lentic habitats. At the upstream ends, where rivers enter reservoirs, fish species characteristic of flowing waters are more abundant. In deeper waters, pelagic species are common. The numerous embayments found in many reservoirs can be more like shallow lakes and have high densities of submerged and emergent macrophytes. The face of the dam may be rock, earth, or concrete. The rocky or concrete dam faces may be unusual benthic habitat for many areas, particularly in reservoirs built in large depositional valleys where rocky habits are rare in natural rivers and lakes.

Restoration of streams and lakes requires consideration of habitat characteristics that lead to successful recovery of fish biodiversity (National Research Council, 1992). The ecological basis for river management should consider what constitutes a natural hydrological regime (Petts et al., 1995; Palmer et al., 2005). Restoring physical structure to stream channels may require wing deflectors, low dams, placement of debris, and other measures (De Jalón, 1995). Regulations on boat traffic may help decrease turbidity and wakes that can interfere with reproduction (Murphy et al., 1995). Habitat restoration or enhancement can improve reproduction and recruitment in lakes and reservoirs. Measures include increasing cover for small prey fishes and juvenile fishes of larger species, and altering the hydrological regime to match spawning requirements of desirable fishes. Addition of old Christmas trees or other large structural material is a common strategy in some areas to increase heterogeneity of reservoir fish habitats.

The diversity of prey species may also alter the diversity of piscivorous species. For example, it is unlikely that the number of predatory species will exceed that of prey species (Fig. 23.4) However, this is a weak relationship and water quality, habitat structure, and large-scale patterns of species dispersal are stronger determinants of fish diversity. Presumably, the more prey species there are, the more species of specialized predators can evolve.

PHYSIOLOGICAL ASPECTS INFLUENCING GROWTH, SURVIVAL, AND REPRODUCTION

Energetics control growth, survival, and reproduction of fishes (or any other organism, for that matter). Environmental extremes can cause excessive energy drains affecting the success of fishes. The quantity of energy required for survival depends on the external environment. Growth requires additional energy, and successful reproduction requires the most energy (Fig. 23.5).

Osmoregulation requires energy and is necessary for survival in freshwaters. Osmoregulation mechanisms vary among freshwater fishes. The lower limit of a fish's

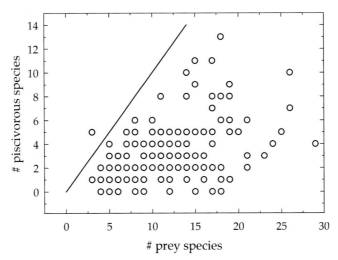

Figure 23.4 Number of piscivorous species as a function of number of prey species from 178 fish assemblages in eastern North America. The line denotes a 1:1 relationship. Note that more predatory than prey species occur in only one case. *After Matthews (1998).*

distribution in an estuary may be a function of its salt tolerance, as may be the response to runoff from road salt additions and hypersalinity in closed-basin lakes. All fishes that enter freshwater must osmoregulate because their internal salt concentrations exceed those found in most freshwaters (they are hypertonic relative to the water). In general, well-developed kidneys excrete excess water. Because water is not limiting, a fish might excrete up to one-third of their body weight per day (Moyle and Cech, 1996). Diadromous species must have versatile osmoregulation strategies to deal with the transitions from one level of salinity to another. Fishes living in highly dilute waters may have ion-specific active transport mechanisms to acquire and retain useful ions. All fishes must regulate the concentration of ions such as sodium and potassium to survive.

Temperature also controls growth, survival, and reproduction of fishes. At low temperatures, metabolic rates are depressed, and swimming requires more energy because of the increased viscosity of water (see Chapter 2); a 22% decrease in viscosity and roughly a doubling in metabolic rate accompany an increase in temperature from 13°C to 23°C. We described general molecular adaptations to temperature extremes in Chapter 15.

High temperatures can be harmful to fishes as well. High enough temperatures will kill fishes, but before that, they become stressed. The ability to stand heat stress is thermal tolerance. Some species can eventually acclimate physiologically to higher temperatures and have plastic thermal responses. With global warming, the ability to

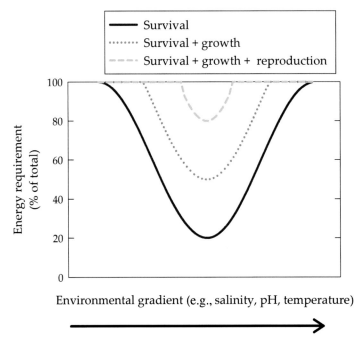

Figure 23.5 Conceptual illustration of energy requirements for survival, growth, and reproduction as a function of an environmental gradient. If 100% or more of the available energy is required for survival, the fish will not survive. If the sum of energy required for survival and growth exceeds 100%, the fish will not grow, and if the sum of energy for survival, growth, and reproduction is less than 100%, the fish can reproduce.

withstand and adapt to higher temperatures is an important feature of the physiology of fishes. Comte and Olden (2017) found that fishes tolerant of the highest temperatures had the lowest physiological plasticity and ability to acclimate to higher temperatures, indicating some upper physiological limit to adaptation. Some taxonomic groups of fishes had greater plasticity in thermal tolerance than others.

Physiological characteristics interact to determine the optimal temperature for growth as mediated by energetic constraints. For example, even though salmonid species seem very similar, they have different temperature requirements for growth and survival (Table 23.1). Temperature influences a variety of fish life history characteristics, so fishes actively seek water that has a temperature near that optimal for growth. Temperature requirements for reproduction may be more stringent than those for survival or growth.

Temperature constraints of fishes in freshwaters and interactions with trophic state have led fisheries managers to consider lakes to be of several general types of thermal regimes with regard to commercial or sport fishes (Fig. 23.6): (1) warm waters dominated by black, white, yellow, and striped basses, sunfish, perch, or catfish; (2) cool

Table 23.1 Temperature requirements for various salmonid species

Species	Maximum growth (°C)	Maximum temperature (°C)	Spawning temperature (°C)
Atlantic salmon (*Salmo salar*)	13—15	16—17	0—8
Brown trout (*Salmo trutta*)	12	19	2—10
Rainbow trout (*Onchorhynchus mykiss*)	14	20—21	4—10
American brook trout (*Salvelinus fontinalis*)	12—14	19	2—10
Grayling (*Thymallus thymallus*)	10—16	20	10

Note how maximum growth rate occurs at a temperature only slightly less than that for maximum survival, and the temperature required for reproduction is much less than that for maximum growth.
Source: After Templeton (1995).

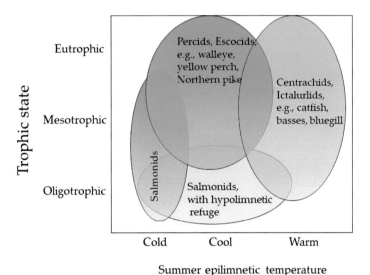

Figure 23.6 Conceptual illustration of fishery types in temperate lakes with summer stratification as a function of temperature and trophic status.

waters with some of the previously mentioned species plus walleye and pike; (3) cold waters dominated by salmonids; and (4) two-story waters in stratified lakes with oxygenated hypolimnia in which cool-water and cold-water species can find refuge from high water temperatures during the summer. During summer months, cold- or cool-water species are hypolimnetic, and warm-water fishes reside in the epilimnion. If thermal refugia are present in streams (e.g., groundwater input), cold-water species can survive in warm streams and rivers. Hypolimnetic release from dams creates cold habitats that allow stocking of salmonids in many rivers where they would not be able to survive otherwise.

Considering temperature optima is necessary to determine how fishes will respond to global warming and other aspects of climate change (Guzzo et al., 2017). Analyses of fish communities in France show shifts toward warmer water species assemblages (Daufresne and Boët, 2007). Additionally, the authors note that impoundments fragment populations and stop migrations of species to cooler waters, further altering the ability of fishes to respond to changes in temperature. The systems mentioned above, even with cool water refugia, may not remain cool enough to maintain cold-water fishes.

Dissolved oxygen is crucial for most fishes. Those that breathe with gills require dissolved O_2. We discussed the problem of fish kills related to O_2 depletion in Chapter 12. The O_2 requirement is greater as temperature increases, and the concentration of O_2 at saturation in warmer water is lower. Fishes can respond to lower O_2 by increasing ventilation rate. Although increased ventilation allows survival under marginal O_2 concentrations, it uses energy and causes stress. Some species of fishes (e.g., lungfish) can use atmospheric O_2, but most others will die if dissolved O_2 concentration becomes too low. Many species can withstand temporary anoxia, but repeated bouts of anoxia can lead to cumulative harmful effects (Hughes, 1981).

Food quantity and quality can alter the survival, reproduction, and growth of fishes, and fishes are selective in what they eat (Hughes, 1997). All fishes require protein, carbohydrates, lipids, vitamins, and minerals. A fish that is piscivorous may be able to obtain reasonable amounts of protein and lipids but the energy content of its food could be the key limiting factor for growth. Zooplanktivorous fishes may have diets that are rich in lipids, and many teleost fishes have a high lipid requirement. Thus, copepods that contain high concentrations of wax esters are an excellent food source for these fish species. Proteins and lipids are more likely to limit herbivorous fishes, and detritivorous fishes are more limited by protein than algivores. Fishes that live in warmer waters can require lipids with a higher melting point.

Herbivorous fishes (such as grass carp) can have special enzymes, extra long guts, or gut bacteria that allow them to use low quality plant materials more efficiently as a source of carbohydrates. If one type of food is of a high quality, it can partially compensate for lower quality food. For example, catfish can use cellulose as a carbon source almost as efficiently as starch if a high proportion of protein is present in their diet (Fig. 23.7). Less is known about the roles of vitamins and minerals in fish diets (Pitcher and Hart, 1982), although a considerable amount of research has gone on with respect to species suitable for aquaculture (Halver and Hardy, 2002).

The rate at which fishes must consume food to allow for growth and reproduction depends on both the requirements imposed by physiological demands of the environment and the actual energy investment in obtaining food. Under stressful conditions, fishes use a greater portion of the food ingested to maintain basic metabolism, and less is

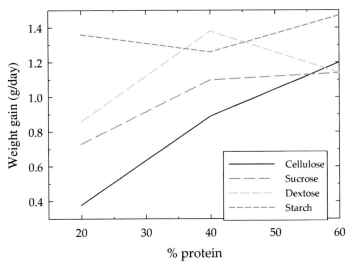

Figure 23.7 Growth of channel catfish (*Ictalurus punctatus*) on four carbohydrate sources as a function of protein content of food. *Data from Simco and Cross (1966).*

available for growth. Determining actual energy requirements can be very complex. For example, in a heterogeneous environment, metabolic costs can vary over space and time, as can the amount of energy required to locate and capture prey. Consider, for instance, the idea that foraging is more difficult when a predator is present (see discussion on optimal foraging in Chapter 20). Ample food may be available in the environment, but when a predator is nearby, a fish must use energy to avoid being eaten (Milinski, 1993) and balance that against its need to forage for food (Fig. 23.8).

Several life history strategies have evolved in fishes given that they must specialize to be successful competitors in different environments. A periodic strategy is one where fishes delay reproduction to grow, and are able to produce very large clutches upon spawning. This allows species to take advantage of spatially or temporally rare favorable habitats. An opportunistic strategy stresses early maturation and frequent reproduction with repeated small clutches. These species are able to recolonize disturbed habitats rapidly and have generally high growth rates. An equilibrium strategy would do well in stable conditions, where density dependent effects are strong. This strategy is to produce large eggs in small clutches, and possibly have parental care (Winemiller and Rose, 1992).

Some interactions of fishes with their environment are not monotonic and a complex series of factors control them. We provide just one example here. In boreal nutrient-poor lakes, colored dissolved organic matter influences yield of brown trout in several ways. The lakes with the lowest dissolved organic matter have lower yields than those with greater concentrations. Perhaps this is because the dissolved matter

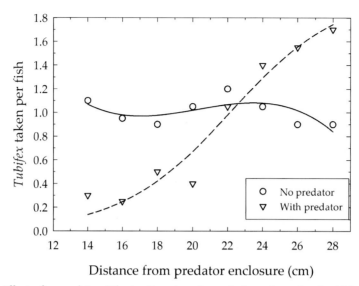

Figure 23.8 Effect of a predator (Tilapia, *Oreochromis mariae*) on the rate at which a stickleback (*Gasterosteus aculeatus*) consumes prey (*Tubifex*). This experiment included an adjacent fish cage that could hold the predator, and worms were placed in tubes at various distances from the adjacent cage. The stickleback had to enter the tube and lose sight of the predator to take a worm. With no predator, the worms were taken from all tubes equally. With the predator, the sticklebacks preferred to feed as far away from the predator as possible. *Modified from Milinski (1993).*

enters the food web or it protects against UV radiation. However, when the organic material increases beyond a point, it starts to cause decreases in fish yield. This could be because it intercepts light, rendering the phytoplankton at the base of the food web less productive (Finstad et al., 2014).

POPULATION DYNAMICS OF FISHES

Fish population dynamics have several key characteristics, including the size of the population (*stock*) and the population growth (*production*). The population can be divided into a number of *age classes* or *size classes*. The number of new fish entering each size or age class (*recruitment*) and the number of fish that are lost while they are in the age or size class (*mortality*) also relate to population dynamics. In addition, the number of eggs females produce (*fecundity*) and the number of larval fish that successfully hatch from those eggs can influence population size.

When sampling a fish population, the first questions that arise usually involve number, size, and age of fish. Most fishes have indeterminate growth, and keep increasing in size as they get older, if their energetic needs are met. However, there is not a strict relationship between age and size, so additional information such as relationships among mass, length, and population structure need to be determined. This information can indicate several important things about the reproduction, growth, and mortality of fishes, and assist with management of populations. This information is dependant on catching the fish and determining their population size (Method 23.1).

The total number and sizes of individuals must be determined after a researcher or manager captures fishes. Length is the easiest to measure but is not a linear function of mass (Fig. 23.9B) because mass is related more closely to volume than length, mass is a logarithmic function of length for most animals, including fishes. The relationship between length and mass correlates to the condition of fishes. A fish that has a high mass per unit length often is considered healthy and in better condition. Researchers have constructed numerous indices to characterize mass—length relationships (Anderson and Neumann, 1996). These indices are useful for assessing the relative condition of individual fish within a habitat or to comparing the condition of fish populations among habitats.

As with many animal populations, small fishes are usually more numerous than large fishes (Fig. 23.9A) because mortality at each age class decreases the number of large individuals in a population. Fig. 23.9 does not include data on the very youngest fish because electrofishing does not sample them effectively. The large numbers of fish of about 80 mm in length are probably 1-year-old. An obvious exception the trend of more small fishes occurs when managers stock adults.

Fisheries managers use many indices to quantify growth, mortality, recruitment, condition, and size structure. We discuss one of these indices based on length data as an example. Relative stock density (RSD) is the percentage of fish in a specific length range relative to the total number of fish above a minimum size estimated to be in the population (Anderson and Neumann, 1996). In other words, it is the proportion of fish in a specific size range relative to all fish sampled effectively. The RSD offers an index that describes the distribution of fish lengths in a population. The index is useful in several specific situations. If the index is restricted to just the largest fish in the population, it can be used to calculate the relative number of trophy fish in a population. If the index considers all fish large enough for anglers to keep, it can indicate the quality of the fishery. The RSD can also indicate if a fishery is balanced (e.g., that there are not too many small fishes, or that competition or some other factor is causing poor condition in a specific size class).

The next aspect of population structure we consider is the size class of each age of fish. If the size class as well as the population and biomass numbers are well known, the rate of growth or the production of the fish population can be calculated. Two of the most commonly used approaches to estimate ages of fishes are length frequency analysis and analysis of anatomical hard parts (Devrie and Frie, 1996).

METHOD 23.1 Sampling fish populations

A wide variety of techniques are available for sampling fish populations. Determining the method to use depends on the physical constraints of the habitat, the species of interest, the information required, and the relative effectiveness of the method. The capture methods can be categorized as passive capture techniques, active netting, and electrofishing. All three are used in freshwaters and require determination of catch per unit effort. If all things are equal, twice the effort should lead to capture of twice as many fishes. Of course, as population density decreases, additional effort to capture individuals will lead to a lower amount of catch per unit effort.

Passive capture techniques work by entanglement, entrapment, or angling with set lines (Hubert, 1996). Gill nets entangle fishes in meshes of specific size. The larger the mesh size, the greater the selectivity for larger fishes. Trammel nets also entangle fishes. They are fine-mesh nets with larger mesh next to the net. When a fish pushes the smaller mesh net through the larger mesh, it forms a pocket in the smaller mesh net that catches the fish. Entrapment gear allows fishes to enter, but not leave, and may use bait to lure the fishes. One type of entrapment gear has cylindrical hoop nets with a series of funnels in which the fishes are able to move into the cylinder but not out. Fyke and other types of trap nets have additional panels of netting at the entrance to a cylindrical trap that guides fishes into the trap. Pot gears also use entrapment. These are rigid traps with funnels or one-way entrances. Minnow traps, crayfish pots, and similar equipment are included in this group. Weirs are barriers built across a stream to divert fishes into a trap; these work well on migratory (usually reproductive) fishes.

Active gears require moving nets through the water to capture organisms (Hayes et al., 1996). Nets can be towed (trawled) at the surface, through the water column, or across the bottom. Alternatively, purse seines can encircle the fishes and confine them into successively smaller space as the net is drawn closed. Other active gears include nets that are thrown or pulled up rapidly from the bottom, spears, dipnets, dynamite, and rotenone (poison).

Electrofishing is a common and effective way to capture fishes. Either AC or DC current alters the behavior of fishes to stun or kill them (Reynolds, 1996). The most common strategy is to stun fishes and then catch them with a net. To stun fishes, the current and its pulsing frequency are controlled at a level that causes paralysis but little injury and death. Water conductivity is an important factor in determining effectiveness of the technique. Low-conductivity waters are poor conductors of electricity and decrease the effectiveness of the method; very high-conductivity waters allow the charge to dissipate too rapidly.

Researchers can count fishes directly when removal is not necessary, commonly with scuba diving or snorkeling. Habitats that are very turbid, with excessive water velocity, or unsafe for human contact are obviously not suitable for such techniques.

Several methods are used to determine total population size, depending on habitat and method of capture. If sampling is highly effective, such as in a small pond with a clean bottom, the sampler captures all the fishes. It is rarely possible to capture all the fishes in a habitat, and several other methods compensate for this fact. A depletion

(Continued)

METHOD 23.1 (Continued)

method repeatedly samples the same area, and the decrease in catch per unit effort is used to estimate the original abundance. Alternatively, mark and recapture techniques are used. In this method, a sample of fishes is taken and marked. Fishes are later recaptured, and the proportion of marked fishes is used to estimate abundance. There are several marking techniques, including fin clipping, external and internal tags, and marks (Guy et al., 1996).

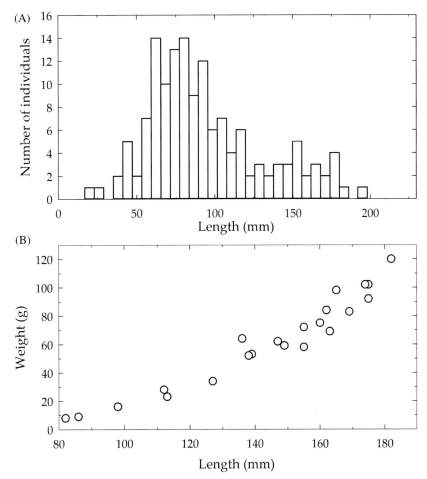

Figure 23.9 (A) the number of bluegill (*Lepomis macrochirus*) at each length and (B) the relationship between length and wet weight for fish taken by electrofishing from Pottawatomie State Fishing Lake II, Kansas, on September 29, 1999.

Length frequency analysis is used where distinct peaks in the distribution of lengths are observable. Population peaks often occur at specific size classes, each corresponding to a reproductive cohort. Such peaks may not be discernable when growth and time of reproduction are variable among individuals and for older fish in the population.

Scales, otoliths, bones, and spines are hard parts used to age fishes. As a fish grows, these hard structures grow in size. Growth marks indicate age as do rings on a tree. Daily marks can sometimes be discerned in young fishes, and where growth has a distinct seasonal pattern as in temperate zones, annual marks can be observed as well (Fig. 23.10). Researchers commonly use scales and fin rays to determine age because sampling them does not kill the fish. Otoliths (calcified structures from the inner ear) yield more reliable aging results but require sacrifice and dissection of the fish.

Recruitment and mortality for each size class can be calculated once age distribution of a population is determined. Reproduction and mortality are influenced by numerous abiotic and biotic factors and can vary tremendously (up to 400-fold) among year classes (Pitcher and Hart, 1982). Explaining this variation is an important part of fish population management and ecology. Life tables such as those used by demographers and population ecologists allow determination of mortality and recruitment rates for each age class in a fish population. Age-specific life tables, in which a cohort moves through age classes through the years, are the most useful. We can construct these tables by calculating the proportion of fish surviving each time interval, generally a year. The recruitment and mortality at each size class often correlates with environmental factors and indicates the influences of natural factors and exploitation on fish populations.

Figure 23.10 A cross section of a fin ray from a river carpsucker (*Carpiodes carpio*) showing growth rings. *Photograph courtesy: Chris Guy.*

If we know the age distribution (age and mass of each year class of fish) then we can calculate the production of the population. Production incorporates both static (biomass and population size) and dynamic (growth, recruitment, and mortality) measures. The mass produced per age class can be of interest to fish managers wanting to know how productive a water body will be for catchable fishes. The total production could be of interest if a species is a food source for sport fishes, used for aquaculture, or is ecologically important. We provide more detailed treatment of calculation of animal production in the ecosystem context in Chapter 24.

Generally, a negative correlation exists between recruitment and growth. If many fish occur in a size class, then significant intraspecific competition can occur. Intense intraspecific competition leads to poor fish yields in the size class. If food is not limiting (rarely the case in natural situations but possible in aquaculture), recruitment and growth may both be high.

REGULATING EXPLOITATION OF FISH STOCKS

Any fishery containing desirable fishes and with unlimited human access fails without management (a phenomenon often referred to as the tragedy of the commons). The ability to predict potential fish yields is central to attempts to manage exploited fish stocks. For many years, managers thought that the maximum sustainable yield (MSY) could be calculated and used to estimate the maximum potential sustainable harvest of any fish population in a particular water body. The MSY was calculated by William Ricker (Biography 23.1) using a curve that plotted the number of reproductive adult fish against the number of reproductive adults they could produce. The Ricker curve has seen much use in fisheries, particularly those in which the numbers of reproductive adults and their offspring can be determined. The MSY concept has several problems, including that it calculates yield without regard to size, that the number of fish produced often does not relate to the number of reproductive adults because each female can produce many eggs, and that survival of eggs and fry is highly variable. Therefore, the actual yield may be highly dependent on unpredictable biotic and abiotic factors, and as such, most freshwater fisheries managers do not calculate MSY. Rather, managers rely on data on growth and recruitment of specific size classes of harvested fishes for recommending management options.

Management of fish stocks usually involves setting regulations on the numbers and sizes of fishes harvested. Such regulations are set depending on the desired type of fishery. Management techniques include closed seasons, limited access, limitation on the number of fishers, regulations on the numbers and sizes of fishes taken by anglers, and regulations on methods used to take fishes (including types of gears and net mesh size, and noodling).

BIOGRAPHY 23.1 William Ricker

Dr. William Ricker (Fig. 23.11) was one of the most renowned fish biologists in the world. He described the stock recruitment curve known as the Ricker curve. He was also one of the first scientists to introduce the concept of chaos into population dynamics. He had more than 100 scientific publications and an equal number of popular articles. Ricker was the recipient of many prestigious awards, including the Eminent Ecologist Award of the Ecological Society of America. He was a fellow of the Royal Society of Canada and of the American Association for Advancement of Science. He was chief scientist of the Fisheries Research Board of Canada.

Ricker was a man of many talents. He identified 80 new species and 46 new genera of stoneflies, wrote a Russian/

Figure 23.11 William Ricker.

English book of fisheries terms, and he played musical instruments. Ricker got started in fisheries in 1938 when he was offered a job for the Fisheries Research Board of Canada. At that time, any job was welcome. The job excited him, and he turned it into a career.

His views were not always popular. In a 1984 article in Sports Illustrated he was quoted as saying "Practically everyone who has ever gone fishing considers himself an expert in fish management and doesn't hesitate to say so. Also, the man who uses any particular type of fishing gear invariably regards all other types as pernicious and destructive; but he can insist, with a straight face, that his kind of fishing couldn't possibly do the stock any harm."

When interviewed for this biography Ricker suggested that it is important to arrange a research career around problems in which you are interested or you will not be successful. His advice was to get hands-on experience in a career before you commit to it so that you will be sure that you will like it. He thought an important part of fisheries management in the future will be to keep people from extirpating fishes. Ricker said that if there are no more fishes, there will be nothing more for fisheries biologists to research.

As discussed previously, a negative correlation generally exists between recruitment and growth. A fishery can be managed either for a large number of small fishes or for fewer large fishes, but generally not both. People often manage sports fisheries for large fishes and commercial fisheries for a high yield of usable fishes (e.g., just above some minimal size). The uncertainties in predicting recruitment and growth make absolute prediction of the effects of exploitation difficult. This uncertainty has led to numerous cases of overexploitation of fish populations (Hilborn, 1996).

Part of the issue is that there are several types of overfishing. If too many adults are taken, then spawning cannot provide enough young fish to replenish the stock. Overfishing can alter food web structure and potentially the ability of the ecosystem to support a sustainable fishery. Economic overfishing considers factors of optimum profitability of a fishery (Hubert and Quist, 2010). Given these complexities and different goals of management, assessments of major fisheries will increasingly include risk analyses. If risk assessment is used, a manager will set limits based in part on the potential risk of long-term harm to production of the fishery. Risk management must be considered in fisheries, and the precautionary principle, (not taking actions until it is certain they will not cause harm), is often necessary (Hilborn et al., 2001).

One of the most common regulations in sports fisheries is *creel limits*, or the numbers of fish that anglers can take per day. Such limits rarely are effective in managing fish populations because most anglers do not catch their limit. However, reaching the limit does provide a degree of satisfaction for the angler.

Setting size limits is another method to ensure that fish reach a size that is acceptable to anglers. Generally, these size limits are set so that several year classes of reproductive fish are less than the catchable size. Thus, in a poor year for recruitment or reproductive failure in a specific age class, the fishery will still have some recruits entering the catchable size range. Given a single length limit and heavy fishing pressure, the size distribution can consist of many fish just below the allowable size. This skewed size distribution can lead to intraspecific competition and low recruitment to fish of allowable size (Noble and Jones, 1999).

Slot limits that do not allow harvest of fish of intermediate size are also used. In this case, surplus young fish can be taken. With removal of small fish, intraspecific competition and recruitment into intermediate size lengths are lower, and growth rates of these fish are high. Slot limits could potentially create high production of large fish that are desirable for anglers.

Managers commonly set seasonal limits. The most common of these is not allowing any taking of fishes while they are spawning. Fishes can be particularly vulnerable to angling pressure at this point in their life cycle. They often feed aggressively to assimilate energy needed for reproduction. Fishes that aggregate to spawn in highly predictable parts of the environment are most susceptible, and regulations to protect spawning fishes can be useful.

STOCKING FOR FISHERIES

Many fisheries use human-produced stock to allow exploitation to occur at desirable rates (Heidinger, 1999). Stocking is done for many reasons, including: (1) introduction of new exploitable species, (2) introduction of new prey species, (3) introduction of biological control agents, (4) provision of fishes to be caught immediately or after they grow, (5) satisfying public pressure to enhance fisheries, (6) reestablishing species where they have been extirpated, and (7) manipulating the size or age class structure of existing fish populations (Heidinger, 1999). Non-native fishes have been introduced worldwide and between 1980 and 2010, rate of introductions increased twofold. About 40% of these introductions were for aquaculture and 17% for improvement of wild stocks (Gozlan et al., 2010). These programs are expensive. For example, managers stock approximately 2.5 billion sport fishes each year in the United States and Canada (Heidinger, 1999). Survey methods and accounting for purchases of fish licenses and fishing gear estimate if revenues and economic stimulation offset the cost of stocking programs.

The stocking of fishes can be a useful management tool. In cases in which reservoirs and ponds are constructed, native stream and river fishes may not survive. Introduction of sport fishes could be desirable (assuming they will not move into rivers and streams and cause problems there or exacerbate the disconnection of streams by the reservoir). In this case, native species are the safest species to stock.

Stocking can also be crucial to the recovery of some endangered fishes. People have stocked many non-native fishes with the intention of improving the fishery. Large-scale introductions can also include inadvertent introductions of other unwanted species or fish diseases. Unfortunately, the ultimate biological effects of such introductions usually have not been carefully considered or researched in advance. Precautionary approaches are best for new species introductions (Bartley, 1996) to avoid negative impacts in the future. These approaches include assuming that a species will have a negative impact until proven otherwise and putting the burden of proof of impact on the agency, company, or individual that releases the fish.

Several problems can arise with fish stocking programs, including competition with or predation on native fish species, hybridization with native fish stocks, lack of genetic diversity in fish populations that may lead to poor fitness, and transfers of parasites and disease (Moyle et al., 1986). We discussed invasive species in Chapter 11, and many of these stocked fishes are non-native and can have large negative consequences for the native fishes. Examples of harm related to intentional stocking include, but are not limited to, salmon and trout introductions in New Zealand endangering the native galaxid populations, carp introductions in North America harming many warm water fishes, and bighead carp in the Danube delta. Invasiveness indices have the potential for assessment of the impacts of intentional introductions (Gozlan et al., 2010).

One example of harm caused by introduction of non-native species is the whirling disease caused by infection by the microscopic cnidarian *Myxobolus cerebralis*. This parasite entered salmonid assemblages of the United States through stocked trout from Europe in 1958. By 2009, the disease had spread through hatcheries and river networks to 25 states. Infection increases vulnerability to predators, and impairs swimming and feeding ability. Initially, the disease was thought to be isolated to just hatcheries. Since then the disease has spread to native trout and salmon populations and led to drastic declines in some populations. This disease can have negative economic impacts and may endanger native fishes. Introduction of whirling disease might have been avoided by careful assessment of stocking programs and more cautious aquaculture procedures. Human activities have introduced numerous other pathogens worldwide from moving sport fishes or fishes used for aquaculture across the globe (Gozlan et al., 2010).

The size and type of fishes managers should stock are often of concern. Large fish survive better, but small fish are less expensive to rear. Eggs are stocked only rarely because of poor survival. Fish stocking is most successful when managers consider when and where they are most likely to survive and grow. Interspecific and intraspecific competition may lead to poor growth if managers stock too many fish. Thus, it is crucial to be aware of growth, mortality, and recruitment characteristics associated with the environment to be stocked, in addition to the ecology and the life history of the species used.

AQUACULTURE

The culture of aquatic organisms is *aquaculture*. Aquaculturists grow fishes, crustaceans, mollusks, and some algae for consumption, stocking, and the aquarium trade. Fish culture may be an important source of protein in developing countries. Freshwaters were the source for 60% of world aquaculture production in 2008 (Bostock et al., 2010). Fishes, crustaceans, and mollusks comprised from 13-17% of the animal protein consumed by humans in 2000 according to the World Health Organization (https://www.who.int/nutrition/topics/3_foodconsumption/en/index5.html downloaded Dec 2018). A billion people rely on fish as their primary source of protein (Allan et al., 2005). World aquaculture production is dominated by production of carps, and the amount of carps produced is about 10-fold greater than that of any other fishes (Fig. 23.12). Most fish production occurs in Southeast Asia.

Culture of fishes and crustaceans is generally more profitable if people use techniques based on an understanding of the ecology of aquatic ecosystems. The trick in aquaculture is to maximize productivity but avoid the negative effects of eutrophication. For example, to grow at maximum rates, fishes need a significant amount of food. Excessive food additions can lead to increased biochemical oxygen demand, anoxia, and death of fish. In addition, aquaculturists grow organisms at high densities, which provides ideal conditions for propagation of diseases and parasites, and attracts predators (Meade, 1989).

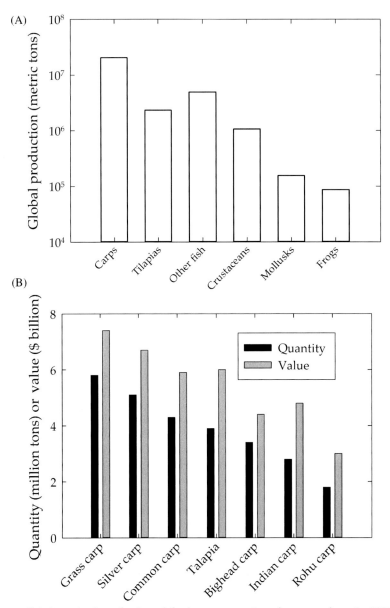

Figure 23.12 Global production of selected freshwater organisms by aquaculture in 1989 (A); note the log scale on the *y* axis (data from Stickney, 1994) and global aquaculture production and value of the top seven species globally in 2012 (B, FAO Yearbook of Fishery and Aquaculture Statistics 2015).

The diseases of cultured aquatic organisms include protozoa, fungi, bacteria, and viruses. Stressed and crowded fish are more prone to infections. Bacteria are more commonly problematic in cold-water species of fish, but antibiotics are generally effective in controlling these infections. Several species of viruses also cause problems for fishes. These diseases cannot be treated once the stock is infected. Rather, disease control requires strict handling practices including use of disease-free stock. Intermediate invertebrate hosts, such as snails and polychaete worms, transmit some invertebrate parasites. In this case, disease prevention may include use of pesticides and pond draining to kill the intermediate hosts.

Perhaps the most ecologically advanced form of aquaculture is the polyculture of carps in Southeast Asia. Trial and error developed this polyculture system over centuries, where fish farmers use close observation of ponds. The first book on aquaculture was written in Chinese in 460 BC (Fichter, 1988). The Chinese developed these systems to use several species with varied food requirements, and the excretion of one species fertilizes the growth of food for the other (Zweig, 1985). The dominant fishes in the Chinese systems are common carp (*Cyprinus carpio*, omnivore, benthic), bighead carp (*Hypophthalmichthys nobilis*, zooplanktivore), grass carp (*Ctenopharyngodon idella*, herbivore), and silver carp (*Hypophthalmichthys molitrix*, phytoplanktivore). In India, the catla (*Catla catla*), rohu (*Labeo rohita*), and mrigal (*Cirrhinus mrigala*) are used. There are about 2.5 million hectares of carp ponds in India and China. In most cases, farmers fertilize these ponds with manure, and feed fish with vegetation or invertebrates. It takes 20 years of training to become adept at all the techniques of disease control, fish feeding, manipulation of reproduction, and fertilization associated with these traditional forms of fish culture (Zweig, 1985).

A rapidly developing area of aquaculture is the use of algal mass for biofuels. This is an attractive route for producing fuel because the algae can also take up nutrient pollution from wastewater. In general, it is difficult to grow pure cultures of algae at industrial scales of production. Therefore, more natural and diverse systems are likely to have applications that are more practical. Sustaining algal production in these settings requires ecological understanding of factors that allow dominance of and production by desired algal species (Smith and Mcbride, 2015).

SUMMARY

1. Biodiversity of fishes relates to factors that operate at a variety of temporal and spatial scales. In general, fish communities are more diverse in tropical areas, in large drainage basins, in areas with high terrestrial productivity, where greater habitat diversity exists, and where more prey species exist.

2. Physiological ecology of fishes considers the energetic requirements of various processes, including osmoregulation, O_2 requirements, responses to temperature, food quality and quantity, and behavioral considerations.

3. Stock size, production, recruitment, and mortality are central aspects of fish populations that people use to manage fish populations.

4. A variety of methods are used to capture fishes depending on their size, the habitat being sampled, and reasons for obtaining data on the fishes.

5. Managers use many indices to assess fish populations and potential yields, including the RSD, which is the percentage of fish in a specified length or age class relative to the total number of fish estimated by population sampling.

6. Managers use a variety of regulations to control overexploitation of fish populations. These regulations require consideration of human dimensions and also generally involve a tradeoff between high numbers of small fish and fewer large fish.

7. Aquaculture of fishes requires understanding of aquatic ecology of fishes. Problems commonly encountered with aquaculture include hypereutrophy of fish culture facilities and diseases.

QUESTIONS FOR THOUGHT

1. Do lakes and streams commonly contain unused niches that can be exploited by fisheries managers wishing to improve sport fisheries?

2. Why are unregulated fisheries overexploited?

3. How might fisheries management for sport fisheries lower diversity of native fish populations?

4. Do the indices used by fisheries managers, such as RSD, have any ecological relevance?

5. Why are herbivorous and detritivorous fishes more often used for food in Asia than in North America?

6. Why should genetic diversity of fish stocks be an important aspect of aquaculture?

CHAPTER 24

Freshwater Ecosystems

Contents

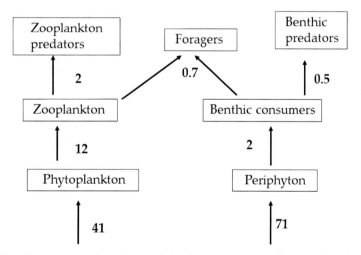

Figure 24.1 Graphic representation of one of the first accountings of energy flux through an ecosystem. Flux rates in g-cal cm^{-2} y^{-1}. *The data are for Cedar Bog Lake (Lindeman (1942)).*

Freshwater Ecology
DOI: https://doi.org/10.1016/B978-0-12-813255-5.00024-7

Raymond Lindeman (1942) described how fluxes of energy could characterize ecosystems in an extremely influential paper published posthumously and summarized in Fig. 24.1, and initiated the ecosystem view of ecological systems. Explicitly accounting for energy fluxes arose from the new ecological idea that an *ecosystem* is the sum of the biotic and abiotic parts of an environment. Covich (2001) provides a more comprehensive, contemporary definition:

> *Ecosystems are thermodynamically open, hierarchically organized communities of producers, consumers, and decomposers together with the abiotic factors that influence species growth, reproduction, and dispersal. These abiotic factors include the flow of energy and the circulation of materials together with the geological, hydrological, and atmospheric forces that influence habitat quality, species distributions, and species abundances. Energy flows through many species, and the way in which this flow affects the persistence of ecosystems is influenced by land-use changes, precipitation, soil erosion, and other physical constraints such as geomorphology.*

This holistic view of ecology is powerful because it allows for analyses of entire systems rather than subsets of systems. However, ecosystems are so complex we still need to reduce them conceptually to units manageable for study. In this chapter, we discuss general methods of approaching ecosystems (including trophic energy transfers, nutrient budgets, and the link between biodiversity and ecosystem function) and then focus on ecosystem properties in groundwaters, rivers, streams, lakes, and wetlands.

GENERAL APPROACHES TO ECOSYSTEMS

Some of the earliest attempts to reduce ecosystems to manageable units revolved around assigning organisms to *trophic levels*. The levels are decomposers, primary producers, primary consumers (herbivores), and higher levels of consumers (secondary, tertiary, etc.). For example, in Fig. 24.1, phytoplankton are primary producers, zooplankton are primary consumers, and zooplanktivores are secondary consumers. Movement of energy through these trophic levels has been a focal point of ecosystem research and can illustrate basic ecosystem concepts.

We discuss ecosystem concepts of energy flow using data from an early ecosystem study on Silver Springs, Florida (Odum, 1957; Odum and Odum, 1959). Biomass of the various trophic levels is considered first (Fig. 24.2A). The *biomass* of primary producers is 10 times greater than that of any other trophic level. This situation, where biomass of each tropic level is less than the level below it, forms what ecologists traditionally called a biomass pyramid because the base of producers has much more biomass than the top-level carnivores. A biomass pyramid with a broad base does not always occur, as production rate per unit biomass also plays a role in ecosystem structure. For example, a pond dominated with macrophytes may have much more

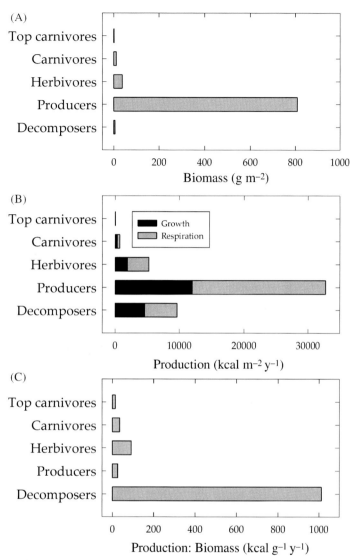

Figure 24.2 (A) Biomass, (B) production, and (C) production per unit biomass of Silver Springs, Florida. Note that production is broken into respiration and growth. *Data from Odum and Odum (1959).*

primary producer biomass than one dominated by phytoplankton, yet both could be equally productive and support higher trophic levels with similar amounts of energy.

The second community characteristic is *production*, or flux of carbon or energy through an ecosystem compartment. It is essential to remember that production is not equivalent to biomass. Biomass is an amount, and production is a rate. A common way to contrast these two values is to use the ratio of production to biomass (P:B). Decomposers in Fig. 24.2 illustrate the problem with assuming that high biomass is equivalent to high production. The biomass of the decomposers is relatively low, but their production is second only to that of the primary producers. This is because metabolic activity per unit biomass (P:B is a ratio that reflects this) is much greater for decomposers than for any of the other trophic levels for Silver Springs. Therefore, P:B can be thought of as an index of relative efficiency.

Net ecosystem production is a gauge of how much carbon all primary producers fix in an ecosystem. As we discussed in Chapter 12, gross ecosystem production—ecosystem respiration = net ecosystem production. Ecosystems can have a positive or negative net production based upon the balance between external carbon imports (*allochthonous*) and primary production within the ecosystem (*autochthonous*), and the relationship between the two is often expressed as the ratio of production to respiration (P:R).

SECONDARY PRODUCTION

As carbon moves up the food web, it is secondary production. Secondary production is tissue elaboration by heterotrophs regardless of its fate (e.g., whether it is consumed by predators, dies, and decomposes, or is molted off). Like primary production, it is a rate usually expressed as mass per unit area per unit time (e.g., $g\ m^{-2}\ year^{-1}$), and it represents actual energy available to higher trophic levels.

Estimates of secondary production can also link animal populations directly to other rates such as nutrient cycling. For example, if a given animal species' dry tissues are 11% nitrogen, and the secondary production of a population of that species is 10 g dry mass $m^{-2}\ year^{-1}$, we know that 1.1 g N $m^{-2}\ year^{-1}$ is cycling through that population $(0.11 \times 10 = 1.1)$. Production estimates are also relevant to population biology because they incorporate densities, biomass, individual growth rates, survivorship, and reproduction (Benke and Whiles, 2011).

Secondary production is the product of a series of ecological efficiencies. After an animal ingests food, it assimilates part of it and egests the rest. The difference between what is ingested and what is egested is *assimilation* (ingestion—egestion = assimilation). One aspect of assimilation is *assimilation efficiency*, which is assimilation/ingestion. Of the portion assimilated, an organism uses some for respiration and loses some through excretion of metabolic wastes. That which is left represents production, which the organism can apply to growth, reproduction, or storage. The efficiency at which assimilated materials are converted to production is the *net production efficiency*; the overall efficiency at which ingested materials are converted to production (e.g., assimilation efficiency × net production efficiency) is *gross production efficiency*.

Secondary production estimates can help assess food availability for managed species such as game fish and waterfowl (e.g., Roell and Orth, 1993), examine ecosystem responses to natural and anthropogenic disturbances (e.g., Nelson et al., 2017b), and quantify roles of consumers in energy flow and nutrient cycling (e.g., Stagliano and Whiles 2002; Evans-White et al., 2003). Production estimates vary considerably as a function of biomass and growth rates. Estimates for single species of freshwater invertebrates range from less than 1 mg dry mass m^{-2} year^{-1} up to extremes of approximately 3000 g dry mass m^{-2} year^{-1} for species with very high biomass and rapid growth rates (Benke, 1993; Huryn and Wallace, 2000; Benke and Huryn, 2010). Likewise, estimates for fishes and amphibians are highly variable depending on biomass and growth rates.

Ecologists estimate secondary production by following changes in density and individual weights of a *cohort*, which is a group of individuals that start life at the same time. The relationship between density and individual weight over time can be plotted as an *Allen curve* (Allen, 1951; Benke and Huryn, 2006), which can be used to actually estimate production and is also useful for conceptualizing production (Fig. 24.3).

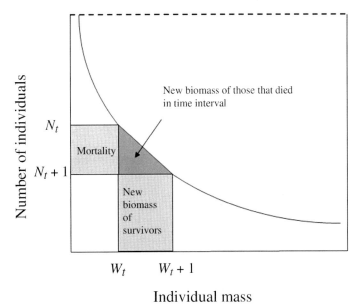

Figure 24.3 An Allen curve showing changes in density and individual weights of a cohort of animals over time. N is number and W is individual weight; subscript t and $t+1$ indicate points in time. Over time, the number of individuals declines (moving down the Y-axis; difference between N_t and N_{t+1} = mortality) and individuals grow (moving left to right on the X-axis). The sum of the new biomass created by survivors (represented by the lower light gray rectangle) and the biomass produced in the interval but lost to mortality (darker gray triangle) represent production for the time interval depicted. Total production for the cohort is the area under the curve; note that the area under the dashed line at the top would represent total production for the cohort if no mortality occurred.

Production for the cohort is the total area under the curve; there are numerous ways to estimate this, some of which we describe here.

One of the simplest ways to estimate production is the *instantaneous growth method*. This method only requires estimates of average biomass and individual growth rates over a time interval, and can be used in situations where a clear cohort cannot be followed, or as a simplified approach when a cohort can be followed. Ecologists create biomass estimates from quantitative field sampling, but estimating growth rates can be more difficult, particularly in cases where rapid growth and/or overlapping cohorts occur. When growth rates cannot be determined through field sampling, researchers can derive them through laboratory or field rearing studies, or mark–recapture procedures. Then they use changes in individual weight over time to estimate instantaneous growth as follows:

$$G = \ln\left(\frac{W_{t+1}}{W_t}\right)$$

where G is the instantaneous growth rate, W_t is the weight at the beginning of the interval, and W_{t+1} is the weight at the end of the interval. The growth rate can be standardized to a daily rate by dividing by the number of days in the interval. Once we obtain G and average biomass for the interval, we estimate interval production as the product of the two. The instantaneous growth method can be used to estimate production using data on a population of stream-dwelling caddisflies (Example 24.1). The increment summation method can give better results when cohorts are clearly

---

Example 24.1 Calculating secondary production with the instantaneous growth method

Using the data in Table 24.1, calculate the secondary production of the caddisfly population. Because a clear cohort structure is present, G can be calculated from the change in average individual weights between sampling intervals. For example, for the first interval of 18 May to 1 June:

$$G = \ln\left(\frac{0.057}{0.021}\right) = \ln(2.714) = 0.998 \text{ interval}^{-1}$$

Using biomass estimates for the same interval (5.94 mg on 18 May and 12.94 mg on 1 June), average biomass is 9.44 mg/m^2 and thus production for the interval is:

$$P = 0.988 \times 9.44 = 9.42 \text{ mg}/m^2$$

While the instantaneous growth method requires less detailed information and is easier to calculate, production estimates using this approach are generally considered less precise than those obtained with cohort procedures such as the method described in Method 24.1. However, in some cases, the instantaneous growth method may be the best option because of unclear cohort structure.

Table 24.1 Data and production calculations for *Brachycentrus spinae*, a stream-dwelling caddisfly with a 1-year life cycle

Date	Days in interval	Density (no./m²), N	Avg. individual weight (mg), W	Biomass (mg/m²), N × W	Weight change ($W_{t+1} - W_t$) Δ weight	Avg. N	Production (mg/m²), Avg. N × Δ weight
18 May	14	283	0.021	5.94	0.036	255.0	9.2
1 June	12	227	0.057	12.94	0.031	204.5	6.3
13 June	16	182	0.088	16.02	0.085	160.5	13.6
29 June	14	139	0.173	24.05	0.179	124.0	22.2
13 July	13	109	0.352	38.37	0.588	98.5	57.9
26 July	22	88	0.940	82.72	0.266	74.0	19.7
17 August	13	60	1.206	72.36	0.590	54.0	31.9
30 August	16	48	1.796	86.21	0.025	42.5	1.1
15 September	18	37	1.821	67.38	1.378	32.0	44.1
3 October	42	27	3.199	86.37	0.358	20.0	7.2
14 November	23	13	3.557	46.24	1.074	11.0	11.8
7 December	51	9	4.631	41.68	2.222	6.5	14.4
27 January	21	4	6.853	27.41	1.624	3.5	5.7
17 February	24	3	8.477	25.43	3.071	2.5	7.7
13 March	45	2	11.548	23.10	3.252	1.0	3.3
27 April		0	14.800	0.00			

Annual average biomass = 41.0   Total production = 256.1 mg m^{-2} $year^{-1}$

Dates are days when samples were collected from the stream; May is used as a start date because this is when newly hatched larvae first appear. Production is calculated using the increment summation method.
Source: Data are from Ross and Wallace (1981).

identifiable (Method 24.1). A third method known as the size-frequency method is now also commonly used. This method involves constructing an "average cohort," somewhat like a life table, and this method applies in situations where a cohort can or cannot be followed. Details of this method are beyond the scope of this text, but Benke and Huryn (2006) present them along with other secondary production methods.

Annual *production to biomass ratios* (P:B) are useful for assessing growth and turnover and explain why biomass is not always directly related to production (e.g., high biomass does not necessarily mean high production). For the example in Table 24.1, P:B is 256.1/41 = 6.2, which is fairly typical for an invertebrate with a 1-year life cycle. Actual *turnover time* (actual biomass turnover rate or replacement rate) can be calculated by dividing the number of days in a year by the annual P:B; 365/6.2 = 59.0 days.

METHOD 24.1 Advanced: calculating secondary production using the increment summation method

Fig. 24.3 shows how production of a cohort for a given time interval is a trapezoid underneath the Allen curve, which is composed of the biomass produced by those that survived and those that died during the time interval. The trapezoid for a given time interval can be calculated from densities and individual weights by multiplying the average number of individuals present during the time interval by the change in individual weight over the interval:

$$\frac{N_t + N_{t+1}}{2}(W_{t+1} - W_t)$$

where N_t is the number of individuals at the beginning of the time interval and N_{t+1} is the number at the end of the interval; W_t is average individual weight at the beginning of the interval and W_{t+1} is average individual weight at the end. Using data for the first interval in Table 24.1 yields:

$$\frac{283 + 227}{2}(0.057 - 0.021) = 255 \times 0.036 = 9.18 \text{ mg m}^{-2} \text{ interval}^{-1}$$

This is production for a given time interval; total production is the sum of the interval values; in this case 256 mg m^{-2} year^{-1}. This is the *increment summation method* (Waters, 1977; Benke and Huryn, 2006), which is accurate, but only applies when researchers can follow a cohort through time. In this case, the estimate from the instantaneous growth method (Method 24.1, 9.42 mg m^{-2} interval^{-1}) and the increment summation method are similar; this is because a clear cohort structure is present, and thus accuracy of both methods is high. A quick test for method suitability is to plot an Allen curve (density vs individual mass) with the data; if density does not decrease and individuals do not get larger through time, the increment summation method will not work.

Annual P:B is a function of growth; faster growing species have high P:B values and annual P:Bs can range from 1—2 for very slow-growing taxa to over 100 for small, rapidly growing species such as some midges and small crustaceans (Benke, 1993). For obvious reasons, biomass can be a poor proxy for secondary production.

ENERGY FLUXES AND NUTRIENT CYCLING

Efficiency relates to the second law of thermodynamics, which roughly states that all processes must lose some energy. However, all processes could be 99.999% efficient, or any other value less than 100%, according to the Second Law. In nature, ecological transformation is considerably less efficient than 100; organisms may be able to turn as much as 75% of the mass of food consumed into biomass or less than 10%, with the remainder lost to respiration. Aquatic herbivorous insects assimilate approximately 30%−60% of the materials they ingest, whereas aquatic predatory insects assimilate anywhere from 15% to 77% (Wiegert and Petersen, 1983).

An organism's efficiency of obtaining and using energy is dependent on the sum off efficiencies of all metabolic pathways, including the amount of energy required to obtain, ingest, and assimilate food, and the energy required for reproduction and survival. Assimilation efficiency is influenced by a variety of factors including food quality (e.g., C:N, % refractory materials such as lignin in plant materials), digestive system morphology and physiology, gut microbiome, and temperature. In general, endothermic organisms are more efficient at assimilating materials than ectotherms, but less efficient at converting materials to production (e.g., production efficiency) because of higher metabolic demands. All of these efficiencies can also vary with abiotic factors over space and time, such as changes in temperature.

The data in Fig. 24.2B from Silver Springs Florida suggest that about half all the carbon taken up by producers is translated into biomass available for the next trophic level, 10% of the decomposer's production becomes biomass, and about 30% of the herbivore's biomass becomes available to predators (Odum and Odum, 1959). Interestingly, this early energetic study on Silver Springs still has relevance; Quinlan et al. (2008) used updated sampling to document a shift in the basic ecosystem energetics related to human influence in the area. The primary production has shifted from dominance by macrophytes to benthic algae as nitrate levels have doubled in the water.

NUTRIENT BUDGETS

Although energy flux makes sense as a primary unit when discussing production of food for secondary consumers and above, it may not be as important for primary producers, herbivores, and microbes, which can be limited by nutrients other than carbon (see Chapter 17). Stoichiometric analysis indicates that many primary food sources (detritus and primary producers) are so poor in nitrogen and phosphorus that primary consumers are not carbon limited (Sterner and Elser, 2002). For example, nitrogen or phosphorus can limit rates of fungal and bacterial activity and growth in streams (Suberkropp, 1995; Tank and Webster, 1998). Also, pelagic bacteria are important consumers of phosphorus and compete for it with phytoplankton (Currie and Kalff, 1984). Hence, consideration of energy limitation alone does not always provide an accurate description of functional ecosystem relationships. Therefore, another branch of ecosystem science is concerned with *nutrient budgets*, which quantify fluxes in nutrient cycles.

We present two examples of nutrient budgets, a flux diagram and a table that tallies total inputs and outputs. A flux diagram allows representation of fluxes that occur within a system. Often, input and output budgets treat the ecosystem as a black box and account for only materials entering and leaving the system.

The nutrient flux diagram is exemplified by the description of a single nutrient budget that was undertaken to describe the importance of nitrogen fixation by a dominant cyanobacterium (*Nostoc*) in a cold-water pond (Fig. 24.4). Nitrogen fixation was responsible for only about 5% of the nitrogen input to the pond. However, this nitrogen input was directly into biota in the pond, whereas much of the nitrogen input that entered in the spring inflow rapidly flowed out of the system. A similar approach demonstrated that nitrogen-fixing organisms could relieve nitrogen limitation in wetlands (Scott et al., 2007). This flux accounting approach can also be used to assess net fate of anthropogenically derived nitrogen inputs to watersheds (Gardner et al., 2016). One interesting feature of Fig. 24.4 is the loss of nitrogen to insect emergence. This flux can be important with respect to nearby terrestrial habitats subsidizing the nutrient input there in both lakes and streams (Gratton and Vander Zanden, 2009).

Isotopic tracers can help delineate nutrient budgets and account for internal nutrient cycling fluxes. Hutchinson and Bowen (1947) pioneered this approach when they added radioactive phosphate to a lake and traced the fate of the materials. They found that the macrophytes in a pond were very good competitors for the added phosphate

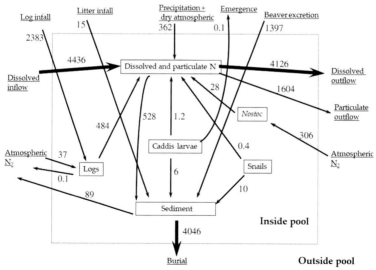

Figure 24.4 Diagram of nitrogen fluxes in a cold-water spring (Mare's Egg Spring, Oregon) dominated by *Nostoc*, a nitrogen-fixing cyanobacterium. Fluxes are in moles N/year. Note: fluxes are measured so budget does not balance. *Redrawn from Dodds and Castenholz (1988).*

relative to the phytoplankton. These approaches are powerful because they allow determination of gross fluxes (internal dynamics) in addition to net fluxes that can be determined by a budgeting approach.

The method of accounting for influx (inputs) and outflux (outputs) of nutrients is demonstrated with a general analysis of nutrient budgets to assess the importance of various sources and losses of nitrogen across ecosystem types (Table 24.2). Such budgets are central to understanding effects of nitrogen contamination and related ecosystem processes. For example, budgets are necessary to determine the efficacy of wetlands at removing nutrient pollution from wastewaters. The data in Table 24.2 suggest that some wetland types are more retentive of nitrogen, and that burial or assimilation in biomass is a significant component of many nitrogen budgets. Such approaches demonstrate that nitrogen fixation rarely accounts for more than 5% of nitrogen inputs in streams (Marcarelli et al., 2008), with the exception of some oligotrophic streams (Kunza and Hall, 2014). In some systems, nitrogen fixation (generally by cyanobacteria) can be a major nitrogen input, and denitrification can be a notable loss. Bratt et al. (2017) also used the approach to estimate nitrogen and phosphorus export from urban watersheds where fall leaves were found to be a significant nutrient source. Wetzel (2001) synopsized a more in-depth analysis of such budgets for streams and lakes for a variety of elements.

BIODIVERSITY AND ECOSYSTEM FUNCTION

The biodiversity of ecosystems can link to rates of ecosystem processes and biological structure of ecosystems in a variety of ways. For example, diverse assemblages of net-spinning caddisflies filter higher portions of suspended particles from the water column than monocultures because of facilitation among different species (Cardinale et al., 2002). However, relationships among biodiversity and ecosystem functions can be variable and complex. A study of functionally and competitively dominant leaf-shredding *Pycnopsyche* caddisflies found a negative relationship between species diversity and leaf litter decomposition because *Pycnopsyche* reduced the presence of other shredder species (strong competitor) and decomposed large amounts of litter and was functionally dominant (Creed et al., 2009).

Given the current biodiversity crisis, particularly in freshwater habitats, efforts to establish relationships between *biodiversity* and *ecosystem function* have been a central issue in freshwater ecology (Schulze and Mooney, 1994). The first problem is how to define ecosystem function. Function can refer to rates of basic processes, such as photosynthesis, respiration, denitrification, or phosphorus retention. Function can be more specific, such as production of plant biomass for herbivores. *Functionally redundant* species may be at the crux of the diversity-function link. Functionally redundant species are those that can be removed from a community without a change in a specific

Table 24.2 Nitrogen budgets of some aquatic systems showing the importance of nitrogen fixation and denitrification

	Inputs				Outputs			Reference
	Inflow	Litter deposition	Atmospheric	N₂ fixation	Outflow	Accumulation	Denitrification	
Sphagnum bog	0	0	1.4 (95)	0.05 (5)	0.2 (15)	0.9 (71)	0.2 (13)	Urban and Eisenreich (1988)
Riparian zone in agricultural area	2.9 (56)	?	1.2 (24)	1.1 (20)	1.3 (33)	5.2 (54)	3.2 (33)	Lowrance et al. (1984)
Groundwater fed fen in agricultural area	2.1 (28)	?	4.2 (56)	1.3 (17)	21 (24)	6.6 (75)	0.1 (1)	Koerselman et al. (1990)
River fed fen in agricultural area	0.7 (14)	?	4.4 (82)	0.2 (4)	1.0 (20)	3.8 (76)	0.1 (3)	Koerselman et al. (1990)
Eutrophic lake (Clear Lake, California)	0.3 (45)	0	0.06 (9)	0.3 (45)	0.4 (55)	0.3 (40)	0.03 (5)	Horne and Goldman (1972, 1994)
Eutrophic lake (Lake Okeechobee, Florida)	3.2 (58)	0	1.7 (31)	0.6 (11)	1.6 (29)	3.0 (53)	1 (18)	Messer and Brezonik (1983)
Spring pool with dominant *Nostoc*	27 (50)	24 (42)	2 (4)	2 (4)	36 (59)	25 (41)	0.6 (1)	Dodds and Castenholz (1988a, 1988b)
Desert stream	0.47 (92)	?	?	04 (8)	49 (89)	05 (11)	?	Grimm and Petrone (1997)
Forest stream	11 (73)	3 (22)	?	1 (5)	11 (75)	4 (25)	?	Triska et al. (1984)
Alpine streams/lakes	0.52 (57)	0	0.39 (43)	?	3 (99)	?	0.025 (1)	Baron and Campbell (1997)

Values in g N m⁻² y⁻¹. % of total input or output in parentheses. Accumulation includes burial and storage in plant parts. When only accumulation is listed, the value includes burial + denitrification.

ecosystem process. Functional redundancy is dependent on ecosystem process, habitat, and time. Many continental habitats have diverse assemblages with considerable redundancy.

We know less about the links between diversity and ecosystem processes driving nutrient cycling because we are just learning about microbial diversity and the degree of redundancy of specific functional groups (Meyer, 1994). Although Hutchinson made early arguments about the functional redundancy of phytoplankton species with regard to the paradox of the plankton (see Chapter 17), the degree of functional separation of different phytoplankton species has not been established (Steinberg and Geller, 1994).

Finlay et al. (1997) argue that microbial diversity is never so impoverished in natural communities that biogeochemical cycling is seriously altered. Delgado-Baquerizo et al. (2016) found evidence for lack of redundancy for bacterial functions in mesocosm experiments, although Lear et al. (2014) found less compelling evidence for functional redundancy in bacteria from freshwater ponds. Diversity is also related to variance in ecosystem properties (France and Duffy, 2006). Experiments with multiple microbial communities assembled differently correlated variation in respiration rate with biodiversity. Respiration rate was similar among highly diverse communities, but varied widely among less diverse communities (Morin and McGrady-Steed, 2004).

Benthic animal diversity can have strong influences on processes of material exchange between the water column and the benthic zone (Covich et al., 1999; Capps et al., 2015a,b). Benthic animals can be important in energy flow and nutrient cycling (Fig. 24.5) and different species alter nutrient flux in different ways (e.g., burrowing, digging tunnels, stirring up sediments, actively pumping oxygenated water into the sediments, and processing different types of benthic materials). A study of a tropical river with a diverse fish assemblage indicated that loss of just one large migratory detritivorous fish species significantly decreased downstream transport of carbon (Taylor et al., 2006). This ingenious study used a fence along the middle of the stream and excluded or included the fish of interest. The data suggest that traits of a single species can alter ecosystem function, even in a diverse system where functional redundancy is high. Another clever study used ^{15}N labeled algae as food for the unionid mussel *Lampsilis siliquoidea* and subsequently returned them to the river they were taken from. The label in different food web compartments was sampled over time after addition, and as much as 74% of the nitrogen came from mussels (Atkinson et al., 2014).

Some of the best data for functional separation of ecosystem processes related to biodiversity derive from studies of two shrimp species that break down leaf litter in Puerto Rican streams (Covich et al., 1999). The two species (*Xiphocaris elongata* and *Atya lanipes*) can both degrade leaf litter, but breakdown is significantly more efficient and the streams are more retentive of organic particles when both species are present.

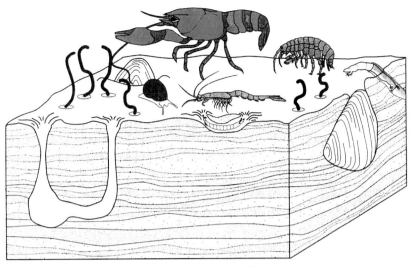

Figure 24.5 Benthic macroinvertebrates that burrow into layered sediments and accelerate nutrient cycling and movement of materials into the water column. Burrowing bivalves, crayfish, tubificid worms, and aquatic insect larvae mix O_2 into the sediments with their activities. Surface-dwelling invertebrates increase turnover of microbial communities and increase sediment suspension. *Reproduced with permission from Covich et al. (1999).*

Atya does not break down intact leaves as rapidly as *Xiphocaris*, but it scrapes microbes from the leaves and filters fine particles from the water column more efficiently. Particulate transport is highest in streams in which both species of shrimp are rare because predatory fishes are present (Pringle et al., 1999). Such a relationship between biodiversity and ecosystem function may occur in mainland streams as well (Jonsson et al., 2001). The Puerto Rican example is but one instance of how facilitation among diverse species can mediate species effects on ecosystem processes, a finding replicated for marsh plants (Bertness and Hacker, 1994; Bertness and Leonard, 1997).

Some ecosystem functions appear not to be replaceable. The catastrophic loss of frogs to chytridiomycosis, a fungal pathogen, in a high elevation Panamanian stream led to significant alterations to ecosystem function (Whiles et al., 2013). The tadpole community in the studied stream was diverse, with up to 20 species inhabiting the stream. Some of the tadpoles were herbivorous, others detritivorous, and yet others mixed feeders. Soon after they disappeared from the system from the disease, algal biomass and detrital stocks increased, and nitrogen uptake decreased by half. The surprise was that no other invertebrates moved in or increased their population sizes to assume the roles the tadpoles were playing, even 6 years after the extirpation of the tadpoles from the system (Rantala et al., 2015). These results indicated that the tadpole biodiversity strongly influenced ecosystem function in unique ways that were not replaceable on ecological timeframes.

A more in-depth meta-analysis of the relationships between diversity and ecosystem function has become possible as researchers publish more studies. A report commissioned by the Ecological Society of America (Hooper et al., 2005) expressed certainty that (1) functional characteristics of species influence ecosystems, (2) human-caused species extinctions and introductions have influenced ecosystems in ways that are costly to humans, (3) effects of species loss or shifts in community structure on ecosystems are contingent upon local conditions, (4) some ecosystem properties are decoupled from diversity because of functional redundancy, and (5) as spatial and temporal variability increases, more species are required to generate a stable supply of ecosystem services. Analysis of 446 terrestrial and aquatic studies found the strongest effects on primary producer and primary consumer abundances, as well as decomposer activity (Balvanera et al., 2006). Analysis of 111 studies indicated that species identity was the single most important determinant of ecosystem effects related to species loss (Cardinale et al., 2006).

GROUNDWATER ECOSYSTEMS

Groundwater ecosystems often rely on organic material derived from surface habitats (Gibert et al., 1994a,b). Alternatively, chemoautotrophic processes, such as sulfide oxidation (e.g., Highlight 14.2), ammonium oxidation, or use of iron or manganese as electron donors, can form the basis of autotrophic production in some systems. Investigation of respiration rates of groundwater sediments suggests that respiration rates decrease with depth (Fig. 24.6). Deep groundwater sediments have the slowest rates of biological activity of any known habitats.

Groundwater ecosystems occur across continua of permeability and average interstitial space. The size and connectivity of the pores or channels through an aquifer can control the transfer of materials through the aquifer and limit the size of the organisms that inhabit the aquifer. Movement of water and organisms has direct effects on energy flux through ecosystems. In aquifers with very fine pore sizes, only dissolved materials, fine particles, bacteria, and very small protozoa can move through the sediments. Thus, only bacterial "producers" and primary consumers are present. Karst aquifers have the largest channels, but there are still few trophic levels relative to streams and lakes; however, there are more trophic levels in karst aquifers than in those with fine pore sizes. Predatory fishes, amphibians, crayfish, or planarians can be the top carnivores; four trophic levels may be the highest number found in karst groundwaters (Culver, 1994), but usually there are less.

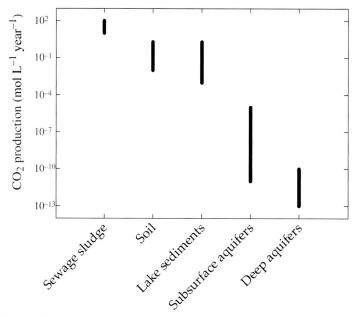

Figure 24.6 Ranges of documented respiration rates from various types of sediments. *Data from Kieft and Phelps (1997).*

Other ways to classify groundwater ecosystems are by their depth or by their degree of connectivity with surface waters and terrestrial habitats. Connection of groundwaters and streams is dependent on the type of substrata and spatial and temporal scales (Harvey and Wagner, 2000). Over the short term (mainly less than a few months) in aquifers with low hydraulic conductivity, only shallow groundwater interacts with lakes, streams, and wetlands. Over longer periods, deep groundwater can have considerable interaction with surface waters.

Groundwater can link lakes across the landscape over years. In northern Wisconsin, lakes occur in sandy glacial outwash, and groundwater links many of them (Kratz et al., 1997). Drought leads to changes in groundwater flow and increases in ion concentrations in some lakes. It can take up to 5 years for drought effects on ion concentrations to move through lakes as influenced by groundwater dynamics (Webster et al., 1996; Webster et al., 2000). Groundwater effects in these lakes can sometimes override the effects of climate, depending on lake position (Baines et al., 2000).

Given the difficulty of sampling groundwater habitats, detailed trophic analyses and energetic or nutrient budgets are rarer than in surface waters. However, knowledge of groundwater habitats increased tremendously in the past two decades. We are just beginning to understand a virtually undescribed group of biomes.

One of the most interesting developments in groundwater ecology is the documentation of deep microbe-dominated habitats. Geological processes give rise to hydrogen gas and this can combine with carbon dioxide to generate methane and energy. We discussed these ecosystems in the chapter on extreme habitats (Chapter 15). These habitats are the only that do not depend upon photosynthesis or the redox gradient ultimately generated when the part of the biosphere that is oxidized by photosynthesis is in interface with the lower redox ions generated in Earth's crust.

STREAMS AND RIVERS

The major concepts associated with streams we consider here are (1) autochthonous versus allochthonous production, (2) inverted biomass pyramids (Allen's paradox), (3) nutrient spiraling, (4) the flood pulse concept, (5) the concept of functional processing zones, and (6) the river continuum concept (RCC). These exemplify some of the key features of stream ecosystem science.

The concept of *allochthonous* (organic material provided from outside the system) versus *autochthonous* (organic material from photosynthetic organisms within the system) production has been stressed in streams because of the potentially strong influence of terrestrially-derived organic material and a substantial standing stock and production of periphyton (Minshall, 1978) and macrophytes (Hill and Webster, 1983) in some systems. The source of organic material is important because different invertebrates specialize on different types of carbon (Cummins, 1973), and varied sources of carbon can alter pathways of carbon transfer through the food web. For example, invertebrates that process leaf litter should provide important routes for energy flux into the food web in a small, forested stream (allochthonous input). Exclusion of litter from forest streams has a profound effect on stream invertebrate communities (Wallace et al., 1999). Wood inputs can be tremendous, but not all wood is available to consumers because few animals are equipped to digest it. In the Queets River, Washington, most of the large wood is less than 50 years old, but some of the buried logs are up to 1,400 years old (Hyatt and Naiman, 2001).

The relative contributions of primary production and external sources of organic carbon can be difficult to identify. In small streams that are heavily wooded, the input of leaves and wood is high, and shading limits primary production; some slow-growing mosses may be abundant. In larger streams, algal biomass is high when sufficient light can reach the bottom. However, dissolved and particulate organic carbon enters the stream from the surrounding terrestrial areas and from upstream. One way to establish the relative importance of internal versus external supplies of carbon is to compare the respiration and photosynthesis occurring in the stream (*stream metabolism*). These metabolic estimates indicate trophic state (Dodds, 2006).

Estimates of stream metabolism can be used to determine the ratio of photosynthesis to respiration (P:R), which serves as an index to the degree of autotrophy (relative autochthonous production) in the system. Researchers have mostly used two methods to make such estimates based mainly on rates of O_2 production and consumption: isolation of shallow benthic substrata in sealed recirculating chambers and measurements of whole-stream diurnal O_2 flux (see Chapter 12). In general, chamber methods have indicated that primary production exceeds respiration in well-lighted streams (Minshall et al., 1983; Naiman, 1983; Bott et al., 1985). Whole-stream estimates suggest that production over a 24-hour period rarely exceeds respiration, and that P:R is usually less than 1, even in streams that receive substantial sunlight (Young and Huryn, 1999; Mulholland et al., 2001; Bernot et al., 2010). The discrepancy between these two methods occurs because the chambers include only the top layer of benthic organisms, whereas the whole-stream methods include significantly more subsurface respiration (the influence of the hyporheic component). Because hyporheic metabolic activity links to instream O_2 dynamics, it makes sense to include it in estimates of whole system metabolic activity.

The trophic dynamics of stream invertebrates and the relationship between standing stocks (biomass) and production have received attention with regard to stream ecosystems. Observations of invertebrate biomass have yielded examples of greater biomass of secondary and higher consumers than of primary consumers. This inverted biomass pyramid has been termed the Allen Paradox following the observation that fishes in a stream required 100 times more benthic prey than was available at any one time (Allen, 1951; Hynes, 1970). The Allen Paradox is an example of problems that can arise when researchers assume that biomass is directly proportional to production. The production of primary consumers per unit biomass can be very high because of high turnover (growth rates) and can support the secondary and tertiary consumer biomass, even when the biomass of primary consumers is relatively low (Allan, 1983; Benke, 1984). Such analyses have become very detailed, including description of carbon flux associated with each species in invertebrate communities. This detailed description involves determination of the *trophic basis of production*, or the relative contributions of individual food sources to the production of each species (Benke and Wallace, 1980).

Materials cycle as they move downstream. Cycling of materials in unidirectional flow environments is termed *nutrient spiraling* (Webster, 1975). Molecules are in the water column for an average amount of time while they move downstream. They are then taken up or adsorbed and move downstream more slowly. Therefore, the nutrient cycle that an ecologist would typically conceptualize as a wheel in lakes becomes a spiral in streams. The spiral length (S) for a nutrient is the sum of the distance that an average molecule travels in the water column in the dissolved form (S_w) and how far it is transported in the primary particulate compartments (algae, microbes, suspended particles, and animals; S_p):

$$S = S_w + S_p$$

The relationship can be described graphically (Fig. 24.7) or in more mathematical detail (Newbold et al., 1981). In practice, modeling spiraling length can be difficult (Stream Solute Workshop, 1990) because understanding the movement of nutrients and environmental contaminants through streams requires detailed descriptions of the processes of dilution, uptake, and remineralization. Description of the influence of the hyporheic zone may be particularly important (Mulholland and DeAngelis, 2000).

Some generalizations are possible with regard to spiral length: length should be greater with greater stream discharge, increased average water velocity, decreased uptake rates per unit area, increased disturbance of the benthic zone, and increased insect drift. Spiral lengths of inorganic nutrients (S_w) that are in high demand are generally short; uptake lengths of phosphate and ammonium are often less than 100 m (Mulholland et al., 1990; Hart et al., 1992; Butturini and Sabater, 1998). The concept of resource spiraling is powerful because it allows comparison of how nutrients are retained by a variety of streams (Elwood et al., 1983), a process that is particularly important in small streams (Peterson et al., 2001; Mulholland et al., 2008).

Two more metrics that are important relate to spiral lengths and can help parse out various factors influencing spiral length. These additional metrics include uptake rate per unit area of stream bottom and average movement of nutrients toward the benthic

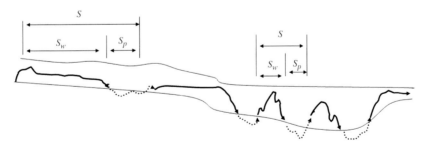

Figure 24.7 A diagram of nutrient spiraling in streams. S is the total spiral length, S_p is the time spent in particulate form in water column or the benthic zone, and S_w is the average time spent dissolved in the water. Average velocity is greater in the riffle on the left, so spiral length is greater than in the pool at the right.

zone (Stream Solute Workshop, 1990). The average rate at which a molecule moves toward the bottom of the stream is uptake velocity (v_f):

$$V_f = \frac{uz}{S_w}$$

where u is the water velocity and z is the average depth. Then V_f can be used to calculate uptake rate in mass per unit area per unit time (U).

$$U = v_f C$$

where C is the concentration of the nutrient being taken up. Baker and Webster (2017) provide methods in more detail.

Understanding how streams process nutrients as they move downstream has become particularly important because of the global increase in episodes of near-shore hypoxia fueled by nutrient enrichment from streams (Diaz and Rosenberg, 2008). The nutrients, most commonly coming from agricultural runoff, enter coastal habitats where they stimulate phytoplankton blooms along the coast. When these blooms sink, they in turn stimulate heterotrophic decomposition, which can deplete O_2 in deeper marine waters. While river networks retain some of the nutrients that enter them (Alexander et al., 2008), retention becomes less efficient as loading increases (O'Brien et al., 2007; Mulholland et al., 2008). Nutrient retention is spatially and temporally variable, and there is some indication that instream biological processes are more important in nitrogen than phosphorus retention (Marti and Sabater, 1996).

The *RCC* (Vannote et al., 1980) has been one of the most influential ideas in stream ecosystem theory. This concept views flowing waters in temperate systems as a connected continuum from small, forested headwater streams to large rivers. It uses the associated gradient in abiotic and riparian characteristics to make specific predictions about the biological community. The concept posits that dense canopy cover and low light found in small streams supply leaf material as the primary carbon source, and shredders dominate the invertebrate community. Empirical data support this idea (e.g., Rosi-Marshall et al., 2016). As the stream increases in width downstream, light increases and leaf input becomes less important. Benthic algal productivity and fine organic material washed from upstream contribute most heavily to production of available carbon, and grazing and collecting invertebrates dominate. In the largest rivers, benthic production is low, suspended particulate material is high, zooplankton and phytoplankton can become established in the water column, and collectors dominate the invertebrate community (Fig. 24.8).

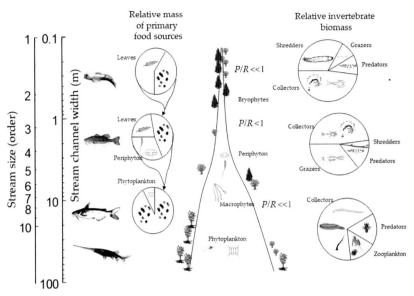

Figure 24.8 Diagram of the river continuum concept. See text and Table 24.3 for explanation. *Modified from Vannote et al. (1980).*

This original model is a simple description of the possible ecosystem parameters that can vary from headwaters to large rivers (Sedell et al., 1989). An expanded view (Table 24.3) considers other abiotic (e.g., temperature and inorganic substrata) and biotic (e.g., woody debris) factors. This model is powerful in part because it encourages consideration of stream ecosystems across landscapes and as influenced by the watershed (including terrestrial and in-stream processes) above each point.

There are clear exceptions to the generalizations of the RCC (as there are to any general ecological model). The original paper acknowledges the specificity of the model for forested streams. Streams that are frequently disturbed may not exhibit spatial trends in functional feeding groups of insects (Winterbourn et al., 1981). Large rivers may be as influenced by carbon input from their side channels as upstream (Sedell et al., 1989). Also, small streams can flow directly into oceans without ever moving into higher order rivers. Further, grassland, desert, and tundra streams and rivers should have limited leaf input in the smallest streams (Gurtz et al., 1988). A study of longitudinal patterns along a North American grassland stream showed different overall patterns than the RCC, but still supported the basic premise that changes in the physical template along stream continua result in predictable changes in consumer groups (e.g., Whiting et al., 2011). Finally, the idea of serial discontinuity (Ward and Stanford, 1983) suggests that dams disrupt the expected natural river continuum as has been well documented for the Colorado

Table 24.3 Summary of an expanded view of the river continuum concept

Feature	Headwaters	Middle reaches	Large rivers
Physical			
Stream order	1–3	4–7	>7
Discharge	Low	Medium	High
Flooding	Flashy, short, unpredictable	Medium	Regular, predictable
Gradient	High	Medium	Low
Temperature	Cool, constant when shaded	Moderate, variable	Warm, constant
Substrata	Rocky, large wood	Intermediate	Silt, sand
Riparian canopy	Dense, covering stream channel	Above stream channel open	Important only in flood zone
Turbidity	Low	Low	High
Light	Low	High	Low
Metabolic			
Photosynthesis (P)	Low	High	Moderate–low
Respiration (R)[a]	?	?	?
P/R	$\ll 1$	<1	$\ll 1$
Organic carbon	Coarse	Intermediate	Fine
CPOM/FPOM ratio[b]	>1	<1	$\lll 1$
Woody debris	Large wood, debris dams	Along margins	Relatively rare, but an important substrate in sandy or silty rivers (Haden et al., 1999)
Producers			
Periphyton	Moderate	High	Low
Phytoplankton	Low	Low	Relatively high
Macrophytes	Low, but mosses may predominate	Moderate	Low except in side pools
Consumer invertebrates			
Shredders	High	Moderate	Low
Filter feeders	Low	Moderate	High
Scrapers/grazers	Moderate	High	Low
Collector gatherers	Moderate	Moderate	High
Predators	Moderate	Moderate	Moderate
Fishes			
Diversity	Low-cool water	Medium	High-warm water
Sight feeders	High	High	Low
Prey	Invertebrates	Invertebrates/fishes	Invertebrates/plankton/fishes

[a]Relative patterns not established.
[b]CPOM, coarse particulate organic matter; FPOM, fine particulate organic matter.

River below Glen Canyon dam (Cross et al., 2013). Dams can cause settling of organic particles and boost populations of zooplankton and phytoplankton down-stream. Exceptions notwithstanding, the conceptual model provides one way for stream ecologists to think critically about streams in their ecosystem context (Cummins, 1977; Cummins et al., 1984).

The *flood pulse concept* promoted a paradigm shift from viewing floods as distur-bances that alter an ecosystem that is otherwise at equilibrium to viewing flooding as a characteristic property of river and stream ecosystem function. A central idea of the flood pulse concept is that lateral habitats, particularly in large rivers, connect with the main channel in times of floods and disconnect during lower flow periods. This means that local processes have strong influences on lateral habitats for much of the year, but they occasionally can strongly influence ecosystems and communities of the main river by upstream processes and local lateral effects (Junk et al., 1989). For example, the length of time of inundation substantially influences sediment respiration (Vallett et al., 2005), and can significantly influence biogeochemical cycles.

Flood pulses can be important in smaller braided rivers as well. These rivers have multiple side channels that are variously connected depending upon discharge, and their degree of connection alters macroinvertebrate diversity in both backwater and main channel habitats (Whiles and Goldowitz, 2005; Malard et al., 2006). Regular flood pulses in tropical rivers lead to higher fish species richness, more stable avian populations, elevated rates of forest production (Jardine et al., 2015), and predictable seasonal changes in consumer–resource interactions (Winemiller et al., 2014). Floodplain areas can serve as important locations for fish reproduction and growth (Jardine et al., 2012). Rivers differ in the responses to flooding of their side channels. Roach et al. (2014) found rivers with high sediment loads had greater rates of heterotrophic respiration in side channels during high water, whereas clear-water rivers had strong net autochthonous production.

It is sometimes difficult for people to view flooding as a natural process in rivers that does not need to be controlled. For example, debris flows in small streams can exceed 10,000 m^3 of mud, logs, rocks, and sediment that sweep through portions of steep watersheds in a matter of minutes. Such flows may appear disastrous, but recov-ery does occur (Lamberti et al., 1991). Flood disturbances in small streams can control primary producers (Biggs, 1995, 2000) and, thus ecosystem function. Consideration of flow and the ramifications of flooding are becoming important to ecosystem and river management plans, such as in efforts to protect the endangered whooping crane (Highlight 24.1).

In contrast to the river continuum and flood pulse concepts, the *riverine productivity model* suggests that in-stream primary production (autochthonous energy) and direct riparian litter inputs (not linked to the flood pulse) can be the main source of energy in some large rivers (Thorp and Delong, 1994). This conceptual model may be most

Highlight 24.1 Management of the Platte River for water quality, water quantity, and species preservation

The Platte River and its tributaries drain a large portion of Nebraska and about one-third of Wyoming and Colorado. The endangered pallid sturgeon (*Scaphirhynchus albus*) inhabits the river. The Platte serves as a major stopover for migrant waterfowl and provides vital habitat for endangered whooping cranes (*Grus americana*), piping plovers (*Charadrius melodus*), and least terns (*Sterna antillarum*). About 80% of all sandhill cranes (*Grus canadensis*) stop there during their migrations. Roughly 70% of the Platte's discharge is diverted by consumptive uses in Colorado, Wyoming, and western Nebraska. There is concern about how this diversion will influence the Platte River ecosystem (US Fish and Wildlife Service, 1981).

A major result of flow reduction and control has been an alteration of channel morphology since the early 1900s. Historically, the river was sandy and braided with a wide, shallow channel, and floods from mountain snow runoff would create a wide shallow river that could be kilometers wide in places. Flow modification has resulted in lack of floods and allowed encroachment by woody species (*Populus* and *Salix*) and loss of much of the initial channel width (Johnson, 1994). Human modifications to the river and its floodplain have also facilitated invasions by salt cedar, purple loosestrife, and other exotic plants. Flow modification has also decreased water in the channel during winter, when seedlings are vulnerable to damage by ice. Lack of spring flooding has further encouraged establishment of vegetation on sandbars. Since the 1960s, the width of the channel has stabilized, vegetation has trapped sediments, and riparian forests have developed. This has necessitated active management efforts such as mechanical and chemical vegetation removal, controlled burns, and sandbar disking to remove vegetation.

Channel modification and flow alteration influence the extent and ecosystem characteristics of riparian wetlands. Cranes obtain much of their nutrition in wet riparian meadows, acquiring needed fat for continued migration. These wet meadows have a unique assemblage of organisms associated with them and they are rapidly disappearing (Whiles et al., 1999; Whiles and Goldowitz, 2001). Areas with narrow channels have fewer associated wetlands and lower than historical usage rates by whooping cranes and other waterfowl (US Fish and Wildlife Service, 1981). Increasing recognition of the importance of the floodplain wetlands in the Platte valley has led to extensive restoration efforts, with some degree of success, but the restoration of natural hydrologic regimes is difficult, as much of the water is appropriated upstream (Meyer et al., 2008; Meyer and Whiles, 2008).

The management of this ecosystem requires knowledge of how hydrology relates to habitat and organisms over long time scales. Vegetation removal to widen the channel may not be advisable because it causes sudden large sediment releases (Johnson, 1997). Mimicking the natural discharge regime to discourage establishment of riparian vegetation may be the preferred alternative. Given the tremendous demand for water from the Platte River basin, it may be difficult to obtain a discharge regime similar to that occurring historically to maintain the desirable biotic features of the Platte.

appropriate for large rivers with constricted channels and limited floodplains, such as the upper Ohio River, where it was originally developed. However, studies of smaller streams also indicate a much greater reliance on autochthonous energy and nutritional sources than previously thought (e.g., Hayden et al., 2016; Neres-Lima et al., 2017). A more nuance view of riverine food webs is obviously important as different parts of the food web can rely on separate sources. For example, Jardine et al. (2017) found that in the Mitchell River of Queensland, Australia, small-bodied fishes relied on autochthonous materials, but larger bodied animals such as predatory fishes and crocodiles were more dependent upon allochthonous dietary sources.

The RCC, flood pulse concept, and riverine productivity model all have obvious merits as conceptual foundations for studying rivers, and the appropriateness of the various components of each varies among and within different types of river systems. The *riverine ecosystem synthesis* model draws on elements of all of these and incorporates patterns in time and space in river networks at multiple scales. For example, due to changes in geomorphology along a river system, there may be reaches where the channel is constricted and there is limited floodplain in contrast to reaches with extensive floodplains and backwater habitats. These different reaches would obviously differ in a variety of physical and biological ways. These, and other sections of the network with different geomorphology and hydrology, can be viewed as *hydrogeomorphic patches*. In turn, these hydrogeomorphic patches give rise to different *functional processing zones* because ecological communities, and thus ecosystem structure and function, differ among patches (Benda et al., 2004; Thorp et al., 2006). In some cases, patches and zones may be discrete, whereas in other cases they only may be indicated by subtle changes along the river network.

A study of the highly impacted Colorado River in the Grand Canyon provides support for geomorphic patches driving river food webs. Cross et al. (2013) found that non-native New Zealand mudsnails and introduced rainbow trout dominated food webs above tributaries. However, diverse food webs of native omnivores occurred below the tributaries where a larger detrital food base was more prevalent.

The issue with these models is understanding how they all link together. Some factors, such as sediment load, might originate high in the watershed; for example, first-order streams drive concentrations of nutrients and other pollutants (Dodds and Oakes, 2008). In contrast, fishes might prefer specific geomorphic areas (e.g., deep pools or confluences). Different ecosystem processes have different effective functional processing zones, and are influenced variably by the river continuum and more local effects. Scaling is a central issue in pulling these concepts together, and we will discuss it more fully in the next chapter.

One effort to synthesize all these models is the *river wave concept*. The overarching hypothesis of this concept is "the location and source of autochthonous production and allochthonous inputs, storage, transformation, and the longitudinal or lateral transport of the material and energy derived from that production are largely a function of the position (ascending or descending limbs, trough, crest) on a river wave, either temporally or spatially" (Humphries et al., 2014). These authors suggest that repeated high flow events function similarly to waves and that these waves couple or decouple habitats and processes depending upon spatial location, and the frequency and intensity of the flow events. This is a promising approach because it more explicitly considers the temporal aspect of how the various models (flood pulse, river continuum, functional processing zones) link.

LAKES AND RESERVOIRS

One general way to classify lake ecosystems is based on lake autotrophic state (Table 24.4). Nutrients limit oligotrophic lakes so they are not very productive, with oxic processes predominating in most of the water column. On the opposite side of the spectrum, eutrophic lakes are prone to cyanobacterial blooms, have anoxic hypolimnia, have high rates of production in the water column, and production tends to be limited by nitrogen (because nitrogen is lost to denitrification) or light. Important exceptions to this classification scheme include dystrophic lakes (with high concentrations of humic compounds) that have low planktonic production but high macrophyte production, limitation by light for the phytoplankton, and heavily anoxic sediments with high rates of denitrification.

The classical view is that a lake ecosystem has cleanly defined boundaries and river inflow and outflow. This view assumes that biomass produced by phytoplankton photosynthesis dominates carbon dynamics, which feeds the zooplankton, which feed the fishes. This view is useful because it allows simple models of lake ecosystems to be constructed. We present a model of this type in Fig. 24.9. The idea that all fluxes can be accounted for in the closed basin is of particular predictive value; models of lake

Table 24.4 Some generalized ecosystem characteristics of different trophic levels in temperate lakes

Type	Productivity (mg C m^{-2} day^{-1})	Anoxic hypolimnion (in deep enough lakes)	Hypolimnetic O_2 depletion rate (mg m^{-2} day^{-1})	Factors limiting production	Relative rates of denitrification and nitrogen fixation
Oligotrophic	<300	No	<250	N and P	Low
Mesotrophic	300–600	Maybe	250–400	N, P, and grazing	Medium
Eutrophic	>600	Yes	>400	N or light	High

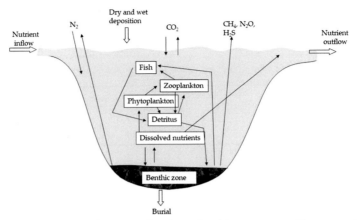

Figure 24.9 A simple diagram of nutrient flux through a lake ecosystem. The system is represented as a two-compartment bioreactor with a pelagic zone and the benthic zone. *Modified from Covich et al. (1999).*

eutrophication (see Chapter 18) can represent material balances and planktonic algal biomass of lakes reasonably well. Of course, as with all ecological constructs, there are exceptions to the simplification.

Many ecological studies have assumed that benthic primary production is not important. In most large, deep lakes, this approximation is probably reasonable. However, small lakes outnumber large lakes, and reservoirs tend to be shallow (see Chapter 7). Consequently, benthic primary production may play a significant role in lake ecosystems (i.e., half or more of the production may be from littoral algae or macrophytes in shallow lakes; Wetzel, 1983; Figs. 24.1 and 24.9). Now, models even suggest that benthic algal production is important in large lakes as well (Vadeboncoeur et al., 2008). Wetzel (2001) defended the importance of benthic processes in small lakes, and data continue to provide increased support for his views.

A carbon budget for Lawrence Lake, Michigan illustrates some of the primary carbon flux pathways in lakes (Fig. 24.10). This lake has significantly greater rates of autochthonous production than allochthonous inputs and primary producers in the system dominate carbon cycling. Macrophytes and associated algae were responsible for about two-thirds of the primary production. A large portion of the macrophyte production ended up as benthic detrital carbon, whereas less than one-third of the phytoplankton production ended up in the sediments. Rates of carbon burial were about half of export via streams, and the lake was a net source of organic carbon to the watershed.

More recent analyses of whole-ecosystem metabolism indicate that many lakes may be net heterotrophic (Dodds and Cole, 2007), giving credence to the concept of separating autotrophic and heterotrophic states of lakes. Rates of heterotrophy exceed

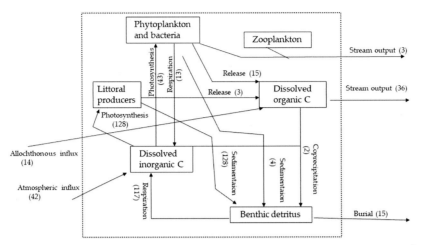

Figure 24.10 Diagram of carbon flux in Lawrence Lake, Michigan. Units are in g C m^{-2} year^{-1}. *Data from Wetzel (1983).*

photosynthetic rates when a broad number of lakes are considered. The evidence for net heterotrophy started with analysis of the degree of saturation of CO_2 in 4,665 lakes throughout the world (Cole et al., 1994). The data showed that 87% of the lakes were supersaturated with CO_2, indicating that respiration rates exceed carbon sequestration and export. These results can be explained most easily if externally derived carbon (*allochthonous* sources) exceeds washout plus burial in the sediments. So, the CO_2 data indicate that lakes are generally net heterotrophic. Of course, this generalization covers a range of lake types. High rates of respiration relative to photosynthesis may be common in more oligotrophic aquatic ecosystems (Duarte and Agustí, 1998). Trophic cascades may alter the relative importance of heterotrophy in lakes (Carpenter et al., 2001). Additional chemical and factors may play roles as well (Lapierre et al., 2017). Little research to date has linked predators to both heterotrophic and autotrophic pathways in food webs (Sitvarin et al., 2016). For example, *Daphnia* can switch to consumption of heterotrophic microbes when algal cells are less available (Mcmeans et al., 2015). However, higher inputs of organic carbon do not necessarily result in greater fish biomass, even if organic carbon enters the food web (Karlsson et al., 2015).

In some large lakes, such as the Great Lakes of North America, photosynthesis is likely high relative to allochthonous organic carbon input. External carbon inputs (in fine and dissolved organic material) may drive other systems when bacteria consume them and zooplankton eat those bacteria. Different sources of terrestrial carbon can influence different parts of the food web (Cole et al., 2006). In general, watersheds influence metabolic characteristics of lakes more than was previously thought (Cole and Caraco, 2001).

Whole lake ecosystem additions of stable isotope tracer ^{13}C indicated that about half the energy for the food web was derived from allochthonous sources in three Wisconsin lakes, two oligotrophic and one dystrophic (Pace et al., 2004; Carpenter et al., 2005). Some more directed inputs, such as terrestrial invertebrates for surface feeding fishes, can have unexpectedly high importance in lakes as well (Cole et al., 2006). Lakes with high concentrations of nonliving suspended particles can support a productive fish community despite very low algal biomass and productivity, and small humic lakes may have high production of bacteria that consume humic substances (Münster et al., 1999).

The idea that benthic primary production and allochthonous carbon provide considerable energy input into the food web complicates the view of energetics of lakes compared to that of a simple model considering phytoplankton—zooplankton—fish linkages. A model incorporating allochthonous inputs and the role of the microbial loop may more accurately characterize lake and reservoir ecosystems. The actual importance of each path of energy flux is context-dependent. If a lake is shallow and clear, macrophytes may dominate, whereas phytoplankton will dominate in a large, deep, clear lake. A lake with high throughput and an extensive littoral zone may function more similarly to a stream and be dominated by allochthonous carbon sources. Benthic periphyton can be important, especially in large oligotrophic lakes or high latitude lakes (Vadeboncoeur et al., 2008).

In Chapter 20, we discussed food webs in lakes and the trophic cascade systems of interacting populations of organisms, but not from the perspective of ecosystem energy flux. An interesting aspect of ecosystem energy flux relates to the fact that nutrients usually limit primary producers, but energy more often limits consumers. Where the switch from nutrient to energy limitation occurs depends on the stoichiometry of the system. The stoichiometry of grazers can feed back and intensify or relieve nutrient limitation (Elser and Urabe, 1999). Consequently, predicting ecosystem energy flux may require knowledge of community structure. For example, large *Daphnia* lower phytoplankton by grazing and intensify phosphorus limitation because of their high phosphorus demand (Elser and Hassett, 1994). Changes in trophic structure that alter *Daphnia* populations can thus affect factors that limit primary production.

Viewing lakes from a regional or landscape perspective can yield important information (Magnuson and Kratz, 1999; Kratz and Frost, 2000). One of the major aspects of groups of lakes is the coherence of lake properties with time (Magnuson et al., 1990). Documenting this coherence allows estimation of how well research results from one lake in an area apply to another. For example, lakes tend to have more similar chemical and biological properties across a landscape when hydrological throughput is high (Soranno et al., 1999). Lakes have also been classified by how well they are linked to other lakes by hydrology and by how far down in the drainage they are (similar to stream ordering). This classification correlates with patterns of species richness, chlorophyll concentrations, and major ion concentration (Riera et al., 2000).

ADVANCED: RESERVOIRS AS UNIQUE ECOSYSTEMS

Reservoirs comprise less surface area (about one-tenth) relative to natural lakes (Lehner and Döll, 2004). Throughout the chapters, we have described a number of features that tend to distinguish reservoirs from other lentic habitats. In general, these are (1) dendritic shape, (2) river influence in upper areas with shallower habitats common in the upstream portions, (3) deepest portion of the water body right next to the dam, (4) possible water releases from deep in the pool, and (5) unnatural flow regimes, with potentially large swings in depth. The fact that humans generally use reservoirs for power generation, flood protection, and water storage mean that the trends of water removal and depth are highly unnatural. Also, reservoirs are often made in areas that do not have many natural lakes, so they can represent an artificial habitat and allow introduction of species that are non-native or naturally rare in a region.

Wetzel (1990) provided a more comprehensive list of generalizations about reservoirs as compared with lakes including (1) more reservoirs are found outside of areas of glaciations, (2) areas of drier climate have a greater relative number of reservoirs, (3) reservoirs tend to have longer, narrower watersheds, (4) reservoirs have greater shoreline development indices and are shallower and less likely to stratify, (5) reservoirs tend to be dominated by larger inflows, irregular outflows, and higher flushing rates, (6) reservoirs experience high sediment loading and more sediment deposition in riverine zones, (7) reservoirs are warmer, (8) variable sediment loading and flushing in reservoirs leads to more variable light extinction, and (9) reservoirs display low diversity of benthic fauna and are characterized by warm water fisheries. Of course, these are generalizations, but they illustrate why reservoirs deserve special consideration.

Most reservoirs are relatively shallow, particularly nearer to the river or stream inputs. Shallower benthic habitat suggests that benthic primary production could be quite important in many reservoirs. The fluctuations in water level may make it difficult for macrophytes to establish; desiccation or freezing harms many macrophyte species. Pelagic production should be more important closer to the dam where the deep open water predominates. Hamre et al. (2017) studied the dinoflagellate *Peridinium* in a reservoir and how its population correlated with position in the river/pelagic continuum from the riverine to the dam portions of a reservoir. They found that physical processes and recruitment from resting stages in the sediments were the most important factors in the riverine section, but that water chemistry was the dominant factor in the pelagic zone.

Nutrient limitation gradients can occur with more nutrient limitation closer to the dam. Storms can create pulses of nutrients influencing productivity, particularly close to river inputs (Vanni et al., 2006). However, turbidity can offset nutrient pulses in zones with greater riverine influence by causing light limitation. In this case, there is a tradeoff between light and nutrient limitation as water progresses from the river into the main reservoir and maximum productivity will occur downstream from the input.

Reservoirs have large watersheds leading to substantial influence of terrestrial processes. This terrestrial influence can have strong effects on production of food webs, particularly if omnivorous organisms that can eat detritus or algae are abundant (Vanni et al., 2005). As with nutrients, the terrestrial materials entering reservoirs should have their greatest influence near the river and stream inputs with attenuation of availability of terrestrial carbon with distance from inflows. Reservoirs are more likely to be net heterotrophic because they interact more with their watershed than do many lakes, particularly in turbid reservoirs where primary producers are light-limited.

Reservoirs can be important zones of carbon and nitrogen retention in river networks. Reservoirs can act to "starve" rivers of organic carbon. By slowing movement of water, large organic particles will sink to the sediments where they are processed. The water leaving reservoirs can be impoverished in large particulate material, including organic carbon that would have subsidized downstream food webs. The larger area of fine anoxic sediments can also stimulate denitrification, leading to significant losses of total nitrogen as water passes through a reservoir.

WETLANDS

Wetland ecosystems can be classified along a hydrologic continuum from those that have very little hydrologic throughput to those that are closely coupled to rivers, estuaries, or lakes. In Chapter 5, we introduced the idea that some wetlands can have high hydrologic throughput (minerotrophic), whereas others receive water mainly by precipitation and have low hydrologic throughput (ombrotrophic). This variation in hydrology has implications for ecosystem function. Minerotrophic wetlands use nutrients from outside and nutrients wash from them easily. Ombrotrophic wetlands must rely more heavily on nutrient input from precipitation and internal nutrient cycling.

Autochthonous production in wetlands is usually extremely high; wetlands are characterized by some of the greatest rates of primary productivity of any habitats on Earth (Fig. 24.11). Production in even temporary wetlands may be important across dry landscapes that usually have very low production (Robarts et al., 1995). Freshwater marshes have very high rates of production; peatlands and deepwater swamps have lower rates of production (Table 24.5). Production of methane appears to follow primary production trends, with more productive wetlands producing

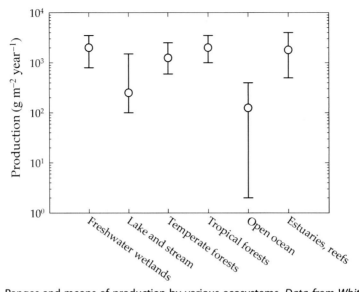

Figure 24.11 Ranges and means of production by various ecosystems. *Data from Whittaker (1975).*

Table 24.5 Ecosystem function in some wetland types

Type	Distribution	Production (g C m^{-2} year^{-1})	Methane production (mg C m^{-2} year^{-1})	Nutrient retention
Freshwater marsh	Worldwide	1,000–6,000	45–285	Sometimes N and P sink
Tidal freshwater marsh	Mid to high latitude, in regions with a broad coastal plain	1,000–3,000	440	N and P sink
Riparian wetland	Worldwide	600–1,300	?	Sometimes N and P sink
Northern wetland	Cold temperate climates of high humidity, generally in Northern Hemisphere	240–1,500	0.1–90	Usually N and P sink, may be an N source
Deepwater swamp	Southeast United States	200–1,700	1–15	

See Table 5.2 for description of wetland types.
Source: After Mitsch and Gosselink (1993). *Wetlands* (2nd ed.).

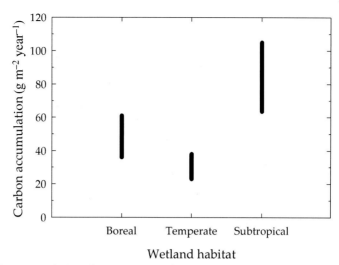

Figure 24.12 Organic sediment deposition rates for wetlands. *Data from Schlesinger (1997).*

greater amounts of methane. Greater production presumably increases the extent of anoxia and leads to greater methane production (Table 24.5).

The majority of primary production by macrophytes is not grazed directly by herbivores in wetlands (Mitsch and Gosselink, 1993); rather, it is deposited as detritus, and much of this organic production can be stored in the sediment (Fig. 24.12) or consumed by microbes and invertebrates. Although less than 4% of the Earth's surface is wetlands, wet soils contain about one-third of all organic matter stored in the world's soils. The vast deposits of coal are remnants of such organic storage from the swamps of the Carboniferous period (because the ability to degrade lignin was not as widespread in microbial communities). Given this potentially great production of carbon, variable rates of storage, respiration, and hydrological throughput, wetlands can serve as either sinks or sources of organic matter in the landscape. Natural wetlands with high hydrologic throughput can be significant sources of organic C in the watershed (Mulholland and Kuenzler, 1979). However, artificial wetlands are used as sewage treatment systems; in this case, the wetland has a net consumption of organic carbon.

Wetlands represent a hybrid between terrestrial and aquatic systems. The carbon flux diagram of Creeping Swamp, North Carolina illustrates some unique features of wetland ecosystems (Fig. 24.13). Trees, followed by algae and small plants, dominated carbon production. The production of carbon in coarse particulate organic material fuels the food web of the wetland. Most of the carbon in ecosystem compartments is in trees and sediments. The majority of the carbon flux into the detrital pool derives from trees in the wetland. The majority of the carbon fixed by photosynthesis is

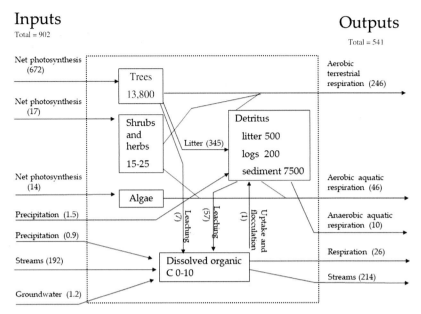

Figure 24.13 Carbon biomass and flux rates for the Creeping Swamp ecosystem. Biomass in g C m^{-2} listed in the compartment boxes, and flux rates in g C m^{-2} y^{-1} are listed in parentheses. *Data from Mulholland (1981).*

released by respiration and burial, but the swamp serves as a net source of organic carbon to the stream water.

Invertebrate communities in wetlands rely heavily on detritus from emergent plants and algae, with macrophyte tissue being less important (Batzer and Wissinger, 1996). In contrast to streams, shredders that are specifically adapted to process plant litter are generally rare, with some notable exceptions (e.g., Whiles et al., 1999). Numerous invertebrates, particularly midge larvae, specialize on algal producers. Availability of algae can limit production of these invertebrates (Batzer and Wissinger, 1996). Insect herbivory on the areal portions of emergent plants can be important (Wissinger, 1999). Top predators in the water are generally insects, crayfish, small fishes, and amphibians (Porter et al., 1999). Vertebrate predators, such as raccoon and birds, rely on the larger animals in the water.

Floods strongly influence wetlands associated with rivers and streams. The flood pulse concept provides a view of the river as a dynamic system connected with its floodplain, and it may provide valuable ways to describe the riparian ecosystem (Lewis et al., 2000). In large rivers, this pulsing is particularly important because such systems have seasonal flooding. Flooding connects the rivers with riparian wetlands. As discussed previously, flooding can be important in the biology of large river fishes (see Highlight 23.1). The flood—riparian connection provides riverine species with food resources and spawning habitat, and riparian lake and wetland species with an avenue for dispersal. The connection is also important in material transport.

Floods inundate riparian areas, which slows water velocity and allows for settling of sediments. Dry, coarse organic debris initially floats and can be moved from the riparian zones into the large channels or moved within the wetland (Molles et al., 1998). Likewise, nutrients move from the river into side pools (Knowlton and Jones, 1997). This process can be important in tropical floodplain rivers, in which lakes and wetlands in the riparian zone become progressively less productive during the dry season as organisms take up nutrients which end up deposited into the sediments. Flooding then provides a new pulse of nutrients that can boost productivity (Hamilton and Lewis, 1990). Flooding in arid-zone rivers also provides nutrients to the riparian zone. This subsidy by flooding can occur even in the absence of overland flow because the flooding can pulse water and nutrients into the hyporheic zone which riparian vegetation can access through their roots (Martí et al., 2000).

Flood pulsing is an essential characteristic of riparian wetland ecosystems. These wetlands are highly endangered by river channelization and modification. Addition of dams and regulation of extreme flow can severely alter their natural cycles. Restoration of riparian wetlands requires management of floods and connectivity to the main river channel (Middleton, 1999). The Pantanal is the largest wetland in the world and is threatened by changes in the hydrologic regime (Highlight 24.2) as are many others globally.

WHOLE-ECOSYSTEM EXPERIMENTS

Although sometimes plagued with statistical issues such as low replication, whole-ecosystem studies have provided important insight into the structure and function, and relationships between the two, of freshwater ecosystems. A classic series of studies in Wisconsin involving whole lake consumer and nutrient manipulations revealed trophic cascades that ultimately determined if lakes were net sources or sinks of CO_2 (Carpenter et al., 2001). Similarly, a long-term phosphorus addition to an Alaskan tundra stream showed responses at multiple trophic levels, from periphyton to predatory fishes (Peterson et al., 1993). The system continued to shift over more than a decade of phosphorus addition; after 8 years, bryophytes replaced diatoms as the dominant primary producers, which in turn altered nutrient uptake rates, primary production, and invertebrate communities (Slavik et al., 2004).

Nitrogen tracers (^{15}N) allow us to examine nutrient cycling at the reach scale in streams and have greatly improved our understanding of nutrient uptake, cycling, and

Highlight 24.2 The Pantanal, the world's largest wetland ecosystem complex

The Pantanal is a vast complex of seasonally flooded wetlands, lakes, and streams along the Paraguay River in Brazil, Bolivia, and Paraguay. The total area of wetlands and savannah includes 140,000 km^2 in Brazil and 100,000 km^2 in Bolivia and Paraguay; the portion in Brazil is larger than the state of New York. During the wet season, about 80% of the area is flooded and the flood pulse moves from the north of the wetland to the south over a period of months. During the dry season, grasses emerge, and the area is grazed heavily by cattle. This wetland is highly dependent upon a wet-season flood pulse and vulnerable to hydrologic alteration (Junk et al., 2006).

Figure 24.14 Scenes from the Pantanal: (A) satellite view of the numerous small patches of surface water, (B) aerial photo of a main channel and adjacent flooded areas, (C) a giant river otter (*Pteronura brasiliensis*), and (D) caimans (*Caiman yacare*) congregate around a pool that remains during the dry season. *Images courtesy Steve Hamilton.*

The wetland undergoes a strong seasonal succession. During the wet season, several meters of water cover all but the highest tree islands. Massive growths of macrophytes occur and aquatic species disperse. During the dry season, the ponds become isolated from the main channel. Fishes cannot escape these ponds and waterbirds and caimans congregate to feed on the trapped fishes (Fig. 24.14). The ponds become eutrophic from nutrients released from the fish carcasses and excreta of the predators (Heckman, 1994).

The Pantanal is a wetland of international importance for ecological conservation. It is habitat for the endangered spotted jaguar (*Panthera onca*), giant anteater (*Myrmecophaga tridactyla*), and giant river otters (*Pteronura brasiliensis*). Numerous waterbirds use the wetland, as does a diverse assemblage of parrots. The Pantanal also has more than 400 species of fishes and attracts large numbers of sport anglers.

The greatest threats from development are channelization and wetland modification that could lower water levels. These threats include a major project to increase the navigability of the Paraguay River. The dredging and channelization project will cost approximately $1 billion, and economic assessments have not carefully considered ecological impacts and the altered hydrology with associated flooding down river (Gottgens et al., 1998). With only a 0.25-m decrease in water level, the inundated area of the wetland will be decreased by more than half in the upper regions of the wetland and 5,790 km^2 overall (Hamilton, 1999). The project will negatively influence the livelihoods of thousands of indigenous people that inhabit the region. Political pressure has caused a decrease in the scale of the plans, but a comprehensive development plan is lacking, and dredging and channelization plans continue. Expansion of agriculture, industry and population all influence the wetland (de Oliviera Neves, 2009). As of 2005, only moderate conservation progress had been made (Harris et al., 2005). If past cases of wetland exploitation are any indication (e.g., the Everglades), the Pantanal ecosystem will suffer greatly as humans develop the area (Gottgens et al., 1998).

---

export from a variety of stream types (Mulholland et al., 2008; Norman et al., 2017; Peterson et al., 2001). The use of ^{13}C tracers has revealed the importance of bacteria to higher trophic levels, such as macroinvertebrates in headwater streams (Hall and Meyer, 1998; Collins et al., 2016a,b). The same tracer allowed determination of allochthonous versus autochthonous inputs into lake food webs (Pace et al., 2004; Carpenter et al., 2005).

Numerous whole system experimental manipulations have taken place at the Coweeta Hydrologic Laboratory in the Appalachian Mountains, and these studies have greatly enhanced our understanding of the ecology of detritus-based headwater streams. Experimental removal of insects from headwater streams at Coweeta resulted in significantly reduced leaf litter decomposition and organic seston export, underscoring the importance of invertebrates in detritus processing and energy flow (Cuffney et al., 1990; Wallace et al., 1991; Whiles et al., 1993). Exclusion of leaf litter inputs to Coweeta streams resulted in greatly reduced invertebrate abundance, biomass, and production, illustrating the tight linkage between headwater streams and riparian forests (Wallace et al., 1997). Multiple years of nitrogen additions to a headwater stream at Coweeta greatly enhanced invertebrate production, but severely reduced amounts of leaf litter in the stream channel, which may have negative long-term consequences for instream energy cycling and food web structure (Cross et al., 2006). Follow-up studies showed how the whole-stream nutrient additions altered food web structure (Manning et al., 2016).

Whole-ecosystem studies (*holistic approaches*) have the advantage of taking into account the entire, intact system. However, difficulties with statistical replication at this scale and the complexity of natural systems sometimes necessitates a *reductionist approach*, whereby researchers examine individual components or subsets of systems. Reductionist approaches range in scale from microcosms in controlled environments, to mesocosms (e.g., replicated ponds, pools, or experimental streams), to field manipulations such as experimental exclusions. Reductionist approaches have the distinct advantage of being more statistically robust and they allow for isolation of specific variables, but results may not always reflect processes and patterns in whole systems. For example, analyses of biodiversity and ecosystem function studies indicate that small-scale manipulations likely underestimate the importance of biodiversity to ecosystem functioning (Duffy, 2009). Both holistic and reductionist approaches have their merits and have greatly contributed to our understanding of freshwater systems; combined approaches have proved particularly valuable.

COMPARISON OF FRESHWATER ECOSYSTEMS

All the generalizations presented in previous chapters apply to classification of ecosystems. The difficult issue is how to weight the importance of the various factors. Keep in mind that classification is mainly a tool for scientists to deal with a very complex world. Primary abiotic differences are associated with hydrologic throughput, the availability of light, the amount of allochthonous input, and the extent of benthic habitat (depth). If we view each of these factors as gradients, where most combinations are possible, in addition to variance associated with each of the factors over spatial and temporal scales, we can describe many essential characteristics of aquatic habitats (Fig. 24.15). Classification of abiotic parameters can provide information about constraints on evolutionary and physiological processes. For example, forested temperate headwater streams are highly variable (hydrologically and with seasons), receive low light, and have a high degree of terrestrial influence (including allochthonous carbon input). In contrast, deep groundwaters are highly stable, receive no light, receive low amounts of allochthonous input and even less autochthonous input, and have very low hydrologic throughput. The RCC is an excellent example of a conceptual model

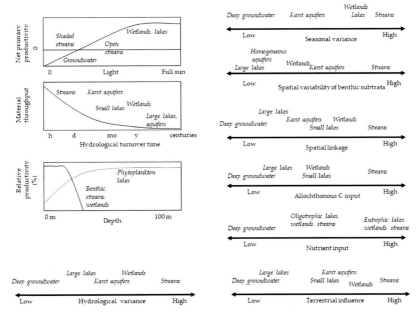

Figure 24.15 Freshwater habitats aligned across gradients that drive ecosystem properties.

that uses classifications of gradients and spatial linkages to make specific predictions about ecosystem function and community structure.

Of course, problems arise when we try to use just a few abiotic gradients to classify ecosystems. Very shallow lakes can occur in two ecological states. They may have very little emergent vegetation or macrophyte production but high cyanobacterial biomass, or alternatively be dominated by macrophytes. Wind, unstable benthic substrata, high grazing, and high nutrients can lead to loss of macrophytes and establish cyanobacterial blooms. Once the bloom is established, it is difficult for the macrophyte assemblage to become reestablished. This is just one example of how the complexity of natural systems precludes our ability to completely classify and make predictions about many of their processes (Dodds, 2009). Perhaps the most interesting ecosystems are those that are the most difficult to classify, and in many cases contingencies based on evolution of unique characteristics of organisms are essential in understanding system behavior.

Abiotic gradients lead to biotic differences in the food webs of various freshwater habitats (Fig. 24.16). Simple planktonic food webs characterize the pelagic zone of lakes and simple food webs based on consumption of biofilms characterize groundwaters. Shallow benthic habitats with heterogeneous substrata (wetlands, streams, and lake littoral zones) can contain diverse assemblages of organisms that process large amounts of detritus in addition to grazers and predators. The vagaries of chance, dispersal, and time have acted in concert with these abiotic gradients to allow evolution

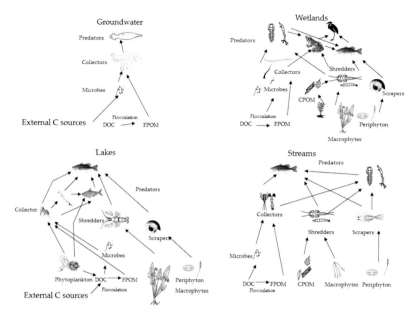

Figure 24.16 Comparison of carbon pathways and trophic complexity of groundwater, wetland, lake, and stream habitats.

and community assembly to produce the diversity of organisms associated with each of the habitats.

We can characterize many of the human-caused disturbances of freshwater habitats by shifts of systems on these axes of classification, particularly those related to habitat modification. Description and understanding of how these complex ecosystems are constrained are essential to describing and mitigating the effects that people have on freshwater, our most valuable resource.

SUMMARY

1. Ecosystems can be represented by trophic levels and the fluxes of carbon, energy, or nutrients between the trophic levels.
2. Biomass is an amount; production is a rate. The two are not the same. In a more general sense, standing stocks are not the same as fluxes.
3. Secondary production is tissue elaboration by heterotrophs, regardless of its fate. It reflects energy available to higher trophic levels and because it is a rate. It can quantitatively link consumers to processes such as nutrient cycling.
4. Biodiversity may have an influence on ecosystem function, particularly in low-diversity systems.

5. Groundwater ecosystems are driven primarily by external inputs of organic carbon. Little is known about energetic fluxes in most groundwater ecosystems, and nutrient budgets are not well characterized for groundwaters.

6. Streams are mainly nonequilibrium ecosystems in which flooding and unidirectional downstream flow provide strong abiotic influences on the ecosystem.

7. Nutrients spiral in streams as they cycle and move downstream.

8. The River Continuum Concept describes a series of physical and biological changes observed when moving from small headwater streams to large lowland streams. Alternative models of stream ecosystem structure and function have been developed, mostly focused on the complexity of large rivers.

9. Planktonic production and consumption can be very important in deep lakes. Allochthonous inputs are minimal, but benthic production can dominate in shallow lakes and reservoirs.

10. Reservoirs have unique ecosystem properties compared to lakes such as being shallower, more terrestrially dominated, and more influenced by riverine inputs.

11. Wetland ecosystems are highly dependent on autochthonous production by plants and algae.

12. Aquatic habitats occur on a continuum of abiotic axes that describe many of the essential parameters that constrain ecosystem function and biotic capacities.

13. Whole system studies have greatly enhanced our understanding of freshwater ecosystems, but statistical limitations and the complexity of systems sometimes dictate a reductionist approach; both are important in freshwater ecology.

QUESTIONS FOR THOUGHT

1. At what point on an Allen curve would you expect secondary production to be highest?

2. Why are wetland ecosystems generally more productive than streams?

3. Why do wetland ecosystems store a greater amount of carbon than lakes or streams?

4. Can a lake ecosystem be described adequately with a two-compartment model, one for the pelagic zone and another for the limnetic zone?

5. Do whole communities evolve over time to optimally exploit ecosystems?

6. If global warming increased the number of freshwater marshes by converting northern peatlands into marshes, what would happen to methane production?

7. How well does an equilibrium model represent stream ecosystems relative to lake and groundwater ecosystems?

8. Should we preserve ecosystems in addition to endangered species?

9. Some people refer to "biotic integrity" and "ecosystem health" in the context of conservation of the environment. What do you think these terms mean and how should we define them?

10. How can some ecosystems have a higher biomass of predators than primary consumers?

11. How would you design a study to look at predator–prey interactions in a pond using a reductionist approach and holistic approach; what about a combined approach?

CHAPTER 25

Scaling, Landscapes, Macroecology, and Macrosystems in Freshwaters

Contents

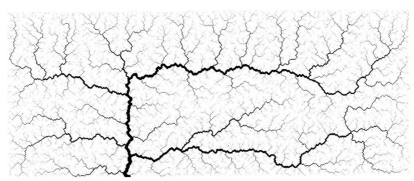

Figure 25.1 A mathematically generated drainage pattern. The image illustrates how mathematical representation of energy dissipation leads to dendritic drainage networks of streams that have fractal scale-invariant properties (Rinaldo et al., 2014). *Image courtesy: Andrea Rinaldo.*

Ecological processes, management, and evolution operate at the very broadest scales, such as watersheds and across continents, though experiments are usually performed at relatively small spatial scales. Some methods allow for large-scale monitoring (e.g., remote sensing), few experiments are done at broad geographical scales. Even whole-watershed or whole-lake experiments usually represent only a tiny fraction of any single type of waterbody across its entire range. Additionally, size of organisms varies about 8 orders of magnitude, from a μm to 100 m, yet they are made of the same biochemical materials. How they vary as a function of size is an avenue to ecological understanding. There is a need for methods that allow us to scale up experiments and observations to whole systems, and ways to make broader generalizations if we are to understand ecological processes at scales relevant to freshwater conservation and management of freshwater habitats. *Landscape ecology* is the study of spatial relationships and ecology. *Macroecology* seeks to identify broad generalizations about organismic biology across levels of biological classification and organization. *Macrosystems* are very large-scale patterns (regional to continental). We discuss these approaches and their relevance to freshwater ecology, but first consider the general topic of scaling in ecology.

SCALING

Considerations of scaling form a major problem in understanding ecology and are one of our biggest challenges (Levin, 1992). Scaling enters into any process where there are compartments or nested structures. Biologists use the levels of biological organization (i.e., molecules, cells, organisms, species, populations, communities, ecosystems) to understand causes and responses of biological systems. We approximate the hierarchical biological organization scaling approach as a useful organizing scheme in this book. Additionally, spatial and temporal scale and hierarchies influence freshwater ecology. We have already discussed the vast range of scales that influence properties of freshwater ecology with respect to factors that influence the movement of water (Fig. 2.16) and habitats for organisms (Table 4.1 and Fig. 6.22).

Studies of biological phenomena can be conducted at any given order of biological organization, but implications of those phenomena are often extrapolated to higher levels of organization. For example, cellular physiological influences the reproductive capacity of an individual fish. The implications of individual fish reproduction are in turn manifested in population dynamics of the fish species. However, this generalization is not always true; single celled bacteria mediate most biogeochemical processes, but their influence radiates to entire watersheds or even global scales with respect to cycling, transport, and quantities of nutrients.

In practice, freshwater ecologists are constrained by the tools they have to approach particular scales (Table 25.1). Thus, if we are to understand properties of entire ecosystems, we must confidently scale from our measurements to the scale of interest for the process we are studying. Think, for example, about how to determine the limiting nutrient in a lake. The easiest approach is to take bottles of water from the lake and add different nutrients and see what nutrient or nutrients stimulate productivity. However, complications related to scaling arise. What light, temperature, and pressure is reasonable to use for the incubations? What response variable should be measured (e.g., chlorophyll)? How do factors in the bottle vary from those in the whole lake?

Table 25.1 Scales of study in freshwater ecology and tools to assess those scales

Scale	Units	Habitats	Methods and comments
Microscale	μm–mm	Silt particles, individual cells	Microscopy, microelectrodes, labels of individual cells
Deciscale	mm–dm	Cobbles, pools, riffles, lake shores, large woody debris, small wetlands, ponds	Common methods of limnology, samples taken and analyzed, mesocosms (note fish researchers often refer to this as microhabitat)
Large scale	dm–km	Stream reaches, small watersheds, larger lakes and wetlands	Whole-system approaches, tracers and large-scale manipulations
Macroscale	km to continents	Larger watersheds, lake districts	Remote sensing, large scale monitoring networks

Are the experiments long enough to account for changes in species composition? Some have argued that small-scale approaches can be very misleading with respect to whole-system responses (e.g., Carpenter et al., 1995; Schindler, 1998).

Ecological systems often have *emergent properties*, characteristics that scientists cannot predict from understanding of individual smaller components of the system or lower levels of biological organization (Odum and Barrett, 1971). This concept is often expressed as the whole being greater than the sum of its parts. If there are emergent properties, then predictive ecology based on small-scale attributes is difficult if not impossible. At the other end of the spectrum, some suggest that if every bit of an ecological system could be sampled, then the whole-system properties could be predicted by summing the individual parts. It is difficult to know if inability to predict larger scales from smaller scale phenomena is due to emergent properties, or incomplete understanding or accounting for smaller scale phenomena. At the deepest level, life is, as far as we currently know, an emergent property; we do not yet understand how to create life from its building blocks. Emergent properties occur in many areas of freshwater ecology. For example, shoaling (schooling) behavior in fishes is an emergent property of individual behavior (Hölker and Breckling, 2005). Self-organization in river meandering is an emergent property of water flowing through valleys.

Emergent properties may alter freshwater system characteristics. Moore et al. (2014) analyzed fisheries, water flows, and temperature on the mostly free-flowing Fraser River in Canada and found that larger watersheds had more stable fisheries catches, water flows, and water temperatures than smaller watersheds. They suggested that these properties of larger systems can buffer them against extreme events. Models of stream diversity also indicate that scaling can control diversity along linear habitats (Holt and Chesson, 2016). Dong et al. (2017) found that self-organization properties influenced spatial patterns of nutrients as a desert stream made a successional transition from an open channel to a wetland dominated condition.

Part of the attraction of approaches that account for spatial and temporal scale is the possibility that there are scale-invariant properties that can predict system characteristics. River and stream geomorphology is one area where such patterns have been observed, as represented by the work of Leopold (1994) on the dynamic equilibrium of stream channels (covered in Chapter 5). In this view, the sinuosity of river channels as they meander is proportional to channel size. That is, the radius of curvature of a meander's loop is a function of the length of the meander. If you have an aerial photograph of a stream channel, it is difficult to tell the size of the channel without a reference object based only on the shape of the meander (Fig. 25.2). We develop this approach to scale more fully in the section below on macroecology.

LANDSCAPE ECOLOGY

According to the International Association for Landscape Ecology, "Landscape ecology is the study of spatial variation in landscapes at a variety of scales. It includes the biophysical and societal causes and consequences of landscape heterogeneity. Above all, it is broadly interdisciplinary." This concept does not necessarily imply broad scale approach, as does the term "landscape" implied from common usage, but in practice, people often assume it means large spatial areas. However, a "landscape" to a microbe might seem relatively small compared to the human-scales of observation. Pickett and Cadenasso (1995) specify that landscape ecology is concerned with spatial dynamics of fluxes of organisms, materials, and energy. This definition means that spatial scale is considered, but not constrained to, the largest scales. The term "landscape ecology" ends up being an unfortunate label for a field of general ecological study, not only because it invokes one specific scale of study with respect to common usage of the word, but also because "land" excludes freshwater and marine habitats, even if the field of study does not. Spatial pattern and population dynamics and community linkages are two areas where landscape ecology has made large scientific advances.

Metapopulations

A *metapopulation* is a group of spatially distributed populations with varying degrees of gene-flow among them (Gilpin and Hanksi, 1991). In metapopulations, multiple populations exchange individuals of the same species through dispersal, and this process enhances and maintains genetic diversity. Distinct metapopulations may eventually diverge into separate species, so the concept is important with respect to evolution and gene-flow. The metapopulation concept is central to conservation because it describes how seemingly separate populations can make up a larger, interactive population of a species. The extinction vortex discussed previously in Chapter 11, can apply to small subpopulations, and without dispersal from outside, some subpopulations are destined for extirpation.

(A)

(B)

(C)

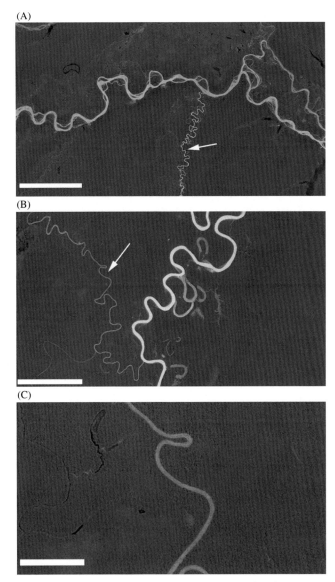

Figure 25.2 River meanders are not distinguishable without scale. Three images from the same area of the Amazon River in the region of the confluence with the Rio Solimões. Bar in (A) is 60 km, the arrow shows the approximate area of image (B). Bar in (B) is 9 km and arrow shows approximate area of image (C). Bar in (C) is 1.8 km. *Images courtesy: Google Earth.*

The concept of metapopulations is important to conservation biology because it explains how maintenance of habitat heterogeneity alongside disruption of connections among patches could be essential to conserving species and explaining species distributions. For example, Dahlgren and Ehrlén (2005) found that macrophyte species in lakes were determined by the degree of connectivity among the lakes, in addition to physical factors such as lake area, noting that emergent species were more dependent upon spatial connections among lakes. Metapopulation models can also be useful for conservation decisions in vernal pools with respect to amphibian populations. Compton et al. (2007) created a spatially explicit model of almost 30,000 wetlands in Massachusetts and used home range estimates for several species of salamanders to identify particular areas of vulnerability. They found that the highest vulnerability to decreased connectivity of temporary wetlands coincided with high rates of suburban development.

The metapopulation concept is entwined with the idea of source-sink population dynamics. A metapopulation is composed of multiple subpopulations, some more successful than others. Some of the subpopulations, termed "sink populations," may be extirpated without input from more successful adjacent populations. Populations that occur in favorable habitats and supply individuals to other populations are source populations.

One example of how spatial environmental heterogeneity influences metapopulation dynamics is in predator–prey systems. In a spatially homogenous environment, a predator will eventually wipe out prey and then go extinct itself if alternative prey is unavailable. In contrast, with high habitat heterogeneity it is possible for prey and predators to coexist (Huffaker, 1958). Similar concepts may apply with disease dynamics. A study of variously connected metapopulations of an endangered aquatic Australian frog, *Litoria raniformis*, found that population connectivity and habitat heterogeneity sustained metapopulations because some populations lived in relatively warm, saline habitats that served as refugia from the fungal disease chytridiomycosis (Heard et al., 2015). This finding is particularly interesting because we often think of population connectivity as a pathway for the spread of diseases; in this case connections with populations inhabiting disease refugia facilitated recolonization of affected areas, which sustained metapopulations. Hence, populations in disease-prone areas could be replenished with individuals from disease-resistant areas.

We can also view lakes from the source-sink framework. Many lakes are connected by streams, and the genetic similarity of organisms among lakes will be constrained by hierarchical structure and factors that influence colonization (Salisbury et al., 2016). Many isolated lakes in mountainous regions are naturally fishless. Fish introductions (often for recreational purposes) into these fishless lakes can harm native organisms. Fish stocking introduces a "new" predator and competitor to communities that have developed without them, and can represent important source populations of aquatic organisms that are intolerant of the presence of fishes (Ventura et al., 2017).

Wetland complexes also can be viewed from a source-sink framework with respect to species conservation and protection. For example, network structure of prairie potholes determined where mallards (*Anas platyrhynchos*) moved. It allowed the ducks to avoid human disturbances (Beatty et al., 2014). Also, some human created wetlands mitigated decreased wetland disconnection and improved functional connectivity for anurans in Nebraska wetlands. However, the wetlands were still way less connected than they were historically before anthropogenic modification (Uden et al., 2014).

The simple model originated by Levins (1969) used to quantify the patchy nature of spatial population distribution is foundational to the field of landscape ecology. In this model, the proportion of patches occupied of those suitable for a species is a function of how often patches are colonized and how often populations are extirpated within a patch.

$$\frac{dp}{dt} = c\,(1-p) * p - e$$

where dp/dt is the change in proportion of patches that are occupied (p) per unit time (t), c is the rate at which occupied patches produce colonists, and e is the rate that occupied patches are extirpated. This framework sets up a situation where there can be sources and sinks of populations; connection of habitat that are conducive to high population density and growth (sources) can provide other areas with less optimal characteristics for survival with immigrants that keep the population going (sinks).

Gotelli and Taylor (1999) tested the predictions of Levins's model to stream fish assemblages. They used 10 years of data from the Cimarron River and found that the simple model did not apply. This was because the probability of extinction and colonization correlated with position in the river; the upstream sites had the greatest probability of extirpation and relatively slower colonization rates. Thus, the assumptions of the simple model that include homogenous patches and constant colonization and extirpation probability were not met. As with any model, more complexity is needed to match conditions in the real world, but even the simplest models can be valuable for conceptualizing ecological processes and generating hypotheses.

We can think of landscape ecology several ways in freshwater ecology. The organisms of interest drive much of the approach. Protists, with global distribution, may be considered a very large-scale group of metapopulations (e.g., Khomich et al., 2017). Fishes, with no life stage that can do well out of water, are much more constrained by the spatial connections of freshwater habitats. Macroinvertebrates with reproductive adults that can move across terrestrial systems are intermediate in the range of vagility. Rivers connect to lakes in different ways depending upon geological history (Fergus et al., 2017). Rivers and streams are largely organized in a dendritic structure with unidirectional transport. Therefore, these rivers are more constrained than many terrestrial landscapes for organisms that are confined to freshwaters. Natural barriers, such as waterfalls, can further constrain connectivity among habitats within a watershed. Human-constructed barriers such as dams and road crossings can disconnect the system as well. Drought can further disconnect stream networks, as can anthropogenic water extraction that leads to stream drying, and point-sources of pollution (Inostroza et al., 2016). Human constructed barriers such as dams and road crossings disconnect systems as well (Mattocks et al., 2017).

Harm to the fish community by network fragmentation is an important large-scale issue. Dewatering is common and obstruction by dams impedes fish movement and reproduction in many river systems worldwide (Liermann et al., 2012; Winemiller et al., 2016). In one example of a landscape perspective on rivers, fragmentation and dewatering of streams in the Great Plains associated with groundwater depletion of the Ogallala Aquifer and reservoir construction had strong effects on native river fishes (Perkin et al., 2014; Perkin et al., 2015). Species that are particularly affected are those that release eggs into the water column that must remain suspended (they are neutrally buoyant) until the larvae hatch. Interruptions to stream flow by dry segments or reservoirs will cause their eggs to sink and fail to hatch. As a result, large fishes have disappeared and smaller fishes remain, resulting in homogenization and simplification of the fish communities and food webs (Perkin et al., 2017). Most broadcast spawning species are extirpated in shorter remaining river fragments (Perkin and Gido, 2011, Fig. 25.3). Bishop-Taylor et al. (2017) documented similar processes of disconnection harming fishes in fish communities of the Murray Darling basin of Australia.

Similar relationships occur with wetlands and anurans. Lehtinen et al. (1999) found that the number of anuran species decreased with increasing distance to the next wetland in Minnesota. They found that habitat fragmentation decreased amphibian diversity. The proportion of urban land nearby had particularly strong negative effects on diversity (Fig. 25.3).

The metapopulation framework was also used to analyze the effects of fires and related ash flowing into streams on native and non-native fishes in the Gila River of New Mexico (Whitney et al., 2015). Their analysis indicated that ash flows increased the probability of extirpation of native fishes from patches more strongly than non-native fishes, and that main-stem habitats served as refugia and source populations for native fishes.

Genetic analyses are an important tool that can allow for examination of population structure and degree of isolation in patches, providing insight in to how metapopulations are structured. Populations that are tightly connected and exchange individuals frequently will not differ significantly in their genetic distribution. In contrast, populations that are rarely connected through movement of reproductive individuals will diverge genetically. For example, Kelson et al. (2015) analyzed brook trout (*Salvelinus fontinalis*) populations in New Hampshire and found that waterfalls led to genetic isolation of populations.

Human activities can also increase physical connectivity of water bodies, often with negative ecological and economic consequences. For example, the spread of sea lamprey (*Petromyzon marinus*) through the North American Great Lakes was facilitated by

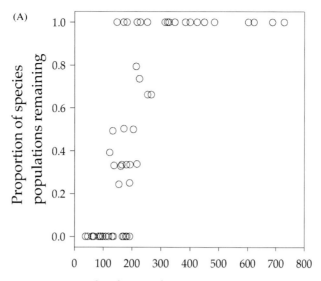

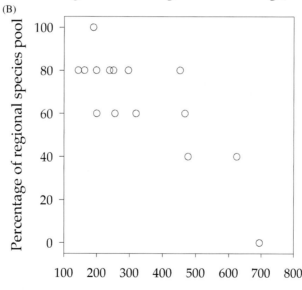

Figure 25.3 Proportion of broadcast-spawning species remaining in Great Plains stream fragments as a function of length of remaining fragment (A), and percentage of regional anuran species pool present as a function of distance to nearest wetland (B). *(A) Data from Perkin and Gido (2011); (B) data replotted from Lehtinen et al. (1999).*

construction and expansion of canals. Sea lamprey have had major ecological impacts in the Great Lakes. These changes have had a large economic impact on the region because of declines in commercially harvested fish stocks and ongoing expensive lamprey control efforts, which ironically include construction of barriers to block their movements into streams where they spawn. Similarly, the zebra mussel spread from Eastern Europe to Western Europe via transportation canals, where it was taken into ships in ballast water and transported to the Great Lakes of North America. From there, it spread through North America, in part through the canal system that increased connection among watersheds.

Metacommunities

In a concept related to metapopulations, *metacommunities* are groups of communities spread across a landscape that interact through dispersal and exchanges of multiple species (Holyoak and Leibold, 2005). A metacommunity is an assemblage of separate communities, with the potential for interactions among all species in the region, connected to various degrees across a spatially heterogeneous landscape (Leibold et al., 2004). Major aspects of metacommunities include species sorting (local processes serve as filters determining community structure), mass effects (where source-sink dynamics override importance of local patches), neutrality (based on random models of colonization and extinction), and patch dynamics (how spatial arrangement of patches influence communities). This view however, may not be nuanced enough to truly describe metacommunity dynamics (Brown et al., 2017).

Neutral models have been used to explain species diversity in communities and explore metacommunity dynamics (Hubbell, 2001). Neutral theory predicts that the diversity of a community is a function of immigration, extinction, and speciation; essentially, all species are considered equal in their ability to exist and reproduce in a community. This is in contrast to the more traditional niche-assembly concept (niche theory), which focuses on resource use by individual species, or community assembly rules where species can have priority or be required before another is established. Neutral models are essentially null hypotheses to traditional niche theory. In a simulation modeling study, White and Rashleigh (2012) implemented a neutral metacommunity model using two stream network types, a widely branched network and a narrowly branched network. The wide branched network had much higher connectivity and shorter distances between stream segments compared to the narrowly branched network. In their simulations, the wide branched system ended up with higher community (alpha) diversity than the narrow branched one. The authors hypothesized that the greater connectivity of the widely branched system facilitated more even spatial dispersal and ultimately higher species diversity. Neutral theory is not universally accepted and has been criticized by some as being simplistic, but it has stimulated extensive research and development of testable hypotheses in community ecology (Leigh, 2007).

The applicability of the metacommunity concept has not received as much study in ecology as metapopulations, for the concept is newer and it is more difficult to establish community structure. Population structure and flow among populations are described with genetic tools, so the metapopulation concept is more straightforward than metacommunities. However, there has been progress. For example, researchers and managers can improve bioassessment by considering community structure of aquatic organisms to assess environmental conditions in light of metacommunity structure (Heino, 2013). Consideration of metacommunity dynamics can also help understand the effectiveness of river and stream restoration (Swan and Brown, 2017). Links among aquatic and riparian habitats that influence β diversity are related to metacommunity dynamics as well (Tonkin et al., 2016).

This view is not as quantitative as the concept of metapopulations because entire communities rarely are extirpated, and parts of communities can colonize others. However, this concept does make clear the idea that similar habitats will have similar community composition if they are connected, but may diverge if not. This approach has led to ways to classify metacommunities based on different criteria (e.g., Leibold and Mikkelson, 2002). Application of this classification scheme to freshwater systems suggested that freshwater communities that tend to organize across ecological gradients, and nested communities (where species-poor sites are subsets of nearby species-rich sites) are most common (Heino et al., 2015).

The metacommunity concept is based on assumptions that patch area, degree of patch isolation, and interactions among species and their environment shape communities in individual patches; these assumptions are similar to those of island biogeography theory discussed in Chapter 11, although variably permeable connections may differ from "islands" that are considered to be very weakly connected to sources of colonists. Parris (2006) tested metacommunity assumptions by examining amphibian communities in urban ponds in Australia, and found that species richness was positively correlated with pond size and negatively correlated with degree of isolation, which was measured as the amount of surrounding area covered by roads. Also consistent with assumptions of metacommunity theory, Parris (2006) found that amphibian species richness in ponds was related to habitat quality.

Lake districts are areas with a relatively isolated cluster of lake types. One example is mountainous areas with many glacial lakes separated from other mountainous areas by warmer and more productive lowland lakes. These districts develop unique communities over evolutionary time with structured metapopulations (Catalán et al., 2009) and presumably metacommunity structure. Even areas with relatively closely situated ponds can have metacommunity structure of zooplankton communities (Cottenie et al., 2003). Lake districts have even developed the basis for modeling human behavior regarding choice of areas and even specific lakes to fish (Carpenter and Brock, 2004).

Metacommunities of bacteria can also occur in river systems (Jackson et al., 2014). These authors analyzed rivers in the Mississippi Basin. They found that sub-basins had differing bacterial communities, but the differences were based on relative abundances, even though many of the same species could be found everywhere. They also demonstrated differences in large scale patterns of particle-bound bacteria and planktonic bacteria. Given how little research has been done on bacterial diversity, the idea of bacterial metacommunities has seen little testing to date.

Lear et al. (2014) found that functional diversity of bacterial communities in a series of sampling sites between 3.5 and 60 m apart in three alpine ponds varied substantially. They used enzyme activities to assess functional characteristics of bacterial communities and DNA fingerprinting to assess community structure. They found that samples more than about 20 m apart mostly drove the spatial variability in the dataset. Muscarella et al. (2016) found that resource gradients among ponds (supply of dissolved organic carbon) could influence bacterial community responses to disturbances (pulses of inorganic nutrients). In a related study, Rofner et al. (2017) found that climate-related changes in terrestrial systems could alter bacterial communities in alpine lakes.

The metacommunity framework was used to investigate bacterial diversity among habitats in Lake Kitkajärvi in Finland (Langenheder et al., 2017). This study showed strong influences of local conditions within a lake in determining species community structure. There was weak evidence for dispersal limitation of microbes in the lake constraining community structure.

The metacommunity framework has also been used to investigate fish disease in a Swiss watershed. Carraro et al. (2017) used eDNA and models of eDNA decay to investigate the transmission of a myxozoan pathogen on salmonids. The pathogen has a complex life cycle; the salmonids are secondary hosts and a bryozoan is the primary host. They were able to create a spatial model of disease dynamics by considering prevalence of salmonids, the parasite, and the bryozoan in addition to spatial linkages among the different communities.

Large-scale and unexpected sources of connectivity can alter community structure. For example, dispersal by water birds of plants, animals, microbes, and fungi is common. This process can mediate gene flow among organisms. It can also be important in the spread of invasive species over large distances (Coughlan et al., 2017). Habitat connectivity along migratory flyways will influence such dispersal.

Network shape and connectivity can also alter metacommunity structure. For example, Seymour et al. (2015) connected microcosms in linear and in dendritic connection topologies. They found that rotifers and protists in dendritic networks had higher α diversity. Beta diversity was initially greater in linear configurations because of dispersal limitations, but became more similar to dendritic networks over time as dispersal progressed. Hanly and Mittelbach (2017) also observed temporal changes in the importance of dispersal ability using 179 plankton taxa in a series of Michigan ponds. Their results suggested that dispersal and network connectivity are primary drivers of alpha and beta diversity dynamics. In contrast, Erős et al. (2017) found niche based mechanisms of species sorting were more important than network structure in fish communities in the Danube River catchment. Vitorino et al. (2016) found dispersal and disturbance had strong influences on the fish community along a 100 km segment of the Upper Tocantins River in the Amazon Basin. Sarremejane et al. (2017) studied stream invertebrate community structure in 241 Finnish streams. They found dispersal limitation was most important in the smallest streams, but in larger rivers the neutral model was more applicable.

MACROECOLOGY

Brown (1995a) proposed *macroecology* to encompass very broad patterns in ecology. This area of research concentrates on understanding the distribution and abundance of species at large spatial scales (Gaston and Blackburn, 2008) in addition to other broad scale patterns, such as metabolism as a function of organism size. Researchers viewing these patterns over large scales generally use graphical and statistical approaches to establish validity of patterns. Patterns are then linked theoretical approaches to ecology. Some major areas explored in the context of macroecology include patterns of species richness, range size of animals, species abundance, and effects of body size on metabolism and food web structure. Mechanism is often difficult to determine in these studies, and Bell (2001) suggests that most of these patterns can be generated assuming species all have the same characteristics, but colonize new habitats based on random dispersal.

The species area curves we discussed in Chapter 11, form the predominant relationship between species richness and area, which was an early macroecological approach to species richness. Some small-scale experiments confirm the predictions of the patterns of greater numbers of species with greater areas, but the relationships appear to scale well to large islands and continents as well. Other generalizations of species richness are more difficult. For example, some have claimed that latitude should be negatively related to species richness. However, this is only true for some freshwater taxa. Salamanders, stoneflies, and caddisflies defy this predicted trend (Heino, 2011). The idea of species diversity has been expanded to include genetic diversity, leading to global-scale maps of the diversity for some aquatic organisms (Miraldo et al., 2016).

One area for which broad generalizations across scales has been particularly successful is metabolic theory, and relating biological properties to the sizes of animals

(Brown et al., 2004). In these approaches, body size describes the essential physical characteristics mass, length, volume, and surface area. Metabolic theory is a component of metabolic scaling theory, which addresses the central role of metabolic rates as related to body size in biological patterns and processes at all biological scales from individual cells to landscapes. It has been an active, if contentious, area of ecological research (e.g., O'Connor et al., 2007). Metabolic theory applies to phytoplankton growth rates (Kremer et al., 2017), longitudinal patterns of fish diversity (Bailly et al., 2014), and excretion rates of fishes and invertebrates (Allgeier et al., 2015).

Broad generalizations about animal range sizes as related to body size is another area of macroecology. There is enormous variation in range size of organisms, with some of the smallest known range sizes belonging to animals that are endemic to a single freshwater spring. In general, range size of fishes and vascular plants tend to be smaller than those of terrestrial animals (Brown et al., 1996). One common generalization is that body size should positively correlate with range size. However, Taylor and Gotelli (1994) analyzed geographical range size of 27 minnow species and found that latitude and longitude were better predictors of geographic range size than was body size, and that ancestral species had larger body sizes. After controlling for latitude, longitude, and phylogeny, they did find that body size was positively related to range size. Pyron (1999) found a positive relationship between body size and range size for North American sunfishes and suckers, and species with smaller ranges also tended to have more specialized or narrow niches. However, Blanchet et al. (2013) found that phylogeny was not related to range size of Canadian fishes; the best predictors of range size were traits that conferred the ability to disperse along river networks.

Other factors can also be important in range size, such as tolerances to temperature and other environmental gradients. Calosi et al. (2010) found that European diving beetles' (*Deronectes*) temperature ranges correlated positively to latitudinal range sizes, as well as dispersal ability, measured as wing length. Briers (2003) found that range sizes of British gastropods positively related to ability to survive under a wide range of calcium concentrations.

Macroecological patterns of abundance could be important because species that are less abundant could be more prone to extinction. The general pattern (which may be considered a law, McGill et al., 2007) is that communities have a few common species and many rare species. Freshwater animal species tend to follow a lognormal distribution of species abundance (Ulrich et al., 2010). Smaller species are more abundant (e.g., there are about 10^6 bacterial cells/mL of lake, stream or wetland water), and have a greater probability of dispersal (Fenchel and Finlay, 2004).

The relationships between metabolism and body size of animals have been noted for over 150 years. Work by Kleiber (1932) found that in homiotherms (warm blooded animals) metabolic rate was roughly related to body mass to the ¾ power across the smallest to the largest animals. As few of the dominant freshwater animals

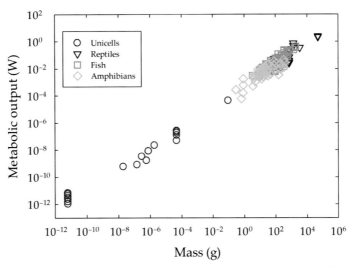

Figure 25.4 Metabolic rates and body size of various aquatic organisms. *Data from Gillooly et al. (2001), kindly provided by Jamie Gillooly.*

are warm blooded, this result is not so useful in freshwater ecology, and it is necessary to correct metabolic rates to a standard temperature before comparisons of metabolism across organism size is made (Fig. 25.4).

Relationships between body mass and metabolic rate are relatively well constrained across the roughly 17 orders of magnitude range of organismal body size. Therefore, it is possible to predict roughly how much energy each sized organism processes. Metabolic rate also varies by temperature, and once the metabolic rates are corrected for body mass, these relationships become clearer. Gillooly et al. (2001) provide relationships to predict lifespan based on body mass for many aquatic organisms, as well as relationships for resting metabolism as a function of temperature for mammals, birds, fish, amphibians, reptiles, unicells, plants, and multicellular invertebrates. These relationships can then be used to compare the ecology of different freshwater organisms, and can be used to understand how organisms control fluxes of materials from local to global scales. However, Ward et al. (2017) note that the size/metabolism relationships may not be so simple. As cells decrease in size, their surface area increases and their nutrient uptake capacity increases, leading to greater potential maximum metabolic rates.

Community structure could also scale by mass. Hatton et al. (2015) note that both aquatic and terrestrial food webs scale with mass. They also found that predator and prey mass and production both scaled similarly across ecosystems, and that highly productive systems should have a much larger prey base relative to predator mass.

MACROSYSTEMS

Researchers have defined very large-scale approaches to ecology as macrosystems biology. As this is a relatively recent field, the approach is still being solidified, but some definitions have been attempted. Heffernan et al. (2014) state, "Macrosystems ecology (MSE) treats the components of regions to continents as a set of interacting parts of a system" and is "...the study of diverse ecological phenomena at the scale of regions to continents and their interactions with phenomena at other scales". Thorp (2014) expands this to "... a hierarchically organized, integrated terrestrial, inland aquatic and/or marine ecological unit of large spatial extent (ca. 10^2-10^6 km^2 or more depending on the types and sizes of ecosystems present) whose temporal interactions within the unit and within regional through global processes are especially significant over periods of decades to millennia." The concept was specifically applied to flowing water by McCluney et al. (2014) as "watershed-scale networks of connected and interacting riverine and upland habitat patches."

It is difficult to separate these concepts from those in the previous section on macroecology. It seems that macroecology has an emphasis on animal biology, or at least specific organisms, and less emphasis on ecosystem processes. In this sense, macrosystems approaches attempt to be broader, but in another sense can be narrower because of an implicit limitation to large-scale spatial processes, whereas macroecology considers patterns across phylogeny as well.

Here, we describe attempts to create a predictive framework across the broadest scales under the concept of macrosystems. Such approaches are just becoming possible as more research becomes available from different parts of the world, and as technology improves in global mapping of aquatic resources (e.g., Verpoorter et al., 2014). In the previous chapter, we discussed abiotic gradients that might occur within a region. However, many aspects of these gradients can occur across larger spatial scales from very large watersheds to continents. Terrestrial ecologists have long recognized that large-scale patterns of climate, as driven by precipitation and temperature, can constrain plant communities and lead to morphologically similar assemblages of plants. This concept of biomes recognizes that high altitude tundra in the tropics bears certain similarities to high latitude tundra in the polar regions, and that plants have convergent adaptations to dry conditions of deserts throughout the world.

The concept of biomes has relevance for freshwater systems. The Stream Biome Gradient Concept is one attempt at such synthesis (Dodds et al., 2015). This scheme uses precipitation and temperature to explain some fundamental properties of streams, including some tied to the terrestrial vegetation (biome) that surrounds streams, and others to ways that climate influences hydrology. The core of the scheme is based on the tradeoffs between temperature (as influenced by altitude and latitude) and precipitation (Fig. 25.5).

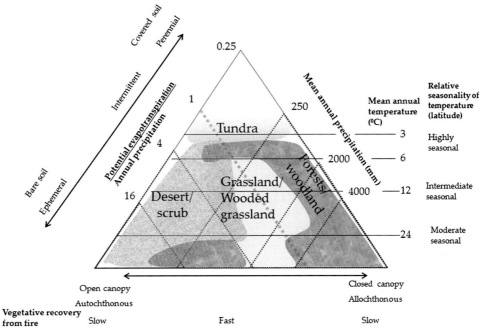

Figure 25.5 The stream biome gradient framework. *Image redrawn from Dodds et al. (2015).*

There are clear relationships between the density of permanent waters (rivers, lakes, and wetlands) and climate. High latitude areas tend to have high densities of lakes and wetlands because evaporation rates are low. Intermittent rivers and lakes are more common in desert areas (Fig. 25.6).

The assumption is that freshwater biomes exist. Consistent with the historical definitions for terrestrial systems, biomes are defined based on functional similarities, not phylogenetic properties. For example, deserts around the world have plants with convergent characteristics even if they are not closely related.

An example of the biome gradient framework is to consider a small stream in the high latitude tundra and one in a temperate deciduous forest. The tundra stream most likely has an open canopy, more primary production, colder water, and complete freezing over a substantial portion of the year, whereas the temperate forested site will have a closed canopy for the summer, warmer temperatures, more reliance on allochthonous material, and significantly less ice cover.

When streams are characterized by vegetation as determined by biomes, most run-off occurs from tropical evergreen forests. Grassland and temperate forests are important sources of runoff as well (Table 25.2). Such classifications can be used to quantify global fluxes of materials such as carbon input to oceans that flows through river networks and that can be linked to terrestrial vegetation types (Meybeck, 1993). These

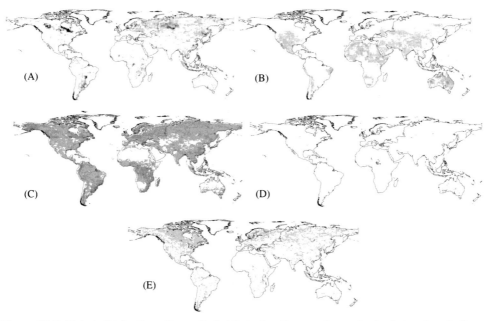

Figure 25.6 Global distribution of aquatic habitats. In all cases, increasingly dark colors indicate a greater density of the type of habitat. (A) Wetlands, (B) intermittent rivers, (C) perennial rivers, (D) intermittent lakes, and (E) perennial lakes. *Data from Cogley (1991).*

vegetation types can alter many things such as the stream channel morphology. For example, replacement of natural riparian forest with pasture can lead to narrower streams because grasses and other pasture vegetation encroach on the stream channel (Davies-Colley, 1997).

Most research on freshwaters has occurred in temperate forested zones, but may not apply everywhere. For example, tropical systems have unique characteristics based on temperature, lack of glaciation, and characteristic seasonality (Ramírez et al., 2015). Drier habitats may have different flow regimes, and those regimes can translate to different food web structure (Sabo et al., 2010).

An important consideration for freshwaters is the point where potential evapotranspiration exceeds actual transpiration. If this is the case, then excess water can fill lakes, bring up groundwater levels, allow for permanent streams, and allow vegetative ground cover and canopy to cover streams. The consequences for ecology, mediated by other abiotic factors, are numerous (Table 25.3). For example, drier areas have more bare soil, which alters erosion patterns, increases sediment transport into freshwaters, concentrates nutrients and salts into runoff, and alters water yield. More sediment in addition to increased nutrients and salts have obvious ecological relevance. For example, in lakes, drier areas are more likely to have higher conductivity and are

Table 25.2 Calculated areal cover of seven vegetative or cover classes and associated runoff, counts of perennial and intermittent rivers, and combined vegetation types

Vegetation class	Coverage		Runoff		Water yield (m/year)	Rivers			
	km²×10⁶	%	km³/year	%		Perennial	% (counts)	Intermittent	% (counts)
Broadleaf evergreen forest	13.4	9.0	14,663	29.8	1.09	16,269	10.4	407	1.1
Grasslands–wooded grasslands	42.2	28.4	13,709	27.9	0.33	39,676	25.4	15,085	39.2
Temperate forests, seasonal forests	28.7	19.3	9,438	19.2	0.33	57,985	37.1	3,069	8.0
Cultivated land	13.3	8.9	4,377	8.9	0.33	14,820	9.5	4,141	10.8
Ice	15.9	10.7	3,838	7.8	0.24	1,446	0.9	0	0.0
Tundra	7.1	4.8	2,235	4.5	0.32	20,142	12.9	6	0.0
Shrub–desert	27.9	18.8	909	1.8	0.03	5,968	3.8	15,747	40.9
Total	148.4	100	49,169	100		156,306	100	38,455	100

The seven classes are ordered by runoff.
Source: From Dodds (1997a). Reprinted by permission of the *Journal of the North American Benthological Society*.

Table 25.3 Features of lotic and lentic waters that are expected to vary from dry to wet habitats

Category of factors	Specific factor	Trend going from dry to wet
Terrestrial	Vegetative cover	Increasing from desert to forest
Physiographic	Intermittence	Decreasing
	Distance between water bodies	Decreasing
	Connectivity of habitats	Increasing
	Watershed size per water yield	Decreasing
	Temperature	Decreasing
	Depth to groundwater	Decreasing
Light	Reaching small water bodies/ littoral	Less (but seasonal in temperate)
Chemistry	Turbidity	Decreasing inorganic turbidity
	Total ions	Decreasing salinity
	Nutrients	Decreasing nutrients
Biological	Trophic state	Increasingly oligotrophic
	GPP/ER	Decreasing
	Desiccation resistant life stages	Decreasing
	Fish and amphibian diversity	Increasing
Anthropogenic	Proportion of natural lakes to reservoirs	Increasing
	Per area effect on water	Less
	Proportion of water abstracted	Less

more frequently saline systems because terminal lakes (those with no outflows) form where evaporation is the main water loss from the lake (Read et al., 2015). The ecological ramifications of salinity were discussed in Chapter 15.

Freshwater vertebrate species richness tends to be greater in wetter biomes. When Dodds et al. (2016) assessed gradients from deserts to forests across all continents but Antarctica, they found lowest diversity of fishes and amphibians in deserts, intermediate levels in grasslands, and the highest diversity in forests (Fig. 25.7). In the Amazon, total fish diversity was greatest and the total amphibian diversity was greatest in Australia. However, diversity may not link to food web structure in ecosystems; Jardine (2014) found latitudinal changes in species in Australia that did not link well with food web structure.

The biome gradient concept can also apply to microbial communities (Fig. 25.8). Battin et al. (2016) describe how stream and hyporheic microbial communities are expected to vary across altitude from high altitude tundra to lower altitude boreal forest habitats. They find considerable differences based on dominate type of substrata and amount of hyporheic habitat (which increases downstream).

Nutrient ecoregions (e.g., Omernik and Griffith, 2014) are another spatial classification method for freshwaters that are based on expected background concentrations of nutrients and was discussed in Chapter 18. One large missing point in the freshwater biome gradient concept is that geology also determines many important parts of

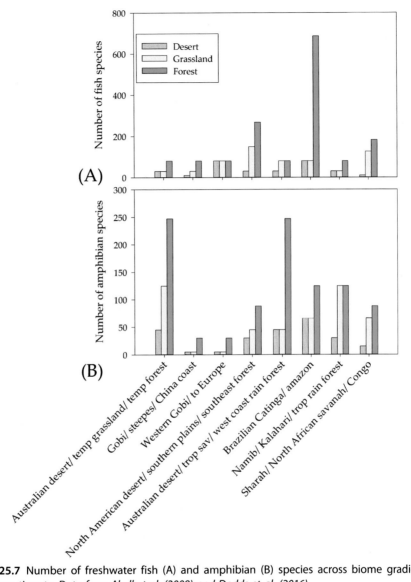

Figure 25.7 Number of freshwater fish (A) and amphibian (B) species across biome gradients on various continents. *Data from Abell et al. (2008) and Dodds et al. (2016).*

aquatic systems, including the types of aquatic habitats that will form, water chemistry, and turbidity. Watersheds cross terrestrial biomes, as well as areas of different geology. This system has been developed for North America (Fig. 25.9), but has not been developed for many other areas of the world.

Taxonomic characterization of ecoregions based on animal phylogeny also exists (Abell et al., 2000). This approach to freshwater ecoregions (Fig. 25.10) is most useful

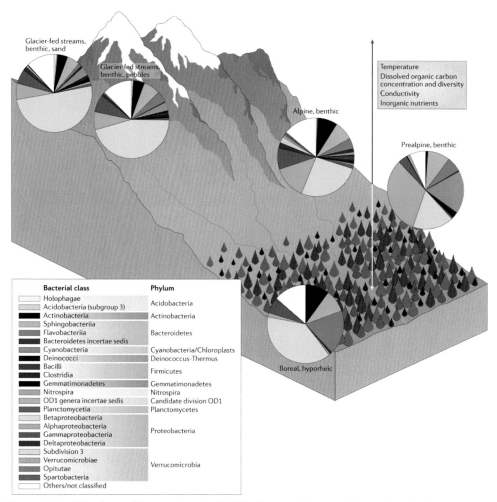

Figure 25.8 Altitude gradients and microbial biodiversity from tundra to boreal forests (Battin et al., 2016).

for conservation. It can be used to identify high diversity unique areas of freshwaters. Then, we can better understand the regions where preservation or restoration are most crucial.

Nonlinear relationships associated with anthropogenic impacts may complicate very large-scale assessment of biological patterns across biomes. Woodward et al. (2012) conducted leaf decomposition studies across Europe. Decomposition rates were low in pristine sites because of nutrient limitation. Decomposition rates reached a maximum in moderate nutrient sites. In heavily nutrient polluted sites, decomposition rates were inhibited.

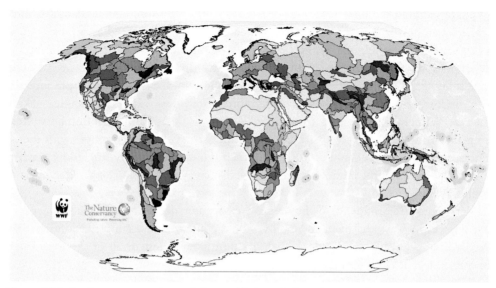

Figure 25.9 Freshwater ecoregions of the world. See Abell et al. (2000) for details. *Image courtesy: World Wildlife Fund and The Nature Conservancy.*

Figure 25.10 USEPA nutrient ecoregions based on geology, land use, and ambient water quality parameters. Ecoregions are overlaid on major river basins. 1, Willamette and Central Valleys; 2, Western Forested Mountains; 3, Xeric West; 4, Great Plains Grass and Shrublands; 5, South Central Cultivated Great Plains; 6, Corn Belt and Northern Great Plains; 7, Mostly Glaciated Dairy Region; 8, Nutrient-poor, Largely Glaciated Upper Midwest and Northeast; 9, Southeastern Temperate Forested Plains and Hills; 10, Texas-Louisiana Coastal and Mississippi Alluvial Plains; 11, Central and Eastern Forested Uplands; 12, Southern Coastal Plain; 13, Southern Florida Coastal Plain; and 14, Eastern Coastal Plain. Areas separated but of the same color are the same Ecoregion. *Image obtained from the USGS and modified.*

Understanding the idea of biomes becomes even more important as we move into a time when humans are impacting freshwaters globally, and given that countries such as Russia, Brazil, the United States, Canada, and China are very large and span multiple biomes. Evolution and water quality and quantity management occur on very large scales, so the large-scale approach has specific relevance to how people interact with freshwaters (Scholes, 2017). Furthermore, most freshwater research has taken place in Western Europe and the United States. Little is known about how well temperate-zone results apply to tropical countries or developing countries with moderate amounts of research and biomes other than those most heavily studied by aquatic ecologists. Still, tropical limnology has received increased attention (e.g., Tundisi and Tundisi, 2012).

Very large-scale influence of humans on freshwater decreases the value of ecosystem services in ways that depend upon regional context. For example, countries that receive little precipitation are more likely to appropriate most or all of the freshwater for human use, leaving little for the biota that inhabits it (Dodds et al., 2013).

While they have not been identified as macrosystems, many researchers working on large-scale patterns of ecosystem rates and characteristics have been characterizing properties of systems at the macroscale. For example, Smith et al. (2003) determined baseline nutrient conditions for the contiguous United States with a model that extrapolated reference conditions regionally. Beaulieu et al. (2011) created a global model to estimate nitrous oxide emissions from streams. Similarly, Harrison et al. (2009) created a model to make regional to global estimates of nitrogen retention by lakes and reservoirs, and Wollheim et al. (2006) made a model to estimate the effect of large rivers on nutrient retention.

Large-scale patterns of freshwater systems are being rapidly refined as people around the world understand the importance and intrinsically fascinating nature of freshwater systems and as the latest technology continues to spread with increasing speed. Consequently, a global framework that addresses how to make comparisons of freshwater systems among biomes is relevant to the future of freshwater science and how it applies to society. We are living in an exciting time where these large-scale approaches can be successful given the rapidly accumulating information on freshwaters globally and concurrent remote sensing, analytical approaches, and computing methods that can provide information at these scales.

SUMMARY

1. Scale is a central issue in ecology in general, as well as in freshwater ecology.
2. While research is relatively easy at smaller scales, applying results of this research to large scales can be difficult and can give misleading results, particularly if larger scale

systems exhibit emergent properties that cannot be predicted from individual parts of the system.

3. Landscape ecology is concerned with the spatial dynamics of fluxes of materials, organisms, and energy in ecology. It seeks to understand issues such as habitat fragmentation, and the relationships of populations and communities of organisms to spatial heterogeneity.

4. The concept of metapopulations is central to landscape ecology and conservation.

5. The metacommunity concept is based on assumptions that habitat patch size, degree of isolation, and quality are drivers of community structure. Neutral models, in contrast to niche-based models, essentially assume all species are equal, and have been used to explore metacommunity dynamics and related patterns of biodiversity.

6. Macroecology is the study of very broad patterns in ecology, which can include patterns across organism types, as well as spatial patterns such as species–area curves.

7. The concepts of macrosystems and macroecology are similar.

8. One macrosystem approach that could be useful to freshwater ecologists is the freshwater biome gradient framework, which explores how climate leads to predictable ecological properties in freshwater systems. However, research in this area is not yet well developed.

QUESTIONS FOR THOUGHT

1. Can you think of any good candidates for emergent properties in freshwater ecological systems?

2. Are lakes, rivers, or wetlands better systems for large, landscape-scale studies?

3. Why would the landscape ecology of karst aquifers vary from that of an unconsolidated sand aquifer?

4. Should a system based on freshwater biomes consider how humans variably alter different biomes?

5. How different is landscape ecology from macroecology?

6. Can you provide specific examples of freshwater habitats where metapopulations could be important?

7. Why is dispersal ability tied to the existence of metapopulation structure?

8. How do you expect annual lake productivity to vary from high latitude to low latitude?

9. How do you expect autochthonous and allochthonous production to vary in streams across a gradient from dry to wet habitats?

CHAPTER 26

Conclusions

We have passed the point at which we can continue to take unabatedly without casting back some comprehensive understanding and wise use in return. Nature is remarkably resilient to human insults. Yet, humans must learn what are nature's dynamic capacities because excessive violation without harmony will only unleash her intolerable vengeance. The very survival of humankind depends on our understanding of our finite freshwater resources.

Robert Wetzel (2001).

Figure 26.1 A stream on Konza Prairie in northeastern Kansas (top), a nearby suburban stream (middle), and an agricultural stream in Indiana (bottom). Agricultural stream photo by N. A. Griffiths.

Freshwater Ecology
DOI: https://doi.org/10.1016/B978-0-12-813255-5.00026-0

Freshwater ecology is a tremendously rich and detailed field of study. We have covered only the most basic aspects of it in this text. We hope that students will have gained some useful insights that can be applied in the making of informed decisions regarding how our aquatic resources are used or studied. Hopefully, some of these students will be inspired to further academic study and careers in aquatic ecology. Regardless, we have tried to make the concepts and applications understandable and interesting to a wide variety of students. We presented details in some cases to illustrate complexity of real systems. Comments from students and instructors that would improve the text are welcome, and we deeply appreciate those many comments we have received so far.

Those of us in developed countries who are able to read (and write) this book live in a golden age. It is an incredible luxury and privilege to pursue academic interests. As the world's population increases, our appetite for resources grows, the global impacts of people on the environment become ever more pronounced, and the existence of unspoiled aquatic habitats will become increasingly rare. A large number of the world's human population is malnourished and impoverished. Of specific relevance to this book, many of these people do not have access to safe and sanitary drinking water. They are exposed to diseases carried by freshwaters (e.g., dysentery) or animals that live in them (e.g., mosquitos). The minority of the people who hold the majority of the economic and political power do so, in part, at the expense of the environment. Neither the rich nor the poor have shown much inclination to conserve our aquatic resources in the past, and it is unlikely that they will do so in the future. The depressing consequence across the Earth is that freshwater biodiversity, quality, and quantity will decrease drastically in our lifetimes; these realities are already well underway. As competition for limited aquatic resources increases, aquatic ecologists will have fewer baselines against which to compare impacted ecosystems; we need to study and appreciate what we have now because there may not be an opportunity to do so later. In addition, a mechanistic understanding of the environment is our only chance to predict a freshwater world that has no analog.

There is reason for hope. Many rivers and streams in developed areas are cleaner than they were in the middle of the 20th century. Knowledge of methods for lowering eutrophication, ways to avoid contamination by sewage, techniques for mitigation of acid precipitation sources, and other technological advances have led to improvements. We are finding ways to manipulate mosquito populations and snails that carry schistosomiasis through understanding in biology. Humanity is applying results from our freshwater science worldwide, and the public has come to realize the importance of aquatic resources as scientific understanding continues to expand. Ecological economics has explicitly tied freshwater ecosystem services to human economies.

Unfortunately, some of the problems associated with human activities are not reversible over a human life span. Extinction is a clear example. There is no known case of complete removal of a pollutant from groundwater. Unwanted species introductions are the same; once they are established, eradication is generally impossible.

Much of the future job of aquatic ecologists may be in "damage control" or restoration. Ecologists will continue to provide information on what is necessary to maintain ecosystem function and preserve desirable species. A more detailed knowledge of aquatic ecology than we currently possess is necessary to provide this information. For example, we have limited understanding of the links between diversity and ecosystem function. We are just beginning to understand microbial diversity, and know little of the consequences of alterations of that diversity. The redundancy of ecosystem services by species (i.e., what is the minimum assemblage of species necessary to maintain productivity and the ability to neutralize pollutants in aquatic ecosystems) is not well documented. We simply cannot predict any but the extreme effects of our impacts on aquatic habitats. Likewise, preserving species requires detailed knowledge of its biology. Such knowledge is sorely lacking for all but the most popular game fishes. Natural history is not currently fashionable, but we need specific details of life history for most species nonetheless. A specific example is the biological species concept; we have no idea if most closely related organisms are able to reproduce sexually or not.

Many aquatic ecologists enter the field because of a love of freshwater and the organisms living in it. This leads to high levels of satisfaction derived from studying aquatic environments. Many scientists with such motivation never directly study environmental problems. However, their contributions may increase our ability to understand aquatic habitats. The graduate study of the first author of this text centered on a spring-fed pool in which the cyanobacterium *Nostoc* grows unusually large (Fig. 26.2). Scientific curiosity motivated this study. However, the Forest Service eventually used the research to develop a management plan to protect the *Nostoc* in that pool. It can be difficult to predict what scientific information will ultimately be useful. The second author of this text has had a life-long fascination with insects, amphibians, and reptiles. His fascination with the communities inhabiting Panamanian streams and curiosity about what would happen when disease eliminated amphibians from the ecosystem led him to work on disease-driven amphibian declines in tropical streams. While inspired by a simple fascination with stream biodiversity, this research ended up providing quantitative insight in to the ecological consequences of declining biodiversity.

Some information presented in this book may seem marginally useful to students taking a class as a requirement for a program specializing in other aspects of aquatic sciences. For example, a fisheries student may have difficulty motivating interest for nutrient cycling, and a student interested in algae may yawn at the thought of studying fishes. However, experience has shown that managers who do not take a holistic approach are doomed to make mistakes. Such mistakes may lead to financial loss and permanent ecosystem damage. For example, people have intentionally introduced many aquatic species to provide benefits, only to later cause unintended problems that

Figure 26.2 Textbook authors and some systems that they are interested in. (A) Tadpoles on rock in a fast moving portion of Rio Maria, El Valle, Panama prior to the disease *Batrachochytrium dendrobatidis* extirpating frogs from this stream. Each tadpole is about 4 cm long. (B) Matt Whiles. (C) The largest documented colony of *Nostoc pruniforme* from Mare's Egg Spring. (D) Walter Dodds.

were not anticipated. More knowledge of how aquatic ecosystems work may decrease the chances of making such mistakes.

If you are reading this, you are a persistent reader! Why read a part of the text unlikely to be on a test? This conclusion is for you. Please take time to reflect on what you have learned from this text, and take with you the valuable parts and make the world a better place. Get your feet wet, enjoy the water.

APPENDIX: EXPERIMENTAL DESIGN IN AQUATIC ECOLOGY

Figure A.1 Sealed chambers with periphyton covered rocks incubated under a range of light intensities to a establish relationship between photosynthesis and irradiance. Each replicate is incubated under a different light level (independent variable) and oxygen production is measured (dependent variable).

Experimental results supplied in a text or course setting are often presented as facts with little information on the scientific process that was used to decide if an effect was "real" or not. Obtaining much of the following information from a statistics course or text would be preferable (e.g., Sokal and Rohlf, 1981; Zar, 1996); at a minimum, some background in statistical methods is recommended. Because ecological knowledge requires formal statistical treatment, the following brief introduction to experimental design is provided for those having little or no statistical experience. We have found that it is common for students to reach the advanced undergraduate level without much understanding about how science is really done with respect to design and analysis of experiments. While this treatment is not specific to freshwater ecology, we hope this will convince even math-phobic students of the necessity of statistical training for the study of aquatic ecology.

Generally, when a scientist approaches a problem, she or he has an idea of what causes a particular phenomenon to occur. We state this initial expectation formally as a hypothesis (though such formal statement may not always occur, and initial hypotheses are often discarded as naive). The hypothesis states that some factor (or factors) has an effect. We can also think of a specific question that is related to a hypothesis. For example, consider the question "Does microbial diversity in lakes decrease invasive ability of unwanted algae?" This can be stated as a hypothesis: lakes with greater microbial diversity are more resistant to invasion by unwanted algae. The job of the scientist then is to test the hypothesis. Other authors have discussed the construction of hypotheses and their testing and usefulness to ecologists (Quinn and Dunham, 1981; Pickett et al., 1994). It will be helpful to know some basic statistical terms before discussion of the methods (Table A.1).

Ecologists use several methods to test hypotheses; some rely on formal experimental methods, and others on observation of natural history or simulation. Although formal experimental methods can provide unequivocal results under the exact laboratory conditions, they are not always possible to apply to the real world or can be limited in scope. All these methods have a place in the aquatic ecologist's toolbox.

NATURAL OBSERVATIONS AND EXPERIMENTS

Historically, much of biology was an observation of patterns of natural history and the use of these patterns to support explanations of the way the natural world works. Many of these approaches are natural experiments. The theory of evolution is an

Table A.1 Some common statistical terms

Term	Description
Mean	Average
Variance	The amount of variation about the mean, often denoted by standard error or standard deviation
N	The size of sample (total replicates)
Normal distribution	The expected distribution when a factor is sampled randomly; most statistical tests rely upon the assumption that the data are sampled from a normal distribution
Nonparametric test	Statistical tests that can be used when data distribution is not normal
P value	The probability that a statistical test shows that the hypothesis is correct
Replication	More than one sample from an independent treatment
Independent variable	A variable that influences the dependent variable
Dependent variable	A variable that changes as a function of the independent variable or variables

excellent example. Prior to advances in genetics and ultimately molecular biology, scientists did not understand the precise mechanisms by which evolution could occur. This notwithstanding, Darwin set forth an impressive array of observations that strongly supported his theory. Some of his observations were of domestication and how selection by breeders could create vastly different looking animals or plants, an experiment of sorts. Biologists continued to amass support for evolution over the years, and today it is the central pillar of biological science. The original theory was proposed and accepted because of the vast weight of natural history observations that supported it. Only relatively recently have experiments been devised to directly test predictions related to evolutionary processes. Currently, much of molecular biology takes a similar approach. Molecular biologists describe genes and molecular interactions (pathways) in much the same way that natural historians describe communities. However, as they fit the pieces together, an understanding of the larger network of controls in cells is emerging. Other fields, such as astronomy, rely heavily upon observation of pattern (Dodds, 2009).

Some facets of ecology and environmental sciences rest mainly on correlation. Correlation is a way to measure the co-occurrence of two variables, and it may provide the only available method to study some systems. For example, we cannot directly test results of the models that predict global warming from carbon dioxide experimentally because there is only one Earth and replication is thus not possible. However, the models continue to predict observed trends of global warming giving stronger and stronger certainty that they are correct. If a polluter contaminates a stream with an unusual organic compound of unknown toxicity and all the fish and aquatic invertebrates die, the most prudent option is to assume that the chemical caused the deaths. Maybe a scientist would start small-scale experiments, but regulators would probably halt releases to the same or other streams, and would not allow experiments to replicate the entire event. Likewise, when large-scale changes are observed in large lakes, river systems, or aquifers there is no feasible way to perform replicated experiments on these systems, so other techniques are required. Correlation is one of these techniques.

The primary problem with correlative approaches lies in the inability to separate cause from effect. A tongue-in-cheek example of this is a plot of the number of churches per town against the number of bars per town for a variety of towns with different populations. The plot will show a positive correlation. The correlation occurs not because bars cause more churches to be built but because larger towns have more churches and more bars. Correlation can support causation, but does not provide a mechanism to support ecological hypotheses.

A classic example of the power of correlation from limnology is the controversy regarding the causes of eutrophication. More phosphorus is certainly positively correlated with algal biomass in lakes. However, not until David Schindler and colleagues conducted whole-lake phosphorus addition experiments (see Fig. 18.1) did polluters' denial that increased phosphorus led to eutrophication become ineffective. After the causation was determined, the correlation was more likely to be used to predict responses of lakes to alterations in phosphorus supply.

We can strengthen natural experiments in several ways (Carpenter, 1989). Time series can provide replication, as described below for whole-system studies. For example, if a lake exhibits one level of fish production for several years, then researchers add nitrogen and fish production increases, we assign a higher probability that the nitrogen addition caused the increase in production. The certainty is greater if researchers discontinue the nitrogen addition and the fish production decreases to its original level. Another way to strengthen such observations is to compare two similar, if not identical, ecosystems.

Even given the potential problems, natural experiments have undeniable benefits. Perhaps the strongest of these benefits is that the observations and correlations occur in entire systems, not in a beaker or a bottle. As such, natural, ecosystem-scale experiments do not require extrapolation. Natural experiments may be most relevant to community- or ecosystem-level processes that operate in the real world (Carpenter et al., 1995), and the results may contradict those extrapolated from small-scale replicated experiments (Schindler, 1998). Collaborations that involve both small scale, manipulative experiments, as well as ecosystem-scale observations and sampling are obviously the most powerful approaches to testing ecological hypotheses.

MULTIVARIATE METHODS

Examination of complex natural observational data sets can require multivariate methods. Many of these are becoming more common in the ecological literature, so at least a passing familiarity with the methods is desirable. Multivariate techniques allow analysis of community or assemblage level data and can be powerful tools for finding patterns in large data sets. We present an overview of some common multivariate methods in Table A.2.

Multiple regression is one of the more commonly used multivariate techniques. While regression examines the relationship between one dependent variable and one independent variable, multiple regression allows for examination of multiple independent variables and a dependent variable. For example, an ecologist may be interested in how nutrient concentrations, turbidity, and substrata composition (independent variables) influence the biotic integrity scores of streams (dependent variable). Multiple regression can predict the relative influence of each independent variable on biotic integrity scores. An important limitation of multiple regression is that correlation among the independent variables is problematic and limits the reliability of the results.

Table A.2 Description of some common multivariate statistical methods

Method	Description	Comments
Multiple regression	Examine influence of multiple independent variables on a dependent variable	Sensitive to correlation among independent variables
Multiple analysis of variance	Examine influence of independent variables on multiple dependent variables	Accounts for correlation among dependent variables
Ordination	Finds patterns in complex data sets	Often used for data exploration and reduction of variables
Cluster analysis	Groups samples based on similarity	Creates groups based on distance measures
Discriminant analysis and classification	Identify variables that define predesignated groups	Classification used to assign samples or sites to groups

As such, careful consideration of which independent variables to include in the analysis is important. In cases where there are numerous potentially correlated variables, ordination techniques described below can be used to reduce the number of variables used or combine some of them.

Multivariate analysis of variance (MANOVA) can indicate responses of multiple dependent variables to independent variables. In analysis of variance (ANOVA), multiple independent variables may be included, but only one dependent variable is considered. For example, an ecologist may want to examine how stream temperature, conductivity, and pH differ among defined geographic regions; the question is best addressed using MANOVA because three dependent variables are being examined. An alternative approach is to run separate ANOVAs for each dependent variable, but this increases the probability of committing a type I error, whereby the chance of finding a significant difference increases with the number of tests, particularly if any of the variables are correlated. MANOVA accounts for potential correlation among the multiple dependent variables.

Researchers often employ ordination and cluster analysis techniques to examine patterns, including similarities and differences, among communities and assemblages. These methods are powerful tools of community ecologists. For example, Fig. A.2, which was generated using nonmetric multidimensional scaling analysis, a common ordination technique, depicts clear differences in macroinvertebrate assemblages associated with different substrata in a river. Patterns like these can be difficult to identify by just looking at data or using standard univariate techniques. Various statistical procedures such as analysis of similarity can test for statistical

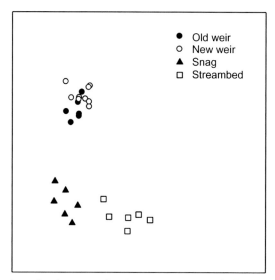

Figure A.2 An ordination plot (generated using nonmetric multidimensional scaling) showing differences in macroinvertebrate assemblages associated with all dominant substrata types in the Cache River, Illinois, except for those on old and new rock weirs. The other substrata are submerged wood (snags) and the ambient scoured clay streambed. Analysis of similarity indicated that all assemblages except for those on old and new weirs were significantly different. *Data from Walther and Whiles (2008).*

significance of patterns (Clarke, 1993). There are also statistical methods for identifying taxa or other variables that drive observed patterns.

There are numerous different ordination techniques, including principle components analysis and canonical correspondence analysis, each with its own set of assumptions, advantages, and limitations; the use of a particular method depends on the specifics of the data set and the research questions asked (Ludwig and Reynolds, 1988). Cluster analysis is similar in that it is often used to group samples, and is based on similarities. Cluster analysis uses measures of distance among samples to construct groups and does not rely on the multivariate normal distribution.

Discriminant analysis and classification techniques can separate samples or sites into groups or allocate samples or sites to groups based on multivariate data. Discriminant function analysis is often used in an exploratory manner to identify variables that define predesignated groups. For example, an ecologist might designate ponds in a region as "impaired" or "unimpaired" based on field observations. Discriminant function analysis based on water quality and biological data collected from the ponds could then identify the variables that separate impaired and unimpaired ponds. Similarly, classification techniques can be used to assign a new pond to a category based on measured variables.

SIMULATION MODELING

Simulation modeling is an additional method that researchers can use to explore possible hypotheses. Perhaps we can consider these models virtual experiments. In this case, researchers build computer models of a system with the desired level of detail and representation of processes, and the system can be perturbed as desired. This is the main approach used to investigate global climate change, atmospheric dynamics, and large-scale physical oceanography. Benefits to modeling are that it is relatively inexpensive, easily replicated, and may indicate critical factors in complex systems that control the observed behavior. More detailed studies targeting these critical factors can occur once the critical factors are determined by modeling. Thus, modeling can improve efficiency of environmental research.

The results of such models are strengthened in several ways. Modelers use *sensitivity analysis* to test systematically the sensitivity of the model to the assumptions used to construct it. For this type of analysis, the modeler compares outputs based in variation of input parameters under different model scenarios. If much uncertainty exists over the value of a parameter that has a great influence over the behavior of a model, the predictions of the model are suspect. The scientist may then design the research program to characterize the uncertain parameter more thoroughly. Another method for validating a model is *hind casting*. In this method, the model predicts a system response to the known response from the past. This is possible in systems in which sufficient data are available from the past but may be difficult where conditions are changing over long periods of time (e.g., under global change).

MANIPULATIVE EXPERIMENTS

Manipulative experiments are often necessary to provide more formal proof of a hypothesis. To test a hypothesis experimentally, all other factors must be held constant, the factor of interest must be varied, and the effect must be noted. There is no way to know which factor caused a response if multiple factors always were varied. In the example given in the previous section, the polluter that released the organic chemical into the stream could argue that the chemical was not necessarily responsible for the fish and invertebrate deaths and that they would have died anyway. Such an argument would hold less water if the fish and invertebrates in nearby, similar streams did not die.

Deciding which is the control and which is the experimental treatment can be difficult because this designation depends on how the hypothesis is stated. For example, if we form the hypothesis that ultraviolet (UV) light lowers fertility of frog eggs, then no UV light is the control, and UV light exposure is the experimental treatment. Conversely, if the hypothesis is that lack of UV increases fertility of frog eggs, then

UV light can be the control and removal of UV the experimental treatment. Neither way is right nor wrong, but often it makes ecological sense to frame a hypothesis in one particular way. In the previous example, ambient UV exposure from sunlight might be a control, and increased UV equal to that expected to result from decreased ozone in the upper atmosphere could be the treatment.

The next problem that arises is the certainty of any particular result. Say that we treated one frog egg with increased UV and it did not hatch, and the control egg treatment with ambient UV did hatch. Have we supported the hypothesis that increased UV will lower frog egg fertility? Perhaps the UV-treated egg was not fertilized properly in the first place. Replicating or repeating the experiment could increase the certainty of the result. If we did the experiment five more times and always had the same result, this would increase the certainty that we had proved the hypothesis. This replication is another key point in experimental design.

Statistics are involved when scientists replicate an experiment. Statistics allow comparisons among experimental and control treatments and formal expression of the degree of certainty that the hypothesis is true (Tables A.1 and A.3). The statistical test is used that matches the experimental design. These tests result in a probability value to express certainty in the trends that researchers observed in their experiment. The research community has adopted the convention that a result is not significant until the investigator is 95% certain. However, this is purely an arbitrary value. When human lives are at stake, 95% is probably not good enough. When publication of an ecological paper is at stake, 95% is probably sufficient.

We present experimental design here in a simplistic fashion. Replication is often a problem for the environmental scientist. On the one hand, many do not understand

Table A.3 Description of some common statistical methods

Method	Description	Comments
T-test	Comparison of one mean and a number or of two means	Common, simple procedure for assessing differences between means
Regression	Fits the best line to the data (minimizes variance)	Can be linear or nonlinear, with one or more independent variables
Correlation	Fit to a simple linear relationship	Does not denote causation; normally used for data exploration
Analysis of variance (ANOVA)	Tests the effects of multiple treatments on one dependent variable	Commonly used in replicated manipulative experiments
Chi square	Compares expected versus observed occurrences in categories	Commonly used to assess whether observed patterns are random or not
Meta-analysis	Compares results from many studies	Arnqvist and Wooster (1995) and Gurevitch and Hedges (1993)

replication; on the other hand, we cannot replicate some experiments or systems, as discussed previously.

Pseudoreplication can be a problem because it can lead to undue faith in results when scientists do not understand replication correctly. Pseudoreplication occurs when "replicate" samples are taken from one treatment (Hurlbert, 1984). For example, if we are testing the effects of nitrogen on water quality in two watersheds, and fertilize one watershed with nitrogen and leave the second untreated, it is inappropriate to call repeated water samples from each watershed statistical replicates. We may be able to strongly infer results, but inference is limited to just the sampled watersheds.

WHOLE-SYSTEM MANIPULATIONS

Whole-system manipulations can be very informative, but are often plagued with statistical problems because of low, or no, replication. Freshwater systems can be hard to replicate (e.g., no two streams are exactly alike) and funding and logistical constraints often preclude replicating ecosystem-scale manipulations even if suitable replicates were available. Nonetheless, some landmark studies in freshwater ecology have involved system-level manipulations; some of these studies were discussed in Chapter 24. Ultimately, combined approaches that use large-scale manipulations to examine "real world" responses, coupled with smaller scale experiments to isolate variables, will be most powerful.

To compensate for statistical limitations, large-scale manipulations generally involve collection of robust, long-term data sets. Ideally, the system to be manipulated and a designated control system are sampled for a long period of time to develop a statistical relationship between them prior to the manipulation. The two systems do not need to be identical, but there needs to be a consistent statistical relationship between them and they should respond similarly to regional climate, etc. The frequency and magnitude of sampling necessary will depend on the variability of the systems and measured variables; highly dynamic systems or variables might require multiple years of monthly sampling to develop statistical relationships. Once a robust premanipulation data set has been developed, the manipulation is applied and sampling continues. Again, the frequency and duration of sampling following the manipulation will depend on the variability of systems, along with the magnitude of the response to the manipulation.

Large-scale manipulations of this type are often analyzed using a before–after, control-impact (BACI) design (Snedcor and Cochran, 1980; Stewart-Oaten et al., 1986; Stewart-Oaten et al., 1992). BACI designs are based on statistical relationships of response variables between control and manipulated systems before and after a manipulation, and whether those relationships change after the manipulation is applied.

Randomized intervention analysis (RIA) is a nonparametric alternative to BACI. That is, RIA and other nonparametric tests do not require that data are normally distributed, as is the case for parametric techniques. RIA has been used to effectively show changes in system-level parameters associated with manipulations in a variety of whole-system lake and stream studies (e.g., Carpenter et al., 1989; Wallace et al., 1997; Connelly et al., 2008). Like BACI, RIA is based on relationships between systems before and after a manipulation, and whether those relationships change after manipulation.

Ultimately, the results of any experimental endeavor are a function of the quality of the experimental design and data. Careful consideration of the questions addressed and associated hypotheses should guide experimental design and selection of specific statistical analyses. Freshwater systems present some significant challenges, and consultation with a statistician during the design and analysis phases of a project is recommended.

SUMMARY

1. There are several general types of experiments: natural experiments, simulation models, and manipulative experiments. Each has its own strengths and weaknesses.
2. Natural observation and experiments may be the most realistic but often are not replicated, and assigning causation to statistical results is difficult.
3. Multivariate techniques can be powerful tools for identifying patterns in large, complex data sets. Ordination techniques can be used to graphically convey patterns in data sets.
4. Simulation models may offer a way to approach intractably large or complex systems and provide insights into key factors in these systems.
5. Manipulative experiments can be replicated and the results subject to statistical determination of certainty of outcome. However, such experiments require replication, and this may lead to small-scale treatments or artificial conditions that make relevance of the results to the system of interest questionable.
6. Whole-system experiments can be powerful for understanding ecosystem responses to disturbances, but they require careful consideration because of limited replication and associated statistical issues.
7. All of the methods are useful to aquatic ecologists. The trick is in asking the important questions and determining the best method to use to obtain satisfactory answers. Consulting with a statistician while planning an experiment and analyzing data is a prudent approach.

Glossary

This glossary explains words used in the text, and also words common in the freshwater literature that do not appear in this text. We hope that this glossary will be useful for students reading the scientific literature in aquatic ecology.

A area

A_d drainage basin area

A_0 surface area of a lake

A_z area at a specific lake depth (z)

Abiotic oxidation a chemical oxidation in the absence of organisms

Abiotic sulfur oxidation reduced forms of sulfur combine with O_2 for a net release of potential energy spontaneously without biological mediation

Absorb to soak up (compare to adsorb)

Acidity the proton ion concentration (pH); also the ability of a solution to donate protons to a base

Active filters organisms that sieve particles from the water column by actively pumping water or creating currents

Acute over short periods

Acute toxicity poisoning with large pulses over short periods

Adaptive radiation relatively rapid evolution of a single species into many species

Additive toxicity when two toxicants together are more toxic than individually

Adhesion sticking to other things

Adsorb to adhere to a surface

Advection movement by flow

Advective transport movement of materials by movement of parcels of water (as opposed to molecular diffusion)

Aeolian process formed by wind action

Aerobic oxic (with oxygen)

Aeration see reaeration

Age class organisms within a specific age range

Aggradation deposition of excess stream load (sediment) to its channel

AHOD (oxygen deficit per unit area) decrease in O_2 in the hypolimnion per unit area surface; typically expressed as mg O_2 cm^{-2} day^{-1} or mg O_2 m^{-2} day^{-1}

Airlift pump pump that uses compressed gas to move water up a tube; used for groundwater sampling and hypolimnetic aeration

Akinete resting cell of cyanobacteria and some green algae, generally resistant to poor growth conditions

Alarm chemical chemical produced by a damaged or stressed prey that alarms other prey

Algae nonvascular organisms capable of oxygenic photosynthesis without sterile cells covering gametangia

Alkaline phosphatase enzyme that hydrolyzes organic phosphorus compounds under basic conditions, rendering some P obtainable as soluble inorganic phosphate; used in some P deficiency assays and cyanobacterial toxin bioassays

Alkalinity the capacity of a solution to neutralize acid; the sum of all titratable bases; usually a function of carbonate, bicarbonate, and hydroxide contents

Allelochemical a chemical produced by one species that alters behavior or growth of another species

Allen curve a plot of individual size versus number of individuals in a cohort through time

Allen's paradox based on studies in New Zealand streams; standing stock of invertebrate prey appears too small to support predatory fish production. Underscores the fact that standing stocks are not production.

Allochthonous originating from outside the system; often refers to organic carbon

Allomone chemical released by one species that affects the behavior of an individual of another species and benefits the species that produces and releases it

Alpha diversity (α diversity), within-habitat diversity

Alternative stable states the idea that ecosystems can exist under different biotic and abiotic conditions that persist in the face of normal environmental variation

Amensalism interspecific interaction in which one species is harmed and the other is not influenced ($-/0$)

Amictic when a lake almost never mixes

Ammonia gas (NH_3) ammonium ion (NH_4^+) is converted to its toxic form, ammonia gas, at basic pH

Ammonification ammonium (NH_4^+) produced from organic nitrogenous compounds by metabolism of living organisms and decomposition of organic matter

Ammonium (NH_4^+) an inorganic N compound; the ionic form of ammonia

Amphidromous regular migration from fresh to saltwater (see diadromous); not always for breeding purposes

Anadromous moving from marine habitats to freshwater for spawning

Anaerobic without oxygen; anoxic

Anaerobic ammonium oxidation conversion of nitrate and ammonium directly to dinitrogen gas under anaerobic conditions

Anammox see anaerobic ammonium oxidation

Anchor ice ice that forms on the bottom of open, flowing streams when temperatures are below $0°C$

Angiosperms true flowering plants

Anoxic without oxygen; anaerobic

Anoxygenic photosynthesis photosynthesis that does not evolve oxygen, as either cyclic photophosphorolization or using sulfide as an electron donor

Antagonism when two toxicants alleviate the influence of each chemical alone

Anthropocene the current geological era originating from human activity

Anthropogenic originating from human activity

Apatite calcium phosphate

Aquaculture the farming of aquatic organisms

Aquiclude a layer that groundwater flows through slowly (unsuitable for a well)

Aquifer permeable deposit that can yield water by a well

Aquifuge an impermeable layer through which groundwater cannot flow

Archaea a domain of organisms, unicellular without organelles; one of three domains (super kingdoms) of organisms; distinguished on the basis of biochemical and genetic differences from the Bacteria and Eukarya

Aerenchymous containing spongy, air-filled tissues; occurs in some wetland plants, and allows for movement of oxygen among parts of the plant, as well as release of methane to the atmosphere

Argillotrophy a mode of obtaining nutrition in turbid systems in which the primary source is organic material associated with clay particles

Artesian well a well with water naturally under pressure so it continuously flows from the wellhead

Ash-free dry mass (AFDM) the mass of material that will combust from a dry sample; used as an estimate of biomass

Assimilation the process of using nutrients to synthesize components of a cell; with regard to consumers, ingested material that is actually digested and absorbed (difference between ingestion and egestion)

Assimilation efficiency the amount of food assimilated divided by the amount ingested

Astatic unstable; refers to water level

Astrobleme a meteor crater; lakes can form in these

Atelomixis unfinished vertical blending of stratified water, combining layers of varied chemical properties without affecting the hypolimnion

Athalassohaline saline water having relative ionic proportions very different from those of seawater

Atmospheric loading nutrient input from dry and wet atmospheric depositions

Attenuation absorption of light as it passes through a medium

Attenuation coefficient a value that indicates how rapidly light is absorbed

Aufwuchs see periphyton

Autecology study of the ecology of a species; population not community or ecosystem ecology

Autochthonous originating within the system (e.g., organic carbon supplied by primary producers in the system)

Autofluorescence natural fluorescence in pigments such as chlorophyll

Autotrophic the capacity to perform primary production; self-feeding; able to use CO_2 as a source of carbon using chemical (chemoautotrophy) or light (photoautotrophy) energy

Autotrophic index chlorophyll:ATP ratio

Autotrophic state a whole ecosystem's CO_2 fixation per unit area

Bacteria a domain of organisms, unicellular without organelles; one of three domains (super kingdoms) of organisms; distinguished on the basis of biochemical differences from the Archaea

Bacterioplankton suspended bacteria

Bailer in groundwater science, a tube with a one-way valve on the bottom (allowing only inflow of water) used to sample well water

Baseflow the level of stream discharge in the absence of recent storms

Basin see watershed or catchment

Bathylimnion the deepest part of a stratified lake

Bathymetric map a topographical map of a lake that indicates the distribution of depths and shape of the bottom

Bed load sediment transported by rolling, sliding, or saltation on or close to the stream or river bed

Beer's law law and associated equations relating light absorption to properties of the medium that light is traveling through; the law states that light absorption is linearly proportional to concentration

Benthic associated with the bottom

Benthos benthic animals

Beta diversity (β diversity) diversity between habitats; sum of species unique to each habitat when examining more than one habitat

Bicarbonate (HCO_3^-) an ionic form of inorganic carbon that dissolves in water

Bicarbonate equilibrium the chemical equilibrium involving the dissolved inorganic forms of carbon dioxide, carbonic acid, bicarbonate, and carbonate; the relative proportions of these ions depends upon pH

Bioaccumulation bioconcentration plus the accumulation of a compound in an organism from food

Bioassessment use of organisms to evaluate environmental quality

Biochemical oxygen demand (BOD) the demand for O_2 created by compounds that can be respired by organisms plus the chemicals that will react with O_2

Bioconcentration ability of a compound to move into an organism from the water

Biodiversity the number of different species, organisms, genotypes, or genes of ecological functional groups in a region (also referred to as biocomplexity)

Biofilm a film of organisms and associated materials attached to a solid surface (substratum)

Biogenic meromixis an increase in the density of the hypolimnion caused by dissolved materials released from sinking organic matter so that mixing cannot occur

Biological assessment see bioassessment

Biologically available phosphorus (BAP) the amount of phosphorus that can be used by organisms; can refer to the instantaneously available phosphate or the P that will become available with long-term decomposition

Biomagnification increase in concentration of a chemical at higher trophic levels of a food web

Biomanipulation to improve water quality by manipulating the trophic structure

Biomass mass of organisms

Biome a large terrestrial area defined by its climate and dominant vegetation; however, it is sometimes used interchangeable with ecoregion for freshwater systems.

Bioremediation cleanup of pollution using organisms

Biotic oxidation oxidation of compounds by organisms

Biotic sulfur oxidation biotic oxidation of sulfur by chemoautotrophic organisms using oxidized sulfur to react with organic carbon and release energy

Bioturbation stirring of sediments by movement and activity of sediment-dwelling organisms

Bittern bromide, magnesium, and chloride blend left when seawater almost completely evaporates

Bivoltine producing two generations annually

Blackfish species that are resistant to deoxygenated conditions and survive in standing floodplain water during dry periods

Blackwater river water colored by dissolved organic matter (humic substances) low in dissolved inorganic and suspended solids and usually with a low pH

Bloom a large-dense population of algae

BOD see biochemical oxygen demand

BOD$_5$ oxygen consumed in 5 days by biological and chemical processes

Bog a wetland in which peat accumulates; with minimal inflow or outflow; supports acid-loving mosses such as *Sphagnum*

Borehole a hole drilled to sample geological material or for well installation

Bottomland lowland along a stream or river that is flooded periodically

Bottom–up control control of system productivity by nutrients or light

BPOM benthic particulate organic matter

Brackish saline water with salinity less than that of seawater, as low as 100 ppm

Braided stream a stream having elaborate, multiple channels rather than one large channel; also referred to as an anastomosing stream

Brine water more saline than seawater

Brook small stream

Brownian motion molecules or particles moving independently at microscopic scales

Brownwater New Zealand term for "blackwater"

Bryophyte mosses, liverworts and other small, nonvascular plants

Budget an accounting of the relative magnitude of the fluxes between compartments in an environment

Buffering the ability of a solution to resist changes in pH (also known as buffering capacity)

Buttress base a common adaption of many wetland tree species (i.e., bald cypress) of root projections to increase stability in saturated soils

Byssal threads (byssus) strong, proteinaceous fibers used by some bivalves to attach to substrata

^{14}**C** a radioactive isotope of carbon often used as a tracer in ecological studies, particularly to measure photosynthetic rates

^{13}**C** a stable isotope of carbon used in some ecological studies; naturally present in the environment at trace levels relative to the more abundant ^{12}C

Caldera a collapsed volcanic crater; some calderas contain lakes

Capacity the amount that a stream can transport when full to the banks

Capillary involving or resulting from surface tension

Capillary action movement of water by capillary forces

Capillary fringe belt of soil above groundwater that contains some water drawn up by capillary action; immediately above the water table

Carbonate ($CO_3{}^{2-}$) an inorganic ion with carbon that forms when CO_2 dissolves in water

Carbon dioxide (CO_2) a gas particularly important as a respiratory byproduct, a reactant for photosynthesis, and as a greenhouse gas

Carbon sequestration atmospheric carbon capture and storage

Carbonic acid (H_2CO_3) the form first assumed by CO_2 when it dissolves in water

Carnivore animal that eats other animals

Catadromous moving from freshwater to saltwater to reproduce

Catchment (watershed) surface area drained by a network of stream channels; although "watershed" is used synonymously, watershed has been defined in European literature as a line that joins the highest points of the perimeter of a catchment

Cellulose complex carbohydrate synthesized by plants

Certainty something that rarely happens in ecology

Chelators compounds that maintain higher concentrations of chemicals in solution than would normally be able to remain dissolved, particularly important with respect to iron dissolved in oxic conditions

Chemical diffusion movement of dissolved materials in water

Chemoautotrophic obtains energy from chemicals other than organic C

Chemocline a steep chemical gradient or pycnocline in a lake; often found at the metalimnion

Chemolithotrophy autotrophy with the energy sources of inorganic chemical bonds and inorganic substances as electron donors (chemosynthesis); also called chemoautotrophy

Chemophobic repelled by a chemical

Chemotactic chemotaxis, attracted to a chemical

Chlorophyll *a* the primary pigment of photosynthesis in cyanobacteria and eukaryotic autotrophs; often used to indicate biomass; absorbs red and blue light

Chronic over long periods

Chronic toxicity toxicity with long-term exposure

Cirque a bowl formed by glacial action at the head of a valley; can contain a lake

Clear water stream water with high transparency, lacking visible suspended material and brown color; ranging from acidic to slightly alkaline in pH

Cline any continuum or gradient

Clinograde distribution showing a gradient; often used to describe temperature or oxygen curves

Clinolimnion part of a lake in which temperature distribution declines exponentially from the epilimnion to the hypolimnion due to turbulence

Cnidocyte (nematocyst) specialized venomous cells of cnidarians used to secure food

Coarse particulate organic matter (CPOM) any carbon containing materials greater than an arbitrary size of 0.5 mm; includes leaves, wood, and aquatic organisms

Coastal wetlands wetlands associated along coasts and influenced by oceans and tidal movement

COD$_5$ oxygen removed by chemical process from water in 5 days

Cohesion sticking together

Cohort a group of individuals that started life at the same time

Cold monomixis one annual total circulation without ice cover; includes cold thereimictic and warm thereimictic lakes

Cold thereimixis circulation in summer at approximately 4°C; cold monomixis that occurs in a polar lake

Collector an organism that makes its living collecting fine particles, either by filtering from the water column (collector-filterer) or feeding on deposited particles (collector-gatherer)

Colloidal particles particles not settled by gravity

Combined nitrogen organic N, nitrate, nitrite, or ammonium (not N_2 gas)

Commensalism interspecific interaction in which one species is influenced positively and the other is not influenced

Community all organisms in an area or a group of species in an area

Community assembly theory the concept that environmentally similar sites with different species assemblages arise from random fluctuations of a common pool of species, different species traits, or the order that they come into the environment

Compartments parts of a budget into which materials are divided; for example, in the global hydrologic cycle, the ocean is a compartment and in the carbon cycle, leaves in a stream are a compartment

Compensation point light level at which O_2 production by photosynthesis equals consumption by respiration, or CO_2 assimilation is equal to production by respiration

Competition an interspecific interaction in which both species harm each other (−, −)

Competitive exclusion principle the idea that only the competitively dominant organism will survive in an equilibrium environment

Conditioning microbial colonization of detritus

Conductivity the ability of water to conduct electricity, a function of the number of dissolved ions in the water; measured in units of mhos (reciprocal ohms) or Siemens per unit distance

Confined aquifer an aquifer between two impermeable layers (aquifuges)

Consumptive use a use of water that does not return it to the stream channel (causes loss by evaporation or to groundwater)

Contest competition a form of intraspecific competition; some individuals of a species get more of a resource to the detriment of other individuals of the same species

Control an experimental condition to which the treatments are compared

Coprophagy ingestion of feces

Core of depression an area around a well where the pumping lowers the level of the groundwater (drawdown)

Coriolis force a force that induces circulation in very large lakes; caused by rotation of the Earth as water or air flows along lines of longitude

Correlation a statistical way to measure relatedness of two variables; correlation does not necessarily indicate causation

Cosine collectors sensors used to measure solar radiation from above

Cosmopolitan distribution having a very large range; found almost everywhere on the planet

CPOM see coarse particulate organic matter

Creel limits numbers of fishes that anglers can take per day

Creek low-order, small stream (crick)

Crenogenic meromixis differences in density between waters of monimolimnion and mixolimnion attributable to below-surface flows of spring or seep saline water

Critical mixing depth mixing depth below which phytoplankton growth does not occur

Cryogenic lake lake located in a thaw basin of permanently frozen ground

Cryoperiphyton (kryoperiphyton) periphyton attached to the bottom of the ice

Cryptic species organisms those are morphologically indistinguishable, but genetically different enough to be considered different species

Cryptorheic concealed drainage with below-surface stream flow, usually in limestone, karstic areas

Cultural eutrophication a human-caused increase in a nutritive factor or factors leading to greater rates of whole-system heterotrophic or autotrophic metabolism

Cumulative a response to the sum of numerous events

Current known as water velocity or the speed of water in a channel

Cycle all of the fluxes of a material that occur in an environment

Cyclic colonization colonization of less permanent habitats by colonists from more permanent habitats

Cyclomorphosis successive emergence of distinctive morphologies in the same species; often observed in microcrustaceans and rotifers

D_L shoreline development index

Dam barrier preventing the flowing of water; impoundment

DAPI a fluorescent dye specific for DNA used to count total bacteria using epifluorescent microscopy or flow cytometry

Darcy's law a relationship used to calculate groundwater flow rates

Decomposition the breakdown of organic materials

Decomposer organism that consumes or breaks down organic materials

Deflation basin a basin formed by action of the wind; can contain a lake or wetland

Degradation erosion of stream channels; opposite of aggradation

Denitrification conversion of nitrate to N_2 gas by microorganisms; a form of respiration that uses nitrate rather than O_2 to oxidize organic carbon

Density current current going along the bottom or through a stratified layer; current of different temperature, ionic strength, or turbidity than the water through which it flows

Density-dependent factors that are a function of population size

Density-independent factors that are not a function of population size

Density-mediated interactions interactions evidenced by changes in population size

Depressional wetlands inland wetlands that occur from topographical depressions

Desalinization removal of salt from water

Detritivores organisms that eat detritus; also called saprophytes

Detritus decaying organic material

Dewatering removing water from a river or stream (also abstraction)

DGGE (denaturing gradient gel electrophoresis) molecular fingerprinting technique that separates polymerase chain reaction (PCR)-generated DNA products

Diadromous migrating between fresh and salt waters

Diagenesis conversion of sediment into rock

Diapause part of the normal life cycle that is stationary, physiologically dormant and often resistant to environmental extremes such as drying or heat

Diatoms single-celled algae with silica shell (frustule) and golden brown coloration (Bacillariophyceae)

Diel 24-h day with a light–dark cycle (as opposed to a period of consistent light)

Diffusion boundary layer the thin aqueous layer near a solid surface where diffusion is dominated by molecular diffusion; its thickness can control metabolic rates of microorganisms

Diffusion coefficient a constant used to describe diffusion of a compound or heat independent of distance and concentration

Diffusion flux the mass of a compound diffusing across an area per unit time

Dimethylsulfide (DMS; C_2H_6S) water insoluble flammable liquid organosulfur compound

Dimictic a lake that mixes twice each year

Dinoflagellates unicellular algae that move by means of flagella (Pyrrhophyta)

Direct interaction occurs between two species and involves no others

Discharge the volume of a fluid flowing past a point per unit time

Dispersion diffusion of material via turbulent transport from the main pulse as it flows downstream

Disproportionation breakdown of a compound that has two atoms of the same element in different redox states to yield energy (for example, acetate to methane and CO_2 or thiosulfate to sulfate and sulfide)

Dissimilatory nitrate reduction to ammonium (DNRA) nitrate reduction to ammonium without assimilation involved; an alternative pathway to oxidize organic carbon or sulfide using nitrate with ammonium as the end product

Dissimilatory process a chemical transformation mediated by organisms that does not involve assimilation

Dissimilatory sulfur reduction using oxidized sulfur as an electron receptor to respire organic carbon or react with reduced inorganic compounds to yield energy

Dissolved inorganic nitrogen (DIN) the sum of ammonium, nitrate, and nitrite

Dissolved materials materials smaller than a particular size (e.g., 0.45 μm) that remain in solution

Dissolved organic carbon (DOC) organic carbon compounds dissolved in solution

Dissolved organic nitrogen (DON) organic nitrogen able to pass through a 0.45-μm filter

Dissolved oxygen (DO) O_2 dissolved in water

Disturbance an event that disrupts ecosystem, community, or population structure

DNA barcoding analysis of short genetic markers in DNA to identify species

DNRA see dissimilatory nitrate reduction to ammonium

Dolina depression caused by dissolution of limestone substrata; sink or swallow hole

DOM see dissolved organic carbon, dissolved organic matter

Drainage area the land area that is drained by all of the tributary streams above a chosen point in the channel

Drawdown lowering a water table by pumping

Drift the material that washes downstream, particularly invertebrates

Droop equation relationship between intracellular nutrient concentrations and growth

Dy sediment of dystrophic lakes, generally with a high organic content and allochthonous origin

Dynamic equilibrium model (or concept) the idea that response of species richness to disturbance is a function of competition intensity in a community and productivity of the system

Dynamic viscosity a constant that describes the intrinsic viscosity of a fluid

Dystrophic a lake that is not productive because it has been influenced by factors that attenuate light or retard photosynthetic organisms; a lake that is high in tannin and lignin is dystrophic

Ebullition release of bubbles

EC_{50} concentration of a toxin which induces a response halfway between baseline and maximum after a given time period

Eckman dredge a dredge that is lowered to the bottom, with a messenger that is sent down its line and triggers jaws on the bottom to shut, taking in sediment

Ecoestrogens organic compounds that mimic natural estrogen and lead to endocrine disruption; estrogens or environmental estrogens

Ecohydrology study of ecology as influenced by hydrology

Ecological state condition of a habitat based on variables of interest

Ecological subsidies movement of nutrients and energy from one habitat to another

Ecoregion an area with a geographically distinct assemblage of communities

Ecosystem all living and nonliving community constituents

Ecosystem engineer organisms that strongly alter their environments

Ecosystem goods products of value to humans from ecosystems

Ecosystem function the rate of specific basic ecosystem processes

Ecosystem respiration (ER) the entire respiration of all living organisms in an ecosystem (see respiration)

Ecosystem services services of value to humans associated with ecosystem processes

Ecotone transitional zone between two habitats, e.g., where there is a change between groundwater and surface water organisms

Ectogenic meromixis meromixis caused by inflow of materials from exterior sources (e.g., saline water entering a dilute lake)

Edaphic relating to the ground or soil, particularly matter influenced by them or from them

Eddy current or small whirlpool moving counter to the main flow

Eddy diffusion diffusion by transport or mixing of a diffusing substance or heat; much faster than molecular diffusion; also called transport diffusion or advective transport

eDNA environmental DNA, used to help detect organisms in their environment

Effective population size the number of individuals in a population that actually contribute offspring

Effective concentration (EC) the concentration of toxic substances that causes some other effect other than death

Effluent the water released from a sewage plant, factory, or other point source

Emergent growing above the water

Emergent properties ecological system characteristics that scientists cannot predict from understanding of individual smaller components of the system or lower levels of biological organization; the whole is greater than the sum of its parts

Endemic species having a distribution that is restricted to a relatively small region

Endocrine disrupting compounds organic compounds that mimic natural metabolic compounds leading to disruption of endocrine function

Endogenic endogenous, created or produced internally; occasionally used instead of autochthonous

Endorheic stream or basin that does not drain to a larger stream or basin (a closed basin)

Endosymbiont an organism living inside another organism

Engulfing predators predators that either swallow or bite off chunks of their prey

Entrainment mixing of part of the hypolimnion into the epilimnion with high winds

Ephemeral not holding surface water most of the year

Ephippium resistant egg case produced by a cladoceran, ephippia

Epifluorescent microscopy microscopy using a light source from above that excites fluorescent molecules and then filters the excitation wavelengths from the observed light, rendering the emitted fluorescence visible

Epigean living in surface water (above the ground or sediment)

Epilimnion the surface layer of a stratified lake, above the metalimnion

Epilithic growing on rocks

Epineuston organisms on the upper part of the water surface film

Epipelion community inhabiting mud surfaces

Epiphytic growing on a plant or macrophyte

Epipsammic growing on sand

Epixylon microbial communities associated with wood

Epizooic growing on an animal

EPT index component of biological assessment that measures the number of Ephemeroptera, Plecoptera, and Trichoptera taxa present (these three groups are generally intolerant of pollution)

Erosional processes movement of particles off land by water or wind

Euphotic zone the region where light is above the compensation point, so net photosynthesis is positive

Eutrophic very productive

Eutrophication an increase in a nutritive factor or factors leading to greater rates of whole-system heterotrophic or autotrophic metabolism

Evaporation to convert into vapor

Evapotranspiration evaporation plus plant transpiration

Evenness a measure of the degree of equal distribution of numbers of each species in a community

Exoenzymes enzymes that can be outside the cell

Exotic species species located outside their natural range

Exploitation interaction that harms one species and helps another

Exploitation competition competition between organisms that are using the same resource; one species reduces the resource or uses it more efficiently, reducing its availability for the other species

External loading supply of nutrients from outside the system

Extinction a condition where no living individuals of a species exist anywhere on Earth, sometimes used to denote more local losses (extirpation)

Extinction coefficient same as the absorption coefficient

Extinction vortex the mutual reinforcement of biotic and abiotic factors that drive small declining populations extinct

Extirpated species that no longer occur in a region (though are not necessarily extinct)

Fall mixing the autumnal period in temperate lakes after summer stratification breaks down

Fault a fracture in rock layers where adjacent layers have moved parallel to the fracture

Fecundity the number of offspring that reproductive females produce

Fen like a bog, with peat accumulation but more input of water from outside

Fermentation organisms extracting energy from organic carbon in the absence of oxygen

Ferric iron ion in oxidized state (Fe^{3+})

Ferric hydroxide flocculent precipitate of ferric ions and hydroxyl ions

Ferrous iron ion in reduced state (Fe^{2+})

Fetch the longest uninterrupted distance on a lake that wind can traverse to create waves

Fick's law a mathematical formula describing the relationship between diffusion flux, distance, and concentration gradients

Filterers organisms that sieve small particles from the water column

Filter feeding filtering fine particles for food, see filterers

Fine particulate organic matter (FPOM) organic material between 0.45 and 500 μm in diameter

Fixed groundwater groundwater trapped in rocks (confined)

Fjord lake a steep glacial valley containing a lake (also spelled fiord)

Flashy having repeated, rapid discharge or floods

Floating attached macrophytes with floating leaves that are rooted in the sediments

Floating marshes wetlands that float on the water (usually found in some coastal areas) whose vegetation forms thick mats of roots and organic matter

Floating unattached free-floating macrophytes

Flood rising and overflowing of a body of water, generally above the banks

Flood pulse concept lateral habitats (especially in large rivers) connect with the main channel in times of floods and disconnect during lower flow periods, and these connections are ecologically important

Floodplain a flat region in the bottom of a valley that is, or historically was, influenced by river flooding

Flow movement of a fluid; generally a velocity, but can be used to mean discharge

Flow boundary layer the zone in which turbulent flow is rare and laminar flow dominates; near solid surfaces

Fluvial produced by action of a stream or river

Fluvial zone the flowing channel of a stream or river

Fluxes movements of materials between pools or compartments in a cycle (e.g., of water through the global hydrologic cycle)

Food chain the simplest view of food webs in which only trophic levels are considered

Food web a network of consumer–resource/predator–prey interactions that occur in an ecological community

Fossil water deep aquifers that have held water for thousand to millions years and is a nonrenewable resource (e.g., the Ogallala in the central United States)

FPOM fine particulate organic material

Free groundwater groundwater not trapped or confined

Freshwater water with less than 1 part per thousand of dissolved salts

Freshwater tidal marshes nonsaline wetlands that are further inland than tidal salt marshes but still influenced by ocean tides, and usually higher in biological diversity than salt marshes

Fringe wetlands wetlands on the edges of other habitats (i.e., coastal wetlands)

Froude number a dimensionless measure relating inertia forces to gravitation effects; important when gravity is dominant (i.e., flow in open channels)

Frustule hard cell walls of diatoms composed mostly of silica

Fulvic acids the humic fraction that does not precipitate when a solution is acidified

Functional feeding group a subset of organisms from a community that feed using similar strategies (e.g., filterers, scrapers, and shredders)

Functional processing zones zones created by hydrogeomorphic conditions that control ecological structure and functions that differ among patches

Functional redundancy the degree to which different species provide the same ecosystem function

Functional response the number of prey eaten per unit prey density; increased consumption rates of predators in response to increasing prey densities

Fungi a kingdom of parasitic and saprophytic organisms

Gaining stream channel is below the water table so it gains water and increases in discharge as it flows down slope

Gape limited size of prey consumed is limited by mouth size

Gas vesicles intracellular protein structures that lend buoyancy to cyanobacteria

Gemmule resistant, asexual buds of sponges

Geotaxis movement with respect to gravity

GIS (Geographic Information System) a computer-based mapping system that can be used for complex spatial and temporal geographic analysis

Glacial relating to or produced by glaciers

Glacier a large body of flowing ice

Global water budget estimated water movement (fluxes) between compartments throughout the world

Glochidia specialized larval stage of freshwater mussels that attach to fishes for dispersal

Graben a depression between two faults when one block slips down relative to two others and creates a basin (see also horst); lakes can form in the depression

Gravitational water water in rocks and soils above groundwater, includes perched aquifers

Gravitoidal particles particles that will settle

Grazer primary consumer; eats algae or plants

Greyfish Australian fishes living in backwater, shoreline vegetation, lake edges, or stagnant channels in dry seasons

Gross ecosystem production (GEP) production of an entire ecosystem (all primary producers and heterotrophs in an area) before accounting for ecosystem respiration (ER).

Gross photosynthesis the total amount of photosynthesis; in bottle experiments, the O_2 production rate in the light bottle minus that in the dark bottle

Gross primary production the total amount of photosynthesis that occurs before accounting for losses to respiration

Gross production efficiency the overall efficiency at which ingested materials are converted to production

Groundwater water in or below the water table

Groundwater recharge replenishment of groundwater

Guild organisms that use a similar resource in a similar manner

Gyttja partially reduced, minute-grained, organic, profundal sediments of eutrophic lakes; copropel, mainly of autochthonous origin

[³H] tritium a radioactive isotope of hydrogen used as a tracer

Habitat diversity beta diversity, between-habitat diversity; alpha diversity, within-habitat diversity

Halophilic salt loving; requires saline water for growth or reproduction

Halophytic (or halophyte) a plant that grows in highly saline waters. (i.e., mangrove trees)

Hardness a characteristic of water that does not allow soap to dissolve; primarily caused by Ca^{2+} and Mg^{2+}

Head the potential energy of water from gravity

Heat balance see heat budget

Heat budget an account of heat incorporated and lost by a water body during a specified time period

Heat capacity amount of energy required to increase the temperature of a specific mass of material by a specific amount

Heat of fusion the number of calories required for a specific mass of substance to transition between solid and liquid states

Herbivore organism that eats primary producers

Hermaphrodite an organism with both male and female sexual reproductive organs

Heterocyst specialized cyanobacterial cell in which nitrogen fixation occurs at high rates

Heterogeneous consisting of dissimilar ingredients or constituents

Heterogeneous aquifer an aquifer with many impermeable layers or areas where water cannot flow evenly

Heterotrophic state a whole ecosystem's CO_2 production per unit area

Heterotrophy metabolic energy and growth from degradation of organic molecules; carnivory, detritivory, herbivory, microbial decay, and omnivory

Hind-casting when a model is used to predict system response when the data from the system are already known

Hochmoor elevated bog with peaty matter higher than the rim of the cavity housing the bog

Holomixis total circulation or blending as in a holomictic lake

Holistic approaches whole system studies often with a multidisciplinary approach

Homeothermy relatively constant temperature, see isothermy

Homogeneous of the same or similar kind of nature

Homogeneous aquifer an aquifer with evenly distributed substrata and even groundwater flows

Homogocene the current global era of homogenization of biodiversity as one consequence of the Anthropocene

Horst where a portion of the Earth's surface pushes upward. Two horst regions on either side of a graben (a lower region) can cause an area where a lake will form

HPLC high-performance liquid chromatography; an analytical system in which a carrier fluid flows through materials that can separate different types of dissolved molecules

Humic acids humic compounds that are precipitated by acid

Humic compounds from decomposition of plant material, includes humic acids, fulvic acids, and humin

Humin the fraction of humic compounds that cannot be extracted by acid or base

Humus high-molecular-weight organic molecules, polymeric, primarily from decayed plants; humic acids with the −COOH radical; humolimnic acids in lake waters and sediments

Hydraulic conductivity the ability of a material to allow water flow (coefficient of permeability)

Hydraulic gradient in groundwater, a line that will connect the level of water in wells or that follows the top of the water table; in streams, the drop in the channel per unit distance

Hydraulic head difference in water elevation between two connected sites

Hydraulic regimes degree and duration of water inundation and depth

Hydric soils soils with characteristics related to constant water inundation

Hydrochory seed dispersal by water

Hydrodynamics temporal and spatial variations in movement and distribution of water

Hydrogen bonding when molecules are polar (i.e., there is an uneven distribution of charge across the molecule), the positive and negative parts of separate molecules are attracted toward each other; particularly important in water

Hydrogeomorphic patches reaches within a network with different geomorphology and hydrology

Hydrograph plot of discharge as a function of time for a stream or river

Hydrologic cycle the fluxes of water across the landscape

Hydroperiod length of time that surface water is present

Hydrophilic water loving

Hydrophobic repelled by water

Hydrophyte an aquatic plant, usually a macroscopic, rooted variety; macrophyte

Hydropsammon interstitial zone in sand below shallow water; sand-dwelling organisms live here

Hydrostatic pressure the pressure that fluids exert (density times depth)

Hygropetric in water on vertical rock surfaces

Hypereutrophic (hypertrophic) extremely productive, with abundant nutrients and very high primary producer biomass

Hypogean existing underground, subterranean, interstitial, or cave-dwelling

Hyponeuston organisms in the lower part of the water surface film

Hyporheic zone a region of groundwater influenced by a nearby stream or river

Hyporheos a region of groundwater influenced by nearby surface water

Hypolimnion the bottom layer of a stratified lake; below the metalimnion

Hypolimnetic aeration a counteractive eutrophication method that provides O_2 to the hypolimnion so phosphate remains in the sediment

Hypothesis initial expectation; statement of cause to be tested by observation or experiment

Hypsography mapping and measuring of elevations and contours

Hysteresis a memory or lag effect

Ichthyochory seed dispersal by fishes

Impermeable surface see impervious surface

Impervious surface a surface that does not allow infiltration of water

Increment summation method a method for calculating secondary production

Indirect interaction interaction between two species mediated by one or more other species

Inertia resistance of a body to a change in its state of motion

Infauna animals that dwell in the substrata or sediments

Infiltrate flows into; permeate

Infiltration gradual movement or permeation (e.g., of water through soils)

Influent flowing in

Instantaneous immediate response to exposure

Instantaneous growth method method to estimate production from estimates of average biomass and individual growth rates over a time interval

Interception the process of plants stopping precipitation before it reaches the ground

Interference competition direct negative competitive effects between species

Intermediate disturbance hypothesis the idea that species diversity is maximal at intermediate levels of natural disturbance

Intermittent usually drying at least once per year

Internal loading availability of nutrients from within the system; often associated with lake mixing

Internal seiche rocking of the hypolimnion while the surface of epilimnion stays still

Interspecific between two species or among numerous species

Interspecific interactions how a species interacts with other species

Interstices voids between sediment particles or in rocks

Interstitial between particles

Intraspecific within a species

Invasive exotic species species outside their natural range that proliferate and potentially cause damage in new habitats

Inverse thermal stratification warm water under cold water in a vertical temperature profile; can occur below an ice cover during winter stratification or when warm saline water sits below cooler, more dilute water

Irradiance radiance flux density on a given surface

Isobath contour line of lake depth; a bathymetric map of a lake is composed of isobaths

Isothermal with the same temperature; homoeothermal

Kairomone chemical produced by a species that benefits another species with adverse consequences for the species that released it

Karst irregular limestone region with sinks, underground streams, and caverns

Kemmerer sampler a tube that closes by gravity when a messenger is sent down a line to it; used to sample water at a specific depth

Kerogen marine and lacustrine residues, not soil humus

Kettles when large blocks of ice melt and leave lakes, ponds, or wetlands

Keystone species species that have a major impact on their community or ecosystem; impact is disproportionately large relative to abundance

Knees a distinctive morphological structure of cypress swamp trees to help increase stability in soft, saturated soils

Labile organic carbon compounds that are easily broken down or used by heterotrophic organisms

Lacustrine shallow lake habitat

Lake very slowly flowing body of water in a depression of ground not in contact with the sea

Laminar flow flow all in one direction, with little lateral mixing (as opposed to turbulent flow)

Landscape ecology the study of spatial relationships and ecology

Langmuir circulation large spiral circulation patterns in lakes induced by wind

Larvae early life form of an animal, usually not capable of reproduction

LC_{50} the concentration of a toxic compound that will kill half of the test organisms

LD_{50} the dose of a toxic compound that will kill half of the test organisms

Leaf pack natural or artificial amassing of leaves in a stream

Leibig's law of the minimum a law that states that the rate of a process is limited by the rate of its slowest subprocess

Lentic still water

Lethal causing death

Lethal dose amount of toxic substances ingested that causes death

Lichen symbiotic mutualistic partnership between fungi and algae

Lignin a complex polymer produced by plants that is resistant to microbial degradation

Limnocrene water from a spring or artesian well forming a pool

Limnology study of continental waters

Lithotrophy primary production or autotrophy with inorganic substances providing electrons; chemolithotrophy, photolithotrophy, chemoautotrophy

Littoral zone shallow, shoreline area of a body of water; often considered the portion of benthos from zero depth to the deepest extent of rooted plants

Logarithmic an exponential relationship

Lotic moving water

Losing stream channel is above the water table and the stream loses water to groundwater; discharge decreases downstream

Luxury consumption uptake of a nutrient in excess of needs

Lysimeter a sampler used to sample soil water

Maar volcanic eruption crater; can contain a lake or wetland

Macrobenthos small (about 1 mm) bottom-dwelling organisms

Macroecology study of identifying broad generalizations about organismic biology across levels of biological classification and organization

Macrophyte large colony of algae visible without magnification, or aquatic plants

Macrosystems large scale patterns (regional to continental)

Madicolous inhabiting thin sheets of water flowing over rock

Magnetotaxis preferential movement toward north or south magnetic lines

Mainstem primary flowing section of a river or stream (as opposed to forks, tributaries, or other divisions)

Marl calcareous sediments, primarily soft

Manipulative experiment all other factors held constant and each factor of interest is varied independently

Mangrove wetlands coastal wetlands that are extensive in along tropical coastlines dominated by mangrove trees in areas with minimal wave action and sediment accumulation

Marl calcareous sediments, primarily soft

Marsh continuously or usually inundated wetland with saturated soils and emergent vegetation

Mass spectrometer an instrument used to determine molecular weight of molecules and elements; used in stable isotope studies and to identify organic pollutants

Maximum density of water 1.000 g/mL; occurs at 3.9°C

Maximum sustainable yield the largest catch or take that can be harvested from a population over an indefinite period

Meander a natural feature of flowing water where S-shaped curves form

MEI morphoedaphic index to fish production; TDS (total dissolved solids) of water divided by the depth of lake; milligrams per liter divided by meters

Meiobenthos bottom-dwelling organisms (about 0.05 mm in size)

Meniscus an upsloping surface where water meets a hydrophilic solid

Meromictic a lake that is permanently stratified or mixes irregularly

Mesolimnion see metalimnion

Mesotrophic moderately productive system

Messenger a weight that attaches to a line and can be dropped to trigger sampling devices

Meta-analysis a statistical technique that tests the significance of combined results of many different studies

Metacommunity two or more communities linked by dispersal of multiple, potentially interacting species; a community of sub-communities

Metalimnion the intermediate zone in a stratified lake in which the temperature change with depth is rapid, below the epilimnion and above the hypolimnion; also known as the thermocline or the mesolimnion

Metapopulation two or more geographically separated populations that interact to some degree

Methanogenesis anoxic bacteria producing methane

Methanotrophy oxidizing methane to obtain energy; a type of methylotrophy

Methylotrophy bacteria harvesting energy by oxidizing chemical compounds with methyl groups

Michaelis—Menten a relationship used to describe the influence of nutrient concentration on uptake rate

Microbenthos bottom-dwelling microorganisms, including bacteria, small algae, ciliates, gastrotrichs, and rotifers

Microbial conditioning colonization of organic materials such as dead leaves with microbes, which generally increases nutritional quality

Microbial loop the part of the food web based on consumption of bacteria and bacteria-sized algae, including protozoa, rotifers, and other microbes; often refers to the flow of carbon through the microbial community

Microbiome the entire microbial community in a specific area

Microphytobenthos periphyton

Microplastics microparticles of plastics (1 μm−5 mm)

Mineralization organism excretion of inorganic nutrients (see remineralization, regeneration)

Minerotrophic a wetland with a high hydrologic throughput

Minimum viable population the minimum number of individuals of a species required for survival in the wild

Mire a peat-accumulating wetland (European definition)

Mixolimnion upper layer that sometimes mixes in meromictic lakes

Mixotrophy capacity to employ both organic and inorganic carbon sources for nourishment (using both autotrophy and heterotrophy)

Molecular diffusion diffusion that occurs by random movement of molecules (Brownian motion)

Monimolimnion layer in meromictic lakes under the thermocline

Monod equation relationship of growth to concentration of nutrients outside the organism

Monomictic a lake that mixes once a year

Moor a peatland; can be raised or a depression

Moraine a wall of material deposited by a glacier (e.g., when the forward movement of a glacier is approximately equal to its backward melting rate a terminal moraine is deposited); lateral moraines are formed at the margins of a glacier; can dam streams to form lakes

Morphometry shape and size of lakes and their watersheds, or the shape and size of any object

Mortality the number of organisms lost between each size or age class

Moss a type of non-vascular spore bearing lower plant; a small filamentous bryophyte

Multiple stressors when aquatic species are impacted by numerous different types of negative environmental, toxic, or other stressors at one time

Multivoltine (polyvoltine) producing many generations annually

Muskeg a large peatland or bog, particularly in northern North America

Mutualism an interspecific interaction in which both species benefit

^{15}**N** a stable isotope of nitrogen used to trace nitrogen flux

^{14}**N** the most common stable isotopic form of N in the environment

Naiad another term for the larva of an aquatic insect

Nanomaterials very small particles of materials, often with unique chemical properties related to their size

Nanoplankton plankton particles between approximately 3 and 50 μm

Nanoplastics small sub μm particles of plastic

Natural abundance amount of a stable isotope found naturally

Natural experiment observation of natural patterns

Nauplius larval stage of copepods

Nekton plankton able to control their position in water columns by swimming

Net ecosystem production (NEP) the photosynthesis that occurs in excess of the respiratory demand of an ecosystem

Net primary production (NPP) the photosynthesis the occurs in excess of the respiratory demand

Net photosynthetic rate the amount of photosynthesis that is used for growth; in bottle experiments, the O_2 concentration in the light bottle minus the initial concentration

Net plankton plankton particles between approximately 50 and 500 μm

Net production efficiency the efficiency at which assimilated materials are converted to production

Neuston microorganisms living at the water surface

New nutrients nutrients from outside the system

Niche position of a species in its environment, including all relationships with biotic and abiotic components

Nitrate (NO_3^-) an oxidized ionic form of inorganic N in natural waters

Nitrification microbial processes that convert nitrate to ammonium, yielding energy

Nitrite (NO_2^-) an ionic form of dissolved inorganic N occasionally found in significant concentrations in natural waters

Nitrogenase an enzyme with molybdenum as a cofactor that reduces N_2 to NH_4^+ (nitrogen fixation); found only in bacteria

Nitrogen fixation the process of converting N_2 to a form of nitrogen used by organisms (to combined nitrogen)

N:P nitrogen to phosphorus ratio; can be expressed per mole or by weight

Nomogram a graphical analog computation device for deriving unknown values based on known values

No net loss regulations that require no net loss of wetland or other habitat area

Nonpoint source a diffuse source of pollution from the landscape, such as that resulting from cultivation, urban lawn spraying, or runoff from parking lots

Numerical response predator densities increasing with prey densities, number of predators per unit prey density

Nutrient element or chemical compound required by organisms for growth

Nutrient budget a quantification of fluxes in a nutrient cycle

Nutrient cycling the transformation of nutrients from organic to inorganic forms and among different oxidation states

Nutrient ecoregions large-scale classification of the natural distribution of nutrients, fundamental characteristics of watersheds that control nutrients, and how human activities alter those distributions

Nutrient limitation control of growth or production by a nutrient or nutrients

Nutrient loading input of nutrients to a system from river or stream inflow, dry or wet deposition, and internal sources

Nutrient regeneration see regeneration

Nutrient sinks compartments that store nutrients over time

Nutrient spiraling alternate uptake and release of a nutrient; another term for nutrient cycling, with "spiraling" referring to downstream motion between uptake and release

Nymph another term for the larva of an aquatic insect

Oligomixis infrequent circulation of water masses

Oligotrophic nutrient-poor system with relatively low primary production

Ombrotrophic system that receives most of its water and minerals from precipitation; ombrotrophic bog; generally nutrient-poor

Operational taxonomic units (OTUs) empirical classification criteria used by researchers for organisms that are closely related rather than species. Usually used for microbial groups.

Optimal foraging evolution through selection leads to maximization of energy gain while foraging

Organic carbon compounds with carbon bonded to hydrogen, usually oxygen, and sometimes other elements

Organism a complex living structure made up of a cell or cells

Organotrophy heterotrophic nutrition in which energy is taken from fermenting or oxidizing organic compounds

ORP oxygen reduction potential; also called redox

Orthofluvial zone the floodplain of a river that does not regularly flood

Orthograde straight distribution as in a vertical oxygen or temperature profile

Orthophosphate phosphate that is free, in the form of a polymer, or bound to organic compounds

Osmoregulation regulation of salt and water balance of the body or cell

Otolith small, hard particles in the inner ear associated with maintaining orientation. In fishes, they can be used to estimate age in much the same way that tree rings are used

Oxbow lake section of an old river channel that has been cut off from the current channel to form a lake or wetland

Oxic with O_2 (aerobic)

Oxidation–reduction potential see redox

Oxidized microzone the small region in sediment between oxic water above and anoxic water below

^{32}P a radioisotope of phosphorus used as a tracer to study phosphorus dynamics

Paleolimnology the study of ancient lakes and their sediments

Palynology the study of pollen grains and spores

PAR see photosynthetically available radiation

Paradox of the plankton the paradox that multiple competing phytoplankton species are able to coexist without competitive exclusion removing less competitive species

Parafluvial zone the portion of the floodplain where the river regularly floods

Parasitism predation where the predator is substantially smaller than the prey

Parthenogenesis reproduction by development of an unfertilized gamete

Particulate large enough to be retained on a filter (often $\geq 0.45\ \mu m$)

Particulate organic carbon (POC) organic carbon particles larger than $0.45\ \mu m$

Particulate organic nitrogen (PON) organic N retained on a filter

Passive filterers organisms that build nets or have morphological features that filter particles out of flowing waters

Paternoster lakes formed by glacial scour in mountain valleys

P:B ratio of production to biomass

PCB (polychlorinated biphenyls) persistent organic pollutants identified as human carcinogens that were once widely used in industry as insulators and coolants

PCR (polymerase chain reaction) molecular technique used to amplify small amounts of DNA

Peatland any wetland that accumulates organic matter

Péclet number ratio of molecular to advective diffusion of heat

Pelagic in open water

Percentage of transmission $\frac{100 \cdot I_2}{I_1} = percentage\ transmission$

Perched water table a small pocket of groundwater held above the main water table by an impermeable layer

Percolate when solvent passes through a permeable substance

Periphyton the mixed assemblage of organisms attached to solid substrates in lighted benthic habitats, including algae, bacteria, protozoa, and invertebrates; a biofilm containing algae; also called aufwuchs and microphytobenthos

Permeability ability of a substance to transmit water

Perennial holding surface water year-round

pH activity of hydrogen ions (which is closely related to concentration); expressed as log10 (moles H^+/L)

Phobic a response away from, such as a photophobic response in which light is avoided

Pholeterous living in crayfish burrows

Phosphatase enzyme that cleaves organic phosphorus compounds to liberate phosphate

Phosphate PO_4^{3-}, dominant ionic form of inorganic phosphorus in natural waters

Photoautotrophs photosynthetic organisms that use light

Photoinhibition the deleterious effects of high light

Photophobic repelled by light

Photosynthesis−irradiance (P−I) describes the effect of light on photosynthetic rate

Photosynthetically active radiation (PAR) light from 400 to 700 nm that is generally available to cyanobacteria and eukaryotic photosynthetic organisms; often expressed in μmole photons/(m^2 s)

Photosynthetic photon flux density (PPFD) photosynthetically active radiation, often used interchangeably with PAR

Photosynthetic quotient moles of oxygen produced divided by moles of CO_2 fixed

Phototactic phototaxis, attracted toward light

Phreatic from the zone below the water table

Phycobilins protein pigments that collect light for photosynthesis in the cyanobacteria and red algae absorbing in the 550−650-nm range

Phycoerythrin a phycobilin pigment similar to those found in cyanobacteria that allow red algae to use blue-green wavelengths of light

Phylogeny the understanding of evolutionary relationships that make up the tree of life

Physiography the natural geography of a region

Phytoperiphyton the photosynthetic organisms in periphyton, usually algae

Phytoplankton suspended algae

Phytotelmata pools of water found in terrestrial plants (such as bromeliads, pitcher plants or tree holes)

Picophytoplankton algal picoplankton

Picoplankton planktonic organisms smaller than 3 μm in diameter

Piercing predators predators that pierce their prey and suck bodily fluids from them

Pieziometer an instrument for measuring pressure or compressibility; used in determining groundwater flow patterns

Plankton suspended organisms

Playa a shallow marsh-like pond, not formed glacially

Pleuston macroorganisms living on the surface of the water

Pneumatophore specialized aerial roots of plants living in saturated soils

Poikilothermy having variable body temperature; also complicated vertical temperature profile

Point bar a depositional area on the inside of a river bend

Point source a clearly defined source of pollution such as a sewage outfall (as opposed to nonpoint source)

Polar lake pond or lake with a surface temperature of 4°C or lower in the warm season

Polar molecule a molecule with unequal distribution of charge

Polymictic almost continuous circulation or many flowing or mixing periods annually

Polyphosphate anionic phosphate polymers used by some organisms in luxury uptake and for phosphorus storage

POM particulate organic matter that is retained by an approximately 0.45-μm membrane filter

Pool a slow-moving portion of a river or stream

Poor fens fens with low nutrient availability and acidic water and are dominated by short vegetation

Porosity the maximum water that can be stored in hydrated sediment

Potamodromous with migratory pattern entirely in freshwater

Potamology older term for stream biology, river ecology, and lotic limnology

Potamon stream or larger water body environment (rhithron)

Potential energy stored energy that can do work

Potential evapotranspiration the maximum possible water loss

Pothole a shallow pond or wetland formed in glacial till

ppb parts per billion (μg/L)

PPFD photosynthetic photon flux density; see photosynthetically active radiation

ppm parts per million (mg/L)

ppt parts per trillion (ng/L)

P:R ratio of gross primary production to community respiration

Prairie pothole a sizeable, rounded, and often water-filled depression in grassland habitat; formed in glacial drift

Precipitation the deposit of hail, mist, fog, rain, sleet, or snow

Predation one organism eating another

Predator an organism that feeds on other organisms

Predator-permanence gradient increasing prevalence of predators with increasing hydroperiod

Press a continuous disturbance

Primary consumer herbivore, grazer

Primary producer autotrophic organism

Primary production rate of carbon fixation by autotrophs (=gross primary production); this minus respiration and maintenance is net primary production

Primary succession disturbance removes all or most of the organisms or conditions cause creation of new habitat

Probability value used to express significance

Production population growth (in fish ecology)

Production flux of carbon or energy through an ecosystem compartment

Production to biomass ratio (P:B) used for assessing growth and turnover and show how biomass is not always directed related to production

Production efficiency the efficiency at which an animal converts assimilated (net production efficiency) or ingested (gross production efficiency) materials into actual production

Profundal zone benthic zone in lakes deep enough that there is not enough light to support photosynthetic organisms

Psammon a habitat in interstitial water between sand grains

Pseudo-persistent chemicals with short half-lives but chronic inputs

Pseudoreplication repeated samples from one treatment incorrectly treated as true replicates

Psychrophilic cold-loving; requires cold temperatures to grow or reproduce

Pulse a single pulsed event

Putrifaction decomposition of organic materials, primarily proteins, resulting in formation of ammonium (NH_4^+) and sulfide (H_2S)

Pycnocline density gradient

Pyrite a complex of sulfide and metal

Quorum sensing compounds secreted by bacteria with biofilms that signal cells to settle and form aggregates when a high enough concentration is reached

Radula hard, rasping mouthpart of gastropods generally used for scraping biofilms

Random by chance

Rarefaction method used by ecologists to estimate the number of species based on sampling effort

Rating curve a plot of measured stream discharge against depth that is used to predict discharge from depth (stage)

Reach a series of pools and riffles in a stream or river

Reaeration process of gas equilibration with the atmosphere

Recalcitrant resistant to degradation

Recharge water entering an aquifer

Recruitment the number of fish entering each size or age class

Redd area on lake or stream bottom where salmon or trout spawn; often a round, cleared depression in gravel

Redfield ratio algal composition under balanced growth, 106:16:1 C:N:P, by moles

Redox or redox potential, the relative number of free electrons in solution; measured as oxidation—reduction potential

Reductionist approach when researchers examine individual components or subsets of systems may involve laboratory experiments or microcosms or mesocosms

Reedswamp a wetland dominated by reeds (e.g., *Phragmites*)

Reflect to turn, throw, or bend off backwards at an angle

Reflex bleeding (autohemorrhaging) exuding hemolymph (blood) containing chemical deterrents to avoid predation

Refraction light bending as it passes through materials of different densities

Regeneration see remineralization

Remineralization organisms excreting nutrients. Also referred to as regeneration or mineralization.

Replication repeating treatments in an experiment

Reset the return of a habitat to preceding (in time) conditions or to conditions upstream (in a continuum); a flood may "reset" a stream to earlier conditions

Residence time amount of time a material spends in an ecosystem compartment

Resilience the ability to bounce back or recover

Resistance the ability to remain unchanged

Resource ratio theory predictive model of species interactions based on their use of shared resources; predicts that species dominance varies with ratios of resource availability

Retention time average time materials spend in an ecosystem compartment

Reynolds number the ratio of inertia to dynamic viscosity; a unitless number that describes the properties of fluids as related to spatial scale and movement

RFLP (restriction fragment length polymorphism) DNA profiling technique that separates restriction fragments of DNA by their length

Rheocrene a spring-fed brook

Rhithron brook environment

Rice paddies natural or constructed wetlands that allow for rice culture

Rich fens fens with higher nutrient availability, neutral or alkaline, and support diverse plant assemblages

Richness see species richness

Riffle a rapidly flowing portion of a river or stream where the influence of the bottom can be seen at the surface

Riparian related to or located near water; often a transition zone between terrestrial and aquatic habitats

Riparian buffer an area adjacent to a stream or other water body where vegetation is left intact to reduce pollution inputs

Riparian wetlands inland wetlands related to or located near water, also called riverine wetlands

Rising limb the portion of the hydrograph that is increasing during or immediately after a storm

River continuum concept an ecosystem-based view of streams and rivers as a continuum from small forested headwater streams to large rivers

River wave concept a concept of stream ecosystems based on temporal patterns following variations in flow

Riverine productivity model in-stream primary production (autochthonous energy) exceeds direct riparian litter inputs as the main source of energy in some large rivers

Riverine wetlands inland wetlands related or located near rivers

rRNA (ribosomal RNA) focus of molecular taxonomy methods because it is highly conserved (low variability)

Run a portion of a river or stream where flow is rapid, but the surface is smooth

Runoff water that eventually flows into rivers and streams

Runoff coefficient the proportion of precipitation falling on a catchment that enters the river or stream, a dimensionless factor that is used to convert the rainfall amounts to runoff

Salinity amount of inorganic salts dissolved in water, stated as %, ‰, g/kg, ppm, mg/L, mg/dm^3; in seawater, closely connected to Cl^- concentration and conductance; salinity of seawater is approximately 3.5% (35,000 ppm)

Saltern saline water with a similar composition to seawater

Sapropel reduced, nondescript sediment of polluted or hypereutrophic lakes; black with FeS, often odorous with H_2S

Saprophytes heterotrophs that decompose organic carbon

Saturating concentration equilibrium concentration when pure water is left in full contact with the atmosphere or in full contact with an undissolved substance

Scalar collector a spherical light collector

Scale to arrange in a graduated series

Scintillation counter an instrument used to quantify radioactive isotopes

Scramble competition a form of intraspecific competition; multiple individuals of the same species compete for a resource and all get equally less than they need

Scraper (grazer) an organism that obtains its food by scraping off and consuming biofilms and periphyton

Scum a layer of floating algae or organic material

Seasonal succession disturbance that occurs on an annual basis because harsh seasons knock back the community that occurs during the growing season

Seasonality index ratio of runoff in wet season to runoff in dry season

Secchi disk a circular disk with black and white quadrants that is lowered into the water to estimate water clarity

Secondary consumer eats primary consumers

Secondary production tissue production by heterotrophic organisms

Secondary succession community response to a disturbance that removes only some of the species or substantially reduces populations

Sediment the particulate matter that settles to the bottom of a liquid

Segment several reaches of a stream together or the length between major tributaries

Sensitivity analysis to systematically test the sensitivity of a model to the assumptions used to construct it

Seiche rocking of a lake or layer of a lake

Seine a long net used to trap fishes

Sequestration carbon deposition

Serial endosymbiosis theory eukaryotic cells organelles were formed from ingested and enveloped endosymbiotic bacterial cells, many accept this theory as an explanation for the origin of chloroplasts and mitochondria.

Sessile attached to bottom

Seston suspended organic particles

Shannon diversity an index of diversity that includes both species richness and evenness (also called Shannon—Wiener or Shannon—Weaver index)

Shear stress tractive force per unit area

Sheet flow a shallower layer of water flowing across the surface of soil

Shelford's law of tolerance concept that species distributions are ultimately limited by their range of tolerance for local environmental factors

Sherwood number ratio of rates of molecular diffusion to transport diffusion

Shoreline development the circumference of a lake divided by the circumference of a circle with the same area; used as an indicator of trophic state (D_L)

Shredder an organism that makes its living shredding organic material for food

Significance an estimate of the statistical certainty

Silica SiO_2

Silicic acid H_4SiO_4

Simulation modeling nonexperimental method used to explore possible hypotheses

Sinkhole surface depression formed in karst area when a subsurface cavity collapses

Sink population among metapopulations, a population that does not produce enough individuals to be self-sustaining; relies on immigration to persist through time

Sinter sediments from mineral springs, including the siliceous geyserite and the calcareous tufa or travertine

Sinuosity the degree of meandering

Size class discrete size range of an organism

Size fractionate separate into size classes

Slough swamp, shallow lake, or slowly flowing marsh

Spates minor floods

Spring mixing during the spring wind can mix the warmer lake surface water (heated by solar energy) with the cooler water below

Solubility the ability to dissolve in a liquid

Soluble reactive phosphorus a chemically determined fraction containing phosphate and other forms of orthophosphate; also called DRP (dissolved reactive phosphorus)

Solvents can dissolve both gasses and ions

Source population among metapopulations, a population that produces enough individuals to sustain itself, and often produces excess individuals to contribute to other populations

Species area relationship relationship between number of species present and geographical area

Species diversity generally measured as the number of species in an area and their evenness (relative abundance)

Species richness number of species in an area

Specific runoff precipitation runoff per unit catchment area

Spectrophotometer an instrument that measures the absorption of light as a function of wavelength

Spherical collector having the form of a sphere or part of a sphere; spherical irradiance collectors are used to estimate total light available to photosynthetic plankton (also called scalar collector)

Spicule small, hard spine-like structures; form the skeleton of sponges

Spiraling length horizontal distance on a river or stream between successive uptake or release events (nutrient spiraling)

Spring mixing lack of differences in temperature with depth allowing wind to mix an entire lake during spring

SRP see soluble reactive phosphorus

Stability of stratification (S) the energy required to blend a body of water to constant density without adding or subtracting heat; expressed as g-cm/cm^2 of lake surface

Stable isotope natural form of an element that does not emit radioactivity

State change when enough state variables in an ecosystem change that a statistically detectable difference from the prior state occurs

Stenothermal able to survive in only a limited range of water temperatures

Stock size of population

Stoke's law law describing sinking rates of small spherical objects

Stoichiometry the ratio of elements to each other

Strahler classification system used to describe locations in dendritic patterns of stream drainage systems; first-order streams are those without tributaries, second-order streams are the result of two first-order streams joining, third-order streams are the result of two second-order streams joining, and so on; a lower order stream entering a higher order stream does not change the numeric designation of the larger stream

Stratification density differences in water that can maintain stable layers

Steam metabolism respiration and photosynthesis occurring in a stream

Streamline the path of a fluid particle relative to a solid body past which the fluid is moving in smooth flow without turbulence

Streamlined shaped to avoid causing turbulence

Stream order classification of streams; the most commonly used system is the Strahler system

Stygobite obligate groundwater dweller; not found in surface waters

Styigophile groundwater organism with adaptations to living in groundwater, but also found in surface waters

Sublethal not causing death

Sublittoral bottom region between littoral and profundal zones

Submersed a macrophyte that grows below the water surface

Submerged see submersed

Succession the sequence of organisms that colonize and inhabit a disturbed or new habitat over time

Sulfate (SO_4^{2-}) anion of sulfur

Sulfide (S^{2-}) sulfur ion

Summer stratification in temperate lakes the hot period in which the lake water is in stable layers of discreet temperatures

Supercooling dropping the temperature of a liquid or gas below the freezing point without it becoming a solid

Supersaturation containing dissolved materials in excess of equilibrium concentrations

Surber sampler an apparatus with a square, fixed-area frame (often 1 ft^2 or 0.1 m^2) hinged to another frame that has a net; the open first frame is located on the stream bed, the net is elevated to a vertical position, and rocks within the open frame are rubbed and shaken to allow stream flow to move organisms into the net

Surface area to volume relationships a geometric ratio that can be used to indicate the ability of an organism to exchange materials and energy with its surroundings

Surface tension hydrogen bonding pulls water into a tight surface

Suspended load fine materials suspended and transported in stream water

Suspended particles material retained by a 0.45-μm filter that stays suspended (seston in streams)

Swamp wetland dominated by woody vegetation (United States) or forested fen or reedswamp (Europe)

Symbiosis two organisms that live in close proximity and interact with one another; does not determine interaction effect; incorrectly used by some to refer to mutualism

Synecology older term for study of communities (as opposed to single species)

Synergistic the sum of the effects of multiple toxins is greater than adding individual toxic effects

Syntrophy a group of organisms with complimentary metabolic capabilities; characteristic of anoxic microbial communities

Systematics the identification of organisms and their relatedness with taxonomy and phylogeny

T_{50} the amount of time necessary for 50% of original dry leaf mass to breakdown in decomposition studies

Tannins derivatives of multimeric gallic acid that leach from bark and leaf litter with properties similar to those of humic substances

Tarn a small mountain lake (often in a cirque)

Taxis movement toward stimuli, tropism

Taxonomy the identification and naming of life on earth

Tectonic basin a basin formed by movement of the earth's crust

Thalassohaline having ionic proportions similar to seawater

Thalweg the part of a stream channel through which the main or most rapid flow travels

Thanatosis feigning death to avoid predation

Thermal bar vertical or horizontal mass of water separating two areas of less dense water

Thermocline see metalimnion

Thermophilic hot-temperature-loving; requires warm temperatures to grow or reproduce

Thiosulfate ($S_2O_3^{2-}$) an oxyanion of sulfur produced by sulfate ions reacting with elemental sulfur in boiling water; the two sulfur atoms are at different redox states

Threshold when a system makes a rapid state change and goes past a tipping point, they are difficult to predict due to their abruptness and can completely change system structure

Throughfall precipitation falling through vegetation

Thymidine uptake a method using [^{3}H] thymidine incorporation into DNA to assess bacterial growth rate

Tidal marshes freshwater or saline coastal (fringe) wetlands

Tidal salt marshes a halophytic grassland or dwarf brush land on riverine sediments influenced by tides or other water fluctuations

Time series temporal replication in experiments

Top carnivores eat consumers but are generally not eaten by larger animals

Top-down control regulation of lower trophic levels or basal resources by higher trophic levels

Torenticole adapted to live in swift water

Tortuosity average length of the flow path between two points

Total dissolved solids (TDS) total mass of material left after a filtered water sample is dried

Total suspended solids (TSS) total weight of particulate material per unit volume of water

Tracer a dye, isotope, or ion used to trace the movement of water or other chemicals through the environment

Trait-mediated interactions interactions evidenced by evolved traits

Transient storage zone regions of slower flow in streams

Transmit to cause or allow spreading (particularly transmission of light in water)

Transport diffusion diffusion with water movement

Treatment method to remove contaminants from water

Troglobitic obligatory cave-dwelling or hypogean organisms

Troglophilic facultative cave-dwelling or hypogean organisms

Trophic basis of production the contribution of individual food sources to production of each species

Trophic cascade influence of consumer organisms on those lower in the food web with alternating effects at each trophic level (top-down control)

Trophic level position in a food web

Trophic state total ecosystem productivity, including heterotrophic and autotrophic pathways

Trophogenic zone area of photosynthetic production

Tropholytic zone area of decomposition

Tropism growth or movement toward a stimulus; similar to taxis (e.g., chemotaxis or chemotropism)

Tufa sedimentary rock formed by precipitation of carbonate minerals from water

Turbidity the quantity of suspended particles in water (TSS) or absorption of light by those particles

Turbulent flow flow with swirls and eddies not in the direction of the main flow

Turnover mixing of a lake, can also refer to calculation of residence time of water in a lake given volume and amounts of water entering and leaving; can also be the rate of flux of any material through a compartment per unit mass in the compartment (e.g., in nutrient budgets)

Ultraoligotrophic exceptionally unproductive system

Unconfined groundwater water not constrained from above

Unit hydrograph the hydrograph resulting from a single storm event

Univoltine producing one generation annually

Unsaturated zone vadose zone

Uptake nutrients taken into cells from the water surrounding them

Uptake length the average distance a nutrient travels in the water column before being taken up; one component of spiral length

Urban stream syndrome half of the world's population lives in urban areas and urbanization is impacting streams including; flashier hydrographs, simplification of stream channels, concrete-lined channels, urban pesticide contamination, runoff of petrochemicals and salt from roadways, increased frequency/intensity of algal blooms, and reduced species richness

Urea a nitrogen-containing organic compound that can be excreted by organisms

V volume of a body of water

Vadose zone above the water table; below the surface soil (also called unsaturated zone)

Van Dorn bottle a tube that can be lowered to a specific depth on a rope; when a messenger is sent down the rope, it triggers two ends to snap onto the tube, trapping the sampled water

Vaporization evaporation

Variable source area hydrology the expansion and contraction of saturated areas in a watershed during wet and dry periods

Veliger a planktonic larval stage of many mollusks

Vernal pool shallow pool or pond that holds water in spring but is usually dry for much of the year

Vicariant event an atmospheric or geophysical event resulting in disturbance or fragmentation of a previously constant distribution and thus leading to allopatric speciation and radiation

Viscosity the resistance of a fluid to change, an internal friction; two types of viscosity can be considered—the dynamic viscosity (intrinsic property of the fluid) and the viscous force (a property of scale)

Volatile suspended solids mass lost to burning following combustion when using bulk chemical parameters to characterize water chemistry or color of a sample

Volcanic related to action of volcanoes

Voltinism relating to the number of generations produced annually by an organism (e.g., bivoltine)

Vortex circular motion associated with turbulent flow

VPOM very fine (ultrafine) particulate organic matter (0.45–53 μm diameter)

Warm monomixis a period of total circulation during the cold time of year in a lake without ice; temperature stratification occurs in summer

Warm thereimixis summer mixing or cold monomixis with water warmer than 4°C

Water abstraction removal of water to be used for drinking water, irrigation, industry, etc. from rivers, lakes, aquifers, and other sources

Watershed area above a point in a stream that catches the water that flows down to that point (in Europe, called catchment, and the watershed is the high point that divides catchments)

Water residence time see retention time

Water table the top of the groundwater or saturated zone

Water velocity speed of water

Water yield the amount of water (depth) per unit time from a specific area

Weathering dissolution of materials from rocks

Weir a device to concentrate all the fluid in a channel into one place and allow for measurement of discharge

Well casing a pipe used to keep a well from collapsing

Wetland areas inundated or saturated by surface or groundwater at a frequency and duration sufficient to support a prevalence of vegetation adapted for life in saturated soil conditions

Wet meadow a meadow without open water but saturated soils

Wet prairie a shallow soil wetland with grass

Whitefish a fish species that migrates to elude severe conditions or, alternately, a type of North American coreogonid fish

Whitewater streams that carry large quantities of suspended solids and appear muddy or silty or large quantities of dissolved inorganic solids and are slightly alkaline or circumneutral; also sections of rivers or streams with extensive rapids and entrained air causing the water to appear white

Wind a natural movement of air of any velocity

Winter stratification the second period during the year when a cold-temperate lake does not mix as long as there is ice cover

Winterbourne upper reach of a chalk stream that channels water in fall after dry summer periods when aquifer levels decrease

Withdrawal taking something out (abstraction)

Yield the amount of fish that are taken per unit time from a fishery

z depth; vertical distance from surface

$\overline{Z}$ mean depth of a body of water; V/A (volume/area)

z_m maximum depth of a body of water

ZSD the maximum depth of Secchi disc visibility; usually expressed in meters

Zoophyte an organism growing on an animal

Zooplankton suspended invertebrates, generally multicellular and smaller than 1 mm

REFERENCES

Aaronson, S., 1981. Chemical Communication at the Microbial Level, vol. I. CRC Press, Inc, Boca Raton, FL.

Abbott, L.L., Bergey, E.A., 2007. Why are there few algae on snail shells? The effects of grazing, nutrients and shell chemistry on the algae on shells of *Helisoma trivolvis*. Freshw. Biol. 52 (11), 2112–2120.

Abe, T., Lawson, T., Weyers, J.D.B., Codd, G.A., 1996. Microcystin-LR inhibits photosynthesis of *Phaseolus vulgaris* primary leaves: implications for current spray irrigation practice. New Phytol. 133, 651–658.

Abebe, E., Decraemer, W., De Ley, P., 2008. Global diversity of nematodes (Nematoda) in freshwater. Hydrobiologia 595 (1), 67–78.

Abell, R.A., Olson, D.M., Dinerstein, E., Hurley, P.T., Diggs, J.T., Eichbaum, W., et al., 2000. Freshwater Ecoregions of North America, a Conservation Assessment. Island Press, Washington, DC.

Abergel, C., Legendre, M., Claverie, J.-M., 2015. The rapidly expanding universe of giant viruses: Mimivirus, Pandoravirus, Pithovirus and Mollivirus. FEMS Microbiol. Rev. 39 (6), 779–796. Available from: https://doi.org/10.1093/femsre/fuv037.

Alcazar, R.M., Becker, L., Hilliard, A.T., Kent, K.R., Fernald, R.D., 2016. Two types of dominant male cichlid fish: behavioral and hormonal characteristics. Biol. Open 5 (8), 1061–1071. Available from: https://doi.org/10.1242/bio.017640.

Alexander, G.G., Allan, J.D., 2007. Ecological success in stream restoration: case studies from the midwestern United States. Environ. Manage. 40 (2), 245–255. Available from: https://doi.org/10.1007/s00267-006-0064-6.

Alexander, R.B., Smith, R.A., Schwarz, G.E., 2000. Effect of stream channel size on the delivery of nitrogen to the Gulf of Mexico. Nature 403, 758–761.

Alexander, R.B., Smith, R.A., Schwarz, G.E., Boyer, E.W., Nolan, J.V., Brakebill, J.W., 2008. Differences in phosphorus and nitrogen delivery to the Gulf of Mexico from the Mississippi River basin. Environ. Sci. Technol. 42 (3), 822–830.

Alexopolus, C.J., Mims, C.W., Blackwell, M., 1996. Introductory Mycology, fourth ed. John Wiley & Sons, Inc., New York, NY.

Alford, R.A., 1999. Ecology resource use, competition, and predation. In: McDiarmid, R.W., Altig, R. (Eds.), Tadpoles: The Biology of Anuran Larvae. The University of Chicago Press, Chicago, IL, pp. 240–278.

Alfreider, A., Krössbacher, M., Psenner, R., 1997. Groundwater samples do not reflect bacterial densities and activity in subsurface systems. Water Res. 31 (4), 832–840.

Allan, J.D., 1983. Food consumption by trout and stoneflies in a Rocky Mountain stream, with comparison to prey standing crop. In: Fontaine, T.D.I., Bartell, S.M. (Eds.), Dynamics of Lotic Ecosystems. Ann Arbor Science Publishers, Ann Arbor, MI, pp. 371–390.

Allan, J.D., 1995. Stream Ecology: Structure and Function of Running Waters. Chapman and Hall, London.

Allan, J.D., Abell, R., Hogan, Z., Revenga, C., Taylor, B.W., Welcomme, R.L., et al., 2005. Overfishing of inland waters. Bioscience 55 (12), 1041–1051.

Allan, J.D., Carillo, M.M., 2007. Stream Ecology: Structure and Function of Running Waters, second ed. Springer, Dordrecht.

Allan, R.J., 1989. Factors affecting source and fate of persistent toxic organic chemicals: examples from the Laurentian Great Lakes. In: Boudou, A., Ribeyre, F. (Eds.), Aquatic Ecotoxicology: Fundamental Concepts and Methodologies. CRC Press, Boca Raton, FL, pp. 219–248.

Alldred, M., Baines, S.B., 2016. Effects of wetland plants on denitrification rates: a meta-analysis. Ecol. Appl. 26 (3), 676–685. Available from: https://doi.org/10.1890/14-1525.

Allen, K.R., 1951. The Horokiwi stream: a study of a trout population. New Zealand Dept. Fish. Bull. 10, 1–238.

Allen, S.K.J., Wattendorf, R.J., 1987. Triploid grass carp: status and management implications. Fisheries 12 (4), 20–24.

Allen, T.F.H., Hoekstra, T.W., 1992. Toward a Unified Ecology. Columbia University Press, New York, NY.

Allgeier, J.E., Wenger, S.J., Rosemond, A.D., Schindler, D.E., Layman, C.A., 2015. Metabolic theory and taxonomic identity predict nutrient recycling in a diverse food web. Proc. Natl. Acad. Sci. 112 (20), E2640–E2647.

Alroy, J., 2015. Current extinction rates of reptiles and amphibians. Proc. Natl. Acad. Sci. 112 (42), 13003–13008. Available from: https://doi.org/10.1073/pnas.1508681112.

Altieri, A.H., Gedan, K.B., 2015. Climate change and dead zones. Glob. Change Biol. 21 (4), 1395–1406.

Alvarez, M., Peckarsky, B.L., 2014. Cascading effects of predatory fish on the composition of benthic algae in high-altitude streams. Oikos 123 (1), 120–128. Available from: https://doi.org/10.1111/j.1600-0706.2013.00397.x.

Ambrosini, R., Musitelli, F., Navarra, F., Tagliaferri, I., Gandolfi, I., Bestetti, G., et al., 2017. Diversity and assembling processes of bacterial communities in cryoconite holes of a Karakoram Glacier. Microb. Ecol. 73 (4), 827–837. Available from: https://doi.org/10.1007/s00248-016-0914-6.

American Rivers Friends of the Earth Trout Unlimited, 1999. Dam removal success stories, Washington, DC. Retrieved from https://www.michigan.gov/documents/dnr/damsuccess_513764_7.pdf.

Amundrud, J.R., Faber, D.J., Keast, A., 1974. Seasonal succession of free-swimming perciform larvae in Lake Opinicon, Ontario. Can. J. Fish. Aquat. Sci. 31, 1661–1665.

Amy, P.S., 1997. Microbiology of the terrestrial deep subsurface. In: Amy, P.S., Haldeman, D.L. (Eds.), Microbial Dominance and Survival in the Subsurface. Lewis Publishers, Boca Raton, FL, pp. 185–203.

Amyot, M., Clayden, M.G., MacMillan, G.A., Perron, T., Arscott-Gauvin, A., 2017. Fate and trophic transfer of rare earth elements in temperate lake food webs. Environ. Sci. Technol. 51 (11), 6009–6017. Available from: https://doi.org/10.1021/acs.est.7b00739.

Anderson, A.F., Banfield, J.F., 2008. Virus population dynamics and acquired virus resistance in natural microbial communities. Science 320, 1047–1050.

Anderson, M.G., 1995a. Interactions between Lythrum salicaria and native organisms: a critical review. Environ. Manage. 19, 225–231.

Anderson, N.H., Wallace, J.B., 1984. Habitat, life history, and behavioral adaptations of aquatic insects. In: Merritt, R.W., Cummins, K.W. (Eds.), An Introduction to the Aquatic Insects of North America. Kendall/Hunt Publishing Co, Dubuque, IA, pp. 38–57.

Anderson, N.J., 1993. Natural versus anthropogenic change in lakes: the role of the sediment record. Trends Ecol. Evol. 8 (10), 356–361.

Anderson, N.J., 1995b. Naturally eutrophic lakes: reality, myth or myopia? Trends Ecol. Evol. 10 (4), 137–138.

Anderson, R.O., Neumann, R.M., 1996. Length, weight, and associated structural indices. In: Murphy, B.R., Willis, D.W. (Eds.), Fisheries Techniques, second ed. American Fisheries Society, Bethesda, MD, pp. 447–482.

Anderson, R.T., Chapelle, F.H., Lovley, D.R., 1998. Evidence against hydrogen-based microbial ecosystems in basalt aquifers. Science 281, 976–977.

Anderson, R.T., Lovley, D.R., 1997. Ecology and biogeochemistry of in situ groundwater bioremediation. Adv. Microb. Ecol. 15, 289–350.

Andresen, C.G., Lara, M.J., Tweedie, C.E., Lougheed, V.L., 2017. Rising plant-mediated methane emissions from arctic wetlands. Glob. Change Biol. 23 (3), 1128–1139. Available from: https://doi.org/10.1111/gcb.13469.

Angermeier, P.L., 1995. Ecological attributes of extinction-prone species: loss of freshwater fishes of Virginia. Conserv. Biol. 9, 143–158.

Angradi, T.R., 1998. Observations of freshwater jellyfish Craspedacusta sowerbyi Lankester (Trachylina: Petasidae) in a West Virginia reservoir. Brimleyana 25, 34–42.

Antwis, R.E., Preziosi, R.F., Harrison, X.A., Garner, T.W.J., 2015. Amphibian symbiotic bacteria do not show a universal ability to inhibit growth of the global panzootic lineage of *Batrachochytrium dendrobatidis*. Appl. Environ. Microbiol. 81 (11), 3706−3711. Available from: https://doi.org/10.1128/aem.00010-15.

Arnell, N.W., Gosling, S.N., 2016. The impacts of climate change on river flood risk at the global scale. Clim. Change 134 (3), 387−401.

Arnnok, P., Singh, R.R., Burakham, R., Pérez-Fuentetaja, A., Aga, D.S., 2017. Selective uptake and bioaccumulation of antidepressants in fish from effluent-impacted Niagara River. Environ. Sci. Technol. 51 (18), 10652−10662. Available from: https://doi.org/10.1021/acs.est.7b02912.

Arnold Jr., C.L., Gibbons, C.J., 1996. Impervious surface coverage: the emergence of a key environmental indicator. J. Am. Plann. Assoc. 62 (2), 243−258.

Arnold, S.F., Klotz, D.M., Collins, B.M., Vonier, P.M., Guillette Jr., L.J., McLachlan, J.A., 1996. Synergistic activation of estrogen receptor with combinations of environmental chemicals. Science 272, 1489−1492.

Arnqvist, G., Wooster, D., 1995. Meta-analysis: synthesizing research findings in ecology and evolution. Trends Ecol. Evol. 10, 236−240.

Arruda, J.A., 1979. A consideration of trophic dynamics in some tallgrass prairie farm ponds. Am. Midl. Nat. 10 (2), 254−262.

Arruda, J.A., Fromm, C.H., 1989. The relationship between taste and odor problems and lake enrichment from Kansas lakes in agricultural watersheds. Lake Reservoir Manage. 5, 45−52.

Arscott, D.B., Bowden, W.B., Finlay, J.C., 1998. Comparison of epilithic algal and bryophyte metabolism in an arctic tundra stream, Alaska. J. N. Am. Benthol. Soc. 17 (2), 210−227.

Arseneau, K.M., Driscoll, C.T., Brager, L.M., Ross, K.A., Cumming, B.F., 2011. Recent evidence of biological recovery from acidification in the Adirondacks (New York, USA): a multiproxy paleolimnological investigation of Big Moose Lake. Can. J. Fish. Aquat. Sci. 68 (4), 575−592.

Arsuffi, T.L., Suberkropp, K., 1984. Leaf processing capabilities of aquatic hyphomycetes: interspecific differences and influence on shredder feeding preferences. Oikos 42, 144−154.

Arsuffi, T.L., Suberkropp, K., 1985. Selective feeding by stream caddisfly (Trichoptera) detritivores on leaves with fungal-colonized patches. Oikos 45, 50−58.

Arsuffi, T.L., Suberkropp, K., 1989. Selective feeding by shredders on leaf colonizing stream fungi: comparison of macroinvertebrate taxa. Oecologia 79, 30−37.

Atkinson, C., Kelly, J., Vaughn, C., 2014. Tracing consumer-derived nitrogen in riverine food webs. Ecosystems 17 (3), 485−496. Available from: https://doi.org/10.1007/s10021-013-9736-2.

Atlas, R.M., Bartha, R., 1998. Microbial Ecology, Fundamentals and Applications, fourth ed. Addison Wesley Longman, Inc., Menlo Park, CA.

Atwood, T.B., Hammill, E., Richardson, J.S., 2014. Trophic-level dependent effects on CO_2 emissions from experimental stream ecosystems. Glob. Change Biol. 20 (11), 3386−3396. Available from: https://doi.org/10.1111/gcb.12516.

Aubeneau, A.F., Hanrahan, B., Bolster, D., Tank, J., 2016. Biofilm growth in gravel bed streams controls solute residence time distributions. J. Geophys. Res.: Biogeosci. 121 (7), 1840−1850. Available from: https://doi.org/10.1002/2016JG003333.

Auer, A., 1991. Qualitative diatom analysis as a tool to diagnose drowning. Am. J. Forensic Med. Pathol. 12, 213−218.

Auer, M.T., Effler, S.W., 1989. Variability in photosynthesis: impact on DO models. J. Environ. Eng. 115, 944−963.

Auerbach, E.A., Seyfried, E.E., McMahon, K.D., 2007. Tetracycline resistance genes in activated sludge wastewater treatment plants. Water Res. 41 (5), 1143−1151.

Auguet, J., Montanié, H., Hartmann, H., Lebaron, P., Casamayor, E., Catala, P., et al., 2009. Potential effect of freshwater virus on the structure and activity of bacterial communities in the Marennes-Oléron Bay (France). Microb. Ecol. 57 (2), 295−306.

Axelrod, H.R., 1973. African Cichlids of Lakes Malawi and Tanganyika. T. F. H. Publications, Inc., Neptune, NJ.

Axler, R.P., Redfield, G.W., Goldman, C.R., 1981. The importance of regenerated nitrogen to phyto-plankton productivity in a subalpine lake. Ecology 62, 345–354.

Aymerich, I., Acuña, V., Barceló, D., García, M.J., Petrovic, M., Poch, M., et al., 2016. Attenuation of pharmaceuticals and their transformation products in a wastewater treatment plant and its receiving river ecosystem. Water Res. 100, 126–136. Available from: https://doi.org/10.1016/j.watres.2016.04.022.

Azevedo, L.B., van Zelm, R., Elshout, P.M.F., Hendriks, A.J., Leuven, R.S.E.W., Struijs, J., et al., 2013. Species richness–phosphorus relationships for lakes and streams worldwide. Glob. Ecol. Biogeogr. 22 (12), 1304–1314. Available from: https://doi.org/10.1111/geb.12080.

Azevedo-Santos, V.M., Vitule, J.R.S., Pelicice, F.M., García-Berthou, E., Simberloff, D., 2016. Nonnative fish to control aedes mosquitoes: a controversial, harmful tool. Bioscience 67 (1), 84–90. Available from: https://doi.org/10.1093/biosci/biw156.

Baar, J., Paradi, I., Lucassen, E.C.H.E.T., Hudson-Edwards, K.A., Redecker, D., Roelofs, J.G.M., et al., 2011. Molecular analysis of AMF diversity in aquatic macrophytes: a comparison of oligotrophic and utra-oligotrophic lakes. Aquat. Bot. 94 (2), 53–61. Available from: https://doi.org/10.1016/j.aquabot.2010.09.006.

Baattrup-Pedersen, A., Riis, T., 1999. Macrophyte diversity and composition in relation to substratum characteristics in regulated and unregulated Danish streams. Freshw. Biol. 42, 375–385.

Bachmann, R.W., Hoyer, M.V., Canfield Jr., D.E., 1999. The restoration of Lake Apopka in relation to alternative stable states. Hydrobiologia 394, 219–232.

Bachmann, R.W., Jones, B.L., Fox, D.D., Hoyer, M., Bull, L.A., Canfield Jr., D.E., 1996. Relations between trophic state indicators and fish in Florida (U.S.A.) lakes. Can. J. Fish. Aquat. Sci. 53, 842–855.

Baer, S.G., 2013. Restoration ecology. Oxford Bibliogr. Available from: https://doi.org/10.1093/obo/9780199830060-0109.

Bagella, S., Gascón, S., Filigheddu, R., Cogoni, A., Boix, D., 2016. Mediterranean temporary ponds: new challenges from a neglected habitat. Hydrobiologia 782 (1), 1–10. Available from: https://doi.org/10.1007/s10750-016-2962-9.

Bagla, P., Kaiser, J., 1996. India's spreading health crisis draws global arsenic experts. Science 274, 174–175.

Bagnoud, A., de Bruijn, I., Andersson, A.F., Diomidis, N., Leupin, O.X., Schwyn, B., et al., 2016. A minimalistic microbial food web in an excavated deep subsurface clay rock. FEMS Microbiol. Ecol. 92 (1). Available from: https://doi.org/10.1093/femsec/fiv138.

Bagshaw, E.A., Wadham, J.L., Tranter, M., Perkins, R., Morgan, A., Williamson, C.J., et al., 2016. Response of Antarctic cryoconite microbial communities to light. FEMS Microbiol. Ecol. 92 (6). Available from: https://doi.org/10.1093/femsec/fiw076.

Bailly, D., Cassemiro, F.A., Agostinho, C.S., Marques, E.E., Agostinho, A.A., 2014. The metabolic theory of ecology convincingly explains the latitudinal diversity gradient of Neotropical freshwater fish. Ecology 95 (2), 553–562.

Baines, S.B., Webster, K.E., Kratz, T.K., Carpenter, S.R., Magnuson, J.J., 2000. Synchronous behavior of temperature, calcium, and chlorophyll in lakes of northern Wisconsin. Ecology 81 (3), 815–825.

Bajer, P.G., Beck, M.W., Cross, T.K., Koch, J.D., Bartodziej, W.M., Sorensen, P.W., 2016. Biological invasion by a benthivorous fish reduced the cover and species richness of aquatic plants in most lakes of a large North American ecoregion. Glob. Change Biol. 22 (12), 3937–3947. Available from: https://doi.org/10.1111/gcb.13377.

Baker, J., Van Sickle, J., Gagen, C., DeWalle, D.R., Sharpe, W., Carline, R., et al., 1996. Episodic acidi-fication of small streams in the northeastern United States: effects on fish populations. Ecol. Appl. 6 (2), 422–437.

Baker, M.A., Webster, J.R., 2017. Conservative and reactive solute dynamics. In: Lamberti, G.A., Hauer, F.H. (Eds.), Methods in Stream Ecology: Vol. 2: Ecosystem Function. Academic Press, p. 129.

Baker, P.D., Humpage, A.R., 1994. Toxicity associated with commonly occurring cyanobacteria in sur-face waters of the Murray-Darling Basin, Australia. Aust. J. Freshw. Res. 45, 773–786.

Bakker, E., Nolet, B., 2014. Experimental evidence for enhanced top-down control of freshwater macrophytes with nutrient enrichment. Oecologia 176 (3), 825–836. Available from: https://doi.org/10.1007/s00442-014-3047-y.

Bakker, E.S., Wood, K.A., Pagès, J.F., Veen, G.F., Christianen, M.J.A., Santamaría, L., et al., 2016. Herbivory on freshwater and marine macrophytes: a review and perspective. Aquat. Bot. 135, 18–36. Available from: https://doi.org/10.1016/j.aquabot.2016.04.008.

Balian, E.V., Segers, H., Lévêque, C., Martens, K., 2008. The freshwater animal diversity assessment: an overview of the results. Hydrobiologia 595 (1), 627–637. Available from: https://doi.org/10.1007/s10750-007-9246-3.

Balkwill, D.L., Boone, D.R., 1997. Identity and diversity of microorganisms cultured from subsurface environments. In: Amy, P.S., Haldeman, D.L. (Eds.), Microbiology of the Terrestrial Deep Subsurface. Lewis Publishers, Boca Raton, FL, pp. 105–117.

Ballen-Segura, M., Felip, M., Catalan, J., 2017. Some mixotrophic flagellate species selectively graze on Archaea. Appl. Environ. Microbiol. 83 (2). Available from: https://doi.org/10.1128/aem.02317-16.

Balvanera, P., Pfisterer, A.B., Buchamnn, N., He, J.-S., Nakashizuka, T., Raffaelli, D., et al., 2006. Quantifying the evidence for biodiversity effects on ecosystem functioning and services. Ecol. Lett. 9, 1146–1156.

Bancroft, B.A., Baker, N.J., Blaustein, A.R., 2007. Effects of UVB radiation on marine and freshwater organisms: a synthesis through meta-analysis. Ecol. Lett. 10 (4), 332–345.

Bancroft, B.A., Baker, N.J., Blaustein, A.R., 2008. A meta-analysis of the effects of ultraviolet B radiation and its synergistic interactions with pH, contaminants, and disease on amphibian survival. Conserv. Biol. 22 (4), 987–996.

Banner, E., Stahl, A., Dodds, W., 2009. Stream discharge and riparian land use influence in-stream concentrations and loads of phosphorus from Central Plains watersheds. Environ. Manage. 44 (3), 552–565.

Barbour, M.T., Gerritsen, J., Snyder, B.D., Stribling, J.B., 1999. Rapid bioassessment protocols for use in wadeable streams and rivers: periphyton, benthic macroinvertebrates, and fish, second ed. (EPA 841-B-99-002), Washington, DC. Retrieved from https://www3.epa.gov/region1/npdes/merrimackstation/pdfs/ar/AR-1164.pdf.

Barel, C.D.N., Dorit, R., Greenwood, P.H., Fryer, G., Hughes, N., Jackson, P.B.N., et al., 1985. Destruction of fisheries in Africa's lakes. Nature 315, 19–20.

Barkley, D., Song, B., Mukund, V., Lemoult, G., Avila, M., Hof, B., 2015. The rise of fully turbulent flow. Nature 526 (7574), 550–553. Available from: https://doi.org/10.1038/nature15701.

Barnes, K.K., Kolpin, D.W., Meyer, M.T., Thurman, E.M., Furlong, E.T., Zaugg, S.D., et al., 2002. Water-quality data for pharmaceuticals, hormones, and other organic wastewater contaminants in U.S. streams, 1999–2000 (Open-File Report 02-94). Retrieved from https://toxics.usgs.gov/pubs/OFR-02-94/.

Barnett, T.P., Pierce, D.W., Hidalgo, H.G., Bonfils, C., Santer, B.D., Das, T., et al., 2008. Human-induced changes in the hydrology of the western United States. Science 319, 1080–1083.

Baron, J.S., Campbell, D.H., 1997. Nitrogen fluxes in a high elevation Colorado Rocky Mountain basin. Hydrol. Processes 11, 783–799.

Barrett, P.R.F., Curnow, J.C., Littlejohn, J.W., 1996. The control of diatom and cyanobacterial blooms in reservoirs using barley straw. Hydrobiologia 340, 307–311.

Barrett, P.R.F., Littlejohn, J.W., Curnow, J., 1999. Long-term algal control in a reservoir using barley straw. Hydrobiologia 415, 309–313.

Barry, R.G., Chorley, R.J., 2009. Atmosphere, Weather and Climate. Routledge.

Bartels, P., Hirsch, P.E., Svanbäck, R., Eklöv, P., 2016. Dissolved organic carbon reduces habitat coupling by top predators in lake ecosystems. Ecosystems 19 (6), 955–967. Available from: https://doi.org/10.1007/s10021-016-9978-x.

Bartley, D.M., 1996. Precautionary approach to the introduction and transfer of aquatic species. Paper Presented at the Technical Consultation on the Precautionary Approach to Capture Fisheries (Including Species Introductions), Lysekil, Sweden, June 613, 1995. Available from http://www.fao.org/3/W1238E08.htm.

Bartley, T.J., Braid, H.E., McCann, K.S., Lester, N.P., Shuter, B.J., Hanner, R.H., 2015. DNA barcoding increases resolution and changes structure in Canadian boreal shield lake food webs. DNA Barcodes 3 (1), 30−43.

Bass, D., 1992. Colonization and succession of benthic macroinvertebrates in Arcadia Lake, a South-Central USA reservoir. Hydrobiologia 242, 123−131.

Batani, G., Pérez, G., Martínez de la Escalera, G., Piccini, C., Fazi, S., 2016. Competition and protist predation are important regulators of riverine bacterial community composition and size distribution. J. Freshw. Ecol. 31 (4), 609−623. Available from: https://doi.org/10.1080/02705060.2016.1209443.

Batt, A.L., Wathen, J.B., Lazorchak, J.M., Olsen, A.R., Kincaid, T.M., 2017. Statistical survey of persistent organic pollutants: risk estimations to humans and wildlife through consumption of fish from U.S. rivers. Environ. Sci. Technol. 51 (5), 3021−3031. Available from: https://doi.org/10.1021/acs.est.6b05162.

Batt, D.J., Anderson, M.G., Anderson, C.D., Caswell, F.D., 1989. The use of prairie potholes by North American ducks. In: van der Valk, A. (Ed.), Northern Prairie Wetlands. Iowa State University Press, Ames, IA, pp. 204−227.

Battaglia, L.L., Foré, S.A., Sharitz, R.R., 2000. Seedling emergence, survival and size in relation to light and water availability in two bottomland hardwood species. J. Ecol. 88 (6), 1041−1050.

Battin, T.J., Besemer, K., Bengtsson, M.M., Romani, A.M., Packmann, A.I., 2016. The ecology and biogeochemistry of stream biofilms. Nat. Rev. Microbiol. 14 (4), 251−263.

Battin, T.J., Sengschmitt, D., 1999. Linking sediment biofilms, hydrodynamics, and river bed clogging: evidence from a large river. Microb. Ecol. 37, 185−196.

Batzer, D.P., Resh, V.H., 1991. Trophic interactions among a beetle predator, a chironomid grazer, and periphyton in a seasonal wetland. Oikos 60, 251−257.

Batzer, D.P., Sharitz, R.R., 2006. Ecology of Freshwater and Estuarine Wetlands. University of California Press, Berkeley.

Batzer, D.P., Wissinger, S.A., 1996. Ecology of insect communities in nontidal wetlands. Annu. Rev. Entomol. 41, 75−100.

Baxter, R.M., 1977. Environmental effects of dams and impoundments. Annu. Rev. Ecol. Syst. 8, 255−283.

Bayley, I.A.E., 1972. Salinity tolerance and osmotic behavior of animals in a thalassic saline and marine hypersaline waters. Annu. Rev. Ecol. Syst. 3, 233−268.

Beardall, J., Young, E., Roberts, S., 2001. Approaches for determining phytoplankton nutrient limitation. Aquat. Sci. 63 (1), 44−69.

Bearon, R.N., Magar, V., 2010. Simple models of the chemical field around swimming plankton. J. Plankton Res. 32 (11), 1599−1608. Available from: https://doi.org/10.1093/plankt/fbq075.

Beatty, W.S., Kesler, D.C., Webb, E.B., Raedeke, A.H., Naylor, L.W., Humburg, D.D., 2014. The role of protected area wetlands in waterfowl habitat conservation: implications for protected area network design. Biol. Conserv. 176, 144−152.

Beaulieu, J.J., McManus, M.G., Nietch, C.T., 2016. Estimates of reservoir methane emissions based on a spatially balanced probabilistic-survey. Limnol. Oceanogr. 61 (S1), S27−S40. Available from: https://doi.org/10.1002/lno.10284.

Beaulieu, J.J., Tank, J.L., Hamilton, S.K., Wollheim, W.M., Hall, R.O., Mulholland, P.J., et al., 2011. Nitrous oxide emission from denitrification in stream and river networks. Proc. Natl. Acad. Sci. 108 (1), 214−219. Available from: https://doi.org/10.1073/pnas.1011464108.

Beaulieu, M., Pick, F., Gregory-Eaves, I., 2013. Nutrients and water temperature are significant predictors of cyanobacterial biomass in a 1147 lakes data set. Limnol. Oceanogr. 58 (5), 1736−1746.

Beaver, J.R., Manis, E.E., Loftin, K.A., Graham, J.L., Pollard, A.I., Mitchell, R.M., 2014. Land use patterns, ecoregion, and microcystin relationships in US lakes and reservoirs: a preliminary evaluation. Harmful. Algae 36, 57−62.

Bebout, B.M., Garcia-Pichel, F., 1995. UV B-induced vertical migrations of cyanobacteria in a microbial mat. Appl. Environ. Microbiol. 61, 4215−4222.

Beck, W.S., Rugenski, A.T., Poff, N.L., 2017. Influence of experimental, environmental, and geographic factors on nutrient-diffusing substrate experiments in running waters. Freshw. Biol. 62 (10), 1667−1680. Available from: https://doi.org/10.1111/fwb.12989.

Beckmann, M., Václavík, T., Manceur, A.M., Šprtová, L., Wehrden, H., Welk, E., et al., 2014. glUV: a global UV-B radiation data set for macroecological studies. Methods Ecol. Evol. 5 (4), 372−383.

Bedford, B.L., Walbridge, M.R., Aldous, A., 1999. Patterns in nutrient availability and plant diversity of temperate North American wetlands. Ecology 80 (7), 2151−2169.

Behrenfeld, M.J., Lean, D.R.S., Lee II, H., 1995. Ultraviolet-B radiation effects on inorganic nitrogen uptake by natural assemblages of oceanic plankton. J. Phycol. 31, 25−36.

Beisner, B.E., Carey, C.C., 2016. Lake name or name lake? The etymology of lake nomenclature in the United States. Freshw. Biol. 61 (9), 1601−1609. Available from: https://doi.org/10.1111/fwb.12795.

Beketov, M.A., Kefford, B.J., Schäfer, R.B., Liess, M., 2013. Pesticides reduce regional biodiversity of stream invertebrates. Proc. Natl. Acad. Sci. 110 (27), 11039−11043. Available from: https://doi.org/10.1073/pnas.1305618110.

Belk, D., 1984. Patterns in Anostracan distribution. In: Jain, S., Moyle, P. (Eds.), Vernal Pools and Intermittent Streams, vol. 28. Institute of Ecology, University of California, Davis, CA, pp. 168−172. (Reprinted from: May 9, 1981).

Bell, G., 2001. Neutral macroecology. Science 293 (5539), 2413−2418.

Belle, S., Baudrot, V., Lami, A., Musazzi, S., Dakos, V., 2017. Rising variance and abrupt shifts of subfossil chironomids due to eutrophication in a deep sub-alpine lake. Aquat. Ecol. 51 (2), 307−319. Available from: https://doi.org/10.1007/s10452-017-9618-3.

Belmont, P., Gran, K.B., Schottler, S.P., Wilcock, P.R., Day, S.S., Jennings, C., et al., 2011. Large shift in source of fine sediment in the Upper Mississippi River. Environ. Sci. Technol. 45 (20), 8804−8810. Available from: https://doi.org/10.1021/es2019109.

Belt, D., 1992. The world's great lake. Natl. Geogr. Mag. 181, 2−39.

Bencala, K.E., Walters, R.A., 1983. Simulation of solute transport in a Mountain Pool-and-Riffle stream—a transient storage model. Water Resour. Res. 19 (3), 718−724.

Benda, L., Poff, N.L., Miller, D., Dunne, T., Reeves, G., Pess, G., et al., 2004. The network dynamics hypothesis: how channel networks structure riverine habitats. Bioscience 54 (5), 413−427.

Benfield, E.F., Fritz, K.M., Tiegs, S.D., 2017. Leaf-litter breakdown. In: Hauer, F.R., Lamberti, G.A. (Eds.), Methods in Stream Ecology, third ed. Academic Press, San Diego, CA, pp. 71−82.

Benke, A., Whiles, M., 2011. Life table vs secondary production analyses-relationships and usage in ecology. J. N. Am. Benthol. Soc. 30 (4), 1024−1032. Available from: https://doi.org/10.1899/11-007.1.

Benke, A.C., 1984. Secondary production of aquatic insects. In: Resh, V.H., Rosenberg, D.M. (Eds.), The Ecology of Aquatic Insects. Praeger, New York, NY, pp. 289−322.

Benke, A.C., 1993. Concepts and patterns of invertebrate production in running waters. Internationalen vereinigung für theoretische und angewandte Limnologie, Verhandlungen 25, 15−38.

Benke, A.C., Huryn, A.D., 2006. Secondary production of macroinvertebrates. In: Hauer, F.R., Lamberti, G.A. (Eds.), Methods in Stream Ecology, second ed. Academic Press, London, UK, pp. 691−710.

Benke, A.C., Huryn, A.D., 2010. Benthic invertebrate production—facilitating answers to ecological riddles in freshwater ecosystems. J. N. Am. Benthol. Soc. 29 (1), 264−285.

Benke, A.C., Wallace, J.B., 1980. Trophic basis of production among net-spinning caddisflies in a southern Appalachian stream. Ecology 61, 108−118.

Benner, R., Lewis, D.L., Hodson, R.E., 1989. Biogeochemical cycling of organic matter in acidic environments: are microbial degradative processes adapted to low pH? In: Rao, S.S. (Ed.), Acid Stress and Aquatic Microbial Interactions. CRC Press, Boca Raton, FL, pp. 34−43.

Bennett, A.M.R., 2008. Aquatic Hymenoptera. In: Merritt, R.W., Cummins, K.W., Berg, M.B. (Eds.), An Introduction to the Aquatic Insects of North America, fourth ed. Kendall-Hunt, Dubuque, IA, pp. 673−686.

Bennett, E.M., Reed-Andersen, T., Houser, J.N., Gabriel, J.R., Carpenter, S.R., 1999. A phosphorus budget for the Lake Mendota watershed. Ecosystems 2, 69−75.

Bennett, P.C., Rogers, J.R., Choi, W.J., 2001. Silicates, silicate weathering, and microbial ecology. Geomicrobiol. J. 18 (1), 3−19.

Benson, A.J., 2000. Documenting over a century of aquatic introductions in the United States. In: Claudi, R., Leach, J.H. (Eds.), Nonindigenous Freshwater Organisms. Lewis Publishers, CRC Press, Boca Ratan, FL, pp. 1−31.

Benstead, J.P., March, J.G., Pringle, C.M., Scatena, F.N., 1999. Effects of a low-head dam and water abstraction on migratory tropical stream biota. Ecol. Appl. 9 (2), 656−668.

Bergeron, M., Vincent, W.F., 1997. Microbial food web responses to phosphorus supply and solar UV radiation in a subarctic lake. Aquat. Microb. Ecol. 12, 239−249.

Bergman, E., Hansson, L.-A., Persson, A., Strand, J., Romare, P., Enell, M., et al., 1999. Synthesis of theoretical and empirical experiences from nutrient and cyprinid reductions in Lake Ringsjön. Hydrobiologia 404, 145−156.

Berman, T., Chava, S., 1999. Algal growth on organic compounds as nitrogen sources. J. Plankton Res. 21 (8), 1423−1437.

Berner, E.K., Berner, R.A., 1987. The Global Water Cycle. Prentice Hall, Englewood Cliffs, NJ.

Bernert, J.A., Eilers, J.M., Elers, B.J., Blok, E., Daggett, S.G., Bierly, K.F., 1999. Recent wetlands trends (1981/82−1994) in the Willamette Valley, Oregon, USA. Wetlands 19 (3), 545−559.

Bernhardt, E.S., Lutz, B.D., King, R.S., Fay, J.P., Carter, C.E., Helton, A.M., et al., 2012. How many mountains can we mine? Assessing the regional degradation of central Appalachian rivers by surface coal mining. Environ. Sci. Technol. 46 (15), 8115−8122. Available from: https://doi.org/10.1021/es301144q.

Bernhardt, E.S., Palmer, M.A., 2007. Restoring streams in an urbanizing world. Freshw. Biol. 52 (4), 738−751.

Bernhardt, E.S., Palmer, M.A., Allan, J.D., Alexander, G., Barnas, K., Brooks, S., et al., 2005. Ecology—synthesizing US river restoration efforts. Science 308 (5722), 636−637. Available from: https://doi.org/10.1126/science.1109769.

Bernot, M.J., Sobota, D.J., Hall, R.O., Mulholland, P.J., Dodds, W.K., Webster, J.R., et al., 2010. Inter-regional comparison of land-use effects on stream metabolism. Freshw. Biol. 55 (9), 1874−1890.

Bernot, R., Brandenburg, M., 2013. Freshwater snail vital rates affected by non-lethal concentrations of silver nanoparticles. Hydrobiologia 714 (1), 25−34. Available from: https://doi.org/10.1007/s10750-013-1509-6.

Bernot, R.J., Dodds, W.K., Quist, M.C., Guy, C.S., 2006. Temperature and kairomone induced life history plasticity in coexisting Daphnia. Aquat. Ecol. 40 (3), 361−372.

Bernot, R.J., Lamberti, G.A., 2008. Indirect effects of a parasite on a benthic community: an experiment with trematodes, snails and periphyton. Freshw. Biol. 53 (2), 322−329.

Bernstein, L., Bosch, P., Canziani, O., Chen, Z., Christ, R., Riahi, K., 2008. IPCC, 2007: climate change 2007: Synthesis report. IPCC.

Bertilsson, S., Burgin, A., Carey, C.C., Fey, S.B., Grossart, H.-P., Grubisic, L.M., et al., 2013. The under-ice microbiome of seasonally frozen lakes. Limnol. Oceanogr. 58 (6), 1998−2012.

Bertness, M.D., Callaway, R., 1994. Positive interactions in communities. Trends Ecol. Evol. 9 (5), 191−193.

Bertness, M.D., Hacker, S.D., 1994. Physical stress and positive associations among marsh plants. Am. Nat. 144 (3), 363−372.

Bertness, M.D., Leonard, G.H., 1997. The role of positive interactions in communities: lessons from intertidal habitats. Ecology 78 (7), 1976−1989.

Bertrand, K., Gido, K., Dodds, W., Murdock, J., Whiles, M., 2009. Disturbance frequency and functional identity mediate ecosystem processes in prairie streams. Oikos 118 (6), 917.

Beschta, R.L., Bilby, R.E., Brown, G.W., Holtby, L.B., Hofstra, T.D., 1987. Stream temperature and aquatic habitat: fisheries and forestry interactions. In: Salo, E.O., Cundy, T.W. (Eds.), Streamside Management: Forestry and Fishery Interactions, vol. 57. Institute of Forest Resources, University of Washington, Seattle, pp. 191−232.

Beveridge, M.C.M., Ross, L.G., Kelly, L.A., 1994. Aquaculture and biodiversity. Ambio 23 (8), 497−502.

Beyer, J.E., Hambright, K.D., 2016. Persistent and delayed effects of toxic cyanobacteria exposure on life history traits of a common zooplankter. Limnol. Oceanogr. 61 (2), 587−595. Available from: https://doi.org/10.1002/lno.10239.

Bhowmick, S., Pramanik, S., Singh, P., Mondal, P., Chatterjee, D., Nriagu, J., 2018. Arsenic in groundwater of West Bengal, India: a review of human health risks and assessment of possible intervention options. Sci. Total Environ. 612, 148−169. Available from: https://doi.org/10.1016/j. scitotenv.2017.08.216.

Biasi, C., Graça, M.A.S., Santos, S., Ferreira, V., 2017. Nutrient enrichment in water more than in leaves affects aquatic microbial litter processing. Oecologia 184 (2), 555−568. Available from: https://doi.org/10.1007/s00442-017-3869-5.

Bidigare, R.R., Ondrusek, M.E., Kennicutt II, M.C., Iturriaga, R., Harvey, H.R., Hoham, R.W., et al., 1993. Evidence for a photoprotective function for secondary carotenoids of snow algae. J. Phycol. 29, 427−434.

Bierman, V.J.J., James, R.T., 1995. A preliminary modeling analysis of water quality in Lake Okeechobee, Florida: diagnostic and sensitivity analyses. Water Res. 29 (12), 2767−2775.

Biggs, B.J.F., 1995. The contribution of flood disturbance, catchment geology and land use to the habitat template of periphyton in stream ecosystems. Freshw. Biol. 33, 419−438.

Biggs, B.J.F., 2000. Eutrophication of streams and rivers: dissolved nutrient-chlorophyll relationships for benthic algae. J. N. Am. Benthol. Soc. 19, 17−31.

Bilby, R.E., Fransen, B.R., Bisson, P.A., 1996. Incorporation of nitrogen and carbon from spawning Coho salmon into the trophic system of small streams: evidence from stable isotopes. Can. J. Fish. Aquat. Sci. 53, 164−173.

Billen, G., 1991. Protein degradation in aquatic environments. In: Chróst, R.J. (Ed.), Microbial Enzymes in Aquatic Environments. Springer-Verlag, New York, NY, pp. 123−143.

Bini, L.M., Landeiro, V.L., Padial, A.A., Siqueira, T., Heino, J., 2014. Nutrient enrichment is related to two facets of beta diversity for stream invertebrates across the United States. Ecology 95 (6), 1569−1578. Available from: https://doi.org/10.1890/13-0656.1.

Birch, S., McCaskie, J., 1999. Shallow urban lakes: a challenge for lake management. The Ecological Bases for Lake and Reservoir Management. Springer, pp. 365−377.

Biron, P.M., Buffin-Bélanger, T., Larocque, M., Choné, G., Cloutier, C.-A., Ouellet, M.-A., et al., 2014. Freedom space for rivers: a sustainable management approach to enhance river resilience. Environ. Manage. 54 (5), 1056−1073.

Bishop-Taylor, R., Tulbure, M.G., Broich, M., 2017. Surface-water dynamics and land use influence landscape connectivity across a major dryland region. Ecol. Appl. 27 (4), 1124−1137. Available from: https://doi.org/10.1002/eap.1507.

Bitton, G., 1994. Wastewater Microbiology. John Wiley & Sons, New York, NY.

Bixby, R.J., Cooper, S.D., Gresswell, R.E., Brown, L.E., Dahm, C.N., Dwire, K.A., 2015. Fire effects on aquatic ecosystems: an assessment of the current state of the science. Freshw. Sci. 34 (4), 1340−1350. Available from: https://doi.org/10.1086/684073.

Bjærke, O., Jonsson, P.R., Alam, A., Selander, E., 2015. Is chain length in phytoplankton regulated to evade predation? J. Plankton Res. 37 (6), 1110−1119. Available from: https://doi.org/10.1093/plankt/fbv076.

Blackburn, N., Fenchel, T., 1999. Influence of bacteria, diffusion and shear on micro-scale nutrient patches, and implications for bacterial chemotaxis. Mar. Ecol. Prog. Ser. 189, 1−7.

Blakelock, G.C., Xenopoulos, M.A., Norman, B.C., Vincent, J.L., Frost, P.C., 2016. Effects of silver nanoparticles on bacterioplankton in a boreal lake. Freshw. Biol. 61 (12), 2211−2220. Available from: https://doi.org/10.1111/fwb.12788.

Blakemore, R.P., 1982. Magnetotactic bacteria. Annu. Rev. Microbiol. 36, 217−238.

Blanchet, S., Reyjol, Y., April, J., Mandrak, N.E., Rodríguez, M.A., Bernatchez, L., et al., 2013. Phenotypic and phylogenetic correlates of geographic range size in Canadian freshwater fishes. Glob. Ecol. Biogeogr. 22 (9), 1083−1094.

Blaustein, A.R., Johnson, P.T., 2003. The complexity of deformed amphibians. Front. Ecol. Environ. 1 (2), 87−94.

Blewett, T.A., Delompré, P.L.M., He, Y., Folkerts, E.J., Flynn, S.L., Alessi, D.S., et al., 2017. Sublethal and reproductive effects of acute and chronic exposure to flowback and produced water from hydraulic fracturing on the water flea *Daphnia magna*. Environ. Sci. Technol. 51 (5), 3032−3039. Available from: https://doi.org/10.1021/acs.est.6b05179.

Blöschl, G., Hall, J., Parajka, J., Perdigão, R.A.P., Merz, B., Arheimer, B., et al., 2017. Changing climate shifts timing of European floods. Science 357 (6351), 588−590. Available from: https://doi.org/10.1126/science.aan2506.

Blossey, B., Skinner, L., Taylor, J., 2001. Impact and management of purple loosestrife (*Lythrum salicaria*) in North America. Biodiversity Conserv. 10 (10), 1787−1807.

Blus, J.B., Henny, C.J., 1997. Field studies on pesticides and birds: unexpected and unique relations. Ecol. Appl. 7 (4), 1125−1132.

Bockstael, N.E., Freeman III, A.M., Kopp, R.J., Portney, P.R., Smith, V.K., 2000. On measuring economic values for nature. Environ. Sci. Technol. 34 (8), 1384−1389.

Boeye, D., Verhagen, B., Haesebroeck, V.V., Verheyen, R.F., 1997. Nutrient limitation in species-rich lowland fens. J. Veg. Sci. 8, 415−424.

Bogan, M.T., Boersma, K.S., Lytle, D.A., 2013. Flow intermittency alters longitudinal patterns of invertebrate diversity and assemblage composition in an arid-land stream network. Freshw. Biol. 58 (5), 1016−1028.

Bohannan, B.J.M., Lenski, R.E., 1997. Effect of resource enrichment on a chemostat community of bacteria and bacteriophage. Ecology 78 (8), 2303−2315.

Böhme, H., 1998. Regulation of nitrogen fixation in heterocyst-forming cyanobacteria. Trends Plant Sci. 3 (9), 346−351.

Bohn, H.F., Federle, W., 2004. Insect aquaplaning: nepenthes pitcher plants capture prey with the peristome, a fully wettable water-lubricated anisotropic surface. Proc. Natl. Acad. Sci. USA 101 (39), 14138−14143.

Bolgovics, Á., Ács, É., Várbíró, G., Görgényi, J., Borics, G., 2016. Species area relationship (SAR) for benthic diatoms: a study on aquatic islands. Hydrobiologia 764 (1), 91−102. Available from: https://doi.org/10.1007/s10750-015-2278-1.

Bolnick, D., 2004. Can intraspecific competition drive disruptive selection? An experimental test in natural populations of sticklebacks. Evolution 58 (3), 608−618.

Boon, P.I., 1992. Antibiotic resistance of aquatic bacteria and its implications for limnological research. Aust. J. Mar. Freshw. Res. 43, 847−859.

Booth, D.B., Roy, A.H., Smith, B., Capps, K.A., 2015. Global perspectives on the urban stream syndrome. Freshw. Sci. 35 (1), 412−420. Available from: https://doi.org/10.1086/684940.

Borman, S., Korth, R., Temte, J., 1997. Through the Looking Glass: A Field Guide to Aquatic Plants. Wisconsin Lakes Partnership, Merill, WI.

Bormans, M., Sherman, B.S., Webster, I.T., 1999. Is buoyancy regulation in cyanobacteria an adaptation to exploit separation of light and nutrients? Mar. Freshw. Res. 50, 897−906.

Bornette, G., Amoros, C., Lamouroux, N., 1998a. Aquatic plant diversity in riverine wetlands: the role of connectivity. Freshw. Biol. 39, 267−283.

Bornette, G., Amoros, C., Piegay, H., Tachet, J., Hein, T., 1998b. Ecological complexity of wetlands within a river landscape. Biol. Conserv. 85, 35−45.

Borowitzka, M.A., 2016. Chemically-mediated interactions in microalgae. In: Borowitzka, M.A., Beardall, J., Raven, J.A. (Eds.), The Physiology of Microalgae. Springer International Publishing, Cham, pp. 321−357.

Borror, D.J., Triplehorn, C.A., Johnson, N.F., 1989. An Introduction to the Study of Insects. Saunders College Publishing, Orlando, FL.

Borsa, A.A., Agnew, D.C., Cayan, D.R., 2014. Ongoing drought-induced uplift in the western United States. Science 345 (6204), 1587−1590. Available from: https://doi.org/10.1126/science.1260279.

Bostock, J., McAndrew, B., Richards, R., Jauncey, K., Telfer, T., Lorenzen, K., et al., 2010. Aquaculture: global status and trends. Philos. Trans. R. Soc. Lond. B Biol. Sci. 365 (1554), 2897−2912.

Bothe, H., 1982. Nitrogen fixation. In: Carr, N.G., Whitton, B.A. (Eds.), The Biology of Cyanobacteria, vol. 19. University of California Press, Berkeley, pp. 87−104.

Bothwell, M.L., Sherbot, D.M.J., Pollock, C.M., 1994. Ecosystem response to solar ultraviolet-B radiation: influence of trophic-level interactions. Science 265, 97−100.

Bothwell, M.L., Taylor, B.W., Kilroy, C., 2014. The Didymo story: the role of low dissolved phosphorus in the formation of Didymosphenia geminata blooms. Diatom Res. 29 (3), 229−236.

Bott, T., Meyer, G.A., Young, E.B., 2008. Nutrient limitation and morphological plasticity of the carnivorous pitcher plant *Sarracenia purpurea* in contrasting wetland environments. New Phytol. 180 (3), 631−641.

Bott, T.L., 1995. Microbes in food webs. Am. Soc. Microbiol. News 61 (11), 580−585.

Bott, T.L., Brock, J.T., Baattrup-Pedersen, A., Chambers, P.A., Dodds, W.K., Himbeault, K.T., et al., 1997. An evaluation of techniques for measuring periphyton metabolism in chambers. Can. J. Fish. Aquat. Sci. 54, 715−725.

Bott, T.L., Brock, J.T., Dunn, C.S., Naiman, R.J., Ovink, R.W., Petersen, R.C., 1985. Benthic community metabolism in four temperate stream systems: an inter-biome comparison and evaluation of the river continuum concept. Hydrobiologia 123, 3−45.

Boulton, A., 2000. The subsurface macrofauna. In: Jones, J.B., Molholland, P.J. (Eds.), Streams and Ground Waters. Academic Press, San Diego, CA, pp. 337−361.

Boulton, A.J., Peterson, C.G., Grimm, N.B., Fisher, S.G., 1992. Stability of an aquatic macroinvertebrate community in a multiyear hydrologic disturbance regime. Ecology 73 (6), 2192−2207.

Bowden, W.B., 1999. Roles of bryophytes in stream ecosystems. J. N. Am. Benthol. Soc. 18 (2), 151−184.

Bowden, W.B., Finlay, J.C., Maloney, P.E., 1994. Long-term effects of PO_4 fertilization on the distribution of bryophytes in an arctic river. Freshw. Biol. 32, 445−454.

Bowen, R., 1986. Groundwater, second ed. Elsevier Applied Science Publishers, New York, NY.

Bracken, M.E.S., Hillebrand, H., Borer, E.T., Seabloom, E.W., Cebrian, J., Cleland, E.E., et al., 2015. Signatures of nutrient limitation and co-limitation: responses of autotroph internal nutrient concentrations to nitrogen and phosphorus additions. Oikos 124 (2), 113−121. Available from: https://doi.org/10.1111/oik.01215.

Bradley, P.M., Journey, C.A., Romanok, K.M., Barber, L.B., Buxton, H.T., Foreman, W.T., et al., 2017. Expanded target-chemical analysis reveals extensive mixed-organic-contaminant exposure in U.S. streams. Environ. Sci. Technol. 51 (9), 4792−4802. Available from: https://doi.org/10.1021/acs.est.7b00012.

Bradshaw, W.E., Creelman, R.A., 1984. Mutualism between the carnivorous purple pitcher plant and its inhabitants. Am. Midl. Nat. 112, 294−304.

Brainerd, S., 2010. European charter on recreational fishing and biodiversity. T-PVS/Inf (2010) 3, pp. 1−21.

Bratbak, G., Thingstad, F., Heldal, M., 1994. Viruses and the microbial loop. Microb. Ecol. 28, 209−221.

Bratt, A.R., Finlay, J.C., Hobbie, S.E., Janke, B.D., Worm, A.C., Kemmitt, K.L., 2017. Contribution of leaf litter to nutrient export during winter months in an urban residential watershed. Environ. Sci. Technol. 51 (6), 3138−3147. Available from: https://doi.org/10.1021/acs.est.6b06299.

Brawand, D., Wagner, C.E., Li, Y.I., Malinsky, M., Keller, I., Fan, S., et al., 2014. The genomic substrate for adaptive radiation in African cichlid fish. Nature 513 (7518), 375−381. Available from: https://doi.org/10.1038/nature13726. Available from: http://www.nature.com/nature/journal/v513/n7518/abs/nature13726.html#supplementary-information.

Bray, J., Harding, J.S., Kilroy, C., Broady, P., Gerbeaux, P., 2016. Physicochemical predictors of the invasive diatom *Didymosphenia geminata* at multiple spatial scales in New Zealand rivers. Aquat. Ecol. 50 (1), 1−14. Available from: https://doi.org/10.1007/s10452-015-9543-2.

Breitburg, D., Levin, L.A., Oschlies, A., Gregoire, M., Chavez, F.P., Conley, D.J., et al., 2018. Declining oxygen in the global ocean and coastal waters. Science 359 (6371), 46−48. Available from: https://doi.org/10.1126/science.aam7240.

Brem, F.M.R., Lips, K.R., 2008. *Batrachochytrium dendrobatidis* infection patterns among *Panamanian amphibian* species, habitats and elevations during epizootic and enzootic. Dis. Aquat. Organ. 81 (3), 189−202.

Brender, J.D., Weyer, P.J., Romitti, P.A., Mohanty, B.P., Shinde, M.U., Vuong, A.M., et al., 2013. Prenatal nitrate intake from drinking water and selected birth defects in offspring of participants in the National Birth Defects Prevention Study. Environ. Health Perspect. 121 (9), 1083.

Brenner, M., Whitmore, T.J., Lasi, M.A., Cable, J.E., Cable, P.H., 1999. A multi-proxy trophic state reconstruction for shallow Orange Lake, Florida, USA: possible influence of macrophytes on limnetic nutrient concentrations. J. Paleolimnol. 21, 215—233.

Brett, M.T., Bunn, S.E., Chandra, S., Galloway, A.W.E., Guo, F., Kainz, M.J., et al., 2017. How important are terrestrial organic carbon inputs for secondary production in freshwater ecosystems? Freshw. Biol. 62 (5), 833—853. Available from: https://doi.org/10.1111/fwb.12909.

Brett, M.T., Goldman, C.R., 1997. Consumer versus resource control in freshwater. Science 275, 384—386.

Brett, M.T., Müller-Navarra, D.C., 1997. The role of highly unsaturated fatty acids in aquatic foodweb processes. Freshw. Biol. 38, 483—499.

Brewer, M.C., Dawidowicz, P., Dodson, S.I., 1999. Interactive effects of fish kairomone and light on *Daphnia* escape behavior. J. Plankton Res. 21 (7), 1317—1335.

Brezonik, P.L., 1994. Chemical Kinetics and Process Dynamics in Aquatic Systems. CRC Press, Inc., Boca Ratan, FL.

Briand, F., 1985. Structural singularities of freshwater food webs. Verhandlungen des Internationalen Verein Limnologie 22, 3356—3364.

Briand, F., Cohen, J.E., 1990. Environmental correlates of food chain length. In: Cohen, J.E., Briand, F., Newman, C.M. (Eds.), Community Food Webs Data and Theory. Springer-Verlag, Berlin, pp. 55—62.

Bridgham, S.D., Johnston, C.A., Pastor, J., Updegraff, K., 1995. Potential feedbacks of northern wetlands on climate change. Bioscience 45 (4), 262—274.

Bridgham, S.D., Updegraff, K., Pastor, J., 1998. Carbon, nitrogen, and phosphorus mineralization in northern wetlands. Ecology 79 (5), 1545—1561.

Briers, R.A., 2003. Range size and environmental calcium requirements of British freshwater gastropods. Glob. Ecol. Biogeogr. 12 (1), 47—51.

Briggs, J.C., 1986. Introduction to the zoogeography of North American fishes. In: Hocutt, C.H., Wiley, E.O. (Eds.), The Zoogeography of North American Freshwater Fishes. John Wiley & Sons, New York, NY, pp. 1—16.

Brinkhurst, R.O., Chua, K.E., Kaushik, N.K., 1972. Interspecific interactions and selective feeding by tubificid oligochaetes. Limnol. Oceanogr. 17, 122—133.

Brinkhurst, R.O., Gelder, S.R., 1991. *Annelida: Oligochaeta* and *Branchiobdellida*. In: Thorp, J.H., Covich, A.P. (Eds.), Ecology and Classification of North American Freshwater Invertebrates. Academic Press, Inc, San Diego, CA, pp. 401—436.

Brinson, M.M., Kruczynski, W., Lee, L.C., Nutter, W.L., Smith, R.D., Whigham, D.F., 1994. Developing an approach for assessing the functions of wetlands. In: Mitsch, W.J. (Ed.), Global Wetlands Old World and New. Elsevier, Amsterdam, The Netherlands, pp. 615—624.

Brix, H., 1994. Constructed wetlands for municipal wastewater treament in Europe. In: Mitsch, W.J. (Ed.), Global Wetlands Old World and New. Elsevier, Amsterdam, The Netherlands, pp. 325—333.

Brock, E.M., 1960. Mutualism between the midge *Cricotopus* and the alga *Nostoc*. Ecology 41 (3), 474—483.

Brock, T.D., 1978. Thermophilic Microorganisms and Life at High Temperatures. Springer-Verlag, New York, NY.

Brönmark, C., 1985. Interactions between macrophytes, epiphytes and herbivores: an experimental approach. Oikos 45, 26—30.

Brönmark, C., Hansson, L.-A., 1998. The Biology of Lakes and Ponds. Oxford University Press, Inc., New York, NY.

Brönmark, C., Hansson, L.-A., 2000. Chemical communication in aquatic systems: an introduction. Oikos 88, 103—109.

Brönmark, C., Pettersson, L.B., Nilsson, P.A., 1999. Predator-induced defense in Crucian Carp. In: Tollrian, E., Harvell, C.D. (Eds.), The Ecology and Evolution of Inducible Defenses. Princeton University Press, Princeton, NJ, pp. 203–217.

Brönmark, C., Rundle, S.D., Erlandsson, A., 1991. Interactions between freshwater snails and tadpoles: competition and facilitation. Oecologia 87, 8–18.

Brönmark, C.S., Klosiewski, P., Stein, R.A., 1992. Indirect effects of predation in a freshwater, benthic food chain. Ecology 73 (5), 1662–1674.

Brooker, R., Maestre, F., Callaway, R., Lortie, C., Cavieres, L., Kunstler, G., et al., 2008. Facilitation in plant communities: the past, the present, and the future. J. Ecol. 96 (1), 18–34.

Brooks, B.W., Lazorchak, J.M., Howard, M.D.A., Johnson, M.-V.V., Morton, S.L., Perkins, D.A.K., et al., 2017. In some places, in some cases, and at some times, harmful algal blooms are the greatest threat to inland water quality. Environ. Toxicol. Chem. 36 (5), 1125–1127. Available from: https://doi.org/10.1002/etc.3801.

Brooks, E.G.E., Holland, R.A., Darwall, W.R.T., Eigenbrod, F., 2016. Global evidence of positive impacts of freshwater biodiversity on fishery yields. Glob. Ecol. Biogeogr. 25 (5), 553–562. Available from: https://doi.org/10.1111/geb.12435.

Brooks, J.L., 1946. Cyclomorphosis in *Daphnia*. Ecol. Monogr. 16 (4), 409–447.

Brooks, J.L., 1950. Speciation in ancient lakes. Q. Rev. Biol. 25 (30–60), 131–176.

Brooks, J.L., Dodson, S.I., 1965. Predation, body size, and composition of plankton. Science 150, 28–35.

Brooks, J.R., Gibson, J.J., Birks, S.J., Weber, M.H., Rodecap, K.D., Stoddard, J.L., 2014. Stable isotope estimates of evaporation: inflow and water residence time for lakes across the United States as a tool for national lake water quality assessments. Limnol. Oceanogr. 59 (6), 2150–2165.

Brooks, R.P., Wardrop, D.H., Cole, C.A., Campbell, D.A., 2005. Are we purveyors of wetland homogeneity? A model of degradation and restoration to improve wetland mitigation performance. Ecol. Eng. 24 (4), 331–340.

Brothers, S., Vadeboncoeur, Y., Sibley, P., 2016. Benthic algae compensate for phytoplankton losses in large aquatic ecosystems. Glob. Change Biol. 22 (12), 3865–3873. Available from: https://doi.org/10.1111/gcb.13306.

Browder, J.A., Gleason, P.J., Swift, D.R., 1994. Periphyton in the Everglades: spatial variation, environmental correlates, and ecological implications. In: Davis, S.M., Ogden, J.C. (Eds.), Everglades, the Ecosystem and Its Restoration. St. Lucie Press, Delray Beach, FL, pp. 379–418.

Brown, B.L., Creed, R.P., Skelton, J., Rollins, M.A., Farrell, K.J., 2012. The fine line between mutualism and parasitism: complex effects in a cleaning symbiosis demonstrated by multiple field experiments. Oecologia 170 (1), 199–207. Available from: https://doi.org/10.1007/s00442-012-2280-5.

Brown, B.L., Sokol, E.R., Skelton, J., Tornwall, B., 2017. Making sense of metacommunities: dispelling the mythology of a metacommunity typology. Oecologia 183 (3), 643–652. Available from: https://doi.org/10.1007/s00442-016-3792-1.

Brown, D.J., Coon, T.G., 1991. Grass carp larvae in the lower Missouri River and its tributaries. N. Am. J. Fish. Manage. 11, 62–66.

Brown, J.H., 1995a. Macroecology. The University of Chicago Press, Chicago, IL.

Brown, J.H., Gillooly, J.F., Allen, A.P., Savage, V.M., West, G.B., 2004. Toward a metabolic theory of ecology. Ecology 85 (7), 1771–1789.

Brown, K.M., 1991. Mollusca: Gastropoda. In: Thorp, J.H., Covich, A.P. (Eds.), Ecology and Classification of North American Freshwater Invertebrates. Academic Press, Inc., San Diego, CA, pp. 285–314.

Brown, L.R., 1995b. Nature's Limits. W.W. Norton & Co, New York, NY.

Brown, T.C., 2000. Projecting U.S. freshwater withdrawals. J. Water Resour. Res. 36 (3), 769–780.

Brown, J.H., Stevens, G.C., Kaufman, D.M., 1996. The geographic range: Size, shape, boundaries, and internal structure. Annu. Rev. Ecol. Syst. 27 (1), 597–623.

Brunke, M., Gonser, T., 1997. The ecological significance of exchange processes between rivers and groundwater. Freshw. Biol. 37, 1–33.

Brutsaert, W., Jirka, G.H., 2013. Gas Transfer at Water Surfaces, vol. 2. Springer Science & Business Media.

Buckley, M., Roberts, R.J., 2006. Reconciling microbial systematics and genomics, Washington, DC. Retrieved from http://www.asmscience.org/content/report/colloquia/colloquia.36.

Budavari, S., O'Neil, M.J., Smith, A., Heckelman, P.E., 1989. The Merck Index, eleventh ed. Merck and Co., Rahaway, NJ.

Bulat, S.A., 2016. Microbiology of the subglacial Lake Vostok: first results of borehole-frozen lake water analysis and prospects for searching for lake inhabitants. Philos. Trans. R. Soc. A: Math., Phys. Eng. Sci. 374 (2059). Available from: https://doi.org/10.1098/rsta.2014.0292.

Bump, J.K., Bergman, B.G., Schrank, A.J., Marcarelli, A.M., Kane, E.S., Risch, A.C., et al., 2017. Nutrient release from moose bioturbation in aquatic ecosystems. Oikos 126 (3), 389−397. Available from: https://doi.org/10.1111/oik.03591.

Bumpers, P.M., Maerz, J.C., Rosemond, A.D., Benstead, J.P., 2015. Salamander growth rates increase along an experimental stream phosphorus gradient. Ecology 96 (11), 2994−3004. Available from: https://doi.org/10.1890/14-1772.1.

Bundschuh, M., Seitz, F., Rosenfeldt, R.R., Schulz, R., 2016. Effects of nanoparticles in fresh waters: risks, mechanisms and interactions. Freshw. Biol. 61 (12), 2185−2196. Available from: https://doi.org/10.1111/fwb.12701.

Burgherr, P., Ward, J., 2001. Longitudinal and seasonal distribution patterns of the benthic fauna of an alpine glacial stream (Val Roseg, Swiss Alps). Freshw. Biol. 46 (12), 1705−1721.

Burgin, A.J., Hamilton, S.K., 2007. Have we overemphasized the role of denitrification in aquatic ecosystems? A review of nitrate removal pathways. Front. Ecol. Environ. 5, 89−96.

Burgmer, T., Reiss, J., Wickham, S.A., Hillebrand, H., 2010. Effects of snail grazers and light on the benthic microbial food web in periphyton communities. Aquat. Microb. Ecol. 61 (2), 163−178.

Burkholder, J.M., 1996. Interactions of benthic algae with their substrata. In: Stevenson, R.J., Bothwell, M.L., Lowe, R.L. (Eds.), Algal Ecology. Academic Press, San Diego, CA, pp. 253−297.

Burkholder, J.M., Glasgow Jr., H.B., 1997. *Pfiesteria piscicida* and other *Pfiesteria*-like dinoflagellates: behavior, impacts, and environmental controls. Limnol. Oceanogr. 42 (5, part 2), 1052−1075.

Burns, A., Walker, K.F., 2000. Effects of water level regulation on algal biofilms in the River Murray, South Austrailia. Regul. Rivers: Res. Manage. 16, 433−444.

Burns, C.W., 1969. Particle size and sedimentation in the feeding behavior of two species of *Daphnia*. Limnol. Oceanogr. 14, 392−402.

Burns, N.M., 1985. Erie: The Lake That Survived. Rowman & Allanheld, Totowa, NJ.

Burton, G.A., Basu, N., Ellis, B.R., Kapo, K.E., Entrekin, S., Nadelhoffer, K., 2014. Hydraulic "Fracking": are surface water impacts an ecological concern? Environ. Toxicol. Chem. 33 (8), 1679−1689.

Bury, N.R., Eddy, F.B., Codd, G.A., 1995. The effects of the cyanobacterium *Microcystis aeruginosa*, the cyanobacterial toxin microcystin-LR, and ammonia on growth rate and ionic regulation of brown trout. J. Fish. Biol. 46, 1042−1054.

Bush, A., Hoskins, A.J., 2017. Does dispersal capacity matter for freshwater biodiversity under climate change? Freshw. Biol. 62 (2), 382−396. Available from: https://doi.org/10.1111/fwb.12874.

Butler, J.N., 1991. Carbon Dioxide Equilibria and Their Applications. Lewis Publishers Inc, Chelsea, MI.

Butman, D., Stackpoole, S., Stets, E., McDonald, C.P., Clow, D.W., Striegl, R.G., 2016. Aquatic carbon cycling in the conterminous United States and implications for terrestrial carbon accounting. Proc. Natl. Acad. Sci. 113 (1), 58−63. Available from: https://doi.org/10.1073/pnas.1512651112.

Butterworth, A.E., 1988. Control of schistosomiasis in man. In: Englund, P.T., Sher, A. (Eds.), The Biology of Parasitism. A Molecular and Immunological Approach, vol. 9. Alan R. Liss, Inc, New York, NY, pp. 43−59. (Marine Biological Laboratory Lectures in Biology).

Butturini, A., Sabater, F., 1998. Ammonium and phosphate retention in a Mediterranean stream: hydrological versus temperature control. Can. J. Fish. Aquat. Sci. 55, 1938−1945.

Buxton, H.T., Kolpin, D.W., 2002. Pharmaceuticals, hormones, and other organic wastewater contaminants in U.S. streams. U.S. Geological Survey Fact Sheet FS-027-02. Retrieved from https://toxics.usgs.gov/pubs/FS-027-02/pdf/FS-027-02.pdf.

Bylak, A., Kukuła, K., Mitka, J., 2014. Beaver impact on stream fish life histories: the role of landscape and local attributes. Can. J. Fish. Aquat. Sci. 71 (11), 1603−1615. Available from: https://doi.org/10.1139/cjfas-2014-0105.

Cael, B.B., Heathcote, A.J., Seekell, D.A., 2017. The volume and mean depth of Earth's lakes. Geophys. Res. Lett. 44 (1), 209−218. Available from: https://doi.org/10.1002/2016GL071378.

Caires, A.M., Chandra, S., Hayford, B.L., Wittmann, M.E., 2013. Four decades of change: dramatic loss of zoobenthos in an oligotrophic lake exhibiting gradual eutrophication. Freshw. Sci. 32 (3), 692−705.

Cairns, J.J., 1982. Freshwater protozoan communities. In: Bull, A.T., Watkinson, A.R.K. (Eds.), Microbial Interactions and Communities, vol. 1. Academic Press, London, pp. 249−285.

Cairns, J.J., 1993. Can microbial species with a cosmopolitan distribution become extinct? Speculations Sci. Technol. 16, 69−73.

Caissie, D., 2006. The thermal regime of rivers: a review. Freshw. Biol. 51 (8), 1389−1406.

Calabrese, E.J., Baldwin, L.A., 1999. Reevaluation of the fundamental dose-response relationship. Bioscience 49 (9), 725−732.

Calero-Cáceres, W., Muniesa, M., 2016. Persistence of naturally occurring antibiotic resistance genes in the bacteria and bacteriophage fractions of wastewater. Water Res. 95, 11−18. Available from: https://doi.org/10.1016/j.watres.2016.03.006.

Calheiros, D., Hamilton, S., 1998. Limnological conditions associated with natural fish kills in the Pantanal wetland of Brazil. Verhandlungen des Internationalen Verein Limnologie 26, 2189−2193.

Callander, R., 1978. River meandering. Annu. Rev. Fluid. Mech. 10 (1), 129−158.

Callaway, R.M., 1995. Positive interactions among plants. Bot. Rev. 61 (4), 306−348.

Callaway, R.M., King, L., 1996. Temperature-driven variation in substrate oxygenation and the balance of competition and facilitation. Ecology 77 (4), 1189−1195.

Callaway, R.M., Walker, L.R., 1997. Competition and facilitation: a synthetic approach to interactions in plant communities. Ecology 78 (7), 1958−1965.

Calosi, P., Bilton, D.T., Spicer, J.I., Votier, S.C., Atfield, A., 2010. What determines a species' geographical range? Thermal biology and latitudinal range size relationships in European diving beetles (Coleoptera: Dytiscidae). J. Anim. Ecol. 79 (1), 194−204.

Camerano, L., 1994. On the equilibrium of living beings by means of reciprocal destruction. In: Levin, S.A. (Ed.), Frontiers in Mathematical Biology. Springer-Verlag, Berlin, pp. 360−379.

Cañedo-Argüelles, M., Hawkins, C.P., Kefford, B.J., Schäfer, R.B., Dyack, B.J., Brucet, S., et al., 2016. Saving freshwater from salts. Science 351 (6276), 914−916. Available from: https://doi.org/10.1126/science.aad3488.

Canfield Jr, D.E., Langeland, K.A., Maceina, M.J., Haller, W.T., Shireman, J.V., Jones, J.R., 1983. Trophic state classification of lakes with aquatic macrophytes. Can. J. Fish. Aquat. Sci. 40 (10), 1713−1718.

Canter-Lund, H., Lund, J.W.G., 1995. Freshwater Algae: Their Microscopic World Explored. Biopress Ltd, Bristol, UK.

Capps, K.A., Atkinson, C.L., Rugenski, A.T., 2015a. Implications of species addition and decline for nutrient dynamics in fresh waters. Freshw. Sci. 34 (2), 485−496. Available from: https://doi.org/10.1086/681095.

Capps, K.A., Bentsen, C.N., Ramírez, A., 2015b. Poverty, urbanization, and environmental degradation: urban streams in the developing world. Freshw. Sci. 35 (1), 429−435. Available from: https://doi.org/10.1086/684945.

Caputo, J., Beier, C.M., Fakhraei, H., Driscoll, C.T., 2017. Impacts of acidification and potential recovery on the expected value of recreational fisheries in Adirondack lakes (USA). Environ. Sci. Technol. 51 (1), 742−750. Available from: https://doi.org/10.1021/acs.est.6b05274.

Caraco, N., Cole, J., 1999. Regional-scale export of C, N, P, and sediment: what river data tell us about key controlling variables. In: Tenhunen, J.D., Kabat, P. (Eds.), Integrating Hydrology, Ecosystem

Dynamics, and Biogeochemistry in Complex Landscapes. John Wiley & Sons Ltd, New York, NY, pp. 239–254.

Caraco, N.F., Cole, J.J., Findlay, S.E.G., Fischer, D.T., Lampman, G.G., Pace, M.L., et al., 2000. Dissolved oxygen declines in the Hudson River associated with the invasion of the zebra mussel (*Dreissena polymorpha*). Environ. Sci. Technol. 34, 1204–1210.

Caravati, E., Callieri, C., Modenutti, B., Corno, G., Balseiro, E., Bertoni, R., et al., 2010. Picocyanobacterial assemblages in ultraoligotrophic Andean lakes reveal high regional microdiversity. J. Plankton Res. 32 (3), 357–366.

Cardenas, M.B., Ford, A.E., Kaufman, M.H., Kessler, A.J., Cook, P.L.M., 2016. Hyporheic flow and dissolved oxygen distribution in fish nests: the effects of open channel velocity, permeability patterns, and groundwater upwelling. J. Geophys. Res.: Biogeosci. 121 (12), 3113–3130. Available from: https://doi.org/10.1002/2016JG003381.

Carder, J.P., Hoagland, K.D., 1998. Combined effects of alachlor and atrazine on benthic algal communities in artificial streams. Environ. Toxicol. Chem. 17 (7), 1415–1420.

Cardinale, B., Palmer, M., Collins, S., 2002. Species diversity enhances ecosystem functioning through interspecific facilitation. Nature 415 (6870), 426–429.

Cardinale, B.J., Srivastava, D.S., Duffy, J.E., Wright, J.P., Downing, A.L., Sankaran, M., et al., 2006. Effects of biodiversity on the functioning of trophic groups and ecosystems. Nature 443, 989.

Carignan, R., Blais, A.-M., Vis, C., 1998. Measurement of primary production and community respiration in oligotrophic lakes using the Winkler method. Can. J. Fish. Aquat. Sci. 55, 1078–1084.

Carlson, R.E., 1977. A trophic state index for lakes. Limnol. Oceanogr. 22 (2), 361–369.

Carlsson, N., Bronmark, C., Hansson, L., 2004. Invading herbivory: the golden apple snail alters ecosystem functioning in Asian wetlands. Ecology 85 (6), 1575–1580.

Carlton, J.T., 1973. Dispersal mechanisms of the Zebra Mussel (*Dreissena polymorpha*). In: Nalepa, T.F., Schloesser, D.W. (Eds.), Zebra Mussels: Biology, Impacts and Control. Lewis Publishers, Boca Ratan, FL, pp. 677–697.

Carmichael, W.W., 1994. The toxins of cyanobacteria. Sci. Am. 270, 78–86.

Carmichael, W.W., 1997. The cyanotoxins. Adv. Bot. Res. 27, 211–240.

Carney, H.J., 1990. A general hypothesis for the strength of food web interactions in relation to trophic state. Verhein Internationale Verein Limnologie 24, 487–492.

Carpenter, S., Brock, W., 2004. Spatial complexity, resilience, and policy diversity: fishing on lake-rich landscapes. Ecol. Soc. 9 (1).

Carpenter, S., Brock, W., 2006. Rising variance: a leading indicator of ecological transition. Ecol. Lett. 9 (3), 311–318.

Carpenter, S.R., 1989. Replication and treatment strength in whole-lake experiments. Ecology 70, 453–462.

Carpenter, S.R., Bolgrien, D., Lathrop, R.C., Stow, C.A., Reed, T., Wilson, M.A., 1998. Ecological and economic analysis of lake eutrophication by nonpoint pollution. Aust. J. Ecol. 23, 68–79.

Carpenter, S.R., Brock, W.A., Cole, J.J., Pace, M.L., 2014. A new approach for rapid detection of nearby thresholds in ecosystem time series. Oikos 123 (3), 290–297.

Carpenter, S.R., Chisholm, S.W., Krebs, C.J., Schindler, D.W., Wright, R.F., 1995. Ecosystem experiments. Science 269, 324–327.

Carpenter, S.R., Cole, J.J., Hodgson, J.R., Kitchell, J.F., Pace, M.L., Bade, D., et al., 2001. Trophic cascades, nutrients, and lake productivity: whole-lake experiments. Ecol. Monogr. 71 (2), 163–186.

Carpenter, S.R., Cole, J.J., Pace, M.L., Batt, R., Brock, W.A., Cline, T., et al., 2011. Early warnings of regime shifts: a whole-ecosystem experiment. Science 332 (6033), 1079–1082. Available from: https://doi.org/10.1126/science.1203672.

Carpenter, S.R., Cole, J.J., Pace, M.L., Van de Bogert, M., Bade, D.L., Bastviken, D., et al., 2005. Ecosystem subsidies: terrestrial support of aquatic food webs from 1^{3C} addition to contrasting lakes. Ecology 86 (10), 2737–2750.

Carpenter, S.R., Frost, T.M., Heisey, D., Kratz, K.K., 1989. Randomized intervention analysis and the interpretation of whole-ecosystem experiments. Ecology 70, 1142–1152.

Carpenter, S.R., Kitchell, J.F., 1987. The temporal scale of variance in limnetic primary production. Am. Nat. 129, 417−433.

Carpenter, S.R., Kitchell, J.F., Hodgson, J.R., Cochran, P.A., Elser, J.J., Elser, M.M., et al., 1987. Regulation of lake primary productivity by food web structure. Ecology 68 (6), 1863−1876.

Carraro, L., Bertuzzo, E., Mari, L., Fontes, I., Hartikainen, H., Strepparava, N., et al., 2017. Integrated field, laboratory, and theoretical study of PKD spread in a Swiss prealpine river. Proc. Natl. Acad. Sci. 114 (45), 11992−11997. Available from: https://doi.org/10.1073/pnas.1713691114.

Carrillo, P., Delgado-Molina, J.A., Medina-Sanchez, J.M., Bullejos, F.J., Villar-Argaiz, M., 2008. Phosphorus inputs unmask negative effects of ultraviolet radiation on algae in a high mountain lake. Glob. Change Biol. 14 (2), 423−439.

Carrizo, S.F., Jähnig, S.C., Bremerich, V., Freyhof, J., Harrison, I., He, F., et al., 2017. Freshwater megafauna: flagships for freshwater biodiversity under threat. Bioscience 67 (10), 919−927. Available from: https://doi.org/10.1093/biosci/bix099.

Carson, R., 1962. Silent Spring. Houghton Mifflin Company, New York, NY.

Carty, S., 2014. Freshwater Dinoflagellates of North America. Cornell University Press.

Castello, L., Arantes, C.C., McGrath, D.G., Stewart, D.J., Sousa, F.S.D., 2015. Understanding fishing-induced extinctions in the Amazon. Aquat. Conserv.: Mar. Freshw. Ecosyst. 25 (5), 587−598. Available from: https://doi.org/10.1002/aqc.2491.

Castello, L., Macedo, M.N., 2016. Large-scale degradation of Amazonian freshwater ecosystems. Glob. Change Biol. 22 (3), 990−1007. Available from: https://doi.org/10.1111/gcb.13173.

Castenholz, R.W., 1984. Composition of hot springs microbial mats: a summary. In: Cohen, Y., Castenholz, R.W., Halvorson, H.O. (Eds.), Ancient Stromatolites and Microbial Mats. Alan R. Liss, Inc., New York, NY, pp. 101−119.

Castenholz, R.W., 2004. Phototrophic bacteria under UV stress. In: Seckbach, J. (Ed.), Origins: Genesis, Evolution and Diversity of Life. Kluwer Academic Publishers, Dordrecht (pp. 447 +).

Catalán, J., Curtis, C., Kernan, M., 2009. Remote European mountain lake ecosystems: regionalisation and ecological status. Freshw. Biol. 54 (12), 2419−2432.

Catania, K., 2014. The shocking predatory strike of the electric eel. Science 346 (6214), 1231−1234. Available from: https://doi.org/10.1126/science.1260807.

Cha, Y., Cho, K.H., Lee, H., Kang, T., Kim, J.H., 2017. The relative importance of water temperature and residence time in predicting cyanobacteria abundance in regulated rivers. Water Res. 124, 11−19. Available from: https://doi.org/10.1016/j.watres.2017.07.040.

Chambers, C.P., Whiles, M.R., Rosi-Marshall, E.J., Tank, J.L., Royer, T.V., Griffiths, N.A., et al., 2010. Responses of stream macroinvertebrates to Bt maize leaf detritus. Ecol. Appl. 20 (7), 1949−1960.

Chambers, P.A., DeWreede, R.E., Irlandi, E.A., Vandermeulen, H., 1999. Management issues in aquatic macrophyte ecology: a Canadian perspective. Can. J. Bot. 77, 471−487.

Chandra, S., Vander Zanden, J.M., Heyvaert, A.C., Richards, B.C., Allen, B.C., Goldman, C.R., 2005. The effects of cultural eutrophication on the coupling between pelagic primary producers and benthic consumers. Limnol. Oceanogr. 50 (5), 1368−1376. Available from: https://doi.org/10.4319/lo.2005.50.5.1368.

Chapnick, S.D., Moore, W.S., Nealson, K.H., 1982. Microbially mediated manganese oxidation in a freshwater lake. Limnol. Oceanogr. 27 (6), 1004−1015.

Chapra, S.C., Boehlert, B., Fant, C., Bierman, V.J., Henderson, J., Mills, D., et al., 2017. Climate change impacts on harmful algal blooms in U.S. freshwaters: a screening-level assessment. Environ. Sci. Technol. 51 (16), 8933−8943. Available from: https://doi.org/10.1021/acs.est.7b01498.

Characklis, W.G., McFeters, G.A., Marshall, K.C., 1990. Physiological ecology in biofilm systems. In: Characklis, W.G., Marshall, K.C. (Eds.), Biofilms. John Wiley & Sons, Inc., New York, NY, pp. 341−394.

Chen, J., Xie, P., 2008. Accumulation of hepatotoxic microcystins in freshwater mussels, aquatic insect larvae and oligochaetes in a large, shallow eutrophic lake (Lake Chaohu) of subtropical China. Fresen. Environ. Bull. 17 (7A), 849−854.

Chen, F., Gong, Z., Kelly, B.C., 2017. Bioaccumulation behavior of pharmaceuticals and personal care products in adult zebrafish (*Danio rerio*): influence of physical-chemical properties and biotransformation. Environ. Sci. Technol. 51 (19), 11085−11095. Available from: https://doi.org/10.1021/acs.est.7b02918.

Chick, J.H., Cosgriff, R.J., Gittinger, L.S., 2003. Fish as potential dispersal agents for floodplain plants: first evidence in North America. Can. J. Fish. Aquat. Sci. 60 (12), 1437−1439.

Chin, K.S., Lento, J., Culp, J.M., Lacelle, D., Kokelj, S.V., 2016. Permafrost thaw and intense thermokarst activity decreases abundance of stream benthic macroinvertebrates. Glob. Change Biol. 22 (8), 2715−2728. Available from: https://doi.org/10.1111/gcb.13225.

Chivian, D., Brodie, E., Alm, E., Culley, D., Dehal, P., DeSantis, T., et al., 2008. Environmental genomics reveals a single-species ecosystem deep within earth. Science 322 (5899), 275.

Christaki, U., Dolan, J.R., Pelegri, S., Rassoulzadegan, F., 1998. Consumption of picoplankton-size particles by marine ciliates: effects of physiological state of the ciliate and particle quality. Limnol. Oceanogr. 43 (3), 458−464.

Christensen, D.L., Herwig, B.R., Schindler, D.E., Carpenter, S.R., 1996. Impacts of lakeshore residential development on coarse woody debris in north temperate lakes. Ecol. Appl. 6 (4), 1143−1149.

Christensen, T.H., Bjerg, P.L., Banwart, S.A., Jakobsen, R., Heron, G., Albrechtsen, H.-J., 2000. Characterization of redox conditions in groundwater contaminant plumes. J. Contam. Hydrol. 45 (3-4), 165−241.

Christensen, M.R., Graham, M.D., Vinebrooke, R.D., Findlay, D.L., Paterson, M.J., Turner, M.A., 2006. Multiple anthropogenic stressors cause ecological surprises in boreal lakes. Glob. Change Biol. 12 (12), 2316−2322.

Christian, N., Whitaker, B.K., Clay, K., 2015. Microbiomes: unifying animal and plant systems through the lens of community ecology theory. Front. Microbiol. 6. Available from: https://doi.org/10.3389/fmicb.2015.00.869.

Christner, B.C., Priscu, J.C., Achberger, A.M., Barbante, C., Carter, S.P., Christianson, K., et al., 2014. A microbial ecosystem beneath the West Antarctic ice sheet. Nature 512 (7514), 310−313. Available from: https://doi.org/10.1038/nature13667. Available from: http://www.nature.com/nature/journal/v512/n7514/abs/nature13667.html#supplementary-information.

Chróst, R.J., 1991. Environmental control of the synthesis and activity of aquatic microbial ectoenzymes. In: Chróst, R.J. (Ed.), Microbial Enzymes in Aquatic Environments. Springer-Verlag, New York, NY, pp. 29−59.

Clarke, K.R., 1993. Non-parametric multivariate analyses of changes in community structure. Aust. J. Ecol. 18, 117−143.

Clasen, J., Rast, W., Ryding, S.-O., 1989. Available techniques for treating eutrophication. In: Ryding, S.-O., Rast, W. (Eds.), The Control of Eutrophication of Lakes and Reservoirs, vol. 1. UNESCO and the Parthenon Publishing Group, Paris, France, pp. 169−212.

Cleckner, L.B., Garrison, P.J., Hurley, J.P., Olson, M.L., Krabbenhoft, D.P., 1998. Trophic transfer of methyl mercury in the northern Florida Everglades. Biogeochemistry 40, 347−361.

Cleckner, L.B., Gilmour, C.C., Hurley, J.P., Drabbenhoft, D.P., 1999. Mercury methylation in periphyton of the Florida Everglades. Limnol. Oceanogr. 44 (7), 1815−1825.

Clements, W.H., Brooks, M.L., Kashian, D.R., Zuellig, R.E., 2008. Changes in dissolved organic material determine exposure of stream benthic communities to UV-B radiation and heavy metals: implications for climate change. Glob. Change Biol. 14 (9), 2201−2214.

Clements, W.H., Stahl, R.G., Landis, R.C., 2015. Ecological effects of biochar on the structure and function of stream benthic communities. Environ. Sci. Technol. 49 (24), 14649−14654. Available from: https://doi.org/10.1021/acs.est.5b04400.

Clivot, H., Danger, M., Pagnout, C., Wagner, P., Rousselle, P., Poupin, P., et al., 2013. Impaired leaf litter processing in acidified streams. Microb. Ecol. 65 (1), 1−11.

Clymo, R.S., Hayward, P.M., 1982. The ecology of *Sphagnum*. In: Smith, A.J.E. (Ed.), Bryophyte Ecology. Chapman and Hall, London, pp. 229−289.

Codd, G.A., 1995. Cyanobacterial toxins: occurrence, properties and biological significance. Water Sci. Technol. 32, 149−156.

Codd, G.A., Bell, S.G., Kaya, K., Ward, C.J., Beattie, K.A., Metcalf, J.S., 1999a. Cyanobacterial toxins, exposure routes and human health. Eur. J. Phycol. 34, 405–415.

Codd, G.A., Ward, C.J., Beattie, K.A., Bell, S.G., 1999b. Widening perceptions of the occurrence and significance of cyanobacterial toxins. In: Peschek, G.A., Löffelhardt, W., Schmetterer, G. (Eds.), The Phototrophic Prokaryotes. Kluwer Academic, New York, NY, pp. 623–632.

Cogley, J., 1991. GGHYDRO—global hydrographic data release 2.0. Trent Climate Note. Department of Geography, Trent University, 91 (1).

Cohen, A.S., 1995a. Paleoecological approaches to the conservation biology of benthos in ancient lakes: a case study from Lake Tanganyika. J. N. Am. Benthol. Soc. 14 (4), 654–668.

Cohen, A.S., Gergurich, E.L., Kraemer, B.M., McGlue, M.M., McIntyre, P.B., Russell, J.M., et al., 2016. Climate warming reduces fish production and benthic habitat in Lake Tanganyika, one of the most biodiverse freshwater ecosystems. Proc. Natl. Acad. Sci. 113 (34), 9563–9568.

Cohen, J.E., 1995b. Population growth and earth's human carrying capacity. Science 269, 341–346.

Colborn, T., vom Saal, F.S., Soto, A.M., 1993. Developmental effects of endocrine-disrupting chemicals in wildlife and humans. Environ. Health Perspect. 101 (5), 378–384.

Cole, G.A., 1994. Textbook of Limnology, fourth ed. Waveland Press, Inc., Prospect Heights, IL.

Cole, J.J., Caraco, N.F., 2001. Carbon in catchments: connecting terrestrial carbon losses with aquatic metabolism. Mar. Freshw. Res. 52, 101–110.

Cole, J.J., Caraco, N.F., Kling, G.W., Kratz, T.K., 1994. Carbon dioxide supersaturation in thesurface waters of lakes. Science 265, 1568–1570.

Cole, J.J., Carpenter, S.R., Pace, M.L., Van de Bogert, M.C., Kitchell, J.L., Hodgson, J.R., 2006. Differential support of lake food webs by three types of terrestrial organic carbon. Ecol. Lett. 9 (5), 558–568.

Coles, B., Coles, J., 1989. People of the Wetlands. Bogs, Bodies and Lake-Dwellers. Thames and Hudson Publishing Co., German Democratic Republic.

Collen, B., Whitton, F., Dyer, E.E., Baillie, J.E.M., Cumberlidge, N., Darwall, W.R.T., et al., 2014. Global patterns of freshwater species diversity, threat and endemism. Glob. Ecol. Biogeogr. 23 (1), 40–51. Available from: https://doi.org/10.1111/geb.12096.

Collins, J.P., Crump, M.L., 2009. Extinction in Our Time: Global Amphibian Decline. Oxford University Press, New York, NY.

Collins, S.F., Nelson, K.A., DeBoom, C.S., Wahl, D.H., 2017. The facilitation of the native bluegill sunfish by the invasive bighead carp. Freshw. Biol. 62 (9), 1645–1654. Available from: https://doi.org/10.1111/fwb.12976.

Collins, S.F., Wahl, D.H., 2017. Invasive planktivores as mediators of organic matter exchanges within and across ecosystems. Oecologia 184 (2), 521–530. Available from: https://doi.org/10.1007/s00442-017-3872-x.

Collins, S.M., Sparks, J.P., Thomas, S.A., Wheatley, S.A., Flecker, A.S., 2016a. Increased light availability reduces the importance of bacterial carbon in headwater stream food webs. Ecosystems 19 (3), 396–410. Available from: https://doi.org/10.1007/s10021-015-9940-3.

Collins, S.M., Thomas, S.A., Heatherly, T., MacNeill, K.L., Leduc, A.O.H.C., López-Sepulcre, A., et al., 2016b. Fish introductions and light modulate food web fluxes in tropical streams: a whole-ecosystem experimental approach. Ecology 97 (11), 3154–3166. Available from: https://doi.org/10.1002/ecy.1530.

Colman, D.R., Jay, Z.J., Inskeep, W.P., de Jennings, R., Maas, K.R., Rusch, D.B., et al., 2016. Novel, deep-branching heterotrophic bacterial populations recovered from thermal spring metagenomes. Front. Microbiol. 7.

Combarros, R.G., Collado, S., Díaz, M., 2016. Toxicity of titanium dioxide nanoparticles on *Pseudomonas putida*. Water Res. 90, 378–386. Available from: https://doi.org/10.1016/j.watres.2015.12.040.

Compton, B.W., McGarigal, K., Cushman, S.A., Gamble, L.R., 2007. A resistant-kernel model of connectivity for amphibians that breed in vernal pools. Conserv. Biol. 21 (3), 788–799.

Comte, L., Olden, J.D., 2017. Evolutionary and environmental determinants of freshwater fish thermal tolerance and plasticity. Glob. Change Biol. 23 (2), 728−736. Available from: https://doi.org/10.1111/gcb.13427.

Conley, D.J., 2000. Biogeochemical nutrient cycles and nutrient management strategies. Hydrobiologia 410, 87−96.

Conley, D.J., Paerl, H.W., Howarth, R.W., Boesch, D.F., Seitzinger, S.P., Havens, K.E., et al., 2009. Controlling eutrophication: nitrogen and phosphorus. Science 323 (5917), 1014−1015.

Conley, J.M., Evans, N., Cardon, M.C., Rosenblum, L., Iwanowicz, L.R., Hartig, P.C., et al., 2017. Occurrence and in vitro bioactivity of estrogen, androgen, and glucocorticoid compounds in a nationwide screen of United States stream waters. Environ. Sci. Technol. 51 (9), 4781−4791. Available from: https://doi.org/10.1021/acs.est.6b06515.

Connell, J., 1980. Diversity and the coevolution of competitors, or the ghost of competition past. Oikos 131−138.

Connell, J.H., 1978. Diversity in tropical rain forests and coral reefs. Science 199, 1302−1310.

Connell, J.H., 1983. On the prevalence and relative importance of interspecific competition: evidence from field experiments. Am. Nat. 122, 661−696.

Connelly, S., Pringle, C.M., Bixby, R.J., Brenes, R., Whiles, M.R., Lips, K.R., et al., 2008. Changes in stream primary producer communities resulting from large-scale catastrophic amphibian declines: can small-scale experiments predict effects of tadpole loss? Ecosystems 11 (8), 1262−1276.

Conrad, C.C., Hilchey, K.G., 2011. A review of citizen science and community-based environmental monitoring: issues and opportunities. Environ. Monit. Assess. 176 (1), 273−291.

Conrad, H.S., Redfearn Jr., P.L., 1979. How to Know the Mosses and Liverworts. Wm. C. Brown Company Publishers, Dubuque, IA.

Contant, J., Pick, F.R., 2013. Picophytoplankton during the ice-free season in five temperate-zone rivers. J. Plankton Res. 35 (3), 553−565. Available from: https://doi.org/10.1093/plankt/fbt013.

Cook, C.D.K., 1996. Aquatic Plant Book. SPB Academic Publishing, Amsterdam, The Netherlands.

Cooke, G.D., Welch, E.B., Peterson, S.A., Newroth, P.R., 1993. Restoration and Management of Lakes and Reservoirs, second ed.) Lewis Publishers, Boca Raton, FL.

Cooper, G.P., Washburn, G.N., 1949. Relation of dissolved oxygen to winter mortality of fish in Michigan lakes. Trans. Am. Fish. Soc. 76, 23−32.

Copetti, D., Finsterle, K., Marziali, L., Stefani, F., Tartari, G., Douglas, G., et al., 2016. Eutrophication management in surface waters using lanthanum modified bentonite: a review. Water Res. 97, 162−174. Available from: https://doi.org/10.1016/j.watres.2015.11.056.

Corallini, C., Gaino, E., 2003. The caddisfly *Ceraclea fulva* and the freshwater sponge *Ephydatia fluviatilis*: a successful relationship. Tissue Cell 35 (1), 1−7. Available from: https://doi.org/10.1016/s0040-8166 (02)00086-1.

Cordoba-Aguilar, A., Uhia, E., Rivera, A.C., 2003. Sperm competition in Odonata (Insecta): the evolution of female sperm storage and rivals' sperm displacement. J. Zool. 261, 381−398. Available from: https://doi.org/10.1017/s0952836903004357.

Corman, J.R., Moody, E.K., Elser, J.J., 2016. Calcium carbonate deposition drives nutrient cycling in a calcareous headwater stream. Ecol. Monogr. 86 (4), 448−461. Available from: https://doi.org/10.1002/ecm.1229.

Correa, S.B., Winemiller, K.O., Lopez-Fernandez, H., Galetti, M., 2007. Evolutionary perspectives on seed consumption and dispersal by fishes. Bioscience 57 (9), 748−756.

Correll, D.L., 1999. Phosphorus: a rate limiting nutrient in surface waters. Poult. Sci. 78, 674−682.

Cory, R.M., Ward, C.P., Crump, B.C., Kling, G.W., 2014. Sunlight controls water column processing of carbon in arctic fresh waters. Science 345 (6199), 925−928. Available from https://doi.org/10.1126/science.1253119.

Costantini, M.L., Sabetta, L., Mancinelli, G., Rossi, L., 2004. Spatial variability of the decomposition rate of *Schoenoplectus tatora* in a polluted area of Lake Titicaca. J. Trop. Ecol. 20, 325−335.

Costanza, R., de Groot, R., Sutton, P., van der Ploeg, S., Anderson, S.J., Kubiszewski, I., et al., 2014. Changes in the global value of ecosystem services. Glob. Environ. Change 26, 152−158.

Costanza, R., Farley, J., 2007. Ecological economics of coastal disasters: introduction to the special issue. Ecol. Econ. 63 (2−3), 249−253.

Cotner, J.B., Gardner, W.S., Johnson, J.R., Sada, R.H., Cavaletto, J.F., Heath, R.T., 1995. Effects of zebra mussels (*Dreissena polymorpha*) on bacterioplankton evidence for both size-selective consumption and growth stimulation. J. Great Lakes Res. 21 (4), 517–528.

Cottenie, K., Michels, E., Nuytten, N., De Meester, L., 2003. Zooplankton metacommunity structure: regional vs. local processes in highly interconnected ponds. Ecology 84 (4), 991–1000.

Cottingham, K.L., Schindler, D.E., 2000. Effects of grazer community structure on phytoplankton response to nutrient pulses. Ecology 81, 183–2000.

Couch, C.A., Meyer, J.L., 1992. Development and composition of the epixylic biofilm in a blackwater river. Freshw. Biol. 27, 43–51.

Coughlan, N.E., Kelly, T.C., Davenport, J., Jansen, M.A.K., 2017. Up, up and away: bird-mediated ectozoochorous dispersal between aquatic environments. Freshw. Biol. 62 (4), 631–648. Available from: https://doi.org/10.1111/fwb.12894.

Coulter, G.W., 1991a. The benthic fish community. In: Coulter, G.W. (Ed.), Lake Tanganyika and Its Life. Oxford University Press, New York, NY, pp. 151–199.

Coulter, G.W., 1991b. Composition of the flora and fauna. In: Coulter, G.W. (Ed.), Lake Tanganyika and Its Life. Oxford University Press, New York, NY, pp. 200–274.

Covich, A.P., 1993. Water and ecosystems. In: Gleick, P.H. (Ed.), Water in Crisis: A Guide to the World's Freshwater Resources. Oxford University Press, New York, NY, pp. 40–55.

Covich, A.P., 2001. Energy Flow and Ecosystems Encyclopedia of Biodiversity. Academic Press, San Diego, CA.

Covich, A.P., Austen, M.C., Barlocher, F., Chauvet, E., Cardinale, B.J., Biles, C.L., et al., 2004. The role of biodiversity in the functioning of freshwater and marine benthic ecosystems. Bioscience 54 (8), 767–775.

Covich, A.P., Palmer, M.A., Crowl, T.A., 1999. The role of benthic invertebrate species in freshwater ecosystems. Bioscience 49 (2), 119–127.

Cowardin, M.F., Golet, C., LaRoe, E.T., 1979. Classification of wetlands and deepwater habitats of the United States. Retrieved from https://www.fws.gov/wetlands/documents/classification-of-wetlands-and-deepwater-habitats-of-the-united-states.pdf.

Cowley, D.E., 2008. Estimating required habitat size for fish conservation in streams. Aquat. Conserv. -Mar. Freshw. Ecosyst. 18 (4), 418–431.

Craft, C., Clough, J., Ehman, J., Joye, S., Park, R., Pennings, S., et al., 2009. Forecasting the effects of accelerated sea-level rise on tidal marsh ecosystem services. Front. Ecol. Environ. 7 (2), 73–78.

Craft, C.B., Vymazal, J., Richardson, C.J., 1995. Response of Everglades plant communities to nitrogen and phosphorus additions. Wetlands 15 (3), 258–271.

Crain, D.A., Guillette Jr., L.J., Pickford, D.B., Percival, H.F., Woodward, A.R., 1998. Sex-steroid and thyroid hormone concentrations in juvenile alligators (*Alligator mississippiensis*) from contaminated and reference lakes in Florida, USA. Environ. Toxicol. Chem. 17 (3), 446–452.

Crawford, A., Holliday, J., Merrick, C., Brayan, J., van Asten, M., Bowling, L., 2017. Use of three monitoring approaches to manage a major *Chrysosporum ovalisporum* bloom in the Murray River, Australia, 2016. Environ. Monit. Assess. 189 (4), 202.

Crawford, J.T., Stanley, E.H., 2016. Controls on methane concentrations and fluxes in streams draining human-dominated landscapes. Ecol. Appl. 26 (5), 1581–1591. Available from: https://doi.org/10.1890/15-1330.

Crawford, J.T., Stanley, E.H., Spawn, S.A., Finlay, J.C., Loken, L.C., Striegl, R.G., 2014. Ebullitive methane emissions from oxygenated wetland streams. Glob. Change Biol. 20 (11), 3408–3422. Available from: https://doi.org/10.1111/gcb.12614.

Creed, R.P., 2000. Is there a new keystone species in North American lakes and rivers? Oikos 91 (2), 405–408.

Creed, R.P., Cherry, R.P., Pflaum, J.R., Wood, C.J., 2009. Dominant species can produce a negative relationship between species diversity and ecosystem function. Oikos 118 (5), 723–732.

Creed, R.P., Lomonaco, J.D., Thomas, M.J., Meeks, A., Brown, B.L., 2015. Reproductive dependence of a branchiobdellidan annelid on its crayfish host: confirmation of a mutualism. Crustaceana 88 (4), 385–396.

Creed, R.P.J., 1994. Direct and indirect effects of crayfish grazing in a stream community. Ecology 75, 2091–2103.

Crews, D., Willingham, E., Skipper, J.K., 2000. Endocrine disruptors: present issues, future directions. Q. Rev. Biol. 75 (3), 243–260.

Cross, W.F., Baxter, C.V., Rosi-Marshall, E.J., Hall, R.O., Kennedy, T.A., Donner, K.C., et al., 2013. Food-web dynamics in a large river discontinuum. Ecol. Monogr. 83 (3), 311–337. Available from: https://doi.org/10.1890/12-1727.1.

Cross, W.F., Benstead, J.P., Rosemond, A.D., Wallace, J.B., 2003. Consumer-resource stoichiometry in detritus-based streams. Ecol. Lett. 6 (8), 721–732.

Cross, W.F., Wallace, J.B., Rosemond, A.D., Eggert, S.L., 2006. Whole-system nutrient enrichment increases secondary production in a detritus-based ecosystem. Ecology 87 (6), 1556–1565.

Crowl, T.A., Covich, A.P., 1990. Predator-induced life history shifts in a freshwater snail. Science 247, 949–951.

Crutzen, P.J., 2006. The "anthropocene" Earth System Science in the Anthropocene. In: Ehlers, E., Krafft, T. (Eds.), Earth System Science in the Anthropocene. Springer, Berlin, Heidelberg, pp. 13–18.

Cuffney, T.F., Wallace, J.B., Lugthart, G.H., 1990. Experimental evidence quantifying the role of benthic invertebates in organic matter dynamics of headwater streams. Freshw. Biol. 23, 281–299.

Cui, P., Dang, C., Zhuang, J.-q, You, Y., Chen, X.-q, Scott, K.M., 2012. Landslide-dammed lake at Tangjiashan, Sichuan province, China (triggered by the Wenchuan Earthquake, May 12, 2008): risk assessment, mitigation strategy, and lessons learned. Environ. Earth Sci. 65 (4), 1055–1065. Available from: https://doi.org/10.1007/s12665-010-0749-2.

Culver, D.C., 1994. Species interactions. In: Gibert, J., Danielopol, D.L., Stanford, J.A. (Eds.), Groundwater Ecology. Academic Press, San Diego, CA, pp. 271–285.

Culver, D.C., Fong, D.W., 1994. Small scale and large scale biogeography of subterranean crustacean faunas of the Virginias. Hydrobiologia 287, 3–9.

Cummins, K.W., 1973. Trophic relations of aquatic insects. Annu. Rev. Entomol. 18, 183–206.

Cummins, K.W., 1974. Structure and function of stream ecosystems. Bioscience 24, 631–641.

Cummins, K.W., 1977. From headwater streams to rivers. Am. Biol. Teach. 305–312.

Cummins, K.W., Klug, M.J., 1979. Feeding ecology of stream invertebrates. Annu. Rev. Ecol. Syst. 10, 147–172.

Cummins, K.W., Minshall, G.W., Sedell, J.R., Cushing, C.E., Petersen, R.C., 1984. Stream ecosystem theory. Verhandlungen des Internationalen Verein Limnologie 22, 1818–1827.

Cunha, D.G.F., Casali, S.P., de Falco, P.B., Thornhill, I., Loiselle, S.A., 2017. The contribution of volunteer-based monitoring data to the assessment of harmful phytoplankton blooms in Brazilian urban streams. Sci. Total Environ. 584, 586–594.

Currie, D.J., Dilworth-Christie, P., Chapleau, F., 1999. Assessing the strength of top-down influences on plankton abundance in unmanipulated lakes. Can. J. Fish. Aquat. Sci. 56, 427–436.

Currie, D.J., Kalff, J., 1984. The relative importance of bacterioplankton and phytoplankton in phosphorus uptake in freshwater. Limnol. Oceanogr. 29, 311–321.

Cursino, L., Oberdá, S.M., Cecílio, R.V., Moreira, R.M., Chartone-Souza, E., Nascimento, A.M.A., 1999. Mercury concentration in the sediment at different gold prospecting sites along the Carmo stream, Minas Gerais, Brazil, and frequency of resistant bacteria in the respective aquatic communities. Hydrobiologia 394, 5–12.

Cushing, C.E., Gaines, W.L., 1989. Thoughts on recolonization of endorheic cold desert spring-streams. J. N. Am. Benthol. Soc. 8, 277–287.

Dahl, T.E., Johnson, C.E., Frayer, W.E., 1991. Wetland losses in the United States, Washington, DC. Retrieved from https://www.fws.gov/wetlands/Documents%5CWetlands-Loss-Since-the-Revolution.pdf.

Dahlgren, J.P., Ehrlén, J., 2005. Distribution patterns of vascular plants in lakes—the role of metapopulation dynamics. Ecography 28 (1), 49–58.

Dakos, V., van Nes, E.H., Donangelo, R., Fort, H., Scheffer, M., 2010. Spatial correlation as leading indicator of catastrophic shifts. Theor. Ecol. 3 (3), 163–174.

Daleo, P., Panjul, E., Casariego, A.M., Silliman, B.R., Bertness, M.D., Iribarne, O., 2007. Ecosystem engineers activate mycorrhizal mutualism in salt marshes. Ecol. Lett. 10 (10), 902—908.

Danielopol, D.L., des Châtelliers, M.C., Moeszlacher, F., Pospisil, P., Popa, R., 1994. Adaptation of crustacea to interstitial habitats: a practical agenda for ecological studies. In: Gibert, J., Danielopol, D.L., Stanford, J.A. (Eds.), Groundwater Ecology. Academic Press, San Diego, CA, pp. 217—243.

Danielson, T.J., 1998. Wetland bioassessment fact sheets. EPA843-F-98-001, Washington, DC. Retrieved from https://nepis.epa.gov/Exe/ZyNET.exe/200053VA.txt?ZyActionD=ZyDocument& Client=EPA&Index=1995%20Thru%201999&Docs=&Query=&Time=&EndTime=&SearchMethod= 1&TocRestrict=n&Toc=&TocEntry=&QField=&QFieldYear=&QFieldMonth=&QFieldDay=& UseQField=&IntQFieldOp=0&ExtQFieldOp=0&XmlQuery=&File=D%3A%5CZYFILES %5CINDEX%20DATA%5C95THRU99%5CTXT%5C00000012%5C200053VA.txt&User= ANONYMOUS&Password=anonymous&SortMethod=h%7C-&MaximumDocuments=1&Fuzzy Degree=0&ImageQuality=r75g8/r75g8/x150y150g16/i425&Display=hpfr&DefSeekPage=x& SearchBack=ZyActionL&Back=ZyActionS&BackDesc=Results%20page&MaximumPages= 1&ZyEntry=1&slide.

Darcy-Hall, T.L., 2006. Relative strengths of benthic algal nutrient and grazer limitation along a lake productivity gradient. Oecologia 148 (4), 660—671.

Dash, M.C., Hota, A.K., 1980. Density effects on the survival, growth rate, and metamorphosis of *Rana tigrina* tadpoles. Ecology 61, 1025—1028.

Daszak, P., Strieby, A., Cunningham, A.A., Longcore, J.E., Brown, C.C., Porter, D., 2004. Experimental evidence that the bullfrog (*Rana catesbeiana*) is a potential carrier of chytridiomycosis, an emerging fungal disease of amphibians. Herpetol. J. 14 (4), 201—207.

Datry, T., Bonada, N., Boulton, A.J., 2017. Intermittent Rivers and Ephemeral Streams: Ecology and Management. Elsevier Science & Technology Books.

Datry, T., Fritz, K., Leigh, C., 2016. Challenges, developments and perspectives in intermittent river ecology. Freshw. Biol. 61 (8), 1171—1180. Available from: https://doi.org/10.1111/fwb.12789.

Datry, T., Larned, S.T., Tockner, K., 2014. Intermittent rivers: a challenge for freshwater ecology. Bioscience 64 (3), 229—235. Available from: https://doi.org/10.1093/biosci/bit027.

Daufresne, M., Boët, P., 2007. Climate change impacts on structure and diversity of fish communities in rivers. Glob. Change Biol. 13 (12), 2467—2478.

Daughton, C.G., Ternes, T.A., 1999. Pharmaceutical and personal care products in the environment: agents of subtle change? Environ. Health Perspect. 107, 907—938.

Davidson, C., 2004. Declining downwind: amphibian population declines in california and historical pesticide use. Ecol. Appl. 14 (6), 1892—1902.

Davies, B.R., Thoms, M.C., Walker, K.F., O'Keefe, J.H., Gore, J.A., 1994. Dryland rivers: their ecology, conservation and management. In: Calow, P., Petts, G.E. (Eds.), The Rivers Handbook. Hydrological and Ecological Principles, vol. 2. Blackwell Scientific Publications, London, pp. 484—511.

Davies, R.W., 1991. *Annelida:* Leeches, Polychaetes, and Acanthobdellids. In: Thorp, J.H., Covich, A.P. (Eds.), Ecology and Classification of North American Freshwater Invertebrates. Academic Press, Inc., San Diego, CA, pp. 437—480.

Davies-Colley, R.J., 1997. Stream channels are narrower in pasture than in forest. New Zealand J. Mar. Freshw. Res. 31, 599—608.

Davis, S.M., 1994. Phosphorus inputs and vegetation sensitivity in the Everglades. In: Davis, S.M., Ogden, J.C. (Eds.), Everglades, the Ecosystem and Its Restoration. St. Lucie Press, Delray Beach, FL, pp. 357—378.

Davis, S.M., Ogden, J.C., 1994. Introduction. In: Davis, S.M., Ogden, J.C. (Eds.), Everglades, the Ecosystem and Its Restoration. St. Lucie Press, Delray Beach, FL, pp. 3—8.

Davis, S.N., Wiest, J.M.D., 1966. Hydrogeology. Wiley, New York, NY.

Davison, A., Blaxter, M., 2005. Ancient origin of glycosyl hydrolase family 9 cellulase genes. Mol. Biol. Evol. 22 (5), 1273—1284.

Davison, W., George, D.G., Edwards, N.J.A., 1995. Controlled reversal of lake acidification by treatment with phosphate fertilizer. Nature 377, 504—507.

Dawidowicz, P., Prejs, A., Engelmayer, A., Martyniak, A., Kozłowski, J., Kufel, L., et al., 2002. Hypolimnetic anoxia hampers top−down food-web manipulation in a eutrophic lake. Freshw. Biol. 47 (12), 2401−2409. Available from: https://doi.org/10.1046/j.1365-2427.2002.01007.x.

De Jalón, D.G., 1995. Management of physical habitat for fish stocks. In: Harper, D.M., Ferguson, A.J.D. (Eds.), The Ecological Basis for River Management. John Wiley & Sons, Ltd, West Sussex, England, pp. 363−374.

De Lange, H.J., Morris, D.P., Williamson, C.E., 2003. Solar ultraviolet photodegradation of DOC may stimulate freshwater food webs. J. Plankton Res. 25, 111−117.

De Lange, H.J., Noordoven, W., Murk, A.J., Lurling, M., Peeters, E., 2006. Behavioural responses of *Gammarus pulex* (Crustacea, Amphipoda) to low concentrations of pharmaceuticals. Aquat. Toxicol. 78 (3), 209−216.

De Meester, L., Weider, L.J., Tollrian, R., 1995. Alternative antipredator defenses and genetic polymorphism in a pelagic predator-prey system. Nature 378, 483−485.

de Oliveira Neves, A., 2009. Conservation of the pantanal wetlands: the definitive moment for decision making. AMBIO: J. Hum. Environ. 38 (2), 127−128.

de Souza Machado, A.A., Zarfl, C., Rehse, S., Kloas, W., 2017. Low-dose effects: nonmonotonic responses for the toxicity of a *Bacillus thuringiensis* Biocide to *Daphnia magna*. Environ. Sci. Technol. 51 (3), 1679−1686. Available from: https://doi.org/10.1021/acs.est.6b03056.

de Szalay, F.A., Resh, V.H., 1997. Responses of wetland invertebrates and plants important in waterfowl diets to burning and mowing of emergent vegetation. Wetlands 17, 149−156.

de Vaate, A.B., 1991. Distribution and aspects of population dynamics of the zebra mussel, *Dreissena polymorpha* (Pallas, 1771), in the lake Ijsselmeer area (The Netherlands). Oecologia 86, 40−50.

Deacon, J.E., Williams, A.E., Williams, C.D., Williams, J.E., 2007. Fueling population growth in Las Vegas: how large-scale groundwater withdrawal could burn regional biodiversity. AIBS Bull. 57 (8), 688−698.

Dean, J.A., 1985. Langes Handbook of Chemistry, thirteenth ed. McGraw-Hill Book Company, New York, NY.

Deborde, D.C., Woessner, W.W., Kiley, Q.T., Ball, P., 1999. Rapid transport of viruses in a floodplain aquifer. Water Res. 33 (10), 2229−2238.

Debroas, D., Domaizon, I., Humbert, J.-F., Jardillier, L., Lepère, C., Oudart, A., et al., 2017. Overview of freshwater microbial eukaryotes diversity: a first analysis of publicly available metabarcoding data. FEMS Microbiol. Ecol. 93 (4). Available from: https://doi.org/10.1093/femsec/fix023fix023-fix023.

Deharveng, L., D'Haese, C.A., Bedos, A., 2008. Global diversity of springtails (Collembola; Hexapoda) in freshwater. Hydrobiologia 595 (1), 329−338. Available from: https://doi.org/10.1007/s10750-007-9116-z.

Deininger, A., Faithfull, C.L., Bergström, A.K., 2017. Phytoplankton response to whole lake inorganic N fertilization along a gradient in dissolved organic carbon. Ecology 98 (4), 982−994. Available from: https://doi.org/10.1002/ecy.1758.

Delgado-Baquerizo, M., Giaramida, L., Reich, P.B., Khachane, A.N., Hamonts, K., Edwards, C., et al., 2016. Lack of functional redundancy in the relationship between microbial diversity and ecosystem functioning. J. Ecol. 104 (4), 936−946.

DeLorenzo, M.E., Scott, G.I., Ross, P.E., 2001. Toxicity of pesticides to aquatic microorganisms: a review. Environ. Toxicol. Chem.: Int. J. 20 (1), 84−98.

Delorme, L.D., 1991. Ostracoda. In: Thorp, J.H., Covich, A.P. (Eds.), Ecology and Classification of North American Freshwater Invertebrates. Academic Press, Inc., San Diego, CA, pp. 691−722.

DelSontro, T., McGinnis, D.F., Wehrli, B., Ostrovsky, I., 2015. Size does matter: importance of large bubbles and small-scale hot spots for methane transport. Environ. Sci. Technol. 49 (3), 1268−1276. Available from: https://doi.org/10.1021/es5054286.

DelVecchia, A.G., Stanford, J.A., Xu, X., 2016. Ancient and methane-derived carbon subsidizes contemporary food webs. Nat. Commun. 7.

Demars, B.O., Gíslason, G.M., Ólafsson, J.S., Manson, J.R., Friberg, N., Hood, J.M., et al., 2016. Impact of warming on CO_2 emissions from streams countered by aquatic photosynthesis. Nat. Geosci. 9 (10), 758−761.

DeMott, W.R., 1995. The influence of prey hardness on Daphnia's selectivity for large prey. Hydrobiologia 307, 127–138.

DeNicola, D.M., 1996. Periphyton responses to temperature at different ecological levels. In: Stevenson, R.J., Bothwell, M.L., Lowe, R.L. (Eds.), Algal Ecology: Freshwater Benthic Ecosystems. Academic Press, San Diego, CA, pp. 150–183.

Denny, M.W., 1993. Air and Water: The Biology and Physics of Lifes Media. Princeton University Press, Princeton, NJ.

Dent, C.L., Grimm, N.B., Fisher, S.G., 2001. Multiscale effects of surface–subsurface exchange on stream water nutrient concentrations. J. N. Am. Benthol. Soc. 20 (2), 162–181.

Derrien, M., Yang, L., Hur, J., 2017. Lipid biomarkers and spectroscopic indices for identifying organic matter sources in aquatic environments: a review. Water Res. 112, 58–71. Available from: https://doi.org/10.1016/j.watres.2017.01.023.

des Châtelliers, M.C., Poinsart, D., Bravard, J.-P., 1994. Geomorphology of alluvial groundwater ecosystems. In: Gibert, J., Danielopol, D.L., Stanford, J.A. (Eds.), Groundwater Ecology. Academic Press, Inc, San Diego, CA, pp. 157–185.

Devrie, D.R., Frie, R.V., 1996. Determination of age and growth. In: Murphy, B.R., Willis, D.W. (Eds.), Fisheries Techniques, second ed. American Fisheries Society, Bethesda, MD, pp. 483–512.

DeYoe, H.R., Lowe, R.L., Marks, J.C., 1992. Effects of nitrogen and phosphorus on the endosymbiont load of *Rhopalodia gibba* and *Epithemia turgida* (Bacillariophyceae). J. Phycol. 28 (6), 773–777.

Di Sabatino, A., Smit, H., Gerecke, R., Goldschmidt, T., Matsumoto, N., Cicolani, B., 2008. Global diversity of water mites (Acari, Hydrachnidia; Arachnida) in freshwater. Hydrobiologia 595 (1), 303–315. Available from: https://doi.org/10.1007/s10750-007-9025-1.

Diaz, R.J., Rosenberg, R., 2008. Spreading dead zones and consequences for marine ecosystems. Science 321, 926–929.

Díaz, S., Pascual, U., Stenseke, M., Martín-López, B., Watson, R.T., Molnár, Z., et al., 2018. Assessing nature's contributions to people. Science 359 (6373), 270–272. Available from: https://doi.org/10.1126/science.aap8826.

Dickman, M., Rao, S.S., 1989. Diatom stratigraphy in acid-stressed lakes in The Netherlands, Canada, and China. In: Rao, S.S. (Ed.), Acid Stress Aquat. Microb. Interact. CRC Press, Boca Raton, FL, pp. 116–138.

Didham, R.K., Tylianakis, J.M., Hutchison, M.A., Ewers, R.M., Gemmell, N.J., 2005. Are invasive species the drivers of ecological change? Trends Ecol. Evol. 20 (9), 470–474.

Dillard, G.E., 1999. Common Freshwater Algae of the United States, an Illustrated Key to the Genera (Excluding the Diatoms). Gebrüder Borntraeger, Berlin, Germany.

Dittmann, E., Wiegand, C., 2006. Cyanobacterial toxins—occurrence, biosynthesis and impact on human affairs. Mol. Nutr. Food Res. 50 (1), 7–17.

Dodds, W., Clements, W., Gido, K., Hilderbrand, R., King, R., 2010. Thresholds, breakpoints, and nonlinearity in freshwaters as related to management. J. N. Am. Benthol. Soc. 29 (3), 988–997.

Dodds, W.K., 1989. Photosynthesis of two morphologies of *Nostoc parmelioides* (Cyanobacteria) as related to current velocities and diffusion patterns. J. Phycol. 25, 258–262.

Dodds, W.K., 1990. Hydrodynamic constraints on evolution of chemically mediated interactions between aquatic organisms in unidirectional flows. J. Chem. Ecol. 16, 1417–1430.

Dodds, W.K., 1991. Community interactions between the filamentous alga *Cladophora glomerata* (L.) Kuetzing, its epiphytes, and epiphyte grazers. Oecologia 85, 572–580.

Dodds, W.K., 1992. A modified fiber-optic light microprobe to measure spherically integrated photosynthetic photon flux density: characterization of periphyton photosynthesis-irradiance patterns. Limnol. Oceanogr. 37, 871–878.

Dodds, W.K., 1993. What controls levels of dissolved phosphate and ammonium in surface waters? Aquat. Sci. 55 (2), 132–142.

Dodds, W.K., 1997a. Distribution of runoff and rivers related to vegetative characteristics, latitude, and slope: a global perspective. J. N. Am. Benthol. Soc. 16, 162–168.

Dodds, W.K., 1997b. Interspecific interactions: constructing a general neutral model for interaction type. Oikos 78, 377–383.

Dodds, W.K., 2003a. Misuse of inorganic N and soluble reactive P concentrations to indicate nutrient status of surface waters. J. N. Am. Benthol. Soc. 22, 171−181.

Dodds, W.K., 2003b. The role of periphyton in phosphorus retention in shallow freshwater aquatic systems. J. Phycol. 39, 840−849.

Dodds, W.K., 2006. Eutrophication and trophic state in rivers and streams. Limnol. Oceanogr. 51 (1), 671−680.

Dodds, W.K., 2007. Trophic state, eutrophication and nutrient criteria in streams. Trends Ecol. Evol. 22 (12), 669−676. Available from: https://doi.org/10.1016/j.tree.2007.07.010.

Dodds, W.K., 2008. Humanity's Footprint: Momentum, Impact, and Our Global Environment. Columbia University Press, New York, NY.

Dodds, W.K., 2009. Laws, Theories, and Patterns in Ecology. University of California Press, Berkeley.

Dodds, W.K., Banks, M.K., Clenan, C.S., Rice, C.W., Sotomayor, D., Strauss, E.A., et al., 1996. Biological properties of soil and subsurface sediments under abandoned pasture and cropland. Soil Biol. Biochem. 28, 837−846.

Dodds, W.K., Bouska, W.W., Eitzmann, J.L., Pilger, T.J., Pitts, K.L., Riley, A.J., et al., 2009. Eutrophication of US freshwaters: analysis of potential economic damages. Environ. Sci. Technol. 43 (1), 12−19. Available from: https://doi.org/10.1021/Es801217q.

Dodds, W.K., Brock, J., 1998. A portable flow chamber for *in situ* determination of benthic metabolism. Freshw. Biol. 39, 49−59.

Dodds, W.K., Carney, E., Angelo, R.T., 2006. Determining ecoregional reference conditions for nutrients, Secchi depth and chlorophyll a in Kansas lakes and reservoirs. Lake Reserv. Manage. 22 (2), 151−159.

Dodds, W.K., Castenholz, R.W., 1988a. Effects of grazing and light on the growth of *Nostoc pruniforme* (Cyanobacteria). Br. Phycol. J. 23 (3), 219−227.

Dodds, W.K., Castenholz, R.W., 1988b. The nitrogen budget of an oligotrophic cold water pond. Arch. Hydrobiol. Suppl. 79, 343−362.

Dodds, W.K., Cole, J.J., 2007. Expanding the concept of trophic state in aquatic ecosystems: it's not just the autotrophs. Aquat. Sci. 69 (4), 427−439.

Dodds, W.K., Collins, S.M., Hamilton, S.K., Tank, J.L., Johnson, S., Webster, J.R., et al., 2014. You are not always what we think you eat: selective assimilation across multiple whole-stream isotopic tracer studies. Ecology 95 (10), 2757−2767. Available from: https://doi.org/10.1890/13-2276.1.

Dodds, W.K., Gido, K., Whiles, M.R., Daniels, M.D., Grudzinski, B.P., 2015. The stream biome gradient concept: factors controlling lotic systems across broad biogeographic scales. Freshw. Sci. 34 (1), 1. Available from: https://doi.org/10.1086/679756.

Dodds, W.K., Gido, K., Whiles, M.R., Fritz, K.M., Matthews, W.J., 2004. Life on the edge: the ecology of great plains prairie streams. Bioscience 54 (3), 205−216.

Dodds, W.K., Gudder, D.A., 1992. The ecology of *Cladophora*. J. Phycol. 28, 415−427.

Dodds, W.K., Gudder, D.A., Mollenhauer, D., 1995. The ecology of *Nostoc*. J. Phycol. 31, 2−18.

Dodds, W.K., Henebry, G.M., 1996. The effect of density dependence on community structure. Ecol. Model. 93, 33−42.

Dodds, W.K., Hutson, R.E., Eichem, A.C., Evans, M.A., Gudder, D.A., Fritz, K.M., et al., 1996. The relationship of floods, drying, flow and light to primary production and producer biomass in a prairie stream. Hydrobiologia 333, 151−159.

Dodds, W.K., Johnson, K.R., Priscu, J.C., 1989. Simultaneous nitrogen and phosphorus deficiency in natural phytoplankton assemblages: theory, empirical evidence, and implications for lake management. Lake Reserv. Manage. 5, 21−26.

Dodds, W.K., Jones, J.R., Welch, E.B., 1998. Suggested classification of stream trophic state: distributions of temperate stream types by chlorophyll, total nitrogen, and phosphorus. Water Res. 32, 1455−1462.

Dodds, W.K., Nelson, J.A., 2006. Redefining the community: a species-based approach. Oikos 112 (2), 464−472.

Dodds, W.K., Oakes, R.M., 2004. A technique for establishing reference nutrient concentrations across watersheds affected by humans. Limnol. Oceanogr. Methods 2, 333−341.

Dodds, W.K., Oakes, R.M., 2008. Headwater influences on downstream water quality. Environ. Manage. 41 (3), 367–377.

Dodds, W.K., Perkin, J.S., Gerken, J.E., 2013. Human impact on freshwater ecosystem services: a global perspective. Environ. Sci. Technol. 47 (16), 9061–9068.

Dodds, W.K., Priscu, J.C., 1989. Ammonium, nitrate, phosphate, and inorganic carbon uptake in an oligotrophic lake: seasonal variation among light response variables. J. Phycol. 25, 699–705.

Dodds, W.K., Priscu, J.C., 1990. A comparison of methods for assessment of nutrient deficiency of phytoplankton in a large oligotrophic lake. Can. J. Fish. Aquat. Sci. 47, 2328–2338.

Dodds, W.K., Priscu, J.C., Ellis, B.K., 1991. Seasonal uptake and regeneration of inorganic nitrogen and phosphorus in a large oligotrophic lake: size-fractionation and antibiotic treatment. J. Plankton Res. 13 (6), 1339–1358.

Dodds, W.K., Smith, V.H., Lohman, K., 2006. Nitrogen and phosphorus relationships to benthic algal biomass in temperate streams. Can. J. Fish. Aquat. Sci. 63 (5), 1190–1191. Available from: https://doi.org/10.1139/F06-040.

Dodds, W.K., Smith, V.H., Zander, B., 1997. Developing nutrient targets to control benthic chlorophyll levels in streams: a case study of the Clark Fork River. Water Res. 31 (7), 1738–1750.

Dodds, W.K., Welch, E.B., 2000. Establishing nutrient criteria in streams. J. N. Am. Benthol. Soc. 19, 186–196.

Dodds, W.K., Wilson, K.C., Rehmeier, R.L., Knight, G.L., Wiggam, S., Falke, J.A., et al., 2008. Comparing ecosystem goods and services provided by restored and native lands. AIBS Bull. 58 (9), 837–845.

Dodson, S.I., Crowl, T.A., Peckarsky, B.L., Kats, L.B., Covich, A.P., Culp, J.M., 1994. Non-visual communication in freshwater benthos: an overview. J. N. Am. Benthol. Soc. 13 (2), 268–282.

Dodson, S.I., Frey, D.G., 1991. Cladocera and other branchiopoda. In: Thorp, J.H., Covich, A.P. (Eds.), Ecology and Classification of North American Freshwater Invertebrates. Academic Press, Inc, San Diego, CA, pp. 725–745.

Doi, H., Inui, R., Akamatsu, Y., Kanno, K., Yamanaka, H., Takahara, T., et al., 2017. Environmental DNA analysis for estimating the abundance and biomass of stream fish. Freshw. Biol. 62 (1), 30–39. Available from: https://doi.org/10.1111/fwb.12846.

Dole-Olivier, M.-J., Marmonier, P., des Châtelliers, M.C., Martin, D., 1994. Interstitial fauna associated with the alluvial floodplains of the Rhône River (France). In: Gibert, J., Danielopol, D.L., Stanford, J.A. (Eds.), Groundwater Ecology. Academic Press, Inc, San Diego, CA, pp. 313–346.

Döll, P., Müller Schmied, H., Schuh, C., Portmann, F.T., Eicker, A., 2014. Global-scale assessment of groundwater depletion and related groundwater abstractions: combining hydrological modeling with information from well observations and GRACE satellites. Water Resour. Res. 50 (7), 5698–5720. Available from: https://doi.org/10.1002/2014WR015595.

Doll, P., Zhang, J., 2010. Impact of climate change on freshwater ecosystems: a global-scale analysis of ecologically relevant river flow alterations. Hydrol. Earth Syst. Sci. 14 (5), 783–799. Available from: https://doi.org/10.5194/hess-14-783-2010.

Donar, C.M., Neely, R.K., Stoermer, E.F., 1996. Diatom succession in an urban reservoir system. J. Paleolimnol. 15, 237–243.

Dong, X., Ruhí, A., Grimm, N.B., 2017. Evidence for self-organization in determining spatial patterns of stream nutrients, despite primacy of the geomorphic template. Proc. Natl. Acad. Sci. 114 (24), E4744–E4752. Available from: https://doi.org/10.1073/pnas.1617571114.

Donohue, I., Hillebrand, H., Montoya, J.M., Petchey, O.L., Pimm, S.L., Fowler, M.S., et al., 2016. Navigating the complexity of ecological stability. Ecol. Lett. 19 (9), 1172–1185. Available from: https://doi.org/10.1111/ele.12648.

Doolittle, W.F., 1999. Phylogenetic classification and the universal tree. Science 284, 2124–2128.

Doren, R.F., Armentano, T.V., Whiteaker, L.D., Jones, R.D., 1996. Marsh vegetation patterns and soil phosphorus gradients in the Everglades ecosystem. Aquat. Bot. 56, 145–163.

Doubek, J.P., Carey, C.C., Cardinale, B.J., 2015. Anthropogenic land use is associated with N-fixing cyanobacterial dominance in lakes across the continental United States. Aquat. Sci. 77 (4), 681–694.

Doughty, M.J., 1991. Mechanism and strategies of photomovement in protozoa. In: Lenci, F., Ghetti, F., Colombetti, G., Häder, D.-P., Song, P.-S. (Eds.), Biophysics of Photoreceptors and Photomovements in Microorganisms. Plenum Press, New York, NY, pp. 73—102.

Douglas, M., Lake, P.S., 1994. Species richness of stream stones: an investigation of the mechanisms generating the species-area relationship. Oikos 69, 387—396.

Downing, A.L., Leibold, M.A., 2010. Species richness facilitates ecosystem resilience in aquatic food webs. Freshw. Biol. 55 (10), 2123—2137. Available from: https://doi.org/10.1111/j.1365-2427.2010.02472.x.

Downing, J.A., Osenberg, C.W., Sarnelle, O., 1999. Meta-analysis of marine nutrient-enrichment experiments: variation in the magnitude of nutrient limitation. Ecology 80, 1157—1167.

Downing, J.A., Watson, S.B., McCauley, E., 2001. Predicting cyanobacteria dominance in lakes. Can. J. Fish. Aquat. Sci. 58, 1905—1908.

Doyle, M.W., Bernhardt, E.S., 2011. What is a stream? Environ. Sci. Technol. 45 (2), 354—359. Available from: https://doi.org/10.1021/es101273f.

Driscoll, C.T., Mason, R.P., Chan, H.M., Jacob, D.J., Pirrone, N., 2013. Mercury as a global pollutant: sources, pathways, and effects. Environ. Sci. Technol. 47 (10), 4967—4983. Available from: https://doi.org/10.1021/es305071v.

Duarte, C.M., Agustí, S., 1998. The CO_2 balance of unproductive aquatic ecosystems. Science 281, 234—236.

Dubey, T., Stephenson, S.L., Edwards, P.J., 1994. Effect of pH on the distribution and occurrence of aquatic fungi in six West Virginia mountain streams. J. Environ. Qual. 23, 1271—1279.

Dudley, R.K., Platania, S.P., 2007. Flow regulation and fragmentation imperil pelagic-spawning riverine fishes. Ecol. Appl. 17 (7), 2074—2086.

Duever, M.J., Meeder, J.F., Meeder, L.C., McCollom, J.M., 1994. The climate of South Florida and its role in shaping the Everglades ecosystem. In: Davis, S.M., Ogden, J.C. (Eds.), Everglades. St. Lucie Press, Delray Beach, FL, pp. 225—248.

Duff, J.H., Triska, F.J., 2000. Nitrogen biogeochemistry and surface-subsurface exchange in streams. In: Jones, J.B., Molholland, P.J. (Eds.), Streams and Ground Waters. Academic Press, San Diego, CA, pp. 197—220.

Duffy, J.E., 2009. Why biodiversity is important to the functioning of real-world ecosystems. Front. Ecol. Environ. 7 (8), 437—444.

Dugan, H.A., Bartlett, S.L., Burke, S.M., Doubek, J.P., Krivak-Tetley, F.E., Skaff, N.K., et al., 2017. Salting our freshwater lakes. Proc. Natl. Acad. Sci. 114 (17), 4453—4458. Available from: https://doi.org/10.1073/pnas.1620211114.

Dugan, P., 1993. Wetlands in Danger: A World Conservation Atlas. Oxford University Press, New York, NY.

Dumont, H.J., 1995. The evolution of groundwater Cladocera. Hydrobiologia 307, 69—74.

Dutton, C.L., Subalusky, A.L., Hamilton, S.K., Rosi, E.J., Post, D.M., 2018. Organic matter loading by hippopotami causes subsidy overload resulting in downstream hypoxia and fish kills. Nat. Commun. 9 (1), 1951. Available from: https://doi.org/10.1038/s41467-018-04391-6.

Dynesius, M., Nilsson, C., 1994. Fragmentation and flow regulation of river systems in the northern third of the world. Science 266, 753—762.

East, A.E., Pess, G.R., Bountry, J.A., Magirl, C.S., Ritchie, A.C., Logan, J.B., et al., 2015. Large-scale dam removal on the Elwha River, Washington, USA: river channel and floodplain geomorphic change. Geomorphology 228, 765—786.

East, T.L., Havens, K.E., Rodusky, A.J., Brady, M.A., 1999. *Daphnia lumholtzi* and *Daphnia ambigua*: population comparisons of an exotic and a native cladoceran in Lake Okeechobee, Florida. J. Plankton Res. 21 (8), 1537—1551.

Eaton, A.D., Clesceri, L.S., Rice, E.W., Greenberg, A.E., Franson, M.A.H., 2005. Standard Methods for the Examination of Water and Wastewater, vol. 21. American Public Health Association, Washington, DC.

Eaton, A.E., Clesceri, L.S., Greenberg, A.E., 1995. Standard Methods for Examination of Water and Wastewater, nineteenth ed. American Public Health Association, Washington, DC.

Eddy, F.B., 1981. Effects of stress on osmotic and ionic regulation in fish. In: Pickering, A.D. (Ed.), Stress and Fish. Academic Press, Inc., London, UK, pp. 77−102.

Eddy, S., Underhill, J.C., 1969. How to Know the Freshwater Fishes. Wm. C. Brown Company Publishers, Dubuque, IA.

Edler, C., Dodds, W.K., 1996. The ecology of a subterranean isopod, *Caecidotea tridentata*. Freshw. Biol. 35, 249−259.

Edler, C., Georgian, T., 2004. Field measurements of particle-capture efficiency and size selection by caddisfly nets and larvae. J. N. Am. Benthol. Soc. 23 (4), 756−770.

Edmondson, W.T., 1991. The Uses of Ecology: Lake Washington and Beyond. The University of Washington Press, Seattle, WA.

Edmonson, W.T., Lehman, J.T., 1981. The effect of changes in the nutrient income on the condition of Lake Washington. Limnol. Oceanogr. 26, 1−29.

Edwards, K.J., Bond, P.L., Gihring, T.M., Banfield, J.F., 2000. An archaeal iron-oxidizing extreme acidophile important in acid mine drainage. Science 287, 1796−1799.

Edwards, P.J., Abivardi, C., 1998. The value of biodiversity: where ecology and economy blend. Biol. Conserv. 83 (3), 239−246.

Eerkes-Medrano, D., Thompson, R.C., Aldridge, D.C., 2015. Microplastics in freshwater systems: a review of the emerging threats, identification of knowledge gaps and prioritisation of research needs. Water Res. 75, 63−82. Available from: https://doi.org/10.1016/j.watres.2015.02.012.

Effler, S.W., Boone, S.R., Sigfired, C., Ashby, S.L., 1998. Dynamics of zebra mussel oxygen demand in Seneca River, New York. Environ. Sci. Technol. 32, 807−812.

Egeland, G.M., Middaugh, J.P., 1997. Balancing fish consumption benefits with mercury exposure. Science 278, 1904−1905.

Egertson, C.J., Kopaska, J.A., Downing, J.A., 2004. A century of change in macrophyte abundance and composition in response to agricultural eutrophication. Hydrobiologia 524 (1), 145−156.

Eggert, S.L., Wallace, J.B., 2007. Wood biofilm as a food resource for stream detritivores. Limnol. Oceanogr. 52 (3), 1239−1245.

Eichem, A.C., Dodds, W.K., Tate, C.M., Edler, C., 1993. Microbial decomposition of elm and oak leaves in a Karst aquifer. Appl. Environ. Biol. 59, 3592−3596.

Eichmiller, J.J., Best, S.E., Sorensen, P.W., 2016. Effects of temperature and trophic state on degradation of environmental DNA in lake water. Environ. Sci. Technol. 50 (4), 1859−1867. Available from: https://doi.org/10.1021/acs.est.5b05672.

Eisenberg, J.N.S., Washburn, J.O., Schreiber, S.J., 2000. Generalist feeding behaviors of *Aedes sierrensis* larvae and their effects on protozoan populations. Ecology 81 (4), 921−935.

Eldakar, O., Dlugos, M., Pepper, J., Wilson, D., 2009. Population structure mediates sexual conflict in water striders. Science 326 (5954), 816.

Elias, S., Banin, E., 2012. Multi-species biofilms: living with friendly neighbors. FEMS Microbiol. Rev. 36 (5), 990−1004.

Eller, G., Deines, P., Grey, J., Richnow, H.-H., Krüger, M., 2005. Methane cycling in lake sediments and its influence on chironomid larval δ13C. FEMS Microbiol. Ecol. 54 (3), 339−350.

Eller, G., Deines, P., Krüger, M., 2007. Possible sources of methane-derived carbon for chironomid larvae. Aquat. Microb. Ecol. 46 (3), 283−293.

Elliott, J.K., Elliott, J.M., Legett, W.C., 1997. Predation by *Hydra* on larval fish: field and laboratory experiments with bluegill (*Lepomis macrochirus*). Limnol. Oceanogr. 42, 1416−1423.

Elliott, J.M., 1981. Some aspects of thermal stress on freshwater teleosts. In: Pickering, A.D. (Ed.), Stress and Fish. Academic Press, Inc., London, UK, pp. 209−245.

Elser, J.J., Bracken, M.E.S., Cleland, E.E., Gruner, D.S., Harpole, W.S., Hillebrand, H., et al., 2007. Global analysis of nitrogen and phosphorus limitation of primary producers in freshwater, marine and terrestrial ecosystems. Ecol. Lett. 10 (12), 1135−1142.

Elser, J.J., Carney, H.J., Goldman, C.R., 1990. The zooplankton-phytoplankton interface in lakes of contrasting trophic status: an experimental comparison. Hydrobiologia 200/201, 69−82.

Elser, J.J., Dobberfuhl, D.R., MacKay, N.A., Schampel, J.H., 1996. Organism size, life history, and N:P stoichiometry, toward a unified view of cellular and ecosystem processes. Bioscience 46 (9), 674–684.

Elser, J.J., Goldman, C.R., 1991. Zooplankton effects on phytoplankton in lakes of contrasting trophic status. Limnol. Oceanogr. 36, 64–90.

Elser, J.J., Hassett, R.P., 1994. A stoichiometric analysis of the zooplankton-phytoplankton interaction in marine and freshwater ecosystems. Nature 370, 211–213.

Elser, J.J., Marzolf, E.R., Goldman, C.R., 1990. Phosphorus and nitrogen limitation of phytoplankton growth in the freshwaters of North America: a review and critique of experimental enrichments. Can. J. Fish. Aquat. Sci. 47, 1468–1477.

Elser, J.J., Urabe, J., 1999. The stoichiometry of consumer-driven nutrient recycling: theory, observations, and consequences. Ecology 80 (3), 735–751.

Elton, C.S., 1958. The Ecology of Invasion by Animals and Plants. Wiley, New York, NY.

Elwood, J.W., Newbold, J.D., O'Neill, R.V., Winkle, W.V., 1983. Resource spiraling: an operational paradigm for analyzing lotic ecosystems. In: Fontaine, T.D.I., Bartell, S.M. (Eds.), Dynamics of Lotic Ecosystems. Ann Arbor Science Publishers, Ann Arbor, MI, pp. 3–27.

Emsens, W.-J., Schoelynck, J., Grootjans, A.P., Struyf, E., van Diggelen, R., 2016. Eutrophication alters Si cycling and litter decomposition in wetlands. Biogeochemistry 130 (3), 289–299. Available from: https://doi.org/10.1007/s10533-016-0257-x.

Encinas Fernández, J., Peeters, F., Hofmann, H., 2016. On the methane paradox: transport from shallow water zones rather than in situ methanogenesis is the major source of CH_4 in the open surface water of lakes. J. Geophys. Res.: Biogeosci. 121 (10), 2717–2726. Available from: https://doi.org/10.1002/2016JG003586.

Engstrom, D.R., Schottler, S.P., Leavitt, P.R., Havens, K.E., 2006. A reevaluation of the cultural eutrophication of Lake Okeechobee using multiproxy sediment records. Ecol. Appl. 16 (3), 1194–1206.

Entrekin, S., Evans-White, M., Johnson, B., Hagenbuch, E., 2011. Rapid expansion of natural gas development poses a threat to surface waters. Front. Ecol. Environ. 9 (9), 503–511.

Eppley, R.W., 1972. Temperature and phytoplankton growth in the sea. Fish. Bull. 70 (4), 1063–1085.

Erhard, D., Gross, E.M., 2006. Allelopathic activity of *Elodea canadensis* and *Elodea nuttallii* against epiphytes and phytoplankton. Aquat. Bot. 85 (3), 203–211.

Ernst, W., 1980. Effects of pesticides and related organic compounds in the sea. Helgoländer Meeresunters 33, 301–312.

Erős, T., Takács, P., Specziár, A., Schmera, D., Sály, P., 2017. Effect of landscape context on fish meta-community structuring in stream networks. Freshw. Biol. 62 (2), 215–228. Available from: https://doi.org/10.1111/fwb.12857.

Etnier, D.A., 1997. Jeopardized southeastern freshwater fishes: a search for causes. In: Benz, G.W., Collins, D.E. (Eds.), Aquatic Fauna in Peril, the Southeastern Perspective, vol. Special Publication 1. Southeast Aquatic Research Institute, Decatur, Georgia, pp. 87–104.

Euliss Jr., N.H., Wrubleski, D.A., Mushet, D.M., 1999. Wetlands of the prairie pothole region: invertebrate species composition, ecology, and management. In: Batzer, D.P., Rader, R.B., Wissinger, S.A. (Eds.), Invertebrates in Freshwater Wetlands of North America, Ecology and Management. John Wiley and Sons, New York, pp. 471–514.

Evanno, G., Castella, E., Antoine, C., Paillat, G., Goudet, J., 2009. Parallel changes in genetic diversity and species diversity following a natural disturbance. Mol. Ecol. 18 (6), 1137–1144.

Evans, W.C., Kling, G.W., Tuttle, M.L., Tanyileke, G., White, L.D., 1993. Gas buildup in Lake Nyos, Cameroon: the recharge process and its consequences. Appl. Geochem. 8, 207–221.

Evans, W.C., White, L.D., Tuttle, M.L., Kling, G.W., Tanyileke, G., Michel, R.L., 1994. Six years of change at Lake Nyos, Cameroon, yield clues to the past and cautions for the future. Geochem. J. 28, 139–162.

Evans-White, M.A., Dodds, W.K., Huggins, D.G., Baker, D.S., 2009. Thresholds in macroinvertebrate biodiversity and stoichiometry across water-quality gradients in Central Plains (USA) streams. J. N. Am. Benthol. Soc. 28 (4), 855–868. Available from: https://doi.org/10.1899/08-113.1.

Evans-White, M.A., Dodds, W.K., Whiles, M.R., 2003. Ecosystem significance of crayfishes and stone-rollers in a prairie stream: functional differences between co-occurring omnivores. J. N. Am. Benthol. Soc. 22, 423−441.

Evans-White, M.A., Lamberti, G.A., 2006. Stoichiometry of consumer-driven nutrient recycling across nutrient regimes in streams. Ecol. Lett. 9 (11), 1186−1197.

Evaristo, J., Jasechko, S., McDonnell, J.J., 2015. Global separation of plant transpiration from groundwater and streamflow. Nature 525 (7567), 91−94. Available from: https://doi.org/10.1038/nature14983.

Everall, N.C., Lees, D.R., 1996. The use of barley-straw to control general and blue-green algal growth in a Derbyshire Reservoir. Water Res. 30, 269−276.

Everitt, D.T., Burkholder, J.M., 1991. Seasonal dynamics of macrophyte communities from a stream flowing over granite flatrock in North Carolina, USA. Hydrobiologia 222, 159−172.

Fagan, W.F., Holmes, E.E., 2006. Quantifying the extinction vortex. Ecol. Lett. 9 (1), 51−60.

Fahnenstiel, G.L., Bridgeman, T.B., Lang, G.A., McCormick, M.J., Nalepa, T.F., 1995a. Phytoplankton productivity in Saginaw Bay, Lake Huron: effects of zebra mussel (Dreissena polymorpha) colonization. J. Great Lakes Res. 21 (4), 465−475.

Fahnenstiel, G.L., Lang, G.A., Nalepa, T.F., Johengen, T.H., 1995b. Effects of zebra mussel (Dreissena polymorpha) colonization on water quality parameters in Saginaw Bay, Lake Huron. J. Great Lakes Res. 21 (4), 435−448.

Fairchild, G.W., Sherman, J.W., 1990. Effects of liming on nutrient limitation of epilithic algae in an acid lake. Water Air Soil Pollut. 52, 133−147.

Falconer, I.R., 1999. An overview of problems caused by toxic blue-green algae (Cyanobacteria) in drinking and recreational water. Environ. Toxicol. 14, 5−12.

Falk, D.A., Richards, C.M., Montalvo, A.M., Knapp, E.E., 2006. Population and ecological genetics in restoration ecology. Found. Restor. Ecol. 14−41.

Falke, J.A., Gido, K.B., 2006. Spatial effects of reservoirs on fish assemblages in Great Plains streams in Kansas, USA. River Res. Appl. 22, 55−68.

Falkenmark, M., 1992. Water scarcity generates environmental stress and potential conflicts. In: James, W., Niemczynowicz, J. (Eds.), Water, Development and the Environment. Lewis Publishers, Ann Arbor, MI, pp. 279−294.

Famiglietti, J.S., 2014. The global groundwater crisis. Nat. Clim. Change 4 (11), 945−948. Available from: https://doi.org/10.1038/nclimate2425.

Fan, Y., Li, H., Miguez-Macho, G., 2013. Global patterns of groundwater table depth. Science 339 (6122), 940−943.

Fantin-Cruz, I., Pedrollo, O., Girard, P., Zeilhofer, P., Hamilton, S.K., 2016. Changes in river water quality caused by a diversion hydropower dam bordering the Pantanal floodplain. Hydrobiologia 768 (1), 223−238. Available from: https://doi.org/10.1007/s10750-015-2550-4.

Faoro, H., 2015. FAO Yearbook: Fisheries and Aquaculture Statistics. Food and Agriculture Organization of the United Nations, Rome.

Farrell, K.J., Creed, R.P., Brown, B.L., 2014. Preventing overexploitation in a mutualism: partner regulation in the crayfish−branchiobdellid symbiosis. Oecologia 174 (2), 501−510.

Fawell, J., Sheahan, D., James, H., Hurst, M., Scott, S., 2001. Oestrogens and oestrogenic activity in raw and treated water in Severn Trent Water. Water Res. 35 (5), 1240−1244.

Fee, E.J., 1979. A relation between lake morphometry and primary productivity and its use in interpreting whole-lake eutrophication experiments. Limnology 24.

Fee, E.J., Hecky, R.E., Regehr, G.W., Hendzel, L.L., Wilkinson, P., 1994. Effects of lake size on nutrient availability in the mixed layer during summer stratification. Can. J. Fish. Aquat. Sci. 51, 2756−2768.

Feinberg, G., 1969. Light Lasers and Light; Readings From Scientific American. W. H. Freeman & Co, San Francisco, CA, pp. 4−13.

Felip, M., Sattler, B., Psenner, R., Catalan, J., 1995. Highly active microbial communities in the ice and snow cover of high mountain lakes. Appl. Environ. Microbiol. 61 (6), 2394−2401.

Fellows, C., Clapcott, J., Udy, J., Bunn, S., Harch, B., Smith, M., et al., 2006. Benthic metabolism as an indicator of stream ecosystem health. Hydrobiologia 572 (1), 71−87.

Feminella, J.W., 1996. Comparison of benthic macroinvertebrate assemblages in small streams along a gradient of flow permanence. J. N. Am. Benthol. Soc. 15 (4), 651−669.

Fenchel, T., 1980. Suspension feeding in ciliated protozoa: functional response and particle size selection. Microb. Ecol. 6, 1−11.

Fenchel, T., Esteban, G.F., Finlay, B.J., 1997. Local versus global diversity of microorganisms: cryptic diversity of ciliated protozoa. Oikos 80, 220−225.

Fenchel, T., Finlay, B.J., 1995. Ecology and Evolution in Anoxic Worlds. Oxford University Press, Oxford, UK.

Fenchel, T., Finlay, B.J., 2004. The ubiquity of small species: patterns of local and global diversity. AIBS Bull. 54 (8), 777−784.

Ferenci, T., 1999. Regulation by nutrient limitation. Curr. Opin. Microbiol. 2, 208−213.

Fergus, C.E., Lapierre, J.F., Oliver, S.K., Skaff, N.K., Cheruvelil, K.S., Webster, K., et al., 2017. The freshwater landscape: lake, wetland, and stream abundance and connectivity at macroscales. Ecosphere 8 (8).

Feris, K., Ramsey, P., Frazar, C., Rillig, M., Moore, J., Gannon, J., et al., 2004. Seasonal dynamics of shallow-hyporheic-zone microbial community structure along a heavy-metal contamination gradient. Appl. Environ. Microbiol. 70 (4), 2323.

Fernández, N., Aguilar, J., Piña-García, C.A., Gershenson, C., 2017. Complexity of lakes in a latitudinal gradient. Ecol. Complex. 31, 1−20. Available from: https://doi.org/10.1016/j.ecocom.2017.02.002.

Ferreira, M.T., Franco, A., Catarino, L., Moreira, I., Sousa, P., 1999. Environmental factors related to the establishment of algal mats in concrete irrigation channels. Hydrobiologia 415, 163−168.

Fichter, G.S., 1988. Underwater Farming. Pineapple Press, Inc., Sarasota, FL.

Field, D.W., Reyer, A.J., Genovese, P.V., Shearer, B.D., 1991. Coastal Wetlands of the United States: An Accounting of a Valuable National Resource: A Special NOAA 20th Anniversary Report. US Department of Commerce, National Oceanic and Atmospheric Administration, Washington, DC.

Figuerola, J., Green, A.J., 2002. Dispersal of aquatic organisms by waterbirds: a review of past research and priorities for future studies. Freshw. Biol. 47 (3), 483−494.

Filippini, M., Buesing, N., Bettarel, Y., Sime-Ngando, T., Gessner, M.O., 2006. Infection paradox: high abundance but low impact of freshwater benthic viruses. Appl. Environ. Microbiol. 72 (7), 4893−4898. Available from: https://doi.org/10.1128/aem.00319-06.

Findlay, S., 2010. Stream microbial ecology. J. N. Am. Benthol. Soc. 29 (1), 170−181. Available from: https://doi.org/10.1899/09-023.1.

Findlay, S., Pace, M.L., Fischer, D.T., 1998. Response of heterotrophic planktonic bacteria to the zebra mussel invasion of the tidal freshwater Hudson River. Microb. Ecol. 36, 131−140.

Findlay, S., Sobczak, W.V., 2000. Microbial communities in hyporheic sediments. In: Jones, J.B., Molholland, P.J. (Eds.), Streams and Ground Waters. Academic Press, San Diego, CA, pp. 287−306.

Finlay, B.J., 2002. Global dispersal of free-living microbial eukaryote species. Science 296, 1061−1063.

Finlay, B.J., Esteban, G.F., 1998. Freshwater protozoa: biodiversity and ecological function. Biodiversity Conserv. 7, 1163−1186.

Finlay, B.J., Maberly, S.C., Cooper, J.I., 1997. Microbial diversity and ecosystem function. Oikos 80, 209−213.

Finlay, J.C., Doucett, R.R., McNeely, C., 2010. Tracing energy flow in stream food webs using stable isotopes of hydrogen. Freshw. Biol. 55 (5), 941−951. Available from: https://doi.org/10.1111/j.1365-2427.2009.02327.x.

Finlay, J.C., Small, G.E., Sterner, R.W., 2013. Human influences on nitrogen removal in lakes. Science 342 (6155), 247−250. Available from: https://doi.org/10.1126/science.1242575.

Finn, D.S., Khamis, K., Milner, A.M., 2013. Loss of small glaciers will diminish beta diversity in Pyrenean streams at two levels of biological organization. Glob. Ecol. Biogeogr. 22 (1), 40−51.

Finney, B.P., Gregory-Eaves, I., Sweetman, J., Douglas, M.S.V., Smol, J.P., 2000. Impacts of climatic change and fishing on Pacific Salmon abundance over the past 300 years. Science 290, 795−799.

Finstad, A.G., Helland, I.P., Ugedal, O., Hesthagen, T., Hessen, D.O., 2014. Unimodal response of fish yield to dissolved organic carbon. Ecol. Lett. 17 (1), 36–43. Available from: https://doi.org/10.1111/ele.12201.

Firth, P., Fisher, S.G. (Eds.), 1992. Global Climate Change and Freshwater Ecosystems. Springer-Verlag, New York, NY.

Fischer, R., Giebel, H.-A., Hillebrand, H., Ptacnik, R., 2017. Importance of mixotrophic bacterivory can be predicted by light and loss rates. Oikos 126 (5), 713–722. Available from: https://doi.org/10.1111/oik.03539.

Fišer, C., Pipan, T., Culver, D.C., 2014. The vertical extent of groundwater metazoans: an ecological and evolutionary perspective. Bioscience 64 (11), 971–979.

Fisher, S.G., Gray, L.J., 1983. Secondary production and organic matter processing by collector macroinvertebrates in a desert stream. Ecology 64 (5), 1217–1224.

Fisher, S.G., Grimm, N.B., 1991. Streams and disturbance: are cross-ecosystem comparisons useful? In: Cole, J., Lovett, G., Findlay, S. (Eds.), Comparative Analyses of Ecosystems. Patterns, Mechanisms, and Theories. Springer-Verlag, New York, NY, pp. 196–221.

Fisher, T.R., Doyle, R.D., Peele, E.R., 1988. Size-fractionated uptake and regeneration of ammonium and phosphate in a tropical lake. Verhein Internationale Verein Limnologie 23, 637–641.

Fitts, C.R., 2002. Groundwater science. Elsevier.

Fitzgerald, D.J., Cunliffe, D.A., Burch, M.D., 1999. Development of health alerts for cyanobacteria and related toxins in drinking water in South Australia. Environ. Toxicol. 14, 203–209.

Fitzhugh, T.W., Richter, B.D., 2004. Quenching urban thirst: growing cities and their impacts on freshwater ecosystems. AIBS Bull. 54 (8), 741–754.

Falconer, I.R., Humpage, A.R., 2006. Cyanobacterial (blue-green algal) toxins in water supplies: Cylindrospermopsins. Environ. Toxicol.: Int. J. 21 (4), 299–304.

Flecker, A.S., 1996. Ecosystem engineering by a dominant detritivore in a diverse tropical stream. Ecology 77 (6), 1845–1854.

Flecker, A.S., Townsend, C.T., 1994. Community-wide consequences of trout introduction in New Zealand streams. Ecol. Appl. 4 (4), 798–807.

Flower, R.J., Battarbee, R.W., 1983. Diatom evidence for recent acidification of two Scottish lochs. Nature 305, 130–133.

Foley, M.M., Warrick, J.A., Ritchie, A., Stevens, A.W., Shafroth, P.B., Duda, J.J., et al., 2017. Coastal habitat and biological community response to dam removal on the Elwha River. Ecol. Monogr. 87 (4), 552–577. Available from: https://doi.org/10.1002/ecm.1268.

Folino-Rorem, N.C., 2015. Phylum Cnidaria. In: Thorp, J.H., Rogers, D.C. (Eds.), fourth ed. Elsevier, pp. 159–179.

Folkers, D., 1999. Pitcher plant wetlands of the southeastern United States. In: Batzer, D.P., Rader, R.B., Wissinger, S.A. (Eds.), Invertebrates in Freshwater Wetlands of North America: Ecology and Management. John Wiley & Sons, Inc., New York, NY, pp. 247–275.

Folkerts, G.W., 1997. State and fate of the world's aquatic fauna. In: Benz, G.W., Collins, D.E. (Eds.), Aquatic Fauna in Peril, the Southeastern Perspective, vol. Special Publication 1. Southeast Aquatic Research Institute, Decatur, Georgia, pp. 1–16.

Follstad Shah, J.J., Kominoski, J.S., Ardón, M., Dodds, W.K., Gessner, M.O., Griffiths, N.A., et al., 2017. Global synthesis of the temperature sensitivity of leaf litter breakdown in streams and rivers. Glob. Change Biol. 23 (8), 3064–3075.

Folt, C.L., Burns, C.W., 1999. Biological drivers of zooplankton patchiness. Trends Ecol. Evol. 14 (8), 300–305.

Fong, D.W., Culver, D.C., 1994. Fine-scale biogeographic differences in the crustacean fauna of a cave system in West Virginia, USA. Hydrobiologia 287, 29–37.

Food_and_Agriculture_Organization, 2014. Yearbook of fishery statistics.

Forman, R.T.T., Alexander, L.E., 1998. Roads and their major ecological effects. Annu. Rev. Ecol. Syst. 29, 207–231.

Forney, L.J., Zhou, X., Brown, C.J., 2004. Molecular microbial ecology: land of the one-eyed king. Curr. Opin. Microbiol. 7 (3), 210–220.

France, K., Duffy, J., 2006. Diversity and dispersal interactively affect predictability of ecosystem function. Nature 441 (7097), 1139–1143.

France, R., 1996. Ontogenetic shift in crayfish ^{13}C as a measure of land-water ecotonal coupling. Oecologia 107, 239–242.

France, R., 1997a. Land-water linkages: influences of riparian deforestation on lake thermocline depth and possible consequences for cold stenotherms. Can. J. Fish. Aquat. Sci. 54 (6), 1299–1305.

France, R.L., 1997b. Stable carbon and nitrogen isotopic evidence for ecotonal coupling between boreal forests and fishes. Ecol. Freshw. Fish 6, 78–83.

Francoeur, S.N., Lowe, R.L., 1998. Effects of ambient ultraviolet radiation on littoral periphyton: biomass accrual and taxon-specific responses. J. Freshw. Ecol. 13, 29–37.

Frank, D.A., McNaughton, S.J., 1991. Stability increases with diversity in plant communities: empirical evidence from the 1988 Yellowstone drought. Oikos 62, 360–362.

Freckman, D., Blackburn, T.H., Brussaard, L., Hutchings, P., Palmer, M.A., Snelgrove, P.V.R., 1997. Linking biodiversity and ecosystem functioning of soils and sediments. Ambio 26 (8), 556–562.

French III, J., 1993. How well can fishes prey on zebra mussels in eastern North America? Fisheries 18 (6), 13–19.

Fretwell, S.D., 1977. The regulation of plant communities by the food chains exploiting them. Perspect. Biol. Med. Winter, 169–185.

Freyer, G., 1980. Acidity and species diversity in freshwater crustacean faunas. Freshw. Biol. 10, 41–45.

Freyer, G., 1993. Variation in acid tolerance of certain freshwater crustaceans in different natural waters. Hydrobiologia 250, 119–125.

Friedl, G., Wüest, A., 2002. Disrupting biogeochemical cycles—consequences of damming. Aquat. Sci. 64, 55–65.

Friedman, J.M., Osterkamp, W.R., Scott, M.L., Auble, G.T., 1998. Downstream effects of dams on channel geometry and bottomland vegetation: regional patterns in the Great Plains. Wetlands 18 (4), 619–633.

Frisch, D., Morton, P.K., Culver, B.W., Edlund, M.B., Jeyasingh, P.D., Weider, L.J., 2017. Paleogenetic records of Daphnia pulicaria in two North American lakes reveal the impact of cultural eutrophication. Glob. Change Biol. 23 (2), 708–718. Available from: https://doi.org/10.1111/gcb.13445.

Frissell, C.A., Liss, W.J., Warren, C.E., Hurley, M.D., 1986. A hierarchical framework for stream habitat classification: viewing streams in a watershed context. Environ. Manage. 10, 199–214.

Fritz, K.A., Kirschman, L.J., McCay, S.D., Trushenski, J.T., Warne, R.W., Whiles, M.R., 2017. Subsidies of essential nutrients from aquatic environments correlate with immune function in terrestrial consumers. Freshw. Sci. 36 (4), 893–900. Available from: https://doi.org/10.1086/694451.

Fritz, K.M., Dodds, W.K., 2005. Harshness: characterisation of intermittent stream habitat over space and time. Mar. Freshw. Res. 56 (1), 13–23. Available from: https://doi.org/10.1071/Mf04244.

Frost, P.C., Benstead, J.P., Cross, W.F., Hillebrand, H., Larson, J.H., Xenopoulos, M.A., et al., 2006. Threshold elemental ratios of carbon and phosphorus in aquatic consumers. Ecol. Lett. 9 (7), 774–779.

Frost, T.M., 1991. Porifera. In: Thorp, J.H., Covich, A.P. (Eds.), Ecology and Classification of North American Freshwater Invertebrates. Academic Press, Inc., San Diego, CA, pp. 95–120.

Fry, B., 1991. Stable isotope diagrams of freshwater food webs. Ecology 72 (6), 2293–2297.

Fry, B., Mumford, P.L., Tam, F., Fox, D.D., Warren, G.L., Havens, K.E., et al., 1999. Trophic position and individual feeding histories of fish from Lake Okeechobee, Florida. Can. J. Fish. Aquat. Sci. 56, 590–600.

Fryer, G., 2006. Evolution in ancient lakes: radiation of Tanganyikan atyid prawns and speciation of pelagic cichlid fishes in Lake Malawi. Hydrobiologia 568 (1), 131–142.

Fryer, G., Iles, T.D., 1972. The Cichlid Fishes of the Great Lakes of Africa. Their Biology and Evolution. Oliver & Boyd, Great Britain.

Fujiwara, A., Matsuhashi, S., Doi, H., Yamamoto, S., Minamoto, T., 2016. Use of environmental DNA to survey the distribution of an invasive submerged plant in ponds. Freshw. Sci. 35 (2), 748–754. Available from: https://doi.org/10.1086/685882.

Fuller, C., Evans-White, M., Entrekin, S., 2015. Growth and stoichiometry of a common aquatic detritivore respond to changes in resource stoichiometry. Oecologia 177 (3), 837–848. Available from: https://doi.org/10.1007/s00442-014-3154-9.

Fuller, M.M., Drake, J.A., 2000. Modeling the invasion process. In: Claudi, R., Leach, J.H. (Eds.), Nonindigenous Freshwater Organisms. Lewis Publishers, CRC Press, Boca Ratan, FL, pp. 411−414.

Funes, A., de Vicente, J., Cruz-Pizarro, L., Álvarez-Manzaneda, I., de Vicente, I., 2016. Magnetic microparticles as a new tool for lake restoration: a microcosm experiment for evaluating the impact on phosphorus fluxes and sedimentary phosphorus pools. Water Res. 89, 366−374. Available from: https://doi.org/10.1016/j.watres.2015.11.067.

Futter, M.N., Valinia, S., Löfgren, S., Köhler, S.J., Fölster, J., 2014. Long-term trends in water chemistry of acid-sensitive Swedish lakes show slow recovery from historic acidification. Ambio 43 (1), 77−90.

Gafner, K., Robinson, C.T., 2007. Nutrient enrichment influences the responses of stream macroinvertebrates to disturbance. J. N. Am. Benthol. Soc. 26 (1), 92−102.

Gaget, V., Humpage, A.R., Huang, Q., Monis, P., Brookes, J.D., 2017. Benthic cyanobacteria: a source of cylindrospermopsin and microcystin in Australian drinking water reservoirs. Water Res. 124, 454−464. Available from: https://doi.org/10.1016/j.watres.2017.07.073.

Galat, D.L., Fredrickson, L.H., Humburg, D.D., Bataille, K.J., et al., 1998. Flooding to restore connectivity of regulated, large-river wetlands. Bioscience 48 (9), 721−734.

Galatowitsch, S.M., van der Valk, A.G., 1995. Natural revegetation during restoration of wetlands in the southern prairie pothole region of North America. In: Wheeler, B.D., Shaw, S.C., Fojt, W.J., Robertson, R.A. (Eds.), Restoration of Temperate Wetlands. John Wiley & Sons, Chichester, United Kingdom, pp. 129−142.

Galatowitsch, S.M., Zedler, J.B., 2014. Wetland restoration. In: Batzer, D.P., Sharitz, R.R. (Eds.), Ecology of Freshwater and Estuarine Wetlands, second ed. University of California Press, California, pp. 225−259.

Galaziy, G.I., 1980. Lake Baikal's ecosystem and the problem of its preservation. Mar. Technol. Soc. J. 14, 31−38.

Gale, M., Edwards, M., Wilson, L., Greig, A., 2014. The boomerang effect: a case study of the Murray-Darling basin plan. Aust. J. Public Adm. 73 (2), 153−163.

Gallardo, B., Clavero, M., Sánchez, M.I., Vilà, M., 2016. Global ecological impacts of invasive species in aquatic ecosystems. Glob. Change Biol. 22 (1), 151−163. Available from: https://doi.org/10.1111/gcb.13004.

Gantar, M.F., 1985. The effect of heterotrophic bacteria on the growth of *Nostoc* sp. (Cyanobacterium). Arch. Hydrobiol. 103 (4), 445−452.

Gao, J., Barzel, B., Barabási, A.-L., 2016. Universal resilience patterns in complex networks. Nature 530 (7590), 307−312. Available from: https://doi.org/10.1038/nature16948. Available from: http://www.nature.com/nature/journal/v530/n7590/abs/nature16948.html#supplementary-information.

Garbett, P., 2005. An investigation into the application of floating reed bed and barley straw techniques for the remediation of eutrophic waters. Water Environ. 19 (3), 174−180.

Garcia, L.C., Ribeiro, D.B., de Oliveira Roque, F., Ochoa-Quintero, J.M., Laurance, W.F., 2017. Brazil's worst mining disaster: corporations must be compelled to pay the actual environmental costs. Ecol. Appl. 27 (1), 5−9. Available from: https://doi.org/10.1002/eap.1461.

Gardner, J.R., Fisher, T.R., Jordan, T.E., Knee, K.L., 2016. Balancing watershed nitrogen budgets: accounting for biogenic gases in streams. Biogeochemistry 127 (2), 231−253. Available from: https://doi.org/10.1007/s10533-015-0177-1.

Gardner, W.S., Cavaletto, J.F., Johengen, T.H., Johnson, J.R., Heath, R.T., Cotner Jr., J.B., 1995. Effects of the zebra mussel, *Dreissena polymorpha*, on community nitrogen dynamics in Saginaw Bay, Lake Huron. J. Great Lakes Res. 21 (4), 529−544.

Garey, J.R., McInnes, S.J., Nichols, P.B., 2008. Global diversity of tardigrades (Tardigrada) in freshwater. Hydrobiologia 595 (1), 101−106. Available from: https://doi.org/10.1007/s10750-007-9123-0.

Garrick, D.E., Hall, J.W., Dobson, A., Damania, R., Grafton, R.Q., Hope, R., et al., 2017. Valuing water for sustainable development. Science 358 (6366), 1003−1005. Available from: https://doi.org/10.1126/science.aao4942.

Garssen, A.G., Baattrup-Pedersen, A., Riis, T., Raven, B.M., Hoffman, C.C., Verhoeven, J.T.A., et al., 2017. Effects of increased flooding on riparian vegetation: field experiments simulating climate change along five European lowland streams. Glob. Change Biol. 23 (8), 3052−3063. Available from: https://doi.org/10.1111/gcb.13687.

Garvey, J.E., Whiles, M., 2016. Trophic Ecology. CRC Press, Boca Raton, FL.

Gasith, A., Gafny, S., 1990. Effects of water level on the structure and function of the littoral zone. In: Tilzer, M.M., Serruya, C. (Eds.), Large Lakes Ecological Structure and Function. Springer-Verlag, New York, NY, pp. 156−172.

Gasol, J.M., Simons, A.M., Kalff, J., 1995. Patterns in the top-down versus bottom-up regulation of heterotrophic nanoflagellates in temperate lakes. J. Plankton Res. 17 (10), 1879−1903.

Gaston, K., Blackburn, T., 2008. Pattern and Process in Macroecology. John Wiley & Sons, Chichester, United Kingdom.

Gatz, J.J., 1983. Do stream fishes forage optimally in nature? In: Fontaine, T.D.I., Bartell, S.M. (Eds.), Dynamics of Lotic Ecosystems. Ann Arbor Science Publishers, Ann Arbor, MI, pp. 391−403.

Gavrieli, I., 1997. Halite deposition from the Dead Sea: 1960−1993. In: Niemi, T.M., Ben-Avraham, Z., Gat, J.R. (Eds.), The Dead Sea. The Lake and Its Setting. Oxford University Press, New York, NY, pp. 161−183.

Gehrke, C., 1998. Effects of enhanced UV-B radiation on production-related properties of a *Sphafnum fuscum* dominated subarctic bog. Funct. Ecol. 12, 940−947.

Geist, J., 2011. Integrative freshwater ecology and biodiversity conservation. Ecol. Indic. 11 (6), 1507−1516. Available from: https://doi.org/10.1016/j.ecolind.2011.04.002.

Gelwick, F.P., Matthews, W.J., 1992. Effects of an algivorous minnow on temperate stream ecosystem properties. Ecology 73 (5), 1630−1645.

Gemmell, B.J., Sheng, J., Buskey, E.J., 2013. Compensatory escape mechanism at low Reynolds number. Proc. Natl. Acad. Sci. 110 (12), 4661−4666. Available from: https://doi.org/10.1073/pnas.1212148110.

Genkai-Kato, M., Carpenter, S.R., 2005. Eutrophication due to phosphorus recycling in relation to lake morphometry, temperature, and macrophytes. Ecology 86 (1), 210−219.

Gensemer, R.W., Playle, R.C., 1999. The bioavailability and toxicity of aluminum in aquatic environments. Crit. Rev. Environ. Sci. Technol. 29 (4), 315−450.

Gerba, C.P., 1987. Transport and fate of viruses in soils: field studies. In: Rao, V.C., Melnick, J.L. (Eds.), Human Viruses in Sediments, Sludges, and Soils. CRC Press, Boca Raton, FL, pp. 141−154.

Gerlanc, N.M., Kaufman, G.A., 2003. Use of bison wallows by anurans on Konza Prairie. Am. Midl. Nat. 150 (1), 158−168.

Gerling, A.B., Munger, Z.W., Doubek, J.P., Hamre, K.D., Gantzer, P.A., Little, J.C., et al., 2016. Whole-catchment manipulations of internal and external loading reveal the sensitivity of a Century-Old reservoir to hypoxia. Ecosystems 19 (3), 555−571. Available from: https://doi.org/10.1007/s10021-015-9951-0.

Gerritsen, J., Carlson, R.E., Dycus, D.L., Faulkner, C., Gibson, G.R., Harcum, J., et al., 1998. Lake and reservoir bioassessment and biocriteria. EPA 841-B-98-007, Washington, DC. Retrieved from https://nepis.epa.gov/Exe/ZyNET.exe/20004ODM.TXT?ZyActionD=ZyDocument&Client=EPA&Index=1995+Thru+1999&Docs=&Query=&Time=&EndTime=&SearchMethod=1&TocRestrict=n&Toc=&TocEntry=&QField=&QFieldYear=&QFieldMonth=&QFieldDay=&IntQFieldOp=0&ExtQFieldOp=0&XmlQuery=&File=D%3A%5Czyfiles%5CIndex%20Data%5C95thru99%5CTxt%5C00000012%5C20004ODM.txt&User=ANONYMOUS&Password=anonymous&SortMethod=h%7C-&MaximumDocuments=1&FuzzyDegree=0&ImageQuality=r75g8/r75g8/x150y150g16/i425&Display=hpfr&DefSeekPage=x&SearchBack=ZyActionL&Back=ZyActionS&BackDesc=Results%20page&MaximumPages=1&ZyEntry=1&SeekPage=x&ZyPURL.

Gessner, M.O., Chauvet, E., 1994. Importance of stream microfungi in controlling breakdown rates of leaf-litter. Ecology 75 (6), 1807−1817.

Gessner, M.O., Chauvet, E., 2002. A case for using litter breakdown to assess functional stream integrity. Ecol. Appl. 12 (2), 498−510.

Geyh, M., 2005. Dating of old groundwater—history, potential, limits and future. In: Aggarwal, P.K., Froehlich, K.F., Gat, J.R. (Eds.), Isotopes in the Water Cycle. Springer, pp. 221−241.

Ghiorse, W.C., 1997. Subterranean life. Science 275, 789−790.

Giberson, D., Hardwick, M.L., 1999. Pitcher plants (*Sarracenia purpurea*) in Eastern Canadian peatlands. In: Batzer, D.P., Rader, R.B., Wissinger, S.A. (Eds.), Invertebrates in Freshwater Wetlands of North America: Ecology and Management. John Wiley & Sons, Inc., New York, NY, pp. 401−422.

Gibert, J., Danielopol, D.L., Stanford, J.A. (Eds.), 1994a. Groundwater Ecology. Academic Press, San Diego, CA.

Gibert, J., Stanford, J.A., Dole-Olivier, M.-J., Ward, J.V., 1994b. Basic attributes of groundwater ecosystems and prospects for research. In: Gibert, J., Danielopol, D.L., Stanford, J.A. (Eds.), Groundwater Ecology. Academic Press, San Diego, CA, pp. 7−40.

Gibson, K.E., 2014. Viral pathogens in water: occurrence, public health impact, and available control strategies. Curr. Opin. Virol. 4, 50−57.

Gido, K., 2002. Interspecific comparisons and the potential importance of nutrient excretion by benthic fishes in a large reservoir. Trans. Am. Fish. Soc. 131 (2), 260−270.

Giersch, J.J., Hotaling, S., Kovach, R.P., Jones, L.A., Muhlfeld, C.C., 2017. Climate-induced glacier and snow loss imperils alpine stream insects. Glob. Change Biol. 23 (7), 2577−2589. Available from: https://doi.org/10.1111/gcb.13565.

Gilbert, J.J., 2018. Morphological variation and its significance in a polymorphic rotifer: environmental, endogenous, and genetic controls. Bioscience 68 (3), 169−181.

Gilbert, D., Amblard, C., Bourdier, G., Francez, A.-J., 1998. The microbial loop at the surface of a peatland: structure, function, and impact of nutrient input. Microb. Ecol. 35, 83−93.

Giles, N., 1994. Tufted duck (*Aythya fuligula*) habitat use and broad survival increases after fish removal from gravel pits. Hydrobiologia 279/280, 387−392.

Giling, D.P., Nejstgaard, J.C., Berger, S.A., Grossart, H.-P., Kirillin, G., Penske, A., et al., 2017. Thermocline deepening boosts ecosystem metabolism: evidence from a large-scale lake enclosure experiment simulating a summer storm. Glob. Change Biol. 23 (4), 1448−1462. Available from: https://doi.org/10.1111/gcb.13512.

Gillesby, B.E., Zacharewski, T.R., 1998. Exoestrogens: mechanisms of action and strategies for identification and assessment. Environ. Toxicol. Chem. 17, 3−14.

Gillespie, R.G., Howarth, F.G., Roderick, G.K., 2001. Adaptive radiation. Encycl. Biodiversity 1, 25−44.

Gillis, A.M., 1995. What's at stake in the Pacific Northwest salmon debate? Bioscience 45, 125−128.

Gillooly, J.F., Brown, J.H., West, G.B., Savage, V.M., Charnov, E.L., 2001. Effects of size and temperature on metabolic rate. Science 293 (5538), 2248−2251.

Gilmour, C.C., Riedel, G.S., Ederington, M.C., Bell, J.T., Benoit, J.M., Gill, G.A., et al., 1998. Methylmercury concentrations and production rates across a trophic gradient in the northern Everglades. Biogeochemistry 40, 327−345.

Gilpin, M.E., Hanksi, I.A., 1991. Metapopulation Dynamics: Empirical and Theoretical Investigations. Academic Press, London.

Gilpin, M.E., Soulé, M.E., 1986. Minimum viable populations: the processes of species extinctions. In: Soulé, M.E. (Ed.), Conservation Biology: The Science of Scarcity and Diversity. Sinauer Associates, Sunderland, MA, pp. 13−34.

Ginzburg, B., Chalifa, I., Gun, J., Dor, I., Hadas, O., Lev, O., 1998. DMS formation by dimethylsulfoniopropionate route in freshwater. Environ. Sci. Technol. 32 (14), 2130−2136.

Giovannoni, S.J., Turner, S., Olsen, G.J., Barns, S., Lane, D.J., Pace, N.R., 1988. Evolutionary relationships among Cyanobacteria and green chloroplasts. J. Bacteriol. 170, 3584−3592.

Giovannoni, S.J., Britschgi, T.B., Moyer, C.L., Field, K.G., 1990. Genetic diversity in Sargasso Sea bacterioplankton. Nature 345, 60−62.

Gladyshev, E., Meselson, M., Arkhipova, I., 2008. Massive horizontal gene transfer in bdelloid rotifers. Science 320 (5880), 1210.

Glagolev, A.N., 1984. Motility and Taxis in Prokaryotes. Harwood Academic Publishers, Cher, Switzerland.

Gleason, P.J., Stone, P., 1994. Age, origin, and landscape evolution of the Everglades peatland. In: Davis, S.M., Ogden, J.C. (Eds.), Everglades. St. Lucie Press, Delray Beach, FL, pp. 149−198.

Gleeson, T., Wada, Y., Bierkens, M.F.P., van Beek, L.P.H., 2012. Water balance of global aquifers revealed by groundwater footprint. Nature 488 (7410), 197−200. http://www.nature.com/nature/journal/v488/n7410/abs/nature11295.html#supplementary-information.

Gleick, P.H., 1998. Water in crisis: paths to sustainable water use. Ecol. Appl. 8 (3), 571−579.

Gleick, P.H. (2008). Water conflict chronology 11/10/08. Retrieved from http://www.worldwater.org/conflict/map/.

Gleick, P.H., Burns, C.G., Chalecki, E.L., Cohen, M., Cushing, K.K., Mann, A., et al., 2004. The World's Water 2002−2003. Island Press, Washington, DC.

Gliwicz, Z.M., 1980. Filtering rates, food size selection, and feeding rates in cladocerans—another aspect of interspecific competition in filter-feeding zooplankton. Paper Presented at the Evolution and Ecology of Zooplankton Communities, Dartmouth College.

Glud, R.N., Fenchel, T., 1999. The importance of ciliates for interstitial solute transport in benthic communities. Mar. Ecol. Prog. Ser. 186, 87−93.

Goldman, C.R., 1960. Molybdenum as a factor limiting primary productivity in Castle Lake, California. Science 132 (3433), 1016−1017.

Goldman, C.R., 1962. A method of studying nutrient limiting factors *in situ* in water columns isolated by polyethylene film. Limnol. Oceanogr. 7, 99−101.

Goldman, C.R., 1972. The role of minor nutrients in limiting the productivity of aquatic ecosystems. Paper Presented at the Symposium on Nutrients and Eutrophication.

Goldman, J.C., Caron, D.A., Dennett, M.R., 1987. Regulation of gross growth efficiency and ammonium regeneration in bacteria by substrate C:N ratio. Limnol. Oceanogr. 32 (6), 1239−1252.

Goldman, C.R., Jassby, A., Powell, T., 1989. Interannual fluctuations in primary production: meteorological forcing at two subalpine lakes. Limnol. Oceanogr. 34, 310−323.

Goldman, C.R., Jassby, A.D., Hackley, S.H., 1993. Decadal, interannual, and seasonal variability in enrichment bioassays at Lake Tahoe, California-Nevada, USA. Can. J. Fish. Aquat. Sci. 50, 1489−1496.

Goldschmidt, T., Witte, F., Wanink, J., 1993. Cascading effects of the introduced Nile Perch on the detritivorous/phytoplanktivorous species in the sublittoral areas of Lake Victoria. Conserv. Biol. 7, 686−700.

Golladay, S.W., Webster, J.R., Benfield, E.F., 1983. Factors affecting food utilization by a leaf shredding aquatic insect: leaf species and conditioning time. Holarctic Ecol. 6, 157−162.

Golterman, H.L., de Oude, N.T., 1991. Eutrophication of lakes, rivers and coastal seas. In: Hutzinger, O. (Ed.), *The Handbook of Environmental Chemistry*, vol. 5. Springer-Verlag, Berlin, pp. 79−124.

Gonzalez, J.M., Suttle, C.A., 1993. Grazing by marine nanoflagellates on viruses and virus-sized particles: ingestion and digestion. Mar. Ecol. Prog. Ser. 94, 1−10.

Goodwin, S., McPherson, J.D., McCombie, W.R., 2016. Coming of age: ten years of next-generation sequencing technologies. Nat. Rev. Genet. 17 (6), 333.

Gopal, B., Goel, U., 1993. Competition and allelopathy in aquatic plant communities. Bot. Rev. 59, 155−193.

Gordon, N.D., McMahon, T.A., Finlayson, B.L., 1992. Stream Hydrology. An Introduction for Ecologists. John Wiley & Sons Ltd, West Sussex, England.

Gore, J.A., Shields Jr., F.D., 1995. Can large rivers be restored? Bioscience 45 (3), 142−152.

Gorham, E., 1991. Northern peatlands: role in the carbon cycle and probable responses to climatic warming. Ecol. Appl. 1 (2), 182−195.

Gotelli, N.J., Taylor, C.M., 1999. Testing metapopulation models with stream-fish assemblages. Evol. Ecol. Res. 1 (7), 835−845.

Gottgens, J.F., Fortney, R.H., Meyer, J., Perry, J.E., Rood, B.E., 1998. The case of the Paraguay-Paraná waterway ("Hidrovia") and its impact on the Pantanal of Brazil: a summary report to the society of wetland scientists. Wetlands Bull. 1998, 12−18.

Goulding, M., 1980. The Fishes and the Forest. University of California Press, Berkeley.

Gounot, A.M., 1994. Microbial ecology of groundwaters. In: Gibert, J., Danielopol, D.L., Stanford, J.A. (Eds.), Groundwater Ecology. Academic Press, San Diego, CA, pp. 189−215.

Govedich, F.R., Blinn, D.W., Hevly, R.H., Keim, P.S., 1999. Cryptic radiation in erpobdellid leeches in xeric landscapes: a molecular analysis of population differentiation. Can. J. Zool. 77, 52−57.

Gozlan, R.E., Britton, J., Cowx, I., Copp, G., 2010. Current knowledge on non-native freshwater fish introductions. J. Fish. Biol. 76 (4), 751−786.

Grabicova, K., Grabic, R., Fedorova, G., Fick, J., Cerveny, D., Kolarova, J., et al., 2017. Bioaccumulation of psychoactive pharmaceuticals in fish in an effluent dominated stream. Water Res. 124, 654—662. Available from: https://doi.org/10.1016/j.watres.2017.08.018.

Grace, J., Wetzel, R., 1981. Habitat partitioning and competitive displacement in cattails (Typha): experimental field studies. Am. Nat. 463—474.

Grace, M.R., Giling, D.P., Hladyz, S., Caron, V., Thompson, R.M., Mac Nally, R., 2015. Fast processing of diel oxygen curves: estimating stream metabolism with BASE (BAyesian Single-station Estimation). Limnol. Oceanogr.: Methods 13 (3), 103—114.

Graczyk, T., Conn, D., Marcogliese, D., Graczyk, H., De Lafontaine, Y., 2003. Accumulation of human waterborne parasites by zebra mussels (*Dreissena polymorpha*) and Asian freshwater clams (*Corbicula fluminea*). Parasitol. Res. 89 (2), 107—112.

Grafe, T.U., Schöner, C.R., Kerth, G., Junaidi, A., Schöner, M.G., 2011. A novel resource—service mutualism between bats and pitcher plants. Biol. Lett. 7 (3), 436—439.

Graham, L.E., Wilcox, L.W., 2000. Algae. Prentice Hall, Upper Saddle River, NJ.

Graham, J.L., Jones, J.R., Jones, S.B., Downing, J.A., Clevenger, T.E., 2004. Environmental factors influencing microcystin distribution and concentration in the Midwestern United States. Water Res. 38 (20), 4395—4404.

Granick, S., Bae, S.C., 2008. A curious antipathy for water. Science 322 (5907), 1477—1478.

Grant, W.D., Gemmell, R.T., McGenity, T.J., 1998. Halophiles. In: Horikoshi, K., Grant, W.D. (Eds.), Extremophiles: Microbial Life in Extreme Environments. Wiley-Liss, Inc, New York, NY, pp. 93—132.

Grant, S.B., Azizian, M., Cook, P., Boano, F., Rippy, M.A., 2018. Factoring stream turbulence into global assessments of nitrogen pollution. Science 359 (6381), 1266—1269. Available from: https://doi.org/10.1126/science.aap8074.

Gratton, C., Zanden, M.J.V., 2009. Flux of aquatic insect productivity to land: comparison of lentic and lotic ecosystems. Ecology 90 (10), 2689—2699. Available from: https://doi.org/10.1890/08-1546.1.

Gray, L.J., 1989. Emergence production and export of aquatic insects from a tallgrass prairie stream. Southwest. Nat. 34, 313—318.

Gray, N.F., 1998. Acid mine drainage composition and the implications for its impact on lotic systems. Water Res. 32 (7), 2122—2134.

Green, W.J., Canfield, D.E., Shensong, Y., Chave, K.E., Ferdelman, T.G., DeLanois, G., 1993. Metal transport and release processes in Lake Vanda: the role of oxide phases. In: Green, W.J., Friedmann, E.I. (Eds.), Physical and Biogeochemical Processes in Antarctic Lakes, vol. 59. American Geophysical Union, Washington, DC, pp. 145—164.

Green, D.S., Boots, B., O'Connor, N.E., Thompson, R., 2017. Microplastics affect the ecological functioning of an important biogenic habitat. Environ. Sci. Technol. 51 (1), 68—77. Available from: https://doi.org/10.1021/acs.est.6b04496.

Greenwood, P.H., 1974. The Cichlid Fishes of Lake Victoria, East Africa: The Biology and Evolution of a Species Flock. Bulletin of the British Museum, London.

Greenwood, J.L., Clason, T.A., Lowe, R.L., Belanger, S.E., 1999. Examination of endopelic and epilithic algal community structure employing scanning electron microscopy. Freshw. Biol. 41, 821—828.

Greenwood, J.L., Rosemond, A.D., Wallace, J.B., Cross, W.F., Weyers, H.S., 2007. Nutrient stimulate leaf breakdown rates and detritivore biomass: bottom-up effects via heterotrophic pathways. Oecologia 151, 637—649.

Greer, M.J.C., Hicks, A.S., Crow, S.K., Closs, G.P., 2017. Effects of mechanical macrophyte control on suspended sediment concentrations in streams. N. Z. J. Mar. Freshw. Res. 51 (2), 254—278. Available from: https://doi.org/10.1080/00288330.2016.1210174.

Gregory, S.V., Swanson, F.J., McKee, W.A., Cummins, K.W., 1991. An ecosystem perspective of riparian zones. Bioscience 41, 540—551.

Greulich, S., Bornette, G., 1999. Competitive abilities and related strategies in four aquatic plant species from an intermediately disturbed habitat. Freshw. Biol. 41, 493—506.

Griffith, G.W., 2012. Do we need a global strategy for microbial conservation? Trends Ecol. Evol. 27 (1), 1—2.

Griffiths, N.A., Tank, J.L., Royer, T.V., Rosi, E.J., Shogren, A.J., Frauendorf, T.C., et al., 2017. Occurrence, leaching, and degradation of Cry1Ab protein fromtransgenic maize detritus in agricultural streams. Sci. Total Environ. 592, 97−105. Available from: https://doi.org/10.1016/j.scitotenv.2017.03.065.

Grimm, N.B., Petrone, D.C., 1997. Nitrogen fixation in a desert stream ecosystem. Biogeochemistry 37, 33−61.

Grippo, M.A., Hlohowskyj, I., Fox, L., Herman, B., Pothoff, J., Yoe, C., et al., 2017. Aquatic nuisance species in the great lakes and Mississippi River basin—a risk assessment in support of GLMRIS. Environ. Manage. 59 (1), 154−173. Available from: https://doi.org/10.1007/s00267-016-0770-7.

Grist, D.H., 1986. Rice. Longman, New York, NY.

Groffman, P.M., Dorsey, A.M., Mayer, P.M., 2005. N processing within the geomorphic structures in urban streams. J. N. Am. Benthol. Soc. 24 (3), 613−625.

Groffman, P.M., Baron, J.S., Blett, T., Gold, A.J., Goodman, I., Gunderson, L.H., et al., 2006. Ecological thresholds: the key to successful environmental management or an important concept with no practical application? Ecosystems 9 (1), 1−13.

Grosjean, H., Oshima, T., 2007. How nucleic acids cope with high temperature. In: Gerday, C., Glandsdorff, N. (Eds.), Physiology and Biochemistry of Extremophiles. ASM Press, Washington, DC, pp. 39−56.

Gross, E.M., 1999. Allelopathy in benthic and littoral areas: case studies on allelochemicals from benthic cyanobacteria and submersed macrophytes. In: Inderjit, K.M.M., Foy, C.L. (Eds.), Principles and Practices in Plant Ecology. CRC Press, Boca Raton, FL, pp. 179−199.

Gross, E., 2003. Allelopathy of aquatic autotrophs. CRC. Crit. Rev. Plant Sci. 22 (3), 313−339.

Gross, E.M, Hilt, S., Lombardo, P., Mulderij, G., 2007. Searching for allelopathic effects of submerged macrophytes on phytoplankton—state of the art and open questions. In: Gulati, R.D., Lammens, E., DePauw, N., Van Donk, E. (Eds.), Shallow Lakes in a Changing World. Springer, pp. 77−88.

Gsell, A.S., Scharfenberger, U., Özkundakci, D., Walters, A., Hansson, L.-A., Janssen, A.B.G., et al., 2016. Evaluating early-warning indicators of critical transitions in natural aquatic ecosystems. Proc. Natl. Acad. Sci. 113 (50), E8089−E8095. Available from: https://doi.org/10.1073/pnas.1608242113.

Guan, B.H., Yao, X., Jiang, J.H., Tian, Z.Q., An, S.Q., Gu, B.H., et al., 2009. Phosphorus removal ability of three inexpensive substrates: physicochemical properties and application. Ecol. Eng. 35 (4), 576−581.

Guégan, J., Lek, S., Oberdorff, T., 1998. Energy availability and habitat heterogeneity predict global riverine fish diversity. Nature 391 (6665), 382−384.

Guiry, M.D., 2012. How many species of algae are there? J. Phycol. 48 (5), 1057−1063. Available from: https://doi.org/10.1111/j.1529-8817.2012.01222.x.

Gulis, V., Rosemond, A.D., Suberkropp, K., Weyers, H.S., Benstead, J.P., 2004. Effects of nutrient enrichment on the decomposition of wood and associated microbial activity in streams. Freshw. Biol. 49 (11), 1437−1447.

Gumbricht, T., Roman-Cuesta, R.M., Verchot, L., Herold, M., Wittmann, F., Householder, E., et al., 2017. An expert system model for mapping tropical wetlands and peatlands reveals South America as the largest contributor. Glob. Change Biol. 23 (9), 3581−3599. Available from: https://doi.org/10.1111/gcb.13689.

Guo, F., Kainz, M.J., Sheldon, F., Bunn, S.E., 2016. The importance of high-quality algal food sources in stream food webs—current status and future perspectives. Freshw. Biol. 61 (6), 815−831. Available from: https://doi.org/10.1111/fwb.12755.

Guo, F., Bunn, S.E., Brett, M.T., Kainz, M.J., 2017. Polyunsaturated fatty acids in stream food webs—high dissimilarity among producers and consumers. Freshw. Biol. 62 (8), 1325−1334. Available from: https://doi.org/10.1111/fwb.12956.

Gurevitch, J., Hedges, L.V., 1993. Meta-analysis: combining the results of independent experiments. In: Scheiner, S.M., Gurevitch, J. (Eds.), Design and Analysis of Ecological Experiments. Chapman & Hall, New York, NY, pp. 378−426.

Gurevitch, J., Padilla, D.K., 2004. Are invasive species a major cause of extinctions? Trends Ecol. Evol. 19, 470—474.

Gurtz, M.E., Marzolf, G.R., Killingbeck, K.T., Smith, D.L., McArthur, J.V., 1988. Hydrologic and riparian influences on the import and storage of coarse particulate organic matter in a prairie stream. Can. J. Fish. Aquat. Sci. 45, 655—665.

Güsewell, S., Bailey, K.M., Roem, W.J., Bedford, B.L., 2005. Nutrient limitation and botanical diversity in wetlands: can fertilisation raise species richness? Oikos 109 (1), 71—80.

Gustafsson, Ö., Gschwend, P.M., 1997. Aquatic colloids: concepts, definitions, and current challenges. Limnol. Oceanogr. 42, 519—528.

Guthrie, M., 1989. Animals of the Surface Film. The Richmond Publishing Co. Ltd, Slough, UK.

Guy, C.S., Blankenship, H.L., Nielsen, L.A., 1996. Tagging and marking. In: Murphy, B.R., Willis, D.W. (Eds.), Fisheries Techniques, second ed. American Fisheries Society, Bethesda, MD, pp. 353—383.

Guzzo, M.M., Blanchfield, P.J., Rennie, M.D., 2017. Behavioral responses to annual temperature variation alter the dominant energy pathway, growth, and condition of a cold-water predator. Proc. Natl. Acad. Sci. 114 (37), 9912—9917. Available from: https://doi.org/10.1073/pnas.1702584114.

Haag, W.R., Warren Jr., M.L., 1999. Mantle displays of freshwater mussels elicit attacks from fish. Freshw. Biol. 42, 35—40.

Haden, G.A., Blinn, D.W., Shannon, J.P., Wilson, K.P., 1999. Driftwood: an alternative habitat for macroinvertebrates in a large desert river. Hydrobiologia 397, 179—186.

Häder, D.-P., 1997. Effects of UV radiation on phytoplankton. In: Jones, J.G. (Ed.), Advances in Microbial Ecology, vol. 15. Plenum Press, New York, NY, pp. 1—26.

Häder, D.-P., Worrest, R.C., Kumar, H.K., Smith, R.C., 1995. Effects of increased solar ultraviolet on aquatic ecosystems. Ambio 24, 174—180.

Haga, H., Nagata, T., Sakamoto, M., 1995. Size-fractionated NH_4^+ regeneration in the pelagic environments of two mesotrophic lakes. Limnol. Oceanogr. 40 (6), 1091—1099.

Hagerthey, S.E., Kerfoot, W.C., 1998. Groundwater flow influences on the biomass and nutrient ratios of epibenthic algae in a north temperate seepage lake. Limnol. Oceanogr. 43 (6), 1227—1242.

Hagiwara, A., Yamamiya, N., de Araujo, A.B., 1998. Effects of water viscosity on the population growth of the rotifer Brachionus plicatilis Müller. Hydrobiologia 387/388, 489—494.

Hairston, N.G.J., 1987. Diapause as a predator-avoidance adaptation. In: Kerfoot, W.C., Sih, A. (Eds.), Predation. Direct and Indirect Impacts on Aquatic Communities. University Press of New England, Hanover, NH, pp. 281—290.

Hairston, N.G.J., 1996. Zooplankton egg banks as biotic reservoirs in changing environments. Limnol. Oceanogr. 41 (5), 1087—1092.

Hairston, N.G.J., Brunt, R.A.V., Kearns, C.M., Engstrom, D.R., 1995. Age and survivorship of diapausing eggs in a sediment egg bank. Ecology 76 (6), 1706—1711.

Hairston, N.G.S., Smith, F.E., Slobodkin, L.B., 1960. Community structure, population control, and competition. Am. Nat. 94, 421—425.

Hakenkamp, C.C., Palmer, M.A., 2000. The ecology of hyporheic meiofauna. In: Jones, J.B., Molholland, P.J. (Eds.), Streams and Ground Waters. Academic Press, San Diego, CA, pp. 307—336.

Hakvoort, J., 1994. Absorption of Light by Surface Water. Delft University of Technology, Delft.

Halbwachs, M., Sabroux, J., Grangeon, J., Kayser, G., Tochon-Danguy, J.-C., Felix, A., et al., 2004. Degassing the "Killer Lakes" Nyos and Monoun, Cameroon. EOS Trans. Am. Geophys. Union 85. Available from: https://doi.org/10.1029/2004EO300001.

Hall, A.S., 2016. Acute artificial light diminishes central texas anuran calling behavior. Am. Midland Nat. 175 (2), 183—193. Available from: https://doi.org/10.1674/0003-0031-175.2.183.

Hall, D.J., Threlkeld, S.T., 1976. The size-efficiency hypothesis and the size structure of zooplankton communities. Annu. Rev. Ecol. Syst. 7, 177—208.

Hall, E., Maixner, F., Franklin, O., Daims, H., Richter, A., Battin, T., 2011. Linking microbial and ecosystem ecology using ecological stoichiometry: a synthesis of conceptual and empirical approaches. Ecosystems 1—13.

Hall Jr, R.O., Hotchkiss, E.R., 2017. Stream metabolism. In: Lamberti, G.A., Hauer, F.H. (Eds.), Methods in Stream Ecology: Vol. 2: Ecosystem Function. Academic Press, p. 219.

Hall, R.I., Leavitt, P.R., Dixit, A.S., Quinlan, R., Smol, J.P., 1999. Limnological succession in reservoirs: a paleolimnological comparison of two methods of reservoir formation. Can. J. Fish. Aquat. Sci. 56, 1109−1121.

Hall, R.O., 1995. Use of a stable carbon-isotope addition to trace bacterial carbon through a stream food-web. J. N. Am. Benthol. Soc. 14 (2), 269−277.

Hall, R.O.J., Meyer, J.L., 1998. The trophic significance of bacteria in an detritus-based stream food web. Ecology 79 (6), 1995−2012.

Hall, R.O.J., Peterson, B.J., Meyer, J.L., 1998. Testing of a nitrogen-cycling model of a forest stream by using a nitrogen-15 tracer addition. Ecosystems 1, 283−298.

Halling-Sorensen, B., Nielsen, S.N., Lanzky, P.F., Ingerslev, F., Lutzhoft, H.C.H., Jorgensen, S.E., 1998. Occurrence, fate and effects of pharmaceutical substances in the environment—a review. Chemosphere 36 (2), 357−394.

Halver, J.E., Hardy, R.W., 2002. Fish Nutrition. Elsevier.

Hambright, K.D., Hairston, N.G., Schaffner, W.R., Howarth, R.W., 2007. Grazer control of nitrogen fixation: synergisms in the feeding ecology of two freshwater crustaceans. Fundam. Appl. Limnol. 170 (2), 89−101.

Hamilton, S.K., 1999. Potential effects of a major navigation project (Paraguay-Paraná Hiedrovía) on inundation in the Pantanal floodplains. Regul. Rivers: Resour. Manage. 15, 289−299.

Hamilton, S.K., Lewis Jr., W.M., 1990. Basin morphology in relation to chemical and ecological characteristics of lakes on the Orinoco River floodplain, Venezuela. Arch. Hydrobiol. 119 (4), 393−425.

Hamner, W.M., Gilmer, R.W., Hamner, P.P., 1982. The physical, chemical, and biological characteristics of a stratified, saline, sulfide lake in Palau. Limnol. Oceanogr. 27 (5), 896−909.

Hampton, S.E., Galloway, A.W.E., Powers, S.M., Ozersky, T., Woo, K.H., Batt, R.D., et al., 2017. Ecology under lake ice. Ecol. Lett. 20 (1), 98−111. Available from: https://doi.org/10.1111/ele.12699.

Hamre, K.D., Gerling, A.B., Munger, Z.W., Doubek, J.P., McClure, R.P., Cottingham, K.L., et al., 2017. Spatial variation in dinoflagellate recruitment along a reservoir ecosystem continuum. J. Plankton Res. 39 (4), 715−728. Available from: https://doi.org/10.1093/plankt/fbx004.

Hanly, P.J., Mittelbach, G.G., 2017. The influence of dispersal on the realized trajectory of a pond metacommunity. Oikos 126 (9), 1269−1280. Available from: https://doi.org/10.1111/oik.03864.

Hansen, A.T., Hondzo, M., Sheng, J., Sadowsky, M.J., 2014. Microscale measurements reveal contrasting effects of photosynthesis and epiphytes on frictional drag on the surfaces of filamentous algae. Freshw. Biol. 59 (2), 312−324. Available from: https://doi.org/10.1111/fwb.12266.

Hanson, P.C., Carpenter, S.R., Cardille, J.A., Coe, M.T., Winslow, L.A., 2007. Small lakes dominate a random sample of regional lake characteristics. Freshw. Biol. 52 (5), 814−822.

Hansson, L.A., Hylander, S., Sommaruga, R., 2007. Escape from UV threats in zooplankton: a cocktail of behavior and protective pigmentation. Ecology 88 (8), 1932−1939.

Hansson, L.-A., Annadotter, H., Bergman, E., Hamrin, S.F., Jeppesen, E., Kairesalo, T., et al., 1998. Biomanipulation as an application of food-chain theory: constraints, synthesis, and recommendations for temperate lakes. Ecosystems 1, 558−574.

Hantke, B., Fleischer, P., Domany, I., Koch, M., Pleb, P., Wiendl, M., et al., 1996. P-release from DOP by phosphatase activity in comparison to P excretion by zooplankton. Studies in hardwater lakes of different trophic levels. Hydrobiologia 317, 151−162.

Hardie, L.A., 1984. Evaporites: marine or non-marine? Am. J. Sci. 284, 193−240.

Hardin, G., 1960. The competitive exclusion principle. Science 131, 1292−1297.

Harke, M.J., Davis, T.W., Watson, S.B., Gobler, C.J., 2016. Nutrient-controlled niche differentiation of western lake erie cyanobacterial populations revealed via metatranscriptomic surveys. Environ. Sci. Technol. 50 (2), 604−615. Available from: https://doi.org/10.1021/acs.est.5b03931.

Harke, M.J., Jankowiak, J.G., Morrell, B.K., Gobler, C.J., 2017. Transcriptomic responses in the bloom-forming cyanobacterium microcystis induced during exposure to zooplankton. Appl. Environ. Microbiol. 83 (5). Available from: https://doi.org/10.1128/aem.02832-16.

Harman, C.J., Ward, A.S., Ball, A., 2016. How does reach-scale stream-hyporheic transport vary with discharge? Insights from rSAS analysis of sequential tracer injections in a headwater mountain stream. Water Resour. Res. 52 (9), 7130−7150. Available from: https://doi.org/10.1002/2016WR018832.

Harris, G.P., 1986. Phytoplankton Ecology, Structure, Function and Fluctuation. Chapman and Hall, Cambridge, UK.

Harris, J.M., 1993. The presence, nature, and role of gut microflora in aquatic invertebrates: a synthesis. Microb. Ecol. 25, 195−231.

Harris, M., Tomas, W., Mourao, G., Da Silva, C., Guimaraes, E., Sonoda, F., et al., 2005. Safeguarding the pantanal wetlands: threats and conservation initiatives. Conserv. Biol. 19 (3), 714.

Harris, S.C., Martin, T.H., Cummins, K.W., 1995. A model for aquatic invertebrate response to Kissimmee River restoration. Restor. Ecol. 3 (3), 181−194.

Harrison, J.A., Maranger, R.J., Alexander, R.B., Giblin, A.E., Jacinthe, P.-A., Mayorga, E., et al., 2009. The regional and global significance of nitrogen removal in lakes and reservoirs. Biogeochemistry 93 (1), 143−157. Available from: https://doi.org/10.1007/s10533-008-9272-x.

Harrison, J.W., Smith, R.E.H., 2009. Effects of ultraviolet radiation on the productivity and composition of freshwater phytoplankton communities. Photochem. Photobiol. Sci. 8 (9), 1218−1232.

Hart, B.T., Freeman, P., McKelvie, I.D., 1992. Whole-stream phosphorus release studies: variation in uptake length with initial phosphorus concentration. Hydrobiologia 235/236, 573−584.

Hart, B.T., Jaher, B., Lawrence, I., 1999. New generation water quality guidelines for ecosystem protection. Freshw. Biol. 41, 347−359.

Hart, D.D., 1992. Community organization in streams: the importance of species interactions, physical factors, and chance. Oecologia 91, 220−228.

Hartman, K.J., Kaller, M.D., Howell, J.W., Sweka, J.A., 2005. How much do valley fills influence headwater streams? Hydrobiologia 532 (1−3), 91.

Hartman, P.E., 1983. Nitrate/nitrite ingestion and gastric cancer mortality. Environ. Mutagen. 5, 111−121.

Harvey, J.W., Wagner, B.J., 2000. Quantifying hydrologic interactions between streams and their subsurface hyporheic zones. In: Jones, J.B., Molholland, P.J. (Eds.), Streams and Ground Waters. Academic Press, San Diego, CA, pp. 3−44.

Harwell, M.A., 1998. Science and environmental decision making in south Florida. Ecol. Appl. 8 (3), 580−590.

Haselkorn, R., Buikema, W.J., 1992. Nitrogen fixation in cyanobacteria. In: Stacey, G., Burris, R. H., Evans, H.J. (Eds.), Biological Nitrogen Fixation. Chapman and Hall, New York, NY, pp. 166−190.

Haslam, S.M., 1978. River Plants: The Macrophytic Vegetation of Water Courses. Cambridge University Press, London, UK.

Hasler, A.D., Scholz, A.T., 1983. Olfactory Imprinting and Homing in Salmon. Springer-Verlag, Berlin, Germany.

Hatton, I.A., McCann, K.S., Fryxell, J.M., Davies, T.J., Smerlak, M., Sinclair, A.R.E., et al., 2015. The predator-prey power law: biomass scaling across terrestrial and aquatic biomes. Science 349 (6252). Available from: https://doi.org/10.1126/science.aac6284.

Havel, J.E., Hebert, P.D.N., 1993. Daphnia lumholtzi in North America: another exotic zooplankter. Limnol. Oceanogr. 38 (8), 1823−1827.

Haveman, S.A., Pedersen, K., 1999. Distribution and metabolic diversity of microorganisms in deep igneous rock aquifers of Finland. Geomicrobiol. J. 16, 277−294.

Havens, K.E., 1991. Fish-induced sediment resuspension: effects on phytoplankton biomass and community structure in a shallow hypereutrophic lake. J. Plankton Res. 13 (6), 1163−1176.

Havens, K.E., 1992a. Acidification effects on the algal-zooplankton interface. Can. J. Fish. Aquat. Sci. 49, 2507−2514.

Havens, K.E., 1992b. Scale and structure in natural food webs. Science 257, 1107−1109.

Havens, K.E., 1993. Effect of scale on food web structure. Science 260, 242−243.

Havens, K.E., 1995. Secondary nitrogen limitation in a subtropical lake impacted by non-point source agricultural pollution. Environ. Pollut. 89 (3), 241−246.

Havens, K.E., Aumen, N.G., James, R.T., Smith, V.H., 1996a. Rapid ecological changes in a large sub-tropical lake undergoing cultural eutrophication. Ambio 25 (3), 150–155.

Havens, K.E., Bull, L.A., Warren, G.L., Crisman, T.L., Philips, E.J., Smith, J.P., 1996b. Food web structure in a subtropical lake ecosystem. Oikos 75, 20–32.

Hawksworth, D., 2000. Freshwater and marine lichen-forming fungi. Fungal Divers. 5, 1–7.

Hayden, B., McWilliam-Hughes, S.M., Cunjak, R.A., 2016. Evidence for limited trophic transfer of allochthonous energy in temperate river food webs. Freshw. Sci. 35 (2), 544–558. Available from: https://doi.org/10.1086/686001.

Hayes, D.B., Ferreri, C.P., Taylor, W.W., 1996. Active fish capture methods. In: Murphy, B.R., Willis, D.W. (Eds.), Fisheries Techniques. American Fisheries Society, Bethesda, MD, pp. 193–220.

He, Y., Flynn, S.L., Folkerts, E.J., Zhang, Y., Ruan, D., Alessi, D.S., et al., 2017. Chemical and toxicological characterizations of hydraulic fracturing flowback and produced water. Water Res. 114, 78–87. Available from: https://doi.org/10.1016/j.watres.2017.02.027.

Healey, F.P., Stewart, W.P.D., 1973. Inorganic nutrient uptake and deficiency in algae. Crit. Rev. Microbiol. 3, 69–113.

Heard, G.W., Thomas, C.D., Hodgson, J.A., Scroggie, M.P., Ramsey, D.S.L., Clemann, N., 2015. Refugia and connectivity sustain amphibian metapopulations afflicted by disease. Ecol. Lett. 18 (8), 853–863. Available from: https://doi.org/10.1111/ele.12463.

Heath, R.T., Fahnenstiel, G.L., Gardner, W.S., Cavaletto, J.F., Hwang, S.-J., 1995. Ecosystem-level effects of zebra mussels (*Dreissena polymorpha*): an enclosure experiment in Saginaw Bay, Lake Huron. J. Great Lakes Res. 21 (4), 501–516.

Heckman, C.W., 1994. The seasonal succession of biotic communities in wetlands of the tropical wet- and dry-climatic zone: I. Physical and chemical causes and biological effects in the Pantanal of Mato Grosso, Brazil. Int. Rev. gest Hydrobiol. 79 (3), 397–421.

Hecky, R.E., Rosenberg, D.M., Campbell, P., 1994. The 25th anniversary of the experimental lakes area and the history of lake 227. Can. J. Fish. Aquat. Sci. 51, 2243–2246.

Hedin, L.O., von Fischer, J.C., Ostrom, N.E., Kennedy, B.P., Brown, M.G., Robertson, G.P., 1998. Thermodynamic constraints on nitrogen transformations and other biogeochemical processes at soil-stream interfaces. Ecology 79 (2), 684–703.

Hedin, R.S., Watzlaf, G.R., Nairn, R.W., 1994. Passive treatment of acid mine drainage with limestone. J. Environ. Qual. 23, 1338–1345.

Heffernan, J.B., Soranno, P.A., Angilletta, M.J., Buckley, L.B., Gruner, D.S., Keitt, T.H., et al., 2014. Macrosystems ecology: understanding ecological patterns and processes at continental scales. Front. Ecol. Environ. 12 (1), 5–14.

Heidinger, R.C., 1999. Stocking for sport fisheries enhancement. In: Kohler, C.C., Hubert, W.A. (Eds.), Inland Fisheries Managament in North America. American Fisheries Society, Bethesda, MD, pp. 375–401.

Heikki, M., Mikko, T., Heino, J., 2017. Environmental degradation results in contrasting changes in the assembly processes of stream bacterial and fungal communities. Oikos 126 (9), 1291–1298. Available from: https://doi.org/10.1111/oik.04133.

Hein, M., 1997. Inorganic carbon limitation of photosynthesis in lake phytoplankton. Freshw. Biol. 37, 545–552.

Heino, J., 2011. A macroecological perspective of diversity patterns in the freshwater realm. Freshw. Biol. 56 (9), 1703–1722.

Heino, J., 2013. The importance of metacommunity ecology for environmental assessment research in the freshwater realm. Biol. Rev. 88 (1), 166–178.

Heino, J., Soininen, J., Alahuhta, J., Lappalainen, J., Virtanen, R., 2015. A comparative analysis of meta-community types in the freshwater realm. Ecol. Evol. 5 (7), 1525–1537.

Heino, J., Virkkala, R., Toivonen, H., 2009. Climate change and freshwater biodiversity: detected patterns, future trends and adaptations in northern regions. Biol. Rev. 84 (1), 39–54.

Helfield, J., Naiman, R., 2006. Keystone interactions: salmon and bear in riparian forests of Alaska. Ecosystems 9 (2), 167–180.

Hemmersbach, R., Volkmann, D., Häder, D.-P., 1999. Graviorientation in protists and plants. J. Plant Physiol. 154, 1–15.

Henriksen, A., Lien, L., Traaen, T.S., Rosseland, B.O., Sevalrud, I.S., 1990. The 1000-lake survey in Norway 1986. In: Mason, B.J. (Ed.), The Surface Waters Acidification Programme. Cambridge University Press, Cambridge, pp. 199−213.

Herb, W.R., Janke, B., Mohseni, O., Stefan, H.G., 2008. Thermal pollution of streams by runoff from paved surfaces. Hydrol. Process. 22 (7), 987−999.

Herdendorf, C.E., 1990. Distribution of the world's largest lakes. In: Tilzer, M.M., Serruya, C. (Eds.), Large Lakes: Ecological Structure and Function. Springer-Verlag, New York, NY, pp. 3−38.

Hernández, D.L., Vallano, D.M., Zavaleta, E.S., Tzankova, Z., Pasari, J.R., Weiss, S., et al., 2016. Nitrogen pollution is linked to US listed species declines. Bioscience 66 (3), 213−222.

Herndl, G.J., Mullerniklas, G., Frick, J., 1993. Major role of ultraviolet-B in controlling bacterioplankton growth in the surface-layer of the ocean. Nature 361 (6414), 717−719.

Herren, C.M., Webert, K.C., Drake, M.D., Jake Vander Zanden, M., Einarsson, Á., Ives, A.R., et al., 2017. Positive feedback between chironomids and algae creates net mutualism between benthic primary consumers and producers. Ecology 98 (2), 447−455. Available from: https://doi.org/10.1002/ecy.1654.

Herrmann, J., Degerman, E., Gerhardt, A., Johansson, C., Lingdell, P., Muniz, I.P., 1993. Acid-stress effects on stream biology. Ambio 22, 298−307.

Hesse, L.W., Chaffin, G.R., Brabander, J., 1989. Missouri River mitigation: a system approach. Fisheries 14, 11−15.

Hessen, D.O., 1990. Niche overlap between herbivorous cladocerans; the role of food quality and habitat homogeneity. Hydrobiologia 190, 61−78.

Hesthagen, T., Fjellheim, A., Schartau, A.K., Wright, R.F., Saksgård, R., Rosseland, B.O., 2011. Chemical and biological recovery of Lake Saudlandsvatn, a formerly highly acidified lake in southernmost Norway, in response to decreased acid deposition. Sci. Total Environ. 409 (15), 2908−2916.

Hettyey, A., Thonhauser, K.E., Bókony, V., Penn, D.J., Hoi, H., Griggio, M., 2016. Naive tadpoles do not recognize recent invasive predatory fishes as dangerous. Ecology 97 (11), 2975−2985. Available from: https://doi.org/10.1002/ecy.1532.

Heuschele, J., Selander, E., 2014. The chemical ecology of copepods. J. Plankton Res. 36 (4), 895−913. Available from: https://doi.org/10.1093/plankt/fbu025.

Hewlett, J.D., Hibbert, A.R., 1967. Factors affecting the response of small watersheds to precipitation in humid areas. In: Sopper, W.E., Lull, H.W. (Eds.), Forest Hydrology. Pergamon Press, Oxford, pp. 275−290.

Hey, D.L., Philippi, N.S., 1995. Flood reduction through wetland restoration: the Upper Mississippi River basin and a case history. Restor. Ecol. 3, 4−17.

Hickman, C.P., Roberts, L.S., 1995. Animal Diversity. Wm.C. Brown Pub., Dubuque, IA.

Hicks, K.A., Fuzzen, M.L.M., McCann, E.K., Arlos, M.J., Bragg, L.M., Kleywegt, S., et al., 2017. Reduction of intersex in a wild fish population in response to major municipal wastewater treatment plant upgrades. Environ. Sci. Technol. 51 (3), 1811−1819. Available from: https://doi.org/10.1021/acs.est.6b05370.

Hiebert, F.K., Bennett, P.C., 1992. Microbial control of silicate weathering in organic-rich ground water. Science 258, 278−281.

Hietala, J., Reinikainen, M., Walls, R., 1995. Variation in life history responses of *Daphnia* to toxic *Microcystis aeruginosa*. J. Plankton Res. 17, 2307−2318.

Higgins, S., Vander Zanden, M., 2010. What a difference a species makes: a meta-analysis of dreissenid mussel impacts on freshwater ecosystems. Ecol. Monogr. 80 (2), 179−196.

Higgins, T., McCutchan, J., Lewis, W., 2008. Nitrogen ebullition in a Colorado plains river. Biogeochemistry 89 (3), 367−377.

Hilborn, E., Carmichael, W., Soares, R., Yuan, M., Servaites, J., Barton, H., et al., 2007. Serologic evaluation of human microcystin exposure. Environ. Toxicol.: Int. J. 22 (5), 459−463.

Hilborn, R., 1996. The development of scientific advice with incomplete information in the context of the precautionary approach. Paper Presented at the Technical Consultation on the Precautionary Approach to Capture Fisheries (Including Species Introductions), Lysekil, Sweden, June 6−13, 1995.

Hilborn, R., Maguire, J., Parma, A., Rosenberg, A., 2001. The precautionary approach and risk management: can they increase the probability of successes in fishery management? Can. J. Fish. Aquat. Sci. 58 (1), 99−107.

Hill, A.M., Lodge, D.M., 1995. Multi-trophic-level impact of sublethal interactions between bass and omnivorous crayfish. J. N. Am. Benthol. Soc. 14, 306−314.

Hill, A.R., Lymburner, D.J., 1998. Hyporheic zone chemistry and stream-subsurface exchange in two groundwater-fed streams. Can. J. Fish. Aquat. Sci. 55, 495−506.

Hill, B.H., Webster, J.R., 1983. Aquatic macrophyte contribution to the new river organic matter budget. In: Fontaine, T.D.I., Bartell, S.M. (Eds.), Dynamics of Lotic Ecosystems. Ann Arbor Science Publishers, Ann Arbor, MI, pp. 273−282.

Hill, W.R., Dimick, S.M., McNamara, A.E., Branson, C.A., 1997. No effects of ambient UV radiation detected in periphyton and grazers. Limnol. Oceanogr. 42, 769−774.

Hilsenhoff, W.L., 1991. Diversity and classification of insects and *Colembola*. In: Thorp, J.H., Covich, A.P. (Eds.), Ecology and Classification of North American Freshwater Invertebrates. Academic Press, Inc, San Diego, CA, pp. 593−664.

Hilsenhoff, W.L., 2017. An improved biotic index of organic stream pollution. The Great Lakes Entomol. 20 (1), 7.

Hilt, S., Brothers, S., Jeppesen, E., Veraart, A.J., Kosten, S., 2017. Translating regime shifts in shallow lakes into changes in ecosystem functions and services. Bioscience 67 (10), 928−936. Available from: https://doi.org/10.1093/biosci/bix106.

Hilt, S., Gross, E.M., 2008. Can allelopathically active submerged macrophytes stabilise clear-water states in shallow lakes? Basic Appl. Ecol. 9 (4), 422−432.

Hinck, J., Blazer, V., Schmitt, C., Papoulias, D., Tillitt, D., 2009. Widespread occurrence of intersex in black basses (*Micropterus* spp.) from US rivers, 1995−2004. Aquat. Toxicol. 95 (1), 60−70.

Hinck, S., Neu, T.R., Lavik, G., Mussmann, M., de Beer, D., Jonkers, H.M., 2007. Physiological adaptation of a nitrate-storing *Beggiatoa* sp. to diel cycling in a phototrophic hypersaline mat. Appl. Environ. Biol. 73 (21), 7013−7022.

Hintz, W.D., Mattes, B.M., Schuler, M.S., Jones, D.K., Stoler, A.B., Lind, L., et al., 2017. Salinization triggers a trophic cascade in experimental freshwater communities with varying food-chain length. Ecol. Appl. 27 (3), 833−844. Available from: https://doi.org/10.1002/eap.1487.

Hinz, H.L., Schwarzländer, M., Gassmann, A., Bourchier, R.S., 2014. Successes we may not have had: a retrospective analysis of selected weed biological control agents in the United States. Invasive Plant Sci. Manage. 7 (4), 565−579.

Hoagland, K.D., Roemer, S.C., Rosowski, J.R., 1982. Colonization and community structure of two periphyton assemblages, with emphasis on the diatoms (*Bacillariophyceae*). Am. J. Bot. 69, 188−213.

Hobbie, J.E., 1992. Microbial control of dissolved organic carbon in lakes: research for the future. Hydrobiologia 229, 169−180.

Hobbie, J.E., Peterson, B.J., Bettez, N., Deegan, L., O'Brien, W.J., Kling, G.W., et al., 1999. Impact of global change on the biogeochemistry and ecology of an Arctic freshwater system. Polar. Res. 18 (2), 207−214.

Hobbs, H.H.I., 1991. Decapoda. In: Thorp, J.H., Covich, A.P. (Eds.), Ecology and Classification of North American Freshwater Invertebrates. Academic Press, Inc, San Diego, CA, pp. 823−858.

Hobbs, R.J., Higgs, E., Harris, J.A., 2009. Novel ecosystems: implications for conservation and restoration. Trends Ecol. Evol. 24 (11), 599−605.

Hodgson, J.R., Hodgson, C.J., Brooks, S.M., 1991. Trophic interaction and competition between largemouth bass (*Micropterus salmoides*) and rainbow trout (*Oncorhynchus mykiss*) in a manipulated lake. Can. J. Fish. Aquat. Sci. 40, 1704−1712.

Hodoki, Y., 2005. Effects of solar ultraviolet radiation on the periphyton community in lotic systems: comparison of attached algae and bacteria during their development. Hydrobiologia 534, 193−204.

Hodson, P.V., 1975. Zinc uptake by Atlantic salmon (*Salmo salar*) exposed to a lethal concentration of zinc at 3, 11, and 19 C. J. Fish. Res. Board Can. 32 (12), 2552−2556.

Hoellein, T.J., McCormick, A.R., Hittie, J., London, M.G., Scott, J.W., Kelly, J.J., 2017. Longitudinal patterns of microplastic concentration and bacterial assemblages in surface and benthic habitats of an urban river. Freshw. Sci. 36 (3), 491−507. Available from https://doi.org/10.1086/693012.

Hoellein, T.J., Zarnoch, C.B., Bruesewitz, D.A., DeMartini, J., 2017. Contributions of freshwater mussels (Unionidae) to nutrient cycling in an urban river: filtration, recycling, storage, and removal. Biogeochemistry 135 (3), 307−324. Available from: https://doi.org/10.1007/s10533-017-0376-z.

Hoham, R.W., 1980. Unicellular chlorophytes-snow algae. In: Cox, E.R. (Ed.), Phytoflagellates: Developments in Marine Biology, vol. 2. Elsevier/North Holland, Amsterdam, Netherlands, pp. 61−84.

Holeck, K.T., Mills, E.L., MacIssac, H.J., Dochoda, M.R., Colautii, R.I., Ricciardi, A., 2004. Bridging troubled waters: biological invasions, transoceanic shipping, and the Laurential Great Lakes. Bioscience 54 (10), 919−929.

Hölker, F., Breckling, B., 2005. A spatiotemporal individual-based fish model to investigate emergent properties at the organismal and the population level. Ecol. Modell. 186 (4), 406−426.

Holland, R.E., Johengen, T.H., Beeton, A.M., 1995. Trends in nutrient concentrations in Hatchery Bay, western Lake Erie, before and after *Dreissena polymorpha*. Can. J. Fish. Aquat. Sci. 52, 1202−1209.

Holloway, M., 2000. The Killing Lakes. Sci. Am. 93−99.

Holt, G., Chesson, P., 2016. Scale-dependent community theory for streams and other linear habitats. Am. Nat. 188 (3), E59−E73. Available from: https://doi.org/10.1086/687525.

Holt, J.G., Krieg, N.R., Sneath, P.H., Staley, J.T., Williams, S.T., 1994. Bergeys Manual of Determinative Bacteriology, nineth ed. Williams and Wilkins, Baltimore, MD.

Holyoak, M., Leibold, M., 2005. Metacommunities: Spatial Dynamics and Ecological Communities. The University of Chicago Press, Chicago, IL.

Hood, G.A., Larson, D.G., 2015. Ecological engineering and aquatic connectivity: a new perspective from beaver-modified wetlands. Freshw. Biol. 60 (1), 198−208. Available from: https://doi.org/10.1111/fwb.12487.

Hooper, D., Chapin Iii, F., Ewel, J., Hector, A., Inchausti, P., Lavorel, S., et al., 2005. Effects of biodiversity on ecosystem functioning: a consensus of current knowledge. Ecol. Monogr. 75 (1), 3−35.

Hori, M., Gashagaza, M.M., Nshombo, M., Kawanabe, H., 1993. Littoral fish communities in Lake Tanganyika: irreplaceable diversity supported by intricate interactions among species. Conserv. Biol. 7, 657−666.

Horne, A.J., Goldman, C.R., 1972. Nitrogen fixation in Clear Lake, California. I. Seasonal variation and the role of heterocysts. Limnol. Oceanogr. 17, 678−692.

Horne, A.J., Goldman, C.R., 1994. Limnology, second ed. McGraw-Hill, Inc, New York, NY.

Hörnström, E., 1999. Long-term phytoplankton changes in acid and limed lakes in SW Sweden. Hydrobiologia 394, 93−102.

Horst, G.P., Sarnelle, O., White, J.D., Hamilton, S.K., Kaul, R.B., Bressie, J.D., 2014. Nitrogen availability increases the toxin quota of a harmful cyanobacterium, *Microcystis aeruginosa*. Water Res. 54, 188−198.

Hörtnagl, P., Pérez, M.T., Zeder, M., Sommaruga, R., 2010. The bacterial community composition of the surface microlayer in a high mountain lake. FEMS Microbiol. Ecol. 73 (3), 458−467. Available from: https://doi.org/10.1111/j.1574-6941.2010.00904.x.

Horton, A.A., Walton, A., Spurgeon, D.J., Lahive, E., Svendsen, C., 2017. Microplastics in freshwater and terrestrial environments: evaluating the current understanding to identify the knowledge gaps and future research priorities. Sci. Total Environ. 586, 127−141. Available from: https://doi.org/10.1016/j.scitotenv.2017.01.190.

Houlton, B.Z., Morford, S.L., Dahlgren, R.A., 2018. Convergent evidence for widespread rock nitrogen sources in Earth's surface environment. Science 360 (6384), 58−62. Available from: https://doi.org/10.1126/science.aan4399.

Howard-Williams, C., Hawes, I., 2007. Ecological processes in Antarctic inland waters: interactions between physical processes and the nitrogen cycle. Antarct. Sci. 19 (2), 205−217.

Howard-Williams, C., Schwarz, A.-M., Hawes, I., Priscu, J.C., 1998. Optical properties of the McMurdo dry valley lakes, Antarctica. In: Priscu, J.C. (Ed.), Ecosystem Dynamics in a Polar Desert, vol. 72. American Geophysical Union, Washington, DC, pp. 189−203.

Howarth, R.W., Cole, J.J., 1985. Molybdenum availability, nitrogen limitation, and phytoplankton growth in natural waters. Science 229, 653−655.

Hrycik, A.R., Almeida, L.Z., Höök, T.O., 2017. Sub-lethal effects on fish provide insight into a biologically-relevant threshold of hypoxia. Oikos 126 (3), 307−317. Available from: https://doi.org/10.1111/oik.03678.

Hu, D.L., Bush, J.W., 2005. Meniscus-climbing insects. Nature 437 (7059), 733.

Hu, D.L., Bush, J.W., 2010. The hydrodynamics of water-walking arthropods. J. Fluid. Mech. 644, 5−33.

Huang, C., Wikfeldt, K.T., Tokushima, T., Nordlund, D., Harada, Y., Bergmann, U., et al., 2009. The inhomogeneous structure of water at ambient conditions. Proc. Natl. Acad. Sci. Available from: https://doi.org/10.1073/pnas.0904743106.

Hubbell, S.P., 2001. The Unified Neutral Theory of Species Abundance and Diversity. Princeton University Press, Princeton, NJ.

Huber, R., Burggraf, S., Mayer, T., Barns, S.M., Rossnagel, P., Stetter, K.O., 1995. Isolation of a hyper-thermophilic archaeum predicted by in situ RNA analysis. Nature 376, 57−58.

Huber, V., Wagner, C., Gerten, D., Adrian, R., 2012. To bloom or not to bloom: contrasting responses of cyanobacteria to recent heat waves explained by critical thresholds of abiotic drivers. Oecologia 169 (1), 245−256. Available from: https://doi.org/10.1007/s00442-011-2186-7.

Hubert, W., Quist, M.C., 2010. Inland Fisheries Management in North America, third ed. American Fisheries Society, Bathesda, MD.

Hubert, W.A., 1996. Passive capture techniques. In: Murphy, B.R., Willis, D.W. (Eds.), Fisheries Techniques, second ed. American Fisheries Society, Bethesda, MD, pp. 157−181.

Hudson, J.J., Taylor, W.D., 1996. Measuring regeneration of dissolved phosphorus in planktonic communities. Limnol. Oceanogr. 41 (7), 1560−1565.

Hudson, J.J., Taylor, W.D., 2005. Rapid estimation of phosphate at picomolar concentrations in freshwater lakes with potential application to P-limited marine systems. Aquat. Sci. 67 (3), 316−325.

Huffaker, C., 1958. Experimental studies on predation: dispersion factors and predator-prey oscillations. Calif. Agric. 27 (14), 343−383.

Hughes, G.M., 1981. Effects of low oxygen and pollution on the respiratory systems of fish. In: Pickreing, A.D. (Ed.), Stress and Fish. Academic Press, New York, NY, pp. 212−246.

Hughes, J.M., Finn, D.S., Monaghan, M.T., Schultheis, A., Sweeney, B.W., 2014. Basic and applied uses of molecular approaches in freshwater ecology. Freshw. Sci. 33 (1), 168−171.

Hughes, R.N., 1997. Diet selection. In: Godin, J.-G.J. (Ed.), Behavioural Ecology of Teleost Fishes. Oxford University Press, New York, NY, pp. 134−162.

Hughes, S.R., Kay, P., Brown, L.E., 2012. Global synthesis and critical evaluation of pharmaceutical data sets collected from river systems. Environ. Sci. Technol. 47 (2), 661−677. Available from: https://doi.org/10.1021/es3030148.

Huisman, J., Weissing, F.J., 1999. Biodiversity of plankton by species oscillations and chaos. Nature 402, 407−410.

Humphries, P., Keckeis, H., Finlayson, B., 2014. The river wave concept: integrating river ecosystem models. Bioscience 64 (10), 870−882.

Hunsaker, C.T., Levine, D.A., 1995. Hierarchical approaches to the study of water quality in rivers. Bioscience 45 (3), 193−203.

Hunter, R.G., Faulkner, S.P., Gibson, K.A., 2008. The importance of hydrology in restoration of bottomland hardwood wetland functions. Wetlands 28 (3), 605−615.

Hunter-Cevera, J.C., 1998. The value of microbial diversity. Curr. Opin. Microbiol. 1, 278−285.

Hurd, C.L., Stevens, C.L., 1997. Flow visualization around single-and multiple-bladed seaweeds with various morphologies. J. Phycol. 33, 360−367.

Hurlbert, S.H., 1984. Pseudoreplication and the design of ecological field experiments. Ecol. Monogr. 54, 187−211.

Hurley, J.P., Krabbenhoft, D.P., Cleckner, L.B., Olson, M.L., Aiken, G.R., Rawlik Jr., P.S., 1998. System controls on the aqueous distribution of mercury in the northern Florida Everglades. Biogeochemistry 40, 293−311.

Huryn, A.D., Wallace, J.B., 2000. Life history and production of stream insects. Annu. Rev. Entomol. 45, 83−110.

Hussner, A., Stiers, I., Verhofstad, M.J.J.M., Bakker, E.S., Grutters, B.M.C., Haury, J., et al., 2017. Management and control methods of invasive alien freshwater aquatic plants: a review. Aquat. Bot. 136, 112−137. Available from: https://doi.org/10.1016/j.aquabot.2016.08.002.

Huston, M.A., 1994. Biological Diversity, the Coexistence of Species on Changing Landscapes. Cambridge University Press, Cambridge, Great Britain.

Huszar, V.L, Caraco, N.F., Roland, F., Cole, J., 2006. Nutrient-chlorophyll relationships in tropical-subtropical lakes: do temperate models fit? In: Martinelli, L.A., Howarth, R.W. (Eds.), Nitrogen Cycling in the Americas: Natural and Anthropogenic Influences and Controls. Springer, pp. 239−250.

Hutchens, E., Radajewski, S., Dumont, M.G., McDonald, I.R., Murrell, J.C., 2004. Analysis of methanotrophic bacteria in Movile Cave by stable isotope probing. Environ. Microbiol. 6 (2), 111−120.

Hutchens Jr., J.J., Wallace, J.B., Grubaugh, J.W., 2017. Transport and storage of FPOM. In: Hauer, F.R., Lamberti, G.A. (Eds.), Methods in Stream Ecology, Vol. 2: Ecosystem Function. Elsevier, New York, NY, pp. 37−53.

Hutchin, P.R., Press, M.C., Lee, J.A., Ashenden, T.W., 1995. Elevated concentrations of CO_2 may double methane emissions from mires. Glob. Change Biol. 1, 125−128.

Hutchinson, G.E., 1957. A Treatise on Limnology. Geography, Physics and Chemistry, vol. 1. John Wiley & Sons, Inc, New York, NY.

Hutchinson, G.E., 1959. Homage to Santa Rosalia or why are there so many kinds of animals? Am. Nat. 93, 145−159.

Hutchinson, G.E., 1961. The paradox of the plankton. Am. Nat. 95 (882), 137−145.

Hutchinson, G.E., 1967. A Treatise on Limnology. An Introduction to Lake Biology and Limnoplankton, vol. 2. John Wiley & Sons, Inc, New York, NY.

Hutchinson, G.E., 1975. A Treatise on Limnology. Limnological Botany, vol. 3. John Wiley & Sons, Inc, New York, NY.

Hutchinson, G.E., 1993. A Treatise on Limnology. The Zoobenthos, vol. 4. John Wiley & Sons, Inc, New York, NY.

Hutchinson, G.E., Bowen, V.T., 1947. A direct demonstration of the phosphorus cycle in a small lake. Proc. Natl. Acad. Sci. USA 33 (5), 148−153.

Hutchinson, G.E., Cowgill, U., 1970. Ianula: An account of the history and development of the Lago di Monterosi, Latium, Italy. The History of the lake: a synthesis. Trans. Am. Philos. Soc. 60 (4), 163−170.

Huttunen, J.T., Lappalainen, K.M., Saarijärvi, E., Väisänen, T., Martikainen, P.J., 2001. A novel sediment gas sampler and a subsurface gas collector used for measurement of the ebullition of methane and carbon dioxide from a eutrophied lake. Sci. Total Environ. 266 (1), 153−158.

Hyatt, T.L., Naiman, R.J., 2001. The residence time of large woody debris in the Queets River, Washington, USA. Ecol. Appl. 11 (1), 191−202.

Hynes, H.B.N., 1960. The Biology of Polluted Waters. Liverpool University Press, Liverpool, UK.

Hynes, H.B.N., 1970. The Ecology of Running Waters. University of Toronto Press, Toronto, Ontario, Canada.

Hynes, H.B.N., 1975. Edgardo Baldi memorial lecture; the stream and its valley. Verhandlungen des Internationalen Verein Limnologie 19, 1−15.

Ibelings, B.W., Bormans, M., Fastner, J., Visser, P.M., 2016. CYANOCOST special issue on cyanobacterial blooms: synopsis—a critical review of the management options for their prevention, control and mitigation. Aquat. Ecol. 50 (3), 595−605. Available from: https://doi.org/10.1007/s10452-016-9596-x.

Ibelings, B.W., Portielje, R., Lammens, E.H.R.R., Noordhuis, R., van den Berg, M.S., Joosse, W., et al., 2007. Resilience of alternative stable states during the recovery of shallow lakes from eutrophication: Lake Veluwe as a case study. Ecosystems 10 (1), 4−16.

Imhof, H.K., Laforsch, C., Wiesheu, A.C., Schmid, J., Anger, P.M., Niessner, R., et al., 2016. Pigments and plastic in limnetic ecosystems: a qualitative and quantitative study on microparticles of different size classes. Water Res. 98, 64−74. Available from: https://doi.org/10.1016/j.watres.2016.03.015.

Immerzeel, W.W., van Beek, L.P.H., Bierkens, M.F.P., 2010. Climate change will affect the Asian water towers. Science 328 (5984), 1382−1385. Available from: https://doi.org/10.1126/science.1183188.

Ingersoll, T.L., Baker, L.A., 1998. Nitrate removal in wetland microcosms. Water Res. 32 (3), 677−684.

Intergovernmental Panel on Climate Change, 2013. The physical science basis. Contribution of working group I to the fifth assessment report of the intergovernmental panel on climate change. Retrieved from: http://www.ipcc.ch/report/ar5/wg1/.

Inostroza, P.A., Vera-Escalona, I., Wicht, A.-J., Krauss, M., Brack, W., Norf, H., 2016. Anthropogenic stressors shape genetic structure: insights from a model freshwater population along a land use gradient. Environ. Sci. Technol. 50 (20), 11346−11356. Available from: https://doi.org/10.1021/acs.est.6b04629.

Ioannou, C.C., Guttal, V., Couzin, I.D., 2012. Predatory fish select for coordinated collective motion in virtual prey. Science 337 (6099), 1212−1215. Available from: https://doi.org/10.1126/science.1218919.

Iredale, R.S., McDonald, A.T., Adams, D.G., 2012. A series of experiments aimed at clarifying the mode of action of barley straw in cyanobacterial growth control. Water Res. 46 (18), 6095−6103. Available from: https://doi.org/10.1016/j.watres.2012.08.040.

Isaak, D.J., Young, M.K., Luce, C.H., Hostetler, S.W., Wenger, S.J., Peterson, E.E., et al., 2016. Slow climate velocities of mountain streams portend their role as refugia for cold-water biodiversity. Proc. Natl. Acad. Sci. 113 (16), 4374−4379. Available from: https://doi.org/10.1073/pnas.1522429113.

Ishii, S., Yan, T., Shively, D.A., Byappanahalli, M.N., Whitman, R.L., Sadowsky, M.J., 2006. *Cladophora* (Chlorophyta) spp. harbor human bacterial pathogens in nearshore water of Lake Michigan. Appl. Environ. Microbiol. 72 (7), 4545−4553.

Islam, F.S., Gault, A.G., Boothman, C., Polya, D.A., Charnock, J.M., Chatterjee, D., et al., 2004. Role of metal-reducing bacteria in arsenic release from Bengal delta sediments. Nature 430 (6995), 68.

Jack, J.D., Fang, W., Thorp, J.H., 2006. Vertical, lateral and longitudinal movement of zooplankton in a large river. Freshw. Biol. 51 (9), 1646−1654.

Jackrel, S.L., Morton, T.C., Wootton, J.T., 2016. Intraspecific leaf chemistry drives locally accelerated ecosystem function in aquatic and terrestrial communities. Ecology 97 (8), 2125−2135. Available from: https://doi.org/10.1890/15-1763.1.

Jackson, C.R., Millar, J.J., Payne, J.T., Ochs, C.A., 2014. Free-living and particle-associated bacterioplankton in large rivers of the Mississippi river basin demonstrate biogeographic patterns. Appl. Environ. Microbiol. 80 (23), 7186. Available from: https://doi.org/10.1128/aem.01844-14.

Jackson, J.K., Sweeney, B.W., Bott, T.L., Newbold, J.D., Kaplan, L.A., 1994. Transport of *Bacillus thuringiensis* var. *israelensis* and its effect on drift and benthic densities of nontarget macroinvertebrates in the susquehanna river, Northern Pennsylvania. Can. J. Fish. Aquat. Sci. 51 (2), 295−314.

Jackson, L.J., Corbett, L., Scrimgeour, G., 2015. Environmental constraints on *Didymosphenia geminata* occurrence and bloom formation in Canadian Rocky Mountain lotic systems. Can. J. Fish. Aquat. Sci. 73 (6), 964−972. Available from: https://doi.org/10.1139/cjfas-2015-0361.

Jackson, M.C., Loewen, C.J.G., Vinebrooke, R.D., Chimimba, C.T., 2016. Net effects of multiple stressors in freshwater ecosystems: a meta-analysis. Glob. Change Biol. 22 (1), 180−189. Available from: https://doi.org/10.1111/gcb.13028.

Jackson, R.B., Vengosh, A., Carey, J.W., Davies, R.J., Darrah, T.H., O'sullivan, F., et al., 2014. The environmental costs and benefits of fracking. Annu. Rev. Environ. Resour. 39, 327−362.

Jacobsen, D., 2008. Low oxygen pressure as a driving factor for the altitudinal decline in taxon richness of stream macroinvertebrates. Oecologia 154 (4), 795−807. Available from: https://doi.org/10.1007/s00442-007-0877-x.

Jacoby, J.M., Collier, D.C., Welch, E.B., Hardy, F.J., Crayton, M., 2000. Environmental factors associated with a toxic bloom of *Microcystis aeruginosa*. Can. J. Fish. Aquat. Sci. 57, 231−240.

Jacomini, A.E., Avelar, W.E.P., Martinez, A.S., Bonato, P.S., 2006. Bioaccumulation of atrazine in freshwater bivalves *Anodontites trapesialis* (Lamarck, 1819) and *Corbicula fluminea* (Muller, 1774). Arch. Environ. Contam. Toxicol. 51 (3), 387−391.

Jaffé, R., Ding, Y., Niggemann, J., Vähätalo, A.V., Stubbins, A., Spencer, R.G., et al., 2013. Global charcoal mobilization from soils via dissolution and riverine transport to the oceans. Science 340 (6130), 345−347.

Jahn, T.L., Bovee, E.C., Jahn, F.F., 1979. How to Know the Protozoa. Wm. C. Brown Company Publishers, Dubuque, IA.

Jake Vander Zanden, M., Fetzer, W.W., 2007. Global patterns of aquatic food chain length. Oikos 116 (8), 1378−1388.

James, D.A., Bothwell, M.L., Chipps, S.R., Carreiro, J., 2015. Use of phosphorus to reduce blooms of the benthic diatom *Didymosphenia geminata* in an oligotrophic stream. Freshw. Sci 34 (4), 1272−1281. Available from: https://doi.org/10.1086/683038.

James, R.T., Havens, K.E., 1996. Algal bloom probability in a large subtropical lake. Water Resour. Bull. 32 (5), 995−1006.

James, R.T., Pollman, C.D., 2011. Sediment and nutrient management solutions to improve the water quality of Lake Okeechobee. Lake Reserv. Manage. 27 (1), 28−40. Available from: https://doi.org/10.1080/07438141.2010.536618.

James, T.Y., Toledo, L.F., Rodder, D., Leite, D.D., Belasen, A.M., Betancourt-Roman, C.M., et al., 2015. Disentangling host, pathogen, and environmental determinants of a recently emerged wildlife disease: lessons from the first 15 years of amphibian chytridiomycosis research. Ecol. Evol. 5 (18), 4079−4097. Available from: https://doi.org/10.1002/ece3.1672.

Jana, B.B., 1994. Ammonification in aquatic environments: a brief review. Limnologica 24 (4), 389−413.

Janssens, L., Op de Beeck, L., Stoks, R., 2017. Stoichiometric responses to an agricultural pesticide are modified by predator cues. Environ. Sci. Technol. 51 (1), 581−588. Available from: https://doi.org/10.1021/acs.est.6b03381.

Jansson, R., Nilsson, C., Dynesius, M., Andersson, E., 2000. Effects of river regulation on river-margin vegetation: a comparison of eight boreal rivers. Ecol. Appl. 10, 203−224.

Jansson, R., Nilsson, C., Malmqvist, B., 2007. Restoring freshwater ecosystems in riverine landscapes: the roles of connectivity and recovery processes. Freshw. Biol. 52 (4), 589−596.

Jaramillo, F., Destouni, G., 2015. Local flow regulation and irrigation raise global human water consumption and footprint. Science 350 (6265), 1248−1251. Available from: https://doi.org/10.1126/science.aad1010.

Jardine, T., Pusey, B., Hamilton, S., Pettit, N., Davies, P., Douglas, M., et al., 2012. Fish mediate high food web connectivity in the lower reaches of a tropical floodplain river. Oecologia 168 (3), 829−838. Available from: https://doi.org/10.1007/s00442-011-2148-0.

Jardine, T.D., 2014. Organic matter sources and size structuring in stream invertebrate food webs across a tropical to temperate gradient. Freshw. Biol. 59 (7), 1509−1521. Available from: https://doi.org/10.1111/fwb.12362.

Jardine, T.D., Bond, N.R., Burford, M.A., Kennard, M.J., Ward, D.P., Bayliss, P., et al., 2015. Does flood rhythm drive ecosystem responses in tropical riverscapes? Ecology 96 (3), 684−692. Available from https://doi.org/10.1890/14-0991.1.

Jardine, T.D., Rayner, T.S., Pettit, N.E., Valdez, D., Ward, D.P., Lindner, G., et al., 2017. Body size drives allochthony in food webs of tropical rivers. Oecologia 183 (2), 505−517. Available from: https://doi.org/10.1007/s00442-016-3786-z.

Jardine, T.D., Woods, R., Marshall, J., Fawcett, J., Lobegeiger, J., Valdez, D., et al., 2015. Reconciling the role of organic matter pathways in aquatic food webs by measuring multiple tracers in individuals. Ecology 96 (12), 3257−3269. Available from: https://doi.org/10.1890/14-2153.1.

Jassby, A.D., Goldman, C.R., Reuter, J.E., 1995. Long-term change in Lake Tahoe (California-Nevada, U.S.A.) and its relation to atmospheric deposition of algal nutrients. Arch. Hydrobiol. 135, 1−21.

Javor, B., 1989. Hypersaline Environments. Microbiology and Biogeochemistry. Springer-Verlag, Berlin, Germany.

Jeffrey, W.H., Pledger, R.J., Aas, P., Hager, S., Corrin, R.B., Haven, R.V., et al., 1996. Diel and depth profiles of DNA photodamage in bacterioplankton exposed to ambient solar ultraviolet radiation. Mar. Ecol. Prog. Ser. 137, 283−291.

Jeffries, M., Mills, D., 1990. Freshwater Ecology: Principles and Applications. Belhaven Press, London.

Jenkins, D.G., Buikema Jr., A.L., 1998. Do similar communities develop in similar sites? A test with zooplankton structure and function. Ecol. Monogr. 68 (3), 421−443.

Jenkins, D.G., Underwood, M.O., 1998. Zooplankton may not disperse readily in wind, rain, or waterfowl. Hydrobiologia 387/388, 15−21.

Jenkins, M., 2003. Prospects for biodiversity. Science 302 (5648), 1175−1177.

Jenny, J.-P., Francus, P., Normandeau, A., Lapointe, F., Perga, M.-E., Ojala, A., et al., 2016. Global spread of hypoxia in freshwater ecosystems during the last three centuries is caused by rising local human pressure. Glob. Change Biol. 22 (4), 1481−1489. Available from: https://doi.org/10.1111/gcb.13193.

Jeppesen, E., Søndergaard, M., Jensen, J.P., Mortensen, E., Hansen, A.-M., Jørgensen, T., 1998. Cascading trophic interactions from fish to bacteria and nutrients after reduced sewage loading: an 18-year study of a shallow hypertrophic lake. Ecosystems 1, 250−267.

Jerde, C.L., Chadderton, W.L., Mahon, A.R., Renshaw, M.A., Corush, J., Budny, M.L., et al., 2013. Detection of Asian carp DNA as part of a Great Lakes basin-wide surveillance program. Can. J. Fish. Aquat. Sci. 70 (4), 522−526.

Jerde, C.L., Olds, B.P., Shogren, A.J., Andruszkiewicz, E.A., Mahon, A.R., Bolster, D., et al., 2016. Influence of stream bottom substrate on retention and transport of vertebrate environmental DNA. Environ. Sci. Technol. 50 (16), 8770−8779. Available from: https://doi.org/10.1021/acs.est.6b01761.

Jernelöv, A., 2010. The threats from oil spills: now, then, and in the future. AMBIO: J. Hum. Environ. 39 (6), 353−366.

Joabsson, A., Christensen, T.R., Wallén, B., 1999. Vascular plant controls on methane emissions from northern peatforming wetlands. Trends Ecol. Evol. 14 (10), 385−388.

Jocque, M., Vanschoenwinkel, B., Brendonck, L.U.C., 2010. Freshwater rock pools: a review of habitat characteristics, faunal diversity and conservation value. Freshw. Biol. 55 (8), 1587−1602. Available from: https://doi.org/10.1111/j.1365-2427.2010.02402.x.

Jöhnk, K.D., Huisman, J., Sharples, J., Sommeijer, B., Visser, P.M., Stroom, J.M., 2008. Summer heatwaves promote blooms of harmful cyanobacteria. Glob. Change Biol. 14 (3), 495−512.

Johns, C., 1995. Contamination of Riparian wetlands from past copper mining and smelting in the headwaters region of the Clark-Fork River, Montana, USA. J. Geochem. Explor. 52 (1−2), 193−203.

Johnson, D., 2007. Physiology and ecology of acidophilic microorganisms. In: Gerday, C., Glandsdorff, N. (Eds.), Physiology and Biochemistry of Extremophiles. ASM Press, Washington, DC, pp. 257−270.

Johnson, L., Tank, J., Dodds, W., 2009. The influence of land use on stream biofilm nutrient limitation across eight North American ecoregions. Can. J. Fish. Aquat. Sci. 66 (7), 1081−1094.

Johnson, P.T.J., Paull, S.H., 2011. The ecology and emergence of diseases in fresh waters. Freshw. Biol. 56 (4), 638−657. Available from: https://doi.org/10.1111/j.1365-2427.2010.02546.x.

Johnson, T.C., Scholz, C.A., Talbot, M.R., Kelts, K., Ricketts, R.D., Ngobi, G., et al., 1996. Late pleistocene desiccation of Lake Victoria and rapid evolution of cichlid fishes. Science 273, 1091−1092.

Johnson, W.C., 1994. Woodland expansion in the Platte River, Nebraska: patterns and causes. Ecol. Monogr. 64, 45−84.

Johnson, W.C., 1997. Equilibrium response of riparian vegetation to flow regulation in the Platte River, Nebraska. Regul. Rivers Res. Manage. 13, 403−415.

Johnston, C.A., 1991. Sediment and nutrient retention by freshwater wetlands: effects on surface water quality. Crit. Rev. Environ. Control 21 (5,6), 491−565.

Jones, H.G., 1999. The ecology of snow-covered systems: a brief overview of nutrient cycling and life in the cold. Hydrol. Process. 13, 2135−2147.

Jones, J.B.J., Holmes, R.M., 1996. Surface-subsurface interactions in stream ecosystems. Trends Ecol. Evol. 11, 239–242.

Jones, J.R., Bachmann, R.W., 1976. Prediction of phosphorus and chlorophyll levels in lakes. J. Water Pollut. Control Fed. 48 (9), 2176–2183.

Jones, R.H., Flynn, K.J., 2005. Nutritional status and diet composition affect the value of diatoms as copepod prey. Science 307 (5714), 1457–1459.

Jones, R.I., Grey, J., 2011. Biogenic methane in freshwater food webs. Freshw. Biol. 56 (2), 213–229. Available from: https://doi.org/10.1111/j.1365-2427.2010.02494.x.

Jones, S., Chiu, C., Kratz, T., Wu, J., Shade, A., McMahon, K., 2008. Typhoons initiate predictable change in aquatic bacterial communities. Limnol. Oceanogr. 53 (4), 1319–1326.

Jonsson, M., Malmqvist, B., Hoffsten, P.O., 2001. Leaf litter breakdown rates in boreal streams: does shredder species richness matter? Freshw. Biol. 46 (2), 161–171.

Jørgensen, B.B., Marais, D.J.D., 1988. Optical properties of benthic photosynthetic communities: fiber-optic studies of cyanobacterial mats. Limnol. Oceanogr. 33, 99–113.

Jouzel, J., Petit, J.R., Souchez, R., Barkov, N.I., Lipenkov, V.Y., Raynaud, D., et al., 1999. More than 200 meters of lake ice above subglacial Lake Vostok, Antarctica. Science 286, 2138–2141.

Joyce, D.A., Lunt, D.H., Bills, R., Turner, G.F., Katongo, C., Duftner, N., et al., 2005. An extant cichlid fish radiation emerged in an extinct Pleistocene lake. Nature 435 (7038), 90.

Juanes, F., 1994. What determines prey size selectivity in piscivorous fishes? In: Stouder, D.J., Fresh, K.L., Feller, R.J. (Eds.), Theory and Application in Fish Feeding Ecology. University of South Carolina Press, Columbia, pp. 80–100.

Juneau, P., Harrison, P.J., 2005. Comparison by PAM fluorometry of photosynthetic activity of nine marine phytoplankton grown under identical conditions. Photochem. Photobiol. 81 (3), 649–653.

Juniper, B.E., Robins, R.J., Joel, D.M., 1989. The Carnivorous Plants. Academic Press, London, UK.

Junk, W.J., An, S., Finlayson, C., Gopal, B., Květ, J., Mitchell, S.A., et al., 2013. Current state of knowledge regarding the world's wetlands and their future under global climate change: a synthesis. Aquat. Sci. 75 (1), 151–167.

Junk, W.J., Bayley, P.B., Sparks, R.E., 1989. The flood pulse concept in river-floodplain systems. Can. Spec. Publ. Fish. Aquat. Sci 106 (1), 110–127.

Junk, W.J., da Cunha, C.N., Wantzen, K.M., Petermann, P., Strüssmann, C., Marques, M.I., et al., 2006. Biodiversity and its conservation in the Pantanal of Mato Grosso, Brazil. Aquat. Sci. 68, 278–309.

Justic, D., Rabalais, N.N., Turner, R.E., 1995a. Stoichiometric nutrient balance and origin of coastal eutrophication. Mar. Pollut. Bull. 30, 41–46.

Justic, D., Rabalais, N.N., Turner, R.E., Dortch, Q., 1995b. Changes in nutrient structure of river-dominated coastal waters: stoichiometric nutrient balance and its consequences. Estuarine, Coastal Shelf Sci. 40, 339–356.

Jüttner, F., Backhaus, D., Matthias, U., Essers, U., Greiner, R., Mahr, B., 1995. Emissions of two- and four-stroke outboard engines—II. Impact on water quality. Water Res. 28 (8), 1983–1987.

Kadlec, R.H., 1994. Wetlands for water polishing: free water surface wetlands. In: Mitsch, W.J. (Ed.), Global Wetlands Old World and New. Elsevier, Amsterdam, The Netherlands, pp. 335–349.

Kalbus, E., Reinstorf, F., Schirmer, M., 2006. Measuring methods for groundwater—surface water interactions: a review. Hydrol. Earth Syst. Sci. 10 (6), 873–887.

Kamjunke, N., Zehrer, R.F., 1999. Direct and indirect effects of strong grazing by Daphnia galeata on bacterial production in an enclosure experiment. J. Plankton Res. 21 (6), 1175–1182.

Kamthonkiat, D., Rodfai, C., Saiwanrungkul, A., Koshimura, S., Matsuoka, M., 2011. Geoinformatics in mangrove monitoring: damage and recovery after the 2004 Indian Ocean tsunami in Phang Nga, Thailand. Nat. Hazards Earth Syst. Sci. 11 (7), 1851.

Kann, J., Smith, V.H., 1999. Estimating the probability of exceeding elevated pH values critical to fish populations in a hypereutrophic lake. Can. J. Fish. Aquat. Sci. 56, 2262–2270.

Kao-Kniffin, J., Freyre, D.S., Balser, T.C., 2010. Methane dynamics across wetland plant species. Aquat. Bot. 93 (2), 107–113.

Kapitsa, A.P., Ridley, J.K., de Q Robin, G., Siegert, M.J., Zotikov, I.A., 1996. A large deep freshwater lake beneath the ice of central East Antarctica. Nature 381, 684−686.

Karatayev, A.Y., Burlakova, L.E., Padilla, D.K., 2015. Zebra versus quagga mussels: a review of their spread, population dynamics, and ecosystem impacts. Hydrobiologia 746 (1), 97−112.

Kardinaal, W.E.A., Janse, I., Kamst-van Agterveld, M., Meima, M., Snoek, J., Mur, L.R., et al., 2007. Microcystis genotype succession in relation to microcystin concentrations in freshwater lakes. Aquat. Microb. Ecol. 48 (1), 1−12.

Kareiva, P., Marvier, M., McClure, M., 2000. Recovery and management options for spring/summer Chinook Salmon in the Columbia River Basin. Science 290, 977.

Karentz, D., Bothwell, M.L., Coffin, R.B., Hanson, A., Herndl, G.J., Kilham, S.S., et al., 1994. Impact of UV-B radiation on pelagic freshwater ecosystems: report of working group on bacteria and phytoplankton. Arch. Hydrobiol. Beih. 43, 31−69.

Karl, D.M., Bird, D.F., Björkman, K., Houlihan, T., Shackelford, R., Tupas, L., 1999. Microorganisms in the accreted ice of Lake Vostok, Antarctica. Science 286, 2144−2147.

Karlsson, J., Bergström, A.-K., Byström, P., Gudasz, C., Rodríguez, P., Hein, C., 2015. Terrestrial organic matter input suppresses biomass production in lake ecosystems. Ecology 96 (11), 2870−2876. Available from: https://doi.org/10.1890/15-0515.1.

Karr, J.R., 1981. Assessment of biotic integrity using fish communities. Fisheries 6, 21−27.

Karr, J.R., 1991. Biological integrity: a long-neglected aspect of water resource management. Ecol. Appl. 1, 66−84.

Kashian, D.R., Zuellig, R.E., Mitchell, K.A., Clements, W.H., 2007. The cost of tolerance: sensitivity of stream benthic communities to UV-B and metals. Ecol. Appl. 17 (2), 365−375.

Kaspari, M., Powers, J.S., 2016. Biogeochemistry and geographical ecology: embracing all twenty-five elements required to build organisms. Am. Nat. 188 (S1), S62−S73. Available from: https://doi.org/10.1086/687576.

Kassaye, Y.A., Skipperud, L., Einset, J., Salbu, B., 2016. Aquatic macrophytes in Ethiopian Rift Valley lakes: their trace elements concentration and use as pollution indicators. Aquat. Bot. 134, 18−25. Available from: https://doi.org/10.1016/j.aquabot.2016.06.004.

Kasza, H., Winohradnik, J., 1986. Development and structure of the Gozalkowice reservoir ecosystem VII. Hydrochemistry. Ekol. Pol. 34 (3), 365−395.

Katz, G.L., Denslow, M.W., Stromberg, J.C., 2012. The Goldilocks effect: intermittent streams sustain more plant species than those with perennial or ephemeral flow. Freshw. Biol. 57 (3), 467−480.

Kaufman, L., 1992. Catastrophic change in species-rich freshwater ecosystems. Bioscience 42 (11), 846−858.

Kaushal, S.S., Groffman, P.M., Likens, G.E., Belt, K.T., Stack, W.P., Kelly, V.R., et al., 2005. Increased salinization of fresh water in the northeastern United States. Proc. Natl. Acad. Sci. USA 102 (38), 13517−13520.

Kaushal, S.S., Likens, G.E., Utz, R.M., Pace, M.L., Grese, M., Yepsen, M., 2013. Increased river alkalinization in the Eastern US. Environ. Sci. Technol. 47 (18), 10302−10311. Available from: https://doi.org/10.1021/es401046s.

Kaye, J.P., Groffman, P.M., Grimm, N.B., Baker, L.A., Pouyat, R.V., 2006. A distinct urban biogeochemistry? Trends Ecol. Evol. 21 (4), 192−199.

Kazamia, E., Helliwell, K.E., Purton, S., Smith, A.G., 2016. How mutualisms arise in phytoplankton communities: building eco-evolutionary principles for aquatic microbes. Ecol. Lett. 19 (7), 810−822. Available from: https://doi.org/10.1111/ele.12615.

Keating, K.I., 1977. Allelopathic influence on blue-green bloom sequence in a eutrophic lake. Science 196, 885−887.

Keating, K.I., 1978. Blue-green algal inhibition of diatom growth: transition from mesotrophic to eutrophic community structure. Science 199, 971−973.

Keck, F., Lepori, F., 2012. Can we predict nutrient limitation in streams and rivers? Freshw. Biol. 57 (7), 1410−1421. Available from: https://doi.org/10.1111/j.1365-2427.2012.02802.x.

Keddy, P.A., 1989. Effects of competition from shrubs on herbaceous wetland plants: a 4-year field experiment. Can. J. Bot. 67, 708−716.

Keddy, P.A., Fraser, L.H., Solomeshch, A.I., Junk, W.J., Campbell, D.R., Arroyo, M.T.K., et al., 2009. Wet and wonderful: the world's largest wetlands are conservation priorities. Bioscience 59 (1), 39−51. Available from: https://doi.org/10.1525/bio.2009.59.1.8.

Keddy, P.A., Twolan-Strutt, L., Wisheu, I.C., 1994. Competitive effect and response rankings in 20 wetland plants: are they consistent across three environments? J. Ecol. 82, 635−643.

Keeler, B.L., Polasky, S., Brauman, K.A., Johnson, K.A., Finlay, J.C., O'Neill, A., et al., 2012. Linking water quality and well-being for improved assessment and valuation of ecosystem services. Proc. Natl. Acad. Sci. 109 (45), 18619−18624. Available from: https://doi.org/10.1073/pnas.1215991109.

Keeling, P.J., Burger, G., Durnford, D.G., Lang, B.F., Lee, R.W., Pearlman, R.E., et al., 2005. The tree of eukaryotes. Trends Ecol. Evol. 20 (12), 670−676. Available from: https://doi.org/10.1016/j.tree.2005.09.005.

Keenan, C.W., Wood, J.H., 1971. General College Chemistry, fourth ed. Harper & Row, Publishers, New York, NY.

Keenleyside, M.H.A., 1991. Parental care. In: Keenleyside, M.H.A. (Ed.), Cichlid Fishes, Behaviour, Ecology and Evolution. Chapman and Hall, Cambridge, Great Britain, pp. 191−208.

Kehew, A.E, Lord, M.L., 1987. Glacial-lake outbursts along the mid-continent margins of the Laurentide ice-sheet. In: Burr, D.M., Carling, P.A., Baker, V.R. (Eds.), Megaflooding on Earth and Mars. Allen & Unwin, Boston, MA, pp. 95−120. (vol. Symposia in Geomorphology).

Kehoe, M.J., Chun, K.P., Baulch, H.M., 2015. Who smells? Forecasting taste and odor in a drinking water reservoir. Environ. Sci. Technol. 49 (18), 10984−10992. Available from: https://doi.org/10.1021/acs.est.5b00979.

Kehoe, T., 1997. Cleaning up the Great Lakes. Northern Illinois University Press, Dekalb, IL.

Kelly, B.C., Ikonomou, M.G., Blair, J.D., Morin, A.E., Gobas, F.A., 2007. Food web-specific biomagnification of persistent organic pollutants. Science 317 (5835), 236−239.

Kelson, S.J., Kapuscinski, A.R., Timmins, D., Ardren, W.R., 2015. Fine-scale genetic structure of brook trout in a dendritic stream network. Conserv. Genet. 16 (1), 31−42.

Kenefick, S.L., Hrudey, S.E., Peterson, H.G., Prepas, E.E., 1993. Toxin release from *Microcystis aeruginosa* after chemical treatment. Water Sci. Technol. 27 (3−4), 433−440.

Kennedy, M.P., Lang, P., Tapia Grimaldo, J., Varandas Martins, S., Bruce, A., Lowe, S., et al., 2016. The zambian macrophyte trophic ranking scheme, ZMTR: a new biomonitoring protocol to assess the trophic status of tropical southern African rivers. Aquat. Bot. 131, 15−27. Available from: https://doi.org/10.1016/j.aquabot.2016.01.006.

Kent, G., 1987. Fish, Food and Hunger. Westview Press, Boulder, CO.

Kerfoot, W.C., Newman, R.M., Hanscom III, Z., 1998. Snail reaction to watercress leaf tissues: reinterpretation of a mutualistic 'alarm' hypothesis. Freshw. Biol. 40, 201−213.

Keshri, J., Pradeep Ram, A.S., Colombet, J., Perriere, F., Thouvenot, A., Sime-Ngando, T., 2017. Differential impact of lytic viruses on the taxonomical resolution of freshwater bacterioplankton community structure. Water Res. 124, 129−138. Available from: https://doi.org/10.1016/j.watres.2017.07.053.

Khomich, M., Kauserud, H., Logares, R., Rasconi, S., Andersen, T., 2017. Planktonic protistan communities in lakes along a large-scale environmental gradient. FEMS Microbiol. Ecol. 93 (4). Available from: https://doi.org/10.1093/femsec/fiw231.

Kidd, K., Schindler, A.D.W., Muir, D.C.G., Lockhart, W.L., Hesslein, R.H., 1995. High concentrations of toxaphene in fishes from a subarctic lake. Science 269, 240−242.

Kieft, T.L., Murphy, E.M., Haldeman, D.L., Amy, P.S., Bjornstad, B.N., McDonald, E.V., et al., 1998. Microbial transport, survival, and succession in a sequence of buried sediments. Microb. Ecol. 36, 336−348.

Kieft, T.L., Phelps, T.J., 1997. Life in the slow lane: activities of microorganmisms in the subsurface. In: Amy, P.S., Haldeman, D.L. (Eds.), Microbiology of the Terrestrial Deep Subsurface. Lewis Publishers, Boca Raton, FL, pp. 137−163.

Kiesecker, J.M., 2002. Synergism between trematode infection and pesticide exposure: a link to amphibian limb deformities in nature? Proc. Natl. Acad. Sci. USA 99, 9900−9904.

Kiesecker, J.M., Blaustein, A.R., Belden, L.K., 2001. Complex causes of amphibian population declines. Nature 410 (6829), 681–684.

Kilham, P., Kilham, S.S., 1990. Endless summer: internal loading processes dominate nutrient cycling in tropical lakes. Freshw. Biol. 23, 379–389.

Kilham, S.S., Theriot, E.C., Fritz, S.C., 1996. Linking planktonic diatoms and climate change in the large lakes of the Yellowstone ecosystem using resource theory. Limnol. Oceanogr. 41, 1052–1062.

Kilroy, C., Biggs, B.J.F., Vyverman, W., 2007. Rules for macroorganisms applied to microorganisms: patterns of endemism in benthic freshwater diatoms. Oikos 116, 550–564.

Kilroy, C., Larned, S.T., 2016. Contrasting effects of low-level phosphorus and nitrogen enrichment on growth of the mat-forming alga *Didymosphenia geminata* in an oligotrophic river. Freshw. Biol. 61 (9), 1550–1567. Available from: https://doi.org/10.1111/fwb.12798.

King, J.L., Simovich, M.A., Brusca, R.C., 1996. Species richness, endemism and ecology of crustacean assemblages in northern California vernal pools. Hydrobiologia 328, 85–116.

King, R.S., Baker, M.E., Kazyak, P.F., Weller, D.E., 2010. How novel is too novel? Stream community thresholds at exceptionally low levels of catchment urbanization. Ecol. Appl. 21 (5), 1659–1678. Available from: https://doi.org/10.1890/10-1357.1.

Kinney, C.A., Furlong, E.T., Kolpin, D.W., Burkhardt, M.R., Zaugg, S.D., Werner, S.L., et al., 2008. Bioaccumulation of pharmaceuticals and other anthropogenic waste indicators in earthworms from agricultural soil amended with biosolid or swine manure. Environ. Sci. Technol. 42 (6), 1863–1870.

Kinzie, R.A.I., Banaszak, A.T., Lesser, M.P., 1998. Effects of ultraviolet radiation on primary productivity in a high altitude tropical lake. Hydrobiologia 385, 23–32.

Kiørboe, T., Saiz, E., Visser, A., 1999. Hydrodynamic signal perception in the copepod *Acartia tonsa*. Mar. Ecol. Prog. Ser. 179, 97–111.

Kiørboe, T., Visser, A.W., 1999. Predator and prey perception in copepods due to hydromechanical signals. Mar. Ecol. Prog. Ser. 179, 81–95.

Kirk, J.T.O., 1994. Light and Photosynthesis in Aquatic Ecosystems, second ed. Cambridge University Press, Cambridge, UK.

Kirk, K.L., Gilbert, J.J., 1990. Suspended clay and the population dynamics of planktonic rotifers and cladocerans. Ecology 71 (5), 1741–1755.

Kirschke, S., Bousquet, P., Ciais, P., Saunois, M., Canadell, J.G., Dlugokencky, E.J., et al., 2013. Three decades of global methane sources and sinks. Nat. Geosci. 6, 813. Available from: https://doi.org/10.1038/ngeo1955. Available from: https://www.nature.com/articles/ngeo1955#supplementary-information.

Kitchell, J.F., Carpenter, S.R., 1992. Summary: accomplishments and new directions of food web management in Lake Mendota. In: Kitchell, J.F. (Ed.), Food Web Management. A Case Study of Lake Mendota. Springer-Verlag, New York, NY, pp. 539–544.

Klaminder, J., Brodin, T., Sundelin, A., Anderson, N.J., Fahlman, J., Jonsson, M., et al., 2015. Long-term persistence of an anxiolytic drug (Oxazepam) in a large freshwater lake. Environ. Sci. Technol. 49 (17), 10406–10412. Available from: https://doi.org/10.1021/acs.est.5b01968.

Klann, A., Levy, G., Lutz, I., Müller, C., Kloas, W., Hildebrandt, J.-P., 2005. Estrogen-like effects of ultraviolet screen 3-(4-methylbenzylidene)-camphor (Eusolex 6300) on cell proliferation and gene induction in mammalian and amphibian cells. Environ. Res. 97 (3), 274–281.

Kleiber, M., 1932. Body size and metabolism. ENE 1, E9.

Klemer, A.R., Cullen, J.J., Mageau, M.T., Hanson, K.M., Sundell, R.A., 1996. Cyanobacterial buoyancy regulation: the paradoxical roles of carbon. J. Phycol. 32, 47–53.

Kleyheeg, E., van Leeuwen, C.H.A., 2015. Regurgitation by waterfowl: an overlooked mechanism for long-distance dispersal of wetland plant seeds. Aquat. Bot. 127, 1–5. Available from: https://doi.org/10.1016/j.aquabot.2015.06.009.

Kline, T.C.J., Goering, J.J., Mathisen, O.A., Poe, P.H., 1990. Recycling of elements transported upstream by runs of Pacific salmon I. ^{15}N and ^{13}C evidence in Sashin Creek, Southeastern Alaska. Can. J. Fish. Aquat. Sci. 47, 136–144.

Kling, G.W., 1987. Seasonal mixing and catastrophic degassing in tropical lakes, Cameroon, West Africa. Science 237, 1022–1024.

Kling, G.W., Clark, M.A., Compton, H.R., Devine, J.D., Evans, W.C., Humphrey, A.M., et al., 1987. The 1986 Lake Nyos gas disaster in Cameroon, West Africa. Science 236, 175−179.

Klose, K., Cooper, S.D., Bennett, D.M., 2015. Effects of wildfire on stream algal abundance, community structure, and nutrient limitation. Freshw. Sci. 34 (4), 1494−1509. Available from: https://doi.org/10.1086/683431.

Knapp, A.K., Blair, J.M., Briggs, J.M., Collins, S.L., Hartnett, D.C., Johnson, L.C., et al., 1999. The keystone role of bison in North American tallgrass prairie: bison increase habitat heterogeneity and alter a broad array of plant, community, and ecosystem processes. Bioscience 49 (1), 39−50.

Knight, T.M., McCoy, M.W., Chase, J.M., McCoy, K.A., Holt, R.D., 2005. Trophic cascades across ecosystems. Nature 437 (7060), 880−883.

Knowlton, M.F., Jones, J.R., 1997. Trophic status of Missouri River floodplain lakes in relation to basin type and connectivity. Wetlands 17 (4), 468−475.

Ko, C.-Y., Lai, C.-C., Hsu, H.-H., Shiah, F.-K., 2017. Decadal phytoplankton dynamics in response to episodic climatic disturbances in a subtropical deep freshwater ecosystem. Water Res. 109, 102−113. Available from: https://doi.org/10.1016/j.watres.2016.11.011.

Koelmans, A.A., Bakir, A., Burton, G.A., Janssen, C.R., 2016. Microplastic as a vector for chemicals in the aquatic environment: critical review and model-supported reinterpretation of empirical studies. Environ. Sci. Technol. 50 (7), 3315−3326. Available from: https://doi.org/10.1021/acs.est.5b06069.

Koerselman, W., Bakker, S.A., Blom, M., 1990. Nitrogen, phosphorus and potassium budgets for two small fens surrounded by heavily fertilized pastures. J. Ecol. 78, 428−442.

Koh, J.-S., Yang, E., Jung, G.-P., Jung, S.-P., Son, J.H., Lee, S.-I., et al., 2015. Jumping on water: surface tension dominated jumping of water striders and robotic insects. Science 349 (6247), 517−521. Available from https://doi.org/10.1126/science.aab1637.

Kolar, C.S., Lodge, D.M., 2001. Progress in invasion biology: predicting invaders. Trends Ecol. Evol. 16 (4), 199−204.

Kolar, C.S., Wahl, D.H., 1998. Daphnid morphology deters fish predators. Oecologia 116, 556−564.

Kolasa, J., 1991. Flatworms: *Turbellaria* and *Nemertea*. In: Thorp, J.H., Covich, A.P. (Eds.), Ecology and Classification of North American Freshwater Invertebrates. Academic Press, Inc., San Diego, CA, pp. 145−172.

Kolmonen, E., Haukka, K., Rantala-Ylinen, A., Rajaniemi-Wacklin, P., Lepistö, L., Sivonen, K., 2011. Bacterioplankton community composition in 67 Finnish lakes differs according to trophic status. Aquat. Microb. Ecol. 62 (3), 241−250.

Kolpin, D.W., Furlong, E.T., Meyer, M.T., Thurman, E.M., Zaugg, S.D., Barber, L.B., et al., 2002. Pharmaceuticals, hormones, and other organic wastewater contaminants in US streams, 1999-2000: a national reconnaissance. Environ. Sci. Technol. 36 (6), 1202−1211.

Kondolf, G.M., Boulton, A.J., O'Daniel, S., Poole, G.C., Rahel, F.J., Stanley, E.H., et al., 2006. Process-based ecological river restoration: visualizing three-dimensional connectivity and dynamic vectors to recover lost linkages. Ecol. Soc. 11 (2).

Kondolf, G.M., Gao, Y., Annandale, G.W., Morris, G.L., Jiang, E., Zhang, J., et al., 2014. Sustainable sediment management in reservoirs and regulated rivers: experiences from five continents. Earth's Future 2 (5), 256−280.

Konikow, L.F., Kendy, E., 2005. Groundwater depletion: a global problem. Hydrogeol. J. 13 (1), 317−320.

Konstantinidis, K.T., Ramette, A., Tiedje, J.M., 2006. The bacterial species definition in the genomic era. Philos. Trans.: Biol. Sci. 361 (1475), 1929−1940.

Koopman, M., Carstens, B., 2011. The microbial biogeography of the carnivorous plant *Sarracenia alata*. Microb. Ecol. 61 (4), 750−758. Available from: https://doi.org/10.1007/s00248-011-9832-9.

Kopf, R.K., Finlayson, C.M., Humphries, P., Sims, N.C., Hladyz, S., 2015. Anthropocene baselines: assessing change and managing biodiversity in human-dominated aquatic ecosystems. Bioscience 65 (8), 798−811.

Koprivnjak, J.-F., Blanchette, J.G., Bourbonniere, R.A., Clair, T.A., Heyes, A., Lum, K.R., et al., 1995. The underestimation of concentrations of dissolved organic carbon in freshwaters. Water Res. 29, 91−94.

Koshland, D.E.J., 1980. Bacterial Chemotaxis as a Model Behavioral System. Raven Press Books, Ltd, New York, NY.

Kosten, S., Huszar, V.L.M., Bécares, E., Costa, L.S., van Donk, E., Hansson, L.-A., et al., 2012. Warmer climates boost cyanobacterial dominance in shallow lakes. Glob. Change Biol. 18 (1), 118−126. Available from: https://doi.org/10.1111/j.1365-2486.2011.02488.x.

Kota, S., Borden, R.C., Barlaz, M.A., 1999. Influence of protozoan grazing on contaminant biodegradation. FEMS Microbiol. Ecol. 29, 179−189.

Kotak, B.G., Hrudey, S.E., Kenefick, S.L., Prepas, E.E., 1993. Toxicity of cyanobacterial blooms in Alberta lakes. Paper Presented at the Proceedings of the 19th Annual Aquatic Toxicity Workshop, Edmonton, Alberta, Canada, October 4−7, 1992.

Kotelnikova, S., Pedersen, K., 1998. Distribution and activity of methanogens and homoacetogens in deep granitic aquifers at Äspö Hard Rock Laboratory, Sweden. FEMS Microbiol. Ecol. 26, 121−134.

Koussoroplis, A.M., Lemarchand, C., Bec, A., Desvilettes, C., Amblard, C., Fournier, C., et al., 2008. From aquatic to terrestrial food webs: decrease of the docosahexaenoic acid/linoleic acid ratio. Lipids 43 (5), 461−466.

Kovalak, W.P., Longton, G.D., Smithee, R.D., 1993. Infestation of power plant water systems by the zebra mussel (*Dreissena polymorpha* Pallas). In: Nalepa, T.F., Schloesser, D.W. (Eds.), Zebra Mussels. Biology, Impacts, and Control. Lewis Publishers, Boca Raton, FL, pp. 359−379.

Kovalenko, K.E., Reavie, E.D., Allan, J.D., Cai, M., Smith, S.D.P., Johnson, L.B., 2017. Pelagic phytoplankton community change-points across nutrient gradients and in response to invasive mussels. Freshw. Biol. 62 (2), 366−381. Available from: https://doi.org/10.1111/fwb.12873.

Koven, C.D., Ringeval, B., Friedlingstein, P., Ciais, P., Cadule, P., Khvorostyanov, D., et al., 2011. Permafrost carbon-climate feedbacks accelerate global warming. Proc. Natl. Acad. Sci. 108 (36), 14769−14774. Available from: https://doi.org/10.1073/pnas.1103910108.

Kozhov, M., 1963. Lake Baikal and Its Life. Dr. W. Junk, Publishers, The Hague.

Kratz, T.K., Frost, T.M., 2000. The ecological organisation of lake districts: general introduction. Freshw. Biol. 43, 297−299.

Kratz, T.K., Webster, K.E., Bowser, C.J., Magnuson, J.J., Benson, B.J., 1997. The influence of landscape position on lakes in northern Wisconsin. Freshw. Biol. 37, 209−217.

Kremer, C.T., Thomas, M.K., Litchman, E., 2017. Temperature-and size-scaling of phytoplankton population growth rates: reconciling the Eppley curve and the metabolic theory of ecology. Limnol. Oceanogr. 62 (4), 1658−1670.

Krienitz, L., Ballot, A., Kotut, K., Wiegand, C., Pütz, S., Metcalf, J.S., et al., 2003. Contribution of hot spring cyanobacteria to the mysterious deaths of Lesser Flamingos at Lake Bogoria, Kenya. FEMS Microbiol. Ecol. 43 (2), 141−148.

Kroeze, C., Dumont, E., Seitzinger, S.P., 2005. New estimates of global emissions of N_2O from rivers and estuaries. Environ. Sci. 2, 159−165.

Kromm, D.E., White, S.E., 1992a. Groundwater problems. In: Kromm, D.E., White, S.E. (Eds.), Groundwater Exploitation in the High Plains. University Press of Kansas, Lawrence, KS, pp. 44−63.

Kromm, D.E., White, S.E., 1992b. The high plains Ogallala region. In: Kromm, D.E., White, S.E. (Eds.), Groundwater Exploitation in the High Plains. University Press of Kansas, Lawrence, KS, pp. 1−27.

Kronvang, B., Grant, R., Larsen, S.E., Svendsen, L.M., Kristensen, P., 1995. Non-point-source nutrient losses to the aquatic environment in Denmark: impact of agriculture. Mar. Freshw. Res. 46, 167−177.

Kruk, C., Devercelli, M., Huszar, V.L.M., Hernández, E., Beamud, G., Diaz, M., et al., 2017. Classification of Reynolds phytoplankton functional groups using individual traits and machine learning techniques. Freshw. Biol. 62 (10), 1681−1692. Available from: https://doi.org/10.1111/fwb.12968.

Krumholz, L.R., McKinley, J.P., Ulrich, G.A., Suflita, J.M., 1997. Confined subsurface microbial communities in Cretaceous rock. Nature 386, 64−66.

Krzyzanek, E., 1986. Development and structure of the Goczalkowice Reservoir ecosystem XIV. Zoobenthos. Ekol. Pol. 34 (3), 491−513.

Kuczynski, A., Auer, M.T., Brooks, C.N., Grimm, A.G., 2016. The Cladophora resurgence in Lake Ontario: characterization and implications for management. Can. J. Fish. Aquat. Sci. 73 (6), 999−1013. Available from: https://doi.org/10.1139/cjfas-2015-0460.

Kuflikowski, T., 1986. Development and structure of the Goczalkowice Reservoir ecosystem X. Macrophytes. Ekol. Pol. 34 (3), 429−445.

Kumar, A., Smith, R.P., Häder, D.-P., 1996. Effect of UV-B on enzymes of nitrogen metabolism in the cyanobacterium *Nostoc calcicola*. J. Plant Physiol. 148, 86−91.

Kumar, S., Arya, S., Nussinov, R., 2007. Temperature-dependent molecular adaptation features in proteins. In: Gerday, C., Glansdorff, N. (Eds.), Physiology and Biochemistry of Extremophiles. ASM Press, Washington, DC, pp. 75−85.

Kunza, L.A., Hall Jr, R.O., 2014. Nitrogen fixation can exceed inorganic nitrogen uptake fluxes in oligotrophic streams. Biogeochemistry 121 (3), 537−549.

Kurilkina, M.I., Zakharova, Y.R., Galachyants, Y.P., Petrova, D.P., Bukin, Y.S., Domysheva, V.M., et al., 2016. Bacterial community composition in the water column of the deepest freshwater Lake Baikal as determined by next-generation sequencing. FEMS Microbiol. Ecol. 92 (7). Available from: https://doi.org/10.1093/femsec/fiw094.

Kutalek, R., Kassa, A., 2005. The use of gyrinids and dyctiscids for stimulating breast growth in East Africa. J. Ethnobiol. 25, 115−128.

la Rivière, J.W.M., 1989. Threats to the worlds water. Sci. Am. 9/89, 80−94.

Laetz, C.A., Baldwin, D.H., Hebert, V.R., Stark, J.D., Scholz, N.L., 2014. Elevated temperatures increase the toxicity of pesticide mixtures to juvenile coho salmon. Aquat. Toxicol. 146, 38−44. Available from: https://doi.org/10.1016/j.aquatox.2013.10.022.

Lagrue, C., Poulin, R., 2015. Bottom−up regulation of parasite population densities in freshwater ecosystems. Oikos 124 (12), 1639−1647. Available from: https://doi.org/10.1111/oik.02164.

Lake, P., 2000. Disturbance, patchiness, and diversity in streams. J. N. Am. Benthol. Soc. 19 (4), 573−592.

Lam, A.K.-Y., Prepas, E., Spink, D., Hrudey, S.E., 1995. Chemical control of hepatotoxic phytoplankton blooms: implications for human health. Water Res. 29 (8), 1845−1854.

Lam, A.K.-Y., Prepas, E.E., 1997. In situ evaluation of options for chemical treatment of hepatotoxic cyanobacterial blooms. Can. J. Fish. Aquat. Sci. 54, 1736−1742.

Lam, B., Baer, A., Alaee, M., Lefebvre, B., Moser, A., Williams, A., et al., 2007. Major structural components in freshwater dissolved organic matter. Environ. Sci. Technol. 41 (24), 8240−8247. Available from: https://doi.org/10.1021/es0713072.

Lambert, M.R., Stoler, A.B., Smylie, M.S., Relyea, R.A., Skelly, D.K., 2016. Interactive effects of road salt and leaf litter on wood frog sex ratios and sexual size dimorphism. Can. J. Fish. Aquat. Sci. 74 (2), 141−146. Available from: https://doi.org/10.1139/cjfas-2016-0324.

Lamberti, G.A., Gregory, S.V., Ashkenas, L.R., Wildman, R.C., Moore, K.M.S., 1991. Stream ecosystem recovery following a catastrophic debris flow. Can. J. Fish. Aquat. Sci. 48, 196−208.

Lamberti, G.A., Moore, J.W., 1984. Aquatic insects and primary consumers. In: Resh, V.H., Rosenberg, D.M. (Eds.), The Ecology of Aquatic Insects. Praeger, New York, NY, pp. 164−195.

Lamouroux, N., Gore, J.A., Lepori, F., Statzner, B., 2015. The ecological restoration of large rivers needs science-based, predictive tools meeting public expectations: an overview of the Rhône project. Freshw. Biol. 60 (6), 1069−1084. Available from: https://doi.org/10.1111/fwb.12553.

Lampert, W., 1997. Zooplankton research: the contribution of limnology to general ecological paradigms. Aquat. Ecol. 31, 19−27.

Lancaster, J., Hildrew, A.G., 1993. Characterization in-stream flow refugia. Can. J. Fish. Aquat. Sci. 50, 1663−1675.

Lane, P.A., 1985. A food web approach to mutualism in lake communities. In: Boucher, D.H. (Ed.), The Biology of Mutualism. Oxford University Press, New York, NY, pp. 344−374.

Lang, J.M., McEwan, R.W., Benbow, M.E., 2015. Abiotic autumnal organic matter deposition and grazing disturbance effects on epilithic biofilm succession. FEMS Microbiol. Ecol. 91 (6). Available from: https://doi.org/10.1093/femsec/fiv060.

Langenheder, S., Wang, J., Karjalainen, S.M., Laamanen, T.M., Tolonen, K.T., Vilmi, A., et al., 2017. Bacterial metacommunity organization in a highly connected aquatic system. FEMS Microbiol. Ecol. 93 (4). Available from: https://doi.org/10.1093/femsec/fiw225fiw225-fiw225.

Langford, T., 1990. Ecological Effects of Thermal Discharges. Springer Science & Business Media.

Lapierre, J.-F., Seekell, D.A., Filstrup, C.T., Collins, S.M., Emi Fergus, C., Soranno, P.A., et al., 2017. Continental-scale variation in controls of summer CO_2 in United States lakes. J. Geophys. Res.: Biogeosci. 122 (4), 875−885. Available from: https://doi.org/10.1002/2016JG003525.

Larkum, A.W.D., Chen, M., Li, Y., Schliep, M., Trampe, E., West, J., et al., 2012. A novel epiphytic chlorophyll d-containing cyanobacterium isolated from a mangrove-associated red alga. J. Phycol. 48 (6), 1320−1327. Available from: https://doi.org/10.1111/j.1529-8817.2012.01233.x.

Larned, S.T., Datry, T., Arscott, D.B., Tockner, K., 2010. Emerging concepts in temporary-river ecology. Freshw. Biol. 55 (4), 717−738.

Larned, S.T., Kilroy, C., 2014. Effects of *Didymosphenia geminata* removal on river macroinvertebrate communities. J. Freshw. Ecol. 29 (3), 345−362. Available from: https://doi.org/10.1080/02705060.2014.898595.

Larsen, B.B., Miller, E.C., Rhodes, M.K., Wiens, J.J., 2017. Inordinate fondness multiplied and redistributed: the number of species on earth and the new pie of life. Q. Rev. Biol. 92 (3), 229−265. Available from: https://doi.org/10.1086/693564.

Larson, C.A., Liu, H., Passy, S.I., 2015. Iron supply constrains producer communities in stream ecosystems. FEMS Microbiol. Ecol. 91 (5). Available from: https://doi.org/10.1093/femsec/fiv041.

Larson, D., 1993. The recovery of Spirit Lake. Am. Sci. 81, 166−177.

Larson, E.R., Renshaw, M.A., Gantz, C.A., Umek, J., Chandra, S., Lodge, D.M., et al., 2017. Environmental DNA (eDNA) detects the invasive crayfishes *Oronectes rusticus* and *Pacifastacus leniusculus* in large lakes of North America. Hydrobiologia 800 (1), 173−185. Available from: https://doi.org/10.1007/s10750-017-3210-7.

Lassen, C., Revsbech, N.P., Pedersen, O., 1997. Macrophyte development and resuspension regulate the photosynthesis and production of benthic microalgae. Hydrobiologia 350, 1−11.

Lathrop, R.C., Carpenter, S.R., Rudstam, L.G., 1996. Water clarity in Lake Mendota since 1900: responses to differing levels of nutrients and herbivory. Can. J. Fish. Aquat. Sci. 53, 2250−2261.

Lathrop, R.C., Carpenter, S.R., Stow, C.A., Soranno, P.A., Panuska, J.C., 1998. Phosphorus loading reductions needed to control blue-green algal blooms in Lake Mendota. Can. J. Fish. Aquat. Sci. 55, 1169−1178.

Laurion, I., Lean, D.R.S., Vincent, W.F., 1998. UVB effects on a plankton community: results from a large-scale enclosure assay. Aquat. Microb. Ecol. 16, 189−198.

Lavoie, C., 2010. Should we care about purple loosestrife? The history of an invasive plant in North America. Biol. Invasions 12 (7), 1967−1999.

Lavoie, R.A., Jardine, T.D., Chumchal, M.M., Kidd, K.A., Campbell, L.M., 2013. Biomagnification of mercury in aquatic food webs: a worldwide meta-analysis. Environ. Sci. Technol. 47 (23), 13385−13394. Available from: https://doi.org/10.1021/es403103t.

Lavrentyev, P.J., Gardner, W.S., Cavaletto, J.F., Beaver, J.R., 1995. Effects of the zebra mussel (*Dreissena polymorpha* Pallas) on protozoa and phytoplankton from Saginaw Bay, Lake Huron. J. Great Lakes Res. 21 (4), 545−557.

Law, A., Jones, K.C., Willby, N.J., 2014. Medium vs. short-term effects of herbivory by Eurasian beaver on aquatic vegetation. Aquat. Bot. 116 (0), 27−34. Available from: https://doi.org/10.1016/j.aquabot.2014.01.004.

Lawler, A., 2016. Iraq confronts a new enemy—the Persian Gulf. Science 351 (6269), 111−112. Available from: https://doi.org/10.1126/science.351.6269.111.

Laws, E.A., 1993. Aquatic Pollution. John Wiley & Sons, New York, NY.

Laws, J., Heppell, K., Sheahan, D., Liu, C.-f, Grey, J., 2016. No such thing as a free meal: organotin transfer across the freshwater−terrestrial interface. Freshw. Biol. 61 (12), 2051−2062. Available from: https://doi.org/10.1111/fwb.12733.

Lawton, J.H., 1991. Are species useful? Oikos 62, 3−4.

Laybourn-Parry, J., Bayliss, P., Ellis-Evans, J.C., 1995. The dynamics of heterotrophic nanoflagellates and bacterioplankton in a large ultra-oligotrophic Antarctic lake. J. Plankton Res. 17 (9), 1835−1850.

Lazarus, E.D., Constantine, J.A., 2013. Generic theory for channel sinuosity. Proc. Natl. Acad. Sci. 110 (21), 8447−8452.

Lean, D.R.S., Pick, F.R., 1981. Photosynthetic response of lake plankton to nutrient enrichment: a test for nutrient limitation. Limnol. Oceanogr. 26, 1001−1019.

Leao, P., Vasconcelos, M., Vasconcelos, V., 2009. Allelopathy in freshwater cyanobacteria. Crit. Rev. Microbiol. 35 (4), 271−282.

Lear, G., Bellamy, J., Case, B.S., Lee, J.E., Buckley, H.L., 2014. Fine-scale spatial patterns in bacterial community composition and function within freshwater ponds. ISME J. 8 (8), 1715.

Lear, L., 1997. Rachel Carson: Witness for Nature. H. Holt & Co, New York, NY.

Leavitt, P.R., Vinebrooke, R.D., Donald, D.B., Smol, J.P., Schindler, D.W., 1997. Past ultraviolet radiation environments in lakes derived from fossil pigments. Nature 388, 457−459.

Lee, B.-G., Griscom, S.B., Lee, J.-S., Choi, H.J., Koh, C.-H., et al., 2000. Influences of dietary uptake and reactive sulfides on metal bioavailability from aquatic sediments. Science 287, 282−284.

Lee, G., Jones, R., 1991. Effects of eutrophication on fisheries. Rev. Aquat. Sci. 5 (3), 287−305.

Lee, S.S., Paspalof, A.M., Snow, D.D., Richmond, E.K., Rosi-Marshall, E.J., Kelly, J.J., 2016. Occurrence and potential biological effects of amphetamine on stream communities. Environ. Sci. Technol. 50 (17), 9727−9735. Available from: https://doi.org/10.1021/acs.est.6b03717.

Leff, L.G., McArthur, J.V., Shimkets, L.J., 1993. Spatial and temporal variability of antibiotic resistance in freshwater bacterial assemblages. FEMS Microbiol. Ecol. 13, 135−144.

Lehman, J.T., Scavia, D., 1982. Microscale nutrient patches produced by zooplankton. Proc. Natl. Acad. Sci. 789, 5001−5005.

Lehmkuhl, D.M., 1979. How to Know the Aquatic Insects. Wm. C. Brown Company Publishers, Dubuque, IA.

Lehner, B., Döll, P., 2004. Development and validation of a global database of lakes, reservoirs and wetlands. J. Hydrol. 296 (1), 1−22.

Lehtinen, R.M., Galatowitsch, S.M., Tester, J.R., 1999. Consequences of habitat loss and fragmentation for wetland amphibian assemblages. Wetlands 19 (1), 1−12. Available from: https://doi.org/10.1007/bf03161728.

Leibold, M.A., Hall, S.R., Smith, V.H., Lytle, D.A., 2017. Herbivory enhances the diversity of primary producers in pond ecosystems. Ecology 98 (1), 48−56. Available from: https://doi.org/10.1002/ecy.1636.

Leibold, M.A., Holyoak, M., Mouquet, N., Amarasekare, P., Chase, J.M., Hoopes, M.F., et al., 2004. The metacommunity concept: a framework for multi-scale community ecology. Ecol. Lett. 7 (7), 601−613.

Leibold, M.A., Mikkelson, G.M., 2002. Coherence, species turnover, and boundary clumping: elements of meta-community structure. Oikos 97 (2), 237−250.

Leigh, E.G., 2007. Neutral theory: a historical perspective. J. Evol. Biol. 20 (6), 2075−2091. Available from: https://doi.org/10.1111/j.1420-9101.2007.01410.x.

Leitchenkov, G.L., Antonov, A.V., Luneov, P.I., Lipenkov, V.Y., 2016. Geology and environments of subglacial Lake Vostok. Philos. Trans. R. Soc. A: Math., Phys. Eng. Sci. 374 (2059). Available from: https://doi.org/10.1098/rsta.2014.0302.

Leopold, L.B., 1994. A View of the River. Harvard University Press, Cambridge, MA.

Leopold, L.B., Davis, K.S., 1996. Water. Time Inc., New York, NY.

Leopold, L.B., Wolman, M.G., Miller, J.P., 1964. Fluvial Processes in Geomorphology. W. H. Freeman and Company, San Francisco, CA.

Lesica, P., Allendorf, F.W., 1999. Ecological genetics and the restoration of plant communities: mix or match? Restor. Ecol. 7 (1), 42−50.

Lever, C., 1994. Naturalized Animals: The Ecology of Successfully Introduced Species. T & AD Poyser Ltd, London.

Levi, P.S., Starnawski, P., Poulsen, B., Baattrup-Pedersen, A., Schramm, A., Riis, T., 2017. Microbial community diversity and composition varies with habitat characteristics and biofilm function in macrophyte-rich streams. Oikos 126 (3), 398−409. Available from: https://doi.org/10.1111/oik.03400.

Levin, S.A., 1992. The problem of pattern and scale in ecology: the Robert H. MacArthur award lecture. Ecology 73 (6), 1943−1967.

Levine, J.M., 2000. Species diversity and biological invasions: relating local process to community pattern. Science 288, 852−854.

Levins, R., 1969. Some demographic and genetic consequences of environmental heterogeneity for biological control. Bull. Entomol. Soc. Am. 15, 237−240.

Lewis Jr, W.M., 2002. Causes for the high frequency of nitrogen limitation in tropical lakes. Internationale Vereinigung für theoretische und angewandte Limnologie: Verhandlungen 28 (1), 210−213.

Lewis, W.M.J., 1986. Evolutionary interpretations of allelochemical interactions in phytoplankton algae. Am. Nat. 127 (2), 184−194.

Lewis, W.M.J., Hamilton, S.K., Lasi, M.A., Rodríguez, M., Saunders III, J.F., 2000. Ecological determinism on the Orinoco floodplain. Bioscience 50, 681−692.

Lewis, W.M.J., Wurtsbaugh, W.A., 2008. Control of lacustrine phytoplankton by nutrients: erosion of the phosphorus paradigm. Int. Rev. gest Hydrobiol. 93 (4−5), 446−465.

Li, H.W., Rossignol, P.A., Castillo, G., 2000. Risk analysis of species introductions: insights from qualitative modeling. In: Claudi, R., Leach, J.H. (Eds.), Nonindigenous Freshwater Organisms. Lewis Publishers, CRC Press, Boca Ratan, FL, pp. 431−447.

Li, M., Luo, Z., Yan, Y., Wang, Z., Chi, Q., Yan, C., et al., 2016. Arsenate accumulation, distribution, and toxicity a sociated with titanium dioxide nanoparticles in *Daphnia magna*. Environ. Sci. Technol. 50 (17), 9636−9643. Available from: https://doi.org/10.1021/acs.est.6b01215.

Li, X., Zhou, L., Yu, Y., Ni, J., Xu, W., Yan, Q., 2017. Composition of gut microbiota in the Gibel Carp (*Carassius auratus gibelio*) varies with host development. Microb. Ecol. 74 (1), 239−249. Available from: https://doi.org/10.1007/s00248-016-0924-4.

Li, Y., Schichtel, B.A., Walker, J.T., Schwede, D.B., Chen, X., Lehmann, C.M.B., et al., 2016. Increasing importance of deposition of reduced nitrogen in the United States. Proc. Natl. Acad. Sci. 113 (21), 5874−5879. Available from: https://doi.org/10.1073/pnas.1525736113.

Liao, J.C., Beal, D.N., Lauder, G.V., Triantafyllou, M.S., 2003. Fish exploiting vortices decrease muscle activity. Science 302 (5650), 1566−1569.

Licht, L.E., 2003. Shedding light on ultraviolet radiaiton and amphibian embryos. Bioscience 53 (6), 551−561.

Liermann, C.R., Nilsson, C., Robertson, J., Ng, R.Y., 2012. Implications of dam obstruction for global freshwater fish diversity. Bioscience 62 (6), 539−548.

Lieth, H.F.H., 1975. Primary production of the major vegetation units of the world. In: Lieth, H.F.H., Whittaker, R.H. (Eds.), Primary Productivity of the Biosphere. Springer-Verlag, New York, NY, pp. 203−215.

Light, S.S., Dineen, J.W., 1994. Water control in the Everglades: a historical perspective. In: Davis, S.M., Ogden, J.C. (Eds.), Everglades. St. Lucie Press, Delray Beach, FL, pp. 47−84.

Likens, G.E., 2001. Biogeochemistry, the watershed approach: some uses and limitations. Mar. Freshw. 52, 5−12.

Likens, G.E., Bormann, F.H., Pierce, R.S., Reiners, W.A., 1978. Recovery of a deforested ecosystem. Science 199, 492−496.

Likens, G.E., Driscoll, C.T., Buso, D.C., 1996. Long-term effects of acid rain: response and recovery of a forest ecosystem. Science 272, 244−246.

Liljedahl, A.K., Boike, J., Daanen, R.P., Fedorov, A.N., Frost, G.V., Grosse, G., et al., 2016. Pan-arctic ice-wedge degradation in warming permafrost and its influence on tundra hydrology. Nat. Geosci. 9 (4), 312−318. Available from: https://doi.org/10.1038/ngeo2674. Available from: http://www.nature.com/ngeo/journal/v9/n4/abs/ngeo2674.html#supplementary-information.

Lillesand, T.M., Johnson, W.L., Deuell, R.L., Lindstrom, O.M., Meisner, D.E., 1983. Use of Landsat data to predict the trophic state of Minnesota lakes. Photogramm. Eng. Remote Sens. 12, 2045−2063.

Lindeman, R.L., 1942. The trophic-dynamic aspect of ecology. Ecology 23 (4), 399−418.

Lindenschmidt, K.-E., Hamblin, P.F., 1997. Hypolimnetic aeration in Lake Tegel, Berlin. Water Res. 31 (7), 1619−1628.

Lips, K.R., Brem, F., Brenes, R., Reeve, J.D., Alford, R.A., Voyles, J., et al., 2006. Emerging infectious disease and the loss of biodiversity in a Neotropical amphibian community. Proc. Natl. Acad. Sci. 103 (9), 3165−3170.

Lips, K.R., Diffendorfer, J., Mendelson, J.R., Sears, M.W., 2008. Riding the wave: reconciling the roles of disease and climate change in amphibian declines. PLoS Biol. 6 (3), 441−454.

Lips, K.R., Reeve, J.D., Witters, L.R., 2003. Ecological traits predicting amphibian population declines in Central America. Conserv. Biol. 17 (4), 1078−1088.

Lipson, S.M., Stotzky, G., 1987. Interactions between clay minerals and viruses. In: Rao, V.C., Melnick, J.L. (Eds.), Human Viruses in Sediments, Sludges, and Soils. CRC Press, Boca Raton, FL, pp. 197−230.

Liston, S.E., Newman, S., Trexler, J.C., 2008. Macroinvertebrate community response to eutrophication in an oligotrophic wetland: an in situ mesocosm experiment. Wetlands 28 (3), 686−694.

Litchman, E., 1998. Population and community responses of phytoplankton to fluctuating light. Oecologia 117, 247−257.

Little, T.J., Hebert, P.D.N., 1996. Endemism and ecological islands: the ostracods from Jamaican bromeliads. Freshw. Biol. 36, 327−338.

Liu, H., Guo, X., Gooneratne, R., Lai, R., Zeng, C., Zhan, F., et al., 2016. The gut microbiome and degradation enzyme activity of wild freshwater fishes influenced by their trophic levels. Sci. Rep. 6.

Liu, K., Brown, M.G., Carter, C., Saykally, R.J., Gregory, J.K., Clary, D.C., 1996. Characterization of a cage form of the water hexameter. Nature 381, 501−503.

Liu, Z.-H., Lu, G.-N., Yin, H., Dang, Z., Rittmann, B., 2015. Removal of natural estrogens and their conjugates in municipal wastewater treatment plants: a critical review. Environ. Sci. Technol. 49 (9), 5288−5300. Available from: https://doi.org/10.1021/acs.est.5b00399.

Lloyd, R., 1960. The toxicity of zinc sulphate to rainbow trout. Ann. Appl. Biol. 48, 84−94.

Locey, K.J., Lennon, J.T., 2016. Scaling laws predict global microbial diversity. Proc. Natl. Acad. Sci. 113 (21), 5970−5975. Available from: https://doi.org/10.1073/pnas.1521291113.

Lodge, D.M., Barko, J.W., Strayer, D., Melack, J.M., Mittelbach, G.G., Howarth, R.W., et al., 1987. Spatial heterogeneity and habitat interactions in lake communities. In: Carpenter, S.R. (Ed.), Complex Interactions in Lake Communities. Springer-Verlag, Berlin, pp. 181−208.

Loeb, S.L., Spacie, A. (Eds.), 1994. Biological Monitoring of Aquatic Systems. Lewis Publishers, Ann Arbor, MI.

Loecke, T.D., Burgin, A.J., Riveros-Iregui, D.A., Ward, A.S., Thomas, S.A., Davis, C.A., et al., 2017. Weather whiplash in agricultural regions drives deterioration of water quality. Biogeochemistry 133 (1), 7−15. Available from: https://doi.org/10.1007/s10533-017-0315-z.

Loehr, R.C., 1974. Characteristics and comparative magnitude of non-point sources. J. Water Pollut. Control Fed. 46 (8), 1849−1872.

Loehr, R.C., Ryding, S.-O., Sonzogni, W.C., 1989. Estimating the nutrient load to a waterbody. In: Ryding, S.-O., Rast, W. (Eds.), The Control of Eutrophication of Lakes and Reservoirs, vol. I. The Parthenon Publishing Group, Paris, France, pp. 115−146. Man and the Biosphere Series.

Loiselle, S.A., Cunha, D.G.F., Shupe, S., Valiente, E., Rocha, L., Heasley, E., et al., 2016. Micro and macroscale drivers of nutrient concentrations in urban streams in South, Central and North America. PLoS One 11 (9), e0162684.

Loman, J., 2004. Density regulation in tadpoles of Rana temporaria: a full pond field experiment. Ecology 85 (6), 1611−1618.

Lomans, B.P., Smolders, A.J.P., Intven, L.M., Pol, A., den Camp, H.J.M.O., van der Drift, C., 1997. Formation of dimethyl sulfide and methanethiol in anoxic freshwater sediments. Appl. Environ. Microbiol. 63 (12), 4741−4747.

Long, S.P., Humphries, S., Falkowski, P.G., 1994. Photoinhibition of photosynthesis in nature. Annu. Rev. Plant Physiol. Plant Mol. Biol. 45, 633−662.

Lopez-Bueno, A., Tamames, J., Velazquez, D., Moya, A., Quesada, A., Alcami, A., 2009. High diversity of the viral community from an Antarctic Lake. Science 326 (5954), 858.

Lorang, M., Hauer, F.R., 2007. Fluvial geomorphological processes. In: Hauer, F.R., Lamberti, G.A. (Eds.), Methods in Stream Ecology, second ed. Academic Press, San Diego, CA, pp. 145–168.

Loreau, M., Naeem, S., Inchausti, P., Bengtsson, J., Grime, J.P., Hector, A., et al., 2001. Ecology—biodiversity and ecosystem functioning: current knowledge and future challenges. Science 294 (5543), 804–808.

Losey, J.E., Raynor, L.S., Carter, M.E., 1999. Transgenic pollen harms monarch larvae. Nature 399, 214.

Loudon, C., Alstad, D.N., 1990. Theoretical mechanics of particle capture: predictions for hydropsychid caddisfly distribution ecology. Am. Nat. 135 (3), 360–381.

Louhi, P., Richardson, J.S., Muotka, T., 2016. Sediment addition reduces the importance of predation on ecosystem functions in experimental stream channels. Can. J. Fish. Aquat. Sci. 74 (1), 32–40. Available from: https://doi.org/10.1139/cjfas-2015-0530.

Louis, V.L.S., Kelly, C.A., Duchemin, É., Rudd, J.W., Rosenberg, D.M., 2000. Reservoir surfaces as sources of greenhouse gases to the atmosphere: a global estimate reservoirs are sources of greenhouse gases to the atmosphere, and their surface areas have increased to the point where they should be included in global inventories of anthropogenic emissions of greenhouse gases. Bioscience 50 (9), 766–775.

Lourenço-Amorim, C., Neres-Lima, V., Moulton, T.P., Sasada-Sato, C.Y., Oliveira-Cunha, P., Zandonà, E., 2014. Control of periphyton standing crop in an Atlantic Forest stream: the relative roles of nutrients, grazers and predators. Freshw. Biol. 59 (11), 2365–2373. Available from: https://doi.org/10.1111/fwb.12441.

Lovley, D.R., Coates, J.D., Blunt-Harris, E.L., Phillips, E.J.P., Woodward, J.C., 1996. Humic substances as electron acceptors for microbial respiration. Nature 382, 445–448.

Lowe, R.L., Pillsbury, R.W., 1995. Shifts in benthic algal community structure and function following the appearance of zebra mussels (Dreissena polymorpha) in Saginaw Bay, Lake Huron. J. Great Lakes Res. 21 (4), 558–566.

Lowrance, R., Todd, R., Fail Jr., J., Hendrickson Jr., O., Leonard, R., Asmussen, L., 1984. Riparian forests as nutrient filters in agricultural watersheds. Bioscience 34, 374–377.

Lucassen, E.C., Smolders, A.P., VanderSalm, A.L., Roelofs, J.G.M., 2004. High groundwater nitrate concentrations inhibit eutrophication of sulphate-rich freshwater wetlands. Biogeochemistry 67 (2), 249–267.

Ludes, B., Coste, M., Tracqui, A., Mangin, P., 1996. Continuous river monitoring of the diatoms in diagnosis of drowning. J. Forensic Sci. 41, 425–428.

Ludwig, J.A., Reynolds, J.F., 1988. Statistical Ecology: A Primer on Methods & Computing. John Wiley & Sons, New York, NY.

Ludyanskiy, M.L., McDonald, D., MacNeill, D., 1993. Impact of the zebra mussel, a bivalve invader. Bioscience 43 (8), 533–544.

Lukens, N.R., Kraemer, B.M., Constant, V., Hamann, E.J., Michel, E., Socci, A.M., et al., 2017. Animals and their epibiota as net autotrophs: size scaling of epibiotic metabolism on snail shells. Freshw. Sci. 36 (2), 307–315. Available from: https://doi.org/10.1086/691438.

Lund, J.W.G., 1964. Primary production and periodicity of phytoplankton. Verhein Internationale Verein Limnologie 15, 37–56.

Lüring, M., 1998. Effect of grazing-associated infochemicals on growth and morphological development in Scededesmus acutus (Chlorophyceae). J. Phycol. 34 (4), 578–586.

Lürling, M., Donk, E.V., 2000. Grazer-induced colony formation in Scenedesmus: are there costs to being colonial? Oikos 88, 111–118.

Luzar, A., Chandler, D., 1996. Hydrogen-bond kinetics in liquid water. Nature 379, 55–57.

Lytle, D.A., Poff, N.L., 2004. Adaptation to natural flow regimes. Trends Ecol. Evol. 19 (2), 94–100.

MacArthur, R., 1955. Fluctuations of animal populations and a measure of community stability. Ecology 36, 533–536.

MacArthur, R.H., Wilson, E.O., 1967. The Theory of Island Biogeography. Princeton University Press, Princeton, NJ.

Macdonald, H.C., Ormerod, S.J., Bruford, M.W., 2017. Enhancing capacity for freshwater conservation at the genetic level: a demonstration using three stream macroinvertebrates. Aquat. Conserv.: Mar. Freshw. Ecosyst. 27 (2), 452–461. Available from: https://doi.org/10.1002/aqc.2691.

Maceina, M.J., Cichra, M.F., Betsill, R.K., Bettoli, P.W., 1992. Limnological changes in a large reservoir following vegetation removal by grass carp. J. Freshw. Ecol. 7, 81–95.

Mack, R.N., Simberloff, D., Lonsdale, W.M., Evans, H., Clout, M., Bazzaz, F.A., 2000. Biotic invasions: causes epidemiology, global consequences and control. Ecol. Appl. 10, 689–710.

Madigan, M.T., Oren, A., 1999. Thermophilic and halophilic extremophiles. Curr. Opin. Microbiol. 2, 265–269.

Madronich, S., McKenzie, R.L., Caldwell, M.M., Bjorn, L.O., 1995. Changes in ultraviolet radiation reaching the earth's surface. Ambio 24, 143–152.

Madsen, E.L., Ghiorse, W.C., 1993. Groundwater microbiology: subsurface ecosystem processes. In: Ford, T.E. (Ed.), Aquatic Microbiology: An Ecological Approach. Blackwell Scientific Publications, Oxford, UK, pp. 167–213.

Madsen, E.L., Sinclair, J.L., Ghiorse, W.C., 1991. In situ biodegradation: microbiological patterns in a contaminated aquifer. Science 252, 830–833.

Magnuson, J.J., Benson, B.J., Kratz, T.K., 1990. Temporal coherence in the limnology of a suite of lakes in Wisconsin, U.S.A. Freshw. Biol. 23, 145–159.

Magnuson, J.J., Kratz, T.K., 1999. Lakes in the landscape: approaches to regional limnology. Verhandlungen des Internationalen Verein Limnologie 27, 1–14.

Magnuson, J.J., Robertson, D.M., Benson, B.J., Wynne, R.H., Livingstone, D.M., Arai, T., et al., 2000. Historical trends in lake and river ice cover in the Northern Hemisphere. Science 289, 1743–1746.

Magnuson, J.J., Webster, K.E., Assel, R.A., Bowser, C.J., Dillon, P.J., Eaton, J.G., et al., 1997. Potential effects of climate changes on aquatic systems: Laurential Great Lakes and precambrian shield region. Proc. Hydrol. 11, 825–871.

Magurran, A.E., 2004. Measuring Biological Diversity. Blackwell, Oxford.

Maier, H.R., Burch, M.D., Bormans, M., 2001. Flow management strategies to control blooms of the cyanobacterium, *Anabaena circinalis*, in the River Murray at Morgan, South Australia. Regul. Rivers: Res. Manage.: Int. J. Devoted River Res. Manage. 17 (6), 637–650.

Maire, R., Pomel, S., 1994. Karst geomorphology and environment. In: Gibert, J., Danielopol, D.L., Stanford, J.A. (Eds.), Groundwater Ecology. Academic Press, Inc., San Diego, CA, pp. 129–155.

Majewski, S.P., Cumming, B.F., 1999. Paleolimnological investigation of the effects of post-1970 reductions of acidic deposition on an acidified Adirondack lake. J. Paleolimnol. 21, 207–213.

Makarewics, J.C., Lewis, T.W., Bertram, P., 1999. Phytoplankton composition and biomass in offshore waters of Lake Erie: pre- and post-*Dreissensa* introduction (1983–1993). J. Great Lakes Res. 25, 135–148.

Mal, T.K., Lovett-Doust, J., Lovett-Doust, L., Mulligan, G.A., 1992. The biology of Canadian weeds. 100. *Lythrum salicaria*. Can. J. Plant Sci. 72, 1305–1330.

Malard, F., Hervant, F., 1999. Oxygen supply and the adaptations of animals in groundwater. Freshw. Biol. 41, 1–30.

Malard, F., Reygrobellet, J.-L., Mathieu, J., Lafont, M., 1994. The use of invertebrate communities to describe groundwater flow and contaminant transport in a fractured rock aquifer. Arch. Hydrobiol. 131, 93–110.

Malard, F., Uehlinger, U., Zah, R., Tochkner, K., 2006. Flood-pulse and riverscape dynamics in a braided glacial river. Ecology 87 (3), 704–716.

Malecki, R.A., Blossey, B., Hight, S.D., Schroeder, D., Kok, L.T., Coulson, J.R., 1993. Biological control of purple loosestrife. Bioscience 43, 680–686.

Malin, G., Kirst, G.O., 1997. Algal production of dimethyl sulfide and its atmospheric role. J. Phycol. 33, 889–896.

Malinsky, M., Challis, R.J., Tyers, A.M., Schiffels, S., Terai, Y., Ngatunga, B.P., et al., 2015. Genomic islands of speciation separate cichlid ecomorphs in an East African crater lake. Science 350 (6267), 1493–1498. Available from: https://doi.org/10.1126/science.aac9927.

Mallory, M.L., McNicol, D.K., Cluis, D.A., Laberge, C., 1998. Chemical trends and status of small lakes near Sudbury, Ontario, 1983–1995: evidence of continued chemical recovery. Can. J. Fish. Aquat. Sci. 55, 63–75.

Malmqvist, B., Wotton, R.S., Zhang, Y.X., 2001. Suspension feeders transform massive amounts of seston in large northern rivers. Oikos 92 (1), 35–43.

Mangin, A., 1994. Karst hydrogeology. In: Gibert, J., Danielopol, D.L., Stanford, J.A. (Eds.), Groundwater Ecology. Academic Press, Inc, San Diego, CA, pp. 43–67.

Manning, D.W.P., Rosemond, A.D., Gulis, V., Benstead, J.P., Kominoski, J.S., Maerz, J.C., 2016. Convergence of detrital stoichiometry predicts thresholds of nutrient-stimulated breakdown in streams. Ecol. Appl. 26 (6), 1745–1757. Available from: https://doi.org/10.1890/15-1217.1.

Maranger, R., Bird, D.F., 1995. Viral abundance in aquatic systems: a comparison between marine and fresh waters. Mar. Ecol. Prog. Ser. 121, 217–226.

Marcarelli, A.M., Baker, M.A., Wurtsbaugh, W.A., 2008. Is in-stream N_2 fixation an important N source for benthic communities and stream ecosystems? J. N. Am. Benthol. Soc. 27, 186–211.

March, J.G., Benstead, J.P., Pringle, C.M., Scatena, F.N., 1998. Migratory drift of larval freshwater shrimps in two tropical streams, Puerto Rico. Freshw. Biol. 40 (2), 261–273.

Marino, N.A.C., Srivastava, D.S., Farjalla, V.F., 2016. Predator kairomones change food web structure and function, regardless of cues from consumed prey. Oikos 125 (7), 1017–1026. Available from: https://doi.org/10.1111/oik.02664.

Markovic, D., Carrizo, S.F., Kärcher, O., Walz, A., David, J.N.W., 2017. Vulnerability of European freshwater catchments to climate change. Glob. Change Biol. 23 (9), 3567–3580. Available from: https://doi.org/10.1111/gcb.13657.

Marleau, J.N., Guichard, F., Loreau, M., 2015. Emergence of nutrient co-limitation through movement in stoichiometric meta-ecosystems. Ecol. Lett. 18 (11), 1163–1173. Available from: https://doi.org/10.1111/ele.12495.

Marmonier, P., Vervier, P., Gibert, J., Dole-Olivier, M.-J., 1993. Biodiversity in ground waters. Trends Ecol. Evol. 8 (11), 392–395.

Marotta, H., Pinho, L., Gudasz, C., Bastviken, D., Tranvik, L.J., Enrich-Prast, A., 2014. Greenhouse gas production in low-latitude lake sediments responds strongly to warming. Nat. Clim. Change 4 (6), 467–470.

Martel, A., Spitzen-van der Sluijs, A., Blooi, M., Bert, W., Ducatelle, R., Fisher, M.C., et al., 2013. *Batrachochytrium salamandrivorans* sp. nov. causes lethal chytridiomycosis in amphibians. Proc. Natl. Acad. Sci. 110 (38), 15325–15329.

Martí, E., Fisher, S.G., Schade, J.D., Grimm, N.B., 2000. Flood frequency and stream-riparian linkages in arid lands. In: Jones, J.B., Molholland, P.J. (Eds.), Streams and Ground Waters. Academic Press, San Diego, CA, pp. 111–136.

Marti, E., Sabater, F., 1996. High variability in temporal and spatial nutrient retention in Mediterranean streams. Ecology 77, 854–869.

Martinez, M.L., Lopez-Barrera, F., 2008. Special issue: restoring and designing ecosystems for a crowded planet. Ecoscience 15 (1), 1–5.

Martinez, N.D., 1991. Artifacts or attributes? Effects of resolution on the Little Rock Lake food web. Ecol. Monogr. 61 (4), 367–392.

Martinez, N.D., 1993. Effect of scale on food web structure. Science 260, 242–243.

Martínez-Cortizas, A., Pontevedra-Pombal, X., Garcia-Rodeja, E., Nóvoa-Muñoz, J.C., Shotyk, W., 1999. Mercury in a Spanish peat bog: archive of climate change and atmospheric metal deposition. Science 284, 939–942.

Marzadri, A., Dee, M.M., Tonina, D., Bellin, A., Tank, J.L., 2017. Role of surface and subsurface processes in scaling N_2O emissions along riverine networks. Proc. Natl. Acad. Sci. 114 (17), 4330–4335. Available from: https://doi.org/10.1073/pnas.1617454114.

Marzolf, E., Mulholland, P.J., Steinman, A.D., 1998. Reply: improvements to the diurnal upstream-downstream dissolved oxygen change technique for determining whole-stream metabolism in small streams. Can. J. Fish. Aquat. Sci. 55, 1786–1787.

Marzolf, E.R., Mulholland, P.J., Steinman, A.D., 1994. Improvement to the diurnal upstream-downstream dissolved oxygen change technique for determining whole-stream metabolism in small streams. Can. J. Fish. Aquat. Sci. 51, 1591–1599.

Mas-Martí, E., Muñoz, I., Oliva, F., Canhoto, C., 2015. Effects of increased water temperature on leaf litter quality and detritivore performance: a whole-reach manipulative experiment. Freshw. Biol. 60 (1), 184–197. Available from: https://doi.org/10.1111/fwb.12485.

Mason, C.F., 1996. Biology of Freshwater Pollution, third ed. Longman Press, Essex, UK.

Matena, J., Simek, K., Fernando, C.H., 1995. Ingestion of suspended bacteria by fish: a modified approach. J. Fish. Biol. 47, 334–336.

Mats, V.D., 1993. The structure and development of the Baikal rift depression. Earth-Sci. Rev. 34, 81–118.

Matthews, W.J., 1998. Patterns in Freshwater Fish Ecology. Kluwer Academic Publishers, Norwell, MS.

Matthews, W.J., Stewart, A.J., Power, M.E., 1987. Grazing fishes as components of North American stream ecosystems: effects of *Campostoma anomalum*. In: Matthews, W.J., Heins, D.C. (Eds.), Community and Evolutionary Ecology of North American Fishes. University of Oklahoma Press, Norman, OK, pp. 128–135.

Mattocks, S., Hall, C.J., Jordaan, A., 2017. Damming, lost connectivity, and the historical role of anadromous fish in freshwater ecosystem dynamics. Bioscience 67 (8), 713–728.

Maul, J.D., Belden, J.B., Schwab, B.A., Whiles, M.R., Spears, B., Farris, J.L., et al., 2006a. Bioaccumulation and trophic transfer of polychlorinated biphenyls by aquatic and terrestrial insects to tree swallows (*Tachycineta bicolor*). Environ. Toxicol. Chem. 25 (4), 1017–1025.

Maul, J.D., Schuler, L.J., Belden, J.B., Whiles, M.R., Lydy, M.J., 2006b. Effects of the antibiotic ciprofloxacin on stream microbial communities and detritivorous macroinvertebrates. Environ. Toxicol. Chem. 25 (6), 1598–1606.

Maxwell, R.M., Condon, L.E., 2016. Connections between groundwater flow and transpiration partitioning. Science 353 (6297), 377–380. Available from: https://doi.org/10.1126/science.aaf7891.

May, R.M., 1972. Will large complex systems be stable? Nature 238, 413–414.

May, R.M., 1988. How many species are there on earth? Science 241, 1441–1448.

Mazak, E.J., MacIsaac, H.J., Servos, M.R., Hesslein, R., 1997. Influence of feeding habits on organochlorine contaminant accumulation in waterfowl on the great lakes. Ecol. Appl. 7 (4), 1133–1143.

Mazumder, A., Taylor, W.D., McQueen, D.J., Lean, D.R.S., 1990. Effects of fish and plankton on lake temperature and mixing depth. Science 247, 312–315.

McAuliffe, J.R., 1983. Competition, colonization patterns, and disturbance in stream benthic communities. In: Barnes, J.R., Minshall, G.W. (Eds.), Stream ecology: application and testing of general ecological theory. Plenum Press, New York, pp. 137–156.

McCabe, D.J., Gotelli, N.J., 2000. Effects of disturbance frequency, intensity, and area on assemblages of stream macroinvertebrates. Oecologia 124, 270–279.

McCafferty, W.P., 1988. Aquatic Entomology, the Fisherman's and Ecologists' Illustrated Guide to Insects and Their Relatives. Jones and Bartlett Publishers, Sudbury, MA.

McClain, M.E., Richey, J.E., Pimentel, T.P., 1994. Groundwater nitrogen dynamics at the terrestrial-lotic interface of a small catchment in the Central Amazon Basin. Biogeochemistry 27, 113–127.

McCluney, K.E., Poff, N.L., Palmer, M.A., Thorp, J.H., Poole, G.C., Williams, B.S., et al., 2014. Riverine macrosystems ecology: sensitivity, resistance, and resilience of whole river basins with human alterations. Front. Ecol. Environ. 12 (1), 48–58.

McCormick, A., Hoellein, T.J., Mason, S.A., Schluep, J., Kelly, J.J., 2014. Microplastic is an abundant and distinct microbial habitat in an urban river. Environ. Sci. Technol. 48 (20), 11863–11871. Available from: https://doi.org/10.1021/es503610r.

McCormick, P.V., Chimney, M.J., Swift, D.R., 1997. Diel oxygen profiles and water column community metabolism in the Florida Everglades, U.S.A. Arch. Hydrobiol. 140, 117–129.

McCrackin, M.L., Jones, H.P., Jones, P.C., Moreno-Mateos, D., 2017. Recovery of lakes and coastal marine ecosystems from eutrophication: a global meta-analysis. Limnol. Oceanogr. 62 (2), 507–518. Available from: https://doi.org/10.1002/lno.10441.

McDonald, R.I., Green, P., Balk, D., Fekete, B.M., Revenga, C., Todd, M., et al., 2011. Urban growth, climate change, and freshwater availability. Proc. Natl. Acad. Sci. 108 (15), 6312–6317. Available from: https://doi.org/10.1073/pnas.1011615108.

McFadden, C.H., Keeton, W.T., 1995. Biology: An Exploration of Life. Norton, New York, NY.

McGee, M.D., Borstein, S.R., Neches, R.Y., Buescher, H.H., Seehausen, O., Wainwright, P.C., 2015. A pharyngeal jaw evolutionary innovation facilitated extinction in Lake Victoria cichlids. Science 350 (6264), 1077−1079. Available from: https://doi.org/10.1126/science.aab0800.

McGill, B.J., Etienne, R.S., Gray, J.S., Alonso, D., Anderson, M.J., Benecha, H.K., et al., 2007. Species abundance distributions: moving beyond single prediction theories to integration within an ecological framework. Ecol. Lett. 10 (10), 995−1015.

McInerney, P.J., Rees, G.N., 2017. Co-invasion hypothesis explains microbial community structure changes in upland streams affected by riparian invader. Freshw. Sci. 36 (2), 297−306. Available from: https://doi.org/10.1086/692068.

McInnes, S.J., Pugh, P.J.A., 1998. Biogeography of limno-terrestrial Tardigrada, with particular reference to the Antarctic fauna. J. Biogeogr. 25, 31−36.

McIntyre, P.B., Jones, L.E., Flecker, A.S., Vanni, M.J., 2007. Fish extinctions alter nutrient recycling in tropical freshwaters. Proc. Natl. Acad. Sci. USA 104 (11), 4461−4466.

McKeon, D.M., Calabrese, J.P., Bissonnette, G.K., 1995. Antibiotic-resistant gram-negative bacteria in rural groundwater supplies. Water Res. 29 (8), 1902−1908.

McKinney, T., Rogers, R.S., Ayers, A.D., Persons, W.R., 1999. Lotic community responses in the Lees Ferry Reach. In: Webb, R.H., Schmidt, J.C., Marzolf, G.R., Valdez, R.A. (Eds.), The Controlled Flood in the Grand Canyon, vol. 110. American Geophysical Union, Washington, DC, pp. 249−258. Geophysical Monograph Series.

McKnight, D.M., Boyer, E.W., Westerhoff, P.K., Doran, P.T., Kulbe, T., Andersen, D.T., 2001. Spectrofluorometric characterization of dissolved organic matter for indication of precursor organic material and aromaticity. Limnol. Oceanogr. 46 (1), 38−48.

McKnight, E., García-Berthou, E., Srean, P., Rius, M., 2017. Global meta-analysis of native and nonindigenous trophic traits in aquatic ecosystems. Glob. Change Biol. 23 (5), 1861−1870. Available from: https://doi.org/10.1111/gcb.13524.

McLachlan, J.A., Arnold, S.F., 1996. Environmental estrogens. Am. Sci. 84, 452−461.

McMahon, R.F., 1991. Mollusca: bivalvia. In: Thorp, J.H., Covich, A.P. (Eds.), Ecology and Classification of North American Freshwater Invertebrates. Academic Press, Inc., San Diego, CA, pp. 315−400.

McMahon, R.F., 2000. Invasive characteristics of the freshwater bivalve *Corbicula fluminea*. In: Claudi, R., Leach, J.H. (Eds.), Nonindigenous Freshwater Organisms. Lewis Publishers, CRC Press, Boca Ratan, FL, pp. 315−343.

McMahon, T.A., Sears, B.F., Venesky, M.D., Bessler, S.M., Brown, J.M., Deutsch, K., et al., 2014. Amphibians acquire resistance to live and dead fungus overcoming fungal immunosuppression. Nature 511 (7508), 224−227. Available from: https://doi.org/10.1038/nature13491. Available from: http://www.nature.com/nature/journal/v511/n7508/abs/nature13491.html#supplementary-information.

McManamay, R.A., Brewer, S.K., Jager, H.I., Troia, M.J., 2016. Organizing environmental flow frameworks to meet hydropower mitigation needs. Environ. Manage. 58 (3), 365−385. Available from: https://doi.org/10.1007/s00267-016-0726-y.

Mcmeans, B.C., Koussoroplis, A.-M., Arts, M.T., Kainz, M.J., 2015. Terrestrial dissolved organic matter supports growth and reproduction of *Daphnia magna* when algae are limiting. J. Plankton Res. 37 (6), 1201−1209. Available from: https://doi.org/10.1093/plankt/fbv083.

Meade, J.W., 1989. Aquaculture Management. Van Nostrand Reinhold, New York, NY.

Mearns, L.O., Gleick, P.H., Schneider, S.H., 1990. Climate forecasting. In: Waggoner, P.E. (Ed.), Report of the American Association for the Advancement of Science Panel on Climatic Variability, Climate Change and the Planning and Management of U. S. Water Resources. John Wiley & Sons, Inc, New York, NY, pp. 87−138.

Meegan, S.K., Perry, S.A., 1996. Periphyton communities in headwater streams of different water chemistry in the central Appalachian Mountains. J. Freshw. Ecol. 11 (3), 247−256.

Meehl, G.A., Stocker, T.F., Collins, W.D., Friedlingstein, P., Gaye, A.T., Gregory, J.M., et al., 2007. Global climate projections. In: Solomon, S., Qin, D., Manning, M., Chen, Z., Marquis, M., Averyt, K.B., et al.,Climate Change 2007: The Physical Science Basis. Contribution of Working Group I to the 4th Assessment Report of the Intergovernmental Panel on Climate Change. Cambridge University Press, United Kingdom and New York, NY.

Meeks, J.C., 1998. Symbiosis between nitrogen-fixing cyanobacteria and plants. Bioscience 48 (4), 266−276.

Megonigal, J.P., Schlesinger, W.H., 1997. Enhanced CH_4 emissions from a wetland soil exposed to elevated CO_2. Biogeochemistry 37, 77−88.

Meijer, M.-L., Jeppesen, E., van Donk, E., Moss, B., Scheffer, M., Lammens, E., et al., 1994. Long-term responses to fish-stock reduction in small shallow lakes: interpretation of five-year results of four biomanipulation cases in the Netherlands and Denmark. Hydrobiologia 275/276, 457−466.

Mellina, E., Rasmussen, J.B., Mills, E.L., 1995. Impact of zebra mussel (Dreissena polymprpha) on phosphorus cycling and chlorophyll in lakes. Can. J. Fish. Aquat. Sci. 52, 2553−2573.

Melone, G., 1998. The rotifer corona by SEM. Hydrobiologia 387/388, 131−134.

Memmott, J., Craze, P.G., Waser, N.M., Price, M.V., 2007. Global warming and the disruption of plant−pollinator interactions. Ecol. Lett. 10 (8), 710−717.

Meriano, M., Eyles, N., Howard, K.W.F., 2009. Hydrogeological impacts of road salt from Canada's busiest highway on a Lake Ontario watershed (Frenchman's Bay) and lagoon, City of Pickering. J. Contam. Hydrol. 107 (1−2), 66−81.

Merritt, R.W., Cummins, K.W., Berg, M.B. (Eds.), 2008. An Introduction to the Aquatic Insects of North America. fourth ed. Kendall/Hunt Pub. Co., Dubuque, IA.

Merritt, R.W., Higgins, M.J., Cummins, K.W., Vandeneeden, B., 1999. The Kissimmee River-riparian marsh ecosystem, Florida. Seasonal differences in invertebrate functional feeding group relationships. In: Batzer, D.P., Rader, R.B., Wissinger, S.A. (Eds.), Invertebrates in Freshwater Wetlands of North America: Ecology and Management. John Wiley & Sons, Inc., New York, NY, pp. 55−79.

Merritt, R.W., Wallace, J.R., 2000. The role of aquatic insects in forensic investigations. Forensic entomology: the utility of arthropods in legal investigations, pp. 177−222.

Messager, M.L., Lehner, B., Grill, G., Nedeva, I., Schmitt, O., 2016. Estimating the volume and age of water stored in global lakes using a geo-statistical approach. Nat. Commun. 7, 13603. Available from: https://doi.org/10.1038/ncomms13603. Available from: http://www.nature.com/articles/ncomms13603#supplementary-information.

Messer, J., Brezonik, P.L., 1983. Comparison of denitrification rate estimation techniques in a large, shallow lake. Water Res. 17, 631−640.

Metson, G.S., Lin, J., Harrison, J.A., Compton, J.E., 2017. Linking terrestrial phosphorus inputs to riverine export across the United States. Water Res. 124, 177−191. Available from: https://doi.org/10.1016/j.watres.2017.07.037.

Meybeck, M., 1982. Carbon, nitrogen, and phosphorus transport by world rivers. Am. J. Sci. 282, 401−450.

Meybeck, M., 1993. Riverine transport of atmospheric carbon: sources, global typology and budget. Water Air Soil Pollut. 70, 443−463.

Meybeck, M., 1995. Global distribution of lakes. In: Lerman, A., Imboden, D.M., Gat, J.R. (Eds.), Physics and Chemistry of Lakes. Springer Verlag, Berlin, Germany, pp. 1−36.

Meybeck, M., Chapman, D.V., Helmer, R., 1989. Global Freshwater Quality. A First Assessment. Published on behalf of the World Health Organization and the United Nations Environment Programme by Blackwell Inc., Cambridge, MA.

Meyer, A., 1993. Phylogenetic relationships and evolutionary processes in East African cichlid fishes. Trends Ecol. Evol. 8 (8), 279−284.

Meyer, C.K., Baer, S.G., Whiles, M.R., 2008. Ecosystem recovery across a chronosequence of restored wetlands in the platte river valley. Ecosystems 11 (2), 193−208.

Meyer, C.K., Whiles, M.R., 2008. Macroinvertebrate communities in restored and natural Platte River slough wetlands. J. N. Am. Benthol. Soc. 27 (3), 626−639.

Meyer, C.K., Whiles, M.R., Baer, S.G., 2010. Plant community recovery following restoration in temporally variable riparian wetlands. Restor. Ecol. 18 (1), 52−64.

Meyer, F.P., 1990. Introduction. In: Meyer, F.P., Barklay, L.A. (Eds.), Field Manual for the Investigation of Fish Kills. United States Department of the Interior, Fish and Wildlife Service, Washington, DC, pp. 1–5. Vol. Resource Publication 177.

Meyer, J.L., 1994a. The microbial loop in flowing waters. Microb. Ecol. 28, 195–199.

Meyer, J.L., Paul, M.J., Taulbee, W.K., 2005. Stream ecosystem function in urbanizing landscapes. J. N. Am. Benthol. Soc. 24 (3), 602–612.

Meyer, L.A., Sullivan, S.M.P., 2013. Bright lights, big city: influences of ecological light pollution on reciprocal stream–riparian invertebrate fluxes. Ecol. Appl. 23 (6), 1322–1330. Available from: https://doi.org/10.1890/12-2007.1.

Meyer, O., 1994. Functional groups of microorganisms. In: Schulze, E.-D., Mooney, H.A. (Eds.), Biodiversity and Ecosystem Function. Springer-Verlag, Berlin, pp. 67–96.

Michael, H.J., Boyle, K.J., Bouchard, R., 1996. Water quality affects property prices: a case study of selected Maine lakes. Miscellaneous Report 398, Orono, ME. Retrieved from https://www.midcoastconservancy.org/wp-content/uploads/2017/01/MR398-Water-Quality-Affects-Property-Prices-A-Case-Study-of-Sel-1.pdf.

Michalak, A.M., Anderson, E.J., Beletsky, D., Boland, S., Bosch, N.S., Bridgeman, T.B., et al., 2013. Record-setting algal bloom in Lake Erie caused by agricultural and meteorological trends consistent with expected future conditions. Proc. Natl. Acad. Sci. 110 (16), 6448–6452. Available from: https://doi.org/10.1073/pnas.1216006110.

Michaletz, P.H., Bonneau, J.L., 2005. Age-0 gizzard shad abundance is reduced in the presence of macrophytes: implications for interactions with bluegills. Trans. Am. Fish. Soc. 134 (1), 149–159. Available from: https://doi.org/10.1577/FT04-011.1.

Michel, E., 1994. Why snails radiate: a review of gastropod evolution in long-lived lakes, both recent and fossil. Arch. Hydrobiol. 44, 285–317.

Middleton, B., 1999. Wetland Restoration, Flood Pulsing, and Disturbance Dynamics. John Wiley & Sons, New York, NY.

Miles, G.R., Mitchell, M.J., Mayer, B., Likens, G., Welker, J., 2012. Long-term analysis of Hubbard Brook stable oxygen isotope ratios of streamwater and precipitation sulfate. Biogeochemistry 111 (1–3), 443–454.

Milinski, M., 1993. Predation risk and feeding behavior. In: Pitcher, T.J. (Ed.), Behavior of Teleost Fishes, second ed. Chapmand and Hall, London, UK, pp. 285–306.

Millenium Ecosystem Assessment, 2003. Ecosystems and Human Well-Being: A Framework for Assessment. Island Press, Washington, DC.

Miller, A.M., Golladay, S.W., 1996. Effects of spates and drying on macroinvertebrate assemblages of an intermittent and a perennial prairie stream. J. N. Am. Benthol. Soc. 15 (4), 670–689.

Miller, B.L., Arntzen, E.V., Goldman, A.E., Richmond, M.C., 2017. Methane ebullition in temperate hydropower reservoirs and implications for US policy on greenhouse gas emissions. Environ. Manage. 60 (4), 615–629. Available from: https://doi.org/10.1007/s00267-017-0909-1.

Miller, E.J., Tomasic, J.J., Barnhart, M.C., 2014. A comparison of freshwater mussels (Unionidae) from a late-Archaic archeological excavation with recently sampled Verdigris River, Kansas, populations. Am. Midl. Nat. 171 (1), 16–26. Available from: https://doi.org/10.1674/0003-0031-171.1.16.

Miller, G.T.J., 1998. Living in the Environment, tenth ed. Wadsworth, Belmont, CA.

Milliman, J.D., Broadus, J.M., Gable, F., 1989. Environmental and economic implications of rising sea level and subsiding deltas: the Nile and Bengal examples. Ambio 18, 340–345.

Mills, E.L., Leach, J.H., Carlton, J.T., Secor, C.L., 1994. Exotic species and the integrity of the Great Lakes. Bioscience 44 (10), 666–676.

Milly, P.C.D., Betancourt, J., Falkenmark, M., Hirsch, R.M., Kundzewicz, Z.W., Lettenmaier, D.P., et al., 2008. Climate change—stationarity is dead: whither water management? Science 319 (5863), 573–574. Available from: https://doi.org/10.1126/science.1151915.

Milner, A.M., Robertson, A.L., Brown, L.E., Sønderland, S.H., McDermott, M., Veal, A.J., 2011. Evolution of a stream ecosystem in recently deglaciated terrain. Ecology 92 (10), 1924–1935. Available from: https://doi.org/10.1890/10-2007.1.

Milvy, P., Cothern, C.R., 1990. Scientific background for the development of regulations for radionuclides in drinking water. In: Cothern, C.R., Rebers, P.A. (Eds.), Radon, Radium and Uranium in Drinking Water. Lewis Publishers, Chelsea, MI, pp. 1−16.

Minshall, G., Brock, J., Varley, J., 1989. Wildfires and Yellowstone's stream ecosystems. Bioscience 707−715.

Minshall, G.W., 1978. Autotrophy in stream ecosystems. Bioscience 28 (12), 767−771.

Minshall, G.W., Petersen, R.C., Cummins, K.W., Bott, T.L., Sedell, J.R., Cushing, C.E., et al., 1983. Interbiome comparison of stream ecosystem dynamics. Ecol. Monogr. 53, 1−25.

Miraldo, A., Li, S., Borregaard, M.K., Flórez-Rodríguez, A., Gopalakrishnan, S., Rizvanovic, M., et al., 2016. An Anthropocene map of genetic diversity. Science 353 (6307), 1532−1535. Available from: https://doi.org/10.1126/science.aaf4381.

Mitsch, W.J., Cronk, J.K., Wu, X., Nairn, R.W., 1995. Phosphorus retention in constructed freshwater riparian marshes. Ecol. Appl. 5 (3), 830−845.

Mitsch, W.J., Gosselink, J.G., 1993. Wetlands, second ed. Van Nostrand Reinhold, New York, NY.

Mitsch, W.J., Gosselink, J.G., 2007. Wetlands, fourth ed. John Wiley & Sons, Hoboken, NJ.

Mittelbach, G., 1988. Competition among refuging sunfishes and effects of fish density on littoral zone invertebrates. Ecology 614−623.

Mittelbach, G.G., Osenberg, C.W., 1994. Using foraging theory to study trophic interactions. In: Stouder, D.J., Fresh, K.L., Feller, R.J. (Eds.), Theory and Application in Fish Feeding Ecology. University of South Carolina Press, Columbia, pp. 45−59.

Moeller, R., Burkholder, J., Wetzel, R., 1988. Significance of sedimentary phosphorus to a rooted submersed macrophyte (*Najas flexilis* (Willd.) Rostk. and Schmidt) and its algal epiphytes. Aquat. Bot. 32 (3), 261−281.

Moelzner, J., Fink, P., 2015. Gastropod grazing on a benthic alga leads to liberation of food-finding infochemicals. Oikos 124 (12), 1603−1608. Available from: https://doi.org/10.1111/oik.02069.

Molles, M.C.J., Crawford, C.S., Ellis, L.M., Valett, H.M., Dahm, C.N., 1998. Managed flooding for riparian ecosystem restoration. Bioscience 48, 749−756.

Montgomery, J.C., Macdonald, F., Baker, C.F., Carton, A.G., 2002. Hydrodynamic contributions to multimodal guidance of prey capture behavior in fish. Brain Behav. Evol. 59 (4), 190−198.

Moore, J.W., Beakes, M.P., Nesbitt, H.K., Yeakel, J.D., Patterson, D.A., Thompson, L.A., et al., 2014. Emergent stability in a large, free-flowing watershed. Ecology 96 (2), 340−347. Available from: https://doi.org/10.1890/14-0326.1.

Moore, K.A., Williams, D.D., 1990. Novel strategies in the complex defense repertoire of a stonefly (*Pteronarcys dorsata*) nymph. Oikos 57, 49−56.

Moore, W.S., 1996. Large groundwater inputs to coastal waters revealed by ^{226}Ra enrichments. Nature 380, 612−614.

Morin, P.J., 1983. Predation, competition, and the composition of larval anuran guilds. Ecol. Monogr. 53, 119−138.

Morin, P.J., McGrady-Steed, J., 2004. Biodiversity and ecosystem functioning in aquatic microbial systems: a new analysis of temporal variationo and species richness-predictability relations. Oikos 104, 458−466.

Morisawa, M., 1968. Streams Their Dynamics and Morphology. McGraw-Hill Book Co, New York, NY.

Morita, R.Y., 1997. Bacteria in Oligotrophic Environments. Chapman and Hall, New York, NY.

Morris, D.P., Lewis Jr., W.M., 1988. Phytoplankton nutrient limitation in Colorado mountain lakes. Freshw. Biol. 20, 315−327.

Morris, D.P., Zagarese, H., Williamson, C.E., Balseiro, E.G., Harbraves, B.R., Modenutti, B., et al., 1995. The attenuation of solar UV radiation in lakes and the role of dissolved organic carbon. Limnol. Oceanogr. 40, 1381−1391.

Morris, J.T., 1991. Effects of nitrogen loading on wetland ecosystems with particular reference to atmospheric deposition. Annu. Rev. Ecol. Syst. 22, 257−279.

Morrison, W., Hay, M., 2011. Induced chemical defenses in a freshwater macrophyte suppress herbivore fitness and the growth of associated microbes. Oecologia 165 (2), 427−436. Available from: https://doi.org/10.1007/s00442-010-1791-1.

Mortimer, C.H., 1941. The exchange of dissolved substances between mud and water in lakes. J. Ecol. 29, 280−329.

Morton, P., Frisch, D., Jeyasingh, P., Weider, L., 2015. Out with the old, in with the new? Younger Daphnia clones are competitively superior over centuries-old ancestors. Hydrobiologia 749 (1), 43−52. Available from: https://doi.org/10.1007/s10750-014-2145-5.

Morton, R., Cunningham, R.B., 1985. Longitudinal profile of salinity in the Murray River. Aust. J. Soil Res. 23, 1−13.

Moss, B., Stephen, D., Balayla, D., Bécares, E., Collings, S., Fernández-Aláez, C., et al., 2004. Continental-scale patterns of nutrient and fish effects on shallow lakes: synthesis of a pan-European mesocosm experiment. Freshw. Biol. 49 (12), 1633−1649.

Moy, N.J., Dodson, J., Tassone, S.J., Bukaveckas, P.A., Bulluck, L.P., 2016. Biotransport of algal toxins to riparian food webs. Environ. Sci. Technol. 50 (18), 10007−10014. Available from: https://doi.org/10.1021/acs.est.6b02760.

Moyle, P.B., Cech Jr., J.J., 1996. Fishes an Introduction to Ichthyology. Prentice Hall, Upper Saddle River, NJ.

Moyle, P.B., Li, H.W., Barton, B.A., 1986. The Frankenstein effect: impact of introduced fishes on native fishes in North America. Paper Presented at the Symposium on the Role of Fish Culture in Fisheries Management, Lake Ozark, MO, March 31−April 3, 1985.

Moyle, P.B., Light, T., 1996. Biological invasions of fresh water: empirical rules and assembly theory. Biol. Conserv. 78, 149−161.

Muggelberg, L.L., Hartz, K.E.H., Nutile, S.A., Harwood, A.D., Heim, J.R., Derby, A.P., et al., 2017. Do pyrethroid-resistant *Hyalella azteca* have greater bioaccumulation potential compared to non-resistant populations? Implications for bioaccumulation in fish. Environ. Pollut. 220, 375−382. Available from: https://doi.org/10.1016/j.envpol.2016.09.073.

Mulholland, P.J., 1981. Organic carbon flow in a swamp-stream ecosystem. Ecol. Monogr. 51 (3), 307−322.

Mulholland, P.J., DeAngelis, D.L., 2000. Surface-subsurface exchange and nutrient spiraling. In: Jones, J.B., Molholland, P.J. (Eds.), Streams and Ground Waters. Academic Press, San Diego, CA, pp. 149−166.

Mulholland, P.J., Fellows, C.S., Tank, J.L., Grimm, N.B., Webster, J.R., Hamilton, S.K., et al., 2001. Inter-biome comparison of factors controlling stream metabolism. Freshw. Biol. 46 (11), 1503−1517.

Mulholland, P.J., Helton, A.M., Poole, G.C., Hall, R.O., Hamilton, S.K., Peterson, B.J., et al., 2008. Stream denitrification across biomes and its response to anthropogenic nitrate loading. Nature 452 (7184), 202−U246. Available from: https://doi.org/10.1038/Nature06686.

Mulholland, P.J., Kuenzler, E.J., 1979. Organic carbon export from upland and forested wetland watersheds. Limnol. Oceanogr. 24 (5), 960−966.

Mulholland, P.J., Sale, M.J., 1998. Impacts of climate change on water resources: findings of the IPCC regional assessment of vulnerability for North America. Water Res. 112, 10−15.

Mulholland, P.J., Steinman, A.D., Elwood, J.W., 1990. Measurement of phosphate uptake rate in streams: comparison of radio tracer and stable PO_4 releases. Can. J. Fish. Aquat. Sci. 47, 2351−2357.

Mulholland, P.J., Tank, J.L., Webster, J.R., Bowden, W.B., Dodds, W.K., Gregory, S.V., et al., 2002. Can uptake length in streams be determined by nutrient addition experiments? Results from an inter-biome comparison study. J. N. Am. Benthol. Soc. 21 (4), 544−560.

Müller, B., Gächter, R., Wüest, A., 2014. Accelerated water quality improvement during oligotrophication in Peri-Alpine Lakes. Environ. Sci. Technol. 48 (12), 6671−6677. Available from: https://doi.org/10.1021/es4040304.

Muller-Navarra, D.C., Brett, M.T., Liston, A.M., Goldman, C.R., 2000. A highly unsaturated fatty acid predicts carbon transfer between primary producers and consumers. Nature 403, 74−77.

Müller-Navarra, D.C., Brett, M.T., Park, S., Chandra, S., Ballantyne, A.P., Zorita, E., et al., 2004. Unsaturated fatty acid content in seston and tropho-dynamic coupling in lakes. Nature 427 (6969), 69.

Münster, U., Haan, H.D., 1998. The role of microbial extracellular enzymes in the transformation of dissolved organic matter in humic waters. Ecol. Stud. 133, 199–257.

Münster, U., Heikkinen, E., Likolammi, M., Järvinen, M., Salonen, K., Haan, H.D., 1999. Utilisation of polymeric and monomeric aromatic and amino acid carbon in a humic boreal forest lake. Arch. Hydrolbiol. Spec. Issues Adv. Limnol. 54, 105–134.

Murdock, J.N., Gido, K.B., Dodds, W.K., Bertrand, K.N., Whiles, M.R., 2010. Consumer return chronology alters recovery trajectory of stream ecosystem structure and function following drought. Ecology 91 (4), 1048–1062.

Murphy, K., Willby, N.J., Eaton, J.W., 1995. Ecological impacts and management of boat traffic on navigable inland waterways. In: Harper, D.M., Ferguson, A.J.D. (Eds.), The Ecological Basis for River Management. John Wiley & Sons, Ltd, West Sussex, England, pp. 427–442.

Murphy, K.J., Barrett, P.R.F., 1990. Chemical control of aquatic weeds. In: Pieterse, A.H., Murphy, K.J. (Eds.), Aquatic Weeds: The Ecology and Management of Nuisance Aquatic Vegetation. Oxford University Press, New York, NY, pp. 136–173.

Murphy, K.J., Pieterse, A.H., 1990. Present status and prospects of integrated control of aquatic weeds. In: Pieterse, A.H., Murphy, K.J. (Eds.), Aquatic Weeds: The Ecology and Management of Nuisance Aquatic Vegetation. Oxford University Press, New York, NY, pp. 222–227.

Murray, A.B., Paola, C., 1994. A cellular model of braided rivers. Nature 371, 54–57.

Murray, A.G., 1995. Phytoplankton exudation: exploitation of the microbial loop as a defense against algal viruses. J. Plankton Res. 17, 1079–1094.

Muscarella, M.E., Jones, S.E., Lennon, J.T., 2016. Species sorting along a subsidy gradient alters bacterial community stability. Ecology 97 (8), 2034–2043. Available from: https://doi.org/10.1890/15-2026.1.

Naiman, R.J., 1983. The annual pattern and spatial distribution of aquatic oxygen metabolism in boreal forest watersheds. Ecol. Monogr. 53, 73–94.

Naiman, R.J., Décamps, H., 1997. The ecology of interfaces: riparian zones. Annu. Rev. Ecol. Syst. 28, 621–658.

Naiman, R.J., Décamps, H., Pollock, M., 1993. The role of riparian corridors in maintaining regional biodiversity. Ecol. Appl. 3 (2), 209–212.

Naiman, R.J., Johnston, C.A., Kelley, J.C., 1988. Alteration of North American streams by beaver. Bioscience 38 (11), 753–762.

Naiman, R.J., Pinay, G., Johnston, C.A., Pastor, J., 1994. Beaver influences on the long-term biogeochemical characteristics of boreal forest drainage networks. Ecology 75 (4), 905–921.

Naiman, R.J., Rogers, K.H., 1997. Large animals and system-level characteristics in river corridors. Bioscience 47 (8), 521–528.

Nakai, K., 1993. Foraging of brood predators restricted by territory of substrate-brooders in a cichlid fish assemblage. In: Kawanabe, H., Cohen, J.E., Iwasaki, K. (Eds.), Mutualism and Community Organization. Behavioural, Theoretical, and Food-Web Approaches. Oxford University Press, New York, NY, pp. 84–108.

Nakai, S., Zhou, S., Hosomi, M., Tominaga, M., 2006. Allelopathic growth inhibition of cyanobacteria by reed. Allelopathy J. 18 (2), 277–285.

Nakano, S., Murakami, M., 2001. Reciprocal subsidies: dynamic interdependence between terrestrial and aquatic food webs. Proc. Natl. Acad. Sci. USA 98 (1), 166–170. Available from: https://doi.org/10.1073/pnas.98.1.166.

Nalepa, T.F., 1994. Decline of native unionid bivalves in Lake St. Clair after infestation by the zebra mussel, Dreissena polymorpha. Can. J. Fish. Aquat. Sci. 51, 2227–2233.

Napolitano, G.E., Cicerone, D.S., 1999. Lipids in water-surface microlayers and foams. In: Arts, M.T., Wainman, B.C. (Eds.), Lipids in Freshwater Ecosystems. Springer-Verlag, New York, NY, pp. 235–262.

Napolitano, G.E., Pollero, R.J., Gayoso, A.M., MacDonald, B.A., Thompson, R.J., 1997. Fatty acids as trophic markers of phytoplankton blooms in the Bahia Blanca Estuary (Buenos Aires, Argentina) and in Trinity Bay (Newfoundland, Canada). Biochem. Syst. Ecol. 25, 739−755.

Nasri, H., El Herry, S., Bouaïcha, N., 2008. First reported case of turtle deaths during a toxic *Microcystis* spp. bloom in Lake Oubeira, Algeria. Ecotoxicol. Environ. Saf. 71 (2), 535−544.

National-Research-Council, 1992. Restoration of Aquatic Ecosystems. Science, Technology, and Public Policy. National Academy Press, Washington, DC.

NRC, 1996. Stemming the Tide. Controlling Introductions of Nonindigenous Species by Ships' Ballast Water. National Academy Press, Washington, DC.

Nedunchezhian, N., Ravindran, K.C., Abadia, A., Abadia, J., Kulandaivelu, J., 1996. Damages of photosynthetic apparatus in *Anacystis nidulans* by ultraviolet-B radiation. Biol. Plant. 38, 53−59.

Nedwell, D.B., 1999. Effect of low temperature on microbial growth: lowered affinity for substrates limits growth at low temperature. FEMS Microbiol. Ecol. 30, 101−111.

Needham, J.G., Needham, P.R., 1975. A Guide to the Study of Fresh-Water Biology. Holden-Day, Inc., San Francisco, CA.

Nelson, D., Benstead, J.P., Huryn, A.D., Cross, W.F., Hood, J.M., Johnson, P.W., et al., 2017a. Experimental whole-stream warming alters community size structure. Glob. Change Biol. 23 (7), 2618−2628. Available from: https://doi.org/10.1111/gcb.13574.

Nelson, D., Benstead, J.P., Huryn, A.D., Cross, W.F., Hood, J.M., Johnson, P.W., et al., 2017b. Shifts in community size structure drive temperature invariance of secondary production in a stream-warming experiment. Ecology 98 (7), 1797−1806. Available from: https://doi.org/10.1002/ecy.1857.

Nelson, D.R., 1991. Tardigrada. In: Thorp, J.H., Covich, A.P. (Eds.), Ecology and Classification of North American Freshwater Invertebrates. Academic Press, Inc, San Diego, CA, pp. 501−522.

Neres-Lima, V., Machado-Silva, F., Baptista, D.F., Oliveira, R.B.S., Andrade, P.M., Oliveira, A.F., et al., 2017. Allochthonous and autochthonous carbon flows in food webs of tropical forest streams. Freshw. Biol. 62 (6), 1012−1023. Available from: https://doi.org/10.1111/fwb.12921.

Neue, H.U., Quijano, C., Senadhira, D., Setter, T., 1998. Strategies for dealing with micronutrient disorders and salinity in lowland rice systems. Field Crops Res. 56, 139−155.

Neves, R.J., Bogan, A.E., Williams, J.D., Ahlstedt, S.A., Hartfield, P.W., 1997. Status of aquatic mollusks in the southeastern United States: a downward spiral of diversity. In: Benz, G.W., Collins, D.E. (Eds.), Aquatic Fauna in Peril, the Southeastern Perspective. Southeast Aquatic Research Institute, Decatur, GA, pp. 43−86. Vol. Special Publication 1.

Newbold, J.D., Elwood, J.W., O'Neill, R.V., Winkle, W.V., 1981. Measuring nutrient spiraling in streams. Can. J. Fish. Aquat. Sci. 38, 860−863.

Newell, D.A., Goldingay, R.L., Brooks, L.O., 2013. Population recovery following decline in an endangered stream-breeding frog (*Mixophyes fleayi*) from subtropical Australia. PLoS One 8 (3), e58559.

Newman, S., Schuette, J., Grace, J.B., Rutchey, K., Fontaine, T., Reddy, K.R., et al., 1998. Factors influencing cattail abundance in the northern Everglades. Aquat. Bot. 60, 265−280.

Ng, F., Liu, S., Mavlyudov, B., Wang, Y., 2007. Climatic control on the peak discharge of glacier outburst floods. Geophys. Res. Lett. 34 (21).

Nielsen, L.T., Asadzadeh, S.S., Dölger, J., Walther, J.H., Kiørboe, T., Andersen, A., 2017. Hydrodynamics of microbial filter feeding. Proc. Natl. Acad. Sci. 114 (35), 9373−9378. Available from: https://doi.org/10.1073/pnas.1708873114.

Nifong, R.L., 2017. Experimental effects of grazers on autotrophic species assemblages across a nitrate gradient in Florida springs. Aquat. Bot. 139, 57−64. Available from: https://doi.org/10.1016/j.aquabot.2017.02.010.

Nikora, V.I., Goring, D.G., Biggs, B.J.F., 1997. On stream periphyton-turbulence interactions. N. Z. J. Mar. Freshw. Res. 31, 435−448.

Nilsson, C., Jansson, R., Zinko, U., 1997. Long-term responses of river-margin vegetation to water-level regulation. Science 276, 798−800.

Nilsson, C., Reidy, C.A., Dynesius, M., Revenga, C., 2005. Fragmentation and flow regulation of the world's large river systems. Science 308 (5720), 405−408.

Nilsson, C., Riis, T., Sarneel, J.M., Svavarsdóttir, K., 2018. Ecological restoration as a means of managing inland flood hazards. Bioscience 68 (2), 89−99.

Nilsson, C., Sarneel, J.M., Palm, D., Gardeström, J., Pilotto, F., Polvi, L.E., et al., 2017. How do biota respond to additional physical restoration of restored streams? Ecosystems 20 (1), 144–162. Available from: https://doi.org/10.1007/s10021-016-0020-0.

Noble, R.L., Jones, T.W., 1999. Managing fisheries with regulations. In: Kohler, C.C., Hubert, W.A. (Eds.), Inland Fisheries Management in North America, second ed. American Fisheries Society, Bethesda, MD, pp. 455–477.

Norman, B.C., Whiles, M.R., Collins, S.M., Flecker, A.S., Hamilton, S.K., Johnson, S.L., et al., 2017. Drivers of nitrogen transfer in stream food webs across continents. Ecology 98 (12), 3044–3055. Available from: https://doi.org/10.1002/ecy.2009.

North, R.L., Guildford, S.J., Smith, R.E.H., Havens, S.M., Twiss, M.R., 2007. Evidence for phosphorus, nitrogen, and iron colimiatino of phytoplankton communities in Lake Erie. Limnol. Oceanogr. 52, 315–328.

Nowell, L.H., Capel, P.D., Dileanis, P.D., 1999. Pesticides in Stream Sediment and Aquatic Biota, Distribution, Trends, and Governing Factors, vol. 4. Lewis Publishers, Boca Ratan, FL.

Nummi, P., Holopainen, S., 2014. Whole-community facilitation by beaver: ecosystem engineer increases waterbird diversity. Aquat. Conserv.: Mar. Freshw. Ecosyst. 24 (5), 623–633. Available from: https://doi.org/10.1002/aqc.2437.

Nurnberg, G.K., 2007. Lake responses to long-term hypolimnetic withdrawal treatments. Lake Reserv. Manage. 23 (4), 388–409.

Nürnberg, G.K., 1996. Trophic state of clear and colored, soft- and hardwater lakes with special consideration of nutrients, anoxia, phytoplankton and fish. J. Lake Reserv. Manage. 12 (4), 432–447.

O'Brien, J., Lessard, J., Plew, D., Graham, S.E., McIntosh, A., 2014. Aquatic macrophytes alter metabolism and nutrient cycling in lowland streams. Ecosystems 17 (3), 405–417. Available from: https://doi.org/10.1007/s10021-013-9730-8.

O'Brien, J.M., Dodds, W.K., Wilson, K.C., Murdock, J.N., Eichmiller, J., 2007. The saturation of N cycling in Central Plains streams: N-15 experiments across a broad gradient of nitrate concentrations. Biogeochemistry 84 (1), 31–49. Available from: https://doi.org/10.1007/s10533-007-9073-7.

O'Connell, T.C., 2017. 'Trophic' and 'source' amino acids in trophic estimation: a likely metabolic explanation. Oecologia 184 (2), 317–326. Available from: https://doi.org/10.1007/s00442-017-3881-9.

O'hanlon, S.J., Rieux, A., Farrer, R.A., Rosa, G.M., Waldman, B., Bataille, A., et al., 2018. Recent Asian origin of chytrid fungi causing global amphibian declines. Science 360 (6389), 621–627.

Oberdorff, T., Hugueny, B., 1995. Global scale patterns of fish species richness in rivers. Ecography 18 (4), 345–352.

Oberemm, A., Becker, J., Codd, G.A., Steinberg, C., 1999. Effects of cyanobacterial toxins and aqueous crude extracts of cyanobacteria on the development of fish and amphibians. Environ. Toxicol. 14, 77–88.

Obermeyer, B.K., Edds, D.R., Miller, E.J., Prophet, C.W., 1997. Range reductions of southeast Kansas unionids. Paper Presented at the Conservation and Management of Freshwater Mussels II. Inititiatives for the Future. Proceedings of a UMCRR Symposium, St. Louis, MO, October 16–18, 1995.

Obeysekera, J., Barnes, J., Nungesser, M., 2015. Climate sensitivity runs and regional hydrologic modeling for predicting the response of the greater Florida Everglades ecosystem to climate change. Environ. Manage. 55 (4), 749–762.

Ochumba, P.B.O., 1990. Massive fish kills within the Nyanza Gulf of Lake Victoria, Kenya. Hydrobiologia 208, 93–99.

O'Connor, M.P., Kemp, S.J., Agosta, S.J., Hansen, F., Sieg, A.E., Wallace, B.P., et al., 2007. Reconsidering the mechanistic basis of the metabolic theory of ecology. Oikos 116 (6), 1058–1072. Available from: https://doi.org/10.1111/j.0030-1299.2007.15534.x.

Odum, E.P., Barrett, G.W., 1971. Fundamentals of Ecology, vol. 3. Saunders, Philadelphia.

Odum, H.T., 1956. Primary production in flowing waters. Limnol. Oceanogr. 1, 102–117.

Odum, H.T., 1957. Trophic structure and productivity of Silver Springs, Florida. Ecol. Monogr. 27, 55–112.

Odum, H.T., Odum, E.P., 1959. Principles and concepts pertaining to energy in ecological systems. In: Odum, E.P., Odum, H.T. (Eds.), Fundamentals of Ecology, second ed. W. B. Saunders Company, Philadelphia, PA, pp. 43—87.

OECD, 1982. Eutrophication of waters. Monitoring assessment and control. Final report, Paris, France. Retrieved from https://onlinelibrary.wiley.com/doi/pdf/10.1002/iroh.19840690206.

O'Hara, S.L., Street-Perrott, F.A., Burt, T.P., 1993. Accelerated soil erosion around a Mexican highland lake caused by prehispanic agriculture. Nature 362, 48—51.

O'Hare, M.T., Clarke, R.T., Bowes, M.J., Cailes, C., Henville, P., Bissett, N., et al., 2010. Eutrophication impacts on a river macrophyte. Aquat. Bot. 92 (3), 173—178.

Okamura, B., Feist, S.W., 2011. Emerging diseases in freshwater systems. Freshw. Biol. 56 (4), 627—637. Available from: https://doi.org/10.1111/j.1365-2427.2011.02578.x.

Oki, T., Kanae, S., 2006. Global hydrological cycles and world water resources. Science 313 (5790), 1068—1072.

Olsen, B.K., Chislock, M.F., Wilson, A.E., 2016. Eutrophication mediates a common off-flavor compound, 2-methylisoborneol, in a drinking water reservoir. Water Res. 92, 228—234. Available from: https://doi.org/10.1016/j.watres.2016.01.058.

Olson, M.H., Hage, M.M., Binkley, M.D., Binder, J.R., 2005. Impact of migratory snow geese on nitrogen and phosphorus dynamics in a freshwater reservoir. Freshw. Biol. 50 (5), 882—890.

Olsson, H., 1991. Phosphatase activity in an acid, limed Swedish lake. In: Chróst, R.J. (Ed.), Microbial Enzymes in Aquatic Environments. Springer-Verlag, New York, NY, pp. 206—219.

Omernik, J., Griffith, G., 2014. Ecoregions of the conterminous United States: evolution of a hierarchical spatial framework. Environ. Manage. 54 (6), 1249—1266. Available from: https://doi.org/10.1007/s00267-014-0364-1.

Omernik, J.M., 1995. Ecoregions: a spatial framework for environmental management. In: Davis, W.S., Simon, T.P. (Eds.), Biological Assessment and Criteria. Tools for Water Resource Planning and Decision Making. Lewis Publishers, Boca Ratan, FL, pp. 49—66.

O'Neill Jr., C.R., 1997. Economic impact of zebra mussels- results of the 1995 National Zebra Mussel Information Clearinghouse Study. Great Lakes Res. Rev 3 (1), 35—44.

O'Neil, J., Davis, T.W., Burford, M.A., Gobler, C., 2012. The rise of harmful cyanobacteria blooms: the potential roles of eutrophication and climate change. Harmful Algae 14, 313—334.

O'Reilly, C.M., Sharma, S., Gray, D.K., Hampton, S.E., Read, J.S., Rowley, R.J., et al., 2015. Rapid and highly variable warming of lake surface waters around the globe. Geophys. Res. Lett. 42 (24), 10773—10781.

Oremland, R.S., Stolz, J.F., 2003. The ecology of arsenic. Science 300 (5621), 939—944.

Orihel, D.M., Bird, D.F., Brylinsky, M., Chen, H., Donald, D.B., Huang, D.Y., et al., 2012. High microcystin concentrations occur only at low nitrogen-to-phosphorus ratios in nutrient-rich Canadian lakes. Can. J. Fish. Aquat. Sci. 69 (9), 1457—1462. Available from: https://doi.org/10.1139/f2012-088.

Orihel, D.M., Schindler, D.W., Ballard, N.C., Wilson, L.R., Vinebrooke, R.D., 2016. Experimental iron amendment suppresses toxic cyanobacteria in a hypereutrophic lake. Ecol. Appl. 26 (5), 1517—1534. Available from: https://doi.org/10.1890/15-1928.

O'Riordan, T., 1999. Economic challenges for lake management. Hydrobiologia 395/396, 13—18.

Orr, C.H., Kroiss, S.J., Rogers, K.L., Stanley, E.H., 2008. Downstream benthic responses to small dam removal in a coldwater stream. River Res. Appl. 24 (6), 804—822.

Osenberg, C., Mittelbach, G., Wainwright, P., 1992. Two-stage life histories in fish: the interaction between juvenile competition and adult performance. Ecology 255—267.

Osman, O.A., Beier, S., Grabherr, M., Bertilsson, S., 2017. Interactions of freshwater cyanobacteria with bacterial antagonists. Appl. Environ. Microbiol. 83 (7). Available from: https://doi.org/10.1128/aem.02634-16.

Otten, T.G., Graham, J.L., Harris, T.D., Dreher, T.W., 2016. Elucidation of taste- and odor-producing bacteria and toxigenic cyanobacteria in a midwestern drinking water supply reservoir by shotgun metagenomic analysis. Appl. Environ. Microbiol. 82 (17), 5410—5420. Available from: https://doi.org/10.1128/aem.01334-16.

Ottinger, R.L., Wooley, D.R., Robinson, N.A., Hodas, D.R., Babb, S.E., 1990. Environmental Costs of Electricity. Oceana Publications, Inc., New York, NY.

Ottman, N., Smidt, H., de Vos, W.M., Belzer, C., 2012. The function of our microbiota: who is out there and what do they do? Front. Cell. Infect. Microbiol. 2. Available from: https://doi.org/10.3389/fcimb.2012.00104.

Ozaki, A., Adams, E., Binh, C.T.T., Tong, T., Gaillard, J.-F., Gray, K.A., et al., 2016. One-time addition of nano-TiO_2 triggers short-term responses in benthic bacterial communities in artificial streams. Microb. Ecol. 71 (2), 266—275. Available from: https://doi.org/10.1007/s00248-015-0646-z.

Ozersky, T., Pastukhov, M.V., Poste, A.E., Deng, X.Y., Moore, M.V., 2017. Long-term and ontogenetic patterns of heavy metal contamination in lake Baikal Seals (*Pusa sibirica*). Environ. Sci. Technol. 51 (18), 10316—10325. Available from: https://doi.org/10.1021/acs.est.7b00995.

Pace, M.L., Batt, R.D., Buelo, C.D., Carpenter, S.R., Cole, J.J., Kurtzweil, J.T., et al., 2017. Reversal of a cyanobacterial bloom in response to early warnings. Proc. Natl. Acad. Sci. 114 (2), 352—357. Available from: https://doi.org/10.1073/pnas.1612424114.

Pace, M.L., Carpenter, S.R., Johnson, R.A., Kurtzweil, J.T., 2013. Zooplankton provide early warnings of a regime shift in a whole lake manipulation. Limnol. Oceanogr. 58 (2), 525—532. Available from: https://doi.org/10.4319/lo.2013.58.2.0525.

Pace, M.L., Cole, J.J., 1994. Comparative and experimental approaches to top down and bottom-up regulation of bacteria. Microb. Ecol. 28, 181—193.

Pace, M.L., Cole, J.J., Carpenter, S.R., 1998. Trophic cascades and compensation: differential responses of microzooplankton in whole-lake experiments. Ecology 79, 138—152.

Pace, M.L., Cole, J.J., Carpenter, S.R., Kitchell, J.F., Hodgson, J.R., Van de Bogert, M.C., et al., 2004. Whole-lake carbon-13 additions reveal terrestrial support of aquatic food webs. Nature 427, 240—243.

Pace, N.R., 1997. A molecular view of microbial diversity and the biosphere. Science 276, 734—740.

Pachauri, R.K., Allen, M.R., Barros, V.R., Broome, J., Cramer, W., Christ, R., et al., 2014. Climate change 2014: synthesis report. Contribution of working groups I, II and III to the 5th assessment report of the Intergovernmental Panel on Climate Change: IPCC.

Paerl, H.W., 1990. Physiological ecology and regulation of N_2 fixation in natural waters. Adv. Microb. Ecol. 11, 305—344.

Paerl, H.W., 2017. Controlling harmful cyanobacterial blooms in a climatically more extreme world: management options and research needs. J. Plankton Res. 39 (5), 763—771. Available from: https://doi.org/10.1093/plankt/fbx042.

Paerl, H.W., Hall, N.S., Calandrino, E.S., 2011. Controlling harmful cyanobacterial blooms in a world experiencing anthropogenic and climatic-induced change. Sci. Total Environ. 409 (10), 1739—1745.

Paerl, H.W., Otten, T.G., 2013. Harmful cyanobacterial blooms: causes, consequences, and controls. Microb. Ecol. 65 (4), 995—1010.

Paerl, H.W., Paul, V.J., 2012. Climate change: links to global expansion of harmful cyanobacteria. Water Res. 46 (5), 1349—1363. Available from: https://doi.org/10.1016/j.watres.2011.08.002.

Paerl, H.W., Pinckney, J.L., 1996. A mini-review of microbial consortia: their roles in aquatic production and biogeochemical cycling. Microb. Ecol. 31, 225—247.

Paerl, H.W., Scott, J.T., McCarthy, M.J., Newell, S.E., Gardner, W.S., Havens, K.E., et al., 2016. It takes two to tango: when and where dual nutrient (N & P) reductions are needed to protect lakes and downstream ecosystems. Environ. Sci. Technol. 50 (20), 10805—10813. Available from: https://doi.org/10.1021/acs.est.6b02575.

Page, L.M., Burr, B.M., 2011. Peterson Field Guide to Freshwater Fishes, second ed. Houghton Mifflin.

Pajak, G., 1986. Development and structure of the Goczalkowice Reservoir ecosystem VIII. Phytoplankton. Ekol. Pol. 34 (3), 397—413.

Palmer, M.A., Bernhardt, E.S., Allan, J.D., Lake, P.S., Alexander, G., Brooks, S., et al., 2005. Standards for ecologically successful river restoration. J. Appl. Ecol. 42 (2), 208—217. Available from: https://doi.org/10.1111/j.1365-2664.2005.01004.x.

Palmer, M.A., Covich, A.P., Finlay, B.J., Gibert, J., Hyde, K.D., Johnson, R.K., et al., 1997. Biodiversity and ecosystem processes in freshwater sediments. Ambio 26 (8), 571—577.

Palmer, M.A., Poff, N.L., 1997. Heterogeneity in streams. the influence of environmental heterogeneity on patterns and processes in streams. J. N. Am. Benthol. Soc. 16, 169–173.

Palumbo, A.V., Mulholland, P.J., Elwood, J.W., 1989. Epilithic microbial populations and leaf decomposition in acid-stressed streams. In: Rao, S.S. (Ed.), Acid Stress and Aquatic Microbial Interactions. CRC Press, Boca Ratan, FL, pp. 70–88.

Pang, L., Close, M.E., 1999. Attenuation and transport of atrazine and picloram in an alluvial gravel aquifer: a tracer test and batch study. N. Z. J. Mar. Freshw. Res. 33, 279–291.

Paragamian, V.I., 1987. Standing stocks of fish in some Iowa streams, with a comparison of channelized and natural stream reaches in the southern Iowa drift plain. Proc. Iowa Acad. Sci. 94 (4), 128–134.

Parker, J.D., Caudill, C.C., Hay, M.E., 2007. Beaver herbivory on aquatic plants. Oecologia 151 (4), 616–625.

Parris, K.M., 2006. Urban amphibian assemblages as metacommunities. J. Anim. Ecol. 75 (3), 757–764.

Parris, M.J., Cornelius, T.O., 2004. Fungal pathogen causes competitive and developmental stress in larval amphibian communities. Ecology 85 (12), 3385–3395.

Patrick, R., 1967. The effect of invasion rate, species pool, and size of area on the structure of diatom community. Proc. Natl. Acad. Sci. 58, 1335–1342.

Patrick, R., Reimer, C.W., 1966. The Diatoms of the United States, vol. 1. Monographs of the Academy of Natural Sciences of Philadelphia, Philadelphia, PA.

Patrick, R., Reimer, C.W., 1975. The Diatoms of the United States, vol. 2. Monographs of the Academy of Natural Sciences of Philadelphia, Philadelphia, PA.

Patten, B.C., 1993. Discussion: promoted coexistence through indirect effects: need for a new ecology of complexity. In: Kawanabe, H., Cohen, J.E., Iwasaki, K. (Eds.), Mutualism and Community Organization. Behavioural, Theoretical, and Food-Web Approaches. Oxford University Press, New York, NY, pp. 323–349.

Patten, D.T., 1998. Riparian ecosystems of semi-arid North America: diversity and human impacts. Wetlands 18 (4), 498–512.

Patterson, D.J., 1999. The diversity of eukaryotes. Am. Nat. 154 (Suppl.), S96–S124.

Paul, M.J., Meyer, J.L., Couch, C.A., 2006. Leaf breakdown in streams differing in catchment land use. Freshw. Biol. 51, 1684–1695.

Paul, V.G., Mormile, M.R., 2017. A case for the protection of saline and hypersaline environments: a microbiological perspective. FEMS Microbiol. Ecol. 93 (8).

Pawlik-Skowronska, B., Toporowska, M., 2016. How to mitigate cyanobacterial blooms and cyanotoxin production in eutrophic water reservoirs? Hydrobiologia 778 (1), 45–59. Available from: https://doi.org/10.1007/s10750-016-2842-3.

Payne, W.J., 1981. Denitrification. John Wiley & Sons, New York, NY.

Peacock, A.D., White, D.C., 2017. Microbial biomass and community composition analysis using phospholipid fatty acids. In: McGenity, T.J., Timmis, K.N., Nogales, B. (Eds.), Hydrocarbon and Lipid Microbiology Protocols: Microbial Quantitation, Community Profiling and Array Approaches. Springer Berlin Heidelberg, Berlin, Heidelberg, pp. 65–76.

Peary, J.A., Castenholz, R.W., 1964. Temperature strains of a thermophilic blue-green alga. Nature 5/64, 720–721.

Peckarsky, B.L., 1982. Aquatic insect predator-prey relations. Bioscience 32 (4), 261–266.

Peckarsky, B.L., Penton, M.A., 1989. Early warning lowers risk of stonefly predation for a vulnerable mayfly. Oikos 54, 301–309.

Peckarsky, B.L., Wilcox, R.S., 1989. Stonefly nymphs use hydrodynamic cues to discriminate between prey. Oecologia 79, 265–270.

Peipoch, M., Brauns, M., Hauer, F.R., Weitere, M., Valett, H.M., 2015. Ecological simplification: human influences on riverscape complexity. Bioscience 65 (11), 1057–1065. Available from: https://doi.org/10.1093/biosci/biv120.

Pejler, B., 1995. Relation to habitat in rotifers. Hydrobiologia 313/314, 267–278.

Pekel, J.-F., Cottam, A., Gorelick, N., Belward, A.S., 2016. High-resolution mapping of global surface water and its long-term changes. Nature 540 (7633), 418–422. Available from: https://doi.org/

10.1038/nature20584. Available from: http://www.nature.com/nature/journal/v540/n7633/abs/nature20584.html#supplementary-information.

Pelicice, F.M., Pompeu, P.S., Agostinho, A.A., 2015. Large reservoirs as ecological barriers to downstream movements of Neotropical migratory fish. Fish Fish. 16 (4), 697−715.

Pelton, D.K., Levine, S., Braner, M., 1998. Measurements of phosphorus uptake by macrophytes and epiphytes from the LaPlatte River (VT) using ^{32}P in stream microcosms. Freshw. Biol. 39, 285−299.

Peng, S., Liao, H., Zhou, T., Peng, S., 2017. Effects of UVB radiation on freshwater biota: a meta-analysis. Glob. Ecol. Biogeogr. 26 (4), 500−510. Available from: https://doi.org/10.1111/geb.12552.

Pennak, R.W., 1978. Fresh-Water Invertebrates of the United States, second ed. John Wiley & Sons, New York, NY.

Pennisi, E., 2014. Water's tough skin. Science 343 (6176), 1194−1197. Available from: https://doi.org/10.1126/science.343.6176.1194.

Peoples, B.K., Frimpong, E.A., 2016a. Biotic interactions and habitat drive positive co-occurrence between facilitating and beneficiary stream fishes. J. Biogeogr. 43 (5), 923−931. Available from: https://doi.org/10.1111/jbi.12699.

Peoples, B.K., Frimpong, E.A., 2016b. Context-dependent outcomes in a reproductive mutualism between two freshwater fish species. Ecol. Evol. 6 (4), 1214−1223. Available from: https://doi.org/10.1002/ece3.1979.

Pérez, C., Muckle, M.T., Zaleski, D.P., Seifert, N.A., Temelso, B., Shields, G.C., et al., 2012. Structures of cage, prism, and book isomers of water hexamer from broadband rotational spectroscopy. Science 336 (6083), 897−901.

Pérez, M.S., 2015. Where the xingu bends and will soon break. Am. Sci. 103 (6), 395.

Perez, R., Richards-Zawacki, C., Krohn, A.R., Robak, M., Griffith, E.J., Ross, H., et al., 2014. Field surveys in Western Panama indicate populations of *Atelopus varius* frogs are persisting in regions where *Batrachochytrium dendrobatidis* is now enzootic.

Perkin, J.S., Gido, K.B., 2011. Stream fragmentation thresholds for a reproductive guild of Great Plains fishes. Fisheries 36 (8), 371−383.

Perkin, J.S., Gido, K.B., Cooper, A.R., Turner, T.F., Osborne, M.J., Johnson, E.R., et al., 2015. Fragmentation and dewatering transform Great Plains stream fish communities. Ecol Monogr. 85 (1), 73−92.

Perkin, J.S., Gido, K.B., Costigan, K.H., Daniels, M.D., Johnson, E.R., 2015. Fragmentation and drying ratchet down Great Plains stream fish diversity. Aquat. Conserv.: Mar. Freshw. Ecosyst. 25 (5), 639−655. Available from: https://doi.org/10.1002/aqc.2501.

Perkin, J.S., Gido, K.B., Falke, J.A., Fausch, K.D., Crockett, H., Johnson, E.R., et al., 2017. Groundwater declines are linked to changes in Great Plains stream fish assemblages. Proc. Natl. Acad. Sci. 114 (28), 7373−7378.

Perron, J.T., Richardson, P.W., Ferrier, K.L., Lapotre, M., 2012. The root of branching river networks. Nature 492 (7427), 100−103. Available from: http://www.nature.com/nature/journal/v492/n7427/abs/nature11672.html#supplementary-information.

Perry, M., 2008. Effects of environmental and occupational pesticide exposure on human sperm: a systematic review. Hum. Reprod. Update 14 (3), 233−242.

Perry, W.L., Lodge, D.M., Lamberti, G.A., 1997. Impact of crayfish predation on exotic zebra mussels and native invertebrates in a lake-outlet stream. Can. J. Fish. Aquat. Sci. 54, 120−125.

Persson, A., 1997. Effects of fish predation and excretion on the configuration of aquatic food webs. Oikos 79, 137−146.

Petersen, E.J., Akkanen, J., Kukkonen, J.V.K., Weber, W.J., 2009. Biological uptake and depuration of carbon nanotubes by *Daphnia magna*. Environ. Sci. Technol. 43 (8), 2969−2975. Available from: https://doi.org/10.1021/es8029363.

Peterson, B.J., 1999. Stable isotopes as tracers of organic matter input and transfer in benthic food webs: a review. Acta Oecol. 20, 479−487.

Peterson, B.J., Bahr, M., Kling, G.W., 1997. A tracer investigation of nitrogen cycling in a pristine tundra river. Can. J. Fish. Aquat. Sci. 54, 2361−2367.

Peterson, B.J., Deegan, L.A., Helfrich, J., Hobbie, J.E., Hullar, M., Moller, B., et al., 1993. Biological response of a tundra river to fertilization. Ecology 74 (3), 653–672.

Peterson, B.J., Fry, B., 1987. Stable isotopes in ecosystem studies. Annu. Rev. Ecol. Syst. 18 (1), 293–320.

Peterson, B.J., Wollheim, W.M., Mulholland, P.J., Webster, J.R., Meyer, J.L., Tank, J.L., et al., 2001. Control of nitrogen export from watersheds by headwater streams. Science 292, 86–90.

Peterson, S.A., Van Sickle, J., Herlihy, A.T., Hughes, R.M., 2007. Mercury concentration in fish from streams and rivers throughout the western United States. Environ. Sci. Technol. 41 (1), 58–65.

Petrie, B., Barden, R., Kasprzyk-Hordern, B., 2015. A review on emerging contaminants in wastewaters and the environment: current knowledge, understudied areas and recommendations for future monitoring. Water Res. 72, 3–27. Available from: https://doi.org/10.1016/j.watres.2014.08.053.

Petruzzella, A., Guariento, R.D., Gripp, Ad.R., Marinho, C.C., Figueiredo-Barros, M.P., Esteves, Fd.A., 2015. Herbivore damage increases methane emission from emergent aquatic macrophytes. Aquat. Bot. 127, 6–11. Available from: https://doi.org/10.1016/j.aquabot.2015.07.003.

Pettit, N.E., Ward, D.P., Adame, M.F., Valdez, D., Bunn, S.E., 2016. Influence of aquatic plant architecture on epiphyte biomass on a tropical river floodplain. Aquat. Bot. 129, 35–43. Available from: https://doi.org/10.1016/j.aquabot.2015.12.001.

Petts, G., Maddock, I., Bickerton, M., Ferguson, A.J.D., 1995. Linking hydrology and ecology: the scientific basis for river management. In: Harper, D.M., Ferguson, A.J.D. (Eds.), The Ecological Basis for River Management. John Wiley & Sons, Ltd, West Sussex, England, pp. 1–16.

Phillips, G., Lipton, J., 1995. Injury to aquatic resources caused by metals in Montana's Clark Fork River basin: historic perspective and overview. Can. J. Fish. Aquat. Sci. 52, 1990–1993.

Phillips, G., Willby, N., Moss, B., 2016. Submerged macrophyte decline in shallow lakes: what have we learnt in the last forty years? Aquat. Bot. 135, 37–45. Available from: https://doi.org/10.1016/j.aquabot.2016.04.004.

Pick, F.R., 2016. Blooming algae: a Canadian perspective on the rise of toxic cyanobacteria. Can. J. Fish. Aquat. Sci. 73 (7), 1149–1158. Available from: https://doi.org/10.1139/cjfas-2015-0470.

Pickett, S.T., Cadenasso, M.L., 1995. Landscape ecology: spatial heterogeneity in ecological systems. Science 269 (5222), 331.

Pickett, S.T.A., Kolasa, J., Jones, C.G., 1994. Ecological Understanding. Academic Press, San Diego, CA.

Pidwirny, M., 2006. Hydrology. Fundamentals of physical geography, second ed. Retrieved from <http://www.physicalgeography.net/fundamentals/contents.html>.

Pielou, E.C., 1977. Mathematical Ecology. John Wiley & Sons, New York, NY.

Piggott, J.J., Salis, R.K., Lear, G., Townsend, C.R., Matthaei, C.D., 2015. Climate warming and agricultural stressors interact to determine stream periphyton community composition. Glob. Change Biol. 21 (1), 206–222. Available from: https://doi.org/10.1111/gcb.12661.

Pijanowska, J., 1997. Alarm signals in Daphnia? Oecologia 112, 12–16.

Pimentel, D., 1996. Green revolution agriculture and chemical hazards. Sci. Total Environ. 188, S86–S98.

Pimentel, D., Acquay, H., Biltonen, M., Rice, P., Silva, M., Nelson, J., et al., 1992. Environmental and economic costs of pesticide use. Bioscience 42, 750–760.

Pimentel, D., Cooperstein, S., Randell, H., Filiberto, D., Sorrentino, S., Kaye, B., et al., 2007. Ecology of increasing diseases: population growth and environmental degradation. Hum. Ecol. 35 (6), 653–668.

Pimentel, D., Harvey, C., Resosudarmo, P., Sinclair, K., Kurz, D., McNair, M., et al., 1995. Environmental and economic costs of soil erosion and conservation benefits. Science 267, 1117–1123.

Pimentel, D., Zuniga, R., Morrison, D., 2005. Update on the environmental and economic costs associated with alien-invasive species in the United States. Ecol. Econ. 52, 273–288.

Pimm, S.L., 1982. Food Webs. Chapman and Hall Ltd., London, GB.

Pimm, S.L., Russell, G.J., Gittleman, J.L., Brooks, T.M., 1995. The future of biodiversity. Science 269, 347–350.

Pina, S., Creus, A., González, N., Gironés, R., Felip, M., Sommaruga, R., 1998. Abundance, morphology and distribution of planktonic virus-like particles in two high-mountain lakes. J. Plankton Res. 20 (12), 2413—2421.

Pineda-Mendoza, R.M., Zúñiga, G., Martínez-Jerónimo, F., 2016. Microcystin production in *Microcystis aeruginosa*: effect of type of strain, environmental factors, nutrient concentrations, and N:P ratio on mcyA gene expression. Aquat. Ecol. 50 (1), 103—119. Available from: https://doi.org/10.1007/s10452-015-9559-7.

Pink, M., Abrahams, M.V., 2015. Temperature and its impact on predation risk within aquatic ecosystems. Can. J. Fish. Aquat. Sci. 73 (6), 869—876. Available from: https://doi.org/10.1139/cjfas-2015-0302.

Piper, G.L., 1996. Biological control of the wetlands weed purple loosestrife (*Lythrum salicaria*) in the Pacific northwestern United States. Hydrobiologia 340, 291—294.

Pirc, U., Vidmar, M., Mozer, A., Kržan, A., 2016. Emissions of microplastic fibers from microfiber fleece during domestic washing. Environ. Sci. Pollut. Res. 23 (21), 22206—22211.

Pitcher, T.J., Hart, P.J.B., 1982. Fisheries Ecology. The Avi Publishing Company, Inc., Westport, CN.

Platts, W.S., Megahan, W.F., Minshall, G.W., 1983. Methods for Evaluating Stream, Riparian, and Biotic Conditions. US Department of Agriculture, Forest Service, Intermountain Forest and Range Experiment Station, Ogden, UT, Gen. Tech. Rep. INT-138, 70 pp. 138.

Podolsky, R.D., 1994. Temperature and water viscosity: physiological versus mechanical effects on suspension feeding. Science 265, 100—103.

Poff, N.L., Allan, J.D., Bain, M.B., Karr, J.R., Prestegaard, K.L., Richter, B.D., et al., 1997. The natural flow regime. Bioscience 47 (11), 769—784.

Poff, N.L., Olden, J.D., Merritt, D.M., Pepin, D.M., 2007. Homogenization of regional river dynamics by dams and global biodiversity implications. Proc. Natl. Acad. Sci. USA 104 (14), 5732—5737.

Poff, N.L., Ward, J.V., 1989. Implications of streamflow variability and predictability for lotic community structure: a regional analysis of streamflow patterns. Can. J. Fish. Aquat. Sci. 46, 1805—1818.

Poinar, G.O.J., 1991. *Nematoda* and *Nematomorpha*. In: Thorp, J.H., Covich, A.P. (Eds.), Ecology and Classification of North American Freshwater Invertebrates. Academic Press, Inc., San Diego, CA, pp. 249—284.

Polis, G.A., Anderson, W.B., Holt, R.D., 1997. Toward an integration of landscape and food web ecology: the dynamics of spatially subsidized food webs. Annu. Rev. Ecol. Syst. 28, 289—316.

Polis, G.A., Power, M.E., Huxel, G.R. (Eds.), 2004. Food Webs at the Landscape Level. The University of Chicago Press, Chicago, IL.

Pollard, A.I., Reed, T., 2004. Benthic invertebrate assemblage change following dam removal in a Wisconsin stream. Hydrobiologia 513 (1—3), 51—58.

Pollock, M.M., Beechie, T.J., Wheaton, J.M., Jordan, C.E., Bouwes, N., Weber, N., et al., 2014. Using beaver dams to restore incised stream ecosystems. Bioscience 64 (4), 279—290.

Pollock, M.M., Naiman, R.J., Hanley, T.A., 1998. Plant species richness in riparian wetlands: a test of biodiversity theory. Ecology 79, 94—105.

Pollux, B.J.A., 2011. The experimental study of seed dispersal by fish (ichthyochory). Freshw. Biol. 56 (2), 197—212. Available from: https://doi.org/10.1111/j.1365-2427.2010.02493.x.

Pompeani, D.P., Abbott, M.B., Steinman, B.A., Bain, D.J., 2013. Lake sediments record prehistoric lead pollution related to early copper production in North America. Environ. Sci. Technol. 47 (11), 5545—5552. Available from: https://doi.org/10.1021/es304499c.

Ponsatí, L., Corcoll, N., Petrović, M., Picó, Y., Ginebreda, A., Tornés, E., et al., 2016. Multiple-stressor effects on river biofilms under different hydrological conditions. Freshw. Biol. 61 (12), 2102—2115. Available from: https://doi.org/10.1111/fwb.12764.

Ponton, F., Otalora-Luna, F., Lefevre, T., Guerin, P.M., Lebarbenchon, C., Duneau, D., et al., 2011. Water-seeking behavior in worm-infected crickets and reversibility of parasitic manipulation. Behav. Ecol. 22 (2), 392—400. Available from: https://doi.org/10.1093/beheco/arq215.

Poor, P.J., Boyle, K.J., Taylor, L.O., Bouchard, R., 2001. Objective versus subjective measures of water clarity in hedonic property value models. Land Econ. 77 (4), 482—493.

Popisil, P., 1994. The groundwater fauna of a Danube aquifer in the "Lobau" wetland in Vienna, Austria. In: Gibert, J., Danielopol, D.L., Stanford, J.A. (Eds.), Groundwater Ecology. Academic Press, Inc., San Diego, CA, pp. 347–390.

Poppe, M., Kail, J., Aroviita, J., Stelmaszczyk, M., Giełczewski, M., Muhar, S., 2016. Assessing restoration effects on hydromorphology in European mid-sized rivers by key hydromorphological parameters. Hydrobiologia 769 (1), 21–40. Available from: https://doi.org/10.1007/s10750-015-2468-x.

Por, F.D., 2007. Ophel: a groundwater biome based on chemoautotrophic resources: the global significance of the Ayyalon cave finds, Israel. Hyhdrobiologia 592, 1–10.

Porter, K.G., 1976. Enhancement of algal growth and productivity by grazing zooplankton. Science 192, 1332–1334.

Porter, K.G., 1977. The plant-animal interface in freshwater ecosystems. Am. Sci. 65, 159–170.

Porter, K.G., Bergstedt, A., Freeman, M.C., 1999. The Okefenokee Swamp. Invertebrate communities and foodwebs. In: Batzer, D.P., Rader, R.B., Wissinger, S.A. (Eds.), Invertebrates in Freshwater Wetlands of North America: Ecology and Management. John Wiley & Sons, Inc, New York, NY, pp. 121–135.

Porter, K.G., Gerritsen, J., Orcutt Jr., J.D., 1982. The effect of food concentration on swimming patterns, feeding behavior, ingestion, assimilation, and respiration by *Daphnia*. Limnol. Oceanogr. 27 (5), 935–949.

Porter, K.G., Sherr, E.B., Sherr, B.F., Pace, M., Sanders, R.W., 1985. Protozoa in planktonic food webs. J. Protozool. 32 (3), 409–415.

Poste, A.E., Hecky, R.E., Guildford, S.J., 2011. Evaluating microcystin exposure risk through fish consumption. Environ. Sci. Technol. 45 (13), 5806–5811. Available from: https://doi.org/10.1021/es200285c.

Postel, S., 1996. Forging a sustainable water strategy. In: Brown, L.R. (Ed.), State of the World. A Worldwatch Institute Report on Progress Toward a Sustainable Society. W. W. Norton & Co, New York, NY, pp. 40–59.

Postel, S.L., Daily, G.C., Ehrlich, P.R., 1996. Human appropriation of renewable fresh water. Science 271, 785–788.

Pough, F.H., Andrews, R.M., Cadle, S.E., Crump, M.L., Savitzky, A.H., Wells, K.D., 1998. Herpetology. Prentice-Hall, Englewood Cliffs, NJ.

Pounds, J.A., Bustamante, M.R., Coloma, L.A., Consuegra, J.A., Fogden, M.P.L., Foster, P.N., et al., 2006. Widespread amphibian extinctions from epidemic disease driven by global warming. Nature 439 (7073), 161–167.

Powell, R., Conant, R., Collins, J.T., 2016. Peterson Field Guide to Reptiles and Amphibians of Eastern and Central North America, fourth ed. Houghton Mifflin.

Power, M.E., 1990a. Effects of fish in river food webs. Science 250, 811–814.

Power, M.E., 1990b. Resource enhancement by indirect effects of grazers: armored catfish, algae, and sediment. Ecology 71 (3), 897–904.

Power, M.E., 1992. Top-down and bottom-up forces in food webs: do plants have primacy? Ecology 73 (3), 733–746.

Power, M.E., Matthews, W.J., 1983. Algae-grazing minnows (*Campostoma anomalum*), piscivorous bass (*Micropterus* spp.), and the distribution of attached algae in a small prairie-margin stream. Oecologia 60, 328–332.

Power, M.E., Sun, A., Parker, G., Dietrich, W.E., Wootton, J.T., 1995. Hydraulic food-chain models: an approach to the study of food-web dynamics in large rivers. Bioscience 45 (3), 159–167.

Power, M.E., Tilman, D., Estes, J.A., Menge, B.A., Bond, W.J., Mills, L.S., et al., 1996. Challenges in the quest for keystones. Bioscience 46 (8), 609–620.

Prepas, E.E., Kotak, B.G., Campbell, L.M., Evans, J.C., Hrudey, S.E., Holmes, C.F.B., 1997. Accumulation and elimination of cyanobacterial hepatotoxins by the freshwater clam *Anodonta grandis simpsoniana*. Can. J. Fish. Aquat. Sci. 54, 41–46.

Prescott, G.W., 1978. How to Know the Freshwater Algae. Wm C. Brown Co., Dubuque, IA.

Prescott, G.W., 1982. Algae of the Western Great Lakes Area. Otto Koeltz Science Publishers, Koenigstein, Germany.

Preston, F.W., 1962. The canonical distribution of commonness and rarity: Part I. Ecology 43 (2), 187–215.

Pringle, C.M., 1997. Exploring how disturbance is transmitted upstream: going against the flow. J. N. Am. Benthol. Soc. 16, 425−438.

Pringle, C.M., Hemphill, N., McDowell, W.H., Bednarek, A., March, J.G., 1999. Linking species and ecosystems: different biotic assemblages cause interstream differences in organic matter. Ecology 80 (6), 1860−1872.

Priscu, J.C., Adams, E.A., Lyons, W.B., Voytek, M.A., Mogk, D.M., Brown, R.L., et al., 1999. Geomicrobiology of subglacial ice above Lake Vostok, Antarctica. Science 286, 2141−2144.

Priscu, J.C., Fritsen, C.H., Adams, E.A., Giovannoni, S.J., Paerl, H.W., McKay, C.P., et al., 1998. Perennial Antarctic lake ice: an oasis for life in a polar desert. Science 280, 1095−2098.

Proctor, V.W., 1959. Dispersal of fresh-water algae by migratory water birds. Science 130, 623−624.

Proctor, V.W., 1966. Dispersal of desmids by waterbirds. Phycologia 5 (4), 227−232.

Proulx, M., Pick, F.R., Mazumdre, A., Hamilton, P.B., Lean, D.R.S., 1996. Experimental evidence for interactive impacts of human activities on lake algal species richness. Oikos 76, 191−195.

Pruden, A., Arabi, M., Storteboom, H.N., 2012. Correlation between upstream human activities and riverine antibiotic resistance genes. Environ. Sci. Technol. 46 (21), 11541−11549. Available from: https://doi.org/10.1021/es302657r.

Pugh, P.J.A., McInnes, S.J., 1998. The origin of Arctic terrestrial and freshwater tardigrades. Polar. Biol. 19, 177−182.

Purcell, L.M., 1977. Life at low Reynolds number. Am. J. Phys. 45, 3−11.

Pyke, G.H., Pulliam, H.R., Charnov, E.L., 1977. Optimal foraging: a selective review of theory and tests. Q. Rev. Biol. 52 (137−154).

Pyne, M.I., Poff, N.L., 2017. Vulnerability of stream community composition and function to projected thermal warming and hydrologic change across ecoregions in the western United States. Glob. Change Biol. 23 (1), 77−93. Available from: https://doi.org/10.1111/gcb.13437.

Pyron, M., 1999. Relationships between geographical range size, body size, local abundance, and habitat breadth in North American suckers and sunfishes. J. Biogeogr. 26 (3), 549−558.

Qin, B., Zhu, G., Gao, G., Zhang, Y., Li, W., Paerl, H.W., et al., 2010. A drinking water crisis in Lake Taihu, China: linkage to climatic variability and lake management. Environ. Manage. 45 (1), 105−112.

Quiblier, C., Wood, W., Echenique-Subiabre, I., Heath, M., Villeneuve, A., Humbert, J., 2013. A review of current knowledge on toxic benthic freshwater cyanobacteria − Ecology, toxin production and risk management. Water Research 47 (15), 5464−5479. Available from: https://doi.org/10.1016/j.watres.2013.06.042.

Quick, A.M., Reeder, W.J., Farrell, T.B., Tonina, D., Feris, K.P., Benner, S.G., 2016. Controls on nitrous oxide emissions from the hyporheic zones of streams. Environ. Sci. Technol. 50 (21), 11491−11500. Available from: https://doi.org/10.1021/acs.est.6b02680.

Quinlan, E.L., Phlips, E.J., Donnelly, K.A., Jett, C.H., Sleszynski, P., Keller, S., 2008. Primary producers and nutrient loading in Silver Springs, FL, USA. Aquat. Bot. 88 (3), 247−255.

Quinn, J.F., Dunham, A.E., 1981. On hypothesis testing in ecology and evolution. Am. Nat. 122 (5), 602−617.

Rabalais, N.N., Turner, R.E., Scavia, D., 2002. Beyond science into policy: Gulf of Mexico Hypoxia and the Mississippi River: nutrient policy development for the Mississippi River watershed reflects the accumulated scientific evidence that the increase in nitrogen loading is the primary factor in the worsening of hypoxia in the northern Gulf of Mexico. AIBS Bull. BioSci. 52 (2), 129−142.

Rader, R.B., 1999. The Florida Everglades, natural variability, invertebrate diversity, and foodweb stability. In: Batzer, D.P., Rader, R.B., Wissinger, S.A. (Eds.), Invertebrates in Freshwater Wetlands of North America: Ecology and Management. John Wiley & Sons, Inc, New York, NY, pp. 25−54.

Rader, R.B., Belish, T.A., 1997a. Effects of ambient and enhanced UV-B radiation on periphyton in a mountain stream. J. Freshw. Ecol. 12 (4), 615−628.

Rader, R.B., Belish, T.A., 1997b. Short-term effects of ambient and enhanced UV-B on moss (Fontinalis neomexicana) in a mountain stream. J. Freshw. Ecol. 12 (3), 395−403.

Ragotzkie, R.A., 1978. Heat budgets of lakes. In: Lerman, A. (Ed.), Lakes: Chemistry, Geology, Physics. Springer New York, New York, NY, pp. 1−19.

Rahel, F.J., 2000. Homogenization of fish faunas across the United States. Science 288, 854−856.

Rahel, F.J., Olden, J.D., 2008. Assessing the effects of climate change on aquatic invasive species. Conserv. Biol. 22 (3), 521–533.

Rainey, P.B., Rainey, K., 2003. Evolution of cooperation and conflict in experimental bacterial populations. Nature 425 (6953), 72.

Ramade, F., 1989. The pollution of the hydrosphere by global contaminants and its effects on aquatic ecosystems. In: Boudou, A., Ribeyre, F. (Eds.), Aquatic Ecotoxicology: Fundamental Concepts and Methodologies, vol. 1. CRC Press, Boca Raton, FL, pp. 152–180.

Ramírez, A., Ardón, M., Douglas, M.M., Graça, M.A.S., 2015. Tropical freshwater sciences: an overview of ongoing tropical research. Freshw. Sci. 34 (2), 606–608. Available from: https://doi.org/10.1086/681257.

Rantala, H., Glover, D., Garvey, J., Phelps, Q., Tripp, S., Herzog, D., et al., 2016. Fish assemblage and ecosystem metabolism responses to reconnection of the bird's point-New Madrid floodway during the 2011 Mississippi River flood. River Res. Appl. 32 (5), 1018–1029.

Rantala, H.M., Nelson, A.M., Fulgoni, J.N., Whiles, M.R., Hall, R.O., Dodds, W.K., et al., 2015. Long-term changes in structure and function of a tropical headwater stream following a disease-driven amphibian decline. Freshw. Biol. 60 (3), 575–589. Available from: https://doi.org/10.1111/fwb.12505.

Ranvestel, A.W., Lips, K.R., Pringle, C.M., Whiles, M.R., Bixby, R.J., 2004. Neotropical tadpoles influence stream benthos: evidence for the ecological consequences of decline in amphibian populations. Freshw. Biol. 49 (3), 274–285.

Räsänen, J., Kauppila, T., Salonen, V.-P., 2006. Sediment-based investigation of naturally or historically eutrophic lakes—implications for lake management. J. Environ. Manage. 79 (3), 253–265.

Raven, J.A., 1992. How benthic macroalgae cope with flowing freshwater: resource acquisition and retention. J. Phycol. 28, 133–146.

Ravishankara, A., Daniel, J.S., Portmann, R.W., 2009. Nitrous oxide (N_2O): the dominant ozone-depleting substance emitted in the 21st century. Science 326 (5949), 123–125.

Raymond, P.A., Cole, J.J., 2001. Gas exchange in rivers and estuaries: choosing a gas transfer velocity. Estuaries Coasts 24 (2), 312–317.

Raymond, P.A., Hartmann, J., Lauerwald, R., Sobek, S., McDonald, C., Hoover, M., et al., 2013. Global carbon dioxide emissions from inland waters. Nature 503 (7476), 355–359. Available from: https://doi.org/10.1038/nature12760. Available from: http://www.nature.com/nature/journal/v503/n7476/abs/nature12760.html#supplementary-information.

Read, E.K., Patil, V.P., Oliver, S.K., Hetherington, A.L., Brentrup, J.A., Zwart, J.A., et al., 2015. The importance of lake-specific characteristics for water quality across the continental United States. Ecol. Appl. 25 (4), 943–955. Available from: https://doi.org/10.1890/14-0935.1.

Reale, J.K., Van Horn, D.J., Condon, K.E., Dahm, C.N., 2015. The effects of catastrophic wildfire on water quality along a river continuum. Freshw. Sci. 34 (4), 1426–1442. Available from: https://doi.org/10.1086/684001.

Reche, I., Pulido-Villena, E., Morales-Baquero, R., Casamayor, E.O., 2005. Does ecosystem size determine bacterial species richness? Ecology 86 (7), 1715–1722. Available from: https://doi.org/10.1890/04-1587.

Redfield, A.C., 1958. The biological control of chemical factors in the environment. Am. Sci. 46, 205–221.

Reemtsma, T., These, A., 2005. Comparative investigation of low-molecular-weight fulvic acids of different origin by SEC-Q-TOF-MS: new insights into structure and formation. Environ. Sci. Technol. 39 (10), 3507–3512.

Regester, K.J., Lips, K.R., Whiles, M.R., 2006. Energy flow and subsidies associated with the complex life cycle of ambystomatid salamanders in ponds and adjacent forest in southern Illinois. Oecologia 147 (2), 303–314.

Regester, K.J., Whiles, M.R., Lips, K.R., 2008. Variation in the trophic basis of production and energy flow associated with emergence of larval salamander assemblages from forest ponds. Freshw. Biol. 53 (9), 1754–1767.

Reid, G.K., Fichter, G.S., 1967. Pond Life, A Guide to Common Plants and Animals of North American Ponds and Lakes. GoldenPress, Western Publishing Company Inc., New York, NY.

Reid, J.W., 1994. Latitudinal diversity patterns of continental benthic copepod species assemblages in the Americas. Hydrobiologia 292/293, 341−349.

Reisinger, A.J., Tank, J.L., Hoellein, T.J., Hall, R.O., 2016. Sediment, water column, and open-channel denitrification in rivers measured using membrane-inlet mass spectrometry. J. Geophys. Res.: Biogeosci. 121 (5), 1258−1274. Available from: https://doi.org/10.1002/2015JG003261.

Rejmankova, E., Macek, P., Epps, K., 2008. Wetland ecosystem changes after three years of phosphorus addition. Wetlands 28 (4), 914−927.

Relyea, R., 2005. The lethal impact of Roundup on aquatic and terrestrial amphibians. Ecol. Appl. 15 (4), 1118−1124.

Resh, V.H., Brown, A.B., Covich, A.P., Gurtz, M.E., Li, H.W., Minshall, G.W., et al., 1988. The role of disturbance in stream ecology. J. N. Am. Benthol. Soc. 7 (4), 433−455.

Resh, V.H., Lévêque, C., Statzner, B., 2004. Long-term, large-scale biomonitoring of the unknown: assessing the effects of insecticides to control river blindness (onchocerciasis) in West Africa. Annu. Rev. Entomol. 49 (1), 115−139.

Revsbech, N.P., Jørgensen, B.B., 1986. Microelectrodes: their use in microbial ecology. In: Marshall, K.C. (Ed.), Advances in Microbial Ecology, vol. 9. Plenum, New York, NY, pp. 293−352.

Reynolds, C., Padisák, J., Sommer, U., 1993. Intermediate disturbance in the ecology of phytoplankton and the maintenance of species diversity: a synthesis. In: Padisák, J., Reynolds, C.S. (Eds.), Intermediate Disturbance Hypothesis in Phytoplankton Ecology. Springer, pp. 183−188.

Reynolds, C.S., 1984. The Ecology of Freshwater Phytoplankton. Cambridge University Press, Cambridge, Great Britain.

Reynolds, C.S., 1994. The role of fluid motion in the dynamics of phytoplankton in lakes and rivers. In: Giller, P.S., Hildrew, A.G., Raffaelli, D.G. (Eds.), Aquatic Ecology. Scale, Pattern and Process. Blackwell Scientific Publications, Oxford, UK, pp. 141−188.

Reynolds, J.B., 1996. Electrofishing. In: Murphy, B.R., Willis, D.W. (Eds.), Fisheries Techniques, second ed. American Fisheries Society, Bethesda, MD, pp. 221−253.

Rheinheimer, G., 1991. Aquatic Microbiology, fourth ed. John Wiley and Sons, Chichester, UK.

Rhoades, C.C., Hubbard, R.M., Elder, K., 2017. A decade of streamwater nitrogen and forest dynamics after a mountain pine beetle outbreak at the Fraser Experimental Forest, Colorado. Ecosystems 20 (2), 380−392. Available from: https://doi.org/10.1007/s10021-016-0027-6.

Ribblett, S.G., Palmer, M.A., Wayne Coats, D., 2005. The importance of bacterivorous protists in the decomposition of stream leaf litter. Freshw. Biol. 50 (3), 516−526.

Ricciardi, A., MacIsaac, H.J., 2000. Recent mass invasion of the North American Great Lakes by Ponto-Caspian species. Trends Ecol. Evol. 15 (2), 62−65.

Rice, E.L., 1984. Allelopathy, second ed. Academic Press, Orlando, FL.

Rice, G., Stedman, K., Snyder, J., Wiedenheft, B., Willits, D., Brumfield, S., et al., 2001. Viruses from extreme thermal environments. Proc. Natl. Acad. Sci. 98 (23), 13341−13345.

Richardson, C.J., King, R.S., Qian, S.S., Vaithiyanthan, P., Qualls, R.G., Stow, C.A., 2007. Estimating ecological thresholds for phosphorus in the Everglades. Environ. Sci. Technol. 41, 8084−8091.

Richardson, C.J., Schwegler, B.R., 1986. Algal bioassay and gross productivity experiments using sewage effluent in a Michigan wetland. Water Resour. Bull. 22, 111−120.

Richardson, J.O., Pérez, C., Lobsiger, S., Reid, A.A., Temelso, B., Shields, G.C., et al., 2016. Concerted hydrogen-bond breaking by quantum tunneling in the water hexamer prism. Science 351 (6279), 1310−1313. Available from: https://doi.org/10.1126/science.aae0012.

Richardson, L.F., 1922. Weather Prediction by Numerical Processes. Cambridge University Press, Cambridge.

Richardson, L.L., Aguilar, C., Nealson, K.H., 1988. Manganese oxidation in pH and O_2 microenvironments produced by phytoplankton. Limnol. Oceanogr. 33 (3), 352−363.

Richardson, L.L., Castenholz, R.W., 1987a. Diel vertical movements of the cyanobacterium *Oscillatoria terebriformis* in a sulfide-rich hot spring microbial mat. Appl. Environ. Microbiol. 53 (9), 2142−2150.

Richardson, L.L., Castenholz, R.W., 1987b. Enhanced survival of the cyanobacterium *Oscillatoria terebriformis* in darkness under anaerobic condition. Appl. Environ. Microbiol. 53 (9), 2151–2158.

Richmond, E.K., Rosi-Marshall, E.J., Lee, S.S., Thompson, R.M., Grace, M.R., 2016. Antidepressants in stream ecosystems: influence of selective serotonin reuptake inhibitors (SSRIs) on algal production and insect emergence. Freshw. Sci. 35 (3), 845–855. Available from: https://doi.org/10.1086/687841.

Richter, J., Martin, L., Beachy, C.K., 2009. Increased larval density induces accelerated metamorphosis independently of growth rate in the frog rana sphenocephala. J. Herpetol. 43 (3), 551–554.

Rickerl, D.H., Sancho, F.O., Ananth, S., 1994. Vesicular-arbuscular endomycorrhizal colonization of wetland plants. J. Environ. Qual. 23, 913–916.

Ridge, I., Walters, J., Street, M., 1999. Algal growth control by terrestrial leaf litter: a realistic tool? Hydrobiologia 395/396, 173–180.

Riemer, D.N., 1984. Introduction to Freshwater Vegetation. AVI Publishing, Westport, UK.

Rier, S.T., Kinek, K.C., Hay, S.E., Francoeur, S.N., 2016. Polyphosphate plays a vital role in the phosphorus dynamics of stream periphyton. Freshw. Sci. 35 (2), 490–502. Available from: https://doi.org/10.1086/685859.

Riera, J.L., Magnuson, J.J., Kratz, T.K., Webster, K.E., 2000. A geomorphic template for the analysis of lake districts applied to the Northern Highland Lake District, Wisconsin, U.S.A. Freshw. Biol. 43, 301–318.

Riess, W., Giere, O., Kohls, O., Sarbu, S.M., 1999. Anoxic thermomineral cave waters and bacterial mats as habitat for freshwater nematodes. Aquat. Microb. Ecol. 18, 157–164.

Rigler, F., 1966. Radiobiological analysis of inorganic phosphorus in lakewater. Verhandlungen des Internationalen Verein Limnologie 16, 465–470.

Rigosi, A., Hanson, P., Hamilton, D.P., Hipsey, M., Rusak, J.A., Bois, J., et al., 2015. Determining the probability of cyanobacterial blooms: the application of Bayesian networks in multiple lake systems. Ecol. Appl. 25 (1), 186–199. Available from: https://doi.org/10.1890/13-1677.1.

Riley, A.J., Dodds, W.K., 2012. Whole-stream metabolism: strategies for measuring and modeling diel trends of dissolved oxygen. Freshw. Sci. 32 (1), 56–69.

Rinaldo, A., Rigon, R., Banavar, J.R., Maritan, A., Rodriguez-Iturbe, I., 2014. Evolution and selection of river networks: statics, dynamics, and complexity. Proc. Natl. Acad. Sci. 111 (7), 2417–2424. Available from: https://doi.org/10.1073/pnas.1322700111.

Ringelberg, J., Gool, E.V., 1998. Do bacteria, not fish, produce 'fish kairomone'? J. Plankton Res. 20 (9), 1847–1852.

Ripley, E.A., Redmann, R.E., Crowder, A.A., 1996. Environmental Effects of Mining. St. Lucie Press, Delray Beach, FL.

Ripple, W., Beschta, R., 2004. Wolves, elk, willows, and trophic cascades in the upper Gallatin Range of Southwestern Montana, USA. For. Ecol. Manage. 200 (1–3), 161–181.

Risgaard-Petersen, N., Kristiansen, M., Frederiksen, R.B., Dittmer, A.L., Bjerg, J.T., Trojan, D., et al., 2015. Cable bacteria in freshwater sediments. Appl. Environ. Microbiol. 81 (17), 6003–6011. Available from: https://doi.org/10.1128/aem.01064-15.

Roach, K.A., Winemiller, K.O., Davis, S.E., 2014. Autochthonous production in shallow littoral zones of five floodplain rivers: effects of flow, turbidity and nutrients. Freshw. Biol. 59 (6), 1278–1293. Available from: https://doi.org/10.1111/fwb.12347.

Roback, S.S., 1974. Insects (Arthropoda: Insecta). In: Hart, C.W.J., Fuller, S.L.H. (Eds.), Pollution Ecology of Freshwater Invertebrates. Academic Press, New York, NY, pp. 313–376.

Robarts, R.D., Donald, D.B., Arts, M.T., 1995. Phytoplankton primary production of three temporary northern prairie wetlands. Can. J. Aquat. Sci. 52, 897–902.

Robb, G.A., Robinson, J.D.F., 1995. Acid drainage from mines. Geogr. J. 161, 47–54.

Roberts, E.C., Laybourn-Parry, J., 1999. Mixotrophic crytophytes and their predators in the dry valley lakes of Antarctica. Freshw. Biol. 41, 737–746.

Robinson, A., Fleischmann, A., Mcpherson, S., Heinrich, V., Gironella, E., Peña, C., 2009. A spectacular new species of *Nepenthes* L. (Nepenthaceae) pitcher plant from central Palawan, Philippines. J. Linn. Soc. 159, 195–202.

Robinson, C., Minshall, G., 1986. Effects of disturbance frequency on stream benthic community structure in relation to canopy cover and season. J. N. Am. Benthol. Soc. 237−248.

Robinson, C., Uehlinger, U., 2008. Experimental floods cause ecosystem regime shift in a regulated river. Ecol. Appl. 18 (2), 511−526.

Robinson, C.T., Thompson, C., Freestone, M., 2014. Ecosystem development of streams lengthened by rapid glacial recession. Fundam. Appl. Limnol./Arch. Hydrobiol. 185 (3−4), 235−246. Available from: https://doi.org/10.1127/fal/2014/0667.

Robson, T.M., Pancott, V.A., Flint, S.D., Ballaré, C.L., Sala, O.E., Scopel, A.L., et al., 2003. Six years of solar UV-B manipulations affect growth of *Sphagnum* and vascular plants in Tierra del Fuego peatland. New Phytol. 160, 379−389.

Rochman, C.M., Kross, S.M., Armstrong, J.B., Bogan, M.T., Darling, E.S., Green, S.J., et al., 2015. Scientific evidence supports a ban on microbeads. Environ. Sci. Technol. 49 (18), 10759−10761. Available from: https://doi.org/10.1021/acs.est.5b03909.

Rodas, V.L., Costas, E., 1999. Preference of mice to consume *Microcystis aeruginosa* (Toxin-producing cyanobacteria): a possible explanation for numerous fatalities of livestock and wildlife. Res. Vet. Sci. 67, 107−110.

Rode, M., Wade, A.J., Cohen, M.J., Hensley, R.T., Bowes, M.J., Kirchner, J.W., et al., 2016. Sensors in the stream: the high-frequency wave of the present. Environ. Sci. Technol. 50 (19), 10297−10307. Available from: https://doi.org/10.1021/acs.est.6b02155.

Röder, G., Mota, M., Turlings, T.C.J., 2017. Host plant location by chemotaxis in an aquatic beetle. Aquat. Sci. 79 (2), 309−318. Available from: https://doi.org/10.1007/s00027-016-0498-8.

Rodrigues, D.F., Tiedje, J.M., 2008. Coping with our cold planet. Appl. Environ. Microbiol. 74 (6), 1677−1686.

Rodríguez-Lado, L., Sun, G., Berg, M., Zhang, Q., Xue, H., Zheng, Q., et al., 2013. Groundwater arsenic contamination throughout China. Science 341 (6148), 866−868. Available from: https://doi.org/10.1126/science.1237484.

Roell, M.J., Orth, D.J., 1993. Trophic basis of production of stream-dwelling smallmouth bass, rock bass, and flathead catfish in relation to invertebrate bait harvest. Trans. Am. Fish. Soc. 122 (1), 46−62.

Roeselers, G., Van Loosdrecht, M., Muyzer, G., 2007. Heterotrophic pioneers facilitate phototrophic biofilm development. Microb. Ecol. 54 (3), 578−585.

Rofner, C., Peter, H., Catalán, N., Drewes, F., Sommaruga, R., Pérez, M.T., 2017. Climate-related changes of soil characteristics affect bacterial community composition and function of high altitude and latitude lakes. Glob. Change Biol. 23 (6), 2331−2344. Available from: https://doi.org/10.1111/gcb.13545.

Rogers, P., 1986. Water: not as cheap as you think. Tech. Rev. 31−43 (11-12/86).

Rogulj, B., Marmonier, P., Lattinger, R., Danielopol, D., 1994. Fine-scale distribution of hypogean Ostracoda in the interstitial habitats of the Rivers Sava and Rhône. Hydrobiologia 287, 19−28.

Root, R.B., 1967. The niche exploitation pattern of the blue-gray gnatcatcher. Ecol. Monogr. 37 (4), 317−350.

Rosemond, A.D., Benstead, J.P., Bumpers, P.M., Gulis, V., Kominoski, J.S., Manning, D.W., et al., 2015. Experimental nutrient additions accelerate terrestrial carbon loss from stream ecosystems. Science 347 (6226), 1142−1145.

Rosemond, A.D., Mulholland, P.J., Elwood, J.W., 1993. Top-down and bottom-up factors varied among biomass and productivity parameters and that top-down and bottom-up effects, alone, were less important than their combined effects. Ecology 74 (4), 1264−1280.

Rosenberg, N.J., Epstein, D.J., Wang, D., Vail, L., Srinivasan, R., Arnold, J.G., 1999. Possible impacts of global warming on the hydrology of the Ogallala aquifer region. Clim. Change 42 (4), 677−692.

Rosenzweig, M.L., 1995. Species Diversity in Space and Time. Cambridge University Press, Cambridge, UK.

Rosenzweig, M.L., 1999. Heeding the warning in biodiversity's basic law. Science 284, 276−277.

Rosenzweig, M.L., 2001. The four questions: what does the introduction of exotic species do to diversity? Evol. Ecol. Res. 3 (3), 361−367.

Rosgen, D.L., 2008. Discussion: "critical evaluation of how the rosgen classification and associated 'natural channel design' methods fail to integrate and quantify fluvial processes and channel responses" by A. Simon, M. Doyle, M. Kondolf, FD Shields Jr., B. Rhoads, and M. McPhillips 1. JAWRA J. Am. Water Resour. Assoc. 44 (3), 782−792.

Roshier, D., Whetton, P., Allan, R., Robertson, A., 2001. Distribution and persistence of temporary wetland habitats in arid Australia in relation to climate. Austral Ecol. 26 (4), 371−384.

Rosi-Marshall, E.J., Royer, T.V., 2012. Pharmaceutical compounds and ecosystem function: an emerging research challenge for aquatic ecologists. Ecosystems 15 (6), 867−880.

Rosi-Marshall, E.J., Tank, J.L., Royer, T.V., Whiles, M.R., Evans-White, M., Chambers, C., et al., 2007. Toxins in transgenic crop byproducts may affect headwater stream ecosystems. Proc. Natl. Acad. Sci. USA 104 (41), 16204−16208.

Rosi-Marshall, E.J., Vallis, K.L., Baxter, C.V., Davis, J.M., 2016. Retesting a prediction of the River Continuum concept: autochthonous versus allochthonous resources in the diets of invertebrates. Freshw. Sci. 35 (2), 534−543. Available from: https://doi.org/10.1086/686302.

Rosińska, J., Rybak, M., Gołdyn, R., 2017. Patterns of macrophyte community recovery as a result of the restoration of a shallow urban lake. Aquat. Bot. 138, 45−52. Available from: https://doi.org/10.1016/j.aquabot.2016.12.005.

Roskov, Y., Abucay, L., Orrell, T., Nicolson, D., Flann, C., Bailly, N., et al., 2016. Species 2000 & ITIS catalogue of life, 2016 annual checklist. Retrieved from: <www.catalogueoflife.org/annual-checklist/2016>.

Ross, D.H., Wallace, J.B., 1981. Production of Brachycentrus spinae Ross (Trichoptera: Brachycentridae) and its role in seston dynamics of a southern Appalachian stream (USA). Environ. Entomol. 10 (2), 240−246.

Roughgarden, J., 1989. The structure and assembly of communities. In: Roughgarden, J., May, R.M., Levin, S.A. (Eds.), Perspectives in Ecological Theory. Princeton University Press, Princeton, NJ, pp. 203−226.

Rovira, A., Alcaraz, C., Trobajo, R., 2016. Effects of plant architecture and water velocity on sediment retention by submerged macrophytes. Freshw. Biol. 61 (5), 758−768. Available from: https://doi.org/10.1111/fwb.12746.

Rüegg, J., Brant, J.D., Larson, D.M., Trentman, M.T., Dodds, W.K., 2015. A portable, modular, self-contained recirculating chamber to measure benthic processes under controlled water velocity. Freshw. Sci. 34 (3), 831−844. Available from: https://doi.org/10.1086/682328.

Rundle, H.D., Nagel, L., Boughman, J.W., Schluter, D., 2000. Natural selection and parallel speciation in sympatric sticklebacks. Science 287, 306−308.

Runkel, R.L., 2002. A new metric for determining the importance of transient storage. J. N. Am. Benthol. Soc. 21 (4), 529−543.

Russell, D.F., Wilkens, L.A., Moss, F., 1999. Use of behavioural stochastic resonance by paddle fish for feeding. Nature 402, 291−294.

Russell, N.J., Hamamoto, T., 1998. Psychrophiles. In: Horikoshi, K., Grant, W.D. (Eds.), Extremophiles. Microbial Life in Extreme Environments. Wiley-Liss, Inc, New York, NY, pp. 2−45.

Ruttner, F., 1963. Fundamentals of Limnology. University of Toronto Press, Toronto, Ontario, Canada.

Ryding, S.-O., Rast, W., 1989. The Control of Eutrophication of Lakes and Reservoirs, vol. 1. UNESCO and Parthenon, Paris and United Kingdom.

Rytter, L., Arveby, A.S., Granhall, U., 1991. Dinitrogen (C_2H_2) fixation in relation to nitrogen fertilization of grey alder [Alnus incana (L.) Moench.] plantations in a peat bog. Biol. Fertil. Soils 10, 233−240.

Saad, J.F., Unrein, F., Tribelli, P.M., López, N., Izaguirre, I., 2016. Influence of lake trophic conditions on the dominant mixotrophic algal assemblages. J. Plankton Res. 38 (4), 818−829. Available from: https://doi.org/10.1093/plankt/fbw029.

Sabo, J.L., Finlay, J.C., Kennedy, T., Post, D.M., 2010. The role of discharge variation in scaling of drainage area and food chain length in rivers. Science 330 (6006), 965−967.

Sahimi, M., 1995. Porous Media and Fractured Rock. Weinham, New York, NY.

Sakai, A., Larcher, W., 1987. Frost Survival of Plants. Springer-Verlag, Berlin, Germany.

Salazar Torres, G., Silva, L.S., Rangel, L., Attayde, J., Huszar, V.M., 2016. Cyanobacteria are controlled by omnivorous filter-feeding fish (*Nile tilapia*) in a tropical eutrophic reservoir. Hydrobiologia 765 (1), 115−129. Available from: https://doi.org/10.1007/s10750-015-2406-y.

Salcher, M.M., Ewert, C., Šimek, K., Kasalický, V., Posch, T., 2016. Interspecific competition and protistan grazing affect the coexistence of freshwater betaproteobacterial strains. FEMS Microbiol. Ecol. 92 (2). Available from: https://doi.org/10.1093/femsec/fiv156.

Salisbury, S.J., McCracken, G.R., Keefe, D., Perry, R., Ruzzante, D.E., 2016. A portrait of a sucker using landscape genetics: how colonization and life history undermine the idealized dendritic metapopulation. Mol. Ecol. 25 (17), 4126−4145.

Salzet, M., 2001. Anticoagulants and inhibitors of platelet aggregation derived from leeches. FEBS Lett. 492 (3), 187−192.

Sánchez-Carrillo, S., Angeler, D., Álvarez-Cobelas, M., Sánchez-Andrés, R., 2010. Freshwater wetland eutrophication. In: Abid, A., Ansari, A.A., Gill, S.S. (Eds.), Eutrophication: Causes, Consequences and Control. Springer, pp. 195−210.

Sanderson, S.L., Cech Jr., J.J., Patterson, M.R., 1991. Fluid dynamics in suspension-feeding blackfish. Science 251, 1346−1348.

Sand-Jensen, K., Riis, T., Markager, S., Vincent, W.F., 1999. Slow growth and decomposition of mosses in Arctic lakes. Can. J. Fish. Aquat. Sci. 56, 388−393.

Sapp, J., 1991. Living together: symbiosis and cytoplasmic inheritance. In: Margulis, L., Fester, R. (Eds.), Symbiosis as a Source of Evolutionary Innovation. MIT Press, Cambridge, MA, pp. 15−25.

Sarbu, S.M., Kane, T.C., Kinkle, B.K., 1996. A chemoautotrophically based cave ecosystem. Science 272, 1953−1995.

Sarneel, J., Huig, N., Veen, G.F., Rip, W., Bakker, E.S., 2014. Herbivores enforce sharp boundaries between terrestrial and aquatic ecosystems. Ecosystems 17 (8), 1426−1438. Available from: https://doi.org/10.1007/s10021-014-9805-1.

Sarnelle, O., Cooper, S.D., Wiseman, S., Mavuti, K.M., 1998. The relationship between nutrients and trophic-level biomass in turbid tropical ponds. Freshw. Biol. 40, 65−75.

Sarremejane, R., Mykrä, H., Bonada, N., Aroviita, J., Muotka, T., 2017. Habitat connectivity and dispersal ability drive the assembly mechanisms of macroinvertebrate communities in river networks. Freshw. Biol. 62 (6), 1073−1082. Available from: https://doi.org/10.1111/fwb.12926.

Sasser, M., 1990. Identification of bacteria by gas chromatography of cellular fatty acids. MIDI Technical Note 101. Newark Microbial ID, Inc.

Sato, T., Egusa, T., Fukushima, K., Oda, T., Ohte, N., Tokuchi, N., et al., 2012. Nematomorph parasites indirectly alter the food web and ecosystem function of streams through behavioural manipulation of their cricket hosts. Ecol. Lett. 15 (8), 786−793. Available from: https://doi.org/10.1111/j.1461-0248.2012.01798.x.

Savage, A.E., Zamudio, K.R., 2011. MHC genotypes associate with resistance to a frog-killing fungus. Proc. Natl. Acad. Sci. 108 (40), 16705−16710.

Scavia, D., Fahnenstiel, G.L., 1984. Small-scale nutrient patchiness: some consequences and a new encounter mechanism. Limnol. Oceanogr. 29, 785−793.

Schaake, J.C., 1990. From climate to flow. In: Waggoner, P.E. (Ed.), Climate Change and U.S. Water Resources. John Wiley and Sons, New York, NY, pp. 177−206.

Schaeffer, D.J., Malpas, P.B., Barton, L.L., 1999. Risk assessment of microcystin in dietary *Aphanizomenon flos-aquae*. Ecotoxicol. Environ. Saf. 44, 73−80.

Schalla, R., Walters, W.H., 1990. Rationale for the design of monitoring well screens and filter pack. In: Nielsen, D.M., Johnson, A.I. (Eds.), Ground Water and Vadose Zone Monitoring. ASTM, Ann Arbor, MI, pp. 64−75.

Scheffer, M., 1998. Ecology of Shallow Lakes. Chapman and Hall, London.

Scheffer, M., Carpenter, S., Foley, J.A., Folke, C., Walker, B., 2001. Catastrophic shifts in ecosystems. Nature 413 (6856), 591−596.

Schelske, C.L., Stoermer, E.F., 1972. Phosphorus, silica, and eutrophication of Lake Michigan. Limnol. Oceanogr., Spec. Symp. 1, 157−171.

Schindler, D.E., Carpenter, S.R., Cole, J.J., Kitchell, J.F., Pace, M.L., 1997. Influence of food web structure on carbon exchange between lakes and the atmosphere. Science 277, 248—251.

Schindler, D.E., Smits, A.P., 2017. Subsidies of aquatic resources in terrestrial ecosystems. Ecosystems 20 (1), 78—93. Available from: https://doi.org/10.1007/s10021-016-0050-7.

Schindler, D.W., 1974. Eutrophication and recovery in experimental lakes: implications for lake management. Science 184, 897—899.

Schindler, D.W., 1998. Replications versus realism: the need for ecosystem-scale experiments. Ecosystems 1, 323—334.

Schindler, D.W., 2001. The cumulative effects of climate warming and other human stresses on Canadian freshwaters in the new millennium. Can. J. Fish. Aquat. Sci. 58 (1), 18—29.

Schindler, D.W., 2012. The dilemma of controlling cultural eutrophication of lakes. Proc. Biol. Sci. 279 (1746), 4322—4333.

Schindler, D.W., Carpenter, S.R., Chapra, S.C., Hecky, R.E., Orihel, D.M., 2016. Reducing phosphorus to curb lake eutrophication is a success. Environ. Sci. Technol. 50 (17), 8923—8929. Available from: https://doi.org/10.1021/acs.est.6b02204.

Schindler, D.W., Curtis, P.J., Parker, B.R., Stainton, M.P., 1996. Consequences of climate warming and lake acidification for UV-B penetration in North American boreal lakes. Nature 22, 705—708.

Schindler, D.W., Hecky, R.E., 2008. Reply to Howarth and Paerl: is control of both nitrogen and phosphorus necessary? Proc. Natl. Acad. Sci. USA 105 (49). Available from: https://doi.org/10.1073/pnas.0809744105, E104-E104.

Schindler, D.W., Hecky, R.E., Findlay, D.L., Stainton, M.P., Parker, B.R., Paterson, M.J., et al., 2008. Eutrophication of lakes cannot be controlled by reducing nitrogen input: results of a 37-year whole-ecosystem experiment. Proc. Natl. Acad. Sci. USA 105 (32), 11254—11258.

Schlesinger, W.H., 1997. Biogeochemistry, an Analysis of Global Change, second ed. Academic Press, San Diego, CA.

Schloegel, L.M., Picco, A.M., Kilpatrick, A.M., Davies, A.J., Hyatt, A.D., Daszak, P., 2009. Magnitude of the US trade in amphibians and presence of *Batrachochytrium dendrobatidis* and ranavirus infection in imported North American bullfrogs (*Rana catesbeiana*). Biol. Conserv. 142 (7), 1420—1426.

Schlumpf, M., Cotton, B., Conscience, M., Haller, V., Steinmann, B., Lichtensteiger, W., 2001. In vitro and in vivo estrogenicity of UV screens. Environ. Health Perspect. 109 (3), 239.

Schmid, R., 2001. Recent advances in the description of the structure of water, the hydrophobic effect, and the like-dissolves-like rule. Monatshefte für Chemie/Chem. Mon. 132 (11), 1295—1326.

Schmitt, W.L., 1965. Crustaceans. The University of Michigan Press, Ann Arbor, MI.

Schneider, S., Melzer, A., 2004. Sediment and water nutrient characteristics in patches of submerged macrophytes in running waters. Hydrobiologia 527 (1), 195—207.

Schoener, T.W., 1983. Field experiments on interspecific competition. Am. Nat. 122, 240—285.

Schoener, T.W., 1987. A brief history of optimal foraging ecology. In: Kamil, A.C., Krebs, J.R., Pulliam, H.R. (Eds.), Foraging Behavior. Plenum Press, New York, NY, pp. 5—68.

Scholes, R.J., 2017. Taking the mumbo out of the jumbo: progress towards a robust basis for ecological scaling. Ecosystems 20 (1), 4—13. Available from: https://doi.org/10.1007/s10021-016-0047-2.

Scholtz, G., Braband, A., Tolley, L., Reimann, A., Mittmann, B., Lukhaup, C., et al., 2003. Ecology—parthenogenesis in an outsider crayfish. Nature 421 (6925), 806—806.

Schooler, S., McEvoy, P., Coombs, E., 2006. Negative per capita effects of purple loosestrife and reed canary grass on plant diversity of wetland communities. Divers. Distrib. 12 (4), 351—363.

Schopf, J.W., 1993. Microfossils of the early Archaen apex Chert: new evidence of the antiquity of life. Science 260, 640—646.

Schrader, K.K., Dennis, M.E., 2005. Cyanobacteria and earthy/musty compounds found in commercial catfish (*Ictalurus punctatus*) ponds in the Mississippi Delta and Mississippi-Alabama Blackland Prairie. Water Res. 39 (13), 2807—2814.

Schreder, E.D., La Guardia, M.J., 2014. Flame retardant transfers from U.S. households (dust and laundry wastewater) to the aquatic environment. Environ. Sci. Technol. 48 (19), 11575—11583. Available from: https://doi.org/10.1021/es502227h.

Schröder, R., Prasse, R., 2013. Do cultivated varieties of native plants have the ability to outperform their wild relatives? PLoS One 8 (8), e71066.

Schueler, T., Simpson, J., 2001. Introduction: why urban lakes are different. Watershed Prot. Tech. 3 (4), 747.

Schulze, E.-D., Mooney, H.A., 1994. Ecosystem function of biodiversity: a summary. In: Schulze, E.-D., Mooney, H.A. (Eds.), Biodiversity and Ecosystem Function. Springer-Verlag, Berlin, pp. 498–510.

Schutz, D., Taborsky, M., Drapela, T., 2007. Air bells of water spiders are an extended phenotype modified in response to gas composition. J. Exp. Zool. Part A (10).

Schuur, E., McGuire, A.D., Schädel, C., Grosse, G., Harden, J., Hayes, D., et al., 2015. Climate change and the permafrost carbon feedback. Nature 520 (7546), 171.

Schwarzenbach, R.P., Escher, B.I., Fenner, K., Hofstetter, T.B., Johnson, C.A., Von Gunten, U., et al., 2006. The challenge of micropollutants in aquatic systems. Science 313 (5790), 1072–1077.

Scott, J., McCarthy, M., 2010. Nitrogen fixation may not balance the nitrogen pool in lakes over timescales relevant to eutrophication management. Limnol. Oceanogr. 55 (3), 1265–1270.

Scott, J.T., Doyle, R.D., Back, J.A., Dworkin, S.I., 2007. The role of N_2 fixatino in alleviating N limitation in wetland metaphyton: enzymatic, iisotopic, and elemental evidence. Biogeochemistry 84, 207–218.

Scott, W., 1923. The diurnal oxygen pulse in Eagle (Winona) Lake. In: Davis, J.J. (Ed.), Proceedings of the Indiana Academy of Science, vol. 33. Wm. B. Burford, Indianapolis, IN, pp. 311–314.

Scrimshaw, S., Kerfoot, W.C., 1987. Chemical defenses of freshwater organisms: beetles and bugs. In: Kerfoot, W.C., Sih, A. (Eds.), Predation. Direct and Indirect Impacts on Aquatic Communities. University Press of New England, Hanover, NH, pp. 240–262.

Sculthorpe, C.D., 1967. The Biology of Aquatic Vascular Plants. Edward Arnold, Ltd, London, UK.

Sedell, J.R., Froggat, J.L., 1984. Importance of streamside forests to large rivers: the isolation of the Willamette River, Oregon, USA, from its floodplain by snagging and streamside forest removal. Verein Internationale Verein Limnologie 22, 1828–1843.

Sedell, J.R., Richey, J.E., Swanson, F.J., 1989. The river continuum concept: a basis for the expected ecosystem behavior of very large rivers? Paper Presented at the Proceedings of the International Large River Symposium.

Seehausen, O., van Alphen, J.J.M., 1999. Can sympatric speciation by disruptive sexual selection explain rapid evolution of cichlid diversity in Lake Victoria? Ecol. Lett. 2, 262–271.

Seehausen, O., van Alphen, J.J.M., Witte, F., 1997. Cichlid fish diversity threatened by eutrophication that curbs sexual selection. Science 277, 1808–1811.

Seely, C.J., Lutnesky, M.M.F., 1998. Odour-induced antipredator behaviour of the water flea Cariodaphnia reticulata, in varying predator and prey densities. Freshw. Biol. 40, 17–24.

Seifert, R.P., Seifert, F.H., 1976. A community matrix analysis of Heliconia insect communities. Am. Midl. Nat. 110 (973), 461–483.

Selosse, M.-A., Charpin, M., Not, F., 2017. Mixotrophy everywhere on land and in water: the grand écart hypothesis. Ecol. Lett. 20 (2), 246–263. Available from: https://doi.org/10.1111/ele.12714.

Semlitsch, R., O'Donnell, K., Thompson III, F., 2014. Abundance, biomass production, nutrient content, and the possible role of terrestrial salamanders in Missouri Ozark forest ecosystems. Can. J. Zool. 92 (12), 997–1004.

Sereda, J.M., Hudson, J.J., Taylor, W.D., Demers, E., 2008. Fish as sources and sinks of nutrients in lakes. Freshw. Biol. 53 (2), 278–289.

Service, R.F., 2016. A debris-dammed lake threatens a flood. Science 353 (6301), 735–736. Available from: https://doi.org/10.1126/science.353.6301.735.

Seybold, H., Rothman, D.H., Kirchner, J.W., 2017. Climate's watermark in the geometry of stream networks. Geophys. Res. Lett. 44 (5), 2272–2280. Available from: https://doi.org/10.1002/2016GL072089.

Seymour, M., Fronhofer, E.A., Altermatt, F., 2015. Dendritic network structure and dispersal affect temporal dynamics of diversity and species persistence. Oikos 124 (7), 908–916. Available from: https://doi.org/10.1111/oik.02354.

Shannon, J.P., Blinn, D.W., McKinney, T., Benenati, E.P., Wilson, K.P., O'Brien, C., 2001. Aquatic food base response to the 1996 test flood below Glen Canyon Dam, Colorado River, Arizona. Ecol. Appl. 11, 672–685.

Shapiro, J., 1979. The importance of trophic-level interactions to the abundance and species composition of algae in lakes. In: Barica, J., Mur, L.R. (Eds.), Hypertrophic Ecosystems Developments in Hydrobiology, vol. 2. Dr. W. Junk, The Netherlands, pp. 105–121.

Shapiro, J., 1997. The role of carbon dioxide in the initiation and maintenance of blue-green dominance in lakes. Freshw. Biol. 37, 307–323.

Sharitz, R.R., Batzer, D.P., 1999. An introduction to freshwater wetlands in North America and their invertebrates. In: Batzer, D.P., Rader, R.B., Wissinger, S.A. (Eds.), Invertebrates in Freshwater Wetlands of North America: Ecology and Management. John Wiley & Sons, Inc, New York, NY, pp. 1–22.

Sharpe, D.M.T., De León, L.F., González, R., Torchin, M.E., 2017. Tropical fish community does not recover 45 years after predator introduction. Ecology 98 (2), 412–424. Available from: https://doi.org/10.1002/ecy.1648.

Shearer, C.A., Descals, E., Kohlmeyer, B., Kohlmeyer, J., Marvanová, L., Padgett, D., et al., 2007. Fungal biodiversity in aquatic habitats. Biodiversity Conserv. 16 (1), 49–67. Available from: https://doi.org/10.1007/s10531-006-9120-z.

Sheath, R.G., Müller, K.M., 1997. Distribution of stream macroalgae in four high arctic drainage basins. Arctic 50, 355–364.

Shelton, A.O., O'Donnell, J.L., Samhouri, J.F., Lowell, N., Williams, G.D., Kelly, R.P., 2016. A framework for inferring biological communities from environmental DNA. Ecol. Appl. 26 (6), 1645–1659. Available from: https://doi.org/10.1890/15-1733.1.

Sherbakov, D.Y., 1999. Molecular phylogenetic studies on the origin of biodiversity in Lake Baikal. Trends Ecol. Evol. 14 (3), 92–95.

Sherbakov, D.Y., Kamaltynov, R.M., Ogarkov, O.B., Verheyen, E., 1998. Patterns of evolutionary changes in Baikalian Gammarids inferred from DNA sequences (Crustacea, Amphipoda). Mol. Phylogenet. Evol. 10 (2), 160–167.

Sherr, B.F., Sherr, E.B., Fallon, R.D., 1987. Use of monodispersed fluorescently labeled bacteria to estimate in situ protozoan bactivory. Appl. Environ. Microbiol. 53, 958–965.

Sherr, E.B., Sherr, B.F., 1994. Bacterivory and herbivory: key roles of phagotrophic protists in pelagic food webs. Microb. Ecol. 28, 223–235.

Shi, W., Wang, M., Guo, W., 2014. Long-term hydrological changes of the Aral Sea observed by satellites. J. Geophys. Res.: Oceans 119 (6), 3313–3326.

Shimizu, S., Ueno, A., Naganuma, T., Kaneko, K., 2015. *Methanosarcina subterranea* sp. nov., a methanogenic archaeon isolated from a deep subsurface diatomaceous shale formation. Int. J. Syst. Evol. Microbiol. 65 (4), 1167–1171.

Shimizu, S., Ueno, A., Tamamura, S., Naganuma, T., Kaneko, K., 2013. *Methanoculleus horonobensis* sp. nov., a methanogenic archaeon isolated from a deep diatomaceous shale formation. Int. J. Syst. Evol. Microbiol. 63 (11), 4320–4323.

Short, F.T., Kosten, S., Morgan, P.A., Malone, S., Moore, G.E., 2016. Impacts of climate change on submerged and emergent wetland plants. Aquat. Bot. 135, 3–17. Available from: https://doi.org/10.1016/j.aquabot.2016.06.006.

Short, S.M., Short, C.M., 2008. Diversity of algal viruses in various North American freshwater environments. Aquat. Microb. Ecol. 51 (1), 13.

Shotyk, W., Weiss, D., Appleby, P.G., Cheburkin, A.K., Frei, R., Gloor, M., et al., 1998. History of atmospheric lead deposition since 12,370 ^{14}C yr BP from a peat bog, Jura Mountains, Switzerland. Science 281, 1635–1640.

Siegert, M.J., Kwok, R., Mayer, C., Hubbard, B., 2000. Water exchange between the subglacial Lake Vostok and the overlying ice sheet. Nature 403, 643–646.

Siegert, M.J., Ross, N., Le Brocq, A.M., 2016. Recent advances in understanding Antarctic subglacial lakes and hydrology. Philos. Trans. R. Soc. A: Math., Phys. Eng. Sci. 374 (2059). Available from: https://doi.org/10.1098/rsta.2014.0306.

Sigee, D., 2005. Freshwater Microbiology. Wiley, Chichester, England.

Simco, B.A., Cross, F.B., 1966. Factors affecting growth and production of channel catfish, *Ictalurus punctatus*. Univ. Kansas Mus. Nat. Hist. Publ. 17, 193−256.

Simek, K., Babenzien, D., Bittl, T., Koschel, R., Macek, M., Nedoma, J., et al., 1998. Microbial food webs in an artificially divided acidic bog lake. Int. Rev. Hydrobiol. 83, 3−18.

Simis, S.G.H., Tijdens, M., Hoogveld, H.L., Gons, H.J., 2005. Optical changes associated with cyanobacterial bloom termination by viral lysis. J. Plankton Res. 27 (9), 937−949.

Simmons, M., Tucker, A., Chadderton, W.L., Jerde, C.L., Mahon, A.R., 2015. Active and passive environmental DNA surveillance of aquatic invasive species. Can. J. Fish. Aquat. Sci. 73 (1), 76−83. Available from: https://doi.org/10.1139/cjfas-2015-0262.

Simon, A., Doyle, M., Kondolf, M., Shields Jr, F., Rhoads, B., McPhillips, M., 2007. Critical evaluation of how the rosgen classification and associated "natural channel design" methods fail to integrate and quantify fluvial processes and channel response 1. JAWRA J. Am. Water Resour. Assoc. 43 (5), 1117−1131.

Simon, A., Doyle, M., Kondolf, M., Shields Jr, F., Rhoads, B., Mcphillips, M., 2008. Reply to discussion 1 by Dave Rosgen 2: "critical evaluation of how the rosgen classification and associated 'natural channel design' methods fail to integrate and quantify fluvial processes and channel responses" 3. JAWRA J. Am. Water Resour. Assoc. 44 (3), 793−802.

Simonich, S.L., Hites, R.A., 1995. Global distribution of persistent organochlorine compounds. Science 269, 1851−1854.

Sinclair, J.L., Ghiorse, W.C., 1989. Distribution of aerobic bacteria, protozoa, algae, and fungi in deep subsurface sediments. Geomicrobiol. J. 7, 5−31.

Sinsabaugh, R.L., Repert, D., Weiland, T., Golladay, S.W., Linkins, A.E., 1991. Exoenzyme accumulation in epilithic biofilms. Hydrobiologia 222, 29−37.

Sinton, L.W., Finlay, R.K., Pang, L., Scott, D.M., 1997. Transport of bacteria and bacteriophages in irrigated effluent into and through an alluvial gravel aquifer. Water Air Soil Pollut. 98, 17−42.

Sirota, J., Baiser, B., Gotelli, N.J., Ellison, A.M., 2013. Organic-matter loading determines regime shifts and alternative states in an aquatic ecosystem. Proc. Natl. Acad. Sci. 110 (19), 7742−7747. Available from: https://doi.org/10.1073/pnas.1221037110.

Sitvarin, M.I., Rypstra, A.L., Harwood, J.D., 2016. Linking the green and brown worlds through nonconsumptive predator effects. Oikos 125 (8), 1057−1068. Available from: https://doi.org/10.1111/oik.03190.

Siver, P.A., Lord, W.D., McCarthy, D.J., 1994. Forensic limnology: the use of freshwater algal community ecology to link suspects to an aquatic crime scene in southern New England. J. Forensic Sci. 39, 847−853.

Skelly, D.K., 1997. Tadpole communities. Am. Sci. 85, 36−45.

Skerratt, L.F., Berger, L., Speare, R., Cashins, S., McDonald, K.R., Phillott, A.D., et al., 2007. Spread of chytridiomycosis has caused the rapid global decline and extinction of frogs. EcoHealth 4, 125−134.

Skubinna, J.P., Coon, T.G., Batterson, T.R., 1995. Increased abundance and depth submersed macrophytes in response to decreased turbidity in Saginaw Bay, Lake Huron. J. Great Lakes Res. 21 (4), 476−488.

Slattery, M., Lesser, M.P., 2017. Allelopathy-mediated competition in microbial mats from Antarctic lakes. FEMS Microbiol. Ecol. 93 (5).

Slavik, K., Peterson, B.J., Deegan, L.A., Bowden, W.B., Hershey, A.E., Hobbie, J.E., 2004. Long-term responses of the Kuparuk River ecosystem to phosphorus fertilization. Ecology 85 (4), 939−954.

Slobodkin, L.E., Bossert, P.E., 1991. The freshwater *Cnidaria*-or *Coelenterates*. In: Thorp, J.H., Covich, A.P. (Eds.), Ecology and Classification of North American Freshwater Invertebrates. Academic Press, Inc, San Diego, CA, pp. 125−143.

Sloey, W.E., Spangler, F.L., Fetter Jr., C.W., 1978. Management of freshwater wetlands for nutrient assimilation. In: Good, R.E., Whigham, D.F., Simpson, R.L. (Eds.), Freshwater Wetlands: Ecological Processes and Management Potential. Academic Press, New York, NY, pp. 321−340.

Small, G.E., Ardón, M., Duff, J.H., Jackman, A.P., Ramírez, A., Triska, F.J., et al., 2016. Phosphorus retention in a lowland Neotropical stream following an eight-year enrichment experiment. Freshw. Sci. 35 (1), 1−11. Available from: https://doi.org/10.1086/684491.

Smith, D.G., 2001. Pennak's Freshwater Invertebrates of the United States, Porifera to Crustacea, fourth ed. John Wiley & Sons, New York, NY.

Smith, G.R., Rettig, J.E., Mittelbach, G.G., Valiulis, J.L., Schaack, S.R., 1999. The effects of fish on assemblages of amphibians in ponds: a field experiment. Freshw. Biol. 41, 829−837.

Smith, I.M., Cook, D.R., 1991. Water mites. In: Thorp, J.H., Covich, A.P. (Eds.), Ecology and Classification of North American Freshwater Invertebrates. Academic Press, Inc, San Diego, CA, pp. 523−592.

Smith, J.P.J., 1977. Vascular Plant Families. Mad River Press, Inc., Eureka, CA.

Smith, K.G., Lips, K.R., Chase, J.M., 2009. Selecting for extinction: nonrandom disease-associated extinction homogenizes amphibian biotas. Ecol. Lett. 12, 1−10.

Smith, R.A., Alexander, R.B., Schwarz, G.E., 2003. Natural background concentrations of nutrients in streams and rivers of the conterminous United States. Environ. Sci. Technol. 37 (14), 2039−3047.

Smith, R.L., Böhlke, J.K., Song, B., Tobias, C.R., 2015. Role of anaerobic ammonium oxidation (anammox) in nitrogen removal from a freshwater aquifer. Environ. Sci. Technol. 49 (20), 12169−12177. Available from: https://doi.org/10.1021/acs.est.5b02488.

Smith, S., Renwick, W., Bartley, J., Buddemeier, R., 2002. Distribution and significance of small, artificial water bodies across the United States landscape. Sci. Total Environ. 299 (1−3), 21−36.

Smith, V.H., 1982. The nitrogen and phosphorus dependence of algal biomass in lakes: an empirical and theoretical analysis. Limnol. Oceanogr. 27 (6), 1101−1112.

Smith, V.H., Mcbride, R.C., 2015. Key ecological challenges in sustainable algal biofuels production. J. Plankton Res. 37 (4), 671−682. Available from: https://doi.org/10.1093/plankt/fbv053.

Smolders, A., Roelofs, J.G.M., 1993. Sulphate-mediated iron limitation and eutrophication in aquatic ecosystems. Aquat. Bot. 46, 247−253.

Smolders, E., Baetens, E., Verbeeck, M., Nawara, S., Diels, J., Verdievel, M., et al., 2017. Internal loading and redox cycling of sediment iron explain reactive phosphorus concentrations in lowland rivers. Environ. Sci. Technol. 51 (5), 2584−2592. Available from: https://doi.org/10.1021/acs.est.6b04337.

Smriga, S., Fernandez, V.I., Mitchell, J.G., Stocker, R., 2016. Chemotaxis toward phytoplankton drives organic matter partitioning among marine bacteria. Proc. Natl. Acad. Sci. 113 (6), 1576−1581. Available from: https://doi.org/10.1073/pnas.1512307113.

Smyth, J.D., Smyth, M.M., 1980. Frogs as Host-Parasite Systems I. The MacMillan Press Ltd, Hong Kong.

Snedcor, G.W., Cochran, W.G., 1980. Statistical Methods, seventh ed. Iowa State University Press, Ames, IA.

Snell, T.W., 1998. Chemical ecology of rotifers. Hydrobiologia 387/388, 267−276.

Snodgrass, J.W., Moore, J., Lev, S.M., Casey, R.E., Ownby, D.R., Flora, R.F., et al., 2017. Influence of modern stormwater management practices on transport of road salt to surface waters. Environ. Sci. Technol. 51 (8), 4165−4172. Available from: https://doi.org/10.1021/acs.est.6b03107.

Sobsey, M.D., Shields, P.A., 1987. Survival and transport of viruses in soils: model studies. In: Rao, V.C., Melnick, J.L. (Eds.), Human Viruses in Sediments, Sludges, and Soils. CRC Press, Boca Raton, FL, pp. 155−177.

Sofaer, H.R., Skagen, S.K., Barsugli, J.J., Rashford, B.S., Reese, G.C., Hoeting, J.A., et al., 2016. Projected wetland densities under climate change: habitat loss but little geographic shift in conservation strategy. Ecol. Appl. 26 (6), 1677−1692. Available from: https://doi.org/10.1890/15-0750.1.

Sokal, R.R., Rohlf, F.J., 1981. Biometry, the Principles and Practice of Statistics. W. H. Freeman and Company, New York, NY.

Sokolow, S.H., Huttinger, E., Jouanard, N., Hsieh, M.H., Lafferty, K.D., Kuris, A.M., et al., 2015. Reduced transmission of human schistosomiasis after restoration of a native river prawn that preys on the snail intermediate host. Proc. Natl. Acad. Sci. 112 (31), 9650−9655. Available from: https://doi.org/10.1073/pnas.1502651112.

Solomon, S., Ivy, D.J., Kinnison, D., Mills, M.J., Neely, R.R., Schmidt, A., 2016. Emergence of healing in the Antarctic ozone layer. Science . Available from: https://doi.org/10.1126/science.aae0061.

Soltero, R.A., Sexton, L.M., Ashley, K.I., McKee, K.O., 1994. Partial and full lift hypolimnetic aeration of Medical Lake, WA to improve water quality. Water Res. 28 (11), 2297−2308.

Sommer, U., 1989. Toward a Darwinian ecology of plankton. In: Sommer, U. (Ed.), Plankton Ecology. Springer-Verlag, New York, NY, pp. 1−8.

Sommer, U., 1999. A comment on the proper use of nutrient ratios in microalgal ecology. Arch. Hydrobiol. 146, 55−64.

Søndergaard, M., 1991. Phototrophic picoplankton in temperate lakes: seasonal abundance and importance along a trophic gradient. Int. Rev. Hydrobiol. 76, 505−522.

Søndergaard, M., Laegaard, S., 1977. Vesicular-arbuscular mycorrhiza in some aquatic vascular plants. Nature 268, 233.

Song, C., Dodds, W.K., Rüegg, J., Argerich, A., Baker, C.L., Bowden, W.B., et al., 2018. Continental-scale decrease in net primary productivity in streams due to climate warming. Nat. Geosci. 11 (6), 415−420. Available from: https://doi.org/10.1038/s41561-018-0125-5.

Sonnenschein, C., Soto, A.M., 1997. An updated review of environmental estrogen and androgen mimics and antagonists. J. Steroid Biochem. Mol. Biol. 65, 43−150.

Sophocleous, M., 2005. Groundwater recharge and sustainability in the High Plains aquifer in Kansas, USA. Hydrogeol. J. 13 (2), 351−365.

Soranno, P.A., Webster, K.E., Riera, J.L., Kratz, T.K., Baron, J.S., Bukaveckas, P.A., et al., 1999. Spatial variation among lakes within landscapes: ecological organization along lake chains. Ecosystems 2, 395−410.

South, G.R., Whittick, A., 1987. Introduction to Phycology. Blackwell Scientific Publications, Oxford, UK.

Spackman, S.C., Hughes, J.W., 1995. Assessment of minimum stream corridor width for biological conservation: species richness and distribution along mid-order streams in Vermont, USA. Biol. Conserv. 71, 325−332.

Spadinger, R., Maier, G., 1999. Selection and diel feeding of the freshwater jellyfish *Craspedacusta sowerbyi*. Freshw. Biol. 41, 567−573.

Sparling, D.W., Fellers, G.M., 2009. Toxicity of two insecticides to California, USA, anurans and its relevance to declining amphibian populations. Environ. Toxicol. Chem. 28 (8), 1696−1703.

Spencer, C.N., Ellis, B.K., 1998. Role of nutrients and zooplankton in regulation of phytoplankton in Flathead Lake (Montana, USA), a large oligotrophic lake. Freshw. Biol. 39, 755−763.

Spencer, C.N., King, D.L., 1984. Role of fish in regulation of plant and animal communities in eutrophic ponds. Can. J. Fish. Aquat. Sci. 41, 1851−1855.

Spencer, C.N., McClelland, B.R., Stanford, J.A., 1991. Shrimp introduction, salmon collapse, and bald eagle displacement: cascading interactions in the food web of a large aquatic ecosystem. Bioscience 41, 14−21.

Sperfeld, E., Raubenheimer, D., Wacker, A., 2016. Bridging factorial and gradient concepts of resource co-limitation: towards a general framework applied to consumers. Ecol. Lett. 19 (2), 201−215. Available from: https://doi.org/10.1111/ele.12554.

Spigel, R.H., Priscu, J.C., 1998. Physical limnology of the McMurdo dry valley lakes. In: Priscu, J.C. (Ed.), Ecosystem Dynamics in a Polar Desert, vol. 72. American Geophysical Union, Washington, DC, pp. 153−186.

Spribille, T., Tuovinen, V., Resl, P., Vanderpool, D., Wolinski, H., Aime, M.C., et al., 2016. Basidiomycete yeasts in the cortex of ascomycete macrolichens. Science 353 (6298), 488−492. Available from: https://doi.org/10.1126/science.aaf8287.

Sprung, M., 1993. The other life: an account of present knowledge of the larval phase of *Dreissena polymorpha*. In: Nalepa, T.F., Schloesser, D.W. (Eds.), Zebra Mussels. Biology, Impacts and Control. Lewis Publishers, Boca Ratan, FL, pp. 39−53.

Sridhar, H., Beauchamp, G., Shanker, K., 2009. Why do birds participate in mixed-species foraging flocks? A large-scale synthesis. Anim. Behav. 78 (2), 337−347. Available from: https://doi.org/10.1016/j.anbehav.2009.05.008.

Srivastava, D.S., 2006. Habitat structure, trophic structure and ecosystem function: interactive effects in a bromeliad−insect community. Oecologia 149 (3), 493−504.

Stagliano, D.M., Whiles, M.R., 2002. Macroinvertebrate production and trophic structure in a tallgrass prairie headwater stream. J. N. Am. Benthol. Soc. 21 (1), 97−113.

Stahlschmidt-Allner, P., Allner, B., Römbke, J., Knacker, T., 1997. Endocrine disrupters in the aquatic environment. Environ. Sci. Pollut. Res. 4 (3), 155−162.

Stander, E.K., Ehrenfeld, J.G., 2009. Rapid assessment of urban wetlands: do hydrogeomorphic classification and reference criteria work? Environ. Manage. 43 (4), 725−742.

Stanford, J.A., Gaufin, A.R., 1974. Hyporheic communities of two Montana rivers. Science 185, 700−702.

Stanford, J.A., Ward, J.V., 1988. The hyporheic habitat of river ecosystems. Nature 335, 64−66.

Stanford, J.A., Ward, J.V., 1993. An ecosystem perspective of alluvial rivers: connectivity and the hyporheic corridor. J. N. Am. Benthol. Soc. 12, 48−60.

Stanish, L.F., Nemergut, D.R., McKnight, D.M., 2011. Hydrologic processes influence diatom community composition in Dry Valley streams. J. N. Am. Benthol. Soc. 30 (4), 1057−1073. Available from: https://doi.org/10.1899/11-008.1.

Stanish, L.F., O'Neill, S.P., Gonzalez, A., Legg, T.M., Knelman, J., McKnight, D.M., et al., 2013. Bacteria and diatom co-occurrence patterns in microbial mats from polar desert streams. Environ. Microbiol. 15 (4), 1115−1131. Available from: https://doi.org/10.1111/j.1462-2920.2012.02872.x.

Stanley, E.H., Casson, N.J., Christel, S.T., Crawford, J.T., Loken, L.C., Oliver, S.K., 2016. The ecology of methane in streams and rivers: patterns, controls, and global significance. Ecol. Monogr. 86 (2), 146−171. Available from: https://doi.org/10.1890/15-1027.

Stanley, E.H., Fisher, S.G., Grimm, N.B., 1997. Ecosystem expansion and contraction in streams. Bioscience 47 (7), 427−435.

Stanley, E.H., Jones, J.B., 2000. Surface-subsurface interactions: past, present, and future. In: Jones, J.B., Molholland, P.J. (Eds.), Streams and Ground Waters. Academic Press, San Diego, CA, pp. 405−417.

Stanley, J.G., Woodard Miley II, W., Sutton, D.L., 1978. Reproductive requirements and likelihood for naturalization of escaped grass carp in the United States. Trans. Am. Fish. Soc. 107, 119−127.

Starmach, J., 1986. Development and structure of the Goczalkowice Reservoir ecosystem XV. Ichthyofauna. Ekol. Pol. 34 (3), 515−521.

Stebbins, R.C., 2003. A Field Guide to Western Reptiles and Amphibians, third ed. Houghton Mifflin Harcourt, Boston.

Stegemeier, J.P., Colman, B.P., Schwab, F., Wiesner, M.R., Lowry, G.V., 2017. Uptake and distribution of silver in the aquatic plant landoltia punctata (duckweed) exposed to silver and silver sulfide nanoparticles. Environ. Sci. Technol. 51 (9), 4936−4943. Available from: https://doi.org/10.1021/acs.est.6b06491.

Stehle, S., Schulz, R., 2015. Agricultural insecticides threaten surface waters at the global scale. Proc. Natl. Acad. Sci. 112 (18), 5750−5755. Available from: https://doi.org/10.1073/pnas.1500232112.

Stein, R.A., DeVries, D.R., Dettmers, J.M., 1995. Food-web regulation by a planktivore: exploring the generality of the trophic cascade hypothesis. Can. J. Fish. Aquat. Sci. 52, 2518−2526.

Steinberg, C.E.W., Geller, W., 1994. Biodiversity and interactions within pelagic nutrient cycling and productivity. In: Schulze, E.-D., Mooney, H.A. (Eds.), Biodiversity and Ecosystem Function. Springer-Verlag, Berlin, pp. 43−64.

Steinman, A.D., 1996. Effects of grazers on freshwater benthic algae. In: Stevenson, R.J., Bothwell, M.L., Lowe, R.L. (Eds.), Algal Ecology: Freshwater Benthic Ecosystems. Academic Press, San Diego, CA, pp. 341−373.

Steinman, A.D., Conklin, J., Bohlen, P.J., Uzarski, D.G., 2003. Influence of cattle grazing and pasture land use on macroinvertebrate communities in freshwater wetlands. Wetlands 23 (4), 877−889.

Steinmetz, J., Soluk, D., Kohler, S., 2008. Facilitation between herons and smallmouth bass foraging on common prey. Environ. Biol. Fish. 81 (1), 51–61.

Stemberger, R., Gilbert, J.J., 1985. Body size, food concentration, and population growth in planktonic rotifers. Ecology 66 (4), 1151–1159.

Stemberger, R.S., Chen, C.Y., 1998. Fish tissue metals and zooplankton assemblages of northeastern U.S. lakes. Can. J. Fish. Aquat. Sci. 55, 339–352.

Stenberg, J., Stenberg, J., 2012. Herbivory limits the yellow water lily in an overgrown lake and in flowing water. Hydrobiologia 691 (1), 81–88. Available from: https://doi.org/10.1007/s10750-012-1035-y.

Sterner, R.W., 1990. The ratio of nitrogen to phosphorus resupplied by herbivores: zooplankton and the algal competitive arena. Am. Nat. 136 (2), 209–229.

Sterner, R.W., 1993. *Daphnia* growth on varying quality of *Scenedesmus*: mineral limitation of zooplankton. Ecology 74, 2351–2360.

Sterner, R.W., Elser, J.J., 2002. Ecological Stoichiometry: The Biology of Elements From Molecules to the Biosphere. Princeton University Press, Princeton, NJ.

Sterner, R.W., Elser, J.J., Hessen, D.O., 1992. Stoichiometric relationships among producers, consumers and nutrient cycling in pelagic ecosystems. Biogeochemistry 17, 49–67.

Sterner, R.W., Hessen, D.O., 1994. Algal nutrient limitation and the nutrition of aquatic herbivores. Annu. Rev. Ecol. Syst. 25, 1–29.

Stetter, K.O., 1998. Hyperthermophiles: isolation, classification, and properties. In: Horikoshi, K., Grant, W.D. (Eds.), Extremophiles. Microbial Life in Extreme Environments. Wiley-Liss, Inc, New York, NY, pp. 1–24.

Stevens, L.E., 1995. Flow regulation, geomorphology, and Colorado River marsh development in the Grand Canyon, Arizona. Ecol. Appl. 5 (4), 1025–1039.

Stevens, L.E., Shannon, J.P., Blinn, D.W., 1997. Colorado River benthic ecology in Grand Canyon, Arizona, USA: dam, tributary and geomorphological influences. Regul. Rivers: Res. Manage. 13, 129–149.

Stevens, M.H.H., Cummins, K.W., 1999. Effects of long-term disturbance on riparian vegetation and instream characteristics. J. Freshw. Ecol. 14, 1–17.

Stevens, T.O., 1997. Subsurface microbiology and the evolution of the biosphere. In: Amy, P.S., Haldeman, D.L. (Eds.), Microbiology of the Terrestrial Deep Subsurface. Lewis Publishers, Boca Raton, FL, pp. 205–223.

Stevens, T.O., McKinley, J.P., 1995. Lithoautotrophic microbial ecosystems in deep basalt aquifers. Science 270, 450–454.

Stevenson, R.J., 1996. The stimulation of drag and current. In: Stevenson, R.J., Bothwell, M.L., Lowe, R.L. (Eds.), Algal Ecology: Freshwater Benthic Ecosystems. Academic Press, San Diego, CA, pp. 321–340.

Stewart-Oaten, A., Bence, J.R., Osenberg, C.W., 1992. Assessing effects of unreplicated perturbations: no simple solutions. Ecology 73, 1396–1404.

Stewart-Oaten, A., Murdoch, W.W., Parker, K.R., 1986. Environmental impact assessment: "Pseudoreplication" in time? Ecology 67, 929–940.

Stickney, R.R., 1994. Principles of Aquaculture. John Wiley & Sons, Inc, New York, NY.

Stockner, J.G., MacIsaac, E.A., 1996. British Columbia lake enrichment programme: two decades of habitat enhancement for Sockeye salmon. Regul. Rivers: Res. Manage. 12, 547–561.

Stoddard, J.L., Jeffries, D.S., Lükewille, A., Clair, T.A., Dillon, P.J., Driscoll, C.T., et al., 1999. Regional trends in aquatic recovery from acidification in North America and Europe. Nature 401, 575–578.

Stoddard, J.L., Van Sickle, J., Herlihy, A.T., Brahney, J., Paulsen, S., Peck, D.V., et al., 2016. Continental-scale increase in lake and stream phosphorus: are oligotrophic systems disappearing in the United States? Environ. Sci. Technol. 50 (7), 3409–3415. Available from: https://doi.org/10.1021/acs.est.5b05950.

Stoks, R., Geerts, A.N., De Meester, L., 2014. Evolutionary and plastic responses of freshwater invertebrates to climate change: realized patterns and future potential. Evol. Appl. 7 (1), 42–55. Available from: https://doi.org/10.1111/eva.12108.

Stølum, H.-H., 1996. River meandering as a self-organization process. Science 271, 1710–1713.

Stone, R., 2007. The last of the leviathans. Science 316, 1684–1688.

Stone, W.W., Gilliom, R.J., Ryberg, K.R., 2014. Pesticides in U.S. streams and rivers: occurrence and trends during 1992–2011. Environ. Sci. Technol. 48 (19), 11025–11030. Available from: https://doi.org/10.1021/es5025367.

Storey, K.B., Storey, J.M., 1988. Freeze tolerance in animals. Physiol. Rev. 68, 27–84.

Strahler, A.N., Strahler, A.H., 1979. Elements of Physical Geography, second ed. John Wiley and Sons, New York, NY.

Strauss, A.T., Shocket, M.S., Civitello, D.J., Hite, J.L., Penczykowski, R.M., Duffy, M.A., et al., 2016. Habitat, predators, and hosts regulate disease in Daphnia through direct and indirect pathways. Ecol. Monogr. 86 (4), 393–411. Available from: https://doi.org/10.1002/ecm.1222.

Strauss, E.A., Dodds, W.K., 1997. Influence of protozoa and nutrient availability on nitrification rates in subsurface sediments. Microb. Ecol. 34 (2), 155–165.

Strauss, E.A., Dodds, W.K., Edler, C.C., 1994. The impact of nutrient pulses on trophic interactions in a farm pond. J. Freshw. Ecol. 9 (3), 217–228.

Strayer, D.L., 1991. Projected distribution of the zebra mussel, Dreissena polymorpha, in North America. Can. J. Fish. Aquat. Sci. 48, 1389–1395.

Strayer, D.L., 1994. Limits to biological distributions in groundwater. In: Gibert, J., Danielopol, D.L., Stanford, J.A. (Eds.), Groundwater Ecology. Academic Press, Inc, San Diego, CA, pp. 287–310.

Strayer, D.L., 1999. Effects of alien species on freshwater mollusks in North America. J. N. Am. Benthol. Soc. 18, 74–98.

Strayer, D.L., 2001. Endangered freshwater invertebrates. In: Levin, S.A. (Ed.), Encyclopedia of Biodiversity, vol. 2. Academic Press, San Diego, CA, pp. 425–439.

Strayer, D.L., 2006. Challenges for freshwater invertebrate conservation. J. N. Am. Benthol. Soc. 25 (2), 271–287.

Strayer, D.L., Caraco, N.F., Cole, J.J., Findlay, S., Pace, M.L., 1999. Transformation of freshwater ecosystems by bivalves. Bioscience 49, 19–27.

Strayer, D.L., Downing, J.A., Haag, W.R., King, T.L., Layzer, J.B., Newton, T.J., et al., 2004. Changing perspectives on pearly mussels, North America's most imperiled animals. Bioscience 54 (5), 429–439.

Strayer, D.L., Dudgeon, D., 2010. Freshwater biodiversity conservation: recent progress and future challenges. J. N. Am. Benthol. Soc. 29 (1), 344–358.

Strayer, D.L., Hummon, W.D., 1991. Gastrotricha. In: Thorp, J.H., Covich, A.P. (Eds.), Ecology and Classification of North American Freshwater Invertebrates. Academic Press, Inc, San Diego, CA, pp. 173–186.

Strayer, D.L., Smith, L.C., Hunter, D.C., 1998. Effects of the zebra mussel (Dreissena polymorpha) invasion on the macrobenthos of the freshwater tidal Hudson River. Can. J. Zool. 76, 419–425.

Stream Solute Workshop, 1990. Concepts and methods for assessing solute dynamics in stream ecosystems. J. N. Am. Benthol. Soc. 9 (2), 95–119.

Strickland, J.D.H., Parsons, T.R., 1972. A Practical Handbook of Sea-Water Analysis, second ed. Bulletin Fisheries Research Board of Canada, Nanaimo, Canada, p. 167.

Strock, K.E., Nelson, S.J., Kahl, J.S., Saros, J.E., McDowell, W.H., 2014. Decadal trends reveal recent acceleration in the rate of recovery from acidification in the northeastern U.S. Environ. Sci. Technol. 48 (9), 4681–4689. Available from: https://doi.org/10.1021/es404772n.

Strong, E.E., Gargominy, O., Ponder, W.F., Bouchet, P., 2008. Global diversity of gastropods (Gastropoda; Mollusca) in freshwater. Hydrobiologia 595 (1), 149–166.

Stuart, S.N., Chanson, J.S., Cox, N.A., Young, B.E., Rodrigues, A.S.L., Fischman, D.L., et al., 2004. Status and trends of amphibian declines and extinctions worldwide. Science 306 (5702), 1783–1786.

Stübing, D., Hagen, W., Schmidt, K., 2003. On the use of lipid biomarkers in marine food web analyses: an experimental case study on the Antarctic krill, Euphausia superba. Limnol. Oceanogr. 48, 1685–1700.

Stumm, W., Morgan, J.J., 1981. Aquatic Chemistry: An Introduction Emphasizing Chemical Equilibria in Natural Waters, second ed. John Wiley and Sons, New York, NY.

Subalusky, A.L., Dutton, C.L., Rosi, E.J., Post, D.M., 2017. Annual mass drownings of the Serengeti wildebeest migration influence nutrient cycling and storage in the Mara River. Proc. Natl. Acad. Sci. 114 (29), 7647−7652. Available from: https://doi.org/10.1073/pnas.1614778114.

Subalusky, A.L., Dutton, C.L., Rosi-Marshall, E.J., Post, D.M., 2015. The hippopotamus conveyor belt: vectors of carbon and nutrients from terrestrial grasslands to aquatic systems in sub-Saharan Africa. Freshw. Biol. 60 (3), 512−525.

Suberkropp, K., 1995. The influence of nutrients on fungal growth, productivity, and sporulation during leaf breakdown in streams. Can. J. Bot. 73, S1361−S1369.

Suberkropp, K., Chauvet, E., 1995. Regulation of leaf breakdown by fungi in streams: influences of water chemistry. Ecology 76 (5), 1433−1445.

Suberkropp, K., Weyers, H., 1996. Application of fungal and bacterial production methodologies to decomposing leaves in streams. Appl. Environ. Microbiol. 62 (5), 1610−1615.

Sugiura, N., Iwami, N., Nishimura, Y.I.O., Sudo, R., 1998. Significance of attached cyanobacteria relevant to the occurrence of musty odor in Lake Kasumigaura. Water Res. 32 (12), 3549−3554.

Sullivan, P.L., Gaiser, E.E., Surratt, D., Rudnick, D.T., Davis, S.E., Sklar, F.H., 2014. Wetland ecosystem response to hydrologic restoration and management: the everglades and its urban-agricultural boundary (FL, USA). Wetlands 34 (1), 1−8. Available from: https://doi.org/10.1007/s13157-014-0525-2.

Suplee, M.W., Watson, V., Dodds, W.K., Shirley, C., 2012. Response of algal biomass to large-scale nutrient controls in the Clark Fork River, Montana, United States. JAWRA J. Am. Water Resour. Assoc. 48 (5), 1008−1021.

Sushchik, N.N., Gladyshev, M.I., Moskvichova, A.V., Makhutova, O.N., Kalachova, G.S., 2003. Comparison of fatty acid composition in major lipid classes of the dominant benthic invertebrates of the Yenisei River. Comp. Biochem. Physiol. 134B, 111−122.

Suttle, C.A., Chan, A.M., Cottrell, M.T., 1990. Infection of phytoplankton by viruses and reduction of primary productivity. Nature 347, 467−469.

Suttle, C.A., Stockner, J.G., Shortreed, K.S., Harrison, P.J., 1988. Time-courses of size-fractionated phosphate uptake: are larger cells better competitors for pulses of phosphate than smaller cells? Oecologia 74, 571−576.

Svensson, J.R., Lindegarth, M., Jonsson, P.R., Pavia, H., 2012. Disturbance−diversity models: what do they really predict and how are they tested? Proc. R. Soc. B 279 (1736), 2163−2170.

Swan, C.M., Brown, B.L., 2017. Metacommunity theory meets restoration: isolation may mediate how ecological communities respond to stream restoration. Ecol. Appl. 27 (7), 2209−2219. Available from: https://doi.org/10.1002/eap.1602.

Swan, C.M., Palmer, M.A., 2006. Composition of speciose leaf litter alters stream detritivore growth, feeding activity and leaf breakdown. Oecologia 147 (3), 469−478.

Syvitski, J.P.M., Vörösmarty, C.J., Kettner, A.J., Green, P., 2005. Impact of humans on the flux of terrestrial sediment to the global coastal ocean. Science 308, 376.

Syvitski, J.P.M., Kettner, A.J., Overeem, I., Hutton, E.W.H., Hannon, M.T., Brakenridge, G.R., et al., 2009. Sinking deltas due to human activities. Nat. Geosci. 2 (10), 681−686. Available from: http://www.nature.com/ngeo/journal/v2/n10/suppinfo/ngeo629_S1.html.

Szuroczki, D., Richardson, J.M.L., 2009. The role of trematode parasites in larval anuran communities: an aquatic ecologist's guide to the major players. Oecologia 161 (2), 371−385.

Takai, K., Nakamura, K., Toki, T., Tsunogai, U., Miyazaki, M., Miyazaki, J., et al., 2008. Cell proliferation at 122°C and isotopically heavy CH_4 production by a hyperthermophilic methanogen under high-pressure cultivation. Proc. Natl. Acad. Sci. 105 (31), 10949−10954. Available from: https://doi.org/10.1073/pnas.0712334105.

Tanaka, T., Kawasaki, K., Daimon, S., Kitagawa, W., Yamamoto, K., Tamaki, H., et al., 2014. A hidden pitfall in the preparation of agar media undermines microorganism cultivability. Appl. Environ. Microbiol. 80 (24), 7659−7666. Available from: https://doi.org/10.1128/aem.02741-14.

Tank, J.L., Martí, E., Riis, T., von Schiller, D., Reisinger, A.J., Dodds, W.K., et al., 2018. Partitioning assimilatory nitrogen uptake in streams: an analysis of stable isotope tracer additions across continents. Ecol. Monogr. 88 (1), 120−138. Available from: https://doi.org/10.1002/ecm.1280.

Tank, J.L., Reisinger, A.J., Rosi, E.J., 2017. Nutrient limitation and uptake. In: Lamberti, G.A., Hauer, F.R. (Eds.), Methods in Stream Ecology, third ed. Elsevier, pp. 147−171.

Tank, J.L., Webster, J.R., 1998. Interaction of substrate and nutrient availability on wood biofilm processes in streams. Ecology 79 (6), 21268−22179.

Tank, S.E., Schindler, D.W., 2004. The role of ultraviolet radiation in structuring epilithic algal communities in Rocky Mountain montane lakes: evidence from pigments and taxonomy. Can. J. Fish. Aquat. Sci. 61 (8), 1461−1474.

Taranu, Z.E., Gregory-Eaves, I., Steele, R.J., Beaulieu, M., Legendre, P., 2017. Predicting microcystin concentrations in lakes and reservoirs at a continental scale: a new framework for modelling an important health risk factor. Glob. Ecol. Biogeogr. 26 (6), 625−637. Available from: https://doi.org/10.1111/geb.12569.

Taranu, Z.E., Zurawell, R.W., Pick, F., Gregory-Eaves, I., 2012. Predicting cyanobacterial dynamics in the face of global change: the importance of scale and environmental context. Glob. Change Biol. 18 (12), 3477−3490. Available from: https://doi.org/10.1111/gcb.12015.

Tatariw, C., Chapman, E.L., Sponseller, R.A., Mortazavi, B., Edmonds, J.W., 2013. Denitrification in a large river: consideration of geomorphic controls on microbial activity and community structure. Ecology 94 (10), 2249−2262. Available from: https://doi.org/10.1890/12-1765.1.

Tavares-Cromar, A.F., Williams, D.D., 1996. The importance of temporal resolution in food web analysis: evidence from a detritus-based stream. Ecol. Monogr. 66, 91−113.

Taylor, B.W., Bothwell, M.L., 2014. The origin of invasive microorganisms matters for science, policy, and management: the case of *Didymosphenia geminata*. Bioscience 64 (6), 531−538.

Taylor, B.W., Flecker, A.S., Hall, R.O., 2006. Loss of a harvested fish species disrupts carbon flow in a diverse tropical river. Science 313 (5788), 833−836.

Taylor, C.A., Warren Jr., M.L., Fitzpatrick Jr., J.F., Hobbs III, H.H., Jezerinac, R.F., Pflieger, W.L., et al., 1996. Conservation status of crayfishes of the United States and Canada. Fisheries 21 (4), 25−38.

Taylor, C.M., Gotelli, N.J., 1994. The macroecology of Cyprinella: correlates of phylogeny, body size, and geographical range. Am. Nat. 144 (4), 549−569.

Taylor, F.J.R., 1999. Morphology (tabulation) and molecular evidence for dinoflagellate phylogeny reinforce each other. Appl. Environ. Microbiol. 35, 1−6.

Taylor, J.M., Back, J.A., Brooks, B.W., King, R.S., 2016. Consumer-mediated nutrient recycling is influenced by interactions between nutrient enrichment and the antimicrobial agent triclosan. Freshw. Sci. 35 (3), 856−872. Available from: https://doi.org/10.1086/687838.

Taylor, T.N., Taylor, E.L., 1993. The Biology and Evolution of Fossil Plants. Prentice Hall, Englewood Cliffs, NJ.

Taylor, W.D., Sanders, R.W., 1991. Protozoa. In: Thorp, J.H., Covich, A.P. (Eds.), Ecology and Classification of North American Freshwater Invertebrates. Academic Press, San Diego, CA, pp. 37−93.

Telford, R.J., Vandvik, V., Birks, H.J.B., 2006. Dispersal limitations matter for microbial morphospecies. Science 312, 1015.

Tellenbach, C., Tardent, N., Pomati, F., Keller, B., Hairston, N.G., Wolinska, J., et al., 2016. Cyanobacteria facilitate parasite epidemics in Daphnia. Ecology 97 (12), 3422−3432. Available from: https://doi.org/10.1002/ecy.1576.

Templeton, R.G., 1995. Freshwater Fisheries Management. Fishing News Books, Osney Mead, Oxford.

Ternes, T.A., 1998. Occurrence of drugs in German sewage treatment plants and rivers. Water Res. 32 (11), 3245−3260.

Tessier, A.J., Consolatti, N.L., 1991. Resource quantity and offspring quality in *Daphnia*. Ecology 72, 468−478.

Tezuka, Y., 1990. Bacterial regeneration of ammonium and phosphate as affected by the carbon: nitrogen:phosphorus ratio of organic substrates. Microb. Ecol. 19, 227−238.

Thackeray, S.J., Sparks, T.H., Frederiksen, M., Burthe, S., Bacon, P.J., Bell, J.R., et al., 2010. Trophic level asynchrony in rates of phenological change for marine, freshwater and terrestrial environments. Glob. Change Biol. 16 (12), 3304−3313.

Thämer, M., De Marco, L., Ramasesha, K., Mandal, A., Tokmakoff, A., 2015. Ultrafast 2D IR spectroscopy of the excess proton in liquid water. Science 350 (6256), 78−82. Available from: https://doi.org/10.1126/science.aab3908.

Thiébaut, G., Muller, S., 1999. A macrophyte communities sequence as an indicator of eutrophication and acidification levels in weakly mineralised streams in north-eastern France. Hydrobiologia 410, 17−24.

Thomas, E.P., Blinn, D.W., Keim, P., 1998. Do xeric landscapes increase genetic divergence in aquatic ecosystems? Freshw. Biol. 40, 587−593.

Thomas, F., Schmidt-Rhaesa, A., Martin, G., Manu, C., Durand, P., Renaud, F., 2002. Do hairworms (Nematomorpha) manipulate the water seeking behaviour of their terrestrial hosts? J. Evol. Biol. 15 (3), 356−361.

Thomas, M.J., Creed, R.P., Skelton, J., Brown, B.L., 2016. Ontogenetic shifts in a freshwater cleaning symbiosis: consequences for hosts and their symbionts. Ecology 97 (6), 1507−1517.

Thomas, W.H., Duval, B., 1995. Sierra Nevada, California, USA, snow-algae: snow albedo changes, algal-bacterial interrelationships, and ultraviolet radiation effects. Arct. Alp. Res. 27 (4), 389−399.

Thomson, J.R., Hart, D.D., Charles, D.F., Nightengale, T.L., Winter, D.M., 2005. Effects of removal of a small dam on downstream macroinvertebrate and algal assemblages in a Pennsylvania stream. J. N. Am. Benthol. Soc. 24, 192−207.

Thorgersen, M.P., Lancaster, W.A., Vaccaro, B.J., Poole, F.L., Rocha, A.M., Mehlhorn, T., et al., 2015. Molybdenum availability is key to nitrate removal in contaminated groundwater environments. Appl. Environ. Microbiol. 81 (15), 4976−4983. Available from: https://doi.org/10.1128/aem.00917-15.

Thorne, R., 1984. Are California's vernal pools uniqueVernal Pools and Intermittent Streams, vol. 28. Institute of Ecology Publication, pp. 1−8.

Thorp, J.H., 2014. Metamorphosis in river ecology: from reaches to macrosystems. Freshw. Biol. 59 (1), 200−210.

Thorp, J.H., Bowes, R.E., 2017. Carbon sources in riverine food webs: new evidence from amino acid isotope techniques. Ecosystems 20 (5), 1029−1041. Available from: https://doi.org/10.1007/s10021-016-0091-y.

Thorp, J.H., Covich, A.P., 1991a. Introduction to freshwater invertebrates. In: Thorp, J.H., Covich, A.P. (Eds.), Ecology and Classification of North American Freshwater Invertebrates. Academic Press, San Diego, CA, pp. 1−15.

Thorp, J.H., Covich, A.P., 1991b. An overview of freshwater habitats. In: Thorp, J.H., Covich, A.P. (Eds.), Ecology and Classification of North American Freshwater Invertebrates. Academic Press, San Diego, CA, pp. 17−36.

Thorp, J.H., Covich, A.P., 2001. Ecology and Classification of North American Freshwater Invertebrates, second ed. Academic Press, San Diego, CA.

Thorp, J.H., Delong, M.D., 1994. The riverine productivity model: an heuristic view of carbon sources and organic processing in large river ecosystems. Oikos 305−308.

Thorp, J.H., Rogers, D.C., 2014. Thorp and Covich's Freshwater Invertebrates: Ecology and General Biology, vol. 1. Elsevier.

Thorp, J.H., Thoms, M.C., Delong, M.D., 2006. The riverine ecosystem synthesis: biocomplexity in river networks across space and time. River Res. Appl. 22 (2), 123−147.

Thouvenot, L., Haury, J., Pottier, G., Thiébaut, G., 2017. Reciprocal indirect facilitation between an invasive macrophyte and an invasive crayfish. Aquat. Bot. 139, 1−7. Available from: https://doi.org/10.1016/j.aquabot.2017.02.002.

Tijdens, M., van deWaal, D.B., Slovackova, H., Hoogveld, H.L., Gons, H.J., 2008. Estimates of bacterial and phytoplankton mortality caused by viral lysis and microzooplankton grazing in a shallow eutrophic lake. Freshw. Biol. 53 (6), 1126−1141.

Tillmanns, A.R., Wilson, A.E., Pick, F.R., Sarnelle, O., 2008. Meta-analysis of cyanobacterial effects on zooplankton population growth rate: species-specific responses. Fundam. Appl. Limnol./Arch. Hydrobiol. 171 (4), 285−295.

Tilman, D., 1982. Resource Competition and Community Structure. Princeton University Press, Princeton, NJ.

Tilman, D., Kilham, S.S., Kilham, P., 1982. Phytoplankton community ecology: the role of limiting nutrients. Annu. Rev. Ecol. Syst. 13, 349–372.

Tilman, D., Naeem, S., Knops, J., Reich, P., Siemann, E., Wedin, D., et al., 1997. Biodiversity and ecosystem properties. Science 278 (5345), 1866–1867.

Tilzer, M.M., Gaedke, U., Schweizer, A., Beese, B., 1991. Interannual variability of phytoplankton productivity and related parameters in Lake Constance: no response to decreased phosphorus loading? J. Plankton Res. 13, 755–777.

Timperman, J., 1969. Medico-legal problems in death by drowning: its diagnosis by the diatom method. J. Forensic Med. 16 (2), 45–73.

Tinbergen, L., 1951. The Study of Instinct. Oxford University Press, New York, NY.

Tobert, H.A., Prior, S.A., Rogers, H.H., Schlesinger, W.H., Mullins, G.L., Runion, G.B., 1996. Elevated atmospheric carbon dioxide in agroecosystems affects groundwater quality. J. Environ. Qual. 25, 720–726.

Tockner, K., Uehlinger, U., Robinson, C., 2009. Rivers of Europe. Academic Press, Amsterdam.

Tockner, K., Ward, J.V., 1999. Biodiversity along riparian corridors. Arch. Hydrobiol. Suppl. 115/3, 293–310.

Todd, D.K., 1970. The Water Encyclopedia. A Compendium of Useful Information on Water Resources. Water Information Center, Port Washington, NY.

Tokmakoff, A., 2007. Shining light on the rapidly evolving structure of water. Science 317 (5834), 54–55. Available from: https://doi.org/10.1126/science.1144515.

Tollrian, R., Dodson, S.I., 1999. Inducible defenses in Cladocera: constraints, costs and multipredator environments. In: Tollrian, E., Harvell, C.D. (Eds.), The Ecology and Evolution of Inducible Defenses. Princeton University Press, Princeton, NJ, pp. 177–202.

Tomassen, H.B., Smolders, A.J., Limpens, J., Lamers, L.P., Roelofs, J.G., 2004. Expansion of invasive species on ombrotrophic bogs: desiccation or high N deposition? J. Appl. Ecol. 41 (1), 139–150.

Tonkin, J.D., Stoll, S., Jähnig, S.C., Haase, P., 2016. Contrasting metacommunity structure and beta diversity in an aquatic-floodplain system. Oikos 125 (5), 686–697. Available from: https://doi.org/10.1111/oik.02717.

Tonn, W.M., Magnuson, J.J., 1982. Patterns in the species composition and richness assemblages in northern Wisconsin lakes. Ecology 63 (4), 1149–1166.

Topping, D.J., Schmidt, J.C., Vierra, L.E. Jr., 2003. Computation and analysis of the instantaneous-discharge for the Colorado River at Lees Ferry, Arizona: May 8, 1921, through September 30, 2000. U.S. Geological Survey Professional Paper 1677, GB1225.A6T67. Retrieved from https://pubs.usgs.gov/pp/pp1677/.

Torres-Ruiz, M., Wehr, J.D., Perrone, A.A., 2007. Trophic relations in a stream food web: importance of fatty acids for macroinvertebrate consumers. J. N. Am. Benthol. Soc. 26 (3), 509–522.

Toth, L.A., 1996. Restoring the hydrogeomorphology of the channelized Kissimmee River. In: Brookes, A., Shields Jr, F.D. (Eds.), River Channel Restoration. Guiding Principles for Sustainable Projects. John Wiley & Sons, Ltd, Chichester, England, pp. 369–383.

Townsend, C.R., 1989. The patch dynamics concept of stream community ecology. J. N. Am. Benthol. Soc. 8, 36–50.

Townsend, C.R., 1996. Invasion biology and ecological impacts of brown trout *Salmo trutta* in New Zealand. Biol. Conserv. 78, 13–22.

Townsend, C.R., Scarsbrook, M.R., Dolédec, S., 1997. The intermediate disturbance hypothesis, refugia, and biodiversity in streams. Limnol. Oceanogr. 42 (5), 938–949.

Traill, L.W., Bradshaw, C.J.A., Brook, B.W., 2007. Minimum viable population size: a meta-analysis of 30 years of published estimates. Biol. Conserv. 139 (1–2), 159–166.

Tricarico, E., Junqueira, A.O.R., Dudgeon, D., 2016. Alien species in aquatic environments: a selective comparison of coastal and inland waters in tropical and temperate latitudes. Aquat. Conserv.: Mar. Freshw. Ecosyst. 26 (5), 872–891. Available from: https://doi.org/10.1002/aqc.2711.

Trimble, S.W., 1999. Decreased rates of alluvial sediment storage in the Coon Creek Basin, Wisconsin, 1975-93. Science 285, 1244–1246.

Triska, F.J., Sedell, J.R., Cromack Jr., K., Gregory, S.V., McCorison, F.M., 1984. Nitrogen budget for a small coniferous forest stream. Ecol. Monogrogr. 54, 119−140.

Troester, M., Brauch, H.-J., Hofmann, T., 2016. Vulnerability of drinking water supplies to engineered nanoparticles. Water Res. 96, 255−279. Available from: https://doi.org/10.1016/j.watres.2016.03.038.

Tsiknia, M., Paranychianakis, N.V., Varouchakis, E.A., Nikolaidis, N.P., 2015. Environmental drivers of the distribution of nitrogen functional genes at a watershed scale. FEMS Microbiol. Ecol. 91 (6). Available from: https://doi.org/10.1093/femsec/fiv052.

Tsui, M.T.K., Finlay, J.C., Nater, E.A., 2009. Mercury bioaccumulation in a stream network. Environ. Sci. Technol. 43 (18), 7016−7022.

Tuchman, N.C., Wetzel, R.G., Rier, S.T., Wahtera, K.A., Teeri, J.A., 2002. Elevated atmospheric CO_2 lowers leaf litter nutritional quality for stream ecosystem food webs. Glob. Change Biol. 8 (2), 163−170.

Tuckett, Q.M., Koetsier, P., 2016. Mid- and long-term effects of wildfire and debris flows on stream ecosystem metabolism. Freshw. Sci. 35 (2), 445−456. Available from: https://doi.org/10.1086/686151.

Tundisi, J.G., Tundisi, T.M., 2012. Limnology. CRC Press, Boca Raton, Fl.

Turetsky, M.R., Kotowska, A., Bubier, J., Dise, N.B., Crill, P., Hornibrook, E.R.C., et al., 2014. A synthesis of methane emissions from 71 northern, temperate, and subtropical wetlands. Glob. Change Biol. 20 (7), 2183−2197. Available from: https://doi.org/10.1111/gcb.12580.

Turner, M.A., Robinson, G.G.C., Townsend, B.E., Hann, B.J., Amaral, J.A., 1995. Ecological effects of blooms of filamentous green algae in the littoral zone of an acid lake. Can. J. Fish. Aquat. Sci. 52, 2264−2275.

Turner, R.E., Rabalais, N.N., 1994. Coastal eutrophication near the Mississippi river delta. Nature 368, 619−621.

Turner, R.E., Rabalais, N.N., Justic, D., Dortch, Q., 2003. Global patterns of dissolved N, P and Si in large rivers. Biogeochemistry 64, 297−317.

Twolan-Strutt, L., Keddy, P.A., 1996. Above- and belowground competition intensity in two contrasting wetland plant communities. Ecology 77, 259−270.

Tyree, M., Clay, N., Polaskey, S., Entrekin, S., 2016. Salt in our streams: even small sodium additions can have negative effects on detritivores. Hydrobiologia 775 (1), 109−122. Available from: https://doi.org/10.1007/s10750-016-2718-6.

U.S. Department of the Interior, U.S. Fish and Wildlife Service, U.S. Department of Commerce, Economics and Statistics Administration Vacant, U.S. Census Bureau, 2011. National survey of fishing, hunting, and wildlife-associated recreation. https://wsfrprograms.fws.gov/subpages/nationalsurvey/natsurveyindex.htm.

Uden, D.R., Hellman, M.L., Angeler, D.G., Allen, C.R., 2014. The role of reserves and anthropogenic habitats for functional connectivity and resilience of ephemeral wetlands. Ecol. Appl. 24 (7), 1569−1582.

Ulrich, W., Ollik, M., Ugland, K.I., 2010. A meta-analysis of species−abundance distributions. Oikos 119 (7), 1149−1155.

United Nations, Department of Economic and Social Affairs, Population Division, 2015. World population prospects: The 2015 revision, key findings and advance tables. Working Paper No. ESA/P/WP.241.

United States Fish and Wildlife Service, 1981. The Platte River ecology study special research report.

Untergasser, U., 1989. Handbook of Fish Diseases. T.F.H. Publications, Inc., Holbokon, NJ.

Urabe, J., Nakanishi, M., Kawabata, K., 1995. Contributions of metazoan plankton to the cycling of nitrogen and phosphorus in Lake Biwa. Limnol. Oceanogr. 40 (2), 232−241.

Urban, N.R., Eisenreich, S.J., 1988. Nitrogen cycling in a forested Minnesota bog. Can. J. Bot. 66, 435−449.

U.S. Air Force, 1960. Handbook of Geophysics, revised ed. McMillan, New York, NY.

USEPA, 1997. EPA national water quality report. Retrieved from: <http://www.epa.gov/watrhome/resources>.

USEPA, 2005. *Bacillus thuringiensis* Cry3Bb1 protein and the genetic material necessary for its production (Vector ZMIR13L) in event MON 863 corn & *Bacillus thuringiensis* Cry1Ab delta-endotoxin and the genetic material necessary for its production in corn (006430, 006484) fact sheet. EPA 730-F-05-001,

Washington, DC. Retrieved from https://archive.epa.gov/pesticides/biopesticides/web/html/frnotices_006430-006484.html.

Vadeboncoeur, Y., Peterson, G., Vander Zanden, M.J., Kalff, J., 2008. Benthic algal production across lake size gradients: interactions among morphometry, nutrients, and light. Ecology 89 (9), 2542−2552.

Vadeboncoeur, Y., Power, M.E., 2017. Attached algae: the cryptic base of inverted trophic pyramids in freshwaters. Annu. Rev. Ecol., Evol., Syst. 48 (1), 255−279. Available from: https://doi.org/10.1146/annurev-ecolsys-121415-032340.

Vajda, A.M., Barber, L.B., Gray, J.L., Lopez, E.M., Woodling, J.D., Norris, D.O., 2008. Reproductive disruption in fish downstream from an Estrogenic wastewater effluent. Environ. Sci. Technol. 42 (9), 3407−3414.

Vallentyne, J.R., 1974. The Algal Bowl, Lakes and Man. Department of the Environment, Fisheries and Marine Service, Ottawa, Canada.

Vallett, H.M., Baker, M.A., Morrice, J.A., Crawford, C.S., Molles Jr., M.C., Dahm, C.N., et al., 2005. Biogeochemical and metabolic responses to the flood pulse in a semiarid floodplain. Ecology 86, 220−234.

Van de Waal, D.B., Smith, V.H., Declerck, S.A.J., Stam, E.C.M., Elser, J.J., 2014. Stoichiometric regulation of phytoplankton toxins. Ecol. Lett. 17 (6), 736−742. Available from: https://doi.org/10.1111/ele.12280.

van der Valk, A., 2012. The Biology of Freshwater Wetlands. Oxford University Press, Oxford, England.

van Dijk, W., van de Lageweg, W., Kleinhans, M., 2012. Experimental meandering river with chute cutoffs. J. Geophys. Res. 117 (F3), F03023.

Van Donk, E., 1989. The role of fungal parasites in phytoplankton succession. In: Sommer, U. (Ed.), Plankton Ecology: Succession in Plankton Communities. Springer-Verlag, New York, NY, pp. 171−194.

Van Donk, E., Ianora, A., Vos, M., 2011. Induced defences in marine and freshwater phytoplankton: a review. Hydrobiologia 668 (1), 3−19. Available from: https://doi.org/10.1007/s10750-010-0395-4.

van Duinen, G.A., Vermonden, K., Bodelier, P.L.E., Hendriks, A.J., Leuven, R.S.E.W., Middelburg, J.J., et al., 2013. Methane as a carbon source for the food web in raised bog pools. Freshw. Sci. 32 (4), 1260−1272. Available from: https://doi.org/10.1899/12-121.1.

Van Duren, I.C., Boeye, D., Grootjans, A.P., 1997. Nutrient limitations in an extant and drained poor fen: implications for restoration. Plant Ecol. 133, 91−100.

van Gerven, L.P.A., de Klein, J.J.M., Gerla, D.J., Kooi, B.W., Kuiper, J.J., Mooij, W.M., 2015. Competition for light and nutrients in layered communities of aquatic plants. Am. Nat. 186 (1), 72−83. Available from: https://doi.org/10.1086/681620.

Van Hannen, E.J., Zwart, G., van Agterveld, M.P., Gons, H.J., Ebert, J., Laanbroek, H.J., 1999. Changes in bacterial and eukaryotic community structure after mass lysis of filamentous cyanobacteria associated with viruses. Appl. Environ. Microbiol. 65 (2), 795−801.

van Rijssel, J.C., Hecky, R.E., Kishe-Machumu, M.A., Meijer, S.E., Pols, J., van Tienderen, K.M., et al., 2016. Climatic variability in combination with eutrophication drives adaptive responses in the gills of Lake Victoria cichlids. Oecologia 182 (4), 1187−1201. Available from: https://doi.org/10.1007/s00442-016-3721-3.

van Wilgen, B.W., Cowling, R.M., Burgers, C.J., 1996. Valuation of ecosystems services. Bioscience 46 (3), 184−189.

Vanni, M., Andrews, J., Renwick, W., Gonzalez, M., Noble, S., 2006. Nutrient and light limitation of reservoir phytoplankton in relation to storm-mediated pulses in stream discharge. Arch. Hydrobiol. 167 (1−4), 421−445.

Vanni, M., Arend, K., Bremigan, M., Bunnell, D., Garvey, J., Gonzalez, M., et al., 2005. Linking landscapes and food webs: effects of omnivorous fish and watersheds on reservoir ecosystems. Bioscience 55 (2), 155−167.

Vanni, M.J., 2002. Nutrient cycling by animals in freshwater ecosystems. Annu. Rev. Ecol. Syst. 33 (1), 341−370.

Vanni, M.J., Bowling, A.M., Dickman, E.M., Hale, R.S., Higgins, K.A., Horgan, M.J., et al., 2006. Nutrient cycling by fish supports relatively more primary production as lake productivity increases. Ecology 87 (7), 1696−1709.

Vanni, M.J., Layne, C.D., 1997. Nutrient recycling and herbivory as mechanisms in the "top-down" effect of fish on algae in lakes. Ecology 78, 21−40.

Vanni, M.J., Layne, C.D., Arnott, S.E., 1997. "Top-down" trophic interactions in lakes: effects of fish on nutrient dynamics. Ecology 78, 1−20.

Vannote, R.L., Minshall, G.W., Cummins, K.W., Sedell, J.R., Cushing, C.E., 1980. The river continuum concept. Can. J. Fish. Aquat. Sci. 37, 130−137.

Vannote, R.L., Sweeney, B.W., 1980. Geographic analysis of thermal equilibria: a conceptual model for evaluating the effect of natural and modified thermal regimes on aquatic insect communities. Am. Nat. 115 (5), 667−695.

Vaughn, C.C., 2010. Biodiversity losses and ecosystem function in freshwaters: emerging conclusions and research directions. Bioscience 60 (1), 25−35. Available from: https://doi.org/10.1525/bio.2010.60.1.7.

Vaughn, C.C., Taylor, C.M., 1999. Impoundments and the decline of freshwater mussels: a case study of an extinction gradient. Conserv. Biol. 13 (4), 912−920.

Vaux, P.D., Paulson, L.J., Axler, R.P., Leavitt, S., 1995. The water quality implications of artificially fertilizing a large desert reservoir for fisheries enhancement. Water Environ. Res. 67 (2), 189−200.

Veach, A., Bernot, M.J., Mitchell, J.K., 2012. The influence of six pharmaceuticals on freshwater sediment microbial growth incubated at different temperatures and UV exposures. Biodegradation 23 (4), 497−507. Available from: https://doi.org/10.1007/s10532-011-9528-3.

Veach, A.M., Stegen, J.C., Brown, S.P., Dodds, W.K., Jumpponen, A., 2016. Spatial and successional dynamics of microbial biofilm communities in a grassland stream ecosystem. Mol. Ecol. 25 (18), 4674−4688.

Veh, G., Korup, O., Roessner, S., Walz, A., 2018. Detecting Himalayan glacial lake outburst floods from Landsat time series. Remote Sens. Environ. 207, 84−97. Available from: https://doi.org/10.1016/j.rse.2017.12.025.

Vellieux, F., Madern, D., Zaccai, G., Ebel, C., 2006. Molecular adaptation to salts. In: Gerday, C., Glansdorff, N. (Eds.), Physiology and Biochemistry of Extremophiles. ASM Press, Washington, DC, pp. 240−253.

Vengosh, A., Jackson, R.B., Warner, N., Darrah, T.H., Kondash, A., 2014. A critical review of the risks to water resources from unconventional shale gas development and hydraulic fracturing in the United States. Environ. Sci. Technol. 48 (15), 8334−8348. Available from: https://doi.org/10.1021/es405118y.

Ventura, M., Tiberti, R., Buchaca, T., Buñay, D., Sabás, I., Miró, A., 2017. Why should we preserve fishless high mountain lakes? High Mountain Conservation in a Changing World. Springer, pp. 181−205.

Verduin, J., 1988. Chemical limnology. Verhein Internationale Verein Limnologie 23, 103−105.

Verhoeven, J.T.A., 1986. Nutrient dynamics in minerotrophic peat mires. Aquat. Bot. 25, 117−137.

Verhoeven, J.T.A., Koerselman, W., Meuleman, A.F.M., 1996. Nitrogen- or phosphorus-limited growth in herbaceous, wet vegetation: relations with atmospheric inputs and management regimes. Trends Ecol. Evol. 11 (12), 494−497.

Verhougstraete, M.P., Martin, S.L., Kendall, A.D., Hyndman, D.W., Rose, J.B., 2015. Linking fecal bacteria in rivers to landscape, geochemical, and hydrologic factors and sources at the basin scale. Proc. Natl. Acad. Sci. 112 (33), 10419−10424. Available from: https://doi.org/10.1073/pnas.1415836112.

Verpoorter, C., Kutser, T., Seekell, D.A., Tranvik, L.J., 2014. A global inventory of lakes based on high-resolution satellite imagery. Geophys. Res. Lett. 41 (18). Available from: https://doi.org/10.1002/2014gl0606412014GL060641.

Vighi, M., Zanin, G., 1994. Agronomic and ecototoxicological aspects of herbicide contamination of groundwater in Italy. In: Bergman, L., Pugh, D.M. (Eds.), Environmental Toxicology, Economics and Institutions. Kluwer Academic Publishers, Dordrecht, The Netherlands, pp. 111−139.

Vincent, W.F., 1988. Microbial Ecosystems of Antarctica. University Press, London, UK.

Vinebrooke, R.D., Leavitt, P.R., 1999. Differential responses of littoral communities to ultraviolet radiation in an alpine lake. Ecology 80, 223−237.

Visser, P.M., Ibelings, B.W., Bormans, M., Huisman, J., 2016a. Artificial mixing to control cyanobacterial blooms: a review. Aquat. Ecol. 50 (3), 423−441. Available from: https://doi.org/10.1007/s10452-015-9537-0.

Visser, P.M., Ibelings, B.W., Veer, B.V.D., Koedood, J., Mur, L.R., 1996. Artificial mixing prevents nuisance blooms of the cyanobacterium *Microcystis* in Lake Nieuwe Meer, the Netherlands. Freshw. Biol. 36, 435−450.

Visser, P.M., Verspagen, J.M., Sandrini, G., Stal, L.J., Matthijs, H.C., Davis, T.W., et al., 2016b. How rising CO_2 and global warming may stimulate harmful cyanobacterial blooms. Harmful Algae 54, 145−159.

Vitorino Júnior, O.B., Fernandes, R., Agostinho, C.S., Pelicice, F.M., 2016. Riverine networks constrain β-diversity patterns among fish assemblages in a large Neotropical river. Freshw. Biol. 61 (10), 1733−1745. Available from: https://doi.org/10.1111/fwb.12813.

Vitousek, P.M., 1994. Beyond global warming: ecology and global change. Ecology 75 (7), 1861−1876.

Vitousek, P.M., Aber, J., Howarth, R.W., Likens, G.E., Matson, P.A., Schindler, D.W., et al., 1997. Human alteration of the global nitrogen cycle: causes and consequences. Issues Ecol. 1, 2−15.

Vitousek, P.M., D'Antonio, C.M., Loope, L.L., Westbrooks, R., 1996. Biological invasions as global environmental change. Am. Sci. 84, 468−478.

Vogel, S., 1994. Life in Moving Fluids, second ed. Princeton University Press, Princeton, NJ.

Vogt, K.A., Gordon, J.C., Wargo, J.P., Vogt, D.J., Asbjornsen, H., Palmiotto, P.A., et al., 1997. Ecosystems. Balancing Science With Management. Springer-Verlag, New York, NY.

Vogt, R.J., St-Gelais, N.F., Bogard, M.J., Beisner, B.E., del Giorgio, P.A., 2017. Surface water CO_2 concentration influences phytoplankton production but not community composition across boreal lakes. Ecol. Lett. 20 (11), 1395−1404. Available from: https://doi.org/10.1111/ele.12835.

Voigt, C., Marushchak, M.E., Lamprecht, R.E., Jackowicz-Korczyński, M., Lindgren, A., Mastepanov, M., et al., 2017. Increased nitrous oxide emissions from Arctic peatlands after permafrost thaw. Proc. Natl. Acad. Sci. 114 (24), 6238−6243. Available from: https://doi.org/10.1073/pnas.1702902114.

Volke, M.A., Scott, M.L., Johnson, W.C., Dixon, M.D., 2015. The ecological significance of emerging deltas in regulated rivers. Bioscience 65 (6), 598−611.

Vollenweider, R.A., 1976. Advances in defining critical loading levels for phosphorus in lake eutrophication. Memorie dell'Instituto Italiano di Idrobiologia 33, 53−83.

von Elert, E., Franck, A., 1999. Colony formation in *Scenedesmus*: grazer-mediated release and chemical features of the infochemical. J. Plankton Res. 21 (4), 789−804.

Vonlanthen, P., Bittner, D., Hudson, A.G., Young, K.A., Muller, R., Lundsgaard-Hansen, B., et al., 2012. Eutrophication causes speciation reversal in whitefish adaptive radiations. Nature 482 (7385), 357−362. Available from: http://www.nature.com/nature/journal/v482/n7385/abs/nature10824.html#supplementary-information.

Vorburger, C., Ribi, G., 1999. Aggression and competition for shelter between a native and an introduced crayfish in Europe. Freshw. Biol. 42, 111−119.

Vörösmarty, C.J., Green, P., Salisbury, J., Lammers, R.B., 2000. Global water resources: vulnerability from climate change and population growth. Science 289, 284−288.

Vorosmarty, C.J., McIntyre, P.B., Gessner, M.O., Dudgeon, D., Prusevich, A., Green, P., et al., 2010. Global threats to human water security and river biodiversity. Nature 467 (7315), 555−561. Available from: https://doi.org/10.1038/nature09440.

Vörösmarty, C.J., Meybeck, M., Pastore, C.L., 2015. Impair-then-repair: a brief history & global-scale hypothesis regarding human-water interactions in the anthropocene. Daedalus 144 (3), 94−109. Available from: https://doi.org/10.1162/DAED_a_00345.

Voshell, J.J., 2002. A Guide to Common Freshwater Invertebrates of North America. McDonald and Woodward, Blacksburg, VA.

Voshell, J.R., Simmons Jr., G.M., 1984. Colonization and succession of benthic macroinvertebrates in a new reservoir. Hydrobiologia 112, 27−39.

Voss, K.A., Bernhardt, E.S., 2017. Effects of mountaintop removal coal mining on the diversity and secondary productivity of Appalachian rivers. Limnol. Oceanogr. 62 (4), 1754−1770. Available from: https://doi.org/10.1002/lno.10531.

Voyles, J., Woodhams, D.C., Saenz, V., Byrne, A.Q., Perez, R., Rios-Sotelo, G., et al., 2018. Shifts in disease dynamics in a tropical amphibian assemblage are not due to pathogen attenuation. Science 359 (6383), 1517−1519. Available from: https://doi.org/10.1126/science.aao4806.

Vrede, T., Tranvik, L.J., 2006. Iron constraints on planktonic primary production in oligotrophic lakes. Ecosystems 9 (7), 1094−1105.

Vrijenhoek, R.C., 1998. Conservation genetics of freshwater fish. J. Fish. Biol. 53, 394−412.

Vuori, K.-M., Joensuu, I., 1996. Impact of forest drainage on the macroinvertebrates of a small boreal headwater stream: do buffer zones protect lotic biodiversity? Biol. Conserv. 77, 87−95.

Vymazal, J., 1995. Algae and Element Cycling in Wetlands. CRC Press, Boca Raton, FL.

Waajen, G., van Oosterhout, F., Douglas, G., Lürling, M., 2016. Geo-engineering experiments in two urban ponds to control eutrophication. Water Res. 97, 69−82. Available from: https://doi.org/10.1016/j.watres.2015.11.070.

Wade, P.M., 1990. Physical control of aquatic weeds. In: Pieterse, A.H., Murphy, K.J. (Eds.), Aquatic Weeds: The Ecology and Management of Nuisance Aquatic Vegetation. Oxford University Press, New York, NY, pp. 93−135.

Wagenhoff, A., Liess, A., Pastor, A., Clapcott, J.E., Goodwin, E.O., Young, R.G., 2017. Thresholds in ecosystem structural and functional responses to agricultural stressors can inform limit setting in streams. Freshw. Sci. 36 (1), 178−194. Available from: https://doi.org/10.1086/690233.

Waggoner, P.E., Schefter, J., 1990. Future water use in the present climate. In: Waggoner, P.E. (Ed.), Climate Change and U.S. Water Resources. Report of the American Association for the Advancement of Science Panel on Climatic Variability, Climate Change and the Planning and Management of U.S. Water Resources. John Wiley & Sons, New York, NY, pp. 19−40.

Wagner, R., 1991. The influence of the diel activity pattern of the larvae of *Sericostoma personatum* (Kirby and Spence) (Trichoptera) on organic matter distribution in stream-bed sediments—a laboratory study. Hydrobiologia 224, 65−70.

Wake, D.B., Vredenburg, V.T., 2008. Are we in the midst of the sixth mass extinction? A view from the world of amphibians. Proc. Natl. Acad. Sci. USA 105, 11466−11473.

Walker, K.F., Boulton, A.M., Thoms, M.C., Sheldon, F., 1994. Effects of water-level changes induced by weirs on the distribution of littoral plants along the River Murray, South Australia. Aust. J. Mar. Freshw. Res. 45, 1421−1438.

Wallace, J.B., Cuffney, T.F., Webster, J.R., Lugthart, G.J., Chung, K., Goldowitz, B.S., 1991. Export of fine organic particles from headwater streams: effects of season, extreme discharges, and invertebrate manipulation. Limnol. Oceanogr. 36 (4), 670−682.

Wallace, J.B., Eggert, S.L., Meyer, J.L., Webster, J.R., 1997. Multiple trophic levels of a forest stream linked to terrestrial litter inputs. Science 277, 102−104.

Wallace, J.B., Eggert, S.L., Meyer, J.L., Webster, J.R., 1999. Effects of resource limitation on a detrital-based ecosystem. Ecol. Monogr. 69, 409−442.

Wallace, J.B., Grubaugh, J.W., Whiles, M.R., 1996. Biotic indices and stream ecosystem processes: results from an experimental study. Ecol. Appl. 6 (1), 140−151.

Wallace, J.B., Merritt, R.W., 1980. Filter-feeding ecology of aquatic insects. Annu. Rev. Entomol. 25 (1), 103−132.

Wallace, J.B., Webster, J.R., 1996. The role of macroinvertebrates in stream ecosystem function. Annu. Rev. Entomol. 41, 115−139.

Wallace, R.T., Snell, T.W., 1991. Rotifera. In: Thorp, J.H., Covich, A.P. (Eds.), Ecology and Classification of North American Freshwater Invertebrates. Academic Press, Inc, San Diego, CA, pp. 187−248.

Wallberg, P., Bergqvist, P.-A., Andersson, A., 1997. Potential importance of protozoan grazing on the accumulation of polychlorinated biphenyls (PCBs) in the pelagic food web. Hydrobiologia 357, 53−62.

Walls, S.C., Ball, L.C., Barichivich, W.J., Dodd, C.K., Enge, K.M., Gorman, T.A., et al., 2017. Overcoming challenges to the recovery of declining amphibian populations in the United States. Bioscience 67 (2), 156−165.

Walsby, A.E., 1994. Gas vesicles. Microbiol. Rev. 58, 94−144.

Walsh, C.J., Roy, A.H., Feminella, J.W., Cottingham, P.D., Groffman, P.M., Morgan II, R.P., 2005. The urban stream syndrome: current knowledge and the search for a cure. J. N. Am. Benthol. Soc. 24 (3), 706–723.

Walter, R.C., Merritts, D.J., 2008. Natural streams and the legacy of water-powered mills. Science 319 (5861), 299–304.

Walters, D.M., Fritz, K.M., Otter, R.R., 2008. The dark side of subsidies: adult stream insects export organic contaminants to riparian predators. Ecol. Appl. 18 (8), 1835–1841. Available from: https://doi.org/10.1890/08-0354.1.

Walters, D.M., Jardine, T.D., Cade, B.S., Kidd, K.A., Muir, D.C.G., Leipzig-Scott, P., 2016. Trophic magnification of organic chemicals: a global synthesis. Environ. Sci. Technol. 50 (9), 4650–4658. Available from: https://doi.org/10.1021/acs.est.6b00201.

Walther, D.A., Whiles, M.R., 2008. Macroinvertebrate responses to constructed riffles in the Cache River, Illinois, USA. Environ. Manage. 41 (4), 516–527.

Wang, H., Zhang, Z., Liang, D., du, H., Pang, Y., Hu, K., et al., 2016. Separation of wind's influence on harmful cyanobacterial blooms. Water Res. 98, 280–292. Available from: https://doi.org/10.1016/j.watres.2016.04.037.

Wang, Y., Hatt, J.K., Tsementzi, D., Rodriguez-R, L.M., Ruiz-Pérez, C.A., Weigand, M.R., et al., 2017. Quantifying the importance of the rare biosphere for microbial community response to organic pollutants in a freshwater ecosystem. Appl. Environ. Microbiol. 83 (8). Available from: https://doi.org/10.1128/aem.03321-16.

Wang, Y., Lin, W., Li, J., Pan, Y., 2013. High diversity of magnetotactic deltaproteobacteria in a freshwater niche. Appl. Environ. Microbiol. 79 (8), 2813–2817.

Wania, F., Mackay, D., 1993. Global fractionation and cold condensation of low volatility organochlorine compounds in polar regions. Ambio 22, 10–18.

Wanjugi, P., Fox, G.A., Harwood, V.J., 2016. The interplay between predation, competition, and nutrient levels influences the survival of *Escherichia coli* in aquatic environments. Microb. Ecol. 72 (3), 526–537. Available from: https://doi.org/10.1007/s00248-016-0825-6.

Ward, A.K., Dahm, C.N., Cummins, K.W., 1985. *Nostoc* (Cyanophyta) productivity in Oregon stream ecosystems: invertebrate influences and differences between morphological types. J. Phycol. 21, 223–227.

Ward, B.A., Marañón, E., Sauterey, B., Rault, J., Claessen, D., 2017. The size dependence of phytoplankton growth rates: a trade-off between nutrient uptake and metabolism. Am. Nat. 189 (2), 170–177. Available from: https://doi.org/10.1086/689992.

Ward, B.B., 1996. Nitrification and denitrification: probing the nitrogen cycle in aquatic environments. Microb. Ecol. 32, 247–261.

Ward, C.J., Codd, G.A., 1999. Comparative toxicity of four microcystins of different hydrophobicities to the protozoan, *Tetrahymena pyriformis*. J. Appl. Microbiol. 86, 874–882.

Ward, D.M., Weller, R., Bateson, M.M., 1990. 16S rRNA sequences reveal numerous uncultured microorgansims in a natural community. Nature 345, 63–65.

Ward, J.V., 1985. Thermal characteristics of running water. Hydrobiologia 125, 31–46.

Ward, J.V., Stanford, J.A., 1979. The Ecology of Regulated Rivers. Plenum Publishing Company, New York, NY.

Ward, J.V., Stanford, J.A., 1983. The serial discontinuity concept of lotic ecosystems. In: Fontaine, T.D. I., Bartell, S.M. (Eds.), Dynamics of Lotic Ecosystems. Ann Arbor Science Publishers, Ann Arbor, MI, pp. 29–42.

Ward, J.V., Voelz, N.J., 1994. Groundwater fauna of the South Platte River system, Colorado. In: Gibert, J., Danielopol, D.L., Stanford, J.A. (Eds.), Groundwater Ecology. Academic Press, Inc, San Diego, CA, pp. 391–423.

Ward-Perkins, J.B., 1970. Monterosi in the Etruscan and Roman periods. Ianula: an account of the history and development of the Lago di Monterosi, Latium, Italy. Trans. Am. Philos. Soc. 60 (4), 10–16.

Warne, R.W., Kirschman, L., Zeglin, L., 2017. Manipulation of gut microbiota reveals shifting community structure shaped by host developmental windows in amphibian larvae. Integr. Comp. Biol. 57 (4), 786−794. Available from: https://doi.org/10.1093/icb/icx100.

Warren, D.R., Collins, S.M., Purvis, E.M., Kaylor, M.J., Bechtold, H.A., 2017. Spatial variability in light yields colimitation of primary production by both light and nutrients in a forested stream ecosystem. Ecosystems 20 (1), 198−210. Available from: https://doi.org/10.1007/s10021-016-0024-9.

Warren, M.L., Burr, B.M., Walsh, S.J., Bart, H.L., Cashner, R.C., Etnier, D.A., et al., 2000. Diversity, distribution, and conservation status of the native freshwater fishes of the southern United States. Fisheries 25 (10), 7−31.

Warren, M.L.J., Burr, B.M., 1994. Status of freshwater fishes of the United States. Fisheries 19, 6−18.

Watanabe, H., Tokuda, G., 2001. Animal cellulases. Cell. Mol. Life Sci. (CMLS) 58 (9), 1167−1178.

Waters, M.N., 2016. A 4700-year history of cyanobacteria toxin production in a shallow subtropical lake. Ecosystems 19 (3), 426−436. Available from: https://doi.org/10.1007/s10021-015-9943-0.

Waters, T.F., 1977. Secondary production in inland waters. Adv. Ecol. Res. 10, 91−164.

Waters, T.F., 1995. Sediment in Streams, Sources, Biological Effects, and Control. American Fisheries Society, Bethesda, MD, Vol. American Fisheries Society Monograph 7).

Watson, V., 1989. Maximum levels of attached algae in the Clark Fork River. Proc. Montana Acad. Sci. 49, 27−35 (Biological Science).

Weast, R.C., 1978. CRC Handbook of Chemistry and Physics. CRC Press, West Palm Beach, FL.

Weaver, D.M., Coghlan, S.M., Zydlewski, J., 2016. Sea lamprey carcasses exert local and variable food web effects in a nutrient-limited Atlantic coastal stream. Can. J. Fish. Aquat. Sci. 73 (11), 1616−1625. Available from: https://doi.org/10.1139/cjfas-2015-0506.

Webster, J.R., 1975. Potassium and Calcium Dynamics in Stream Ecosystems on Three Southern Appalachian Watersheds of Contrasting Vegetation (Dissertation). University of Georgia, Athens, GA.

Webster, J.R., Benfield, E.F., Ehrman, T.P., Schaeffer, M.A., Tank, J.L., Hutchens, J.J., et al., 1999. What happens to allochthonous material that falls into streams? A synthesis of new and published information from Coweeta. Freshw. Biol. 41, 687−705.

Webster, J.R., Ehrman, R.P., 1996. Solute dynamics. In: Hauer, F.R., Lamberti, G.A. (Eds.), Methods in Stream Ecology. Academic Press, San Diego, CA, pp. 145−160.

Webster, K.E., Kratz, T.K., Bowser, C.J., Magnuson, J.J., Rose, W.J., 1996. The influence of landscape position on lake chemical responses to drought in northern Wisconsin. Limnol. Oceanogr. 41 (5), 977−984.

Webster, K.E., Soranno, P.A., Baines, S.B., Kratz, T.K., Bowser, C.J., Dillon, P.J., et al., 2000. Structuring features of lake districts: landscape controls on lake chemical responses to drought. Freshw. Biol. 43, 499−515.

Wehr, J., Sheath, R., 2003. Freshwater Algae of North America: Ecology and Classification. Academic Press.

Wehr, J.D., Sheath, R.G., 2002. Freshwater Algae of North America: Ecology and Classification. Elsevier, New York, NY.

Wehr, J.D., Sheath, R.G., Kociolek, J.P., 2015. Freshwater Algae of North America: Ecology and Classification. Elsevier.

Welch, D.M., Meselson, M., 2000. Evidence for the evolution of Bdelloid rotifers without sexual reproduction or genetic exchange. Science 288, 1211−1215.

Welcomme, R.L., 1979. Fisheries Ecology of Floodplain Rivers [Tropics]. Longman.

Weldon, C., Du Preez, L.H., Hyatt, A.D., Muller, R., Speare, R., 2004. Origin of the amphibian chytrid fungus. Emerg. Infect. Dis. 10, 2100−2105.

Weller, M.W., 1995. Use of tow waterbird guilds as evaluation tools for the Kissimmee River restoration. Restor. Ecol. 3 (3), 211−224.

Wenger, S.J., Roy, A.H., Jackson, C.R., Bernhardt, E.S., Carter, T.L., Filoso, S., et al., 2009. Twenty-six key research questions in urban stream ecology: an assessment of the state of the science. J. N. Am. Benthol. Soc. 28 (4), 1080−1098.

Werner, E.E., Hall, D.J., 1974. Optimal foraging and the size selection of prey by the bluegill sunfish (*Lepomis macrochirus*). Ecology 55, 1042−1052.

Wernet, P., Nordlund, D., Bergmann, U., Cavalleri, M., Odelius, M., Ogasawara, H., et al., 2004. The structure of the first coordination shell in liquid water. Science 304 (5673), 995−999.

Westermann, P., 1993. Wetland and swamp microbiology. In: Ford, T.E. (Ed.), Aquatic Microbiology: An Ecological Approach. Blackwell Scientific Publications, Oxford, UK, pp. 215−238.

Weston, D.P., Poynton, H.C., Wellborn, G.A., Lydy, M.J., Blalock, B.J., Sepulveda, M.S., et al., 2013. Multiple origins of pyrethroid insecticide resistance across the species complex of a nontarget aquatic crustacean, *Hyalella azteca*. Proc. Natl. Acad. Sci. USA 110 (41), 16532−16537. Available from: https://doi.org/10.1073/pnas.1302023110.

Wetlands Committee on Characterization of Wetlands, 1995. Commission on Geosciences, Environment, and Resources. Wetlands Characteristics and Boundaries. National Research Council, Washington, DC.

Wetzel, R., 1990. Reservoir ecosystems: conclusions and speculations. Reserv. Limnol.: Ecol. Perspect. 227−238.

Wetzel, R.G., 1983. Limnology, second ed. Saunders College Publishing, Orlando, FL.

Wetzel, R.G., 1991. Extracellular enzymatic interactions: storage, redistribution, and interspecific communication. In: Chróst, R.J. (Ed.), Microbial Enzymes in Aquatic Environments. Springer-Verlag, New York, NY, pp. 6−28.

Wetzel, R.G., 2001. Limnology, third ed. Academic Press, San Diego, CA.

Wetzel, R.G., Likens, G.E., 1991. Limnological Analyses, second ed. Springer-Verlag, New York, NY.

Weyhenmeyer, G.A., Kortelainen, P., Sobek, S., Müller, R., Rantakari, M., 2012. Carbon dioxide in boreal surface waters: a comparison of lakes and streams. Ecosystems 15 (8), 1295−1307.

Whiles, M.R., Dodds, W.K., 2002. Relationships between stream size, suspended particles, and filter-feeding macroinvertebrates in a Great Plains drainage network. J. Environ. Qual. 31 (5), 1589−1600.

Whiles, M.R., Gladyshev, M.I., Sushchik, N.N., Makhutova, O.N., Kalachova, G.S., Peterson, S.D., et al., 2010. Fatty acid analyses reveal high degrees of omnivory and dietary plasticity in pond-dwelling tadpoles. Freshw. Biol. 55 (7), 1533−1547.

Whiles, M.R., Goldowitz, B.S., 2001. Hydrologic influences on insect emergence production from central Platte River wetlands. Ecol. Appl. 11 (6), 1829−1842.

Whiles, M.R., Goldowitz, B.S., 2005. Macroinvertebrate communities in Central Platte River wetlands: patterns across a hydrologic gradient. Wetlands 25 (2), 462−472.

Whiles, M.R., Goldowitz, B.S., Charlton, R.E., 1999. Life history and production of a semi-terrestrial limnephilid caddisfly in an intermittent Platte River wetland. J. N. Am. Benthol. Soc. 18 (4), 533−544.

Whiles, M.R., Hall, R.O., Dodds, W.K., Verburg, P., Huryn, A.D., Pringle, C.M., et al., 2013. Disease-driven amphibian declines alter ecosystem processes in a tropical stream. Ecosystems 16 (1), 146−157. Available from: https://doi.org/10.1007/s10021-012-9602-7.

Whiles, M.R., Lips, K.R., Pringle, C.M., Kilham, S.S., Bixby, R.J., Brenes, R., et al., 2006. The effects of amphibian population declines on the structure and function of Neotropical stream ecosystems. Front. Ecol. Environ. 4 (1), 27−34.

Whiles, M.R., Wallace, J.B., Chung, K., 1993. The influence of Lepidostoma (Trichoptera: Lepidostomatidae) on recovery of leaf-litter processing in disturbed headwater streams. Am. Midl. Nat. 130, 356−363.

White, D., Rashleigh, B., 2012. Effects of stream topology on ecological community results from neutral models. Ecol. Modell. 231, 20−24. Available from: https://doi.org/10.1016/j.ecolmodel.2012.01.022.

White, D.A., Visser, J.M., 2016. Water quality change in the Mississippi River, including a warming river, explains decades of wetland plant biomass change within its Balize delta. Aquat. Bot. 132, 5−11. Available from: https://doi.org/10.1016/j.aquabot.2016.02.007.

White, D.C., 1995. Chemical ecology: possible linkage between macro- and microbial ecology. Oikos 74, 177−184.

White, D.S., Hendricks, S.P., 2000. Lotic macrophytes and surface-subsurface exchange processes. In: Jones, J.B., Molholland, P.J. (Eds.), Streams and Ground Waters. Academic Press, San Diego, CA, pp. 363−379.

White, I., Macdonald, B.C.T., Somerville, P.D., Wasson, R., 2009. Evaluation of salt sources and loads in the upland areas of the Murray-Darling Basin, Australia. Hydrol. Process. 23 (17), 2485−2495.

White, J.D., Sarnelle, O., Hamilton, S.K., 2017. Unexpected population response to increasing temperature in the context of a strong species interaction. Ecol. Appl. 27 (5), 1657−1665. Available from: https://doi.org/10.1002/eap.1558.

White, J.W.C., Ciais, P., Figge, R.A., Kenny, R., Markgraf, V., 1994. A high-resolution record of atmospheric CO_2 content from carbon isotopes in peat. Nature 367, 153–156.

White, P.S., Pickett, S.T.A., 1985. Natural disturbance and patch dynamics: an introduction. In: Pickett, S.T.A., White, P.S. (Eds.), The Ecology of Natural Disturbance and Patch Dynamics. Academic Press, Inc, Orlando, FL, pp. 3–13.

White, W.B., Culver, D.C., Herman, J.S., Kane, T.C., Mylroie, J.E., 1995. Karst lands. Am. Sci. 83, 450–459.

Whitehead, P.G., Wilby, R.L., Battarbee, R.W., Kernan, M., Wade, A.J., 2009. A review of the potential impacts of climate change on surface water quality. Hydrol. Sci. J. 54 (1), 101–123.

Whiteley, M., Diggle, S.P., Greenberg, E.P., 2017. Progress in and promise of bacterial quorum sensing research. Nature 551, 313. Available from: https://doi.org/10.1038/nature24624.

Whitfield, J., 2005. Biogeography: is everything everywhere? Researchers have dug up some surprising evidence casting doubt on the long-held belief that microbes are impervious to geographic constraints. Science 310 (5750), 960–962.

Whiting, D.P., Whiles, M.R., Stone, M.L., 2011. Patterns of macroinvertebrate production, trophic structure, and energy flow along a tallgrass prairie stream continuum. Limnol. Oceanogr. 56, 887–898.

Whitman, W.B., Coleman, D.C., Wiebe, W.J., 1998. Prokaryotes: the unseen majority. Proc. Natl. Acad. Sci. 95, 6578–6583.

Whitney, J.E., Gido, K.B., Pilger, T.J., Propst, D.L., Turner, T.F., 2015. Consecutive wildfires affect stream biota in cold-and warmwater dryland river networks. Freshw. Sci. 34 (4), 1510–1526.

Whittaker, R.H., 1975. Communities and Ecosystems, second ed. MacMillan Publishing Co., Inc, New York, NY.

Wickstrom, C.E., Castenholz, R.W., 1978. Association of *Pleurocapsa* and *Calothrix* (Cyanophyta) in a thermal stream. J. Phycol. 14, 84–88.

Wickstrom, C.E., Castenholz, R.W., 1985. Dynamics of cyanobacterial and ostracod interactions in an Oregon hot spring. Ecology 66 (3), 1024–1041.

Wiegert, R., Petersen, C., 1983. Energy transfer in insects. Annu. Rev. Entomol. 28 (1), 455–486.

Wigand, C., Andersen, F.Ø., Christensen, K.K., Holmer, M., Jensen, H.S., 1998. Endomycorrhizae of isoetids along a biogeochemical gradient. Limnol. Oceanogr. 43 (3), 508–515.

Wiggins, G.B., 1995. Trichoptera. In: Merritt, R.W., Cummins, K.W. (Eds.), An Introduction to the Aquatic Insects of North America, third ed. Kendall/Hunt Publishing Company, Dubuque, IA, pp. 271–347.

Wigington Jr, P., DeWalle, D.R., Murdoch, P.S., Kretser, W., Simonin, H., Van Sickle, J., et al., 1996. Episodic acidification of small streams in the northeastern United States: ionic controls of episodes. Ecol. Appl. 6 (2), 389–407.

Wilber, C.G., 1983. Turbidity in the Aquatic Environment: An Environmental Factor in Fresh and Oceanic Waters. Charles C. Thomas Publisher, Springfield, IL.

Wilbur, H.M., 1987. Regulation of structure in complex systems: experimental temporary pond communities. Ecology 68 (5), 1437–1452.

Wilbur, H.M., 1997. Experimental ecology of food webs: complex systems in temporary ponds. Ecology 78 (8), 2279–2302.

Wilhelm, F.M., Hudson, J.J., Schindler, D.W., 1999. Contribution of *Gammarus lacustris* to phosphorus recycling in a fishless alpine lake. Can. J. Fish. Aquat. Sci. 56, 1679–1686.

Wilkinson, J., Maeck, A., Alshboul, Z., Lorke, A., 2015. Continuous seasonal river ebullition measurements linked to sediment methane formation. Environ. Sci. Technol. 49 (22), 13121–13129. Available from: https://doi.org/10.1021/acs.est.5b01525.

Willett, S.D., McCoy, S.W., Perron, J.T., Goren, L., Chen, C.-Y., 2014. Dynamic reorganization of river basins. Science 343 (6175). Available from: https://doi.org/10.1126/science.1248765.

Williams, D., Febria, C., Schriever, T., 2009. Structure and mechanics of intermittent wetland communities: bacteria to anacondas. International Wetlands: Ecology, Conservation, and Restoration (pp. 17–55). Nova, New York.

Williams, D.D., 1996. Environmental constraints in temporary fresh waters and their consequences for the insect fauna. J. N. Am. Benthol. Soc. 15 (4), 634—650.

Williams, D.M., Embley, T.M., 1996. Microbial diversity: domains and kingdoms. Annu. Rev. Ecol. Syst. 27, 569—595.

Williams, W.D., 1993. Conservation of salt lakes. Hydrobiologia 267, 291—306.

Williamson, C.E., 1991. Copepoda. In: Thorp, J.H., Covich, A.P. (Eds.), Ecology and Classification of North American Freshwater Invertebrates. Academic Press, Inc, San Diego, CA, pp. 787—822.

Williamson, C.E., Zagarese, H.E., Schulze, P.C., Hargreaves, B.R., Seva, J., 1994. The impact of short-term exposure to UV-B radiation on zooplankton communities in north temperate lakes. J. Plankton Res. 16 (3), 205—218.

Willson, M.F., Gende, S.M., Marston, B.H., 1998. Fishes and the forest. Bioscience 48 (6), 455—462.

Wilson, E.O., 2016. Half-Earth: Our Planet's Fight for Life. WW Norton & Company.

Wilson, L.G., 1990. Methods for sampling fluids in the vadose zone. In: Nielsen, D.M., Johnson, A.I. (Eds.), Ground Water and Vadose Zone Monitoring. ASTM, Ann Arbor, MI, pp. 7—24.

Wilson, M.A., Carpenter, S.R., 1999. Economic valuation of freshwater ecosystem services in the United States: 1971—1997. Ecol. Appl. 9 (3), 772—783.

Wilson, N., 2018. Nanoparticles: environmental problems or problem solvers? Bioscience 68 (4), 241—246.

Wilson, S.C., Duarte-Davidson, R., Jones, K.C., 1996. Screening the environmental fate of organic contaminants in sewage sludges applied to agricultural soils: 1. The potential for downward movement to groundwaters. Sci. Total Environ. 185, 45—57.

Winemiller, K., McIntyre, P., Castello, L., Fluet-Chouinard, E., Giarrizzo, T., Nam, S., et al., 2016. Balancing hydropower and biodiversity in the Amazon, Congo, and Mekong. Science 351 (6269), 128—129.

Winemiller, K.O., Montaña, C.G., Roelke, D.L., Cotner, J.B., Montoya, J.V., Sanchez, L., et al., 2014. Pulsing hydrology determines top-down control of basal resources in a tropical river—floodplain ecosystem. Ecol. Monogr. 84 (4), 621—635. Available from: https://doi.org/10.1890/13-1822.1.

Winemiller, K.O., Rose, K.A., 1992. Patterns of life-history diversification in North American fishes: implications for population regulation. Can. J. Fish. Aquat. Sci. 49 (10), 2196—2218.

Wingham, D.J., Siegert, M.J., Shepherd, A., Muir, A.S., 2006. Rapid discharge connects Antarctic subglacial lakes. Nature 440 (7087), 1033.

Winkworth-Lawrence, C., Lange, K., 2016. Antibiotic resistance genes in freshwater biofilms may reflect influences from high-intensity agriculture. Microb. Ecol. 72 (4), 763—772. Available from: https://doi.org/10.1007/s00248-016-0740-x.

Winterbourn, M.J., 1990. Interactions among nutrients, algae, and invertebrates in a New Zealand mountain stream. Freshw. Biol. 23, 463—474.

Winterbourn, M.J., Rounick, J.S., Cowie, B., 1981. Are New Zealand stream ecosystems really different? N. Z. J. Mar. Freshw. Res. 15, 321—328.

Winton, R.S., Richardson, C.J., 2017. Top-down control of methane emission and nitrogen cycling by waterfowl. Ecology 98 (1), 265—277. Available from: https://doi.org/10.1002/ecy.1640.

Wissinger, S.A., 1999. Ecology of wetland invertebrates. synthesis and applications for conservation and management. In: Batzer, D.P., Rader, R.B., Wissinger, S.A. (Eds.), Invertebrates in Freshwater Wetlands of North America: Ecology and Management. John Wiley & Sons, Inc, New York, NY, pp. 1043—1086.

Wissinger, S.A., Whiteman, H.H., Sparks, G.B., Rouse, G.L., Brown, W.S., 1999. Foraging trade-offs along a predator-permanence gradient in subalpine wetlands. Ecology 80 (6), 2102—2116.

Witte, F., Witte-Maas, E.L.M., 1980. Haplochromine cleaner fishes: a taxonomic and eco-morphological description of two new species. Neth. J. Zool. 31 (1), 203—230. Available from: https://doi.org/10.1163/002829680X00249.

Wnorowski, A., 1992. Tastes and odours in the aquatic environment: a review. Water SA 18 (3), 203—214.

Woese, C.R., Kandler, O., Wheelis, M.L., 1990. Towards a natural system of organisms: proposal for the domains Archae, Bacteria, and Eucarya. Proc. Natl. Acad. Sci. 87, 4576–4579.

Wohl, E., Hall, R.O., Lininger, K.B., Sutfin, N.A., Walters, D.M., 2017. Carbon dynamics of river corridors and the effects of human alterations. Ecol. Monogr. 87 (3), 379–409. Available from: https://doi.org/10.1002/ecm.1261.

Wolfe, B.E., Weishampel, P.A., Klironomos, J.N., 2006. Arbuscular mycorrhizal fungi and water table affect wetland plant community composition. J. Ecol. 94 (5), 905–914.

Wolfe, G.V., Steinke, M., Kirst, G.O., 1997. Grazing-activated chemical defense in a unicellular marine alga. Nature 387, 203–214.

Wolke, C.T., Fournier, J.A., Dzugan, L.C., Fagiani, M.R., Odbadrakh, T.T., Knorke, H., et al., 2016. Spectroscopic snapshots of the proton-transfer mechanism in water. Science 354 (6316), 1131–1135. Available from: https://doi.org/10.1126/science.aaf8425.

Woller-Skar, M.M., Jones, D.N., Luttenton, M.R., Russell, A.L., 2015. Microcystin Detected in Little Brown Bats (*Myotis lucifugus*). Am. Midl. Nat. 174 (2), 331–334.

Wollheim, W.M., Vörösmarty, C., Peterson, B.J., Seitzinger, S.P., Hopkinson, C.S., 2006. Relationship between river size and nutrient removal. Geophys. Res. Lett. 33 (6).

Woltemade, C., 2000. Ability of restored wetlands to reduce nitrogen and phosphorus concentrations in agricultural drainage water. J. Soil Water Conserv. 55 (3), 303–309.

Wommack, K.E., Hill, R.T., Muller, T.A., Colwell, R.R., 1996. Effects of sunlight on bacteriophage viability and structure. Appl. Environ. Microbiol. 62, 1342–1346.

Wong, M.K.M., Goh, T.-K., Hodgkiss, I.J., Hyde, K.D., Ranghoo, V.M., Tsui, C.K.M., et al., 1998. Role of fungi in freshwater ecosystems. Biodiversity Conserv. 7, 1187–1206.

Wong, P.P., Losada, I.J., Gattuso, J.-P., Hinkel, J., Khattabi, A., McInnes, K.L., et al., 2014. Coastal systems and low-lying areas. Clim. Change 2104, 361–409.

Wood, S.A., Atalah, J., Wagenhoff, A., Brown, L., Doehring, K., Young, R.G., et al., 2017a. Effect of river flow, temperature, and water chemistry on proliferations of the benthic anatoxin-producing cyanobacterium *Phormidium*. Freshw. Sci. 36 (1), 63–76. Available from: https://doi.org/10.1086/690114.

Wood, S.A., Borges, H., Puddick, J., Biessy, L., Atalah, J., Hawes, I., et al., 2017b. Contrasting cyanobacterial communities and microcystin concentrations in summers with extreme weather events: insights into potential effects of climate change. Hydrobiologia 785 (1), 71–89. Available from: https://doi.org/10.1007/s10750-016-2904-6.

Wood, S.A., Kuhajek, J.M., de Winton, M., Phillips, N.R., 2012. Species composition and cyanotoxin production in periphyton mats from three lakes of varying trophic status. FEMS Microbiol. Ecol. 79 (2), 312–326. Available from: https://doi.org/10.1111/j.1574-6941.2011.01217.x.

Wood, T.S., 1991. Bryozoans. In: Thorp, J.H., Covich, A.P. (Eds.), Ecology and Classification of North American Freshwater Invertebrates. Academic Press, Inc, San Diego, CA, pp. 481–500.

Woodward, C., Shulmeister, J., Larsen, J., Jacobsen, G.E., Zawadzki, A., 2014. The hydrological legacy of deforestation on global wetlands. Science 346 (6211), 844–847. Available from: https://doi.org/10.1126/science.1260510.

Woodward, G., Gessner, M.O., Giller, P.S., Gulis, V., Hladyz, S., Lecerf, A., et al., 2012. Continental-scale effects of nutrient pollution on stream ecosystem functioning. Science 336 (6087), 1438–1440.

Woodward, G., Dybkjær, J.B., Ólafsson, J.S., Gíslason, G.M., Hannesdóttir, E.R., Friberg, N., 2010. Sentinel systems on the razor's edge: effects of warming on Arctic geothermal stream ecosystems. Glob. Change Biol. 16 (7), 1979–1991. Available from: https://doi.org/10.1111/j.1365-2486.2009.02052.x.

Wootton, J.T., 1994. The nature and consequences of indirect effects in ecological communities. Annu. Rev. Ecol. Syst. 25, 443–466.

Wootton, K.L., 2017. Omnivory and stability in freshwater habitats: does theory match reality? Freshw. Biol. 62 (5), 821–832. Available from: https://doi.org/10.1111/fwb.12908.

Worthington, E.B., 1931. Vertical movements of fresh-water macroplankton. Internationale Revue der gesamten Hydrobiologie und Hydrographie 25, 394–436.

Wrona, F.J., Prowse, T.D., Reist, J.D., Hobbie, J.E., Lévesque, L.M.J., Vincent, W.F., 2006. Climate change effects on aquatic biota, ecosystem structure and function. AMBIO: J. Hum. Environ. 35 (7), 359−369.

Wu, L., Culver, D.A., 1991. Zooplankton grazing and phytoplankton abundance: an assessment before and after invasion of *Dreissena polymorpha*. J. Great Lakes Res. 17 (4), 425−436.

Wu, S., Wang, G., Angert, E.R., Wang, W., Li, W., Zou, H., 2012. Composition, diversity, and origin of the bacterial community in grass carp intestine. PLoS One 7 (2), e30440. Available from: https://doi.org/10.1371/journal.pone.0030440.

Wurtsbaugh, W.A., 1992. Food-web modification by an invertebrate predator in the Great Salt Lake (USA). Oecologia 89, 168−175.

Wurzbacher, C.M., Bärlocher, F., Grossart, H.-P., 2010. Fungi in lake ecosystems. Aquat. Microb. Ecol. 59, 125−149.

Xenopoulos, M.A., Lodge, D.M., Alcamo, J., Märker, M., Schulze, K., Van Vuuren, D.P., 2005. Scenarios of freshwater fish extinctions from climate change and water withdrawal. Glob. Change Biol. 11 (10), 1557−1564.

Xenopoulos, M.A., Schindler, D.W., 2003. Differential responses to UVR by bacterioplankton and phytoplankton from the surface and the base of the mixed layer. Freshw. Biol. 48, 108−122.

Xia, K., Bhandari, A., Das, K., Pillar, G., 2005. Occurrence and fate of pharmaceuticals and personal care products (PPCPs) in biosolids. J. Environ. Qual. 34 (1), 91−104.

Xia, X., Jia, Z., Liu, T., Zhang, S., Zhang, L., 2017. Coupled nitrification-denitrification caused by suspended sediment (SPS) in rivers: importance of SPS size and composition. Environ. Sci. Technol. 51 (1), 212−221. Available from: https://doi.org/10.1021/acs.est.6b03886.

Xu, H., Paerl, H.W., Zhu, G., Qin, B., Hall, N.S., Zhu, M., 2017. Long-term nutrient trends and harmful cyanobacterial bloom potential in hypertrophic Lake Taihu, China. Hydrobiologia 787 (1), 229−242. Available from: https://doi.org/10.1007/s10750-016-2967-4.

Yamamoto, Y., Kouchiwa, T., Hodoki, Y., Hotta, K., Uchida, H., Harada, K.-I., 1998. Distribution and identification of actinomycetes lysing cyanobacteria in a eutrophic lake. J. Appl. Phycol. 10, 391−397.

Yan, N.D., Keller, W., Scully, N.M., Lean, D.R.S., Dillon, P.J., 1996. Increased UV-B penetration in a lake owing to drought-induced acidification. Nature 381, 141−143.

Yan, Z., Han, W., Peñuelas, J., Sardans, J., Elser, J.J., Du, E., et al., 2016. Phosphorus accumulates faster than nitrogen globally in freshwater ecosystems under anthropogenic impacts. Ecol. Lett. 19 (10), 1237−1246. Available from: https://doi.org/10.1111/ele.12658.

Yanagita, T., 1990. Natural Microbial Communities. Ecological and Physiological Features. Japan Scientific Societies Press, Tokyo, Japan.

Yang, Z., Kong, F., Shi, X., Cao, H., 2006. Morphological response of *Microcystis aeruginosa* to grazing by different sorts of zooplankton. Hydrobiologia 563 (1), 225−230.

Yao, M., Henny, C., Maresca, J.A., 2016. Freshwater bacteria release methane as a by-product of phosphorus acquisition. Appl. Environ. Microbiol. 82 (23), 6994−7003. Available from: https://doi.org/10.1128/aem.02399-16.

Ye, H., Sugihara, G., 2016. Information leverage in interconnected ecosystems: Overcoming the curse of dimensionality. Science 353 (6302), 922−925.

Yard, M.D., Bennett, G.E., Mietz, S.N., Coggins, L.G., Stevens, L.E., Hueftle, S., et al., 2005. Influence of topographic complexity on solar insolation estimates for the Colorado River, Grand Canyon, AZ. Ecol. Modell. 183 (2), 157−172.

Yim, K.J., Kwon, J., Cha, I.-T., Oh, K.-S., Song, H.S., Lee, H.-W., et al., 2015. Occurrence of viable, red-pigmented haloarchaea in the plumage of captive flamingoes. Sci. Rep. 5, 16425. Available from: https://doi.org/10.1038/srep16425. Available from: https://www.nature.com/articles/srep16425#-supplementary-information.

Yoch, D.C., Carraway, R.H., Friedman, R., Kulkarni, N., 2001. Dimethylsulfide (DMS) production from dimethylsulfoniopropionate by freshwater river sediments: phylogeny of Gram-positive DMS-producing isolates. FEMS Microbiol. Ecol. 37 (1), 31−37.

Yokota, K., Waterfield, H., Hastings, C., Davidson, E., Kwietniewski, E., Wells, B., 2017. Finding the missing piece of the aquatic plastic pollution puzzle: interaction between primary producers and microplastics. Limnol. Oceanogr. Lett. 2, 91–104. Available from: https://doi.org/10.1002/lol2.10040.

Yoon, H.-S., Golden, J.W., 1998. Heterocyst pattern formation controlled by a diffusible peptide. Science 30, 935–938.

Yoon, I., Williams, R., Levine, E., Yoon, S., Dunne, J., Martinez, N., 2004. Webs on the Web (WOW): 3D visualization of ecological networks on the WWW for collaborative research and education. Proceedings of the IS&T/SPIE Symposium on Electronic Imaging, Visuzliation and Data Analysis, vol. 5295, pp. 124–132.

Young, J.P.W., 1992. Phylogenetic classification of nitrogen-fixing organisms. In: Stacey, G., Burris, R.H., Evans, H.J. (Eds.), Biological Nitrogen Fixation. Chapman & Hall, New York, NY, pp. 43–86.

Young, P., 1996. Safe drinking water: a call for global action. Am. Soc. Microbiol. News 62 (7), 349–352.

Young, P., 1997. Major microbial diversity initiative recommended. Am. Soc. Microbiol. News 63, 417–421.

Young, R.G., Huryn, A.D., 1999. Effects of land use on stream metabolism and organic matter turnover. Ecol. Appl. 9 (4), 1359–1376.

Young, R.G., Matthaei, C.D., Townsend, C.R., 2008. Organic matter breakdown and ecosystem metabolism: functional indicators for assessing river ecosystem health. J. N. Am. Benthol. Soc. 27 (3), 605–625.

Young, T.P., Petersen, D., Clary, J., 2005. The ecology of restoration: historical links, emerging issues and unexplored realms. Ecol. Lett. 8 (6), 662–673.

Yuan, A.H., Hochschild, A., 2017. A bacterial global regulator forms a prion. Science 355 (6321), 198–201. Available from: https://doi.org/10.1126/science.aai7776.

Yuan, L., Pollard, A., 2015. Classifying lakes to quantify relationships between epilimnetic chlorophyll *a* and hypoxia. Environ. Manage. 55 (3), 578–587. Available from: https://doi.org/10.1007/s00267-014-0412-x.

Yuan, L.L., Pollard, A.I., 2017. Using national-scale data to develop nutrient–microcystin relationships that guide management decisions. Environ. Sci. Technol. 51 (12), 6972–6980. Available from: https://doi.org/10.1021/acs.est.7b01410.

Yuan, L.L., Pollard, A.I., Pather, S., Oliver, J.L., D'Anglada, L., 2014. Managing microcystin: identifying national-scale thresholds for total nitrogen and chlorophyll *a*. Freshw. Biol. 59 (9), 1970–1981. Available from: https://doi.org/10.1111/fwb.12400.

Yuen, T.K., Hyde, K.D., Hodgkiss, I.J., 1999. Interspecific interactions among tropical and subtropical freshwater fungi. Microb. Ecol. 37, 257–262.

Yurista, P.M., 2000. Cyclomorphosis in *Daphnia lumholtzi* induced by temperature. Freshw. Biol. 43, 207–213.

Zalasiewicz, J., Williams, M., Haywood, A., Ellis, M., 2011. The Anthropocene: a new epoch of geological time? Philos. Trans. R. Soc. Lond. A: Math., Phys. Eng. Sci. 369 (1938), 835–841.

Zar, J.H., 1996. Biostatistical Analysis, vol. 55, fourth ed., p. 75309.

Zaret, T.M., 1980. Predation and Freshwater Communities. Yale University Press, New Haven, CT.

Zedler, J.B., Doherty, J.M., Miller, N.A., 2012. Shifting restoration policy to address landscape change, novel ecosystems, and monitoring. Ecol. Soc. 17 (4). Available from: https://doi.org/10.5751/es-05197-170436.

Zedler, J.B., Kercher, S., 2005. Wetland resources: status, trends, ecosystem services, and restorability. Annu. Rev. Environ. Resour. 30, 39–74.

Zeglin, L.H., Dahm, C.N., Barrett, J.E., Gooseff, M.N., Fitpatrick, S.K., Takacs-Vesbach, C.D., 2011. Bacterial community structure along moisture gradients in the parafluvial sediments of two ephemeral desert streams. Microb. Ecol. 61 (3), 543–556.

Zevenboom, W., de Vaate, A.B., Mur, L.R., 1982. Assessment of factors limiting growth rate of *Oscillatoria agardhii* in hypertrophic Lake Wolderqijd, 1978, by use of physiological indicators. Limnol. Oceanogr. 27, 39–52.

Zhang, C., Hu, Z., Deng, B., 2016. Silver nanoparticles in aquatic environments: physiochemical behavior and antimicrobial mechanisms. Water Res. 88, 403−427. Available from: https://doi.org/10.1016/j.watres.2015.10.025.

Zhang, F., Lee, J., Liang, S., Shum, C., 2015. Cyanobacteria blooms and non-alcoholic liver disease: evidence from a county level ecological study in the United States. Environ. Health 14 (1), 41.

Zhang, X.-X., Fu, Z., Zhang, Z., Miao, C., Xu, P., Wang, T., et al., 2012. Microcystin-LR promotes melanoma cell invasion and enhances matrix metalloproteinase-2/-9 expression mediated by NF-κB activation. Environ. Sci. Technol. 46 (20), 11319−11326. Available from: https://doi.org/10.1021/es3024989.

Zhang, Y., Ma, R., Hu, M., Luo, J., Li, J., Liang, Q., 2017a. Combining citizen science and land use data to identify drivers of eutrophication in the Huangpu River system. Sci. Total Environ. 584, 651−664.

Zhang, Z., Zimmermann, N.E., Stenke, A., Li, X., Hodson, E.L., Zhu, G., et al., 2017b. Emerging role of wetland methane emissions in driving 21st century climate change. Proc. Natl. Acad. Sci. 114 (36), 9647−9652. Available from: https://doi.org/10.1073/pnas.1618765114.

Zhao, J., Cao, X., Wang, Z., Dai, Y., Xing, B., 2017. Mechanistic understanding toward the toxicity of graphene-family materials to freshwater algae. Water Res. 111, 18−27. Available from: https://doi.org/10.1016/j.watres.2016.12.037.

Zheng, G., Xu, R., Chang, X., Hilt, S., Wu, C., 2013. Cyanobacteria can allelopathically inhibit submerged macrophytes: effects of *Microcystis aeruginosa* extracts and exudates on *Potamogeton malaianus*. Aquat. Bot. 109 (0), 1−7. Available from: https://doi.org/10.1016/j.aquabot.2013.02.004.

Zimba, P.V., 1998. The use of nutrient enrichment bioassays to test for limiting factors affecting epiphytic growth in Lake Okeechobee, Florida: confirmation of nitrogen and silica limitation. Arch. Hydrobiol. 141 (4), 459−468.

Zlotnik, I., Dubinsky, Z., 1989. The effect of light and temperature on DOC excretion by phytoplankton. Limnol. Oceanogr. 34 (5), 831−839.

Zwahlen, C., Hilbeck, A., Gugerli, P., Nentwig, W., 2003. Degradation of the Cry1Ab protein within transgenic *Bacillus thuringiensis* corn tissue in the field. Mol. Ecol. 12, 765−775.

Zwart, G., Crump, B.C., Kamst-van Agterveld, M.P., Hagen, F., Han, S.-K., 2002. Typical freshwater bacteria: an analysis of available 16S rRNA gene sequences from plankton of lakes and rivers. Aquat. Microb. Ecol. 28, 141−155.

Zweerde, V.D., 1990. Biological control of aquatic weeds by means of phytophagous fish. In: Pieterse, A.H., Murphy, K.J. (Eds.), Aquatic Weeds. The Ecology and Management of Nuisance Aquatic Vegetation. Oxford University Press, New York, NY, pp. 201−221.

Zweig, R.D., 1985. Freshwater aquaculture management for survival. Ambio 14 (2), 66−74.

TAXONOMIC INDEX

Note: Page numbers followed by "*f*" and "*t*" refer to figures and tables, respectively.

A

GEOGRAPHIC INDEX

Note: Page numbers followed by "*f*" and "*t*" refer to figures and tables, respectively.

SUBJECT INDEX

Note: Page numbers followed by "*f*" and "*t*" refer to figures and tables, respectively.

A

Acetogenesis, 384*t*, 385–386
Acid mine drainage, 429, 479, 485, 500
Acid precipitation, 11–12, 339, 476–485,
    546–547, 792
    aluminum, 500–501
    biological effects, 479–483
    chemical effects, 496*t*
    mollusks, effects on, 375
    sources, 477–479
Acid rain. *See* Acid precipitation
Acidity, 241, 339–340, 375, 478–479
Acute exposure or toxicity, 456
Adaptation to extreme environments, 426–430
Adaptive radiation, 300–301
Advective transport, 50, 52, 70–71, 685
Aeolian lakes, 167
Aerenchymous, 664
Aerobic, 348. *See also* Oxic; Oxygen; Redox
Agriculture, 5, 12–13, 400–401, 490–491
    nutrient runoff, 144–145, 573–574
    pollution, 562
Akinete, 221
Alarm chemicals, 632, 634
Algae, 65, 206–207, 248, 505. *See also*
                Periphyton; Phytoplankton
    toxins, 224, 544, 605
Alkalinity, 339–340, 375, 477. *See also*
                Bicarbonate
Allelochemicals, 612
    in microbes, 612
    in streams, 613–614
Allen curve, 727–728, 727*f*, 730
Allochthonous, 601, 625, 726, 739, 749–751,
    760–761
Allomone, 632
Alpha diversity. *See* Diversity
Alternative stable states, 689–692
Aluminum, 26, 338–339, 500–501
Amensalism, 205–206, 590, 668
Amictic lake, 362

Ammonia, 397, 403–404, 477, 563
Ammonia oxidation (nitrification), 403–404
Ammonification, 401
Ammonium, 216–217, 343–345, 396–398,
                400–402, 404, 419, 422, 477–478, 510,
                522–524, 741
Amphibians, 54–55, 108–109, 112, 119, 219,
                271, 291, 299, 327, 329, 354, 784, 793
    global declines, 330
Amphidromous, 286
Anadromous, 142, 290
Anaerobic. *See* Anoxic (anaerobic)
Anaerobic ammonium oxidation, 404
Anammox. *See* Anaerobic ammonium
                oxidation
Anoxic (anaerobic), 105–106, 240, 362, 383–384,
                386, 433–434, 642
    carbon, 421–422, 542
    fish kills, 362, 365–366, 544
    groundwaters, 366
    lakes, 365, 407, 562
    redox, 400
    streams, 562
    wetlands, 407
Antarctic Lakes, 270, 437–439
Anthropocene, 15–17
Anthropogenic, 9–10, 98, 485–488, 498–499,
                727, 786
    eutrophication (cultural), 547–552
    global change, 117–118
    hydrological perturbations, 695
    loss of biodiversity, 330
    nutrient pollution, 538–539
    toxic pollution, 454
Antibiotic resistance, 475, 614
Apatite (calcium phosphate), 412–413, 513
Aquaculture, 11, 212–213, 285–286, 401, 695,
                699
Aquifer, 85–89, 542, 772
Arctic lakes, 439
Arsenic, 485, 490–491